Magnesium Technology 2011

TMS2011

140th Annual Meeting & Exhibition

Magnesium Technology 2011

Proceedings of a symposium sponsored by the
Magnesium Committee of the Light Metals Division of
The Minerals, Metals & Materials Society (TMS)

Held during
TMS 2011 Annual Meeting & Exhibition
San Diego, California, USA
February 27-March 3, 2011

Edited by

Wim H. Sillekens
Sean R. Agnew
Neale R. Neelameggham
Suveen N. Mathaudhu

Editors
Wim H. Sillekens
Sean R. Agnew
Neale R. Neelameggham
Suveen N. Mathaudhu

ISBN 978-3-319-48568-3
ISBN 978-3-319-48223-1 (eBook)
DOI 10.1007/978-3-319-48223-1

Chemistry and Materials Science: Professional

Published by Springer International Publishers, Switzerland, 2016
Reprint of the original edition published by John Wiley & Sons, Inc., 2011, 978-1-11802-936-7

Printed on acid-free paper

This Springer imprint is published by Springer Nature
The registered company is Springer International Publishing AG
The registered company address is: Gewerbestrasse 11, 6330 Cham, Switzerland

TABLE OF CONTENTS
Magnesium Technology 2011

Magnesium Technology 2011

Opening Session

Primary Production; Characterization and Mechanical Performance

Casting and Solidification

Alloy Design/Development; Grain Refinement and Severe Plastic Deformation

High-Temperature Alloys; High-Strength Alloys; Precipitation

Deformation Mechanisms I

Deformation Mechanisms II: Formability and Forming

New Applications (Biomedical and Other)

Advanced Materials and Processing

Corrosion and Coatings

Poster Session

Appendix

PREFACE

The world of today faces a number of immense challenges relating to such diverse issues as sustainability (environmental concerns, long-term availability of energy and mineral resources), security and quality of life. Most interestingly, magnesium can – and likely will – increasingly contribute to the resolution of several of these issues. Weight saving by introducing magnesium alloy components in vehicles is a recognized means of enhancing fuel efficiency and thus of reducing energy consumption and greenhouse gas emissions. Different from several other metals, magnesium is virtually inexhaustibly available from natural resources. By using attributes of magnesium-based materials other than low density (such as impact resistance, biocompatibility and chemical affinity), a variety of new applications relating to ballistic armor, biomedical implants and hydrogen storage rises at the horizon.

It is against this background that the Magnesium Committee of the Light Metals Division of TMS has organized and sponsored its 12th edition of the Magnesium Technology Symposium, held at the TMS Annual Meeting in San Diego, CA (USA) from February 28 to March 3, 2011. The symposium was organized in an opening session, a poster session and nine technical oral sessions, covering a broad range of topics. This included primary production and characterization, casting and solidification, alloy design, high-temperature and high-strength alloys, deformation mechanisms, formability and forming, new applications, advanced materials and processing, and corrosion and coatings.

The volume at hand represents the Proceedings of this Symposium. Like in previous years, contributions come from countries around the globe that are active in magnesium research and development and reflect the latest advancements in the field. To ensure TMS standards, all papers were peer reviewed by a pool of volunteers acting on behalf of the Magnesium Committee. In addition to these Proceedings, some Symposium contributions on biomedical applications will be published in full in an upcoming JOM Special Issue sponsored by the Magnesium Committee and entitled “Biomedical Applications of Magnesium” (issue 63/4, April 2011).

The bar chart on the next page visualizes the development of the TMS Magnesium Technology Symposium since its inception in the year 2000 in terms of size and international participation. In the initial years, the number of contributions was quite steady, but as of 2005 when parallel sessions were introduced this number rose steeply – although variations between the successive years exist. Contributing countries can roughly be divided into two categories, depending on if their market emphasis is on magnesium supply (e.g., China and Israel) or consumption (e.g., Germany, Japan and the USA). The largest share of the Proceedings publications comes from the listed eight countries while the remaining category “others” comprises more than 20 other countries from Europe and Asia (with the United Kingdom and Norway being the most prominent of these). Overall, the USA accounts for roughly a quarter of all abstracts and papers to date, followed by Canada (14%), China (11%) and Germany (10%). Notably, the chart also reflects the market changes that the magnesium sector has seen over the last decade, the most pronounced example of this being the large increase in Chinese contributions along with the country’s development of primary and alloy production during the more recent years.

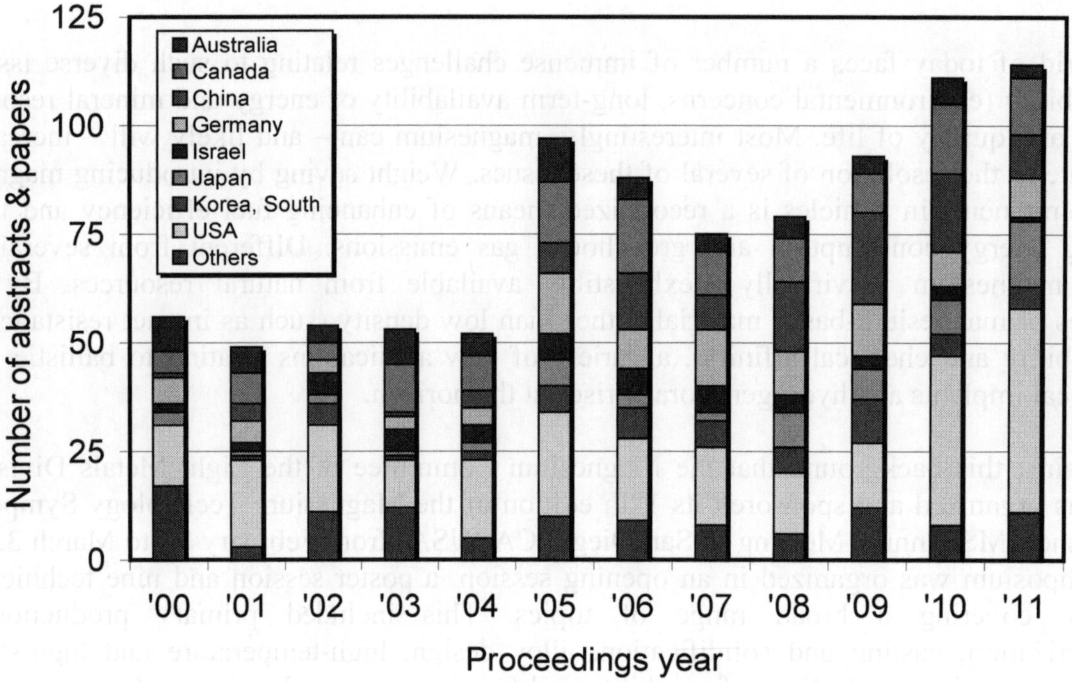

Keystone data of the TMS Magnesium Technology Proceedings (country of origin of each contribution is based on the affiliation of the main author)

The organization of the Magnesium Technology Symposium and the realization of its Proceedings would not have been possible without the support of numerous engaged volunteers. Hence this is the place to acknowledge the contribution of all authors, reviewers and session chairs that have been instrumental in making this happen. Further, the continued support by TMS staff has definitely facilitated the job and is well-appreciated.

While Symposium Proceedings traditionally reflect the state-of-the-art and spirit of the age, may this volume become a valuable part of your reference library for the years to come and in retrospect mark a memorable period in advancing the field of magnesium technology.

Wim H. Sillekens
Sean R. Agnew
Neale R. Neelameggham
Suveen N. Mathaudhu

ABOUT THE EDITOR

WIM SILLEKENS
MAGNESIUM TECHNOLOGY 2011 EDITOR

Wim H. Sillekens (1963) is a Senior Scientist at the Netherlands Organization for Applied Scientific Research (TNO), where he is involved in national and European research projects. He obtained his Ph.D. from the University of Technology Eindhoven, Netherlands, on a subject relating to metal-forming technology. Since he has been engaged in light-metals research (aluminum and magnesium), amongst others on (hydro-mechanical) forming, (hydrostatic) extrusion, forging, recycling/refining, and more recently on biomedical applications. His professional career includes positions as post-doc researcher at his alma mater and as a research scientist / project leader at TNO. International working experience covers a placement as a research fellow at the Mechanical Engineering Laboratory (AIST-MITI) in Tsukuba, Japan, and – more recently – shorter stays as a visiting scientist at GKSS in Geesthacht, Germany, and at PNNL in Richland WA, USA. He has co-authored a variety of journal and conference papers (about 50 entries to date). Current research interests are in the physical and mechanical metallurgy of light metals in general and magnesium wrought alloys in particular.

ABOUT THE ORGANIZERS

Sean R. Agnew is the Heinz and Doris Wilsdorf Distinguished Research Chair and Associate Professor in the Department of Materials Science and Engineering at the University of Virginia. He earned his Ph.D. in Materials Science and Engineering at Northwestern University in 1998. His dissertation focused on fatigue behavior and texture of ultrafine grain copper produced by severe plastic deformation. He first began working on magnesium as a Wigner post-doctoral fellow at the Oak Ridge National Laboratory in 1999, with research into the mechanical behavior of both cast and wrought alloys. Since moving to UVA in 2001, his magnesium research has focused on the development of constitutive models that account for dislocation- and twinning-based deformation mechanisms; modeling and control of texture development; measurement and modeling of the hot workability and sheet metal formability; and alloy design. His research group also conducts research on fatigue, creep, diffraction-based characterization, and non-destructive testing of a variety of metallic alloys.

Neale R. Neelameggham is the Technical Development Scientist for US Magnesium LLC. He has 38 years of expertise in magnesium production technology, having been with the plant from its startup company NL magnesium. Dr. Neelameggham's expertise includes all aspects of the magnesium process, from solar ponds through the cast house including solvent extraction, spray drying, molten salt chlorination, electrolytic cell and furnace designs, lithium ion battery chemicals and by-product chemical processing. In addition, he has an in-depth and detailed knowledge of alloy development as well as all competing technologies of magnesium production, both electrolytic and thermal processes worldwide. Dr. Neelameggham holds 13 patents and has several technical papers to his credit. As a member of TMS, AIChE, and a former member of American Ceramics Society he is well versed in energy engineering, bio-fuels and related processes. Dr. Neelameggham has served in the Magnesium Committee of LMD since its inception in 2000, chaired it in 2005, and has been a co-organizer of the Magnesium Symposium since 2004. In 2007 he was made a Permanent Co-organizer for the Magnesium Symposium. He has been a member of the Reactive Metals Committee, Recycling Committee and Programming Committee Representative of LMD. In 2008, LMD and EPD created the Energy Committee following the symposium on CO_2 Reduction Metallurgy Symposium initiated by him. Dr. Neelameggham was selected as the inaugural Chair for the Energy Committee with a two-year term. He is also a member of LMD council. Dr. Neelameggham holds a doctorate in extractive metallurgy from the University of Utah.

Suveen N. Mathaudhu is a Program Manager with the US Army Research Office (ARO), Materials Science Division. Dr. Mathaudhu also concurrently serves as an Adjunct Assistant Professor in the Department of Materials Science and Engineering at North Carolina State University. Dr. Mathaudhu received his B.S.E. from Walla Walla College and his Ph.D. in Mechanical Engineering from Texas A&M University. Upon graduating in 2006, he accepted a post-doctoral fellowship, and subsequently a civil servant position at the US Army Research Laboratory with the purpose of establishing deformation-processing laboratories for research on advanced metallic and composite materials. Since joining ARO in 2010, he manages programs which focus on the use of innovative approaches for processing high performance structural materials reliably and at lower costs. Dr. Mathaudhu's current research interests include: ultrafine-grained and nanostructured materials by severe plastic deformation, microstructural optimization and homogenization, consolidation of metastable particulate materials and processing-microstructure-property relationships of refractory metals and lightweight metals, integrated computational materials engineering, and thermally stable nanocrystalline materials. He has co-authored over 40 technical publications in these areas. He is also an active member of TMS where he is the primary organizer of the Ultrafine-Grained Material Symposium, and also serves on the Magnesium Technology Committee, and the Nanomaterials Committee.

Magnesium Technology 2011

Opening Session

Session Chairs:

Wim H. Sillekens
(TNO Science and Industry, Netherlands)
Suveen N. Mathaudhu
(US Army Research Office, USA)

Magnesium Technology 2011
Edited by: Wim H. Sillekens, Sean R. Agnew, Neale R. Neelameggham, and Suveen N. Mathaudhu
TMS (The Minerals, Metals & Materials Society), 2011

MAGNESIUM IN NORTH AMERICA: A CHANGING LANDSCAPE

Susan Slade
US Magnesium LLC; 238 North 2200 West; Salt Lake City, UT 84116, USA

Keywords: magnesium market, North America

Abstract

The changing landscape of North American manufacturing in the context of global competition is impacting the market of all raw materials, including magnesium. Current automotive fuel economy legislation and pending legislation on the emissions of greenhouse gases are impacting magnesium's largest consuming industries, such as aluminum, automotive components, steel and transition metals. These industries are all considering innovative ways to efficiently incorporate the needed raw materials into their processes. The North American magnesium market differs from other regions based on maturity, supply streams, changing manufacturing capabilities and trade cases, combined with the transformation of North American manufacturing.

The impact of these factors on the supply/demand dynamics of the North American magnesium market in both the short and long-term will be reviewed. The influence of new applications, products, and legislative changes are considered in the equation.

Magnesium Technology 2011
Edited by: Wim H. Sillekens, Sean R. Agnew, Neale R. Neelameggham, and Suveen N. Mathaudhu
TMS (The Minerals, Metals & Materials Society), 2011

GLOBAL MAGNESIUM RESEARCH: STATE-OF-THE-ART AND WHAT'S NEXT?

Karl Ulrich Kainer

Helmholtz Zentrum Geesthacht, Magnesium Innovation Center; Max-Planck-Straße 1, 21502 Geesthacht, Germany

Keywords: magnesium research and development, global situation, issues and challenges

Abstract

In recent years magnesium and its alloys have been successfully introduced into weight-saving applications in the transportation industries in order to reduce fuel consumption and greenhouse gas emissions as well as to increase the performance of modern cars. Besides advantages, e.g. superior specific strength and excellent processability, applications of magnesium alloys are limited due to their inferior properties at elevated temperatures, e.g. low creep resistance and reduced corrosion behavior, especially when in galvanic contact with other metallic materials.

Current developments are revealing possibilities to improve these properties by using modern alloys and processing routes. While the majority of industrial applications utilize cast products, the use of wrought magnesium alloys is still at an early stage. Within the framework of ongoing research and development, the corrosion behavior of both cast and wrought magnesium materials in standalone uses or in galvanic couples with other metallic materials is gaining increasing attention. New coating systems tailored to selected applications will have to be developed in order to increase the usage of magnesium alloys in the transportation industries in the future. This work also needs to be coordinated with new processes for joining magnesium alloys with similar and dissimilar metals and alloys, to achieve a broad spectrum of materials that fulfill the requirements given by the applications.

After years of intensive research in Europe, North America, Australia and Asia, an increase in these activities has taken place in recent years, in particular in China and Korea. The magnesium industry has to face new challenges with regard to market issues, the breakdown of the Western magnesium industry and finally the carbon footprint discussion of the life-cycle assessment of components for the transportation industry.

This presentation will first address these issues and challenges, then discuss new developments and finally show some examples of new applications. In the conclusions, gaps and challenges will be analyzed and recommendations for sustainable research and development will be given.

Magnesium Technology 2011
Edited by: Wim H. Sillekens, Sean R. Agnew, Neale R. Neelameggham, and Suveen N. Mathaudhu
TMS (The Minerals, Metals & Materials Society), 2011

ENVIRONMENTAL CHALLENGES FOR THE MAGNESIUM INDUSTRY

Robert E. Brown
Magnesium Assistance Group Inc; 226 Deer Trace; Prattville, Alabama 36067, USA

Keywords: magnesium production, environment, recycling

Abstract

The subject of environmental concerns with magnesium production and magnesium processing first started showing up in technical analysis of problems about the time of Life Cycle Analysis articles. Magnesium is produced and processed in relatively small quantities throughout the world. Annual magnesium production has been around 500-700,000 metric tons per year. This compares to aluminum which is produced in annual amounts up to 35 million metric tons.

There have been some excellent review papers done, but a great amount of the work related to electrolytic magnesium production which was the predominant method of production. That situation has changed totally in the past 10 years and now 85% of the world's magnesium is produced by thermal processes and most of that is in China.

Comparison papers have been written on the environmental impacts of the two main magnesium production processes. As the measurement technology improves and as the total information references are better understood the environmental challenges can be more clearly identified. This paper reviews the situation and suggests some forward looking steps that might need to be taken.

Introduction

Most of the major environmental problem areas can be divided into the electrolytic magnesium and the thermal magnesium production methods. And these areas can again be subdivided into the preparation of the magnesium containing feed materials and the actual processing of this material to produce magnesium metal. All of these areas are impacted by secondary processes which impact the magnesium production such as the production of process power and necessary reduction materials such as 75% ferrosilicon for the Pidgeon thermal reduction process.

The first LCA papers were prompted by the environmental work that was done to get new construction permits for newer electrolytic magnesium plants planned for Canada and Australia in the late 1980's. The work was well done and focused on the GHG (Greenhouse Gas) effect. Some of the work was made part of investigations on the advantage of a lighter weight magnesium car in regard to emissions. For the first time, the advantage of SF_6 as a cover was questioned because of its global warming potential. SF_6 was found to have a global warming potential of 23,900 times that of Carbon Dioxide (CO_2) [1]. This problem was immediately addressed with a partnership between the U.S. Environmental Protection Agency (EPA) and the magnesium community. There were 16 members along with the International Magnesium Association that joined in this work [2]. At that time, the magnesium usage of SF_6 amounted to about 7% of the total world usage. Initially this was estimated to be about 4 kg/metric ton of magnesium product. Later it was found to be closer to 2kg/mt [2].

Electrolytic Magnesium Production

Most electrolytic magnesium production plants have been closed. Some of the original LCA studies were done on the Norsk Hydro plants in Porsgrunn, Norway and Becancour, Québec, Canada [3]. The primary feed for Porsgrunn was sea water and dolomite. Feed for Becancour is magnesite. Total emissions from the two plants were included in the study by Albright and Haagensen. At the time, it was said, "Separate data for emissions from Porsgrunn and Becancour sites during the period 1994 to 1996 were charted. See Table I. For each of the emissions, the 1996 level at Porsgrunn is above the 1995 level at the Becancour plant. This difference arises from the fact that the Becancour plant was constructed with new, improved technology, whereas the older Porsgrunn facility is being continually improved from substantially higher emission levels a decade ago."

Table I. Emissions from Norsk Hydro Production Sites (1995-1996)

	CHC to Water, kg		Dioxins to air, grams		Dioxins to water, grams		SO_2 to air, tons	
Site	1996	1995	1996	1995	1996	1995	1996	1995
Pors-grunn	3.0	3.2	1.3	3.5	2.3	1.6	276	190
Becan-cour	0.4	0.8	0.42	0.51	0.05	0.02	25	3

Albright and Haagensen did not list the production capacities of the plants, although the published estimates for each of these plants were about 40,000 mtpy of capacity [4]. Chlorinated hydrocarbon (CHC) emissions were actually reduced by half at both plants. Dioxins decreased substantially. The authors pointed out, "that in 1997 the permitted values of total dioxin emissions for Porsgrunn are but two grams per year to atmosphere and one gram per year to the water" [5].

Further work on the electrolytic process was discussed in work done by scientists at the Commonwealth Scientific and Industrial Research Organization (CSIRO) in Australia [6]. Unfortunately they used the Australian Magnesium Corporation technology in their calculations. This process was run in a 1500 mtpy pilot operation, but the commercial plant was not constructed. Magnesite was used for the process feed.

The initial calculations were part of a study to assess the environmental impact of a magnesium engine block supply chain. This study also was broadened to include an engine block made from magnesium alloy ingots produced at Becancour and secondary ingots produced in the USA. The study concluded that "the use stage of a passenger car contributes significantly to the total GHG impact of magnesium components over their entire life cycle. A significant reduction in the GHG impact during the use stage, and hence over the entire life cycle, may be achieved from a significant mass reduction of a car by using magnesium components" [7].

A second study was run using a torque converter housing as the automobile part produced [8]. The results were very similar to the engine block study and showed a significant reduction only if there was the substitution of a cover gas which had a lower GWP than SF_6 for the processing. A simple substitution of magnesium without a large reduction in GHG is not competitive environmentally with other metals such as demonstrated in the case of Chinese magnesium.

The development of the new electrolytic reduction plant designs, required environmental permitting and such items as chlorine emissions and dioxin generation were addressed in great detail. In the case of the work by Albright and Haagensen a LCI (Life Cycle Inventory) was developed for Norsk Hydro magnesium plants [3]. Clear tables were developed for both pure magnesium production and for magnesium alloy production. See Table II below.

In regard to the emissions and wastes at Norsk Hydro plants, it was reported that "The solid wastes from Hydro's plants consists mostly of inorganic salts and minerals from the raw materials. This waste is disposed of at authorized sites and does not cause any negative environmental impact. The treatment of gas and waste water on site, along with incineration of residues containing chlorinated hydrocarbons, form a portion of the (LCI) analysis as well."

Table II. Life Cycle Inventory for Magnesium Production [3]

	Pure Mg	Mg Alloy AZ91
Total Energy MJ/kg metal	144	151
Global warming effect kg CO_{2eq}/kg	19	19
Acidification kg/kg metal	0.02	0.025
Winter Smog kg/kg metal	0.015	0.017
Solid Waste kg/kg metal	0.5	0.5
Dioxins to air µg/kg metal	0.24	0.21
CHC to air mg/kg metal	13.7	12.4

The subject of chlorinated hydrocarbons, dioxins and chlorine emissions to air is continually addressed in the design and operation of electrolytic magnesium reduction plants [11]. Newer and better scrubbing technology plus process modifications have made great steps in reducing the emissions. Although the construction of electrolytic plants continues to be slow, the research and development of the new equipment and methods should be encouraged on a global basis.

Thermal Reduction Processes

The silicothermic process has been investigated very carefully in several excellent papers [9, 10]. The basis for most of the work is the Pidgeon process as operated in most Chinese plants. Results of these studies indicate that direct comparison with electrolytically produced magnesium shows the GCI (Global Climate Impact) is much greater for the silicothermic process.

The papers qualify their calculations by noting that the Chinese production is continually being upgraded and the effects of these changes have greatly improved the GCI over the past ten years. One inherent advantage of the Chinese casting operations is that they use very small amounts of SF_6 in the ingot casting operations. This has mainly occurred because the majority of the first plants built in China were very rudimentary and used very simple melting and casting operations with sulfur powder directly dusted on the cast ingots to prevent burning during solidification. As the plants increased in size, most of the operations retained the sulfur dusting for convenience and simplicity.

The Chinese magnesium and ferrosilicon operations were impacted by the 2008 Olympic Games that were held in Beijing. A central approach was taken to drastically reduce emissions during this period. Many of the older and smaller Pidgeon process plants with inefficient collection equipment were forced to close to meet the strict guidelines issued. It is predicted that the central approach to control emissions and to encourage greater energy efficiency will continue. This has recently been seen by new rules that have been issued regarding the size and design of magnesium plants that would be given permits [12].

China, in its 12th Five-Year-Plan, is planning to control its annual national output of magnesium, aluminum, copper, nickel, lead, zinc, tin, antimony, titanium and mercury, with total production not exceeding 41 million mt/year until 2015, according to China Nonferrous Metals Industry Association [12].

The government plans to eliminate magnesium smelters with capacities below15,000 mt/year and requires existing plants to have a minimum smelting capacity of 15,000 mt/year, while new entrants must have a minimum capacity of 20,000 mt/year, Ministry of Industry and Information Technology, or MITT, officials said at a magnesium industry conference in Ningxia in October 2010 [12].

In Beijing, a source with China Magnesium Association (CMA) said, "The government's stricter regulations will be good to the domestic magnesium sector as these will help improve industry management, which is in line with the state's macro goal of energy saving and emissions reduction."

According to MITT, the proposed new regulations will require existing smelters to consume a maximum of 5.5-6.0 mt of coal for one ton of magnesium smelting, and for the new smelters, a maximum of 5 mt of coal. Chinese magnesium smelters, over the past few decades, have been consuming as much as 11-18 mt of coal for each ton of magnesium smelting, although some smelters in the last two years were able to cut the usage to around 5-6 mt coal, following the state's advocating a low-carbon economy, according to CMA [12].

The environmental impacts of the products must be compared throughout their entire life-cycle i.e. from "cradle-to-grave", and this is done by Life-cycle Assessment (LCA), also known as Life-cycle Analysis. LCA allows impacts to be assessed throughout all stages of production and use to final disposal.

A newer study has a focus on the global warming impact; a cradle-to-grave life cycle study is conducted using averaged data for magnesium production in China. Calculations show that the cradle-to-grave global warming impact of Chinese magnesium ingots is 42 kg CO_2 eq/kg Mg ingot, within an uncertain range of 37–47 kg CO_2 eq/kg Mg ingot. The value of impact for the magnesium produced in China is ~60% higher than the global warming impact of aluminum, a competing material that is also produced in China in abundance [13].

The results of other recent studies show that the direct emission of fuel combustion in the process is the major contributor to the pollutants emission of magnesium production. Global warming potential and acidification potential make the main contribution to the accumulative environmental impact. The different fuel use strategies in the practice of magnesium production cause much different impacts on the environmental performance. The accumulative environmental impact of coal burned directly is the highest, and that of producer-gas comes to the next, while that of coke-oven gas is the lowest [13, 14].

Much of the fuel-based pollution will be addressed under the new rules and natural gas is being designed into many of the newer plants where it is available.

Discussion

We know that magnesium is the lightest structural metal and it is basically available in limitless quantities. The earth is literally covered with magnesium as one cubic mile of seawater contains 6,000,000 tons of magnesium and there are 330,000,000 cubic miles of seawater according to The Scientific American. There is magnesium in all of the major salt lakes in the world and it is found in deposits in various forms on all continents.

Unfortunately, it is not easy to recover economically due to some basic quirks of chemistry and the mechanical properties of the final metal produced have some serious limitations. The lightweight that makes magnesium as attractive as a material of construction is also a penalty in the modern method of calculation. Plants are rated in tons per day or tons per year. Extrusion presses are rated by tonnage. Rolling mills are rated by tonnage. Environmental emissions are rated by tonnage. One ton of magnesium is much larger than a ton of lead, hence it takes more material and more processing to produce the end product.

This phenomenon was discussed by Albright and Haagensen [3] where they say, "The energy consumption (at the primary plant site only) to produce magnesium and aluminum has been estimated as 35 and 30 kWh/kg. The comparative number for steel is about 11kWhr/kg. However, when these data are considered on a unit volume basis, which is the common design concern (that is, components must be designed or packaged into a given space or volume), the life cycle advantage of magnesium is more clear. On this basis, the above numbers correspond to 63, 81, and 87 kWh/l for magnesium, aluminum and steel, respectively. For recycled materials, the relative energy used is even more favorable for magnesium; with recycled magnesium energy consumption is only about 1kWh/l."

Cherubini et al. [15] conclude that "World magnesium production contributes to the Global Warming Potential with an emission of about 25.5Mt of CO2 eq. per year. This will be lowered as the substitutions and elimination of SF_6 is taken into account with improved energy efficiency.

China has the higher emissions, but it is important to notice that while it supplies the Magnesium market with about 77% of total production, in terms of emissions it accounts for about 89.7% of the total: Mg industrial production in China is thus more polluting than in other countries, although, by an economic point of view, it is cheaper. This aspect is even more clearly shown by the Acidification Potential (where the contribution of Chinese Mg to the total AP is equal to 93.1%), because of the relatively high sulfur content of the low-grade coal used in China [13].

These concerns are being addressed by several approaches in China and in the many major magnesium production research centers that they have recently established. The newest developments in China indicate that the major environmental challenges are being aggressively addressed by both private and governmental strategies.

The price of magnesium on a world wide basis is still quite high. Most of the small Chinese Pidgeon process plants that used coal for firing and large quantities of hand labor are slowly disappearing. Larger, more efficient plants, with advanced technology for the process and equipment are being built. All areas of the Chinese metals and magnesium community has become very aware of the global climate situation and are working very hard to become cleaner and more energy efficient.

The Chinese are approaching the problem of addressing GWI by improving the production plants and working hard in downstream processing to produce lightweight magnesium products for use in bicycles, motorcycles, cars, trains and airplanes.

The China Magnesium Association (CMA) review of 2009 [16] said, in part: "China's magnesium industry still faces huge challenges of adjusting structure and transforming development mode in 2010. The capability of independent innovation and core competitiveness need to be promoted. It's a brand new topic and task to revive the increasing of magnesium output and improve economic benefits, to expand the domestic demand and guarantee the growth, and to make more breakthroughs in domestic consumption." CMA tried to outline a development layout for China's magnesium industry and bring forward some ideas in the face of weakening global demand for Chinese magnesium.

Developing more new technologies of conserving energy and reducing emissions, developing low-carbon magnesium industry

- ✧ Developing and using high-efficient and energy-saving MELZ double-chamber shaft kiln, direct heat-storage U-kiln and annular shaft kiln;
- ✧ Developing and using reduction furnace clean production system with vertical retort in which briquette is loading from the top and slag is dripping out of the bottom;
- ✧ Developing and using new electric Magnetherm technology;
- ✧ Developing electrolytic method of magnesium smelting;
- ✧ Recycling the waste heat of magnesium residues;
- ✧ Recycling carbon dioxide;
- ✧ Exploiting new smelting technologies using olivine, serpentine and brucite.

Expanding magnesium application and boosting domestic consumption, trying to build innovative strategic alliances

- ✧ Chery, partnered with North University of China and Shanxi Yinguang Huasheng Magnesium Co. Ltd, built a magnesium innovative strategic alliance in the aim of promoting magnesium alloy wheels' application on automobiles;
- ✧ CHANA Automobile built an innovative strategic alliance cooperated with Chongqing Boao Mg-Al Manufacturing Co. Ltd. And Chongqing University in the aim of promoting magnesium alloy's mass application on automobiles;
- ✧ Promoting magnesium alloy's application on trains.

China will continue industrial restructuring for the purpose of optimizing magnesium industry and upgrading competitiveness.

Targets will be to achieve low carbon economy, symbolized with low energy consumption, low pollution and low emission, they are now orienting and guiding the business of industrial restructuring. China efforts should be stepped up to adjust the structure of product, industry and industrial distribution in terms of the Planning for Adjusting and Reviving of the Non-ferrous Industry.

- ✧ Main smelters are gradually congregated in resource- and energy-intensive areas while processing companies are located in consumer-intensive areas;
- ✧ More products of high-quality, high value-added, high technology content and high competitiveness are manufactured;
- ✧ The development mode of energy-saving, low emission, clean, safe and recycling is widely adopted.

Boosting China magnesium's quality and profitability by transforming development mode

- ✧ To set up magnesium industry model bases and strength their guide role by enhancing their ability of independent innovation, extending and maturing industry chain as well as expanding industrial scale;
- ✧ To eliminate the under-developed capacity through competition. It is urgent requirement of transforming development mode and enhancing economic growth quality and profitability. It's also requirement of dealing effectively with financial crisis. Similarly, it is requirement of promoting energy-saving & emission-reducing and tackling global climate changes.

The Chinese review concluded by saying "In the era following the financial crisis of 2008, we believe that the global magnesium industry will have a bright and promising future only if international coordination and cooperation are being increased, only if the idea of win-win is being truly carried out, only if common challenges are being tackled jointly. "

The US Government is also moving more into taking control of the emissions of several areas including magnesium. The reporting of greenhouse gas emissions by major sources of these pollutants is gaining momentum [17].

The U.S. Environmental Protection Agency (EPA) is finalizing requirements under its national mandatory greenhouse gas (GHG) reporting program for underground coal mines, industrial wastewater treatment systems, industrial waste landfills and magnesium production facilities. The data from these sectors will provide a better understanding of GHG emissions and will help EPA and businesses develop effective policies and programs to reduce them.

It is encouraging that the total Global Warming Potential of the magnesium operations is being addressed. It has been reported that carmakers in the EU are cutting their fleets' carbon dioxide emissions faster than expected and will reach the European Union targets ahead of time [18].

Europe has instituted the REACH program which is also covering magnesium. The program is named for the description of Registration, Evaluation, Authorization and restriction of Chemicals. A group of companies with similar interests and handling the same class of materials can form a consortium. This has been done by many of the companies directly importing, using, or producing magnesium metal in Europe. A Substance Information Exchange Forum (SIEF) has been established as a joint registration forum.

Over three months have passed since the first survey carried out by Magnesium REACH Consortium (MAREC) for its members' registration intention. The questionnaire covers the check of substance sameness, indication of envisaged registration deadline, registration intention, nomination of Lead Registrant (LR) and SIEF Formation Facilitator (SFF) and so on. Magnesium Elektron UK won the LR nomination, replacing previous ECKA Granulate Essen GmbH.

Magnesium (EC No.231-104-6) SIEF is a large SIEF with up to 2,441 members. However, it is unexpected that so far only 152 companies have given feedbacks. Furthermore, the number of SIEF members has sharply decreased to 104 now, which greatly increases individual costs. For this reason, MAREC has planned to launch another survey, which is the final survey for registration intentions of SIEF members [19].

In the meantime, it is reported that the use of SF_6 gas has been eliminated in magnesium processing in Europe [20].

Conclusions

Magnesium as an industry has recognized the major problems with its position in Global Warming Potential. Many various companies, researchers, universities have responded rapidly and effectively to identify and address the problems. The largest problem areas have all had actions taken to quantify and address the problems.

China is the world leader in magnesium production and is making great efforts to mobilize an effective force to improve their environment. New magnesium production methods are being investigated and tested for favourable impact on the GWP. They have also started a REACH program and are cooperating with several European groups to get help in establishing their program. Magnesium was not specifically mentioned.

Europe continues to serve as a leader in its aggressive approach to overall analysis and addressing of the problems, including the use of Life Cycle Analysis (LCA) on the newer car designs.

The US EPA actively and positively stepped into the SF_6 cover gas problem as have several producers of more environmentally friendly materials. The Australian research community is also working on cover gas development.

References

1. H. Gjestland, H. Westengen, S. Phahte, "Use of SF_6 in the Magnesium Industry: An Environmental Challenge", Proceedings of the Third International Magnesium Conference, The Institute of Materials, London, 1997 pp 33-41.
2. S.C. Bartos, "Building a Bridge for Climate Protection: U. S. EPA and the Magnesium Industry", Proceedings of the 59th Annual World Magnesium Conference (IMA), Montreal, Canada May 2002 pp 22-24.
3. D. Albright, J. Haagensen, "Life Cycle Inventory of Magnesium", Proceedings of the 54th Annual World Magnesium Conference (IMA), Toronto, Canada, June 1997 pp 32-37.

4. C. Burstow et al "A Strategic Analysis of the Global Magnesium Industry", Metal Bulletin Research, London, UK, December 1999.
5. Albright and Haagensen, p. 33.
6. S. Ramakrishnan, P. Koltun, "A Comparison of the Greenhouse Impacts of Magnesium Produced by Electrolytic and Pidgeon Processes" Magnesium Technology 2004, Presented at 133rd TMS Annual Meeting, Charlotte, NC, March 2004 pp. 173-178.
7. A. Tharumarajah and P. Koltun, "Life Cycle Assessment of Magnesium Component Supply Chain" Proceedings of the 62nd World Magnesium Conference (IMA), Berlin, Germany May 2005 pp. 67-73.
8. P. Koltun, A. Tharumarajah, S. Ramakrishnan, "Life Cycle Environmental Impact of Magnesium Automotive Components", Magnesium Technology 2005, Presented at 134th TMS Annual Meeting, San Francisco, CA, February 2005 pp. 43-48.
9. E. Aghion, S. Bartos, "Comparative Review of Primary Magnesium Production Technologies as Related to Global Climate Change", Proceedings of the 65th Annual World Magnesium Conference (IMA), Warsaw, Poland, May 2008. pp. 97-104.
10. S. Ehrenberger, S. Schmid, S. Song, H. Friedrich, "Status and Potentials of Magnesium Production in China: Life cycle Analysis Focusing on CO_{2eq} Emissions" Proceedings of the 65th Annual World Magnesium Conference (IMA), Warsaw, Poland, May 2008 pp. 105-112.
11. R. E. Brown, "M-Cell Modernization Improves US Magnesium Process and Environmental Performance", Light Metal Age, June 2003 pp. 6-12.
12. J. Leung, A. Yee, "Chinese Magnesium Rules to Eliminate Small Smelters", Platt's Metals Week, October 25, 2010 p 1.
13. S. Ramakrishnan, P. Koltun , "Global Warming Impact of the Magnesium Produced in China Using the Pidgeon Process", Resources, Conservation and Recycling, Volume 42, Issue 1, August 2004, pp. 49-64.
14. F. Gao, Z. Nie, Z. Wang, X. Gong, T. Zuo, "Assessing Environmental Impact of Magnesium Production using Pidgeon Process in China", Transactions of Nonferrous Metals Society of China, Volume 18, Issue 3, June 2008, pp. 749-754.
15. F. Cherubini, M. Raugei, S. Ulgiati, "LCA of Magnesium Production: Technological Overview and Worldwide Estimation of Environmental Burdens", Resources, Conservation and Recycling, Volume 52, Issues 8-9, July 2008, pp. 1093-1100.
16. M. Shukun,W. Xiuming, X. Jinxiang, "China Magnesium Development Report in 2009", Proceedings of 67th Annual World Magnesium Conference (IMA), Hong Kong China, May 2010.
17. R. Greenway, Environmental News Network, July 2, 2010.
18. Associated Press News in Montgomery, Alabama Advertiser, Nov 1, 2010.
19. Bulletin of REACH 24H Compliance Services, Ireland Oct 2010.
20. F. D'Errico, S. Fare, G. Graces, "The Next Generation of Magnesium Based Material to sustain the Intergovernmental Panel on Climate Change Policy" – *this proceedings volume*.

Magnesium Technology 2011
Edited by: Wim H. Sillekens, Sean R. Agnew, Neale R. Neelameggham, and Suveen N. Mathaudhu
TMS (The Minerals, Metals & Materials Society), 2011

PREDICTING MG STRENGTH FROM FIRST-PRINCIPLES: SOLID-SOLUTION STRENGTHENING, SOFTENING, AND CROSS-SLIP

Dallas R. Trinkle[1], Joseph A. Yasi[2], and Louis G. Hector, Jr.[3]

[1]Department of Materials Science and Engineering, Univ. Illinois at Urbana-Champaign, Urbana, IL 61801, USA
[2]Department of Physics, Univ. Illinois at Urbana-Champaign, Urbana, IL 61801, USA
[3]General Motors R&D Center, 30500 Mound Road, Warren, MI 48090, USA

Keywords: magnesium alloys; dislocations; plastic deformation; density functional theory

Abstract

Predictive modeling of strength from first-principles electronic structure methods offers great promise to inform Mg alloy design. Simulating the mechanical behavior for new alloys requires an understanding of mechanisms for deformation at atomic-length scales, with accurate chemistry, extended to larger length- and time-scales. To design ductile Mg alloys, we identify solutes that strengthen basal slip and increase cross-slip. First-principles modeling of dislocations predict dislocation motion under stress through a field of solutes at a finite temperature. First-principles flexible boundary conditions compute accurate core structures of basal and prismatic dislocations, and dislocation/solute interactions. We develop new models to predict the solute-strengthening for basal dislocations; cross-slip from basal- to prismatic-slip for *a*-type screw dislocations; and cross-slip stress with solutes. First-principles data provides insight into the response of dislocations to solutes and quantitative data to build new predictive models.

Introduction

Decreasing energy consumption in the transportation sector requires weight reduction. Accomplishing this is no small task, and requires new light-weight structural alloys that can be formed into parts efficiently. Magnesium—whose density is two-thirds that of aluminum—is now finding its way into automotive applications. Use of Mg alloys in future automobile designs for structural and body panel materials could lead to weight and fuel savings, which in turn translate to a decrease in emissions. However its wide-spread use is greatly hampered by its anisotropic plastic response which limits its formability at room and warm ($\lesssim 300°C$) temperatures. This comes partly from the hexagonal-closed packed crystal-structure, where slip in the basal plane is nearly one-hundred times *easier* than in perpendicular planes. Magnesium also twins at fairly low strains. New alloy development involves new solute additions and deformation to affect mechanical properties. Physics-based modeling offers great promise to aid the development of current and new structural alloys, but ultimately relies on insight from computational and experimental studies. Models for submicron length-scales describe the motion of individual dislocations, but they cannot address the effects of the complex chemistry of alloys. Chemical effects in dislocations and twin boundaries are of special importance, as are computational techniques that address chemical effects on mechanical properties. Recent advances in first-principles electronic-structure methods to study dislocation/solute interactions with accurate treatment of both metallic bonding and the long-range strain field of dislocations sets the stage to incorporate new data into a constitutive finite-element model for Mg alloys.

Results

New computational studies showcase the predictive power of density-functional theory and atomistic-scale methods to predict dislocation behavior in magnesium and magnesium alloys. These calculations are now available thanks to state-of-the-art first-principles calculations[1, 2, 3, 4, 5] with flexible boundary condition methods[6, 7], which have already proven successful for Mo[8] and Al[9], and now Mg[10, 11]. Flexible boundary condition methods accurately couple the dislocation core to the long-range bulk response; this reproduces the correct long-range strain field with a computationally tractable number of atoms. Moreover, by using first-principles methods, chemical identity can be changed to compute the interaction of solutes with dislocations: changes in energies and geometry and forces to resist glide. For magnesium, we have now predicted the equilibrium dislocation core geometry[10] as well as the interaction energy and forces for basal dislocations with solutes[11]. This latter work was combined with a simple solid-solution strengthening model for dilute concentrations and low temperatures to connect direct interactions with predictions of strength. The end result is a simple map connecting solute characteristics to strengthening "potency"[11].

Predicting basal strengthening from first-principles is the first part of reducing slip anisotropy in Mg at room temperature; a simultaneous reduction in prismatic Peierls stress is also required. For example, there is experimental evidence that Al simultaneously strengthens basal and softens prismatic slip[12, 13, 14]. Slip on the prismatic slip is a thermally activated process at low temperatures, involving

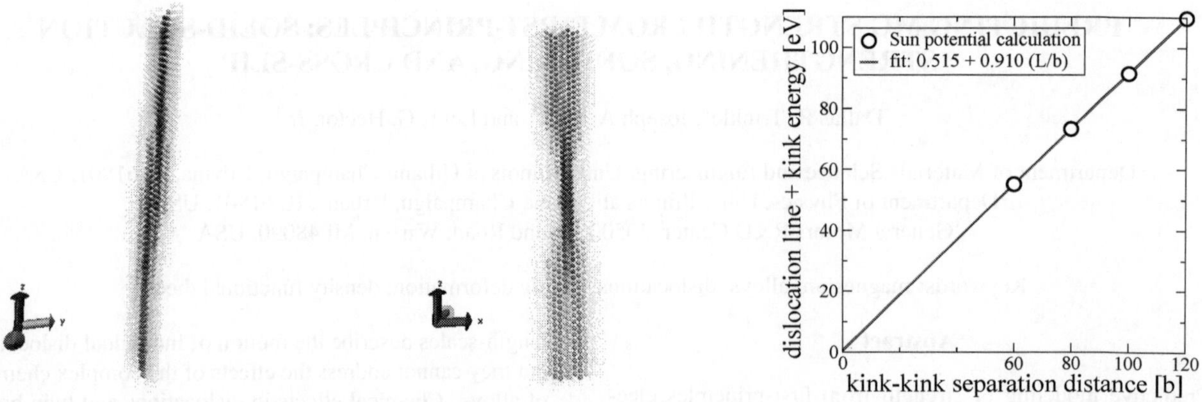

Figure 1: The geometry of a kink for an *a*-type screw dislocation, and the computation of formation energy. Atoms are colored from transparent to blue based on their local energy; the spreading on the basal plane (XZ in the above plot) is evident, along with the narrowing to constrict on the prismatic (YZ) plane. To account for kink-kink interaction energies, as well as the line-energy of the basal dislocation, we compute the total energies for different line lengths with a single kink; the fit of total energy to a linear function of line length gives the energy of a single kink (intercept) plus the line energy of a basal dislocation.

the nucleation of kinks along a screw dislocation allowing cross-slip onto the prismatic plane. Softening of prismatic slip, as in other solid-solution softened alloys with thermal activation[8], requires an attractive interaction with a solute and a dislocation; this reduces the kink-nucleation energy in the presence of a solute, increasing the cross-slip rate at a fixed stress and temperature. Modeling cross-slip for quantitative predictions requires: (1) determining energies for kinks on the prismatic plane, and (2) interaction energies of prismatic dislocations with solutes.

Figure 1 shows the relaxed geometry of a kink including the computation of the formation energy. For the computation of kinks, we use an atomistic classical potential parametrized by Sun *et al.*[15]. This potential has already proven to reasonably reproduce dislocation geometries in pure magnesium[10]; here, we construct a basal screw dislocation with a "step" of height c in the prismatic plane with a long threading periodicity. After we relax this geometry, it forms a kink in the prismatic plane. We perform this relaxation for several different line lengths; in order to compute a kink formation energy, we must remove the line energy of the basal dislocation line and kink-kink interaction energies. The total energy can be fit to a linear function of total line length (or kink-kink separation for our geometry); the intercept is then the energy to create a single kink, while the slope is the line energy of the dislocation. For the Sun potential, we predict a kink formation energy of 0.515eV with this method.

Figure 2 shows the interaction energy of an Al atom with two types of prismatic *a*-type dislocations. In both cases, Al atoms are substituted for Mg atoms in the dislocation core, and the geometry relaxed; after multiple sites are computed, the energy map for a solute at any site around the dislocation—including the far-field from continuum elasticity—is known. The prismatic edge dislocation has both attractive and repulsive interactions due to the volumetric strain in an edge dislocation core. Normally, the *a*-type screw dislocation splits into partials on the basal slip plane at zero stress. As a first approximation to a cross-slipped prismatic screw dislocation, we relax a perfect *a*-type screw dislocation but allow only displacements along the threading direction. This constraint produces a constricted dislocation core with no splitting, and a small amount of spreading on the prismatic plane. This constrained core has a large attractive interaction with an Al atom, and falls off quickly away from the core. The attractive interaction of prismatic dislocations with Al is a necessary ingredient for a reduction in kink nucleation energy in proximity to an Al atom; this is expected to promote cross-slip, and lead to more isotropic strength response for Mg alloys.

Conclusion

The future direction for this work is to use atomistic-scale data to build a predictive model for cross-slip stress as a function of temperature for pure Mg. The interaction energies of solutes with prismatic screw dislocations is then used to predict local changes in kink formation energy, and hence the change in cross-slip stress with solute chemistry and concentration. Combining the softening results for prismatic slip with the hardening results for basal slip gives an overall reduction in the strength anisotropy. In addition, the basic framework presented here is directly applicable to any sub-

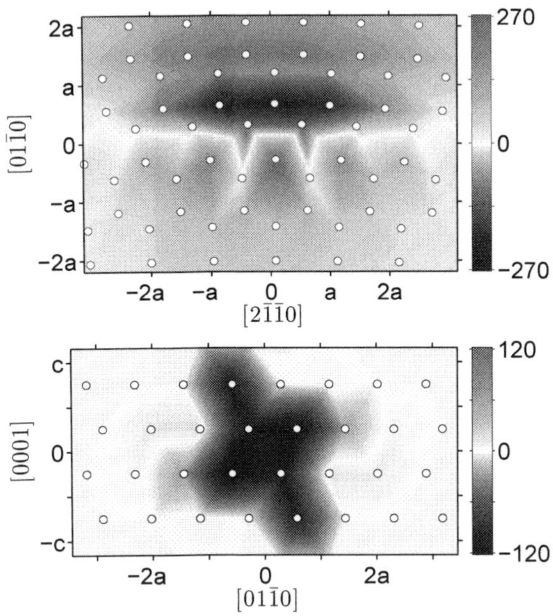

Figure 2: Interaction energy for Al with a prismatic a-type edge dislocation, and a constrained a-type screw dislocation. Each white circle is a dislocation site where an Al atom can substitute for Mg. In both cases, the interaction energy is computed using density-functional theory; the orange represents a repulsive (positive) interaction energy, and blue represents an attractive (negative) interaction energy. Both dislocations show an attractive interaction of dislocations to Al; this is expected to reduce the energy to form a kink in the presence of an Al solute, which should hence aid in the cross-slip of a screw dislocation in Mg.

stitutional or interstitial solute, allowing for first-principles-based modeling of strength, and incorporation into the design of new alloys.

Acknowledgments

This research was sponsored by NSF through the GOALI program, grant 0825961, and with the support of General Motors, LLC. This research was supported in part by the National Science Foundation through TeraGrid resources provided by NCSA and TACC, and with a donation from Intel.

References

[1] G. Kresse and J. Hafner, "Ab initio molecular dynamics for liquid metals," *Phys. Rev. B*, 47 (1993), RC558.

[2] G. Kresse and J. Furthmüller, "Efficient iterative schemes for ab initio total-energy calculations using a plane-wave basis set," *Phys. Rev. B*, 54 (1996), 11169.

[3] J. P. Perdew and Y. Wang, "Accurate and simple analytic representation of the electron-gas correlation energy," *Phys. Rev. B*, 45 (1992), 13244.

[4] D. Vanderbilt, "Soft self-consistent pseudopotentials in a generalized eigenvalue formalism," *Phys. Rev. B*, 41 (1990), 7892.

[5] G. Kresse and J. Hafner, "Norm-conserving and ultrasoft pseudopotentials for first-row and transition elements," *J. Phys.: Cond. Mat.*, 6 (1994), 8245.

[6] S. Rao *et al.*, "Green's function boundary conditions in two-dimensional and three-dimensional atomistic simulations of dislocations," *Philos. Mag. A*, 77 (1998), 231.

[7] D. R. Trinkle, "Lattice Green function for extended defect calculations: Computation and error estimation with long-range forces," *Phys. Rev. B*, 78 (2008), 014110.

[8] D. R. Trinkle and C. Woodward, "The Chemistry of Deformation: How Solutes Soften Pure Metals," *Science*, 310 (2005), 1665.

[9] C. Woodward, D. R. Trinkle, L. G. Hector, and D. L. Olmsted, "Prediction of dislocation cores in aluminum from density functional theory," *Phys. Rev. Lett.*, 100 (2008), 045507.

[10] J. A. Yasi *et al.*, "Basal and prism dislocation cores in magnesium: comparison of first-principles and embedded-atom-potential methods predictions," *Model. Simul. Mater. Sci. Eng.*, 17 (2009), 055012.

[11] J. A. Yasi, L. G. Hector, and D. R. Trinkle, "First-principles data for solid-solution strengthening of magnesium: From geometry and chemistry to properties," *Acta mater.*, 58 (2010), 5704.

[12] A. Akhtar and E. Teghtsoonian, "Solid solution strengthening of magnesium single crystals—I alloying behavior in basal slip," *Acta Metall.*, 17 (1969), 1339.

[13] A. Akhtar and E. Teghtsoonian, "Substitutional Solution Hardening of Magnesium Single Crystals," *Philos. Mag.*, 25 (1972), 897.

[14] C. H. Cáceres and D. M. Rovera, "Solid solution strengthening in concentrated Mg-Al alloys," *J. Light Metals*, 1 (2001), 151.

[15] D. Y. Sun *et al.*, "Crystal-melt interfacial free energies in hcp metals: A molecular dynamics study of Mg," *Phys. Rev. B*, 73 (2006), 024116.

Magnesium Technology 2011
Edited by: Wim H. Sillekens, Sean R. Agnew, Neale R. Neelameggham, and Suveen N. Mathaudhu
TMS (The Minerals, Metals & Materials Society), 2011

BIODEGRADABLE MAGNESIUM IMPLANTS – HOW DO THEY CORRODE IN-VIVO?

Frank Witte[1], Norbert Hort[2], Frank Feyerabend[2]
[1]Hannover Medical School; Anna-von-Borries-Straße 1-7; 30625 Hannover, Germany
[2]Helmholtz Zentrum Geesthacht, Magnesium Innovation Center; Max-Planck-Straße 1, 21502 Geesthacht, Germany

Keywords: biodegradable magnesium implants, in-vivo corrosion

Abstract

Biodegradable magnesium implants are currently breaking the paradigm of designing and producing only corrosion resistant metallic biomaterials. The academic and industrial interest in this novel class of biomaterials is increasing dramatically in the recent years. First biodegradable metal implants have been realized as vascular stents and bone screws. However, the knowledge of the underlying degradation mechanism of these metal implants remains mainly undiscovered.

This lecture will summarize the current published knowledge and recent advances in elucidating the in-vivo corrosion processes of these novel biodegradable magnesium implants [1].

Reference

[1] F. Witte, N. Hort, F. Feyerabend, "Biodegradable magnesium implants – How do they corrode in-vivo", *JOM* 63/4 (2011), *to be published.*

Magnesium Technology 2011
Edited by: Wim H. Sillekens, Sean R. Agnew, Neale R. Neelameggham, and Suveen N. Mathaudhu
TMS (The Minerals, Metals & Materials Society), 2011

THE NEXT GENERATION OF MAGNESIUM BASED MATERIAL TO SUSTAIN THE INTERGOVERNMENTAL PANEL ON CLIMATE CHANGE POLICY

F.D'Errico[1], S.Farè[1], G.Garces[2]
[1]Politecnico di Milano, Department of Mechanical Engineering, Via La Masa 34, 20156 Milan, Italy
[2]CENIM, Department of Physical Metallurgy, Av. De Gregorio del Amo 8, 28040 Madrid

Keywords: Magnesium alloys, ultrafine microstructure, sustainability, CO_2 reduction

Abstract

Current Mg alloys have several drawbacks that limit wide and profitable utilization in the industrial sector. From an environmental point of view, lighter metals like magnesium are currently considered unclean products as they require energy-intensive. But they have been proven to be "clean" in the transport sector, as they can reduce fuel consumption. Here the potential of magnesium based materials is addressed through double-tasking: a) establish innovative lean-manufacturing processes, avoid the classic melting step to substantially reduce carbon footprint of the magnesium products; b) encourage the using of no-melt processes, realizing high-resistant ultra-fined microstructures. The "Green Metallurgy 2020", a project funded by European Community in the LIFE+ 2009 Program, started in September 2010, coordinated by Politecnico di Milano (ITA) aims to scale to industrial route such impressive results experienced by CENIM (SPA) for some ultrafine bi-phase Mg –Zr (-Y) produced by no-melting route that achieved up to 400 MPa UTS and elongation capability of about 13%.

Introduction

While there is considerable debate about when global oil and natural gas production is likely to peak, there is no debate that fossil fuels like oil, coal, and natural gas constitute a non-renewable, finite resource. A resource that has powered cars, trucks, power plants and factories, and caused the recent relative and dramatic build up of greenhouse gases. As a basic requirement for the strategy to cut CO_2 levels in the transport sector, the IPCC report claims more fuel-efficient vehicles are needed in the future [1]. These would include advanced electric and hybrid vehicles with more powerful and reliable batteries, cleaner diesel vehicles, bio-fuels, modal shifts from road transport to rail and public transport systems, non-motorized transport (cycling, walking), and higher efficiency aircraft. In Europe, 12% of man-made CO_2 comes from passenger cars, with overall transport producing 26%. Between 1995 and 2003, motorists in the EU-25 increased their annual mileage by 16.4%. While a major challenge, it is clear that vehicle emissions are part of a much larger jigsaw [1]. The new industrial modern era is facing a difficult challenge: on the one hand there is the problem of increasing power energy, coupled with the dramatic increase of world energy consumption related to the growing economies in ASIA and AFRICA. On the other hand, there is an urgent and strong need to decrease the GHGs produced as a byproduct of this growth. These opposing forces bring an environmental imperative to the economic accountability of the actual cost of power production and need to include all of the environmental perimeters of our energy use. While fossil fuels have been considered to be relatively inexpensive in the past, a more modern approach starts with the realization that the atmosphere's assimilation of the byproducts produced by fossil fuel combustion is not without consequences. Today their economic impact on environmental and health must also be considered. For this reason a more modern approach focuses on supply chains and total product life cycles which involve minerals, energy resource processing, manufacturing steps, logistics, re-use, and recycling routes. Among these, Life Cycle Assessment (LCA) concerns over the limitations of raw materials and energy resources sparked interest in finding ways to cumulatively account for energy use and to project future resource supplies and use. LCA analysis adopts a schematic method of assessing the entire product life cycle, which is usually divided into the following stages: a) cradle-to-entry gate (raw material extraction and refining); b) entry-gate-to exit gate (product manufacture) and c) exit gate-to-grave (product use, recycling, and disposal). The risk, however, may be that reduced fuel consumption, i.e. the cut CO_2 in an internal combustion engine (ICE) vehicle, cannot balance the equivalent carbon dioxide[1] produced during the cradle-to-exit gate.

Achieving a Net Positive Balance of CO_2 Emission in Transport

The ICE is expected to continue to be the dominant technology for the short-medium term, as the ICE can maintain similar performance standards by utilizing more lightweight designs that directly can reduce CO_2 emissions, and promote further engine downsizing. For the long-term, more lightweight hybrid, fully electric or hydrogen cars that can travel long distances will surely be a key-factor to saving energy. According to the European Automobile Manufacturers Associations (ACEA), automakers will have to reduce CO_2 emissions[2] in new cars to 130 grams per kilometer by 2012/15, with an additional 10 gram reduction coming from 'complementary measures' including a greater use of bio-fuels. And a new objective of 95 grams per kilometer was established for 2020 [2]. From 2012 onwards, manufacturers who do not meet their targets must pay an "excess emissions penalty"[3], which may become law as early as 2012, may become an incentive for automakers to reduce CO_2 emissions by levying fines on automakers who don't respect emission thresholds (stated on 130gCO_2/Km).

---

[1] The kgCO_2 equivalent emissions is a parameter used for directly accounting of Global Warming; in fact it is also refereed as Global Warming Potential (GWP).

[2] In 2009, the market share of cars emitting 120 gCO_2/km had risen to 25%. Cars with emission above 160 gCO_2/km accounted for 23% of the market, compared to 39% in 2006 and to 80% in 1995 (source: ACEA).

[3] Excess emission penalty shall be calculated as the excess emissions (over EU targets), multiplied by the number of new passenger cars (yearly basis) multiplied by the excess emissions premium that shall vary accordingly to the calendar year (for the year 2012, 20 Euros, for the year 2013, 35 Euros; for the year 2014, 60 Euros; 2015 and subsequent calendar years, 95 Euros).

If forced to pay a fine on cars that do not meet EU emission standards[4], automakers, especially those in the medium-high quality area, would be encouraged to look for new short-term emission reducing strategies, and reducing the weight of each car may be the most practical, and least invasive way of reaching that goal. In fact, alternative solutions like cutting-performance or eliminating accessories would affect client's perceptions and expectations of the product.

The Magnesium Perspective for Vehicles: Superlight Structural Components Target Environmental Problems

Thanks to their low density, magnesium alloys offer one of the best opportunities. With a density of 1.81 kg/dm^3, they are one lowest structural metals and are excellent alternative to heavier materials like steels and even aluminum. Current environmental points of view state that while lighter metals like magnesium are currently unclean products to make, they could be "very clean to use". This can be attributed to the substantial weight reduction they offer and the resulting reduction of fuel consumption and CO_2 on ICE vehicles. With that in mind, consider the following key points regarding internal combustion engine transportation taken from [3]:

•2.85 kg CO_2eq is the average emissions per 1 liter of gasoline consumption;

•10.83 kg CO_2 per 1000 km is the calculated amount of reduced emissions for every 100 kg mass saved on passenger vehicles;

•if 200,000 km is the total life distance of a common passenger vehicle, a total amount of 2,166 kg CO_2eq can be saved through 100 kg weight reduction.

Although magnesium density is less than 25% of that of steel, its lower mechanical resistance is an impediment to achieving a 1 to 1 substitution ratio. Conservatively speaking, 1 to 2.65 is the (volumetric) substitution ratio to guarantee similar resistance of components (Table I).

Table I. Total Average Pollution Emissions Over Medium-size Vehicle Life.

	Total amount	Metrics
Car mass	1,300	Kg
Emissions	2.85	kg CO_{2eq}/liter
Average consumption	8.5	liter / 100 km
Total CO_2 emission over 200,000 km	48,450	kg $CO_{2\,eq}$ per car

Such a statement comes from the mechanical resistance of conventional Mg alloys, that is about 35% of common medium-carbon steels (50% for very low carbon steels). Substituting medium carbon steel with conventional magnesium alloys on structural components will result in an increase of the sections and volume of the components. On the other hand, thanks to magnesium's lower density, an overall weight reduction of approximately 40% can be expected using a magnesium component rather than a medium-carbon steel component (Table II).

Table II. Mg-alloy Design and Weight Saving by Substituting Medium Carbon Steel with Conventional (wrought) Mg-based Alloys.

Features	Mg	Steel
Density (kg/dm^3)	1.81	7.87
Resistance limit (MPa)	170	450
Usage ratio	2.65	1.0
Volume (mm^3)	265	100
Mass total (kg)	4.8	7.9
Total weight saving if 100 kg steel is substituted with magnesium conventional alloy (kg)	39	

To evaluate the impact of magnesium usage on medium-size passenger cars in terms of final net reduction of kgCO_2eq over the vehicle's travel-life distance, it is necessary to calculate the negative impact on the environment. The impact can be expressed in terms of kgCO_2eq, as measure of the Global Warming Potential (GWP) of magnesium-based products manufactured[5]. In the LCA of magnesium product to use onboard, the net GWP shall be calculated starting from the GWP of the cradle-to-exit gate path, but subtracting the reduction of CO_2 emissions obtained by reduced fuel consumption for the reduced total weight onboard. The importance of properly calculate the net GWP of magnesium product is due to the fact all magnesium processes are energy intensive and expensive, in comparison with steel-making[6]. The availability of primary minerals, the energy sources used for basic extracting process route, the power generation method employed for electricity in case of electrolytic process route[7], drive the entire GWP process, specifically the carbon-footprint of the production of raw materials.

[4] The European Commission calculates it will cost on average EUR 1500 per car to comply with new EU CO_2 targets. In addition, there will be the costs for the 10gCO_2/km 'complementary measures' (gear-shift indicators, tire-pressure monitoring, low-rolling resistance tires, air-conditioning and bio-fuels) (source: Proposal for a REGULATION OF THE EUROPEAN PARLIAMENT AND OF THE COUNCIL, "Setting emission performance standards for new passenger cars as part of the Community's integrated approach to reduce CO_2 emissions from light-duty vehicles", presented by Commission, Brussels, 19 December 2007).

[5] Great impact is ascribable to the use of sulfur hexafluoride (SF_6) as a cover gas, yet it is used in various low-cost plants. SF_6 is used to protect molten magnesium against oxidation both in primary magnesium smelters and in secondary magnesium production (alloy making). However, in Europe SF6 is banned while in the United States the use is optional for industry. The SF6 (used often in the Chinese magnesium industry) GWP impact has been calculated to be 22,800 kg CO_2eq per kg of SF_6 used (in other words, 22,800 times greater than 1 kg of CO_2 emitted). Usually 1 kg of SF_6 is required as cover gas per ton of melting magnesium, resulting in a 22,800 kg CO_2 equivalent per kilogram of melt magnesium.

[6] The GWP average value for steel production by the integrated route is around 2.3 kg CO_2eq/kg steel.

[7] Actually, the emissions from the electrolytic route strongly depend on energy sources for electricity generation (or energy-mix adopted) in the producer Country.

As raw materials are supplied, manufacturing and assembly steps include various conventional steps that also contribute to the carbon footprint of the final magnesium-based product[8]: a) transportation of ingots of the purchased alloy to the casting plant; b) ingot preheating, melting, and holding; c) casting into precision dies; d) finishing (i.e., fettling and grinding); e) quality control, with an average product reject rate of 3%; f) and transportation and assembly [4].

Table III shows data comparing different raw magnesium extraction processes for a simplified calculation of GWP per kg of magnesium less costly cast product.

Table III. Environmental Impact of Usual Magnesium Component ("Cradle-to-Gate" path) Expressed in GWP per kg of Magnesium Manufactured.

Plants	Primary Mg Production Manufacturing and Total and Alloy Making (*) [kg CO_2eq /kg]	Manufacturing and assembly (**) [kg CO_2eq /kg]	Total cradle-to-gate [kg CO_2eq /kg]	References
Hydro Magnesium (I)	16.1	19.4	35.5	Tharumarajah et al.
Pigdeon (China) (II)	43.3	19.4	62.7	Tharumarajah et al.
AM (Australia) (III)	27.9	19.4	47.3	Ramakrishnan et al.
Bolzano (Brazil) (IV)	13.8	19.4	33.2	Cherubini et al.
Magnetherm (France) (V)	17.6	19.4	37.0	Cherubini et al.

Footnotes:
(I) Recycling ratio: 70% primary: 30% secondary. All upstream stages included in alloy making. Use of SF_6 as cover gas.
(II) Mg ingots produced in China. For manufacturing and assembly stages: Mg ingots transported to North America or Europe for AZ91D alloy making.
(III) Electrolytic process employed in Australia (now ceased, but an interesting case for comparison). HFC-134a used as cover gas instead of SF6. Alloy making by AM-Cast.
(IV) Current mix for electricity and natural gas for plant operations in Brazil. Alloy making as for the AM-Cast process.
(IV) Thermal process formerly used in France up until 2001. Alloy making as for AM-Cast process.
(*) For comparison: GWP values for steel and aluminum production (not comprehensive of alloy making) are 2.3 kg CO_2 eq / kg steel and 22.4 kg CO_2 eq / kg aluminum respectively.
(**) GWP of common manufacturing steps derived from magnesium product LCA [4]; manufacturing steps include preheating of magnesium alloy ingots, melting, precision sand casting, fettling in a common open-loop recycling cycle.

[8] In material flow, 80% of the recycling material has been assumed to be from scraps, rejects, and fettling waste.

Table IV shows the calculation of the CO2 cut over the exit-gate-to-grave path, based on the general assumptions reported in Table II.

Table IV. Percentage Reduction of CO2 Estimated with the Increasing Usage of Conventional Magnesium-alloys for the Manufacturing of Components Substituting Heavy Materials on Medium-size Passenger Vehicles.

Features	Scenarios			
	I	II	III	IV
Magnesium usage (kg)	93	125	219	256
Weight reduction (kg)	60	80	140	165
Total GWP of Bolzano route (kgCO_{2eq})	-926	-176	2,004	3,405
Total percentage of CO_2 cut$^{(*)}$	*-1.9%*	*-0,4%*	*4%*	*7%*

(*) The percentage is calculated over the total amount of 48,450 KgCO_2 estimated for 200,000 km average travel-life distance.

As Friedrich and Schumann [5,6] stated in their early study on the potential uses of magnesium components for Volkswagen vehicles, a total average usage of around 150 kg could potentially be achieved in upcoming years. This value was actually close to the 154 kg targeted by a report issued in 2004 by the United States Automotive Materials Partnership (USAMP), the Automotive Metals Division (AMD) for the United States Council for Automotive Research (USCAR) entitled "Magnesium Vision 2020: A North American Automotive Strategic Vision for Magnesium" [7].

Taking into consideration the "unclean side" of the product cycle, around 150 kg of magnesium usage would be not sufficient for substantial benefits in terms of CO_2 percentage reduction, as shown in Table IV.

The challenge suggests the implementation of a double correlated target:
a) from the alloy-design side, it is mandatory to reduce the substitution ratio (namely, usage ratio, refer to Table 2) from 1 to 2.64, switching to a 1 to 1.8, in order to reduce mass part of magnesium component substituting heavy materials;
b) and from the process-design side, it is mandatory to drastically reduced the carbon-footprint of the unclean route steps needed in the manufacturing process of magnesium components.

While we think these two statements should be considered fundamental to the realization of novel, eco-sustainable magnesium-based alloy development [8,9], in the following we mainly focus on the alloy-design side issue.

The Alloy-Design Approach: The Environmental Problem Targeted

As previously explained and illustrated in Table IV, it is possible to refine the carbon-footprint of the entire life product cycle and decrease the amount of CO_2 by reducing the use of heavy materials. This assumption is tied to the strategy of alloy-design that should lead to a substantial reduction of the substitution ratio for low-carbon steel (namely, usage ratio) from 1: 2.65, as shown in Table 2, to a more effective 1: 1.8 value, calculated in Table V.

Table V. Mg-alloy Design and Weight Saving by Substitution of Medium Carbon Steel with High-Performance Novel Mg-Based Alloys.

Features	Mg	Steel
Density (kg/dm^3)	1.81	7.87
Resistance limit (MPa)	250	450
Usage ratio	1.8	1.0
Volume (mm^3)	180	100
Mass total (kg)	3.3	7.9
Total weight saving if 100 kg steel is substituted with magnesium High-Performance Novel Mg-Based Alloys alloy (kg)	39	

(*) Mg-alloy Design and Weight Saving by Substitution of Medium Carbon Steel with High-Performance Novel Mg-Based Alloys.

Based on the LCA of the entire magnesium product manufacturing cycle, a positive balance of emissions over cradle-to-grave is achievable using a highly environmentally friendly and energy-saving program. There can be a drastic cut of $kgCO_2eq$ emitted during the manufacturing steps by simply switching from 19.4 $kgCO_2eq$ in Table 3 to a value close to 1-2 $kgCO_2eq$ per kg of processed magnesium, primarily attributed to the use of a SF_6-free route and net-finishing shaping that eliminates additional machining and finishing. The total GWP could be decreased even further by using fully-electrical machines powered by renewable sources which is produced on-site or centralized, and by increasing the market share of the recycled magnesium. As a cleaner manufacturing cycle could be adopted, the GWP of cradle-to-exit gate cycle is lowered considerably. If approximately 175 kg of high performance materials are used onboard, around 11% of CO_2 can be saved on ICE vehicles, as shown in Table VI.

Table VI. – Percentage Reduction of CO_2 Estimated with the Increasing Usage of High-Performance Magnesium-Alloys Produced with Lower Environmental Impact Manufacturing Cycle.

Features	Scenarios			
	I	II	III	IV
Magnesium usage (kg)	63	85	149	175
Weight reduction (kg)	59	121	211	247
Total GWP of Bolzano route (kgCO2eq)	1,260	3,556	6,891	5,592
Total percentage of CO2 cut(*)	2.6%	7.3%	14.2%	11.5%

(*) The percentage is calculated over the total amount of 48,450 $KgCO_2$ estimated for 200,000 km average travel-life distance.

Using this premise, the way to reduce CO_2 is to evaluate the performance of magnesium-based alloys and target its most effective usage, and do it in the most environmentally friendly process possible. Electric or hybrid motors can also benefit from the use of magnesium structural components. The potential quantity of magnesium that can be used in electric cars is high because they do not reach the high temperatures that you find on ICE motors. This is a very interesting perspective as a relevant reduction in weight would greatly enhance the distance that vehicles powered with electric batteries can travel, which is currently the main objection to their wide spread use.

Integration of Process-Design and Alloy-Design Strategies for Lowering Carbon Footprint in Transport

Reducing the grain-size of metals is the key to improve mechanical properties of metal alloys in general. Nanostructured materials in particular (i.e. less than 1 micron grain size) achieve the highest performances allowed for a metallic material, with increased strength, toughness, corrosion resistance, and creep resistance. Various laboratory-scale methods exist to produce ultrafine structures through top-down routes, namely, solidification grain refining, severe plastic deformation of billet, etc., but nanostructured materials cannot be produced for large parts and no examples exist of industrial process lines for the fabrication of cost-effective components made of nanostructured metallic materials.

Currently, there are two approaches to produce bulk nanostructured materials:

- TOP-DOWN, consists of breaking down the bulk material into particles of nanometer size grain (in metals, e.g.: mechanical alloying, equal channel angular pressing, high-energy ball milling, accumulative roll bonding, high-pressure torsion).
- BOTTOM-UP, in which individual atoms or clusters are put together to form required nanoparticles (e.i: gas-phase synthesis, inert gas condensation, physical/chemical vapor condensation, plasma synthesis, etc.).

Extensive research has been conducted at CENIM [10-15] on ultrafine and nano-grain size magnesium-base alloys directly produced from a solid state (from Rapidly Solidified Powders, RSP), proving the effectiveness of Mg-based direct-powder extrusion. The effect of extrusion temperatures on the final microstructure and mechanical behavior for various Mg-based bi-phase systems has been extensively investigated. It was found that warm (below 0.5 Tm) extrusion of Mg-based alloys produced using the RSP route showed that the lower the compacting extrusion temperature, the higher the yield stress of monolithic product.

Several Mg-based bi- and ternary-phases systems were studied and optimized at CENIM. Generally speaking, microstructures provide high strength as RSPs are compacted by the extrusion process at around 250 °C. This is the temperature at which the inner shearing realized during extrusion is sufficient to finely distribute second phase particles. As very fine particles homogeneously dispersed in the Mg-matrix, a further strengthening mechanism begins, reinforcing the grain-refinement strengthening mechanism. For example, the MgZn3.3Y0.43 alloy showed a yield stress of 410 MPa with 12% elongation [10]; this effect agrees with the general Hall-Petch equation valid for metals associated with the metallurgical phenomenon that produces mechanical properties according to grain-size refinement. Such high grain-refining leads to further benefits in terms of increased physical properties related to corrosion and creep resistance, and formability. At medium-low temperatures RSPs compacted by warm-extrusion exhibit superplastic behavior. These temperatures can vary from 200°C to 400°C, depending on the metallurgical Mg-system and the targeted results, as well as at high strain rates, attaining elongations to failure around 500% [15]. Generally speaking, impressive mechanical behavior was obtained at CENIM for this new process of magnesium alloy, as reported in Table VII.

Table VII. – Main key-features relative to ultra-fine powder warm extrusion of dual-phase magnesium alloys [10] versus average values for conventional steels.

Mg-based High Performance Novel Alloys from PWD metallurgy	Low-carbon steels	Medium carbon steels
up to 350 MPa YS	from 300 – 400 YS MPa	from 550 – 700 YS MPa
about 400 MPa UTS	from 500 – 600 UTS MPa	from 600 – 850 UTS MPa
elongation about 13%	elongation about 20%	elongation about 15%

Basing on such previous experience, the authors are currently involved in the European Project titled "Green Metallurgy 2020"[9], a 2 ML€ project financed by European Community as one of the successful proposals submitted to the EU LIFE+ 2009 Program. This 3-years project officially started on the 1th of September 2010 and it aims to overcome main historic barriers that currently prevent magnesium from widely being applied as substitute ultra-lightweight material for the transport sector. The main scope of this project is to industrially scale to the production plant such impressive results obtained in laboratories, in order to commercially introduce a no-melting route that will be characterized by very low carbon footprint. This outcome shall be realized through application of specific technological steps tested for several years in CENIM's laboratories and engineered with the support of the Politecnico di Milano's Advanced Material Research Group in order to figure out how the basic process steps can be integrated in a single or in a double-step plant machinery. This fact is actually fundamental to greatly reduce the high costs related to magnesium part manufacturing using conventional melting processes. According to preliminary experiments, grain coarsening can be avoided. This prevents decaying of final mechanical properties of finished components thanks to maintain the ultrafine grain-size of the precursor material. As the historic barriers for high-performance usage of magnesium-base materials are removed, the magnesium industry will be able to increase their market share by capturing segments of the market now dominated by heavier materials like aluminum alloys and steels.

References

1. IPCC Fourth Assessment Report: Climate Change 2007 (AR4)

2. ACEA, European Automobile Industry Report, 2009/10 (http://www.acea.be/index.php)

3. L. Ridge, EUCAR - Automotive LCA Guidelines - Phase 2, In: Total Life Cycle Conference and Exposition; Graz, SAE paper 982185, 193-204 (1998)

4. A. Tharumarajah, and P. Koltun, "Is There an Environmental Advantage of Using Magnesium Components for Light-Weighting Cars?" *Journal of Cleaner Production*, 15 (2007), 1007–1013

5. H. Friedrich, S. Schumann, "The second age of magnesium" (Paper presented at the Second Israeli International Conference, Sdom, Israel, February 2000)

6. H. Friedrich, S. Schumann, "Research for a new age of magnesium in the automotive industry", Journal of Material Process Technology, 117 (2001), 276–281

7. Source USAMP report on website of USCAR (last check June 2010:http://www.uscar.org/commands/files_download.php?files_id=99.)

8. F. D'Errico, B. Rivolta, R. Gerosa, and G. Perricone, "Thixomolded Magnesium Alloys: Strategic Product Innovation in Automobiles", *Journal of the Minerals, Metals and Material Society*, 2008, no 11: 70-76

9. F. D'Errico, G.Perricone, and R.Oppio, "A new integrated lean manufacturing model for magnesium products", *Journal of the Minerals, Metals and Material Society*, 2009, no 4: 14-18

10. E. Mora, G. Garces, E. Onõrbe, P. Perez and P. Adeva, "High-strength Mg–Zn–Y alloys produced by powder metallurgy", *Scripta Materialia*, Volume 60, 2009, 776–779

11. P. Perez, G. Garces, F. Sommer, and P. Adeva, "Mechanical properties of amorphous and crystallised Mg-35 wt.% Ni", *Journal of Alloys and Compounds*, Volume 396, Issues 1-2, 2005, 175-181

12. P. Perez, S. Gonzalez, and G. Garces, "Influence of partial replacement of cerium-rich mischmetal (CeMM) by yttrium on the crystallization and mechanical properties of amorphous $Mg_{80}Ni_{10}CeMM_{10}$ alloy", *Intermetallics*, Volume 15, Issue 3, 2007, 315-326

13. P. Perez, M. Eddahbi, G. Garces, F. Sommer, and P. Adeva, "Mechanical properties of crystallised amorphous Mg-23.5Ni(wt.%) alloy", *Scripta Materialia*, Volume 50, Issue 7, 2004, 1039-1043

14. M. Eddahbi, P. Perez, M.A. Monge, G. Garces, R. Pareja, and P. Adeva, "Microstructural characterization of an extruded Mg-Ni-Y-RE alloy processed by equal channel angular extrusion", *Journal of Alloys and Compounds*, Volume 473, Issues 1-2, 2009, 79-86

15. P. Perez, S. Gonzalez, G. Garces, G. Caruana, and P. Adeva, "High-strength extruded $Mg_{96}Ni_2Y_1RE_1$ alloy exhibiting superplastic behavior", *Materials Science and Engineering: A*, Volume 485, Issues 1-2, 2008, 194-199

[9] This project is in accordance bases on the project proposal finally approved on 31 May 2010 for the co-financing by the EU Commission, Directorate of the LIFE+ Program.

Magnesium Technology 2011
Edited by: Wim H. Sillekens, Sean R. Agnew, Neale R. Neelameggham, and Suveen N. Mathaudhu
TMS (The Minerals, Metals & Materials Society), 2011

FRACTURE MECHANISM AND TOUGHNESS IN FINE- AND COARSE-GRAINED MAGNESIUM ALLOYS

Hidetoshi Somekawa, Alok Singh, Toshiji Mukai
Structural Metals Center, National Institute for Materials Science, 1-2-1 Sengen, Tsukuba, Ibaraki 305-0047 Japan

Keywords: Toughness, Magnesium, Fracture mechanism, Twins

Abstract

The fracture mechanisms in the extruded magnesium alloys with two different grain sizes, 2 and 50 μm, were investigated by SEM, TEM and EBSD microstructural observations. The coarse-grained alloy showed that the {10-12} type deformation twins formed at the beginning of test, and the crack was propagated into the boundaries between twins and matrix. On the other hand, the fine-grained alloy showed that the sub-grain boundaries formed instead of the deformation twins. No formation of twins at the early deformation stage causes a crack-tip blunting, and thus, the fracture toughness has a high value.

Introduction

Magnesium alloys have a high potential for application as structural materials because of being the lightest among all structural alloys in use. For use in structural applications, their mechanical properties must satisfy both reliability and safety requirements. A method of ensuring that is to investigate their fracture toughness. Several reports exist on the fracture toughness of magnesium [1] and magnesium alloys [2,3], and the wrought magnesium alloys have higher fracture toughness compared to the cast magnesium alloys [4]. However, the fracture toughness in magnesium alloys has generally been reported to be lower than that in aluminum alloys [4].

The fracture toughness in magnesium and magnesium alloys is reported to be affected by several microstructural features, such as texture [5] and dispersion of particle morphologies (particle shapes [6], sizes [6,7], volume fraction [8] and interfaces [4,9]). The fracture toughness is also influenced by the grain size [1,10]. Above all things, since the grain refinement enhances not only the fracture toughness but also changing the fracture features, the control of grain size is known to be the most effective method to improve the mechanical properties. One of the reasons for the enhancement fracture toughness is that the fine-grained magnesium alloys tend to have a decrease in the formation of deformation twins, which is the origin of fracture [11]; the dominant plastic deformation mechanism changes from twins to dislocation slip by the refinement of the grain structures [12,13]. However, the detailed mechanisms during the fracture toughness test in fine- and coarse-grained magnesium alloys have not been understood yet. Therefore, in this study, the deformed microstructure of fracture toughness tested fine- and coarse-grained magnesium alloys were investigated using scanning electron microscopy (SEM), transmission electron microscopy (TEM) observations combined with the focused ion beam (FIB) and electron backscattered diffraction (EBSD).

Experimental Procedure

The materials used in the present study were extruded Mg-6.2wt.%Zn and Mg-3wt.%Zn-1wt.%Al (AZ31) alloys. The grain sizes of Mg-Zn alloy and AZ31 alloy were 2 and 50 μm, respectively, which consisted of high-angle grain boundaries. The detail initialed microstructures and mechanical properties in each alloy have been already reported elsewhere [8,14].

The specimen was a three point bending sample with a width of 10 mm, a thickness of 5 mm and a length of 44 mm, based on ASTM E399 [15]. The specimen was machined directly from the extruded bar with parallel to extrusion direction. The fracture toughness tests were discontinued, when the load reached around half of the maximum value or the maximum value, in order to obtain the deformed samples. The deformed microstructures were examined at the crack-tip region by EBSD in the coarse-grained alloy and TEM observation in the fine-grained alloy, respectively. The TEM samples were picked up by the FIB at the crack-tip region of 5 × 10 μm. The fracture surfaces after the fracture toughness tests were also observed by SEM in the both alloys.

Results and Discussion

Coarse-grained Alloy

Typical EBSD analysis near the crack-tip region in the coarse-grained alloy is shown in Figure 1: (a) the sample where the test was stopped at around half of the maximum load and (b) the sample where the test was stopped at the maximum load. Figure 1 shows that many deformation twins were formed near the crack-tip region even before the crack-propagation, i.e., before the crack blunted. The deformation twins are the {10-12} type, <*c*>-axis tensile deformation twins. The same type of deformation twins has also been observed in samples, deformed by the tensile and the compression tests [16,17]. One of the possible reasons for the easy formation of {10-12} deformation twins compared to the other types of deformation twins, i.e., {10-11} and {10-13} deformation twins (<*c*>-axis compression deformation twins) is due to the lower critical resolved shear stress (CRSS) [18,19]. Figure 1(b) shows that, in the sample where the test was stopped at the maximum load, the crack propagates into the boundaries between the deformation twins and the matrix, which is marked by the arrows.

Fine-grained Alloy

A typical TEM micrograph at the crack-tip region in the fine-grained alloy is shown in Figure 2: (a) around half of the maximum load and (b) maximum load in the fracture toughness tests. Selected area electron diffraction (SAED) patterns inset in each image show streaked or split spots, suggesting a local variation in orientation and the presence of sub-grain structures. The microstructural features of the deformed sample are very different from that before the testing. Since the as-extruded alloy is a fully recrystallized structure with high-angle grain boundaries [8], the sub-grain structures are found to form during the crack-blunting. No deformation twins are found in Figure 2(a). On the other hand, Figure 2(b) shows not only the existence of many fine sub-grain structures, but also the {10-12} deformation twins, which is a well known type and similar to Figure 1. This observed deformation twins are nano-scale size.

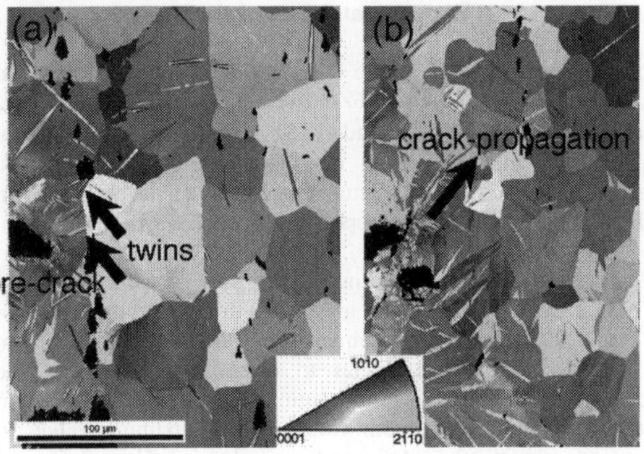

Figure 1. Typical EBSD images in the coarse-grained alloy: (a) half of maximum load and (b) maximum load in the fracture toughness tests.

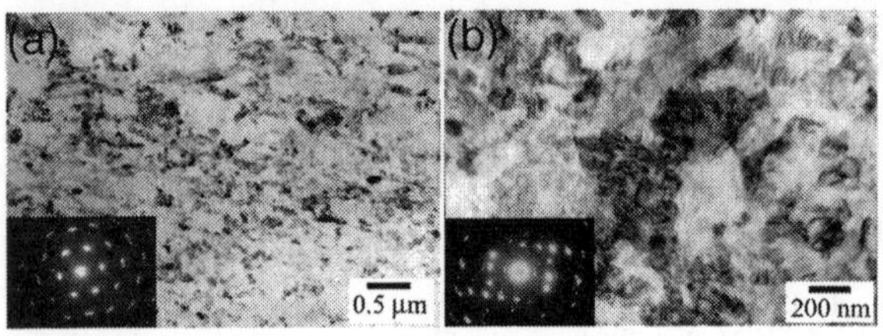

Figure 2. Typical TEM micrograph in the fine-grained alloy: (a) half of maximum load and (b) maximum load in the fracture toughness tests.

Fracture Mechanisms

Typical SEM micrographs of the fracture surface after the fracture toughness test are shown in Figure 3: (a) coarse- and (b) fine-grained alloys, respectively. The pre-crack direction is marked in this figure. The coarse-grained alloy shows that the fracture surface consists of a few parts with ductile patterns but is mainly comprised of brittle-like patterns related to the twin boundary fracture, marked by white arrows. On the other hand, the fine-grained alloy shows many dimple patterns, which is a typical trace of ductile fracture.

The effect of grain size on the fracture mechanism is discussed hereafter. Simple illustrations are shown in Figure 4: (a) coarse- and (b) fine-grained alloys. On applying a stress, a plastic zone is created ahead of the crack-tip, because the stress concentration occurs at the crack-tip. When the grain size is the coarse or medium ($\geq \sim 10$ μm), the stress to form the twins is smaller than that for dislocation slip: the dominant deformation mechanism is twinning. Thus, the deformation twins quite easily form at the very beginning of the fracture toughness test. The twin boundary plays the role of strain accumulation [14]; however, since the accumulated plastic strain by the twins is smaller than that of the grain boundary migration by the dislocation slip, the twins are associated with the crack propagation route. Therefore, as soon as the deformation twins are formed ahead of the crack-tip, the crack is propagated along the twins without blunting, i.e., no formation of a void, and the fracture morphology is of a brittle feature (shown in Figure 3(a)).

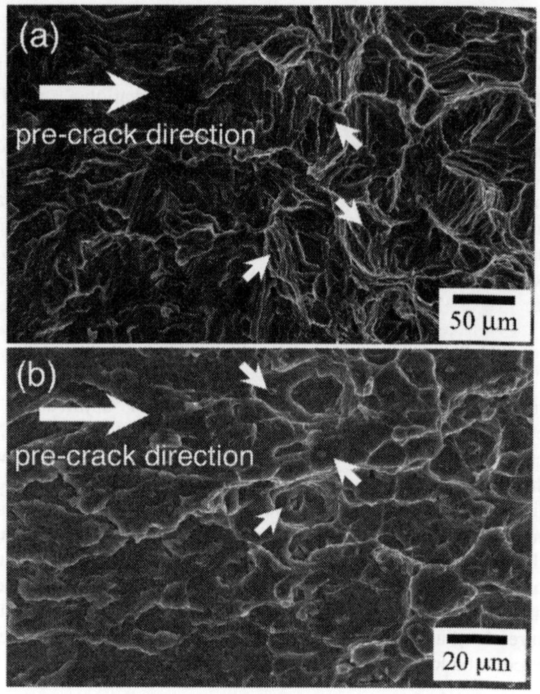

Figure 3. Typical SEM micrographs of the fracture surface after the fracture toughness test: (a) coarse- and (b) fine-grained alloys.

On the other hand, in fine-grained materials, since the volume fraction of the grain boundary increases with grain refinement, the deformation behavior is affected by the grain boundary characteristics. Non-basal slip systems are activated even at room temperature due to the operation of compatibility stress at grain boundary [20]. In addition, the grain boundary sliding occurs due to high diffusion rate along grain boundaries [21]. These deformation mechanisms make up for the lack of slip systems at the beginning of the test. As the fracture toughness test proceeds, the crack-tip blunts sufficiently and causes high fracture toughness and a ductile fracture (shown in Figure 3(b)), because of absence of the crack-propagation sites, such as the twin boundaries. On applying a stress, since a large stress operates at the crack-tip, the twins form as nano-scale size. The deformation twins of a micron-scale size in metallic materials are the origin of crack-nucleation; however, the existence of nano-scale twins is reported to contribute to the enhancement of strength and ductility [22]. The nanocrystalline Ni alloys with a grain size of ~ 10 nm are also observed within the intergranular fracture [23], the fracture surface by the SEM observation of these alloys appears a void-like fracture, i.e., ductile feature [23,24]. Therefore, the size of the deformation twins is assumed to be too fine not only to affect the fracture mechanism but also for the brittle fracture to occur.

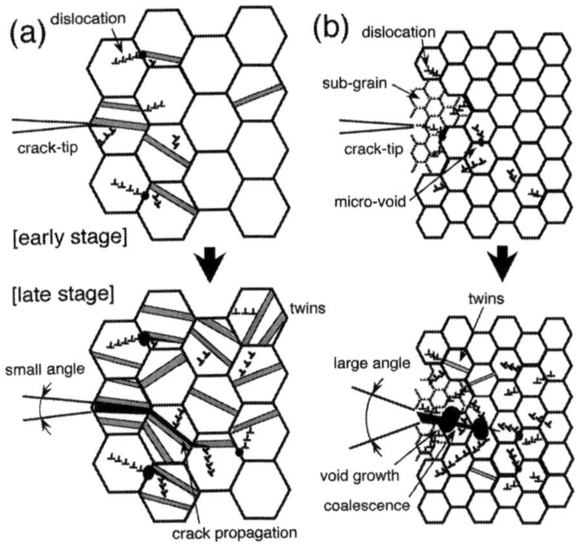

Figure 4. Illustrations of fracture mechanism: (a) coarse- and (b) fine-grained alloys.

Summary

The effect of grain size on fracture mechanisms was investigated by SEM, TEM and EBSD microstructural observations. The following results were obtained.

1) The coarse-grained alloy showed that the {10-12} type deformation twins formed at the beginning of test. On applying a strain, the crack propagated into the boundaries between twins and matrix.

2) The fine-grained alloy showed that many fine sub-grains formed instead of deformation twins. Non formation of deformation twins at the beginning of test caused sufficient crack-tip blunting; therefore, the fracture toughness had a high value.

Acknowledgement

This work was partly supported by JSPS Grant-in-Aid for Young Scientists (B) Grant No.21760564.

References

1. H. Somekawa, T. Mukai, "Effect of grain refinement on fracture toughness in extruded pure magnesium", *Scripta Mater.* 53 (2005), 1059.

2. T. Sasaki, H. Somekawa, A. Takara, Y. Nishikawa, K. Higashi, "Plane-strain fracture toughness on thin AZ31 wrought magnesium alloy sheet", *Mater. Trans.* 44 (2003), 986.

3. ASM Specialty Handbook, Magnesium and magnesium alloys, (Materials Park, OH, ASM International, 1999).

4. H. Somekawa, A. Singh, T. Mukai, "High fracture toughness of extruded Mg-Zn-Y alloy by the synergitic effect of grain refinement and dispersion of quasicrystalline phase", *Scripta Mater.* 56 (2007), 1091.

5. H. Somekawa, T. Mukai, "Effect of texture on fracture toughness in extruded AZ31 magnesium alloy", *Scripta Mater.* 53 (2005), 541.

6. H. Somekawa, A. Singh, T. Mukai, "Effect of precipitate shapes on fracture toughness in extruded Mg-Zn-Zr magnesium alloys", *J. Mater. Res.* 22 (2007), 965.

7. H. Somekawa, T. Mukai, "High strength and fracture toughness balance on the extruded Mg-Ca-Zn alloy", *Mater. Sci. Eng.* A459 (2007), 366.

8. H. Somekawa, Y. Osawa, A. Singh, T. Mukai, "Effect of precipitate volume fraction on fracture toughness of extruded Mg-Zn alloys", *J. Mater. Res.* 23 (2008), 1128.

9. H. Somekawa, A. Singh, Y. Osawa, T. Mukai, "High strength and fracture toughness balances in extruded Mg-Zn-RE alloys by dispersion of quasicrystalline phase particles", *Mater. Trans.* 49 (2008), 1947.

10. H. Somekawa, H. S. Kim, A. Singh, T. Mukai, "Fracture toughness in direct extruded Mg-Al-Zn alloys", *J. Mater. Res.* 22 (2007), 2598.

11. H. Somekawa, K. Nakajima, A. Singh, T. Mukai, "Ductile fracture mechanism in fine-grained magnesium alloy", *Philo. Mag. Lett.* 90 (2010), 831.

12. M. A. Meyers, O. Vohringer, V. A. Lubarda, "The onset of twinning in metals; A constitutive description", *Acta Mater.* 49 (2001), 4025.

13. M. R. Barnett, "A rationale for the strong dependence of mechanical twinning on grain size", *Scripta Mater.* 59 (2008), 696.

14. H. Somekawa, A. Singh, T. Mukai, "Fracture mechanism of a coase-grained magnesium alloy during fracture toughness testing", *Philo. Mag. Lett.* 89 (2009), 2.

15. ASTM E399, Standard Test Method for Plane-Strain Fracture Toughness of Metallic Materials, American Society for Testing and Materials (West Conshohocken, PA).

16. J. Koike, "Enhanced deformation mechanisms by anisotropic plasticity in polycrystalline Mg alloys at room temperature", *Metall. Mater.* 36 (2005), 1689.

17. S. R. Agnew, O. Duygulu, "Plastic anistotropy and the role of non-basal slip in magnesium alloy AZ31B", *Int. J. Plast.* 21 (2005), 1161.

18. R. E. Reed-Hill, W. D. Robertson, "The crystallographic characteristic of fracture in magnesium single crystals", *Acta Mater.* 5 (1957), 728.

19. H. Yoshinaga, R. Horiuchi, "Deformation mechanisms in magnesium single crystals compressed in the direction parallel to hexagonal axis", *Mater. Trans.* 4 (1963), 1.

20. J. Koike, T. Kobayashi, T. Mukai, H. Watanabe, M. Suzuki, K. Maruyama, K. Higashi, "The activity of non-basal slip system and dynamic recovery at room temperature in fine-grained AZ31B magnesium alloys", *Acta Mater.* 51 (2003), 2055.

21. H. Somekawa, T. Mukai, "Nanoindentation creep behavior of grain boundary in pure magnesium", *Phil. Mag. Lett.* 90 (2010), 883.

22. L. Lu, Y. Shen, X. Chen, L. Qian, K. Lu, "Ultrahigh strength and high electrical conductivity in copper", *Science*, 304 (2004), 422.

23. H. Li, F. Ebrahimi, "Ductile-to-brittle transition in nanocrystalline metals", *Adv. Mater.* 17 (2005), 1969.

24. A. Hasnaoui, H. V. Swygenhoven, P. M. Derlet, "Dimples on nanocrystalline fracture surfaces as evidence for shear plane formation", *Science*, 300 (2003), 1550.

Magnesium Technology 2011

Primary Production; Characterization and Mechanical Performance

Session Chairs:

Neale R. Neelameggham
(US Magnesium LLC, USA)

Adam C. Powell
(Metal Oxygen Separation Technologies Inc., USA)

Magnesium Technology 2011
Edited by: Wim H. Sillekens, Sean R. Agnew, Neale R. Neelameggham, and Suveen N. Mathaudhu
TMS (The Minerals, Metals & Materials Society), 2011

THE DEVELOPMENT OF THE MULTIPOLAR MAGNESIUM CELL: A CASE HISTORY OF INTERNATIONAL COOPERATION IN A COMPETITIVE WORLD

Dr. Olivo G. Sivilotti
Sigma Tau Technologies Inc. 185-1108 Ontario St., Kingston, Ontario, K7L 2Y7, Canada

Keywords: Magnesium, Electrolysis, Multipolar

Abstract

The author conceived the first multipolar magnesium electrolytic cell in the late 1970s, in order to offset the impact of an 80kA monopolar cell developed in cooperation with Osaka Titanium Company (OTC) being licensed by Alcan to a major competitor of OTC. During the 1980s the commercial value of the multipolar magnesium cell technology was established. The 1990s saw significant further progress by OTC, while attempts by Alcan to commercialize it in magnesium production plants met with technical difficulties and failure, due primarily to the lack of magnesium chloride feed of adequate quality. Now the author plans to commercialize a new multipolar cell design with a target productivity of 8-10 tons/day and a unit power consumption of 8.5-10 KWHR/kg of magnesium.

Introduction

In 2002 Byron Clow, a retired executive vice-president of the International Magnesium Association, reviewed the history of world magnesium production [1]. He listed the electrolytic magnesium plants being closed in the previous ten years due to a twenty-fold increase in production out of China based on the Pidgeon process. Today, only a handful of electrolytic plants are still in operation. While this evidence is compelling, the magnesium industry has seen dramatic swings in production volumes and technologies before, due to the strategic importance of magnesium for military uses and to alternative process challenges. More recently, for the years 2007, 2008 and 2009, the world economic downturn is reflected in primary annual magnesium production worldwide of 792, 722 and 615 kilotons, of which 627, 559 and 470 kilotons originated from China [2].

Bob Brown illustrates the technology issues in a well researched discussion of the techno-economic scenarios [3]. Having introduced Mr. Brown to a ninety-year-old Dr. Lloyd Pidgeon, I still recall how fascinated I was by the quick and passionate responses of the elder scientist to Mr. Brown's questions. Clearly, if you get the magnesium bug at an early stage in your life, it stays with you forever.

As a freshly minted engineer from Italy, in 1957 I was assigned to trouble-shoot technological problems in an under-performing electrolytic magnesium plant at Alcan in Arvida, Quebec, Canada. In less than a year, the 32kA electrolytic cells were performing well and I was asked to estimate how long it would take to fix the chlorinators next door. On my estimate that it would take at least five years, Alcan decided to strike a deal with Dow. Its budding magnesium operation in Arvida was shut down and Alcan became the agent for Dow, selling Dow magnesium all over the world for many years.

Soon afterwards the opportunity arose to continue development activities to improve the Alcan cell design, which as the first diaphragm-less cell was an offspring of the classical I. G. Farben cell, with North American connotations. The activity was carried out *sub-rosa,* without an official budget and without time-sheets. The skunk works consisted of *ad-hoc* meetings at Place Ville Marie with two friends, John Taylor and Meine Vandermeulen, whenever I was in Montreal for other business. We were called the Magnesium Mafia (Tom Anderson was the Godfather, I was the *consigliere* and John and Meine the hit men). As the world was struck by the first oil shock in the seventies, the stage was set for dramatic improvements to occur in cell productivity and energy efficiency. The Alcan multipolar cell was the product of these converging realities and of the cooperation with a Japanese company interested in recovering magnesium and chlorine from the production of titanium sponge by the Kroll process

Enter Osaka Titanium Company

In 1960, eight of the fifty four 32kA Alcan cells from Arvida found a new home in Japan at the Amagasaki plant of the Osaka Titanium Co. (OTC), a young company in the Sumitomo Group. I went to help in the start up of the magnesium plant at the end of 1961 and I experienced firsthand the technical challenges of a plant start-up together with the hospitality and friendship of the Japanese people. Two persons who stood out were Dr. Toshio Noda, a ceramic art collector at heart and the mover and shaker of the Japanese team, and Junkichi Iseki, a young process engineer who was learning the trade and who was to become the Strait of Gibraltar, through which all subsequent technical developments had to pass. These friends and this first plant start-up experience had a lasting effect on the rest of my life.

After two years of monthly reports from Japan and review notes from Canada, the Japanese team was on their own. By the early seventies, they had managed to expand production by increasing the cell amperage to 40kA and by adding a row of new cells supplied by a second rectifier, but housed in the same cellroom. However, power consumption had also increased to 20 KWHR/kg of magnesium and this meant big trouble for the bottom line of OTC when the *oil-shokku* hit Japan.

Meanwhile, I provided technical support to engineering activities and to another magnesium plant start-up in 1967 by Oregon Metallurgical in Albany, Oregon, a new licensee of the Alcan cell. Following that event, physical and mathematical modeling suggested to us that the power consumption could be reduced to 13.5-14.5 KWHR/kg by increasing the amperage to between 80 and 120kA and by tapering the anodes [4]. A new 100kA cell design was prepared for a Greenfield magnesium project under study by Shell/Billington in the Netherlands, based on deep-well brines discovered in Friesland. The Dutch project never materialized, and the new cell design was presented to Norsk Hydro, who intended to revamp their I. G. Farben technology in Posgrun, Norway. A license deal with Norsk Hydro was never sealed.

On a courtesy visit to OTC, I noticed the two 40kA bussbar loops in their expanded cellroom and proposed to OTC to join these loops together in order to provide a site for a scaled-down 80kA tapered-anode pilot cell. This was done successfully and in 1975 the cell produced magnesium, as per design, at less than 14 KWHR/kg [5]. Other improvements, such as the introduction of a heat exchanger, made its operation easier to control [6].

OTC was building a 100kA cellroom when Timet, their main competitor in titanium sponge production, asked Alcan to license the new 80kA cell to upgrade their WWII-era cellroom in Henderson, Nevada, which was still operating with the original 20kA I.G. Farben cells. In January 1976 I took the Timet team through the Amagasaki plant to show them the 80kA pilot cell in operation. They decided that this design was what they wanted but, as expected, OTC was reluctant to cooperate, fearing the loss of their competitive advantage. During a sleepless jet-lagged night at the Royal Osaka Hotel, I drew up the first sketch for a multipolar cell as shown in Figure 1. In a private session with OTC the next morning, I proposed to work together with them on a new cell based on this idea. It is a sign of trust and of courage that Dr. Noda and his team accepted the proposal and released their support to Alcan for licensing the tapered anode cell to Timet.

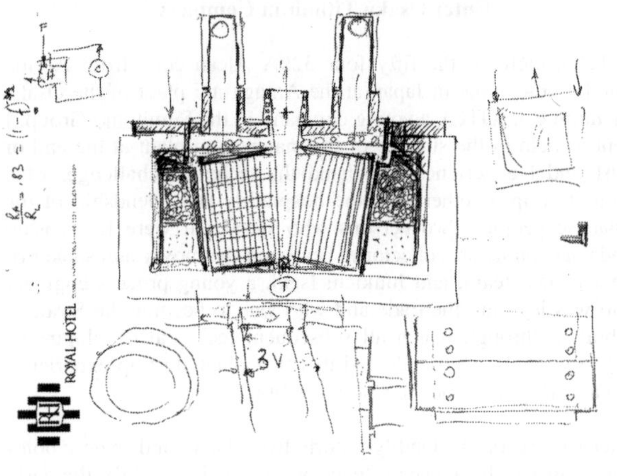

Figure 1. First sketch of a multipolar magnesium cell.

It took until July 1979 to successfully negotiate an agreement on the multipolar cell, but things moved very quickly after the agreement was signed. The events are described in detail elsewhere [7-8]. The design and development activity is documented in several patents [9-13]. The first multipolar cell was started in October 1980 as a retrofitted 40kA cell (AMC1). Twelve other experimental cells at 40kA and one at 100kA were run in 1981 and the 100kA pilot cell was operated while a new magnesium plant rated at 100kA was being built. The new plant adopted the multipolar cell design (AMC3) and started in February-March 1982. The results were a dramatic improvement over the performance obtained with the tapered anode cell, as shown in Table I. In November 1984 the test period was terminated, the objectives of the multipolar cell program having been fully met or surpassed [8].

Table I. Comparative Performance of Magnesium Electrolytic Cells.

Cell type	Power consumption KWHR/kg	Productivity of cell t/day	Campaign Life of cell years
Monopolar			
Alcan 40kA	20	0.4	2
Alcan Tapered	14	0.8-1.1	3
Norsk Hydro	13	4.0	4
MagCorp	12.6	2.8	N/A
UKTMP	13.2	1.8	N/A
Multipolar			
AMC1	16	0.8	N/A
AMC3	10.5	3	N/A
SMC3*	8.5-10	8-10	>5

*Estimated

A joint development program started again in 1989 to develop with OTC a new multipolar cell with removable electrodes [14]. A cell design [15] was finally agreed upon and a full scale pilot test was organized by retrofitting one of the 100kA, AMC3 cells. The first experiment failed because of shorting problems and a second almost had the same fate, as one of the electrode assemblies refused to carry any current. I remember, on the weekend after the start up, going to a Buddhist shrine with Makoto Yamaguchi, a young process engineer in charge of the start up of the second pilot cell, and praying with him for a solution to the mysterious problem (which turned out to be very simple indeed, consisting in raising the electrolyte level to eliminate a freeze-up plug). The new design is still in operation.

In 1995, the Alcan Multipolar Magnesium Cell was awarded the International Magnesium Association Award for Process Excellence, an honor that Mr. Iseki and I accepted in Berlin, Germany, on behalf of our companies.

There is life after retirement

In 1995, at the time of my retirement from Alcan, MagCorp (now US Magnesium LLC) started a revamp project [16]. This revamp was certainly successful and is one of the reasons why US Magnesium LLC is the sole survivor in North America of the onslaught by the silico-thermic magnesium production from China. Nevertheless, it is obvious from the reasons mentioned in [16] that the step from the I.G. Farben technology to the Alcan Multipolar Cell was too risky for MagCorp to take at that time.

My main preoccupation during retirement has been for the Magnola project. After supervising the design of the "slice" for the Valleyfield pilot plant, my offers of advice (centered on the need to develop a probe to monitor on-stream the quality of the enriched electrolyte in its return path to the cell and being at the intersection between the Alcan and the Noranda technologies) found no takers.

My recommendation to Noranda to continue the operation of the Valleyfield pilot plant while the Danville, Quebec, facility was being built fell on deaf ears. Alcan was mindful that a failure of the Alcan Multipolar Cell in their backyard would be embarrassing, and could compromise any future licensing activity. As the events unfolded, the operation of the plant in Danville proved to be unsatisfactory. Based on information published by Noranda [17], the performance of the Magnola electrolysis cells can be summarized as follows.

As of Q3-2002, some 27 months after equipment commissioning was scheduled to be completed, the plant was producing at a rate of 28,000 tons/yr, against a rated capacity of 63,000 tons/yr, with 20-22 of 24 cells in operation at 65-70% of nominal current. Over the extended start-up period, progress was slowed by numerous problems involving the electrolysis cells, including a fire, electrolyte leakages, premature cell failures and damaged electrodes. A news release has also reported issues with the production of chlorinated hydro-carbons (CHCs) in excess of the levels expected, a telltale sign that the quality of the feed supplied to the cells was below the required standard.

Elsewhere, the operation of a full scale Alcan Multipolar Cell in the pilot plant of the Australian Magnesium Project was reportedly successful. Unfortunately, this project ran into financial difficulties while in the construction phase of the production plant and was suspended. I had the opportunity to meet the team at the Queensland Metals Corp. (Ian Howard-Smith, Ray Koenig) and at the CSIRO (Dr. Malcolm Frost and Bill Kreuse): great engineers and scientists, full of life, talent and vision.

Calendar time is a precious resource to mature one's thoughts and to revisit one's past experiences. Time and again, partially fulfilled objectives can be reviewed and the reasons for some limited successes can be examined. If one keeps the focus on first principles and on the vision of a logical outcome, it is very rewarding to try and solve problems without constraints of project deadlines. This challenge has kept me amused over the last fifteen years, whenever I was thinking magnesium.

The first problem I tackled was to make the electrode assemblies truly removable and replaceable, without emptying the cell. To do that, the curtain wall became a hood attached to a removable cover [18]. Cell campaign life was to be significantly extended by this method. The second challenge was to deal with the issue of ingress of ambient air via the porous graphite anodes. The proposed solution was to insert triangular anodes through the back of the cell, below the electrolyte level [19]. It is well-known that oxidation of the hygroscopic electrolyte leads to poor metal coalescence. The recent paper by Dr. Boyd Davis's team [20] and the video of their lab experiments clearly show the cathodes affected by poor metal coalescence. The trouble is that the solution of the problem proposed in [19] rendered the first objective utterly impossible to achieve.

After several tentative designs, the two conflicting objectives have been finally met in a satisfactory manner. Mathematical modeling has guided the evolution of the cell geometry from the one described in reference [14] to a new cell design, the SMC3, whose estimated performance is reported in the last row of Table I.

A new patent application covering the novel features of this cell design has just been filed as a measure of protection in support of commercialization activities now underway in collaboration with engineering and technology sale partners.

The main claim of the patent application reads: Claim 1. A process for the production of molten magnesium metal by electrolysis in a cell with an electrolysis chamber and a metal collection chamber, said process comprising electrolyzing in said electrolysis chamber an electrolyte containing a fused salt of said metal to produce said metal, said electrolyte having a greater density than said metal, and at least one electrode assembly that comprises a cathode defining within it a cavity, an anode disposed within the cavity and at least one intermediate bipolar electrode disposed between the anode and the cathode within said cavity, said electrolytic cell being characterized by a primary barrier wall that can be separately removed and replaced while the cell is in full operation, as it is for other modular and durable components, such as the cover of the metal collecting chamber and the equipment assembled to it, while the cover of the electrolysis chamber and the electrode assemblies are replaceable by shutting down the cell operation, but without the need to empty the cell, so that the campaign life of the cell and of each consumable component inside it can be extended to their individual useful limit.

Another claim reads: Claim 8. An electrolytic process as claimed in claim 1, characterized by means to increase cell productivity and power efficiency by improvements in metal coalescence obtained by sealing said cell to prevent any ingress of ambient air during normal cell operation into the chlorine room and into the front compartment.

Critical details on how the new cell is going to be built are being designed and tested experimentally to prove their operating reliability in the field.

What will happen next?

Magnesium chloride sources are practically inexhaustible from natural brines and bitterns from solar evaporation of sea water in natural basins or man-made ponds. Oxidic ores are natural products of the reaction of chlorides with CO_2 in the atmosphere and they can be easily converted back to artificial brines. The electrolysis of molecules into simpler constituents has been a foundation for two centuries of industrial inorganic chemistry. The main obstacle to electrolytic magnesium is the hygroscopic nature of the magnesium chloride cell feed. This is where the breakthrough will occur, once we focus on this long-term grand challenge. Unfortunately, in many ways we still require "at least five years to fix the chlorinators next door".

The Chinese are keenly interested in reaching the one-million-ton-a-year production target but the Pidgeon process route will become more and more a handicap for them as their labor rates normalize.

So, the door is open for those who are ready to come through. Young technologists, for example Dr. Boyd Davis of Kingston Process Metallurgy and others with new ideas and who are not afraid to invest their time and attention on developing the missing link, and engineers with competence in light metals technology will find support either from established industrial enterprises or from entrepreneurs and venture capitalists who can see the long-term potential and the opportunity that lays ahead. Chances are good for the future millionth ton of magnesium to be produced electrolytically.

In what direction is the technology wind blowing? The Kroll process will see the magnesium/chlorine recovery circuit integrated with the titanium reduction step. The production of electrolytic magnesium for sale will see the last step of the feed preparation process done just-in-time, to recover surplus heat from the cell and to avoid any moisture pick up by the feed while in storage. But the two processes will have to sing along carrying the same tune, as, if the quality of the feed is inadequate, the cell will be unforgiving. And, as they say, at the end of the day the process will tell.

Conclusions

Luke 13.12-30 describes how difficult it is to enter the Kingdom of God and how those left out will be weeping and gnashing their teeth. An old magnesium process engineer, after listening to the sermon, approached the preacher saying: "Preacher, but I have no teeth left." To which the preacher replied: "Don't worry; teeth will be provided." For some people, the electrolytic magnesium technology has lost its teeth in the struggle mounted from China by the Pidgeon process. My hope is that they will re-examine their lack of faith or teeth, and find their way to Damascus.

Superficially, my effort may be interpreted as a personal odyssey, but I have attempted to shine some light on the tall people I have met in my journey, who were instrumental in the achievement of the technological progress made over the many years, and to summon the help of some new talent to bring out the full potential of the electrolytic route to produce magnesium.

The human content of a development effort is rarely appreciated by the decision makers, who are more interested in the financial outcome of a project and fail to recognize that the two are intimately linked. Any financially successful project starts with trust and courage and evolves with the tenacity, patience and perseverance of a few individuals who can see the light at the end of the tunnel and are not afraid to be run over by an incoming train. Experience has shown, however, that even if one is on the right track, he may be run over if he does not keep his ear to the ground (see the Magnola project).

Cooperation and competition are the Yin Yang of all business ventures, and the magnesium business has seen both in various times and places, as pointed out in this paper.

Acknowledgments

The history of the multipolar magnesium cell reported here could never have been realized without the efforts of many individuals who soldiered over the years to make things happen. I was fortunate to have met them in the trenches when the technology battles were fought. To them, too numerous to mention, and to those I mentioned and to those who shared data with me in their private communications for this paper goes my deep appreciation.

References

1. Clow, Byron, B. History primary magnesium since WWII. Magnesium Technology 2002. TMS Annual Meeting, Seattle, Washington, 2002.

2. Industry Statistics. The International Magnesium Association, on-line.

3. Brown, Bob. History of magnesium production. Latest Databank Articles on-line.

4. Sivilotti, Olivo. Procedures and apparatus for electrolytic production of metals. US Patent 4,055,474. Oct. 1977.

5. Sivilotti, O. G., et al., Recent developments in the Alcan-type magnesium cell. AIME Annual Meeting, Las Vegas, 1976

6. Sivilotti, O. G., Iseki J., Electrolytic method and cell for metal production. US Patent 4,420,381. Dec. 1983.

7. Sivilotti, Olivo. Developments in electrolytic cells for magnesium production. The International Magnesium Association. The 41st Annual World Magnesium Conference, London, England. June 1984.

8. Sivilotti, Olivo. Operating performance of the Alcan multipolar magnesium cell. Light Metals, AIME Annual Meeting, Phoenix, 1988.

9. Sivilotti, Olivo. Metal production by electrolysis of a molten electrolyte. US Patent 4,514,269. Apr. 1985.

10. Sivilotti, Olivo. Apparatus for metal production by electrolysis of a molten electrolyte. US Patent 4,518,475. May 1985.

11. Sivilotti, Olivo. Metal production by electrolysis of a molten electrolyte. US Patent 4,560,449. Dec. 1985.

12. Sivilotti, Olivo. Electrolysis cell for a molten electrolyte. US Patent 4,604,177. Aug. 1986.

13. Sivilotti, Olivo. et al. Method for magnesium production. US Patent 4, 613,414. Sep. 1986.

14. Sivilotti, Olivo. Electrolytic cell for the production of a metal. US Patent 4,960,501. Oct. 1990.

15. Sivilotti, Olivo, et al. Multi-polar cell for the recovery of a metal by electrolysis of a molten electrolyte. US Patent 5,935,394. Aug. 1999.

16. Thayer, R.L., Neelameggham, R., Improving the electrolytic process for magnesium production. JOM, p.15-17, Aug. 2001.

17. Noranda Inc., 1999 Annual Report. Canada Newswire, Noranda Quarterly Results. Press Releases Q1, Q2, Q3, 2002. Canada Newswire, Noranda Press Release – Progress Report on Magnola, Nov. 22, 2002. Canada Newswire, Noranda Press Release – Magnola presents first results for its organo chloride monitoring program, Oct. 10, 2002.

18. Sivilotti, Olivo. Method and apparatus for electrolysing light metals. US Patent 5,660,710. Aug. 1997.

19. Sivilotti, Olivo. Method and apparatus for electrolysing light metals. US Patent 5,855,757. Jan. 1999.

20. McLean, Kevin; et al., Cathode wetting studies in magnesium electrolysis. Magnesium Technology 2009. TMS Annual Meeting, San Francisco, 2009.

Magnesium Technology 2011
Edited by: Wim H. Sillekens, Sean R. Agnew, Neale R. Neelameggham, and Suveen N. Mathaudhu
TMS (The Minerals, Metals & Materials Society), 2011

Effect of KCl on Liquidus of LiF-MgF_2 Molten Salts

Shaohua Yang[1],Fengli Yang [1], Xianwei Hu[2], Zhaowen Wang[2], Zhongning Shi[2] ,Bingliang Gao[2]

1. School of Materials and Chemistry, Jiangxi University of Science and Technology
2. School of Materials and Metallurgy117#, Northeastern University.

Keywords: Molten salt; Liquidus temperature; Cooling curve; Electrolyte

Abstract

Liquidus temperature of KCl-LiF-MgF_2 molten salts was determined based on cooling curve. The method of cooling curve was reliable and accurate. Experimental accuracy has been measured by the determined liquidus temperature of KCl. The measured value of KCl liquidus temperature of 769.27 °C is near and comparable to the documented value 769.5 °C. The difference value is 0.23 °C, and the relative error was 0.03%. The results show that effect of KCl on liquidus of molten salt MgF_2-LiF electrolyte was great. The liquidus temperatures lowered with increasing of content of KCl in electrolyte. The liquidus temperature could be reduced 20 °C with content of KCl in electrolyte increased to 50wt%. The experiments showed that the compound addition of KCl into LiF-MgF_2 electrolyte not only reduced the liquidus temperature of electrolyte effectively, but also decreased the aluminum-magnesium production cost and increased the economic benefits.

Introduction

As the lightest of all the commonly used metals, magnesium is very attractive for applications in transportation. It also has other advantageous features, such as good ductility, better damping characteristics than other metals and excellent castability. So, it has been widely used for aerospace industry, car manufacture, electronic technology, precise machines and so on. As the secondary element of aluminum alloy, increasing use of magnesium has been about 10-15% every year. Al-Mg alloys are widely used in construction and ship building industry[1-3].

However, the high cost of preparation of magnesium and magnesium alloys is a main factor to prevent them from wider use. For many years, many researchers[4-8] hope to prepare magnesium or its alloys in cell directly from magnesium oxide, and their focus is to cut down the loss of energy, shorten the process of the preparation and reduce the loss of metal oxidation. Preparation of Al-Mg alloys from MgO by molten salt electrolysis method is very popular in this research field. Some studies have attained exciting results. Aluminum-magnesium alloys could be prepared in 200 ampere with 20w%MgF_2-30w%LiF-50w%KCl as electrolyte, liquid aluminum as cathode, graphite as anode and magnesium oxide as raw material. The electrolytic process was stable, range of variation for cell voltage was narrow. The difference between the highest cell voltage and the lowest cell voltage was in 0.6V. Content of magnesium in alloy was not even because of construction of cell, the highest and the lowest were 20 w %, 6w%, respectively. The content of magnesium in alloys was 8.56 w% after refusion, and current efficiency was about 82%[9-14].

Liquidus temperature is one of important properties of electrolyte[15-17]. If liquidus temperature of electrolyte can be decreased, the electrolysis temperature can be reduced correspondingly. Thus, it has many advantages for low temperature electrolysis, such as to reduce energy consumption, decrease volatilization of electrolyte and increase current efficiency.

In this paper, the effect of KCl on liquidus of LiF-MgF_2 molten salts was studied by the cooling curve method. With increasing of KCl content in different electrolytes liquidus of the electrolytes were measured.

Experiment

Experimental apparatus and agents

KCl (Analytic agent), LiF (Analytic agent), MgF_2 (Analytic agent), all regents were dried at 400 °C for 2 hours to remove the water before experiments. Resistance furnace(2kw);Temperature controller(DWK-702); Thermal couple.

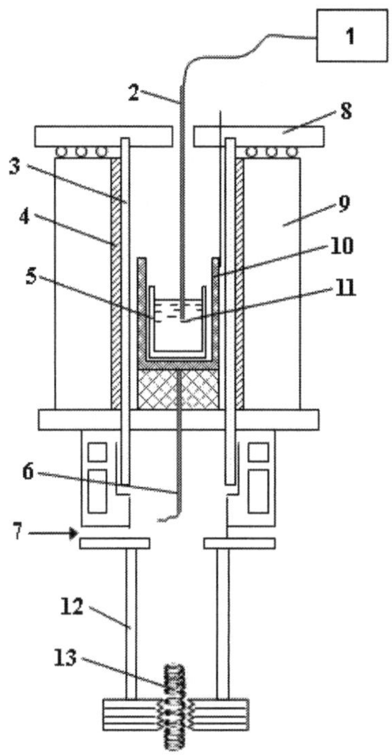

Figure 1. Experimental installation to measure liquidus of molten salt

1-Potential difference meter; 2-Thermal couple; 3- Alundum tube; 4-Heater; 5-Platinum crucible; 6- Thermal couple; 7-Pipe for argon; 8-Cooling water; 9-Resistance furnace; 10-Steel crucible; 11-Platinum ball; 12-Supporter; 13-Elevater;

Experimental process

Firstly, thermocouple was calibrated, and its accuracy was attained. Secondly, experimental system was measured, and accuracy of the method was determined. Amounts of dried KCl was fed into alundum tube container, KCl was melted in resistance furnace. The thermocouple was dipped into KCl molten salt. The temperature of furnace decreased slowly. A cooling curve could be shown on computer, and liquidus temperature of KCl could be estimated from the curve. The experimental results were compared with documented value. Thirdly, liquidus temperatures of electrolytes with different composition were measured by this method.

Accurate of system

The cooling curve of KCl was shown as figure 2, and the error analysis is shown in table 1:

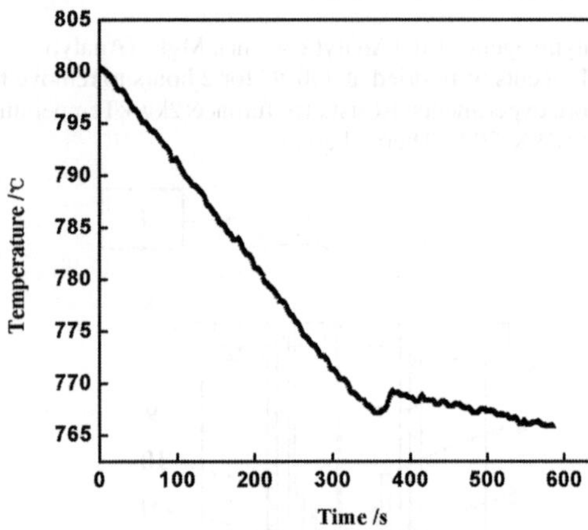

Figure 2. Cooling curve for potassium chloride

Table 1 Relative error analysis

Mass	Experimental value(° C)	Documental value(° C)	Error value(° C)	Relative error(%)
KCl	769.27	769.5	0.23	0.03

Figure 2 and table 1 can show that inflection liquidus temperature was clear, and the relative error was little, so the method was accurate.

Components of electrolyte were given as table 2:

Table 2 Components of electrolyte

Number	Components
1	0w%KCl-30w%LiF-70w%MgF_2
2	10w%KCl-30w%LiF-60w%MgF_2
3	20w%KCl-30w%LiF-50w%MgF_2
4	30w%KCl-30w%LiF-40w%MgF_2
5	40w%KCl-30w%LiF-30w%MgF_2
6	50w%KCl-30w%LiF-20w%MgF_2

Results and discussion

Experiments were carried out at different temperatures with different electrolytes. The results are shown as figure 3:

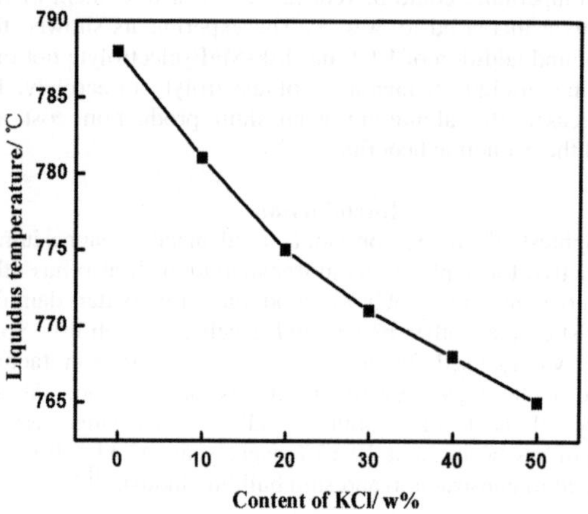

Figure 3 Relationship between content of KCl and liquidus temperature

Figure 3 shows that liquidus temperature decreased with content of KCl in electrolyte, the decreased value was about 20 ° C as content of KCl increased from 0w% to 50w%. After 20w% KCl in electrolyte, the decreased in values were reduced gradually. In other words, too much content of KCl in electrolyte had no contribution to decreasing liquidus temperature. The effect of KCl in electrolyte on liquidus temperature was clear if content of KCl was under 20%. In addition, increasing content of KCl and decreasing content of MgF_2 should be considered together for decreasing liquidus temperature. The decreased content of MgF_2 may have had some effects on decreasing liquidus temperature, not only liquidus temperature decreased by increasing of content KCl in electrolyte.

Conclusions

The method of cooling curve was accurate to measure liquidus temperature of electrolyte, and the relative error was 0.03%.

The effect of KCl on liquidus of molten salt was great for MgF_2-LiF electrolyte. The liquidus temperature could be reduced 20° C with increasing content of KCl in electrolyte from 0wt% to 50wt%. Increasing content of KCl and decreasing content of should be considered together for decreasing liquidus temperature. The decreased content of MgF_2 may have had some effects on decreasing liquidus temperature.

Acknowledgements

The authors wound like to acknowledge the financial supported by "The Education Department of Jiangxi Province (GJJ09513)", and express great thanks to Pr. Zhaowen Wang for his help in experiments.

References

[1] A.K. Dahle, D.H. StJohn, G.L. Dunlop, Developments and challenges in the utilization of magnesium alloys, Mater. Forum 24 (2000):167-182.

[2] E.Aghion, B.Bronfin, D.Eliezer. The role of magnesium industry in protecting the environment[J], Journal of materials processing technology2001(117):381-385.

[3] Ranmkrishnan S, Koltun P. Global warming impact of the magnesium produced in China using the Pidgeon process[J]. Resources Conservation and Recycling,2004(42):49-64.

[4] Qiu Z X, Zhang M J, Wang J, et al. Preparation of aluminum-magnesium master alloys by electrolysis of magnesium oxide in fluoride melts[C]Light Metals. Warrendale: Minerals, Metals & Materials Soc, 1991: 349-355.

[5] Ram A. Sharma. Method for producing magnesium metal from magnesium oxide [P].USA. 5279716,1994.1.18

[6] Ram A. Sharma. An electrolytic process for magnesium and its alloys production [C]. Light Metals, WTMS, 1996: 1113~1122.

[7] Ram A. Sharma. A new electrolytic magnesium production process[J]. JOM, 10(1996):39~43

[8] Qiu Z X, Zhang M J. Preparation of aluminum master alloy by electrolysis in molten cryolite[J]. Aluminum, 1990,66 (6): 560~564.

[9] Shaohua Yang, Bingliang Gao, Zhaowen Wang, et al. Preparation of aluminum-magnesium alloys starting from magnesium oxide [C]Light Metals. Warrendale: Minerals, Metals & Materials Soc, 2007:283~286.

[10] Shaohua Yang, Zhaowen Wang Ying Nie, Jidong Li, Xianwei Hu.Preparation of Al-Mg alloys from MgO through molten salt electrolysis method[A].Light Metals [C]. Warrendale : Minerals, Metals & Materials Soc, 2008:17-20

[11] Shaohua Yang, Zhaowen Wang, Ying Nie, Xingliang Zhao.Electrodeposition of Magnesium from BaF_2-LiF-MgF_2 electrolyte[A].Light Metals [C]. Warrendale : Minerals, Metals & Materials Soc, 2008:13-15

[12] Shaohua Yang, Fengli Yang, et al. STUDY ON ELECTROLYSIS OF MAGNESIUM OXIDE ON 200A SCALE[A].Light Metals[C]. Warrendale : Minerals, Metals & Materials Soc, 2009:57-59

[13] Shaohua Yang, Fengli Yang, Xianwei Hu. Preparation of Al-Mg alloys from MgO by molten salt electrolysis method[A].Light Metals. Warrendale[C] : Minerals, Metals & Materials Soc, 2010:107-110

[14] Shaohua Yang, Fengli Yang, Guocheng Wang. Effect of KCl on conductivity of BaF_2-LiF-MgF_2 molten salts[A]. light Metals. Warrendale[C] : Minerals, Metals & Materials Soc, 2010:111-114

[15] Ray D P, Alton T T. Liquidus curves for the cryolite AlF_3-CaF_2-Al_2O_3 system in aluminium cell electrolytes [C]. Warrendale: Minerals, Metals & Materials Soc, 1987, 383-388.

[16] Solheim A, Rolseth S, Skybakmoen E, et al. Liquidus temperature and alumina solubility in the system Na_3AlF_6-AlF_3-LiF-CaF_2-MgF_2 [C]. Warrendale: Minerals, Metals & Materials Soc, 1995, 451-456.

[17] Rolseth S, Verstreken P, Kobbeltvedt O. Liquidus temperature determination in the molten salts [C]. Warrendale: Minerals, Metals & Materials Soc, 1998, 359-366

Magnesium Technology 2011
Edited by: Wim H. Sillekens, Sean R. Agnew, Neale R. Neelameggham, and Suveen N. Mathaudhu
TMS (The Minerals, Metals & Materials Society), 2011

EFFICIENCY AND STABILITY OF SOLID OXIDE MEMBRANE ELECTROLYZERS FOR MAGNESIUM PRODUCTION

E. Gratz[1], S. Pati[1,2,3], J. Milshtein[2], A. Powell[3] and U. Pal[1,2]
[1] Boston University, Division of Materials Science and Engineering, 15 Saint Mary's St. Boston, MA, 02215
[2]Boston University, Department of Mechanical Engineering, 101Cumminton St. Rm 101 Boston MA 02215
[3]Metal Oxygen Separation Technologies, Inc.11 Michigan Dr. Natick, MA 01760-1334

Keywords: Magnesium, SOM, Reduction, Membrane Stability

Abstract

Solid oxide membrane (SOM) process has been successfully employed for the production of magnesium directly from its oxide. The process involves dissolving MgO in a fluoride based ionic flux and electrochemically pumping out the oxygen ions from the flux via an oxygen-ion-conducting SOM to the anode where they are oxidized, while reducing magnesium ions at the cathode. Understanding the long-term stability of the SOM in the flux is critical for the commercial success of this technology. In this study long term SOM stability is investigated under potentio-static conditions. Additionally, study utilizes electrochemical techniques such as impedance spectroscopy and linear sweep voltammetry to investigate key concepts related to MgO dissociation and current efficiency. Results show that the dissociation potential of MgO is dependent on the partial pressures at which magnesium is generated and the membrane stability is likely related to the current efficiency.

Introduction

Magnesium is a leading candidate material to replace steel and aluminum in automobiles for reducing vehicle weight to increase fuel efficiency. Since magnesium is 36% lighter per unit volume than aluminum and 78% lighter than steel estimates show that 22.5 kg of mass reduction would improve fuel efficiency by around 1% [1]. Currently world magnesium production is around 400,000 tonnes per annum and demand for magnesium is expected to grow by 7% per annum [2]. To meet this future demand for primary magnesium, the current technologies have to be made more efficient and new technologies must be explored.

One promising new technology is the solid oxide membrane (SOM) electrolyzer that has the potential to produce primary magnesium at a fraction of its current cost with the added benefit of being environmentally friendly [3, 4]. However, one challenge for the SOM electrolyzer is the long term stability of SOM in a magnesium oxide containing molten flux at high temperatures (1100-1200°C). Previous studies show that fluoride based ionic flux is suitable for the SOM electrolysis [3]. However, long term stability of SOM in the molten flux for more than 2000 hrs is needed before the SOM electrolyzer can be commercially used for magnesium production.

We believe the SOM can degrade in two possible ways: 1) Leaching of yttria from the SOM into the molten flux, and 2) Electronic current in the cell. Some initial studies show that leaching of yttria from the SOM into the molten flux can be prevented by adding yttrium fluoride in the flux to balance the activity of yttrium in the membrane and yttrium in the flux [5]. The second pathway for SOM degradation is due to electronic current in the system, and this is evaluated in this article. Suput et. al. [6] employed the SOM electrolyzer to produce titanium and showed that the multi-valence state of Ti in the flux resulted in electronic current in the SOM process. The electronic current degraded the SOM and the efficiency was significantly lower. In that study it was also reported that by using an electron blocker, the degradation of the SOM decreased significantly and increased the overall efficiency of the SOM electrolyzer.

The advantage of the SOM electrolyzer for magnesium production is that the magnesium is not multivalent as titanium. Therefore, the flux is essentially ionic and acts as an electron blocker. However, Krishnan et. Al [5] observed a small current prior to the dissociation potential of MgO during the potentiodynamic scan and attributed that to the oxygen impurities in the cell. This so-called "leakage current" was also reported in the SOM process for calcium production and it increased with electrolysis time [7]. The increase in leakage current was attributed to electronic current resulting from solubility of metallic calcium in the fluoride based flux [8]. As mentioned before, when the flux becomes electronic the efficiency decreases and also leads to degradation of the membrane. Therefore it is critical to study the nature of the current in the SOM electrolyzer. In this paper we characterize the SOM electrolyzer electrochemically to understand the origins of the leakage current. This is very essential to formulate strategies for reducing or eliminating leakage current and thereby increase process efficiency.

Experimental

<u>*Experimental set-up*</u>: Figure 1 shows the design of SOM electrolyzer used in the present experiments. It consists of an upper electrolysis chamber, heated to 1200°C and a lower condensation chamber kept at a temperature range of 1150°C-500°C. The set-up is fabricated using SS-304 (Grade 304 stainless steel) and heated as mentioned above in a reducing atmosphere (5%H_2-Ar). The electrolysis chamber contains a 0.75" outer diameter yttria stabilized zirconia tube (McDaniel Ceramics) immersed in a molten flux (55 % MgF_2-45%CaF_2-10% MgO).

The stainless steel crucible wall of the electrolysis chamber served as the cathode and silver inside the YSZ tube served as the anode. An alumina spacer was used to insulate the YSZ membrane from the steel which acted as the cathode. A ¼" molybdenum tube (Nanmac) served as the anodic current collector. When the electric potential applied across the cathode and the anode exceeded the dissociation potential of magnesium oxide, magnesium vapor is produced on the stainless steel cathode. It should be noted that during electrolysis the cathodic chamber of the cell is purged with Ultra High Purity Argon (Airgas) at

300cc/min through the annulus between a 1 ¼" inner diameter SS-304 tube and 0.75" outer diameter YSZ tube to lower the partial pressure of magnesium vapor that is produced over the cathode. This is done in order to prevent the reduction of YSZ with magnesium vapor. The magnesium vapor produced on the cathodic side is transported to the condensation chamber by two ¼" outer diameter and 0.12" inside diameter SS-304 tubes. The inlets of these two tubes extend above the flux to prevent any molten flux from entering the condenser. On the anodic side, hydrogen was bubbled in the silver through the molybdenum tube at rate of 60cc/min prevent the molybdenum current collector from oxidizing. Part of the hydrogen also acts as a reductant and the overall cell reaction is given as:

$$MgO + H_2(g) = Mg + H_2O(g) \quad (1)$$

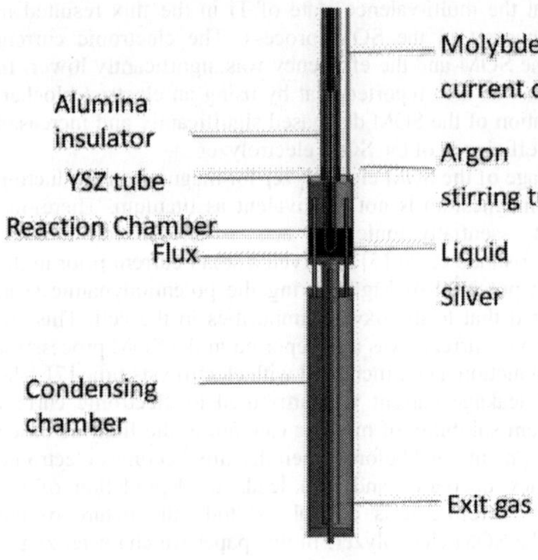

Figure 1. Design of experimental setup for SOM process.

All electrochemical measurements were performed using a Princeton Applied Research Potentiostat model 263A and a Solartron Frequency Response analyzer model 1250. Electrolysis of magnesium was carried out using an Agilent Technologies N5743A Power Supply.

Results and Discussion

Electrochemical measurements: When the SOM electrolyzer was at the desired temperature, pre-electrolysis was performed at an applied potential less than the dissociation potential of magnesium oxide to remove any dissolved oxygen in the flux. Previous studies report that the dissociation potential of MgO in a SOM electrolyzer with hydrogen at the anode is around 0.7 V [3]. Therefore, pre-electrolysis was performed at a potential 0.5 V. Figure 2 shows that after 15 minutes, the current drops to around 0.01 A, which suggests that most of the dissolved oxygen in the flux has been removed. The initial current is most likely due to a small amount of oxygen that is not removed during the purging process.

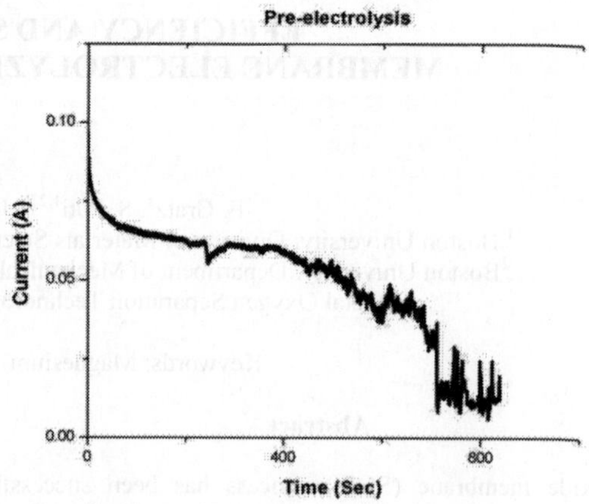

Figure 2. Pre-Electrolysis potentio-static hold at 0.5V

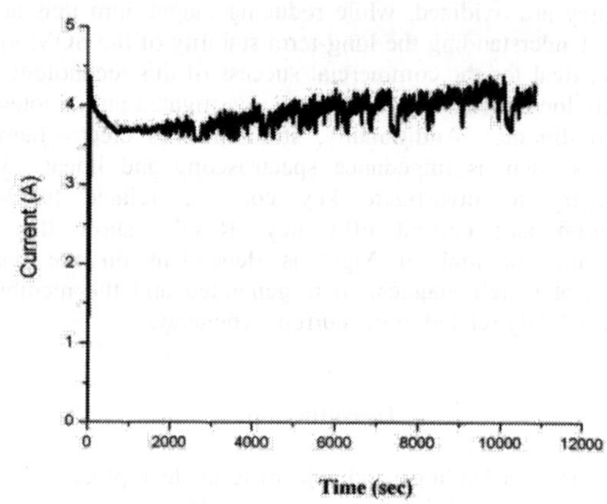

Figure 3. Current-voltage characteristics during electrolysis at 2.6V

Figure 3 shows the current-time characteristics when electrolysis was performed at 2.6 V. It shows initially that the current dropped from 4.4 A to 3.7 A during the first 1000 seconds. Then the current increased slowly and was around 4.0 A after 11000 seconds of electrolysis. This is not expected because with increase in duration of electrolysis the amount MgO in the flux decreases and therefore the mass transfer resistance of oxygen ions in the flux is expected to increase and the current should decrease. Potentiodynamic scans performed before and after electrolysis further show that during electrolysis the contribution of leakage current increases (Figure 4). The possible reason for this could either be a leak of H_2O from the anodic side or electronic current due to dissolution into the flux of metallic magnesium and/or calcium produced during electrolysis.

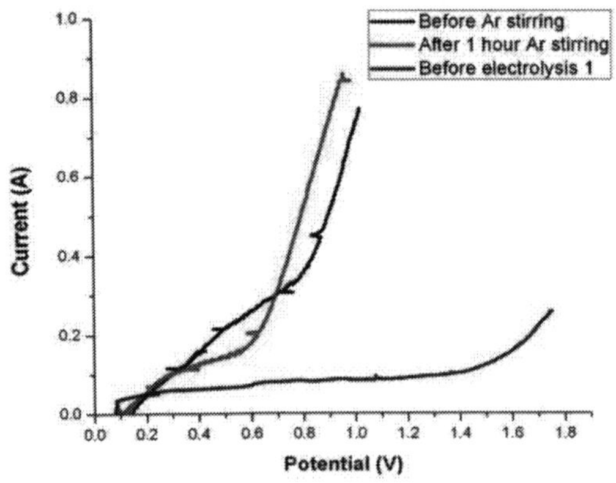

Figure 4. The effect of argon stirring on the leakage current

1) *H_2O leak from anodic side*: It has been shown previously that even a P_{O2} of 10^{-14} atm will lead to noticeable leakage current [5]. In that study the SOM electrolyzer experiments relied on a metal to ceramic seal at a high temperatures. In this study the design was changed so that all of the ceramic- metal sealing was moved to the room temperature region on the top of the set-up using Swagelok UltraTorr fittings. This significantly reduces the potential for H_2O leak from the anodic side and therefore also the leakage current. Although leakage current of this nature decreases the efficiency, it is not expected to degrade the membrane.

2) Metal Solubility: Another potential reason for the leakage current is the dissolution of metallic magnesium and/or calcium in the flux. It is widely known that calcium has a very high solubility in calcium halides [8]. Dissolution of metals in the flux which is originally ionic in nature can cause the flux to start conducting electronically. This can be very detrimental to the SOM electrolyzer as it could potentially reduce the zirconia in the membrane and ultimately degrade the membrane, reducing its lifespan.

It has been shown that magnesium metal does have some solubility in chloride based ionic salts [9]. In that study metallic magnesium and magnesium chloride were melted together and samples from the melt were cooled to room temperature. These samples were then dissolved in dilute acid to produce hydrogen. From the amount of hydrogen evolved the amount of dissolved magnesium was calculated.

In order to determine whether there is any dissolved metal in this flux after electrolysis, dilute acid was added to the flux in a closed container and the increase in pressure due to the formation of hydrogen gas measured. The dissolved metal, Mg and/or Ca in the flux after electrolysis, was estimated to be between 0.02-0.05% by weight. For comparison, a similar procedure was performed on the fluoride flux prior to electrolysis and no hydrogen evolution was detected indicating that prior to electrolysis the flux did not have any metal in solution.

Although the presence of soluble metal in the flux was detected after electrolysis it could not be determined whether the metal was magnesium or calcium. Since the flux contains both calcium and magnesium ions along with oxygen ions, there exist finite activities of both CaO and MgO in the flux. Therefore the metal in solution could be either Mg or Ca or both.

It is possible to lower the activity of the dissolved metal in the flux by stirring the flux with an inert gas such as argon. Figure 4 compares the potentiodynamic scans with and without argon stirring in the flux. This shows that the leakage current can be lowered by stirring argon for an hour but it could not be eliminated.

Although stirring argon did not eliminate the leakage current, it did decrease the dissociation potential of magnesium oxide which in turn decreases the energy requirements of the SOM process. The dissociation potential decreased from 1.4 volts without argon stirring to 0.7 V [Figure 5] with it. The reason for decrease in the the dissociation potential is because magnesium evolves at a lower partial pressure when argon is stirred during electrolysis. The Nernst equation for reaction 1 is given by equation 2.

$$E = E^0 + \frac{RT}{nF} \ln \frac{a_{MgO(l)} \cdot p_{H_2(g)}}{P_{Mg(g)} \cdot P_{H_2O(g)}} \tag{2}$$

Since E^0 is negative, |E| decreases when $P_{Mg(g)}$ decreases. It is believed that the $P_{Ar(g)}$ coming out of the stirring tube is very near 1atm and therefore $P_{Mg(g)}$ must be very low. Since the stirring tube also acts as a cathode it provides a location for magnesium oxide reduction at a lower potential.

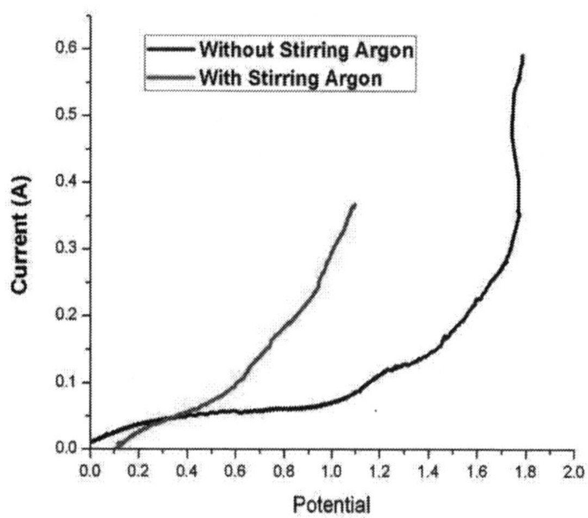

Figure 5. Comparison of potentiodynamic scans with and without argon showing change in dissociation potential

Conclusion

The SOM process is a method for producing magnesium directly from its oxide. The current increased during electrolysis and this was primarily attributed to dissolution of magnesium and/or calcium in the flux that is produced during electrolysis. The leakage current not only lowers the efficiency but also accelerates the degradation of SOM. The leakage current was lowered by stirring argon in the flux. The dissociation potential depends on the partial pressure at which the magnesium is produced, and also decreased with stirring argon in the flux.

Acknowledgements

The authors thank Metal Oxygen Separation Technologies, National Science Foundation for Award No. IIP-0912743 and the

Department of Energy UT-Batelle, LLC c/o Oak Ridge NationalLaboratory for Subcontract Number 4000091722.

References

[1] P. Pollock, "Weight Loss with Magnesium Alloys," *Science,* 328, May 2010, 986-987.

[2] B. Mordike, and T. Ebert, "Magnesium Properties – Applications –Potential," *Materials Science and Engineering,* A302 (2001) 37-45.

[3] A. Krishnan, U. b. Pal and X. G. Lu "Solid Oxide Membrane Process for Magnesium Production Directly from Magnesium Oxide," *Metallurgical and Materials Transactions B,* 36 (4) (2005), 463-473.

[4] D. Sujit, "Primary Magnesium Production Costs for Automotive Applications," *Journal of the Minerals, Metals and Materials Society,* 60 (11) (2008), 63-69.

[5] A. Krishnan, "Solid Oxide Membrane Process for the Direct Reduction of Magnesium from Magnesium Oxide" (Ph.D. thesis, Boston University, 2006.

[6] M. Suput, R. Delucas, S. Pati, G. Ye, U. Pal, A.C. Powell IV, "Solid oxide Membrane Tecnology for Enviromentally Sound Production of Titanium," *Mineral Processsing and Extractive Metallurgy,* 117 (2) 2008, 118.

[7] S. Pati, M. Suput, R. Delucas, and P. Uday, "Solid Oxide Membrane Process for Calcium Production Directly From its Oxide" *TMS EPD Congress,* 2008.

[8] Y. Ochifuji, F. Tsukihashi and N. Sano, "The Activity of Calcium in Calcium-Metal-Flourine Fluxes," *Metallurgical and Materials Transactions B,* 26 (1995), 789-794.

[9] J. Wypartowicz, T. Ostvold, and H. Oye, "The Solubility of Magnesium Metal and the Recombination Reaction in the Industrial Magnesium Electrolysis" *Electrochimica ACTA*, (25) 1980, 151-156.

Magnesium Technology 2011
Edited by: Wim H. Sillekens, Sean R. Agnew, Neale R. Neelameggham, and Suveen N. Mathaudhu
TMS (The Minerals, Metals & Materials Society), 2011

MAGNESIUM PRODUCTION BY VACUUM ALUMINOTHERMIC REDUCTION OF A MIXTURE OF CALCINED DOLOMITE AND CALCINED MAGNESITE

Wen-xin Hu, Nai-xiang Feng, Yao-wu Wang, Zhi-hui Wang
School of Materials and Metallurgy, Northeastern University, Shenyang, Liaoning 110004, China

Keywords: vacuum aluminothermic reduction; magnesium; $CaO{\cdot}2Al_2O_3$; alumina leaching

Abstract

A new method of magnesium production was proposed that using a mixture of calcined dolomite and calcined magnesite as raw materials with the molar ratio of MgO to CaO was 6:1 by vacuum aluminothermic reduction. The reduction process was studied by thermodynamic analysis and X-ray diffraction analysis of reduction residue. The reaction of reduction process was $CaO+6MgO+4Al=CaO{\cdot}2Al_2O_3+6Mg$. The effect of briquetting pressure, reduction temperature, time and CaF_2 (MgF_2) on reduction ratio of MgO was investigated. And the reduction residue that the main phase of $CaO{\cdot}2Al_2O_3$ was leached in alkaline solution for producing sodium aluminate-the raw material for special alumina. The results show that the reduction ratio is increased with increasing of the temperature, time, briquetting pressure in range from 40 to 100 MPa and addition of CaF_2 or MgF_2 in range from 0 to 3%. The alumina leaching ratio of reduction residue reached 88% at the conditions of leaching temperature 95 ℃ and the concentration ratio of Na_2CO_3 to NaOH was 100:75 in leaching solution.

Introduction

Magnesium is the lightest metal of the commonly used metals and can be applied widely. For example, there are metallurgical applications, chemical applications, structural applications, and automobile parts production [1,2]. It means huge demands for magnesium in the world. The raw materials of magnesium production such as magnesite and dolomite are widespread and high purity magnesium can be produced under low cost, simple and flexible conditions of vacuum thermal reduction process versus electrolytic production.

The Pidgeon process is a widely used process in vacuum thermal reduction process, and this technology has been in stage of maturity [3-5]. But because of the low reactivity of ferrosilicon alloy, higher reduction temperature (>1170 ℃) and longer reduction time (>7 hours) are required, which in order to reach high reduction ratio of MgO. The reduction ratio of MgO was only 73.9% at the conditions of temperature 1443 K, time 120 min [6]. More energy is needed for heating, the service life of Ni-Cr alloy reactors decrease greatly under vacuum and the conditions mentioned above, make higher cost of the Pidgeon process eventually; the residue after reduction (main phase is γ-$2CaO{\cdot}SiO_2$) is useless to other fields, such as cement industry with a reason that unreacted MgO has negative influence on the performance of cement [7,8]. According to the above information, there are two problems should be solved, firstly, the service life of reactors and the high cost of process; secondly, the emission of CO_2 and the waste of reduction residue.

A new vacuum thermal reduction method is proposed for solving the two problems about the Pidgeon process in the present paper. The new reduction process is aluminothermic reduction of a mixture of calcined dolomite and calcined magnesite with the molar ratio of MgO to CaO was 6:1. It is studied by thermodynamic analysis and the effects of reduction temperature, reduction time, vacuum of reduction system, briquetting pressure and addition of CaF_2 or MgF_2 on the reduction ratio of MgO, which for ensuring the optimal conditions of reduction process. Using alkaline solution to leach alumina from $CaO{\cdot}2Al_2O_3$, the main phase of reduction residue. The new magnesium production method in the present paper is in order to create a green, lower cost but higher economic benefit process.

Thermodynamic Analysis

Figure 1 is the X-ray diffraction analysis of reduction residue, the reduction condition of experiment is reduction temperature 1413 K, reduction time 120 min, pellets forming pressure of 50 MPa and vacuum of 4 Pa. It is noticed that $CaO{\cdot}2Al_2O_3$ phase with higher intensity; the other phases are from raw materials. Thus, the main reaction of reduction process occurred in reduction is

$$CaO\,(s) + 6MgO\,(s) + 4Al\,(l) = CaO{\cdot}2Al_2O_3\,(s) + 6Mg\,(g) \quad (1)$$

$$\Delta G^{o} = 1013.7 - 0.615T \quad \text{(kJ/mol)} \quad (2)$$

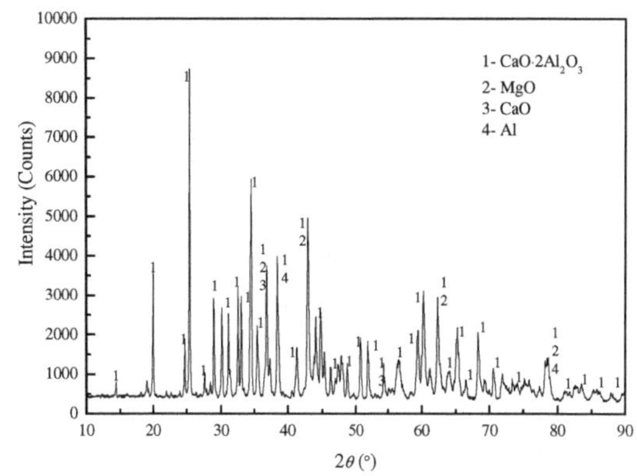

Figure 1. The XRD pattern of reduction residue.

The relationship between the equilibrium magnesium partial pressure (P_{Mg}) and reduction temperature is given by Equation 3.

$$\Delta G = 1013.7 - 0.615T + RT\ln\left[\frac{(\frac{P_{Mg}}{P^{o}})^{6}}{(\frac{\alpha_{Al}}{\alpha^{o}})^{4}}\right] \quad \text{(kJ/mol)} \quad (3)$$

Where α_{Al} is the activity of aluminum; α^{o} and P^{o} are values of standard state; R is the gas constant.

The equilibrium temperature of Equation 1 is 1648.3 K at reduction system pressure of 1 atm. But the reaction temperature decreases to 904.4 K at the experimental vacuum of 4 Pa. Therefore, the reaction Equation 1 could occur at temperatures of 1223, 1273, 1323, 1373, 1413, 1443 and 1473 K at vacuum of 4 Pa (during the reduction process, P_{Mg} is approximately equal to the pressure of vacuum reduction system [9], which displays on vacuometer).

Experimental

Materials

Dolomite and magnesite were obtained from Liaoning, China; aluminum powder (purity 99%), CaF_2, MgF_2, NaOH and Na_2CO_3 were supplied by Tian Jin Kermal Chemical Reagent CO., LTD. Table I and II show the chemical compositions of dolomite and magnesite using chemical analysis.

Table I. Chemical composition of dolomite.

Component	MgO	CaO	SiO_2	Al_2O_3
Content (%, ω)	23.62	29.86	0.22	0.10

Table II. Chemical composition of magnesite.

Component	MgO	CaO	SiO_2	Al_2O_3
Content (%, ω)	47.34	0.67	0.13	0.08

Apparatus

Figure 2 schematically shows the experimental apparatus for reduction. A furnace using silicon carbide rods was used to heat the reactor which was made of high-temperature alloyed steel; temperature control system was used to control the heating ratio of reduction system and shown the temperature which was measured by NiCr-NiSi thermocouple; vacuum system was made up of ZJP-70 roots vacuum pump and ZX-15 rotary-vane vacuum pump; condenser, the upper part of reactor was used to condense magnesium vapor and collected magnesium.

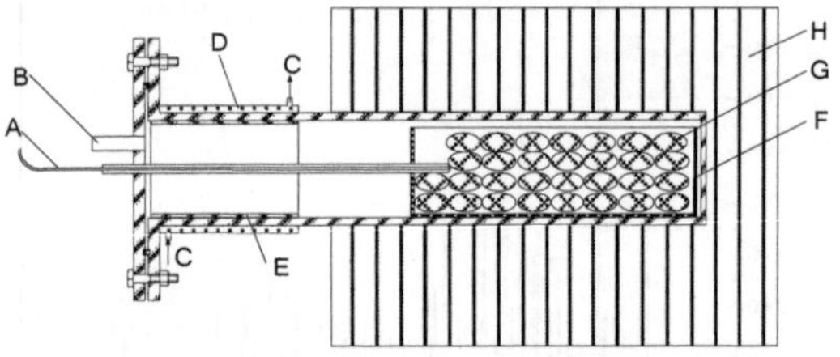

A- thermocouple B- vacuum tube

C- circulating water D- water-cooled jacket

E- condenser F- briquettes bucket

G- briquettes H- furnace

Figure 2. Schematic of experimental apparatus.

Procedures

The vacuum thermal reduction experiments were carried out by using a mixture of calcined dolomite and calcined magnesite. According to the experimental data of calcinations, dolomite and magnesite should be calcined at 1060 °C and 830 °C for 90 min respectively for ensuring the reactivity of MgO·CaO and MgO. The calcined dolomite and calcined magnesite were pulverized and sieved 150~180 μm out, mixed with the molar ratio of MgO, CaO to Al 6:1:4. The mixed charges were formed into pellets under different pressure, having a diameter of 25 mm and a thickness of about 10 mm. The pellets were charged into the reactor for reduction experiments. The magnesium was reduced by aluminum in the conditions of vacuum and high temperature. The magnesium was collected on the condenser. The reduction residue was leached under the conditions of leaching temperature 95 °C and the concentration ratio of Na_2CO_3 to NaOH was 100:75 by calculation of Equation 6. The flow sheet of process show is shown in Figure 3. After reduction and reactor cooling, the condensed magnesium product and the residue obtained were weighed and analyzed.

The reduction ratio of MgO (R) is calculated by the following formula:

$$R = \frac{m_r}{m_0} \times 100\% \tag{4}$$

Where m_0 is the initial quantity of magnesium in the raw materials and m_r is the quantity of magnesium on condenser.

The leaching ratio of Al_2O_3 according to the following formula:

$$\eta_{Al_2O_3} = \frac{\omega_l}{\omega_0} \times 100\% \tag{5}$$

Where ω_0 is the quantity of Al_2O_3 (elemental aluminum in reduction residue is convert to Al_2O_3), ω_l is the quantity of Al_2O_3 in leaching solution.

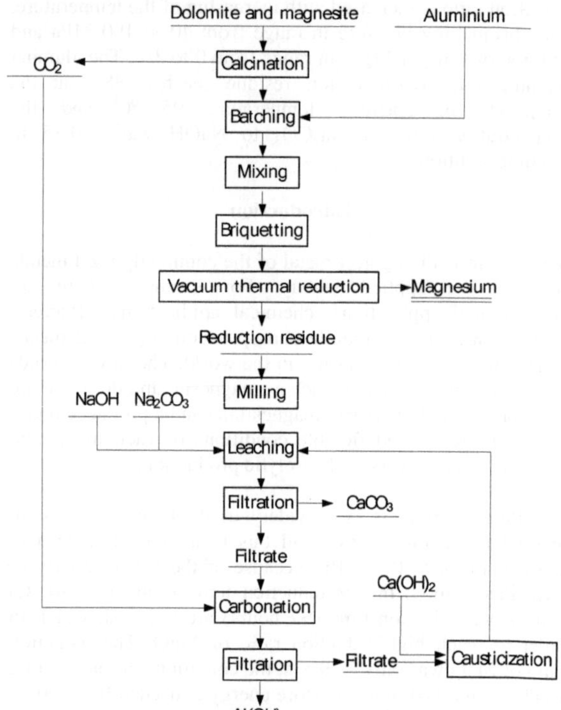

Figure 3. The flow sheet of reduction and leaching process.

Analysis of Experimental Results

Effect of Reduction Temperature on Reduction Ratio of MgO

Figure 4 shows the changes in reduction ratio of MgO at different temperatures. At the conditions of briquetting pressure of 50 MPa, reduction time 120 min and vacuum 4 Pa. With increasing temperature from 1223 K to 1413 K, the reduction ratio of MgO increased remarkably, from 44.47% to 82.06%. The reduction ratio of MgO increased slowly at temperature range from 1413 K to 1443 K. The final reduction ratio was 91.35% at 1473 K.

Reduction temperature is an important factor to the reaction rate; the reaction ratio was accelerated with temperature increasing. But at the stage of 1443 K, the reduction is close to the end which the reason of reduction ratio increased slowly. The high reaction rate and good results of reduction ratio of MgO needed high temperature. But the reasonable reduction temperature is chosen 1413 K by experimental data. The reason that decreasing the reduction temperature for saving cost of high-temperature alloy is one of the principal objectives of the present reduction process. At 1413K, the reduction ratio of MgO is higher than Pidgeon process at 1473 K [6]. The reduction ratio of MgO should increase by change other factors, such as reduction time.

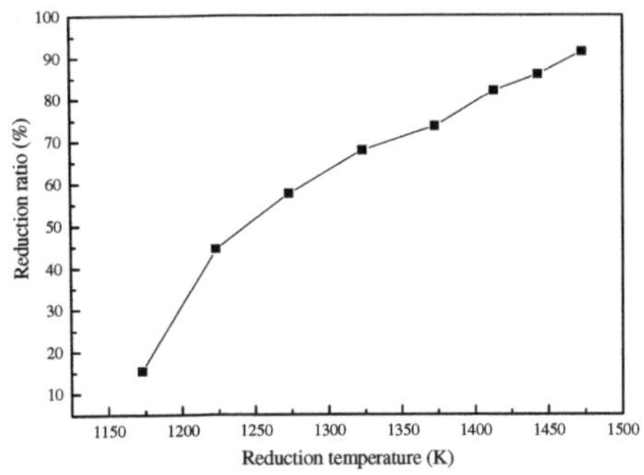

Figure 4. Effect of reduction temperature on reduction ratio of MgO.

Effect of Reduction Time on Reduction Ratio of MgO

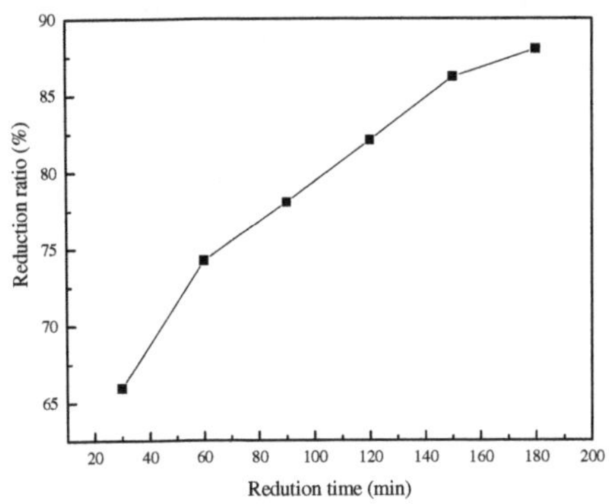

Figure 5. Effect of reduction time on reduction ratio of MgO (the initial reduction ratios according to experimental data are 40.9% in temperature-rise period from room-temperature to 1413 K).

The effect of reduction time on reduction ratio of MgO is presented in Figure 5. The reduction temperature was 1413 K, briquetting pressure 50 MPa and vacuum 4 Pa. The reduction ratio increased obviously at the initial reaction stage from 30 min to 150 min, the reduction ratio reached 86.18% at 150 min. After reduction time of 150 min, the reduction ratio of MgO increased slowly that only increasing 1.81% at 180 min. The reduction ratio increased almost no more. It should be the meaning of the end of reduction.

The more energy will lead to high reaction rate of reduction process. The energy which was supported to pellets by heat transfer from furnace to pellets by thermal radiation almost, and then from surface into the core of pellets with reduction time extended. In present paper, time of 150 min was not the best reduction time, but the time which the corresponding reduction ratio of MgO was closed to the maximum. It is one of key links for saving the cost of energy, and enhanced the effects of briquetting pressure and addition of CaF_2 or MgF_2 on reduction ratio of MgO.

Effect of Briquetting Pressure on Reduction Ratio of MgO

Figure 6 shows the effect of briquetting pressure on reduction ratio of MgO. The reduction temperature was 1413 K, reduction time 120 min and vacuum 4 Pa. For the briquetting pressure of 100 MPa, the reduction ratio proceeded the fastest and the reduction ratio was 88.96%. For the briquetting pressure of 50 MPa or 200 MPa, the reduction ratio was decreased remarkably. The reason of this phenomenon is from 50 MPa to 100 MPa, increasing briquetting pressure can increase the contact area between MgO-CaO and reduction agent; from 100 MPa to 200 MPa, the briquetting pressure make the gap is too small between MgO-CaO and reduction agent that magnesium vapor is difficult to escape from the pellets. Thus the appropriate briquetting pressure could balance progresses of contact between MgO-CaO and reduction agent, and the escape of magnesium vapor.

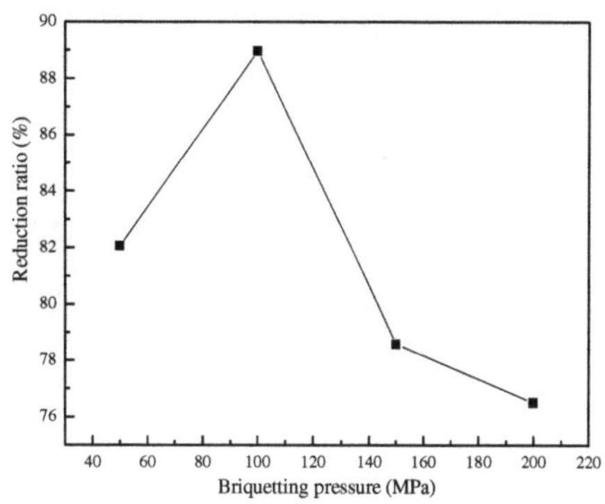

Figure 6. Effect of briquetting pressure on reduction ratio of MgO.

Effect of CaF_2 or MgF_2 on Reduction Ratio of MgO

Figure 7 illustrates the effect of CaF_2 or MgF_2 on reduction ratio of MgO. The reduction temperature 1413 K, briquetting pressure 50 MPa and vacuum 4 Pa. At the addition of CaF_2 and MgF_2 1% of total charge mass respectively, the reduction rates were 87.07% and 84.97%. The reduction rates reached the maximum at the addition of CaF_2 (MgF_2) 3%. With the additions of MgF_2 5%, the reduction rates slightly decreased that 95.59% and 94.95%. The addition of CaF_2 (MgF_2) could activate the surface of MgO-CaO and the reactivity of oxygen ion was increased which enhanced the reaction between MgO-CaO and aluminum.

The addition of CaF_2 (MgF_2) was an effective way to accelerate reduction progress, but the mass of addition should be appropriate as 3% of total charge mass according to the considerate of cost and the high reduction ratio of magnesium. The accelerating effect of MgF_2 to reduction was better than CaF_2 of addition of 3% or above.

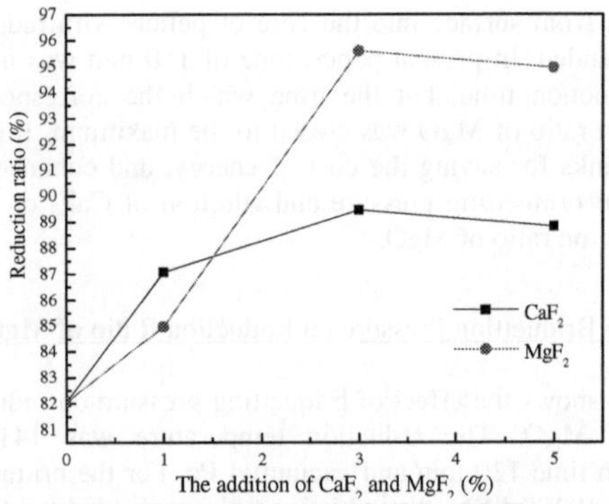

Figure 7. Effect of the addition of CaF_2 or MgF_2 on reduction ratio of MgO.

The reduction ratio of MgO was 97.01% at temperature 1413 K, time 150 min, briquetting pressure 100 MPa and 3% MgF_2.The magnesium produced in reduction as shown in Figure 8, the ratio of purity beyond 99.17% by chemical analysis.

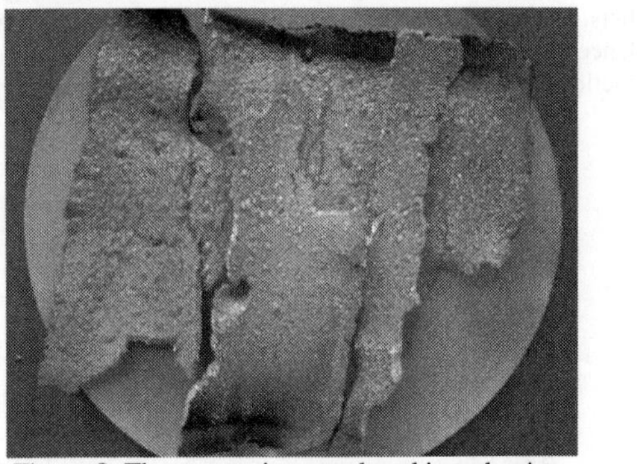

Figure 8. The magnesium produced in reduction.

Alumina Leaching from Reduction Residue

According to the XRD pattern shown in Figure 1, $CaO{\cdot}2Al_2O_3$ was the major phase in reduction residue. Using alkaline solution of a mixture of Na_2CO_3 and NaOH with the concentration ratio of 100:75 to leach $NaAl(OH)_4$ from $CaO{\cdot}2Al_2O_3$ at temperature 95 °C and time 120 min. The leaching ratio of Al_2O_3 reached 88%. $NaAl(OH)_4$ was converted to $Al(OH)_3$ by carbonation. CO_2 was collected at the stage of dolomite and magnesite calcinations. This step is contributed to reduce cost and reduced the emission of CO_2. The product of $Al(OH)_3$ is shown in Figure 9.

The phase of residue after leaching was shown in Figure 10. $CaCO_3$ could be isolated firstly and then be the raw material of CaO in cement industry. The leaching process is the following

$$CaO{\cdot}2Al_2O_3(s) + 2NaOH(l) + Na_2CO_3(l) + 7H_2O = CaCO_3(s) + 4NaAl(OH)_4(l) \quad (6)$$

and the reaction during the carbonation of $NaAl(OH)_4$ is

$$2NaAl(OH)_4(l) + CO_2(g) = 2Al(OH)_3(s) + Na_2CO_3(l) + H_2O \quad (7)$$

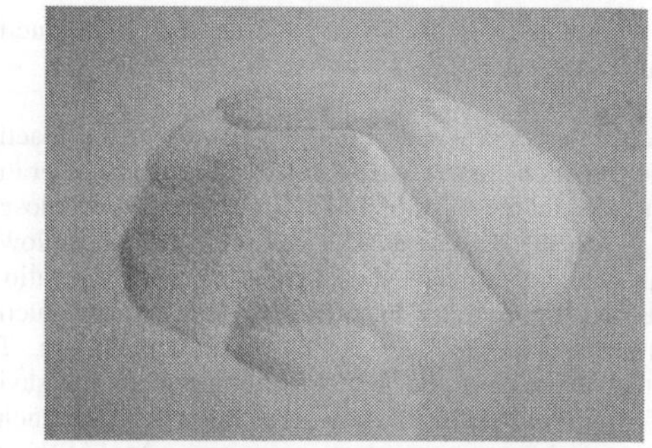

Figure 9. The product of $Al(OH)_3$.

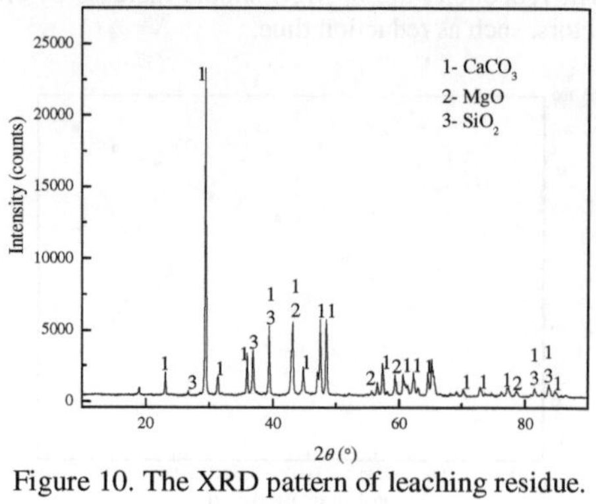

Figure 10. The XRD pattern of leaching residue.

Conclusions

(1) The reduction temperature, reduction time, briquetting pressure and the addition of CaF_2 (MgF_2) are important factors to reduction process. With increasing the above factors increased the reduction ratio of MgO. But briquetting pressure and the addition of CaF_2 (MgF_2) should be in the optimum ranges.

(2) According to the analysis of experimental results, the possible industrial operating conditions of reduction process are reduction temperature 1413 K, reduction time 150 min, vacuum 4 Pa, pressure for briquetting of raw materials 100 MPa and the addition of 3% MgF_2.

(3) The main phase of reduction residue is $CaO{\cdot}2Al_2O_3$ that can extract sodium aluminate and produce $Al(OH)_3$ by carbonation. The leaching ratio reached 88% under leaching temperature 95 °C of 120 min and the concentration ratio of Na_2CO_3 to NaOH was 100:75.

(4) Compare to the Pidgeon process, the new reduction technology presents in this paper can accelerate reduction process, $Al(OH)_3$ can be produced from reduction residue, CO_2 is collected at the stage of dolomite and magnesite calcinations and $CaCO_3$ in the leaching residue also can be used, which mean the new reduction technology reduce production cost effectively and almost none waste be discharged into environment.

References

1. D. Eliezer, E. Aghion, F.H. Froes, "Magnesium Science, Technology and Applications", *Advanced Performance Materials*, 5 (1998), 201-212.

2. G. Hanko, H. Antrekowitsch, P. Ebner, "Recycling Automotive Magnesium Scrap", *JOM*, 2 (2002), 51-54.

3. I.M. Morsi, K.A. Elbarawy, M.B. Morsi et al., "Silicothermic Reduction of Dolomite Ore Under Inert Atmosphere", *Canadian Metallurgical Quarterly*, 1 (2002), 15-28.

4. D. Minic, D. Manasijevic, J. Dokic et al., "Silicothermic reduction process in magnesium production", *Journal of Thermal Analysis and Calorimetry*, 2 (2008), 411-415.

5. G. C. Holywell, "Magnesium: The First Quarter Millennium", *JOM,* 7 (2005), 26-33.

6. W.-X. Hu, J. Liu, N.-X. Feng, "Analysis of New Vacuum Reduction Process in Magnesium Production", *Non-ferrous Mining and Metallurgy*, 4 (2010), 40-42.

7. F. Gao, Z.-R. Nie, Z.-H. Wang et al., "Environmental Assessment of Energy Usage Strategies for Magnesium Production Using the Pidgeon Process", *Journal of Beijing University of Technology*, 6 (2008), 646-650.

8. S.-Y. Yan, "The current status and analysis of making magnesium by means of Pidgeon Process in China", *Light Metals,* 6 (2005), 37-40.

9. J.R Wynnyckyj, L. M. Pidgeon, "Equilibria in the Silicothermic Reduction of Calcined Dolomite", *Metallurgical transactions B*, 2 (1971), 979-986.

Magnesium Technology 2011
Edited by: Wim H. Sillekens, Sean R. Agnew, Neale R. Neelameggham, and Suveen N. Mathaudhu
TMS (The Minerals, Metals & Materials Society), 2011

MULTIPHASE DIFFUSION STUDY FOR Mg-Al BINARY ALLOY SYSTEM

Young-Min Kim[1], Sazol Kumar Das[1], Manas Paliwal[1] and In-Ho Jung[1]
[1]Department of Mining and Materials Engineering, McGill University, 3610 Rue University, Montreal, Quebec, Canada, H3A 2B2

Keywords: Mg-Al binary alloys, Annealing, Multiphase Diffusion

Abstract

Multiphase diffusion simulation and annealing experiments have been performed for Mg-Al binary alloys at various temperatures. Annealing experiments of Mg–3wt% Al and Mg–6wt% Al alloys were carried out at 330 and 400 °C for various times and the change of concentration profiles of Al in grains were measured by Electron Probe Micro Analyzer (EPMA). In order to simulate this annealing process and understand the diffusion of Mg-Al alloys, diffusion model was developed by using Finite Difference Method (FDM) coded in FORTRAN. In the diffusion simulations, composition-independent inter-diffusion coefficients were used and the intermetallic phases were assumed to have equilibrium compositions.

Introduction

Magnesium has the lowest density (two-third that of aluminum) and therefore it is a prime candidate material for use in automobiles [1]. In spite of a plenty of scopes to apply magnesium-based component in cars, the current usage is quite limited due to poor room temperature formability of Mg. Numerous researches are being carried out to improve the formability from the view point of process optimization and alloy design. Most of wrought Mg alloys are casted via ingot casting or twin roll casting route; then they are homogenized, rolled and annealed to produce Mg sheets at relative high temperature. Although these high temperature processes always involve the diffusion of alloying elements, the diffusion in multi-component Mg alloys has not been well studied. The diffusion is also important to understand creep mechanism of Mg alloys [2]. Extensive research has been performed to develop diffusion models and databases for Fe- and Ni-based alloys [3-9]. However, no systematic study has been conducted to model diffusion in Mg-based alloys. One of the reasons might be the asymmetric diffusivity in hcp-Mg. Diffusion in hcp-Mg is faster along the c-axis than the a-axis.

In order to understand and optimize the high temperature diffusion control processes of Mg alloys, fundamental and systematic diffusion studies for Mg alloys are urgently needed. In the present study, we have developed a multiphase diffusion model for Mg-Al binary alloy using the Finite Difference Method (FDM) coded in FORTRAN. Composition-independent and symmetric inter-diffusion coefficients were used for all phases and the intermetallic phases were assumed to have equilibrium compositions throughout the whole diffusion process. Annealing experiments of as-cast Mg–3wt% Al and Mg–6wt% Al alloys were carried out at 330 and 400 °C for various times and the change of Al concentration profiles in grains were measured. The diffusion of Al in randomly oriented Mg matrix was simulated using the diffusion model.

Methodology

Experimental method

Two compositions (Mg–3wt% Al and Mg–6wt% Al) were casted to observe the Al concentration profile in the hcp-Mg and β-$Mg_{17}Al_{12}$ phases. The casting was performed using a Cu plate mold (14 mm × 140 mm × 370 mm) with a cooling rate of about 80 °C/s. The casted Mg-Al binary alloy samples were then annealed at 330 and 400 °C for 1, 2, 4 and 8 h, respectively, and Al concentration profiles were examined using the Electron Probe Micro Analyzer (EPMA).

Diffusion model

A simple diffusion model was developed based on several assumptions: 1) the phase geometry is planar (one dimensional diffusion); 2) the composition at the interface is in equilibrium state; 3) the inter-diffusion coefficient of each phase is constant and 4) the diffusion in hcp-Mg is symmetric. It is well known that the diffusion in hcp-Mg is asymmetric but unfortunately no diffusivity of Al along the a- and c-axis of hcp-Mg have been determined. In the present study, as-cast dendrites (grains) showed random orientation (discussed in the Results section), so the averaged symmetric diffusivity of Al in hcp-Mg was used for the sake of simplicity.

The Finite Difference Method (FDM) is used to solve the diffusion equation, which is coded in FORTRAN. The general equation for diffusion using FDM is as follows:

$$C_i^{m,j+1} = C_i^{m,j} + \tilde{D}_p^m \frac{(\Delta t)}{(\Delta x)^2}(C_{i+1}^{m,j} - 2C_i^{m,j} + C_{i-1}^{m,j}) \qquad (1)$$

where $C_{i,j}{}^m$ is the concentration of component m at time step j in node i (for the binary Mg-Al system, m = Mg and Al), Δt is the time step, Δx is the distance step, and $\tilde{D}_p^m$ is the inter-diffusion coefficient for component m in the p phase. Eq. (1) is the numerical solution for the diffusion equation. For the stability of the FDM approximation, Δt must be chosen with respect to the constant present in that equation. To make the FDM approximation stable, Δt is chosen according to Eq. (2):

$$2\tilde{D}_p^m \frac{\Delta t}{(\Delta x)^2} \le 1 \qquad (2)$$

In this study, we performed two kinds of simulations: single phase diffusion, to simulate the annealing process of as-cast Mg-Al alloys, and multiphase diffusion, to simulate the diffusion couple at the Mg-Al junction. In multiphase diffusion, the most important thing is to solve the moving boundary problem with consideration of the phase transformation and nucleation at the interface. Among many different ways to solve the moving boundary problem, the Murray-Landis transformation [10] is adopted in this study. The concentration gradient was regarded as zero (zero mass transfer) at both boundaries of simulated samples. The interface compositions were determined from the Mg-Al binary equilibrium phase diagram [11] and the inter-diffusion coefficients of the various phases present in the Mg-Al system were obtained from the literature [12-16].

Results and Discussion

Annealing experiments

The variation of the Al concentration profiles between cores of two grains was examined with annealing time and temperature using EPMA. Area mapping was performed for the as-cast Mg-Al plate sample to observe local in-homogeneities throughout the whole plate in the thickness direction and also to find the overall composition of Al inside the plate. The results are reported in Fig. 1 where we can see that more scattering occurred for the Mg-6wt% Al samples than the Mg-3wt% Al ones. This is due to the fact that larger β-$Mg_{17}Al_{12}$ grains can be formed with higher Al concentration. Despite the scatter observed in the Al concentration data, the concentration of Al is constant throughout the samples, which means there is no significant segregation during solidification.

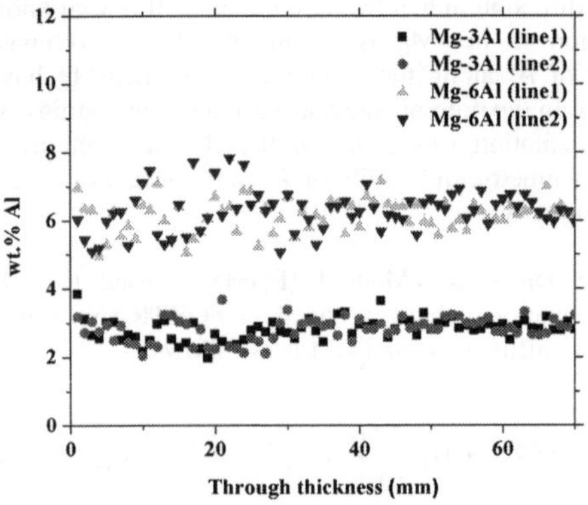

Figure 1. Area mapping results for Al concentration in Mg-3wt% Al and Mg-6wt% Al as-cast plate samples.

In order to confirm that the dendrites in as-cast samples have no preferred orientation, X-ray pole-figures were taken and presented in Fig. 2. A Bruker D8 X-ray diffraction system equipped with 2-dimentional HI-STAR detector was used to measure the texture of as-cast Mg-6wt% Al alloy. Texture measurements were done for the three main peak intensity planes for Mg which is Basal, Prismatic and Pyramidal (see Fig. 2). As seen from the pole figures, the as-cast alloy shows a random texture with no preferred orientation. As well-known, the diffusivity in hcp structure is asymmetric. To consider this asymmetric diffusion of Mg, we are studying the directional diffusivity of Mg with Mg single- and poly-crystals with different grain sizes and we will extend our diffusion model with Mg directional diffusivity in the future. For now, the present study deals with randomly oriented Mg grains (dendrites) and the averaged symmetric diffusion coefficient for Al in hcp-Mg can reasonably explain the diffusion behaviour of Al.

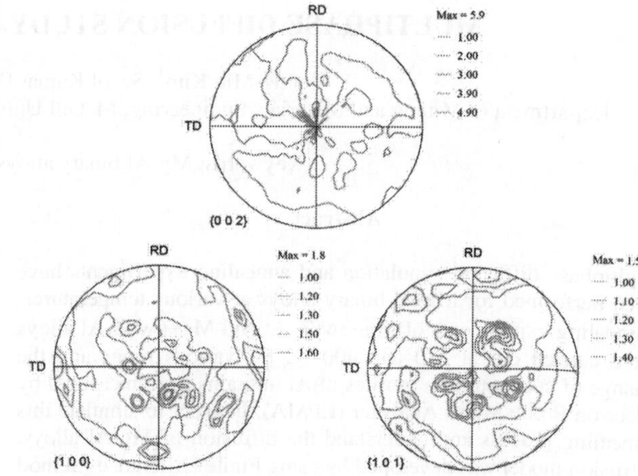

Figure 2. Pole figures of as-cast Mg-6wt%Al sample. a) Basal - (0001), b) Prism - {10 $\overline{1}$ 0} and c) Pyramidal - {10 $\overline{1}$ 1}.

The as-cast microstructures (SEM image) of Mg-3wt% Al and Mg-6wt% Al alloys are presented in Fig. 3. As it can be seen, the size of the dendrites (grains) is larger than 20 μm and β-$Mg_{17}Al_{12}$ is dispersed around the dendrites in both alloys. In the case of Mg-3wt% Al, the size of the β phase is small (~3 μm) compared to the one (up to ~30 μm) in Mg-6wt% Al. The large β phase can be at the origin of the large scattering in the Al concentrations depicted in Fig. 1.

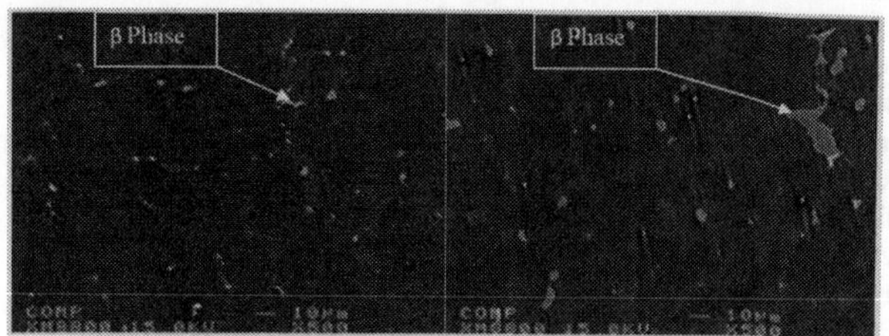

(a) (b)

Figure 3. As-cast microstructures (SEM) of (a) Mg-3wt% Al and (b) Mg-6wt% Al. β phases were confirmed by EDS analysis.

As expected, the dendrites in as-cast samples have quite significant cored microstructure. For the Mg-6wt% Al alloy, the Al concentration at the dendrite core is about 2 wt% and reach about 8 wt% near the boundary. In order to investigate the homogenization of Al concentration in the grain, annealing of the samples was carried out at 330 and 400 °C for up to 8 h. The microstructural changes due to this annealing are presented in Fig. 4. For the Mg-3wt% Al alloy, the sample was almost fully homogenized after 8 h at 330 and 400 °C and the β phase almost vanished. In the case of Mg-6wt% Al alloy, the microstructure is almost fully homogenized at 400 °C after 8 h but not at 330 °C even after 8 h. However, some segregated area can still be observed in the Mg-6wt% Al alloy after 8 h as it can be seen in Fig. 4(d). To investigate the change of Al concentration profile across the grain during the annealing process, EPMA line-scans of Al and Mg were carried out for the as-cast and annealed samples.

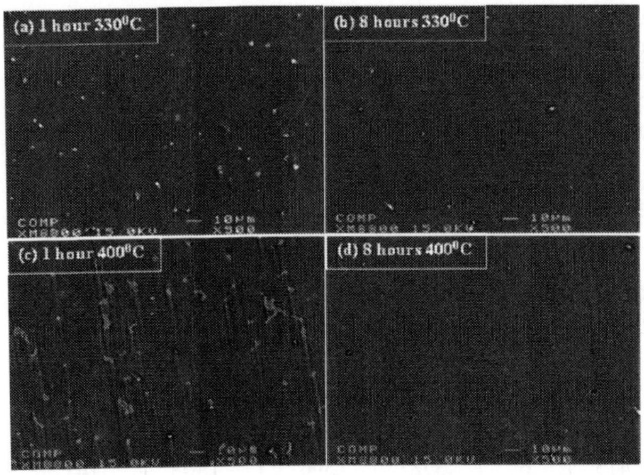

Figure 4. Microstructures of (a) and (b) Mg-3wt% Al, and (c) and (d) Mg-6wt% Al, after annealing.

To avoid the sudden change of the Al profile due to the presence of the β phase at the grain boundary, the line-scans were carefully positioned between the cores of grains through the grain boundary without crossing the β phase as shown by the bold lines in Fig. 5. In each sample, 15 randomly selected lines were scanned. The measured Al concentration profiles for Mg-6wt% Al alloy annealed at 330 and 400 °C are presented in Fig. 6. The as-cast profile shows very strong segregation of Al near the grain (dendrite) boundary. As mentioned above, the Al concentration near the grain boundary is about 8 wt%. Annealing of the sample at 330 °C for 8 h was not enough to homogenize the Al concentration in the grain. On the other hand, the Al concentration was found to be fully homogenized at 400 °C after 8h.

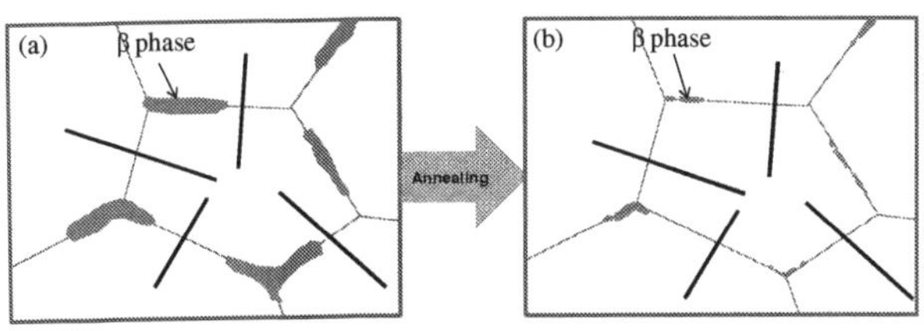

Figure 5. Schematic representation of Mg-Al binary alloy microstructure; (a) as-cast and (b) as-annealed sample (bold lines represent the position of the elemental line scans performed in the present study).

Diffusion simulation

Diffusivity of Al in hcp-Mg has been determined by several researchers [12-14] and the values are slightly different. In order to find out the reliability/accuracy of the available diffusivity data, we performed diffusion simulation for the annealing process shown in Fig. 6 using available inter-diffusion coefficients. The results are reported in Fig. 7. The as-cast concentration profile measured in the present study was used as an input for the simulation. It was also assumed that the Al profile is not influenced by the grain boundary. As the diffusivity of Abe and Onodera [13] is quite similar to that of Moreau et al. [12], we do not present their results here.

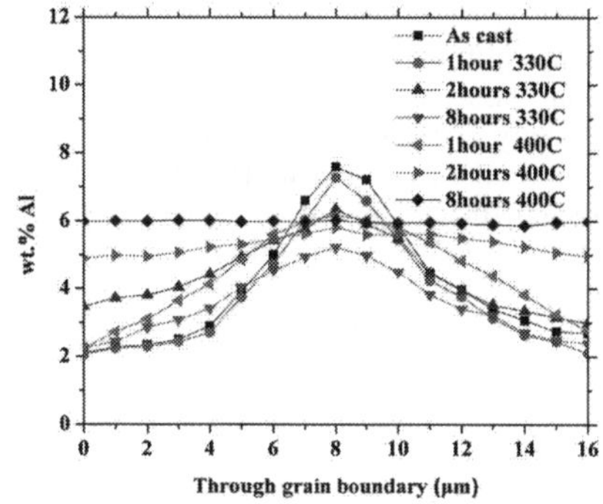

Figure 6. Change of the Al concentration profiles of Mg-6wt% Al between cores of grains (dendrites) through grain boundary with annealing time.

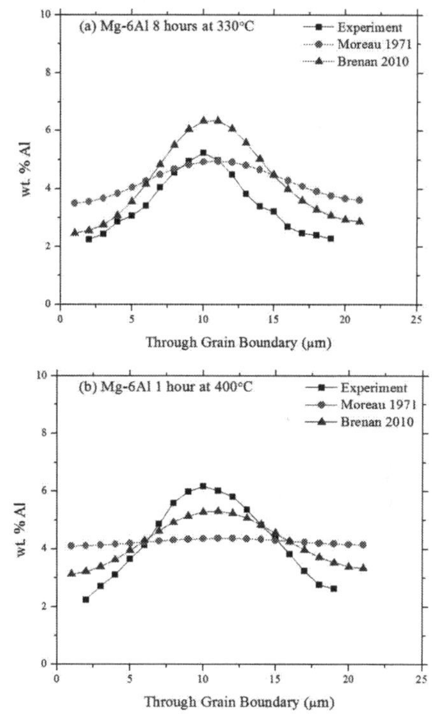

Figure 7. Calculated Al concentration profile in hcp-Mg using two available diffusion coefficients of Al from literature [12,14] in comparison with the present experimental annealing data.

After comparison of the simulation results with the experimental ones, it was found that the diffusivity reported by Moreau et al. [12] at 400 °C is too high to reproduce the experimental results (Fig. 7(b)). The simulation with the diffusivity by Moreau et al. already reached homogenization of the Al concentration at 400 °C in 1 h. On the other hand, the simulated Al profile with the diffusivity value of Brennan et al. [14] is similar to the expeirmental results at 400 °C in 1 h. The difference in the diffusivity values of Al at 330 °C from Moreau et al. [12] and Brennan et al. [14] is smaller than that at 400 °C (Fig. 7(a)), but the diffusivity from Moreau et al. [12] is still higher than that of

Brennan et al. [14]. The simulated Al profile at 330 °C is similar to each oher but still the results with Brennan et al. are in better agreement with the experimental annealing data. Even when the spherical geometry of the grains was considered in the diffusion simulation, the results were similar. Therefore, we found that the inter-diffusion coefficient of Al from the recent study of Brennan et al. [14] can more accurately reproduce the diffusion process for Mg-Al alloys. Using the diffusion coefficient of Al in hcp-Mg from Brennan et al. [14] and the diffusion coefficients for other phases listed in Table I, the multiphase diffusion simulation at 415 and 420 °C for 6 and 10 days, respectively, were performed and the results are presented in Fig. 8.

Table I. Inter-diffusion coefficient values for each phase in the Mg-Al binary system used in the present study.

$$\tilde{D} = D_o \exp(-Q/RT)$$

Elements		D_o (m/s^2)	Q (J)	Reference
Al	HCP	4.13e-3	157696	Brenan 2010
	HCP	1.20e-3	143449	Moreau 1971*
	FCC	1.41e-5	126719	Yao 2008
$Mg_{17}Al_{12}$		7.90e-5	117458	Funamizu 1972
Mg_2Al_3		3.00e-8	56848	

** This value is not used for Mg-Al multiphase diffusion simulation in the present study.*

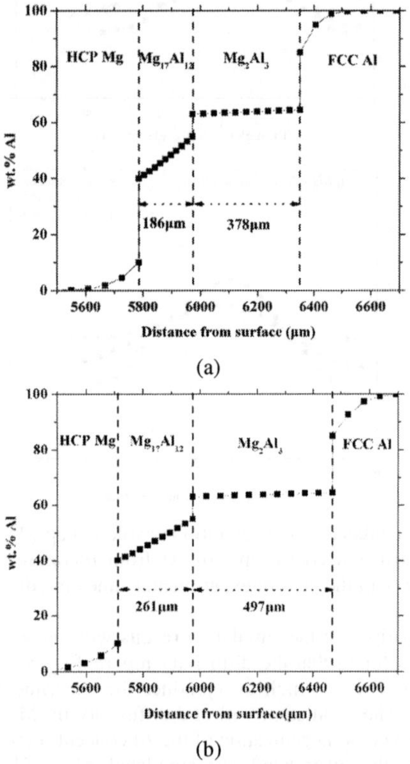

(b)

Figure 8. Calculated concentration profile of Al for Mg-Al binary diffusion couple. (a) at 415 °C for 6 days, (b) at 420 °C for 10 days.

The calculated thicknesses of intermetallic layers are compared with experimental results [17] in Table II. The calculated thicknesses of the $Mg_{17}Al_{12}$ and Mg_2Al_3 layers as well as the Al profiles in hcp and fcc phases are well reproduced by the present simulation with the selected diffusivity data in Table I. The Al concentration profile is clearly changed with diffusion time. Both calculated and experimental results show that the layer of Mg_2Al_3 is thicker than that of $Mg_{17}Al_{12}$, which results from the higher inter-diffusion coefficient for Mg_2Al_3 compared to that for $Mg_{17}Al_{12}$.

Table II. Calculated thickness of diffusion layers for the β and γ phase compared with experimental results [17] obtained with diffusion couple.

T (°C)	Time (days)	Thickness of diffusion layer (μm)			
		Mg_2Al_3 (γ phase)		$Mg_{17}Al_{12}$ (β phase)	
		Present work	Exp.	Present work	Exp.
415	6	378	396	186	141
420	10	497	532	261	229

Summary

A multiphase diffusion simulation model is developed for Mg-Al binary alloy system by using FDM. The model is validated by simple annealing and diffusion couple experiments. In this study, simulation results are in good agreement with experimental results. The growth rate of the Mg_2Al_3 phase is faster than that of the $Mg_{17}Al_{12}$ phase. The present diffusion simulation model is generally applicable to binary diffusion couple. The results of multiphase diffusion simulation for Mg-Al diffusion couple are promising and will lead this study further to the multi-component system.

Acknowledgement

Financial support from General Motors of Canada Ltd. and the CRD research programs of the Natural Sciences and Engineering Research Council of Canada is gratefully acknowledged.

References

1. E. Baril, P. Labelle, and M.O. Pekguleryuz, "Elevated temperature Mg-Al-Sr: Creep resistance, mechanical properties, and microstructure," *JOM-Journal of Minerals, Metals and Materials Society,* 55 (11) (2003), 34-39.
2. M.O. Pekguleryuz, and A.A. Kaya, "Creep Resistant Magnesium Alloys for Powertrain Applications," *Advanced Engineering Materials*, 5(12) (2003), 866-878.
3. J.A. Nesbitt, "COSIM—A Finite-Difference Computer Model to Predict Ternary Concentration Profiles Associated With Oxidation and Interdiffusion of Overlay-Coated Substrates" (Technical report no. NASA/TM-2000-209271, National Aeronautics and Space Administration, Glenn Research Center, 2000)
4. M.S.A. Karunaratne et al., "A multicomponent diffusion model for prediction of microstructural evolution in coated Ni based superalloy systems," *Materials Science and Tecnology*, 25(2) (2009), 287-299.
5. K.V. Dahl, J. Hald, and A. Horsewell, "Interdiffusion Between Ni-Based Superalloy and MCrAlY Coating," *Defect and Diffusion Forum*, 258-260 (2006), 73-78.
6. C.J. O'Brien et al., "VisiMat©–Educational Tool for Multicomponent Diffusion in 2 and 3 Dimensions," *Defect and Diffusion Forum*, 266 (2007), 199-207.

7. B.-J. Lee, "Computer Simulation of Multicomponent and Multiphase Diffusion," *Proceedings of the 2nd Pacific Rim International Conference on Advanced Materials and Processing*, (1995), 829-835.
8. A. Engstrom, L. Hoglund, and J. Agren, "Computer Simulation of Diffusion in Multiphase Systems," *Metallurgical and Materials Transactions: A*, 25A (1994), 1127-1134.
9. A.Engstrom, J.E. Morral, and J. Agren, "Computer Simulations of Ni-Cr-Al Multiphase Diffusion Couples," *Acta Materialia*, 45(3) (1997), 1189-1199.
10. W.D. Murray, and F.Landis, "Numerical and Machine Solutions of Transient Heat Conduction Problems involving Phase Change," Trans. *ASME. J. Heat Transfer*, 81(C) (1959), 106-112.
11. C.W. Bale et al., "FactSage Thermochemical Software and Databases - Recent Developments," *CALPHAD: Computer Coupling of Phase Diagrams and Thermochemistry,* 33(2), 295-311.
12. G. Moreau, J.A. Cornet, and D. Calais, "Acceleration de la diffusion chimique sous irradiation dans le systeme aluminium-magnesium," *Journal of Nuclear Materials*, 38 (1971), 197-202.
13. T. Abe, and H. Onodera, "Solute atmosphere round an edge dislocation in Al-Cu-X and Fe-C-X ternary alloys," *Journal of Phase Equlibria*, 22(4) (2001), 491-497.
14. S. Brennan et al., " Impurity and Tracer Diffusion Studies in Magnesium and its Alloys," *Magnesium Technology 2010, TMS* (2010), 537-538.
15. Y. Funamizu, and K. Watanabe, "Interdiffusion in the Al-Mg system," *Transactions of the Japan Institute of Metals*, 13 (1972), 278-283.
16. J. Yao et al., "Diffusional mobility for fcc phase of Al-Mg-Zn system and its applications," *CALPHAD*, 32 (3) (2008), 602-607.
17. C. Brubaker, and Z.K. Liu, "Diffusion Couple Study of the Mg-Al System," *Magnesium Technology 2004, TMS* (2004), 229-234.

Magnesium Technology 2011
Edited by: Wim H. Sillekens, Sean R. Agnew, Neale R. Neelameggham, and Suveen N. Mathaudhu
TMS (The Minerals, Metals & Materials Society), 2011

Experiments and Modeling of Fatigue Damage in Extruded Mg AZ61 Alloy

J.B. Jordon[1], J.B. Gibson[2], M.F. Horstemeyer[2,3]

[1]Department of Mechanical Engineering, The University of Alabama, Tuscaloosa, Al 35487
[2]Center for Advanced Vehicular Systems (CAVS), Mississippi State University, Mississippi State, MS 39762
[3]Department of Mechanical Engineering, Mississippi State University, Mississippi State, MS 39762

Keywords: Fatigue, Structure-Property Relations; Magnesium alloys

Abstract

In this study, structure-property relations with respect to fatigue of an extruded AZ61 magnesium alloy were experimentally quantified. Strain-life experiments were conducted in the extruded and transverse orientations under low and high cycle conditions. The cyclic behavior of this alloy displayed varying degrees of cyclic hardening depending on the strain amplitude and the specimen orientation. The fracture surfaces of the fatigued specimens were analyzed using a scanning electron microscope in order to quantify structure-property relations with respect to number of cycles to failure. Intermetallic particles were found to be the source of fatigue initiation on a majority of fracture surfaces. Finally, a multistage fatigue model based on the relative microstructural sensitive features quantified in this study was employed to capture the anisotropic fatigue damage of the AZ61 magnesium alloy.

Introduction

Given the recent push for energy efficient design, interest in magnesium alloys for automotive structural applications has greatly expanded in recent years. This heightened interest has pushed the materials community to investigate not only cast magnesium but also wrought products for their high strength-to-weight ratio and fatigue resistance. As these wrought products begin to take hold in the automotive community, the need to quantify and make high integrity fatigue predictions of the material performance is of increased importance.

Limited work has been performed recently towards characterizing wrought magnesium cyclic behavior. Cyclic experiments involving extruded AM30, AZ31B, and AZ31 magnesium alloys have been employed to characterize the cyclic response of these materials. These materials were found to have strong cyclic hardening characteristics. In addition, with an increase in total strain amplitude, both plastic strain amplitude and mean stress increased for these alloys. Furthermore, the well known Coffin-Manson law and Basquin's equation were found to correlate with the experimental fatigue results of these alloys quite well [1-6].

The current body of literature on fatigue of wrought magnesium alloys suggests that the mechanisms of fatigue damage evolution are consistent with the well established stages of fatigue for steel and aluminum alloys [7]: incubation, microstructurally small crack growth; physically small crack growth; and long crack growth. Much like wrought aluminum alloys, inclusions can cause fatigue crack incubation in extruded magnesium alloys [8]. Other studies [1-5] on fatigue of wrought magnesium alloys did not explicitly report on the sources of fatigue incubation. However, persistent slip bands and grain boundaries are also thought to be sources of fatigue initiation [1,2,4,9-11]. The mechanisms of twinning on fatigue incubation are unclear, but this is an area of ongoing research. A small level of interest has been shown to the issue of small crack fatigue in wrought magnesium alloys. The inapplicability of the well established linear elastic fracture mechanics (LEFM) parameters (*i.e.,* stress intensity factors) to characterize the growth of fatigue cracks on the order of microns in length is established in wrought magnesium alloys [12,13]. On the other hand, fatigue cracks on the order of millimeters in length and larger are characterized adequately by LEFM parameters [14]. The transition from "small" crack growth to long crack growth has not been established for extruded magnesium alloys. However, the small crack region for cast magnesium alloys appears to be in the range of 60-100 μm [10], which is six to ten times the dendrite cell size.

In regards to the effect of texture on fatigue life in magnesium alloys, the crystallographic orientation appears to have some influence on final failure [9]. Work on other HCP metals was performed on titanium alloys to capture the effects of texture on the fatigue properties [15,16]. Rolling or extruding processes induce strong anisotropy in metals due to preferred crystallographic orientation. This crystallographic orientation has a large effect on twinning and slip in the HCP, therefore affecting the material cyclic behavior. Like the titanium alloys, the effects of texture in wrought magnesium alloys were found to be very important in the cyclic response [8,13,17].

The purpose of this paper is to experimentally quantify the fatigue behavior and the structure-property of an extruded AZ61 magnesium alloy. In addition, a microstructurally-sensitive fatigue model has been developed for wrought alloys based on the experimental results and observations presented here. This model, the MultiStage Fatigue (MSF) model, first proposed by McDowell *et al.* [14], captures the effects of microstructural features as reflected in the incubation and growth stages for both low cycle fatigue (LCF) and high cycle fatigue (HCF). The MSF model was first developed for cast aluminum alloys and was employed to predict the fatigue performance based variation in porosity, grain size, dendrite cell size, and inclusion size and type. The model was later extended to cast magnesium by capturing the network of porosity for AE44 [18], AM50 [19], and AZ91 [20], and was based upon the *in-situ,* scanning electron microscope (SEM) small crack analysis [9-11]. The MSF model was further modified to predict fatigue life in a high strength wrought aluminum alloy by capturing the fatigue crack incubation and growth resulting from iron rich intermetallic particles [18, 21]. These modeling efforts on the wrought aluminum alloy specifically addressed local constrained cyclic microplasticity and crystallographic orientation

[18]. This updated model has also been used for fatigue life predictions of Laser Engineered Net Shaping (LENS) processed steel [22] for A356 and A380-F aluminum alloys [23, 24].

Materials and Experiments

The material used in the study is an extruded AZ61 magnesium alloy. The as-received air-quenched extruded alloy is in the form of an automotive crash rail. The crash rail was extruded at an average temperature of 500 °C and an extrusion exit speed of 4 ft/min. The as-received material was examined with an optical microscope to quantify the initial microstructure in both a transverse material orientation and an extrusion or longitudinal orientation. X-ray diffraction (XRD) was also performed to quantify texture of the as-received material.

Rectangular cross-section tensile specimens were machined in the transverse and longitudinal orientations. The specimens were designed from a scaled ASTM-E8 standard to accommodate the size of the extrudate. The nominal specimens had a 15 mm gage length with a width of 4.35 mm and an as-extruded thickness of 2.5 mm. This design was used for both quasi-static and cyclic tests with the exception that the specimens used for cyclic loads were hand-ground with silicon carbide paper to remove residual stresses and to reduce the machining surface finish effects, which affect the fatigue characteristics.

Monotonic tensile tests were performed using an MTS servo-hydraulic load frame with a capacity of 25 kN. Flat tensile specimens, previously mentioned, were tested in both the longitudinal and transverse orientations in strain control at a rate of 0.01/s in an ambient laboratory environment.

Fully reversed cyclic tests were also performed on a 25 kN Servo-hydraulic MTS load frame in an ambient laboratory environment with relative humidity level of 45%. The fatigue specimens were tested to failure (50% load drop) for both transverse and longitudinal orientations under strain-controlled conditions at 5 Hz and the following strain amplitudes: 0.3%, 0.4%, 0.5%, and 0.6%. The specimens tested at 0.2% and 0.275% strain amplitudes were fatigued at 5 Hz to 10,000 cycles and then the tests were stopped and resumed in load control at 30 Hz to failure.

Fracture surfaces from the fatigued specimens were mounted and Au-Pl sputter-coated to improve imaging quality. The surfaces were then examined in a scanning electron microscope to determine sources of fatigue crack incubation as well as determine regions of small and long crack growth stages. Additionally, chemical analysis was used to determine the composition of areas of interest on the fracture surfaces.

Fatigue Model

The fatigue model employed in this research is the MultiStage Fatigue (MSF) model originally proposed by McDowell et al [14] and later modified by Xue et al [18] and Jordon et al. [23]. The basis of the MSF model is that it comprises fatigue into three distinct regimes: crack incubation, microstructurally small crack (MSC) and physically small crack (PSC) growth, and long crack (LC) growth as shown in Eq. (1).

$$N_{Total} = N_{Inc} + N_{MSC/PSC} + N_{LC}. \quad (1)$$

For brevity's sake, the complete model equations are not shown here, but can be found in [23].

Results and Discussion

The longitudinal orientation was found to have equiaxed grains averaging 37.3 μm, an average particle size of 8.4 μm, an average nearest neighbor distance of 88.5 μm, and a particle area fraction of 0.038 (Figure 1a). The transverse orientation was found to also have equiaxed grains that averaged 48.4 μm, particles that averaged 9.4 μm, a nearest neighbor distance of 104.7 μm, and a particle area fraction of 0.034 (Figure 1b). EDAX chemical analysis of the particles indicates that these particles are Al-Mn based. While not shown here, pole figures of the longitudinal and transverse orientation from XRD results suggest that the AZ61 magnesium alloys are not strongly textured; however, the basal plane is aligned with the longitudinal direction. For further analysis, the comparison of the calculated average Taylor factor for both orientations were made, where the longitudinal direction had an average calculated Taylor value of 2.07 and the transverse direction had an average calculated Taylor value of 2.04. The slight difference in grain size and Taylor factor of both orientations are attributed to the slow extrusion rate that accommodates recrystallization.

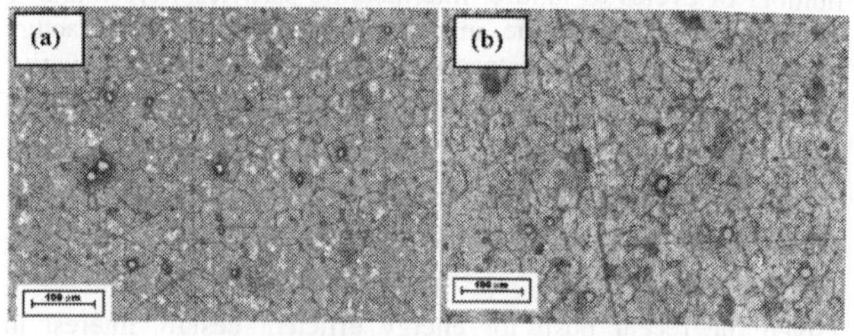

Figure 1. Optical micrographs of as-received extruded AZ61 magnesium alloy polished and etched: (a) longitudinal material orientation with a 37.3μm average grain size (GS), a 8.4μm average particle size (PS), a 0.038 particle area fraction, and a 88.5μm average nearest neighbor distance (NND). (b) Transverse material orientation with a 48.4μm GS, a 9.4μm PS, a 0.034 particle area fraction, and a 104.7μm NND.

Hysteresis and Stress Response

Figure 2 shows the first cycle hysteresis loops for the longitudinal and transverse orientations. The increasing presence of twinning in the first cycles of fatigue loading as the total applied strain amplitude increased was observed for the 0.4% (Figure 2b) and 0.6% (Figure 2c) strain amplitudes. Twinning can accommodate deformation without an increase in stress, which causes the hysteresis loop to have an asymmetric sigmoidal shape to varying degrees in either compression or tension depending on the orientation. A first cycle hysteresis loop for a total applied strain amplitude of 0.2% had a very symmetrical hysteresis loop for both specimen orientations (Figure 2a) while the first cycle hysteresis loop for a total applied strain amplitude of 0.6% had an asymmetric sigmoidal cyclic stress-strain response in tension for the transverse orientation and in compression for the longitudinal specimen orientation due to twin deformation accommodation.

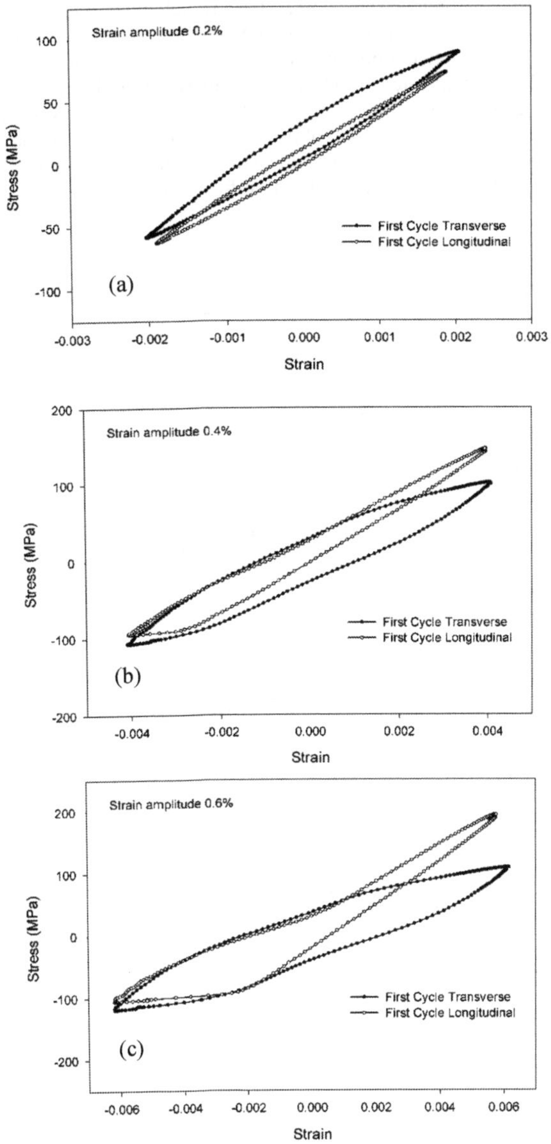

Figure 2. First and half-life cycle hysteresis loops for AZ61 magnesium alloy in the longitudinal and transverse directions with an applied strain amplitude of (a) 0.2%, (b) 0.4%, and (c) 0.6%. Note the stress asymmetry and change in character in the stress-strain response in the 0.4% and 0.6% applied strain amplitudes due to the twinning; these test strain amplitudes were above the cyclic yield stress.

Figure 3. shows the stress amplitude for the longitudinal and transverse orientations versus number of cycles. As the slip becomes the primary deformation mechanism, dislocations begin to pile-up and the material shows a strong hardening response with increased total applied strain amplitude for both orientations as shown in Figure 3a and 3b. This strong observable cyclic hardening is due to the increased plastic strain [1-5]. As the applied strain amplitude decreases, the plastic strain amplitude decreases which in turn diminishes twinning and cyclic hardening alike [25].

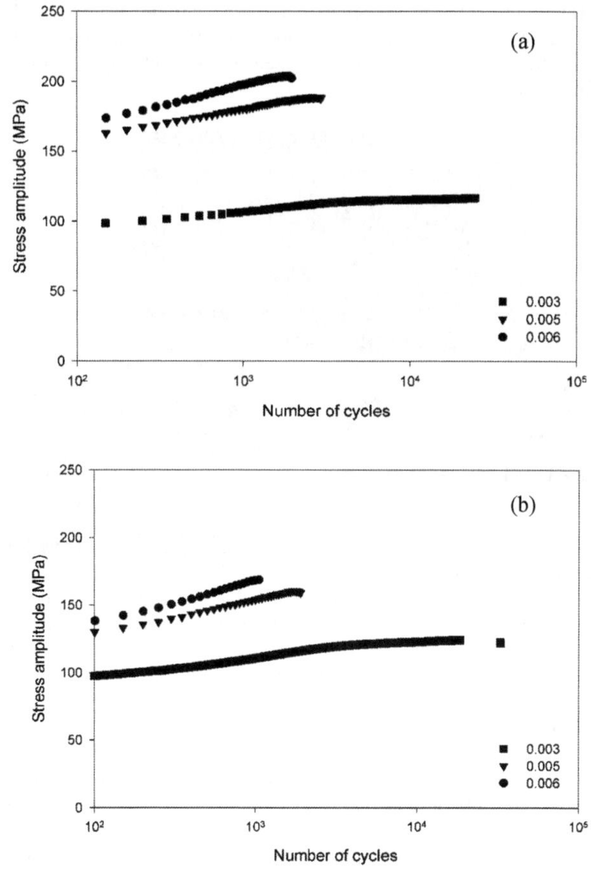

Figure 3. Stress amplitude versus number of cycles for varying applied total strain amplitude of (a) longitudinal and (b) transverse oriented extruded AZ61 magnesium alloy.

Fractography

Extensive fractography was performed on the fatigue fracture surfaces using an SEM to quantify particle sizes that initiated fatigue cracks and the composition of the particles. Fractured particles near the free surface of the specimen were found to be the sites of initiation on most fracture surfaces as shown in Figure 4. Chemical analysis was performed on these particles, and they were found to be Al-Mn based. These particles on the longitudinally oriented specimens had an average equivalent size (derived from the square root of the particle area) of 8.4 μm and a 9.4 μm for the transverse orientation.

SEM fractography conducted in this study resulted in characterization of the distinct stages of fatigue damage: incubation, MSC/PSC, and LC. For low cycle fatigue, the first few hundred cycles of fatigue loading results in particle fracture, which in turn results in incubation of the fatigue crack. The number of cycles for incubation of fatigue cracks is estimated to be approximately 30% based on Bernard et al [26]. SEM analysis of the fracture surfaces showed that a particle near the surface fractured and no debonding was observed. Next, based on comparative textural roughness of the fracture surfaces, the transition from MSC/PSC to long crack growth was estimated to be approximately 1 mm for both the longitudinal and transverse orientation. Due to the small size of the fatigue specimens, the LC growth was considered to comprise a very small amount of the total life and as such is not included in the modeling portion of this work.

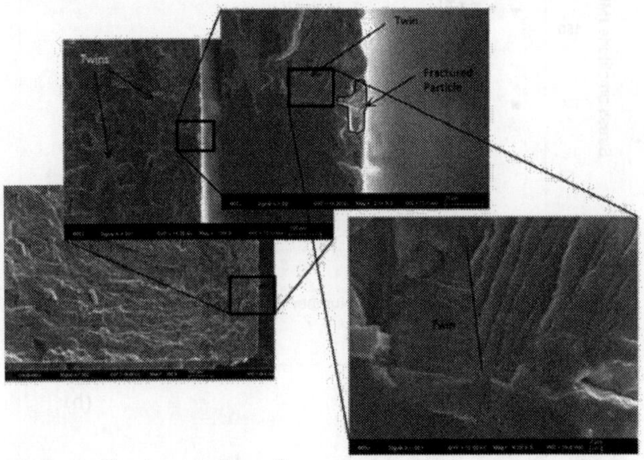

Figure 4. Fracture surface of an extruded AZ61 magnesium alloy longitudinally oriented specimen tested at 0.3% strain amplitude. The scale bar magnitudes from left to right are 1 mm, 200 μm, 100 μm, 20 μm, and 2 μm. Note the twinning on the surface and the particle of initiation (90μm).

Model Correlation

Figure 5 shows the correlations of the MSF model with both the transverse orientation and the longitudinal orientation to experimental data. The difference in the two directions is due to the different grain sizes, particle sizes, yield stresses, and cyclic hardening behavior. All of the other material constants were consistent for both orientation. This is important because it illustrates the sensitivity of the MSF model to capture the effects of microstructural influences on fatigue life. In this particular case, the fatigue life was found to be most sensitive to cyclic hardening parameters related to incubation where the greater the cyclic hardening, the longer the fatigue life.

Figure 6 shows the monotonic stress strain behavior compared to the cyclic stress-strain behavior calculated via companion samples [27]. Maximum tensile stress of the stabilized hysteresis loops was plotted against their respective applied strain amplitudes. Cyclic strain hardening is shown for both the longitudinal and transverse orientations. Cyclic hardening affects the incubation fatigue life by determining the local plastic zone size due to presence of the inclusion. Thus, the differences of the cyclic yield of the two orientations leads to differing local strain fields at the Al-Mn particles responsible for incubating the fatigue cracks.

In an attempt to capture the stochastic nature of the fatigue data, the maximum and minimum particle sizes were employed to obtain error bands for the longitudinal (Figure 7a) and transverse (Figure 7b) directions. In the longitudinal direction, a maximum particle size of 26.4 μm and a minimum of 11 μm were the parameters altered to obtain the bounds of the MSF model for the longitudinal direction. A maximum and minimum particle size of 22 μm and 8.6 μm respectively were used for the transverse orientation bounds. The choices for the selection of the model upper and lower bounds were based on the extreme observable particle sizes taken from the optical analysis and represents the variation of the microstructure. As illustrated in Figure 7, the model captures the upper bounds and lower of the fatigue results, which demonstrates the robustness of the combination of incubation and small crack growth of the MSF model as influenced by varying microstructural features

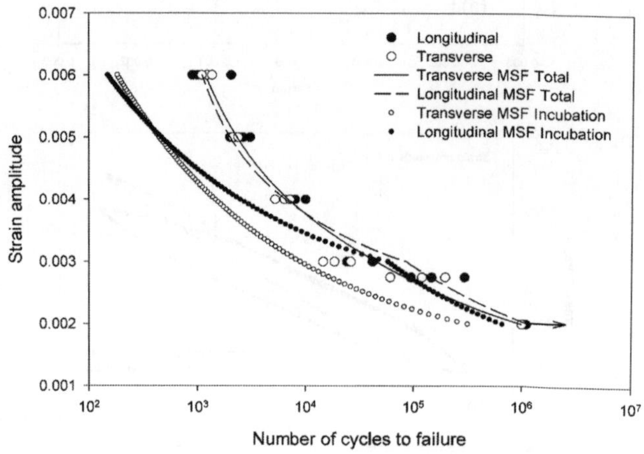

Figure 5. MSF model total life and incubation life comparison with experimental data for extruded AZ61 magnesium alloy.

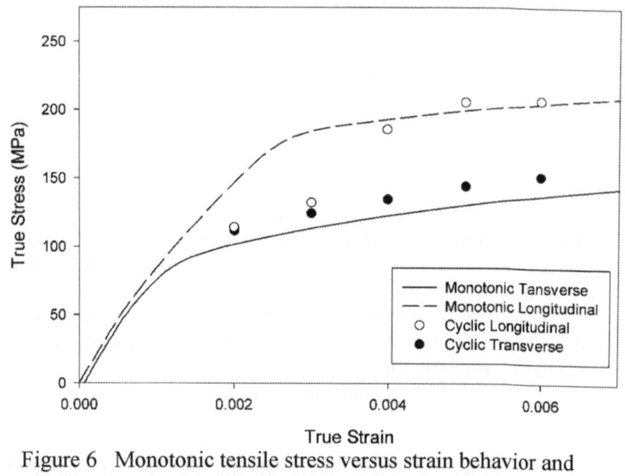

Figure 6 Monotonic tensile stress versus strain behavior and cyclic stress versus strain behavior for fully reversed experiments at half lifetime for an extruded AZ61 magnesium alloy in the transverse and longitudinal material orientations.

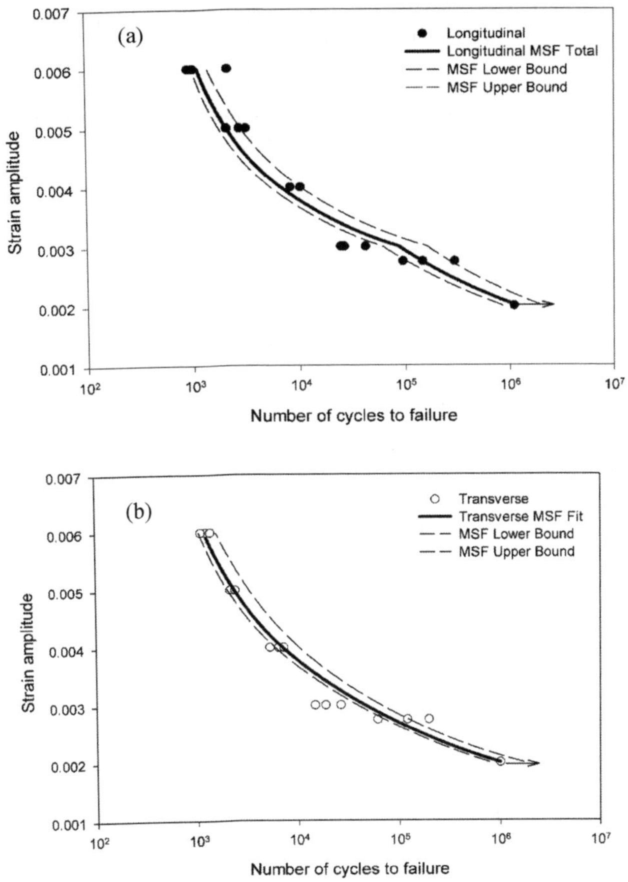

Figure 7 Upper and lower bounds for the MSF model for the (a) longitudinal and (b) transverse directions. These bounds were developed by only varying the particle size by the maximum and minimum of that found on the fracture surfaces.

Conclusions

The monotonic and cyclic behavior of extruded AZ61 magnesium alloy was quantified along with the microstructural characteristics that drove the uncertainty in the mechanical responses. Further, the MultiStage Fatigue (MSF) model was used to capture the strain-life behavior and illustrate the structure-property relationships. The following conclusions were realized from this study.

1) Compressive twinning/detwinning was observed in the longitudinal hysteresis loops due to a tensile stress being induced along the c-axis under compressive loading. Conversely, the transverse hysteresis loops showed tensile twinning/detwinning. Both orientations had a significant Bauschinger effect and were both symmetric and stabilized at the half-life.
2) Fatigue cracks initiated from fractured particles near the free surface.
3) Differences in the longitudinal and transverse orientations were observed on the number of cycles to failure in the high cycle fatigue regime. The longitudinal orientation exhibited greater fatigue resistance compare to the transverse orientation.
4) The MSF model incorporated the differences in grain orientation, grain sizes, particle sizes, and cyclic hardening parameters in order to capture the differences in the longitudinal and transverse fatigue behavior of an extruded AZ61 magnesium alloy. We note that the other material constants were not changed for the incubation and MSC regimes. Also, the MSF model captured the upper and lower bounds on the strain-life curve by simply employing the maximum and minimum particle sizes found in the as-received material. Thus, illustrating the effects of the microstructural content on the fatigue properties.

Acknowledgments - The authors would like to recognize Richard Osborne, James Quinn, Xuming Su, John Allison, Robert McCune, Don Penrod, and Matthew Castanier for their encouragement of this study. This material is based upon work supported by the Department of Energy and the National Energy Technology Laboratory under Award Number No. DE-FC26-02OR22910, and the U.S. Army TACOM Life Cycle Command under Contract No. W56HZV-08-C-0236 through a subcontract with Mississippi State University, and was performed for the Simulation Based Reliability and Safety (SimBRS) research program. This report was prepared as an account of work sponsored by an agency of the United States Government. Neither the United States Government nor any agency thereof, nor any of their employees, makes any warranty, express or implied, or assumes any legal liability or responsibility for the accuracy, completeness, or usefulness of any information, apparatus, product, or process disclosed, or represents that its use would not infringe privately owned rights. Reference herein to any specific commercial product, process, or service by trade name, trademark, manufacturer, or otherwise does not necessarily constitute or imply its endorsement, recommendation, or favoring by the United States Government or any agency thereof. The views and opinions of authors expressed herein do not necessarily state or reflect those of the United States Government or any agency thereof. Such support does not constitute an endorsement by the Department of Energy of the work or the views expressed herein. UNCLASSIFIED: Dist A. Approved for public release.

References

1. Begum S, Chen DL, Xu S, Luo Alan A. Strain-controlled low-cycle fatigue properties of a newly developed extruded magnesium alloy. Metall Mater Trans A 2008;39A:3014-26.

2. Begum S, Chen DL, Xu S, Luo Alan A. Low cycle fatigue properties of an extruded AZ31 magnesium alloy. Int J Fatigue 2009;31:726-35.

3. Begum S, Chen DL, Xu S, Luo Alan A. Low cycle fatigue properties of an extruded AZ31 magnesium alloy. Int J Fatigue 2009;31:726-35.

4. Lin XZ, Chen DL. Strain controlled cyclic deformation behavior of an extruded magnesium alloy. Mater Sci Eng A 2008;496:106-13.

5. Lin XZ, Chen DL. Strain hardening and strain-rate sensitivity of an extruded magnesium alloy. J Mater Eng Perform 2008;17:894-901.

6. Hasegawa S, Tsuchida Y, Yano H, Matsui M. Evaluation of low cycle fatigue life in AZ31 magnesium alloy. Int J Fatigue 2007;29:1839-45.

7. Suresh S. Fatigue of materials. United Kingdom: Cambridge University Press; 1998

8. Bernard JD, Jordon JB, Horstemeyer MF, El Kadiri H, Baird J, Lamb D, Luo AA. Structure-property relations of cyclic damage in a wrought magnesium alloy. Scr Mater. In Press, 2010.

9. Gall K, Biallas G, Maier HJ, Gullett P, Horstemeyer MF, McDowell DL. In-Situ observations of low-cycle fatigue damage in cast AM60B magnesium in an environmental scanning electron microscope. *Metall Mater Trans A* 2004a;35:321-31.

10. Gall K, Biallas G, Maier HJ, Gullett P, Horstemeyer MF, McDowell DL, Fan J. In-Situ observations of high cycle fatigue mechanisms in cast AM60B magnesium in vacuum and water vapor environments. Int J Fatigue 2004b;26:59-70.

11. Gall K, Biallas G, Maier HJ, Horstemeyer MF, McDowell DL. Environmentally influenced microstructurally small fatigue crack growth in cast magnesium. Mater Sci Eng A 2005;396:143-54.

12. Tokaji K, Kamakura M, Ishiizumi Y, Hasegawa N. Fatigue behavior and fracture mechanisim of a rolled AZ31 magnesium alloy. Int J Fatigue 2004;26:1217-24.

13. Sajuri ZB, Miyashita Y, Hosokai Y, Mutoh Y. Effects of Mn content and texture on fatigue properties of as-cast and extruded AZ61 magnesium alloys. Int J Mech Sci 2006;48:198-209.

14. McDowell DL, Gall K, Horstemeyer MF, Fan J. Microstructure-based fatigue modeling of cast A356-T6 alloy. Eng Fract Mech 2003;70:49-80.

15. Bache MR, Evans WJ, Randle V, Wilson RJ. Characterization of mechanical anisotropy in titanium alloys. Mater Sci Eng A 1998;A257:139-44.

16. Whittaker MT, Evans WJ, Lancaster R, Harrison W, Webster PS. The effect of microstructure and texture on mechanical properties of Ti6-4. Int J Fatigue 2009;31:2022-30.

17. Chamos AN, Pantelakis SG, Haidemenopoulos GN, Kamoutsi E. Tensile and fatigue behavior of wrought magnesium alloys AZ31 and AZ61. Fatigue Fract Eng Mater Struct 2008;31:812-21.

18. Xue Y, McDowell DL, Horstemeyer MF, Dale MH, Jordon JB. Microstructure-based multistage fatigue modeling of aluminum alloy 7075-T651. Eng Fract Mech 2007;74:2810-23.

19. El Kadiri H, Xue Y, Horstemeyer MF, Jordon JB, Wang PT. Identification and modeling of fatigue crack growth mechanisms in a die-cast AM50 magnesium alloy. *Acta Mater* 2006;54:5061-76.

20. Horstemeyer MF, Yang N, Gall KA, McDowell DL, Fan J, Gullett P. High cycle fatigue on a die cast AZ91E-T4 magnesium alloy Acta Mater 2004;52:1327-36.

21. Xue Y, Horstemeyer MF, McDowell DL, El Kadiri H, Fan J. Microstructure-based multistage fatigue modeling of a cast AE44 magnesium alloy. Int J Fatigue 2007;29:666-76

22. Xue Y, Pascu A, Horstemeyer MF, Wang L, Wang PT. Microporosity effects on cyclic plasticity and fatigue of LENS-processed steel. Acta Mater 2010;58:4029-38.

23. Jordon JB, Horstemeyer MF, Yang N, Major JF, Gall KA, Fan J. Microstructural inclusion influence on fatigue of a cast A356 aluminum alloy. Metall Mater Trans A 2010;41A:356-363

24. Xue Y, Burton CL, Horstemeyer MF, McDowell DL, Berry JT. Multistage fatigue modeling of cast A356-T6 and A380-F aluminum alloys. Metall Mater Trans B 2007;38B:601-6.

25. El Kadiri H, Oppedal AL. A crystal plasticity theory for latent hardening by glide twinning through dislocation transmutation and twin accommodation effects. J Mech Phys Solids 2010;58:613–24

26. Bernard JD, Jordon JB, Horstemeyer MF. Small fatigue crack growth observations in a wrought magnesium alloy, TMS 2011 Annual Meeting and Exhibition, [submitted].

27. Bannantine JA, Comer JJ, Handrock JL. Fundamentals of metal fatigue analysis. USA: Prentice Hall; 1990.

Magnesium Technology 2011
Edited by: Wim H. Sillekens, Sean R. Agnew, Neale R. Neelameggham, and Suveen N. Mathaudhu
TMS (The Minerals, Metals & Materials Society), 2011

LOW-CYCLE FATIGUE BEHAVIOR OF DIE-CAST Mg ALLOY AZ91

Luke Rettberg[1], Warwick Anderson[1], J. Wayne Jones[1]

[1]University of Michigan, Gerstacker Building Rm. 1021, 2200 Bonisteel Blvd; Ann Arbor, MI, 48109, USA

Keywords: Low-Cycle Fatigue, Die-Cast AZ91, Strain-Life, Hysteresis Loops, Incremental Step Test, Cyclic Stress-Strain, Microstructure, Fractography, Porosity

Abstract

An investigation has been conducted on the influence of microstructure and artificial aging response (T6) on the low-cycle fatigue behavior of super vacuum die-cast (SVDC) AZ91. Fatigue lifetimes were determined from total strain-controlled fatigue tests for strain amplitudes of 0.2%, 0.4% and 0.6%, under fully reversed loading at a frequency of 5 Hz. Cyclic stress-strain behavior was determined using incremental step test (IST) methods. Two locations in a prototype casting with different thicknesses and, therefore, solidification rates, microstructure and porosity, were examined. In general, at all total strain amplitudes fatigue life was unaffected by microstructure refinement and was attributed to significant levels of porosity. Cyclic softening and a subsequent increased cyclic hardening rate, compared to monotonic tests, were observed, independent of microstructure. These results, fractography and damage accumulation processes, determined from metallographic sectioning, are discussed.

Introduction

In recent years, magnesium alloys have gained renewed interest due to their low density and high specific strength. For automotive applications, reducing weight is important to improve fuel economy. For example, replacing steel and aluminum with magnesium in heavy components could result in a $0.25L$ and $0.1L$ per $100km$ reduction in fuel consumption, respectively [1]. North American automotive manufacturers plan to substitute $340lbs.$ of magnesium components for $630lbs.$ of current ferrous and aluminum parts by 2020 [2].

The ternary Mg-Al-Zn alloy AZ91 has excellent castability, mechanical properties at ambient temperature, corrosion resistance and is the most commonly used magnesium die-casting alloy [3]. However, before AZ91 can be used for structural components in automobiles, fatigue properties must be further characterized. Fatigue properties are relevant for such components where cyclic loading rather then constant stress limits service life [4]. For fatigue modeling, the cyclic stress-strain properties need to be characterized and the impact of heat treatment determined. Some work has already been done in this area [5–9] and the goal of this paper is to further examine the effects of microstructure refinement and aging on the low-cycle fatigue and cyclic stress-strain properties of die-cast AZ91. The influence of microstructure on damage accumulation and crack initiation was also examined.

Experimental Procedure

Fatigue specimens were machined from two locations on a prototype die-cast AZ91 shock tower. The two locations (L1 & L2) had the greatest difference in thickness ($4.7mm$ & $3.0mm$, respectively) and thus were expected to have significant differences in solidification time, cell size (secondary dendrite arm spacing) and porosity.

An aging treatment (T6) was performed according to modified ASTM B661 specifications [10]. Specimens were encapsulated in argon and solution treated at 413 °C for $20h$. Following a water quench, specimens were aged for $50h$ at 413 °C.

Axial fatigue tests were performed using a MTS servo-hydraulic test machine with a FlexTest SE controller and MTS TestSuite control software. All tests were performed in total strain control with a strain ratio of $R(\epsilon_{min}/\epsilon_{max}) = -1$ and a constant frequency of 5 Hz. Specimens were tested in the as-cast and T6 conditions from L1 and L2 at total strain amplitudes of 0.2%, 0.4% and 0.6%. A minimum of four specimens was tested for each location/condition/total strain amplitude. Failure was defined as complete specimen separation or 20% tensile load drop. Tests were performed in laboratory air at 25 °C and according to ASTM E606 [11].

Incremental step tests (IST) had a total strain amplitude of 1.0%, 60 cycles per block, constant strain rate of $0.01sec^{-1}$ and $R = -1$. A total strain amplitude of 1.0% was chosen to prevent failure before a stable state was established. Specimens were tested until complete separation, normally six cycles. One specimen was tested for each location/condition. Tests were performed in laboratory air at 25 °C. An example IST

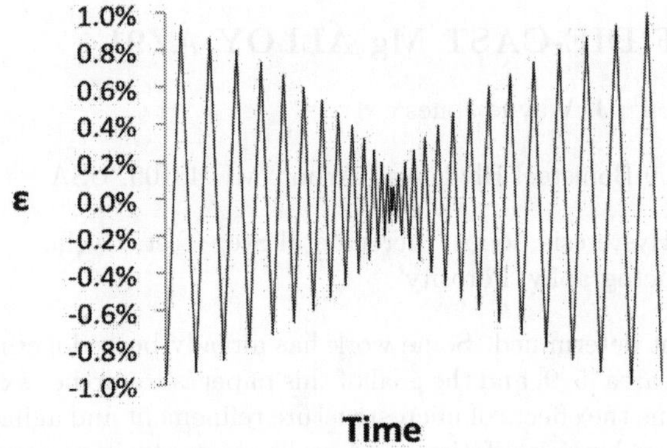

Figure 1: Waveform of a simple IST. In this example there are 30 cycles per block. Strain amplitude changes linearly between cycles.

waveform is shown in Figure 1. An IST allows the determination of the cyclic stress-strain curve from a single specimen [12].

Metallographic preparation involved grinding and polishing to 1 μm particle size. Water was determined to have little effect on the resulting microstructure and was not explicitly avoided during polishing. Samples were ultrasonically cleaned after the final polishing step. The microstructure was examined in the as-polished condition using a Phillips XL30 SEM with a backscatter electron detector. Cell size was measured according to ASTM E112 [13].

Results and Discussion

Microstructure

The microstructure of die-cast AZ91 is shown in Figure 2 and consists primarily of α-Mg. In the as-cast condition, the β phase ($Mg_{17}Al_{12}$) and various intermetallic phases are present in the interdentritic regions. As a result of aging, a eutectic α-Mg and β structure formed along with β precipitates. Intermetallic particles (MnAl) were also present at both conditions. Comparing L1 and L2, a slight difference in average cell size and porosity area fraction was measured (Table I). As expected, L2 had a more refined microstructure with less porosity because of a faster solidification time. L2 also had smaller pores than L1. Heat treatment had negligible effect on porosity. Based on these results, it was expected that L2 would have improved fatigue life because of a smaller cell size and less porosity.

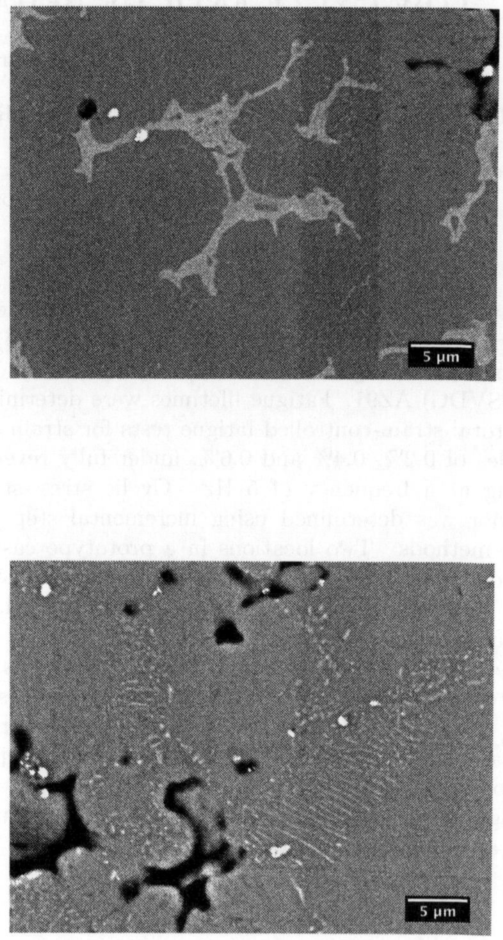

Figure 2: Representative micrographs of as-cast (top) and heat treated (T6) AZ91 (bottom) showing the α-Mg phase (dark) and the β phase ($Mg_{17}Al_{12}$) (light).

Table I: Cell size and porosity for L1 and L2. STDev (Standard Deviation) and STE (Standard Error) are also listed.

	L1	L2
Equivalent Diameter (μm)	4.99	4.17
STDev (μm)	2.26	1.51
STE (μm)	0.05	0.02
Porosity Area Fraction (%)	4.18	1.75

Table II: Low-cycle fatigue test data. N_f^{avg} is the average fatigue-life.

		L1		L2	
	$\Delta\epsilon/2$	As-Cast	T6	As-Cast	T6
N_f^{avg}	0.002	165388	30066	33234	41748
	0.004	6219	5435	5326	4276
	0.006	825	1157	1053	609
STDev	0.002	197977	16944	16092	23144
	0.004	1731	3658	3030	2923
	0.006	557	225	320	382

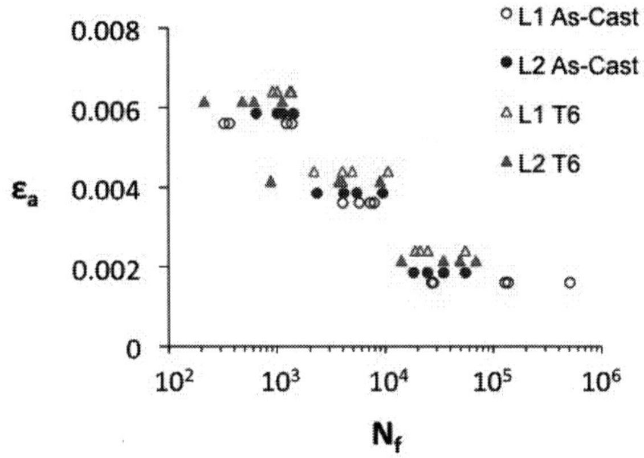

Figure 3: Strain-Life plot for L1 and L2 in the as-cast and T6 conditions. Total strain levels were slightly offset to improve clarity.

Axial Fatigue Behavior

The strain-life plot for L1 and L2 in the as-cast and T6 conditions showed a significant amount of scatter at all strain amplitudes (Figure 3). Table II lists the strain-life results in more detail. Results compared favorably with recent literature on die-cast AZ91 [5, 7, 14]. Fatigue-life was an order of magnitude shorter for each strain amplitude when compared with sand cast AZ91 in the T6 condition [8]. Fatigue parameters (Table III) were determined by separating the total strain amplitude into elastic and plastic components and then using Basquin (1) and Coffin-Manson (2) laws, respectively. The fatigue-life relationship based on the total strain can be found by adding equations (1) & (2), resulting in equation (3). For all locations and conditions the fatigue-life data was well described by this approach (Figure 4). Plastic strain was measured from the width of the half-life hysteresis loops at the mean stress. A slightly negative mean stress was observed in most cases due to compression-tension asymmetry (Figure 5).

$$\frac{\Delta\epsilon_e}{2} = \frac{\sigma'_f}{E}(2N_f)^{-b} \tag{1}$$

$$\frac{\Delta\epsilon_p}{2} = \epsilon'_f(2N_f)^{-c} \tag{2}$$

$$\frac{\Delta\epsilon}{2} = \frac{\sigma'_f}{E}(2N_f)^{-b} + \frac{\epsilon'_f}{E}(2N_f)^{-c} \tag{3}$$

where:
$\Delta\epsilon_e$ = true elastic strain range,
$\Delta\epsilon_p$ = true plastic strain range,
$\Delta\epsilon$ = true strain range,
$2N_f$ = reversals to failure,
b = fatigue strength exponent,
c = fatigue ductility exponent,
σ'_f = fatigue strength coefficient,
ϵ'_f = fatigue ductility coefficient, and
E = Youngs modulus (modulus of elasticity).

Table III: Low-cycle fatigue parameters. N_t is the transition life.

	L1		L2	
	As-Cast	T6	As-Cast	T6
b	0.154	0.243	0.249	0.17
c	0.605	0.771	0.809	0.628
$\sigma'_f(MPa)$	607.5	1323	1386	634.5
ϵ'_f	0.119	0.507	0.785	0.124
N_t	125.4	219.8	324.7	115.8

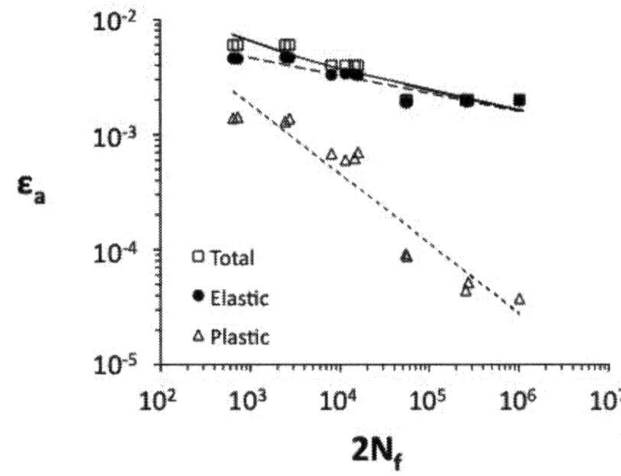

Figure 4: Total, elastic and plastic strain amplitudes vs. number of reversals to failure from L1 in the as-cast condition. Log-log linear regression fits were used to determine the fatigue parameters.

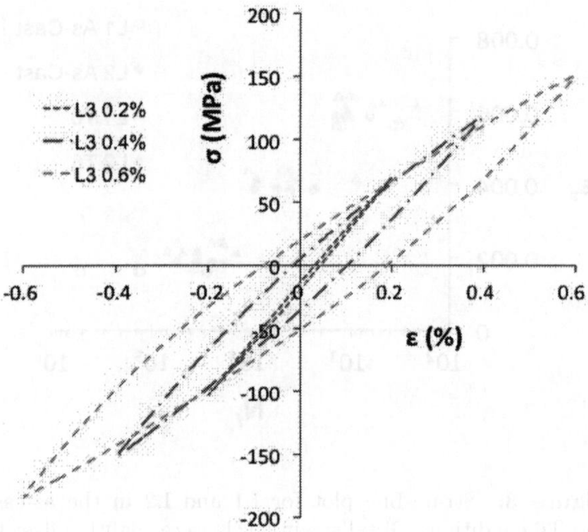

Figure 5: Three representative hysteresis loops from L1 in the as-cast condition corresponding to all three strain amplitudes. Notice the clockwise rotation as the strain amplitude increases and the development of a negative mean stress.

Fractography

Distinct crack growth and fast fracture regions were observed for most failed specimens. Some fracture surfaces experienced severe rubbing during testing and could not be examined. Crack growth regions had a smooth appearance while fast fracture regions showed high contrast regions and cracking around the β phase, similar to fracture surfaces observed from tensile tests [7, 15]. All cracks initiated at either surface or subsurface pores (Figure 6). However, in sand cast AZ91-T6, Goodenberger, et. al. found that cracks initiated primarily at the surface [8]. In samples with shorter fatigue life, fatal cracks initiated from larger pores, similar to results reported by Horstemeyer, et. al. [16]. Pores, present at all locations and conditions, controlled the fatigue life rather than microstructural features.

Fatigue Damage Accumulation

Further examination of crack initiation behavior was done by examining the cross-section on the plane parallel to the stress axis of failed fatigue specimens. Small cracks initiated at pores close to the primary crack for each location/condition (Figure 7). Crack growth appeared to be unaffected by interdentritic and eutectic structures. In some cases, cracks grew parallel to the axis of stress in order to link nearby pores (Figure 8).

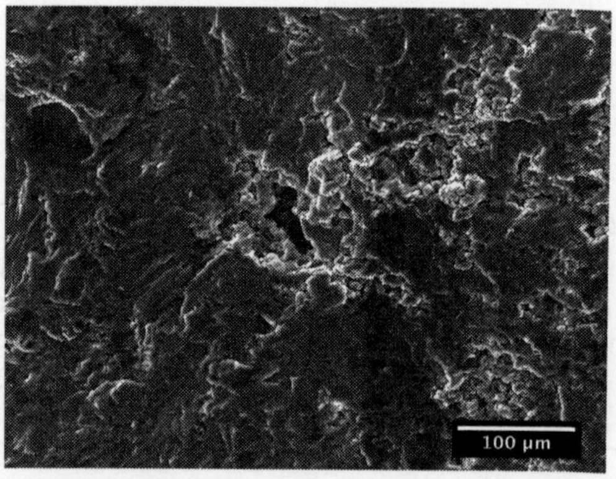

Figure 6: A representative SEM fractograph showing a typical crack initiation pore from L1 in the as-cast condition. This particular specimen had the longest fatigue life ($N_f = 506791$) and a crack initiation pore size of 72 μm.

There did not appear to be general or wide spread crack initiation and growth away from the fracture plane, indicating the dominant role of a single pore or a cluster of closely spaced pores in damage accumulation.

Cyclic Stress-Strain Behavior

The monotonic and cyclic stress-strain parameters (Table IV) were characterized using two power-law fits, equations (4) & (5), respectively. Methods were similar to those described previously to determine low-cycle fatigue parameters. Comparing monotonic and cyclic loading conditions, the strain-hardening rate is higher in cyclic loading and the yield strength is lower, indicating cyclic softening. Heat treatment had a negligible effect on the strain-hardening rate but did increase the yield strength for both locations under cyclic loading.

$$\frac{\Delta\sigma}{2} = K\left(\frac{\Delta\epsilon_p}{2}\right)^n \quad (4)$$

$$\frac{\Delta\sigma}{2} = K'\left(\frac{\Delta\epsilon_p}{2}\right)^{n'} \quad (5)$$

where:
$\Delta\sigma$ = true stress range,
$\Delta\epsilon_p$ = true plastic strain range,
n = monotonic strain hardening exponent,
n' = cyclic strain hardening exponent,
K = monotonic strength coefficient, and
K' = cyclic strength coefficient.

Table IV: Monotonic(M)/cyclic(C) stress-strain parameters.

	L1				L2			
	As-Cast		T6		As-Cast		T6	
	M	C	M	C	M	C	M	C
n/n'	0.185	0.417	0.246	0.400	0.140	0.481	0.233	0.419
$K/K'(MPa)$	482	2140	737	2153	393	3561	671	2334
$E(GPa)$	43.1	43.9	44.7	44.8	42.0	42.7	44.9	43.7
$Y_s(MPa)$	155	136	160	151	166	138	155	142

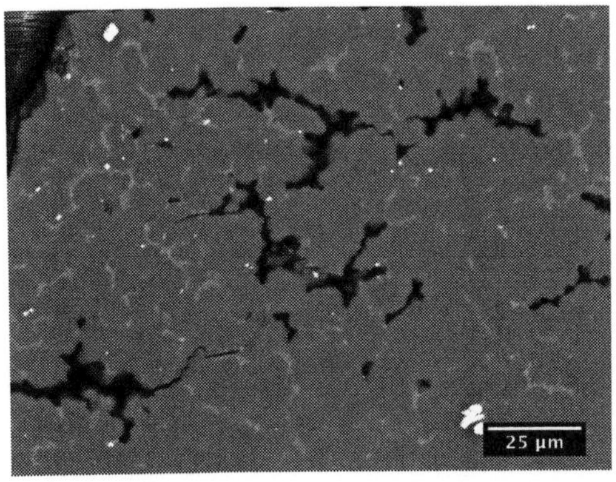

Figure 7: Micrograph from L1 in the as-cast condition showing cracks initiating at pores near the primary crack. Stress axis is parallel to the long direction of the page. Part of the primary crack (fracture surface) can be seen in the upper left corner.

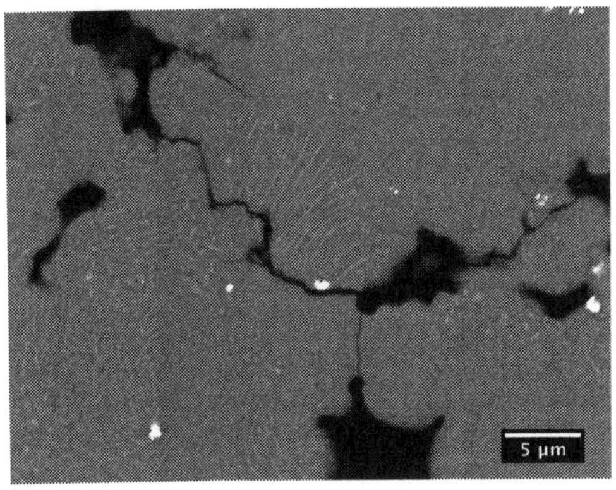

Figure 8: Micrograph from L2 in the T6 condition. A crack in this figure grew parallel to the stress axis in order to link two nearby pores. Stress axis is parallel to the long direction of the page.

Conclusions

1. Fatigue-life was controlled by porosity at both locations/conditions, for all strain amplitudes and was independent of microstructure and heat treatment. In general the larger the initiation pore the shorter the fatigue life.
2. Crack growth from pores was uninhibited by interdentritic and eutectic structures. The crack growth direction was typically perpendicular to the stress axis, however, the driving factor for cracks to link with nearby pores caused some cracks to grow parallel to the stress axis.
3. Cyclic loading caused an increased hardening rate and reduced the yield strength. The cyclic yield strength can be improved with a T6 heat treatment.

Acknowledgments

The authors would like to thank USAMP for funding Project AMD703 and members (specifically Jason Geathers and Garret Huff) of the Fatigue Lab at the University of Michigan for their assistance and guidance.

References

1. E. Aghion and B. Bronfin, "Magnesium Alloys Development towards the 21st Century," *Materials Science Forum*, vol. 350-351, pp. 19–30, 2000.
2. "Magnesium Vision 2020: A North American Automotive Strategic Vision for Magnesium, U.S. Automotive Materials Partnership, (USAMP) U.S. Council for Automotive Research (USCAR)," Sept. 2010. `www.uscar.org`.
3. J. F. King, "Development of magnesium diecasting alloys," in *Magnesium Alloys and their Applications*, pp. 37–47, Werkstoff-Infomationsgesellschaft, 1998.

4. H. R. Mayer, H. Lipowsky, M. Papakyriacou, R. Rösch, A. Stich, and S. E. Stanzl-Tschegg, "Application of ultrasound for fatigue testing of lightweight alloys," *Fatigue Fracture of Engineering Materials and Structures*, vol. 22, pp. 591–599, July 1999.

5. L. J. Chen, J. Shen, W. Wu, F. Li, Y. Wang, and Z. Liu, "Low-Cycle Fatigue Behavior of Magnesium Alloy AZ91," *Materials Science Forum*, vol. 488-489, pp. 725–728, 2005.

6. G. Eisenmeier, B. Holzwarth, H. Hoppel, and H. Mughrabi, "Cyclic deformation and fatigue behaviour of the magnesium alloy AZ91," *Materials Science and Engineering A*, vol. 319-321, pp. 578–582, Dec. 2001.

7. H. El Kadiri, M. Horstemeyer, J. Jordon, and Y. Xue, "Fatigue Crack Growth Mechanisms in High-Pressure Die-Cast Magnesium Alloys," *Metallurgical and Materials Transactions A*, vol. 39, pp. 190–205, Jan. 2008.

8. D. L. Goodenberger and R. I. Stephens, "Fatigue of AZ91E-T6 Cast Magnesium Alloy," *Journal of Engineering Materials and Technology*, vol. 115, no. 4, p. 391, 1993.

9. F. Li, Y. Wang, L. Chen, Z. Liu, and J. Zhou, "Low-cycle fatigue behavior of two magnesium alloys," *Journal of Materials Science*, vol. 40, pp. 1529–1531, Mar. 2005.

10. ASTM B661-06, "Standard practice for heat treatment of magnesium alloys," *ASTM*, 2006.

11. ASTM E606-04, "Standard practice for strain-controlled fatigue testing," *ASTM*, 2005.

12. R. W. Landgraf, J. Morrow, and T. Endo, "Determination of the Cyclic Stress-Strain Curve," *Journal of Materials*, vol. 4, no. 1, pp. 176–188, 1969.

13. ASTM E112-96, "Standard test methods for determining average grain size," *ASTM*, 2006.

14. J. P. Li, G. W. Lorimer, J. D. Robson, and B. Davis, "The Microstructures of As-Cast Mg-Zr and Mg-Mn Alloys," *Materials Science Forum*, vol. 488-489, pp. 329–332, 2005.

15. H. Patel, D. Chen, S. Bhole, and K. Sadayappan, "Microstructure and tensile properties of thixomolded magnesium alloys," *Journal of Alloys and Compounds*, vol. 496, pp. 140–148, Apr. 2010.

16. M. Horstemeyer, "High cycle fatigue of a die cast AZ91E-T4 magnesium alloy," *Acta Materialia*, vol. 52, pp. 1327–1336, Mar. 2004.

Magnesium Technology 2011
Edited by: Wim H. Sillekens, Sean R. Agnew, Neale R. Neelameggham, and Suveen N. Mathaudhu
TMS (The Minerals, Metals & Materials Society), 2011

Small Fatigue Crack Growth Observations in an Extruded Magnesium Alloy

J.D. Bernard[1], J.B. Jordon[2], M.F. Horstemeyer[1,3]
[1]Center for Advanced Vehicular Systems (CAVS), Mississippi State University, Mississippi State, MS 39762
[3]Department of Mechanical Engineering, The University of Alabama, Tuscaloosa, AL 35487
[2]Department of Mechanical Engineering, Mississippi State University, Mississippi State, MS 39762

Keywords: fatigue, replica, AZ61 magnesium alloy, fatigue modeling

Abstract

The purpose of this paper is to quantify the microstructurally small/physically small crack growth behavior in an extruded AZ61 magnesium alloy. Fully-reversed, interrupted load control tests were conducted on notched specimens that were taken from a magnesium alloy extrusion. In order to measure crack growth, replicas of the notch surface were made using a two-part silicon-rubber compound at periodic cyclic intervals. Scanning electron microscopy analysis of the replica surfaces revealed multi site crack initiation and subsequent crack coalescence. The crack growth behavior of the small fatigue cracks was shown to have a strong dependence on the material microstructure as the crack was submitted to a tortuous growth path along grain boundaries and crystallographic slip planes. A microstructurally dependent crack growth model that was previously developed for FCC metals was further extended here to HCP metals.

Introduction

Magnesium alloys are currently being targeted as a potential lightweight metal for widespread use in automotive structural components for the purpose of reducing vehicle weight while maintaining equivalent performance. Although the majority of magnesium alloys that have been researched for use within the automotive industry are cast, extruded magnesium alloys have been shown to exhibit better properties with regards to strength [1]. This is largely because cast products contain microstructural defects like gas pores, pore shrinkage clusters, and oxide films. As such, these microstructural defects can serve as fatigue crack initiation sites and thus reduce the overall fatigue resistance of that alloy. Due to the processing nature, wrought materials lack these defects [2], but they can exhibit anisotropic characteristics and behavior based on loading orientation relative to the process direction (i.e. rolling direction, extrusion, etc) [3]. Since magnesium alloys are naturally anisotropic due to their hexagonal close-packed (HCP) systems which limit the number of potential slip systems that are readily activated [4], extruded magnesium alloys typically have a strong dependence upon loading orientation with respect to the extrusion direction. With regards to characterizing the fatigue behavior of the AZ61 magnesium alloy, previous research has been fairly limited to studying the effects on fatigue by the manganese content, texture, and environmental aspects [5-7]. Furthermore, to the authors' knowledge, studies on the small fatigue crack characteristics on the AZ61 alloy does not exist. Suresh [8] acknowledged the importance of studying small fatigue cracks because a microstructurally small fatigue crack does not necessarily adhere to the more classical approaches of fatigue that are based upon continuum mechanics and linear elastic fracture mechanics [8]. Regarding past studies on other magnesium alloys, Gall *et al.* [9-11] conducted in-situ fatigue tests on cast AM60 magnesium specimens in order to quantify small crack growth behavior. These tests revealed that the material microstructure had a large impact on fatigue crack growth, including that of the small crack nature. They found that fatigue cracks initiated from inclusions, grain boundaries, and persistent slip bands. In the AM60 alloy, the fatigue cracks grew along grain boundaries that contributed to the tortuous crack path. The grain boundaries, in affect, contributed to a varying driving force that resulted in accelerated and retarded crack growth rates.

Initially developed for cast aluminum alloys [12,13], the MultiStage Fatigue (MSF) model was further expanded by Xue *et al.* [14-16] to capture the fatigue characteristics and behavior of LENS shaped steels, wrought aluminum alloys, as well as cast magnesium alloys. However, the small crack formulation of the MSF model has not been verified for wrought magnesium alloys. Thus, the purpose of this paper is quantify the small fatigue crack growth, through the use of the replica method and correlate it to the small crack stage in the MSF model.

Materials and Experimental Methods

The material used in the study is an extruded AZ61 magnesium alloy. The as-received air-quenched extruded alloy is in the form of an automotive crash rail with the cross-section shown in Figure 1a and the specimen layout in Figure 1b. The crash rail was extruded at an average temperature of 500 ºC and an extrusion exit speed of 1.2 m/min. The weight percent composition of the AZ61 magnesium alloy is given in Table 1. The as-received material was examined with an optical microscope to quantify the initial microstructure in both a transverse material orientation (Figure 2a) and an extrusion or longitudinal orientation (Figure 2b). X-ray diffraction (XRD) was also performed to quantify texture of the as-received material and the results are shown in Figure 3.

Table 1. AZ61 Chemical Composition by Weight Percent

Al	Zn	Fe	Ni	Cu	Si	Mn
5.85	0.83	0.023	0.0019	0.0057	0.027	0.326

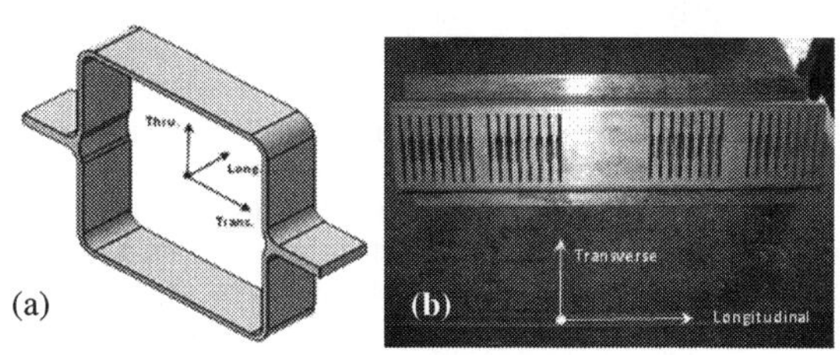

Figure 1. (a) Geometry of the as-received, extruded AZ61 magnesium alloy crash rail and (b) specimen layout rail.

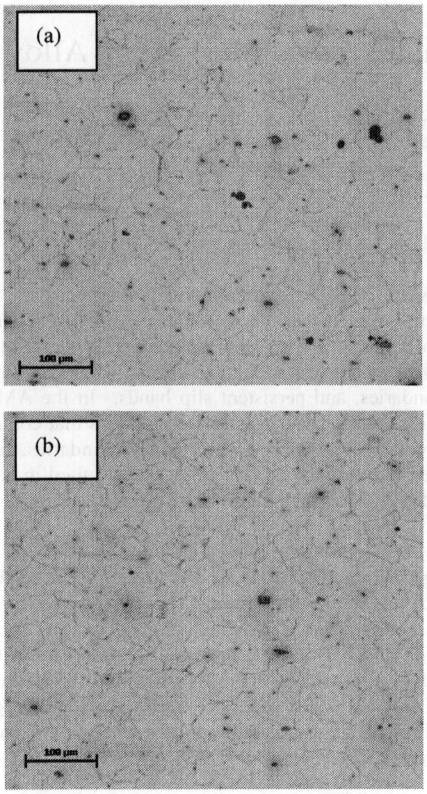

Figure 2. Optical micrographs of the as-received extruded AZ61 magnesium alloy polished and etched: (a) longitudinal orientation (b) transverse orientation.

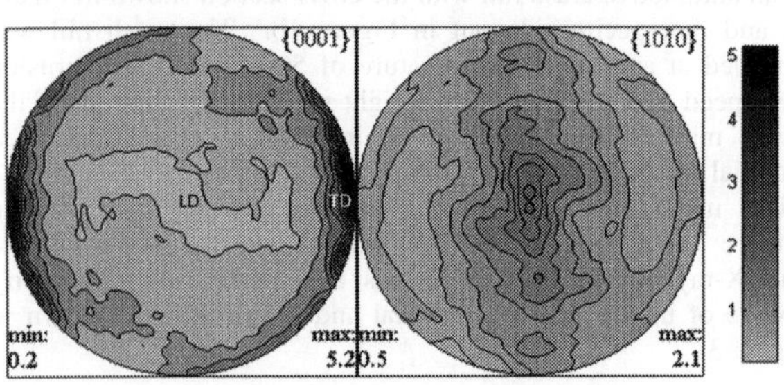

Figure 3. Pole figures of the longitudinal orientation calculated from X-ray diffraction for an AZ61 magnesium alloy.

As previously stated, rectangular cross-sectional fatigue specimens were machined from the crash rail in the transverse orientation as shown in Figure 1b. The specimen design and dimensions are shown in Figure 4. In order to make small fatigue crack growth measurements, a location favorable for crack initiation was required; therefore, a 35mm radius was machined to a depth of 300μm onto the specimen surface according to similar methods found in literature [17]. The shoulder, gage, and notch surfaces of each specimen were hand ground using 800grit sandpaper parallel to the loading direction. In addition to sanding with 800grit sandpaper, several specimens were also polished with a series of 1200grit, 9μm, 6μm, 1μm, and 0.25μm diamond pastes and then polished with OPS, a silicon dioxide (SiO_2) suspension that also mildly etched the specimen surfaces similar to [9-11]. All specimens were then degreased using ethanol. Fully-reversed, load control tests were then conducted to failure on these notched specimens. In addition, several specimens that were ground were fatigue tested but were interrupted at set intervals in order to take replicas of the notch bore. The replicas were made by using a dual-part silicone based compound, called Repliset®. The Repliset® compound has a resolution 0.1μm, but must be sputter coated with gold palladium for optimal SEM image observation. All replicas were sputter coated for a period of 30 seconds. Upon failure, the fracture surfaces were cut off from the specimen and mounted for SEM examination for further fractography analysis.

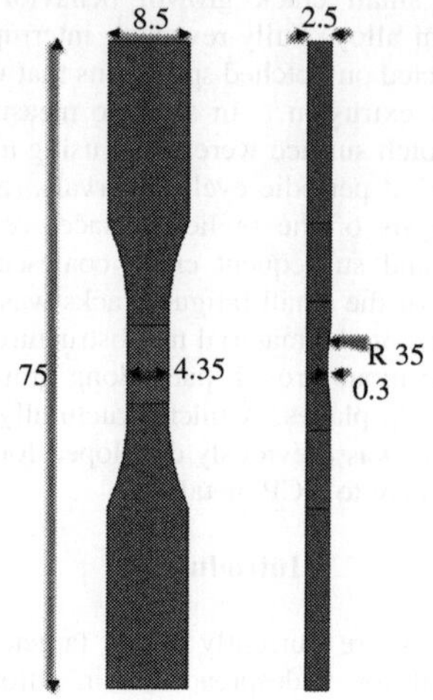

Figure 4. Fatigue specimen geometry, all dimensions are in mm.

Fatigue Life Results of Notched Specimens

The stress-life values were plotted for the failed specimens in order to determine what significance the surface preparation and replication had upon the overall fatigue life. The results are shown in Figure 5.

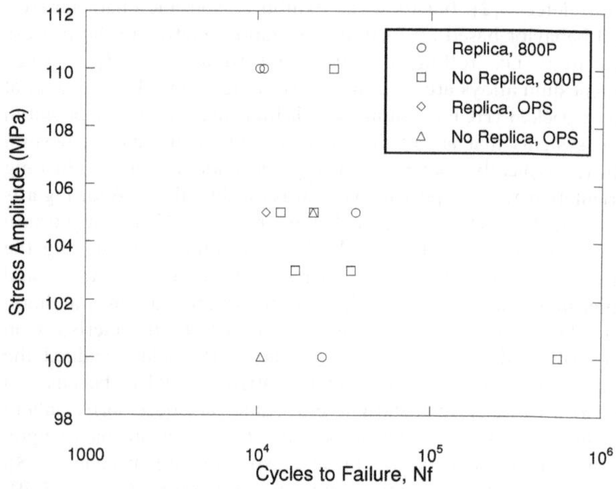

Figure 5. Stress amplitude versus cycles to failure for the specimens tested.

Note that in Figure 5, the circles indicate specimens prepared by sanding with 800grit sandpaper and had replicas taken of the notch. The boxes indicate specimens that were sanded with 800grit, but with no replicas taken. The diamonds indicates OPS specimens with no replicas. Although there is some scatter in the above plot, neither replicating the surface of the specimen nor preparing them with OPS had a profound or consistent impact upon the fatigue life of the specimens.

Small Fatigue Crack Behavior

Using SEM imaging of the replicas, the fatigue crack initiation and growth on the notch surface were experimentally observed. Each replica was observed in the SEM starting with the last replica prior to failure. Once the dominate fatigue crack was located and measured, each preceding replica was examined and the length of the dominate crack was measured until the crack was no longer visible. Figure 6 presents the crack length versus number of cycles. For this particular specimen, the load amplitude was 105MPa (R=-1) and replicas were taken at intervals of 1,000 cycles. Also indicated in Figures 6, 7, and 8 are the points A, B, and C which correspond to cycles 13,000, 19,000, and 24,000, respectively.

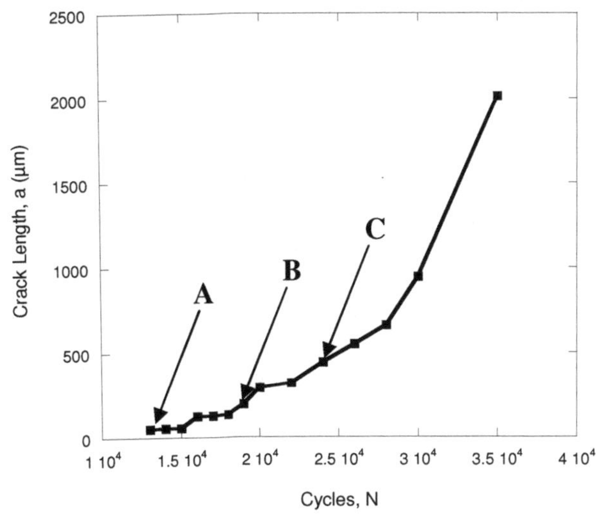

Figure 6. Crack length versus cycles for a stress amplitude of 105MPa

Upon plotting the crack growth rate versus crack length, shown in Figure 7, there is a marked acceleration/deceleration between points A and B which results in the corresponding crack growth from 13,000 cycles to 19,000 cycles. This is likely due to grain boundary blocking where the crack, outlined in Figure 8, grew through and around grain boundaries. Note that this specimen surface was prepared with 800 grit sandpaper and the roughness shown in Figure 8 is from the grinding process. Between points B and C, however, driving force increased such that the growth rate become more constant. This is likely because as the crack grew in length, the driving force increased such that the crack path became less dependent on the crystallographic orientation, and thus rendered it less sensitive to the material microstructure.

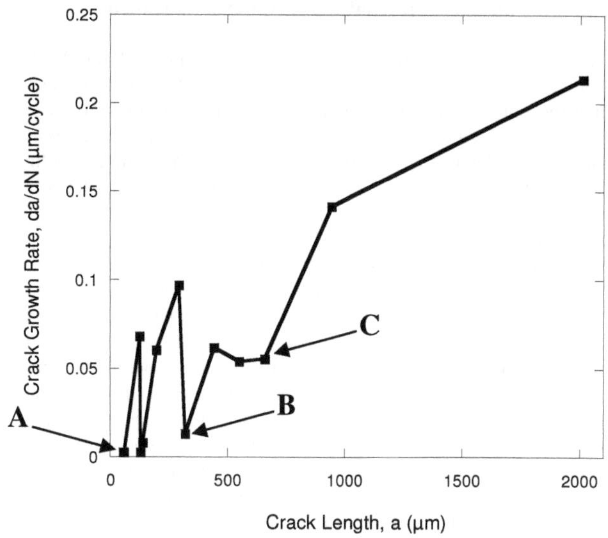

Figure 7. Crack growth rate versus crack length for a stress amplitude of 105MPa.

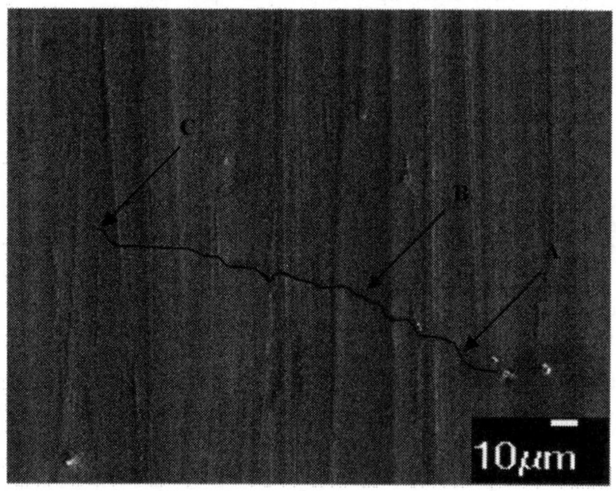

Figure 8. Replica image illustrating the path of the crack shown in Figures 6 and 7, along with the locations of the indicated points.

Previous research conducted by Gall *et al.* [9-11] on cast magnesium showed that the microstructurally small cracks had a strong dependence on the material microstructure. Features such as dendrites and particle laden interdendritic regions demonstrated a significant impact on the small crack growth rate. Although the extruded magnesium alloy examined in this paper had a different microstructure than the cast AM60B previously researched, parallels between the two studies can still be drawn. Small fatigue cracks were shown to preferentially propagate through favorably oriented dendrites in cast magnesium and grains in the extruded AZ61 magnesium alloy. In addition, the smaller fatigue cracks displayed a stronger dependence upon the material microstructure than longer fatigue cracks in both the cast and extruded magnesium materials.

Finite Element Analysis of Notched Specimen

In order to correlate the small fatigue crack stage of the MSF model to crack growth data, the local stress state was needed. Since the notched specimen creates a slightly non-uniaxial stress state, finite element analysis (FEA) of the specimen was conducted to determine the stress state of the notch under cyclic loading. As such, a non-linear finite element material model was employed. The material model used in the FEA is the plasticity-damage internal state variable (ISV) model first developed by Bammann *et al.* [18-20] and later modified by Horstemeyer *et al.* [21,22]. For a complete description of the ISV model, see [18-22]. The material model constants were determined based on monotonic tensile and compression tests. Three-dimensional finite element simulations of the entire notched specimen were performed using the aforementioned constitutive model. The boundary conditions employed in this study were identical to the ones employed in the experiment. In order to satisfy continuum mechanics conditions, the 8-node brick element used in the notch bore region was less than ten times the size of the average grain size of the magnesium AZ61 alloy.

A relationship between equivalent stress at the notch root and remote loading were determined from the FEA, as shown in Figure 9. The relationship between local equivalent stress became non-linear due to local plastic deformation in the notch.

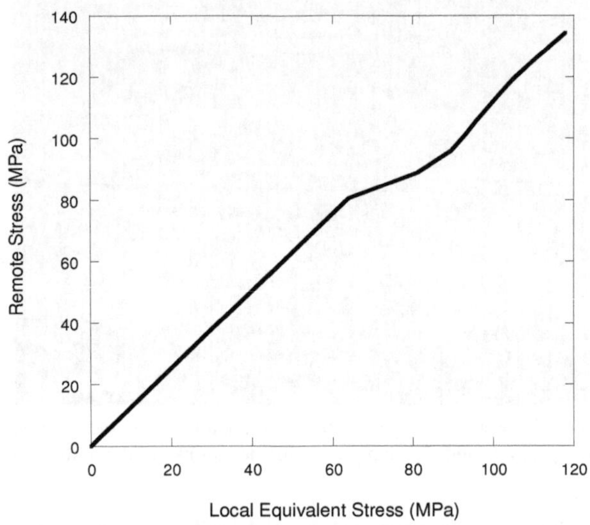

Figure 9. Local equivalent stress as a function of the remote stress for the notched specimens.

Small Fatigue Crack Model

To provide context for the small crack fatigue stage of the model, an overview of the MSF model is presented. The MSF model [12] divides the total fatigue life into three distinct regions: incubation, microstructurally small/physically small crack growth, and long crack growth, as shown in Eq. 1.

$$N_{Total} = N_{Inc} + N_{MSC/PSC} + N_{LC} \tag{1}$$

where N_{Total} is the total fatigue life. N_{Inc} is the number of cycles required to incubate a crack from an inclusion. In the case of the magnesium alloy, the inclusion is an intermetallic particle. The incubated crack then expands from the inclusion into the matrix. $N_{MSC/PSC}$ is the number of cycles required to propagate a crack through the microstructurally small/physically small crack regime. Finally, N_{LC} is number of cycles of long crack (LC) propagation to final failure, which depends on the amplitude of loading and the corresponding extent of plasticity ahead of the crack tip.

The MSC/PSC model is a function of the crack tip displacement (Eq. 2) and portrays the small crack growth regime, where ΔCTD is the crack tip displacement range, χ is a material constant, and ΔCTD_{th} is the range of the threshold for crack tip displacement. χ is 0.32 for aluminum alloys and has been taken as this value for the given material.

$$\left(\frac{da}{dN}\right)_{msc} = \chi\left(\Delta CTD - \Delta CTD_{th}\right) \tag{2}$$

The range of the crack tip displacement, shown in Eq. (3), is a function of remote loading where C_I, C_{II}, ξ, ξ', ω, ω', and ζ are constants based on small crack growth experiments.

$$\Delta CTD = C_{II}\left(\frac{GS}{GS_0}\right)^{\varpi}\left(\frac{GO}{GO_0}\right)^{\xi}\left[\frac{U\Delta\hat{\sigma}}{S_{ut}}\right]^{\zeta} a + C_I\left(\frac{GS}{GS_0}\right)^{\varpi'}\left(\frac{GO}{GO_0}\right)^{\xi'}\left(\left.\frac{\Delta\gamma^{p}_{max}}{2}\right|_{macro}\right)^{2} \tag{3}$$

The ratio of grain size to the reference grain size of the material, shown as $\left(\frac{GS}{GS_0}\right)$, represents the effect of grain size distribution upon the small crack growth. The ratio of the Taylor factor for the orientation of the grain at the crack tip with the Taylor factor of the typical rolling texture, $\left(\frac{GO}{GO_0}\right)$, encompasses the effect of texture upon the MSC growth regime. The nonlocal maximum plastic shear strain amplitude term, represented by $\left(\frac{\Delta\gamma^{p}max}{2}\right)$ in Equation 3, can be found by Equation 4,

$$C_{inc}N_{inc}^{\alpha} = \beta = \frac{\Delta\gamma^{P}_{max}}{2} \tag{4}$$

where N_{inc} is the number of cycles required to incubate a crack, α is a material constant that is based on the macroscopic Coffin-Manson law, and C_{inc} can be solved by using the following equation. The constants related to the Coffin-Manson law were based on strain-life experiments of magnesium AZ61 alloy of Gibson *et al.* [23].

Small Fatigue Crack Model Correlation

The small crack equations presented here were used to make correlations to the crack growth rate measured in the interrupted tests. The small crack model was plotted as a function of crack length and subsequently compared to the crack growth data, shown in Figure 10. It is important to note that the small crack model (Equation 2) is intended to model crack growth for a through crack of crack length, a. In this study, the experimental small crack measurements were taken from surface crack growth. For the correlation process, we assume that the crack grew in a symmetric semi-circular shape and the surface crack length (c) is equal to twice the depth of the crack length (2a). Thus, we correlated the small crack model to ½ of the surface crack length, c, in which we measured the crack growth rate to the left of the initiation site as shown in Figure 7. Although the model captured the mean trend of the crack growth rate, the scatter in the data is quite significant when the crack is less than 500 μm in length. While not shown here, the small crack model can be used to predict the upper and lower bounds of the crack growth rate data.

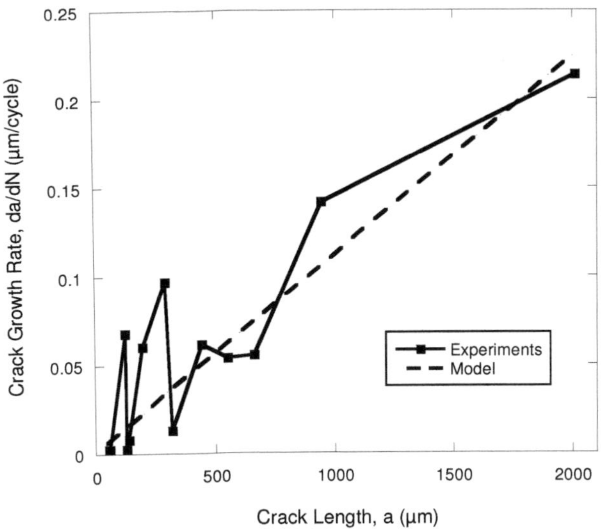

Figure 10. Crack growth rate versus crack length for a stress amplitude of 105MPa with the model approximations also shown.

Summary

Based on the small fatigue crack experiments and modeling presented here, we present the following summary:

1. Surface preparation and replicas taken of the notch surface had no notable impact upon the fatigue life of the material.
2. The replica method was able to capture the mean crack growth of the AZ61 material, particularly within the microstructurally small/physically small crack regime.
3. The small crack equations of the multistage fatigue model were correlated to the small crack growth behavior of the AZ61 alloy.
4. Small fatigue cracks were shown to preferentially propagate through favorably oriented grains in the extruded AZ61 magnesium alloy.
5. When the crack was small, (less than 500μm), the fatigue crack displayed a stronger dependence upon the material microstructure compared to when it was longer in length.

Acknowledgments

The authors would like to recognize Richard Osborne, James Quinn, Xuming Su, John Allison, Robert McCune, Don Penrod, and Matthew Castanier for their encouragement of this study. This material is based upon work supported by the Department of Energy and the National Energy Technology Laboratory through a subcontract with Mississippi State University, and was performed for the Simulation Based Reliability and Safety (SimBRS) research program. This report was prepared as an account of work sponsored by an agency of the United States Government. Neither the United States Government nor any agency thereof, nor any of their employees, makes any warranty, express or implied, or assumes any legal liability or responsibility for the accuracy, completeness, or usefulness of any information, apparatus, product, or process disclosed, or represents that its use would not infringe privately owned rights. Reference herein to any specific commercial product, process, or service by trade name, trademark, manufacturer, or otherwise does not necessarily constitute or imply its endorsement, recommendation, or favoring by the United States Government or any agency thereof. The views and opinions of authors expressed herein do not necessarily state or reflect those of the United States Government or any agency thereof. Such support does not constitute an endorsement by the Department of Energy of the work or the views expressed herein. UNCLASSIFIED: Dist A. Approved for public release.

References

1. S. Begum, D.L. Chen, S. Xu, and Alan A. Luo, "Strain-Controlled Low-Cycle Fatigue Properties of a Newly Developed Extruded Magnesium Alloy," *Metallurgical and Materials Transactions A,* 39A (2008), 3014-3026.

2. F. Yang, S.M. Yin, S.X. Li, and Z.F. Zhang, "Crack initiation mechanism of extruded AZ31 magnesium alloy in the very high cycle fatigue regime," *Materials Science and Engineering A*, 491 (2008), 131-136.

3. J.B. Jordon, M.F. Horstemeyer, K. Solanki, J.D. Bernard, J.T. Berry, T.N. Williams, "Damage characterization and modeling of a 7075-T651 aluminum plate," *Materials Science and Engineering A*, 527 (2009), 169-178.

4. S. Begum, D.L. Chen, S. Xu, Alan A. Luo, "Effect of strain ratio and strain rate on low cycle fatigue behavior of AZ31 wrought magnesium alloy," *Materials Science and Engineering A*, 517 (2009) 334-343.

5. Z.B. Sajuri, Y. Miyashita, Y. Hosokai, Y. Mutoh, "Effects of Mn content and texture on fatigue properties of as-cast and extruded AZ61 magnesium alloys," *International Journal of Mechanical Sciences*, 48 (2) (2006) 198-209.

6. K. Tokaji, M. Nakajima, Y. Uematsu, "Fatigue crack propagation and fracture mechanisms of wrought magnesium alloys in different environments," *International Journal of Fatigue*, 31 (7) (2009) 1137-1143.

7. Z.B. Sajuri, Y. Miyashita, Y. Mutoh, "Effects of humidity and temperature on the fatigue behaviour of an

extruded AZ61 magnesium alloy," *Fatigue and Fracture of Engineering Materials and Structures*, 28, 4 (2005) 373-379.

8. S. Suresh, "Fatigue of Metals," Cambridge University Press, Cambridge, 1998.
9. K. Gall, G. Biallas, Hans J. Maier, P. Gullett, M. F. Horstemeyer, D. L. McDowell, and J. Fan, "In-situ observations of high cycle fatigue mechanisms in cast AM60B magnesium in vacuum and water vapor environments," *International Journal of Fatigue*, 26 (2004) 59-70.
10. K. Gall, G. Biallas, H. J. Maier, M. F. Horstemeyer, and D. L. McDowell, "Environmentally influenced microstructurally small fatigue crack growth in cast magnesium," *Materials Science and Engineering A*, 396 (2004) 143-154.
11. K. Gall, G. Biallas, H. J. Maier, P. Gullett, M. F. Horstemeyer, and D. L. McDowell, "In-Situ Observations of Low-Cycle Fatigue Damage in Cast AM60B Magnesium in an Environmental Scanning Electron Microscope," *Metallurgical and Materials Transactions* A, 35 (2004) 321-331.
12. D.L. McDowell, K. Gall, M.F. Horstemeyer, and J. Fan, "Microstructure-based fatigue modeling of cast A356-T6 alloy," *Engineering Fracture Mechanics,* 70 (1) (2003), 49-80.
13. J.B. Jordon, M.F. Horstemeyer, N. Yang, J.F. Major, K.A. Gall, J. Fan, D.L. McDowell, "Microstructural inclusion influence on fatigue of a cast A356 aluminum alloy," *Metallurgical and Materials Transactions* A, 41A (2) (2010), 356-363.
14. Y. Xue, A. Pascu, M.F. Horstemeyer, L. Wang, P.T. Wang, "Microporosity effects on cyclic plasticity and fatigue of LENS™-processed steel," Acta Materialia 58 (11) (2010), 4029-4038.
15. Y. Xue, D.L. McDowell, M.F. Horstemeyer, M.H. Dale, J.B. Jordon, "Microstructure-based multistage fatigue modeling of aluminum alloy 7075-T651," *Engineering Fracture* Mechanics, 74 (17) (2007), 2810-2823.
16. Y. Xue, M.F. Horstemeyer, D.L. McDowell, H. El Kadiri, J. Fan, "Microstructure-based multistage fatigue modeling of a cast AE44 magnesium alloy," *International Journal of Fatigue,* 29 (4) (2007), 666-676.
17. K. Tokaji, M. Kamakura, Y. Ishiizumi, N. Hasegawa, "Fatigue behaviour and fracture mechanism of a rolled AZ31 magnesium alloy," *International Journal of* Fatigue, 26 (11) (2004), 1217-1224.
18. D.J. Bammann, "Internal variable model of viscoplasticity," *International Journal of Engineering Science*, 22 (8-10) (1983), 1041-1053.
19. D.J. Bammann, E.C. Aifantis, "Model for finite-deformation plasticity," Acta *Mechanica*, 69 (1-4) (1987), 97-117.
20. D.J. Bammann, E.C. Aifantis, "A damage model for ductile metals," *Nuclear Engineering and Design*, 116 (3) (1989), 355-362.
21. M.F. Horstemeyer, A.M. Gokhale, "Void-crack nucleation model for ductile metals," *International Journal of Solids and Structures*, 36 (33) (1999), 5029-5055.
22. M.F. Horstemeyer, J. Lathrop, A.M. Gokhale, M. Dighe, "Modeling stress state dependent damage evolution in a cast Al-Si-Mg aluminum alloy," *Theoretical and Applied Fracture Mechanics*, 33 (1) (2000), 31-47.
23. J.B. Jordon, J.B. Gibson, M.F. Horstemeyer, "Experiments and Modeling of Fatigue Damage in Extruded Mg AZ61 Alloy," *TMS Annual Meeting and 2011*, Submitted.

Magnesium Technology 2011
Edited by: Wim H. Sillekens, Sean R. Agnew, Neale R. Neelameggham, and Suveen N. Mathaudhu
TMS (The Minerals, Metals & Materials Society), 2011

APPLICABILITY OF Mg -Zn-(Y, Gd) ALLOYS FOR ENGINE PISTONS

Kazutaka Okamoto[1], Masato Sasaki[2], Norikazu Takahashi[2], Qudong Wang[3], Yan Gao[3], Dongdi Yin[3], Changjiang Chen[3]
[1]Hitachi, Ltd.; Hitachi, Ibaraki 319-1292, Japan
[2]Hitachi Automotive Systems, Ltd.; Atsugi, Kanagawa 243-8510, Japan
[3]Shanghai Jiao Tong Univ.; Shanghai 200240, China

Keywords: Gravity casting, Heat treatment, Tensile strength, Creep strain, Fatigue strength

Abstract

Commercial magnesium alloys have a great potential for structural applications in automotive due to their significant weight saving. However, they have poor creep resistance at temperature over 125°C, thus making them inadequate for power train applications such as engine pistons, which are operated at temperature up to 300°C. Recently, creep resistant magnesium alloys with rare-earth elements and Zn have been developed, hence the applicability of Mg-Zn-(Y, Gd) alloys for engine pistons was investigated in this paper. Gravity casting was performed with Mg-Zn-(Y, Gd)-Zr alloy, followed by T6 treatment. Effects of the amount of alloying elements on the mechanical properties of tensile strength and creep strain were evaluated. Nominal composition of Mg-2Zn-11Y-5Gd-0.5Zr was selected for the actual piston cast trial and its high cycle fatigue test was conducted comparing to the current aluminum cast alloy of A336 (JIS AC8A) for pistons. At room temperature, the fatigue strength is 27% lower than A336, while it is 35% higher at 300°C. It is suggested that Mg-2Zn-11Y-5Gd-0.5Zr alloy shows attractive high temperature mechanical properties higher than A336, hence it is promising as a candidate material for the engine piston application.

Introduction

Magnesium alloys have a great potential for structural applications in automotive and aerospace industries due to their significant weight savings, thus improving fuel economy and lessening environmental impact. The most significant magnesium applications are in castings, such as instrument panel, transfer cases, valve/cam covers, various housings and brackets, and steering components in automobiles, with commercial magnesium alloys of AZ91, AM50 and AM60. These alloys offer an excellent combination of mechanical properties, corrosion resistance and castability.

On the other hand, power train components are made of heavy parts, hence the specific weight of magnesium, which is 2/3 of aluminum, is very attractive for weight reduction of transmission case or engine block. Although aluminum engine piston is not so weighty, generally about 0.3kg, it is reciprocated at higher velocity over 10m/s, so that the reciprocating mass can be reduced. Furthermore, weight reduction of circumference parts such as connecting rod, less NVH (Noise, Vibration and Harshness) issues, higher performance of engine are expected. However, these components are operated at higher temperatures above 175°C, engine block up to 200°C, and engine pistons up to 300°C, thus making commercial magnesium alloys inadequate for major power-train applications because of their poor creep resistance at temperatures above 125°C.

The requirements for high temperature creep resistance and weight savings have made it imperative to develop new magnesium alloys to decrease use of conventional aluminum and iron castings in vehicles. Both diffusion controlled dislocation climb and grain boundary sliding/shearing have been reported as creep mechanisms in magnesium alloys, depending on the alloy system, microstructure, and stress and temperature regimes. In magnesium alloys, aluminum is often added to obtain a good combination of strength, ductility and castabilty. However, the poor thermal stability of $Mg_{17}Al_{12}$ intermetallics (β phase, with a eutectic temperature of 437°C) and its discontinuous precipitation can result in substantial grain boundary sliding/shearing at elevated temperatures. Moreover, the accelerated diffusion of aluminum solute in magnesium matrix and the self-diffusion of magnesium at elevated temperatures contribute to creep deformation in Mg-Al based alloys. Based on the above arguments and alloy strengthening mechanisms, possible approaches to improve creep resistance in magnesium alloys include solution strengthening, precipitation strengthening and dispersoid strengthening. Most of magnesium alloy systems were developed using these approaches to obtain good elevated/high-temperature creep resistance, namely Mg-Al-Sr (AJ), Mg-Al-Ca(-Sr) (AX(J)), Mg-Al-RE (AE), Mg-Y-Nd (WE), etc. Among those alloys, the enhancement of creep resistance by rare earth (RE) elements is particularly common. RE has a quite large solubility, resulting in solid solution strengthening, decreasing stacking fault energy, thus splitting basal dislocation into partials and increasing resistance of dislocation glide and climb [1-3]. Furthermore, Zn addition significantly enhances the age hardening response and the creep resistance with uniform and dense distribution of basal precipitate plates [4] and/or long period stacking order (LPSO) [5-13], namely Mg-Zn-RE alloy.

In this research, high temperature properties of Mg-Zn-(Y, Gd) cast alloy were evaluated so as to investigate the applicability of the cast to engine pistons, considering those of the current aluminum cast alloy of A336 (JIS AC8A). Conventional gravity casting was carried out, followed by T6 heat treatment, and both creep strain and high cycle fatigue tests as well as tensile test at room and elevated temperature were conducted. Our development goal was set as summarized in Table 1.

Table 1 Development goal for the mechanical properties of Mg-Zn-(Y, Gd) cast alloy

Property and Test Conditions	Goal
Ultimate Tensile Strength	
RT	> 230MPa
300°C	> 150MPa
Creep Strain	
250°C , 80MPa for 20hrs	< 0.30%
300°C , 50MPa for 20hrs	< 0.40%
Fatigue Strength	
300°C, 10^7cycles	> 65MPa

Experimental

Materials to be Evaluated

The chemical compositions of the magnesium alloys are summarized in Table 2. The actual chemical composition of the alloys were evaluated by an inductively couples plasma (ICP).

All these alloy master ingots were prepared with high purity Mg (99.95%) and Zn (99.95%), Mg-25mass%Gd, Mg-25mass%Y and Mg-30mass%Zr alloys. After smelting using an electric resistance furnace under the mixed cover gas of CO_2 and SF_6, the molten metal was pored into a preheated (200°C) cast iron block mold with a wall thickness of 20mm, see Fig.1.

Table 2 Nominal and actual chemical composition of Mg-Zn-(Y, Gd)-Zr cast alloy (mass %)

Nominal comp.	Actual comp.				
	Zn	Y	Gd	Zr	Mg
Mg-10Y-5Gd-0.5Zr	-	8.9	5.4	0.4	bal.
Mg-2Zn-**10Y-5Gd**-0.5Zr	2.1	10.5	4.6	0.4	bal.
Mg-2Zn-**5Y-15Gd**-0.5Zr	1.9	3.6	12.5	0.4	bal.
Mg-2Zn-**5Y-5Gd**-0.5Zr	2.3	4.1	5.0	0.4	bal.
Mg-2Zn-**11Y**-5Gd-0.5Zr	2.2	10.7	5.6	0.4	bal.
Mg-2Zn-**12Y**-5Gd-0.5Zr	2.0	11.9	5.6	0.4	bal.
Mg-2Zn-**14Y**-5Gd-0.5Zr	2.2	14.4	4.9	0.5	bal.
Mg-**0.5Zn**-10Y-5Gd-0.5Zr	0.6	7.2	4.1	0.4	bal.
Mg-**2Zn**-10Y-5Gd-0.5Zr	2.1	10.5	4.6	0.4	bal.
Mg-**3Zn**-10Y-5Gd-0.5Zr	3.5	9.1	4.8	0.5	bal

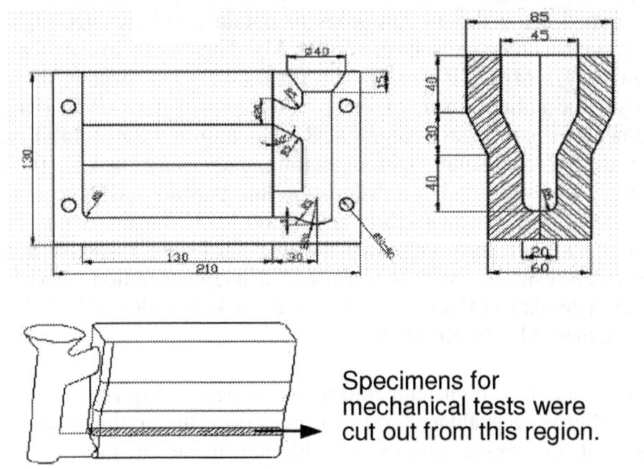

Fig. 1 Schematic drawings of cast iron block mold

Microstructure and Hardness

Microstructure observation was carried out using an optical microscope (OM) and a scanning electron microscope equipped with an energy dispersive X-ray spectrometer (SEM-EDS). Specimens were cut and polished, then etched with 5mass% HNO_3 ethanol. Differential scanning calorimetry (DSC) as well as Vickers hardness tests with an indentation load of 49N were also performed in order to decide the heat treatment condition.

Mechanical Properties

Mechanical properties were evaluated with the specimens shown in Fig. 2, cut from the cast magnesium alloy. Tensile test was performed at both room and elevated temperatures with an initial strain rate of 0.5mm/min. Creep strain was evaluated at 200°C, 250°C and 300°C with an initial load of 120MPa, 80MPa and 50MPa, respectively. Furthermore, high cycle tensile-tensile fatigue test was carried out at room temperature, 250°C and 300°C, up to 10^7 cycles. The stress ratio and frequency were set to R=0 and 30Hz, respectively.

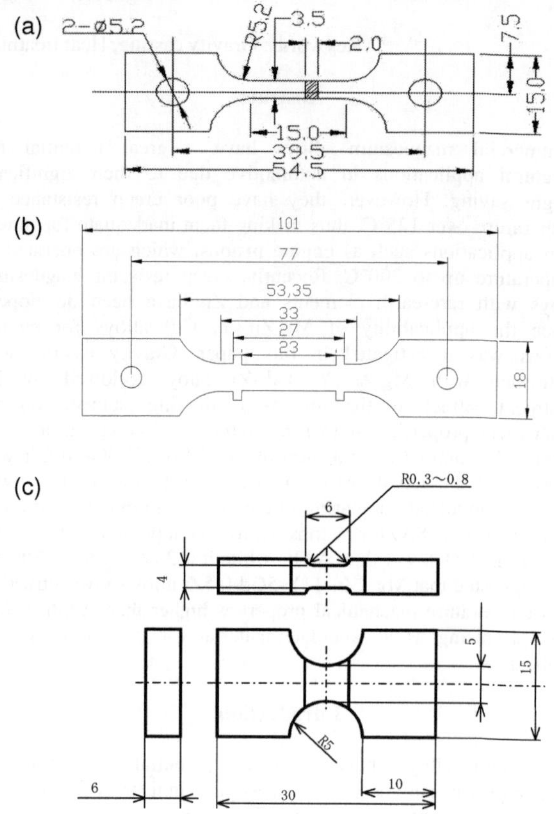

Fig. 2 Schematic drawings of mechanical test specimens for (a) tensile test, (b) creep test and (c) fatigue test

Results and Discussions

Heat Treatment

DSC was conducted for both Mg-10Y-5Gd-0.5Zr and Mg-2Zn-10Y-5Gd-0.5Zr alloys as shown in Fig. 3, and the peak temperatures were measured at 567°C and 540°C, respectively. In order to apply a same heat treatment condition to both alloys, temperature for solution heat treatment was set to 535°C, just below the eutectic temperature of Mg-2Zn-10Y-5Gd-0.5Zr alloy.

Figure 4 shows as-cast and as-T4 microstructure of both alloys. According to the previous research [14], by adding Zn, the resultant microstructure is remarkably changed with long period stacking order formation (seen as a lamellar like morphology in Fig. 4). After 16hrs at 535°C, the eutectic intermetallics of Mg-(Y, Gd) were almost solutionized in α–Mg matrix of Mg-10Y-5Gd-0.5Zr alloy, while the remnant of the eutectic intermetallics was still observed in Mg-2Zn-10Y-5Gd-0.5Zr alloy so that Zn addition decreases the solubility of Y and Gd into α-Mg matrix. The average grain size was approximately 100μm.

After the solution heat treatment, aging treatment was applied at various temperature of 225, 250 and 300°C. Figure 5 shows the age hardening response of Mg-10Y-5Gd-0.5Zr alloy. The peak hardness decreased with increasing aging temperature. At 225°C, the hardness increased rapidly by holding 1hr, and the peak hardness of around135Hv was obtained at about 16-24hrs. After reaching the peak hardness, a plateau region was observed up to 64hrs, then the hardness gradually decreased.

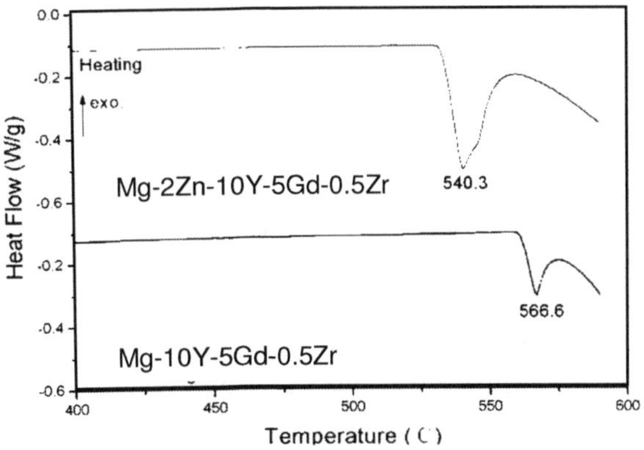

Fig.3 DCS curves for the magnesium alloys

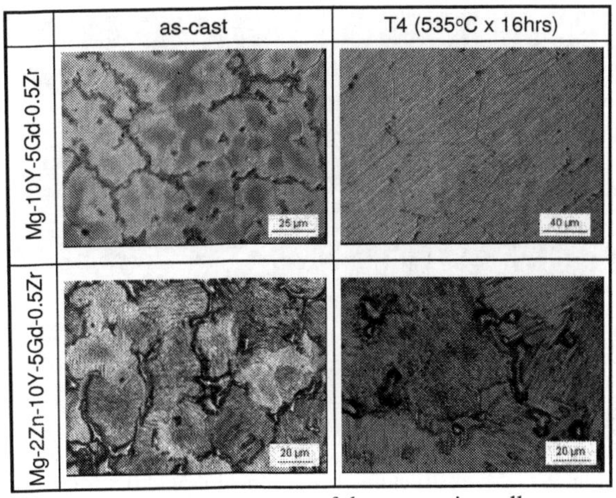

Fig.4 Microstructure of the magnesium alloys

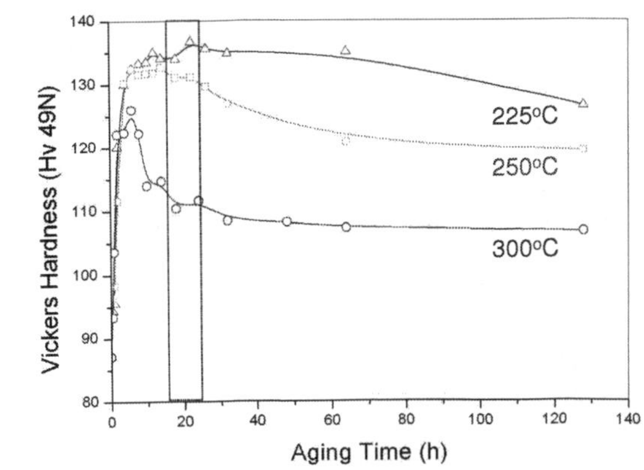

Fig.5 Age hardening response of Mg-10Y-5Gd-0.5Zr alloy

Tensile Strength

In order to investigate the effects of the total amount of RE elements, effects of Y and Gd content on the tensile properties were evaluated. Table 3 summarizes the tensile properties of Mg-2Zn-(Y, Gd)-0.5Zr alloys. It is clearly shown that both 0.2% proof stress ($\sigma_{0.2}$) and ultimate tensile strength (UTS) increased with increasing the total amount of (Y, Gd) while the elongation (ε) remarkably decreased down to less than 1%. However, the melt loss is quite significant for Mg-2Zn-5Y-15Gd-0.5Zr alloy, so that the amount of Gd was selected to 5mass %. Figure 6 shows stress-strain curves of Mg-2Zn-10Y-5Gd-0.5Zr alloy at 300°C and 350°C, respectively. Even at 300°C, UTS is higher than 200MPa, which meets our development goal. However, UTS remarkably decreased over 300°C with significant increase in the elongation up to 28%.

Table 3 Effect of (Y, Gd) on the tensile properties

	$\sigma_{0.2}$ / MPa	UTS / MPa	ε (%)
Mg-2Zn-**10Y-5Gd**-0.5Zr	248	273	0.6
Mg-2Zn-**5Y-15Gd**-0.5Zr	236	305	0.9
Mg-2Zn-**5Y-5Gd**-0.5Zr	126	251	12

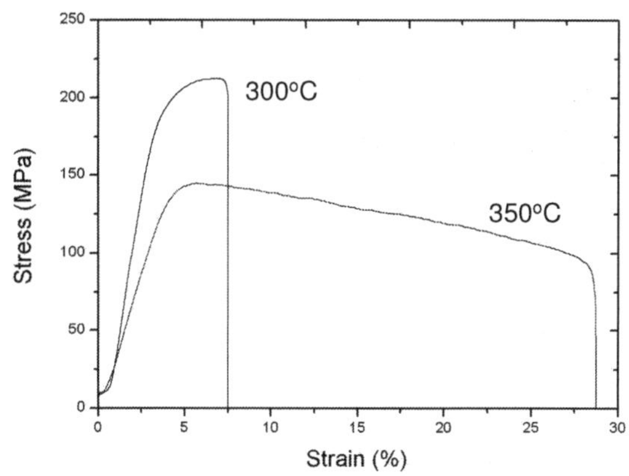

Fig. 6 Stress-strain curves of Mg-2Zn-10Y-5Gd-0.5Zr alloy at elevated temperatures

Creep Strain

In order to determine the composition range, effects of Zn and Y content on the creep properties were investigated.

It was found that the addition of Zn to Mg-10Y-5Gd-0.5Zr alloy remarkably improves the creep resistance, as shown in Table 4, summarizing the creep strain and secondary creep rate at 250°C, 80MPa. Furthermore, Zn content of 2mass % is most affective for the improvement of creep resistance.

Figure 7 shows creep curves of the magnesium alloys at 300°C, 50MPa. Amount of Y, a primary alloying element for these magnesium alloys, has a strong influence on the creep strain. Creep stain is minimized and our goal was achieved for the alloys with Y content of 11-12mass %. However higher Y was demonstrated to be not beneficial for the improvement of creep resistance and the material cost will increase. According to Mg-Y binary diagram, the maximum solubility of Y into α-Mg matrix at the eutectic temperature is about 12.5mass %, hence over high Y addition results in the existence of large amount of eutectic phase, which can not be dissolved into the matrix completely even after the solution heat treatment. These thermal unstable intemetallic compounds may deteriorate the creep resistance at α-Mg grain boundaries.

Figure 8 shows the microstructure evolution of Mg-2Zn-11Y-5Gd-0.5Zr alloy during the creep test at 300°C, 50MPa for 0-100hrs. It can be seen that the lamellar like morphology is developed as increasing the testing time and maybe the structure causes higher creep resistance.

Table 4 Effect of Zn on the creep properties at 250°C, 80MPa

	Creep strain for 20hrs (%)	Secondary creep rate / s^{-1}
Mg-10Y-5Gd-0.5Zr	0.36	44.6×10^{-9}
Mg-**0.5Zn**-10Y-5Gd-0.5Zr	0.16	6.19×10^{-9}
Mg-**2Zn**-10Y-5Gd-0.5Zr	0.12	3.89×10^{-9}
Mg-**3Zn**-10Y-5Gd-0.5Zr	0.28	11.8×10^{-9}

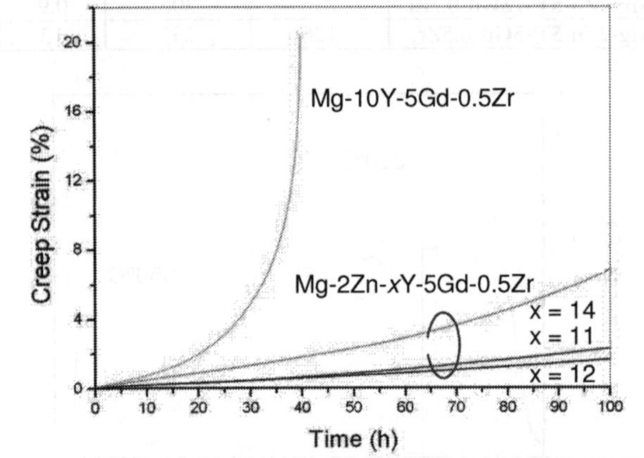

Fig. 7 Creep curves of the magnesium alloys at 300°C, 50MPa

Fig.8 Microstructure evolution in Mg-2Zn-11Y-5Gd-0.5Zr alloy during the creep test at 300°C for (a) 0hr, (b) 20hrs, (c) 50hrs and (d) 100hrs

Fatigue Strength

Based on the mechanical properties of various Mg-Zn-(Y, Gd)-Zr alloys above mentioned, Mg-2Zn-11Y-5Gd-0.5Zr alloy was selected for further investigations. Table 5 summarizes the high cycle tensile-tensile fatigue test at room temperature, 250°C and 300°C, comparing to those of A336-T6. Although the fatigue strength of the magnesium alloy at room temperature and 250°C is 30-40% lower than A336, it is about 85% of A336 at 300°C.

Microstructure evolution was observed after the tests at 250°C and 300°C, within the α-Mg grain interior near the boundaries as described in Fig. 9, which may strengthen the grain boundaries and prevent its deformation.

Figure 10 shows the fracture surface of Mg-2Zn-11Y-5Gd-0.5Zr alloy. It is observed that cleavage like cracks are predominant and facet size of the cleavage plane is almost the same as the grain size. Furthermore, fatigue crack propagates perpendicular to the small steps, so that it is considered that the LPSO along with the basal plane of hexagonal α-Mg results in the cleavage cracks.

Table 5 High cycle fatigue strength of Mg-2Zn-11Y-5Gd-0.5Zr and A336 aluminum alloys

	RT	250°C	300°C
Mg-2Zn-11Y-5Gd-0.5Zr	70MPa	59MPa	53MPa
A336	103MPa	96MPa	63MPa

Fig. 9 Microstructure of Mg-2Zn-11Y-5Gd-0.5Zr alloy after testing for 9.5×10^{6} cycles at 55MPa, 300°C

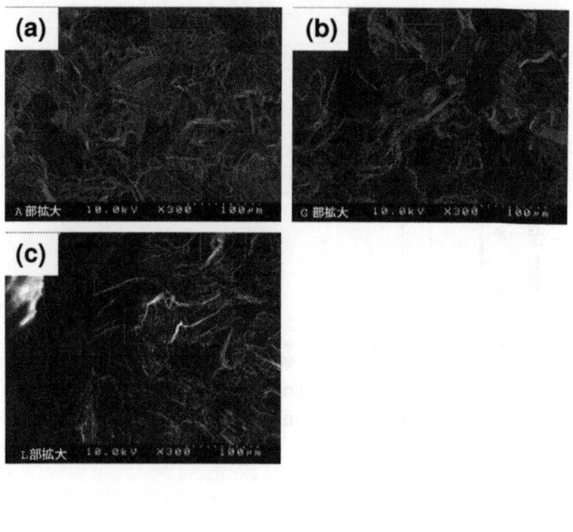

Fig. 10 Fracture surface of fatigue specimens of Mg-2Zn-11Y-5Gd-0.5Zr alloy tested at (a) RT, (b) 250°C and (c) 300°C

<u>Piston Cast Trial</u>

Gravity casting was performed using a piston mold illustrated in Fig. 11. Specimens were cut out from the crown surface for metallurgical and mechanical investigations. The actual chemical composition is analyzed to Mg-2.1Zn-10.9Y-4.6Gd-0.4Zr (mass %). T6 of 535°Cx16-20hrs and 225°Cx24hrs was applied.

Figure 12 show both as-cast and as-T6 microstructures. In case of as-cast, Gd is enriched in α-Mg matrix, whose average grain size is about 100μm, while it is grown to about 150μm via T6 treatment. Zn and Y are detected in the eutectic intermetallics at α-Mg grain boundaries. Furthermore, lamellar like morphology is observed along the α-Mg grain boundaries, which suggests LPSO is formed originated from Y element [6-9]. Yamasaki et al. [9] reported that RE can be categorized into two types; type-I additives such as Y form LPSO during solidification, and type-II additives such as Gd form during high temperature exposure. In case of as-T6, it is suggested that Gd plays an important role forming LPSO. Furthermore, Yamada [11, 12] and Honma [13] reported that 0.3-1.0at. % Zn containing Mg-Gd-Y-Zr alloy shows room temperature tensile strength over 400MPa with 14H-LPSO at α-Mg grain boundaries and metastable β' phase, via solution heat treatment at 500°C followed by aging treatment at 225°C. Therefore, it is considered that the similar type of microstructure evolution can be expected for this alloy.

Figure 13 shows the tensile properties of Mg-2Zn-11Y-5Gd-0.5Zr alloy. It is observed that UTS shows an inverse temperature dependence up to 200°C, and it decreases over the temperature. As mentioned above, the aging treatment at 225°C might form metastable β' phase, so that further investigations of microstructure evolution and phase stability of the resultant microstructure are required. Table 6 summarized the creep properties of Mg-2Zn-11Y-5Gd-0.5Zr alloy at elevated temperatures. It can be said that the creep resistance of the piston specimen is almost comparable to that of block specimen.

Table 6 Creep properties of Mg-2Zn-11Y-5Gd-0.5Zr alloy at elevated temperatures

Temperature / °C	Creep strain for 20hrs (%)	Secondary creep rate / s^{-1}
200	0.17	0.97×10^{-9}
250	0.32	3.65×10^{-9}
300	0.33	4.72×10^{-9}

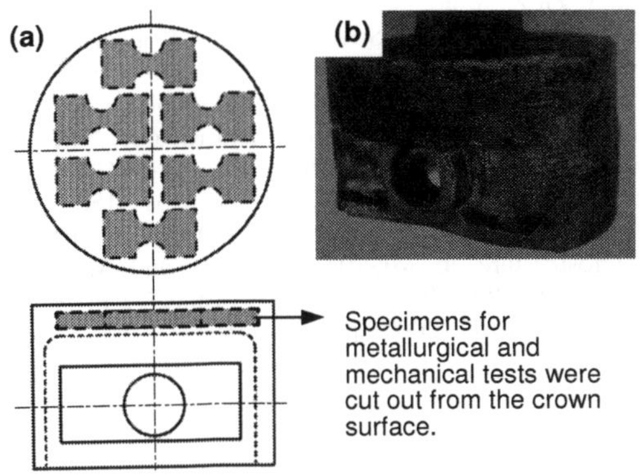

Fig.11 Actual piston cast trial
(a) schematic drawing of piston, (b) as-cast piston

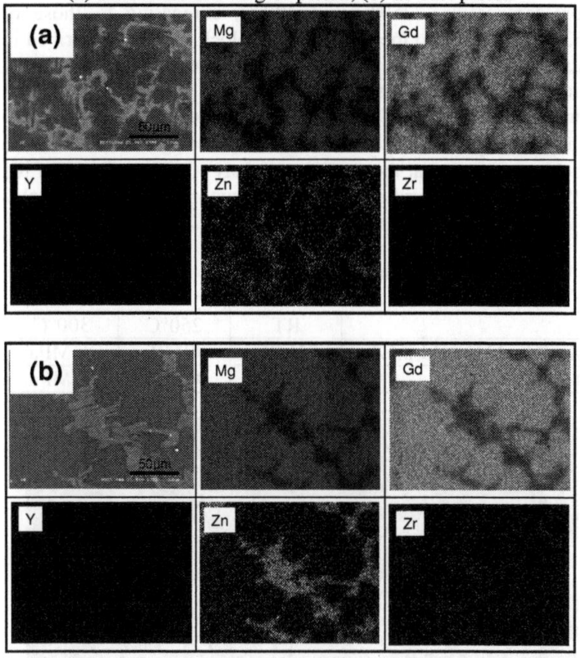

Fig.12 Microstructure of Mg-2Zn-11Y-5Gd-0.5Zr alloy
(a) as-cast and (b) as-T6

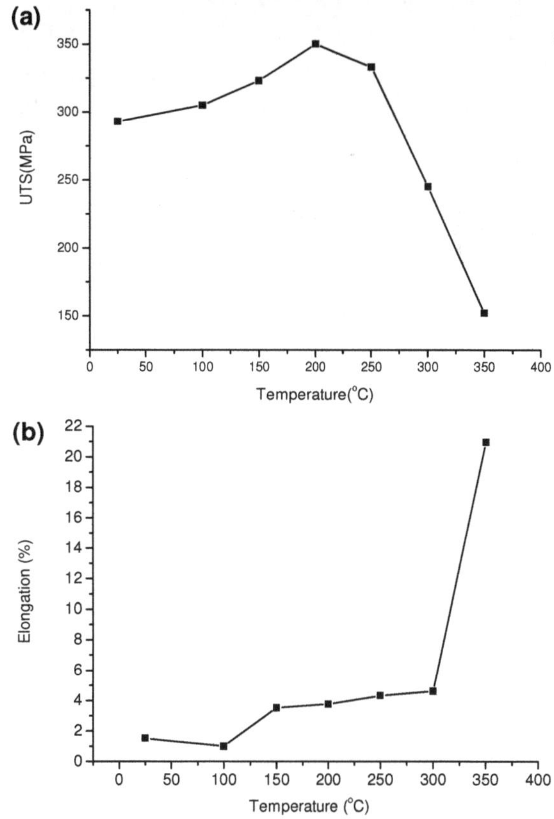

Fig.13 Tensile properties of Mg-2Zn-11Y-5Gd-0.5Zr alloy
(a) UTS and (b) elongation

Table 7 summarizes the high cycle tensile-tensile fatigue test at room temperature, 250°C and 300°C, comparing to those of A336-T6. Similar to the results of the block samples described in Table 5, the fatigue strength at room temperature is 27% lower than A336, while almost the same or even 35% higher strength were obtained at 250°C or 300°C. It is thought that the average grain size of piston crown surface is 150µm, larger than that of the block sample, so that the cooling rate may significantly affect the LPSO formation. Furthermore, microstructure evolution was also observed during the tests at 250°C and 300°C as seen in Fig. 14.

Table 7 High cycle fatigue strength of Mg-2Zn-11Y-5Gd-0.5Zr and A336 aluminum alloys

	RT	250°C	300°C
Mg-2Zn-11Y-5Gd-0.5Zr	75MPa	90MPa	85MPa
A336	103MPa	96MPa	63MPa

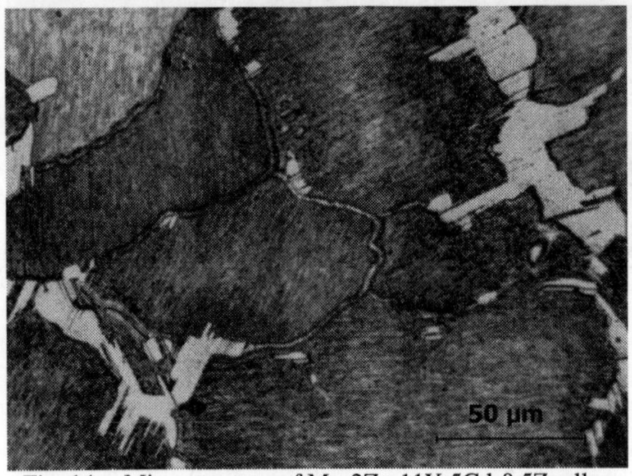

Fig. 14 Microstructure of Mg-2Zn-11Y-5Gd-0.5Zr alloy after testing for $3.3x10^6$ cycles at 90MPa, 300°C

Conclusions

Mechanical properties of Mg-Zn-(Y, Gd) alloys were evaluated by using gravity cast and T6 treated specimens and its applicability for engine pistons was investigated. The results obtained are summarized as follows:

- Effect of the amount of the alloying elements, Zn and Y of Mg-*x*Zn-*y*Y-5Gd-0.5Zr (mass %) alloy on the creep property at elevated temperature were studied. It is suggested that the addition of 2mass %Zn and 11mass %Y minimizes the creep stain at elevated temperatures.
- Mg-2Zn-11Y-5Gd-0.5Zr alloy was selected for high cycle fatigue test. Fatigue strength at room temperature and 250°C is 30-40% lower than A336, while it is about 85% of A336 at 300°C.
- Actual pistons of Mg-2.1Zn-10.9Y-4.6Gd-0.4Zr (mass %) were gravity cast. In the comparison to A336, almost the same or even 35% higher fatigue strength were obtained at 250°C or 300°C.

Mg-2Zn-11Y-5Gd-0.5Zr cast alloy exhibits excellent high temperature mechanical properties such as tensile strength, creep resistance and fatigue strength, equivalent or even higher in the comparison with the current aluminum alloy for piston. Therefore, it can be concluded that this magnesium alloy is very promising as a candidate material for piston applications, since specific strength design for this alloy piston is not required. However, there are still manufacturing related issues remained, such as optimization of casting design, heat treatment schedule for the stability of microstructure at elevated temperatures, wear resistant surface treatment strategy, and so on. Therefore further investigations are on going via piston trial manufacturing to pursue technical possibility of Mg-Zn-(Y, Gd) alloys for engine pistons.

References

1. I.A.Anyanwu, S.Kamado and Y.Kojima: "Aging characteristics and high temperature tensile properties of Mg-Gd-Y-Zr alloys", Mater. Trans. 42 (2001), pp. 1206-1211
2. I.A.Anyanwu, S.Kamado and Y.Kojima: "Creep properties of Mg-Gd-Y-Zr alloys", Mater. Trans. 42 (2001), pp. 1212-1218
3. Y.Gao, Q.Wang, J.Gu, Y.Zhao, Y.Tong and J.Kaneda: "Effects of heat treatments on microstructure and mechanical properties of Mg-15Gd-5Y-0.5Zr alloy", J. of Rare Earth 26 (2008), pp. 298-302
4. J.F.Nie, X. Gao and S.M. Zhu: "Enhanced age hardening response and creep resistance of Mg-Gd alloys containing Zn", Scripta Mater. 53 (2005), pp. 1049–1053
5. Y.Kawamura, K. Hayashi, A. Inoue and T. Masumoto: "Rapidly solidified powder metallurgy Mg97Zn1Y2 alloys with excellent tensile yield strength above 600MPa", Mater. Trans. 42 (2001), pp. 1172-1176
6. E.Abe, Y.Kawamura, K.Hayashi and A.Inoue: ""Long-period ordered structure in a high-strenght manocrystalline Mg-1at%Zn-2at%Y alloy studied by atomic-resolution Z-contrast STEM", Acta Mater. 50 (2002), pp. 3845-3857
7. T.Itoi, T.Seimiya, Y.Kawamura and M.Hirohashi: "Long period stacking structures observed in Mg97Zn1Y2 alloy", Scr. Mater. 51 (2004), pp. 107-111
8. M.Yamasaki, T.Anan, S.Yoshimoto and Y.Kawamura: "Mechanical properties of warm-extruded Mg-Zn-Gd alloy with coherent 14H long periodic stacking ordered structure precipitate", Scr. Mater. 53 (2005), pp. 799-803
9. M.Yamasaki, M.Sasaki, M.Nishijima, K.Hiraga and Y.Kawamura: "Formation of 14H long period stacking ordered structure and profuse stacking faults in Mg-Zn-Gd alloys during isothermal aging at high temperature", Acta Mater. 55 (2007), pp. 6798-6805
10. Y.Kawamura and M.Yamasaki: " Formation and mechanical properties of Mg97Zn1RE2 alloys with long-period stacking ordered structure", Mater. Trans. 48 (2007), pp. 2986-2992
11. K.Yamada, Y.Okubo, M.Shiono, H.Watanabe, S.Kamado and Y.Kojima: "Alloy development of high toughness Mg-Gd-Y-Zn Zr alloys", Mater. Trans. 47 (2006), pp. 1066-1070
12. K.Yamada, Y.Okubo, S.Kamado and Y.Kojima: "Precipitate microstructures of high strength Mg-Gd-Y-Zn-Zr alloys", Advanced Materials Research 11-12 (2006), pp. 417-420
13. T.Honma, T.Ohkubo, S.Kamado and K.Hono: "Effect of Zn additions on the age-hardening of Mg-2.0Gd-1.2Y-0.2Zr alloys", Acta Materialia 55 (2007), pp. 4137-4150
14. Y.Gao, Q.Wang, J.Gu, Y.Zhao and D.Yin: "Comparison of microstructure in Mg-10Y-5Gd-0.5Zr and Mg-10Y-5Gd-2Zn-0.5Zr alloys by conventional casting", J of Alloys and Compounds 477 (2009), pp. 374-378

Magnesium Technology 2011
Edited by: Wim H. Sillekens, Sean R. Agnew, Neale R. Neelameggham, and Suveen N. Mathaudhu
TMS (The Minerals, Metals & Materials Society), 2011

COMPRESSIVE CREEP BEHAVIOUR OF EXTRUDED Mg ALLOYS AT 150 °C

M. Fletcher[1], L. Bichler[1], D. Sediako[2] and R. Klassen[3]
[1]University of British Columbia – Okanagan, School of Engineering, Kelowna, Canada
[2]Canadian Neutron Beam Centre, Chalk River, Canada
[3]University of Western Ontario, London, Canada

Keywords: Creep, Magnesium Alloys, Nano-Indentation

Abstract

Wrought magnesium alloy bars, sections and tubes have been extensively used in the aerospace, electronics and automotive industries, where component weight is of concern. The operating temperature of these components is typically limited to below 100°C, since appreciable creep relaxation of the wrought alloys takes place above this temperature.

The objective of this study was to investigate the high temperature creep performance of two wrought magnesium alloys (AE42 and EZ33) developed for elevated temperature applications. Compressive creep behavior of extruded rods was studied at room temperature and at 150°C using the nano-indentation creep technique (on the microscale) and neutron diffraction (on the macroscale). Measurements were performed in the extrusion and radial directions to observe the effect of texture on the creep resistance, hardness and elastic modulus of the alloys. Microscopic examination of the alloys revealed that the distribution of second phases along the grain boundaries was critical to the alloy's creep resistance.

Introduction

Magnesium (Mg) is one of the lightest structural metals available to design engineers. Mg alloys have high specific strength, are easy to machine and are potentially recyclable. These attributes make them particularly attractive for applications in the transportation industry [1,2]. However, current Mg alloys do not possess the required creep resistance above 125 °C [1]. Creep deformation in Mg alloys has been generally contributed to grain boundary sliding and plastic deformation leading to intergranular failure [**Error! Bookmark not defined.**].

Available literature suggests that zinc (Zn) and aluminum (Al) have been effectively used to increase the room temperature mechanical properties of Mg-based wrought alloys via solid-solution strengthening and dispersion strengthening with intermetallic compounds. Zn has been observed to raise the eutectic temperature thus slightly enhancing the creep resistance of some Mg alloy systems. Zn addition, however, also increased alloy's susceptibility to microporosity and embrittlement. [**Error! Bookmark not defined.**].

Mechanical properties of Al-containing Mg alloys have been extensively studied, due to their commercial relevance. The rapid loss of strength of Al-containing alloys above 120°C has been generally attributed to the presence of β-$Mg_{17}Al_{12}$ phase (with incipient melting temperature of ~430°C) at the grain triple-points. To retard formation of the β phase, addition of rare earths (REs) to Mg-Al systems was implemented. Rare earths were seen to bind aluminum and form Al_xRE_y thermally stable intermetallics. The role of cerium, lanthanum, neodymium and praseodymium on improving creep resistance, however, remains the subject of extensive research.

In the case of RE addition to wrought Mg-Zn alloy systems, the rare earths generally reduced the tendency for microporosity and embrittlement. In these alloy systems, REs have combined to form additional Al-Zn-RE intermetallics. These compounds effectively pinned the grain boundaries above 120°C and the maximum effect from RE addition was seen in the 2-6 wt% level [3,4].

One of the most advanced Al-RE-Mg alloy systems is the AE42, with 4wt% Al and 2wt%RE. Research has shown that at 150°C, the creep resistance of the AE42 alloy begins to significantly deteriorate, because lamellar $Al_{11}RE_3$ intermetallic typically located in the vicinity of the grain boundaries, begins to dissolve and the Al_2RE and β-$Mg_{17}Al_{12}$ phases appear. The $Mg_{17}Al_{12}$ phase with its low incipient melting temperature subsequently facilitates creep-induced grain boundary movement [3].

Research focusing on EZ33 alloy, with 3wt%Re and 3wt%Zn, suggests that formation of intermetallics (and their stoichiometry) in this system is complex and strongly depends on the constitutive elements of the Rare Earth mischmetal (mostly forming T-Mg_9(Ce,La,Pr,Nd) phase. In this alloy, T-phase is stable at temperatures as high as 420°C [5]. Thus, available literature suggests that at 150°C, the EZ33 alloy should have a higher creep resistance than AE42.

Experimental Setup

The alloy samples used in this research were produced by Timminco Corporation, Canada. The alloy compositions can be found in Table I. The alloys were cast using a proprietary controlled-cooling static casting process, followed by hot extrusion. Both alloys were cast and formed without cracking or other processing defects. Figure 1 shows a schematic of the torus-like extruded specimens produced for this research. The outer diameter of the torus was 25mm, inner diameter was 7mm and thickness 9mm. The torus was sectioned to reveal the cross section (i.e., surface parallel with the extrusion direction) and the radial section (i.e., surface perpendicular to the extrusion direction).

Table I. Alloy Composition

Constituents (wt%)	Alloy	
	AE42	EZ33
Al	3.5-4.5	≤0.05
Zn	≤0.20	2.0-3.1
Mn	0.20-0.50	0.20-0.50
Fe	≤0.01	≤0.01
Ni	≤0.005	≤0.005
Cu	≤0.05	≤0.05
Si	≤0.05	≤0.05
RE	1.2-3.0	2.5-3.5
Zr	0.45-0.70	

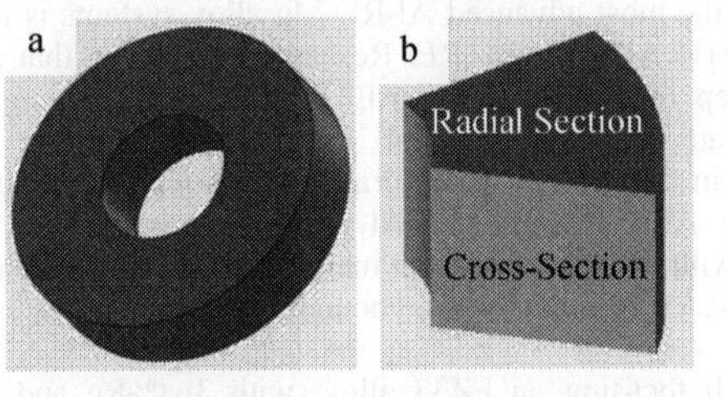

Figure 1. Alloy samples a) Extruded torus and b) Sample surfaces examined.

Nano-indentation Creep Tests

Nano-indentation tests were carried using a sapphire Berkovich indenter. Creep indentation measurements were conducted at 25°C and 150°C (±1°C) on both alloys and both sample surfaces (i.e., cross-section and radial section). For each alloy, six independent 1-hour constant load tests were performed to validate the repeatability of results. From these tests a plot of indentation depth versus time was generated and used for comparison of creep resistances of the samples.

Six indentation tests to a depth of 10 μm were performed for all samples. Based on these experiments, the applied force, *F*, was plotted as a function of the total indentation depth, thus enabling calculation of the elastic (h_e) and plastic (h_p) deformation of the materials. These tests were also used to calculate sample hardness, *H*, using Equation 1. The indenter was assumed to be a perfect pyramid and no material pile-up or sink-in occurred during indentation testing.

(Equation 1)

Creep-induced Residual Strain Measurement

High-temperature compressive creep residual strain measurements were performed at the Canadian Neutron Beam Centre in Chalk River, Canada. Detailed description of these experiments is available in earlier papers [6,7]. Creep-induced residual strain was calculated using the peak-shift method (Equation 2), where the lattice spacing of a sample, *d*, was obtained from Bragg's Law (Equation 3), with the wavelength of the neutron beam, λ, at 0.237nm, first order diffraction (n=1) and scattering angle, θ, measured with a wire angle detector.

(Equation 2)

$$n*\lambda = 2*d*\sin(\theta) \quad \text{(Equation 3)}$$

Creep-induced residual strain measurements for crystallographic plane were made in a direction parallel to the extrusion direction (i.e., principal strain direction). The applied load during the neutron diffraction creep tests was 50MPa at a test temperature of 150 °C for 120 hours.

Microscopic Evaluation

Metallographic samples were prepared using standard techniques for Mg alloys. Zeiss AxioObserver A1m microscope with differential interference contrast (DIC) capability enabled imaging of the grain structure of the as-extruded specimens without the need for chemical etching. Scanning electron microscopy was conducted using JEOL-6410 SEM with XEDS capability.

Results and Discussion

Neutron Diffraction Results

The solidus temperatures, T_s, for the AE42 and EZ33 alloys are 590°C and 545°C, respectively [8,9]. Thus, in relation to the alloy's solidus, the test temperatures corresponded only to $0.27T_s$ and $0.25T_s$, respectively. However, as Figure 2 illustrates, ND measurements revealed that the residual creep-induced strain in the AE42 alloy was significantly greater than that of EZ33. AE42 exhibited primary creep deformation up to ~3% (27 hrs), with steady-state creep deformation thereafter, reaching 6% at the end of the test. In contrast, EZ33 exhibited primary creep up to 0.05% (7 hrs) and subsequently steady-state creep of ~0.06% remained for the remainder of the test.

It is worth noting, that the ND method of evaluating creep resistance of the alloy provides strain data on the bulk deformation of the specimen. Thus, the contributions of the solid-solution matrix and the intermetallics located on the grain boundaries are considered together. This is particularly relevant for comparison with the following nano-indentation results.

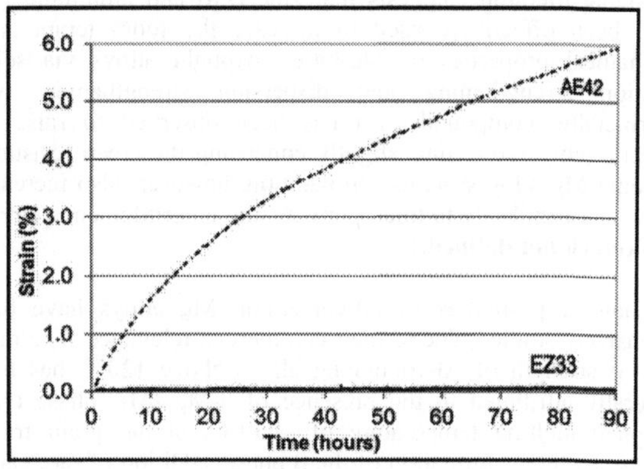

Figure 2. Compression-induced creep strain for (10 0) planes.

Nano-Indentation Results

The results of nano-indentation creep experiments are plotted in Figure 3 and Figure 4. The results confirm the same behavior as observed with ND experiments: EZ33 alloy has a higher creep resistance than AE42. However, the nano-indentation data also demonstrate that the extruded material was anisotropic, since the strain in the cross-section and radial directions varied significantly. This is in agreement with an earlier study which revealed strong tendency for the basal planes to align with the extrusion axis [10]. The pole figure (Figure 5), obtained from ND texture experiments with the AE42 and EZ33 alloy demonstrate this fact. At 150°C, Figure 3**Error! Reference source not found.**b shows that AE42 alloy began to exhibit anisotropic response, since the radial and cross-section strains continuously diverged. In the case of EZ33, the creep indentation strains remained comparable for the duration of the test.

The nano-indentation results in Figure 4 also suggest that the strain in the extrusion direction was significantly more sensitive to increase in temperature than the strain in the radial direction. This was attributed to the significantly different microstructures for these samples. As will be discussed in the following sections, the extrusion process generated high-aspect ratio grains in the extrusion direction, with non-uniform dispersion of intermetallic compounds, while the radial direction had spheroidal grains and uniformly dispersed intermetallics.

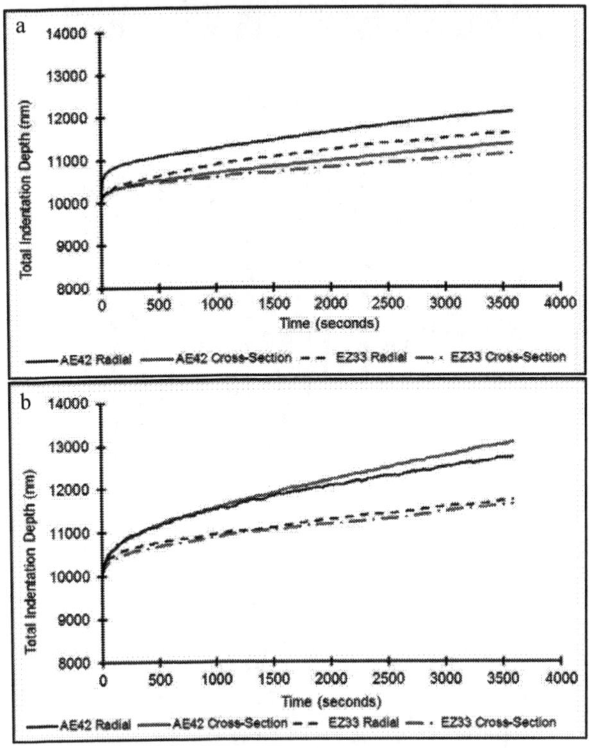

Figure 3. Nano-indentation compressive creep measurements in the radial and cross sections at a) 25°C and b) 150°C.

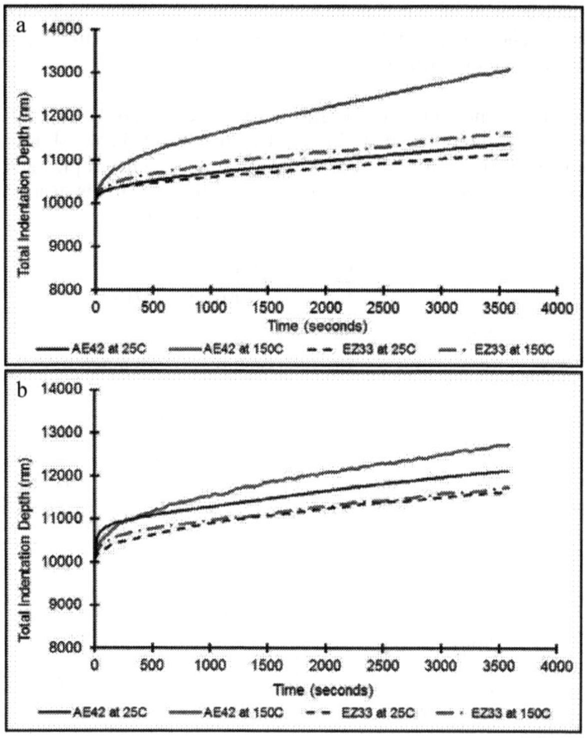

Figure 4. Nano-indentation compressive creep measurements at 25°C and 150° of the a) cross-section and b) radial section.

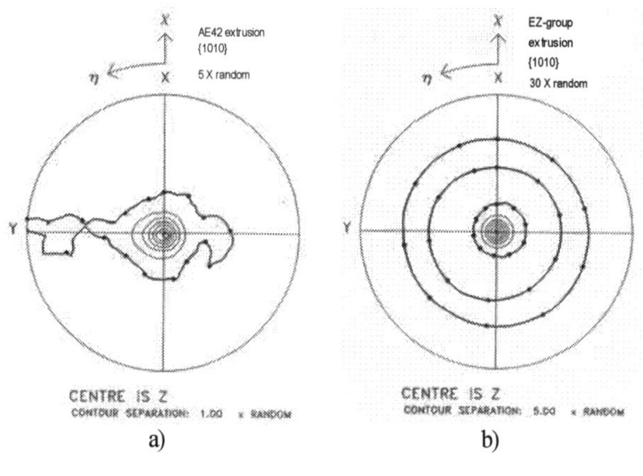

Figure 5. Pole figure for {1010} planes: a) AE42 alloy, b) EZ33 alloy.

Hardness and Elastic Modulus

The nano-indentation hardness data (Table I) directly support the creep results. On average, the EZ33 alloy was ~35% harder than the AE42 alloy. Increasing the test temperature from 25°C to 150°C resulted in ~35% and ~20% loss of strength for the EZ33 and AE42 alloys, respectively. For both alloys, the cross-section (i.e., plane parallel to extrusion direction) exhibited higher hardness than the radial plane. This is possibly the result of the presence of coherent intermetallic compounds which form on the basal planes of the Mg matrix. Consequently, the hardness values were higher suggesting that dislocation glide along the basal planes was more difficult.

The elastic modulus data reveal that on average AE42 had 30-43% lower modulus than EZ33. The results in Table II also confirm that AE42 was significantly more anisotropic and temperature sensitive. This is possibly indicative of dislocation glide being a factor in creep deformation of this alloy.

Table I. Alloy Hardness

Alloy		Hardness (MPa)	
		25 °C test temperature	150 °C test temperature
AE42	Cross section	571 ± 30	437 ± 41
	Radial section	334 ± 16	302 ± 25
EZ33	Cross section	781 ± 50	563 ± 40
	Radial section	682 ± 83	503 ± 39

Table II. Alloy Elastic Modulus

Alloy		Elastic modulus (GPa)	
		25 °C test temperature	150 °C test temperature
AE42	Cross section	28 ± 4	33 ± 3
	Radial section	14 ± 1	19 ± 2
EZ33	Cross section	40 ± 2	40 ± 7
	Radial section	38 ± 4	39 ± 10

<ins>Alloy Microstructure</ins>

As can be seen in Figure 6, the microstructure of the alloys was highly anisotropic for the cross-section and radial directions. The hot extrusion process caused deformation of the grain boundary regions resulting in banding of intermetallics along the extrusion direction (vertical in Figure 6a), as expected. The same microstructural characteristics were observed for EZ33 alloy.

Light optical microscopy with DIC prism revealed a fundamental difference between the AE42 and EZ33 alloys in terms of the distribution of the intermetallic phases. As Figure 7 shows, the AE42 alloy contained intermetallics (predominantly acicular $Al_{11}RE_3$ and globular Al_2RE) dispersed throughout the grain boundaries as well as the matrix. In contrast, EZ33 contained intermetallics (globular $Mg_xZn_yRE_z$) exclusively on the grain boundaries. Thus, the intermetallics in the EZ33 were significantly more effective in pining the grains during high temperature deformation. Further, as Figure 8 suggests, the cuboid intermetallics in the AE42 alloy had relatively flat interface with the matrix, while in the EZ33 alloy the intermetallics were finer and with serrated edges, thereby more effectively anchoring neighboring grains.

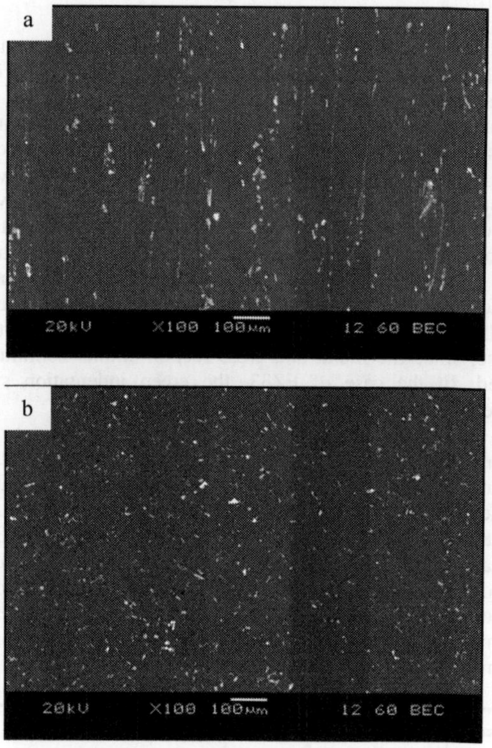

Figure 6. Microstructure of AE42 alloy (SEM image) a) Cross-section and b) Radial section.

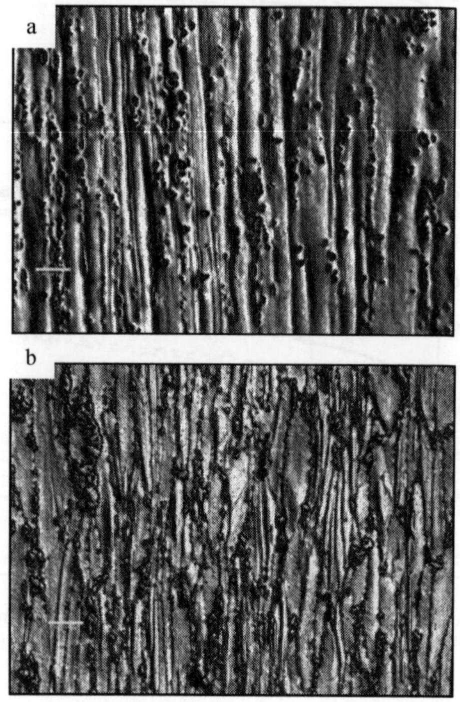

Figure 7. Microstructures for cross section surfaces. 100x; scale = 50μm (DIC polarization) a) AE42 and b) EZ33

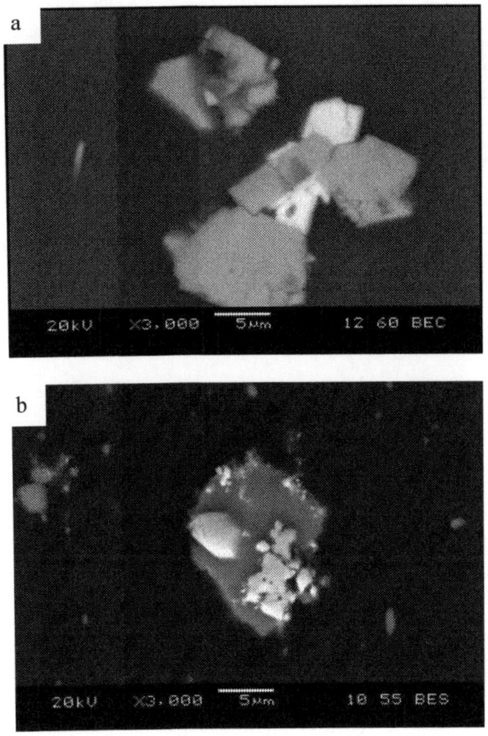

Figure 8. Morphology of intermetallics (SEM image) of a) AE42 and b) EZ33

Conclusions

This research demonstrated that neutron diffraction creep-induced residual strain measurements and nano-indentation creep measurements can be used to provide valuable data on creep performance of metallic alloys. The results of this research provided for the following conclusions:

1. AE42 was seen to be highly isotropic with properties highly sensitive to temperature variation.
2. On average, the EZ33 alloy was ~35% harder and had a 30-43% higher elastic modulus than the AE42 alloy.
3. Residual creep-induced strain in the AE42 alloy was significantly greater than that of EZ33.
4. The microstructure of the alloys was highly anisotropic for the cross-section and radial directions. Banding of intermetallics along the extrusion direction was observed in both alloys.
5. AE42 alloy contained acicular and globular intermetallics dispersed throughout the grain boundaries as well as the matrix.
6. EZ33 contained fine and irregular intermetallics exclusively on the grain boundaries, which significantly enhanced the creep-resistance of the alloy.

Acknowledgement

The authors would like to acknowledge the financial support of the Natural Sciences and Engineering Research Council of Canada for financial support and Mr. S. Shook for material donation.

References

1. M. Avedesian and H. Baker, *ASM Specialty Handbook; Magnesium and Magnesium Alloys,* (Materials Park: ASM International, 1999), 3-239.

2. E. Emley, *Principles of Magnesium technology* (New York: Pergamon Press, 1966), 42.

3. B. R. Powell et al., "Microstructure and Creep Behaviour in AE42 Magnesium Die-Casting Alloy," *JOM,* 54 (8) (2002), 34-38.

4. L. Peng, F. Yang, J. Nie, and J.C.M. Li, "Impression creep of a Mg-8Zn-4Al-0.5Ca alloy," *Materials Science and Engineering A,* 410-411 (2005), 42-47.

5. T. Rypaev et al., "Microstructure of superplastic QE22 and EZ33 magnesium alloys," *Materials Letters,* 62 (2008), 4041-4043.

6. D. Sediako and M. Gharghouri, "Neutron Diffraction Measurements of Residual Stress in Creep Resistant Mg Alloys," *Magnesium Technology 2008*, TMS, 2008, 407-409.

7. D. Sediako and S. Shook, "Application of Neutron Diffraction in In-Situ Studies of Stress Evolution in High Temperature Creep Testing of Creep-resistant Mg Alloys," *Magnesium Technology 2009*, TMS, 2009, 255-259.

8. K. Mills, *Recommended Values of Thermophysical Properties for Selected Commercial Alloys,* (Cambridge, England: ASM International, 2002), 143-157.

9. International Magnesium Association. *Melting point/range for Magnesium.* Accessed: September 14, 2010: http://www.intlmag.org/phys07.html

10. S. Shook, and D. Sediako, "Analysis Of Stress Evolution in High Temperature Creep Testing of Creep-resistant Mg Alloys," *IMI – 2009*, San Francisco, 43-51.

Magnesium Technology 2011
Edited by: Wim H. Sillekens, Sean R. Agnew, Neale R. Neelameggham, and Suveen N. Mathaudhu
TMS (The Minerals, Metals & Materials Society), 2011

THE EFFECT OF THERMOMECHANICAL PROCESSING ON THE CREEP BEHAVIOR AND FRACTURE TOUGHNESS OF THIXOMOLDED® AM60 ALLOY

Z. Chen[1], J. Huang[2], R. Decker[2], S. LeBeau[2], and C.J. Boehlert[1]

[1]Michigan State University, East Lansing, Michigan, 48824
[2]Thixomat, Incorporated, Ann Arbor, Michigan, 48108

Keywords: Magnesium, Lightweight Alloys, Creep, Fracture Toughness

Abstract

Creep and fracture toughness experiments were performed on a commercially available magnesium-aluminum alloy (AM60) after three processing treatments: (1) As-Thixomolded® (as-molded), (2) Thixomolded® then thermomechanically processed (TTMP), and (3) Thixomolded® then TTMP then annealed (annealed). The conventional tensile-creep experiments were performed at applied stresses ranging between 20-75MPa and temperatures between 373-473K (100-200°C). *In-situ* tensile-creep tests were performed on selected samples. The as-molded material exhibited creep resistance superior to the thermomechanically processed materials. Creep experiments indicated grain boundary cracking, and grain size was expected to be an important microstructural parameter that affected the creep behavior. Fracture toughness experiments were performed at room temperature (RT) on single edge notched tension (SENT) samples. The TTMP and annealed materials exhibited fracture toughness values almost twice that of the as-molded material.

Introduction

With increasing fuel cost and global environmental awareness, the demand for lightweight structural materials to increase fuel efficiency is paramount in the automotive industry. Magnesium (Mg) alloys have low density and exhibit high specific strengths. Thus, they serve as attractive candidate for replacing the Al alloys widely used in automotive industry. Previous work by our group was conducted to investigate the RT and elevated-temperature tensile and fatigue properties of a Thixomolded® AM60 alloy processed under different conditions [1]. This study focused on the effect of thermomechanical processing on the creep behavior and fracture toughness of the Thixomolded® AM60 alloy.

Experimental

The measured bulk chemical composition of the as-molded material is shown in Table I. The AM60 alloy was thixomolded into 3.2mm thick plates which represented the as-molded condition [2]. Some of the plates were then thermomechanically processed in order to induce dynamic recrystallization and this represented the TTMP condition. Several of the thermomechanically-processed plates were then annealed and this represented the annealed condition.

Conventional tensile-creep tests were performed using vertical load frames manufactured by Applied Test System, Incorporated (Butler, PA). Flat dogbone-shaped samples, containing a gage length of 25mm and a gage width of 12mm, were electro-discharge machined, and polished using 600 grit silicon carbide paper to remove the recast layers before testing. Test temperatures were ranged between 373-473K (100-200°C), and targeted test temperatures were maintained within ±3K. Applied stresses ranged between 20-75MPa. Strain was measured and documented throughout the test, by a linear variable differential transformer on a high temperature extensometer. After a minimum strain rate was achieved, either the load or temperature was increased or the test was stopped. Upon termination of the creep experiments, the samples remained under load while the temperature was decreased in order to maintain the deformed state of the sample. Several of the deformed gage sections were cut and polished using silicon carbide paper and diamond paste to a final finish of 0.25μm with ethanol as a lubricant, and then observed using a JEOL 6500 Field Emission scanning electron microscope (SEM).

Table I. Chemical composition in wt% of the studied AM60 alloy.

Al	Mn	Zn	Fe	Si	Ni	Be	Mg
6.29	0.28	0.05	0.001	0.02	0.0009	0.0007	93.34

In-situ tensile-creep experiments were conducted on the annealed material. Flat dogbone-shaped samples, with gage dimensions of 3mm wide by 2.5mm thick by 10mm long, were EDM cut. The specimens were glued to a metallic platen and polished to a 1μm finish using an automatic polishing machine and ethanol as a polishing lubricant. These experiments were performed using a screw-driven tensile stage (built by Ernest F. Fullam, Incorporated, which was purchased by MTI, Inc.) placed inside an Xl-30 FEG FEI SEM. Temperature was controlled using a constant-voltage power supply to a 6mm diameter tungsten-based heater located just below the gage section of the sample. An open-bath, closed-loop chiller was used to circulate distilled water at RT through copper tubes to prevent the tensile stage from overheating. A fine-gage K-type thermocouple was spot-welded to the gage section of each sample. After the sample's gage-section temperature reached the desired creep temperature, a 30 minute period was given to stabilize the thermal stress prior to applying load. The load, which was measured using a 4,448 N load cell, was applied at 3.7N/s until reaching the desired creep stress. The tests were considered constant load where the stress fluctuation varied ±3MPa. The displacement data acquired during the experiments included that of the sample as well as the gripping fixtures. Thus the displacement values reported do not represent the sole displacement of the reduced gage section of the sample. Secondary electron SEM images were taken before loading and at periodic displacements throughout the creep experiments without interrupting the experiment. The pressure in the SEM chamber never exceeded 10^{-6} torr, and therefore oxidation did not detrimentally affect the SEM imaging. Further details of this apparatus and testing technique can be found elsewhere [3-5].

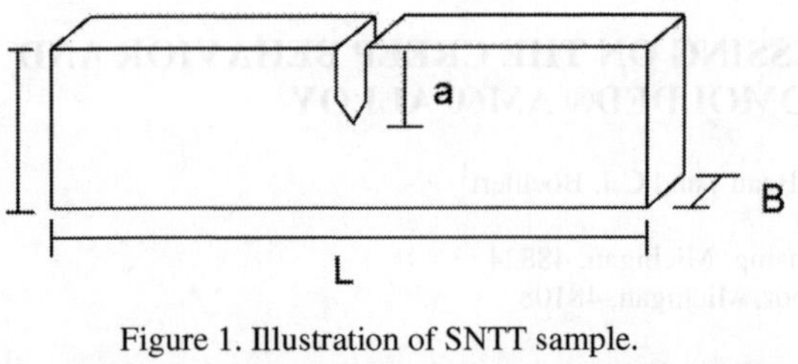

Figure 1. Illustration of SNTT sample.

Fracture toughness experiments were also performed at RT using a servohydraulic system [6]. SENT samples were EDM machined and the geometry of the sample is shown in Figure 1, where width W=12.5mm; crack length a=7mm; thickness B=3.2mm for as-molded material, and B=1.6mm for the TTMP and annealed materials. The longitudinal direction of the specimen was parallel to the longitudinal direction of the plates. The specimen surfaces were polished using silicon carbide paper to reduce the effects of surface defects. No fatigue precrack was introduced and the initial precrack (introduced through EDM) length a was measured using digital imaging. The samples were loaded in tension at a 75N/s loading rate for the TTMP and annealed materials, and at 150N/s for the as-molded material. The crack opening displacement was measured using an extensometer attached to the specimen. Two experiments were performed for each material.

Results

Previous study showed that the as-molded material exhibited an equiaxed α grain morphology where the $Mg_{17}Al_{12}$ β phase decorated the grain boundary [1]. The thermomechanical processing induced a finer grain size and strong basal texture to the TTMP and annealed material. The TTMP material exhibited the highest strength while the annealed material exhibited an intermediate strength but the highest elongation-to-failure (ε_f) More information can be found in reference [1].

Creep

The creep behavior resembled that typical for pure metals indicating three stages of creep. Creep strain versus time curves for the studied AM60 alloys tested under three different applied stress levels at 423K (150°C) are shown in Figure 2. The minimum creep rate versus stress plots for the studied AM60 alloys are shown in Figure 3. The creep stress exponents (n) can be calculated based on these plots, and the n values at 423K (150°C) were 4.9, 5.8, and 4.5 for the as-molded, TTMP, and annealed materials, respectively. Figure 4 shows the temperature dependence of the minimum creep rate under 20MPa applied stress, and the apparent activation energy (Q_{app}) was calculated based on this plot. The plot showed a two-stage linear relationship. Between 373-423K (100-150°C), the Q_{app} for the as-molded, TTMP, and annealed materials were 48kJ/mol, 64kJ/mol and 67kJ/mol, respectively. Between 423-473K (150-200°C), the Q_{app} for the as-molded, TTMP, and annealed materials were 128kJ/mol, 174kJ/mol, and 126kJ/mol, respectively. Grain boundaries served as crack nucleation sites in creep, see Figure 5.

The *in-situ* creep experiments highlighted the surface grain boundary cracking, see Figures 6 and 7. It appeared that the grain

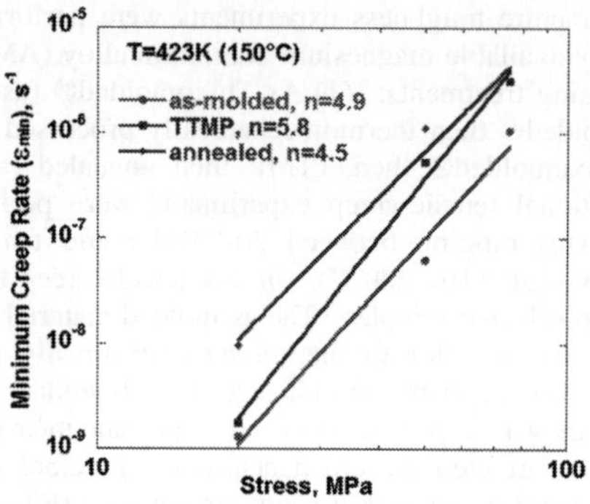

Figure 3. Minimum creep rate versus stress plot at T=423K (150°C) for the studied AM60 alloys. The creep stress exponent (n) values are shown.

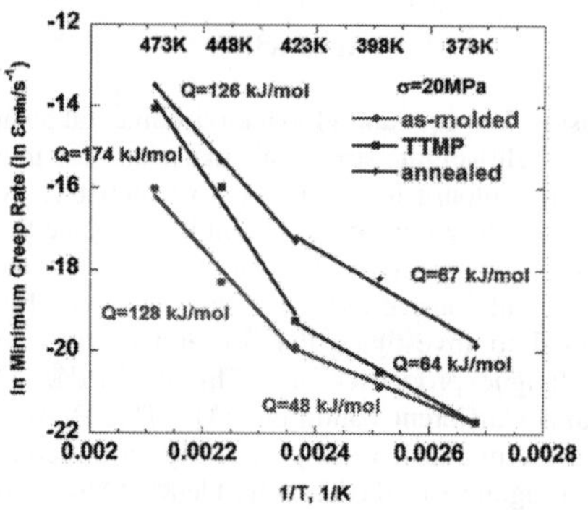

Figure 4. ln minimum creep rate versus reciprocal of temperature plot for the studied AM60 alloys at σ=20MPa. The apparent creep activation energy (Q_{app}) values are shown.

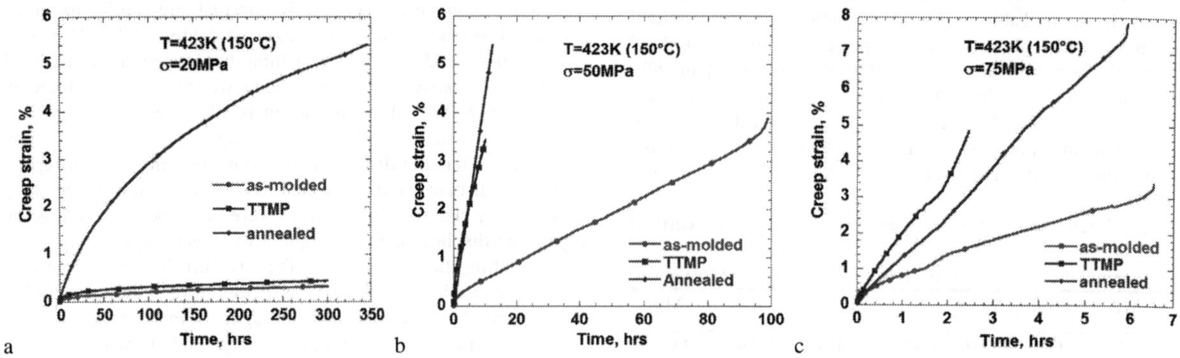

Figure 2. Creep strain versus time plot of the studied materials at 423K (150°C): (a) σ=20MPa, (b) σ=50MPa, and (c) σ=75MPa.

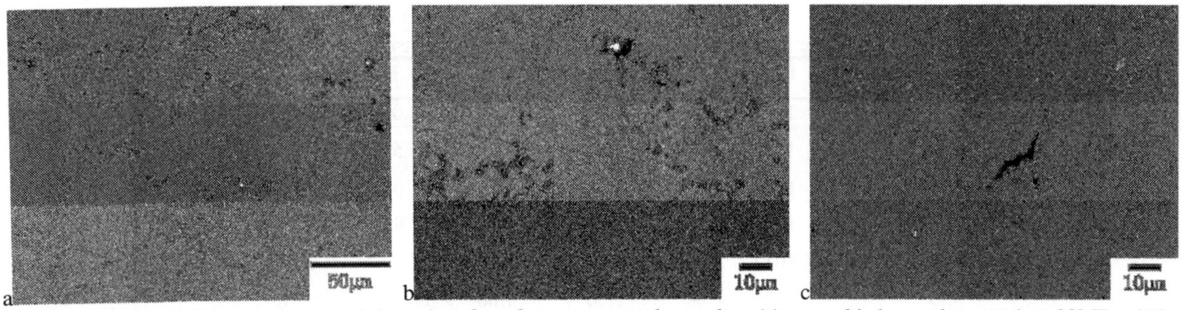

Figure 5. Backscattered electron images of the subsurface for creep tested samples: (a) as-molded sample, tested at 20MPa, 373-473K (100-200°C), total strain was 2.3%, (b) TTMP sample, tested at 75MPa, 423K (150°C), total strain was 4.8%, (c) annealed sample, tested at 20MPa, 373-473K (100-200°C), total strain was 10.5%

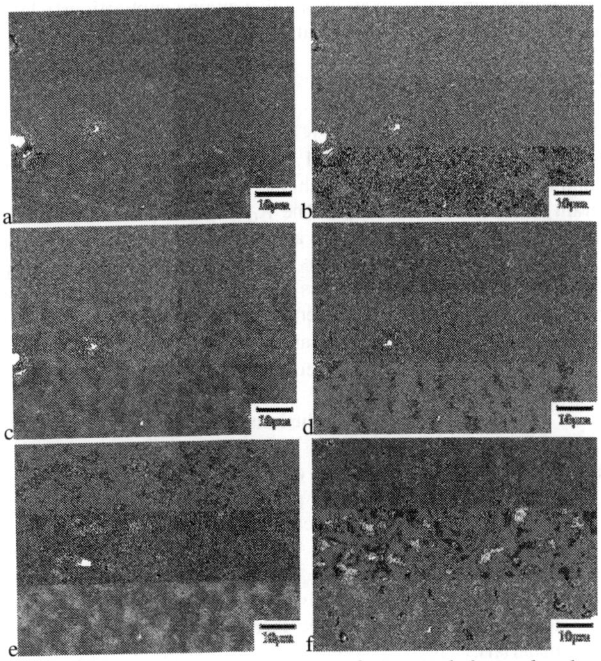

Figure 6. Secondary electron images of an annealed sample taken during an *in-situ* creep test at 50MPa and 423K (150°C). The displacements were (a) 0.168mm, (b) 0.328mm, (c) 0.579mm, (d) 0.945mm, (e) 1.730mm, (f) 2.553mm. A significant amount of grain boundary cracking was evident.

Figure 7. Secondary electron images of the surface of an annealed sample after *in-situ* creep testing at 423K (150°C) and 75MPa. The final displacement was 3.127mm.

boundary cracking was associated with grain boundary sliding as surface fiducial scratches were jogged at grain boundary locations, see Figure 7. Thus grain boundary cracking may have been accommodating the grain boundary sliding.

Fracture Toughness

The fracture toughness was calculated according to the following equations [7]:

$$f\left(\frac{a}{W}\right)=\frac{K_Q B\sqrt{W}}{P_Q}, \tag{1}$$

where for the SENT geometry:

$$f\left(\frac{a}{W}\right)=\frac{\sqrt{2\tan\frac{\pi a}{2W}}}{\cos\frac{\pi a}{2W}}\left[0.752+2.02\left(\frac{a}{W}\right)+0.37\left(1-\sin\frac{\pi a}{2W}\right)^3\right] \tag{2}$$

P_Q is the critical load when fracture occurred and was determined from the load-displacement relationship according to ASTM E 399 [8]. Table II summarizes the fracture toughness data. Although the P_{max}/P_Q ratio exceeded 1.10, the conditional fracture toughness K_Q was calculated based on the P_Q value. The K_Q value was then used to estimate the minimum thickness of the sample necessary for a valid plane-strain fracture toughness test, based on the relationship suggested by ASTM E 399:

$$B, a, W-a > 2.5\left(\frac{K_Q}{\sigma_{ys}}\right) \tag{3}$$

As shown in Table II, the actual sample thickness was significantly smaller than that necessary for achieving plane-strain conditions.

Discussion

Creep

The as-molded material exhibited the lowest creep strain rates, see Figures 2-4. The reason for this might be that the TTMP and annealed materials both had smaller grain size than the as-molded material. The TTMP material exhibited a lower minimum creep rate than the annealed material at applied stresses of 20MPa and 50MPa. However this was not the case at an applied stress of 75MPa. The reason for this crossover is not clear and will be the subject of further investigation.

Table II. Summary of the fracture toughness measurements.

Condition	Specimen ID	P_Q,N	Pmax,N	Pmax/P_Q	f(a/w)	K_Q,MPam$^{0.5}$	B_{min} for valid K_{IC}, mm
As-molded	A	1128	2276	2.01	4.61	15.0	32.6
	B	1322	2295	1.74	4.64	17.5	44.7
TTMP	A	1111	1288	1.16	4.76	31.8	23.1
	B	1012	1328	1.31	4.61	27.9	17.8
Annealed	A	1012	1614	1.59	4.59	27.4	36.4
	B	1021	1596	1.56	4.66	28.1	38.2

The measured creep exponents imply dislocation climb to be the dominant mechanism controlling the secondary creep rate. Similar creep exponents have been reported for QE22 (Mg-2Ag-2Nd) [9], Mg-Y alloys [10], Mg-Zn-Zr alloys [11], and die cast AZ91D and AS21 [12]. In both ingot and die cast AZ91, creep mechanisms based on dislocation motion (on basal and non-basal planes) were proposed [13, 14], where the ingot exhibited a creep rate one order of magnitude lower than the die cast alloy which was proposed to be due to the larger grain size in the ingot microstructure. During creep of Mg-5.6Y-0.04Zn(wt.%) (Mg-1.6mol%Y-0.015mol%Zn), bowed out dislocations were observed to trail straight dislocation segments parallel to the trace of basal planes [10]. The bowed-out dislocations were moving on prismatic planes. The deformation observations in the current work suggest grain boundary sliding also contributed significantly to the strain rates, see Figures 5-7, where grain boundary cracking may have been accommodating the grain boundary sliding. Thus, these two creep deformation processes may be competing and the measured activation energy suggest that temperature may have an influence on this competition.

For temperatures above 423K (150°C), the measured activation energy resembled that for lattice self diffusion [15], while for temperatures less than 423K (150°C), the activation energies were roughly half that for lattice self diffusion. This suggests that grain boundary diffusion may be dominant at lower temperatures. The measured Q_{app} values in low-temperature regime were in good agreement with previous measurements of particle strengthened Mg alloys [11, 12]. Dargusch and Dunlop [12] reported Q_{app} values between 36-44kJ/mol for creep of AZ91D and AS21 and related it to the grain boundary sliding mechanism promoted by discontinuous precipitation of $Mg_{17}Al_{12}$ for which the activation energy is 30kJ/mol [16]. For the high-temperature creep regime (423-473K (150-200°C)), the measured Q_{app} values were close to the 125kJ/mol measured for high-temperature creep of pure Mg, and in the range of 92-135 kJ/mol reported for the activation energy for dislocation glide in basal planes in pure Mg which is also the self-diffusion activation energy in the Mg lattice [15]. In addition, others have measured Q_{app} values between 120-143kJ/mol in the high-temperature regime [17]. The grain boundary cracking and grain boundary sliding observations were made at the transition temperature (423K (150°C)) between the high-temperature and low-temperature regimes. It is expected that the equiaxed α grain size is an important microstructural feature, especially at temperatures of 423K (150°C) and below. The refinement caused by the TMP process would be expected to decrease the creep resistance. Thus grain size appeared to have an influence on the creep behavior. This is a strong consideration for the implementation of TMP Thixomolded® Mg alloys in creep-driven applications, such as automotive engine applications which are subjected to elevated temperatures.

Fracture Toughness

According to Table II, the SENT sample thicknesses were well below those required for valid plane-strain fracture toughness condition. Thus the K_Q values should not be interpreted as plane-strain fracture toughness K_{IC}. The K_Q values for the TTMP and annealed materials were similar, and they were roughly twice that of the as-molded materials. Thus the TMP treatment significantly improved the fracture toughness and this was considered to be related to the significantly higher tensile strength achieved in the TTMP material and the significantly higher ε_f value achieved in the annealed material as compared to the as-molded material [1].

Conclusions

1. The creep resistance of the as-molded material was superior to the thermomechanically processed materials. The creep experiments indicated cracking preferentially occurred at grain boundaries. Grain size was expected to be an important microstructural parameter, and this partially explains why the creep resistance of the as-molded material was superior to that for the thermomechanically processed materials.

2. Through thermomechanical treatment of AM60, the fracture toughness can be almost doubled.

Acknowledgement

This research was conducted in part through the Oak Ridge National Laboratory's High Temperature Materials Laboratory User Program, which is sponsored by the U. S. Department of Energy, Office of Energy Efficiency and Renewable Energy, Vehicle Technologies Program, and through the Oak Ridge National Laboratory's SHaRE User Facility, which is sponsored by the Division of Scientific User Facilities, Office of Basic Energy Sciences, U.S. Department of Energy. A portion of this work was supported by the Faculty and Student Teams (FAST) Program, which is a cooperative program between the Department of Energy Office of Science and the National Science Foundation. The authors are grateful to Mr. Bryan Kuhr and Mr. Alex Ritter of Michigan State University for their technical assistance with the SEM, and *in-situ* deformation characterization

References

1. Chen, Z., et al., *The Effect of Thermomechanical Processing on the Tensile and Fatigue Behavior of Thixomolded(R) AM60*, in *Magnesium Technology 2010*, S.R. Agnew, et al., Editors. 2010, TMS: Seattle. p. 495-500.
2. Decker, R.F., et al., *Magnesium Semi-Solid Metal Forming*. Advanced Materials & Processes, 1996. **149**(2): p. 41-42.
3. Boehlert, C.J., et al., *In situ Scanning Electron Microscopy Observations of Tensile Deformation in a Boron-Modified Ti-6Al-4V Alloy*. Scripta Materialia, 2006. **55**(5): p. 465-468.
4. Cowen, C.J. and C.J. Boehlert, *Comparison of the Microstructure, Tensile, and Creep Behavior for Ti-22Al-26Nb (at. pct) and Ti-22Al-26Nb-5B (at. pct)*. Metallurgical and Materials Transactions a-Physical Metallurgy and Materials Science, 2007. **38A**(1): p. 26-34.
5. Quast, J.P. and C.J. Boehlert, *Comparison of the Microstructure, Tensile, and Creep Behavior for Ti-24Al-17Nb-0.66Mo (atomic percent) and Ti-24Al-17Nb-2.3Mo (atomic percent) Alloys*. Metallurgical and Materials Transactions a-Physical Metallurgy and Materials Science, 2007. **38A**(3): p. 529-536.
6. Hartman, G.A. and S.M. Russ, *Techniques for Mechanical and Thermal Testing of Ti3Al/SCS-6 Metal Matrix Composites*, in *Metal Matrix Composites: Testing, Analysis and Failure Modes*, W.S. Johnson, Editor. 1989, American Society for Testing and Materials: Philadelphia, PA. p. 43-53.
7. Anderson, T.L., *Fracture Mechanics: Fundamentals and Applications*. 1991, Boca Raton, FL: CRC Press, Inc. 793.
8. *Standard Test Method for Plane-Strain Fracture Toughness of Metallic Materials*, in *Annual Book of ASTM Standards, vol. 03.01*. 2000, American Society for Testing and Materials: Philadelphia, PA. p. 431-461.
9. Mordike, R.L. and P. Lukac. in *Proceedings of the 3rd International Magnesium Conference*. 1997. London, England: The Institute of Metals.
10. Maruyama, K., M. Suzuki, and H. Sato, *Creep Strength of Magnesium-Based Alloys*. Metallurgical and Materials Transactions a-Physical Metallurgy and Materials Science, 2002. **33**(3): p. 875-882.
11. Boehlert, C.J., *The Tensile and Creep Behavior of Mg-Zn Alloys with and without Y and Zr as Ternary Elements*. Journal of Materials Science, 2007. **42**(10): p. 3675-3684.
12. Dargusch, M.S. and G.L. Dunlop, in *Magnesium Alloys and Their Applications*, B.L. Mordike and K.U. Kainer, Editors. 1998, Werkstoff-Informationsgesellschaft: Frankfurt, Germany. p. 277-282.
13. Regev, M., et al., *Creep Studies of Coarse-Grained AZ91D Magnesium Castings*. Materials Science and Engineering a-Structural Materials Properties Microstructure and Processing, 1998. **252**(1): p. 6-16.
14. Regev, M., E. Aghion, and A. Rosen, *Creep Studies of AZ91D Pressure Die Casting*. Materials Science and Engineering a-Structural Materials Properties Microstructure and Processing, 1997. **234**: p. 123-126.
15. Dieter, G.E., *Mechanical Metallurgy*. 1986, New York: McGraw-Hill.
16. Uchida, H. and T. Shinya, J. Jpn. Inst. Light Metals, 1995. **45**(10): p. 572.
17. Vagarali, S.S. and T.G. Langdon, *Deformation Mechanisms in Hcp Metals at Elevated-Temperatures .2. Creep-Behavior of a Mg-0.8-Percent Al Solid-Solution Alloy*. Acta Metallurgica, 1982. **30**(6): p. 1157-1170.

Magnesium Technology 2011

Casting and Solidification

Session Chairs:

Baicheng Liu
(Tsinghua University, China)
Elhachmi Essadiqi
(CANMET, Canada)

Magnesium Technology 2011
Edited by: Wim H. Sillekens, Sean R. Agnew, Neale R. Neelameggham, and Suveen N. Mathaudhu
TMS (The Minerals, Metals & Materials Society), 2011

SIMULATION OF POROSITY AND HOT TEARS IN A SQUEEZE CAST MAGNESIUM CONTROL ARM

K.D. Carlson[1], C. Beckermann[1], J. Jekl[2], R. Berkmortel[2]
[1]Department of Mechanical and Industrial Engineering, University of Iowa, Iowa City, IA 52242, USA
[2]Meridian Lightweight Technologies, Strathroy, Ontario, Canada

Keywords: Magnesium Alloys, Casting, Shrinkage Porosity, Hot Tears, Modeling

Abstract

Simulations are performed for the squeeze casting of AM60 and AZ91 automotive control arms. Advanced feeding flow and stress models are used within commercial casting simulation software to predict shrinkage porosity and hot tears. The simulations are validated through comparisons with observations made on experimental castings. Generally good agreement is obtained between the measured and predicted defect locations and extents. Design and process changes are introduced to mitigate the shrinkage and hot tear problems in these castings. The comparisons in the present study establish considerable confidence in the ability of casting simulation to predict shrinkage and hot tears in squeeze casting of magnesium alloys.

Introduction

Both shrinkage porosity and hot tears are common defects that occur during solidification of magnesium alloy castings. Shrinkage porosity forms when feeding flow becomes limited to a casting region containing liquid metal. Hot tears form when tensile strains create "volume deficits" in the mushy zone (semi-solid region). These volume deficits develop into hot tears if the local solid fraction is large enough that the deficits cannot be fed by the remaining liquid.

Predicting shrinkage porosity and hot tears with casting simulation software is a difficult task. Accurate prediction requires accurate modeling of casting solidification (including feeding flow velocities and shrinkage porosity formation), as well as accurate modeling of the evolution of stresses and strains throughout the solidifying casting. Two simulation models have recently been developed that can predict the relevant phenomena involved in shrinkage porosity formation and hot tearing. The first model is an advanced feeding model that predicts melt pressure, feeding flow, and shrinkage porosity formation and growth during casting solidification. This model solves a pressure equation that is derived by combining the multiphase mass and momentum conservation equations. During solidification, melt pressure and feeding velocity are calculated throughout the casting cavity. Shrinkage porosity forms in solidifying metal when the local melt pressure drops sufficiently low, and then this porosity grows until solidification is complete [1-2]. The second model is a viscoplastic deformation model that predicts stresses, strains, deformations and porous damage evolution during casting solidification and subsequent cooling. This model employs a viscoplastic constitutive model for the mushy zone. The total strain is taken as the sum of the thermal, elastic and viscoplastic strains. Porous damage is computed by integrating the volumetric portion of the viscoplastic strain rate over time, beginning when the feeding flow is cut off. Hot tears occur in regions containing porous damage [3-5]. Two recent studies have been performed utilizing the new viscoplastic strain model; one study successfully predicted hot tears in simple experimental AZ91D castings [3], and the other study used forces measured during binary Mg-Al alloy solidification to calibrate mechanical properties of magnesium alloys [5]. Due to space limitations, these new models are not presented here; see Refs. [1-5] for model descriptions and details. The present investigation focuses on the application of these models, utilizing them to simulate squeeze cast magnesium alloy control arms.

Control Arm Casting Trials

This study considers two preliminary designs of a magnesium alloy control arm (see Fig. 1). These control arms were produced as squeeze castings, using the steel die depicted in Fig. 2. Squeeze casting employs quiescent, laminar filling through a relatively thick ingate. High pressurization (up to 950 bar in the present process) is applied to the solidifying casting after filling is

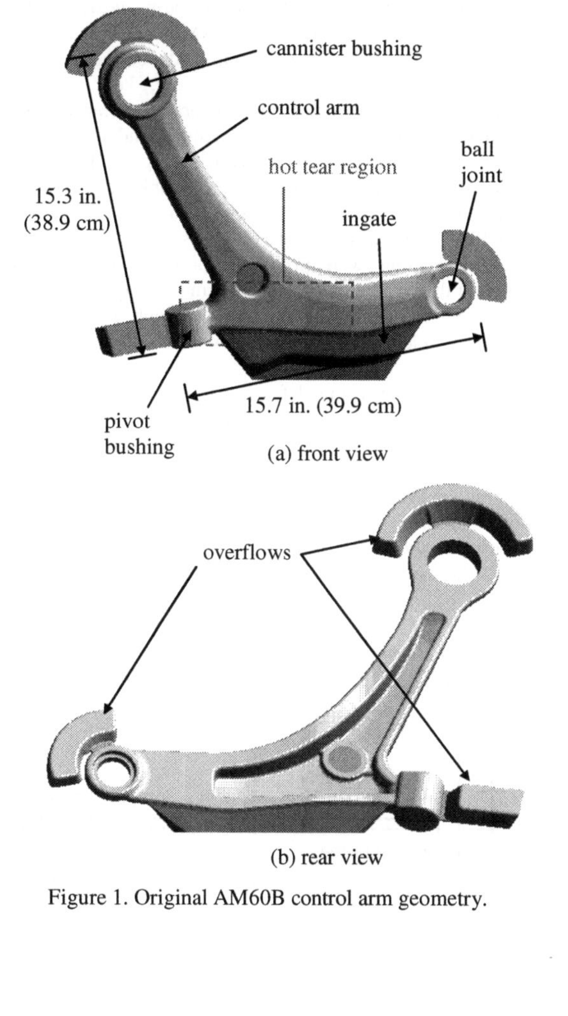

Figure 1. Original AM60B control arm geometry.

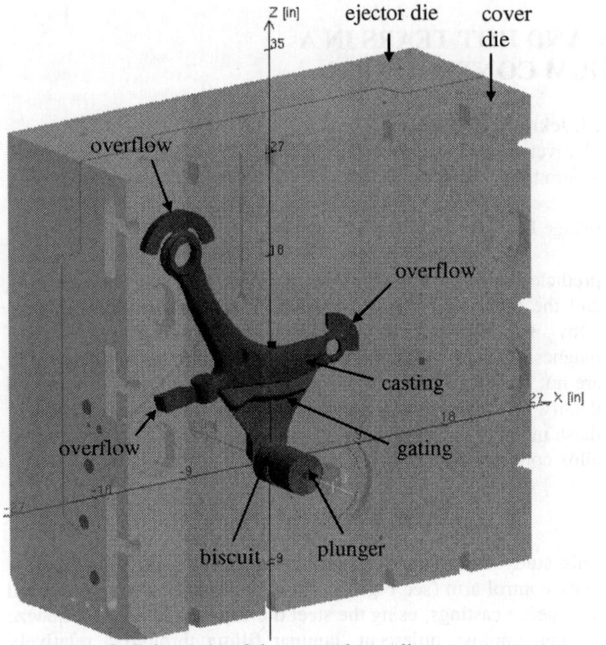

Figure 2. Schematic of the control arm die arrangement.

complete, and this pressure is maintained until shortly before the die is opened. The original control arm design, shown schematically in Fig. 1, was cast in AM60B. The metal was injected at 665°C. A series of heating and cooling lines were placed near the casting cavity in the die, kept at various temperatures selected to promote die temperatures near the casting surface that were favorable for producing sound castings. The time line for the casting cycle is given in Fig. 3. The castings produced with this process had excessive shrinkage porosity in a few locations, most notably in the pivot bushing of the control arm (see Fig. 1), because this was a hot spot in the casting. In addition, the castings were prone to hot tears near the ingate connection at the bottom of the control arm (see Fig. 1). Images of the radiographs showing porosity and photographs of the hot tears will be presented later, in comparison with simulation results.

In an effort to remedy the defects resulting from the original design, the casting process was revised. The alloy was changed to AZ91D, and the control arm casting geometry was slightly modified as shown in Fig. 4. The control arm pivot bushing was made hollow, in order to remove the hot spot. Also, the ribs on the back of the control arm were re-designed for structural and process optimization. The new AZ91D control arm design was cast using the same squeeze process as the original design; the

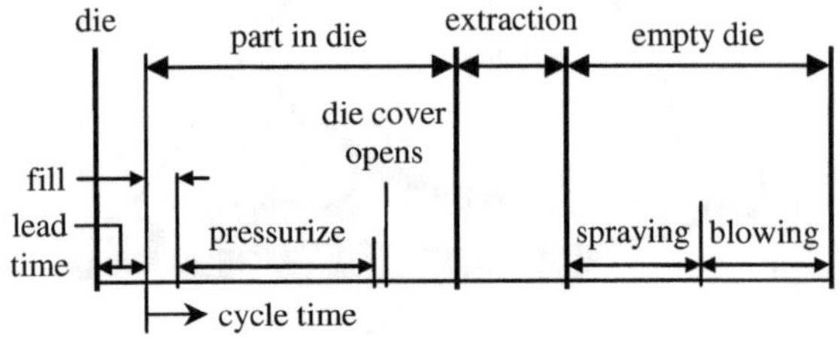

Figure 3. Casting cycle time line for the control arm.

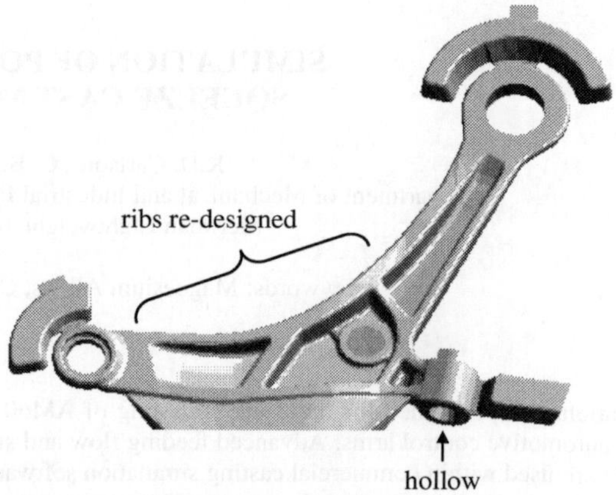

Figure 4. Revised AZ91D control arm geometry.

heating and cooling line temperatures, the die preheat, the metal temperature at injection, the pressurization schedule and the casting cycle time line (Fig. 3) all remained the same as for the AM60B control arms. The primary goal of the casting re-design was achieved: hot tears did not occur in the revised AZ91D control arms. Some shrinkage porosity was still evident in the AZ91D control arms, but to a lesser extent than in the AM60B control arms.

Simulation Properties and Settings

The solidification path and thermophysical property data required to simulate casting solidification and cooling for AM60B and AZ91D were computed with the thermodynamic simulation software package JMatPro [6]. The mechanical properties required to perform the stress simulations for these alloys were determined as described in Refs. [3-5]. The control arm castings were simulated using the new advanced feeding and viscoplastic deformation models described in the introduction. Both of these new models have been implemented as special models within the commercial casting simulation software package MAGMAsoft [7]. The simulation parameters for both the original AM60B and the revised AZ91D castings were identical, except for the control arm geometry and the magnesium alloy utilized. The die set-up utilized for the simulations is shown in Fig. 2, and the casting cycle time line is given in Fig. 3. Automatic grid generation was used, resulting in about 400,000 computational cells in the casting cavities of the die. The die was modeled with the STEEL database provided in MAGMA. The metal injection temperature was set to 665°C for both alloys. The cooling and heating lines were all set to the reported temperatures used in the casting trials (the same values were used for both alloys). The casting/die interfacial heat transfer coefficient (IHTC) for each alloy was modeled with temperature-dependent curves provided in the MAGMA database, namely AM60B-HPDC and AZ91-HPDC. The casting/die contact boundary conditions necessary for stress calculations were defined such that the steel die was completely rigid, and the casting was deformable. Nine warm-up cycles were simulated to reach a relatively steady-state, and then the tenth cycle was simulated for the present analysis, utilizing the new advanced feeding and viscoplastic deformation models.

Representative Simulation Results

An example of the melt pressure predicted during solidification for the AM60B control arm is shown in Fig. 5. The melt pressure contours (Fig. 5a) and liquid fraction contours (Fig. 5b) are shown 5.1 s into the cycle, during pressurization. Note that because the advanced feeding algorithm calculates the melt pressure, the squeeze casting pressurization (950 bar) is taken into account. Fig. 5b shows that the liquid fraction in the part of the control arm leading to the top ring is low (less than 30%). Feed metal moving up the control arm toward the top ring through partially solidified metal results in a large pressure drop from the pressurized value of 950 bar near the biscuit. Fig. 5b indicates two hot spots in the AM60B control arm; one in the pivot bushing and one just below the top ring of the control arm. Shrinkage porosity is expected to form in these locations because substantial liquid metal remains after feeding becomes difficult, which results in the melt pressure dropping low enough for porosity to form. The more liquid metal that is present when porosity forms, the more porosity will form to feed the remaining local solidification shrinkage.

Fig. 6 shows simulated feeding velocity results 2.9 s into the cycle, at a cross-section of the AM60B control arm near the top ring. Fig. 6a shows the liquid fraction contours at this time, and Fig. 6b shows velocity contours and vectors. The vector length indicates the relative magnitude of the velocity, as does the contour color scale. At the bottom right of Fig. 6b, feeding flow is seen moving up the control arm, through a channel in the arm with a high liquid fraction (see Fig. 6a). The velocity decreases once the feed metal reaches the hot spot because there is more high-liquid-fraction space in which the feed metal can move. As the feed metal flows into the upper ring, the feed metal velocity increases once again because of the small high-liquid-fraction channel that remains around the upper ring.

The final porosity predictions are given in Fig. 7a for the AM60B control arm, and in Fig. 7b for the AZ91D control arm. Fig. 7 shows a cross-sectional view that cuts through the upper part of

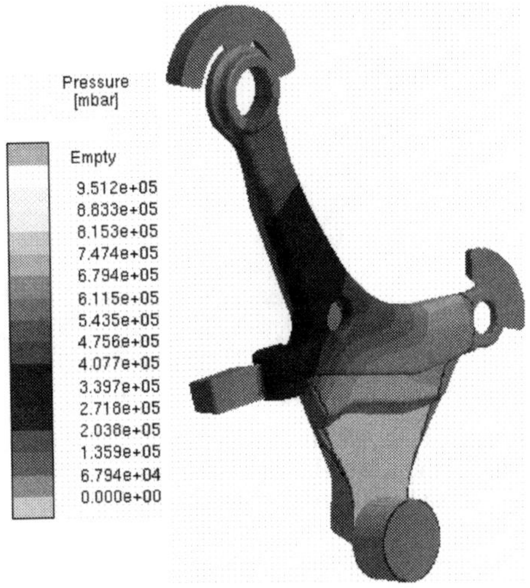

(a) melt pressure

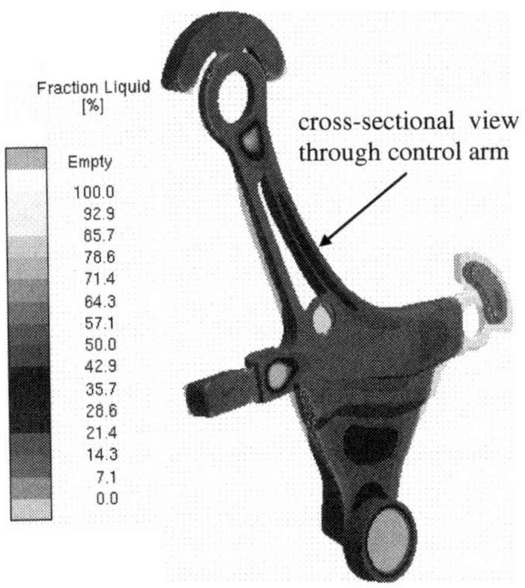

(b) liquid fraction

Figure 5. Contours showing (a) melt pressure and (b) liquid fraction during pressurization of the AM60B control arm casting.

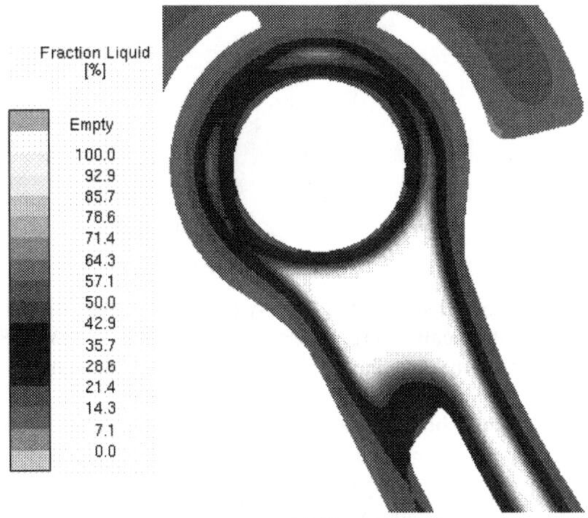

(a) liquid fraction

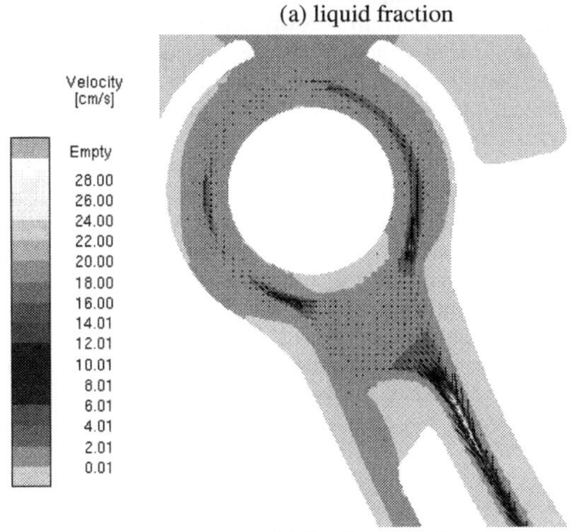

(b) feeding velocity

Figure 6. Contours showing (a) liquid fraction and (b) feeding flow velocity at a cross-section near the top of the AM60B control arm during solidification.

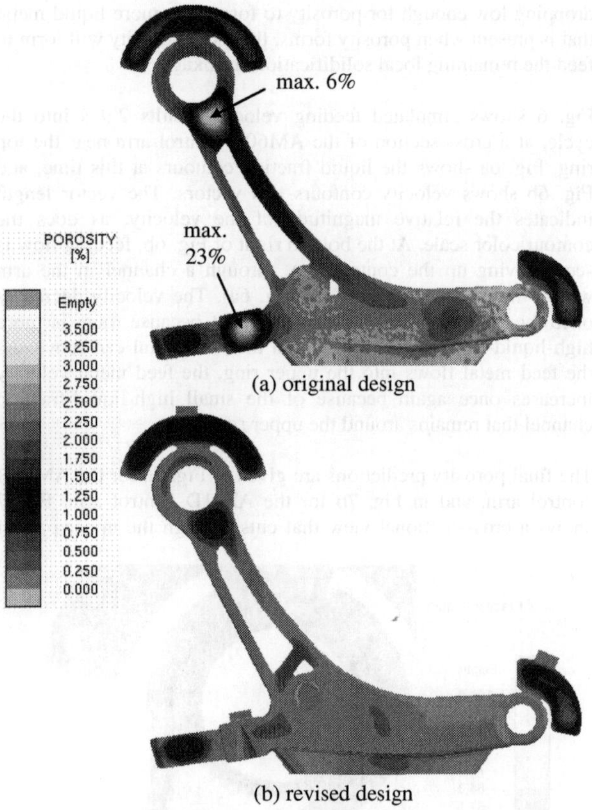

Figure 7. Predicted shrinkage porosity contours shown at a cross-section that reveals internal porosity for (a) original AM60B control arm design, and (b) revised AZ91D design.

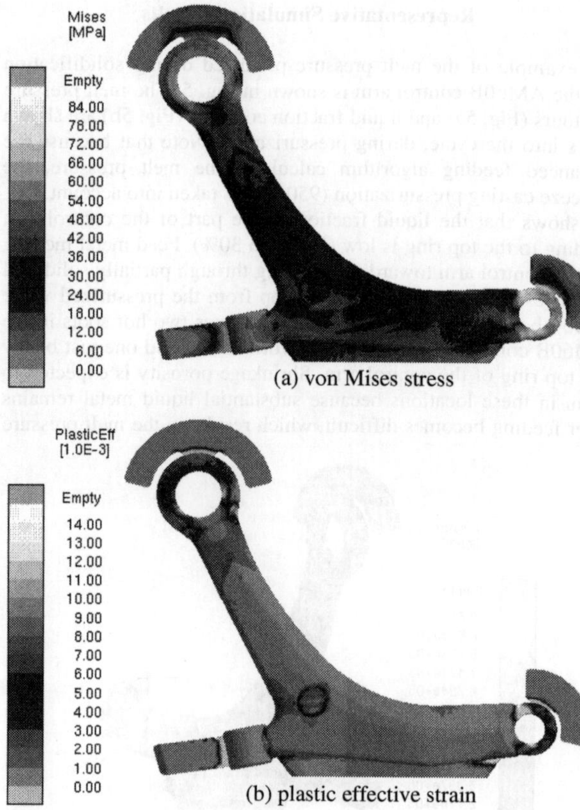

Figure 8. Simulated (a) von Mises stress and (b) plastic effective strain results for the AM60B control arm 20 s into the casting cycle, immediately prior to ejection from the die.

the control arms, showing the porosity in the two hot spots in the AM60B castings mentioned above; the control arms were not oriented in the simulation in a way that allows results to be shown on the casting mid-plane, which would be optimal. However, the mid-plane (i.e., maximum) porosity contour plots in each region of interest in both control arms are provided later, when the simulated porosity is compared to radiographs. Fig. 7a shows that, as expected, the two hot spots in the AM60B casting contain significant amounts of shrinkage porosity. In addition, smaller amounts of porosity are predicted in the upper part of the control arm below the hot spot. Comparing the revised AZ91D control arm porosity prediction in Fig. 7b to the AM60B prediction in Fig. 7a, it is clear that changing from AM60B to AZ91D and modifying the geometry significantly reduces the amount of shrinkage porosity predicted. The porosity in the hot spot near the upper ring is less severe in Fig. 7b than in Fig. 7a, and the porosity in the arm directly below this hot spot is significantly reduced as well. Also, hollowing out the pivot bushing in the revised geometry has alleviated the large porosity indication in that location.

Sample stress-strain results are provided for the AM60B control arm in Fig. 8. These results are shown immediately prior to the part being ejected from the die. Fig. 8a shows the von Mises stress in the control arm, and Fig. 8b shows the plastic effective strain. The control arm is completely solidified at this time. The stresses seen in Fig. 8a are an instantaneous result; the high stress regions around the rings (i.e., the cannister bushing and ball joint) are due to the rings straining against the rigid die, which is preventing the control arm from freely contracting as it cools. On the other hand, the plastic strains in Fig. 8b are a cumulative result, because of the plastic (irreversible) nature of the strain.

The final damage and distortion predictions for the control arms are given in Fig. 9. The magnitude of the porous damage scale is not important; only the relative amounts of damage are relevant. The AM60B control arm prediction is given in Fig. 9a. The shadow shows the original control arm shape, and the distortion is magnified by a factor of twenty to make it more noticeable. Note the distortion around the two rings, where the die prevented the casting from freely contracting as it solidified and cooled. The contour colors indicate the predicted level of porous damage. The highest damage predictions shown in Fig. 9a occur around the pivot bushing. Smaller amounts of damage are predicted just below the top ring of the control arm, and along the junction between the bottom of the control arm and the ingate; the latter indication is difficult to see in this view, but it will be shown more clearly when the simulated porous damage results are compared to the experimental hot tear pictures. Fig. 9b shows the final distortion and damage prediction for the AZ91D control arm. No significant damage is predicted in the revised control arm, which agrees with the finding that hot tears did not occur in the AZ91D control arms. Comparing Figs. 9a and 9b, it is also evident that the AZ91D control arm undergoes significantly less distortion than the AM60B control arm.

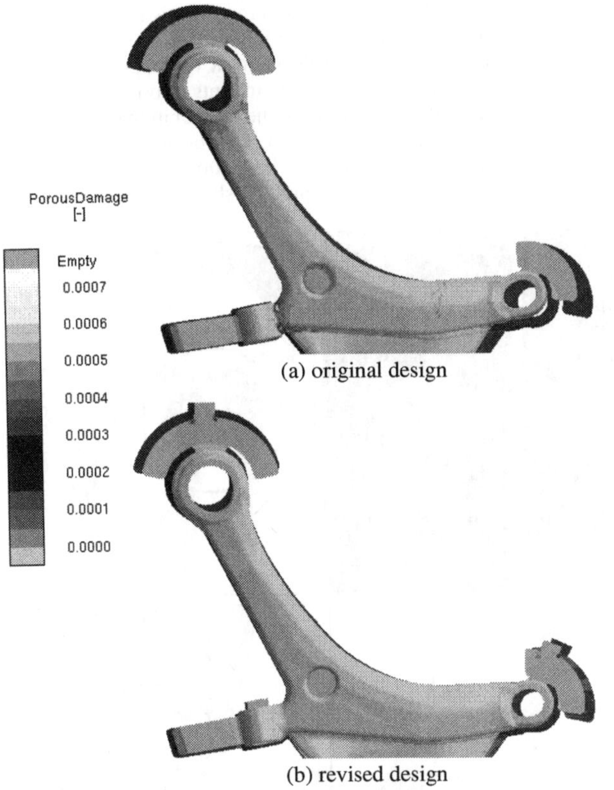

Figure 9. Predicted distortion and damage at end of casting cycle, for (a) original AM60B design, and (b) revised AZ91D design. Contours show porous damage, with the shape indicating distortion (magnified 20x, original shape shown as a shadow).

As explained in the introduction, porous damage is determined by integrating the volumetric portion of the plastic strain rate over time, beginning when feeding flow is cut off. This can be understood by considering the AM60B simulation results. For example, high strain is seen around the lower right ring in Fig. 8b. However, the porous damage result in Fig. 9a indicates that no damage is evident near this ring. This can be explained by considering Fig. 7a, which shows that almost no shrinkage porosity forms in this region, indicating that it is well-fed until solidification is complete. So although the strain in the area is high, no porous damage forms because the region is well fed throughout solidification. Another example is the region where the pivot bushing connects to the control arm. Fig. 8b shows elevated strain indications in this region. Furthermore, Fig. 7a shows that this location contains a significant amount of shrinkage porosity, indicating that there was a substantial amount of liquid remaining in the pivot bushing when feeding was cut off (as shown in Fig. 5b). Therefore, shrinkage porosity was readily available in this region for tensile strains to stretch and enlarge, creating the predicted porous damage.

Comparison with Casting Results

The simulated AM60B porosity results are compared to radiographs and micrographs of these control arm castings in Fig. 10. The micrographs in the top row of Fig. 10 illustrate the dendritic nature of the porosity (with the possible exception of region 4), indicating that it is indeed solidification shrinkage seen in the radiographs. Regions 1 and 2, where hot spots are located in the casting, show good agreement between experiment and simulation; porosity is predicted where visible indications are seen in the radiographs of these regions. The radiograph of region 3 shows very minor indications, observed better in the micrograph above the region 3 radiograph. Correspondingly, small amounts of porosity (0.25 – 0.5%) are predicted in that region. The

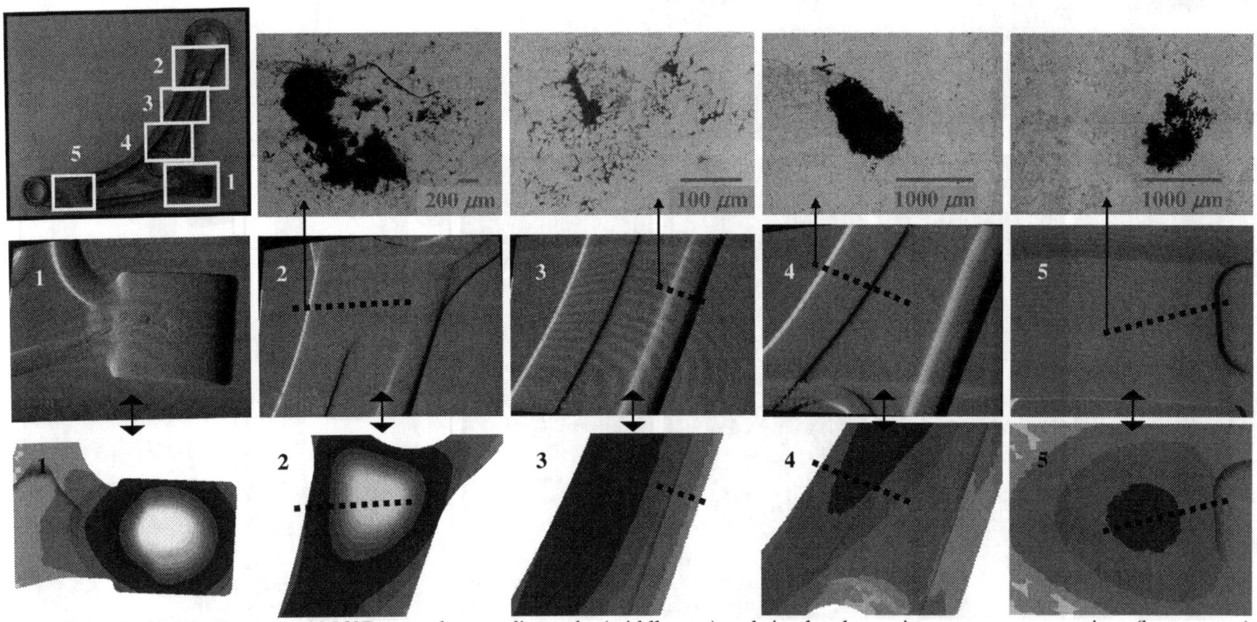

Figure 10. Comparison between AM60B control arm radiographs (middle row) and simulated porosity contour cross-sections (bottom row) at five locations indicated on the upper left photograph. The reminder of the top row contains micrographs at the locations indicated by the dotted lines. The porosity scale, given in Fig. 7, ranges from 0 – 3.5%.

radiographs and micrographs for regions 4 and 5 also show evidence of porosity, and the simulation results for those regions have porosity predictions of 0.75 – 1%. Overall, Fig. 10 shows excellent agreement between regions where porosity is predicted and visible radiographic shrinkage indications.

Similarly, the AZ91D porosity simulation results are compared to casting radiographs in Fig. 11. Circles encompass visible porosity in the radiographs. The top row of this figure compares simulated and measured shrinkage in region 1, the upper portion of the control arm (see the schematic at the upper left of Fig. 11). The simulation predicts small amounts of porosity (note the porosity scale at the upper right of the figure) in the hot spot and circumferentially in the top half of the ring in this region, and the radiographs show shrinkage indications in these areas as well. Region 2, containing the pivot bushing, is compared in the

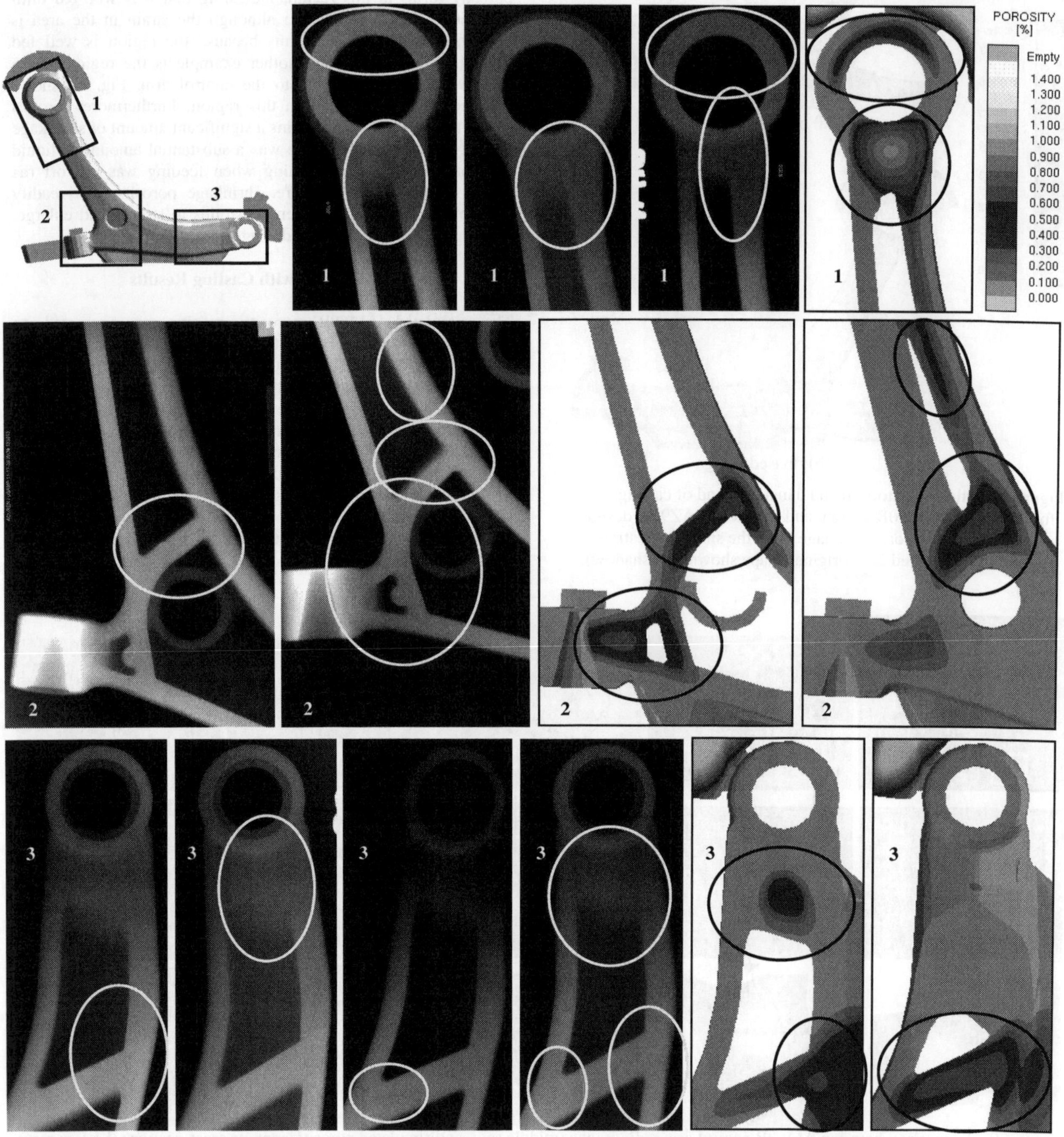

Figure 11. Comparison between radiographs and simulated porosity contours for the revised AZ91D control arm at three locations (one location per row), with locations indicated on the upper left schematic of the control arm. The porosity scale is given in the upper right.

middle row. Two simulated cross-sections are shown because a single cross-section does not characterize all the porosity. Again, excellent agreement is found between porosity visible in the radiographs in the ribs and near the pivot bushing, and regions where porosity is predicted in the simulation. Finally, region 3, containing the other ring, is compared in the bottom row (rotated 90° counterclockwise). Again, two simulation cross-sections are given to show all the relevant porosity. Once more, the locations near the ring and in the ribs, where indications are visible on the radiographs, are in excellent agreement with the regions where porosity is predicted in the simulation. In summary, both the AM60B and the AZ91D porosity predictions show excellent qualitative agreement with the locations of visible shrinkage shown in the radiographs and micrographs of the experimental castings.

Figs. 12 and 13 provide a comparison between the hot tears observed in the AM60B control arm castings (see Fig. 1 for location) and the simulated damage for these castings. Fig. 12a is a photograph of a hot tear that runs along the junction between the control arm and the ingate (indicated by a dotted line), and Fig. 12b shows the damage prediction in this region. Some degree of damage is predicted all along the length of the hot tear in the photograph. Fig. 13 focuses on the region where the pivot bushing connects to the control arm (i.e., left side of Fig. 12a). The photograph in the upper left indicates four regions where additional hot tear pictures are shown. The pictures from all four of these regions show evidence of hot tearing, and the simulation predicts relatively high porous damage in all of these regions.

The simulated damage prediction for the AM60B control arms did produce one false positive result. A small spot of relatively high damage was predicted on the bottom of the control arm, on the far right edge of the ingate where it meets the control arm; the location corresponds to the region of high plastic strain at the

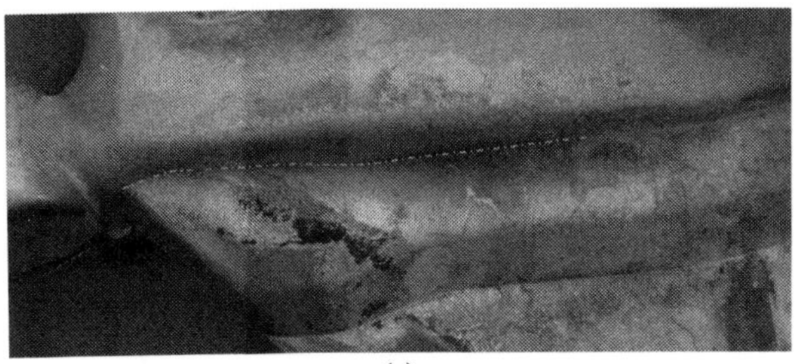

(a)

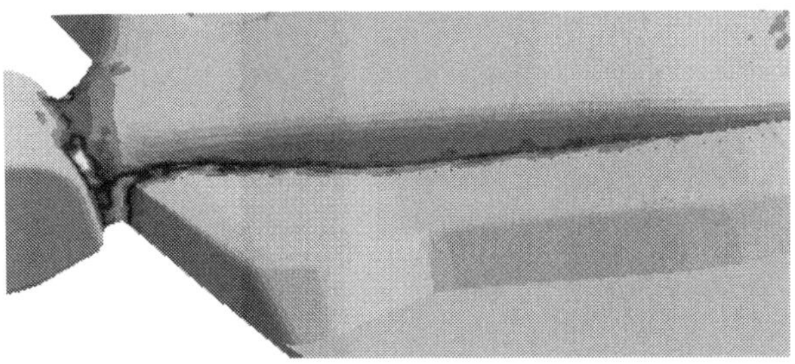

(b)

Figure 12. Comparison between (a) experimental AM60B casting hot-tear (indicated by dotted line), and (b) simulated porous damage. The porous damage scale is given in Fig. 9.

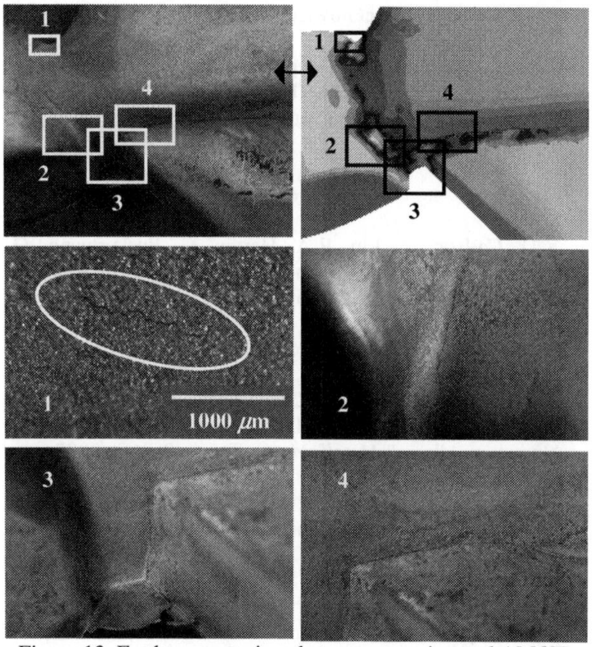

Figure 13. Further comparison between experimental AM60B casting hot-tears and simulated porous damage (top row), along with four close-up pictures showing hot tearing. The porous damage scale is given in Fig. 9.

top right corner of the ingate in Fig. 8b. Because the damage indication is on the bottom surface of the control arm, this spot is not visible in Fig. 9a. Although damage was predicted here, no hot tearing was seen in this region. However, aside from this, the correlation between locations with predicted damage and the occurrence of hot tears in both the AM60B and the AZ91D control arm castings is excellent. In general, porous damage predictions indicate potential initiation sites for hot tears. The results for the AM60B casting imply that the tears likely initiated near the pivot bushing, where the highest damage is predicted.

Conclusions

An experimental study was performed involving two preliminary designs of squeeze cast magnesium alloy control arms. The first design utilized AM60B, and resulted in control arms with significant porosity and hot tears. After this, the geometry was re-designed and the alloy was changed to AZ91D. Control arms cast with these revisions exhibited much less shrinkage porosity, and did not show evidence of hot tears. The casting of both designs was simulated using a new advanced feeding model and a new viscoplastic deformation model. For both the AM60B and AZ91D control arms, excellent agreement was seen between regions where porosity was predicted and where visible shrinkage was seen on radiographs, as well as between regions where porous damage was predicted and regions containing hot tears. The excellent qualitative agreement observed in both the porosity and hot tear predictions indicates that both new models are useful predictive tools for the squeeze casting process used to produce these castings.

Acknowledgements

This work was supported, in part, under the High Integrity Magnesium Cast Components (HIMAC) project, sponsored by the U.S. Department of Energy under Award Number DE-FC26-02OR22910.

References

1. K.D. Carlson, Z. Lin, R.A. Hardin, C. Beckermann, G. Mazurkevich, and M.C. Schneider, "Modeling of Porosity Formation and Feeding Flow in Steel Casting," *Modeling of Casting, Welding and Advanced Solidification Processes - X*, ed. D.M. Stefanescu et al., (Warrendale, PA: TMS, 2003), pp. 295-302.

2. K.D. Carlson, Z. Lin and C. Beckermann, "Modeling the Effect of Finite-Rate Hydrogen Diffusion on Porosity Formation in Aluminum Alloys," *Metall. Mater. Trans. B*, 36B (2007), pp. 541-555.

3. M.G. Pokorny, C.A. Monroe, C. Beckermann, L. Bichler, and C. Ravindran, "Prediction of Hot Tear Formation in a Mg Alloy Permanent Mold Casting," *Int. J. Metalcasting*, 2 (2008), pp. 41-53.

4. C.A. Monroe, C. Beckermann, and J. Klinkhammer, "Simulation of Deformation and Hot Tears Using a Visco-Plastic Model with Damage," *Modeling of Casting, Welding, and Advanced Solidification Processes - XII*, ed. S.L. Cockcroft and D.M. Maijer (Warrendale, PA: TMS, 2009) pp. 313-320.

5. M.G. Pokorny, C.A. Monroe, C. Beckermann, Z. Zhen, and N. Hort, "Simulation of Stresses During Casting of Binary Magnesium-Aluminum Alloys," *Metall. Mater. Trans. A*, (in press, 2010).

6. JMatPro v.4.0, Sente Software Ltd., Surrey Technology Centre, Surrey GU2 7YG, United Kingdom.

7. MAGMAsoft v4.5, Magma GmbH, Aachen, Germany.

Magnesium Technology 2011
Edited by: Wim H. Sillekens, Sean R. Agnew, Neale R. Neelameggham, and Suveen N. Mathaudhu
TMS (The Minerals, Metals & Materials Society), 2011

Dendritic Microstructure in Directional Solidification of Magnesium Alloys

Morteza Amoorezaei, Sebastian Gurevich, Nikolas Provatas

Department of Materials Science and Engineering, McMaster University, 1280 Main Street West; Hamilton, Ontario, L8S4L7, Canada

Keywords: directional solidification, magnesium, dendrite, transition

Abstract

We demonstrate morphological transitions in Mg-Al alloy dendritic microstructure as the cooling conditions change during steady state and transient directional solidification. The effect of temperature gradient on the transition is investigated numerically using two-dimensional phase field simulations. The six-fold symmetry of Mg alloys leads to very different dendrite morphologies than those encountered in alloys exhibiting four-fold surface tension anisotropy. In particular, we find that at high temperature gradients primary dendrites become columnar in the direction of thermal gradient. In contrast, in the regions where surface energy anisotropy is dominant, primary stalks cross at 60-degree angles that characterize hexagonal crystal structure. Our modelling observations are compared to new Mg-Al experiments.

Introduction

Predicting and controlling the as-cast microstructure during solidification of alloys is crucial to obtain desired properties. Specifically in emerging technologies such as strip casting, where few thermomechanical processes are applied, the microstructure established at the time of solidification strongly correlates with the mechanical properties of the alloy.

Columnar dendritic arrays are an important class of solidification microstructures that have been studied through directional solidification, where the temperature gradient is established in the direction that heat is extracted while the temperature is essentially uniform in the transverse direction. Directional solidification experiments are carried out under constant (Bridgeman-type) or transient (industrial-type) growth conditions. Industrial experiments usually occur under transient growth conditions where front velocity (V) and temperature gradient (G) are coupled. In contrast, V and G can independently vary in Bridgeman-type directional solidification.

Columnar dendritic microstructure has mostly been studied in alloys with cubic crystal structure [1, 2, 3, 4], where their four-fold symmetry allows studying the dynamical spacing selection between dendrites as the growth conditions vary dynamically. However, very little is known about dendritic microstructure of alloys with hexagonal crystal structure such as Mg-based alloys. Mg is gaining importance in transportation industries due to the high strength-weight ratio, which accordingly leads to lower fuel consumption in the transportation sector. The six-fold symmetry of these alloys induces a competition over growth direction between the anisotropy of surface energy and temperature gradient imposed by the external heat flux. This leads to various dendritic morphologies as the growth conditions and orientation change, which consequently introduces a transition from anisotropy-dominant to temperature gradient-dominant microstructure.

In this paper we investigate the anisotropy-heat flux competition in Mg-Al alloys directionally solidified under transient conditions relevant to industrial applications. We present new 2-D phase field simulations implemented on an adaptive mesh scheme. For brevity, we present simulations for a single grain where interaction between the grains is neglected. Our multigrain simulations and more qualitative studies on the morphological transition will be presented in upcoming publications.

Methods

Experimental Procedure

As received Mg-$0.5wt\%Al$ alloys were employed in this study. Upward transient directional solidification was achieved utilizing a cylindrical stainless steel crucible being water jet cooled from below. To prevent heat extraction from the walls, the crucible was shielded by cylindrical alumina insulation with a thickness of about 10 mm. To measure the temperature at different height of the melt, twelve K-type thermocouples were aligned at the centre of the crucible 1 mm separated from each other and starting at 1 mm from the chilling surface. More details about the experimental set up are found in [1]. We applied air and water with different pressures as the coolant to obtain a wide range of cooling rates. The obtained temperature gradient at liquidus temperature covered the range of 0.6 to 4.45 K/mm and the growth velocity varied from 0.43 to 3.7 mm/sec.

To reveal the microstructure, the specimens were polished down to 0.05 μm and etched in a solution of 20-ml Water, 20-ml acetic acid, 60-ml ethylene glycol and 1-ml HNO_3 for about 5 minutes.

Phase field method

We consider unidirectional solidification of Mg-$0.5wt\%Al$. In the dilute limit, the solidus and liquidus lines can be approximated by straight lines of slopes m/k and m, respectively, leading to the partition relation $c_s = kc_l$, where c_s (c_l) is the concentration of impurities at the solid (liquid) side of the interface, and k is the partition coefficient. The simulations are two-dimensional, corresponding to a longitudinal section through a row of dendrites coinciding with the basal plane. Given that thermal diffusion is orders of magnitude faster than solute diffusion, latent heat production is neglected, resulting in the frozen temperature approximation $T(z,t) = T_0 + G(t)(z - z_0 - \int_0^t V_p(t')dt')$, where $T(z_0, 0) = T_0$ is a reference temperature, while $G(t)$ and $V_p(t)$ are the local thermal gradient and pulling speed, respectively. Fixing the reference concentration as the impurity concentration on the liquid side of an advancing steady-state planar interface $c_l^0 = c_0/k$, where c_0 is the nominal alloy composition, we obtain the following sharp interface solidification equations:

$$\partial_t c = D\nabla^2 c \tag{1}$$

$$c_l(1-k)v_n = -D\partial_n c|_l \tag{2}$$

$$c_l/c_l^0 = 1 - (1-k)\kappa d_0 \gamma(\theta) - (1-k)\left(z - \int_0^t V_p(t')dt'\right)/l_T - (1-k)\beta v_n \tag{3}$$

where $d_0 = \Gamma/\Delta T_0$ is the solutal capillary length, $\Delta T_0 = |m|(1-k)c_l^0$ the freezing range, $\gamma(\theta)$ the orientation dependent interface stiffness according to the underlying crystalline structure, $l_T = \Delta T_0/G$ the thermal length, D the diffusivity of solute in the liquid, and $\beta = 1/(\mu_k \Delta T_0)$ the kinetic coefficient.

Our phase-field formulation of equations (1)-(3) is based on the model developed by Karma and co-workers and detailed extensively elsewhere [5, 6]. The corresponding evolution equations are given in terms of a generalization of the field $\tilde{U} = (c - c_l^0)/(c_l^0(1-k))$, which represents the local supersaturation with respect to the point (c_l^0, T_0):

$$U = \frac{1}{1-k}\left(\frac{c/c_l^0}{(1-\phi)/2 + k(1+\phi)/2} - 1\right) \tag{4}$$

with the phase-field variable valued at $\phi = 1$ (-1) in the solid (liquid). Their explicit form is given by:

$$\tau(\hat{n})\left(1 - (1-k)\frac{(z - z_{\text{int}})}{l_T}\right)\frac{\partial \phi}{\partial t} = w_0^2 \vec{\nabla}\left[a(\hat{n})^2 \vec{\nabla}\phi\right] + \phi - \phi^3 - \lambda(1-\phi^2)^2\left(U + \frac{z - z_{\text{int}}}{l_T}\right) \tag{5}$$

$$\left(\frac{1+k}{2} - \frac{1-k}{2}\phi\right)\frac{\partial U}{\partial t} = \vec{\nabla}\left[q(\phi)D\vec{\nabla}U - \frac{w_0}{2\sqrt{2}}\left(1 + (1-k)U\right)\hat{n}\frac{\partial \phi}{\partial t}\right] + \left(\frac{1 + (1-k)U}{2}\right)\frac{\partial \phi}{\partial t} \tag{6}$$

where $z_{\text{int}} \equiv \int_0^t V_p\, dt'$ is the interface position, $\tau(\hat{n}) = \tau_0 \cdot a^2(\hat{n})$ the relaxation time and $\hat{n} \equiv -(\vec{\nabla}\phi)/(|\vec{\nabla}\phi|)$ the unit vector normal to the interface. The interpolation function $q(\phi) = (1-\phi)/2$ governs diffusivity across the interface.

A sixfold symmetry is implemented through the anisotropy function $a(\hat{n}) \equiv a(\theta) = 1 + \epsilon_0 + \epsilon_6 cos[6(\theta - \theta_0]$, where θ is the angle between the normal to the interface and the underlying crystalline axis of the HCP structure corresponding to the $< 11\bar{2}0 >$ direction in the basal plane, and θ_0 is the angle between that crystalline axis and the direction of the thermal gradient, defining the orientation of the crystalline structure as a whole, meaning $\theta_0 = 0$ corresponds to the $< 11\bar{2}0 >$ direction of the crystal coinciding with that of the thermal-gradient. This anisotropy function is the projection in the basal plane of the spherical harmonics representing the space group of the HCP crystal lattice: $a_{3d} = 1 + \epsilon_{20} y_{20} + \epsilon_{66} y_{66}$, where ϵ_{20} and ϵ_{66} are constant coefficients weighting the contribution of each of the spherical harmonic functions $y_{20} = \sqrt{5/16\pi}[3cos^2(\Theta) - 1]$ and $y_{20} = \sqrt{6006/64\sqrt{\pi}}[sin^6(\Theta)cos(\Phi)]$ while Θ and Φ are the inclination (or elevation) and azimuth spherical coordinate angles, respectively. The applied anisotropy function has been used by Eiken et al [7] and has been validated by the experimental findings of Pettersen et al [8] and Zhang et al [9]. From the molecular dynamics study of Sun et al ([10]) the anisotropy coefficients are given by $\epsilon_0 = 0.084$ and $\epsilon_6 = 0.03$, thus we approximate $1 + \epsilon_0 \sim 1$ and define $\epsilon_6 \equiv \epsilon = 0.03$.

In order to promote sidebranching, thermal noise induced concentration fluctuations are also included as described in [11]. Their explicit inclusion in the evolution equations is omitted here for clarity and brevity. Further details can be found at [1].

The material parameters are presented in table 1. Kinetic effects are neglected (i.e. $\beta = 0$), at least to first order, as shown in [6]. The phase-field equations are simulated using the adaptive-mesh-refinement (AMR) scheme developed by Provatas and co-workers, details of which can be found in [12, 13].

$\lvert m \rvert$ $(K/wt\%)$	5.50
c_0 $(wt\%)$	0.50
k	0.40
D $(\mu m^2/s)$	1800
Γ $(K \cdot \mu m)$	0.62
ϵ	0.03

Table 1: Material parameters defining the MgAl system. m is the liquidus slope, c_0 the alloy composition, k the partition coefficient, D the diffusivity of impurities in the liquid, Γ the Gibbs-Thomson constant and ϵ the anisotropy strength.

Two different growth conditions are simulated, on relatively small two dimensional domains, of width 1 and 2 mm. In the first case, direct thermocouple data from unidirectional solidification experiments are used to extract the local thermal gradient across the solid-liquid interface and the effective front velocity. These were then fitted to provide the functions representing $G(t)$ and $V_p(t)$. Since the pulling speed was modelled after a fit of the experimental front velocity, and the simulated interface is initially positioned at T_L, the simulated front velocity is systematically lower than the experimental front velocity used to determine the pulling speed, with the discrepancy decreasing as the system evolves. In the second case, the thermal gradient and pulling speed are kept constant. This report only presents simulations using constant growth conditions.

Results and discussions

We simulated a single grain with maximum misorientation with respect to temperature gradient under constant growth conditions of $v_p = 1\ mm/sec$ and $G = 2\ K/mm$ as shown in figure 1. These constant parameters being very close to the average transient growth conditions in our experiments. A hexagonal structure dominates at the onset, but it eventually becomes columnar. This transition can be attributed to the orientation of temperature gradient, which is selected to be different from the anisotropy direction, to be dominant. The boundary conditions at the sidewalls may also be playing a role. A detailed study of this effect, which requires simulating larger domains, is in progress. Since the thermal gradient and the surface tension compete during the whole solidification period, we can expect that a more intense thermal gradient will induce the transition earlier. To investigate the effect of temperature gradient, for the same pulling speed and initial orientation, we varied temperature gradients to $4G$, $8G$ and $16G$ as presented in figure 2 . As expected, higher temperature gradients make the transition manifest itself earlier.

Figure 1: Single grain morphology at different stages of its evolution. The width of the simulation domain corresponds to 1 mm. Grid lines map out the structure of the adaptive mesh.

As opposed to the more commonly studied cubic structures, at the earlier stages of evolution, before the hexagonal-to-columnar transition, there can be regions where there is no clear distinction between primary and secondary arms. This is illustrated in the left hand frames of figure 3. The various orientations in these structures, and their relative change during hexagonal-to-columnar transitions, were analyzed here using a two-dimensional power spectral analysis algorithm developed by Kuchnio et al [14], which registers the main wavelength at different angular directions with respect to the centre of an image. The right hand frames of figure 3 show an overall shift in the distribution of orientations towards the y-directions, indicative of the columnar transition shown in the left hand frames.

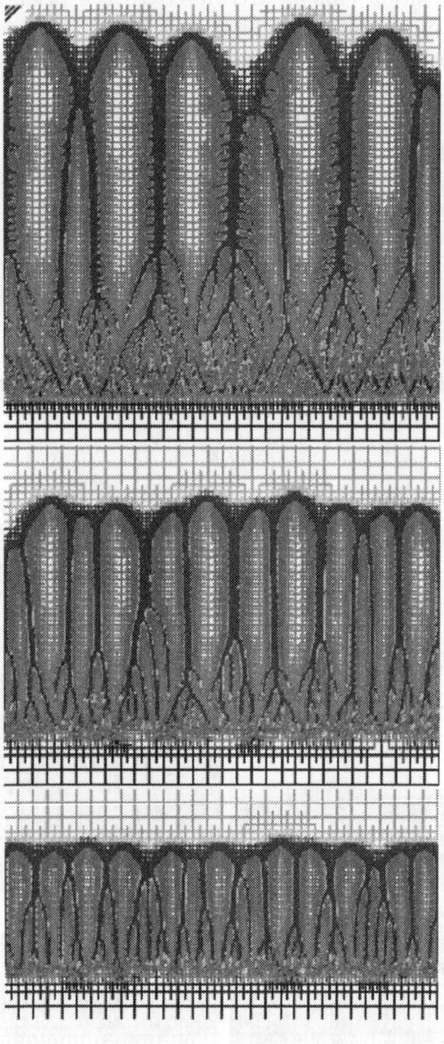

Figure 2: The hexagonal-to-columnar transition as it manifest itself earlier for more intense thermal gradient in each image, from top to bottom, increased by a factor of 4. All other parameters are the same as for figure 1.

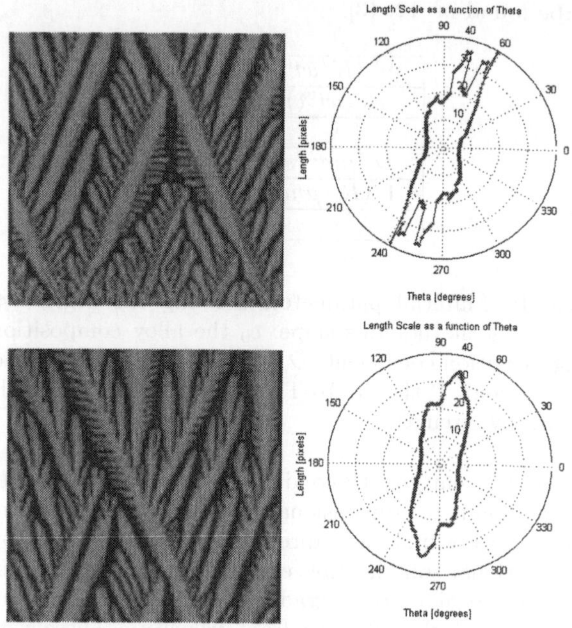

Figure 3: Main wavelength of a power-spectrum analysis performed at different angular directions with respect to the centre of a subdomain of the final structure presented in figure 1. The subdomain centred closer to the onset of the hexagonal-to-columnar transition present a closer alignment with the direction of the thermal gradient.

The hexagonal-to-columnar transition was also investigated experimentally under transient growth conditions. Growth velocity and temperature gradient are coupled in the experiments and decrease as the solidification front moves away from the chilling wall. Under these conditions, we expected to observe the transition at earlier times when the temperature gradient is highest.

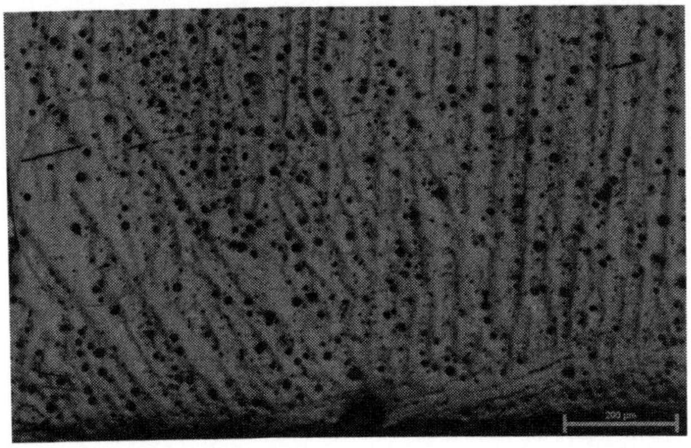

Figure 4: Different dendritic orientation showcased in a single grain obtained from a water-jet cooled sample. The red oval drawn close to the grain boundary shows the transition region from hexagonal to columnar microstructure.

However, high interface growth velocity opposes the orientation change since higher velocity means less effective time for dendrites to transition from hexagonal to columnar structure. Figure 4 shows an experimental image of a dendritic microstructure initialized from a chilling wall. Dendrites with different initial orientations exhibit both six-fold and columnar microstructure in a single grain. A hexagonal to columnar transition is observed very close to the left boundary. It is not clear to what effect this is caused by the finite size effect (i.e. due to interaction with another grain) or due to the competitive effect of thermal gradient and anisotropy, as indicated can exist from our simulations.

References

[1] Amoorezaei M., Gurevich S., Provatas N. *Spacing Characterization in Al-Cu Alloys Directionally Solidified Under Transient Growth Conditions* Acta Mater., **58**, n18, 6115-6124 (2010)

[2] Trivedi R., *Theory of dendritic growth during the directional solidification of binary alloys.*, Journal of Crystal Growth, **49**, n2, 219-32 (1980)

[3] Bouchard D. and Kirkaldy J. S., *Prediction of dendrite arm spacings in unsteady-and steady-state heat flow of unidirectionally solidified binary alloys.*, Metallurgical and Materials Transactions B, **27B**, n1, 101-13 (1996)

[4] Hunt J. D., Solidification and casting of metals, *Cellular and primary dendrite spacings.*, The metals society, London, 3 (1979)

[5] Karma A.,*Phase-field formulation for quantitative modelling of alloy solidification*, Phys. Rev. Lett. **87**, 115701/1-4 (2001)

[6] Echebarria B., Folch R., Karma A., Plapp M.,*Quantitative phase-field model of alloy solidification*, Phys. Rev. E **70**, 061604 (2004)

[7] Eiken J., *Dendritic growth texture evolution in Mg-based alloys investigated by phase-field simulation.*, Int. J. of Cast Materials Research,**22**, 1-4 (2009)

[8] Pettersen K., Lohne O., Ryum N., *Dendritic solidification of magnesium alloy AZ91.*, Mettal. Trans. A.,**21A**, 221-230 (1990)

[9] Zhang C., Ma D., Wu K.-S., Cao H.-B., Cao G.-P., Kou S., Chang Y. A., Yan X.-Y., *Microstructure and microsegregation in directionally solidified Mg-4Al alloy.*, Mettal. Trans. A.,**21A**, 221-230 (2007)

[10] *Crystal-melt interfacial free energies in hcp metals: A molecular dynamics study of Mg* D. Y. Sun, M. I. Mendelev, C. A. Becker, K. Kudin, T. Haxhimali, M. Asta, J. Hoyt, A. Karma and D. J. Srolovitz: Phy. Rev. B, **73**, 024116 (2006)

[11] Echebarria B., Karma A., Gurevich S. *Onset of sidebranching in directional solidification.*, Phys. Rev. E **81**, 021608 (2010)

[12] Provatas N., Goldenfeld N., Dantzig J.,*Efficient computation of dendritic microstructures using adaptive mesh refinement.*,Phys. Rev. Lett. **80**, n15, 3308 (1998)

[13] Athreya B. P., Goldenfeld N., Dantzig J. A., Greenwood M., Provatas N., *Adaptive mesh computation of polycrystalline pattern formation using a renormalization-group reduction of the phase-field crystal model.*, Phys. Rev. E,**76**, n5, 056706/1-14 (2007)

[14] Kuchnio P., Tetervak A., Watt C., Henein H., Provatas N., *Quantification of rapidly solidified micrpstructure of Al-Fe droplets using correlation length analysis.*, Metalurgical and Materials transactions A, **40**, n1, 196-203 (2009)

Magnesium Technology 2011
Edited by: Wim H. Sillekens, Sean R. Agnew, Neale R. Neelameggham, and Suveen N. Mathaudhu
TMS (The Minerals, Metals & Materials Society), 2011

Microstructures and Casting Defects of Magnesium Alloy Made By A New Type of Semisolid Injection Process

Yuichiro Murakami[1], Naoki Omura[1], Mingjun Li[1], Takuya Tamura[1], Shuji Tada[1], Kenji Miwa[1]
[1]Material Research Institute for Sustainable Development, National Institute of Advanced Industrial Science and Technology (AIST)
2266-98 Anagahora, Shimo-Shidami, Moriyama-ku, Nagoya 463-8560, Japan

Keywords: magnesium, semisolid, injection speed, fraction solid, microstructure, casting defect

Abstract

We have developed a new type of semisolid injection process that allows magnesium alloys to be formed in high material yields approximating 90%. In this process, generic magnesium billets are heated into their semisolid temperature range in an injection cylinder, without cover gas, and then the material is injected into a mold.

In this study, several billets were precision-heated in the cylinder to obtain a desired fraction solid. Plate specimens were produced by injecting the material at different injection speeds. Microstructures were observed by optical microscopy, and casting defects were detected on an X-ray computed tomography scanner. As injection speed was increased, the size and shape of α-Mg solid particles became smaller and more spherical, and the defect volume fraction increased. In contrast, as the fraction solid was increased, the defect volume fraction decreased. Spheroidization and miniaturization of solid particles were attributed to shear stress at the nozzle, and defects were affected by viscosity.

Introduction

In recent years, the development of environmental protection programs and green technology has become increasingly important. In particular, reducing carbon dioxide emissions and fuel efficiency improvement are urgent issues for the automobile industry, and vehicle weight reduction is one highly effective means of improving fuel efficiency. Hence, extensive research has been conducted on reducing vehicle weight. Toward this end, one approach is the expanded use of light metals such as aluminum and magnesium. Magnesium in particular is considered to be a promising lightweight structural material because it has the lowest density among applicable metallic materials, and has excellent specific strength. In recent years, magnesium alloy consumption has increased markedly, and it is expected that this growth trend will continue.

The die-casting process is among the most commonly used methods for forming magnesium alloys. However, components molded by die-casting exhibit low engineering performance due to the existence of inherent defect such as porosity, hot cracks and oxide inclusions. Additionally, die-casting of magnesium alloys requires the use of cover gas (e.g., SF_6) because magnesium alloys are easy combustible in the liquid state; However, SF_6 gas has a high global warming potential of about 24000, and thus its use should be avoided.

Recently, semisolid processes have been developed to fabricate high-quality aluminum alloy products [1- 3]. The semisolid injection process uses a semisolid billet that has a higher viscosity than liquid metal; consequently, casting defects in the final components can be reduced. This process is useful for magnesium alloys because processing temperatures lower than conventional casting processes result in decreased combustibility of the magnesium alloy.

We have developed a new type of semisolid injection process that allows magnesium alloys form in high material yield of about 90% [4, 5]. In this process, generic magnesium billets are heated to the semisolid temperature range in an injection cylinder. Thus, the semisolid magnesium alloy is not exposed to air. For this reason, this process requires no cover gas usage.

In this study, an apparatus was developed for preparing specimens using the semisolid injection process. Plate specimens were prepared by injecting AZ91D semisolid billet into a permanent mold through a narrow nozzle. The effects of the volume fraction solid and injection speed on microstructures and casting defects were investigated.

Experimental procedure

A diagram of the semisolid forming apparatus is shown in Fig. 1. This experimental setup has a vertical injection system. The injection cylinder is able to heat the magnesium alloy billets to the semisolid temperature, as well as to maintain this temperature. The semisolid billet can then be injected into a permanent mold by a piston. The injection cylinder has an inner diameter of 25 mm, an outer diameter of 60 mm and a length of 54 mm; the

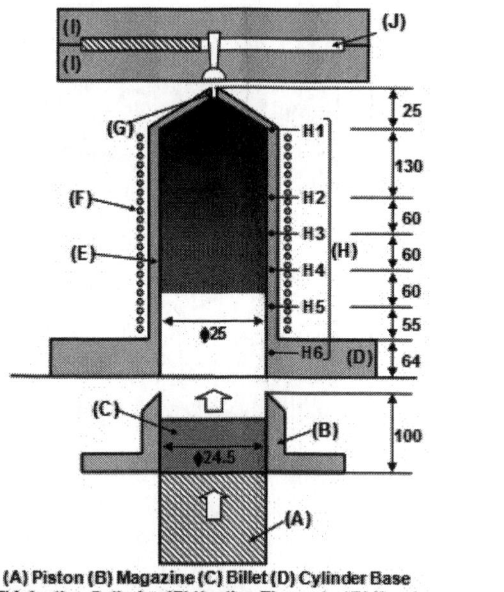

(A) Piston (B) Magazine (C) Billet (D) Cylinder Base
(E) Injection Cylinder (F) Heating Elements (G) Nozzle
(H) Thermocouples (I) Permanent mold (J) Cavity

Fig. 1 Schematic representation of semisolid forming apparatus

nozzle is 3 mm in diameter. In this setup, the extrusion ratio is about 70. The injection cylinder is always filled with billets from the nozzle to a 320 mm height in the apparatus, and the billet in the uppermost part of the cylinder is heated to the semisolid temperature range.

The injection cylinder is equipped with six heaters on the surface of the outer wall; these heaters can control the temperature of the billet precisely. The six heaters are controlled independently on the basis of measurements from six thermocouples (H1 to H6), each of which is inserted at the location of a heater. The thermocouples measure the temperature of the cylinder wall, but this value is different from the actual temperature of the billet inside the cylinder. Therefore, the actual temperature of the billet was measured by inserting a thermocouple into the billet directly from the nozzle, in order to calibrate the heaters and thermocouples. Then, the temperature of the billet could be controlled precisely to obtain the optimal temperature balance from heaters H1 to H6. The temperature of the billet in the uppermost part of the injection cylinder (from the nozzle to 130 mm height) was set to a temperature in the semisolid range, namely, 591°C, 586°C or 581°C. The fraction solid f_s was 0.3, 0.4 or 0.5, respectively, at each of these temperatures. Meanwhile, the temperature of the billet was set lower, nearer to the bottom of the injection cylinder. The bottom of the billet was below the solidus temperature. Thus, the semisolid billet was not exposed to air; therefore, cover gas was not used in this experiment.

The billet was inserted into the magazine and moved directly under the cylinder by an air piston. The billet was then inserted into the injection cylinder by a hydraulic piston. The injection cylinder was continuously filled with billets, thus forcing semisolid billet from the uppermost part of the heated cylinder through the nozzle into the permanent mold. The injection speed was set to 220, 300 or 400 mm/s. The permanent mold had two plate cavities of 20 mm in width, 100 mm in length and 5 mm in thickness. In this experiment, one plate specimen was made per injection because one cavity was filled by dummy.

The specimens were polished by grinding with SiC paper, followed by polishing with diamond paste. Then, the specimens were etched in a solution of 75 ml ethylene glycol, 1 ml nitric acid and 24 ml distilled water. The microstructures of these specimens were observed by optical microscopy. Also, the specimens were analyzed on an X-ray computerized tomography (CT) scanner. The casting defects were detected from 3D volume images of specimens made from these X-ray images.

Results and discussion

Effects of fraction solid and injection speed on microstructure

Fig. 2 shows typical microstructures of AZ91D magnesium alloys injected under various conditions. These micrographs show that the AZ91D magnesium alloys had a homogeneous microstructure with a uniform dispersion of primary α-Mg in the matrix (eutectic α-Mg and β-$Mg_{17}Al_{12}$). We feel that the α-Mg particles were the solid phase and the matrix was the liquid phase when the slurry was injected. With increasing fraction solid, the size of primary solid particles increased and the primary solid particle shape changed from irregular to spherical.

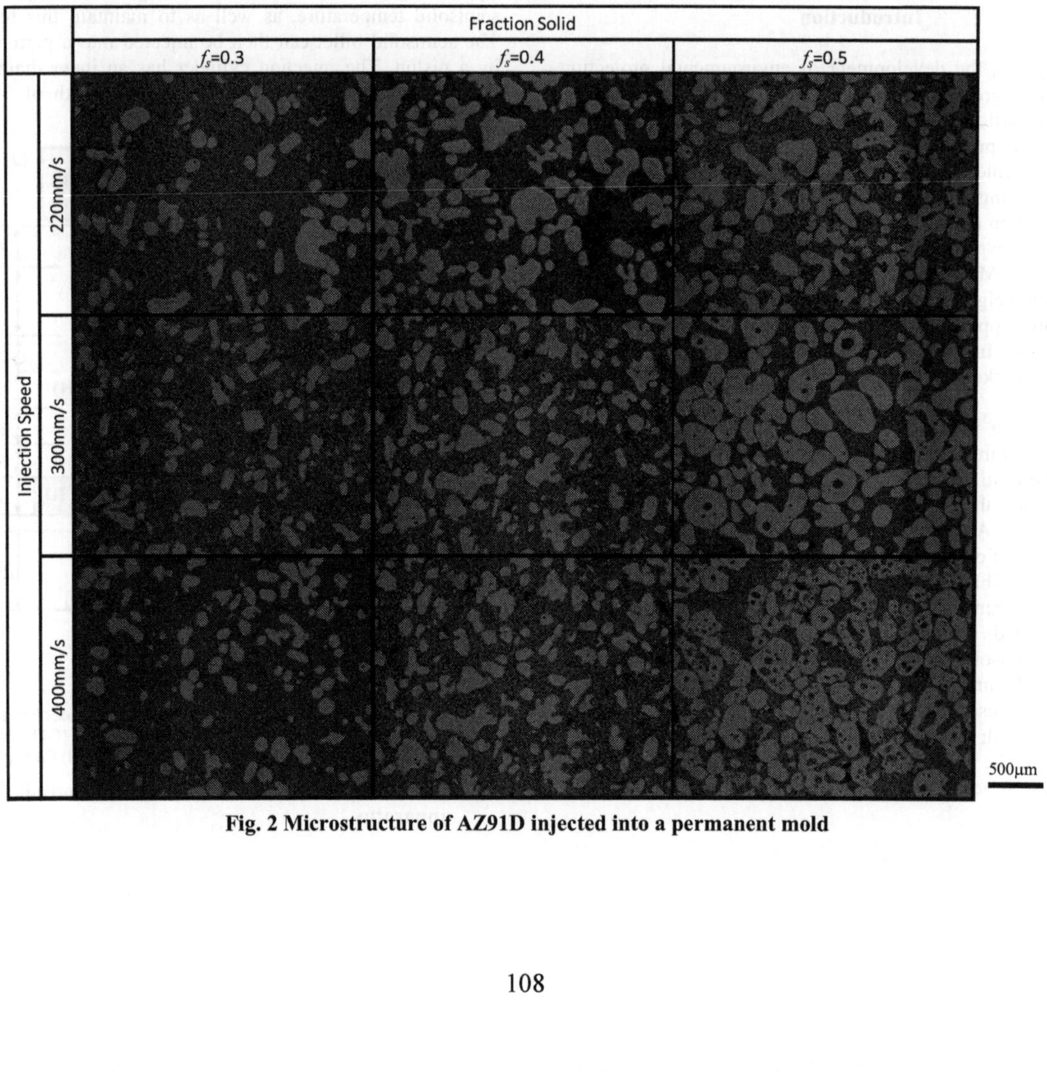

Fig. 2 Microstructure of AZ91D injected into a permanent mold

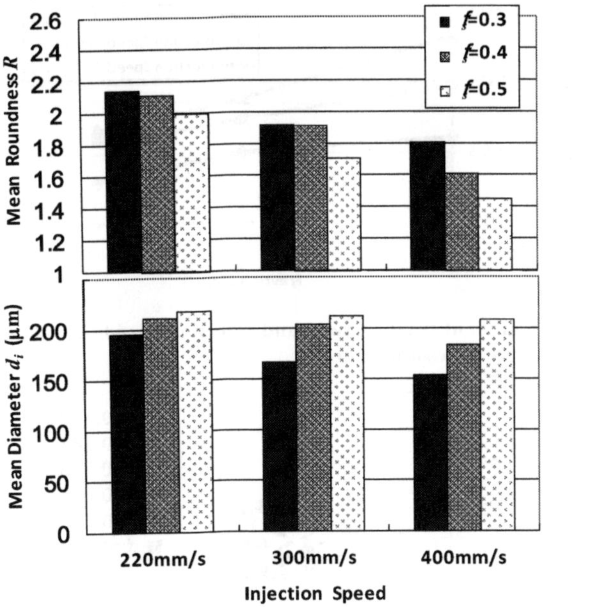

Fig. 3 Effect of injection speed and fraction solid on the mean roundness and diameter of primary α-Mg particle

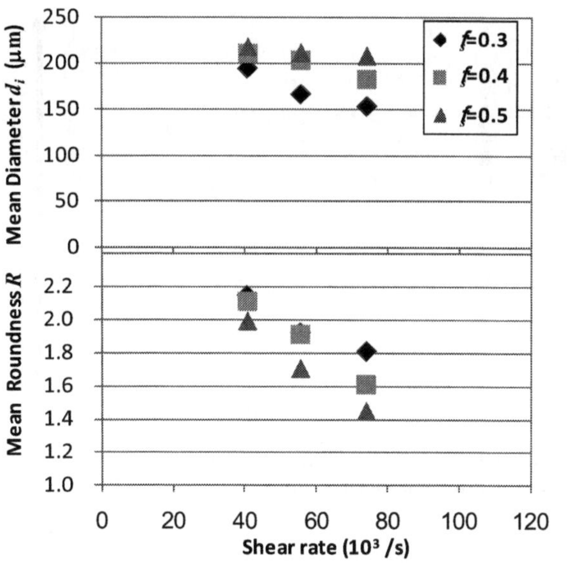

Fig. 4 Relations of mean diameter and roundness of primary α-Mg particles versus shear rate at the nozzle

The α-Mg particle diameter d [μm] (equivalent circle diameter) and the roundness R was measured at five locations of a specimen by image analysis.

Roundness was calculated as follows:

$$R = L^2/(4\pi A),$$

where L [μm] and A [μm^2] are the boundary length and area of an α-Mg particle. When R = 1, the particle is a true circle. Also, the area-weighted mean diameter d_s [μm] and area-weighted mean roundness R_s were calculated using the following equations:

$$d_s = \Sigma d_i A_i / \Sigma A_i$$

and

$$R_s = \Sigma R_i A_i / \Sigma A_i,$$

where d_i [μm], A_i [μm^2] and R_i are the diameter, area and roundness, respectively, of an α-Mg particle. Fig. 3 shows the weighted mean diameter d_s [μm] and area-weighted mean roundness R_s calculated as described above. These results show that increasing fraction solid or increasing injection speed decreased the mean roundness and increased the mean diameter of the α-Mg particles.

The shear rate and viscosity of slurry strongly affect the shear stress at the nozzle. Increasing the shear rate or viscosity increases shear stress. The shear rate γ [1/s] can be calculated from the injection speed S [m/s] and the dimensions of the nozzle and cylinder [6]:

$$\gamma = \frac{4Q}{\pi r_1^3} \quad (1)$$

,

where Q [m^3/s] is the flow rate through the nozzle given by $Q = \pi r_2^2 S$, r_1 [m] is the diameter of the nozzle, and r_2 [m] is the diameter of the cylinder.

Fig. 4 shows the relationships between mean particle roundness and shear rate, and between mean particle diameter and shear rate. The roundness and particle diameter decreased with increasing shear rate. In addition, for the same shear rate at the nozzle, particles became more spherical as the fraction solid increased. This result suggests that the primary solids were sheared through the nozzle during injection, and as a result became smaller and more spherical at high injection speeds. On the other hand, a high fraction solid increases shear stress, but particle size is affected by both shear stress and fraction solid. For this reason, particles became spherical with increasing fraction solid only at the same shear rate.

Effects of fraction solid and injection speed on casting defects

Fig. 5 shows the distribution of casting defects in the specimen prepared at an injection speed of 400 mm/s, as observed from X-ray CT images. In the case of zero fraction solid $f_s = 0$, many casting defects were dispersed throughout the specimen. It is considered that air in the mold was trapped in the specimen because this process is different from the die-casting process in that the mold has no overflows and runner. On the other hand, when semisolid material was injected, casting defects were significantly decreased. For fraction solid $f_s = 0.3$ and 0.4, the casting defects tended to be generated only in the middle of the specimen in the thickness direction. Additionally, many casting defects were dispersed throughout the specimens in the length direction and some of these had a crescent shape. In the semisolid injection process, it is known that the liquid-rich phase with higher fluidity is injected from nozzle to the end of the cavity first. After that, the solid-rich phase is injected into the cavity [7].

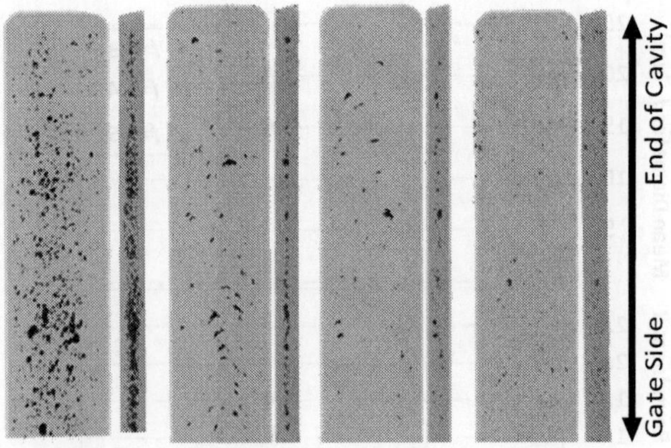

f_s=0.0 f_s=0.3 f_s=0.4 f_s=0.5

Fig. 5 Distribution of casting defects in the specimen prepared at the injection speed of 400 mm/s

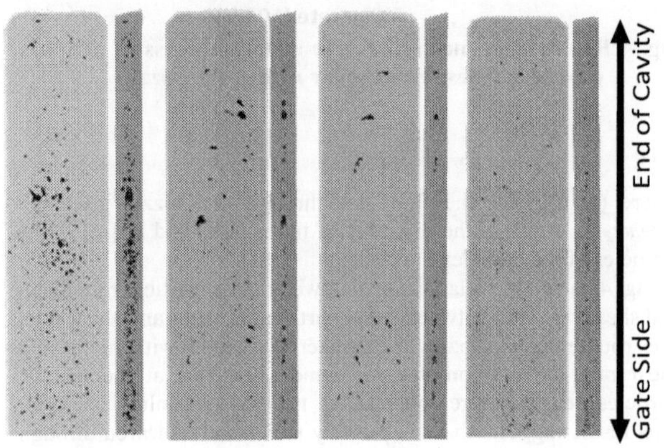

f_s=0.0 f_s=0.3 f_s=0.4 f_s=0.5

Fig. 6 Distribution of casting defects in the specimen prepared at the injection speed of 300 mm/s

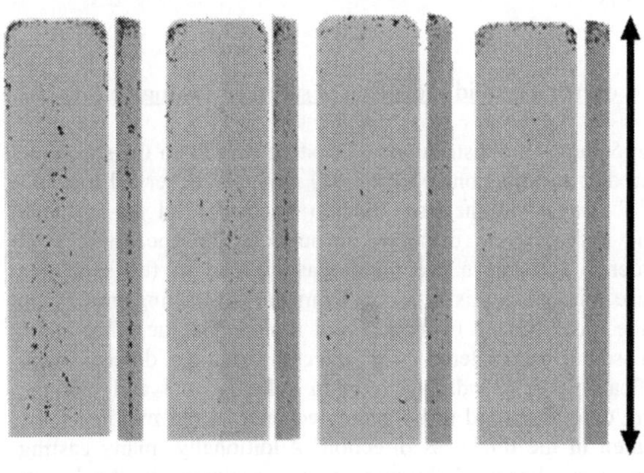

Fig. 7 Distribution of casting defects in the specimen prepared at the injection speed of 220 mm/s

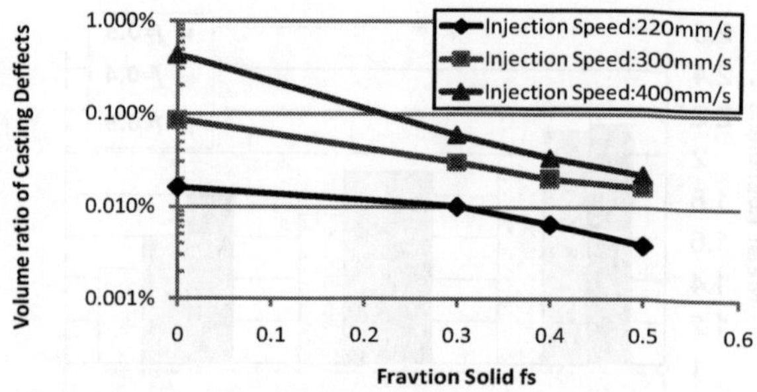

Fig. 8 Effects of fraction solid and injection speed on volume content of casting defects

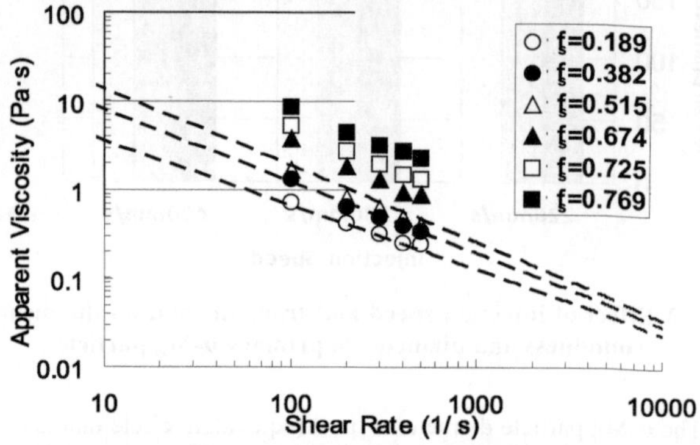

Fig. 9 Comparison of steady-state apparent viscosity with shear rate at various fraction solids for AZ91 [Error! Bookmark not defined.]. **Dashed lines show approximate values calculated using Eq. (2).**

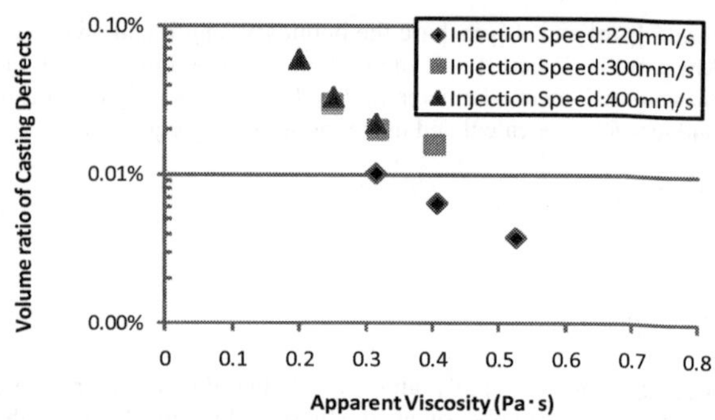

Fig. 10 Relationship between steady-state apparent viscosity in cavity and volume of casting defects

When the fraction solid is low, it is thought that the liquid-rich phase filled the air vents. For this reason, air in the mold was trapped in the specimen at the gap between the liquid-rich phase injected first and the solid-rich phase injected second.

Fig. 6 and Fig. 7 show the distribution of casting defects in the specimens prepared at injection speeds of 300 and 220 mm/s, respectively. At the injection speed of 300 mm/s, there were fewer casting defects in comparison with the injection speed of 400 mm/s for a given fraction solid. In addition, the shape of the defects exhibited the same tendency as in the case of an injection speed if 400 mm/s. It appears that air was trapped in specimens prepared at 300 and 400 mm/s for the same reason. On the other hand, for the specimen prepared at an injection speed of 220 mm/s, casting defects in the center part of the specimen were significantly diminished. However, the number of small casting defects at the end of the cavity increased. It appears that the slurry fluidity was insufficient to fill the entire cavity when the injection speed was 220 mm/s.

Error! Reference source not found. Fig. 8 shows the defect volume content in the specimens in relation to their fraction solid. The volume ratio of the defects decreased with increasing fraction solid or decreasing injection speed. This seems to be attributable to the fact that the viscosity of the slurry increased as the fraction solid increased [8].

Ghosh et al. [9] have reported the effects of the shear rate and the fraction solid on the apparent viscosity of AZ91D at steady state. In their study, the apparent viscosity was measured with a coaxial-cylinder rheometer. At $f_s \leq 0.55$, the apparent viscosity η [Pa·s] of AZ91D in the steady state was approximated by the following equation:

$$\eta = \frac{7.10\exp(5.17 f_s)}{\gamma^{(0.43 f_s + 0.64)}}, \qquad (2)$$

and the shear rate γ in the cavity can be calculated as

$$\gamma = \frac{6Q}{L^2 D}, \qquad (3)$$

where Q [m^3/s] is the flow rate through the nozzle given by $Q = \pi r^2 S$, r [m] is the diameter of the cylinder, and L [m] and D [m] are the width and thickness of the cavity, respectively.

Fig. 10 shows the relationship between the apparent viscosity in the cavity and the volume of casting defects. At injection speeds of 300 and 400 mm/s, the volume of casting defects similarly decreased with increasing the apparent viscosity. On the other hand, at the injection speed of 220 mm/s, the volume of casting defects significantly diminished in comparison to the injection speeds of 300 and 400 mm/s for the same viscosity. If the slurry does not reach a steady state, the apparent viscosity of slurry will be affected not only by the fraction solid and shear rate but also by particle morphology and particle agglomeration [10]. Fig. 3 showed that α-Mg particles in the slurry injected at 220 mm/s were not spherical. Thus, it appears that the viscosity of the slurry was less than that in the steady state. Hence, the volume of casting defects was significantly diminished. Additionally, for the injection speed of 220 mm/s, it was thought that the shear stress at the nozzle was insufficient for improving fluidity.

Conclusions

In order to investigate the effects of the fraction solid and injection speed on microstructure and casting defects, experiments on the semisolid forming of AZ91D magnesium were performed. The primary α-Mg particles became large and spherical with increasing fraction solid. Furthermore, with increasing injection speed, the size of the primary solid particle decreased and the shape of the primary solid particle became spherical. By calculating the shear rate at the nozzle, it was revealed that the roundness and diameter of α-Mg particles decreased with increasing shear stress. However, the fraction solid affected the diameter of α-Mg particles more strongly than shear stress.

The amount of casting defects decreased with increasing fraction solid or decreasing injection speed. For injection speeds of 300 and 400 mm/s, casting defects from trapped air were distributed throughout the specimen. On the other hand, for the injection speed of 220 mm/s, many casting defects were generated at the end of cavity because the fluidity of slurry was insufficient. For the injection speeds of 300 and 400 mm/s, the volume of casting defects was similarly decreased with increasing apparent viscosity. However, at the injection speed of 220 mm/s, the volume of casting defects was substantially reduced in comparison to the injection speeds of 300 and 400 mm/s for a given viscosity. The reason that the α-Mg particles in the slurry injected at 220 mm/s were not spherical is that shear stress at the nozzle was insufficient.

References

1. W.G. Cho and C.G. Kang, "Mechanical properties and their microstructure evaluation in the thixoforming process of semi-solid aluminum alloys," *Journal of Materials Processing Technology*, 105 (2000), 269-277

2. S. Nafisi and R. Ghomashchi, "Grain refining of conventional and semi-solid A356 Al–Si alloy," *Journal of Materials Processing Technology*, 174 (2006), 371-383

3. H.K. Jung and C.G. Kang, "Induction heating process of an Al–Si aluminum alloy for semi-solid die casting and its resulting microstructure," *Journal of Materials Processing Technology*, 174 (2006), 355-364

4. K. Miwa, R. Rachmat, T. Tamura and Y. Sakaguchi, "Relationship between volume fraction solid and casting defect on semi solid injection in AZ91D magnesium alloy," *Journal of Japan Foundry Engineering Society*, 78 (2006), 187-193 (in Japanese)

5. N. Omura, Y. Murakami, M.G. LI, T. Tamura and K. Miwa, "Effect of Volume Fraction Solid and Injection Speed on Mechanical Properties in New Type Semi-solid Injection Process," *Solid State Phenomena*, 141-143 (2008) 761-766

6. The Society of Polymer Science Japan, ed., *Plastic Processing Handbook* (Nikkan Kogyo Shimbun Ltd., 1995) 1401 (in Japanese)

7. R. Rachmat, T. Tamura and K. Miwa, "Fluidity and Microstructures Characteristics of AZ91D by using New Type Semi-solid Injection Process," *Solid State Phenomena*, 116-117 (2006) 534-537

8. D. B. Spencer, R. Mehrabian and M. C. Flemings, "Rheological behavior of Sn-15 Pct Pb in the Crystallization Range," Metallurgical Transactions, 3 (1972) 1925-1932

9. D. Ghosh, R. Fan and C. VanSchilt, "Thixotropic roperties of Semi-Solid Magnesium Alloys AZ91D and AM50," *Proceedings of The 3rd International Conference on Semi-Solid Processing of Alloys and Composites*, (1994) 85-94

10. M. C. Flemings, "Behavior of metal alloys in the semisolid state," *Metallurgical Transactions B*, 22B (1991) 269-293

Magnesium Technology 2011
Edited by: Wim H. Sillekens, Sean R. Agnew, Neale R. Neelameggham, and Suveen N. Mathaudhu
TMS (The Minerals, Metals & Materials Society), 2011

Macrostructure evolution in directionally solidified Mg-RE alloys

M.A. Salgado-Ordorica, W. Punessen, S. Yi, J. Bohlen, K.U. Kainer and N. Hort

GKSS-Forschungszentrum, Max-Planck-Strasse 1, 21502 Geesthacht, Germany.

Keywords: Magnesium alloys, rare earths, directional solidification, microstructure.

Abstract

The use of Rare-Earths (RE) to develop new cast- and wrought-magnesium alloys has acquired increased interest in recent years. The good mechanical properties of Mg-RE alloys at room temperature, and in particular their high strength at relatively high temperatures are at present well-known facts that make them very promising materials for transport applications. In this context, it is necessary to achieve a better understanding of the macro and microstructure evolution of cast Mg-metals directionally solidified. To this end, binary Mg-RE alloys (where RE = Gd, Nd and Y) were cast by permanent mould direct chill casting. This process was performed in a specially optimized laboratory-scale installation in order to ensure the obtention of "clean" ingots, with homogeneous composition and free of porosity and inclusions. A set of different processing conditions was evaluated in order to better control the final microstructure, mainly in terms of grain size, orientation and distribution. The grain selection mechanisms operating during the solidification of these specimens, namely texturization and Columnar to Equiaxed Transition (CET), were characterized and put into relation with the initial composition of the alloy and the imposed cooling conditions.

Introduction

The characteristics of Rare Earths (RE) and their potential effects when added to Mg melts have been characterized during the last 60 years (1). However, due to new technological developments on RE extraction and separation, it has been only until recently that Mg-RE alloys haver acquired new interest for well defined applications, in particular in transport industry and bio-medical developments. The use of RE in Mg-alloys increase notably their properties at relatively high temperatures (2; 3) and due to their promising role in automotive and aircraft applications, it is important to know at which extent their microstructure evolution can be controlled either in as-cast products or wrought materials.

Chill Casting (CC) of Mg-alloys present a number of complications related to the melt preparation and atmosphere protection. In particular during melting and pouring into the mould, the molten metal can react with air and burn or create oxide layers that will eventually be detrimental to the homogeneity of the final product. Due to volume limitations on the mould size, permanent mould CC has only be applied at a laboratory scale, but the solidified ingots can be further processed to fullfill other requirements.

Other than in die-casting processes, the as-cast microstructure of simple Mg-RE alloys has been poorly investigated and focus has been mainly directed to understand the effect of these alloying elements in further processing stages. In the context of Mg-alloys directionally solidified, some work has been devoted to measure and understand the dendrite growth directions and texture development, in particular in AZ and AM alloys. These studies have shown that there is a strong dependence of the microstructure on the processing conditions, i.e., imposed thermal gradient G and solidification velocity v_s, and the intrinsic properties of the alloy, in particular the solid-liquid interfacial energy $\gamma_{s\ell}$ anisotropy. A pioneer effort on characterizing the microstructure of an AZ91 alloy was performed by Pettersen et al (4; 5). They found that in these alloys, a $\langle 11\bar{2}0\rangle$ texture evolves in specimens directionally solidified at a low G/v_s ratio ($\approx 48/0.32 = 150$ K/s), whereas at $G/v_s \approx 10/0.053 \approx 190$ K/s the texture evolves along a $\langle 22\bar{4}5\rangle$ direction, which is of course not contained within the basal plane. In the first case, the columnar trunks appear bonded by three secondary arms growing at 60 deg. from the trunk, while in the second case two arms belong also to $\langle 22\bar{4}5\rangle$ directions, at 35 deg. from the trunk, and a third one grow along a $\langle 11\bar{2}0\rangle$ direction, at 54 deg. from the trunk. Mirkovic and Fetzer (6) performed also Bridgman solidification experiments and measured primary and secondary arm spacings of AZ31 and AM50 dendritic structures. They also found the growth of a primary trunk with three secondary arms growing at 60 deg. from it. These later observations were further investigated by Eiken et al. and compared to numerical predictions of dendrite growth using the phase-field method (7). In order to compare simulations and experimental observations, not only the processing conditions were taken into account, but also an appropriate form for the anisotropy of the solid-liquid interfacial energy $\gamma_{s\ell}(\mathbf{n})$ for an hexagonal system was considered. The dependency of $\gamma_{s\ell}$ with respect to the crystallographic orientation, expressed as a vector **n** normal to the solid liquid interface in the crystallographic reference frame, was constructed through mathematical functions, called spherical harmonics, that included the two- and six-fold symmetries of an hexagonal system. These simulations reproduced well the texture evolution for dendrites growing along $\langle 11\bar{2}0 \rangle$ directions only in the range of $G/v_s \leq 150K/s$. However, it has been stated in this and other works (8; 9) that for non-basal growth directions, the use of a more complex function of $\gamma_{s\ell}(\mathbf{n})$, might be necessary.

The present work first describes a reliable process to cast Mg-RE ingots of very good quality. A comparison of different solidification conditions for these alloys is presented in order to establish a range of parameters that will allow to control the final microstructure. Finally, an analysis of texture evolution is presented, in particular for Mg-10wt.%Gd alloys.

Experiments and Methods

Alloys preparation

Different binary Mg-RE alloys, where RE = 4wt.%Y, 10wt.%Gd and 2wt.%Nd, were prepared from their pure elements (purity of 99.99%) in a resistances furnace able to heat up a stainless steel crucible of 8 kg capacity. Mg ingots of about 750 g were first molten in the furnace at 720°C, then the corresponding alloying element was added in small quantities not bigger than a few hundreds of grams. After a waiting time of 5 minutes, a six-blade boronitride coated stainless steel propeller, turning at 150 rpm during 30 min, was used to stir the melt and ensure complete mixing. A constant flux of an Ar-SF_6 mixture (in a ratio 5:1) was introduced to the furnace during the whole melting and stirring times, in order to reduce oxidation and burning of the melt. After stirring, oxides remaining on the upper surface of the melt were cleaned out with a boronitride coated stainless steel paddle. The temperature before pouring into the permanent mould was of 700°C.

Permanent mould chill casting

This process has been developed in order to produce, at a laboratory scale, Mg-RE ingots of good quality, free of porosity, with homogeneous composition and controlled microstructure. A schematic representation of the experimental set-up can be observed in Fig. 1a. A three-zones resistances furnace having a tubular shape is bounded in the upper and bottom parts by well-insulated opening hatches. The permanent mould, fabricated with stainless steel and coated in its inner surface with a thin layer of boronitride, was preheated into the tubular furnace at a temperature of 650°C during 15 to 20 min. After pre-heating, the mould was extracted from the tubular furnace through the upper hatch and fixed to a support where the melting furnace could directly fill it. A smooth flux of Ar was passed through the furnace nozzle in order to avoid oxidation and burning of the molten metal during this stage. After pouring, the mould was introduced again into the tubular furnace through the upper hatch and it was covered with a thin steel deck. A secondary flux of an Ar-SF_6 mixture was flowing through the deck in order to keep the atmosphere above the molten metal free of oxygen. Additionally, a 1 mm diameter thermocouple was introduced into a boronitride coated stainless steel capillary tube (3 mm diameter), that was in direct contact with the melt, at about 50 mm from the bottom of the mould.

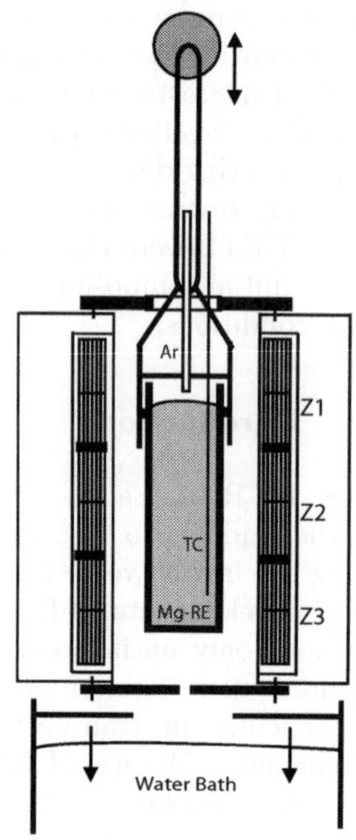

Figure 1: Schematic representation of the three-zones tubular furnace used to produce Mg-RE ingots.

The mould and melt were hold alternatively during 60

min at temperatures of 650°C or 750°C, depending on the experiment. After this holding time, the thermocouple was extracted from the melt and the bottom hatch was opened. A y-axis movement inductor was specially designed in order to achieve different pulling velocities of the mould into a water bath located at about 5 cm below the tubular furnace. The pulling velocity v_p ranged in between 8 to 20 cm/min (this corresponds to a range from 1.33×10^{-3} to 3.33×10^{-3} m/s). The different mould geometries used in this study, i.e., rectangular, insulated and non-insulated pipe-shape, can be observed in Fig. 2. The insulated pipe-shape mould was simply surrounded by an external pipe with an inter-wall spacing of 5 mm. Only the Mg-10wt.%Gd ingots were obtained in the three different geometries, whereas the other two alloys were only processed in the insulated and non-insulated pipe-shape moulds.

Figure 2: Images of the different mould geometries used in this work: a) rectangular, b) insulated pipe, and c) non-insulated pipe

After solidification, the ingots were extracted from the moulds. The upper and bottom regions of the ingot (about 1.5 cm in length along the main ingot axis) were directly sectioned prior to further analysis of the specimens. In the case of the cylindrical ingots solidified in the insulated and non-insulated pipe-shape moulds, the ingot was sectioned in two halves along its main axis and then sections at 4 different heights were extracted. In order to evaluate the composition homogeneity in the whole volume, chemical analysis was performed by means of an X-ray fluorescense analyzer XRF (Bruker S4explorer) for the Mg-Gd specimens, and through spark emission spectrum analysis in a Spectrolab2003 (Sprectro Analytical Instruments GmbH and Co.) for the Mg-Y and Mg-Nd ingots.

In order to perform macrostructure analyses, a 1 cm thick slice parallel to the main axis of each ingot (rectangular or cylindrical) was extracted. These slices were prepared according to the standard procedure developed by Kree et al. (12), although no etching was necessary to reveal the grain structure of the Mg-10wt.%Gd alloys. In particular the study of the macrostructure in the rectangular Mg-Gd ingots allowed to identify the Columnar-to-Equiaxed Transition (CET) and put it into relation with the solidification conditions.

In order to investigate the texture of the columnar grains observed in some of the specimens, small sections of about 2×2 cm^2 were mechanically polished to mirror quality and then electro-polished in a standard Struers AC2 solution cooled down to -20°C (33 V during 90 s). Then, EBSD measurements were performed in an electron microscope SEM Zeiss Ultra 55.

Results and Discussion

Macrostructure

The macrographs in Fig. 3a, b and c, show the different grain structures observed for each alloy solidified in the insulated pipe-shape permanent moulds. These ingots were solidified by pulling down the mould at 12 cm/min only to contact with the water bath. Then the tubular furnace was shut down while the mould was maintained at this position during 20 min, meaning that except for the bottom 3 cm of the ingot, the rest of the specimen remained inside of the tubular furnace. As the bottom part of the ingot was previously sectioned, only a fully developed columnar region can be observed, i.e., the region containing only equiaxed grains near the chill bottom surface does not appear in the macrographs. In the upper part of the ingot in Fig. 3a, a Columnar to Equiaxed Transition (CET) took place, but the equiaxed grains only occupied the upper 3 cm of the ingot length. The other two alloys did not exhibited a CET. The average chemical composition at different locations in the ingot are indicated in each macrograph. None of the specimens here obtained showed any macrosegregated regions. Below each macrograph, optical micrographs of each alloy can be observed. The Mg-10wt.%Gd alloys exhibit a well-defined dendritic structure, while the other ones, due to low content of solute, present more likely a columnar cellular morphology with some microsegregation channels of a phase rich in Y or Nd in between the cells.

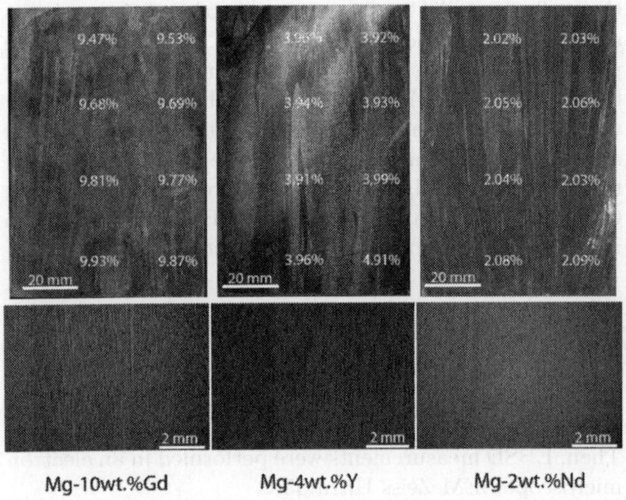

Figure 3: Macro and micrographs of a) Mg-10wt.%Gd, b) Mg-4wt.%Y, and c) Mg-2wt.%Nd alloys solidified in the insulated pipe-shape mould pulled down at v_p = 12 cm/min to contact with the water bath, then cooled down directly in the furnace. The composition measurements at different regions of the ingots are clearly indicated.

Another casting was obtained with the same type of insulated mould, but by keeping v_p constant until the 95% of the ingot volume was inside the water bath. In this case, the CET was present for alloys. This CET occurred much earlier during solidification and relatively big equiaxed grains occupied more than a half of the ingot volume. Indeed, although the external insulation helps to keep the heat inside the mould when it is pulled down to the water bath, the melt temperature still decreases when the ingot is inside the water bath, and therefore equiaxed grains can nucleate ahead of the columnar front. This was confirmed by the macrostructure observed in the ingot solidified in the non-insulated pipe-shape mould pulled down at a speed of 12 cm/min. In this case only in the bottom and near the lateral walls a few columnar grains developed along the first half of the ingot. The rest of the volume was conformed by equiaxed grains.

Further characterization of the grain distribution can be better observed in Fig. 4, which shows the macrostructure observed in Mg-10wt.%Gd alloys solidified in the rectangular mould at different pulling velocities. Due to the symmetry along the vertical axis of the ingot, only a half of a 2D projection of these ingots is shown, in the upper row for the ingots where the initial melt temperature T_{melt} (i.e., before pulling downwards into the water bath) was of 650°C, and in the bottom row for those with T_{melt} = 700°C. The effect of the pulling velocity v_p on the grain orientation with respect to the casting direction can be clearly observed. At the lowest pulling velocity, columnar grains evolve slightly inclined towards the centre of the specimen. This means that the isotherms are nearly perpendicular to the casting direction. As v_p is increased, the depth of the mushy zone increases as one moves towards the centre of the specimen and the isotherms acquire a nearly U-shape. Therefore the columnar grains grow rather towards the centre of the specimen, at an angle α with respect to the mould wall that increases with v_p. In the centre of each of the ingots when $v_p \geq 8$ cm/min, a region containing equiaxed grains can be observed. These grains appeared always slightly larger in a precise direction, along the main axis of the ingot when $v_p \approx 12$ cm/min, perpendicular to it when $v_p \geq 12$ cm/min. The region containing the equiaxed grains is larger in volume as v_p and T_{melt} increase, but the size of these grains gets considerably reduced. The averaged equiaxed grain size ϕ for each specimen is also indicated in Fig. 4.

Texture

The texture of the as-cast specimens was investigated through EBSD measurements performed on the fully developed columnar region of a Mg-10wt.%Gd ingot solidified in the insulated cylindrical mould at an initial pulling speed of 12 cm/min, until the mould bottom made contact with the water bath, and then cooled down inside the tubular furnace. Under these directional solidification conditions, the texture measurement can be directly related to the primary trunk growth direction. Very simply expressed, the probability of finding the dendrite growth direction at an angle θ from the thermal gradient is proportional to a function $\sin\theta$ (10; 11). At the beginning of solidification, in the chill surface, the probability of finding the specific growth direction of a dendrite perfectly aligned with G is small, but it increases to a maximum as the grown distance from the chill also increases. Naturally, the angle between the growth direction and G also decreases to a value near, but different to zero, and the distribution of possible grain orientations around this maximum gets also reduced.

Figure 5b presents a 2×4 mm^2 surface containing

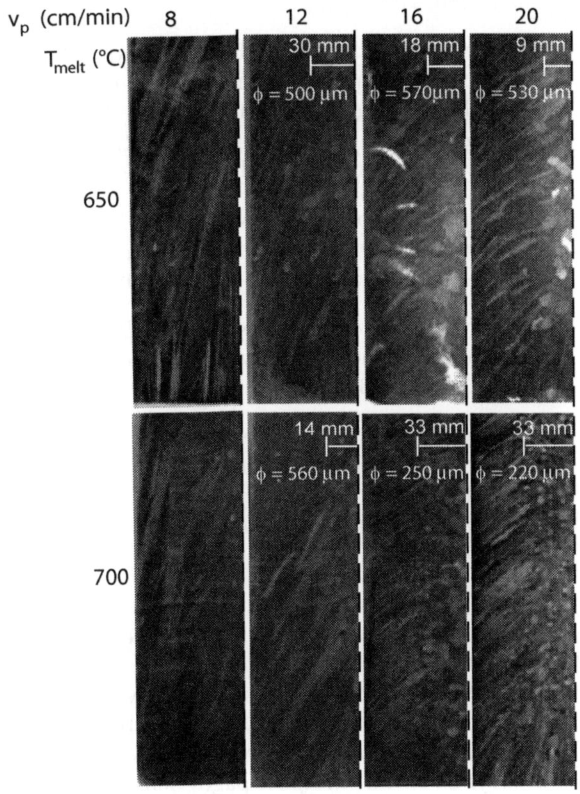

Figure 4: Effect of pulling velocity on the macrostructure of Mg-10wt.%Gd alloys solidified in the rectangular mould. The average equiaxed grain size ϕ is indicated in each ingot.

about 20 grains with different crystallographic orientations. Note that the grey shades refer to differences in the crystallographic orientation which will be discussed with respect to the texture below. The coloured grains are marked in order to reveal their specific orientation in the respective pole figure. The $\langle 0001 \rangle$ pole figure in Fig. 5a clearly shows that a texture along a well defined direction exists in the grains contained in the examined surface, i.e., the $\langle 0001 \rangle$ directions of the crystals in these grains rotate all around a single axis. Due to the distribution of the $\langle 0001 \rangle$ directions of the crystals inside these grains, the rotation axis is clearly not lying within the basal plane. This means that the texturization along a $\langle 11\bar{2}0 \rangle$ direction observed in previous works for Mg-Al alloys is not present in the Mg-Gd system. The pole figures presented in Fig. 5c

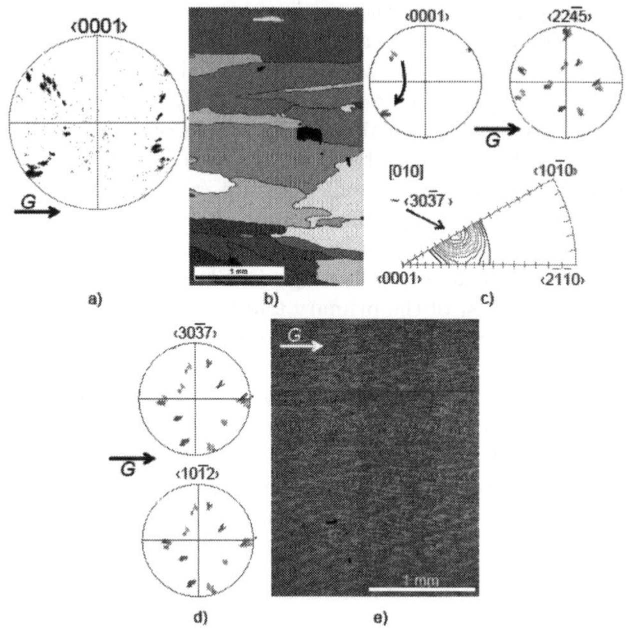

Figure 5: EBSD texture analysis of a directionally solidified Mg-10wt.%Gd alloy. a) $\langle 0001 \rangle$ pole figure drawn from the crystallographic orientation of the Mg-10wt.%Gd directionally solidified grains shown in b); c) $\langle 0001 \rangle$ and $\langle 22\bar{4}5 \rangle$ pole figures of selected grains (top) and inverse pole figure of the whole analyzed surface (bottom); d) $\langle 30\bar{3}7 \rangle$ and $\langle 10\bar{1}2 \rangle$ pole figures of selected grains; e) microstructure of the specimen inside the analyzed surface.

were drawn to find precisely the rotation axis along which the texture direction could be contained. In order to better perform this analysis, only the orientation of four selected grains are shown. Blue and red grains have similar orientations, different from the also similar dark and light green grains. Therefore only two well localized spots with their corresponding colours can be identified in the $\langle 0001 \rangle$ pole figure. The other type of texture reported in literature for Mg-alloys is along $\langle 22\bar{4}5 \rangle$ directions. In the corresponding $\langle 22\bar{4}5 \rangle$ pole figure none of the six equivalent directions from each of the selected grains is aligned with the thermal gradient, and actually these are distributed randomly in the pole figure.

In order to identify the crystallographic direction better aligned with G, an inverse pole figure was drawn from the crystallographic orientation of each grain. This is shown in the bottom part of Fig. 5c. The [010] index indicated in the upper-left corner means that the inverse pole figure was drawn by taking the specimen as seen from a plane perpendicular to the direction of G. A well texturized grain distribution can be identified in this inverse pole figure, with a maximum near to a $\langle 30\bar{3}7 \rangle$ direction. As can be seen in the remaining pole figures of Fig. 5d, the $\langle 30\bar{3}7 \rangle$ and $\langle 10\bar{1}2 \rangle$ directions are almost equivalent. The main texture direction, which should correspond to the dendrite trunk growth direction, has been indicated in these two pole figures. The dendritic microstructure inside the grains analyzed through EBSD can be observed in Fig. 5e. Most of the primary trunks are aligned nearly parallel to G, which confirms the assumption that the texture is a measure of the dendrite trunk growth direction. Nevertheless, it is not possible to make any assumption on the specific secondary arms growth directions, since these are not always lying in the observed section. Further studies of these microstructures through X-ray tomography still need to be performed.

Conclusion

An experimental technique has been developed in order to produce ingots of homogeneous composition and free of porosity and inclusions. A set of parameters were imposed in order to better control the macrostructure. A columnar structure can be obtained when $v_p < 12$ cm/min and the occurrence of the CET can be retarded by stopping the pulling and leaving the specimen to cool down inside the furnace. A well texturized structure along a $\approx \langle 10\bar{1}2 \rangle$ direction develops in directionally solidified Mg-10wt.%Gd alloys. This is different from previous observations on the texture evolution in AZ and AM alloys. The growth of columnar well texturized grains in this binary system is a promising field, in particular for the production of single-grain components made of Mg-alloys.

Acknowledgments

The authors would like to thank Sabine Schubert and Volker Kree for their technical support in the chemical analysis and EBSD measurements, respectively.

References

[1] L.L. Rokhlin. *Magnesium alloys containing rare earth metals.* Taylor and Francis, 2003.

[2] H. Dieringa, N. Hort, and K.U. Kainer. Investigation of minimum creep rates and stress exponents calculated from tensile and compressive creep data of magnesium alloy ae42. *Mat. Sci. and Eng. A*, 510-511:382–386, 2009.

[3] M.A. Gibson, S. Zhu, M.A. Easton, and J.-F. Nie. Microstructure, tensile properties and creep resistance of binary Mg-Rare Earths alloys. *Magnesium Technology*, Edited by S.R. Agnew, N.R. Neelameggham, A. Yberg and W.H. Sillekens:233–238, 2010.

[4] K. Pettersen, O. Lohne, and N. Ryum. Crystallography of directionally solidified magnesium alloy AZ91. *Metall. Mater. Trans.*, 20 A:847–852, 1989.

[5] K. Pettersen, O. Lohne, and N. Ryum. Dendritic solidification of magnesium alloy AZ91. *Metall. Mater. Trans.*, 21A:221–230, 1990.

[6] D. Mirkovic and R. Schmid-Fetzer. Directional solidification of mg-al alloys and microsegregation study of mg alloys az31 and am50: Part ii. comparison between az31 and am50. *Metall. Mater. Trans.*, 40 A:974–980, 2009.

[7] J. Eiken, G. Klaus, D. Mirkovic, I. Steinbach, A. Bührig-Polaczek, and R. Schmid-Fetzer. Numerical and experimental investigation of dendritic growth texture evolution in mg-al alloys with hcp lattice anisotropy. *Modeling of Casting, Welding and Advanced Solidification Processes XII (Edited by S.L. Cockcroft and D.M. Maijer)*, pages 553–560, 2009.

[8] A. Mariaux. *Texture Formation in Hot-Dip Galvanized Coatings: Nucleation and Growth of Anisotropic Grains in a Confined Geometry.* PhD thesis, EPFL No. 4646, 2010.

[9] A. Mariaux, T.-V.-Putte, and M. Rappaz. Modeling nucleation and growth of zinc grains in hot-dip galvanized coatings. *Modellling of Casting, Welding and Advanced Solidification Processes. Edited by S. Crockcroft and D. Maijer*, XII:667–674, 2009.

[10] C.A. Gandin, M. Rappaz, D. West, and B.L. Adams. Grain texture evolution during columnar growth of dendritic alloys. *Metall. Mater. Trans. A*, 26 A:1543–1551, 1995.

[11] F. Gonzales and M. Rappaz. Grain selection and texture evolution in directionally solidified Al-Zn alloys. *Metall. Mater. Trans. A*, 39 A (9):2148–2160, 2008.

[12] V. Kree, J. Bohlen, D. Letzig and K.-U. Kainer. The Metallographical Examination of Magnesium Alloys. *Prakt. Metall.*, 41 (5):233–246, 2004.

Magnesium Technology 2011
Edited by: Wim H. Sillekens, Sean R. Agnew, Neale R. Neelameggham, and Suveen N. Mathaudhu
TMS (The Minerals, Metals & Materials Society), 2011

Microstructure and Mechanical Behavior of Cast Mg AZ31B Alloy Produced by the Magnetic Suspension Melting Process

N.W. Rimkus, M.L. Weaver, N. El-Kaddah
The University of Alabama, Tuscaloosa, Alabama, USA

Keywords: Containerless Melting, Casting, Magnesium AZ31B alloy, Microstructure, Fracture

Abstract

Magnesium is the lightest of all structural metals and offers significant weight savings compared to traditional automotive materials. This paper describes the macrostructure and the microstructure of Mg AZ31B alloy produced via the Magnetic Suspension Melting (MSM) technique at a low superheat of 5°C. It was found that casting at this low superheat produced a fine globular grain structure in comparison to a dendritic structure in conventionally cast alloys. The intermetallic phases were analyzed in detail and compared with the conventionally cast alloy. In the MSM cast alloy, the $Mg_{17}Al_{12}$ phase formed mainly at the grain boundaries, in contrast to typical dendritic entrapment of this phase within the grains in conventional castings. The formation of the Al-rich secondary-α phase during solidification was investigated. The effects of this morphology change on mechanical and fracture behavior of this material are presented. These results are discussed relative to conventionally cast Mg alloys.

Introduction

With increasing demand for more fuel efficient automobiles, recent years have brought considerable interest in using cast magnesium components in vehicles. Magnesium will reduce weight, improving the fuel efficiency of automobiles [1,2]. Traditional AZ and AM series magnesium alloys are currently being used in low-strength applications such as steering wheels, instrument panels, air intake systems, and tank covers [1]. However, their use in structural applications such as: cross car beams, seat frames, steering column brackets, and vehicle front end structures have been rather limited due to contamination of the alloys during processing [1,3,4], macrosegregation and microsegregation [1,3], and the precipitation of intermetallic phases such as β-$Mg_{17}Al_{12}$ in the matrix and at the grain boundaries, all of which are known to be detrimental to the mechanical properties of these alloys [1,2,5,6].

Grain refinement in magnesium alloys has been a large focus in research in an effort to minimize the adverse effects of segregation and intermetallic precipitates on the mechanical properties. The addition of various grain refiners such as carbon and zirconium on the final grain structures have been extensively investigated [6,7,8]. These studies showed that the use of grain refiners can effectively reduce the grain size to about 20 μm, which is generally desirable to improve mechanical performance of the cast material [6,7]. For typical solidification rates in permanent mold and sand castings the grains are essentially dendritic, leading to entrapment of the aluminum-rich secondary-α phase between secondary dendrite arms [5,7]. Further improvement in the mechanical properties of the castings could be achieved by avoiding dendritic entrapment of the secondary-α phase by changing the solidification morphology to a globular one by way of another grain refinement method such as casting at low superheat [9,10,11,12].

The high reactivity of magnesium alloys makes contamination of the metal during melting in conventional melting systems unavoidable. The use of a protective SF_6 or CO_2-SF_6 gas to prevent magnesium burning only minimizes oxidation of the melt [13]. Furthermore, melting in a steel crucible leads to contamination with iron, which adversely affects grain refinement [8]. Therefore, the development of new melting technology that completely eliminates contamination would be highly desirable for casting magnesium structural automotive components.

The Magnetic Suspension Melting process (MSM), an integrated containerless induction melting and casting process, completely eliminates oxidation and contamination of the metal during melting. This is done by electromagnetically supporting the liquid metal in space. Previous work on melting and casting Al-Li alloys using this process at low superheat has yielded homogeneous, oxide free castings with a fine globular grain structures [10,11,12]. This paper examines low superheat castings of AZ31B produced by the MSM process. The microstructural features of MSM cast alloy are compared to those obtained at high superheat and the Charpy impact behavior of these castings are presented.

Experimental Technique

Figure 1 shows a schematic drawing of the experimental MSM system used in this study. It consists of two components: 1) an induction melting unit and 2) a casting chamber. The melting unit is basically an induction coil surrounding a silica tube with an inner diameter of 80 mm mounted on a stainless steel plate. The top of the tube is sealed using a brass flange fitted with ports for gas input and output and for a thermocouple. The charge material, which is a round billet 3" in diameter and 2" long, is placed on a water-cooled stainless steel chill block inside the silica tube. In this system, the induction coil is not stationary and moves during melting to maintain the containment force needed to support the molten metal. The motion of the coil is controlled using the laser tracking system.

The casting chamber is a cylindrical stainless steel tube attached to the chill plate of the melting unit. The casting mold is placed on the bottom plate of the casting chamber. In this work the mold was designed to allow unidirectional solidification of the melt. The mold is essentially an insulating refractory tube around a stainless steel chill plate located on the bottom. The refractory tube has a 3" ID and 3.5" OD, and a height of 6" made by Zircar Refractory. The stainless steel chill plate is 3" in diameter and 2" long. A thermocouple holder is placed on top of the refractory tube to allow measurements of the cooling rate along the mold during solidification. Thermocouples are connected to an Omega TC-08 data acquisition system.

The entire system is sealed using O-rings, and is connected to a gas flow system which includes an argon gas cylinder and a vacuum pump. The induction coil, which surrounds the silica tube, has 10 turns and is connected to an Inductotherm 125 VIP power supply.

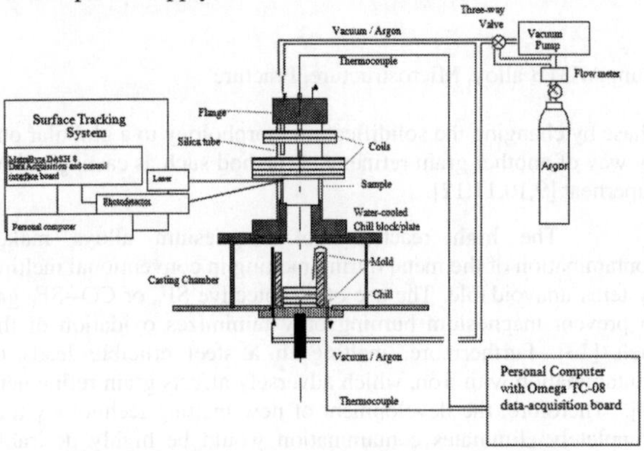

Figure 1. Schematic of the MSM experimental setup.

Casting experiments at high superheat were performed using conventional melting techniques. In these experiments the magnesium alloy was remelted in a steel crucible in a 1320 W resistance furnace under cover of CO_2-SF_6 gas. The molten metal was poured at 700 °C in the same mold used in the MSM experiments. The cooling curves along the mold during solidification were also measured using the Omega TC-08 data acquisition unit.

The melting and casting experiments were carried out using AZ31B magnesium alloy. The chemical composition of the alloy is shown is Table 1. Characterization of the castings include optical microscopy, scanning electron microscopy (SEM) and chemical analyses. The optical analysis was conducted on a Nikon Epiphot 200 Microscope with a SPOT Insight 2 camera attachment. SEM spectroscopy was conducted on a JEOL 7000 FEG SEM equipped with Oxford EDS and WDS systems. Microprobe analysis was conducted on a JEOL 8600 electron microprobe. This instrument was also equipped with a Si drift EDS system for rapid chemical mapping.

Impact toughness was assessed by means of room temperature Charpy impact tests. The specimens for impact testing were standard Charpy V-notch test bars with dimensions of 10 mm × 10 mm × 55 mm (V-notch depth = 2 mm) with their longitudinal directions parallel to the bottom of the mold.

Results and Discussion

This section presents results of characterization of unidirectional solidified AZ31B magnesium specimens prepared using the MSM process at low superheat of 5°C. The corresponding results for castings produced by conventional casting techniques at high superheat of 70°C are also presented.

Solidification Morphology and Grain Structure

Figures 2(a) and 2(b) show typical optical micrographs of the grain structure taken from the center of the ingot for the MSM and conventional castings respectively. These figures show that the grains of MSM cast alloy at 5°C superheat are much smaller than those obtained by conventional casting. The average grain size of the MSM cast alloy is 84μm while it is about 334 μm for the casting produced at high superheat of 70°C. It should be mentioned that there is very little variation in the grain size of the MSM cast ingot, despite the decrease in the solidification rate along the mold, away from the chill plate. Figure 2(a) also shows that the grain boundaries of the MSM casting are characterized by sharp interfaces, in contrast to the irregular grain boundaries observed in the conventional casting, Figure 2(b). The observed faceted interfaces in the MSM cast AZ31B suggests a plane front solidification morphology, which has been observed in Al alloys cast at such a low superheat [10,11].

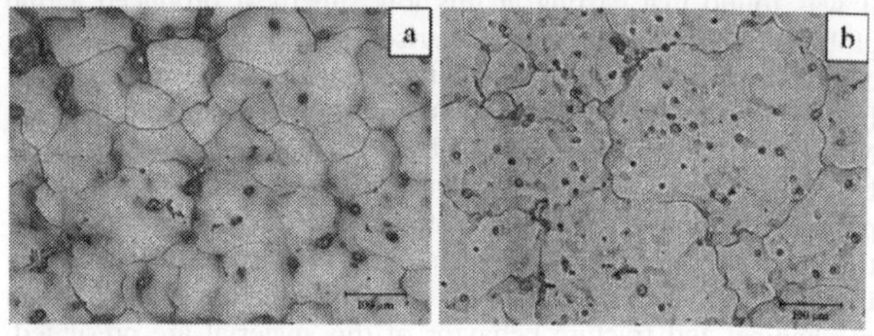

Figure 2. Optical micrographs of MSM (a) and conventionally cast (b) alloys etched using a Citric Acid solution .

Figures 3(a) and 3(b) show the morphologies of the grains produced at low and high superheat, respectively. The bright area corresponds to the primary α-Mg phase while the aluminum-rich secondary α-Mg phase appears dark. These figures show that the grains in the low superheat (i.e., MSM cast) alloy essentially consist of a primary-α phase with the secondary-α in the interstices between the grains; which is indicative of a globular solidification morphology, Figure 3(a), while it is dendritic at high superheat, Figure 3(b). At the solidification rate used in this investigation, the dendritic grains are equiaxed with the secondary-α phase between dendrite arms. The pronounced segregation of Al and other alloying elements within the equiaxed dendritic grains is the main source for the formation of the intermetallic phases observed within the matrix as shown in Figure 2(b). EDS mapping was conducted to confirm the segregation of aluminum in the grains. Figures 3(c) and 3(d) show the EDS aluminum maps for the grains produced at low and high superheat, respectively. Grain boundaries have been added in white showing that the aluminum rich secondary-α phase is found in the matrix between dendrite arms in the high superheat casting, Figure 3(d). Comparatively, the MSM low superheat casting, Figure 3(c), confirms the secondary-α phase is found along the grain boundaries only. This segregation explains the low superheat's lack of intermetallic phases appearing in the matrix.

The significant difference in the grain size and solidification morphology can be attributed to the high nucleation potential of the melt in the MSM process due to casting at low superheat [10]. This also acts to enhance the rate of grain growth and reduces the temperature gradient at the solidification front,

Table 1 Chemical Composition of AZ31B Ingot used in experiments

Magnesium	Aluminum	Silicon	Zinc	Nickel	Copper	Iron	Calcium	Manganese	OET
95.444%	3.140%	0.051%	0.661%	0.001%	0.001%	0.005%	0.020%	0.377%	0.300%

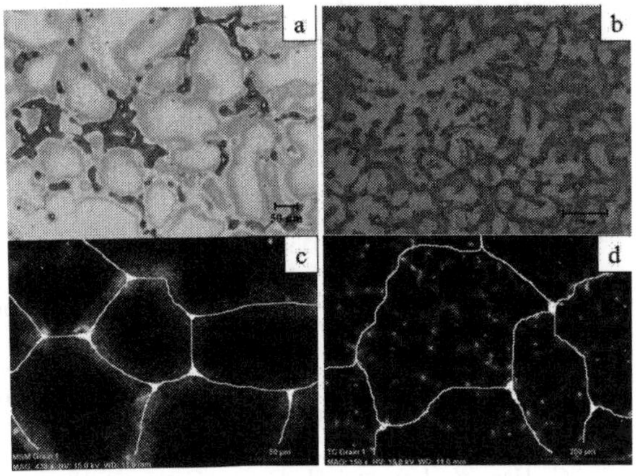

Figure 3. (a)Optical micrograph of the MSM casting showing a globular structure. (b) Optical micrograph of the conventional casting showing a dendritic structure. (c) EDS Map of Al for the MSM casting showing a globular structure. (d) EDS Map of Al for the conventional casting showing a dendritic structure.

thus favoring the transition from dendritic to globular solidification [15].

Intermetallics

The grain structure of the conventionally cast ingot, Figure 3(b), shows typical formation of the β-intermetallic particles in both the matrix and along the grain boundaries [5]. The distribution of the intermetallic precipitates shown in Figure 3 depends on the solidification morphology. For globular solidification morphology, corresponding to casting at low superheat in the MSM process, the intermetallic precipitates are found only at the grain boundaries. For dendritic solidification the precipitates are formed both within the grains as well as along the grain boundaries. This feature is attributed to the difference in the microsegregration of aluminum during solidification, Figure 3. For the globular solidification morphology, aluminum-rich liquid remains ahead of the solidification front rather becoming entrapped between the primary and secondary dendrite arms in dendritic solidification morphology.

The two main intermetallic phases formed in cast AZ31B alloys are β-$Mg_{17}Al_{12}$ and Al_8Mn_5. The β-$Mg_{17}Al_{12}$ phase is the most predominant precipitate in the castings. For casting at high superheat, the β-$Mg_{17}Al_{12}$ precipitates are found in two forms: (1) within the grains the β phase is formed as divorced or partially divorced particles, Figure 4(b); while (2) along the grain boundaries, in addition to the divorced and partially divorced particles, β-$Mg_{17}Al_{12}$ is present in a lamellar-type form. The formation of lamellar β phase occurs during the final stages of solidification

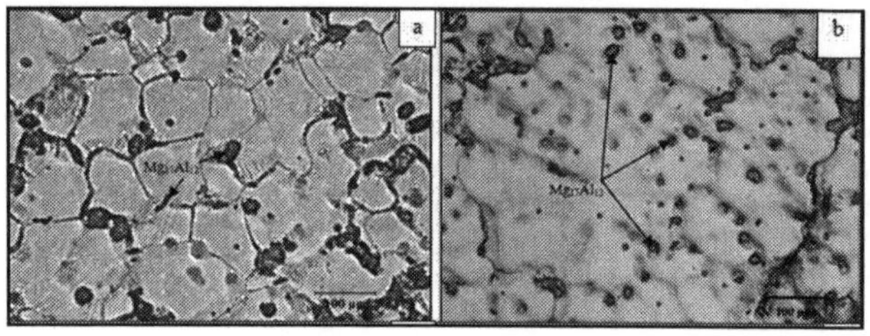

Figure 4. (a) Optical micrograph of the MSM casting showing $Mg_{17}Al_{12}$. (b) Optical micrograph of the conventional casting showing $Mg_{17}Al_{12}$

and is generated by grain boundary movement. In the MSM cast alloy, the β precipitates are mainly of the lamellar type and are found exclusively along the grain boundaries. The lamellae size and spacing in globular solidification is much larger than that for dendritic solidification. This may be attributed to the relatively large volume of the liquid between impinging equiaxed globular grains, and the low thermal gradient present in the liquid.

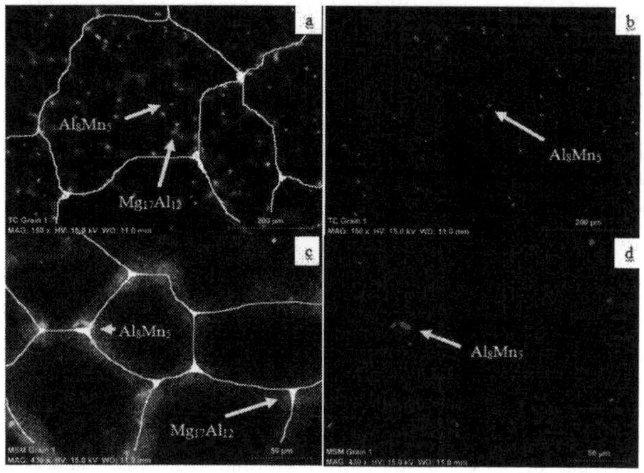

Figure 5. Al and Mn EDS Maps for (a,b) conventionally cast AM31B showing $Mg_{17}Al_{12}$ and Al_8Mn_5 phases. (c,d) EDS maps for MSM cast AM31B showing $Mg_{17}Al_{12}$ and Al_8Mn_5 phases along grain boundaries

Figure 5 shows EDS maps of aluminum (a,c) and manganese (b,d) for high and low superheats. These maps show overlapping Mn and Al dots where Al_8Mn_5 intermetallics are located. The EDS maps for the high superheat casting, Figures 5(a,b), show that the Al_8Mn_5 intermetallic is found both in the matrix as well as along the grain boundaries. The EDS maps for the low superheat casting, Figures 5(c,d), show that these intermetallics are only located along the grain boundaries. The EDS maps further support the occurrence of plane front solidification in the low superheat MSM castings.

Fracture Behavior

Charpy impact energies of 9.5 and 10 J were measured respectively for conventionally cast and MSM cast alloys. These values are equivalent to those for common Mg-Al-Zn alloys (i.e., 3 to 9 J) [16,17]. Figure 6 shows fracture surface images for the MSM and conventionally cast alloys. At low magnifications, the fracture surfaces for both castings exhibited large regions of quasi-cleavage decohesion with sizes corresponding to the grain sizes observed in each casting. Superimposed upon these regions were small cleavage facets and microshrinkage voids, arrowed in Figure 6(a), and ductile microvoids located within some of the Mg-rich solid solution regions. The small cleavage facets, which were associated with $Mg_{17}Al_{12}$ particles, and the microshrinkage voids appeared to act as fracture initiation sites within the conventional castings. Some similar features were observed in the

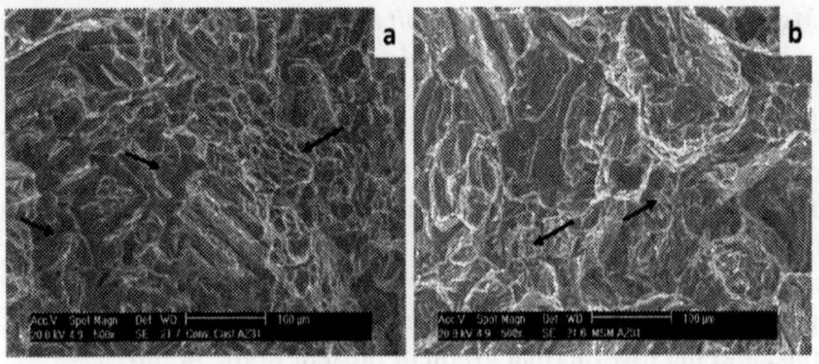

Figure 6 Fractographs (SEM) of (a) Conventionally Cast and (b) MSM cast impact toughness specimens.

MSM casting; however, the MSM casting contained no microshrinkage pores. Furthermore, the regions containing small cleavage facets were smaller and less numerous, and the percentage of microvoids was higher, suggesting a higher potential for plastic deformation.

Conclusions

Experiments were performed to investigate the solidification morphology and structure of AZ31B alloy cast at low superheats using the Magnetic Suspension Melting process. Casting at a superheat of 5°C was found to produce fine globular grains that were three times smaller than those obtained at a high superheat of 70°C. In the MSM casting, SEM/EDS analysis showed that the intermetallic precipitates were found exclusively at the grain boundaries. The main intermetallic precipitates in the cast ingot were β-$Mg_{17}Al_{12}$ and Al_8Mn_5. These results suggest that the MSM process may prove to be a viable technique to improve the properties for casting high strength magnesium automotive components.

Acknowledgments

The authors graciously thank the NSF for funding of this project under grant number CMMI-0856320

References

[1] N. Li, R. Osborne, B. Cox and D. Penrod, Magnesium Engine Cradle: The USCAR Structure Cast Magnesium Development Project: ASE paper, (2005) p. 337

[2] J.P. Thomson, S. Xu, M. Sadayappan, P.D. Newcombe, L. Millette, and M.Sahoo, Low Pressure Casting of Magnesium Alloys AZ91 and AM50: AFS Transactions, (2004) p. 1-10

[3] Soon Gi Lee, G.R. Patel, and A.M. Gokhale, Characterization of the effects of process parameters on macrosegregation in a high-pressure die-cast Magnesium alloy: Materials Characterization, (2005), V.55, p. 219-224

[4] Y. Chino, T. Furuta, M. hakamada, and M. Mabuchi "Influence of distribution of oxide contaminants on fatigue behavior in AZ31 Mg alloy recycled by solid-state processing," Materials Science and Engineering: A, V424 (2006) p. 355-360

[5] D.G. Leo Prakash, Doris Regener, "Quantitative characterization of $Mg_{17}Al_{12}$ phase and grain size in HPDC AZ91 Magnesium Alloy," Journal of Alloys and Compounds, V461 (2007) p. 139-146

[6] Q. Jin, J.-P. Eom, S.-G. Lim, W.-W. Park and B.-S. You, Grain Refining Mechanism of a Carbon Addition Method in a Mg-Al Magnesium Alloy: Scripta Materialia 49 (2003) 1129-1132.

[7] G. Han, X. Liu, and H. Ding, "Grain refinement of AZ31 magnesium alloy by new Al-Ti-C master alloys," Transactions of Nonferrous Metals Society of China, V19 (2009) p.1057-1064

[8] P. Cao, M. Qian, and D. H. StJohn, "Native grain refinement of magnesium alloys," Scripta Materialia, V53 (2005) p.841-844

[9] Frank Czerwinski, Near-liquidus molding of Mg-Al and Mg-Al-Zn alloys: Acta Materialia, (2005), V.53, p. 1973-1984

[10] J. Adams and N. Ei-Kaddah, Containerless processing and characterization of high purity aluminum alloys: in "Proceedings of the 1st International Conference on Processing Materials for Properties" (TMS, Warrendale, OH, 1993) p. 905-908.

[11] D. J. Reynolds, M. Shamsuzzoha and N. El-Kaddah, Characterization of Al-Li Castings Produced by the Magnetic Suspension Melting Process: in "Proceedings of the 4th Decennial International Conference on Solidification Processing", edited by J. Beech and H. Jones (Department of Engineering Materials, University of Sheffield, Sheffield, 1997) p. 45-48.

[12] C. Mahato, M. Shamsuzzoha and N. El-Kaddah, Solidification Morphology and Structure of Cast Al-Li 2090 Alloy at Low Superheats: in "Solidification of Aluminum Alloys", edited by M. G. Chu, D. A. Granger and Q. Han (TMS, Warrendale, OH, 2004) p. 321-328.

[13] H. Proffitt, "Magnesium and Magnesium Alloys," in "ASM Handbook on Casting" (ASM International, Materials Park, OH, 1988) p. 798.

[14] D.H. Kang, Manas Paliwal, Elhachmi Essadiqi, and In-Ho Jung, "Experimental Studies on the As-Cast Microstructure of Mg-Al Binary Alloys with Various Solidification Rates and Compositions," Magnesium Technology TMS, (2010) p.533-536

[15] Doru Michael Stefanescu, *Science and Engineering of Casting Solidification*, KA/PP, 2002, p.145-148

[16] H. W. Wagener, J. Hosse-Hartmann and R. Friz, "Deep Drawing and Impact Extrusion of Magnesium Alloys at Room Temparature," *Advanced Engineering Materials* 5 (2003) 237-242.

[17] J. Liao, M. Hotta, K. Kaneko and K. Kondoh, "Enhanced impact toughness of magnesium alloy by grain refinement," *Scripta Materialia* 61 (2009) 208-211.

Magnesium Technology 2011
Edited by: Wim H. Sillekens, Sean R. Agnew, Neale R. Neelameggham, and Suveen N. Mathaudhu
TMS (The Minerals, Metals & Materials Society), 2011

INVESTIGATIONS ON HOT TEARING OF MG-ZN-(AL) ALLOYS

Le Zhou[1,2], Yuanding Huang[1], Pingli Mao[2], Karl Ulrich Kainer[1], Zheng Liu[2], Norbert Hort[1]

[1]MagIC-Magnesium Innovation Center, GKSS Research Center, Max-Planck-Strasse 1, D-21502, Geesthacht, Germany
[2] School of Materials Science and Engineering, Shenyang University of Technology, No.111, Shenliao West Road, Economic & Technological Development Zone, Shenyang,110178, China

Keywords: Mg-Zn alloy, casting, hot tearing, thermodynamic calculation, solidification

Abstract

Mg-Zn alloys are widely used as wrought alloys such as ZM, ZK and ZE series. They are reportedly to be prone to hot tearing due to the presence of Zn. The present work first evaluates the hot tearing susceptibility (HTS) of binary Mg-Zn alloys using quantitative experimental methods and thermodynamic simulations based on Clyne's model, and then further investigate the addition of aluminum on the HTS in the ternary alloys Mg-Zn-Al. The results show that the curve of the HTS vs. the content of Zn has a typical "λ" shape. With increasing the content of Zn, the HTS increases firstly, reaches the maximum at 1.5% Zn and then decreases again. The addition of Al in Mg-Zn alloys influences the HTS. In the Mg-Zn-Al ternary system, the HTS decreases with the increase of Al content. The curve of the HTS as a function of Zn content in the ternary Mg-Zn-Al system is a little different from that observed in the binary Mg-Zn alloys. Two peaks are obtained: one is approximately at 1.0 to 1.5 wt.% Zn, another at 3.0 wt.% Zn.

Introduction

The hot tearing, also called hot cracking, is often a major defect in casting of alloys. It occurs during solidification due to obstructed contraction of the solidifying alloy, often at hot spots where the casting finishes solidifying or at locations with a sudden change in cross sections. Hot tearing is a complex solidification phenomenon that is still not fully understood, though various mechanisms and criterion have been proposed [1-4]. The factors dominating the formation and susceptibility of hot tears include alloying elements, freezing range, amount of eutectic phases and initial mold temperature.

For hot tearing susceptibility of magnesium alloys, most of previous inspections were performed only for Mg-Al series. They surveyed the effect of alloying elements such as RE, Ca and Sr on the hot tearing susceptibility of Mg-Al alloys [5-8]. The additions of these alloying elements more or less deteriorate the castability and increase the tendency for hot tearing. Mg-Zn alloys are most widely used as the wrought magnesium alloys, which is a basic composition of ZK series commercial alloys. Previous investigations were mainly performed on their corrosion behavior, age-hardening behavior and microstructure, but there are very limited investigations on their hot tearing. Clyne et al. carried out the preliminary investigations on the hot tearing of Mg-Zn alloys and concluded that the largest hot tearing susceptibility can be observed with 2% Zn [3]. The present work investigates the influence of Zn content and mold temperature on the hot tearing susceptibility for Mg-Zn binary alloys and Mg-Zn-Al ternary alloys using thermodynamic modeling and quantitative experimental measurements.

Experimental procedures

Thermodynamic calculations based on Clyne's model

The hot tearing criterion proposed by Clyne and Davies [2, 3] is based on the assumption that the strain applied during the earlier stage of liquid and mass feeding is accommodated without problem by the casting. When the dendrites are no longer free to move easily, the liquid mass feeding can not accommodate the strains developed during this stage. The last stage of freezing is considered the most susceptible to hot tearing in this criterion. The cracking susceptibility coefficient (CSC) is defined by the ratio of t_V, the vulnerable time period where hot tearing may develop, and t_R, the time available for the stress-relief process where mass feeding and liquid feeding occur. The CSC reads:

$$CSC = \frac{t_V}{t_R} = \frac{t_{f_L=0.01} - t_{f_L=0.1}}{t_{f_L=0.1} - t_{f_L=0.6}} \qquad \text{Equation 1}$$

where $t_{f_L} = 0.01$ is the time when the liquid fraction, f_L, is 0.01, $t_{f_L} = 0.1$ is the time when f_L is 0.1 and $t_{f_L} = 0.6$ is the time when f_L is 0.6. The CSC values were calculated using Pandat and PanEngine thermodynamic software based on the thermodynamic database PanMg 8.0. The detailed description about the calculation of CSC can be found in the previous paper [9].

Melting and quantitative setup to characterize hot tearing

A cylindrical mild steel crucible coated with BN was used for melting in an electrical resistance furnace. Pure magnesium (99.9 wt. %), zinc (99.6 wt. %) and aluminum (99.6 wt. %) were used as staring materials. About 700g of pure magnesium was melted in the crucible under a protective gas of high pure Ar + 0.2% SF6 to minimize the formation of oxide films on the melt surface. When the temperature of melt reached about 700°C, pure zinc and aluminum were added to the melt. The melt was stirred manually for 2 minutes and then held at the pouring temperature for about 5 minutes before pouring. The pouring temperature was set at 80°C above the liquidus temperature. The melt was poured into a permanent mold with a thin layer of BN. For a certain pouring temperature for each Mg-Zn alloy, five different initial mold temperatures ranging from 200 to 450°C were selected.

To detect the initiation of hot tearing, a quantitative method based on the contraction stress measurement principle was developed [10-12]. The system consists of a constrained rod casting (CRC) mold, a contraction force measurement system with a load cell, a data logging unit and a data recording program. During solidification, the contraction of the horizontal bar is restrained by both the sprue and the flange. This restraint can cause hot tears to form. The hot tears always occurred at the junction of the sprue and the horizontal bar. Once the casting started, the force measurement system was activated. The force and temperatures of the mold at different positions and temperature at the hot spot area were recorded by computer during casting. A computer-based data acquisition system, including ALTHEN load cell (ADBBPS), was used to record the temperature and force during solidification. The force curve (force vs. time) and cooling curve (temperature vs. time) were used for analyzing the hot tearing.

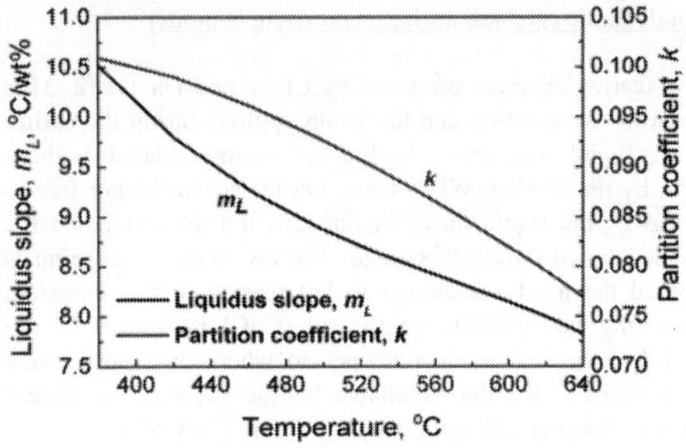

Figure 1. Calculated values m_L and k as a function of temperature.

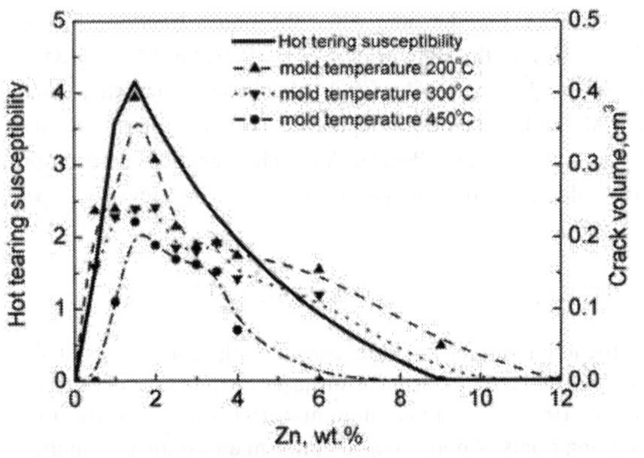

Figure 2. Comparison of hot tearing susceptibility obtained by thermodynamic calculations and crack volume by experimental measurement for Mg-Zn alloys at different mold temperatures.

Crack volume measurement

In the previous investigations, the hot crack size was normally used as the index of hot tearing susceptibility. The larger the crack size, the higher the hot tearing susceptibility. Besides that, either the total size or the width of crack was quantified instead of the crack size. In fact, due to the complexity of crack pattern and the fact that the depth of crack was not considered, the large errors become inevitable when using these methods. In order to reduce the system error and increase the precise of evaluation, the present work takes the volume of total cracks as an index to quantitatively describe hot tearing susceptibility. The detailed descriptions about the measurement of the crack volume can be found in Ref. [10-12].

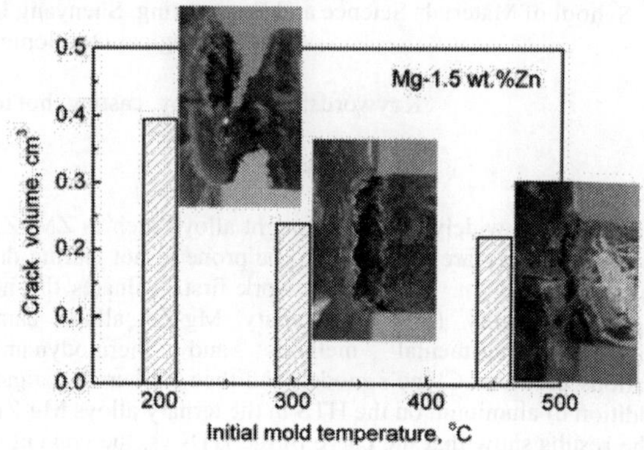

Figure 3. Influence of the initial mold temperature on the hot tearing susceptibility of Mg-1.5Zn alloy.

Results and discussion

Hot tearing susceptibility of Mg-Zn binary alloys

Influence of the Zn content on hot tearing susceptibility. In the present thermodynamic calculations, some modifications have been done for the parameters "liquidus slope m_L" and "partition coefficient k". These two parameters were considered to be constant in the previous Clyne's method. In fact, they vary with temperature, as shown in Figure 1. The liquidus slope m_L increases from 7.8 to 10.5 as the temperature decreases from 648°C (melting point of Mg) to 340°C (eutectic temperature of binary Mg-Zn alloys). The partition coefficient k also changes from 0.076 to 0.1 in the temperature range from 640°C to 340°C.

Figure 2 shows the prediction of CSC as a function of the content of Zn for the binary Mg-Zn alloys. The curve shows a typical "λ-shape". The CSC value first increases with the content of Zn, reaches the maximum at about 1.5 wt.% Zn, and then decreases again with further increasing Zn content. The measurements of crack volume for the hot tearing samples show the same tendency in the variation of hot tearing susceptibility with Zn content (Figure 2). It also indicates that the maximum hot tearing susceptibility occurs at about 1.5 wt%. Zn. Good agreement is obtained between the current thermodynamic predictions and the experimental measurements of crack volume. However, it is noticed that the present thermodynamic calculations cannot predict the effects of mold temperature on the hot tearing susceptibility (Figure 2), although they can give a satisfied result in predicting the influence of the content of Zn on the hot tearing susceptibility. Actually, with the mold temperature increasing, the hot tearing susceptibility decreases (More details are given later). To optimize the alloy composition from the considering of hot tearing susceptibility, the thermodynamic calculations based on Clyne and Davies' model would still be helpful and valuable.

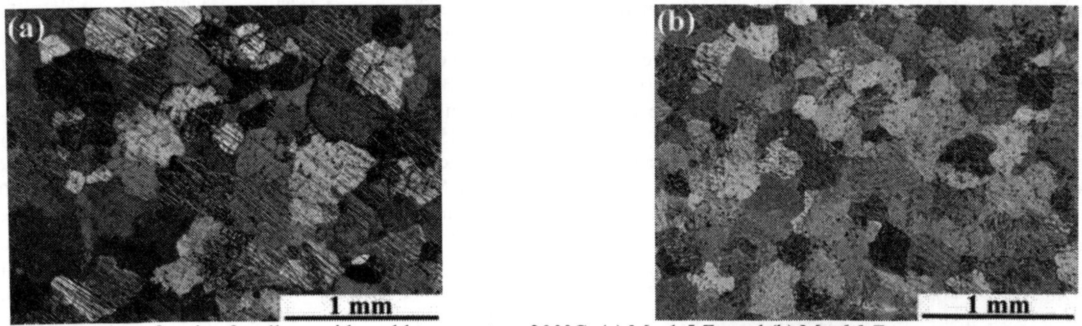

Figure 4. Morphologies of grains for alloys with mold temperature 300°C, (a) Mg-1.5 Zn and (b) Mg-6.0 Zn.

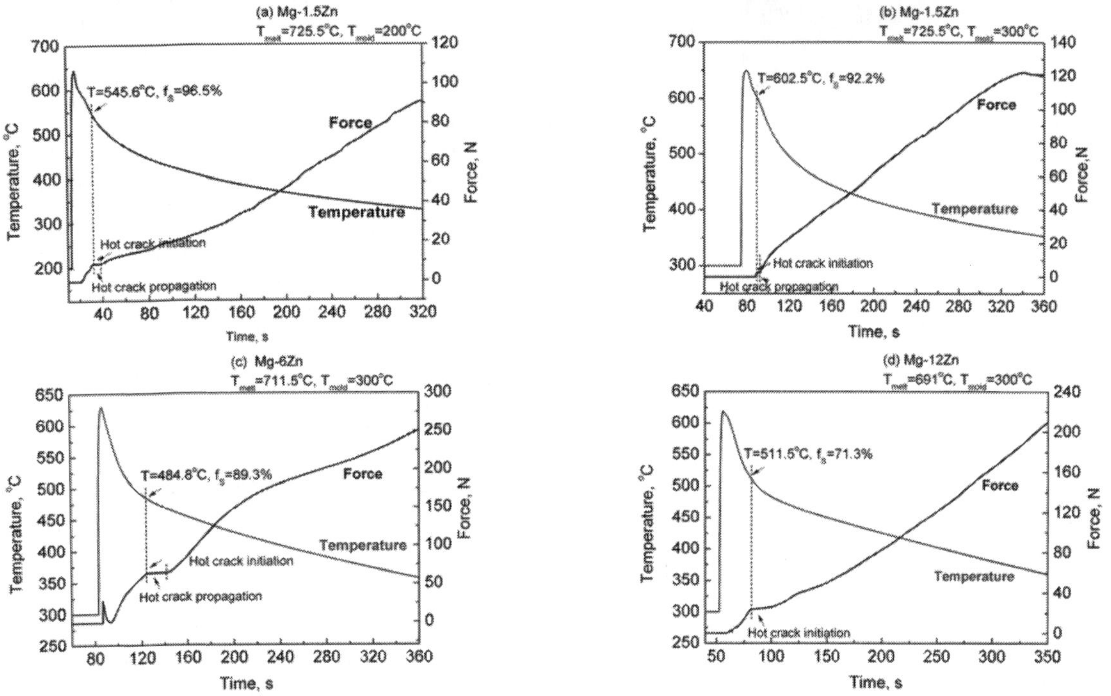

Figure 5. Typical curves of contraction force as a function of solidification time, (a) Mg-1.5Zn, mold temperature 200°C; (b) Mg-1.5Zn, mold temperature 300°C; (c) Mg-6Zn, mold temperature 300°C; and (d) Mg-12Zn, mold temperature 300°C.

Although the detailed mechanism about the effect of primary alloying elements on the hot tearing is unclear, the following explanations are more or less helpful to understand the role of primary alloying elements. The resultant microstructure such as the dendrites and the amount of eutectic phases are associated with the content of primary alloying elements. At a lower content of Zn, the amount of eutectic phases is less. With increasing Zn content, the amount of Zn-containing intermetallics increases due to the non-equilibrium solidification. This may lead increasing tendency of hot tearing when the Zn content increases. At higher Zn content, the dendrite size decreases [13]. On the other hand, the amount of low melting point eutectic phases increases, and that improves the fluidity of liquid. The refilling of the hot cracks by these liquids may proceed at the later stages of solidification, and the tendency of hot tearing is alleviated. Due to the complexity of hot tearing [4], there is still much work to do to clarify the mechanism. For example, at the later stages of solidification, the segregation of impurities at the dendritic and grain boundaries may also influence the hot tearing susceptibility [4].

Influence of mold temperature. The cooling rate significantly affects the solidification process, and has a considerable influence on the hot tearing susceptibility. It is easy to obtain different cooling rates by changing the mold temperatures. Figure 3 illustrates the crack volume in castings prepared under different initial mold temperatures from 200 to 500°C for the Mg-1.5 wt.% Zn alloy. It is clearly shown that the increment in the initial mold temperature decreases the hot tearing susceptibility. The crack volume of the sample with the initial mold temperature at 450°C is smaller than that at an initial mold temperature of 200°C. More data for the other alloys is shown in Figure 2. For each alloy, both the hot tearing susceptibility and measured crack size increase with decreasing initial mold temperature. In addition, the critical value of the initial mold temperature to suppress the occurrence of the hot tearing is different when the Zn content is different (Figure 2). For example, the crack is hardly observed when the initial mold temperature is 450°C for the Mg-6wt.% Zn alloy. With further increases in Zn content to 12 wt.%, no visible crack is observed even with an initial mold temperature at 200°C.

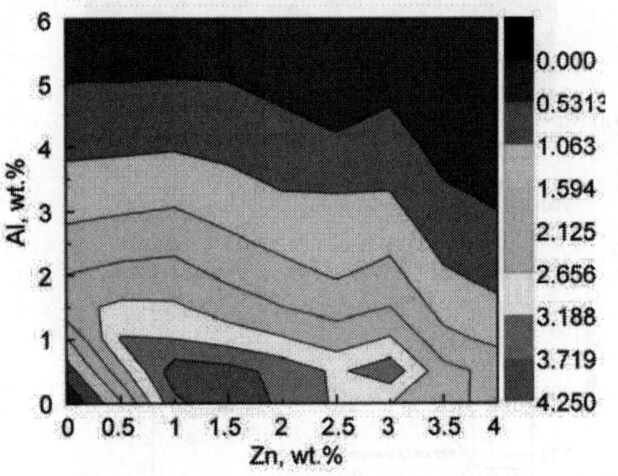

Figure 6. Influence of the content of Zn and Al on the hot tearing susceptibility of Mg-Zn-Al ternary alloys.

Different initial mold temperatures can lead to different cooling rates in the casting. Hot tearing is a defect normally formed in the hot spot areas, where the lattermost solidification takes place. For an irregularly shaped casting, a higher cooling rate will generate larger temperature gradients, thus resulting in more severe hot spots. At the same time, the larger temperature gradient leads to higher thermal stresses in the casting [4]. Therefore, the hot tearing susceptibility is higher when the cooling rate is high, i.e. the initial mold temperature is low. The fact that the critical initial mold temperature to suppress the hot tearing decreases with the increment in the content of Zn, is attributed to the following possible reasons:

(1) With increasing Zn content, the liquidus temperature of Mg-Zn alloys reduces.
(2) The fluidity of liquid improves due to the increment in the content of Zn.
(3) The size of dendrites and grains decreases when the content of Zn increases (Figure 4).

Initiation and propagation of hot cracks. Some typical experimental curves for Mg-Zn binary alloys with different initial mold temperatures are summarized in Figure 5. The analyses of these curves can satisfactory explain the hot tearing response that occurred during solidification [10-12]. The sudden drop was observed on the force curves. The peak point of the drop corresponds to the initiation of hot tearing. These peak positions are affected by both the initial mold temperature and the Zn content. As shown in Figure 5(a), the hot crack initiated at 545.6°C for Mg-1.5 wt.%Zn alloy with the initial mold temperature 200°C. This corresponds to a solid fraction of 96.5%, according to the thermodynamic calculation by Pandat software using the Scheil model. This solid fraction value is close to the well-established criterion that hot tearing normally occurs at the late stages of solidification with the solid fraction in the range from 85% to 95% [1, 14, 15]. With the initial mold temperature increasing from 200°C to 300°C, the onset temperature increases from 545.6°C to 602.5°C (Figure 5(b)). This corresponds to a solid fraction decreasing from 96.5% to 92.2%. For Mg-6 wt.% Zn and Mg-12 wt.% Zn with the mold temperature 300°C, the onset temperature is 484.8°C and 511.5°C, respectively (Figure 5(c),(d)), corresponding to a solid fraction of 89.3% and 71.3%, respectively. Therefore, with increasing Zn content, the solid fraction decreases at which the hot tearing happens during solidification. The cause is the same as that mentioned above, i.e. the higher Zn content leads to a higher amount of eutectic phases and improves the liquid fluidity. Besides the information about the initiation of hot tearing, more information such as the propagation of hot crack can also be obtained by further analyzing the experimental curves. During the propagation of a hot crack, the force releases as marked in Figure 5(a), (b) and (c) with a dashed line. In Figure 5(d), no apparent drop is observed on the curves, indicating that no visible hot cracks exist in Mg-12 wt.% Zn alloy with the initial mold temperature of 300°C. This is in good agreement with the experimental observation of hot crack (Figure 2). However, it should be noted that, although the apparent force drops are not observed, the curve slope greatly reduces after some time (see the position of the dashed line in Figure 5(d)), indicating that the accumulated force is released in some ways. This may proceed with the formation of micro-hot tears [9]. The formation of these micro-hot tears decreases the rate of force accumulation, and in the subsequent solidification they could be healed by liquid refilling because this sample has a higher Zn content and the amount of eutectic phases is higher.

Hot tearing susceptibility of Mg-Zn-Al ternary alloys

Thermodynamic calculations. The contour plot of hot tearing susceptibility for Mg-Zn-Al ternary alloys with thermodynamic calculation is shown in Figure 6. The maximum hot tearing susceptibility occurs at the composition of around 1.5 wt.% Zn and <0.5 wt.%Al. The data from the two sections in this contour plot are compared with the experimental data in Figure 7. The hot tearing susceptibility at the section with a fixed content of Zn (1.5 wt.%), i.e. the Mg-1.5Zn-yAl (y=0.5,3,6) alloys, decreases with increasing Al content (Figure 7 (a)). Both the thermodynamic calculations and experimental measurement of crack volume show that the hot tearing susceptibility decreases when the content of Al increases in Mg-1.5 Zn-y Al ternary alloys. This indicates that the thermodynamic calculations on the hot tearing susceptibility based on Clyne and Davies' model are also suitable for the composition design of ternary alloy system from the consideration of castability. When the content of Al is fixed at 0.5 wt.% Al, i.e. the Mg-xZn-0.5Al (x=0.5, 1.5, 2.5, 4) alloys, good agreement for the hot tearing susceptibility as a function of Zn content is also observed between the thermodynamic calculations and experimental measurements (Figure 7(b)). As shown in Figure 7(b), in the investigated range of Zn content two peaks are observed: one is at about 1.0 to 1.5 wt.% Zn, another at about 3.0 wt.%.

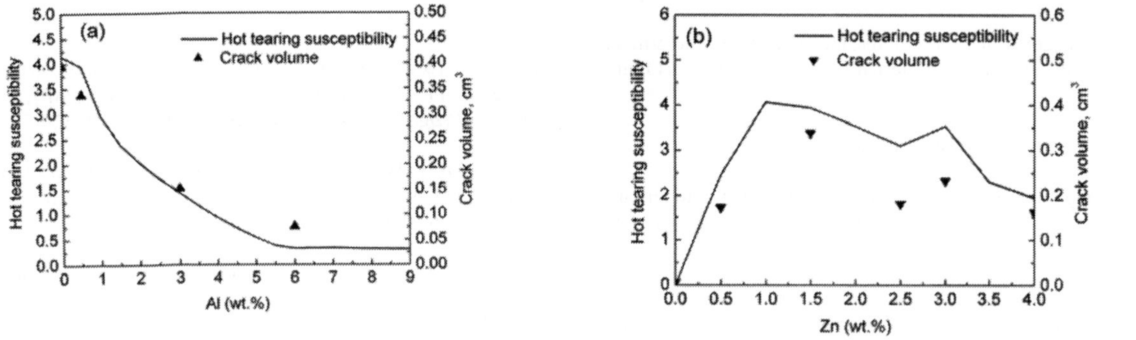

Figure 7. Comparison of the hot tearing susceptibility obtained by thermodynamic calculations and the crack volume by experimental measurement which is proportional to the hot tearing susceptibility for Mg-Zn-Al ternary system, (a) with a fixed value of 1.5 wt.% Zn; and (b) with a fixed value of 0.5 wt.% Al.

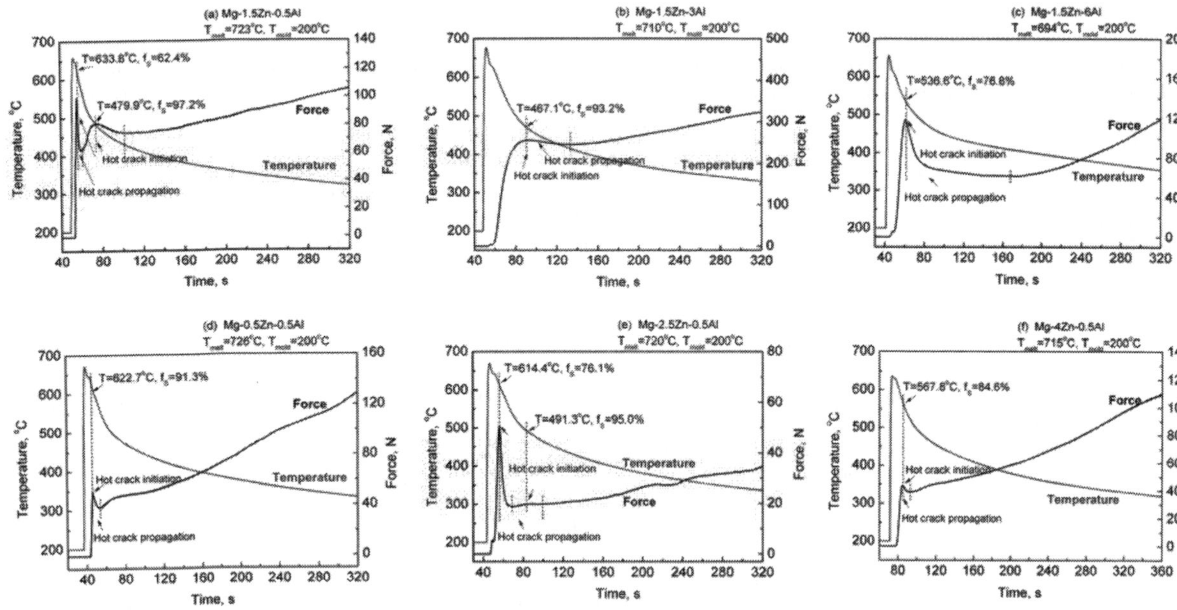

Figure 8. Typical curves of contraction force as a function of solidification time at mold temperature 200°C, (a) Mg-1.5Zn-0.5Al; (b) Mg-1.5Zn-3Al; (c) Mg-1.5Zn-6Al; (d) Mg-0.5Zn-0.5Al; (e) Mg-2.5Zn-0.5Al and (f) Mg-4.0 Zn-0.5Al.

Figure 8 shows typical curves of contraction forces as a function of solidification time for Mg-1.5Zn-yAl and Mg-xZn-0.5Al alloys at a mold temperature of 200°C. A clear drop in the contraction force is observed in all curves, indicating that the hot tears form in all these samples. For Mg-1.5Zn-0.5Al alloy, two peaks exist on the curve (Figure 8(a)). The first peak occurs at temperature 633.8°C which corresponds to a solid fraction of 62.4%. The second peak is located at 479.9°C which corresponds to a solid fraction of 97.2%. This sample contains 1.5 wt.% Zn and 0.5 wt.% Al. The investigations of the binary Mg-Zn alloys (Figure 2) and Mg-Al alloys [11] illustrate that when the content of Zn is 1.5 wt.% and Al 0.5 wt.% the hot tearing susceptibility is very large. The formation of two peaks on the contraction force curve may also depict that this alloy has a large hot tearing susceptibility. When the first hot tear occurs at 633.8°C, the solid fraction is relatively low with a value of 62.4%. A large amount of liquid still remains and the formed dendritic network does not so densely contact each other. The liquid still has a chance to flow into the first formed hot tears and repair them. With the solidification proceeding, the force increases again. When it reaches a critical value, new hot tearing occurs and the force drops again (second peak in Figure 8(a)). The same situation is also observed for alloy Mg-2.5Zn-0.5Al, indicating that this alloy has a large hot tearing tendency.

As observed for the binary Mg-Zn alloys, the initiation of hot tears and their propagation can also be reflected by the curves of contraction force for the ternary Mg-Zn-Al alloys (Figure 8). Based on the peak position and its corresponding temperature, the critical solid fraction at which the hot tear occurs can be calculated with the thermodynamic calculations. The hot tearing occurs at a higher sold fraction similar to that observed for the binary Mg-Zn alloys, normally more than 90%. Due to the complexity of hot tearing, at this moment it is not easy to explain the recorded curves completely and clearly both for the binary and ternary alloys. More investigations are needed in the future. In spite of this, it seems that with increasing Al content, the critical solid fraction decreases for the ternary Mg-Zn-Al alloys (Figure 8).

Conclusions

The hot tearing susceptibility of binary Mg-Zn system and ternary Mg-Zn-Al system has been investigated by thermodynamic calculations and experimental methods. The conclusions are summarized as follow:

(1) Both the measurements of hot crack volume and thermodynamic calculations show that the hot tearing susceptibility as a function of Zn content follows the "λ" shape. The HTS first increases with Zn content, reaches the maximum at about 1.5 wt.% Zn and then decreases with further increasing the Zn content. The thermodynamic calculations on HTS, based on Clyne and Davies' model, are helpful and valuable for the optimal design of compositions for magnesium alloys from the considerations of hot tearing susceptibility. However, this model cannot predict the influence of other parameters on HTS, such as the initial mold temperature.

(2) With increasing the initial mold temperature, the hot tearing susceptibility decreases, because the higher initial mold temperature results in a lower cooling rate.

(3) In the Mg-Zn-Al ternary system, the HTS decreases with increasing Al content. The curve of the HTS as a function of Zn content in the ternary Mg-Zn-Al system is a little different from that observed in the binary Mg-Zn alloys. Two peaks are obtained: one is at about 1.0 to 1.5 wt.% Zn, another at about 3.0 wt.%.

(4) The recorded contraction force-temperature-time curves can supply some useful information about the initiation of hot tears, the corresponding solid fraction, propagation of hot tears, and load release by the formation of macrocracks or microcracks.

Acknowledgements

The authors would like to thank Mr. W. Punessen and Mr. G. Meister for their technical assistance on setting up the equipment and performing the casting experiments. One of the authors (Le Zhou) also gives thanks to Helmholtz-CSC Scholarship for supporting her stay in the GKSS Research Centre.

References

1. J. Campbell, *Castings* (Oxford: Butterworth-Heinemann, 2003), 205-231.

2. T.W. Clyne, and G.J. Davies, "Comparision between Experimental Data and Theoretical Predictions Relating to Dependence of Solidification Cracking on Composition", Solidification and casting of metals, edited by Sheffield, UK, 1977, 275-278.

3. T.W. Clyne, and G.J. Davies, "The Influence of Composition on Solidification Cracking in Binary Alloy Systems", *Brit. Foundryman*, 74 (1981), 65-73.

4. D.G. Eskin, Suyitno, and L. Katgerman, "Mechanical Properties in the Semi-Solid State and Hot Tearing of Aluminium Alloys", *Progress in Materials Science*, 49 (2004), 629-711.

5. G. Cao, I. Haygood, and S. Kou, "Onset of Hot Tearing in Ternary Mg-Al-Sr Alloy Castings", *Metallurgical and Materials Transactions A*, 41 (2010), 2139-2150.

6. G. Cao, and S. Kou, "Hot Tearing of Ternary Mg-Al-Ca Alloy Castings", *Metallurgical and Materials Transactions*, 37A (2006), 3647-3663.

7. Y.S. Wang, Q.D. Wang, C.J. Ma, W.J. Ding, and Y.P. Zhu, "Effects of Zn and RE Additions on the Solidification Behavior of Mg-9Al Magnesium Alloy", *Materials Science and Engineering A*, 342 (2003), 178-182.

8. W. Zheng, S. Li, B. Tang, D. Zeng, and X. Guo, "Effect of Rare Earths on Hot Cracking Resistant Property of Mg-Al Alloys", *Journal of Rare Earths*, 24 (2006), 346-351.

9. L. Zhou, Y.D. Huang, P.L. Mao, K.U. Kainer, Z. Liu, and N. Hort, "Influence of Composition on Hot Tearing in Binary Mg-Zn Alloys", Proceedings of the 8th Pacific Rim International Conference on Modeling of Casting and Solidification Processes, edited by J. K. Choi, H. Y. Hwang and J. T. Kim, Incheon, Korea, 2010, 97-104.

10. Z.S. Zhen, N. Hort, Y.D. Huang, N. Petri, O. Utke, and K.U. Kainer, "Quantitative Determination on Hot Tearing in Mg-Al Binary Alloys", *Light Metals Technology 2009*, 618-619 (2009), 533-540.

11. Z.S. Zhen, N. Hort, Y.D. Huang, O. Utke, N. Petri, and K.U. Kainer, "Hot Tearing Behaviour of Binary Mg-1Al Alloy Using a Contraction Force Measuring Method", *International Journal of Cast Metals Research*, 22 (2009), 331-334.

12. Z.S. Zhen, N. Hort, O. Utke, Y.D. Huang, N. Petri, and K.U. Kainer, "Investigations on Hot Tearing of Mg-Al Binary Alloys by Using a New Quantitative Method", *Magnesium Technology 2009*, (2009), 105-110.

13. D.H. StJohn, M. Qian, M.A. Easton, P. Cao, and Z. Hildebrand, "Grain Refinement of Magnesium Alloys", *Metallurgical and Materials Transactions*, 36A (2005), 1669-1679.

14. D. Eskin, and L. Katgerman, "A Quest for a New Hot Tearing Criterion", *Metallurgical and Materials Transactions A*, 38 (2007), 1511-1519.

15. Suyitno, V. Savran, L. Katgerman, and D. Eskin, "Effects of Alloy Composition and Casting Speed on Structure Formation and Hot Tearing During Direct-Chill Casting of Al-Cu Alloys", *Metallurgical and Materials Transactions A*, 35 (2004), 3551-3561.

Magnesium Technology 2011
Edited by: Wim H. Sillekens, Sean R. Agnew, Neale R. Neelameggham, and Suveen N. Mathaudhu
TMS (The Minerals, Metals & Materials Society), 2011

PROPORTIONAL STRENGTH-DUCTILITY RELATIONSHIP OF NON-SF6 DIECAST AZ91D ECO-MG ALLOYS

Shae K. Kim
Casting Research Center, Korea Institute of Industrial Technology,
7-47 Songdo-dong, Yeonsu-gu, Incheon 406-840, Korea

Keywords: Mg alloys, Diecasting, CaO, Non-SF_6, Strength, Ductility

Abstract

SF_6 gas has been generally used for Mg alloys during melting and casting as a cover gas. The use of SF_6 gas, however, will be limited owing to its significant impact on global warming. Non-SF_6 process during melting and casting in diecasting industry has been proved with Eco-Mg alloys by a simple addition of small amount of CaO into AZ91D and AM60B Mg alloys. This paper will show non-SF_6 diecasting procedures for AZ91D Eco-Mg alloys. Cold-chamber and hot-chamber diecasting mass productions were performed by using a Toshiba 135-ton cold chamber and a Frech 200-ton hot chamber diecasting machines under CO_2 atmosphere without SF_6 gas. An emphasis will be on proportional strength and ductility relationship of Eco-Mg alloys, in part, due to a high-quality melt, refined grain size and Al_2Ca second phase strengthening. Microstructures and mechanical properties of AZ91D Eco-Mg alloys will be evaluated in comparison with those of conventional AZ91D Mg alloy.

Introduction

Molten Mg alloys react with ambient atmosphere without the melt protective gases during melting and casting process due to their chemical unstability [1]. Oxides and inclusions formed by this reaction decrease the quality of melt and safety of work. SF_6 gas is generally used for Mg alloys during melting and casting as a cover gas and has proved to be a successful inhibitor. The use of SF_6 gas, however, will be limited owing to its significant impact on global warming. SF_6 gas has recognized as a very potent greenhouse gas with a high global warming potential (GWP) value of 23,900 compared with CO_2 and has a long atmospheric lifetime of 3,200 years [2]. For the expansion of Mg diecasting industry, non-SF_6 process is necessary. The one simple and effective way of non-SF_6 process is to improve oxidation resistance of Mg alloys without change of original Mg alloys properties. The automotive industry accounts for 90% of the casting demand, and diecasting is the dominant commercial fabrication process. The high-pressure die-cast (HPDC) Mg-alloys contain a considerable amount of casting defects such as oxides, inclusions and micro-porosity which adversely affect the mechanical properties of cast Mg alloys [3-5]. Although the SF_6 gas is used to protect the melt from oxidation in diecasting process, the surface of Mg alloys melts is only protected from oxidation in SF_6 gas affected zone such as melting furnace. The diecasting defects, in spite of using SF_6 gas, still exist and decrease the quality of diecast Mg alloys. So, the mechanical properties of diecast Mg alloys will be able to increase by improvement of oxidation resistance of Mg alloys.

The total vehicle weight can be reduced by using magnesium alloy components. The automotive part that may be subjected to deformation during crash must withstand a certain degree of deformation and absorb a certain amount of energy without fracture for safety. AZ91D alloys are the most widely used diecasting magnesium alloy in automotive applications due to their excellent castability and good room-temperature mechanical properties. But AZ91D alloys have poor elongation due to presence of brittle β-$Mg_{17}Al_{12}$. The enhancement of the ductility and strength of Mg alloys by alloying elements is very difficult but required for increased application of Mg alloys in automotive parts. Recently, the non-SF_6 process during melting and casting in diecasting industry and enhancement of the ductility and strength of Mg alloys have been proved with Eco-Mg alloys, a simple addition of small amount of CaO into AZ91D and AM60B Mg alloys. CaO added into the Mg melt does not exist as CaO itself in the solidified state. By the reactive phase formation, Mg_2Ca (C14) in pure Mg and Al_2Ca (C15) or $(Mg, Al)_2Ca$ (C36) in Al-bearing Mg alloys are formed only in the grain or sub-grain boundary region. The distinguished difference is that CaO does not form α-phase (solid solution phase) in pure Mg regardless of content and process conditions. The effect of CaO on Mg alloys has been already demonstrated in previous studies. Eco-Mg is characterized by (1) non-SF_6 processing, (2) Be elimination, (3) improved melt cleanliness, (4) ensuring original process abilities for casting, forming, joining as well as surface treatment, (5) improved mechanical properties by grain refinement and internal soundness, (6) ensured safety during manufacturing and application by raising oxidation and ignition resistances, (7) improved recyclability, and (8) cost reduction [8-14].

In this paper, the effect of CaO addition to AZ91D Mg alloy on non-SF_6 diecasting procedures and tensile properties has been investigated, with focus on proportional strength and ductility relationship. Cold-chamber and hot-chamber diecasting mass productions were performed by using a Toshiba 135-ton cold chamber and a Frech 200-ton hot chamber diecasting machines under CO_2 atmosphere without SF_6 gas. Microstructures and mechanical properties of AZ91D Eco-Mg alloys will be evaluated in comparison with those of conventional AZ91D Mg alloy.

Experimental Procedures

For diecasting procedure, AZ91D Mg alloy ingot was melted under SF_6 and CO_2 gases; on the other hand, The pre-made 0.3wt.%CaO added AZ91D Mg alloys and 0.7wt.%CaO added AZ91D Mg alloys ingot was melted respectively under CO_2 atmosphere without SF_6 gas in a steel crucible heated to 720°C. The diecasting was performed by using a Toshiba 135-ton cold chamber and a Frech 200-ton hot chamber diecasting machines as shown in Fig. 1. The metallic mold was designed for cellular phone case and preheated at 200°C. The metallic mold for tensile test specimen was prepared specifically. Fig. 2 (a) and Fig. 2 (b) show the hot chamber diecasting process for the cell phone casing and tensile test specimen under nitrogen atmosphere without protective gas, respectively. The temperature of melt was 690°C. Chemical composition analysis was performed by ICP-mass spectrometer. In order to observe the change of the microstructures by CaO addition, the cast billets were polished and observed using an Olympus PME3 microscopy. Tensile test was carried out at room temperature. Cross-head movement speed was determined as 1 mm/min. The obtained values of tensile properties were based on the average of 3 times. Tensile specimen had a gauge length of 30mm and a diameter of 6mm as shown in Fig. 7.

Figure 1. Photograph of Frech 200-ton hot chamber diecasting machine.

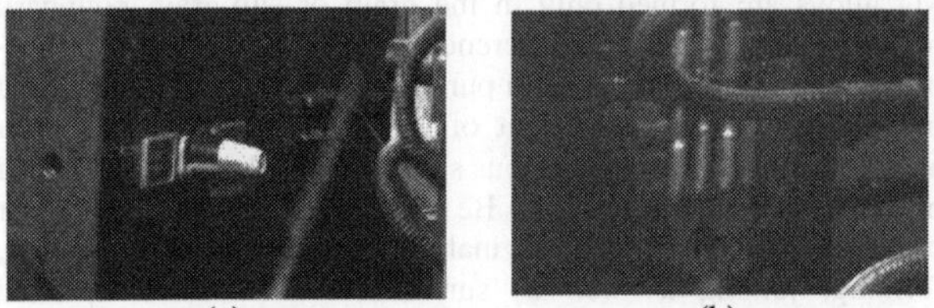

(a) (b)

Figure 2. Diecastings of CaO added AZ91D Mg alloy under nitrogen atmosphere without protective gas; (a) cell phone case and (b) tensile test specimen.

Results and Discussions

The results of tests to demonstrate the effect of small amount of CaO on non-SF_6 processing for AZ91D Mg alloys are shown in Fig. 3. The each melt, under the same melting conditions, is poured into large circular mold though slide for vigorous reaction with ambient atmosphere. Fig. 3 (a) shows the surfaces of AZ91D Mg alloy melt during casting and solidification under an ambient atmosphere without protective gas. Terrible oxidation began and the melt surface became tarnished in 10 seconds just after being poured. Severe burning started in 30 seconds and continued even with white flame to the point until the melt was consumed by burning. It is well known that continuous oxidation and resultant burning of Mg alloy at high temperature over 500℃ is due to the porous surface MgO oxide films that do not act as a protective layer to prevent further oxidation and burning. The experiment with 0.1wt.%CaO added AZ91D Mg alloy was carried out to investigated the minimum amount of CaO for non-SF_6 processing. The oxidation and burning occurred at melt surface a little. The result showed that 0.1wt.%CaO addition was not enough to be processed in an ambient atmosphere without protective gas. Fig. 3 (c) shows the change of surfaces of 0.3wt.%CaO added AZ91D Mg alloy during casting and solidification under an ambient atmosphere without a protective gas. No oxidation and burning appeared and shiny surfaces were maintained until the solidification completely finished. The small addition of 0.3wt.%CaO can prevent the oxidation and burning of AZ91D Mg alloys melt without any protective gas.

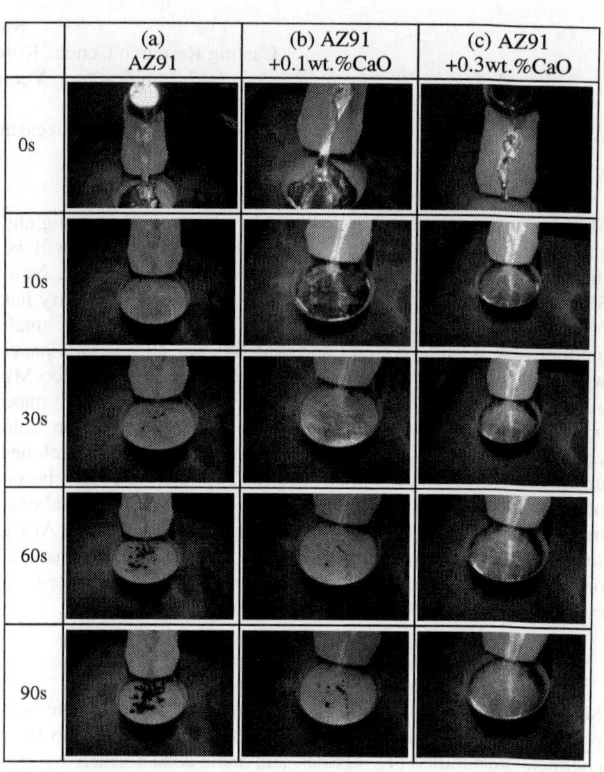

Figure 3. The melt surface change of the (a) AZ91, (b) AZ91+0.1wt.%CaO, and (c) AZ91D+0.3wt.%CaO alloys during casting and solidification under an ambient atmosphere without protective gas.

Fig. 4 shows the non-SF_6 diecasting process for 0.3wt.%CaO added AZ91D under nitrogen atmosphere without protective gas in the furnace. Fig. 4 (b) shows clean surface of 0.3wt.%CaO added AZ91D Mg alloy melt. There is no visible oxidation and burning on melt surface. The result showed that 0.3wt.%CaO was enough to be processed in non-SF_6 atmosphere.

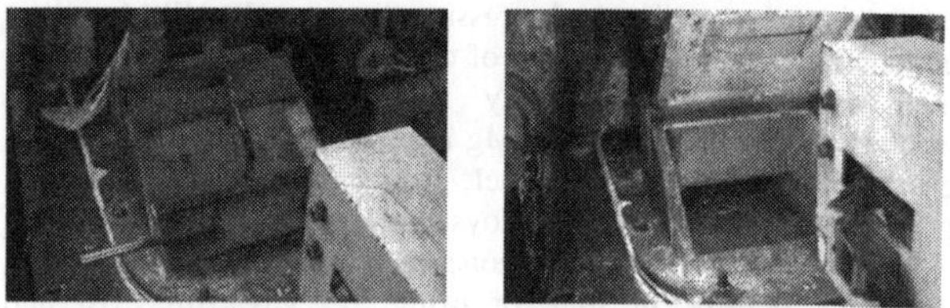

(a) (b)

Figure 4. Non-SF_6 diecasting furnace for 0.3wt.%CaO added AZ91D Mg alloy under nitrogen atmosphere without protective gas; (a) closed furnace door and (b) melt surface in furnace.

Fig. 5 explains the reason of the non-SF_6 process of Eco-Mg schematically. The surface oxide film of Eco-Mg is denser than that of current Mg alloys because it consists of oxides mixture of MgO and CaO. The CaO existing in surface layer is not CaO added into Mg melt itself. That is reproduced CaO after reduction in Mg melt. Due to the dense complex surface oxide film, the casting and solidification of Mg alloys is possible without any protective gas.

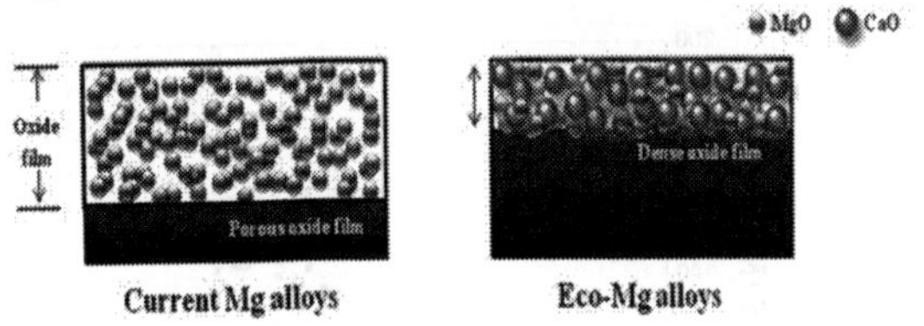

(a) (b)

Figure 5. Schematic illustration of surface oxide film for (a) current Mg alloys and (b) Eco-Mg alloys.

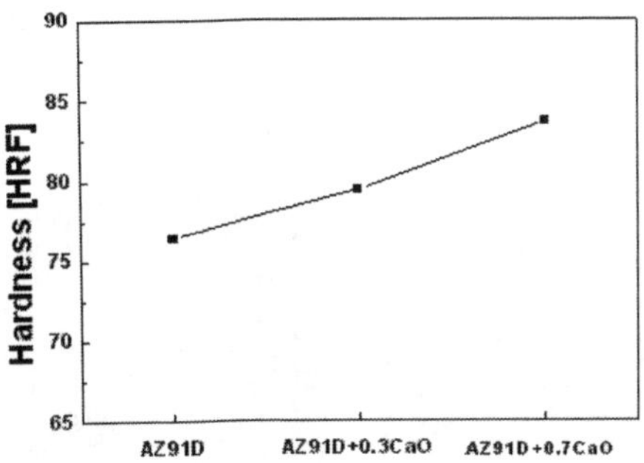

Figure 6. Rockwell hardness values of cellular phone case produced by cold chamber diecasting of AZ91D under SF_6 and CaO added AZ91D Mg alloys without SF_6 gas.

Fig. 6 shows the hardness values of cellular phone case produced by cold chamber diecasting AZ91D under SF6, 0.3wt.%CaO added AZ91D Mg alloys without SF6 gas and 0.7wt.%CaO added AZ91D Mg alloys without SF6 gas. Hardness tests were conducted using the Rockwell hardness tester. The Rockwell hardness values of AZ91D, 0.3wt.%CaO added AZ91D Mg alloys and 0.7wt.%CaO added AZ91D Mg alloys were 76.4, 79.4 and 83.6, respectively. The Rockwell hardness value of 0.3wt.%CaO added AZ91D Mg alloy was little higher than that of AZ91D Mg alloy. The Rockwell hardness value of 0.7wt.%CaO added AZ91D Mg alloy was highest.

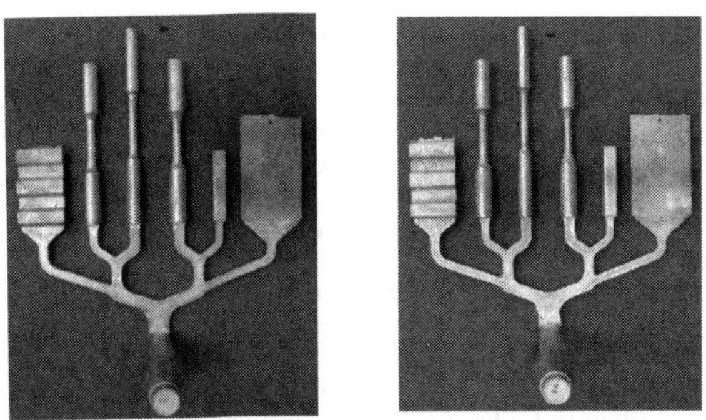

(a) (b)

Figure 7. Tensile test specimen produced by hot chamber diecasting of (a) AZ91D under SF_6 gas and (b) CaO added AZ91D without SF_6 gas

Fig. 7 (a) shows the tensile test specimen produced by hot chamber diecasting of AZ91D Mg alloy under SF_6 and CO_2 protective gas. Fig. 7 (b) shows the mechanical property specimen produced by hot chamber diecasting of CaO added AZ91D Mg alloy under only nitrogen atmospheres without protective gas.

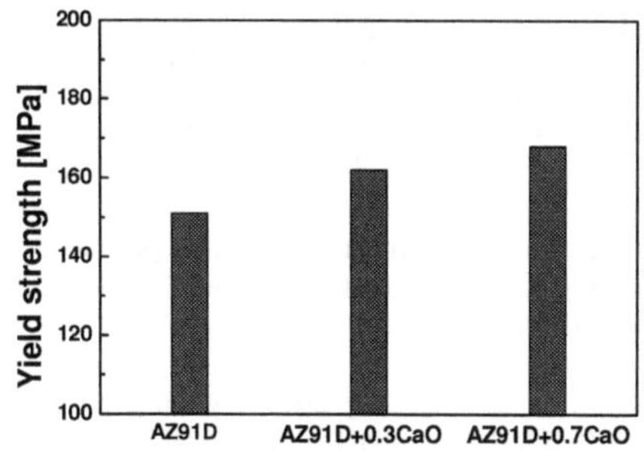

Figure 8. Yield strength of tensile test specimen produced hot-chamber diecasting of AZ91D under SF_6 gas and CaO added AZ91D Mg alloys without SF_6 gas.

Fig. 8 shows the yield strength of tensile test specimen produced by hot chamber diecasting of AZ91D under SF_6 and CO_2 protective gas and CaO added AZ91D Mg alloys without SF_6 gas. The yield strength of AZ91D, 0.3wt.%CaO added AZ91D Mg alloys and 0.7wt.%CaO added AZ91D Mg alloys were 151 MPa, 162 MPa and 168 MPa respectively. The yield strength of CaO added AZ91D Mg alloy, in spite of non-SF_6 process, was higher than that of AZ91D Mg alloy.

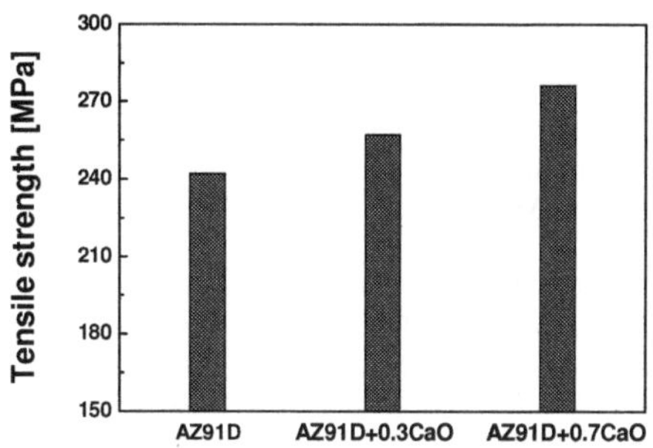

Figure 9. Tensile strength of tensile test specimen produced by hot-chamber diecasting of AZ91D under SF_6 and CaO added AZ91D Mg alloys without SF_6 gas.

Fig. 9 shows the tensile strength of tensile test specimen produced by hot chamber diecasting of AZ91D under SF_6 and CO_2 protective gas and CaO added AZ91D Mg alloys without SF_6 gas. The tensile strength of AZ91D, 0.3wt.%CaO added AZ91D Mg alloys and 0.7wt.%CaO added AZ91D Mg alloys were 242 MPa, 257 MPa and 276 MPa, respectively. The tensile strength of CaO added AZ91D Mg alloy was higher than that of AZ91D Mg alloy.

Fig. 10 shows the elongation of tensile test specimen produced by hot chamber diecasting of AZ91D under SF_6 and CO_2 protective

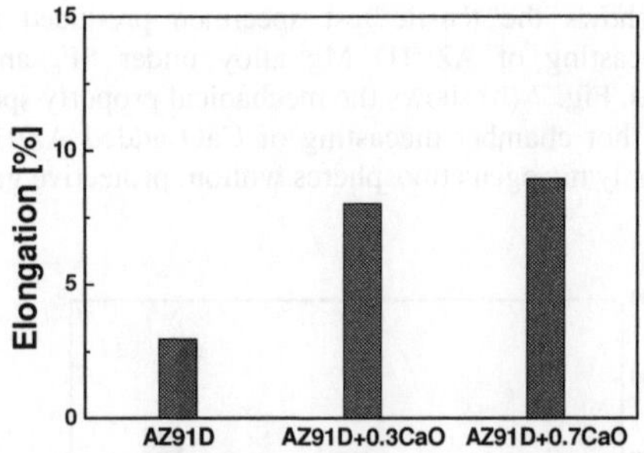

Figure 10. Elongation of tensile test specimen produced by hot-chamber diecasting of AZ91D under SF_6 and CaO added AZ91D Mg alloys without SF_6 gas.

gas and CaO added AZ91D Mg alloys without SF_6 gas. The elongation of AZ91D, 0.3wt.%CaO added AZ91D Mg alloys and 0.7wt.%CaO added AZ91D Mg alloys were 2.96 %, 8 % and 8.91 %, respectively. The elongation of CaO added AZ91D Mg alloy, in spite of non-SF_6 process, was much higher than that of AZ91D Mg alloy.

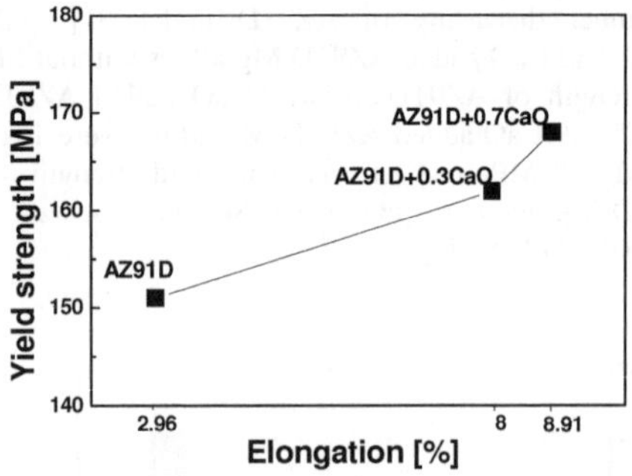

Figure 11. Proportional strength-ductility relationship of non- SF_6 Diecast AZ91D Eco-Mg Alloys.

This leads to a conclusion that additions of CaO to the AZ91D Mg alloys, in spite of non-SF_6 process, improve the ductility as well as UTS and TYS. It is usually experienced that optimizing one property by alloying reaches the expense of one or more other properties. The relationship between yield strength and ductility is a typical example. In order to increase yield strength significantly, some ductility has to be sacrificed. However, tensile yield strength and ductility are all increased by CaO addition. The proportional strength-ductility relation results from high-quality melt, refined grain size and Al_2Ca second phase strengthening. This is illustrated in Fig. 11. Fig. 12 shows another example of the proportional strength-ductility relationship of non-SF_6 diecast Eco-Mg alloys. As the CaO content added into Mg alloys increases from 0.25 to 0.65 wt.% the strength and ductility of Mg alloys improve at the same time proportionally. The strength - ductility relationship of conventional Mg-Al alloys such as AM and AZ series Mg alloys, however, is in inversely proportional with joining Al contents [15].

The alloying element Al increases the strength and castability of Mg alloys but the ductility decreases sharply due to brittle β-$Mg_{17}Al_{12}$ phase formation. In order to significantly improve the

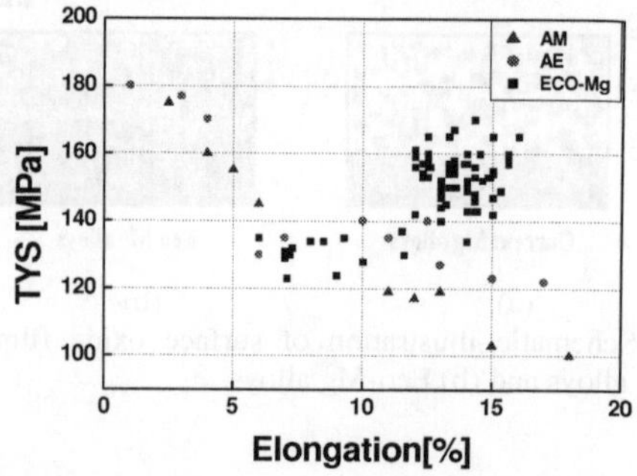

Figure 12. Relationship of strength and elongation of Eco-Mg alloys, compared with those of conventional Mg-Al alloys.

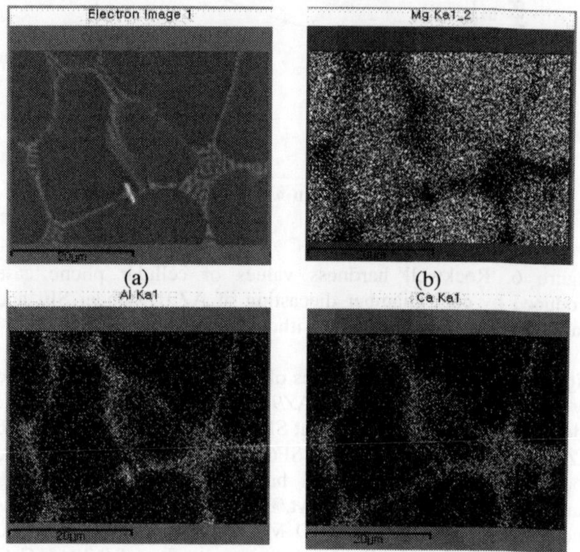

(a) (b) (c) (d)

Figure 13. Comparison of elemental mapping images of (a) BSE image, (b) Mg, (c) Al, and (d) Ca by EDS for CaO added AZ91D Mg alloy .

ductility of Mg-Al based diecasting Mg alloys with high Al contents, the brittle effect of the β-$Mg_{17}Al_{12}$ phase needs to be reduced.

Fig. 13 shows comparison of elemental mapping images of (b) Mg, (c) Al, and (d) Ca by EDS for CaO added AZ91D Mg alloy. Fig. 13 (a) shows BSE (back scattered electron) image for CaO added AZ91D Mg alloy. Ca was distributed on the grain boundary. But O was not detected. CaO did not exist as CaO itself in the solidified state. Ca reduced from CaO forms intermetallic compound with Al. By reactive phase formation, Al_2Ca phase was formed mostly in the grain or sub-grain boundary region. The distinguished difference is that CaO did not forms solid solution phase. Although Ca has a strong segregation tendency, Ca added Mg alloys solidified first as solid solution phase and then as divorce eutectic phase. The brittle β-$Mg_{17}Al_{12}$ phase formation grows down due to Al_2Ca phase formation. The Al_2Ca phase has a lamellar structure with a relatively large extension from the grain boundaries. The formation of Al_2Ca suppresses the formation of brittle β-$Mg_{17}Al_{12}$ by binding up Al. The effect of second phase strenghening increases due to Al_2Ca phase in the grain or sub-grain boundary region. Also, the strength is improved due to grain refinement by Al_2Ca phase formation. So, the ductility and strength are enhanced at the same time by CaO addition.

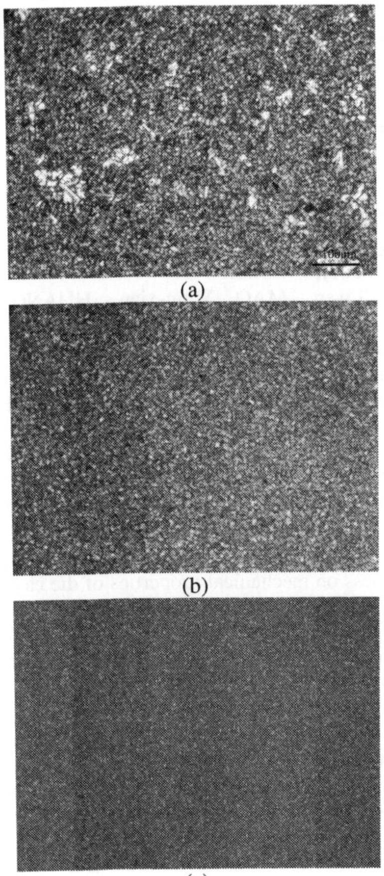

(a) (b) (c)

Figure 14. Microstructures of a cell phone case produced by cold chamber diecasting of (a) AZ91D under SF_6, (b) 0.3wt.%CaO added AZ91D without SF_6 gas and (c) 0.7wt.%CaO added AZ91D without SF_6 gas.

Fig. 14 shows the microstructures of a cell phone case produced by cold chamber diecasting of AZ91D under SF_6 gas, 0.3wt.%CaO added AZ91D Mg alloys without SF_6 gas and 0.7wt.%CaO added AZ91D Mg alloys without SF_6 gas. The microstructure of AZ91D has porosity and coarse grain as shown in Fig. 14 (a). The phase distribution of AZ91D is irregular. Figure 14 (b) shows the microstructure of 0.3wt.%CaO added AZ91D Mg alloy. The grain of 0.3wt.%CaO added AZ91D Mg alloy is refined with little porosity in spite of non-SF_6 process. This tendency is more prominent in 0.7wt.%CaO added AZ91D Mg alloy as seen in Fig. 14 (c). The microstructure of Eco-Mg alloys is refined by Al_2Ca phase formation by CaO addition. The Al_2Ca phase is distributed uniformly. The grain refinement and regular distribution of phase contribute to simultaneous enhancement of ductility and strength of AZ91D Mg alloys.

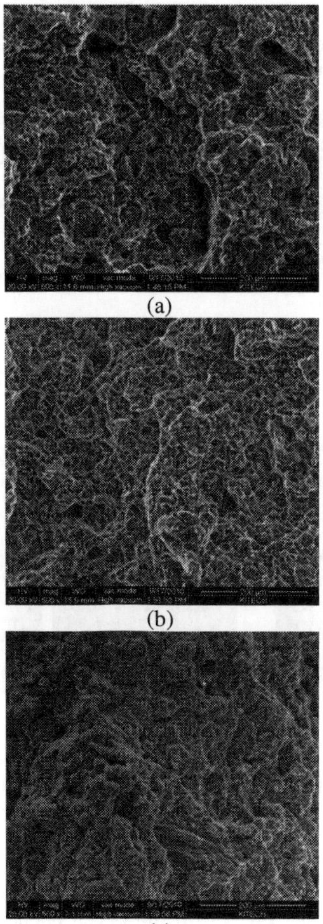

(a) (b) (c)

Figure 15. SEM images of the fracture surface of a tensile test specimen of (a) AZ91D under SF_6, (b) 0.3wt.%CaO added AZ91D without SF_6 gas and (c) 0.7wt.%CaO added AZ91D without SF_6 gas.

There is a close correlation between the tensile ductility and quality of castings such as the area fraction of the porosity and inclusions. The fracture beginning preferentially goes through the regions containing concentrated pores or inclusions [16].

Fig. 15 shows the fracture surfaces of tensile bars of hot chamber diecasting of AZ91D under SF_6 gas and CaO added AZ91D Mg alloys without SF_6 gas. In Fig. 15 (a) corresponding to brittle AZ91D alloys, the cleavage planes are prominently observed inside the grains that also contain evidence of void formation. In CaO added AZ91D Mg alloys samples, on the other hand, cleavage is rarely seen. The dimples on the fracture surface of the CaO added AZ91D Mg alloys sample are demarcated by intense slip at the boundary region. The area fraction of porosity in the fracture surface is high in brittle AZ91D alloys. But in

0.7wt.%CaO added AZ91D Mg alloys, it decreased sharply as shown in Fig. 15 (c).

Fig. 16 (a) shows the small cell phone case produced by hot chamber diecasting of AZ91D Mg alloy under SF_6 and CO_2 atmospheres. Fig. 16 (b) shows the cell phone case produced by hot chamber diecasting of 0.3wt.%CaO added AZ91D Mg alloy under nitrogen atmospheres without protective gas.

(a) (b)

Figure 16. Photograph of a cell phone case produced by hot chamber diecasting of (a) AZ91D under SF_6 and (b) 0.3wt.%CaO added AZ91D without SF_6 gas.

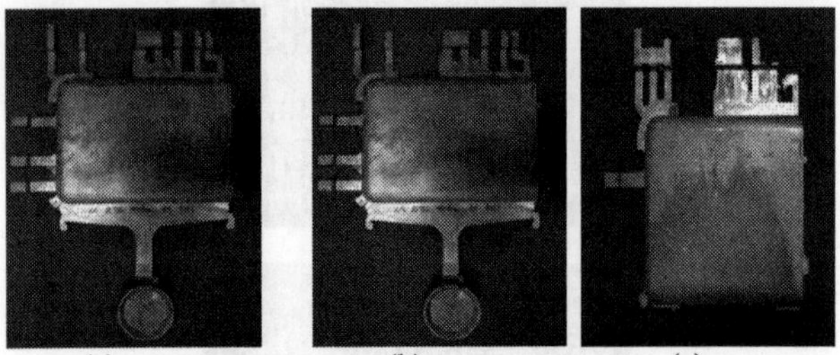

(a) (b) (c)

Figure 17. Photograph of cell phone case produced by cold chamber diecasting of (a) AZ91D under SF6 , (b) 0.3wt.%CaO added AZ91D without SF6 gas and (c) 0.7wt.%CaO added AZ91D.

Fig. 17 shows another larger cell phone cases produced by cold chamber diecasting. Fig. 17 (a) shows cell phone cases produced by cold chamber diecasting of AZ91D under SF_6 and Fig. 17 (b) and (c) show cell phone cases produced by cold chamber diecasting of 0.3wt.%CaO added AZ91D and 0.7wt.%CaO added AZ91D without SF6 gas, respectively. In spite of non-SF_6 process, CaO added AZ91D are well produced by hot chamber diecasting and cold chamber diecasting. It means that non-SF_6 CaO added AZ91D has good diecastability. A possible explanation of the good diecastability of Eco-Mg alloys can be the suppression of the formation of β-$Mg_{17}Al_{12}$, combined with reduced casting defects like inclusions and porosity.

Conclusions

Without SF_6 gas, the 0.3wt.%CaO added AZ91D Mg alloy and 0.7wt.%CaO added AZ91D Mg alloy could be well manufactured by hot chamber diecasting and cold chamber diecasting. This is because oxidation resistance of Mg alloys is improved by dense complex surface oxide film which consists of oxides mixture of MgO and CaO. That lead to melt soundness and improvement of mechanical properties. The Rockwell hardness value of AZ91D Mg alloy increases with CaO contents up. Yield strength and tensile strength of CaO added AZ91D Mg alloy was higher than that of AZ91D Mg alloy. The elongation of CaO added AZ91D Mg alloy was much higher than that of AZ91D Mg alloy. It is remarkable that in spite of ductility improvement, the tensile yield strength is improved and ultimate tensile strength is increased. Addition of CaO results in the formation of Al_2Ca phase (C15 phase) on the grain boundaries. This phase formed by the red uction of CaO leads to the refinement of grain structure and dispersion strengthening. As a result, strength and ductility improve proportionally. The suppression of formation of brittle β-$Mg_{17}Al_{12}$ phase and decrease of fracture surface porosity and other casting defect caused by melt soundness also contribute to enhancement of ductility and strength of AZ91D Mg alloy. The proportional strength-ductility relationship results from high-quality melt, refined grain size and Al_2Ca second phase strengthening.

Reference

1. R. Schmid-Fetzer et al., "Focused Development of Magnesium Alloys Using the Calphad Approach," *Advanced Engineering Materials*, 3 (2001), 947-961.
2. ZENG Yi-wen, MAO Ming-xian, HUANG Zhi-qiang. Research on magnesium melting protected by cover gas mixtures containing HFC-32 [J]. *Foundry*, 2006, 55(8): 776–779.
3. K. Pettersen, P. Bakke, D. Albright, " Magnesium Die Casting Alloy Design", *Magnesium Technology 2002*, Ed.:H.I. Kaplan, TMS-AIME, 2002
4. S. Sannes, H. Westengen, "The influence of process conditions on the microstructure and mechanical properties of magnesium die castings", *Magnesium Alloys and their Applications*, Wolfsburg, Germany, April 28-30, 1998, pp.223-229
5. P. Bakke, K. Pettersen, S. Guldberg, S. Sannes, "The impact of metal cleanliness on mechanical properties of die cast magnesium alloy AM50", *Magnesium Alloys and their Applications*, Ed.: K.U. Kainer, Wiley-VCH, Weinheim, Germany, 2000, pp.739-745
6. S. K. Kim et al., "The behavior of CaO in Mg alloys," *Proc. Magnesium Technology* 2005, (2005), 285-289.
7. J. K. Lee et al., "Development of CaO added wrought Mg alloy for cleaner production," *Proc. Magnesium Technology* 2006, (2006), 185-189.
8. J. K. Lee et al., "Effect of CaO addition on ignition behavior in molten AZ31 and AZ91D Magnesium alloys." *Rare Matals*, 25 (2006) 155-159.
9. S. H. Ha et al., "The behavior of CaO and Ca in pure Mg", *Rare Matals*, 25 (2006) 150-154.
10. S. H. Ha et al., "Effect of CaO on oxidation resistance and microstructure of pure Mg", *Materials Transactions, JIM*, 49 (2008), 1081-1083.
11. D. I. Jang et al., "Effect of CaO on AZ31 Mg Strip Castings", *Materials Transactions, JIM*, 49 (2008), 976-979.
12. J. K. Lee et al., "Development of Novel Environment-friendly magnesium alloys", *Advanced Materials Research*, 47-50 (2008) 940-943.
13. Per Bakke, Ketil Pettersen, Håkon Westengen, "ENHANCED DUCTILITY AND STRENGTH THROUGH RE ADDITION TO MAGNESIUM DIE CASTING ALLOYS,"*Proc. Magnesium Technology* 2003, (2003), 171-176.
14. S. G. Leeal., "Variability in the tensile ductility of high-pressure die-cast AM50 Mg-alloy", Scripta Materialia 53 (2005) 851–856.

Magnesium Technology 2011
Edited by: Wim H. Sillekens, Sean R. Agnew, Neale R. Neelameggham, and Suveen N. Mathaudhu
TMS (The Minerals, Metals & Materials Society), 2011

ESTIMATION OF HEAT TRANSFER COEFFICIENT IN SQUEEZE CASTING OF MAGNESIUM ALLOY AM60 BY EXPERIMENTAL POLYNOMIAL EXTRAPOLATION METHOD

Zhizhong Sun[1], Xiaoping Niu[2], Henry Hu[1]
[1]Dept. of Mechanical, Automotive & Materials Engineering, University of Windsor; 401 Sunset Ave.; Windsor, Ontario N9B 3P4, Canada
[2]Promatek Research Centre, Cosma International; Brampton, Ontario L6T 5R3, Canada

Keywords: Heat transfer coefficient, Squeeze casting, Magnesium alloy, Inverse method

Abstract

In this work, a different wall-thickness 5-step (with thicknesses as 3, 5, 8, 12, 20 mm) casting mold was designed, and squeeze casting of magnesium alloy AM60 was performed in a hydraulic press. The casting-die interfacial heat transfer coefficients (IHTC) in 5-step casting were determined based on experimental thermal histories data throughout the die and inside the casting which were recorded by fine type-K thermocouples. With measured temperatures, heat flux and IHTC were evaluated using the polynomial curve fitting method. The results show that the wall thickness affects IHTC peak values significantly. The IHTC value for the thick step is higher than that for the thin steps.

Introduction

The squeeze casting process with high applied pressures is a promising solution for magnesium castings. Compared to other conventional casting processes, the most attractive features of squeeze casting (SC) are slow filling velocities and the pressurized solidification. Before the solid fraction of the casting is high enough, the applied pressure squeezes liquid metal feed into the air or shrinkage porosities effectively. Therefore, squeeze casting can make castings virtually free of porosity and usually have excellent as-cast quality, and are heat treatable, which is difficult to achieve with other conventional casting processes [1]. Although many research activities on squeeze casting process, some fundamental questions still need to be answered and the process must be optimized so as to expand its application, especially for emerging magnesium alloys.

Numerical simulation improved the productivity and optimized casting process greatly in the last decade. Beside the correct thermophysical property data, the estimating of the interfacial heat transfer coefficients (IHTCs) at the metal-mold interface is also necessary to simulate the solidification process accurately. IHTCs are usually very roughly set in the available FEM/FDM commercial codes. An accurate prediction of the boundary conditions is required to determine temperature distribution, solidification path, formation of shrinkage porosity, microstructure development, and residual stress. The pressure-transfer path is affected by applied hydraulic pressures, pouring and die initial temperatures, alloy and die materials, and casting orientation. Thermal barriers include coating applied on the die surface and air gap caused by shrinkage. The process parameters, such as the applied hydraulic and local pressures, pouring temperatures, and die initial temperatures, have an influence on the formation of pressure-transfer path, which consequently affects heat transfer at the metal-mold interface and the finial quality of squeeze castings [2,3]. In various casting process, the contact between the liquid metal and mold die is imperfect because of the coating applied on the die surface and air gap caused by shrinkage [4]. These thermal barriers may decrease the heat transfer between metal and die and cooling rate of the casting surface, which affect microstructure and quality of the casting significantly. Hence, precise determination of heat transfer coefficients at the metal-mold interface is a critical consideration to simulate the solidification process and model the microstructure of die castings accurately [5-10]. Especially, for thin-wall castings, the evaluation of IHTC becomes vital due to very limited solidification time.

However, many studies only focused on the simple shape die casting [11-14]. Little attention has been paid to variation of casting thicknesses and hydraulic pressures. Actually, in the die casting practice, various section thicknesses at different locations of castings result in significant variation of the local heat transfer coefficients. Therefore, it would be essential to investigate the influence of casting thickness, pressure value, and process parameters on the IHTC. In this study, a special 5-step squeeze casting was designed for understanding casting thickness-dependant IHTC, and the temperature measuring units and the pressure transducers were employed to accurately measure the temperatures and the local pressures during squeeze casting of magnesium alloy AM60.

Experimental Setup

5-Step Casting

A 5-step shape casting was designed special for this study. Figure 1 shows the 3-D model of 5-step casting, which consists of 5 steps (from top to bottom designated as steps 1 to 5) with dimensions of 100x30x3 mm, 100x30x5 mm, 100x30x8 mm, 100x30x12 mm, 100x30x20 mm accordingly. The molten metal fills the cavity from the bottom cylindrical shape sleeve with diameter 100 mm.

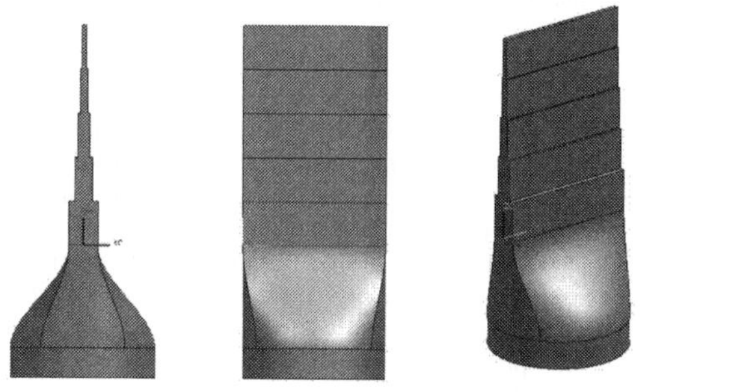

Figure 1. 3-D model of 5-step casting with the round-shape gating system. (A) XZ view; (B) YZ view; (C) isometric view.

Die Assembly

The integrated system included a laboratory hydraulic press, upper-lower die, an electric resistance furnace and a data acquisition system. As Figure 2(a) shows, the mold assembly was composed of three parts. The two upper die of casting cavity split along the center. The bottom sleeve has a diameter of 0.1016 m and a height of 0.127 m. The chill vent was located on the top of the step casting, which can discharge the gas inside the upper die cavity. Both the upper die and the bottom sleeve were heated by cartridge heaters, in which the temperatures were separately controlled by Shinko Temperature Controllers.

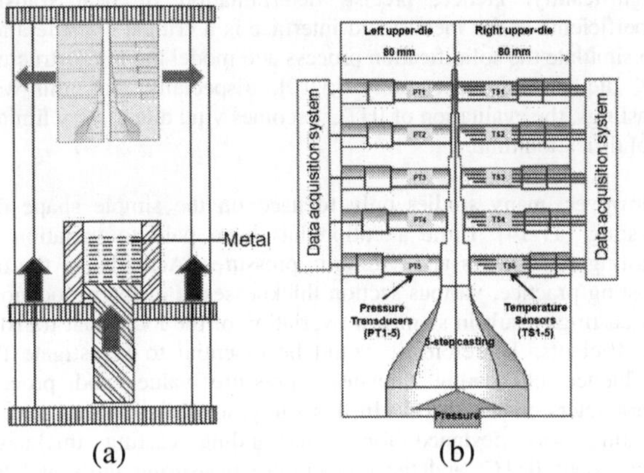

Figure 2. Schematic diagram of (a) squeeze casting machine and (b) upper-die configuration.

To measure the temperatures and pressures at the casting-die interface accurately and effectively in the 5-step squeeze casting, a special thermocouple holder was designed and developed to enable the proper placement of the thermocouples in the upper die. The thermocouple holder was manufactured using the same material as the die to ensure that the heat transfer process would not be distorted. Figure 2 illustrates schematically the configuration of the upper die (left and right parts) mounted on the top ceiling of the press machine. It also included the installation of pressure transducers, thermocouple holders and thermocouples. Local pressures within the die cavity were measured using Kistler pressure transducers 6175A2 with operating temperature 850^{0}C and pressures up to 200MPa.

As Figure 2 shows, the pressure transducers and temperature thermocouples were located opposite each other so that measurements from each sensor could be directly correlated. Five pressure transducers and temperature measuring unit were designated as PT1 through PT5, TS1 through TS5, respectively. Each sensor unit was adjusted into the die until the front wall of the sensor approached the cavity surface. The geometry shape of the temperature measuring unit was purposely designed to be the same as the pressure transducer, so that they could be exchangeable at different positions.

Casting Process

The 75- ton heavy duty hydraulic press made by Technical Machine Products (TMP, Cleveland, Ohio, USA) used in the experimental study. The die material was P20 steel. Commercial magnesium alloy AM60 was used in the experiment. The chemical composition of AM60 is shown in Table I. Table II gives the thermal properties of the related materials in this study.

Table I. Chemical composition of magnesium alloy AM60.

Mg	Al(%)	Mn(%)	Si(%)	Cu(%)	Zn(%)
balance	5.5-6.5	0.13	0.5	0.35	0.22

Table II. Thermophysical properties of magnesium alloy AM60.

Properties	Mg Alloy AM60	
	Solid	Liquid
Thermal Conductivity (W/m K)	62	90
Specific Heat (J/kg K)	1020	1180
Density (kg/m^3)	1790	1730
Latent Heat (KJ/kg)	373	
Liquidus Temperature at 0 MPa (^{0}C)	615	
Solidus Temperature at 0 MPa (^{0}C)		540

Before the pouring, the dies were pre-heated to 275^{0}C using four heating cartridges installed inside the dies. The experimental procedure included pouring molten magnesium alloy AM60 into the bottom sleeve with a pouring temperature 720^{0}C, closing the dies, cavity filling, squeezing solidification with applied pressure, lowering the sleeve die, splitting the two parts of the upper die, and finally the 5-step casting was shaken out from the cavity. The temperatures inside the die and casting were measured by Omega KTSS-116U thermocouples with response time below 10 ms, which could be applied in the squeeze casting process properly. Real-time in-cavity local pressures and temperature data were recorded by a LabVIEW- based data acquisition system. Figure 3 shows a typical 5-step casting poured under above mentioned process condition with applied hydraulic pressure of 30 MPa.

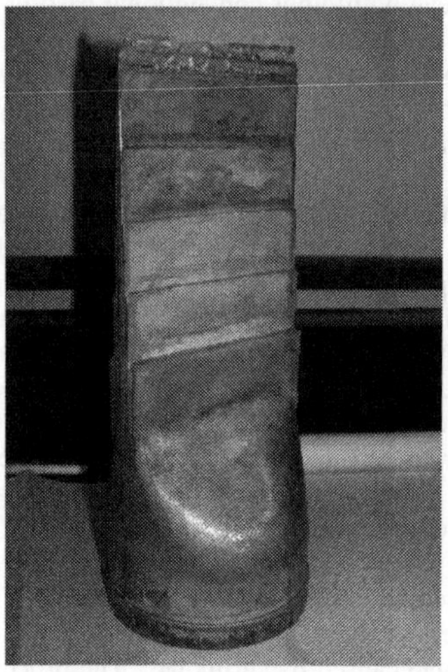

Figure 3. A 5-step casting solidifying under applied pressure 30 MPa.

Determination of IHTC

Based on the principle of heat transfer, the interfacial heat transfer coefficients (IHTC) between metal and die surface can be determined by Equation 1:

$$h(t) = \frac{q(t)}{T_{cs} - T_{ds}} \quad (1)$$

where h is IHTC; q is heat flux at the metal-die interface; T_{cs} and T_{ds} are the casting surface temperature and die surface temperature, respectively; and t is the solidification time. With the known boundary conditions in the form of temperatures or the heat fluxes, the temperature field inside the die or casting can be obtained by the direct heat conduction method.

But, it is almost impossible to measure T_{cs} and T_{ds} because the insertion of thermocouples of finite mass at the interface may distort the temperature field at the interface. Further, the heat flow at the interface may not be unidirectional due to the complex geometry. Therefore, determination of IHTC using measurements of T_{cs}, T_{ds}, and $q(t)$ directly is difficult. As a result, a polynomial curve fitting method needs to be employed to determine the IHTC based on the temperatures measured inside the die or casting [10,12,15]. The direct heat transfer modeling also was involved to calculate heat flux at the casting-die interface, which requires numerical or analytical methods to be solved.

From the measured interior temperature histories, the transient metal-die interface heat flux and temperature distribution were estimated by the polynomial extrapolation method, coupled with the finite difference method (FDM).

Because solidification of magnesium alloy AM60 during squeeze casting involves phase change and its thermal properties are temperature-dependent, the inverse heat conduction is a non-linear problem. To evaluate the IHTC effectively as a function of solidification time in the squeeze casting process, the finite difference method (FDM) was employed as follows based on the heat transfer equations [16].

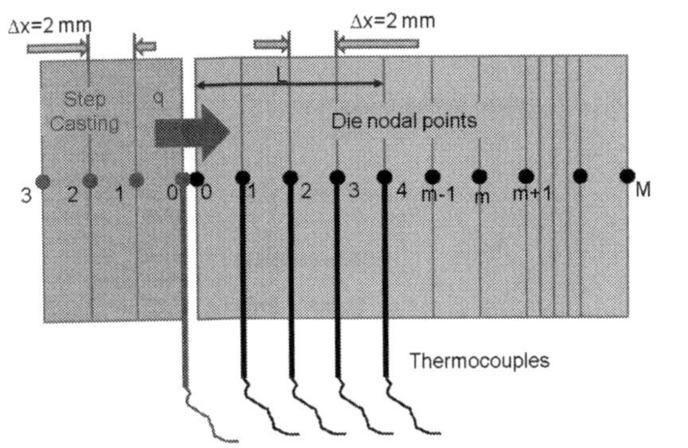

Figure 4. One-dimensional heat transfer at the interface between the casting and die, where temperature measurements were performed.

Since the thickness of each step is much smaller than the width or length of the step, it can be assumed that the heat transfer at each step is one-dimensional. The heat transfer across the nodal points of the step casting and die is shown in Figure 4. The temperatures were measured at 2, 4, 6, 8 mm beneath die surface and the heat flux transferred to the die mould can be evaluated by heat transfer equations.

Heat Transfer Model

The heat transfer inside the die at each step is transient conduction through one-dimensional body which can be described by Equation 2:

$$\rho c \frac{\partial T(x,t)}{\partial t} = \frac{\partial}{\partial x}\left(k \frac{\partial T(x,t)}{\partial x} \right) \quad (2)$$

where ρ is density of conducting die, T is the temperature, t is the time and x is the distance from the die surface to the node point; c, k are specific heat capacity and thermal conductivity of the die.

The initial and boundary conditions are described by the following equations 3 to 5:

$$T(x,0) = T_i(x) \quad (3)$$

$$q(0,t) = -k(T)\frac{\partial T}{\partial x}\bigg|_{x=0} \quad (4)$$

$$T(L,t) = Y(L,t) \quad (5)$$

Where T_i is the initial temperature of the die; q is the heat flux at the casting-die interface; L is the distance from the last temperature measurement point to the die surface; Y is the measured temperature at distance L from die surface.

The heat flux for both the casting and die interface can be calculated from the temperature gradient at the surface and sub-surface nodes by Equation 6:

$$q(t) = -k\frac{dT}{dx} = -k\frac{T_m^t - T_{m-1}^t}{\Delta x} \quad (6)$$

where k is thermal conductivity of the casting or die materials. The superscript t is solidification time. The subscript m means the number of the discrete nodal points. With the heat flux value, the segregated IHTC value can be evaluated from Equation 1.

For the surface node of the die, Equation 2 can be rearranged as Equation 7a:

$$(1 + 2F_0)T_0^{p+1} - 2F_0T_1^{p+1} = 2F_0\frac{\Delta x}{k}q_0 + T_0^p \quad (7a)$$

For any interior node of the die, equation 3 can be solved as Equation 7b:

$$(1 + 2F_0)T_m^{p+1} - F_0(T_{m-1}^{p+1} + T_{m+1}^{p+1}) = T_m^p \quad (7b)$$

where the superscript p is used to denote the time dependence of T. F_0 is a finite different form of the Fourier number:

$$F_0 = \frac{a\,\Delta t}{(\Delta x)^2} = \frac{k}{c\rho}\frac{\Delta t}{(\Delta x)^2} \quad (7c)$$

The heat flux at the casting-die interface (q) at each time step was obtained by applying Equation 7a and 7b. Thus, with T_{ds} estimated by the polynomial curve fitting method, the IHTC values were evaluated by Equation 1.

Polynomial Curve Fitting Method

For example, beneath the step 4 die surface, as Figure 4 showed, thermocouples were positioned at X1 = 2mm, X2 = 4mm, X3 = 6mm, and X4 = 8mm away from the die surface. From the temperature versus time curves obtained at each position inside the die, the temperature at the die surface (X0 = 0mm) can be extrapolated by using a polynomial curve fitting.

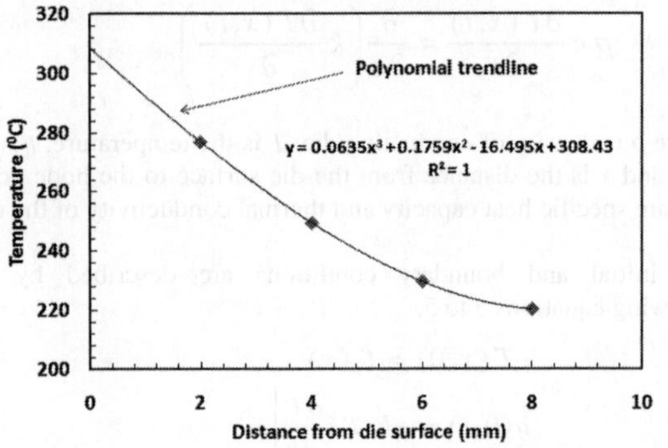

Figure 5. Polynomial curve with various measured temperatures at a time of 4.1 seconds after pressurized solidification.

By selecting a particular time of solidification process, for example t = *4.1 seconds*, the values of temperatures were read from the temperature-time data at position X1, X2, X3, and X4. Figure 5 shows the temperature values against distance X which were fitted by the polynomial trendline. The temperature at the die surface (*$T0=308.43\ ^0C$*) was determined by substituting the value of x=0 in the polynomial curve fitting Equation 6 obtained from the temperature values at various distances inside the die at a chosen time of 4.1 seconds after pressurized solidification.

$$y = 0.0635\,x^3 + 0.1759\,x^2 - 16.495\,x + 308.43 \tag{6}$$

This procedure was repeated for a number of time increments to get series of such temperatures with corresponding times. As Figure 6 shows, for the step 4 under pressure 30 MPa, the temperature curve versus time at the die surface(X0 = 0mm) was extrapolated as "Die-surf-T0-0mm-polynomial" based on the experimental data T1(X1=2mm), T2(X2=4mm), T3(X3=6mm), and T4(X4=8mm) beneath the die surface. By extrapolation, the evaluated peak temperature value of the die surface is 333.39^{0}C at the solidification time t=*6.1* seconds.

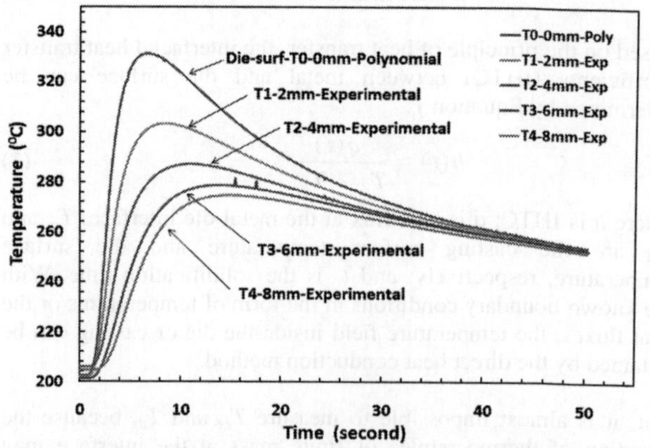

Figure 6. Extrapolated temperature curve at the die surface (T0) by the polynomial curve fitting method with applied pressure 30 MPa.

Results and Discussion

Experimental Cooling Curve

Figure 6 and 7 show typical temperatures versus time curves at the metal-die interface of Step 4 for solidifying magnesium alloy AM60 and the steel die respectively with an applied hydraulic pressure of 30 MPa. The measured locations are described in Figure 4, which include casting surface temperature (Metal-surface-Experimental), T1, T2, T3, and T4 inside the die. Since molten metal filled the cavity from the bottom, pre-solidification occurred upon the completion of cavity filling. No die surface temperatures exceeded 350^{0}C, and metal surface temperature (532.97^{0}C) was also lower than the liquidus temperature of the melt. From Figure 7, it can be observed that the temperature curve at casting surface increases abruptly and drops faster than the temperature measurements obtained at different depths under the die surface. The curves indicated the dynamic temperature change at the metal-die interface.

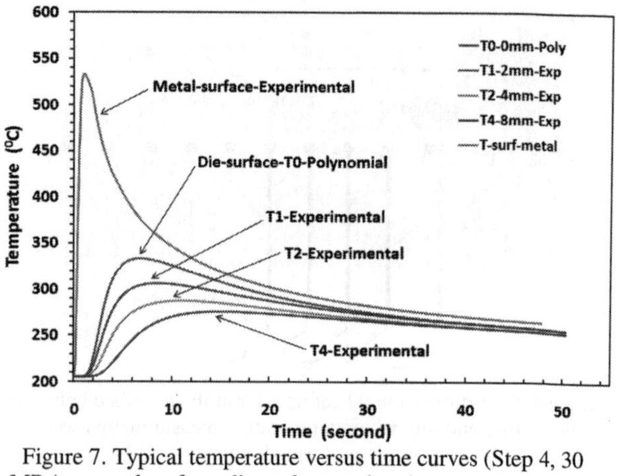

Figure 7. Typical temperature versus time curves (Step 4, 30 MPa) at metal surface, die surface, and various positions inside the die.

Typical Heat flux(q) & IHTC(h) Curves

Substituting the estimated die surface temperature (*T0*) and the measured temperature at T1=2 mm to Equation 2, the interfacial heat flux (*q*) was calculated. Figure 8 shows the interfacial heat flux (*q*) and the heat transfer coefficient (IHTC) versus solidification time of step 4 with applied pressure 30 MPa. The curves were estimated by extrapolated fitting method based on the data in Figure 7. For step 4, the peak heat flux value was 3.4E+05 W/m^2, and the peak value of IHTC was 6,450 W/ m^{2}K. From Figure 8, it can be observed that the heat flux (q) curve reached its peak value abruptly within 2.3 seconds and decreased rapidly to a lower level (5.0E+04 W/m^2) after 20 seconds. While the heat transfer coefficient (h) curve reached its peak value gradually at 12.3 seconds and vibrated around that peak value for about 6.5 seconds, then decreased slowly to the level 3,000 W/ m^{2}K after 28 seconds. Notably, the uncertainty and error of the polynomial extrapolated method should be responsible for the significant variation presented in the heat flux & IHTC curve in Figure 8.

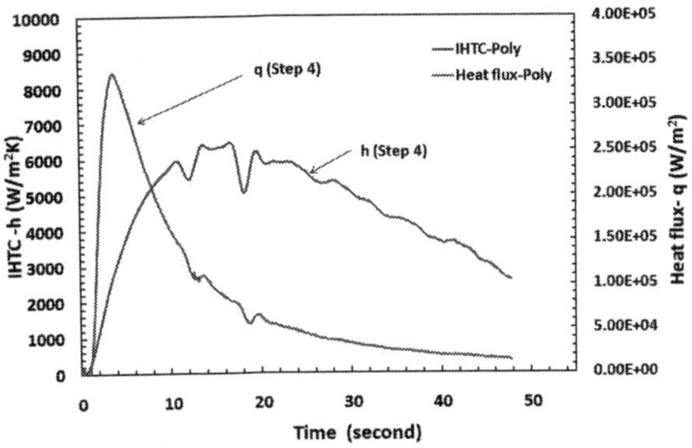

Figure 8. Interfacial heat flux (q) and the heat transfer coefficient (IHTC) curves for step 4 with applied pressure 30 MPa.

Figure 9 shows the heat flux (q) versus solidification time of step 3, step 4, and step 5 with an applied pressure of 30 MPa. The curves were estimated by extrapolation of the experimental data. For steps 3, 4, and 5, the peak heat flux values were 1.8E+05 W/m^2, 3.4E+05 W/m^2, 5.25E+05 W/m^2, respectively. From step 3 to step 5, the heat flux (q) curves reached to their peak value abruptly between 2.4, to 3.8 second and decreased rapidly to the lower level (5.0E+04 W/m^2) at 16, 28, and 42 seconds, respectively.

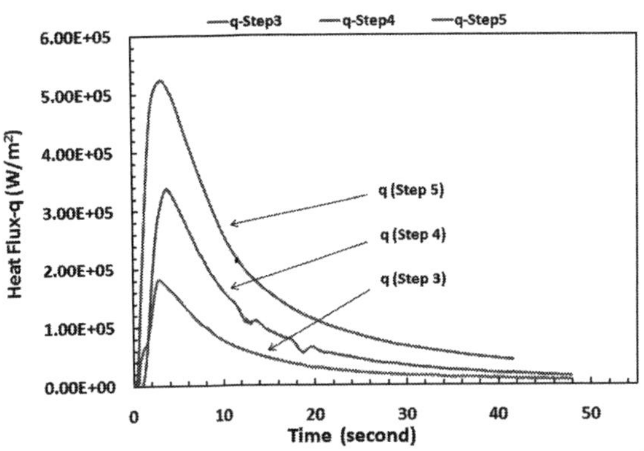

Figure 9. Heat flux (q) curves for step 3,4,5 estimated by the extrapolated fitting method.

Figure 10 shows that the heat transfer coefficient (IHTC) curves of steps 3,4,5 estimated by the extrapolated fitting. For step 3, IHTC began increasing, and reached its peak value of 3,200 W/ m^{2}K at 12.5 seconds, maintained that value for about 6 seconds, then decreased slowly to the level 1,600 W/ m^{2}K at 48 seconds. For step 4, IHTC value increased and reached its peak value (6,450 W/ m^{2}K) at about 12.3 seconds, remained at that value for about 6.5 seconds, then decreased slowly to the level 3,000 W/ m^{2}K at 48 seconds. For step 5, IHTC curve increased sharply to the peak value of 7,850 W/ m^{2}K at 8.2 seconds and then decreased to the lower level 5,500 W/ m^{2}K at 20.5 seconds. Finally IHTC increased again to high value. The up swinging tail of the IHTC curve for step 5 may be attributed to the stability of data collection, which needs to be further verified.

From steps 3 to 5, the peak IHTC value varied from 3,200 W/m^{2}K to 7,850 W/m^{2}K. Therefore, the wall thickness affects IHTC peak values significantly. The peak IHTC value decreased from the bottom to the top of the step casting as the step thickness reduced.

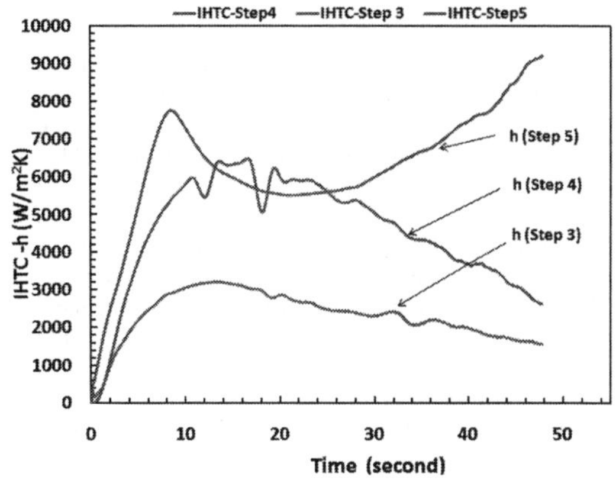

Figure 10. Heat transfer coefficient (IHTC) curves for step 3,4,5 estimated by the extrapolated fitting method.

Conclusions

1. The heat flux and IHTC at metal-die interface in squeeze casting were determined based on an extrapolation method.
2. For all steps, IHTC increased first, and reached its peak value, then dropped gradually until it arrived at a low value.
3. For steps 3, 4, and 5, the peak heat flux values were 1.8E+05 W/m^2, 3.4E+05 W/m^2, 5.25E+05 W/m^2, respectively.
4. For step 3, with a section thickness of 8mm, IHTC began with a increasing stage, and reached its peak value of 3,200 W/ m^{2}K at 12.5 seconds, maintained that value for about 6 seconds, then decreased slowly to the level 1,600 W/m^{2}K at 48 seconds. For step 4 of the thickness 12mm, IHTC value increased and reached its peak value (6,450 W/m^{2}K) at about 12.3 seconds, remained at that value for about 6.5 seconds, then decreased slowly to the level 3,000 W/m^{2}K at 48 seconds. From step 3 to 5, the peak IHTC value varied from 3,200 W/m^{2}K to 7,850 W/m^{2}K.
5. The wall thickness of squeeze cast magnesium alloy AM60 affected IHTC peak values significantly. The peak IHTC value decreased in a direction from the bottom to top as the step thickness reduced.

Acknowledgements

The authors would like to thank the Natural Sciences and Engineering Research Council of Canada, and University of Windsor for supporting this work.

References

1. I. Cho, C. Hong, "Evaluation of heat-transfer coefficients at the casting/die interface in squeeze casting", Int. J. Cast Metals Res, 1996, Vol.9, 227-232.

2. J. Aweda, M. Adeyemi, "Experimental determination of heat transfer coefficients during squeeze casting of aluminum", J. of Materials and Processing Technology, 2009, vol.209, 1477-1483.

3. Z. Guo, S. Xiong, B. Liu, M. Li, J. Allison, "Effect of Process Parameters, Casting Thickness, and Alloys on the Interfacial Heat-transfer Coefficient in the High-pressure Die Casting Process", Metallurgical and Materials Transactions, Vol.39A, 2008, 2896-2905.

4. M. Trovant, S. Argyropoulos, "Finding boundary conditions: A coupling strategy for the modeling of metal casting processes: Part I. Experimental study and correlation development", Metallurgical and Materials Transactions B, Vol.31B, 2000, 75-86.

5. Alfred Yu, "Mathematical Modeling and experimental study of squeeze casting of magnesium alloy AM50A and aluminum alloy A356", Ph.D. dissertation, Department of Mechanical, Automotive & Materials Engineering, University of Windsor, 2007.

6. D. Browne, D. O'Mahoney, "Interface heat transfer in investment casting of aluminum", Metall. Mater. Trans. A, 32A, 2001, 3055-3063.

7. G. Dour, M. Dargusch, C. Davidson, and A. Nef, "Development of a non-intrusive heat transfer coefficient gauge and its application to high pressure die casting", J. of Materials and Processing Technology, 2005, vol.169, 223-233.

8. A. Hamasaiid, M. Dargusch, C. Davidson, S. Tovar, T. Loulou, F. Rezai-aria, and G. Dour, "Effect of mold coating materials and thickness on heat transfer in permanent mold casting of aluminum alloys", Metallurgical and Materials Transactions A, Vol.38A, 2007, 1303-1316.

9. M. Dargusch, A. Hamasaiid, G. Dour, T. Loulou, C. Davidson, and D. StJohn, "The accurate determination of heat transfer coefficient and its evolution with time during high pressure die casting of Al-9%si-3%Cu and Mg-9%Al-1%Zn alloys", Advanced Engineering Materials, 2007, Vol.9, No.11, 995-999.

10. J. Taler, W. Zima, "Solution of inverse heat conduction problems using control volume approach", Int. J. Heat Mass Transf., 1999, Vol.42, 1123-1140.

11. R. Rajaraman, R. Velraj, "Comparison of interfacial heat transfer coefficient estimated by two different techniques during solidification of cylindrical aluminum alloy casting", Heat Mass Transfer, 2008, Vol.44, 1025-1034.

12. J. Su, F. Geoffrey, "Inverse heat conduction problem of estimating time varing heat transfer coefficients", Numer Heat Transf Part A, 2004, Vol.45, 777-789.

13. D. Sui, Z. Cui, "Regularized determination of interfacial heat transfer coefficient during ZL102 solidification process", Transactions of Nonferrous Metals Society of China, 2008, Vol.18, 399-404.

14. S. Broucaret, A. Michrafy, G. Dour, "Heat transfer and thermo-mechanical stresses in a gravity casting die influence of process parameters", J. of Materials Processing Technology, 2001, vol.110, 211-217.

15. J. Beck, "Nonlinear estimation applied to the nonlinear inverse heat conduction problem", Int. J. Heat Mass Transfer, 1970, Vol.13, 703-715.

16. J. Beck, B. Blackwell, A. Haji-sheikh, "Comparison of some inverse heat conduction methods using experimental data", Int. J. Heat Mass Transfer, 1996, Vol.39, 3649-3657.

Magnesium Technology 2011
Edited by: Wim H. Sillekens, Sean R. Agnew, Neale R. Neelameggham, and Suveen N. Mathaudhu
TMS (The Minerals, Metals & Materials Society), 2011

WIDE STRIP CASTING TECHNOLOGY OF MAGNESIUM ALLOYS

W.-J. Park, J.J. Kim, I.J. Kim, D. Choo
RIST (Research Institute of Industrial Science and Technology), San 32, Hyoja-dong, Nam-gu, Pohang, Kyungbuk 790-784, Korea

Keywords: Strip casting, Twin roll casting, Magnesium sheet, Magnesium coil

Abstract

Extensive investigations relating to the production of high performance and low cost magnesium sheet by strip casting have been performed for the application to automotive parts and electronic devices. Research on magnesium sheet production technology started in 2004 by *Research Institute of Industrial Science and Technology* (RIST) with support of *Pohang Iron and Steel Company* (POSCO). POSCO has completed the world's first plant to manufacture magnesium coil. Another big project in order to develop wide strip casting technology for the automotive applications of magnesium sheets was started in succession.

In this paper, the recent results for wide magnesium strip casting will be presented. Magnesium coils of ~16 tons with 2,000 mm width, 2,500 mm diameter and 4~8 mm thickness were successfully manufactured by the strip casting process. Surface defects like inverse segregations could be minimized by adequate process control. As-cast cross-sectional microstructures of magnesium strips consist of a surface chill zone, a dendritic columnar structure and a central region.

Introduction

Magnesium alloys have high potentials as structural materials due to their excellent properties such as lightweight, high specific strength and excellent damping capacity. Recently, request for low cost magnesium sheet-typed materials has been drastically increased for applications to automotive and electronic industries. The research project in order to satisfy these demands has been begun from 2004 by RIST with support of POSCO. The purpose of that project was to confirm which high quality magnesium sheet coil was just able to be produced through twin roll casting technology. For that reason, the maximum width of the as-strip cast strips was limited to about 600 mm, which strip has been tried to be apply to the narrow parts such as electronic parts and kitchen goods. The new integrated process of strip casting and warm coil rolling have been established at the pilot plant scale, and the successful research results have led the construction of POSCO Magnesium coil plant, which has commenced magnesium coil production since summer 2007 in the world' first.

It is necessary to secure competitiveness against aluminum alloys for expanding the demand of magnesium sheets. This is able to be achieved by applying magnesium sheets into large scaled markets such as automotive industry. The process development to produce the wider sheets is requested thus, because the sheet of 600 mm in width was too narrow to be utilized for the main parts of vehicles. A big project relative to magnesium sheet production accordingly has been planned since the completion of POSCO plant in succession. This new one was focused on broadening the width of magnesium strip coil for entering into automotive markets. The project in order to comply with this purpose was finally launched from 2008. The core in the project is the establishment of strip casting technology to be able to produce highly graded magnesium sheet coils with 2,000 mm width for automotive parts.

In this paper, the current results for magnesium 'wide strip casting technology' – being able to produce the strip with much wider width than that in 2004 project – will be introduced. In addition, the as-cast microstructures of different magnesium alloys produced via this process are discussed.

Wide Strip Casting

Strip casting is a well-known economical production technology to manufacture thin sheets of various metallic materials directly from its melt. Recently, a similar technology has been applied to magnesium thin strip production [1-9]. In strip casting, molten metal is fed into gap between a pair of water-cooled rolls, where it is solidified and rolled, then produced to sheet directly from melt.

There are mainly two types of defects: inverse segregation and inclusions. The inclusions, which are divided into metallic and non-metallic, can be controlled by prevention of magnesium oxidation and melt cleanliness. Evolution of inverse segregation, however, is closely correlated with solidification mechanism during strip casting process. The word of inverse segregation means the segregation formed in the opposite side of the position to have to be segregated metallurgically. In as-cast magnesium strip, the inverse segregation is just observed on both surfaces which the solidification begins, as follows; firstly, highly concentrated liquid phase is trapped into central region of strip (Figure 1a) and then the trapped liquid outflow from the center region to strip surface by pushing of a pair of casting rolls through weak inter-dendrite (Figure 1b).

(a)

(b)

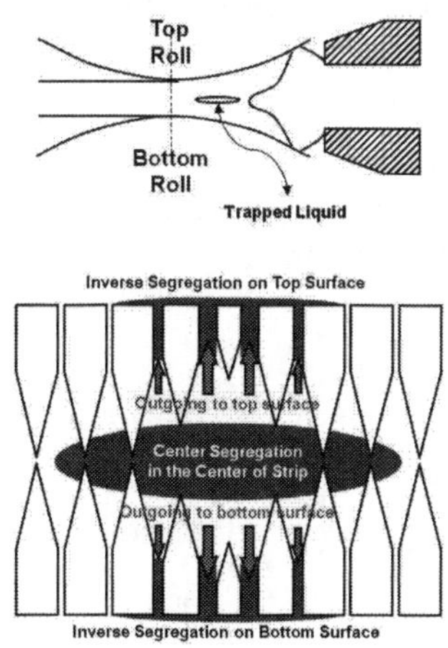

Figure 1. Schematic diagrams for the formation of inverse segregations; (a) trapping of liquid phase into central region, and (b) outgoing of the trapped liquid phase from center to each surface of strip.

Following aspects in this process are important control parameters for decreasing the defect like inverse segregation; casting speed, melt temperature, roll separation force, roll gap and setback distance, etc. In addition, the following factors are especially important to control the quality of final products for magnesium alloys; prevention of magnesium oxidation and melt cleanness.

Width of as-cast strip is able to be extended by broadening the exit slit of nozzle tip that feed melt into roll gap in horizontal strip casting. The extension of nozzle tip makes the optimization of melt flow difficult in general. The broadness of the strip width therefore results in irregular state during strip casting and the decrease of strip quality. High graded wide strip is able to be manufactured by tight control for the above process parameters and a great improvement of nozzle tip design as well.

As shown in Figure 2, wide magnesium coil of about 16 tons with 2,000 mm width, 2,500 mm diameter and 4~8 mm thickness were successfully manufactured during more than 10hrs by the accurate combination of the above-mentioned key parameters in the strip casting process. Magnesium coil in Figure 2 was side-trimmed directly during producing in wide strip caster of our laboratory.

Figure 2. As-cast strip AZ31 alloy coil of about 16 tons with 2,000 mm width, 2,500 mm diameter and 4~8 mm thickness.

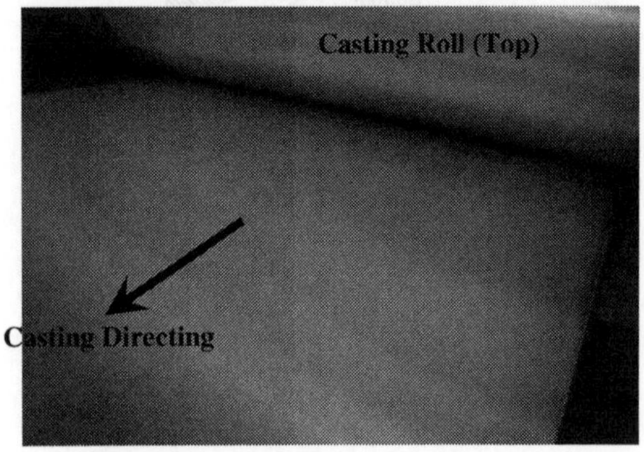

Figure 3. Strip casting process that magnesium strip is produced through the gap of a pair of casting rolls.

Figure 3 shows twin roll casting process that magnesium wide strip was produced through the gap of a pair of casting rolls. As-cast strip had clean surface with silver color and smooth edge shape without the deviation of the width. There were no defects such as inverse segregations and oxide inclusions on the surface of as-cast strip.

Strip Cast Microstructure

Figure 4 shows a typical microstructure of as-cast strip AZ31 alloy produced by wide twin roll casting system with 2,000 mm width named as Supercaster Plus® of Fata Hunter. This microstructure consists of surface chill zone near top and bottom surfaces, central equi-axed zone in the center of strip thickness, and dendritic columnar zone between surface chill zone and central equi-axed zone. The dendritic columnar zone is especially inclined against casting direction as shown in Figure 4. The microstructure was naturally similar to the one of strip cast AZ31 sheet with 600 mm width [6].

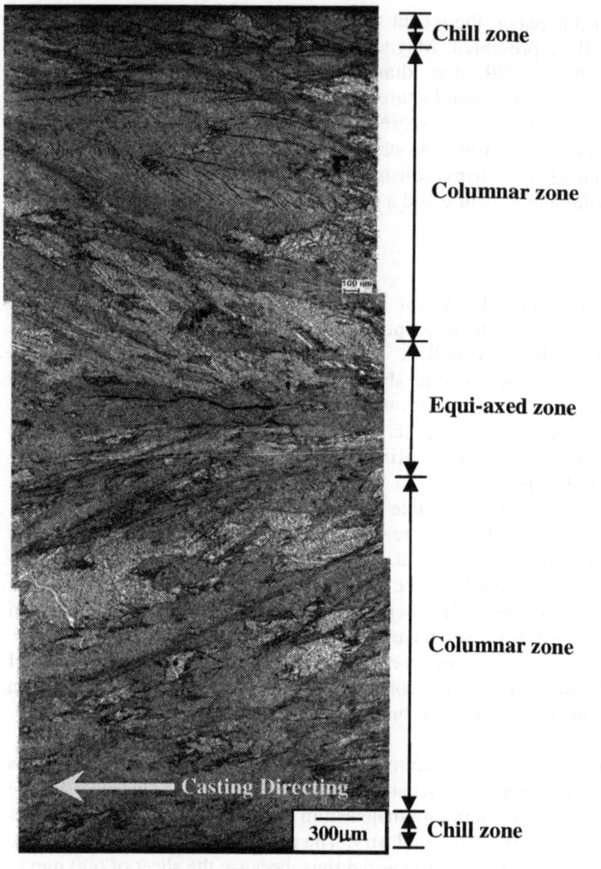

Figure 4. Cross-sectional microstructure of as-cast strip AZ31 alloy of 5.9 mm thickness.

Comparing wide strip cast microstructure of Figure 4 with that of 600 mm width [6], the most important difference in view of as-cast microstructure is the decrease in volume of center segregation in central equi-axed zone. As shown in Figure 1, the trapped liquid phase which induces the center segregation is the source of the principal surface defect of as-cast strip like inverse segregation. High quality of the wide strip overviewed in this paper consequently is due to the reduction in quantity of center segregation.

AZ61 alloy sheet coils were also produced using the same strip caster on the basis of the process condition that was able to manufacture AZ31 strip like Figure 4. As shown in Figure 5, segregation in AZ series alloys was the mixture of β-$Mg_{17}Al_{12}$ phase and α-Mg phase with a small amount of Al-Mn intermetallics and Mg_2Si phase. The Mg_2Si phase was especially supposed to be formed from impurity Si in the magnesium matrix. As the total quantity of alloy element increase, the volume of the segregation increase in AZ series alloys. It is more difficult to make high quality sheet with AZ61 alloy than AZ31 alloy.

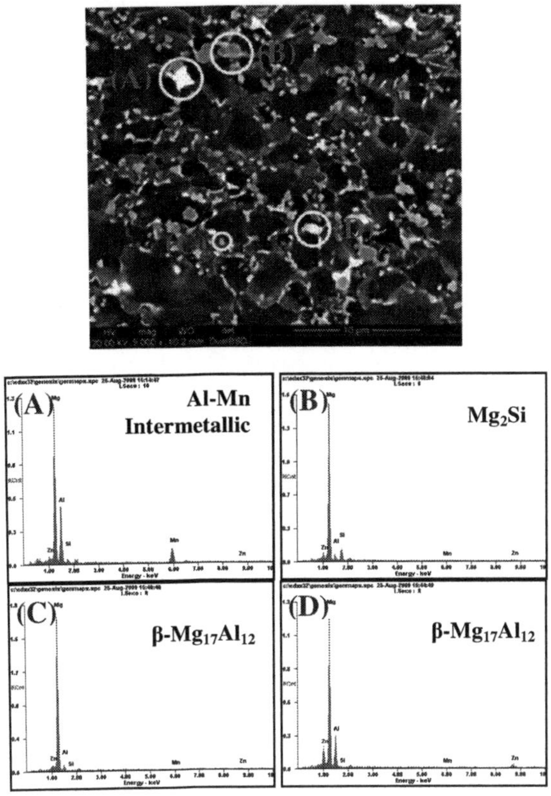

Figure 5. SEM/EDS results for the phases in segregation in as-cast strip AZ31 alloy.

Figure 6 shows as-cast strip microstructure of AZ61 alloy with more quantity of aluminum. Cross-sectional microstructure of as-cast AZ61 alloy consists of surface chill zone, central equi-axed zone and dendritic columnar zone too. In comparison of the cross sectional microstructure of AZ31 strip, the proportion of columnar zone in AZ61 microstructure was reduced and the ratio of surface chill zone and central equi-axed zone is increased. As mentioned above, inverse segregation of each surface originates from trapped liquid in central region. As the fraction of columnar structure decrease and the proportion of equi-axed structure increase, the inter-dendrite route from center to surface is wavier and longer. AZ61 sheet coil without surface defect in Figure 6 was due to the extension of the area with equi-axed dendrites and the decrease of columnar structure, even though the AZ61 alloy is containing more aluminum than the AZ31 alloy.

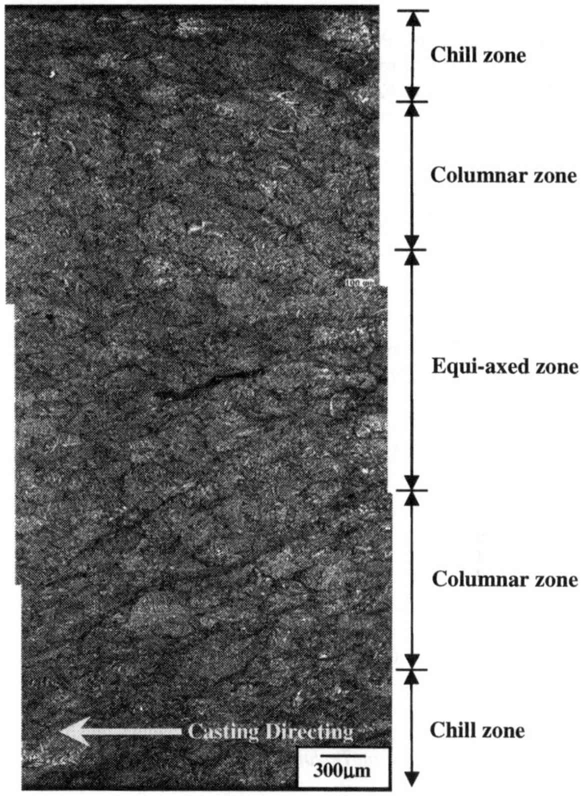

Figure 6. Cross-sectional microstructure of as-cast strip AZ61 alloy of 5.4 mm thickness.

Conclusion

POSCO has started the world's first magnesium coil production since summer 2007 based on the successful results of the projects performed by our institute from 2004. A successive big project for developing wide strip casting technology has commenced in 2008.

Wide magnesium coil of about 16 tons with 2,000 mm width, 2,500 mm diameter and 4~8 mm thickness were successfully manufactured during more than 10 hrs through the accurate combination of key parameters in strip casting process.

As-cast strips show defect-free surface with silver color and smooth edge shape without the deviation of the width. As-cast strip microstructures of AZ31 and AZ61 alloy consists of surface chill zone, central equi-axed zone and dendritic columnar zone. Segregation-free surface of AZ31 strip was estimated to be due to the reduction of central segregation. Improvement of surface quality in AZ61 strip is deduced to be caused by extensive increase of the area with equi-axed dendrites.

Coil rolling test for the as-cast strip coils is going to be performed through the company with magnesium coil roller in future. The mechanical properties for the as-cast and rolled sheets will be evaluated in succession. And then more extensive researches for automotive applications of wide magnesium strip, such as forming, surface treatment, welding, are planning to be executed.

Acknowledgements

Financial support from POSCO is gratefully acknowledged. The authors also thank to Prof. N.J. Kim (POSTECH), Prof. S. Ahn (POSTECH), and Prof. K.S. Shin (Seoul National University) for their helpful advice and encouragement.

References

1. S.S. Park, W.-J. Park, C.H. Kim, B.S. You, and N.J. Kim, *JOM* (2009), pp. 14-18.

2. S.S. Park, Y.S. Park and N.J. Kim, *LiMat-2001* (2001), p. 225.

3. R.V. Allen, D.R. East, T.J. Johnson, W.E. Borbidge and D. Liang, *Magnesium Technology 2001* (2001), pp. 75-79.

4. L. Löchte, H. Westengen and J. Rodseth, *Magnesium Technology 2005* (2005), pp. 247-252.

5. B. Engl, *62th Annual World Magnesium Conference* (2005), pp. 29-36.

6. I.-H. Jung, W. Bang, I.J. Kim, H.-J. Sung, W.-J. Park, D. Choo and S. Ahn, *Magnesium Technology 2007* (2007), p. 85-88.

7. L.K. Fan, K.M. Peng, R. Wang, J. Dong, and W.J. Ding, *Magnesium Technology 2009* (2009), pp. 69-72.

8. O. Duygulu, S. Ucuncuoglu, G. Oktay, D. S. Temur, O. Yucel, and A. A. Kaya, *Magnesium Technology 2009* (2009), pp. 85-92.

9. H. Watari et al., *Journal of Achievements in Materials and Manufacturing Engineering*, 20 (1-2) (2007), pp. 515-518.

Magnesium Technology 2011
Edited by: Wim H. Sillekens, Sean R. Agnew, Neale R. Neelameggham, and Suveen N. Mathaudhu
TMS (The Minerals, Metals & Materials Society), 2011

MICROSTRUCTURAL ANALYSIS OF SEGREGATED AREA IN TWIN ROLL CAST MAGNESIUM ALLOY SHEET

Jae Joong Kim, Woo-Jin Park and D. Choo
Research Institute of Industrial Science and Technology (RIST), San 32, Hyoja-dong, Nam-ku, Pohang, 790-600, Korea

Keywords: Twin roll cast, Magnesium alloy sheet, Segregation, Phase identification

Abstract

Twin roll casting process of magnesium alloy has been studied by various institutes since POSTECH started at the world first time. RIST has done the research and development program of twin roll casting and reverse warm milling of magnesium alloy sheet with 600 mm width. Presently, RIST is working on twin roll casting program of magnesium alloy sheet with 2,000 mm width. Twin roll cast magnesium alloy sheet includes segregated area which is important due to controllable quality of final product. Here we present the microstructural analysis of segregated area in twin roll cast magnesium alloy sheet to understand segregation phenomena. Especially, we studied on phase identification in segregated area, comparing the theoretically calculated simulation. Scanning electron microscopy, transmission electron microscopy and energy dispersive spectrometry were used for the phase identification. Our result shows that two kinds of segregation morphologies are observed in the segregated area, not depending upon the location in the sheet. Phase identification shows that segregated area includes densely continuous or dispersed beta and phi phase, and dispersed Al_8Mn_5 and Mg_2Si.

Introduction

Die-cast magnesium parts have been used widely for weight reduction of automobile so far, and then wrought magnesium part would be applied in the near future [1]. Especially, magnesium alloy sheet is approaching the market with competitive cost and new advanced technologies, such as strip casting process. Twin roll strip casting has been developed competitively by Universities, Research Institutes, and Industries [2-6], respectively, according to two issues. Firstly, twin roll strip casting is cost-competitive process for magnesium alloy sheet due to process consolidation. Secondly, demands of lightweight increase due to moving efficiency.

Market needs to improve the surface quality of magnesium alloy sheet and increase its width up to 2,000 mm to extend the application. Dent defect should be removed on the surface of as-strip cast magnesium alloy sheet. However, segregation could be included in the as-strip cast magnesium alloy sheet during solidification. The segregation is classified into center segregation and inverse segregation. It is well known that the center segregation could be reasonably included during solidification in twin roll casting. The inverse segregation might be possibly formed as well by roll pressure during solidification. Controlling the inverse segregation is an important issue to improve the surface quality. Therefore, it is necessary to analyze the microstructure of segregation area in as-cast sheet to understand segregation phenomena. Additionally, phase identification in segregation area is also an important issue to accomplish the post process [7-9]. This study is focused on understanding segregation phenomena in as-twin roll cast magnesium alloy sheet. Here we present the microstructural analysis of segregation morphology in as-cast magnesium alloy sheet and phase identification of second particles in the segregation area.

Experimental

The magnesium alloy sheet with 600 mm width was manufactured by twin roll casting in RIST. The alloy was AZ31 (Mg-3wt%Al-1wt%Zn). Figure 1 shows the schematic drawing of the twin roll casting process and the manufactured alloy sheet with directions (rolling direction, transverse direction and normal direction). The transverse direction (TD) of the alloy sheet was observed mainly to analyze the segregation morphology. The microstructural analysis and phase identification were conducted by optical microscopy (OM), scanning electron microscopy (SEM), transmission electron microscopy (TEM) and energy dispersive spectrometry (EDS). TEM specimen was prepared by focused ion beam (FIB) to conduct the phase identification of the local area which was segregated in the sheet. EDS analysis of the TEM specimen was performed in scanning transmission electron microscopy (STEM) mode to increase the accuracy.

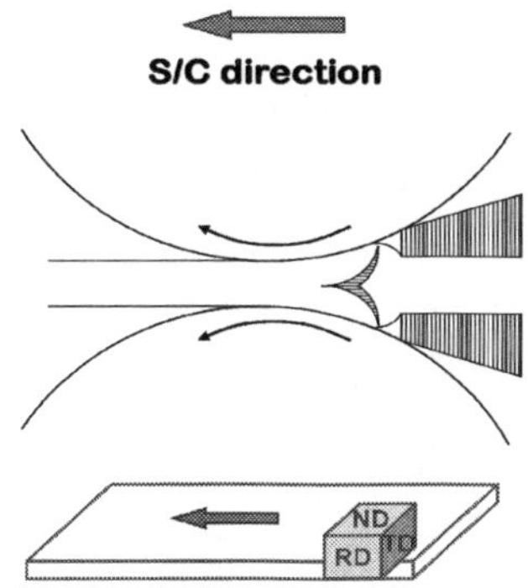

Figure 1. Schematic drawing of twin roll cast process for magnesium alloy sheet.

Figure 2. TEM specimen prepared by FIB.

Results and Discussion

Segregation microstructure

As shown in Figure 1, center segregation can be easily observed in the central area in twin roll cast AZ31 alloy sheet. Figure 3 (a) shows the center segregation observed at TD direction. Figure 3 (b) shows center segregation and inverse segregation which is also observed in an area, having different contrast on the alloy sheet surface. Minimizing inverse segregation is the key issue in twin roll casting process to improve the surface quality of magnesium alloy sheet because the inverse segregation is located near the sheet surface as shown in Figure 4. The inverse segregation is originated from center segregation and formed by roll pressure because it is vertically connected with center segregation and surface.

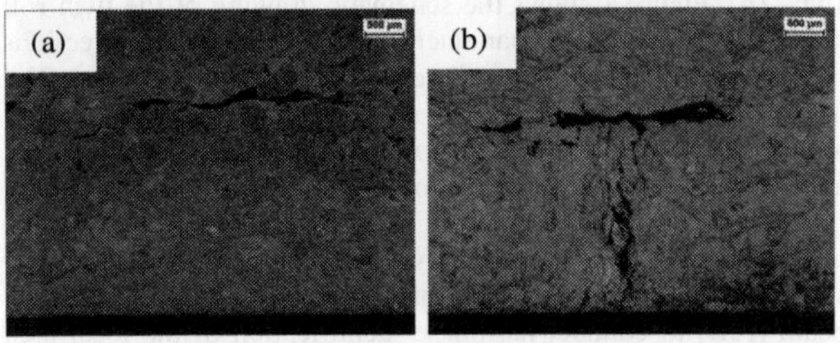

Figure 3. Optical images of (a) center segregation and (b) center and inverse segregation (TD direction).

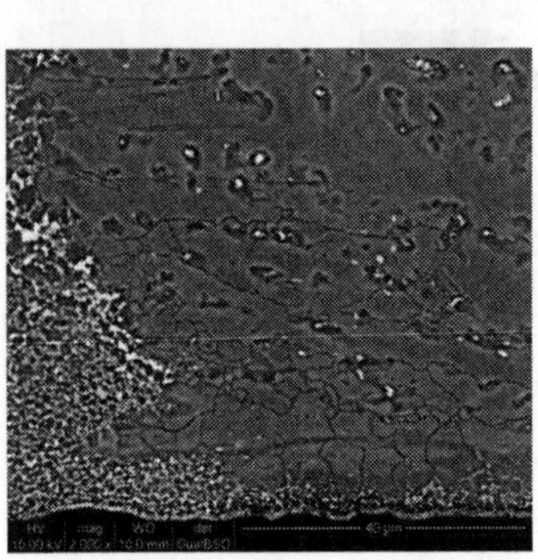

Figure 4. SEM image of inverse segregation near sheet surface.

The segregation area includes second particles with α-Mg matrix. Dispersed second phases and continuous phases are observed in the segregation area. Figure 5 shows magnified microstructures of segregation area. Figure 5 (a) shows the center segregation and the inverse segregation. Two kinds of segregation morphologies are observed in the segregated area, not depending upon the location in the sheet (dispersed phases and continuous phases). Figure 5 (b) shows that the dispersed phases consist of $Mg_{17}Al_{12}$, Al-Mn and Mg_2Si particles according to EDS analysis (as shown in Figure 6 (a) and (b)). Dispersed phases are mixed with continuous phases as shown in Figure 5 (c). Figure 5 (d) shows the continuous phases consist of beta phase and phi-like phase according to EDS analysis (as shown in Figure 6 (c) and (d)). The phi-like phase is brighter phase in Figure 5 (d). The phi phase is known as the last solidified phase according to calculated simulation result based on Scheil cooling of AZ31 alloy [9].

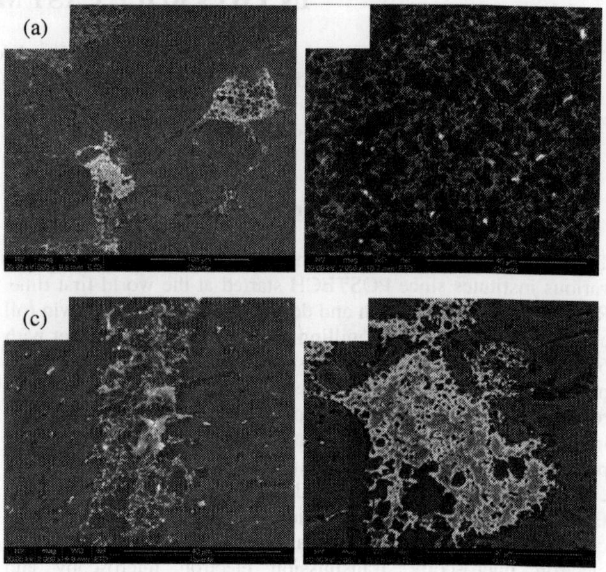

Figure 5. SEM images of segregation area showing dispersed phases and continuous phases.

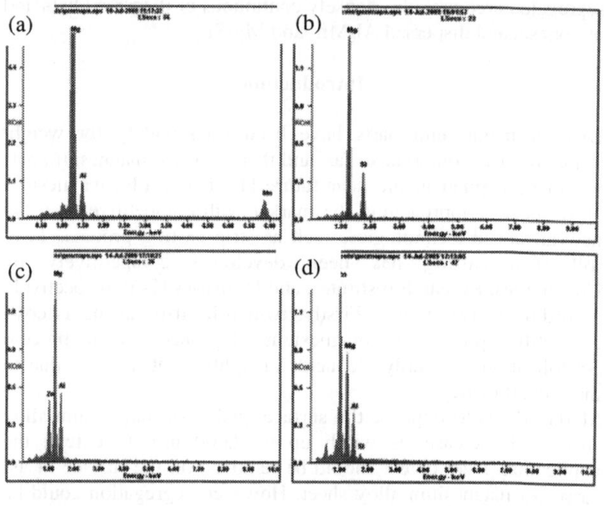

Figure 6. EDS qualitative analysis of (a) and (b) dispersed, and (c) and (d) continuous second phases in Figure 5 (b) and (d), respectively.

Phase identification

TEM works were conducted to identify the second phases observed in dispersed and continuous segregation area. Figure 7 shows TEM image of dispersed phase segregation area. According to EDS quantitative analysis, dispersed segregation area consists of beta phase, Al_8Mn_5, Mg_2Si and phi phase (as shown in Table I). The particle shape of phi phase is similar to Al_8Mn_5 phase, bright particle. Most of the second phases in the dispersed segregation area are beta phases identified as $Mg_{17}Al_{12}$, including Zn element.

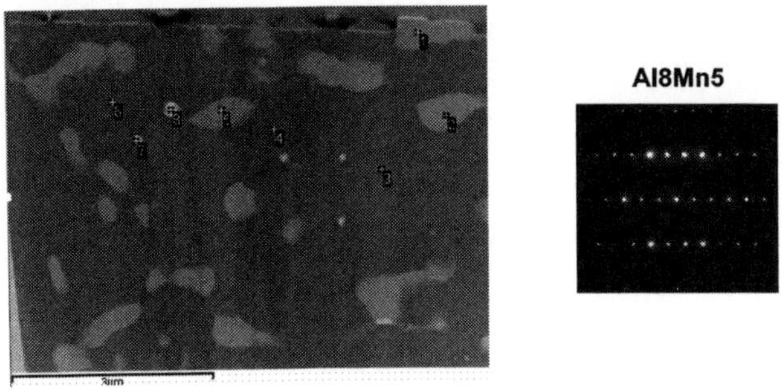

Figure 7. TEM image of dispersed phase segregation area.

Table I. EDS quantitative analysis of second phases in Figure 7.

(Wt. %)

Spectrum	O	Mg	Al	Si	Mn	Zn
1	2.75	48.29	33.88			15.08
2	3.26	46.27	36.51			13.95
3	3.55	90.87	5.57			
4	4.54	91.34	4.12			
5	3.01	48.06	34.69			14.24
6	4.09	92.41	3.49			
7	3.50	37.75	15.68			43.07
8	2.46	7.37	52.94	1.09	36.14	
9			40.84	1.90	57.26	

(At. %)

Spectrum	O	Mg	Al	Si	Mn	Zn
1	4.72	54.50	34.45			6.33
2	5.54	51.81	36.84			5.81
3	5.33	89.71	4.96			
4	6.76	89.59	3.64			
5	5.13	53.89	35.05			5.94
6	6.11	90.79	3.09			
7	7.27	51.56	19.29			21.88
8	4.94	9.73	62.97	1.25	21.12	
9			57.69	2.58	39.73	

Figure 8 shows TEM image of continuous phase segregation area. According to EDS quantitative analysis, continuous segregation area consists of beta phase and phi phase (as shown in Table II). The phi phase solidifies in the eutectic area after solidification of beta phase. The phi phase is brighter than beta phase in the micrograph, and identified as $Mg_{11}Al_5Zn_4$ according to diffraction pattern and EDS quantitative analysis.

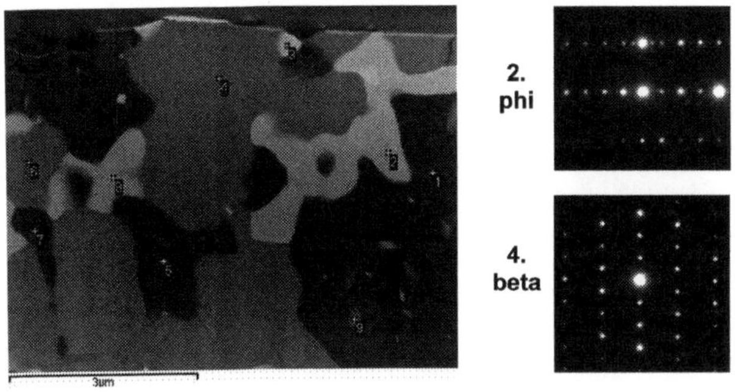

Figure 8. TEM image of continuous phase segregation area.

Table II. EDS quantitative analysis of second phases in Figure 8.

Spectrum	O	Mg	Al	Zn	(Wt. %)
1	3.61	90.22	6.17		
2	2.14	32.81	21.10	43.95	
3	2.46	34.72	19.98	42.83	
4	2.97	47.89	28.62	20.52	
5	6.39	87.48	6.13		
6	4.42	45.97	31.39	18.22	
7	5.32	91.88	2.80		
8	3.17	34.61	20.23	42.00	
9	2.65	51.06	31.63	14.66	

Spectrum	O	Mg	Al	Zn	(At. %)
1	5.41	89.10	5.49		
2	4.55	45.94	26.62	22.89	
3	5.17	47.96	24.87	22.00	
4	5.26	55.80	30.05	8.89	
5	9.45	85.17	5.37		
6	7.65	52.39	32.23	7.72	
7	7.88	89.65	2.46		
8	6.57	47.24	24.88	21.32	
9	4.52	57.34	32.01	6.13	

Summary

As-twin roll cast AZ31 alloy sheet includes center segregation, or could include center segregation and inverse segregation under unsuitable casting condition. The segregation area can be classified into disperse phase segregation and continuous phase segregation. According to EDS quantitative analysis, dispersed segregation area consists of beta phase, Al_8Mn_5, Mg_2Si and phi phase, and continuous segregation area consists of beta phase and phi phase. The beta phases is identified as $Mg_{17}Al_{12}$, including Zn element and the phi phase is identified as $Mg_{11}Al_5Zn_4$ according to this study. The continuous phase inverse segregation should be removed during twin roll casting process to improve the surface quality of magnesium alloy sheet.

Acknowledgements

A financial support from Magnesium Business Department in POSCO is greatly appreciated. Authors also thank to Prof. S.H. Ahn (POSTECH) for his helpful discussion.

References

1. J.J. Kim et al., "Recent Development and Applications of Magnesium Alloys in the Hyundai and Kia Motors Corporation", *Materials Transactions 49* (2008), 894-897.

2. S.S. Park et. al., "The Twin-Roll Casting of Magnesium Alloys", *JOM 61* (2009), 14-18.

3. Y. Nakaura, A. Watanabe and K. Ohori, "Microstructure and Mechanical Properties of AZ31 Magnesium Alloy Strip Produced by Twin Roll Casting", *Materials Transactions 47* (2006), 1743-1749.

4. S.S. Park et. al., "Microstructural Evolution in Twin-roll Strip Cast Mg-Zn-Mn-Al", *Materials Science & Engineering A 449-451* (2007), 352-355.

5. X. Gong et. al., "Enhanced Plasticity of Twin-roll Cast ZK60 Magnesium Alloy Through Differential Speed Rolling", *Materials and Design 30* (2009), 3345-3350.

6. P. Ding et. al., "Twin-roll Strip Casting of Magnesium Alloys in China", Transactions of Nonferrous Metals Society of China 18 (2008) s7-s11.

7. Y.P. Ren et. al., "The α-Mg Solvus and Isothermal Section of Mg-rich Corner in the Mg-Zn-Al Ternary System at 320°C", *Journal of Alloys and Compounds 481* (2009), 176-181.

8. L. Bourgeois, B.C. Muddle and J.F. Nie, "The Crystal Structure of the Equilibrium Φ Phase in Mg-Zn-Al Casting Alloys", *Acta Materialia 49* (2001), 2701-2711.
9. I.H. Jung, "Calculation analysis of Mg-Al-Zn ternary system" in McGill University (2007).

Magnesium Technology 2011
Edited by: Wim H. Sillekens, Sean R. Agnew, Neale R. Neelameggham, and Suveen N. Mathaudhu
TMS (The Minerals, Metals & Materials Society), 2011

Development of the Electromagnetic Continuous Casting Technology for of Magnesium Alloys

Joon-Pyo Park, Myoung-Gyun Kim, Jong-Ho Kim, Gyu-Chang Lee
RIST(Research Institute of industrial Science & Technology); #32, Hyoja-Dong, Nam-Gu, Pohang, Korea

Keywords: magnesium billet, electromagnetic, continuous casting, EMC, EMS, engine pulley

Abstract

Currently, magnesium billets produced by ingot casting or direct chill casting process, result in low-quality surfaces and low productivity, Continuous casting technology to solve these problem has not only high-quality surface billets with fine-grained and homogeneous microstructure but also cost down. The latent heat of fusion per weight (J/g) of magnesium is similar to other metals, however, considering the heat emitted to the mold surface during continuous casting in meniscus region and converting it to the latent heat of fusion per volume, magnesium will be rapidly solidified in the mold during continuous casting, which induces subsequent surface defect formation. In this study, electromagnetic casting and stirring (EMC and EMS) techniques are proposed to control solidification process conveniently by compensating the low latent heat of solidification by volume and to fabricate magnesium billet with high-quality surface. This technique was extended to large scale billets up to 300 mm diameter and continuous casting was successfully conducted. Then magnesium billet was used for the fabrication of prototype automobile pulley.

Introduction

The use of magnesium alloys for structural components is attractive due to their excellent specific strength to weight ratio [1]. Die casting is the typical process for magnesium component production, and wrought magnesium alloys have gained a great attention due to its potential use for vehicle mass reduction and performance improvement. However, the cost of wrought magnesium products remains still expensive because of the low quality of casting billets. Currently, important issues for improving the quality of magnesium alloy billets are these such as surface defects, segregation, and inclusions. Especially, the latent heat of magnesium alloy is low compared with aluminum alloys, resulting in surface defect, during continuous casting [2-4].
Continuous casting in conjunction with electromagnetic field is very effective not only in improving productivity, but also in refining grain size, eliminating surface defect and segregation, stabilizing the temperature distribution during solidification and enhancing the casting speed. In previous study, high quality aluminum billets were successfully fabricated by applying electromagnetic casting and stirring (EMC and EMS) technique. In this study, the electromagnetic casting and stirring techniques is proposed to fabricate magnesium alloy billets of high-quality surface. The casting of magnesium billet was scaled up to large billets up to 300 mm diameter. Then magnesium billet was forged for the fabrication of prototype automotive engine pulley.

Experimental

The equipment for the continuous casting of magnesium alloy billet with electromagnetic field consists of a melting furnace capable of producing about 40 kg of molten metal for magnesium, a tundish to supply the molten metal into the mold, a mold, an electromagnetic casting and stirring devices. After melting the magnesium alloy in a melting furnace, the furnace was tilted and the molten metal was poured into the tundish. A stopper was used to control the flow of molten metal from the tundish into the mold so that a fixed amount would be supplied. The continuous casting experiment was carried out by changing the EMC and EMS conditions when the supplied molten metal reached a certain height.
In the previous study of aluminum electromagnetic continuous casting, it was possible to improve the casting speed and quality characteristics by spraying cooling water directly on the billet surface that comes out of the mold. Analogous technique was applied to electromagnetic continuous casting process of the magnesium billets; however, for the molten metal of magnesium alloy, there are difficulties in the process because a strong and violent reaction occurs when molten metal comes in direct contact with water. To solve this problem, the surface of the molten metal was solidified completely while the magnesium was cast continuously so that explosive reaction was avoided even if the billet surface came in direct contact with water at the mold separation area. Two types of mold: 150 and 300 mm in diameter, was used in this study.
The schematic diagram of experimental apparatus is described in Fig 1. Through direct cooling using a cooling spray on the billet surface in the mold separation area, the cooling effect on the billet surface area was doubled. The EMC coil located at the upper part of the mold can apply a high frequency electromagnetic field to the molten metal inside the mold. The EMS system was installed at the lower part of the mold so that the microstructure in the billet can be improved by electromagnetic stirring during casting process. The experimental conditions of this study are as shown in Table 1. To prevent oxidation and ignition of the molten magnesium during casting, a mixture of CO_2 and SF_6 gases were supplied to the melting furnace, tundish, and the molten metal of the mold. The billet was forged into automotive engine pulley.

Table I. Parameters of experiment

Parameter	Conditions
Alloy	AZ31, AZ31+1%Ca
Mold	Copper, Length 150mm , Diameter 150 mm, 300 mm,
EMC	Frequency 20 kHz, 0~1400 A
EMS	Frequency 15 Hz, 0~150A
Casting Speed	5~40 cm/min.

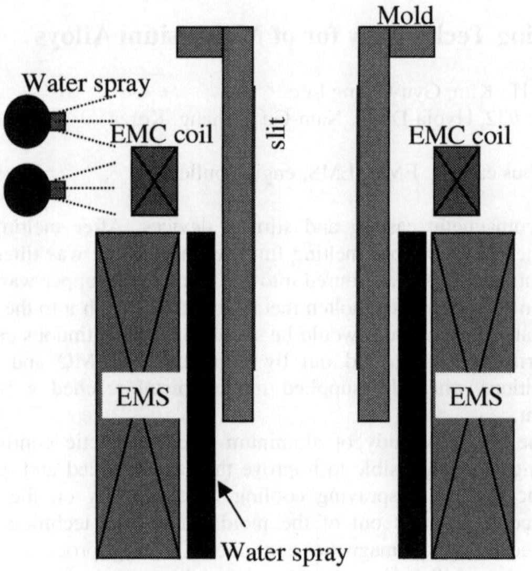

Figure 1. Schematic diagram of continuous casting apparatus combined with electromagnetic field for magnesium alloy billet

Results and discussions

The temperature distributions of molten metal surface when only EMC was applied and when both EMC and EMS were applied simultaneously are shown in Fig 2. The initial part describes the temperature change when only EMC was applied. The temperature difference between different positions which were separated by 10 mm and 50 mm from the mold surface, was very large as 20~30°C. However, the difference was greatly reduced to about 10□ while EMC and EMS were applied simultaneously, as shown in Fig 2. This indicates the temperature of the molten metal becomes uniform because the molten metal is mixed homogeneously by the electromagnetic stirring. It was expected that a uniform temperature distribution of the molten metal plays an important role in reducing the cracks inside the slab by suppressing the creation of dendrite structure and promoting the creation of fine-grained structure. In addition, it was possible to increase casting speed up to 0.4 m/min with a billet diameter of 100 mm and up to 0.2 m/min for a diameter of 150 mm.

Occasionally, the breakout of magnesium billet was observed during continuous casting. Fig 3(a) is applied 600A currents on EMC, 0.2 m/min casting speed with 50L/min. Fig 3(b) is applied 650A currents on EMC, 0.3 m/min casting speed with 20L/min. And fig 3(c) is applied 920~1134A currents on EMC, 0.3~0.4 m/min casting speed with 20L/min. The example of breakout billet is shown in Fig 3(a). If the high power EMS is applied to molten metal, the stirring is excessive so that the circulation of molten metal affect on the solidifying surface and shell. It is important to find optimized process condition for EMS and EMC. The lack of cooling water also causes the breakout of magnesium billet. As the casting speed is increased, the more cooling water is required. If the cooling water is not sufficient, the breakout of billets can occurs. The amount of cooling also can be an important factor for the production of billets. By optimizing these conditions, 1.2m-length magnesium billet with high casting speed was successfully cast as shown in Fig 3(b) and (c).

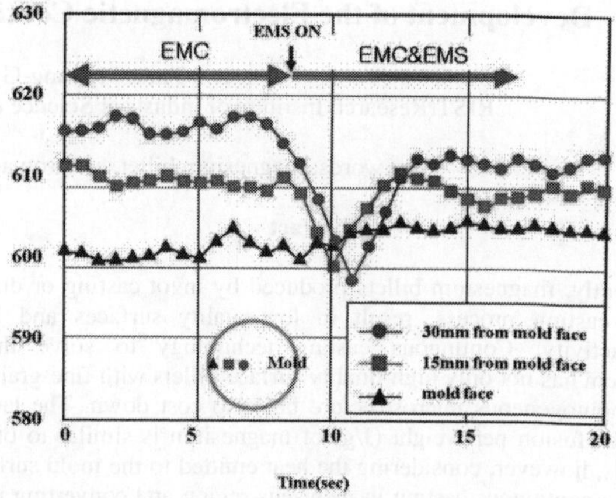

Figure 2. Temperature distribution on the molten magnesium during EMC and EMC+EMS casting

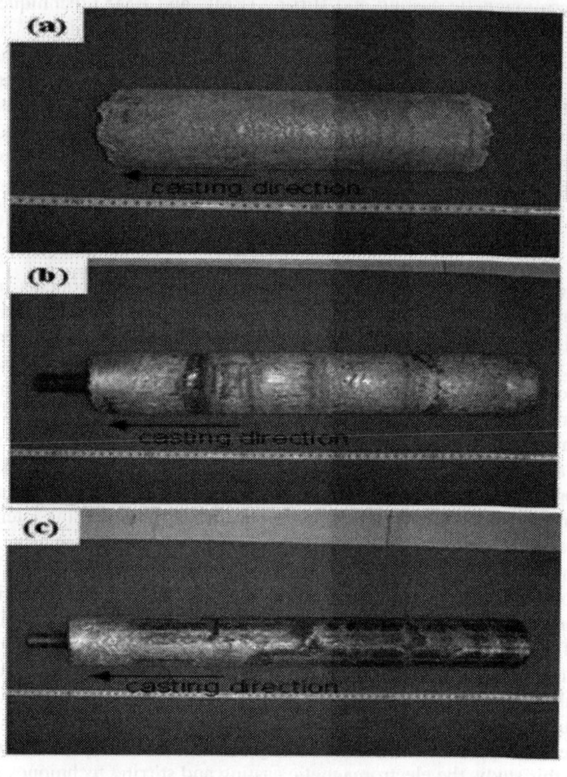

Figure 3. Appearance of magnesium billet with various experimental conditions; (a) applied 600A currents on EMC, 0.2 m/min casting speed with 50L/min (b) applied 650A currents on EMC, 0.3 m/min casting speed with 20L/min (c) applied 920~1134A currents on EMC, 0.3~0.4 m/min casting speed with 20L/min

Fig 4 shows the microstructure of the AZ31+1wt% Ca alloy billet of applying EMC only and applying both EMC and EMS. The grain size of only EMC applied specimen is about 48-60 μm, while it is about 40-50 μm for the specimen of applying EMC and EMS. It indicates that the grain size has been decreased as both EMC and EMS are applied. It is noticeable that the grain side has become significantly smaller and the microstructure refinement was achieved by applying EMC and EMS. The microstructure refinement may be due to the combination of effects of direct cooling, electromagnetic stirring, and the addition of Ca.

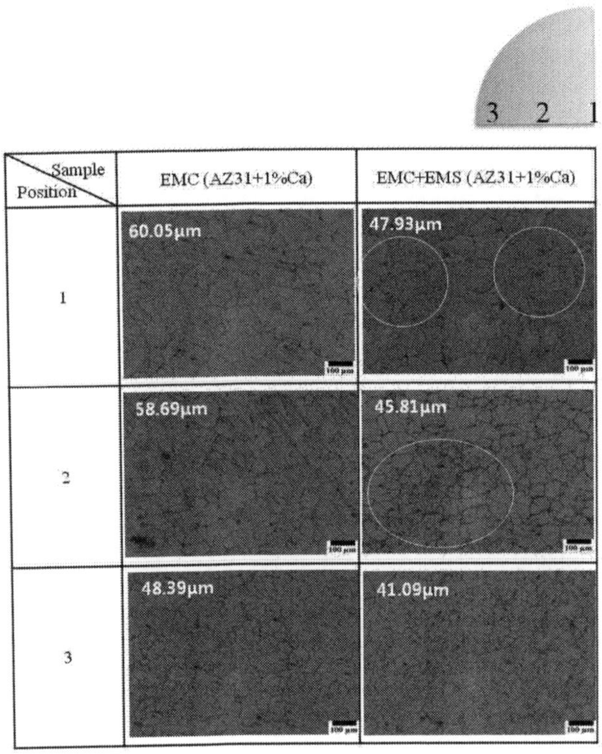

Figure 4. Microstructures of the 1 wt. % Ca-AZ31 billet processed by EMC and EMC+EMS

The casting equipment was scaled up to 300 mm diameter, which enables the production of large scale products. The size of mold was sized up to 300 mm and following apparatus, such as cooling water system, EMC coil, and EMS module were modified for the large scale equipment. Fig 5 shows the magnesium billets of 300 mm diameter made by continuous casting with EMC and EMS technique. The equipment for the fabrication of large scale billets was successfully implemented and operated which means that EMC/EMS can be applied for large scale production.

Fabricated magnesium billets are wrought alloy so that post-processing is required for the final product. In this study, we conducted hot forging process to fabricated automotive engine pulley. An automotive engine pulley made by magnesium billet has a less weight in comparison to aluminum or steel pulley. Forging process was conducted in the temperature range of 230 to 320°C. In the low temperature, the forging process could be finished but the quality of pulley surface is not sufficient. At the high temperature processing with 320°C, there is no crack or defect on the surface of final product. The fabrication process of automotive engine pulley is shown in Fig 6. In the Fig6, second product is forged one and last product is surface-treated one.

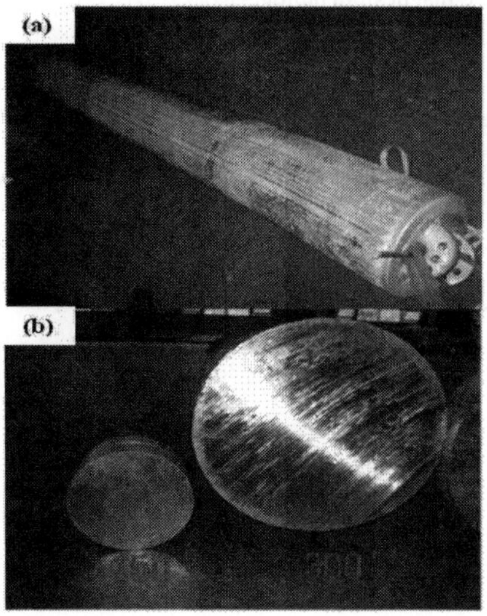

Figure 5. Magnesium billets of 300 mm diameter made by continuous casting

Figure 6. Fabrication of automotive engine pulley with magnesium billets.

Conclusions

The following conclusions were obtained in this study;

(1) High quality magnesium billets can be produced by continuous casting with electromagnetic casting and stirring techniques;

(2) The casting speed can be increased up to 0.4 m/min for a billet diameter of 100 mm, and 0.2 m/min for a billet diameter of 150 mm.;

(3) The microstructure of magnesium billet can be refined by applying the electromagnetic stirring technique;

(4) The casting equipment was modified to produce 300 mm diameter billets and successfully demonstrated;

(5) Magnesium billet was hot forged and made into the prototype of automotive magnesium pulley.

References

1. I.J. Polmear, *Materials Science and Technology*, 10 (1994), 1-16.
2. P. Fu, et al., *Journal of Materials Processing Technology*, 205 (1-3) (2008), 224-234.
3. S. Guo, et al., *Materials Science and Engineering*, 2005, A404: 323-329.
4. JP 2003-2765569, Sumitomo Metal INC LTD, Japan

Magnesium Technology 2011

Alloy Design/Development; Grain Refinement and Severe Plastic Deformation

Session Chairs:

Matthew R. Barnett
(Deakin University, Australia)
Suveen N. Mathaudhu
(US Army Research Office, USA)

Magnesium Technology 2011
Edited by: Wim H. Sillekens, Sean R. Agnew, Neale R. Neelameggham, and Suveen N. Mathaudhu
TMS (The Minerals, Metals & Materials Society), 2011

EFFECT OF Zn/Gd RATIO ON PHASE CONSTITUTIONS IN Mg-Zn-Gd ALLOYS

S. Zhang[1,2], G.Y. Yuan[1,2], C. Lu[1,2], W.J. Ding[1,2]

[1]National Engineering Research Center of Light Alloy Net Forming, Shanghai Jiao Tong University, 800 Dongchuan Road, Shanghai 200240, PR China

[2]State Key Laboratory of Metal Matrix Composite, Shanghai Jiao Tong University, 800 Dongchuan Road, Shanghai 200240, PR China

Keywords: Mg-Zn-Gd alloys, long period stacking ordered (LPSO) structure, I phase

Abstract

This paper discusses the influence of Zn/Gd ratio on the phase constitutions of as-cast Mg-Zn-Gd alloys in the Mg-rich corner within the range of 0.5-3.0 at.% for Zn content and 0.5-3.0 at.% for Gd content. The critical Zn/Gd ratio for the formation of icosahedral quasicrystal phase (I phase) and long period stacking ordered (LPSO) structure in the Mg-Zn-Gd system has been confirmed. LPSO structure and $(Mg,Zn)_3Gd$ phase are formed in Mg-Zn-Gd alloys in the range of Zn/Gd ratio $\leq$1.0. However, only $(Mg,Zn)_3Gd$ phase is observed in the alloys in the range of 1.0< Zn/Gd ratio <1.5. And if Zn/Gd ratio is $\geq$1.5, I phase and $(Mg,Zn)_3Gd$ phase in the alloys begin to form gradually.

Introduction

The great interest in magnesium and its alloys as new light constructional materials for the applications in automobiles and aircrafts has led to numerous publications in this field. Recently, a lot of studies have been focused on Mg-Zn-Gd alloys reinforced by icosahedral quasicrystal phase (I-phase) [1-3], which show a good combination of strength and ductility because of the strong interface between I phase and α-Mg matrix [4]. In addition, Mg-Zn-Gd alloys with 14H-LPSO structure precipitate [5-8] have attracted increasing interest in the past years due to the high mechanical strength and good heat-resistance. Both I phase and LPSO structure are hopeful strengthening phases in Mg alloys, but the formation range of two phases in Mg-Zn-Gd systems are not clear yet. In this study, the critical Zn/Gd ratio for the formation of I phase and LPSO structure is investigated in Mg-Zn-Gd alloys in the Mg-rich corner by conventional cast process.

Experimental Procedure

Twenty Mg-Zn-Gd alloys in the range of 0.5-3.0 at.% for Zn content and 0.5-3.0 at.% for Gd content were prepared by melting the mixture of high purity Mg, Zn and Mg–25 wt.% Gd master alloys using high-frequency induction under protection of Ar gas, then cast into preheated mild steel mould with diameter size of 10mm for further investigations. The microstructure of the specimens was observed by an optical microscope (OM), a LEO1450 scanning electron microscope (SEM) and a JEOL-2010 transmission electron microscope (TEM) operating at 200kV. X-ray diffraction (XRD) measurements were performed for the phase analysis of all specimens.

Results and Discussion

All Mg-Zn-Gd specimens were divided into three groups, (a-c) the specimens in the range of Zn/Gd ratio $\leq$1.0, (d-f) the specimens in the range of 1.0< Zn/Gd ratio <1.5, and (g-i) the specimens in the range of Zn/Gd ratio $\geq$1.5. Figure 1 shows optical microstructures of as-cast Mg-Zn-Gd alloys. It is observed that the volume fraction of secondary phase increased with increasing content of Zn and Gd in the three groups, but their microstructures are similar.

As shown in Figure 1, fine lamellae are observed near grain boundaries in the specimens with Zn/Gd ratio $\leq$1.0. These lamellae precipitates are typically observed in as-cast Mg-Zn-Gd alloys with the content that Zn%<Gd% [5-8]. Wu et al. [7,8] firstly observed fine lamellae precipitates in the as-cast Mg-Zn-Gd alloy and confirmed it has a 14H-LPSO structure. In our study, to further confirm the existence of the LPSO structured lamellae, SEM and TEM observation was carried out for the Mg-1.5Zn-1.5Gd alloy.

Figure1. Optical microstructures of (a) Mg-1.0Zn-1.0Gd, (b) Mg-1.5Zn-1.5Gd, (c) Mg-2.0Zn-2.0Gd, (d) Mg-1.2Zn-1.0Gd, (e) Mg-2.0Zn-1.5Gd, (f) Mg-2.5Zn-2.0Gd, (g)Mg-1.5Zn-1.0Gd, (h) Mg-2.5Zn-1.5Gd and (i) Mg-3.0Zn-2.0Gd alloys.

Figure 2 shows a typical SEM image and TEM graphs $(B//[110]_{\alpha\text{-}Mg})$ of fine lamellae in the Mg-1.5Zn-1.5Gd alloy. The selected area electron diffraction (SAED) pattern of fine lamellae shows that there exist 13 extra spots equally spaced between the central spot and the $(002)_{Mg}$ spots, implying that it has a 14H-type LPSO structure. It also can be observed that these lamellae have a specific relationship orientation with the matrix. The 14H-LPSO structured lamellae are formed along c*axle-[001] of the 2H-Mg. Zhu et al. [9,10] found that the stacking sequence of the 14H-LPSO structure in the Mg alloys is ABABCACACACBABA consisting of 2 ABCA stacking sequence structural units.

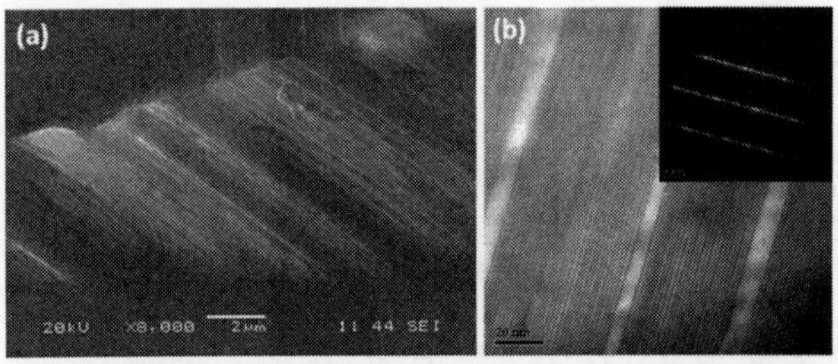

Figure 2. SEM (a) and TEM (b) images of fine lamellae in as-cast Mg-1.5Zn-1.5Gd alloy, the eclectic beam is parallel to [110]α-Mg.

Figure 3 shows XRD analysis of the specimens, which can prove the secondary phase in Mg-Zn-Gd alloys is $(Mg,Zn)_3Gd$ phase. Yamasaki et al. [6] firstly reported that the secondary phase in Mg-Zn-Gd alloy was a Mg_3Gd-type compound with a DO_3-type structure and lattice constant of a=0.720nm. In previous studies [2, 11] which focused on Mg-Zn-Gd alloys reinforced by I phase, $(Mg,Zn)_3Gd$ phase in Mg-Zn-Gd alloys was usually denoted as W phase. The peaks of $(Mg,Zn)_3Gd$ phase are observed a little deviation from the standard with increasing of Zn/Gd ratio. With the increasing of Zn/Gd ratio, the peaks of $(Mg,Zn)_3Gd$ phase deviate to high degrees a little, which means the lattice parameter of $(Mg,Zn)_3Gd$ phase is variable with the change of Zn/Gd ratio. The different lattice parameters of $(Mg,Zn)_3Gd$ phase in the three groups is calculated as a=0.710 (Zn/Gd ratio $\leq$1.0), a=0.701 (1.0< Zn/Gd ratio <1.5) and a=0.693 (Zn/Gd ratio $\geq$1.5). These results suggest that the lattice parameter of $(Mg,Zn)_3Gd$ phase is variable with the composition variety of Zn%-Gd% in the alloys.

As shown in Figure 3(a), only the peaks of α-Mg and $(Mg,Zn)_3Gd$ phase can be observed in the specimens with Zn/Gd ratio $\leq$1.0. But 14H-LPSO structure is not detected by XRD analysis because of their tiny volume fraction. Similarly, with the increasing of Zn/Gd ratio, only the peaks of α-Mg and $(Mg,Zn)_3Gd$ phase can be observed in the specimens with 1.0< Zn/Gd ratio <1.5, as shown in Figure 3b. In Figure 3c, the additional peaks in the specimens with Zn/Gd ratio $\geq$1.5 are identified to be I phase. Quasilattice parameter of I phase is calculated as a=0.520nm by using Elser's indexing method [12].

Figure 4 shows a typical SEM and TEM graphs of I phase in as-cast Mg-2.5Zn-1.5Gd alloy. The diffraction pattern reveals a 5-fold symmetry which is typical icosahedral quasicrystalline structure.

It has been reported [13] that the formation range of I phase in Mg-Zn-Gd system was about in the Zn/Gd ratio range of 1.5-40 in atomic percent. However, no researches appear concerning the relationship between the formation of LPSO structure and Zn/Gd ratio in Mg-Zn-Gd alloys, and whether or not there is a Mg-Zn-Gd alloy containing both I phase and LPSO structure. The formation range of I phase in Mg-Zn-Gd alloys in our studies is consistent with previous research. Furthermore, the formation range of LPSO structure in Mg-Zn-Gd alloy in the range of 0.5-3.0 at.% for Zn content and 0.5-3.0 at.% for Gd content is investigated, and expressed by Zn/Gd ratio. LPSO structured lamellae and I phase in the as-cast Mg-Zn-Gd alloys are formed in the range of Zn/Gd ratio $\leq$1.0 and Zn/Gd ratio $\geq$1.5, respectively. This provided a guide for the preparation and application of Mg-Zn-Gd alloys strengthening by LPSO structure or I phase. Zn/Gd ratio has to be chosen carefully with regard to the requirements of an application.

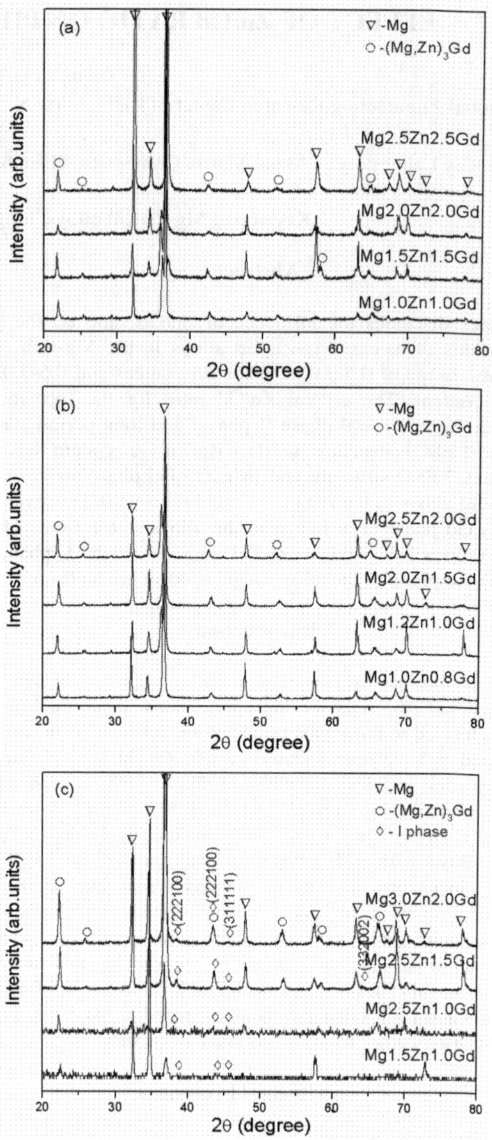

Figure 3. XRD analysis of (a) Mg-1.0Zn-1.0Gd, Mg-1.5Zn-1.5Gd, Mg-2.0Zn-2.0Gd and Mg-2.5Zn-2.5Gd alloys, (b) Mg-1.0Zn-0.8Gd, Mg-1.2Zn-1.0Gd, Mg-2.0Zn-1.5Gd and Mg-2.5Zn-2.0Gd alloys, (c) Mg-1.5Zn-1.0Gd, Mg-2.5Zn-1.0Gd, Mg-2.5Zn-1.5Gd and Mg-3.0Zn-2.0Gd alloys.

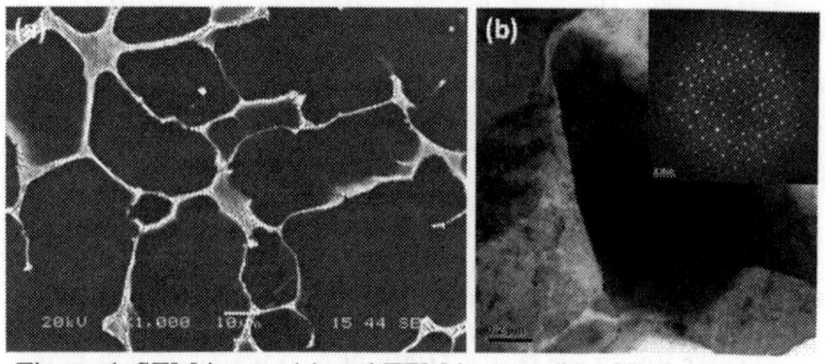

Figure 4. SEM image (a) and TEM images (b) of I-phase in Mg-2.5Zn-1.5Gd alloy, which exhibits 5-fold axe SAED pattern.

Summary

In this study, the formation range of I phase and LPSO structure in as-cast Mg-Zn-Gd alloys within the range of 0.5-3.0 at.% for Zn content and 0.5-3.0 at.% for Gd content was investigated by using XRD, OM, SEM and TEM. LPSO structure and $(Mg,Zn)_3Gd$ phase are formed in Mg-Zn-Gd alloys in the range of Zn/Gd ratio ≤ 1.0. However, there is only $(Mg,Zn)_3Gd$ phase observed in the alloys in the range of $1.0 <$ Zn/Gd ratio < 1.5. If Zn/Gd ratio is ≥ 1.5, I phase and $(Mg,Zn)_3Gd$ phase in the alloys began to form gradually. In general, the phase constitutions of Mg-Zn-Gd alloys under investigation are influenced by Zn/Gd ratio. These findings will be helpful to the alloy design to meet diverse requirements.

Acknowledgements

This work was supported by the National Nature Science Foundation of China (No.:50771063), and Science and Technology Commission of Shanghai Municipality (08JC1412200, 10JC1407400) and Jiangsu Key Laboratory of Advanced Metallic Materials (AMM200903). The help with the TEM observations received from the instrumental analysis center, Shanghai Jiao Tong University is also appreciated.

References

1. G.Y. Yuan, H. Kato, K. Amiya, A.Inoue, "Excellent creep properties of Mg-Zn-Cu-Gd-based alloy strengthened by quasicrystals and Laves phases," *Journal of Materials Research*, 20 (2005), 1278-1286.

2. G.Y. Yuan , Y. Liu , C. Lu, W.J. Ding, "Effect of quasicrystal and Laves phases on strength and ductility of as-extruded and heat treated Mg–Zn–Gd-based alloys," *Materials Science and Engineering A*, 472 (2008), 75-82.

3. Y. Liu, G.Y. Yuan, C. Lu, W.J. Ding, "Stable icosahedral phase in Mg-Zn-Gd alloy," *Scripta Materialia*, 55 (2006), 919-922.

4. A. Singh, H.Somekawa, T. Mukai, "Compressive strength and yield asymmetry in extruded Mg-Zn-Ho alloys containing quasicrystal phase," *Scripta Materialia*, 56 (2007), 935-938.

5. M. Yamasaki, T. Anan, S. Yoshimoto, Y. Kawamura, "Mechanical properties of warm-extruded Mg–Zn–Gd alloy with coherent 14H long periodic stacking ordered structure precipitate," *Scripta Materialia*, 53 (2005), 799–803.

6. M. Yamasaki, M. Sasaki, M. Nishijima, K. Hiraga, Y. Kawamura, "Formation of 14H long period stacking ordered structure and profuse stacking faults in Mg-Zn-Gd alloys during isothermal aging at high temperature," *Acta Materialia*, 55 (2007), 6798-6805.

7. Y.J.Wu, X.Q. Zeng, D.L. Lin, L.M. Peng, W.J. Ding, "The microstructure evolution with lamellar 14H-type LPSO structure in an Mg96.5Gd2.5Zn1 alloy during solid solution heat treatment at 773 K," *Journal of Alloys and Compounds*, 477 (2009), 193-197.

8. Y.J.Wu, D.L. Lin, X.Q. Zeng, L.M. Peng, W.J. Ding, "Formation of a lamellar 14H-type long period stacking ordered structure in an as-cast Mg–Gd–Zn–Zr alloy," *Journal of Materials Science*, 44 (2009), 1607-1612.

9. Y.M. Zhu, A.J. Morton, J.F. Nie, "The 18R and 14H long-period stacking ordered structures in Mg-Y-Zn alloys," *Acta Materialia*, 58 (2010), 2936-2947.

10. Y.M. Zhu, M. Weyland, A.J. Morton, K. Oh-ishi, K. Honod, J.F. Nie, "The building block of long-period structures in Mg-RE-Zn alloys," *Scripta Materialia*, 60 (2009), 980-983.

11. Z.P. Luo, H. Hashimoto, "High-resolution electron microscopy observation of a new crystalline approximant W' of Mg–Zn–Y icosahedral quasicrystal," *Micron*, 31 (2000), 487-492.

12. V.Elser, C.L. Henley, "Crystal and quasicrystal structures in Al-Mn-Si alloys," *Physical Review Letters*, 55 (1985), 2883-2886.

13. Y. Liu, G.Y. Yuan, S. Zhang, X.P. Zhang, C. Lu, W.J. Ding, "Formation range of Icosahedral phase of Mg-Zn-Gd system in the Mg-rich section," *Materials Transactions*, 49 (2008), 941-944.

Magnesium Technology 2011
Edited by: Wim H. Sillekens, Sean R. Agnew, Neale R. Neelameggham, and Suveen N. Mathaudhu
TMS (The Minerals, Metals & Materials Society), 2011

OPTIMIZATION OF MAGNESIUM-ALUMINUM-TIN ALLOYS FOR AS-CAST MICROSTRUCTURE AND MECHANICAL PROPERTIES

Xiaoyu Kang[1], Alan A. Luo[2], Penghuai Fu[1], Zhenzhen Li[1], Tianyu Zhu[1], Liming Peng[1], Wenjiang Ding[1]
[1] National Engineering Research Center of Light Alloys Net Forming and State Key Laboratory of Metal Matrix Composites, Shanghai Jiao Tong University, 200030, Shanghai, P. R. China
[2] General Motors Global Research & Development, Warren, MI, 48090, USA

Keywords: magnesium alloy development, mechanical properties, microstructure, computational material design.

Abstract

The microstructure and mechanical properties of as-cast Mg–Al-Sn alloys have been investigated using computational alloy design and experimental approaches. The as-cast microstructure of Mg-Al-Sn alloys consists of α-Mg, $Mg_{17}Al_{12}$ and Mg_2Sn phases. The volume fractions of $Mg_{17}Al_{12}$ and Mg_2Sn phases increase with increasing Al/Sn contents, and show good agreement between computational thermodynamics modeling and the experimental results. Generally, the yield strength of as-cast alloys increases with Al/Sn alloying contents, while the ductility decreases. This study has confirmed an earlier development of Mg-7Al-2Sn alloy and led to a promising new Mg-7Al-5Sn alloy with significantly improved strength and ductility compared to commercial alloy AZ91 (Mg-9Al-1Zn).

Introduction

The low density of magnesium alloys makes them an attractive material for lightweight components in the automotive industry. There are three main alloy systems in magnesium alloy applications: Mg-Al, Mg-Zn and Mg-RE based alloys. Mg-RE based alloys have the best mechanical properties among these three alloy systems, but their costs are generally too high for high-volume automotive applications. Mg-Zn based alloys are usually used as wrought alloys due to their poor castability, such as the tendency for hot tearing during casting. The most widely used Mg alloys are Mg-Al based alloys [1], as Al improves the castability and mechanical properties at room temperature. Currently, the application of commercially available Mg-Al based alloys is limited to room- or near room-temperature applications due to their poor creep resistance at elevated temperatures. This is mainly caused by the grain boundary $Mg_{17}Al_{12}$ phase, which softens at temperatures exceeding 120°C [2]. Additions of Sn to Mg-Al alloys were found to improve the strength [3] and creep-resistance [4, 5] of the alloys. Also, the Mg-Sn system is known as an age-hardening alloy system due to the precipitation of Mg_2Sn, a face-centered cubic (fcc) phase with a high melting temperature of 770°C [6]. Despite those interesting findings, there has been no systematic investigation and composition optimization of Mg-Al-Sn system for microstructure and resultant mechanical properties.

In the present study, the microstructure and mechanical properties of as-cast Mg-Al-Sn alloys have been investigated using a computational alloy design approach supported by experimental results. While the binary phase diagrams of Mg-Sn, Al-Sn and Mg-Al systems are well established [7-9], the thermodynamic data of the Mg-Al-Sn system was only made available recently [10]. Therefore, thermodynamic calculations of Mg-Al-Sn system were carried out using a software package Pandat developed by CompuTherm, Madison, WI [11]. The calculated phase diagrams, phase constituents, and solidification paths were then used to design the alloy compositions with balanced mechanical properties.

Thermodynamic Calculations and Alloy Design

Phase Constituents and Alloy Design

Fig. 1 shows the calculated liquidus projection of the Mg-Al-Sn system in the Mg-rich corner. No ternary stable phase was discovered, and no ternary solubilities of the binary intermetallic phases were found in this calculation, which is consistent with a previous study [10]. Two binary phases, Mg_2Sn and $Mg_{17}Al_{12}$, are important to this alloy system. Mg_2Sn phase is reportedly to provide significant strengthening [3], while $Mg_{17}Al_{12}$ phase improves strength [12], corrosion and castability [13], but reduces ductility and creep resistance [2]. In designing the new Mg-Al-Sn alloys for automotive structural applications, mechanical properties and corrosion resistance are primary considerations. Fig. 1 suggests that the equilibrium phase constituents (Mg_2Sn and $Mg_{17}Al_{12}$) of Mg-Al-Sn alloys are determined by the Al and Sn contents. Table 1 lists the nominal compositions of the experimental alloys studied in this investigation, along with a baseline alloy AZ91.

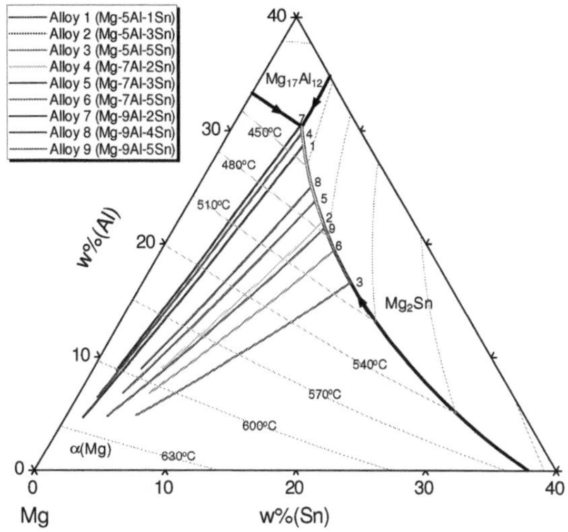

Fig. 1. Calculated Mg-Al-Sn liquidus projection and the solidification paths of the experimental alloys.

Table 1. Chemical compositions (in wt.%) of experimental Mg-Al-Sn alloys and commercial AZ91 alloy

Alloy #	Nominal composition	Al	Sn	Zn	Mg
1	Mg-5Al-1Sn	4.67	1.32	-	Bal.
2	Mg-5Al-3Sn	4.77	3.13	-	Bal.
3	Mg-5Al-5Sn	4.82	5.33	-	Bal.
4	Mg-7Al-2Sn	6.64	1.54	-	Bal.
5	Mg-7Al-3Sn	6.83	3.34	-	Bal.
6	Mg-7Al-5Sn	6.81	5.37	-	Bal.
7	Mg-9Al-2Sn	8.84	1.85	-	Bal.
8	Mg-9Al-4Sn	8.95	3.70	-	Bal.
9	Mg-9Al-5Sn	9.01	5.26	-	Bal.
AZ91	Mg-9Al-1Zn	9.14	-	1.33	Bal.

Scheil Simulation Results

Based on the assumption of complete mixing in the liquid but no diffusion in the solid, Scheil solidification model available in the Pandat package is much closer to the actual casting conditions compared to the equilibrium model based on infinite diffusion in both liquid and solid states. The solidification paths of the nine experimental Mg-Al-Sn alloys are also shown in Fig. 1, and they were calculated using the Scheil model. Table 2 summarizes the Scheil simulation results of the experimental Mg-Al-Sn alloys using Pandat, including the volume fractions of Mg_2Sn and $Mg_{17}Al_{12}$ phases in the alloys, as well as the liquidus and solidus temperatures. The simulation results suggest the following solidification sequence for the alloys upon cooling:

(1) Nucleation of primary magnesium (from liquidus):
$L \rightarrow L + \alpha\text{-Mg}$
(2) Binary eutectic reaction (439°C):
$L \rightarrow L + \alpha\text{-Mg} + Mg_2Sn$
(3) Ternary eutectic reaction (430°C):
$L \rightarrow L + \alpha\text{-Mg} + Mg_2Sn + Mg_{17}Al_{12}$

Table 2. Scheil simulation results of experimental Mg-Al-Sn alloys using Pandat

Alloy #	Nominal composition	Mg_2Sn, vol. %	$Mg_{17}Al_{12}$, vol. %	Liquidus, °C	Solidus, °C
1	Mg-5Al-1Sn	0.24	4.00	610	430
2	Mg-5Al-3Sn	0.72	4.38	606	430
3	Mg-5Al-5Sn	1.47	4.65	602	430
4	Mg-7Al-2Sn	0.33	6.46	605	430
5	Mg-7Al-3Sn	0.88	7.27	600	430
6	Mg-7Al-5Sn	1.62	7.64	596	430
7	Mg-9Al-2Sn	0.45	10.32	599	430
8	Mg-9Al-4Sn	1.08	10.82	595	430
9	Mg-9Al-5Sn	1.72	11.12	590	430

The nucleation of primary magnesium (α-Mg) starts from the liquidus temperature (610-590°C as shown in Table 2), and its volume fraction increases until the melt reaches the binary eutectic temperature of 439°C when Mg_2Sn phase is formed as part of the (α-Mg + Mg_2Sn) eutectic structure. Furthermore, $Mg_{17}Al_{12}$ is formed at 430°C as result of the ternary eutectic reaction (α-Mg + Mg_2Sn + $Mg_{17}Al_{12}$), which completes the solidification at this temperature (solidus in Table 2).

Experimental Procedures

Materials and Casting

The experimental alloys listed in Table 1 were prepared by melting high-purity Mg, Al and Sn ingots in an electrical resistance furnace under the protection of mixture gas of SF_6, CO_2 and air, and then cast into a permanent mold at a pouring temperature of 710±5°C and mold temperature of 150±5°C. Commercial alloy AZ91 (Mg-9Al-1Zn) was also prepared and cast in the same conditions for comparison. The actual chemical compositions of the alloys were analyzed using inductively coupled plasma analyzer (ICP) technique and the results are also listed in Table 1.

Microstructural Analysis and Mechanical Testing

Specimens for microstructural analysis were etched in a 4 vol.% nital or in a solution of 20 ml acetic acid + 60 ml ethanol + 1 ml nitric acid + 19 ml water. Microstructure was examined using an optical microscope (OM) and a JSM-6460 scanning electron microscope (SEM). The identification of various secondary phases was carried out using a D/MAX-IIIA X-ray diffractometer (XRD). Quantitative image analysis technique was used to measure the area fractions of secondary phases from the SEM images of the as-cast alloys. Since the optical images do not clearly reveal grain boundaries, the average grain sizes of the alloys were measured from the solution-treated (16 h at 420°C) samples, using a linear intercept method.

Tensile test samples were cut into flat tensile specimens with dimensions of 3.5 mm width, 2 mm thickness and 10 mm gage length using an electric-sparking wire-cutting machine. Tensile testing was carried out on a Zwick/Roell-20 kN material test machine at a cross head speed of 1 mm/min at room temperature. Three tensile samples were tested for each alloy. Fracture surfaces of the tensile test samples were investigated using an SEM.

Experimental Results and Discussion

Microstructure

Fig. 2 shows the SEM images of as-cast microstructure of Mg-Al-Sn alloys, all of which is comprised of dendrites of magnesium matrix surrounded by eutectic compounds. With increasing Al and/or Sn contents, the volume fractions of the eutectic compounds increase. The XRD patterns in Fig. 3 confirm the two compounds of $Mg_{17}Al_{12}$ and Mg_2Sn in the as-cast microstructure as calculated in this study. The two eutectic phases have different morphology in the SEM images, as shown in Fig. 4(a) which is the as-cast microstructure of the Mg-9Al-5Sn alloy, compared to that of AZ91 alloy, Fig. 4(b). The quantitative EDS analysis results of the different phases in Fig. 4(a) are shown in Table 3. Point A in Fig. 4(a) contains Mg and Al, with the composition ranges indicating the $Mg_{17}Al_{12}$ phase; while Point B contains Mg and Sn, suggesting the Mg_2Sn phase. Table 4 compares the measured area fractions and the calculated volume fractions of $Mg_{17}Al_{12}$ and Mg_2Sn phases in the Mg-Al-Sn alloys, and shows a reasonable agreement between the measurements and the Scheil simulation results.

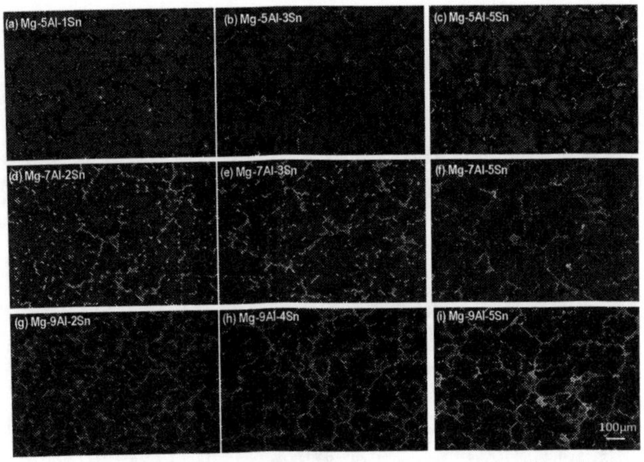

Fig. 2. SEM micrographs showing the as-cast microstructure of Mg-Al-Sn alloys.

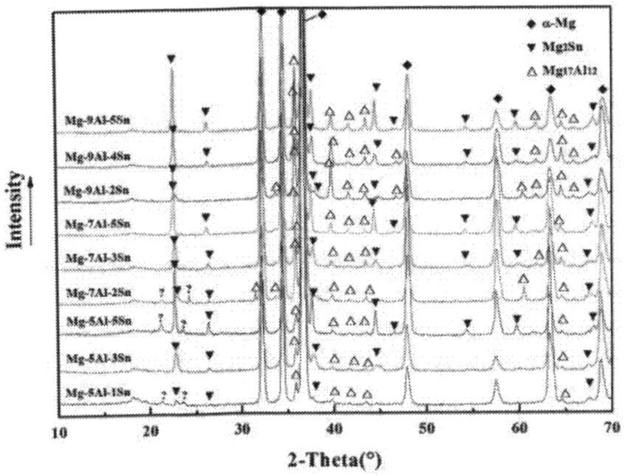

Fig. 3. XRD patterns of as-cast Mg-Al-Sn alloys.

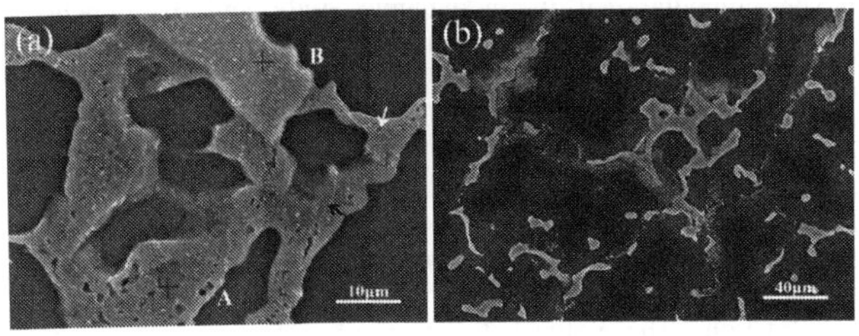

Fig. 4. SEM micrographs showing the as-cast microstructure of (a) Mg-9Al-5Sn alloy; and (b) AZ91 (Mg-9Al-1Zn) alloy.

Table 3. Quantitative EDS analysis results of Point A and Point B in Fig. 4(a)

Element (at.%)	Mg	Al	Sn
Point A	69.96	29.29	00.75
Point B	65.90	04.51	29.59

Fig. 5 shows the grain structure of solution-treated (16 h at 420°C) Mg-Al-Sn alloys. The grain sizes of Mg-Al-Sn alloys are summarized in Table 4. Generally, with increasing Al and/or Sn contents, the average grain size decreases due to the fact that more eutectic structures form during solidification and restrict the growth of α-Mg grains. However, Mg-5Al-3Sn and Mg-9Al-4Sn alloys have slightly finer grains compared to Mg-5Al-5Sn and Mg-9Al-5Sn alloys, respectively. The Mg-5Al-1Sn alloy has the coarsest grains, as shown in Fig. 5(a), about 300 μm, and the average grain size decreases to about 150 μm when the Al content is increased from 5% to 7% or the Sn content from 2% to 4%. Further increases of Al and/or Sn contents lead to finer grains, and the Mg-9Al-4Sn alloy has the finest average grain size of about 74 μm. Compared to the AZ91 alloy, with an average grain size of about 180 μm, all Mg-Al-Sn alloys but the Mg-5Al-1Sn alloy have finer grains (Table 4).

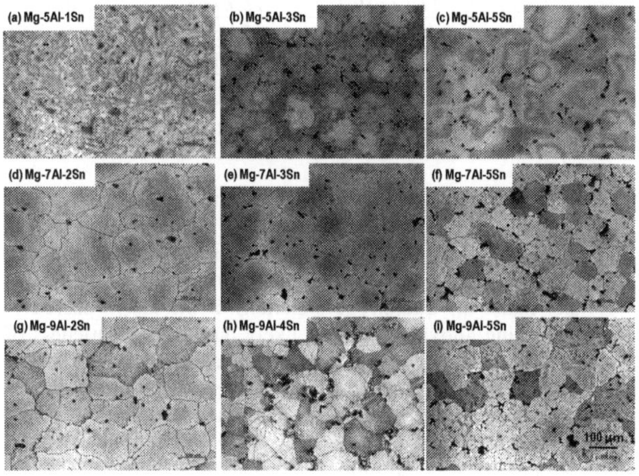

Fig. 5. Optical micrographs showing the grain structure of solution-treated (16 h at 420°C) Mg-Al-Sn alloys.

Table 4. The average grain sizes and second phase contents of Mg-Al-Sn alloys

Alloy	Average grain size (μm)	$Mg_{17}Al_{12}$		Mg_2Sn	
		Calculated (volume %)	Measured (area %)	Calculated (volume %)	Measured (area %)
Mg-5Al-1Sn	295	4.00	3.62	0.24	0.18
Mg-5Al-3Sn	149	4.38	4.36	0.72	1.06
Mg-5Al-5Sn	135	4.65	4.60	1.47	1.49
Mg-7Al-2Sn	147	6.46	6.22	0.33	0.41
Mg-7Al-3Sn	121	7.27	6.79	0.88	0.98
Mg-7Al-5Sn	81	7.64	7.04	1.62	2.07
Mg-9Al-2Sn	119	10.32	11.57	0.45	0.55
Mg-9Al-4Sn	74	10.82	11.30	1.08	1.17
Mg-9Al-5Sn	91	11.12	11.46	1.72	1.84

Tensile Properties

Table 5 lists the tensile properties of as-cast Mg-Al-Sn alloys at room temperature. With increasing Al and/or Sn contents, the yield strength (YS) of the alloys generally increase; while the ultimate tensile strength (UTS) and elongation decrease. Compared to the commercial AZ91 alloy, the properties of Mg-Al-Sn alloys are summarized as follows:

(1) Mg-7Al-2Sn and Mg-7Al-5Sn alloys have better mechanical properties (strength and ductility);
(2) Mg-7Al-3Sn alloy has similar properties;

(3) Mg-5Al-1Sn, Mg-5Al-3Sn and Mg-5Al-5Sn alloys have better UTS and elongation but lower YS; and
(4) Mg-9Al-2Sn, Mg-9Al-4Sn and Mg-9Al-5Sn alloys have better YS but lower UTS and elongation.

Table 5. Tensile properties of Mg-Al-Sn alloys

Alloy	Yield strength (MPa)	Ultimate tensile strength (MPa)	Elongation (%)
Mg-5Al-1Sn	64.7±1.4	179.1±17.0	8.32±0.78
Mg-5Al-3Sn	75.7±3.0	177.5±9.9	7.49±2.19
Mg-5Al-5Sn	82.5±4.0	161.1±5.6	4.88±0.72
Mg-7Al-2Sn	***91.9±2.1***	***175.5±6.2***	***5.34±1.04***
Mg-7Al-3Sn	88.8±3.6	151.5±2.7	3.11±0.24
Mg-7Al-5Sn	***110.1±10.1***	***160.7±5.5***	***3.18±0.64***
Mg-9Al-2Sn	102.5±3.3	154.8±2.0	2.05±0.20
Mg-9Al-4Sn	120.4±6.9	160.1±7.0	1.46±0.11
Mg-9Al-5Sn	127±1.3	149.8±2.1	1.08±0.49
AZ91 (Mg-9Al-1Zn)	89.3±4.4	150.8±11.6	2.37±0.49

The results in Tables 4 and 5 suggest that the yield strength of the Mg-Al-Sn alloys increases with increasing alloying elements (Al and Sn), due to the increased volume fraction of the eutectic compounds and the reduced grain size in the microstructure. On the other hand, the ductility (elongation) of the alloys decreases with increasing alloying contents, likely due to the brittleness of the $Mg_{17}Al_{12}$ and Mg_2Sn intermetallic phases [3]. The UTS of alloys is generally a function of YS and elongation, supposedly increasing with yield strength but restrained by the elongation. Therefore, the UTS of the Mg-Al-Sn alloys do not show a clear trend with the alloy compositions in this study.

Fig. 6 is the SEM fractographs of the Mg-Al-Sn alloy tensile specimens. These fracture surfaces show two types of features, the cleavage planes indicating brittle facture in samples of Mg-5Al-1Sn, Mg-5Al-3Sn and Mg-7Al-2Sn alloys; and the ruptured eutectics in all the fracture samples. Cleavage planes generally lead to transgranular fractures as shown in Fig. 6 (b) and (d). All samples show a mixture of intergranular and transgranular fractures. With increasing Al/Sn contents, the samples have more intergranular fractures resulting from the ruptured eutectics, which is evident in all fracture surfaces.

Alloy Optimization

Thermodynamic calculations of Mg-Al-Sn alloy system suggest two binary intermetallic phases $Mg_{17}Al_{12}$ and Mg_2Sn, whose volume fractions in the composition ranges investigated in this study would provide strengthening. The strengthening effects of $Mg_{17}Al_{12}$ and Mg_2Sn have been previously reported [3, 12], and confirmed in this paper. The liquidus and solidus temperatures of the Mg-Al-Sn alloys (Table 2) are similar to those of AZ91 alloy (600°C and 435°C, respectively [14]), indicating possible good castability, which will be verified in future work. Generally, the yield strength of as-cast alloys increases with the Al/Sn contents, while the ductility decreases. Among the alloys studied, the following two alloys show improved mechanical properties compared to AZ91 alloy.

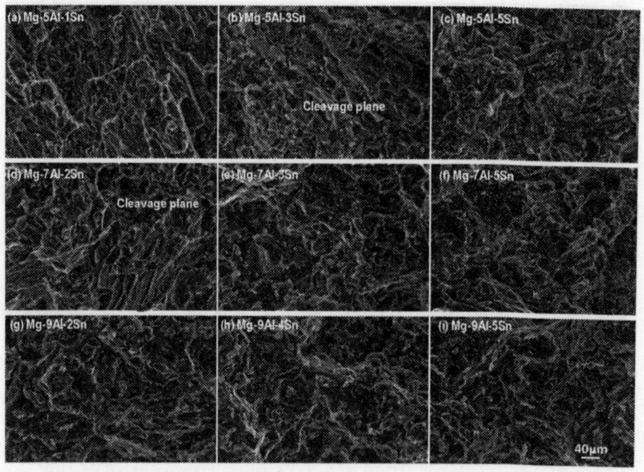

Fig. 6. SEM fractographs showing the fracture surfaces of tensile specimens of Mg-Al-Sn alloys.

(1) The Mg-7Al-2Sn alloy has similar YS, but significant higher UTS (175.5 MPa, 16% increase) and elongation (5.34%, 1.25 times); and
(2) The Mg-7Al-5Sn alloy shows improvements YS (110.1 MPa, 23% increase), UTS (160.7 MPa, 7% increase) and elongation (3.18%, 34% increase)

The identification of Mg-7Al-2Sn alloy in this study confirms the earlier development of AT72 alloy of the same composition for structural applications [3], a high-ductility alloy with similar strength as the AZ91 alloy. It is also promising that the Mg-7Al-5Sn alloy can be developed as a high-strength alloy also with improved ductility. These alloys are presently being studied for die-castability, heat treatment and corrosion resistance.

Conclusions

1. The computational thermodynamic study and experimental investigation of Mg-Al-Sn alloy system have confirmed an earlier development of Mg-7Al-2Sn alloy and led to a promising new Mg-7Al-5Sn alloy with significantly improved strength and ductility compared to commercial AZ91 alloy.

2. The Mg-Al-Sn alloys consist of α-Mg, $Mg_{17}Al_{12}$ and Mg_2Sn phases in the as-cast microstructure, as predicted by computational thermodynamic modeling and experimental validation. The volume fractions of $Mg_{17}Al_{12}$ and Mg_2Sn phases increase with the Al and Sn contents, and show good agreement between the thermodynamic calculations using Scheil solidification model and the experimental measurements.

3. The grain size of as-cast Mg-Al-Sn alloys generally decreases with increasing Al and Sn alloying contents due to the formation of more eutectic structures, restricting the growth of α-Mg grains during solidification. Generally, the yield strength of as-cast Mg-Al-Sn alloys increases with alloying contents, while the ductility decreases, due to the increased volume fractions of the eutectic compounds and the reduced grain size in the microstructure.

Acknowledgments

This work was carried out as a collaborative research project supported by General Motors Global Research and Development (GM R&D), Warren, MI, USA, and Shanghai Jiao Tong University (SJTU), Shanghai, China. Dr. Anil Sachdev of GM R&D and Prof. Xiaoqin Zeng of SJTU are gratefully acknowledged for technical discussions.

References

1. B.L. Mordike, T. Ebert, "Magnesium: Properties-Applications-Potential", Mater. Sci. Eng. A 302 (2001), 37-45.
2. A.A. Luo, "Recent Magnesium Alloy Development for Elevated Temperature Applications", Int. Mater. Rev., 49 (2004), 13-30.
3. A.A. Luo and A.K. Sachdev, "Microstructure and Mechanical Properties of Mg-Al-Mn and Mg-Al-Sn Alloys", in: Magnesium Technology 2009, edited by E.A. Nyberg, S.R. Agnew, N.R. Neelameggham and M.O. Pekguleryuz, TMS, 2009, 437-443.
4. C.L. Mendis, C.J. Bettles, M.A. Gibson, C.R. Hutchinson, "An Enhanced Age Hardening Response in Mg-Sn Based Alloys Containing Zn", Mater. Sci. Eng. A 435-436 (2006), 163-171.
5. S. Avraham, A. Katsman, M. Bamberger, "Strengthening Mechanisms in Mg-Al-Sn Based Alloys", in: Magnesium Technology 2009, edited by E.A. Nyberg, S.R. Agnew, N.R. Neelameggham and M.O. Pekguleryuz, TMS, 2009, 471-475.
6. H. Liu, Y. Chen, Y. Tang, S. Wei, G. Niu, "The Microstructure, Tensile Properties, and Creep Behavior of As-cast Mg-(1-10)%Sn Alloys", J. Alloy Compd. 440 (2007), 122-126.
7. T.B. Massalski, Binary Alloy Phase Diagrams, ASM International, 1990.
8. S.G. Fries and H.L. Lukas, "Optimization of the Mg-Sn System", J Chem. Phys., 90 (1993), 181-187.
9. P. Liang, H.-L. Su, P. Donnadieu, M.G. Harmelin, A. Quivy, P. Ochin, G. Effenberg, H.J. Seifert, H.L. Lukas, and F. Aldinger, "Experimental Investigation and Thermodynamic Calculation of the Central Part of the Mg-Al Phase Diagram", Z. Metallkde., 89 (8) (1998), 536-540.
10. E. Doernberg, A. Kozlov and R. Schmid-Fetzer, "Experimental Investigation and Thermodynamic Calculation of Mg-Al-Sn Phase Equilibria and Solidification Microstructures", Journal of Phase Equilibria and Diffusion, 28 (2007), 523-535.
11. CompuTherm LLC, "Pandat 8.0 - Phase Diagram Calculation Software for Multi-component Systems", CompuTherm LLC, Madison, WI, 53719, USA, 2008.
12. A.A. Luo and A.K. Sachdev: "Microstructure and Mechanical Properties of Magnesium-Aluminum-Manganese Cast Alloys", International Journal of Metalcasting, 4 (4) (2010), 51-58.
13. D. Sachdeva, S. Tiwari, S. Sundarraj and A.A. Luo, "Microstructure and Corrosion Characterization of Squeeze Cast AM50 Magnesium Alloys", Metallurgical and Materials Transactions B, DOI: 10.1007/s11663-010-9433-x.
14. A. Luo, "Understanding the Solidification of Magnesium Alloys" in Proc. of the Third International Magnesium Conference, Ed., G.W. Lorimer, The Institute of Materials, London, UK, 1996, 449-464.

Magnesium Technology 2011
Edited by: Wim H. Sillekens, Sean R. Agnew, Neale R. Neelameggham, and Suveen N. Mathaudhu
TMS (The Minerals, Metals & Materials Society), 2011

THERMODYNAMIC ANALYSIS OF AS-CAST AND HEAT-TREATED MICROSTRUCTURES OF Mg-Ce-Nd ALLOYS

Mark A. Easton[1], Suming Zhu[1], Mark A. Gibson[2], Jian-Feng Nie[1], Joachim Gröbner[3], Artem Kozlov[3], Rainer Schmid-Fetzer[3]

[1]CAST CRC, Monash University; Victoria 3800, Australia
[2]CAST CRC, CSIRO; Clayton, Victoria 3169, Australia
[3]Clausthal University of Technology; Robert-Koch-Straße 42, Clausthal-Zellerfeld, D-38678, Germany

Keywords: Rare earth, CALPHAD, Casting, Isothermal heat treatments

Abstract

The phase relationships in the Mg-rich corner of the Mg-Ce-Nd system have been investigated through the evaluation of selected compositions in the as-cast and heat-treated condition. Consistent thermodynamic CALPHAD-type assessments have also been generated for the Mg-Ce-Nd system [1].

It is shown that this system reveals a significant degree of meta-stability under technologically relevant solidification conditions (i.e., permanent-mould or high pressure die casting). This is simulated in thermodynamic calculations by suppression of the RE_5Mg_{41} phase and reasonable agreement is found with the as-cast microstructures. After heat treatment these microstructures transform, depending on alloy composition, into phase assemblies consistent with the calculated stable equilibrium phase diagram. It is the elucidation of such meta-stable phase formation and the subsequent transformation from the as-cast to the heat treated state that is a particular strength of the thermodynamic approach and which makes it a powerful tool for alloy development.

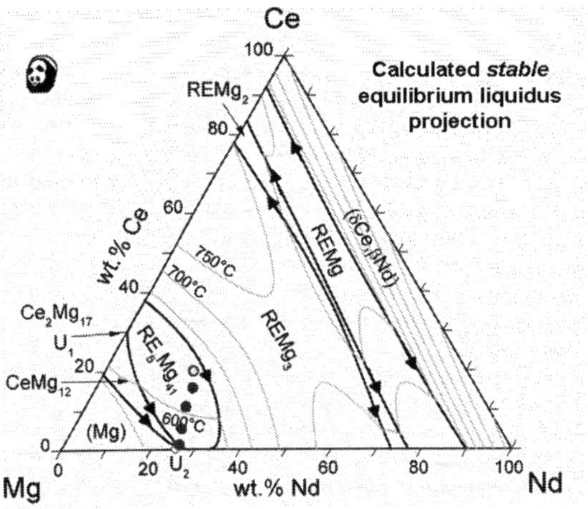

Figure 1. Mg-Ce-Nd Liquidus surface.

The solid dots in Figure 1 indicate alloys exhibiting experimentally evidenced primary precipitation of the $CeMg_{12}$ phase in the as-cast condition, whereas the stable equilibrium phase diagram suggests RE_5Mg_{41} to be the primary phase. The thermodynamic calculation, however, reveals that a relatively small undercooling will result in the (metastable) primary $CeMg_{12}$ phase precipitation in all these alloys, as observed after casting. More results and the phase transformations towards equilibrium after heat treatment are given in detail by Gröbner et al. [1].

References

[1] J. Gröbner, A. Kozlov, R. Schmid-Fetzer, M. A. Easton, S. Zhu, M. A. Gibson, J.-F. Nie, *Acta Materialia*, 2010, doi:10.1016/j.actamat.2010.09.066 (in press).

Magnesium Technology 2011
Edited by: Wim H. Sillekens, Sean R. Agnew, Neale R. Neelameggham, and Suveen N. Mathaudhu
TMS (The Minerals, Metals & Materials Society), 2011

COMPRESSIVE STRENGTH AND HOT DEFORMATION BEHAVIOR OF TX32 MAGNESIUM ALLOY WITH 0.4% Al AND 0.4% Si ADDITIONS

K.P. Rao[1], Y.V.R.K. Prasad[2], K. Suresh[1], C. Dharmendra[1], N. Hort[3], K.U. Kainer[3]

[1]Department of Manufacturing Engineering and Engineering Management
City University of Hong Kong, 83 Tat Chee Avenue Kowloon, Hong Kong SAR, China
[2] Processingmaps.com (formerly at City University of Hong Kong)
[3]GKSS Research Centre Geesthacht GmbH, Max-Planck Str. 1, Geesthact 21502, Germany

Keywords: Mg-Sn-Ca-Al-Si Alloy, Compressive Strength, Hot Workability, Processing Maps, Kinetic Analysis

Abstract

Mg-3Sn-2Ca (TX32) alloy has good corrosion and creep resistance although its strength is inferior to AZ31 alloy. In this paper, the influence of additions of 0.4%Al and 0.4%Si on the compressive strength and hot working characteristics of TX32 is reported. Although the room temperature compressive strength improved marginally with the alloying additions, the drop in higher-temperature strength is significant. By comparing with the alloy having only 0.4% Al, it is clear that the Si addition is responsible for this deterioration. The hot working behavior as characterized by processing maps revealed that TX32 exhibits two domains of dynamic recrystallization occurring in the temperature and strain rate ranges: (1) 300 – 350 °C and 0.0003 – 0.001 s^{-1} and (2) 390 – 500 °C and 0.005 – 0.6 s^{-1}. In Al and Si containing TX32, both the domains moved to higher temperatures and the flow instability is reduced improving the hot workability. In both the domains, the apparent activation energy is higher than that for self-diffusion in magnesium implying that there is a significant contribution from the back stress generated by the hard particles present in the matrix.

Introduction

Their light weight is the main reason that makes Mg based alloys attractive for components in aerospace and automobile applications. However, many of the wrought alloys like Mg-3Al-1Zn have inferior corrosion and creep properties and to improve these two properties newer Mg-Sn-Ca alloys (TX series) are being developed [1-3]. In this system, Sn forms a solid solution with Mg and imparts corrosion resistance while Ca enhances high temperature creep strength by forming CaMgSn intermetallic particles. The first and foremost alloy in this system is Mg-3Sn-1Ca (TX31), which was characterized as regards its corrosion resistance and processability [4-6]. The creep resistance of this alloys can be further improved by increasing the Ca content to 2% but with a small compromise of corrosion properties [3,7]. In the microstructure of Mg-3Sn-2Ca [3], intermetallic particles of CaMgSn form mostly in the matrix and Mg_2Ca phase forms at the grain boundaries both of which are beneficial in enhancing the creep strength. However, there is a need to strengthen the alloy further to make it a candidate material for structural applications. For this purpose, alloying with aluminum and silicon is considered to be attractive since Al causes solid solution strengthening of Mg as there is a significant difference in the atomic diameters of Mg and Al, while Si forms intermetallic particles that can enhance the creep strength. With this in view, Mg-3Sn-2Ca-0.4Al-0.4Si alloy has been designed and chosen for this investigation. The aim of the present investigation is to evaluate the compressive strength and hot working characteristics of this alloy with a view to understand the effect of the combined additions of Al and Si to TX32. For this purpose, the compressive strength of Mg-3Sn-2Ca-0.4Al-0.4Si is measured in the temperature range 25 – 250 °C and the hot working behavior is evaluated in the temperature range 300 – 500 °C. It is proposed to compare the behavior of this alloy with those of Mg-3Sn-2Ca and Mg-3Sn-2Ca-0.4Al.

The hot working behavior has been characterized with the help of the standard kinetic approach [8] as well as processing maps [9,10]. The standard kinetic rate equation relating the steady state flow stress (σ) to strain rate ($\dot{\varepsilon}$) and temperature (T) is given by [8]:

$$\dot{\varepsilon} = A\sigma^n \exp\left[-\frac{Q}{RT}\right] \tag{1}$$

where A = constant, n = stress exponent, Q = activation energy, and R = gas constant. The rate-controlling mechanisms are identified on the basis of the activation parameters n and Q.

Processing maps are based on the Dynamic Materials Model [11], the principles of which were described in earlier publications [9,12]. Briefly, the work-piece undergoing hot deformation is considered to be a dissipator of power and the strain rate sensitivity (m) of flow stress is the factor that partitions power between deformation heat and microstructural changes. The efficiency of power dissipation occurring through microstructural changes during deformation is derived by comparing the non-linear power dissipation occurring instantaneously in the work-piece with that of a linear dissipater for which the m value is unity, and is given by:

$$\eta = 2m/(m+1) \tag{2}$$

The variation of efficiency of power dissipation with temperature and strain rate represents a *power dissipation map* which is generally viewed as an iso-efficiency contour map. Further, the extremum principles of irreversible thermodynamics as applied to continuum mechanics of large plastic flow [13] are explored to define a criterion for the onset of flow instability given by the equation for the instability parameter $\xi(\dot{\varepsilon})$:

$$\xi(\dot{\varepsilon}) = \frac{\partial \ln[m/(m+1]}{\partial \ln \dot{\varepsilon}} + m \le 0 \tag{3}$$

The variation of the instability parameter as a function of temperature and strain rate represents an *instability map* which delineates regimes of instability where $\xi(\dot{\varepsilon})$ is negative. A superimposition of the instability map on the power dissipation map gives a *processing map* which reveals *domains* (efficiency contours converging towards a peak efficiency) where individual microstructural processes occur and the limiting conditions for the *regimes* (bounded by a contour for $\xi(\dot{\varepsilon}) = 0$) of flow instability. Processing maps help in identifying temperature – strain rate windows for hot working where the intrinsic workability of the material is maximum (e.g. dynamic recrystallization (DRX) or superplasticity) and also in avoiding the regimes of flow instabilities (e.g. adiabatic shear bands or flow localization).

Experimental

Mg – 3 wt.% Sn – 2 wt.% Ca – 0.4 wt.% Al – 0.4 wt.% Si alloy was prepared using 99.99% pure Mg, 99.96% pure Sn, 98.5% pure Ca, 99.9% Al and 99.9% Si. The alloy, molten at about 720 °C, was kept under a protective cover of Ar+3% SF_6 gas before casting in a pre-heated permanent mold to obtain cylindrical billets of 100 mm diameter and 350 mm length.

Cylindrical specimens of 10 mm diameter and 15 mm height were machined from the as-cast billet for compression testing. For inserting a thermocouple to measure the specimen temperature as well as the adiabatic temperature rise during deformation, the specimens were provided with a 1 mm diameter hole machined at mid height to reach the centre of the specimen.

For the purpose of evaluating the compressive strength of the alloy, compression tests were conducted at a strain rate of 0.0001 s^{-1} and in the temperature range 25–250 °C using a computer-controlled servo-hydraulic testing machine. The data for developing processing maps were obtained in isothermal uniaxial compression tests were conducted at constant true strain rates in the range 0.0003 – 10 s^{-1} and temperature range 300 – 500 °C. Details of the test set-up and procedure are described in earlier publication [14]. Constant true strain rates during the tests were achieved using an exponential decay of actuator speed in the servo hydraulic machine. Graphite powder mixed with grease was used as the lubricant in all the experiments. The specimens were deformed up to a true strain of about 1.0 and then quenched in water. The load – stroke data were converted into true stress – true strain curves using standard equations. The flow stress values were corrected for the adiabatic temperature rise at different temperatures and strain rates and this correction was less important at lower strain rates and higher temperatures. The deformed specimens were sectioned in the center parallel to the compression axis and the cut surface was mounted, polished and etched for metallographic examination. All the specimens were etched with an aqueous solution containing 3% picric acid.

Results and Discussion

The microstructure of the starting material in cast condition is dendritic and consisted of very large grains (about 500 μm diameter). While the CaMgSn particles are present both in the matrix and at the grain boundaries, Mg_2Ca phase is essentially present at the grain boundaries. Other intermetallics phases including Mg_2Si are also present in the microstructure.

High Temperature Strength

The variation of the ultimate compressive strength of Mg-3Sn-2Ca-0.4Al-0.4Si alloy with temperature in the range 25 – 250 °C obtained at a strain rate of 0.0001 s^{-1} is shown in Figure 1. A strain rate slower than those used for generating processing maps is selected for obtaining the strength values since the material does not have much ductility in this temperature range [15]. The strength drops fast with temperature and reaches a near plateau beyond about 150 °C up to 200 °C beyond which it drops precipitously. The values for the base alloy Mg-3Sn-2Ca and the one with 0.4%Al addition are also included in this figure for the purpose of comparison and to evaluate the effects of additions of Al alone and of Al+Si. The alloy with only 0.4 wt% Al has shown increased strength compared to their TX32 base (Al free alloy) at almost all the test temperatures. Aluminum improves room temperature mechanical properties of Mg by solid solution strengthening due to large atomic size difference (α-Mg) and also precipitation of (β-$Mg_{17}Al_{12}$) intermetallic phase [16]. However, with the addition of 0.4% Si to this alloy decreases the higher temperature strength much more significantly between 100 and 200 °C. This reduction is likely due to the formation of Mg_2Si particles. It is the only stable compound that forms in the Mg-Si System. The Si addition often promotes the precipitation of Mg_2Si particles which form during solidification and are quite detrimental to the mechanical properties of Mg alloys [17,18]. It is reported [19] that the addition of Si to AZ91 alloy does not change the quantity as well as the morphology of the $Mg_{17}Al_{12}$ phase. It may be due to negligible solid solubility of Si in Mg and the impossibility of the formation of any other compound containing Al and Si. On the other hand, Nanjunda Swamy et al. [20] observed that, in cast Mg-Al-Si composites, ultimate tensile strength (UTS) decreases with increasing Mg_2Si content possibly due to increased casting defects such as porosity and segregation which may be counteracting the strengthening effect of increasing Mg_2Si content. Asl et al. [16] have indicated that addition of Si to some aluminum containing magnesium alloys improves creep resistance while reducing the room temperature mechanical properties. With such differing observations, the role of Si on mechanical and creep properties of such alloy systems is not fully understood. The present results clearly show that unlike the Al addition, Si addition to the extent of 0.4% is not useful for improving the strength property of the base alloy.

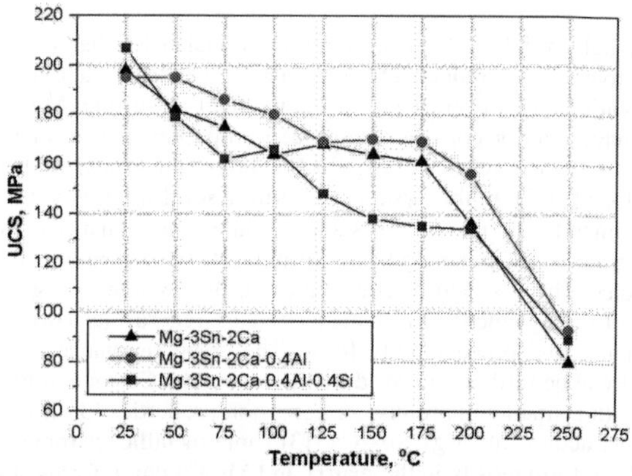

Figure 1. Variation of ultimate compressive strength (UCS) of as-cast Mg-3Sn-2Ca, Mg-3Sn-2Ca-0.4Al, and Mg-3Sn-2Ca-0.4Al-0.4Si alloys with temperature.

Hot Working Behavior

The processing map obtained at a strain of 0.5 (near steady-state flow) covering a temperature range 300 – 500 °C and strain rate range 0.0003 – 10 s^{-1} for the alloy Mg-3Sn-2Ca-0.4Al-0.4Si is shown in Figure 2. The map exhibits two domains occurring in the temperature and strain rate ranges given as follows:

Domain #1: 300 – 410 °C and 0.0003 – 0.003 s^{-1} with a peak efficiency of 38% occurring at 350 °C and 0.0003 s^{-1}.
Domain #2: 420 °C – 500 °C and 0.003 – 3 s^{-1} with a peak efficiency of 39% occurring at 500 °C and 0.1 s^{-1}.
In addition, two flow instability regimes occur: (1) at temperatures lower than 360°C and strain rates higher than 0.01 s^{-1}, (2) at higher temperatures 430 – 490 °C and strain rate of 10 s^{-1}. The second regime covers only a small region and may be ignored.

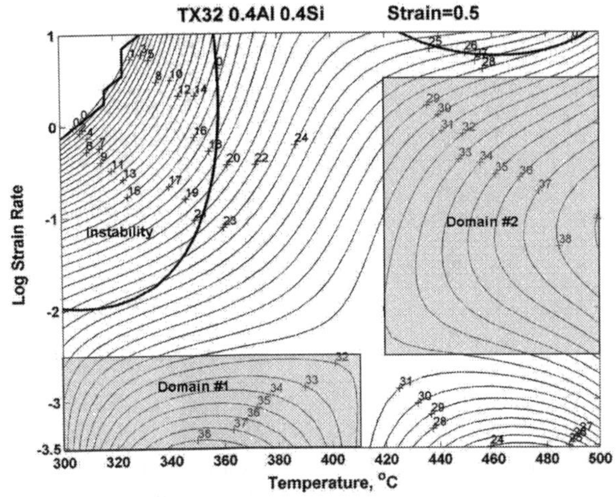

Figure 2. Processing map for Mg-3Sn-2Ca-0.4Al-0.4Si alloy. The numbers against the contours represent efficiency of power dissipation in percent. The dark line separates the flow instability regimes. The blank region at left hand top corner represents cracking of the specimens.

The microstructures recorded at 350 °C/0.0003 s^{-1} (Domain #1) and at 500 °C/0.1 s^{-1} (Domain #2), are shown in Figure 3 (a) and (b). Both these microstructures indicate that the as-cast microstructure is converted into an equiaxed wrought microstructure through a mechanism involving dynamic recrystallization (DRX). It is interesting to note that there are two DRX domains separated at about 410 °C. In terms of dissipative energy which will be low when the efficiency of power dissipation is high, this change-over region may be viewed as a saddle point configuration [12]. It is possible that such a configuration has formed due to the difference in the operating slip systems. Below 400 °C, the flow is dominated by prismatic slip and basal slip, in which case the recovery mechanism is going to be climb of edge dislocations. In this case, cross-slip is ruled out since the stacking fault energy on basal slip systems is lower [21]. It is well known that at temperatures higher than 400 °C, the pyramidal slip systems dominate the flow and the recovery mechanism will be cross-slip since the stacking fault energy on the pyramidal slip systems will be high [22]. Processing at temperatures close to the saddle point is not desirable since it may lead to microstructural instability.

Using the standard kinetic rate equation (Eq. 1) to represent the steady-state flow under hot working conditions [8], the activation parameters – stress exponent (n) and apparent activation energy (Q) – are evaluated. A plot of log(flow stress) versus log(strain rate) at different test temperatures, is shown in Figure 4. All the data at the temperature of 500 °C (Domain #2) exhibited a good linear fit with a slope (m) of 0.225 or stress exponent (n) of 4.44. With decreasing temperatures, the linear fit got progressively limited to lower strain rates (Domain #1) and the m value is 0.2 or stress exponent (n) is 5.0. At higher strain rates and lower temperatures, where the rate equation (Eq. 1) is not obeyed, flow instability occurs.

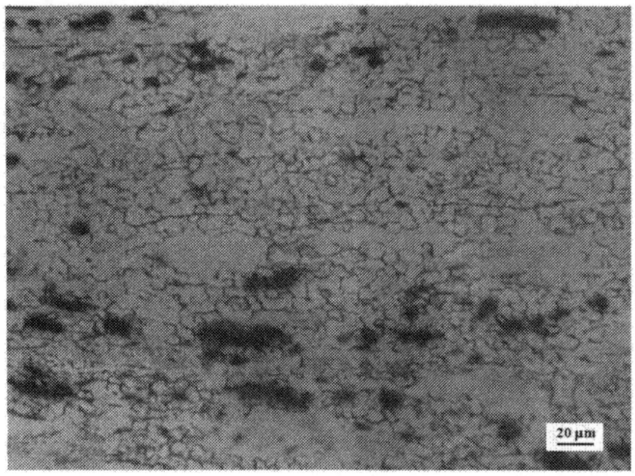

Figure 3(a). Microstructure of Mg-3Sn-2Ca-0.4Al-0.4Si alloy deformed at 350 °C and 0.0003 s^{-1} (Domain #1) exhibiting dynamic recrystallization of as-cast microstructure.

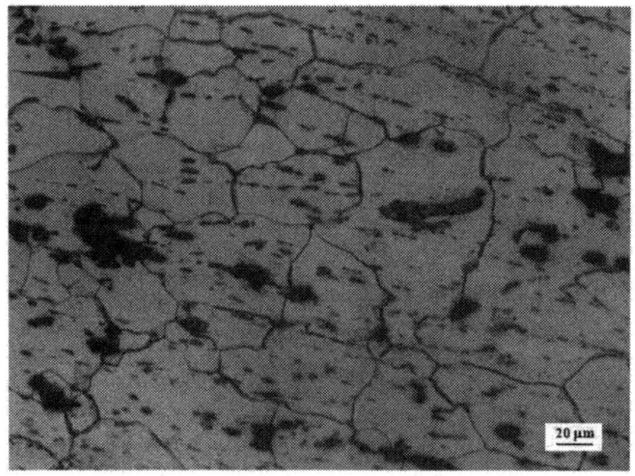

Figure 3(b). Microstructure of Mg-3Sn-2Ca-0.4Al-0.4Si alloy deformed at 500 °C and 0.1 s^{-1} (Domain #2) exhibiting dynamic recrystallization of as-cast microstructure.

Arrhenius plot relating normalized flow stress [ln (σ/μ)] to ($1/T$) (μ = shear modulus of Mg) is shown in Figure 5. The data at strain rates of 0.0003 s^{-1} and 0.001 s^{-1} exhibited a linear fit giving an apparent activation energy of 177 kJ/mole. On the other hand, the data at strain rates of 0.01 s^{-1} to 1 s^{-1} gave a linear fit progressively at higher temperatures and also yielded an apparent activation energy of 177 kJ/mole.

The apparent activation energy values obtained in the two different regimes are larger than that for lattice self diffusion in Mg (135 kJ/mole) [23]. This could be due to the large back stress generated by the large volume fraction of CaMgSn intermetallic particles in the matrix.

For hot working, dynamic recrystallization (DRX) domain is preferred since it helps in converting the as-cast dendritic microstructure by an equi-axed wrought microstructure. While processing in the lower strain rate domain (Domain #1) is too slow from industrial view point, hot working in Domain #2 is recommended for this alloy, the optimum parameters being 500 °C/0.1 s^{-1}. However the hot working window is wide suggesting that this alloy is highly workable.

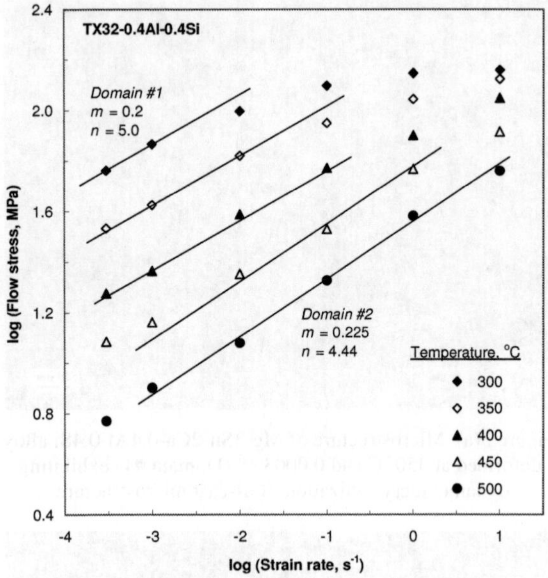

Figure 4. Variation of flow stress (at a strain of 0.5) of Mg-3Sn-2Ca-0.4Al-0.4Si alloy with strain rate on a log-log scale at different test temperatures.

Comparison with Mg-3Sn-2Ca and Mg-3Sn-2Ca-0.4Al Alloys

The processing maps for the base alloy Mg-3Sn-2Ca is shown in Figure 6, which also exhibits two domains of dynamic recrystallization in the temperature and strain rate ranges given as follows: (1) 300 – 350 °C and 0.0003 – 0.001 s^{-1} with a peak efficiency of 42% occurring at 300 °C and 0.0003 s^{-1}, and (2) 390 – 500 °C and 0.005 – 0.6 s^{-1} with a peak efficiency of 42% occurring at about 450 °C and 0.03 s^{-1}. A comparison of the two processing maps (Figures 2 and 6) reveals that both the DRX domains have shifted to higher temperatures by the addition of Al and Si. Domain #1 has expanded from 350 °C to 410 °C and Domain #2 has commenced at a higher temperature of 420 °C by the addition of Al and Si compared to 390 °C for the base TX32 alloy. Also, the strain rate range for domain #1 has slightly increased whereas that for domain #2 is significantly broadened with a range of 0.003 - 3 s^{-1} compared to 0.005 - 0.6 s^{-1} for TX32 alloy. In addition, the flow instability regime is reduced with the addition of Al and Si. Both these changes are favorable for the hot workability since the workability windows are expanded by these two additions. This effect is essentially due to increased volume fraction of intermetallic particles due to the formation of Mg_2Si in the Si containing alloy. Higher temperatures are required to overcome the back stress before the recovery processes including climb or cross-slip to occur. Another noteworthy feature is that the saddle point configuration has not changed by the addition of Al and Si. This is in support of the earlier interpretation in terms of the change in the dominant slip mechanisms from basal+prismatic slip to pyramidal slip, which is not influenced by minor alloying additions. The reduced flow instability at higher temperatures may be interpreted in terms of the promotion of cross-slip by the Mg_2Si particles in addition to CaMgSn particles.

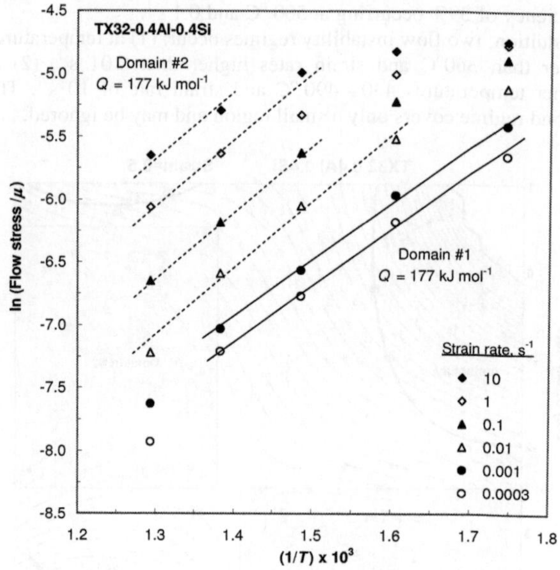

Figure 5. Arrhenius plot showing the variation of normalized flow stress of Mg-3Sn-2Ca-0.4Al-0.4Si with inverse of temperature at different strain rates.

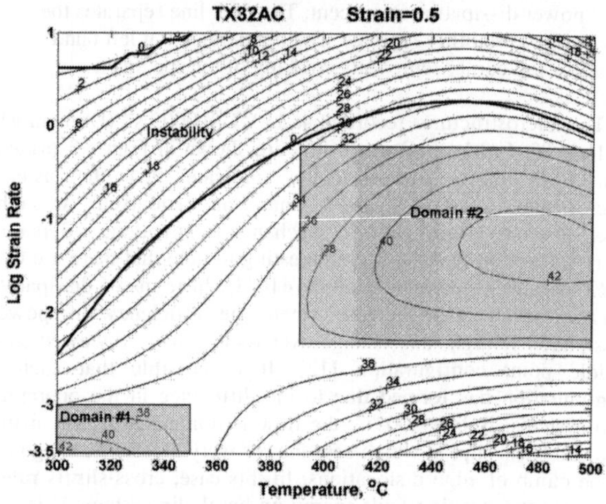

Figure 6. Processing map for Mg-3Sn-2Ca as-cast alloy. The numbers against the contours represent efficiency of power dissipation in percent. The dark line separates the flow instability regimes. The blank region at left hand top corner represents cracking of the specimens.

The activation parameters evaluated in the two domains of all the alloys are shown in Table 1. The values are nearly unchanged due to Al additions [24] while the activation energy value for Domain #2 decreased due to Si addition. This further supports the interpretation that Mg_2Si particles promote the recovery by cross-slip in this domain by increasing the barriers to dislocation movement. However, the kinetics of the climb process controlling Domain #1 is not much influenced by these particles.

Table 1. Activation parameters for Mg-3Sn-2Ca, Mg-3Sn-2Ca-0.4Al, and Mg-3Sn-2Ca-0.4Al-0.4Si alloys

Alloy (as-cast)	Domain #1		Domain #2	
	n	Q kJ/mole	n	Q kJ/mole
Mg-3Sn-2Ca	4.54	177	4.54	197
Mg-3Sn-2Ca-0.4Al	4.54	175	4.54	195
Mg-3Sn-2Ca-0.4Al-0.4Si	5.0	177	4.44	177

Summary and Conclusions

The compressive strength of Mg-3Sn-2Ca-0.4Al-0.4Si alloy in as-cast condition has been evaluated in the temperature range 25 – 250°C and compared with that of the base alloy Mg-3Sn-2Ca and Mg-3Sn-2Ca-0.4Al alloy with a view to assess the effect of Al and Si additions. The hot deformation behavior has also been studied using hot compression tests in the temperature range 300 – 500 °C and strain rate range 0.0003 – 10 s^{-1}. The flow stress data have been analyzed using processing maps and standard kinetic analysis. The effect of Al and Si additions to the hot working behavior has also been assessed by comparing the behaviors. The following conclusions are drawn from this investigation.

(1) The compressive strength of Mg-3Sn-2Ca alloy has increased due to the addition of 0.4 Al but deteriorated by the Si addition, particularly at higher temperatures.
(2) The deterioration in strength may have been caused by the formation of Mg_2Si particles, although the exact reason is unclear.
(3) The processing map for Mg-3Sn-2Ca-0.4Al-0.4Si alloy exhibited two domains in the temperature and strain rate ranges given by: (1) 300 – 410 °C and 0.0003 – 0.003 s^{-1} with a peak efficiency of 38% occurring at 350 °C and 0.0003 s^{-1} and (2) 420 °C – 500°C and 0.003 – 3 s^{-1} with a peak efficiency of 39% occurring at 500 °C and 0.1 s^{-1}.
(4) Comparison of processing map of Mg-3Sn-2Ca-0.4Al-0.4Si with that of the base alloy (Mg-3Sn-2Ca) revealed that both the domains have moved to higher temperatures and the flow instability regime reduced particularly at higher temperatures due to the addition of Al and Si. The basic features however remained unchanged.
(5) Both the domains represent the process of dynamic recrystallization. It is suggested that the rate controlling mechanism in the lower strain rate domain is climb while it is cross-slip in the higher temperature domain.
(6) The apparent activation energy obtained by kinetic analysis in both the domains of the map for Mg-3Sn-2Ca-0.4Al-0.4Si is 177 kJ/mole, which is higher than that for self diffusion in Mg.
(7) The apparent activation energy in the lower strain rate domain is similar to that obtained for the base alloy while it is lower in the second domain, suggesting that cross-slip is promoted by Mg_2Si particles.
(8) The addition of Si is favorable for enhancing the hot workability since it widens the processing window.

Acknowledgement

This work was supported by a grant (Project #115108) from the Research Grants Council of the Hong Kong Special Administrative Region, China.

References

1. T. Abu Leil et al., "Corrosion Behavior and Microstructure of a Broad Range of Mg-Sn-X Alloys," *Magnesium Technology 2006*, ed. A.A. Luo, N.R. Neelameggham and R.S. Beals (Warrendale, PA: The Minerals, Metals and Materials Society, 2006), 281-286.

2. T. Abu Leil et al., "Effect of Heat Treatment on the Microstructure and Creep Behavior of Mg-Sn-Ca Alloys," *Mater. Sci. Forum*, 546-549 (2007), 69-72.

3. T. Abu Leil et al., "Microstructure, Corrosion and Creep of As-cast Magnesium Alloys Mg-2Sn-2Ca and Mg-4Sn-2Ca," *Magnesium Technology 2007*, ed. R.S. Beals, A.A. Luo, N.R. Neelameggham, and M.O. Pekguleryuz (Warrendale, PA: The Minerals, Metals and Materials Society, 2007), 257-262.

4. Y.V.R.K. Prasad et al., "Hot Working Parameters and Mechanisms in As-Cast Mg-3Sn-1Ca Alloy," *Mater. Letters*, 62 (2008), 4207-4209.

5. Y.V.R.K. Prasad et al., "Hot Workability Characteristics of Cast and Homogenized Mg-3Sn-1Ca Alloy," *Jour. Mater. Proc. Tech.*, 201 (2008), 359-363.

6. Y.V.R.K. Prasad et al., "Optimum Parameters and Rate-Controlling Mechanisms for Hot Working as Extruded Mg-3Sn-1Ca Alloy," *Mater. Sci. Eng., A* 502 (2009), 25-31.

7. N. Hort et al., "Creep and Hot Working Behavior of a New Magnesium Alloy Mg-3Sn-2Ca," *Magnesium Technology 2008*, ed. M.O. Pekguleryuz, N.R. Neelameggham, R.S. Beals, and E.A. Nyberg (Warrendale, PA: The Minerals, Metals and Materials Society, 2008), 401-406.

8. J.J. Jonas, C.M. Sellars, and W.J. McG. Tegart, "Strength and Structure under Hot-Working Conditions," *Metall. Rev.*, 14 (1969), 1-24.

9. Y.V.R.K. Prasad and T. Seshacharyulu, "Modeling of Hot Deformation for Microstructural Control," *Inter. Mater. Rev.*, 43 (1998), 243-258.

10. Y.V.R.K. Prasad and S. Sasidhara, *Hot Working Guide: A Compendium of Processing Maps* (Materials Park OH: ASM International, 1997).

11. Y.V.R.K. Prasad et al., "Modeling of Dynamic Material Behavior in Hot Deformation: Forging of Ti-6242," *Metall. Trans.*, 15A (1984), 1883-1892.

12. Y.V.R.K. Prasad, "Processing Maps – A Status Report," *Jour. Mater. Eng. Perf.*, 12 (2003), 638-645.

13. H. Ziegler, "Some Extremum Principles in Irreversible Thermodynamics with Applications to Continuum Mechanics," *Progress in Solid Mechanics*, ed. I.N. Sneddon and R. Hill, Vol.4 (New York: Wiley, 1965), 91-193.

14. Y.V.R.K. Prasad, and K.P. Rao, "Processing Maps for Hot Deformation of Rolled AZ31 Magnesiuim Alloy Plate:

Anisotropy of Hot Workability," *Mater. Sci. Eng., A* 487 (2008), 316-327.

15. K.P. Rao et al., "Effect of Minor Additions of Al and Si on the Mechanical Properties of Cast Mg-3Sn-2Ca Alloys in Low Temperature Range," *Mater. Sci. Forum,* 654-656 (2010), 635-638.

16. K.M. Asl, A. Tari, and F. Khomamizadeh, "The Effect of Different Content of Al, RE and Si Element on the Microstructure, Mechanical and Creep Properties of Mg-Al Alloys," *Mater. Sci. Eng., A* 523 (2009), 1-6.

17. D.H. Kang, S.S. Park, and N.J. Kim, "Development of Creep Resistant Die Cast Mg-Sn-Al-Si Alloy," *Mater. Sci. Eng., A* 413-414 (2005), 555-560.

18. M.S. Dargusch et al., "The Effect of Silicon Content on the Microstructure and Creep Behavior in Die-Cast Magnesium AS Alloys," *Metall. Mater. Trans.*, 35A (2004), 1905-1909.

19. A. Srinivasan, U.T.S. Pillai, and B.C. Pai, "Microstructure and Mechanical Properties of Si and Sb Added AZ91 Magnesium Alloy," *Metall. Mater. Trans.*, 36A (2005), 2235-2243.

20. H.M. Nanjunda Swamy, S.K. Nath, and S. Ray, "Tensile and Fracture Properties of Cast and Forged Composite Synthesized by Addition of Al-Si Alloy to Magnesium," *Metall. Mater. Trans.*, 40A (2009), 3284-3293.

21. D.H. Sastry, Y.V.R.K. Prasad, and K.I. Vasu, "On the Stacking Fault Energies of Some Close-Packed Hexagonal Metals," *Scripta Metall., 3(1969)927-929.*

22. J.R. Morris et al., "Prediction of a {1122} hcp Stacking Fault Using a Modified Generalized Stacking-Fault Calculation," *Philos. Mag., A76* (1997), 1065-1077.

23. H.J. Frost, and M.F. Ashby, *Deformation-Mechanism Maps: The Plasticity and Creep of Metals and Ceramics* (Oxford: Pergamon Press, 1982), 44.

24. K.P. Rao et al., "Effect of Aluminum Addition on the Strengthening and High Temperature Deformation Behavior of Mg-3Sn-2Ca Alloy," *Magnesium Technology 2010,* ed. S.R. Agnew, N.R. Neelameggham, E.A. Nyberg, and W.H. Sillekens (Warrendale, PA: The Minerals, Metals and Materials Society, 2010) 201-205.

AN ANALYSIS OF THE GRAIN REFINEMENT OF MAGNESIUM BY ZIRCONIUM

P. Saha and S. Viswanathan
The University of Alabama, Tuscaloosa, AL 35487

Keywords: Magnesium, Zirconium, Grain refinement, TEM, Particle size distribution, Faceted particle

Abstract

A Design of Experiments approach was used to conduct a systematic study of the grain refinement of magnesium by zirconium; variables included the amount of zirconium, the pouring temperature, and the settling time prior to casting. Optical microscopy was used to measure the grain size. TEM was used to identify particles that are likely nucleation sites. The TEM results show that a range of particle sizes are likely substrates and that particles which are likely nucleation sites are faceted. Sample dissolution followed by SEM was used to characterize particle size and morphology while an AccuSizer 770 was used to measure particle size distributions, both in the master alloy and grain refined samples. While less than 2% of the total particles serve as nucleation sites, a comparison of the grain density vs. faceted particle density shows close agreement, suggesting that only faceted particles are likely nucleation sites.

Introduction

Zirconium is well known as an excellent grain refiner for magnesium alloys that do not contain aluminum, since aluminum poisons the grain refining ability of zirconium. [1] However, zirconium is a heavy element with a density almost four times that of magnesium (ρ_{Mg} = 1740 kg/m^3, ρ_{Zr} = 6520 kg/m^3). As a result, the addition of zirconium into molten magnesium results in the loss of zirconium by settling, since during melting zirconium particles settle to the bottom of the crucible and are wasted as sludge. As zirconium is expensive, this represents a significant additional cost as well as wastage of material.

The magnesium rich region of the binary Mg-Zr phase diagram [2] (see Figure 1) shows that grain refinement of magnesium can occur in theory by a peritectic reaction at 654°C, where α-zirconium reacts with liquid and forms an α-magnesium phase. Currently, approximately 1 wt% zirconium is currently added to ensure the peritectic reaction and grain refinement. However, in a systematic study of the grain refinement of magnesium, Saha et al. [3] showed that grain refinement was possible at additions less than the solubility limit of 0.5 wt% zirconium. (Note: All compositions reported in this work are in wt%.)

The goal of this study was to understand the mechanism of grain refinement of magnesium by zirconium and identify the requirements for a more efficient grain refiner. A permanent mold based "hockey puck" experimental setup was used to quantify the grain refinement efficacy of zirconium in pure magnesium [3]. A Design of Experiments (DOE) approach was used to conduct a systematic study of the grain refinement behavior of magnesium by zirconium; variables included temperature, settling time, and amount of zirconium addition prior to casting. Optical microscopy was used to measure grain size. Transmission electron microscopy (TEM) was used to explore and establish the orientation relationship between zirconium particles and the parent magnesium matrix. Sample dissolution followed by scanning electron microscopy (SEM) was used to characterize particle size and morphology while an AccuSizer 770 was used to measure particle size distributions, both in the master alloy and grain refined samples. The results from the above tasks were used to understand the mechanism of grain refinement of magnesium by zirconium and develop insights for engineering efficient zirconium-based grain refiners for magnesium alloys.

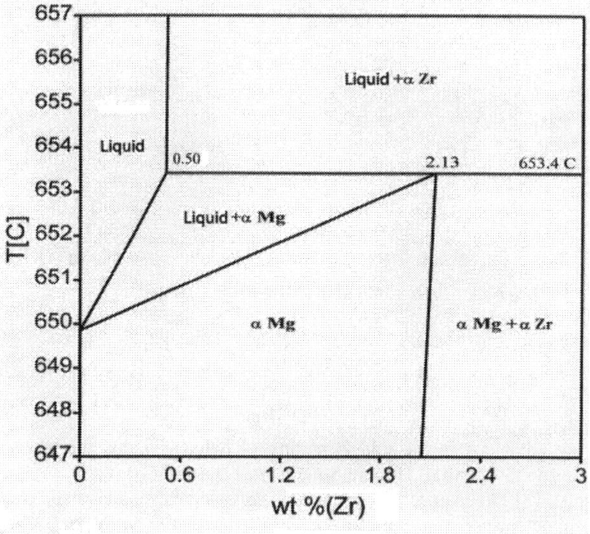

Figure 1. Relevant portion of magnesium-zirconium phase diagram [2] showing peritectic reaction.

Background

Microstructure development in castings is a nucleation and growth process; nucleation is typically heterogeneous. Grain refinement is the introduction of extraneous nucleating agents (grain refiners) to a liquid melt. From classical nucleation theory, the critical radius r^*_{het} for heterogeneous nucleation is calculated as [4]

$$r^*_{het} = -\frac{2\gamma_{SL}}{\Delta G_v} \tag{1}$$

Where γ_{SL} = solid-liquid interfacial energy, and
ΔG_v = Volume free energy change during solidification (i.e. driving force),

and
$$\Delta G_v = \frac{\Delta H \Delta T}{T_m} \tag{2}$$

where ΔT is undercooling, defined as the difference between the actual temperature and the equilibrium melting point or liquidus temperature. Undercooling must be negative for solidification to occur.

For magnesium, [5]
T_m = melting point of magnesium = 922.8K,
ΔH = change in enthalpy during solidification
= 5.898×10^8 J/m^3, and
$\gamma_{SL} = 115 \times 10^{-3}$ J/m^2.

Then, the critical radius r^*_{het} is given by

$$r^*_{het} = -\frac{0.36}{\Delta T} \quad \text{or} \quad d^*_{het} = -\frac{0.72}{\Delta T} \qquad (3)$$

where d^*_{het} is the diameter or size of the critical radius.

Figure 2 is a plot of Eqn. 3 and shows the undercooling required to nucleate a magnesium grain as a function of zirconium particle size.

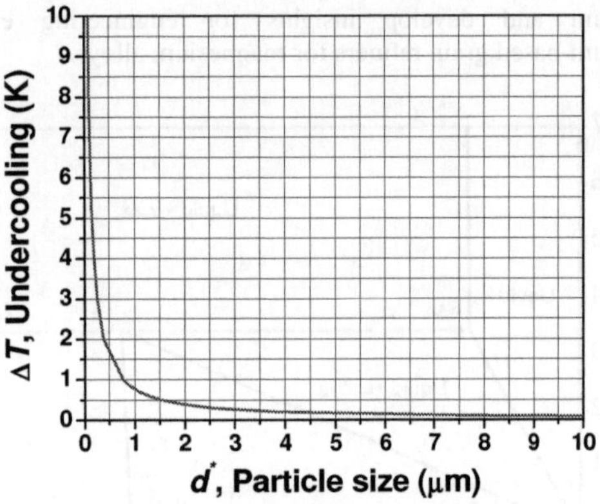

Figure 2. Undercooling required for nucleation of magnesium grain as a function of zirconium particle size.

Nucleation and Growth

Key advances in the understanding of heterogeneous nucleation are due to Turnbull [6] and Sundquist, [7] and of growth to Greer et al. [8] Greer et al. [8] proposed that grain formation is growth limited rather than nucleation limited. They suggested that solidification first occurs on the largest substrates at the lowest undercooling. As undercooling is increased, progressively smaller particles become centers for free growth, which occurs as soon as the required undercooling is reached.

Grain Refinement of Magnesium By Zirconium

Zirconium satisfies all of the requirements of a good grain refiner required by theory, [9] as:

1. Zirconium and magnesium have the same crystal structure (hcp) and the lattice disregistry between them is very small (Zr: a = 0.323 nm, c = 0.514 nm and Mg: a = 0.320 nm, c = 0.520 nm). [10]
2. A peritectic reaction occurs close to 0.5% zirconium at 654ºC. A zirconium-rich phase precipitates from the liquid and reacts with magnesium to form an α-magnesium phase.
3. Zirconium has the highest growth restriction factor among all elements that promote grain refinement in magnesium. [10]

While zirconium is an effective grain refiner, it suffers from two drawbacks. First, zirconium is an expensive addition, and excessive zirconium (1%) is added to achieve optimum grain refinement. Second, the zirconium is allowed to settle, as the large zirconium particles and particle clusters present in the grain refiner master alloy are detrimental to mechanical properties, and this is wasted as sludge.

Qian et al., [11-14] conducted a series of studies on the grain refinement of magnesium. They added 1% zirconium in the form of Zirmax® (Mg-33.3% Zr) master alloy to pure magnesium melt and found undissolved zirconium particles or large clusters (≥ 10 μm). These clusters did not take part in the grain refinement process and settled to the bottom of the crucible. They suggested that only particles between 1 and 5 μm in size acted as nuclei. They also suggested that the majority of grain refinement is achieved from soluble zirconium, although insoluble zirconium particles also contribute to the grain refinement.

DOE Study

Saha et al. [3] conducted experiments with pure magnesium (99.9% Mg) and magnesium-15% zirconium grain refiner master alloy. Three different factors were varied to study grain refinement efficacy – total zirconium addition (0.25, 0.5, and 1%), pouring temperature (705 and 815ºC, or 1300 and 1500ºF), and settling time (0 and 30 min). A total of 12 experiments were carried out, comprising a full factorial DOE matrix. Figs. 4, 5, and 6 show the variation of the average grain size with total zirconium addition, pouring temperature, and settling time [3].

Doe Study Results

Grain size decreased with increasing zirconium addition for both pouring temperatures and settling times. Also, grain refinement was obtained at 0.25% zirconium, which is less than the solubility limit and less than that required for the peritectic reaction. This shows that zirconium additions less than that needed for the peritectic reaction can produce grain refinement, providing significant new opportunities for increasing zirconium efficiency. A smaller grain size is generally obtained at the higher pouring temperature. Finally, a settling time of 30 min generally resulted in a larger grain size. [3]

TEM Study

TEM analysis was used to identify zirconium particles in grain refined samples that are likely nucleation sites. TEM bright field (BF) microscopy and selected area composite diffraction patterns (SADP) were used to analyze the particle-matrix orientation relationship. For particles which were not electron transparent, high resolution TEM (HRTEM) was used to determine orientation relationships at the particle-matrix interface. The bright field micrographs and diffraction patterns were taken using a Tecnai F-20 TEM operating at 200 keV.

Various particles from a grain refined sample (magnesium-0.5% zirconium, 815ºC, 30 min) were observed under TEM. Fig. 3(a) shows a small faceted zirconium particle 0.23 μm in size at the center of a magnesium grain. The SADP of the particle and the matrix illustrated in Fig. 3(b) shows that the particle and the matrix have the same crystallographic orientation. The diffraction spots were indexed based on the $[2\bar{1}\bar{1}0]$ direction of the hcp crystal, and all the diffraction spots from the matrix and the particle were exactly superimposed on each other, i.e. the planar disregistry between the low-index planes was very small. The orientation relationship between the particle and the matrix was established as $(0001)_{Mg} \| (0001)_{Zr}$ and $[2\bar{1}\bar{1}0]_{Mg} \| [2\bar{1}\bar{1}0]_{Zr}$, i.e. the basal planes of the particle and matrix had the same orientation. The similarity in the crystal structure and similar lattice spacings between magnesium and zirconium supports the hypothesis that zirconium particles which are likely nucleation sites will show a parallel axis orientation with the matrix.

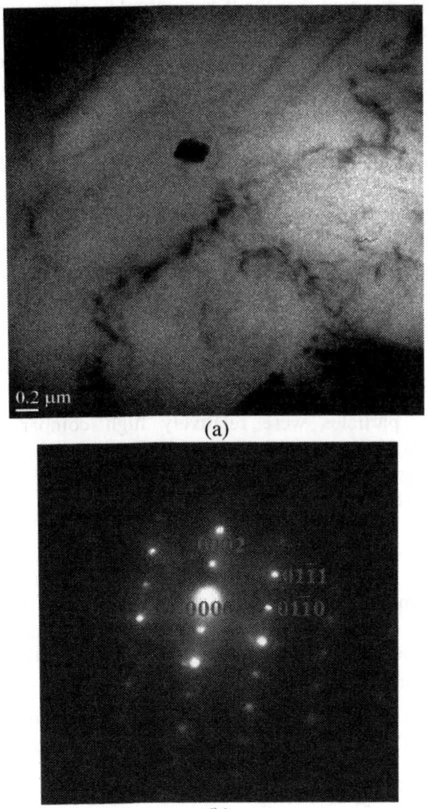

Figure 3. (a) TEM bright field micrograph showing a 0.23 µm faceted zirconium particle in the magnesium matrix, and (b) Composite diffraction pattern of particle and matrix showing that diffraction spots are superimposed.

Fig. 4(a) shows a large faceted zirconium particle of 1.6 µm length and 1 µm width) in the matrix at the center of a single grain from the same sample (magnesium-0.5 % zirconium 815°C-30 min grain refined analyzed in Fig. 7. The bright field micrograph suggests that the particle is faceted in nature. The composite selected area diffraction pattern shown in Fig. 4(b) for the particle and matrix reveals that a parallel axis orientation relationship exists between the particle and matrix. The diffraction spots were indexed from the standard $[2\bar{1}\bar{1}0]$ dffraction pattern of an hcp magnesium crystal.

HRTEM was performed at the particle/matrix interface indicated by an arrow in Fig. 4(a). The lattice fringes in Fig. 4(c) are parallel and continuous across the particle-matrix interface and suggest that the particle is coherent with the matrix and there is no atomic misfit or dislocation at the interface. The result suggests that the magnesium grain has grown epitaxially on the zirconium crystal.

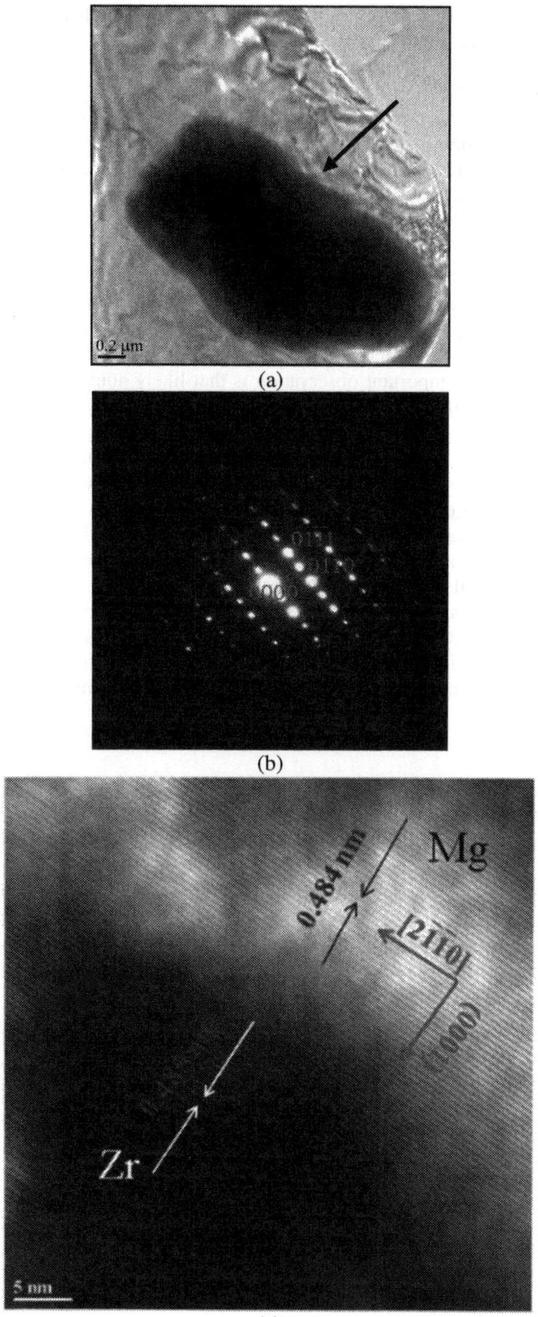

Figure 4. (a) Bright Field TEM micrograph of a zirconium particle (1.6 µm length and 1 µm width) in the matrix, (b) composite diffraction pattern denoting parallel axis orientation of particle and matrix, and (c) HRTEM lattice fringe image at the particle/matrix interface showing parallel and continuous lattice fringes suggesting epitaxial growth of magnesium on zirconium.

Several other particles, both faceted and non-faceted, were found at grain boundaries and did not show any orientation relationship with the matrix, suggesting that they were like not nuclei. This also suggests that not all particles are selected as nucleation sites and cause grain refinement.

TEM Study: Discussion

The results of this section show that TEM with SADP or HRTEM is an excellent tool for identifying particles that are likely nucleation sites. The TEM study from the magnesium-0.5% zirconium, 815°C pouring temperature, 30 min settling time, grain refined sample showed that a range of zirconium particle sizes (0.23 and 1.6 μm) are likely nucleation sites. (Note: a third particle 0.5 μm in size that showed a parallel axis orientation relationship with the matrix was also found). This supports Greer's free growth theory that nucleating substrates of different size will become active when the required undercooling is achieved. An important observation is that likely nucleation sites, i.e. particles that show parallel axis orientation relationship with small planar disregistry (0.1 to 0.2%) appear to be faceted. This is reasonable as zirconium particles which are faceted in nature are likely to expose a suitable nucleating substrate to the matrix. In particles that exhibited a parallel axis orientation with the matrix, the planes that were parallel were generally either the basal or prismatic planes. HRTEM analysis indicated that in particles that show a parallel axis orientation relationship with the matrix, the lattice fringes at the particle matrix interface are parallel and continuous across the particle-matrix interface and suggest that the particle is coherent with the matrix, suggesting that the magnesium grain has grown epitaxially on the zirconium crystal.

Particle Size and Morphology

The TEM study outlined in the previous section suggested that only faceted particles were likely nucleation sites. In an effort to characterize and quantify particle size, particle size distribution, and particle morphology, samples of zirconium particles were prepared from grain refined samples. Samples were dissolved into 1% diluted hydrochloric acid (0.1 g/10 ml basis) and kept for 72 hrs to digest the sample completely. One ml of the solution was transferred into an EPPENDORF tube and centrifuged at 15000 rpm for 15 min for 3 cycles. After each cycle the liquid was decanted, and de-ionized water was added to the mixture, and the tube was sonicated for 10 min. This allowed magnesium chloride, which is highly soluble in water, to be washed away and for the zirconium particles to be collected at the bottom. Zirconium particle size distributions were determined using an Accusizer model 770 particle size analyzer that uses the light obscuration/photozone method (Particle Technology Labs, Downers Grove, IL, USA).

The final centrifuged sample was ultra-sonicated for 30 min and one drop of the solution was evenly distributed on a silicon wafer. The wafer was dried and kept under vacuum to prevent it from being contaminated. The silicon wafer with the zirconium particles was characterized by SEM at an accelerating voltage of 15 kV, working distance of 10 mm and spot size of 7 nm. Faceted zirconium particles evident from the SEM micrographs were counted using the automated threshold picking tool in NIS-Elements image analysis software. Particles which showed a clear facet and did not have any rounded or curved edges or surfaces were designated as faceted particles. The total number of particles was also counted to determine the percent faceted particles. Faceted particle density was estimated from the particle size distribution data.

Particle Size Distribution

The Accusizer model 770 was only capable of measuring particles greater than 0.5 μm. Also, the particle size analysis data showed that particles greater than 5 μm were present in extremely small numbers. Accordingly, only particle sizes ranging between 0.5 and 5 μm are plotted. Fig. 5 shows the number of particles present in the magnesium-0.25% zirconium grain refined sample at 0 and 30 min. It is observed that the 0 min sample exhibits a wide size distribution and a large number of particles (total count was 83,212 in 1 ml of the dissolution sample) while the 30 min sample has a narrow distribution and a drastically smaller number of particles (total count was 6170 in 1 ml of the dissolution sample) . While it might be expected that at 30 min, particles larger than 2 μm particles will settle and be unavailable for grain refinement, the drastic decrease in particles smaller than 2 μm suggests that smaller particles likely coalesced and settled, providing a far smaller particle count at 30 min than expected. The phenomena of particle settling and a consequent increase in grain size is commonly observed in industrial practice and is termed grain refiner fade.

A similar trend was also observed in the magnesium-1% zirconium sample and is shown in Fig. 6. However, the total numbers of particles were relatively high compared to the magnesium-0.25% zirconium sample. This explains why the increase in grain size at 30 min is much smaller at the 1% zirconium level compared to 0.25% zirconium.

Total Particle Efficiency

The fraction of zirconium particles responsible for nucleating grains in each grain refined sample was estimated as the ratio of grain density to particle density. This assumes that each nucleant particle only nucleates a single grain. It was assumed that the total particle density N_v^P (mm^{-3}) was equal to the number of particles present at the start of the melt, i.e. the particle density for the 0 min sample. The grain density N_v^G was calculated as [8]

$$N_v^G = \frac{0.5}{\bar{l}^{3}}$$

where $\bar{l}$ is the mean lineal intercept of grains.

Table 1 lists the grain size, grain density, particle density, and the nucleant particle efficiency for the magnesium-0.25 and 1% zirconium grain refined samples. It is clear that total particle efficiency decreases with increasing settling time. Importantly, in all cases, the fraction of active zirconium particles is less than 2%, suggesting that only a few of the total particles take part in grain refinement. . Interestingly, Greer et al. [8] in their work on aluminum grain refinement with TIBOR, found that only 1% of TiB_2 particles were effective as nuclei.

Faceted Particle Efficiency

Table 2 shows a comparison of the faceted particle density in grain refined samples. It is observed that approximately 2 to 5% of particles present are faceted. Table 3 also shows that the percent of faceted particles increased after 30 min of settling time. The increase in the number of faceted particles at 30 min might be due to the fact that the particles may have dissolved partially and exposed facets in the melt.

Table 3 shows a comparison of the grain density with the faceted particle density. While the total particle efficiency was less than

2% (Table 1), the faceted particle efficiency is around 50% (and varies from 36 to 63%), i.e. a ratio of the grain density to the faceted particle density is approximately half. It is clear that the faceted particle density shows an excellent match with the grain density. This result suggests that only faceted particles are likely active nucleation sites for magnesium grains.

Summary Discussion and Suggestions For Future Grain Refiner Development

Current grain refiners are extremely inefficient; a comparison of grain density with total particle density suggests less than 2% of all particles are active nucleation centers. However, the ratio of the grain density to the faceted particle density is approximately half, suggesting that only faceted particles are active as nucleation centers for magnesium grains. TEM with SADP or HRTEM is an excellent tool for identifying particles in grain refined samples that are likely nucleation sites. Likely nucleation sites are faceted particles that show a parallel axis orientation relationship and small planar disregistry (0.1 to 0.2%) with the matrix. This is reasonable as zirconium particles which are faceted in nature are likely to expose a suitable nucleating site to the matrix. However, only 2 to 5% of all particles are faceted, suggesting that new methods for manufacturing grain refiners need to be developed as the current salt replacement technique is seriously flawed.

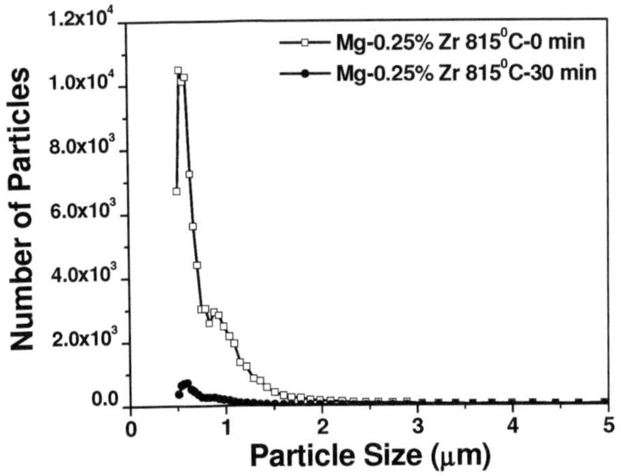

Figure 5. Size distribution of zirconium particles in magnesium-0.25% zirconium grain refined samples at 0 and 30 min. The drastically lower number of particles at 30 min suggests particle agglomeration and settling with time (termed fading).

TEM of grain refined samples showed that a range of zirconium particles act as nucleation sites. This supports Greer's free growth theory that nucleating substrates of various sizes will become active when the required undercooling is achieved, i.e. grain refiners should consist of particles of a range of sizes (e.g. 0.5 to 2 μm).

The dissolution study of grain refined samples suggests that due to the likely coalescence of particles combined with settling, far fewer particles are present at 30 min than would be expected due to settling alone. The reduction in the number of particles at 30 min explains the substantially larger grain size in the 30 min grain refined samples. Thus, the current industrial practice of allowing a melt to settle to allow larger particles to settle out, wastes a lot more zirconium than previously imagined. A far better approach would be to ensure that the maximum particle size is limited to around 2 μm, to limit any deterioration in mechanical properties, and introduce grain refiners as late as possible (Note: late addition is already current practice in aluminum alloy casting, inoculation of iron, etc. but this work provides quantitative data and further confirms why this is important).

The dissolution study of the grain refined samples also showed that more faceted particles were present in the 30 min sample. It is likely that particles partially dissolved and exposed facet in the melt. This explains why grain size was smaller at the higher pouring temperature, as the higher temperature likely increased the dissolution rate of particles.

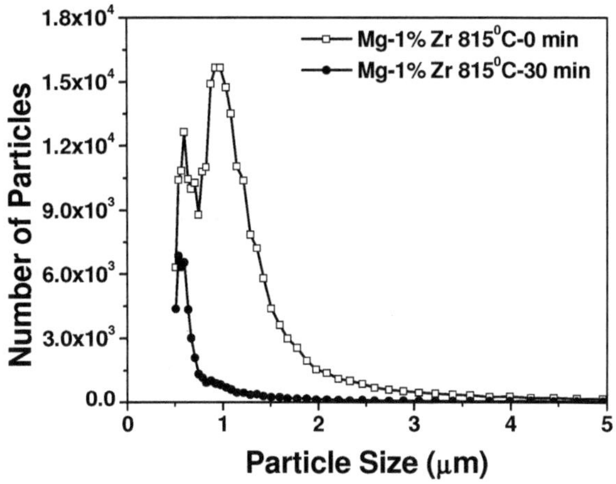

Figure 6. Size distribution of zirconium particles in magnesium-1.0% zirconium grain refined samples at 0 and 30 min.

Table 1. Comparison of nucleant particle efficiency in grain refined samples.

Sample ID	Grain size (μm)	Grain density (mm^{-3})	Total particle density (mm^{-3})	Nucleant particle efficiency (%)
Mg-0.25% Zr 815°C-0 min	109	386	56987	**0.68**
Mg-0.25% Zr 815°C-30 min	230	41	56987	**0.07**
Mg-1.0% Zr 815°C-0 min	51	3769	266667	**1.41**
Mg-1.0% Zr 815°C-30 min	60	2315	266667	**0.87**

Table 2. Fraction of faceted zirconium particles present in grain refined samples.

Sample ID	No. of total particles	No. of faceted particles	Fraction of faceted particles (%)
Mg-0.25% Zr 815°C-0 min	599	15	**2.5**
Mg-0.25% Zr 815°C-30 min	628	33	**5.3**
Mg-1.0% Zr 815°C-0 min	354	8	**2.3**
Mg-1.0% Zr 815°C-30 min	348	16	**4.6**

Table 3. Comparison of faceted particle efficiency in grain refined samples.

Sample ID	Grain size (μm)	Grain density (mm^{-3})	Faceted particle density (mm^{-3})	Faceted particle efficiency (%)
Mg-0.25% Zr 815°C-0 min	109	386	952	**40.5**
Mg-0.25% Zr 815°C-30 min	230	41	114	**36.0**
Mg-1.0% Zr 815°C-0 min	51	3769	5973	**63.1**
Mg-1.0% Zr 815°C-30 min	60	2315	6029	**38.4**

Acknowledgements

This work was supported by The University of Alabama (UA) and the UA Central Analytical Facility and was carried out as part of the High Integrity Magnesium Automotive Component (HIMAC) project sponsored by the United States Automotive Materials Partnership (USAMP). This material is based upon work supported by the Department of Energy National Energy Technology Laboratory under Award Number DE-FC26-02OR22910. This report was prepared as an account of work sponsored by an agency of the United States Government. Neither the United States Government nor any agency thereof, nor any of their employees, makes any warranty, express or implied, or assumes any legal liability or responsibility for the accuracy, completeness, or usefulness of any information, apparatus, product, or process disclosed, or represents that its use would not infringe privately owned rights. Reference herein to any specific commercial product, process, or service by trade name, trademark, manufacturer, or otherwise does not necessarily constitute or imply its endorsement, recommendation, or favoring by the United States Government or any agency thereof. The views and opinions of authors expressed herein do not necessarily state or reflect those of the United States Government or any agency thereof.

References

1. E. F. Emley, *Principles of Magnesium Technology.* (Pergamon Press, Oxford, United Kingdom, 1966), pp. 126-156, 254-260.
2. Pandat Software with PanMagnesium Database, Computherm LLC, Madison Wisconsin, USA.
3. P. Saha, K. Lolies, S. Viswanathan, R. G. Batson, and A. M. Gokhale, "A Systematic Study of the Grain Refinement of Magnesium by Zirconium," *Magnesium Technology 2010* (ISBN: 978-0-87339-746-9, The Minerals, Metals, and Materials Society, Warrendale, PA, 2010), p 425.
4. J. H. Perepezko, "Nucleation Kinetics and Grain Refinement," ASM Handbook: Vol. 15, *Casting* (ASM International, Materials Park, OH, Dec 2008), pp. 278.
5. M. Qian, "Heterogeneous nucleation on potent spherical substrates during solidification," *Acta Materialia* (2007), 55[3] 943-953.
6. D. Turnbull, "Theory of catalysis of nucleation by surface patches," *Acta Metallurgica* (1953), vol. 1, pp. 8-14.
7. B. E. Sundquist, "On "Nucleation catalysis in supercooled liquid tin"," *Acta Metallurgica*, vol. 11, pp. 630-632, 1963.
8. A. L. Greer, A. M. Bunn, A. Tronche, P. V. Evans, and D. J. Bristow, "Modelling of inoculation of metallic melts: application to grain refinement of aluminium by Al-Ti-B," *Acta Materialia* (2000), 48[11] 2823-2835.
9. J. H. Perepezko, "Nucleation Kinetics and Grain Refinement," ASM Handbook: Vol. 15, *Casting* (ASM International, Materials Park, OH, Dec 2008), pp. 281-282.
10. Y. C. Lee, A. Dahle, and D. H. StJohn, "The role of solute in grain refinement of magnesium," *Metallurgical and Materials Transactions A*, 31[11] 2895-2906 (2000).
11. Q. Ma, D. H. StJohn, and M. T. Frost, "Characteristic zirconium-rich coring structures in Mg-Zr alloys," *Scripta Materialia* (2002), vol. 46, pp. 649-654.
12. M. Qian, D. H. StJohn, and M. T. Frost, "Heterogeneous nuclei size in magnesium-zirconium alloys," *Scripta Materialia* (2004), vol. 50, pp. 1115-1119.
13. M. Qian, L. Zheng, D. Graham, M. T. Frost, and D. H. StJohn, "Settling of undissolved zirconium particles in pure magnesium melts," *Journal of Light Metals* (2001), vol. 1, pp. 157-165.
14. M. Qian, D. H. StJohn, and M. T. Frost, "Effect of soluble and insoluble zirconium on the grain refinement of magnesium alloys," *Material Science Forum* (2002), vol. 419-422, pp. 593-598.

Magnesium Technology 2011
Edited by: Wim H. Sillekens, Sean R. Agnew, Neale R. Neelameggham, and Suveen N. Mathaudhu
TMS (The Minerals, Metals & Materials Society), 2011

STUDY ON THE GRAIN REFINEMENT BEHAVIOR OF Mg-Zr MASTER ALLOY AND Zr CONTAINING COMPOUNDS IN Mg-10Gd-3Y MAGNESIUM ALLOY

Guohua Wu[1, 2], Ming Sun[1, 2], Jichun Dai[1, 2], Wenjiang Ding[1, 2]
[1]National Engineering Research Center of Light Alloy Net Forming; Shanghai Jiaotong University; Shanghai, 200240, China
[2]State Key Laboratory of Metal Matrix Composites; Shanghai Jiaotong University; Shanghai, 200240, China

Keywords: Grain refinement, Magnesium alloy, Zirconium (Zr)

Abstract

The effects of Mg-Zr master alloy and a potassium fluozirconate (K_2ZrF_6) salt mixture (KSM) on the grain refinement behavior of Mg-10Gd-3Y magnesium alloy were studied. The results show that the Mg-10Gd-3Y alloy is well refined by Mg-Zr or KSM. The characteristic microstructure feature of the alloy refined by Mg-Zr master alloy is the Zr-rich cores that exist in most grains, while the Zr-rich cores are not observed in the alloy refined by KSM. It is suggested that the grain refinement mechanisms of zirconium in the two cases are different: the Zr released from Mg-Zr master alloy works by adding heterogeneous nucleants, while the Zr produced from the in-situ reaction between Mg melt and K_2ZrF_6 works by restricting grain growth. Compared with the Mg-30.wt%Zr master alloy, the KSM refiner shows much longer fading time during melting.

Introduction

The recently developed Mg-Gd-Y-Zr alloys are becoming one of the most promising heat-treatable magnesium alloys, because they exhibit excellent specific strength at both room and elevated temperatures, and its creep resistance is markedly better than that of WE54, which until now has been the most creep resistant commercially available magnesium alloy [1-4].

In Mg-RE alloys, the powerful grain refiner, Zr, leads to a fine and equiaxed microstructure, which is needed for structural uniformity and consistency in performance of Mg-products [5]. Hitherto, many papers have studied the grain refinement behavior of adding Zr into magnesium alloys [6-16]. Over the decades various approaches to introducing Zr into molten magnesium alloy have been explored, including alloying with different forms of Zr metal, alloying with Zr sponge, alloying with Zn-Zr, with ZrO_2, with Zr-salts, or with Mg-Zr master alloys [10, 16]. Although, in principle, zirconium may be introduced into magnesium by a number of approaches, it appears that only Mg-Zr type master alloys have been in widespread use since 1960 because of the disadvantages of other alloying approaches [10, 11, 15]. However, during the alloy casting process, the utilization rate of Mg-Zr master alloy is not satisfactory (low alloying efficiency [8]) because the undissolved Zr particles spontaneously settle due to the much higher density of Zr (~6.52 g/cm^3) than liquid Mg (~1.7 g/cm^3) [12], and that large zirconium particles (>5μm) are inactive as nucleation centers [6]. Consequently, an excess of Mg-Zr master alloy needs to be introduced in order to achieve the satisfactory zirconium content for optimum grain refinement, which greatly increases the cost [8]. Therefore, work reconsiders the relatively cheap, convenient and easily handled Zr-salt method, and experiments related to the grain refinement effect of K_2ZrF_6 on magnesium alloy have been performed and reported in our recent paper [17].

Moreover, most of the published papers concerning the grain refinement effect of Zr on magnesium alloys only focus on alloying with Mg-Zr master alloy, which may not be too comprehensive for understanding the precise mechanism by which Zr works. The grain refinement mechanism of Zr is still the subject of investigation. So, this study also furthers the knowledge of the grain refinement by Zr addition into magnesium.

Experimental Procedures

The Mg-10Gd-3Y-0.5Zr (GW103K) alloy was prepared from high purity Mg (99.95wt.%) and master alloys of Mg-25wt.%Gd and Mg-25wt.%Y in an electric resistance furnace under a mixed flowing protective gas of CO_2 and SF_6 with the volume ratio of 100:1. Two forms of Zr grain refiner including Mg-Zr master alloy (microstructure as shown in Figure 2) and a potassium fluozirconate (K_2ZrF_6) salt mixture (KSM, composition of 50wt.%K_2ZrF_6-25wt.%NaCl-25wt.%KCl) were used respectively to compare their grain refinement behaviors. For the former, the addition order was Mg→Mg-25%Gd→Mg-25%Y→Mg-30%Zr. However, for the latter, the addition order was Mg→Mg-25%Gd→KSM→Mg-25%Y, where KSM is added before melting Mg-Y because Y can preferentially react with the K_2ZrF_6 causing loss of Y. After the designed holding time, it was cast at 750 °C into metallic molds preheated to 150 °C.

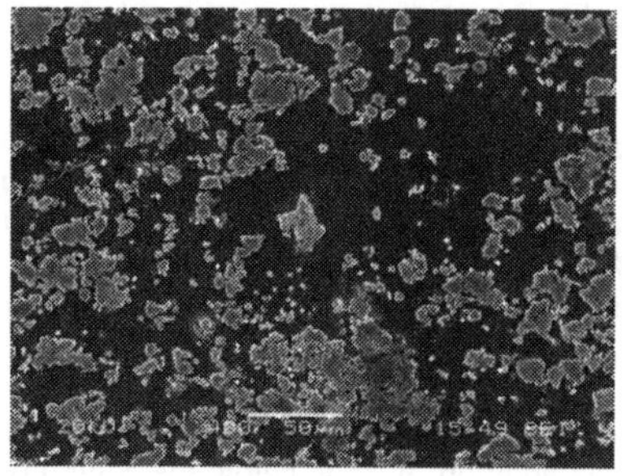

Figure 1. SEM microstructure of as-received Mg-30wt.%Zr master alloy.

A combination of optical microscopy (OM, OLYMPUS BX51M), scanning electron microscopy (SEM, JEOL JSM–6460), electron probe micro-analyzer (EPMA, JXA-8800R), transmission electron microscope (TEM, JEOL 2010), energy dispersive X-ray spectrometry (EDS, Oxford INCA) was carried out to characterize as-cast microstructure. Specimens for microstructure observation were etched in a 4 vol.% nital solution.

Results and Discussion

Grain Refinement Efficiency

The connection between average grain size (AGS) and Zr content is shown in Figure 2 [17, 18]. It can be seen that the AGS reduces with an increase of Zr content regardless of the difference in the two refiners. The AGS of alloy with 0.93% Zr refined by Mg-Zr master alloy is the finest (22μm), while the alloy with 0.56% Zr refined by KSM is the finest (21μm). Hence, it is possible that the grain refinement efficiency of in-situ Zr chemically reduced from the K_2ZrF_6 salts is much better than that of the ex-situ Zr released from the Mg-Zr master alloy.

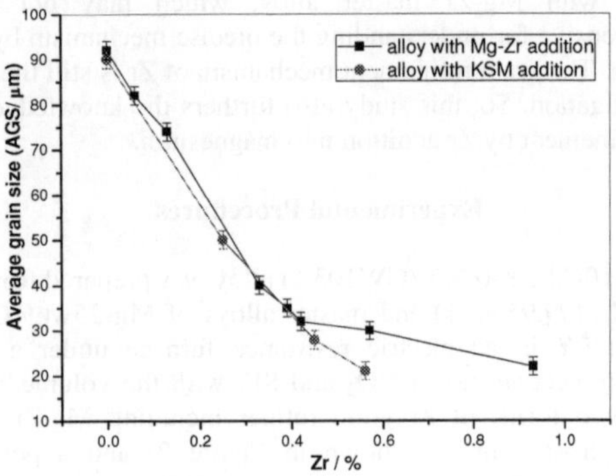

Figure 2. Relationship between Zr content and average grain size (AGS) of Mg-10Gd-3Y-Zr alloys.

Microstructures

Figures 3 and 4 show the microstructures of the alloy refined by Mg-Zr master alloy and the alloy refined by KSM, respectively. They all consist of primary α-Mg matrix and the β-Mg_{24}(Gd, $Y)_5$ eutectic phase distributed along the grain boundaries [17, 18].

The microstructures of alloys with different Zr contents, which are refined by Mg-Zr, are shown in Fig. 3. It can be seen that the AGS reduces with an increasing Zr content. Because the solubility of Zr in molten magnesium at the peritectic temperature (653.6°C) is only ~0.445% and increases to ~0.6% at 780 °C [13], and there are no intermetallics in Mg-Zr binary phase diagram, some undissolved Zr particles are present in the alloy with 0.93% Zr content. Previous report showed that magnesium melts inoculated with Zr particles always contain some undissolved Zr particles even though the addition of Zr is below the solubility limit [13]. Therefore, Figure 3 (c) shows that undissolved Zr particles may also caused grain refinement effect. This is in good agreement with Ma's study [13] which shows that nucleation of magnesium grains occurred on two types of zirconium nucleants, where the first type were formed by precipitation, and the other type were undissolved particles. The conclusion is further testified by Figure 3 (d) which shows that a sample with 3.67%Zr taken from the bottom of the melt has very fine grain size (10μm). Moreover, it is recently shown by Tamura et al. that undissolved Zr particles actually play an important role in grain refinement [19], and more recent work [12, 13, 20] has confirmed this observation and demonstrated that grain refinement of magnesium alloys by Zr is dictated by both soluble and insoluble Zr contents.

An obvious distinction between the Zr-rich cores (the halo structures formed in the melt, which are Zr-rich magnesium crystallites [13]) in Figure 3 (b) and Figure 3 (c) is that the former is petal-like while the latter is spherical. Ma [13] indicates that the major factor noticed affecting the growth patterns of Zr-rich halo is the dissolved Zr content: the magnesium halo structures in a low Zr content magnesium melt is much more prone to dendritic growth than those formed in a high Zr content magnesium melt.

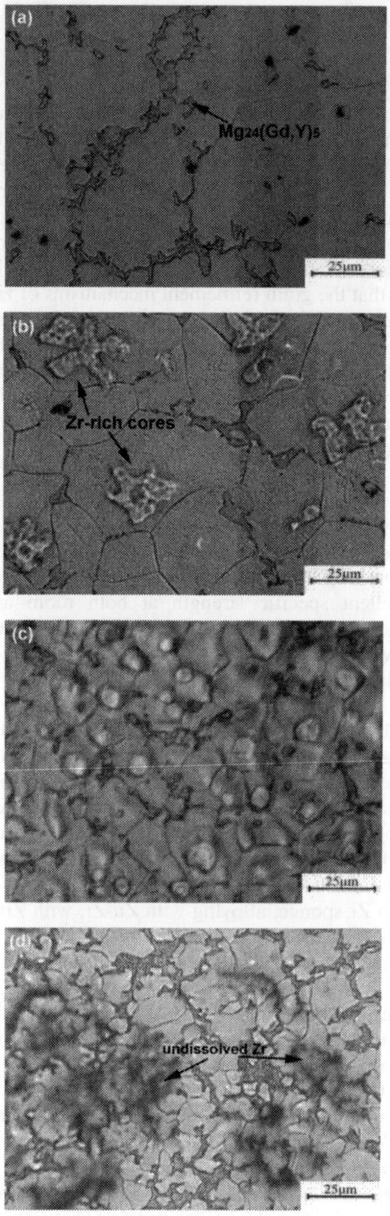

Figure 3. Optical microstructures of as-cast alloys with different Zr contents (wt.%), (a) 0; (b) 0.42; (c) 0.93, and (d) 3.67%, a sludge sample taken from the bottom of the melt has very fine grain size (~10μm).

Figure 4 shows the influence of KSM addition level on the microstructure of GW103 alloy. It is clear that with increasing addition level from 0% to 12% (wt.%), the AGS of GW103K alloy decreases gradually. However, the addition of the salt mixture causes inclusions whose compositions are composed of fluorides, chlorides and oxides which result from the chemical reduction reaction: $K_2ZrF_6+2Mg=Zr+2MgF_2+2KF$. Our previous paper [17] has clarified that the reason for the presence of inclusions whose compositions are mainly O, F, Cl, Na, K, Mg, Gd, Y and Zr is possibly that the reaction products of KF and MgF_2 combined with the residual KSM to form inclusion clusters.

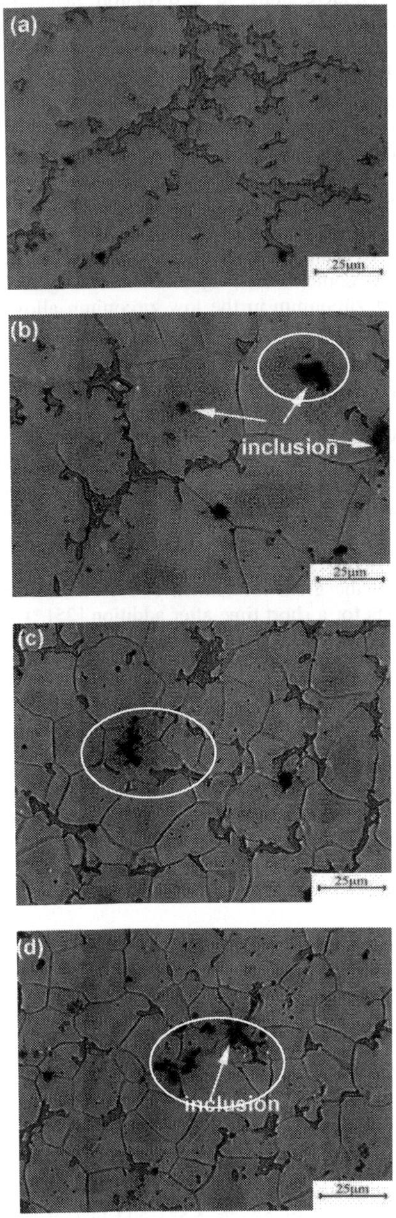

Figure 4. Microstructure of GW103K alloy with different KSM addition level: (a) 0, (b) 4%, (c) 8%, (d) 12%.

Different Existing Feature of Zr

Although both of the two agents can well refine the Mg-10Gd-3Y alloys, the presence of Zr in the alloys is different. As shown in Figure 3, the Zr-rich cores or undissolved Zr particles are obvious in the alloys refined by Mg-Zr master alloy, similar to the previous literatures [1, 9, 17]. The characteristic Zr-rich cores in Figure 5 illustrate that some small white Zr particles located at the centre of the Zr-rich cores which then form Zr clusters and act as nucleates.

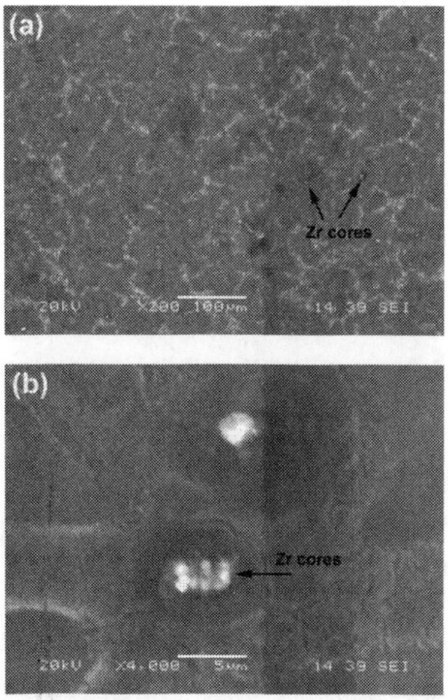

Figure 5. SEM morphology of the Zr cores in the model alloy Mg-10-Gd-3Y-0.52Zr alloy at: (a) low magnification and (b) high magnification.

Figure 6 shows TEM images and corresponding EDS results of the particles in the alloy refined by Mg-Zr, further indicating that Zr is meant as α-Zr, which is in good agreement with previous observations [1, 10].

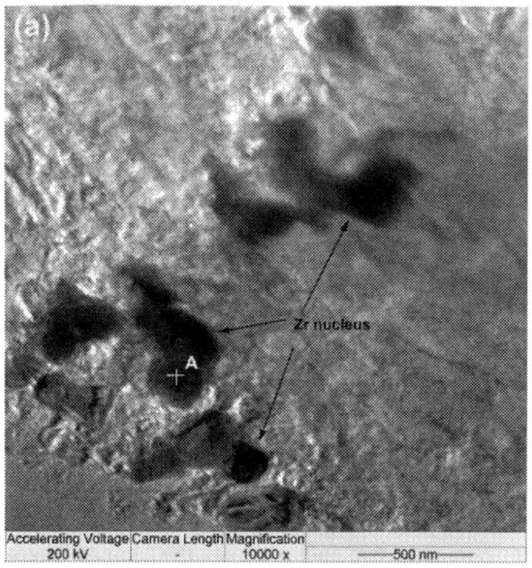

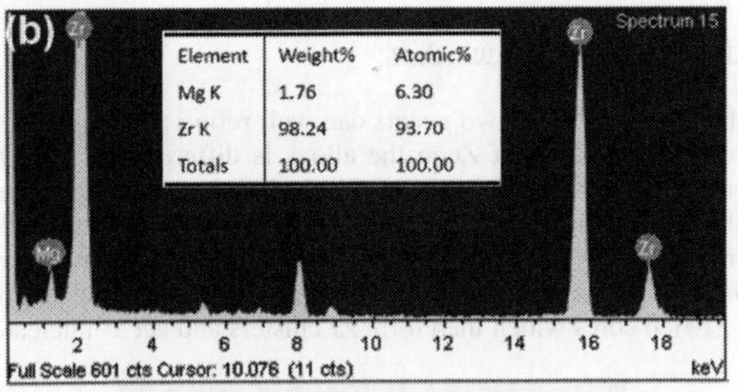

Figure 6. (a) TEM image, (b) EDS of Zr nucleus.

However, the Zr-rich cores are not observed in the alloy refined by K_2ZrF_6 as shown in Fig. 3, and the Zr element distribution is uniform detected by EPMA scanning as shown in Figure 7. So it is possible to believe that the in-situ tiny Zr well dissolves into the magnesium melt and then causes the grain refinement effect.

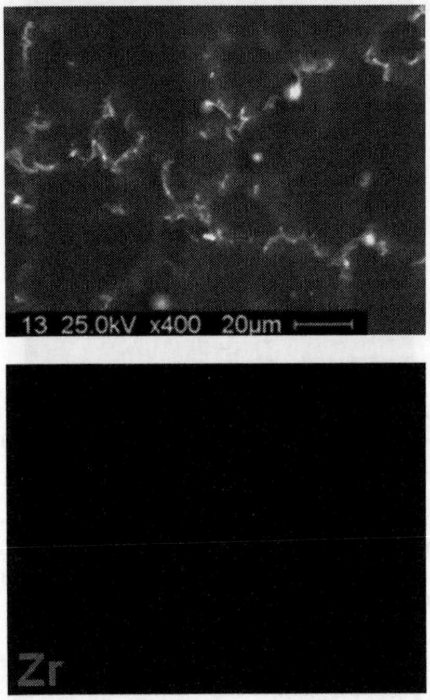

Figure 7. EPMA surface distribution of the element Zr in the 8% KSM processed sample.

<u>Discussion</u>

Although the microstructural feature of Zr distribution in the two alloys is different, the grains are both well refined. The mechanism by which Zr works may be different. For the alloy refined by Mg-Zr, the grain refinement mechanism may be explained by the most popular peritectic mechanism proposed by Emley [21]: since α-Zr and α-Mg have the same crystal structure (*hcp*) and nearly same lattice parameters (Zr: a_{Zr}=0.323nm, c_{Zr}=0.514nm; Mg: a_{Mg}=0.320nm, c_{Mg}=0.520nm), suitably sized α-Zr particles may act as effective nucleant sites for magnesium. The most characteristic feature of the microstructure of a magnesium alloy containing more than a few tenths per cent soluble Zr is the Zr-rich cores that exist in most grains [7]. They are believed to be the products of peritectic solidification. However, for the alloy refined by K_2ZrF_6, where the in-situ Zr dissolves well in the Mg matrix, the grain refinement mechanism can be understood from a growth restriction perspective. According to Lee's calculations [22], the solute Zr has the strongest growth restriction factor (GRF) among the common alloying elements in Mg, such as Ca, Si, Zn, Cu, Al, Sc, Sr, Ce, Y, etc, which makes Zr have powerful grain refinement effect on Mg. Here, the GRF theory means that the solute elements generate constitutional undercooling in a diffusion layer ahead of the advancing solid/liquid interface and then restrict grain growth [22-24]. Consequently, for the alloy refined by K_2ZrF_6, the growth restriction effect of Zr is well established, in which the K_2ZrF_6 salts approach introduced into magnesium alloy produces a large number of in situ fresh Zr particles, and then the dissolved Zr facilitates the grain refinement from a growth restriction perspective. Similarly, Peng [23] finds that, in the Mg-9Gd-4Y-Zr alloys, there is at least one zirconium rich core in almost each grain in alloys with high zirconium content, whereas the characteristic zirconium rich cores are not found in the alloy with low zirconium content. He suggests that the grain refinement mechanism of zirconium in the low zirconium alloy is different from that in the high zirconium alloys: the zirconium works mainly by generating nucleants in the high zirconium alloys, and by restricting grain growth in the low zirconium alloy.

The fading time of the two studied grain refiners is compared in Figure 8. It can be seen that by extending the holding time to 120 min, the Zr content in the alloy refined by KSM almost keeps stable. However, the Zr content in the alloy refined by Mg-30wt.%Zr master alloy gradually fades with increasing holding time. In particular, it decreases significantly after 120 min, indicating that the grain refinement effect of the Mg-Zr master alloy only lasts for a short time after addition [25]. Therefore, the K_2ZrF_6 salt mixture refiner has a much longer fading time comparing that of Mg-Zr master alloy. This is similar with previous study on refining effect of salts containing Ti and B elements in purity aluminum compared with that of an Al-Ti-B master alloy [26].

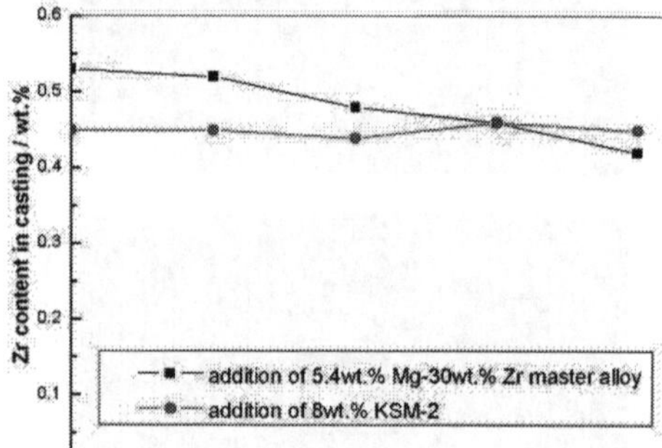

Figure 8. Comparison of the fading time of grain refinement between the addition of 8wt.% KSM and that of 5.4wt.% Mg-30wt.%Zr master alloy. Industrially the latter can yield 0.4~0.6wt.% Zr content in magnesium alloys.

Furthermore, because the total Zr content detected by ICP-AES in the alloy refined by KSM consists of well dissolved Zr content and the inclusion's Zr content, it is reasonable to deduce that, a relatively lower dissolved Zr content may be enough for a satisfactory grain refinement effect. In other words, if the Zr recovery of KSM could be improved by technical methods, the KSM approach has more potential. Based on the above analysis, further work of melting and purification will be investigated to improve the K_2ZrF_6 approach in order to increase the Zr recovery and to remove the subsequent inclusions.

Conclusions

The Mg-10Gd-3Y alloy is grain refined by a Mg-Zr master alloy and KSM. The characteristic microstructural feature of the alloy refined by the Mg-Zr master alloy is Zr-rich cores that exist in most grains, while the Zr-rich cores are not observed in the alloy refined by KSM. It is suggested that the grain refinement mechanisms of zirconium in the two cases are different: the Zr released from Mg-Zr master alloy works by adding heterogeneous nucleation, while the Zr reduced from the in-situ reaction between the Mg melt and K_2ZrF_6 works by restricting grain growth. Compared with the Mg-30.wt%Zr master alloy, the KSM refiner shows much longer fading time during melting.

Acknowledgements

The present study is funded by the National Basic Research Program of China (No. 2007CB613701), Program of Shanghai Subject Chief Scientist (No. 08XD14020) and Aerospace Science and Technology Innovation Fund of China Aerospace Science and Technology Corporation (No. 0502).

References

1. S.M. He, X.Q. Zeng, L.M. Peng, X. Gao, J.F. Nie, W.J. Ding, "Microstructure and Strengthening Mechanism of High Strength Mg-10Gd-2Y-0.5Zr Alloy," *J. Alloys Compd,.* 427 (2007), 316-323.

2. Y.C. Guo, J.P. Li, J.S. Li, Z. Y, J. Zhao, F. Xia, M.X. Liang, "Mg-Gd-Y System Phase Diagram Calculation and Experimental Clarification," *J. Alloys Compd.,* 450 (2008), 446-451.

3. W. Wang, Y.Huang, G. Wu, Q. Wang, M. Sun, W. Ding, "Influence of Flux Containing YCl_3 Additions on Purifying Effectiveness and Properties of Mg-10Gd-3Y-0.5Zr Alloy," *J. Alloys Compd.,* 480 (2009), 386-391.

4. T. Honma, T. Ohkubo, S. Kamado, K. Hono, "Effect of Zn Additions on the Age-Hardening of Mg-2.0Gd-1.2Y-0.2Zr Alloys," *Acta Mater.,* 55 (2007), 4137-4150.

5. Ma Qian, Z.C.G. Hildebrand, D.H. StJohn, "The Loss of Dissolved Zirconium in Zirconium-Refined Magnesium Alloys after Remelting," *Metall. Mater. Trans. A,* 40A (2009), 2470-2479.

6. Ma Qian, D. StJohn, M.T. Frost, "Heterogeneous Nuclei Size in Magnesium-Zirconium Alloys," *Scripta Mater.,* 50 (2004), 1115-1119.

7. Ma Qian, D. StJohn, M.T. Frost, "Characteristic Zirconium-Rich Coring Structures in Mg-Zr Alloys," *Scripta Mater.,* 46 (2002), 649-654.

8. Ma Qian, A. Das, "Grain Refinement of Magnesium Alloys by Zirconium: Formation of Equiaxed Grains," *Scripta Mater.,* 54 (2006), 881-886.

9. P. Cao, Ma Qian, D.H. StJohn, M.T. Frost, "Uptake of Iron and its Effect on Grain Refinement of Pure Magnesium by Zirconium," *Mater. Sci. Technol.,* 20 (2004), 585-592.

10. Ma Qian, D. StJohn, M.T. Frost, "Zirconium Alloying and Grain Refinement of Magnesium Alloys," *Magnesium Technology 2003,* Ed. H. Kaplan, pp. 209-214, San Diego, USA: TMS.

11. Ma Qian, D. StJohn, M.T. Frost, M. R. Barnett, "Grain Refinement of Pure Magnesium Using Rolled Zirmax-Master Alloy (Mg-33.3Zr)," *Magnesium Technology 2003,* Ed. H. Kaplan, pp. 215-220, San Diego, USA: TMS.

12. Ma Qian, L. Zheng, D. Graham, M.T. Frost, D.H. StJohn, "Settling of Undissolved Zirconium Particles in Pure Magnesium Melts," *J. Light Met.,* 1 (2001), 157-165.

13. Ma Qian, D.H. StJohn, "Grain Nucleation and Formation in Mg-Zr Alloys," *International Journal of Cast Metals Research,* 22 (2009), 256-259.

14. Ma Qian, "Heterogeneous Nucleation on Potent Spherical Substrates during Solidification," *Acta Mater.,* 55 (2007), 943-953.

15. Ma Qian, D. Graham, L. Zheng, D. H. StJohn, M. T. Frost, "Alloying of Pure Magnesium with Mg-33.3wt%Zr Master Alloy," *Mater. Sci. Technol.,* 19 (2003), 156-162.

16. Ma Qian, D. StJohn, M.T. Frost, "Magnesium Zirconium Alloying," US Patent 0 161 121 A1, 2005.

17. M. Sun, G.H. Wu, J.C. Dai, W. Wang, W.J. Ding, "Grain Refinement Behavior of Potassium Fluozirconate (K_2zrf_6) Salts Mixture Introduced into Mg-10Gd-3Y Magnesium Alloy," *J. Alloys Compd.,* 494 (2010), 426-433.

18. M. Sun, G.H. Wu, W. Wang, W.J. Ding, "Effect of Zr on the Microstructure, Mechanical Properties and Corrosion Resistance of Mg-10Gd-3Y Magnesium Alloy," *Mater. Sci. Eng. A,* 523 (2009), 145-151.

19. Y. Tamura, N. Kono, T. Motegi, E. Sato, "Grain Refining Mechanism and Casting Structure of Mg-Zr Alloy," *Journal of Japan Institute of Light Metals,* 48 (1998), 185-189.

20. D.H. StJohn, A.K. Dahle, T. Abbott, M.D. Nave, Ma Qian, "Solidification of Cast Magnesium Alloys," *Magnesium Technology 2003,* Ed. H. Kaplan, pp. 95-100, San Diego, USA: TMS.

21. E. F. Emley, *Principles of Magnesium Technology* (Oxford: Pergamon Press, 1966), 257-261.

22. Y. C. Lee, A. K. Dahle, D. H. StJohn, "The role of solute in grain refinement of magnesium," *Metall. Mater. Trans. A,* 31A (2000), 2895-2906.

23. Z.K. Peng, X.M. Zhang, J.M. Chen, Y. Xiao, H. Jiang, "Grain Refining Mechanism in Mg-9Gd-4Y Alloys by Zirconium," *Mater. Sci. Technol.,* 21 (2005), 722-726.

24. D.H. StJohn, Ma Qian, M. Easton, P. Cao, Z. Hildebrand, "Grain Refinement of Magnesium Alloys," *Metall. Mater. Trans. A,* 36A (2005), 1669-1679.

25. H.E. Friedrich, and B.L. Mordike, *Magnesium Technology: Metallurgy, Design Data, Application* (Berlin Heidelberg: Springer-Verlag Berlin Heidelberg, 2006), 128-135.

26. H.H. Zhang, X. Tang, G.J. Shao, L.P. Xu, "Refining Mechanism of Salts Containing Ti And B Elements in Purity Aluminum," *J. Mater. Process. Technol.,* 180 (2006), 60-65.

Magnesium Technology 2011
Edited by: Wim H. Sillekens, Sean R. Agnew, Neale R. Neelameggham, and Suveen N. Mathaudhu
TMS (The Minerals, Metals & Materials Society), 2011

THE EFFECT OF RARE EARTH ELEMENTS ON THE TEXTURE AND FORMABILITY OF SHEAR ROLLED MAGNESIUM SHEET

D Randman[1], B Davis[1], ML Alderman[1], G Muralidharan[2], TR Muth[2], WH Peter[2], TR Watkins[2], OB Cavin[3]
[1]Magnesium Elektron North America; 1001 College St.; P.O. Box 258; Madison; Illinois 62060; USA
[2]Materials Science and Technology Division; Oak Ridge National Laboratory; 1 Bethel Valley Rd; Oak Ridge; Tennessee 37831; USA
[3]Center for Materials Processing; University of Tennessee, Knoxville; Tennessee 37996; USA

Keywords: Magnesium, Asymmetric Rolling, Texture, Formability

Abstract

The lower relative formability of magnesium alloy sheet can be a restrictive factor when designing light weight engineering structures. Standard symmetric rolling introduces a strong basal texture that decreases the formability; however, asymmetric rolling has been put forward as a possible route to produce sheet with weaker basal texture and greater ductility. It has also been shown in recent work that weaker textures can be produced through the addition of rare earth elements to magnesium alloys. Therefore, this study has been carried out to investigate the effect of rare earth additions on the texture changes during asymmetric rolling. Two alloys have been studied (AZ31B and ZEK100), in which a significant difference is the presence of rare earths in the ZEK100 but not in AZ31B. The differences in texture, microstructure and mechanical properties will be discussed.

Introduction

Magnesium alloys are the lowest density engineering materials that are commonly available and have received significant interest from the automotive and aerospace industries in recent times. Aluminum alloys are often used for sheet forming applications where a light weight solution is required, such as car door panels or aircraft seat backs. Although magnesium is approximately 2/3 the density of aluminum, a barrier to the use of magnesium in these types of applications is the reduced formability at lower temperatures. In general magnesium sheet only attains acceptable levels of formability at temperatures above 200°C due to the activation of additional slip planes to accommodate deformation. This creates problems in industry because these temperatures require the use of solid lubricants, such as boron nitride, that must be removed after forming. This removal process is much harder and more energy intensive than the removal of oils that can be used below 200°C. Therefore, if the temperature of formability could be reduced, the use of magnesium sheet for forming applications would become much more attractive.

The lack of formability in magnesium sheet originates from the rolling texture present in the material. During rolling, magnesium is well known to develop a 'basal texture' in which the c-axes of the grains align strongly with the normal direction of rolling. This places the basal planes, the most common slip plane in magnesium, into a poor orientation for further deformation. Another common deformation mechanism that would allow strain accommodation during forming is the twinning system. In magnesium, this acts as a c-axis extension twin and thus is not likely to be active during forming of the textured material, when thickness reduction is required. Therefore, grains are in a poor orientation for both slip and twinning, which leads to a reduced ductility. By either weakening the basal texture or by tilting it away from the normal direction it should be possible to produce magnesium sheet with greater formability.

Several methods have been put forward that change the basal texture of magnesium alloys and thus improve the formability: examples are Equal Channel Angular Processing (ECAP)[1, 2], Equal Channel Angular Rolling (ECAR)[3, 4], 'cross-roll rolling'[5], and asymmetric or shear rolling[6-8]. An important target for the magnesium industry is to improve the formability of magnesium sheet. Not all of these methods are suitable for commercial scale sheet production. ECAP is an extrusion process that on its own cannot be used to produce sheet. ECAR combines rolling with ECAP to produce a shear deformation immediately after rolling. This produces suitably textured sheet but requires a radical change to standard rolling practices and thus is unlikely to be easily scaled up to commercial production. Cross-roll rolling was described by Chino *et al.*[5] as a rolling process whereby the rolls are twisted horizontally from each other. Some improvements in texture and formability were achieved, but this method could not be used to produce useful sheet as the crossed rolls would produce a thickness variation across the width of the sheet. Asymmetric rolling uses work rolls travelling at different speeds to introduce shear into the material during the rolling process. This has been found to reduce the strength of the basal texture in magnesium and to cause a tilt towards the rolling direction. Although this reduction in texture is less than by using some of the other techniques, and despite a small curvature introduced in the sheet, shear rolling is thought to be the only process suitable for scaling up to larger commercial production of rolled sheet.

The formability and deformation behavior of the final sheet is strongly dependent on the microstructure and texture that are developed during the shear rolling process. These vary according to the parameters used during rolling, i.e. temperature, rolling speed and reduction per pass (strain rate). Temperature is one of the most important conditions involved as the deformation mechanisms are strongly affected by this. It is well known that twinning is more prevalent at lower temperatures, while a greater number of slip systems are active at higher temperatures in magnesium alloys[9]. However, it is as yet unknown how these different mechanisms will respond to the different stress states experienced in shear rolling compared to standard symmetric rolling. Temperature also affects the recrystallization behavior of the material, both in dynamic recrystallization during rolling and static recrystallization during both interpass pre-heat cycles and post deformation annealing. Dynamic recovery and recrystallization have been observed in various studies[7, 10-12], often causing grain refinement. It has been seen that asymmetric rolling undergoes higher levels of dynamic recrystallization than symmetric rolling under the same conditions[11], which would be expected due to the higher strains experienced. Recrystallization can also be expected to affect the texture and has previously been observed to decrease the intensity of the basal texture in shear rolled sheet[7, 13, 14].

Different roll speed ratios have been used in the literature, ranging from 1.1 to 3[10, 13, 15]. Effects on the mechanical properties of the sheet have been seen at all ratios, although stronger changes were generally observed at higher roll speed ratios. Despite the general increase in the effectiveness of shear rolling as the roll speed ratio increases, it is thought that an upper limit should be reached due to slipping of the material. This limit will be a function of the friction between the material and the work rolls. Lower than the upper limit for friction, it is thought that there will be a ratio at which the applied shear stress is adequate to produce the necessary changes in microstructure and which it is not necessary to exceed.

The reduction applied per pass affects the final microstructure and properties of the material. This is the result of the amount of strain introduced and retained work in the material, as well as the distribution of the deformation through the thickness. Xia *et al.*[13] stated that flow localization is more likely at larger reductions and that a more homogeneous microstructure, with more dynamic recrystallization, is achieved using smaller reductions in a multi-pass schedule.

Shear rolling has been carried out in the current work using a mill with a speed ratio of 3:1 on both AZ31B and ZEK100 alloys and the differences in behavior have been studied. One set of rolling conditions has been studied in detail to provide an initial analysis of these differences; a commercially viable process route was used, with symmetric rolling down to a low gauge, followed by multiple shear rolling passes to the final gauge.

Scope of the Project

This project has been set up by Oak Ridge National Laboratory in collaboration with Magnesium Elektron North America. The overall aim of the project is to develop more formable magnesium sheet for use in automotive applications, in order to increase the light weighting of vehicles and save fuel usage. The project has been split into 5 different tasks in order to achieve this final goal.

Task 1: Design, fabricate and install shear rolling mill

An important part of this project is the installation of a specially designed shear rolling mill, to advance the understanding of asymmetric rolling of magnesium sheet on a pilot scale, above that possible on laboratory scale mills The shear mill will be capable of rolling strips up to 250mm wide with very accurate control of temperature, via heated work rolls and roller tables, and shear rolling ratios, with independently driven work rolls.

Task 2: Alloy Selection

Four different alloys were initially studied to develop an understanding of their different behaviors under shear rolling conditions. These alloys were AZ31B (Mg-3wt%Al-1wt%Zn), K1A (Mg-<1wt%Zr), ZK10 (Mg-1wt%Zn-<1wt%Zr) and ZEK100 (Mg-1wt%Zn-<1wt%Rare Earths-<1wt%Zr). It was found in preliminary trials that K1A and ZK10 did not perform as well when processed under a fixed set of parameters. These sheets were more prone to cracking than AZ31B and ZEK100. As such the breadth of the study was limited to future processing of AZ31B and ZEK100. Results for the K1A and ZK10 will not be reported here.

Task 3: Develop Shear Rolling Process

Task 3 has been set out to find ideal processing parameters for AZ31B and ZEK100 alloys in order to gain the greatest benefit from shear rolling and impart the highest formability. The vast majority of this work will be carried out on the dedicated shear mill once it is installed, but preliminary results from initial trials are presented here. The research focuses on the microstructural and textural changes from temperature, reduction per pass and roll speed ratio.

Tasks 4 and 5: Demonstrate shear rolling in an industrial process environment and fabricate components

Once optimum processing conditions have been established for the two alloys in question the concept will be scaled up and implemented on an industrial size mill at Magnesium Elektron. The proof of concept will be carried out to form industrially relevant parts for the automotive industry through OEM partnerships.

Experimental Materials and Procedure

Magnesium Elektron supplied the material for these investigations in the form of commercially rolled sheet. The AZ31B and ZEK100 were both direct chill (DC) cast as slabs that were then rolled down to the final gauges of 2.36mm and 2.18mm respectively. This sheet formed the rolling feed stock for the shear rolling program.

The laboratory scale shear mill employed during these experiments had a bottom roll with a diameter of 229mm and a top roll diameter of 76mm, giving a ratio of size and circumferential speed of 3:1. To counteract the curling of the sheet as it exited the mill, a stripper plate arrangement was fitted to the mill, forming a channel through which the curled sheet travelled to remove some of the curve. Two inch wide strips were rolled at different temperatures and reductions per pass.

Table 1 – Table showing the temperature, roll ratio and thickness of the sheet with each pass. AZ31B and ZEK100 underwent 7 and 8 passes respectively.

			Thickness after pass (mm)	
Pass	Temp.	Roll Ratio	AZ31B	ZEK100
1	225°C	3:1	2.29	2.01
2	225°C	3:1	2.13	1.85
3	225°C	3:1	2.03	1.75
4	225°C	3:1	1.96	1.63
5	225°C	3:1	1.70	1.50
6	225°C	3:1	1.55	1.37
7	225°C	3:1	1.47	1.27
8	225°C	3:1		1.19

Samples for optical microscopy were cut from the rolled sheet and mounted and polished using standard metallographic techniques. An acetic-picral etch was used to reveal the grain structure in conjunction with an Olympus GX51 microscope equipped with polarized light optics. The mean linear intercept method was used for grain size analysis.

Samples for X-ray diffraction (XRD) analysis were cut from the rolled sheet and polished down to the centreline for texture analysis. A PTS Tube unit was used with a cobalt x-ray source, and pole figures were measured for the (0002) peaks.

Longitudinal samples for tensile testing were removed from the rolled strips. Two tensile bars were kept as-rolled and two were annealed at 249°C for 2 hours. The microstructures were studied optically and the mechanical properties were compared. Tensile testing was carried out to ASTM E8 specifications.

Results

Microstructures

Figure 1 and Figure 2 show the starting material prior to shear rolling. The microstructures were fully recrystallized and homogeneous through the full thickness. It can also be seen that the grain size differs between the two materials; ZEK100 had a grain size that was approximately twice that observed in the AZ31B sheet, see Table 2.

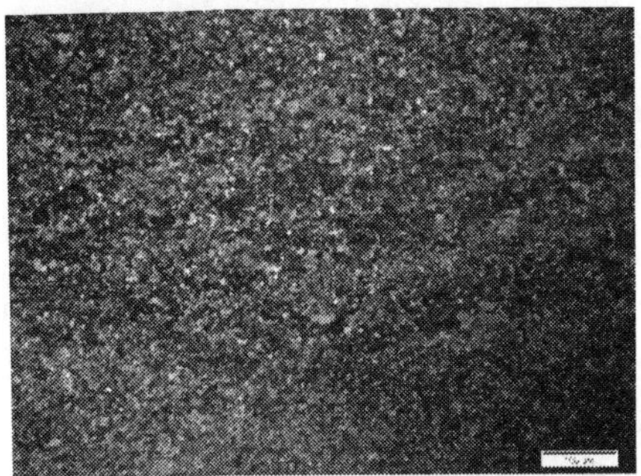

Figure 1 – As received AZ31B feed stock material that was symmetrically rolled to the starting gauge of 2.36mm for asymmetric rolling.

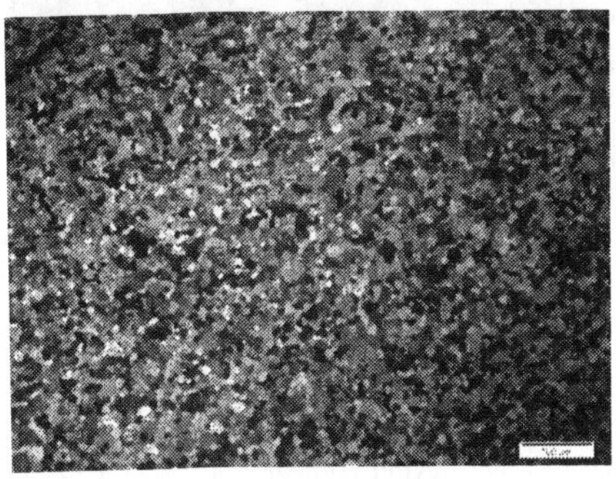

Figure 2 – As received ZEK100 feed stock material that was symmetrically rolled to the starting gauge of 2.18mm for asymmetric rolling.

The shear rolling conditions chosen for the current work were multiple passes at 225°C to a large total strain, as shown in Table 1. Micrographs of the material following rolling are shown in Figure 3 and Figure 4.

Table 2 shows the grain sizes measured in the starting and post-shear rolling conditions of both alloys. It can be seen that there was little change in the grain size of the AZ31B, although this was hard to measure in the deformed specimen as there was a high density of twins and slip bands present. It can be seen in Figure 3 that although there were a lot of other features present, the grain boundaries still exhibited a roughly equiaxed shape. In Figure 4, the ZEK100 sample showed a small amount of grain refinement with shear rolling, although it can be seen that a large number of the grains stayed the same size as in the starting material and showed high levels of deformation, particularly in the form of twins. The recrystallized grains were observed primarily at the grain boundaries of the deformed grains. Some of the recrystallized grains appeared to be undeformed, without evidence of a deformation structure.

Table 2 – Grain sizes of the starting and rolled materials

		Ave. Grain Size (μm)
AZ31B	Starting	4.09±0.17
	After shear rolling	3.98±0.25
ZEK100	Starting	8.59±0.33
	After shear rolling	7.16±0.39

A.

B.

C.

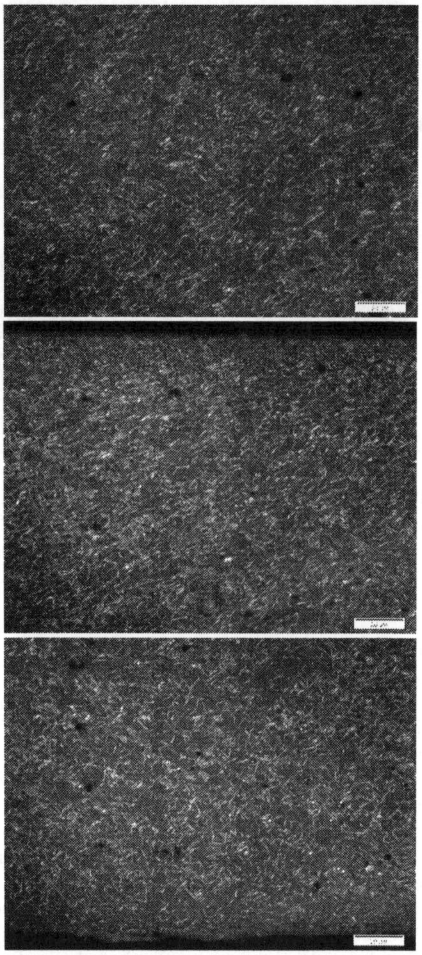

Figure 3 – Micrograph showing the grain structure at the A. centre line, B. slow roll surface and C. fast roll surface of AZ31B sheet after 7 passes of asymmetric rolling to a total strain of 48%. In A., the slow roll surface is towards the top of the micrograph and the fast roll surface is towards the bottom. The rolling feed direction is left to right.

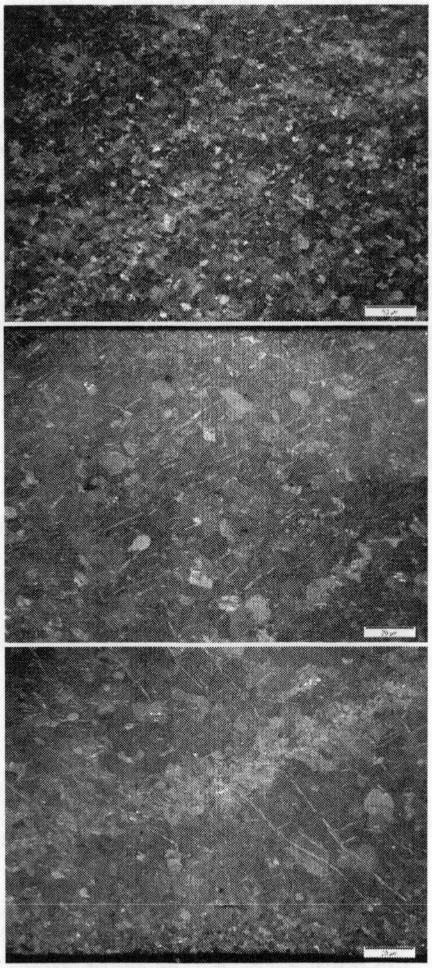

A.
B.
C.

Figure 4 – Micrograph showing the grain structure at A. the centre line, B. slow roll surface, and C. fast roll surface of ZEK100 sheet after 8 passes of asymmetric rolling to a total strain of 65%. In A., the slow roll surface is towards the top of the micrograph and the fast roll is towards the bottom. The rolling feed direction is left to right.

Whilst the bulk of the samples showed homogeneous microstructures, there were significant differences in a thin layer at each surface, generally less than 20μm thick. In the AZ31B sample, finer grains were observed at the slow roll surface, while little change was seen at the fast roll surface. In the ZEK100 sample, a band of fine grains was observed at both surfaces with smaller grains seen at the slow roll surface. It is thought that these grains were formed by recrystallization of an intensely deformed region. It is thought that the bottom surface of the material did not move three times faster than the top surface. The fast work roll was assumed to be moving faster than the material and equally the slow roll slower than the material. This caused slippage during shearing that created a 'smearing effect' at the surfaces of the material with a very high level of strain and induced extensive recrystallization in those regions.

Microstructure after annealing

Following an annealing treatment at 249°C for 2 hours the microstructure was again studied, see Figure 5 and Figure 6. The grain sizes of the annealed material are given in Table 3. It can be seen that the post-annealed AZ31B grain size was only slightly larger than the as shear rolled metal, although the number of twin boundaries and slip bands within the grains appeared to have decreased. The grain size increased significantly in the ZEK100 material, by approximately 50%, and it can be seen from Figure 6 that this was mostly achieved by the large growth of a small number of grains. It is thought that this was grain growth of dynamically recrystallized grains present in the microstructure just after rolling. The remaining grains still show high levels of cold work in the form of twin boundaries and slip bands. This microstructure would not be expected to give good ductility.

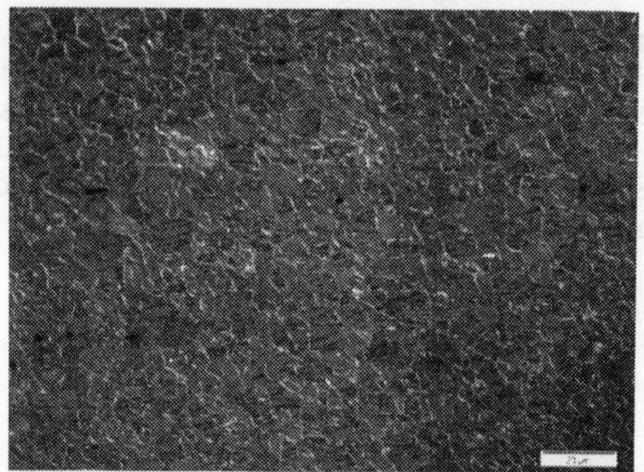

Figure 5 – Micrograph showing the grain structure at the centre line of AZ31B sheet after 7 passes of asymmetric rolling, to a total strain of 48%, then annealed at 249°C for 2 hours.

Figure 6 – Micrograph showing the grain structure at the centre line of ZEK100 material after 8 passes of asymmetric rolling, to a total strain of 65%, then annealed at 249°C for 2 hours.

Table 3 – Grain sizes of the samples following annealing

		Av. Grain Size (μm)
AZ31B	After Anneal	4.34±0.29
ZEK100	After Anneal	10.78±0.72

Texture

The bulk textures on the centerlines of the shear rolled samples were measured with XRD and the {0002} pole figures are presented in Figure 7. These pole figures show that the basal texture is tilted away from the normal direction and towards the rolling direction in each sample.

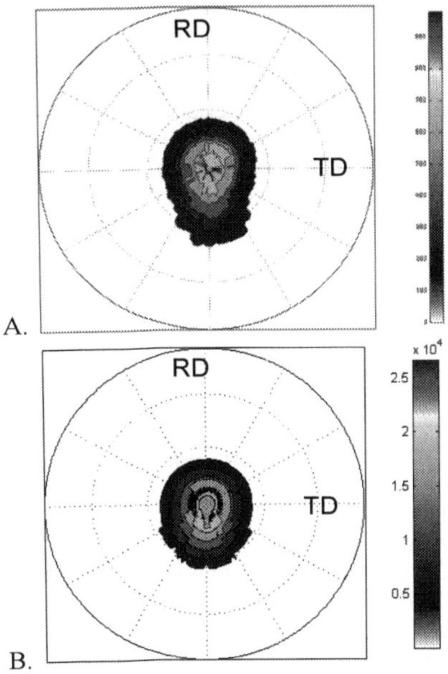

Figure 7 – {0002} pole figure measured at the centerline of A. AZ31B sheet after 7 passes of asymmetric rolling, to a total strain of 48% and B. ZEK100 sheet after 8 passes of asymmetric rolling, to a total strain of 65%.

Mechanical Properties

Table 4 – Mechanical properties of the symmetrically rolled and asymmetrically rolled sheets in the longitudinal direction

Alloy	Processing	Yield (MPa)	UTS (MPa)	Elongation (%)
AZ31B	Symmetric	210	288	10.4
	Symmetric plus anneal	174	265	17.5
	Asymmetric	236	290	6.0
	Asymmetric plus anneal	148	256	8.0
ZEK100	Symmetric plus anneal	212	263	10.4
	Asymmetric	205	249	7.0
	Asymmetric plus anneal	137	232	11.0

While the tensile ductility studied here does not give a direct link to formability, the indication to final forming properties is still useful. It can be seen that shear rolling under the conditions used here degraded the ductility of AZ31B when compared to symmetrically rolled sheet in both the as-rolled and annealed conditions. However, the ductility of the ZEK100 was improved slightly by shear rolling plus annealing.

Discussion

Figure 1 and Figure 2 show that the starting grain sizes were different in the two different materials. It is thought that this is due to the variation in process route as the AZ31B was cast into a much larger slab and therefore underwent a greater reduction to reach this gauge.

Shear rolling of AZ31B did not cause a significant change in grain size or morphology. The grains were approximately equiaxed after a total reduction of 48%, see Figure 3. Under conventional rolling conditions it would be expected that at this level of rolling reduction the grains would become elongated. The presence of equiaxed grains is a strong indication of recrystallization. It can also be seen that the majority of grains contained some evidence of work in the form of twins and/or slip bands. This indicates that deformation continued after the recrystallization event, suggesting that recrystallization was not dynamic but static during the interpass holding times.

Figure 4 showed that there were some smaller grains formed through recrystallization in the ZEK100 during shear rolling. Many of these do not show any further signs of work so it can be assumed that they formed by dynamic recrystallization near to the end of deformation. This change in recrystallization mechanism is a significant difference between the two alloys and could be expected to affect both the shear rolling behavior and the post-rolling properties.

The recovery and recrystallization behavior during post-deformation annealing was also strongly affected by the alloying elements. AZ31B underwent some recovery and recrystallization but little grain growth, whereas ZEK100 underwent significant grain growth but little further recrystallization. The recrystallization seen in AZ31B was not complete as there was still evidence of deformation within many of the grains and thus a longer heat treatment is required to maximize the ductility. This is thought to be the reason the tensile ductility shown in Table 4 was lower for the shear rolled and annealed AZ31B relative to the symmetrically produced and annealed material. It is also thought that the shear rolling process can introduce more flow localization than symmetric rolling and consequently is more likely to induce cracking in those regions, reducing ductility. There are various ways in which the flow localization can be reduced, including lower speeds, higher temperatures, or lower reductions on each pass, and these factors will be looked at in future investigations to improve the shear rolling response.

ZEK100 showed a very small improvement in ductility from symmetrically rolled material, although this could be due simply to experimental variation. The annealed sample of ZEK100 showed grain growth, thought to be primarily from the dynamically recrystallized grains.

Despite this lack of improvement in ductility, it should be remembered that formability is only indirectly related to tensile ductility. Due to the nature of tensile testing, the extension in one direction can be accommodated by contraction in either of the two remaining directions, whichever is easiest. For rolled magnesium tested in the longitudinal direction, the contraction is easiest in the transverse direction, and vice versa. However, in biaxial forming, such as deepdrawing, there is extension in both the longitudinal and transverse directions and thus the contraction occurs in the thickness direction. This is not tested with tensile tests and thus can be very different; true formability testing requires biaxial deformation and this will be analyzed in later studies by the current authors.

The pole figures in Figure 7 showed that the texture had been tilted away from pure basal by shear rolling in both alloys. It can be seen that the main intensity peak was tilted further in ZEK100 than AZ31B, which indicates that shear rolling had a greater effect in ZEK100. Weaker textures have been seen previously in the rolling of rare earth containing magnesium alloys[16] and it is thought that the rare earth content of ZEK100 has allowed a greater tilting of the texture during shear rolling.

The current work has shown that ZEK100, containing rare earths, gives a greater ductility after shear rolling than AZ31B. This is associated with the modification of the texture, as well as the amount of recrystallization that occurred, as discussed above.

Summary

- AZ31B and ZEK100 magnesium sheets have been produced by shear rolling at 225°C.
- Following shear rolling, the microstructures show high levels of cold work in the form of twins and slip bands within grains.
- The texture of the material is altered during shear rolling to give a tilt of the basal pole towards the rolling direction.
- Following annealing, shear rolled AZ31B underwent recrystallization to reduce the deformation inherent within the grains.
- Shear rolled ZEK100 underwent grain growth during annealing.

Acknowledgements

The materials processing research was carried out in collaboration with Oak Ridge National Laboratory under sub contract 4000092046 as part of the American Recovery and Reinvestment Act through the U. S. Department of Energy, Office of Energy Efficiency and Renewable Energy, Industrial Technologies Program. The materials characterization research through the Oak Ridge National Laboratory's High Temperature Materials Laboratory User Program was sponsored by the U. S. Department of Energy, Office of Energy Efficiency and Renewable Energy, Vehicle Technologies Program. Thanks goes to those staff members at Oak Ridge National Laboratory that contributed to the research program.

References

1. Agnew, S.R., et al., *Enhanced ductility in strongly textured magnesium produced by equal channel angular processing.* Scripta Materialia, 2004. **50**(3): p. 377-381.
2. Mukai, T., et al., *Ductility enhancement in AZ 31 magnesium alloy by controlling its grain structure.* Scripta Materialia(USA), 2001. **45**(1): p. 89-94.
3. Cheng, Y.Q., et al., *Improvement of Drawability at Room Temperature in AZ31 Magnesium Alloy Sheets Processed by Equal Channel Angular Rolling.* Journal of Materials Engineering and Performance, 2008. **17**(1): p. 15-19.
4. Kim, W.J., et al., *Grain size and texture control of Mg-3Al-1Zn alloy sheet using a combination of equal-channel angular rolling and high-speed-ratio differential speed-rolling processes.* Scripta Materialia, 2009. **60**(10): p. 897-900.
5. Chino, Y., et al., *Enhanced formability at elevated temperature of a cross-rolled magnesium alloy sheet.* Materials Science and Engineering: A, 2006. **441**(1-2): p. 349-356.
6. Watanabe, H., T. Mukai, and K. Ishikawa, *Differential speed rolling of an AZ31 magnesium alloy and the resulting mechanical properties.* Journal of materials science, 2004. **39**(4): p. 1477-1480.
7. Kim, S.H., et al., *Texture and microstructure changes in asymmetrically hot rolled AZ31 magnesium alloy sheets.* Materials Letters, 2005. **59**(29-30): p. 3876-3880.
8. Huang, X., et al., *Microstructure and texture of Mg-Al-Zn alloy processed by differential speed rolling.* Journal of Alloys and Compounds, 2008. **457**(1-2): p. 408-412.
9. Emley, E.F., *Principles of magnesium technology.* 1966, Oxford: Pergamon. 1032p.
10. Ji, Y.H., J.J. Park, and W.J. Kim, *Finite element analysis of severe deformation in Mg-3Al-1Zn sheets through differential-speed rolling with a high speed ratio.* Materials Science and Engineering: A, 2007. **454**: p. 570-574.
11. Gong, X., et al., *Microstructure and mechanical properties of twin-roll cast Mg-4.5 Al-1.0 Zn sheets processed by differential speed rolling.* Materials & Design, 2010. **31**(3): p. 1581-1587.
12. Huang, X., et al., *Microstructural and textural evolution of AZ31 magnesium alloy during differential speed rolling.* Journal of Alloys and Compounds, 2009. **479**(1-2): p. 726-731.
13. Xia, W., et al., *Microstructure and mechanical properties of AZ31 magnesium alloy sheets produced by differential speed rolling.* Journal of Materials Processing Technology, 2009. **209**(1): p. 26-31.
14. Huang, X., K. Suzuki, and N. Saito, *Textures and stretch formability of Mg-6Al-1Zn magnesium alloy sheets rolled at high temperatures up to 793 K.* Scripta materialia, 2009. **60**(8): p. 651-654.
15. Gong, X., et al., *Enhanced plasticity of twin-roll cast ZK60 magnesium alloy through differential speed rolling.* Materials & Design, 2009. **30**(9): p. 3345-3350.
16. Hantzsche, K., et al., *Effect of rare earth additions on microstructure and texture development of magnesium alloy sheets.* Scripta Materialia, 2010. **63**(7): p. 725-730.

References

1. Agnew, S.R., et al., *Enhanced ductility in strongly textured magnesium produced by equal channel angular processing.* Scripta Materialia, 2004. **50**(3): p. 377-381.
2. Mukai, T., et al., *Ductility enhancement in AZ 31 magnesium alloy by controlling its grain structure.* Scripta Materialia(USA), 2001. **45**(1): p. 89-94.
3. Cheng, Y.Q., et al., *Improvement of Drawability at Room Temperature in AZ31 Magnesium Alloy Sheets Processed by Equal Channel Angular Rolling.* Journal of Materials Engineering and Performance, 2008. **17**(1): p. 15-19.
4. Kim, W.J., et al., *Grain size and texture control of Mg-3Al-1Zn alloy sheet using a combination of equal-channel angular rolling and high-speed-ratio differential speed-rolling processes.* Scripta Materialia, 2009. **60**(10): p. 897-900.
5. Chino, Y., et al., *Enhanced formability at elevated temperature of a cross-rolled magnesium alloy sheet.* Materials Science and Engineering: A, 2006. **441**(1-2): p. 349-356.
6. Watanabe, H., T. Mukai, and K. Ishikawa, *Differential speed rolling of an AZ31 magnesium alloy and the resulting mechanical properties.* Journal of materials science, 2004. **39**(4): p. 1477-1480.
7. Kim, S.H., et al., *Texture and microstructure changes in asymmetrically hot rolled AZ31 magnesium alloy sheets.* Materials Letters, 2005. **59**(29-30): p. 3876-3880.
8. Huang, X., et al., *Microstructure and texture of Mg-Al-Zn alloy processed by differential speed rolling.* Journal of Alloys and Compounds, 2008. **457**(1-2): p. 408-412.
9. Emley, E.F., *Principles of magnesium technology.* 1966, Oxford: Pergamon. 1032p.
10. Ji, Y.H., J.J. Park, and W.J. Kim, *Finite element analysis of severe deformation in Mg-3Al-1Zn sheets through differential-speed rolling with a high speed ratio.* Materials Science and Engineering: A, 2007. **454**: p. 570-574.
11. Gong, X., et al., *Microstructure and mechanical properties of twin-roll cast Mg-4.5 Al-1.0 Zn sheets processed by differential speed rolling.* Materials & Design, 2010. **31**(3): p. 1581-1587.
12. Huang, X., et al., *Microstructural and textural evolution of AZ31 magnesium alloy during differential speed rolling.* Journal of Alloys and Compounds, 2009. **479**(1-2): p. 726-731.
13. Xia, W., et al., *Microstructure and mechanical properties of AZ31 magnesium alloy sheets produced by differential speed rolling.* Journal of Materials Processing Technology, 2009. **209**(1): p. 26-31.
14. Huang, X., K. Suzuki, and N. Saito, *Textures and stretch formability of Mg-6Al-1Zn magnesium alloy sheets rolled at high temperatures up to 793 K.* Scripta materialia, 2009. **60**(8): p. 651-654.
15. Gong, X., et al., *Enhanced plasticity of twin-roll cast ZK60 magnesium alloy through differential speed rolling.* Materials & Design, 2009. **30**(9): p. 3345-3350.
16. Hantzsche, K., et al., *Effect of rare earth additions on microstructure and texture development of magnesium alloy sheets.* Scripta Materialia, 2010. **63**(7): p. 725-730.

Magnesium Technology 2011
Edited by: Wim H. Sillekens, Sean R. Agnew, Neale R. Neelameggham, and Suveen N. Mathaudhu
TMS (The Minerals, Metals & Materials Society), 2011

IMPROVEMENT OF STRENGTH AND DUCTILITY OF MG-ZN-CA-MN ALLOY BY EQUAL CHANNEL ANGULAR PRESSING

L.B. Tong [1, 2], M.Y. Zheng [1, *], S.W. Xu [2], P. Song [1], K. Wu [1], S. Kamado [2]
[1] School of Materials Science and Engineering, Harbin Institute of Technology, Harbin 150001, PR China
[2] Department of Mechanical Engineering, Nagaoka University of Technology, Nagaoka 940-2188, Japan

Keywords: Mg–Zn–Ca–Mn alloy; Equal channel angular pressing; Microstructure; Mechanical properties

Abstract

An ultrafine-grained (UFG) Mg–5.25 wt.% Zn–0.6 wt.% Ca–0.3 wt.% Mn alloy was produced by subjecting the as-extruded alloy to equal channel angular pressing (ECAP) for 4 passes at 250 and 300 °C, respectively. ECAP resulted in a remarkable grain refinement. After ECAP processing at 250°C, basal planes are oriented both parallel and inclined about 45° to the extrusion direction, while most the basal planes are oriented parallel to the extrusion direction after ECAP processing at 300 °C. Both yield strength and elongation were increased after ECAP processing. The yield strength was higher in the ECAPed alloy processed at 300 °C with larger grain size, indicating that the texture strengthening effect was dominant over the strengthening from grain refinement in the ECAPed alloy. The grain refinement may lead to dislocation slip on non-basal plane and grain boundary sliding, which improved the ductility of the ECAPed Mg–Zn–Ca–Mn alloy.

Introduction

Magnesium and its alloy are especially attractive for applications in aeronautics and automobile components due to their low density, high specific strength and good castability [1]. However, the poor ductility of Mg alloys at room temperature due to the limited number of slip systems available in the hcp crystal structure [2], restricts their widespread application. Grain refinement is an effective way to increase the strength and ductility of Mg alloys at ambient temperature [3]. During the last decade, equal channel angular pressing (ECAP) has emerged as a widely-known procedure for the fabrication of UFG metals and alloys through introducing intensive strain during processing [4].

In addition to the grain refinement, texture modification has a great influence on the mechanical properties in Mg alloys [5]. It has been reported that the yield stress of Mg alloys was decreased after ECAP processing in despite of grain refinement, exhibiting an inverse Hall-Petch relationship, because of the intensive development of texture after ECAP processing [6].

Alloying Mg with Ca increases the strength and corrosion resistance [7], while the presence of Zn in the binary Mg-Ca alloys enhances the precipitation hardening response [8], and Mn addition can improve the strength of Mg alloy effectively through grain refinement [9]. Recently, Mg–Zn–Ca alloys with high hardness and good creep resistance have been developed, due to the existence of fine stable precipitates in the alloys [10, 11], which are considered as $Ca_2Mg_6Zn_3$ [11, 12]. However, the investigation on the possibility of improving the mechanical properties in Mg–Zn–Ca–Mn alloy by using ECAP processing has not been investigated yet.

The purpose of the current research is to investigate the influence of the ECAP processing on the microstructure and mechanical properties of the as-extruded Mg–Zn–Ca–Mn alloy, with the main focus on the effect of grain size and texture on the mechanical properties.

Experimental Procedures

The alloy with a chemical composition of Mg - 5.25 wt.% Zn - 0.6 wt.% Ca - 0.3 wt.% Mn was prepared from pure Mg (99.99%), Zinc (99.98%), Ca (99.98%) and Mg–1.18 wt.% Mn master alloy, using electric resistance heating furnace in a SF_6 and CO_2 protective atmosphere. The melting alloy was held at 720 °C for 10 min and then cast into a steel mold.

The as-cast ingot was extruded at 300 °C with a reduction ratio of 10. The as-extruded alloy was machined into billets for ECAP with cross-sectional dimensions of 10 mm × 10 mm and a length of 70 mm. The ECAP die had characteristic angles $\Phi = 90°$ and $\psi = 37°$. The ECAP processing was conducted at 250 and 300 °C for 4 passes (corresponding to a total strain of $\varepsilon \approx 4.2$), using processing route Bc (the sample was rotated by 90° in the same direction between each pass) with constant displacement rate of 10 mm/s. The samples for microstructure observation and tensile tests were cut along y-plane (parallel to ECAP direction and normal direction). EBSD data were obtained from a JEOL FESEM JSM-7000F scanning electron microscopy equipped with TSL MSC-2200. The tensile tests were performed using Instron 5569 tensile machine with an initial strain rate of 1 mm/min at room temperature.

Results and discussion

Microstructure

Fig. 1 shows the microstructure of the as-extruded and as-ECAPed Mg–Zn–Ca–Mn alloy from EBSD analysis. As shown in Fig. 1a, in addition to some elongated grains, fine equiaxed grains with well-defined grain boundaries was observed, with average grain size of 4 μm, which indicated that dynamic recrystallization process occurred during hot extrusion, but it was not completed. After ECAP processing for 4 passes, the homogenous structure was observed, and grain size was decreased substantially due to the dynamic rescrystallization. The average grain size of the as-ECAPed alloys prepared at 250 and 300 °C was 1.0 and 1.2 μm, respectively, as shown in Fig.1 (b) and (c). Some low angle grain boundaries (LAGBs) were observed in the as-ECAPed alloy, which resulted from the dynamic recovery during the repeating ECAP process.

Texture

Fig. 2 shows the texture of the as-extruded and as-ECAPed alloy. A basal texture was observed in the as-extruded alloy, but some basal planes were inclined about 15° to extrusion direction, which

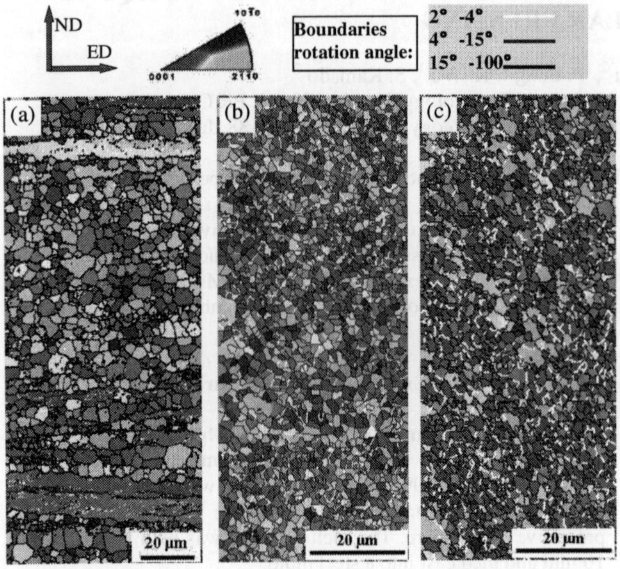

Fig. 1 The orientation imaging microscopy (OIM) maps of Mg-Zn-Ca alloy after (a) extrusion, (b) ECAP for 4 passes at 250 °C and (c) at 300 °C.

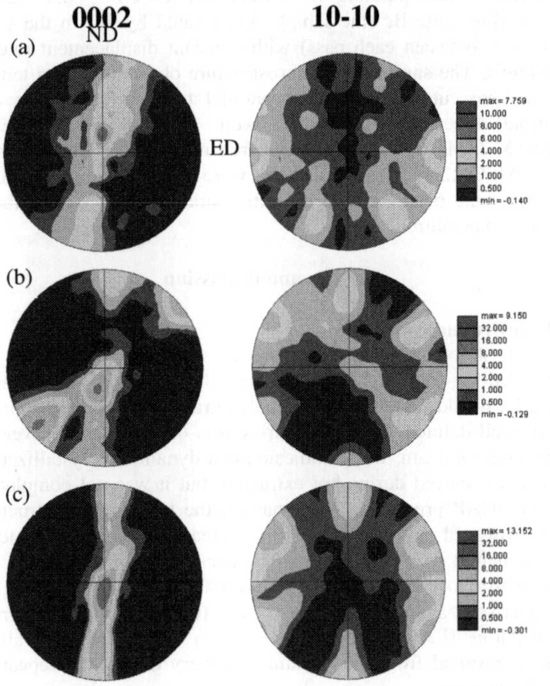

Fig. 2 Pole figures of the Mg-Zn-Ca alloy after (a) extrusion, (b) ECAP for 4 passes at 250 °C and (c) ECAP at 300 °C.

was derived from the recrystallization during the hot extrusion [13]. After ECAP for 4 passes at 250 °C, most of basal planes were inclined about 45° to extrusion direction, in addition, part of the basal planes are oriented parallel to the extrusion direction. While the ECAP was carried out at 300 °C, a distinct texture was observed with basal plane parallel to ED, which was similar with that in the as-extruded alloy. Similar texture evolution was observed in the extruded AZ31 Mg alloy after ECAP processing for 1 pass at 250 and 300 °C. The texture in the specimen processed at 250 °C, may be dominated by tensile twin and basal slip, while in the specimen processed at 300 °C, c+a pyramidal slip occurred at initial stage of deformation, the following crystal rotation dominated by basal slip and grain boundary sliding. The different textures formed during ECAE processing at the different processing temperature perhaps caused by the deformation mechanism at the initial stage of the deformation [14].

Tensile Properties

Nominal stress vs. nominal strain curves of Mg-Zn-Ca alloys in the tensile tests at room temperature were shown in Fig. 6. The ultimate tensile strength (UTS), tensile yield strength (TYS) and elongation of the as-extruded and as-ECAPed alloy were shown in Table 1. After ECAP for 4 passes at 300 °C, both the strength, and elongation were increased, while after ECAP for 4 passes at 250°C, yield strength was decreased and elongation to failure was increased.

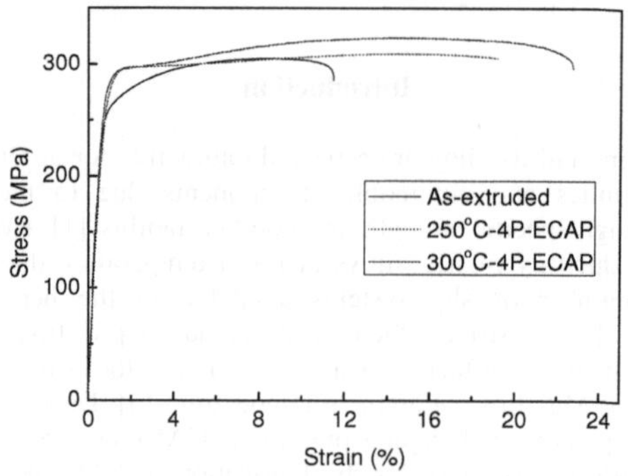

Fig. 3 Tensile stress vs. strain curves of the Mg-Zn-Ca alloy at room temperature after extrusion and ECAP processing.

Table 1 Tensile properties of the as-extruded and as-ECAPed Mg-Zn-Ca alloy.

	TYS (MPa)	UTS (MPa)	Elongation (%)
As-extruded	242	305	11.4
ECAP at 250 °C	228	309	19.1
ECAP at 300 °C	258	324	22.7

In general, the yield stress of polycrystalline materials varied with grain size, the relationship usually following the Hall-Petch equation [15]:

$$\sigma_y = \sigma_0 + kd^{-1/2} \quad (1)$$

where d was a measure of the grain diameter and σ_0 and k were experimentally derived constants. As shown in Fig.1, the alloy processed by ECAP at 300°C has larger grain size than that at 250 °C. However, the tensile strength was higher in the alloy processed by ECAP at 300°C. This indicated that such grain size dependency of yield strength did not exist in the present study. The variation of yield strength was considered to be derived from crystallographic texture modification during ECAP processing at different temperatures.

The Schmid factor for basal plane slip varied with the basal plane orientation, according to the definition of Schmid factor m:

$$m = \tau_c / c_s = \cos\lambda\cos\varphi \quad (2)$$

where λ and φ were the angles between the stress axis and the slip direction and slip plane normal, respectively, τc was the critical resolved shear stress, σ_s was the applied stress. As shown in Fig. 4, the average Schmid factor of basal slip system was decreased to 0.136 in the as-extruded alloy processed at 300 °C.

The TYS of the Mg alloys with hexagonal close packed structure was thought to be dependent on both the grain size and the texture. Although the grain size of the Mg alloy was significantly reduced after ECAP at lower temperature, the TYS of the as-ECAPed alloy was lower than that of the as-extruded alloy. In the alloy processed by ECAP at 250°C, the basal plane was more favorably oriented for basal slip. This indicated that texture softening dominated over the strengthening from the grain refinement in the alloy processed by ECAP at 250°C, leading to a lower yield stress. While after ECAP for 4 passes at 300 °C, both Schmid factor and average grain size were decreased, therefore, both the strengthening effect from grain refinement and texture modification contribute to the increase of the TYS (258 MPa).

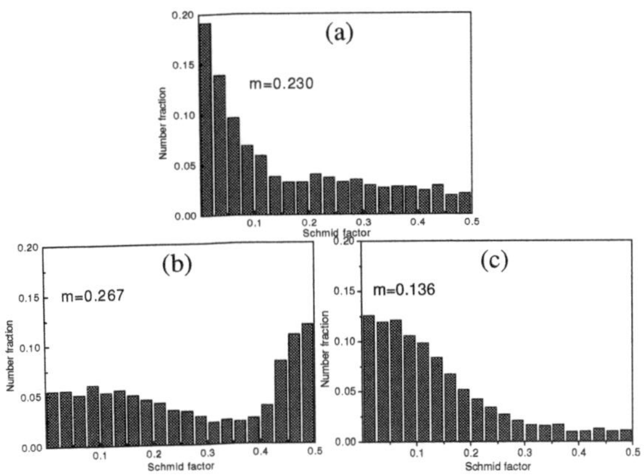

Fig. 4 Schmid factor for basal slip system of the Mg-Zn-Ca alloy after (a) extrusion, (b) ECAP at 250 °C and (c) ECAP at 300 °C.

Ductility

The increased elongation was obtained in the ECAPed Mg-Zn-Ca alloy. It was reported that non-basal slip was induced in fine grained Mg alloys by compatibility stress that operates to maintain continuity at grain boundary [16]. Thus, the activity of non-basal slip near grain boundaries would be helpful for the improvement of ductility of the present ECAPed Mg alloy.

Grain boundary sliding was observed to occur at room temperature in AZ31 alloy with average grain size of 8 μm, and the ratio of the strain by grain boundaries (GBs) to total strain is about 8% at room temperature [17]. The UFG structure in the current study might enhance the grain boundaries sliding at high strain, leading to the improved ductility.

The improved tensile elongation in the ECAPed Mg-Zn-Ca alloy may be resulted from non-basal slip and grain boundary sliding, in addition to basal slip of dislocations.

Conclusion

1. The obvious grain refinement of the as-extruded Mg-Zn-Ca alloy was obtained through ECAP processing at 250 and 300 °C, respectively.
2. After ECAP for 4 passes at 250 °C, most of basal planes were inclined about 45° to ED, but the basal planes in the as-ECAPed alloy processed at 300 °C were parallel to ED.
3. The yield strength was higher in the alloy processed by ECAP at high temperature than that that at low temperature, which indicated that the texture strengthening was dominant over the strengthening from grain refinement.
4. Dislocation slip on non-basal plane and grain boundary sliding may lead to the improved the elongation of the as-ECAPed alloy with fine grains.

Acknowledgements

This work is supported by the National Natural Science Foundation of China (No.50201005 and No.50571031) and the Program for New Century Excellent Talents in University (NCET-08-0169), International S&T Cooperation Key Project of Ministry of Science and Technology of China (2010DFB50210) and International S&T Cooperation Key Project of Heilongjiang Province, China (WB09A202).

References

[1] S.E. Ion, F.J. Humphreys, S.H. White, Acta Metall. 30 (1982) 1909-1919
[2] K. Matsubara, Y. Miyahara, Z. Horita, T.G. Langdon, Acta Mater. 51 (2003) 3073-3084
[3] M.Y. Zheng, S.W. Xu, X.G. Qiao, K. Wu, S. Kamado, Y. Kojima, Mater. Sci. Eng. A 483-484 (2008) 564-567
[4] R.Z. Valiev, T.G. Langdon, Prog. Mater. Sci. 51 (2006) 881-981
[5] W.M. Gan, M.Y. Zheng, H. Chang, X.J. Wang, X.G. Qiao, K.Wu, B. Schwebke, H. G. Brokmeier, J. Alloys Compd. 470 (2009) 256.
[6] W.J. Kim, S.I. Hong, Y.S. Kim, S.H. Min, H.T. Jeong, J.D. Lee, Acta Mater. 51 (2003) 3293-3307.
[7] Erlin Zhang, Lei Yang, Mater. Sci. Eng. A 497 (2008) 111–118.
[8] J.F. Nie, B.C. Muddle, Scripta Mater. 37 (1997) 1475–1481.
[9] S. A. Khan, Y. Miyashita, Y. Mutoh, Z. B. Sajuri, Mater. Sci. Eng. A 420 (2006) 315–321.
[10] J.C. Oh, T. Ohkubo, T. Mukai, K. Hono, Scripta Mater. 53 (2005) 675-679
[11] J.F. Nie, B.C. Muddle, Scripta Mater. 37 (1997) 1475-1481
[12] G. Levi, S. Avraham, A. Zilberov, M. Bamberger, Acta Mater. 54 (2006) 523-530
[13] M. Shahzad, L. Wagner, Mater. Sci. Eng. A 506 (2009) 141–147.
[14] Yu Yoshida, Lawrence Cisar, Shigeharu Kamado, Jun-ichi Koike, Yo Kojima, Mater. Sci. Forum 419–422 (2003) 533–538.

[15] N.J. Petch, J. Iron Steel Inst. 174 (1953) 25–28.
[16] J. Koike, T. Kobayashi, T. Mukai, H. Watanabe, M. Suzuki, K. Maruyama, K. Higashi, Acta Mater. 51 (2003) 2055–2065.
[17] J. Koike, R. Ohyama, T. Kobayashi, M. Suzuki, K. Maruyama, Mater. Trans. 44 (2003) 445-451.

Magnesium Technology 2011
Edited by: Wim H. Sillekens, Sean R. Agnew, Neale R. Neelameggham, and Suveen N. Mathaudhu
TMS (The Minerals, Metals & Materials Society), 2011

DEFORMATION BEHAVIOR OF FRICTION STIR PROCESSED MAGNESIUM ALLOYS

Q. Yang[1], S. Mironov[2], Y.S. Sato[2], K. Okamoto[3]
[1]Research and Development Division, Hitachi America, Ltd., Farmington Hills, MI 48335, USA
[2]Department of Materials Processing, Tohoku University, Sendai 980-8579, JAPAN
[3]Hitachi Research Laboratory, Hitachi Ltd., Ibaraki 319-1292, JAPAN

Keywords: Mg alloy, Friction stir processing, Anisotropy of mechanical properties, texture

Abstract

The present study explores the feasibility of using friction stir processing to produce high-performance Mg alloy sheet. Deformation behavior of friction stir processed Mg alloys is addressed. Analyses of grain size and crystallographic texture are made to interpret the characteristics of strength and ductility of friction stir processed Mg. It is found that friction stir processing causes variation of texture orientation in the processed sheet. The non-uniform texture distribution leads to significant in-plane anisotropy in ductility as well as localized deformation behavior of the processed sheet. Furthermore, the inhomogeneity of high-intensity texture would not be eliminated by hot compression performed along the normal of the processed sheet.

Introduction

Wrought Mg alloys in the forms of sheet and extrusion have been extensively investigated with a target for lightweight structural applications [1,2]. Ultra-fine grained (UFG) Mg alloys produced through severe plastic deformation processing exhibit excellent attributes such as higher yield/tensile strength and formability as well as less strength anisotropy, compared to coarse grained material [3-5]. In this study, the feasibility of using friction stir processing (FSP) to produce high-performance Mg alloy sheet is explored.

FSP, derived from friction stir welding (FSW), is enforced by a non-consumable tool which consists of a circular shoulder and a profiled pin [6]. The tool pin is fully penetrated in the bulk material with the shoulder in contact with the upper plate surface while the rotating tool is translated. Frictional heat and large-strain shear deformation mainly resulted from the rotation of the tool pin enable the processed material to flow around the tool and finally deposit at the back of the tool. The tool is shifted in the transverse direction by a certain distance after each single pass such that there is an overlap between the neighboring beads, as schematically illustrated in Fig. 1(a). Repeat the process until the base material is entirely processed. The heavily deformed material undergoes dynamic recrystallization during process and subsequent static recrystallization, and would be transformed into uniform ultra-fine grained material. FSP has been applied to refining grain structure in various Al alloys and hence developing superplastic forming [7,8].

Mg alloy has a hexagonal close-packed (HCP) crystal structure, and mechanical properties of wrought Mg exhibit high texture-dependence. The present study addresses the characteristics of strength and ductility of friction stir processed (FSPed) Mg alloy. Analyses of grain size and texture are first conducted on a single-pass processed bead, and the deformability of a fully processed Mg is then briefly discussed.

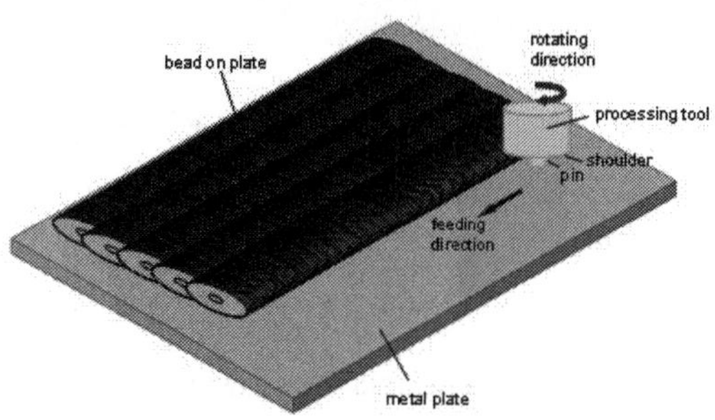

(a)

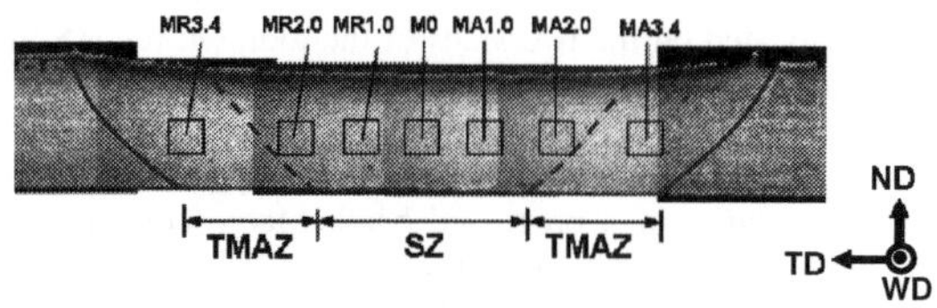

(b)

Fig. 1 (a) Illustration of FSP; (b) Cross sectional macrostructure of a single-pass processed bead

Experiment

Hot rolled AZ31 Mg alloy sheet 2.0mm thick was chosen for the present study. The processing tool consisted of a 10mm-diameter shoulder and a 4mm-diameter threaded pin which was 1.75mm long. FSP was conducted at selected rotational speeds of 1500, 2000 and 3000rpm and a selected travel speeds of 100 mm/min. The tool translation direction (WD) was perpendicular to the sheet rolling direction (RD).

Microstructural examination was made on a plane containing the sheet normal direction (ND) and RD for the base alloy, and on a plane containing the ND and transverse direction (TD) for FSPed beads. The TD is perpendicular to the WD. All specimens were subjected to mechanical grinding and polishing. The specimens are etched with an acetic-picral solution. Orientation image microscopy (OIM) was conducted for texture analysis, with each individual scanning area being 600μm x 600μm.

Mechanical properties of the base metal and the FSPed beads along the WD and TD were investigated by uni-axial tension. Test specimens with gauge width and length of 4 and 20 mm, 6 and 30 mm, and 10 and 55 mm were cut from FSPed beads along the WD with the longitudinal centerlines of tensile specimens and processed beads coincident, as exemplified in Fig. 2. Test specimens cut from the base metal along the RD and from the

FSPed beads along the TD had 6mm gauge width and 30mm gauge length. Sectioned specimens from the processed beads were subject to grinding from the top processing surface and the bottom surface by about 0.2mm deep prior to testing. Specimens were tested at an initial strain rate of $8.33\times10^{-4}s^{-1}$.

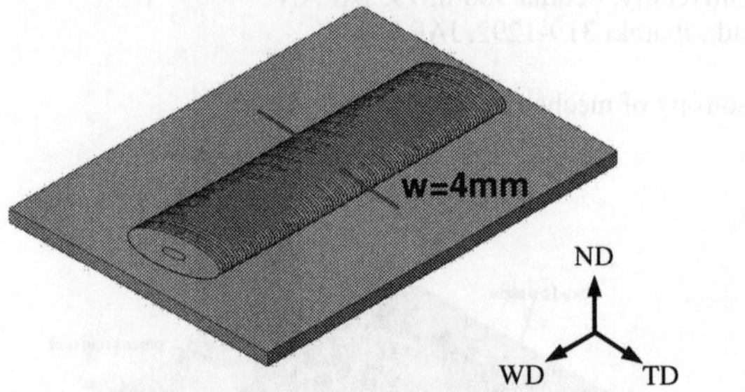

Fig. 2. Illustration of specimen sectioning from a FSP bead along the WD

Results

Microstructure and mechanical property of single-pass FSPed beads

A single bead has an advancing side (AS) and a retreating side (RS) that are located on the two sides of the tool. On the AS, the shear stress induced by the pin rotation has a similar direction to that induced by the forward tool pin translation, while on the RS, the directions of these two shear stresses are opposite. Fig. 1(b) shows macrostructure of a processed bead. Macroscopically, an individual bead can be partitioned into a central stir zone (SZ) and two thermo-mechanical zones (TMAZ) that are covered by the SZ's of the prior and subsequent beads during FSP.

Fig. 3 shows microstructure variation in a bead which is made at 3000rpm. In the center of the bead, a bimodal grain structure is present near the top surface (Fig. 3(a)). The microstructure then becomes quite uniform but coarser through the mid-layer (Fig. 3(d)) towards the bottom surface (Fig. 3(g)). In the mid-layer (Fig. 3(b-f)), microstructure uniformity increases from the TMAZ to the SZ center. Large and elongated grains are present in the TMAZ's. The locations of the mid-layer microstructures are indicated on the micrographs. For example, MR3.4 and MA3.4 represent the locations 3.4mm from the bead center on the RS and AS, respectively.

Fig. 4(b-d) shows the effect of rotation speed on microstructure in the SZ center. The microstructure of the base metal is also included for comparison (Fig. 4(a)). The base metal has a bimodal grain structure with an average grain size of 7.9μm. Grain size and uniformity of FSPed material are affected by process condition. As shown, grain size increases with rotation speed. The average grain sizes in the SZ's are 2.6μm, 7.6μm and 12.0μm at 1500, 2000 and 3000rpm, respectively. The SZ shows a banded structure composed of fine and coarse grain band at 1500rpm, while it shows a nearly uniform microstructure at a rotation speed of 3000rpm.

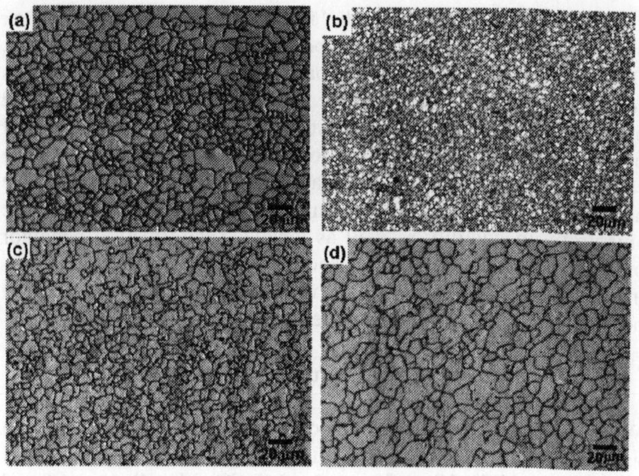

Fig. 4 (a) Microstructure of the base metal; (b-d) Microstructures of the SZ at 1500, 2000, and 3000rpm, respectively.

The rolled base metal has a texture in which the (0002) basal poles are aligned with the ND. FSP develops very strong shear texture in the weld region. Texture mapping along the thickness in the bead center shows: (1) the top surface region about 0.3mm deep has a basal texture tending to be aligned with the surface (*i.e.*, nearly [0002]//ND). This indicates that the tool shoulder dominates plastic deformation in this region; (2) the rotation of the tool pin dominates plastic deformation inside the metal sheet, leading to a single basal fiber in which the (0002) basal poles are roughly aligned with the WD (*i.e.*, the basal planes are parallel to the pin surface). Therefore, texture in the mid-layer across the bead is only shown here. The measurement locations as indicated are consistent with those marked in Fig. 1(b).

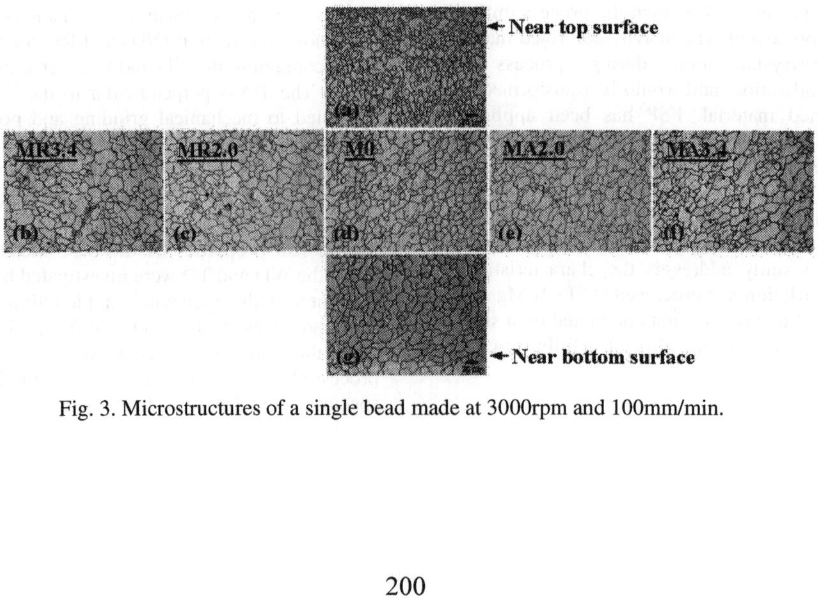

Fig. 3. Microstructures of a single bead made at 3000rpm and 100mm/min.

As shown in Fig. 5, the location 'M0' is the bead center which has a single basal fiber texture (*i.e.*, nearly [0002]//WD). The intensity of basal texture is expressed as times random. On the RS, the location 'MR2.0' is the stir zone extremity where the basal texture splits into two components of similar intensities. In one component, the basal planes are approximately parallel to the pin surface but have a rotation around the plate normal direction (ND). In the other component, the basal planes show a further rotation around the WD from the orientation of the former component. The location 'MR3.4' is immediately close to the base material, and shows a dominant texture component close to the texture of the base metal. Therefore, from the base material to the bead center on the RS, the basal planes aligned with the sheet surface, experiences two rotations around the WD and ND in sequence, to reach the orientation aligned with the pin surface. On the AS, the texture undergoes similar evolution from the base material towards the weld center. It is worth noting that in the location 'MA2.0', the texture splits into two components as well. The component where the basal planes are tilted from the ND by about 40° might be induced by the upward flow of the plasticized material due to the tool translation while this material is dragged into the rotational shear zone around the tool pin due to the tool rotation. The material is subsequently aligned with the pin surface due to the shearing effect, as indicated by nearly (0002) // the pin surface (*i.e.*, the other component).

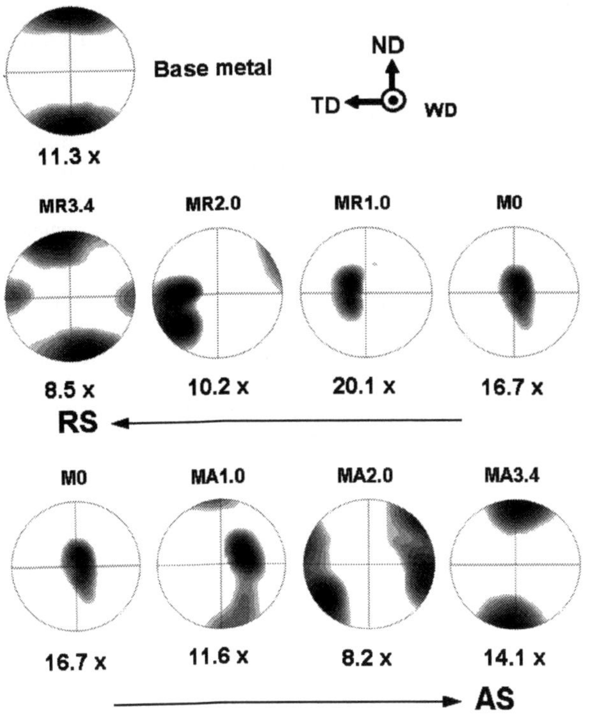

Fig. 5. {0002} pole figures of as received and FSPed AZ31 Mg alloy.

A lower rotation speed of 1500rpm results in cavity in the SZ of the bead near the bottom surface, and mechanical properties at this process condition are therefore not evaluated. Fig. 6(a) shows mechanical properties of defect-free beads along the TD. The base metal has yield strength of 171MPa, ultimate tensile strength of 265MPa, and elongation of 20.1%. The processed beads have yield strengths of about 75MPa and elongations of 3.5%-4.7%; the tensile strength is increased from 196MPa to 225MPa when the rotation speed is raised from 2000rpm to 3000rpm.

Fig. 6(b) indirectly examines the variation with the location of mechanical properties of the processed material in a bead along the longitudinal direction (LD, same as WD). Take the bead made at 3000rpm for example. Since the processed region in a bead is in the shape of a basin, the test specimen 10mm wide includes undisturbed material which is positioned out of the TMAZ's. As shown, increasing specimen width increases yield strength but decreases ductility; it does not result in a prominent increase in ultimate tensile strength. Corresponding to the specimen widths of 4, 6 and 10mm, the yield strengths are 62, 67 and 114MPa, respectively; the tensile strengths are 225, 225 and 243MPa, respectively. The ductility decreases from 36.5% to 26.8% to 20%.

Fig. 6(c) shows the effect of rotation speed on mechanical properties of processed bead along the LD. Test specimen has a gauge width of 4mm. Increasing rotation speed (and therefore grain size) slightly increases both yield strength (from 52Mpa to 62MPa) and tensile strength (from 195MPa to 225MPa), but decreases ductility from 47% to 36.5%.

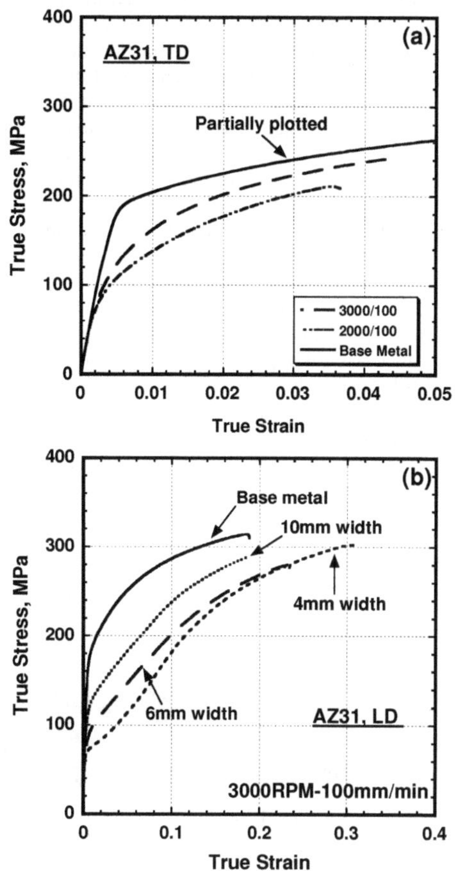

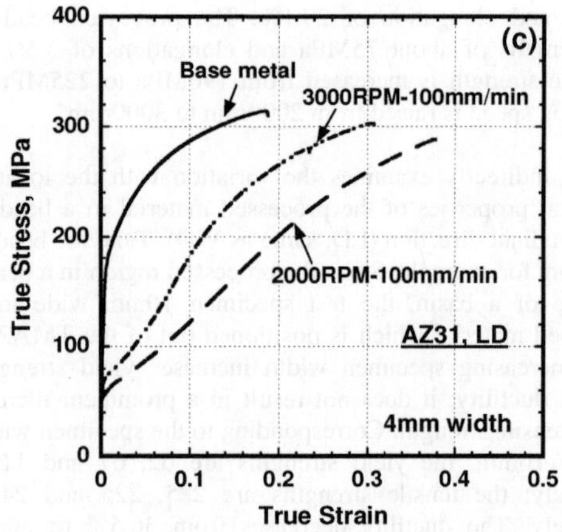

Fig. 6. Mechanical properties of single-pass FSPed beads along the TD and LD

Deformability of multiple-pass FSPed Mg

Fig. 7(a) shows the geometry of a failed tensile specimen. The specimen is cut along the TD from a single-pass AZ31 Mg bead which is made at 3000rpm. As shown, the specimen fails ~2mm from the bead center on the AS after increased local deformation over there (distance measured from the mid-layer). Significant necking is also found at the corresponding location on the RS. Further examination of the geometrical change of the circular grid pattern marked on the tensile specimen shows that the material in the center of the bead is hardly deformed, and that the material near the regions of deformation localization and the fracture surface undergoes appreciable deformation and rotation.

Fig. 7(b) shows the geometry of a failed tensile specimen which is cut from a fully FSPed AZ61 Mg along the TD. The AZ61 Mg alloy is first friction stir processed and then compressed at 300°C with more than 50% thickness reduction. Mechanical testing shows that the step of hot compression increases the ductility of the FSPed material along the TD. However deformation localization still remains. Figs. 6 and 7 indicate that FSP develops strong in-plane anisotropy in deformability of Mg alloys. In other words, FSP enhances the ductility of Mg alloy along the tool travel direction (WD). However it deteriorates the ductility in the perpendicular direction (TD) by developing deformation localization.

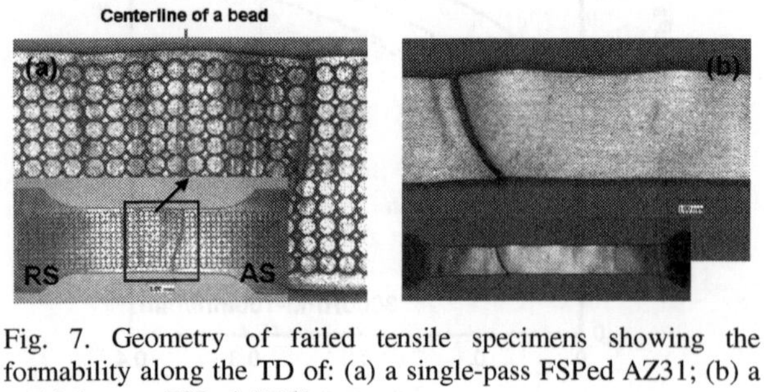

Fig. 7. Geometry of failed tensile specimens showing the formability along the TD of: (a) a single-pass FSPed AZ31; (b) a multiple-pass FSPed AZ61.

Discussion

Friction stir processing, like other severe plastic deformation processes such as ECAE and high pressure torsion, develops a strong ***B***-fiber texture in Mg alloys where the basal plane (0002) is aligned with the substantive shear plane [9-11]. For FSP, the shear plane is mainly the tool pin surface, and therefore leading to (0002)//ND nearly in the SZ (Fig. 5). Across the TMAZ's, the basal plane undergoes two rotations, first around the WD and then around the ND, to achieve the reorientation of crystallographic texture from the base material to the processed SZ. The texture evolution discloses material flow driven by the tool on the two sides of the processing tool.

Mg alloys have a limited number of slip systems at low temperatures due to their HCP crystal structure. The potential mechanisms for deforming Mg at such temperatures mainly include the basal **<a>**, the prismatic **<a>**, and the tensile $\{10\bar{1}2\}$ twinning [12-14]. Each individual deformation mode has its own critical resolved shear stress (τ_{CRSS}). With a designated texture orientation with respect to the external loading direction, each deformation mode corresponds to a yield stress σ_y according to the Schmid's law

$$\sigma_y = \tau_{CRSS}/m \qquad (1)$$

, where m is an average Schmid factor for such a deformation mode. The deformation mode having the smallest yield stress is activated.

The single-pass FSPed material exhibits texture orientations in which the (0002) basal planes are considerably tilted from the sheet plane (Fig. 5). In addition, the basal <a> slip system has a τ_{CRSS} significantly lower than those for the other two deformation modes. Therefore, the basal **<a>** slip would expectably dominate plastic flow at the early stage of deformation.

Consider the effect of distributed texture on deformation behavior of single-pass FSPed material. Calculation shows that the locations 'MA2.0' and 'MR2.0' have the largest m values for the basal **<a>** slip when a bead is loaded along the TD. From Eq. (1), plastic deformation in a Mg bead is initiated from these two locations. The basal poles tilted in the tensile direction (*i.e.*, the TD) enhances the occurrence of mechanical twinning and disfavors the activation of the prismatic slip [2,12,15]. A preliminary microstructure analysis of the failed TD specimens also indicates that the regions 'MA2.0' and 'MR2.0' are severely twinned, and the central area 'M0' and other regions are much less deformed. In Mg, mechanical twins are barriers to dislocation motion, and the induced stress concentration is generally unlikely to be relaxed by other slips, kinking or twinning at an ambient temperature. Therefore, the localized deformation accelerates the failure of the FSPed material with a relatively low strength (Fig. 6(a)).

When the loading direction is along the WD, the m values for the basal **<a>** slip are the largest in the locations 'MA1.0' and 'MR1.0', leading to lower yield strength of the processed material compared to the base metal (Fig. 6(b) and (c)). The extensive twinning takes place in the processed material as well. These two factors result in an increase in strain hardening rate which is the essential reason for ductility enhancement. The rate-sensitivity of the processed material is significantly lower than that of the base material. The present study shows that the grain size effect is not as important as the texture effect on mechanical properties of Mg when different grain sizes are in a comparable range.

Tensile testing of a FSPed and hot pressed Mg alloy along the TD shows evenly-spaced deformation localization (Fig. 7(b)). This indicates that the variation of text orientation caused by FSP would not be eliminated by hot compression. The texture evolution in the FSPed material during hot compression is under investigation to better understand its effect on the deformation behavior of 'modified' FSPed material. The present study shows that eliminating the variation of texture orientation is the key to enabling FSP to produce high-performance wrought Mg sheet.

Conclusions

(1) Friction stir processing results in significant reduction of yield strength and ultimate tensile strength, and develops strong anisotropy of ductility in Mg alloy. The reorientation of crystallographic texture by friction stir processing dominates these changes of mechanical properties.
(2) The variation of texture orientation caused by FSP would not be eliminated by hot pressing.
(3) Eliminating the variation of texture orientation is the key to enabling friction stir processing to produce high-performance wrought Mg sheet.

References

[1] J. Bohlen, J. Swiostek, D. Letzig, K.U. Kainer, Magnesium technology 2009 (2009), 225-230.
[2] J. Bohlen, M.R. Nurnberg, J. Senn, D. Letzig, S.R. Agnew, Acta Materialia 55 (2007) 2101-2112.
[3] Q. Yang, A.K. Ghosh, Acta Materialia 54 (2006) 5147-5158.
[4] Q. Yang, A.K. Ghosh, Acta Materialia 54 (2006) 5159-5170.
[5] Q. Yang, A.K. Ghosh, Metall. Mater. Trans. A 39A (2008) 2781-2796.
[6] W.M. Thomas, E.D. Nicholas, J.C. Needham, M.G. Murch, P. Templesmith, C.J. Dawes, G.B. Patent Application No. 9125978.8, Dec. 1991.
[7] Z.Y. Ma, R.S. Mishra, M.W. Mahoney, Acta Materialia, 50 (2002) 4419-4430.
[8] Z.Y. Ma, R.S. Mishra, M.W. Mahoney, R. Grimes, Metall. Mater. Trans. A 36A (2005) 1447-1458.
[9] Y.S. Sato, H. Kokawa, K. Ikeda, M. Enomoto, S. Jogan, T. Hashimoto, Metall. Mater. Trans. A 32A (2001) 941-948.
[10] S.H.C. Park, Y.S. Sato, H. Kokawa, Metall. Mater. Trans. A 34A (2003) 987-994.
[11] B Beausir, L.S. Toth, K.W. Neale, Acta Mater 55 (2007) 2695.
[12] D.L. Yin, J.T. Wang, J.Q. Liu, X. Zhao, Magnesium technology 2009 (2009), 235-240.
[13] A. Staroselsky, L. Anand, International Journal of plasticity 19 (2003) 1843-1864.
[14] S.R. Agnew, C.N. Tome, D.W. Brown, Scripta Materialia 48 (2003) 1003-1008.
[15] X. Huang, K. Suzuki, A. Watazu, I. Shigematsu, N. Saito, Mater. Sci. Eng. A 488 (2008) 214-220.

Magnesium Technology 2011
Edited by: Wim H. Sillekens, Sean R. Agnew, Neale R. Neelameggham, and Suveen N. Mathaudhu
TMS (The Minerals, Metals & Materials Society), 2011

EFFECT OF HEAT INDEX ON MICROSTRUCTURE AND MECHANICAL BEHAVIOR OF FRICTION STIR PROCESSED AZ31

Wei Yuan, Rajiv S. Mishra
Center for Friction Stir Processing, Department of Materials Science and Engineering
Missouri University of Science and Technology, Rolla, MO 65409, USA

Keywords: friction stir processing, heat index, tensile behavior, anisotropy, work hardening

Abstract

Friction stir processing modifies the microstructure and properties of metals through intense plastic deformation. The frictional heat input affects the microstructure evolution and resulting mechanical properties. 2 mm thick commercial AZ31B-H24 Mg alloy was friction stir processed under various process parameter combinations to investigate the effect of heat index on microstructure and properties. Recrystallized grain structure in the nugget region was observed for all processing conditions with decrease in hardness. Results indicate a reduced tensile yield strength and ultimate tensile strength compared to the as-received material in H-temper, but with an improved hardening capacity. The strain hardening behavior of friction stir processed material is discussed.

Introduction

Friction stir processing (FSP), a derivate from friction stir welding [1], has emerged as an effective tool for microstructural modification. During FSP, a non-consumable rotating tool plunges into the workpiece and traverses along a pre-determined direction. FSP locally modifies the microstructure by refining the grain structure and homogenizing any precipitate particles as a result of severe plastic deformation and dynamic recrystallization [2-3]. The tool rotation rate (ω) and traverse speed (υ) play a key role in frictional heat generation and afterward heat dissipation, which dominate the microstructure evolution. It is generally accepted that higher tool rotation rate or a lower tool traverse speed produces more frictional heat. A higher tool traverse speed is beneficial for achieving higher heat dissipation, i.e., higher cooling rate [4].

It has been shown that reducing the frictional heat and increasing the heat dissipation rate are beneficial for grain refinement during FSP of magnesium alloys [5-8]. However, it is difficult to quantify the heat input during FSP due to the parameter variables and materials to be processed. Arbegast and Hartley [9] established an empirical relationship for aluminum alloys between the nugget temperature and the processing parameters based on experimental observations, the pseudo heat index (HI) $\omega^2/\upsilon*10^{-4}$, expressed by $T/T_m=K(\omega^2/\upsilon*10^{-4})^\alpha$, where T_m is melting point, K is a constant between 0.65 and 0.75, and α varies from 0.04 to 0.06.

The present work seeks to investigate the effect of heat index on microstructure evolution and mechanical properties of FSPed magnesium alloy AZ31. Tensile testing was performed to characterize the anisotropy and work hardening behavior of FSPed AZ31 in nugget region.

Experimental Procedure

A commercial grade AZ31-H24 magnesium alloy having a thickness of 2.0 mm was used. To make friction stir passes, a tool with 12 mm diameter concave shoulder and 1.5 mm conical pin was employed. FSP of AZ31 was carried out at various process parameters as shown in Table 1. Tensile properties were characterized by using mini-tensile specimens having a gage length of 1.3 mm, a width of 1.0 mm, and a thickness of 0.6 mm. Samples were machined along the processing direction (PD) and transverse direction (TD) from the center of the nugget as shown in Figure 1, and were polished with 1 μm finish and tested at room temperature at a strain rate of 1 x 10^{-3} s^{-1}. Microhardness measurements were performed at the middle of the nugget center line with a 0.5 kgf load and 10 s dwell time. Five hardness values were obtained for each heat index condition. Microstructure of FSP samples, cross-section perpendicular to the PD, was examined using optical microscopy. Samples were mechanically polished to 1μm and etched using an acetic picral solution (4.2 g picric acid, 10 ml acetic acid, 10 ml water, and 70 ml ethanol). The linear intercept method was used to obtain an average grain size.

Table I. A summary of process parameters were used in this study.

ω (rpm*100)	5	6	6	6	6	6	6	7	8	9
υ (ipm)	2	2	4	6	8	10	12	2	2	2

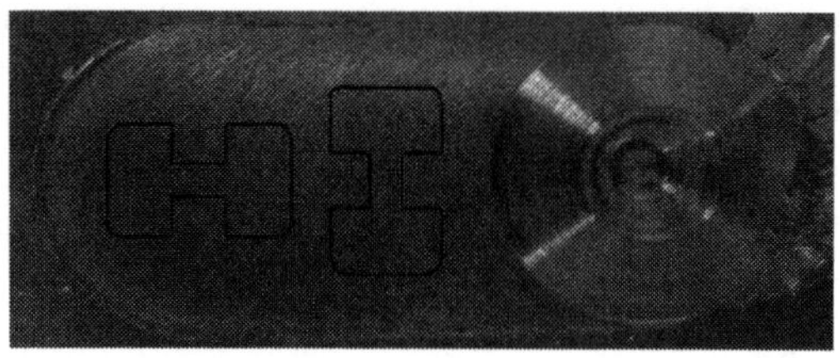

Figure 1. Friction stir processed AZ31 with mini-tensile specimen arrangement.

Results and Discussion

Microstructure Evolution

Figure 2(a) presents the microstructure of as-received AZ31 alloy. The most dominant feature of this microstructure is heavily twinned grains which is consistent with the H temper of as-received material. Figures 2(b) to 2(d) show the nugget microstructures of FSPed AZ31 at various processing conditions. Compared with the as-received material, more uniform grain structures were achieved because of dynamic recrystallization and grain growth. The variation in grain size among the three processing conditions is obvious. A fine grain structure with an average grain size of 3.8±0.6 μm was achieved at a tool rotation rate of 600 revolutions per minute (rpm) and a tool traverse speed of 8 inches per minute (ipm). However, a relatively coarse grain structure was produced at 900 rpm and 2 ipm with an average grain size of 10.2±2.2 μm. The grain size of the nugget at 600 rpm and 2 ipm was 6.0±1.5 μm.

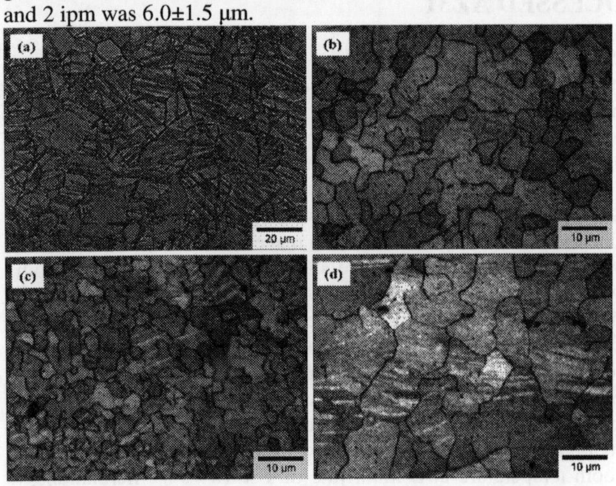

Figure 2. Microstructure of AZ31 alloy in (a) as-received condition, (b) FSPed at 600 rpm and 2 ipm, (c) FSPed at 600 rpm and 8 ipm, and (d) FSPed at 900 rpm and 2 ipm.

The grain size of nugget varied with processing condition or the heat index. The grain structure evolution of magnesium alloys during FSP is complex and generally accepted as a result of dynamic recrystallization [10-12]. The average grain size in the nugget is plotted against heat index in Figure 3. For this particular alloy and tool used, the grain size appears to vary linearly with the heat index; grain size increases with the heat index. A finest grain structure with an average grain size of 3.1 μm was achieved at a heat index of 3.6. On the other hand, a coarsest grain structure with an average grain size of 10.2 μm was achieved at a heat index of 40.5. It has been reported that an increase in tool rotation rate or a decrease in tool traverse speed leads to an increase in grain size [13].

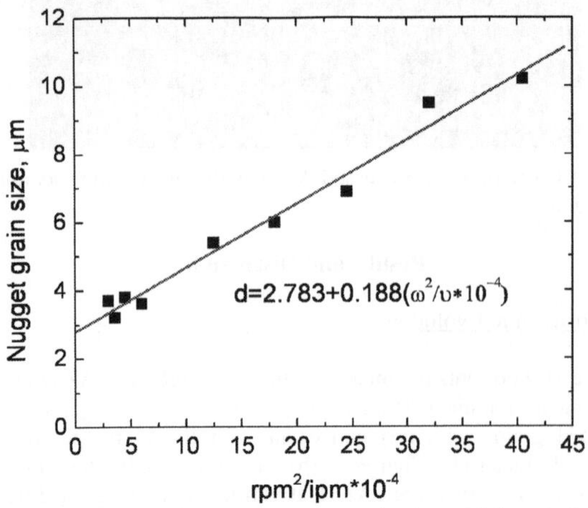

Figure 3. Variation of average nugget grain size with heat index.

Mechanical Behavior

Figure 4(a) shows the plot of microhardness at middle region of the nugget. The hardness of nugget decreases after FSP as compared to the as-received material, which is ~70 HV. Within the heat indices used, the hardness first decreases from about 62 HV to 53 HV as the heat index increases from 4.5 to 12.5, and then stays almost constant. The hardness of the nugget is dominated by grain size and dislocation density, since AZ31 magnesium alloy does not have significant level of precipitates for strengthening. The as-received AZ31 was in work-hardened condition. The stored dislocations and twins contribute to the relatively higher hardness value. On the other hand, the FSPed AZ31 exhibits a fully recrystallized microstructure. The hardness variation indicates that the grain size plays an important role when the average size is below 5 μm. Yang et al. [14] reported that the hardness values of friction stir welded AZ31 did not change much when the grain size was larger than 10 μm. Figure 4(b) presents the relationship between microhardness and average grain size in the nugget region. The Hall-Petch relationship between the hardness and the nugget grain size is observed when the nugget grain size is larger than 5 μm, however, data points with grain sizes less than 4 μm deviate from this expression. Combined with the previous results of FSPed fine grain structure with grain size in the range of 1 to 3 μm, the highest hardness values for the present work fit the Hall-Petch equation reported as $HV=33.3+58.6d^{-1/2}$ [7]. Though Chang et al. [15] reported a single slope for Hall-Petch relationship for FSPed AZ31 with a wide grain size range (3 to 100 μm), the current results indicate an inconsistency when the grain size is between 3 and 5 μm.

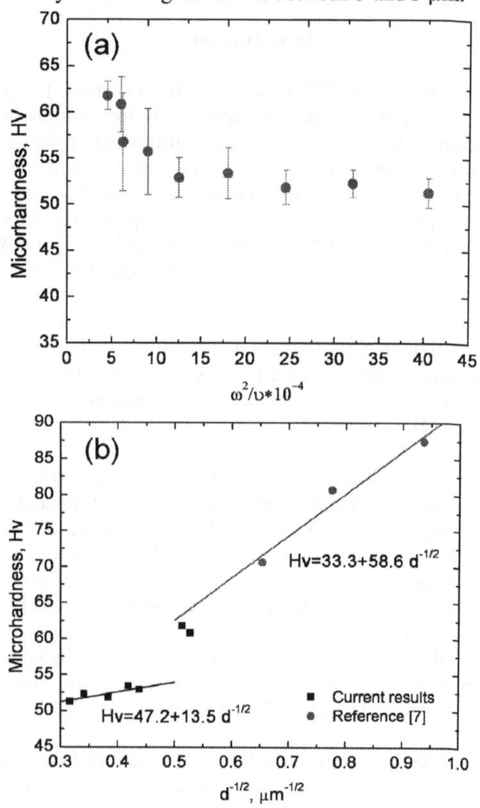

Figure 4. Microhardness variation as a function of (a) heat index, and (b) nugget grain size.

Figure 5 presents room temperature tensile stress-strain curves at a strain rate of $1x10^{-3}$ s^{-1} for the as-received and FSPed AZ31 alloy at four different heat indices 4.5, 18 and 40.5, along two orthogonal directions; the RD for as-received AZ31alloy or the PD for FSPed AZ31 alloy, and the TD for all. Mini-tensile results of the as-received AZ31 alloy exhibited minor anisotropic behavior evidenced by slightly higher strength in the TD. The FSPed AZ31 alloy exhibited significant anisotropic behavior with higher strength but lower ductility in the TD as compared to the PD. The yield strength of the sample tested in the PD decreased from 243 MPa to about 110 MPa, 74 MPa and 50 MPa for heat index at 4.5, 18 and 40.5, respectively. For samples tested in the TD, the yield strength decreased from 274 MPa to about 233 MPa, 213 MPa and 170 MPa, respectively. The specimens tested in PD showed higher ductility and uniform elongation than those of TD. These results are different from the previous results for FSPed ultra-fine grained AZ31, which showed both higher strength and ductility in TD than PD [8].

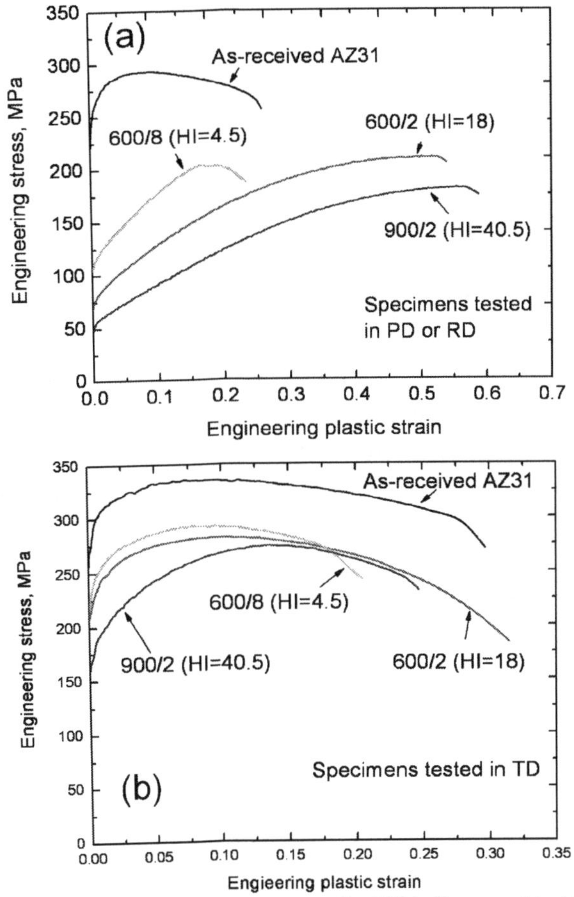

Figure 5. Stress - plastic strain curves for AZ31 alloy tested in (a) processing or rolling direction and (b) transverse direction.

The significant anisotropy in tensile behavior was related to the unique fiber texture generation in the nugget during FSP, where the tilt of basal poles was depth dependent. The tilt of basal poles in the center line of the nugget increased as the depth increased. At a location of 75% of the nugget depth, the basal poles were roughly parallel to the PD [8]. For the current results, the basal poles are expected to have a similar character, which favors the basal plane slip during tensile test when the loading axis is along the PD, indicated by much lower tensile yield strength in Figure 5(a). The difference in the yield strength among various heat indices is related to the variation in the grain size. As the heat index decreases, the average grain size becomes lower, so a higher stress is required for plastic deformation to occur for a similar texture. The yield strength shows similar trend when specimens were tested along TD, it increases as the indices decreases. The stress-strain curves also indicate that the uniform elongation increases as the grain size increases from 3.8 μm to 10.2 μm, which could be related to the twinning effect [16].

Figure 6 shows the degree of anisotropy in yield strength between TD and PD or (RD for as-received AZ31 alloy). The as-received AZ31 alloy exhibited low anisotropy with a difference of about 32 MPa in the yield strength. However, the degree of anisotropy for FSPed AZ31 alloy was significant and a difference of about 120 MPa in yield strength was noted. A ratio of difference in yield strength to the maximum tensile yield strength was also plotted. The ratio decreases as the heat index decreases, i.e., the ratio decreases with the grain size. As the grain size decreases to a critical scale, non-basal slip systems can be activated even at room temperature [17,18], which should reduce the anisotropy.

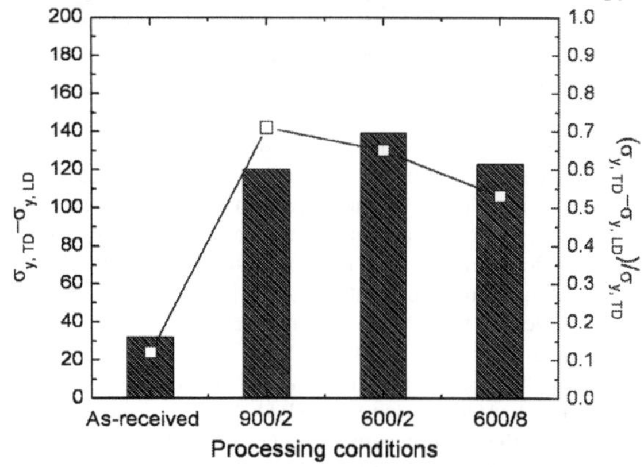

Figure 6. Degree of anisotropy at various processing conditions.

The hardening capacity, H_c, defined as a ratio of the ultimate tensile strength to the yield strength, signifies the capacity to which the material can be strain-hardened [19]. H_c is plotted for AZ31 alloy for various heat indices and compared to the as-received condition in Figure 7. A significant increase in hardening capacity can be observed for the specimens tested in PD. Similar results were reported by Afrin et al. [20] and Chowdhury et al. [21]. On the other hand the increase in hardening capacity is not obvious for TD. As defined, the hardening capacity of a material is directly related to its yield strength and the following strain hardening contributed by dislocation-dislocation interaction and interaction of dislocation with other microstructural features, such as twin boundaries. The as-received AZ31 alloy was in cold-worked and partially annealed condition. With the pre-existence dislocations and twinning, the yield strength was high and the hardening capacity low. However, FSPed AZ31 shows a dynamic recrystallized grain structure with a relatively lower dislocation density [14]. An increase in heat index or grain size decreases the tensile yield strength, but increases the ability of dislocation storage in a grain [19]. This explains the relatively higher hardening capacity at a higher HI. Texture also plays a role in hardening capacity; the higher hardening capacity is also related to the activation of easier basal slip system. Compared with the as-received AZ31, which has a harder texture (low Schmid factor) for basal slip for in-plane tensile tests, the specimens tested in PD for FSPed AZ31 have a softer texture (high Schmid factor) for basal slip. The texture effect on hardening capacity can be seen from the different behaviors between TD and PD. A much reduced hardening capacity is observed for TD due to the harder texture for basal slip, which requires higher stress to activate the non-basal slip systems or twinning.

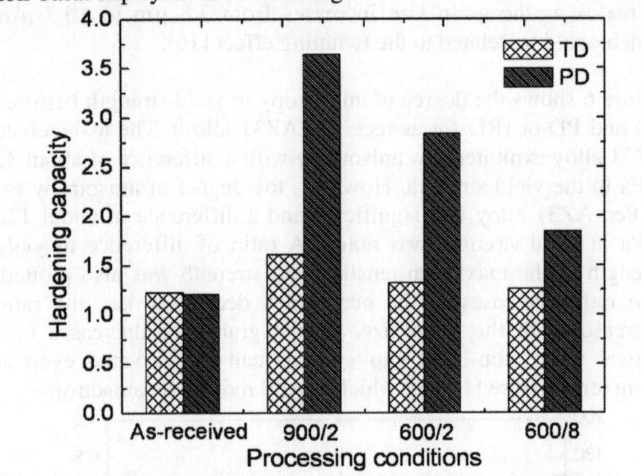

Figure 7. Hardening capacity at various processing conditions.

Figure 8 plots the work hardening rate derived from true stress vs. true strain as a function of net flow stress for various heat indices for both PD and TD. For the specimens tested in PD or RD (Figure 8(a)), the work hardening rate decreases steadily with the increase of flow stress during the initial stage due to the balance between storage and annihilation of dislocations, which is the stage III hardening range [22]. The as-received AZ31 alloy shows a higher work hardening rate than FSPed AZ31 during the initial period and extend of work hardening period is short. The initial drop in work hardening rate for FSPed AZ31 is followed by an increase when the net flow stress exceeds 30 MPa and subsequent decrease again. Similar work hardening phenomenon has been reported by Lee et al. [23]. They suggested that in addition to the texture effect, the increased work hardening rate should be correlated with the deformation-twin induced grain refinement and intense stress concentration as a result of interaction between dislocation and twin. The current results also indicate that the grain size influences the work hardening rate. At same net flow stress, the work hardening rate decreases as the heat index or grain size increases. The increase in work hardening rate after stage III is more prominent for larger grain size material, which may due to easier twinning in larger grains [16]. Figure 8(b) exhibits the work hardening rate for specimens tested in TD. A much higher initial work hardening rate is observed for TD compared to PD for FSPed AZ31 alloy. The difference in hardening behavior could be related to different deformation mechanisms introduced by texture. A key difference is that the intermediate stage of increase in work hardening rate is not observed in TD specimens. The FSPed finer grain AZ31 shows higher work hardening rate at initial hardening stage, but lower work hardening rate at higher net flow stress region. Sinclair et al. [24] and Kovacs et al. [25] reported that grain size strongly influenced strain hardening at lower strains, and the influence diminishes at higher strains due to dislocation screening and dynamic recovery effects at grain boundaries. Recently, de Valle et al. [26] reported that decreasing the grain size resulted in a decrease in work hardening rate in rolled and annealed AZ31 (grain size range of 2 to 55 μm). However, current results indicate that grain size influences work hardening differently when texture effect is superimposed.

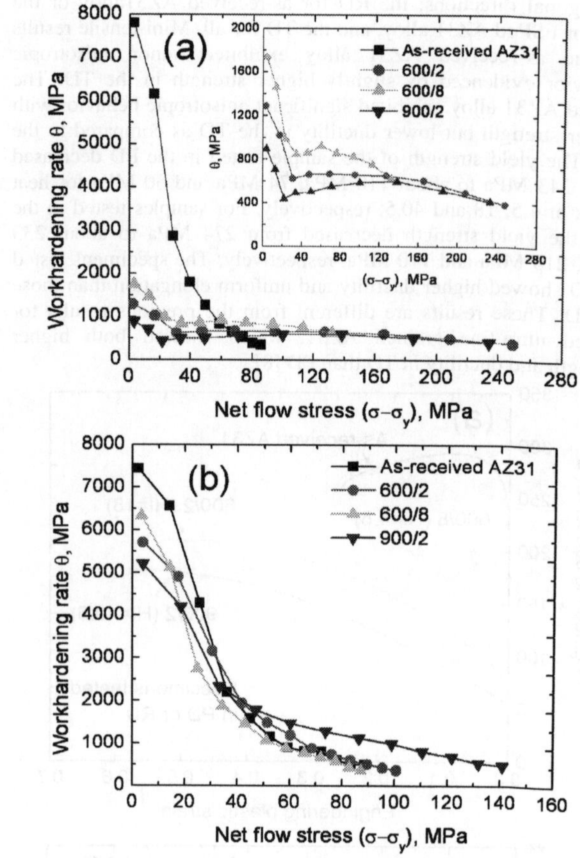

Figure 8. Work hardening rate derived from true stress vs. true strain (considering only uniform elongation), for various processing conditions, tested in (a) PD and (b) TD.

Conclusions

Friction stir processing modifies the microstructure and properties of metals through intense plastic deformation. The frictional heat index exhibits a great effect on the dynamically recrystallized grain structure in the nugget and resulting mechanical properties. Results indicate that nugget grain size increases linearly with the heat index, which in turn reduces the nugget hardness, tensile yield strength, but improves the work hardening capacity. The anisotropy in tensile behavior in transverse direction and processing direction is related to the unique fiber texture generated in the nugget. Both the texture and grain size have a great effect on work hardening behavior. For specimens tested in PD, the finer the grain size, the higher the work hardening rate; however, for specimens tested in TD, the finer grain size exhibits higher work hardening rate only at lower net flow stress stage, but lower work hardening rate at higher net flow stress region.

Acknowledgements

The authors gratefully acknowledge the support of the National Science Foundation through grant NSF-EEC-0531019.

References

[1] W.M. Thomas, E.D. Nicholas, J.C. Needham, M.G. Murch, P. Templesmith, C. J. Dawes, G.B. Patent 9125978.8 (1991).
[2] R.S. Mishra, M.W. Mahoney, S.X. McFadden, N.A. Mara, A.K. Mukherjee, *Scr. Mater.*, 42 (1999) 163-168.
[3] R.S. Mishra and Z.Y. Ma, "Friction stir welding and processing," *Mater. Sci. Eng. R: Reports*, 50 (2005) 1-78.
[4] L. Cui et al., "Friction stir welding of a high carbon steel," *Scr. Mater.*, 56 (2007), 637-640.
[5] C.I. Chang, X.H. Du, J.C. Huang, "Achieving ultrafine grain size in Mg-Al-Zn alloy by friction stir processing," *Scr. Mater.*, 57 (2007) 209-212.
[6] C.I. Chang et al., "Producing nanograined microstructure in Mg–Al–Zn alloy by two-step friction stir processing," *Scr. Mater.*, 59 (2008), 356-359.
[7] G. Bhargava et al., "Influence of texture on mechanical behavior of friction-stir-processed magnesium alloy," *Metall. Mater. Trans.*, 41 (2010) 13-17.
[8] W. Yuan et al., "Effect of texture on the mechanical behavior of ultra-fine grained magnesium Alloy," to be submitted to Scr. Mater..
[9] W.J. Arbegast, P.J. Hartley, in: Proceedings of the Fifth International Conference of Trends in Welding Research, Pine Mountain, GA, June 1–5, 1998, p. 541.
[10] J.A. Esparza et al.," Friction-stir welding of magnesium alloy AZ31B," *J. Mater. Sci. Lett.*, 21 (2002) 917-920.
[11] S.H.C. Park, Y.S. Sato, H. Kokawa, "Microstructural evolution and its effect on Hall-Petch relationship in friction stir welding of thixomolded Mg alloy AZ91D," J. *Mater. Sci.*, 38 (2003) 4379-4383.
[12] W.B. Lee, Y.M. Yeon, S.B. Jung, "Joint properties of friction stir welded AZ31B - H24 magnesium alloy," Mater. Sci. Technol., 19 (2003) 785-790.
[13] L. Commin et al., "Friction stir welding of AZ31 magnesium alloy rolled sheets: Influence of processing parameters," *Acta Mater.*, 57 (2009) 326-334.
[14] J. Yang et al., "Effects of heat input on tensile properties and fracture behavior of friction stir welded Mg-3Al-1Zn alloy," *Mater. Sci. Eng. A*, 527 (2010) 708-714.
[15] C.I. Chang et al., "Relationship between grain size and Zener–Holloman parameter during friction stir processing in AZ31 Mg alloys," *Scr. Mater.*, 51 (2004), 509-514.
[16] M.R. Barnett et al., "Influence of grain size on the compressive deformation of wrought Mg-3Al-1Zn," *Acta Mater.*, 52 (2004) 5093-5103.
[17] J. Koike et al., "The activity of non-basal slip systems and dynamic recovery at room temperature in fine-grained AZ31B magnesium alloys," *Acta Mater.*, 51 (2003) 2055-2065.
[18] S.R. Agnew and O. Duygulu, "Plastic anisotropy and the role of non-basal slip in magnesium alloy AZ31B," *Int. J. Plast.*, 21 (2005) 1161-1193.
[19] J. Luo et al., "Diminishing of work hardening in electroformed polycrystalline copper with nano-sized and uf-sized twins," *Mater. Sci. Eng. A*, 441 (2006) 282-290.
[20] N. Afrin et al., "Strain hardening behavior of a friction stir welded magnesium alloy," *Scr. Mater.*, 57 (2007) 1004-1007.
[21] S.M. Chowdhury et al., "Tensile properties and strain-hardening behavior of double-sided arc welded and friction stir welded AZ31B magnesium alloy," *Mater. Sci. Eng. A*, 527 (2010) 2951-2961.
[22] U.F. Kocks and H. Mecking, "Physics and phenomenology of strain hardening: the FCC case," *Prog. Mater. Sci.*, 48 (2003) 171-273.
[23] H.-W. Lee et al., "Studies on the improvement of tensile ductility of hot-extrusion AZ31 alloy by subsequent friction stir process," *J. Alloys Compd.*, 475 (2009) 139-144.
[24] C.W. Sinclair et al., "A model for the grain size dependent work hardening of copper," *Scr. Mater.*, 55 (2006), 739-742.
[25] I. Kovacs et al., "Grain size dependence of the work hardening process in Al99.99," Phys. Status Solidi A 194 (2002) 3-18.
[26] J.A. del Valle et al., "Influence of texture and grain size on work hardening and ductility in magnesium-based alloys processed by ECAP and rolling," *Acta Mater.*, 54 (2006) 4247-4295.

Magnesium Technology 2011
Edited by: Wim H. Sillekens, Sean R. Agnew, Neale R. Neelameggham, and Suveen N. Mathaudhu
TMS (The Minerals, Metals & Materials Society), 2011

STRENGTHENING MG-AL-ZN ALLOY BY REPETITIVE OBLIQUE SHEAR STRAIN

Toshiji Mukai [1], Hidetoshi Somekawa [1], Alok Singh [1], Tadanobu Inoue [1]

[1]National Institute for Materials Science, 1-2-1 Sengen, Tsukuba, Ibaraki 305-0047 Japan

Keywords: magnesium, grain refinement, shear strain, texture, caliber rolling

Abstract

Grain refinement is one of the possible ways to enhance the strength of magnesium without losing the ductility and/or toughness. In this study, severe plastic working by caliber rolling has been demonstrated to refine the grain structure of a commercial AZ31 Mg-Al-Zn alloy at a commercial processing speed. As a result, ultra-fine-grain structure with sub-grains in a sub-micro-meter scale was obtained. A simultaneous operation of oblique shear strain weakened the basal texture compared to that of the initial as-extruded alloy, and resulted in tensile ductility comparable to that of the commercially extruded alloy, and showed a higher asymmetry ratio of yield stress in compression/tension than that of the as-extruded alloy.

Grain Refinement of Mg for Strengthening

Magnesium alloys are being widely investigated throughout the world for the weight reduction of structural components due to their low density. In structural applications, the strength of magnesium alloys should be increased along with their ductility to improve the relatively low stiffness, so that the alloys can be formed into complex shapes. One of the major microstructural factors for strengthening of metals is grain refinement. It is even more effective in increasing the strength of magnesium alloys because of the relatively large Taylor factor in its HCP structure compared to that of the other lightweight metal, aluminum [1]. The tensile elongation also increases with grain refinement due to the enhanced activity of grain boundary plasticity [2]. Another important factor for enhancing the ductility is the distribution of the basal plane orientation, *i.e.*, the basal texture. The basal texture can be modified by changing the applied shear loading direction during severe plastic working [3], and/or by changing the recrystallization behavior by a combination of distributing nano-scale particles and plastic working [4]. Wrought magnesium alloys are generally known to show yield asymmetry [5]; the yield stress in tension is higher than in compression. The asymmetry of yield stress can be weakened due to the texture modification. Wang *et al.* demonstrated that the asymmetry ratio of yield stress (yield stress in compression / yield stress in tension) increases with grain refinement and approaches 0.9 in the AZ31 alloy [6]. Thus, the combination of grain refinement and the randomization or the weakening of the basal texture may enhance the ductility in high-strength Mg alloys.

There are some reports that the application of repetitive shear loading could refine the grain structure [7-9] and modify the texture of Mg alloys, *i.e.*, repetitive processing of equal-channel-angular-extrusion (or -pressing) (ECAE or ECAP) could incline the basal pole at certain degrees against the extrusion direction, while a strong basal texture is formed almost parallel to the extruded direction by conventional extrusion [10-13]. In our earlier work, when the basal pole was inclined about 45 degrees against the extrusion direction, the tensile yield stress was reduced to just half of that in a directly extruded alloy but the tensile elongation increased to more than double [3]. Koike *et al.* revealed by transmission electron microscopy that an inclined basal pole activates non-basal <a> dislocations as well as basal dislocations even at room temperature [14]. On the other hand, Agnew *et al.* demonstrated by using in-site neutron diffraction that the polycrystalline modeling validated the elastoplastic deformation in the ECAE AZ31 alloy [15]. They pointed out that more than 50 % of the plastic deformation was dominated by the basal slip even after macroscopic yielding, while the prismatic or pyramidal <a> dislocation slip became more active than the basal dislocation in a conventional AZ31 extrusion. Other procedures have been applied to the AZ31 alloy to refine the grain structure and to modify the texture, *e.g.*, Kim *et al.* demonstrated that differential speed rolling (DSR) at a relatively low temperature of 413 K enhanced the strength of AZ31 effectively by means of the ultra-fine grained structure [16]. They also reported that the weakened intensity of the (0002) basal pole could increase the tensile elongation. Xing *et al.* demonstrated that the grain refinement of AZ31 could increase the tensile strength by multi-directional forging by decreasing the processing temperature for each forging [17]. Yang and Ghosh conducted a grain refinement of the AZ31 alloy by weakening the basal texture using an alternate biaxial reverse corrugation pressing [18]. The above mentioned studies suggest that the grain refinement along with the weakened basal texture can activate the basal and/or non-basal dislocations rather than deformation twins and result in enhancing the strength-ductility balance at room temperature.

Repetitive Shear Straining by Caliber Rolling

In this study, grain refinement with a weakened texture is demonstrated by applying a heavy oblique shear strain using caliber rolling, which can operate continuously at a commercial processing speed and commercial scale. Caliber rolling has been used mostly in the research of steels to control the cross sectional shape of products [19] and/or strengthening the material with toughness [20]. Aoki and Yanagimoto reported the effect of caliber shape on the strain distribution in the cross section using the plasticine models. [19]. Inoue *et al.* have demonstrated that the accumulated strain in the cross section could be effectively increased using caliber rolling by numerical simulation [21]. In this study, we have paid attention to the effect of oblique shear strain on the texture evolution. A schematic illustration of the caliber rolling is shown in Fig. 1 [22]. The upper roll has a channel of identical dimensions with the counterpart bottom roll. However, the bottom portion of the caliber has a smaller roll diameter than that of the top portion of the caliber roll. Therefore the outer rolling speed at the bottom (V_{btm}) is slower than that at the top (V_{top}). When a billet is subjected to rolling, the material is deformed plastically during a certain duration. The cross sectional area of the billet is reduced at a fixed ratio, while the discrepancy of V_{top} and V_{btm} creates a certain shear strain simultaneously after passing the roll, as illustrated in Fig. 1(a). After rotating the rolled billet 90 degrees clockwise around the rolling axis, the billet is subjected to rolling in the same way as the former pass. The cross section is reduced again with the same reduction ratio, but the shear direction is the opposite from that of the former pass, as illustrated in Fig. 1(b). Thus the resultant texture after the secondary pass is weakened as compared to that of the formerly rolled billet, although the accumulated strain in the secondary processed billet is much higher than that in the former one. Since the actual deformation may be more complicated due to a lateral extension *etc.* throughout the processing, a detailed numerical simulation by the finite element method has been carried out to understand the deformation sequence [23].

Caliber Rolling for AZ31 alloy

In this experimental work, a commercially extruded AZ31 alloy (Mg-2.9 Al-0.8 Zn-0.42 Mn, by mass %) was used as the starting material. The commercial extrusion was conducted at a temperature of 623 K with an extrusion ratio of 16:1 by Osaka Fuji Kogyo, Japan. The grain structure of the initial material was inspected by optical microscopy, and the average grain size was measured to be ~ 25 μm. A cylindrical billet was machined from the extruded bar with a diameter of 42 mm and a length of 100 mm. The billet was kept at 473 K in a furnace in air for one hour and then subjected to severe plastic working by the caliber rolling facility at room temperature [22]. The outer speed of the roller was fixed at 0.5 m/s. The reduction ratio of the cross sectional area was designated to be -18% for each caliber with an analogy shape as shown in Fig. 1. The billet was rolled repetitively by 18 passes with rotating the rolled bar 90° for each pass. In the final pass, the rolled bar was rotated 90° and rolled with the same caliber as that of the former pass. The cumulative reduction in the area was estimated to be about 95% after 18 passes, corresponding to an equivalent strain of 3.1 estimated from the reduction in area. To avoid abnormal grain growth, reheating was not conducted after the passes. The rolled billet was subjected to subsequent rolling within a few seconds. The surface temperature on the rolled specimen after 18 passes was measured to be ~ 433 K, thus the material was kept in a temperature range of 433 to 473 K during the repetitive rolling.

Effect of Repetitive Shear Straining

The microstructure at the mid-layer of the caliber rolled alloy after 15 passes was inspected by electron backscatter diffraction (EBSD) equipped in a field emission scanning electron microscope (JEOL F7400) to investigate the texture evolution. The orientation map of the grain structure in the caliber rolled alloy after 15 passes was shown in Fig. 2. The grain structure of high angle grain boundaries (> 15 degrees) and low angle grain boundaries are shown by black and gray lines in Fig. 2, respectively. The grain structure was found to be refined to an average grain diameter of 2.5 μm by caliber rolling. (0002) pole figure representing the basal plane was also measured by the Schulz reflection method at α-angles of 20° to 90° [22]. The

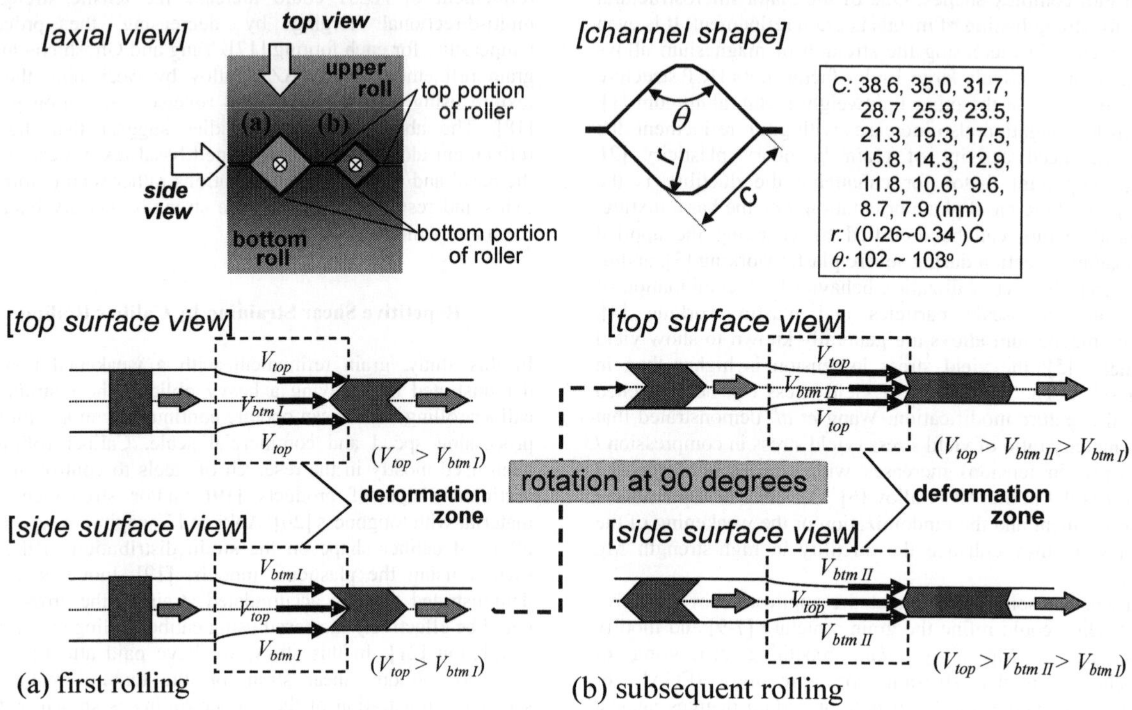

Figure 1 Schematic illustration of the caliber rolling process [22].

XRD measurement revealed that (i) the intensity of the basal pole in the caliber rolled alloy was shifted about 10 degrees from the center toward the transverse direction, (ii) the asymmetrical distribution of the basal plane and the weakening of the maximum intensity of the basal pole were formed in the caliber rolled alloy, although the material was subjected to repetitive severe plastic working.

Mechanical properties were examined using specimens machined from the initial extrusion and rolled bars having a gauge diameter of 3 mm and a gauge length of 15 mm for tensile tests and a diameter of 4 mm and a height of 8 mm for compression tests. The axial direction of each specimen was parallel to the rolling or extrusion direction. The tensile and compression tests were conducted at a strain rate of $1 \times 10^{-3}\ s^{-1}$ at room temperature by an Instron tester. Nominal stress-strain relations in tension and compression for the caliber rolled alloy and the as-extruded alloy are shown in Fig. 3. The tensile yield stress (σ_{tys}) in the caliber rolled alloy being approximately twice as high as that of the as-extruded alloy, although the tensile elongation exhibits a similar value. The curve of caliber rolled alloy in compression exhibit a similar trend with a concave shape similar to that of the as-extruded alloy. The plateau region of the flow stress after yielding increases due to the grain refinement, conforming to the experimental results of the as-extruded AZ31 alloy reported by Barnett *et al.* [24]. Although the yield stress in compression (σ_{cys}) also increases after the caliber rolling, the magnitude of increment in σ_{cys} is noticeably greater than that in σ_{tys}. The asymmetry ratio of yield stress in the caliber rolled alloy is noticeably higher than that in the as-extruded alloy, which suggests the weakened basal texture after caliber rolling. The tensile yield stress-elongation balance of the caliber rolled alloy is shown in Fig. 4. The reference data for the as-extruded alloy and AZ31 wrought alloys using other procedures reported in the literature [13,15,16,25,26] are also shown. The yield stress of the caliber rolled alloy is significantly higher than that of the reference alloys. A direct comparison with the yield stress of a fine-grained AZ31 (d: 1.4 μm) processed by DSR [16] suggests that an additional strengthening factor exists for the high strength caliber rolled alloy, because the caliber rolled alloy has grains of similar sizes (d: 1.5 μm) with high angle boundaries. The EBSD observation in Fig. 2 shows color gradation in several grains which suggests formation of a substructure in the grains. Therefore, the microstructure was inspected by

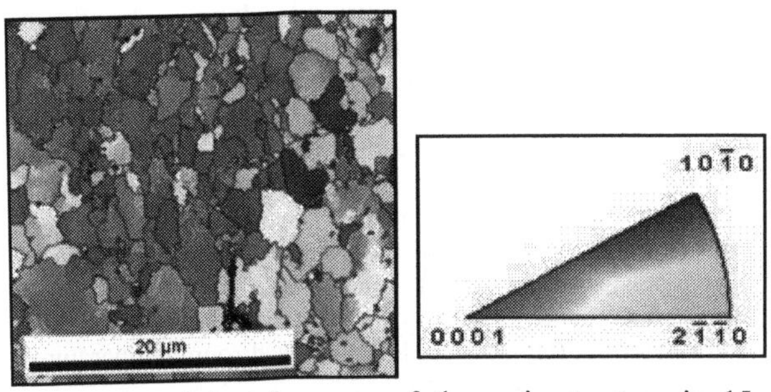

Figure 2: Orientation map of the grain structure in 15 passed alloy observed by EBSD. The grain structure of high angle grain boundaries (> 15 degrees) and low angle grain boundaries are shown by black and gray lines, respectively.

strengthening factor. A typical microstructure of the caliber rolled alloy is shown in Fig. 5. The micrograph shows the formation of fine sub-grains with dense dislocations in the interior. The average size of the sub-grains was measured to be about 0.3 μm in diameter. Kim *et al.* estimated the strengthening effect by grain refinement using the well-known Hall-Petch (H-P) relation in AZ31; the slope (k) of the relation was 336 and 394 $MPa/(\mu m)^{-1/2}$ in the conventionally rolled and DSR alloy, respectively. By assuming that the slope of the H-P relation in the caliber rolled alloy is in the range of 336 ~ 394 $MPa/(\mu m)^{-1/2}$, the additional strength from the reduction of the grain size from 1.4 μm to 0.3 μm was estimated to be 330 ~ 387 MPa. The apparent discrepancy between the tensile yield stress of the caliber rolled alloy and the DSR alloy is 92 MPa, which is smaller than the estimated additional strength. The result suggests that since the caliber rolled alloy has a weakened texture and is composed of fine sub-grains with low angle boundaries, which act as a weaker barrier against dislocation motion than high angle boundaries.

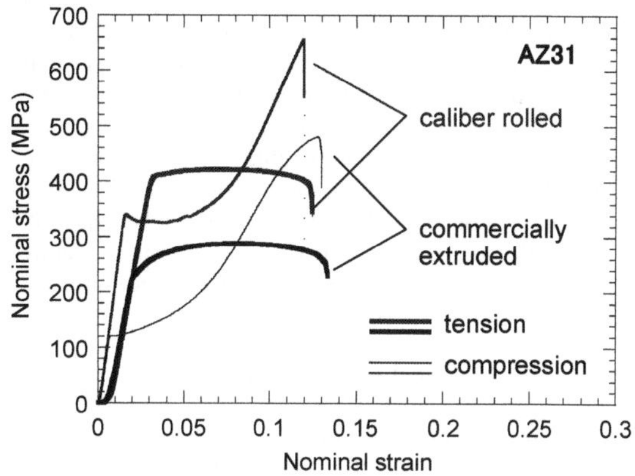

Figure 3: Nominal stress-strain relations in tension and compression.

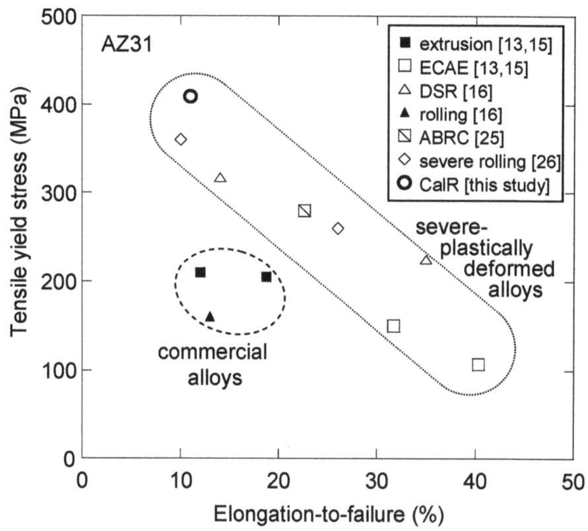

Figure 4: Tensile yield stress-elongation balance in the present alloy exhibits a superior combination [13, 15, 16, 25, 26].

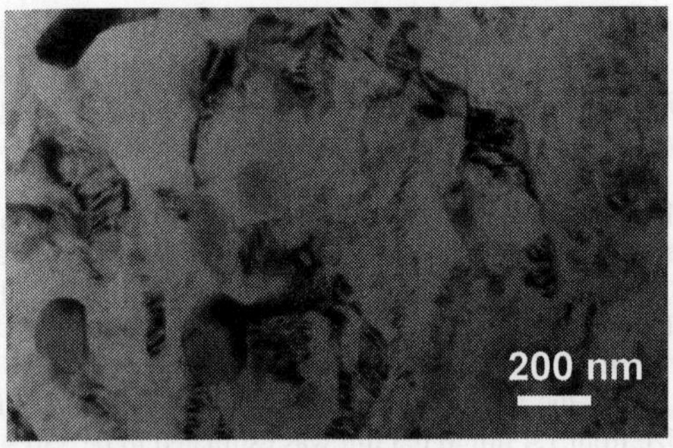

Figure 5: Typical microstructure after caliber oblique shear rolling for 18 passes.

Summary

Severe plastic working by caliber rolling refined the grain structure effectively at a commercial processing speed with the formation of fine sub-grains in sub-micro-meter scale and resulted in a high yield stress of over 400 MPa. A simultaneous operation of oblique shear strain weakened the basal texture from that of the initial as-extruded alloy and resulted in a certain tensile ductility comparable to the commercially extruded alloy and a higher asymmetry ratio of yield stress in compression/tension than that of the as-extruded alloy. It was summarized that the combination of refining grain structure and weakening texture was a possible procedure to improve the mechanical properties in Mg alloys.

Acknowledgement

This work was partly supported by JSPS Grant-in-Aid for Scientific Research (B) No. 21360347 (T.M) and JSPS Grant-in-Aid for Young Scientists (B) No.21760564 (H.S).

References

[1] K. Kubota, M. Mabuchi, K. Higashi, "Processing and mechanical properties of fine-grained magnesium alloys", J Mater Sci 34(1999)2255.

[2] J. Koike, R. Ohyama, T. Kobayashi, M. Suzuki, K. Maruyama, "Grain-Boundary Sliding in AZ31 Magnesium Alloys at Room Temperature to 523K", Mater Trans, 44 (2004) 445.

[3] T. Mukai, M. Yamanoi, H. Watanabe, K. Higashi, "Ductility enhancement in AZ31 magnesium alloy by controlling its grain structure", Scr. Mater., 45 (2001) 89.

[4] A. Singh, H. Somekawa, T. Mukai, "Compressive Strength and Yield Asymmetry in Extruded Mg-Zn-Ho Alloys Containing Quasicrystal Phase", Scripta Mater. 56 (2007) 935.

[5] ASM specialty handbook 'Magnesium and Magnesium Alloys', ASM International, Materials Park, HO, (1999), p166.

[6] J.T. Wang, D.L. Yin, J.Q. Liu, J. Tao, Y.L Su, X. Zhao, "Effect of grain size on mechanical property of Mg–3Al–1Zn alloy", Scr. Mater. 59 (2008) 63.

[7] A. Yamashita, Z. Horita, T.G. Langdon, Mater. Sci. Eng., A 300 (2001) 142.

[8] W.J. Kim, C.W. An, Y.S. Kim, S.I. Hong, Scr. Mater. 47 (2002) 39.

[9] Y. Yoshida, L. Cisar, S. Kamado, Y. Kojima, "Effect of Microstructural Factors on Tensile Properties of an ECAE-Processed AZ31 Magnesium Alloy", Mater. Trans., 44 (2003) 468.

[10] S.R. Agnew, J.A. Horton, T.M. Lillo, D.W. Brown, "Enhanced ductility in strongly textured magnesium produced by equal channel angular processing", Scr. Mater., 50 (2004) 377.

[11] H. Watanabe, A. Takara, H. Somekawa, T. Mukai, K. Higashi, "Effect of texture on tensile properties at elevated temperatures in an AZ31 magnesium alloy", Scr. Mater. 52 (2005) 449.

[12] S.R. Agnew, P. Mehrotra, T.M. Lollo, G.M. Stoica, P.K. Liaw, "Texture evolution of five wrought magnesium alloys during route A equal channel angular extrusion: Experiments and simulations", Acta Mater., 53 (2005) 3135.

[13] H. Somekawa, T. Mukai, "Fracture toughness in Mg–Al–Zn alloy processed by equal-channel-angular extrusion", Scr. Mater. 54 (2006) 633.

[14] J. Koike, et al., "The activity of non-basal slip system and dynamic recovery at room temperature in fine-grained AZ31B magnesium alloys", Acta Mater., 51 (2003) 2055.

[15] S.R. Agnew, D.W. Brown, C.N. Tome, "Validating a polycrystal model for the elastoplastic response of magnesium alloy AZ31 using in situ neutron diffraction", Acta Mater., 54 (2006) 4841.

[16] W.J. Kim, J.B. Lee, W.Y. Kim, H.T. Jeong, H.G. Jeong, "Microstructure and mechanical properties of Mg–Al–Zn alloy sheets severely deformed by asymmetrical rolling", Scr. Mater. 56 (2007) 309.

[17] J. Xing, X. Yang, H. Miura, T. Sakai, "Mechanical Properties of Magnesium Alloy AZ31 after Severe Plastic Deformation", Mater. Trans., 49 (2008) 69.

[18] Q. Yang and A.K. Ghosh, "Production of ultrafine-grain microstructure in Mg alloy by alternate biaxial reverse corrugation", Acta Mater., 54 (2006) 5147.

[19] I. Aoki, S. Yanagimoto, J Jpn Soc Technol Plast, 9 (1968) 597 (in Japanese).

[20] Y. Kimura, T. Inoue, F. Yin, K. Tsuzaki, "Inverse Temperature Dependence of Toughness in an Ultrafine Grain-Structure Steel", Science, 320 (2008) 1057.

[21] T. Inoue, F. Yin, Y. Kimura, "Strain distribution and microstructural evolution in multi-pass warm caliber rolling", Mater Sci Eng, A466 (2007) 114.

[22] T. Mukai, H. Somekawa, T. Inoue, A. Singh, "Strengthening Mg–Al–Zn alloy by repetitive oblique shear strain with caliber roll", Scripta Mater., 62(2010)113.

[23] T. Inoue, H. Somekawa, T. Mukai, "Hardness Variation and Strain Distribution in Magnesium Alloy AZ31 Processed by Multi-pass Caliber Rolling", Adv. Eng. Mater., 11(2009)654.

[24] M.R. Barnett, Z. Keshavarz, A.G. Beer, D. Atwell, "Influence of grain size on the compressive deformation of wrought Mg-3Al-1Zn", Acta Mater. 52 (2004) 5093.

[25] Q. Yang and A.K. Ghosh, "Deformation behavior of ultrafine-grain (UFG) AZ31B Mg alloy at room temperature", Acta Mater., 54 (2006) 5159.

[26] N. Stanford, M.R. Barnett, "Fine grained AZ31 produced by conventional thermo-mechanical processing", J. Alloy Com. 466 (2008) 182

Magnesium Technology 2011

High-Temperature Alloys; High-Strength Alloys; Precipitation

Session Chairs:

Alan A. Luo
(General Motors, USA)
Alok Singh
(National Institute for Materials Science, Japan)

Magnesium Technology 2011
Edited by: Wim H. Sillekens, Sean R. Agnew, Neale R. Neelameggham, and Suveen N. Mathaudhu
TMS (The Minerals, Metals & Materials Society), 2011

Creep and elemental partitioning behavior of Mg-Al-Ca-Sn alloys with the addition of Sr

Jessica R. TerBush[1,2], Olivia H. Chen[1], J. Wayne Jones[1] and Tresa M. Pollock[1,3]
[1] Materials Science and Engineering, University of Michigan; 2300 Hayward St, Ann Arbor, MI 48109, USA
[2] Department of Materials Engineering, Monash University; Clayton, VIC 3800, Australia
[3] Materials Department, University of California; Santa Barbara, CA 93106, USA

Keywords: Creep; Solute Strengthening; Precipitation; Elemental Partitioning

Abstract

The partitioning of elements during solidification can strongly affect precipitation and solute strengthening during creep, especially in alloys where creep is controlled by viscous glide of dislocations. Elemental partitioning during solidification and its influence on creep behavior has been examined for Mg-6.5Al-2.25Ca-0.8Sn with and without 0.25wt% Sr additions. The alloy containing Sr experienced increased partitioning of Al and Ca to the α-Mg phase. However, the creep behavior in compression at 110 MPa and 180°C was not significantly different from that of the alloy without Sr. These observations are discussed in terms of current models for solid solution and precipitation strengthening. In addition, it was determined that the elemental partitioning profiles of permanent mold cast specimens were representative of the grain interiors of die-cast specimens. The partitioning profiles may be less representative for interdendritic regions of the microstructure, however.

Introduction

In the automotive industry, the use of magnesium alloys is aiding efforts to reduce vehicle mass. Many new applications are in automotive powertrain components that experience elevated temperatures, which require alloys with good creep resistance above 150°C [1]. Cast alloys are generally used for powertrain applications, and creep properties in these materials are sensitive to both the solidification behavior and therefore the casting process parameters. These alloys are usually used in the as-cast state, making it important to understand the elemental partitioning that occurs during solidification, as it can ultimately affect high temperature strengthening mechanisms.

Solute redistribution occurs during solidification, resulting in compositional variation from dendrite centers to interdendritic regions [2, 3]. In Mg alloys, this means that the α-Mg that forms at the beginning of solidification (i.e. in the cell interiors) will have a lower solute concentration than the solid that forms in the near-interdendritic regions later in solidification. Solute enrichment in the interdendritic region at the end of solidification leads to the precipitation of intermetallic phases. If no appreciable back diffusion occurs during solidification, the solute will remain inhomogeneously distributed across the microstructure [3]. In the case of no back diffusion in the solid and perfect mixing in the liquid, the Scheil equation is often used to describe the elemental distribution under non-equilibrium conditions:

$$C_s = kC_0(1-f_s)^{(k-1)} \qquad (1).$$

Here, C_s is the concentration of solute in the solid, C_0 is the bulk average concentration, f_s is the fraction solid and k is the partitioning coefficient, which is also defined as (C_s / C_L) where C_L is the concentration of solute in the liquid [4, 5]. Elements with $k > 1$ will preferentially segregate to the dendrite centers, while those with $k < 1$ will segregate to the interdendritic region during solidification. Elements with $k \approx 1$ will not segregate strongly.

Mg-Al-Ca-based alloys have received considerable attention for powertrain applications [1, 6-14]. In our previous work, elemental partitioning has been examined for four Mg-Al-Ca-based alloys [15, 16] where it was discovered that MRI230D (Mg-6.5Al-2.25Ca-0.25Sr-0.8Sn) had significantly increased partitioning of Al and Ca to the α-Mg phase during solidification. It was hypothesized that the presence of Sn in this alloy led to the increased partitioning compared to the other alloys studied, which did not contain Sn. This motivated an investigation of elemental partitioning in quaternary Mg-5Al-3Ca-xSn alloys, described elsewhere [17]. The increased partitioning to the α-Mg observed in MRI230D could not be explained by Sn additions alone. However, MRI230D also contains 0.25 wt% Sr, which was reported to affect the elemental partitioning during solidification of a Mg-5Al-3(Ca,Sr) alloy [8]. Thus, an investigation on the effect of Sr additions on elemental partitioning during solidification has been conducted for this Mg-Al-Ca-Sn alloy.

Of particular interest is how differences in elemental partitioning translate to differences in creep resistance. For example, die-cast MRI153M (Mg-8Al-1Ca-0.25Sr), which has higher Al partitioning and lower Ca partitioning to the α-Mg phase, is less creep resistant than MRI230D and AXJ530 when tested at 110 MPa and temperatures between 100 and 180°C [18]. However, different nominal compositions in these alloys complicates the analysis. Increased solute in the α-Mg can potentially increase solute and precipitation strengthening at elevated temperatures [10], although supersaturation of solute (e.g. Al) has also been linked to decreased creep resistance in some alloys, especially where β-$Mg_{17}Al_{12}$ precipitates form during high temperature exposure [19, 20].

The current investigation examines the elemental partitioning and creep behavior for an Mg-6.5Al-2.5Ca-0.8Sn alloy with and without additions of Sr. The effect of systematic additions of Sn on partitioning of Ca and Al and creep behavior has also been examined.

Experimental Procedure

An alloy with nominal composition of Mg-6.5Al-2.25Ca-0.8Sn (in wt%) was prepared from a master alloy of Mg-4Al-4Ca with additions of pure (99.9%) Mg ingot, (99.5%) Al lump and (99.99%) Sn ingot. The starting materials were melted in a low-carbon steel crucible under a cover gas of 1% SF_6 in Ar using an induction heater. The melt was stirred for several minutes prior to pouring onto an unheated steel plate. A cooling rate of several degrees Celsius per second was obtained for 25 g samples.

Similar casting conditions were used to prepare 25g samples of Mg-6.5Al-2.25Ca-0.25Sr-0.8Sn (MRI230D). For this alloy, the starting material consisted of die-cast specimens of MRI230D, made from original ingots produced by Dead Sea Magnesium. The starting material (die-cast specimens) had an average cell size of ~11 μm, while the slower cooling rate in the current study produced an average cell size of ~20 μm, which facilitated determination of composition profiles.

The as-cast samples for both alloys were sectioned and cold mounted in resin and prepared using standard metallographic procedures. The final polishing step consisted of 0.25 μm diamond paste on a neoprene pad with ethanol lubrication. Samples were not etched prior to microstructural examination in a Philips XL30 SEM operating in backscattered electron (BSE) mode.

Elemental partitioning in as-cast alloys was investigated using a Cameca SX100 electron microprobe. The same sample preparation steps used for SEM analysis were used for the microprobe studies. WDS standards consisted of pure metal and mineral samples, with one exception. The Mg standard used was a sample of Mg-4wt%Al that had been homogenized for 26 days at 400°C. Compositional mapping used a procedure similar to that of Flemings et al. [21], Gungor [22], Tin et al. [23] and Huang et al. [24]. Area scans consisting of at least 400 points were collected from multiple regions of the sample, with an average step size of 10 μm between each point. A voltage of 20 kV and beam current of 15 nA were used for each scan, resulting in an electron beam spot size of approximately 2 μm. Only points with total concentrations between 99 and 101 wt% were used for subsequent analysis. Totals outside this range could be caused by sampling more than one phase simultaneously or by oxide films or inclusions. The resulting data was sorted according to decreasing Mg concentration as described in our previous work [15-17].

Compressive creep tests were conducted on parallelepiped specimens measuring approximately 4x4x8 mm^3 that were sectioned from the induction melted samples. Specimens were ground using 400 grit SiC paper with water lubrication until all sides were flat and parallel. For creep testing, a compression cage was inserted into an Arc-weld constant-load tensile creep frame fitted with a clam-shell furnace. An LVDT attached to an extensometer was used to monitor strain, while temperature was monitored with a K-type thermocouple in contact with the specimen. Testing was carried out at 110 MPa at a temperature of 180°C. Tests were generally interrupted after strain accumulation of 0.1, and specimens were cooled to room temperature under load.

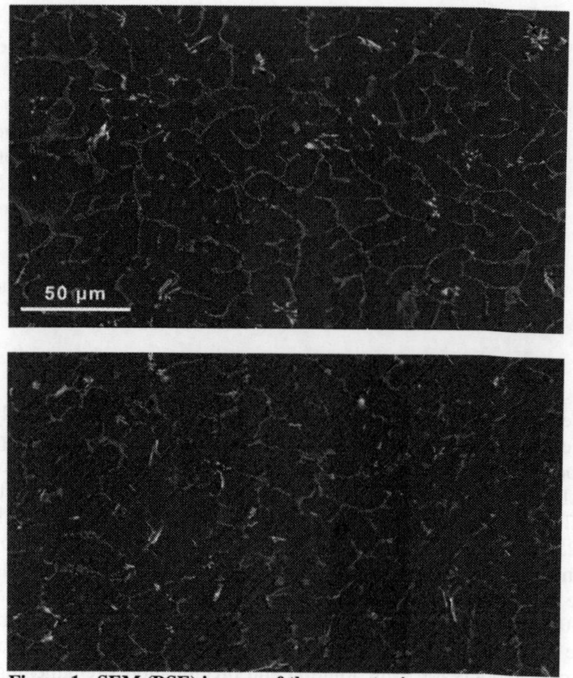

Figure 1 - SEM (BSE) images of the as-cast microstructure of a) MRI230D and b) Mg-6.5Al-2.25Ca-0.8Sn.

For selected crept specimens, dislocation substructures were examined after creep using TEM. Discs 3 mm in diameter were punched from slices made perpendicular to the loading direction, and then thinned by hand to 0.11-0.12 mm thickness. Final thinning to electron transparency was done via twin-jet polishing using a solution of 8 vol% perchloric acid in methanol at -30°C. A Philips CM12 transmission electron microscope was used for the investigation, with an operating voltage of 120 kV.

Results and Discussion

The as-cast microstructures for the alloy with and without Sr are shown in Fig. 1, and are very similar morphologically. The major phase is α-Mg, surrounded by a eutectic microstructure consisting mainly of C36 $(Mg,Al)_2Ca$ and α-Mg. A low volume fraction of a Sn-containing phase, likely CaMgSn ($Ca_{2-x}Mg_xSn$) [25, 26], is also observed in both alloys. A low volume fraction of a Sr-rich phase is also present in the Sr-containing MRI230D. Scheil calculations using the Pandat thermodynamic database predict that β-$Mg_{17}Al_{12}$ will also be present in both alloys, with a higher volume fraction expected in the quaternary Mg-6.5Al-2.25Ca-0.8Sn alloy.

In Fig. 2, the elemental partitioning behavior during solidification is compared for the two alloys for both Al and Ca. The curves shown in this figure are least-mean-square (LMS) fits to the data, presented for ease of comparison. Although not shown here, some scatter does exist in the experimental data, a consequence of the data sorting method used [27]. Based on the shape of the partitioning profiles, MRI230D has higher Al and Ca partitioning to the α-Mg phase than the quaternary alloy without Sr. The Al and Ca partitioning curves are fairly similar for both alloys at the start of solidification and near the end, with the greatest differences observed for fractions solid between 0.5 and 0.8. This suggests that Sr additions may cause differences in the local

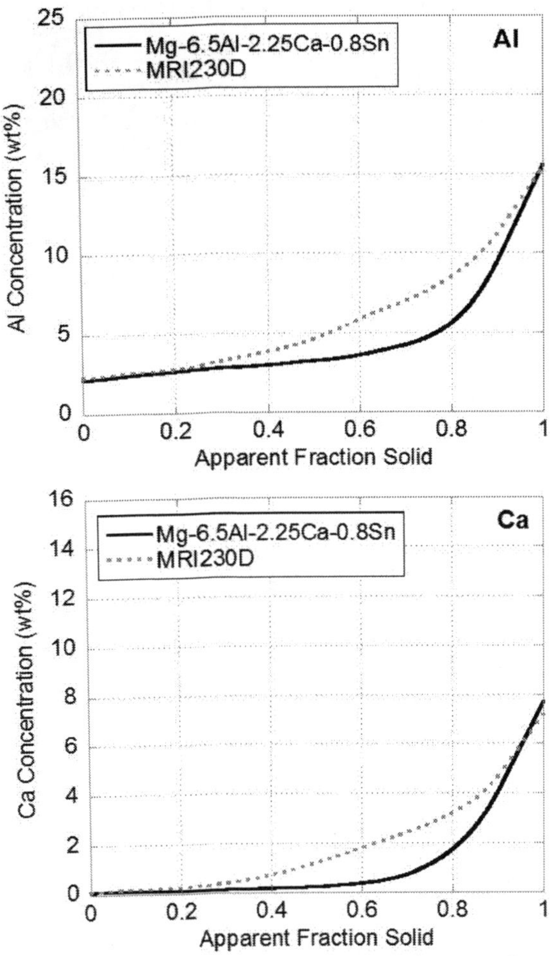

Figure 3 - Comparison of elemental partitioning curves for MRI230D and Mg-6.5Al-2.25Ca-0.8Sn: a) Al and b) Ca profiles.

Table I - Calculated partitioning coefficients (*k*)

Alloy	Mg	Al	Ca	Sn	Sr
AXJ530 [31]	1.03	0.38	0.06	-	*
Mg-5Al-3Ca-0.75Sn	1.08 ±0.03	0.28 ±0.05	0.09 ±0.03	0.05 ±0.02	-
Mg-6.5Al-2.25Ca-0.8Sn	1.07 ±0.03	0.32 ±0.05	0.06 ±0.03	0.07 ±0.04	-
Mg-6.5Al-2.25Ca-0.25Sr	1.14 ±0.03	0.27 ±0.04	0.02 ±0.01	-	0.04 ±0.03
Mg-6.5Al-2.25Ca-0.8Sr	1.11 ±0.03	0.30 ±0.05	0.03 ±0.02	-	0.05 ±0.03
MRI230D [16]	1.10 ±0.02	0.42 ±0.08	0.17 ±0.07	0.08 ±0.04	0.08 ±0.05

* Not reported

concentration, especially in the near-interdendritic region of the microstructure.

To quantify this difference, a Scheil equation has been fit to the experimental data for 10-60% solid. The partitioning coefficients calculated are listed in Table I. For comparison, calculated coefficients are also included for AXJ530 (Mg-5Al-3Ca-0.15Sr), Mg-5Al-3Ca-0.75Sn, Mg-6.5Al-2.25Ca-0.25Sr and Mg-6.5Al-2.25Ca-0.8Sr. MRI230D has the highest partitioning of Al and Ca to the α-Mg phase. Comparing the coefficients for AXJ530 and Mg-5Al-3Ca-0.75Sn, Sn additions appear to increase Ca partitioning to the α-Mg phase. However, the influence of Sn on Al partitioning is less clear; the Mg-5Al-3Ca-0.75Sn alloy has lower Al partitioning to the α-Mg phase than AXJ530, while Mg-6.5Al-2.25Ca-0.8Sn has increased Al partitioning compared to the Mg-6.5Al-2.25Ca-0.8Sr alloy. From comparison of the Sr-containing quaternary alloys, Sr appears to increase Al partitioning to the α-Mg, but does not affect Ca partitioning. Thus, the increased Al and Ca partitioning to the α-Mg phase observed in MRI230D is likely due to the combination of Sn and Sr present in this alloy.

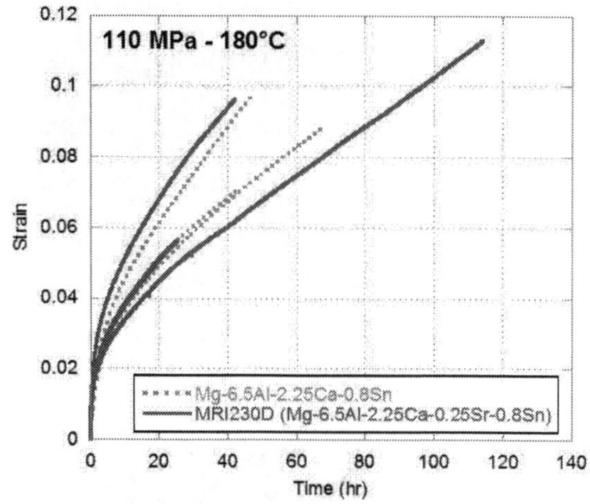

Figure 2 - Compressive creep curves for specimens tested at 110 MPa and 180°C.

Table II - Compressive creep rates at 110 MPa and 180°C

Alloy	Minimum Creep Rate (x 10^{-7} s^{-1})	Avg. Minimum Creep Rate (x 10^{-7} s^{-1})
Mg-6.5Al-2.25Ca-0.8Sn	3.30	2.51
	2.12	
	2.10	
MRI230D	1.97	2.51
	3.01	
	2.54	

To determine how the difference in partitioning with Sr addition affects creep resistance, compressive creep tests were performed at 110 MPa and 180°C, Fig. 3, and minimum creep rates are listed in Table II. The average minimum creep rate for both MRI230D and Mg-6.5Al-2.25Ca-0.8Sn was 2.51×10^{-7} s^{-1}, indicating that Sr does not affect creep resistance in Mg-Al-Ca alloys that also contain Sn. In contrast, minor additions of Sr have been reported to increase creep resistance in Mg-Al-Ca alloys without Sn additions [6, 11, 28]. Further experiments are needed to fully understand the role of Sr in creep resistance.

The dislocation substructures were examined for the crept samples. An example is shown in Fig. 4 for MRI230D. The

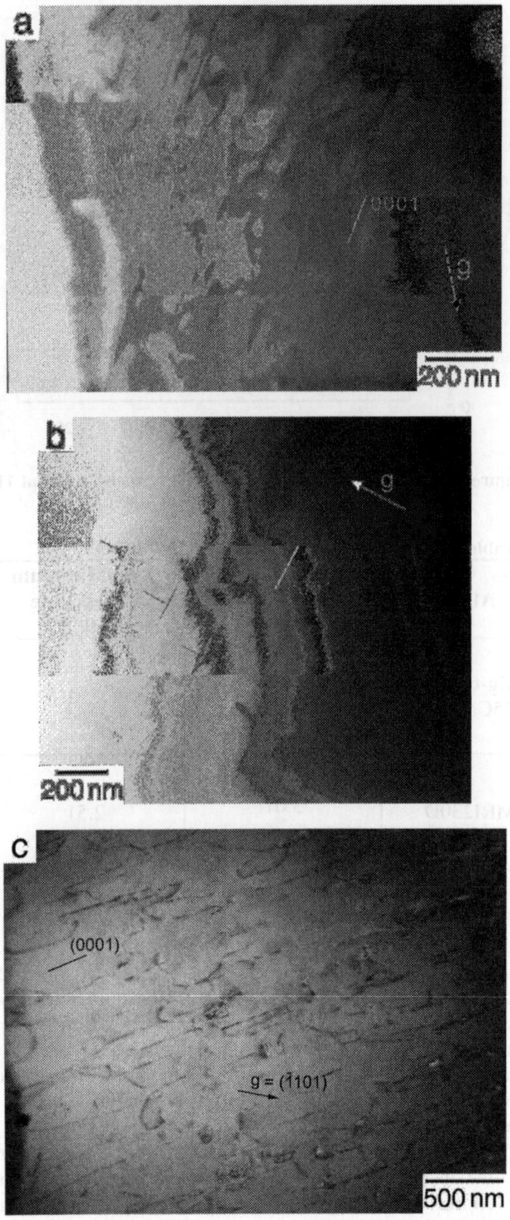

Figure 4 - BF TEM images of a compressively crept sample of MRI230D, taken from a ZA of $[11\bar{2}0]$. a) g = $(1\bar{1}01)$ and b) g = (0002). Precipitates are indicated by the small arrows. c) Crept sample of die-cast MRI230D [18], tested in tension at 110 MPa and 180°C.

Table I - Calculated values of (e^2c/D)

Alloy	Al	Ca	Al + Ca
Mg-6.5Al-2.25Ca-0.8Sn	1.23×10^{12}	1.66×10^{12}	2.89×10^{12}
MRI230D	1.63×10^{12}	6.66×10^{12}	8.29×10^{12}

images were taken from a zone axis of $[11\bar{2}0]$. The majority of the dislocations visible have been identified as basal <a> dislocations, which are invisible under the conditions used in Fig. 4b. The small arrows in Fig. 4b indicate precipitates in the α-Mg matrix. Similar dislocation substructures were identified previously in die-cast samples of MRI230D tested in tension under similar stress and temperature conditions [18], Fig. 4c, with a limited number of non-basal <a> and <c+a> dislocations also present. As in the die-cast samples, viscous glide of basal <a> dislocations was once again identified as the major deformation mechanism during creep.

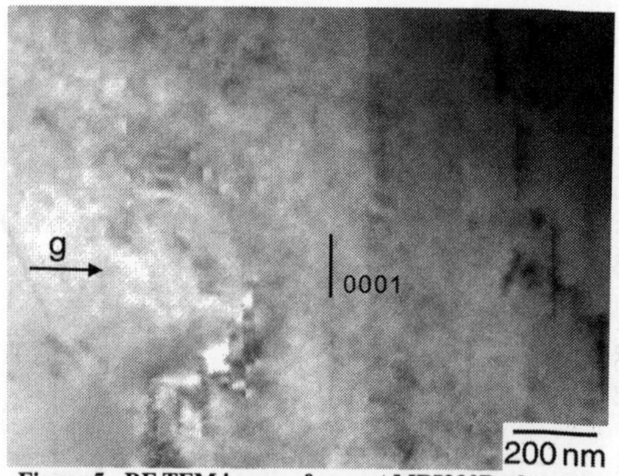

Figure 5 - BF TEM image of as-cast MRI230D, showing precipitates present in the α-Mg. ZA = $[11\bar{2}0]$ and g = (0002).

For alloys where creep is controlled by viscous glide of dislocations, creep resistance can be improved by inhibiting dislocation motion through the α-Mg. This can be accomplished by increasing solute in the matrix, or by the formation of precipitate phases that interact with the moving dislocations. For the alloys currently under investigation, both mechanisms are likely active. Mohamed and Langdon [29] developed an expression relating the creep rate ($\dot{\varepsilon}$) in solute strengthened alloys to solute concentration (c), atomic misfit of the solute atom (e), and diffusivity ($\tilde{D}$):

$$\dot{\varepsilon} \approx \frac{\pi(1-\nu)kT\tilde{D}}{6e^2cb^5G}\left(\frac{\sigma}{G}\right)^3 \quad (2),$$

Where ν is Poisson's ratio, k is Boltzmann's constant, σ is the applied stress, T is the absolute temperature, b is the Burgers vector and G is the shear modulus. A large contribution from the alloy-sensitive ($e^2c/\tilde{D}$) term would result in a low creep rate, assuming all other terms are similar for the alloys. The average concentration measured using the electron microprobe was determined for solid fractions between 0 and 0.7 and used for c in the calculations. The average Al concentration was 2.66 at% for Mg-6.5Al-2.25Ca-0.8Sn and 3.51 at% for MRI230D, and Ca concentrations of 0.12 at% and 0.48 at%, respectively. For e, values of 0.124 and 0.204 were used for Al and Ca, respectively. The diffusion coefficient in these complex alloys is not well known. Approximating the diffusivity as 3.32×10^{-16} cm²/s at 180°C [30] for Al in Mg, and as one order of magnitude less for Ca (~3×10^{-17} cm²/s), the calculated values of ($e^2c/\tilde{D}$) are listed in Table III for both alloys. As would be expected from the higher average concentrations of solute in the α-Mg, the model predicts that MRI230D should have the lower creep rate. However, as described earlier, this was not observed experimentally.

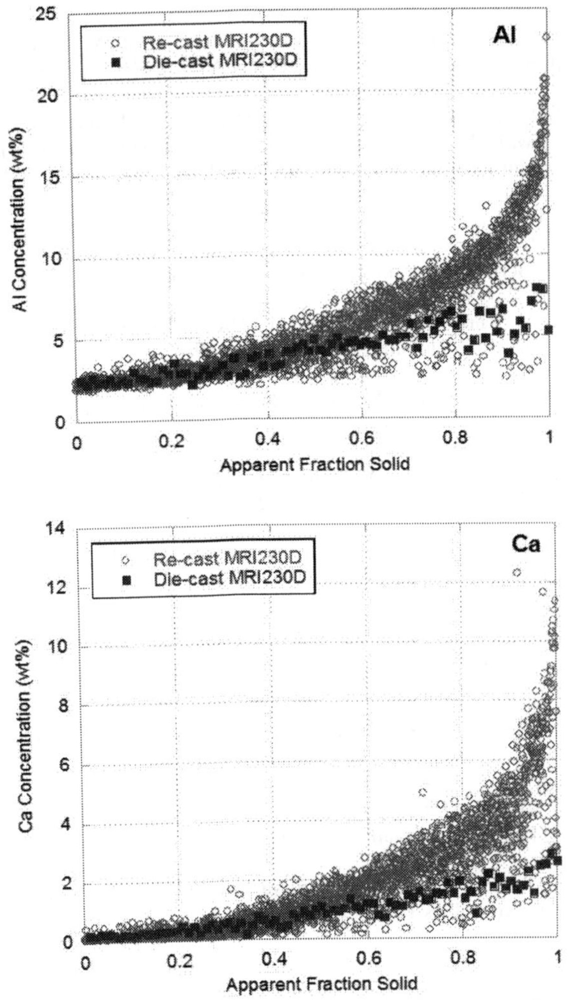

Figure 4 - Comparison of the elemental partitioning curves for die-cast and permanent-mold-type specimens of MRI230D: a) Al and b) Ca profiles.

In the previous calculation, it was assumed that the entire measured concentration of Al and Ca was in the form of solute in the Mg matrix. However, the electron microprobe does not distinguish between solute and nanoscale precipitates when measuring elemental concentrations. As shown in Fig. 5 for MRI230D, precipitates are present even in the as-cast condition. These precipitates lie parallel to the basal plane of the α-Mg matrix, and are likely C15 Al_2Ca as observed in other Mg-Al-Ca alloys, e.g. AXJ530 [10]. These precipitates are also present in the crept alloys, Fig. 4b. According to Pandat equilibrium calculations, similar volume fractions of the C15 phase are expected for both MRI230D and Mg-6.5Al-2.25Ca-0.8Sn. Although not very effective obstacles for basal dislocations, these precipitates are fairly effective obstacles for non-basal dislocations [10]. It has been suggested that non-basal dislocations in the near-grain boundary region account for the localized damage that occurs after the minimum creep rate is reached in samples of permanent mold cast AXJ530 [31]. Moreno et al. [32] and Mori et al. [33] have both suggested that the composition of the near-boundary region may have significant effect on creep resistance of Mg alloys. The increased Al and Ca concentration measured in this region of the microstructure in the current investigation suggests that further study may be needed of the near-boundary region of the α-Mg.

Many automotive applications involve components that are high pressure die cast, and EMPA area and line scans were also conducted on a die-cast sample of MRI230D to determine if segregation was comparable to that observed in permanent mold cast alloys reported here or in previous studies [15-17]. Step sizes of 3-5 μm were used, and the line scans extended across several cells. The die-cast material had an average cell size of ~10 μm, which increased the frequency of sampling multiple phases when both the cell interiors and neighboring interdendritic regions were scanned simultaneously. This was especially true in the near-interdendritic region, and in turn lead to a large number of points that fell outside the acceptable composition range (99-101 total wt%). As before, these points were disregarded during the subsequent analysis. However, enough points were generated for a rough comparison to the segregation behavior of induction melted samples.

Elemental partitioning profiles are shown for both casting conditions in Fig. 6. These profiles show the experimental data and the inherent scatter. The Al and Ca profiles in the die cast material overlap with those in the induction melted samples at apparent fractions solid of approximately 0.5 or less. Thus, the partitioning data presented earlier for the induction melted samples are also representative of the grain interiors of die-cast specimens of the same alloy. However, greater differences may exist in the near-interdendritic region, since lower Al and Ca concentrations were measured in this region for the die-cast specimen. Again, this analysis was somewhat complicated by the small cell size of the die-cast alloys; the electron microprobe technique employed in this study is not ideal for materials with smaller microstructural features (e.g. die-cast alloys). Further investigation is needed, perhaps using higher resolution STEM techniques, to determine whether these differences in local composition in the near-interdendritic region are an artifact of the measurement technique used or reflect elemental partitioning variations with casting process and/or cooling rate.

Summary

1. Increased Al and Ca solute are present in the α-Mg phase of MRI230D and this is attributed to the combined influence of Sr and Sn. Compared with the quaternary alloy without Sr, the largest difference in the Al and Ca concentration profile of MRI230D was observed in the near-interdendritic region.

2. Sr additions do not significantly affect creep resistance in Mg-Al-Ca alloys that also contain Sn, although improvements have been reported in Mg-Al-Ca alloys without Sn. Further study is necessary to mechanistically explain the effect of Sr on creep resistance.

3. Elemental partitioning profiles measured for more slowly cooled permanent-mold cast specimens are representative of the profiles for die-cast specimens for fractions solid of ~0.5 or less (i.e. the grain interiors).

Acknowledgements

The authors acknowledge the support of the National Science Foundation FRG, Grants No. F015408 and DMR-0309468. We also thank B.R. Powell (General Motors), C. Torbet (UC – Santa Barbara) R.R. Adharapurapu (University of Michigan) and C. Henderson (University of Michigan EMAL) for their advice and assistance.

References

[1] A. A. Luo, International Materials Reviews 49 (2004) 13-30.

[2] Y. M. Won, B. G. Thomas, Metallurgical and Materials Transactions 32A (2001) 1755-1767.

[3] D. A. Porter, K. E. Easterling, Phase Transformations in Metals and Alloys, Nelson Thornes Ltd, Cheltenham, 1992, pp. 185-262.

[4] H. D. Brody, M. C. Flemings, Transactions of the Metallurgical Society of AIME 236 (1966) 615-624.

[5] T. F. Bower, H. D. Brody, M. C. Flemings, Transactions of the Metallurgical Society of AIME 236 (1966) 624-634.

[6] A. A. Luo, M. P. Balogh, B. R. Powell, Metallurgical and Materials Transactions 33A (2002) 567-574.

[7] S. M. Zhu, B. L. Mordike, J. F. Nie, Materials Science and Engineering A 483-484 (2008) 583-586.

[8] A. Suzuki, N. D. Saddock, L. Riester, E. Lara-Curzio, J. W. Jones, T. M. Pollock, Metallurgical and Materials Transactions A 38A (2007) 420-427.

[9] A. Suzuki, N. D. Saddock, J. R. TerBush, B. R. Powell, J. W. Jones, T. M. Pollock, SAE Technical Paper 2007-01-1025, SAE International, Detroit, MI, 2007.

[10] A. Suzuki, N. D. Saddock, J. R. TerBush, B. R. Powell, J. W. Jones, T. M. Pollock, Metallurgical and Materials Transactions A 39A (2008) 696-702.

[11] N. D. Saddock, A. Suzuki, J. R. TerBush, J. W. Jones, T. M. Pollock, J. E. Zindel, J. E. Allison, SAE Technical Paper 2007-01-1027, SAE International, Detroit, MI, 2007.

[12] S. M. Zhu, B. L. Mordike, J. F. Nie, Metallurgical and Materials Transactions A 37A (2006) 1221-1229.

[13] Y. Terada, N. Ishimatsu, T. Sato, Mat Trans 48 (2007) 2329-2335.

[14] J. A. Hines, R. C. McCune, J. E. Allison, B. R. Powell, L. J. Ouimet, W. L. Miller, R. Beals, L. Kopka, P. P. Ried, SAE Technical Paper 2006-01-0522, SAE International, Detroit, MI, 2006.

[15] J. R. TerBush, R. R. Adharapurapu, J. W. Jones, T. M. Pollock, in: E. A. Nyberg, S. R. Agnew, N. R. Neelameggham, M. O. Pekguleryuz (Eds.), Magnesium Technology 2009, TMS, San Francisco, CA, 2009, pp. 161-165.

[16] J. R. TerBush, N. D. Saddock, J. W. Jones, T. M. Pollock, Metallurgical and Materials Transactions A 41 (2010) 2435-2442.

[17] J. R. TerBush, O. H. Chen, J. W. Jones, T. M. Pollock, in: S. R. Agnew, E. Nyberg, W. Sillekens, N. R. Neelameggham (Eds.), Magnesium Technology 2010, TMS, Seattle, WA, 2010, pp. 607-611.

[18] J. R. TerBush, A. Suzuki, N. D. Saddock, J. W. Jones, T. M. Pollock, Scripta Materialia 58 (2008) 914-917.

[19] S. M. Zhu, M. A. Gibson, J. F. Nie, M. A. Easton, T. B. Abbott, Scripta Materialia 58 (2008) 477-480.

[20] M. S. Dargusch, S. M. Zhu, J. F. Nie, G. L. Dunlop, Scripta Materialia 60 (2009) 116-119.

[21] M. C. Flemings, D. R. Poirier, R. V. Barone, H. D. Brody, Journal of the Iron and Steel Institute 208 (1970) 371-381.

[22] M. Gungor, Metallurgical Transactions A 20 (1989) 2529-2533.

[23] S. Tin, T. M. Pollock, W. Murphy, Metallurgical and Materials Transactions A 32 (2001) 1743-1753.

[24] S. C. Huang, L. Peluso, D. Backman, in: W. H. Hofmeister, J. R. Rogers, N. B. Singh, S. P. Marsh, P. W. Vorhees (Eds.), Solidification, TMS, 1999, pp. 163-172.

[25] A. Kozlov, M. Ohno, T. Abu Leil, N. Hort, K. U. Kainer, R. Schmid-Fetzer, Intermetallics 16 (2008) 316-321.

[26] A. Kozlov, M. Ohno, R. Arroyave, Z. K. Liu, R. Schmid-Fetzer, Intermetallics 16 (2008) 299-315.

[27] M. Ganesan, D. Dye, P. D. Lee, Metallurgical and Materials Transactions 36A (2005) 2191-2204.

[28] B. R. Powell, V. Rezhets, A. A. Luo, B. L. Tiwari, US Patent 6,264,763, 2001.

[29] F. A. Mohamed, T. G. Langdon, Acta Metall. 22 (1974) 779-788.

[30] G. Moreau, J. A. Cornet, D. Calais, Journal of Nuclear Materials 38 (1971) 197-202.

[31] N. D. Saddock, PhD Thesis, University of Michigan, Ann Arbor, MI, 2007.

[32] I. P. Moreno, T. K. Nandy, J. W. Jones, J. E. Allison, T. M. Pollock, Scripta Materialia 48 (2003) 1029-1034.

[33] Y. Mori, Y. Terada, T. Sato, Materials Transactions 46 (2005) 1747-1752.

Magnesium Technology 2011
Edited by: Wim H. Sillekens, Sean R. Agnew, Neale R. Neelameggham, and Suveen N. Mathaudhu
TMS (The Minerals, Metals & Materials Society), 2011

EFFECT OF Mn ADDITION ON CREEP PROPERTY IN Mg-2Al-2Ca SYSTEMS

T. Homma[1], S. Nakawaki[1], K. Oh-ishi[2], K. Hono[2], S. Kamado[1]
[1]Department of Mechanical Engineering, Nagaoka University of Technology, Nagaoka 940-2188, Japan
[2]National Institute for Materials Science, 1-2-1 Sengen, Tsukuba 305-0047, Japan

Keywords: Magnesium alloy, Creep, Transmission electron microscopy, Atom probe tomography

Abstract

Planar Al_2Ca phase forms on the basal plane in the as-cast AX22 and AXM220 alloys. In the AXM220 alloy, Mn is detected in the Al_2Ca phase. The number density of the Al_2Ca phase does not alter, irrespective of the creep deformation and Mn addition. Fine and planar precipitates appear during the creep deformation. The number density increases by the addition of Mn. The size of the precipitate is reduced by the Mn addition. The Mn addition can improve the creep property.

Introduction

Rare-earth-containing Mg alloys have widely been investigated since they demonstrate remarkable creep properties [1-6]. The latest papers suggest that the improvement in the creep properties is related to dispersion of fine precipitates [4-6]. However, the addition of the rare-earth element into the Mg alloys is indeed costly. Therefore, the development in new-type of magnesium alloys containing "ubiquitous elements" is required.

Al- and Ca-containing Mg alloys attract a lot of attention due to their cost performance and high temperature properties [7-12]. In particular, the use of the materials is expected in automobile industries [13-15]. The advantage in utilizing the Mg-Al-Ca-based alloys is due to the formation of $(Mg,Al)_2Ca$ phases at the grain boundaries and the interior of the grain, leading to preventing grain boundary sliding [8-10,16]. Another possible strengthening mechanism may be dispersion strengthening [17]. Jing [11] and Suzuki et al. [12] reported that Al_2Ca phases precipitate, and they improve the creep properties. Indeed, it is found that application of age hardening treatments to the Mg-Al-Ca alloys can improve the creep properties [12].

The addition of Mn in commercial Mg alloys is usually conducted so as to improve the corrosion resistance [18]. It is reported that the Mn addition can stimulate the formation of Al-Mn type particles which can sweep down iron, resulting in improving the purity in the melt. Nevertheless, it is found that the Mn addition can actually improve the creep property when the creep properties between the Mn-free and Mn-contianing Mg-Al-Ca alloys are compared.

In this study, creep properties in Mg-2Al-2Ca and Mg-2Al-2Ca-0.3Mn alloys were compared. The fine microstructures were analyzed by transmission electron microscopy (TEM) and atom probe tomography (APT).

Experimental procedures

Two types of alloys were investigated in this study. The alloy compositions are listed in Table 1 in mass and mol percent. In this investigation, Al, Ca and Mn are denoted as A, X and M, respectively, following the ASTM designation. Melting was conducted at 720 °C using Mg of 99.95% purity, and then Al and Al-10Mn master alloy were remelted by an electric furnace in

Table 1. Chemical composition (mass% and mol%).

Alloy	Element	Mg	Al	Ca	Mn
AX22	mass%	bal.	2.5	1.8	–
	mol%	bal.	2.3	1.1	
AXM220	mass%	bal.	2.2	2	0.3
	mol%	bal.	2.0	1.2	0.1

a mixed-gas atmosphere of SF_6 and CO_2, and finally Ca was added to the melt. After stirring for about 1 min with the aid of Ar bubbling, the melt was poured into an iron mold held at 250-300 °C. The obtained cast dimensions were 150 × 120 × 40 mm. From the sample, creep test specimens and samples for microstructural observations were cut.

Flat specimens with a gauge length of 50 mm and reduced cross-sectional dimension of 10 × 5 mm were used for a tensile creep test using an A&D CP-6L-500 machine at temperature of 175 °C and an applied stress of 50 MPa. The temperature was maintained at ± 2 °C. No heat treatment was applied to the samples prior to the creep test.

The microstructures were observed using an Olympus BX60M optical microscope. Thin foils for the TEM observations were prepared by low energy beam ion thinning. Conventional TEM observations were made using a JEOL JEM-2010 microscope operated at 200 kV. The specimen thickness and structures of dispersed precipitates were measured using a convergent beam electron diffraction (CBED) technique, and the detailed methods can be found elsewhere [19].

Specimens for atom probe analyses were prepared by mechanical grinding and micro-electropolishing. These specimens were observed by a field ion microscope, and then analyzed by a locally built three dimensional atom probe (3DAP) equipped with a CAMECA fast delay line detector. Field evaporation was assisted by femtosecond laser pulses [20] with the following conditions; wavelength (UV: ultraviolet): 343 nm, pulse power: ~0.5 μJ/pulse; and pulse frequency 100 kHz. The measured temperature and vacuum conditions were 30 K and an ultrahigh vacuum of ~1×10^{-8} Pa, respectively.

Results and Discussion

Creep curves

Figure 1 shows creep curves obtained in the AX22 and AXM220 alloys crept at 175 ºC. The Mn-containing alloy shows a substantially improved (slow) creep rate. It is interesting to note that there was no report on the improved creep properties by the addition of Mn, although most of commercial Mg alloys contained Mn. The minimum creep rates ($\dot{\varepsilon}$) at 175 °C are 5.6×10^{-9} and 2.8×10^{-10} s^{-1} in the AX22 and AXM220 alloys, respectively. $\dot{\varepsilon}$ obtained in the AXM220 alloy is one order of magnitude lower

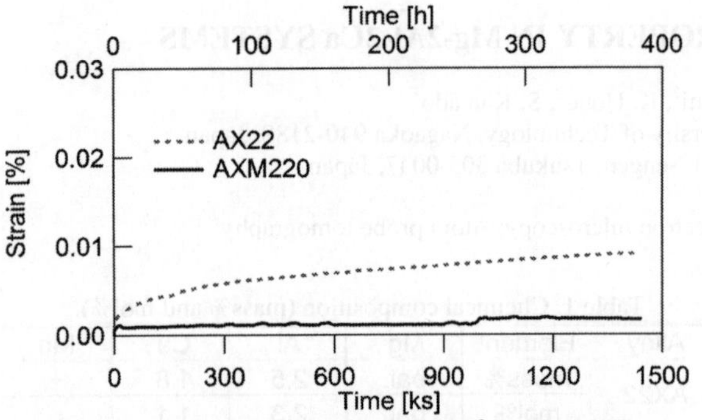

Figure 1. Creep curves obtained at 175 ºC and 50 MPa.

than that in the AX22 alloy. Thus, it is obvious that the Mn addition can improve the creep property of the Mg-Al-Ca system.

Grain boundary structures

Figure 2 indicates optical microscope images obtained in the as-cast samples. Coarse and network-shaped compounds appear at the grain boundaries and in the grains. The compounds were identified based on selected-area electron diffraction patterns, and it was found that the phases are mainly $(Mg,Al)_2Ca$. Mg_2Ca phases were also found. These results are well consistent with previous reports [8-11]. As the other authors reported, the compounds may prevent the grain boundary sliding during the creep deformation, resulting in improving the creep properties. Nevertheless, the Mn addition does not alter the morphology and structure of the compounds.

Intragranular precipitates in the as-cast samples

As shown in Fig. 2, some intragranular precipitates can be confirmed in the interior of the grains in both the alloys. The detailed morphology was observed using the TEM bright-field (BF) images as indicated in Fig. 3. The morphology of the precipitates is a plate shape, and the planar precipitates appear on the basal plane. This precipitate is consistent with Al_2Ca phase [11,12]. In contrast to results reported by Suzuki et al., our samples reveal the presence of the planar precipitate in the interior of the grains even after the as-cast state [12]. This may be due to the difference in the cooling rate between the permanent mold casting and die casting.

Based on foil thickness measurements using a CBED technique, the number densities of the planar precipitates were measured. The number densities of the AX22 and AXM220 alloys are 1.7×10^{19} and 1.8×10^{19} /m^3, respectively. Thus, the Mn

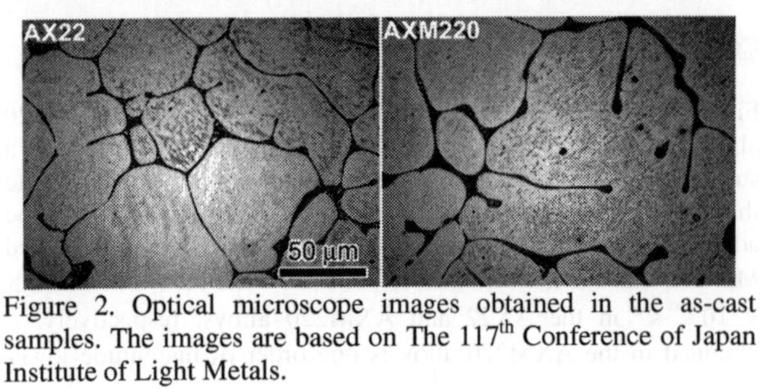

Figure 2. Optical microscope images obtained in the as-cast samples. The images are based on The 117th Conference of Japan Institute of Light Metals.

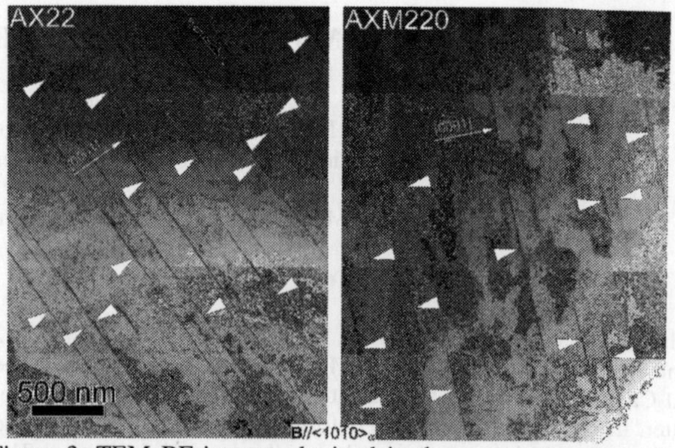

Figure 3. TEM BF images obtained in the as-cast samples. The white arrowheads indicate planar precipitates.

addition does not influence the number density of the Al_2Ca phase in the as-cast conditions.

3DAP elemental maps of Al, Ca and Mn are obtained from the as-cast samples, and the results are given in Fig. 4. As can be seen in the Fig. 4, planar precipitates are detected. Based on the morphology, the planar precipitates are identified as the Al_2Ca phase. Note that the measured composition from the planar precipitates is lower than the stoichiometric composition of the Al_2Ca. This may be due to the fact that the compositional fluctuation may occur in the planar precipitates since the probe direction is parallel to the broad faces of the Al_2Ca phases: as a result, Mg concentration in the phases may be increased, resulting in lower concentration of Al, Ca and Mn. Although the concentration of the planar precipitates is low, CBED patterns reveal a C15 structure (a = 0.804 nm, space group $Fd\bar{3}m$ [21]). Hence, it is certain that the planar precipitates observed in the 3DAP elemental maps shown in Fig. 4 correspond to the Al_2Ca phase.

It should be noted that the partitioning behavior of Mn into the Al_2Ca phase is confirmed. However, the morphology and structure of the Al_2Ca as shown in Fig. 3 do not alter significantly. Segregation of Mn in the planar precipitates has been reported preciously [22]. This may affect the dislocation motion such as dislocation climb or dislocation glide. Hence, the Mn addition influences the creep property.

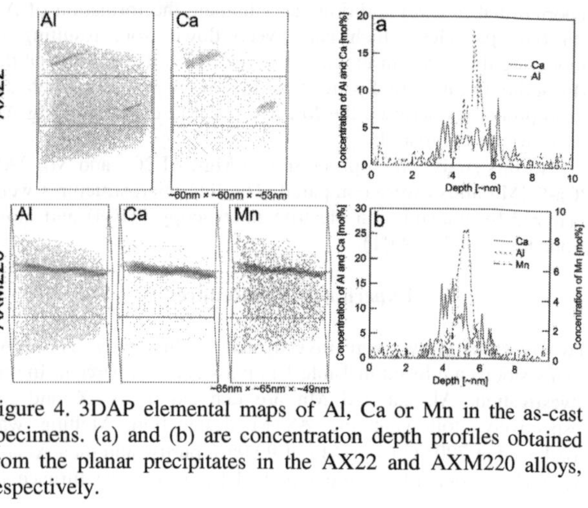

Figure 4. 3DAP elemental maps of Al, Ca or Mn in the as-cast specimens. (a) and (b) are concentration depth profiles obtained from the planar precipitates in the AX22 and AXM220 alloys, respectively.

Intraglanular precipitates in the crept specimens

Microstructures after the creep deformation were observed by conventional BF images, as indicated in Fig. 5. The specimens of the AX22 and AXM220 alloys were crept to the minimum creep rates, and the creep conditions were for 2051 h at 175 °C. The images were taken near $\boldsymbol{B} = [10\bar{1}0]_{\alpha}$ with three different g vectors ($\boldsymbol{g}$); i.e., $\boldsymbol{g} = \bar{2}112$, $\bar{2}110$ and 0002; Fig. 5 is obtained using $\boldsymbol{g} = \bar{2}110$. In both the alloys, the Al_2Ca phase precipitates on the basal plane of the matrix. The number densities were again measured based on the foil thickness measurement, and those in the AX22 and AXM220 alloys were 1.3×10^{19} and 1.9×10^{19} /m^3, respectively. Thus, the number densities before and after the creep test do not alter significantly at the minimum creep rate. In addition, the Mn addition does not influence the number density of the Al_2Ca phase in the crept specimens in spite of the partitioning behavior of Mn.

Despite the fact that the Al_2Ca phases do not influence the creep property in view of the number densities, some fine contrasts newly appear in the matrix between the planar Al_2Ca phases. These fine contrast can be seen when **g** is aligned to the $[\bar{2}110]$ direction. The coarse and planar precipitates are the Al_2Ca phase. As can be seen, small plate-like precipitates are observed with clear strain contrast as indicated by the white arrows. The plates are on the basal plane of the α-Mg matrix.

The measured number densities of the fine precipitates in the AX22 and AXM220 alloys are 4.4×10^{20} and more than 5.5×10^{20} /m^3, respectively. Since the fine precipitates observed in the AXM220 alloy is too small, some of them could not be counted. Hence, the number density in the AXM220 alloy is indeed underestimated. The average diameters of the fine precipitates along the direction perpendicular to $[0001]_{\alpha}$ are 19.4 and 14.3 nm in the AX22 and AXM220 alloys, respectively. Thus, the number density of the fine precipitates increases, while the diameter of the fine precipitates decreases by the Mn addition. In addition, it is found that the fine precipitates dynamically precipitate during the creep, regardless of the Mn addition.

Effect of fine precipitates on the creep property

The effect of dispersion strengthening on creep properties has widely been proposed [4-6, 23-27]. A continuous precipitation during creep test has been confirmed by TEM in a 2024PM (powder metallurgy) Al alloy [23]. The reported particle size is very small (~60 nm). They proposed that the dispersion strengthening may arise due to three different processes: (1) the bowing of dislocations between particles; (2) the back stress associated with the local climb of dislocations over particles; or (3) the stress associated with detaching a dislocation from a particle. While Wang et al. [24] and Lumley et al. [25] reported the effect of dispersion of Ω phase in high temperature resistant Al-Cu-Mg-Ag alloys on the creep property. They found that the dispersion of the Ω phase can enhance the creep properties. By applying a T6 treatment before the creep test, the creep properties are improved [25].

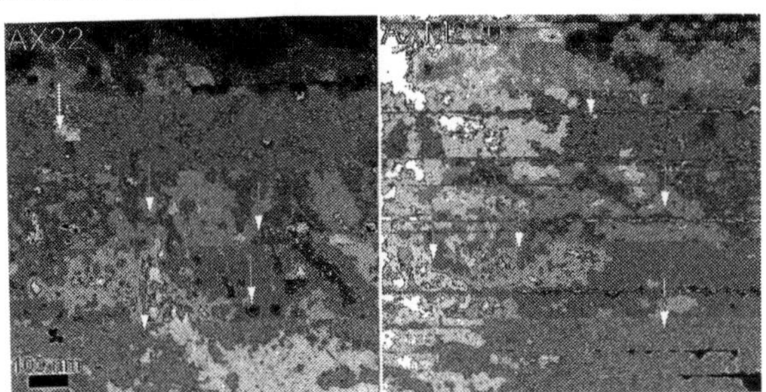

Figure 5. TEM BF images obtained in the crept specimens. The white arrows indicate fine and planar precipitates.

A similar effect was reported by Suzuki et al. in an AXJ530 die cast alloy [12], although they applied the aging treatment without solution treatment. The creep properties are improved by dispersing the fine Al_2Ca phases before the creep test as they impede the nonbasal slip of the a-type dislocations during creep deformation.

The measured number density of the planar precipitates in both the crept AX22 and AXM220 alloys is, in fact, comparable to that peak aged Mg alloys [19,28,29]. The fine and planar precipitates observed in our alloys dynamically precipitate on the basal plane during the creep deformation. The Mn addition can further enhance the fine and dense precipitation. This may interfere the movement of the dislocations, leading to high creep resistance.

It should be emphasized that fine precipitations cannot be observed in the as-cast AXM220 alloy. This result is largely different from that reported preciously in an AXM6305 alloy [22]; fine and spherical Al-Mn precipitates appear in the as-cast state. It is reported that the enthalpy of mixing between Al and Mn is negatively large [30]. This means that aggregation or ordering of Al and Mn is the most stable state. Since the AXM220 alloy processes relatively low Al composition compared to the AXM6305 alloy, the formation of Al-Mn type precipitation may not easily occur; as a result, the precipitation behavior of the fine and planar precipitate is similar to the Mn-free alloy.

Conclusions

The effect of Mn addition on the creep property of the AX22 alloy has been investigated. The following conclusions were obtained. Coarse and network-shaped compounds form in the as-cast AX22 and AXM220 alloys. They may prevent the grain boundary sliding. The Mn addition does not influence the morphology and structure of the compounds. The planar Al_2Ca phase forms on the basal plane even in the as-cast samples. In the AXM220 alloy, Mn is detected in the Al_2Ca phase. This may cause strain fields around the phase, resulting in disturbing the dislocation motion. The number density of the Al_2Ca phase does not alter, regardless of the creep deformation and Mn addition. Fine and planar precipitates appear during the creep deformation. The number density is much higher in the AXM220 alloy than that in the AX22 alloy. The size of the precipitate is reduced by addition of Mn. The Mn addition can improve the creep property.

Acknowledgements

This work was supported in part by the Grant-in-Aid Scientific Research (A) 22246094, 2010. Mr. R. Asakawa of Nagaoka University of Technology, Japan is gratefully acknowledged for his support.

References

1. B.L. Mordike and W. Henning, "Creep and high temperature properties of magnesium based alloys," Magnesium Technology, The Inst. Met. London, (1987), 54-59.

2. H. Harimzadeh, et al., "Tensile and creep fracture of a Mg-Y-RE alloy," Magnesium Technology, (London, The Institute of Metals, 1987), 138-141.
3. M. Ahmed et al., "Creep fracture in a Mg-Y-RE alloy," Magnesium Alloys and Their Application, ed. B.L. Mordike and F. Hehmann (Oberursel, DGM Informationsgesellschaft m.b.H., 1992), 251-257.
4. M. Suzuki et al., "Creep behavior and deformation microstructures of Mg-Y alloys at 550 K," Materials Science and Engineering A, 252 (1998), 248-255.
5. I.A. Anyanwu et al., "Creep properties of Mg-Gd-Y-Zr alloys," Materials Transactions, 42 (2001), 1212-1218.
6. K. Maruyama, M. Suzuki and H. Sato, "Creep strength of magnesium-based alloys," Metallurgical and Materials Transactions A, 33A (2002), 875-882.
7. R. Ninomiya, T. Ojiro and K. Kubota, "Improved heat resistance of Mg-Al alloys by the Ca addition," Acta Metallurgica et Materialia, 43 (1995), 669-674.
8. A.A. Luo, M.P. Balogh and B.R. Powell, "Creep and microstructure of magnesium-alminum-calcium based alloys," Metallurgical and Materials Transactions A, 33A (2002), 567-574.
9. Y. Terada et al, "A thousandfold creep strengthening by Ca addition in die-cast AM50 magnesium alloy," Metallurgical and Materials Transactions A, 35A (2004), 3029-3032.
10. S.M. Zhu, B.L. Mordike and J.F. Nie, "Creep and rapture properties of a squeeze-cast Mg-Al-Ca alloy," Metallurgical and Materials Transactions A, 37A (2006), 1221-1229.
11. B. Jing et al., "Effect of extrusion on microstructures, and mechanical and creep properties of Mg-Al-Sr and Mg-Al-Sr-Ca alloys," Scripta Materialia, 55 (2006), 1163-1166.
12. A. Suzuki et al., "Precipitation strengthening of a Mg-Al-Ca-based AXJ530 die-cast alloy," Metallurgical and Materials Transactions A, 39A (2008), 696-702.
13. A. Luo and T. Shinoda, "Development of a creep-resistant magnesium alloy for die casting applications," Magnesium Alloys and their Applications, ed. B.L. Mordike and K.U. Kainer, (Frankfurt, Wekstoff-Informationsgesellschaft, 1998), 151-156.
14. F.v. Buch et al., "Development of a low-cost, temperature- and creep-resistant magnesium die-casting alloy," Magnesium Alloys and their Applications, ed. K.U. Kainer, (Darmstadt, WILEY-VCH, 2000), 23-28.
15. S. Koike et al., "Development of lightweight oil pans made of a heat-resistant magnesium alloy," HONDA R&D Technical Review, 12 (2000), 167-174.
16. A. Suzuki et al., "Structure and transition of eutectic $(Mg,Al)_2Ca$ Laves phase in a die-cast Mg-Al-Ca base alloy," Scripta Materialia, 51 (2004), 1005-1010.
17. M.E. Kassner, Fundamentals of Creep in Metals and Alloys (Oxford, Elsevier, 2009).
18. E.F. Emley, Principle of Magnesium Technology (New York, Pergamon Press, 1966).
19. T. Homma et al., "Effect of Zr addition on the mechanical properties of as-extruded Mg-Zn-Ca-Zr alloys," Materials Science and Engineering A, 527 (2010), 2356-2362.
20. T. Ohkubo et al., "Laser-assisted atom probe analysis of bulk insulating ceramics," MRS Proc. (2009), Fall Meeting.
21. P. Villars and L.D. Calvert, Pearson's Handbook of Crystallographic Data for Intermetallic Phases, second ed. (Materials Park, OH, ASM International, 1991).
22. T. Homma, S. Nakawaki and S. Kamado, "Improvement in creep property of a cast Mg-6Al-3Ca alloy by Mn addition," Scripta Materialia, (2010), in press.
23. L. Kloc et al., "Significance of continuous precipitation during creep of a powder metallurgy aluminum alloy," Materials Science and Engineering A, 216 (1996), 161-168.
24. J. Wang, X. Wu and K. Xia, "Creep behavior at elevated temperatures of an Al-Cu-Mg-Ag alloy," Materials Science and Engineering A, 234-236 (1997), 287-290.
25. R.N. Lumley, A.J. Morton and I.J. Polmear, "Enhanced creep performance in an Al-Cu-Mg-Ag alloy through underageing," Acta Materialia, 50 (2002), 3597-3608.
26. C.R. Hutchinson, P. Cornall and M. Gouné , "On the origin of the enhanced creep resistance in underaged Al-Cu based alloys," Materials Science Forum, 519-521 (2006), 1029-1034.
27. C.K. Sudbrack et al., "Temporal evolution of the nanostructure and phase compositions in a model Ni-Al-Cr alloy," Acta Materialia, 54 (2006), 3199-3210.
28. C.L. Mendis et al., "Enhanced age hardening in a Mg-2.4at.%Zn alloy by trace additions of Ag and Ca," Scripta Materialia, 57 (2007) 485-488.
29. C.L. Mendis et al., "Precipitation-hardenable Mg-2.4Zn-0.1Ag-0.1Ca-0.16Zr (at.%) wrought magnesium alloy," Acta Materialia, 57 (2009), 749-760.
30. A.R. Miedema, F.R. de Boer and R. Boom, "Model predictions for the enthalpy of formation of transition metal alloys," CALPHAD, 1 (1977), 341-359.

Magnesium Technology 2011
Edited by: Wim H. Sillekens, Sean R. Agnew, Neale R. Neelameggham, and Suveen N. Mathaudhu
TMS (The Minerals, Metals & Materials Society), 2011

THE EFFECT OF PRECIPITATE STATE ON THE CREEP PERFORMANCE OF Mg-Sn ALLOYS

Mark Gibson[1], Xi-Ya Fang[2], Colleen Bettles[3], Christopher Hutchinson[3]

[1]CAST CRC and CSIRO Manufacturing and Infrastructure Technology; Private Bag 33, Clayton, Vic, 3169, Australia
[2]Monash Centre for Electron Microscopy; Monash University, Clayton, Vic 3800, Australia
[3]ARC Centre of Excellence for design in Light Metals; Department of Materials Engineering, Monash University, Clayton, Australia

Keywords: creep, precipitation, micro-alloying

Extended Abstract

The Mg-Sn system has recently received attention as a potential basis for the development of creep resistant alloys. This is an age hardenable system with the potential to form ~10 vol. % of Mg_2Sn. The Mg_2Sn precipitate has a high melting temperature (~770 °C) and alloys based on this system might be thought to show some promise for use in applications requiring elevated temperature creep resistance [1]. The age hardening response of the binary alloy is only modest, despite the large volume fraction of precipitates that may be formed [2]. The reason is that the precipitates form with a relatively coarse distribution, in laths parallel to the basal planes of the magnesium matrix. This configuration is relatively inefficient at impeding dislocation motion [3].

Mendis *et al.* [3,4] have recently shown two methods for improving the room temperature precipitation hardening response of Mg-Sn based alloys. The first approach involved using micro-alloying additions to catalyze a greatly refined distribution of precipitates. The basic idea is to create heterogeneities in the matrix at which Mg_2Sn may precipitate. A qualitative thermo-kinetic criterion for choosing micro-alloying additions was proposed and, using additions of Na, a two orders of magnitude increase in the number density of precipitates and an augmentation of the hardening increment of 270% were achieved [3]. In a second, more ambitious approach, Mendis *et al.* used Zn additions to attempt to manipulate the shape of the Mg_2Sn precipitates [4]. The premise was to manipulate the lattice misfit normal to the basal plane of the magnesium matrix in an effort to change the shape of the Mg_2Sn precipitates. This was also shown to be successful and a substantial increase in age hardening was realized. As a final step, the two approaches were used simultaneously (Na+Zn additions) realizing a further increase in the precipitation hardening response. In all cases the precipitate remained almost exclusively Mg_2Sn. The age hardening responses of the binary alloy, the alloy containing Zn, and the alloy containing Zn+Na have been reported by Mendis *et al.* [4].

However, designing microstructures for high strength is not the same as designing materials for creep resistance. Creep resistance also requires stability of microstructure. Under conditions where the creep strain is accommodated mainly by dislocation motion, stability in the precipitate state is crucial for creep resistance. Approaches to strengthening that involve an increase in precipitate number density suffer from a potential disadvantage for applications requiring creep resistance because for a given volume fraction of precipitates, increases in the number density necessarily correspond to a decrease in the mean particle size. The driving force for coarsening is the excess energy stored in the particle/matrix interfaces and the coarsening rate of a particle distribution scales as R^{-2}. Thus, the increase in strength of precipitate strengthened alloys achieved through refinements in scale may come at the cost of a decrease in the thermal stability of the precipitate distribution.

The Mg-Sn-Zn-Na alloy examined by Mendis *et al.* exhibited a two orders of magnitude increase in the precipitate number density. A surprising result was that the long-term aging behavior of the alloy did not result in significant softening of the alloy, even for times approaching 1000 h at 200 °C. Such stability in properties indicates stability in microstructure and suggests that this alloy may have very interesting creep properties.

In this work, the tensile creep properties of a binary Mg-Sn alloy, a ternary Mg-Sn-Zn alloy and an alloy containing both Zn and Na additions were examined. Sample preparation is outlined in [1]. Each of the three alloys was aged to peak strength at 200 °C. Peak strength for the binary, ternary and quaternary alloys corresponded to ~1000 h, ~200 h and ~7 h, respectively. The average grain size of the alloys was ~150-200 μm. Tensile creep tests were performed in oil at 177 °C with a load of 60 MPa. Selected tests at 150 °C or with loads of 35 MPa were also performed.

The tensile creep strain as a function of loading time at 177 °C under a load of 60 MPa is shown in Fig. 1a for the three alloys. The creep performance of the binary alloy is quite poor and this is consistent with previous reports and expectations based on the low static strength of this alloy. The performance of the ternary alloy containing Zn is much improved compared with the binary alloy but the performance of the Zn+Na containing quaternary alloy is rather striking. The creep performance is orders of magnitude better than both the binary and ternary alloys. The creep performance of the Zn+Na containing quaternary alloy is plotted in Fig. 1b with an extended time axis to illustrate the creep lifetime. Lifetimes in excess of 400 h were obtained under tensile loading conditions at both 177 °C and 150 °C. The tests were stopped soon after 400 h.

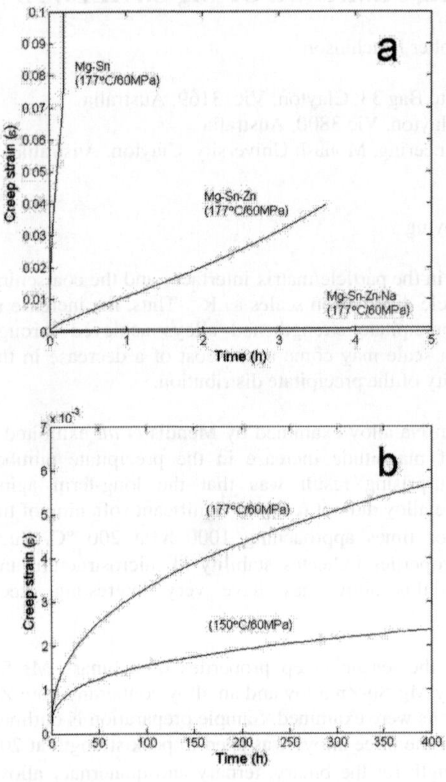

Figure 1. Tensile creep curves for a) the Mg-Sn, Mg-Sn-Zn and Mg-Sn-Zn-Na alloys at 177 °C under a load of 60 MPa and b) the Mg-Sn-Zn-Na alloys at 177 °C and 150 °C under a load of 60 MPa. These tests were stopped after ~ 400 h.

The quaternary Zn+Na containing alloy has a creep rate approximately five orders of magnitude lower than the binary alloy under loads of 60 MPa.

Transmission electron micrographs of the peak-aged microstructures and as crept microstructures of the Mg-Sn-Zn-Na alloy are reproduced in [1,3,4]. The as-crept microstructure of the Mg-Sn-Zn-Na alloy shows very little change from the peak-aged structure, confirming the thermal stability of the precipitate structure. The excellent creep resistance of the Mg-Sn-Zn-Na alloy is linked to the excellent thermal stability of this microstructure.

What is the origin of the thermal stability of the precipitates in the Mg-Sn-Zn-Na alloy?

This thermal stability is not exhibited by the alloy containing Zn additions only so it is tempting to look to the Na for an explanation. Na additions were chosen because they were expected to cluster on the magnesium lattice during aging and hence catalyze the nucleation of Mg_2Sn. Since the Na additions are expected to be intimately linked with the Mg_2Sn precipitates, it may be expected that the Na is either incorporated into the Mg_2Sn precipitate as it grows or may be located at the precipitate/matrix interface. In either case, it would not be surprising for such segregation to affect the lattice misfit between the precipitate and matrix, and therefore the interfacial energy which drives coarsening processes, or in the case of incorporation of the Na into the Mg_2Sn precipitates, the intrinsic thermodynamic stability of the Mg_2Sn precipitates. The obvious next step is to experimentally identify the exact location of the Na additions in the microstructure, and any possible association with the Zn additions. This information would aid in the optimization of the micro-alloying additions for the most beneficial impact on the mechanical properties.

This work illustrates that, through the judicious selection of alloying additions, it is possible to very substantially affect the size and shape of Mg_2Sn precipitates in Mg-Sn based alloys. These modifications are beneficial for the tensile creep performance of these alloys. In particular, The Mg-Sn-Zn-Na alloy exhibits excellent thermal stability of the precipitate structure leading to tensile creep properties that are of practical significance and have a five orders of magnitude improvement in secondary creep rate compared to the binary Mg-Sn alloy.

Acknowledgements

The CAST Co-operative Research Centre was established under, and is funded in part by, the Australian Government's Co-operative Research Centres Scheme. The Australian Research Council is gratefully acknowledged for funding the ARC Centre of Excellence for Design in Light Metals.

References

1. Gibson M., Fang X., Bettles C.J., Hutchinson C.R., *Scripta Materialia*, 63 (2010), 899-902.
2. Van der Planken J., *Journal of Materials Science Letters*, 4 (1969), 927.
3. Mendis C.L., Bettles C.J., Gibson M.A., Gorsse S., Hutchinson C.R., *Philosophical Magazine Letters*, 86 (2006), 443.
4. Mendis C.L., Bettles C.J., Gibson M.A., Hutchinson C.R., *Materials Science and Engineering A*, 435-436 (2006), 163.

Magnesium Technology 2011
Edited by: Wim H. Sillekens, Sean R. Agnew, Neale R. Neelameggham, and Suveen N. Mathaudhu
TMS (The Minerals, Metals & Materials Society), 2011

MICROSTRUCTURE AND MECHANICAL PROPERTIES OF Mg-Zn-Y-M (M: MIXED RE) ALLOYS WITH LPSO PHASE

Jonghyun Kim[1] and Yoshihito Kawamura[2]
[1]Kumamoto Technology and Industrial Foundation; 2080-10 Tabaru, Mashiki-machi, Kamimashiki-gun, Kumamoto, 861-2202, JAPAN
[2]Department of Materials Science, Kumamoto University; 2-39-1, Kurokami, Kumamoto, 860-8555, JAPAN

Keywords: Mg alloy, LPSO, Mechanical properties

Abstract

$Mg_{96}Zn_2Y_{1.9}M_{0.1}$ (M=Monazite base mixed Rare Earth (RE) ($Ce_{52.4}La_{26.0}Nd_{21.6}$) in at. %) alloys were prepared by high-frequency induction melting in an Ar atmosphere. The microstructure and mechanical properties of the extruded alloys were investigated. The microstructure contained very small sized RE intermetallic compound particles within the α-Mg matrix and between the LPSO intergranular phase in the $Mg_{96}Zn_2Y_{1.9}M_{0.1}$ alloys. The presence of the RE compound in the $Mg_{96}Zn_2Y_{1.9}M_{0.1}$ alloy was responsible for its improved tensile and creep properties over those of the $Mg_{96}Zn_2Y_2$ alloy. Excellent tensile yield strength was obtained in the extruded $Mg_{96}Zn_2Y_{1.9}M_{0.1}$ alloys, with a yield strength of 379 MPa at room temperature and 312 MPa at 523K, respectively. Considerable improvement was observed in the creep behavior of the $Mg_{96}Zn_2Y_{1.9}M_{0.1}$ alloys at 473K. Extended creep life and reduced minimum creep rate were also observed.

Introduction

Several wrought magnesium alloys such as AM- and AZ- series are structural materials which are suitable for use in the computer, mobile phone and automobile industries mainly because of their low density [1,2]. However, the number of commercially available Mg alloys is still limited especially for application at room and elevated temperatures. For example, the above-mentioned alloys of AM- and AZ-series cannot be applied to power train parts operating at temperatures higher than 130℃ due to their poor elevated temperature mechanical properties [3].

Rare earth (RE) elements have been used in magnesium alloys and they can improve the casting characteristics, mechanical properties at room and elevated temperatures and corrosion resistance [4,5]. Some investigations have been done on the morphology, microstructure and strengthening mechanisms when Res are added to Mg-Zn and Mg-Al alloys [6,7]. Recently, Kawamura *et al.* developed rapidly solidified powder metallurgy (RS P/M) Mg-Zn-Y alloys with superior mechanical properties including a tensile yield strength of 610 MPa and an elongation of 5 % at room temperature [8]. The extruded $Mg_{97}Zn_1Y_2$ and $Mg_{96}Zn_2Y_2$ alloys also exhibited excellent mechanical properties. The tensile yield strength of $Mg_{96}Zn_2Y_2$ alloy was higher than that of $Mg_{97}Zn_1Y_2$ alloy at room temperature [9,10]. The $Mg_{97}Zn_1Y_2$ alloy consisted of fine α-Mg grains and the long period staking ordered (LPSO) phase and the improvement of mechanical properties in this alloy is consider to originate from the high dispersion of the bent LPSO phase and the refinement of those structures and the α-Mg grains. In contrast, the $Mg_{96}Zn_2Y_2$ alloy consisted of 3 phases with fine α-Mg, LPSO phase and the intermetallic RE containing compound. The volume fraction of intergranular phases (LPSO phase and the intermetallic RE compound) of $Mg_{96}Zn_2Y_2$ alloy was higher than that of the $Mg_{97}Zn_1Y_2$ alloy [10]. Therefore, to improve of mechanical properties of the Mg-Zn-Y alloys, the phases control is necessary to obtain an appropriate volume fraction of the α-Mg, LPSO phase and intermetallic RE compound.

In the present study, the investigation reported here focused on the influence of mixed RE, which was also effective in strengthening Mg-Zn-Y alloys at elevated temperature, on the microstructure and mechanical properties of the Mg-Zn-Y-M (M: Mixed RE) alloys.

Experimental procedure

The alloys, with nominal composition of $Mg_{96}Zn_2Y_{1.9}M_{0.1}$ (M=Monazite base mixed RE ($Ce_{52.4}La_{26.0}Nd_{21.6}$) in at. %), were prepared by high-frequency induction melting in an Ar atmosphere and cast in a cylindrical steel mould (31 mm of diameter). Rods of 70 mm height and 29 mm length were cut from the initial ingots. The rods of as-cast alloy were extruded with an extrusion ratio of 10 at 623K and a ram speed of 2.5 mm/s. The phase structures of as-cast and extruded alloys were investigated by X-ray diffractometry (XRD), optical microscopy (OM), scanning electron microscopy (SEM) and transmission electron microscopy (TEM). The mechanical properties of the investigated alloys were determined using tensile test specimens with a gage section of Ø 2.5 mm × 15 mm.

Results and discussion

Figure 1 shows SEM micrographs of the as-cast $Mg_{96}Zn_2Y_2$ and $Mg_{96}Zn_2Y_{1.9}M_{0.1}$ alloys. As it is shown in the picture, the secondary phase with fine lamellar contrasts was observed in the grain boundary and dendrite arm boundaries for $Mg_{96}Zn_2Y_2$ and $Mg_{96}Zn_2Y_{1.9}M_{0.1}$ alloys. The secondary phase of the $Mg_{96}Zn_2Y_{1.9}M_{0.1}$ alloys had a sharp and smooth interface. According to the XRD results and EDS analysis, it can be confirmed that the secondary phases are LPSO and/or the Mg-RE compound in the $Mg_{96}Zn_2Y_{1.9}M_{0.1}$ alloys. In the $Mg_{96}Zn_2Y_2$ and $Mg_{96}Zn_2Y_{1.9}M_{0.1}$ alloys, continuous net-worked LPSO and intermetallic RE compound form a coarse interdendrite structure. The volume fractions of compound in $Mg_{96}Zn_2Y_{1.9}M_{0.1}$ alloys (3.1 %) were higher than that of the $Mg_{96}Zn_2Y_2$ alloys (0.9 %).

The room and elevated temperature tensile properties of the two kinds of alloy after extrusion are shown in Fig. 2. The yield strength, ultimate tensile strength and elongation of the $Mg_{96}Zn_2Y_2$ and $Mg_{96}Zn_2Y_{1.9}M_{0.1}$ extruded alloys at room temperature were 376 MPa, 425 MPa and 5.1 %, and 379 MPa, 428 MPa and 5.0 %, respectively. The mechanical properties of the $Mg_{96}Zn_2Y_{1.9}M_{0.1}$ are slightly higher than that of the ternary $Mg_{96}Zn_2Y_2$ alloys.

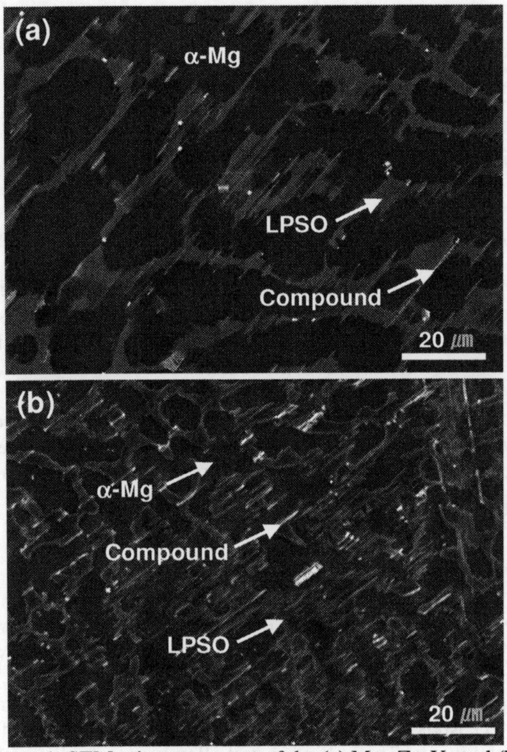

Figure 1. SEM microstructures of the (a) $Mg_{96}Zn_2Y_2$ and (b) $Mg_{96}Zn_2Y_{1.9}M_{0.1}$ alloys

The yield strength, ultimate tensile strength and elongation at elevated temperature (523 K) of the $Mg_{96}Zn_2Y_2$ alloy were 255 MPa, 265 MPa and 19 %, respectively. The tensile strength of the $Mg_{96}Zn_2Y_{1.9}M_{0.1}$ alloys at elevated temperature (523K) is higher than the tensile strength of $Mg_{96}Zn_2Y_2$ alloy.

The creep tests were carried out at 473 K and at applied stresses between 90 and 210 MPa. The 300 h creep curves for $Mg_{96}Zn_2Y_2$ and $Mg_{96}Zn_2Y_{1.9}M_{0.1}$ tested at different stresses are shown in Fig. 3. The 300 h creep strain at 90 MPa was almost the same for both the $Mg_{96}Zn_2Y_2$ and $Mg_{96}Zn_2Y_{1.9}M_{0.1}$ alloys. The creep curves at 150 and 210 MPa exhibit a well-defined primary stage and a secondary stage for both the $Mg_{96}Zn_2Y_2$ and $Mg_{96}Zn_2Y_{1.9}M_{0.1}$ alloys. The creep resistance of both alloys investigated at a pressure of 210 MPa was poor.

However, the creep resistance and tensile strength at elevated temperature for the $Mg_{96}Zn_2Y_{1.9}M_{0.1}$ alloy were both higher than that of the $Mg_{96}Zn_2Y_2$ alloy at a pressure of 210 MPa. Therefore, intermetallic RE compound's effect on the mechanical properties at elevated temperature needs to be investigated. To do this, a heat treatment experiment was designed.

Figure 4 shows the microstructures of heat treated $Mg_{96}Zn_2Y_2$ and $Mg_{96}Zn_2Y_{1.9}RE_{0.1}$ alloys at 573 to 673K for 1h. The average α-Mg grain size of $Mg_{96}Zn_2Y_2$ and $Mg_{96}Zn_2Y_{1.9}RE_{0.1}$ alloys was increased with increasing heat treatment temperature. While the heat treatment conditions were 573K for 1h, no remarkable changes were observed in the microstructure of heat treated $Mg_{96}Zn_2Y_2$ and $Mg_{96}Zn_2Y_{1.9}RE_{0.1}$ alloys. However, the average α-Mg grain size of $Mg_{96}Zn_2Y_2$ alloy was 3.4 μm at 623K and the average α-Mg grain size of $Mg_{96}Zn_2Y_{1.9}RE_{0.1}$ alloys was 1.8 μm at 623K.

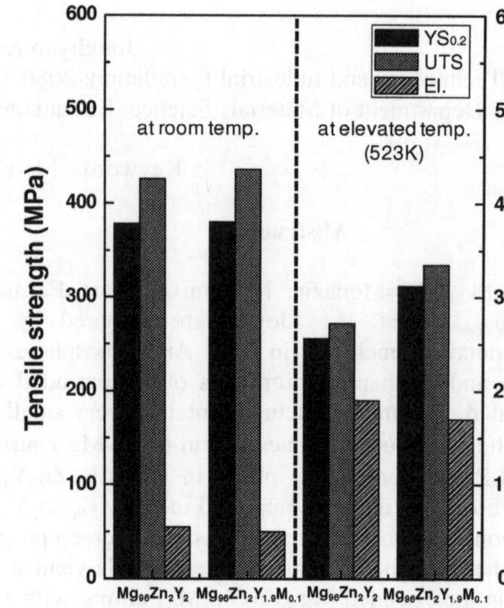

Figure 2. Mechanical properties of $Mg_{96}Zn_2Y_2$ and $Mg_{96}Zn_2Y_{1.9}M_{0.1}$ alloys at room and elevated temperature

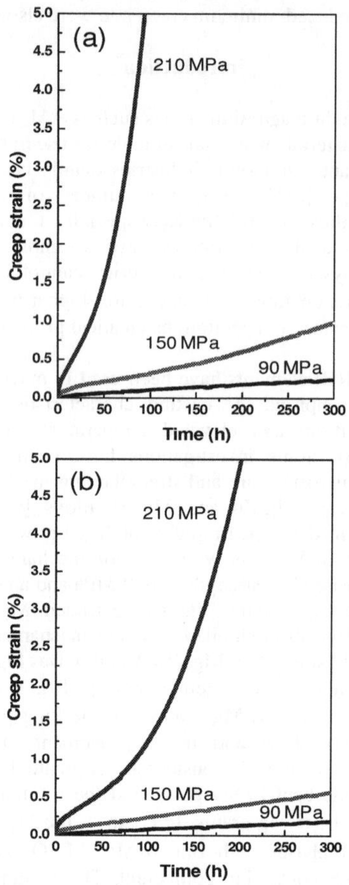

Figure 3. Creep curves of (a) $Mg_{96}Zn_2Y_2$ and (b) $Mg_{96}Zn_2Y_{1.9}M_{0.1}$ alloys at 473 K. the applied stress were 90, 150 and 210 MPa

Figure 5 shows the optical microstructures of extruded $Mg_{96}Zn_2Y_2$ and $Mg_{96}Zn_2Y_{1.9}M_{0.1}$ alloys. The $Mg_{96}Zn_2Y_2$ and $Mg_{96}Zn_2Y_{1.9}M_{0.1}$ alloys consist of α-Mg grains, the internetallic RE compound and the LPSO phase. This Mg-RE compound exists within the α-Mg matrix and between the LPSO phases in the $Mg_{96}Zn_2Y_2$ and $Mg_{96}Zn_2Y_{1.9}M_{0.1}$ alloys. The volume fraction of intermetallic RE compound in $Mg_{96}Zn_2Y_{1.9}M_{0.1}$ alloys (3.1 %) was higher than that of the $Mg_{96}Zn_2Y_2$ alloys (0.9 %). These results indicate that the strengthening of the quaternary $Mg_{96}Zn_2Y_{1.9}RE_{0.1}$ alloy system at elevated temperature is depended on the LPSO phase, fine-grained α-Mg matrix grains and especially, the highly dispersed compounds and the amount of the compounds.

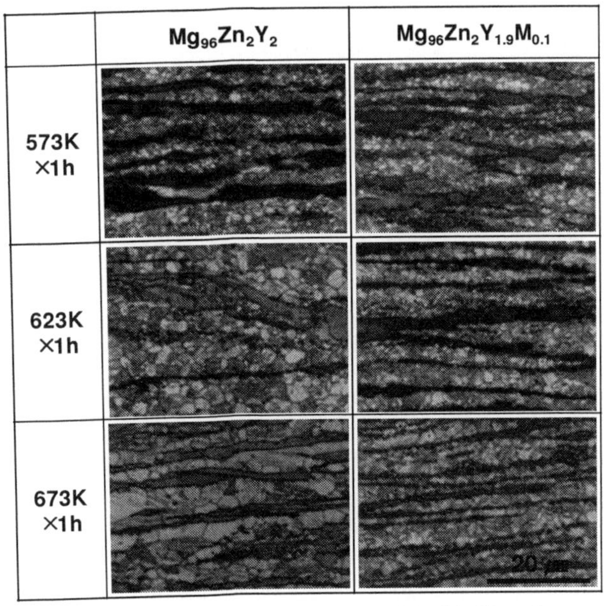

Figure 4. Optical microstructures of $Mg_{96}Zn_2Y_2$ and $Mg_{96}Zn_2Y_{1.9}M_{0.1}$ alloys after heat treatment

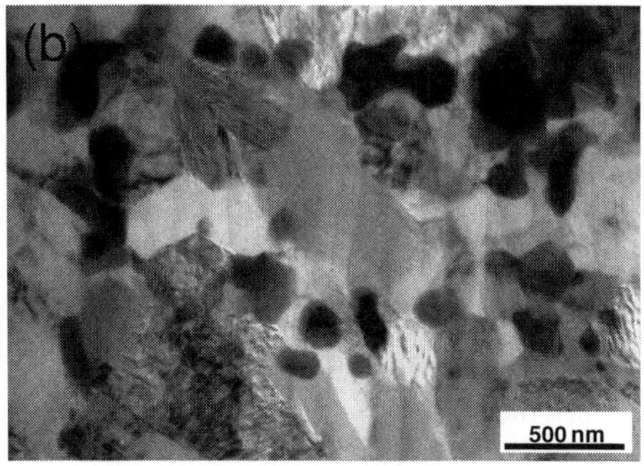

Figure 5. TEM microstructures of the extruded (a) $Mg_{96}Zn_2Y_2$ and (b) $Mg_{96}Zn_2Y_{1.9}M_{0.1}$ alloys

Conclusions

(1) Excellent tensile yield strength was obtained in the extruded $Mg_{96}Zn_2Y_{1.9}M_{0.1}$ alloys, with a yield strength of 379 MPa at room temperature and 312 MPa at 523K.

(2) The very small sized intermetallic RE compound exists in the α-Mg matrix and between the LPSO phases in the $Mg_{96}Zn_2Y_{1.9}M_{0.1}$ alloys. The presence of intermetallic RE compound in $Mg_{96}Zn_2Y_{1.9}M_{0.1}$ alloy was responsible for its improved tensile and creep properties over those of $Mg_{96}Zn_2Y_2$ alloy.

(3) Considerable improvement in the creep behavior of the $Mg_{96}Zn_2Y_{1.9}M_{0.1}$ alloys at 473K was observed. Extended creep life and a reduction in the minimum creep rate were also observed.

Acknowledgements

This work has been financially supported by the Development of Key Technology for Next-generation Heat-resistant Magnesium alloys, Kumamoto Prefecture Collaboration of Regional Entities for the advancement of Technology Excellence, Japan Science and Technology Agency.

References

1. A. A. Luo, "Recent magnesium alloy development for automotive powertrain applications," *Materials Science Forum* , 419-422 (2003) 57-66.
2. Magnesium & Magnesium alloys, ed. M. M. Avedesian & H. Baker: ASM International, Materials Park, OH (1999), 3-6.
3. A. Luo, M.O. Pekguleryuz, "Cast magnesium alloys for elevated temperature applications," *Journal of Materials Science*, 29 (1994) 5259-5271.
4. J.F. Nie and B.C. Muddle, "Precipitation in magnesium alloy WE54 during isothermal ageing at 250°C," *Scripta Materialia*, 40 (1999) 1089-1094.
5. J.F. Nie and B.C. Muddle, "Characterisation of strengthening precipitate phases in a Mg–Y–Nd alloy," *Acta Materialia*, 48 (2000) 1691-1703.

6. J. Bohlen, M.R. Nurnberg, J.W. Senn, D. Letzig and S.R. Agnew, "The texture and anisotropy of magnesium–zinc–rare earth alloy sheets," *Acta Materialia*. 55 (2007) 2101-2112.
7. X. Tian, L.M. Wang, J.L. Wang, Y.B. Liu, J. An and Z.Y. Cao, "The microstructure and mechanical properties of Mg-3Al-3RE alloys," *Journal of Alloys and Compounds*. 465 (2008) 412-416.
8. Y. Kawamura, K. Hayashi and A. Inoue, "Rapidly solidified powder metallurgy $Mg_{97}Zn_1Y_2$ alloys with excellent tensile yield strength above 600 MPa," *Materials Transactions*. 42 (2001) 1171-1174.
9. Y. Kawamura and M. Yamasaki, "Formation and Mechanical Properties of $Mg_{97}Zn_1RE_2$ Alloys with Long Period Stacking Ordered Structure," *Materials Transactions*. 48 (2007) 2986-2992.
10. S. Yoshimoto, M. Yamasaki and Y. Kawamura, "Microstructure and Mechanical Properties of Extruded Mg-Zn-Y Alloys with 14H Long Period Ordered Structure," *Materials Transactions*. 47 (2006) 959-965.

Magnesium Technology 2011
Edited by: Wim H. Sillekens, Sean R. Agnew, Neale R. Neelameggham, and Suveen N. Mathaudhu
TMS (The Minerals, Metals & Materials Society), 2011

APPLICATION OF NEUTRON DIFFRACTION IN CHARACTERIZATION OF TEXTURE EVOLUTION DURING HIGH-TEMPERATURE CREEP IN MAGNESIUM ALLOYS

D. Sediako[1], S. Shook[2], S. Vogel[3], and A. Sediako[4]
[1] Canadian Neutron Beam Centre, National Research Council, Chalk River Laboratories, ON, Canada
[2] Applied Magnesium International, Denver, CO, USA
[3] Los Alamos Neutron Science Center, Los Alamos, NM, USA
[4] McGill University, Montreal, PQ, Canada

Keywords: High temperature creep, magnesium alloy, texture, neutron diffraction

Abstract

A good combination of room-temperature and elevated temperature strength and ductility, good salt-spray corrosion resistance and excellent diecastability are frequently among the main characteristics considered when developing a new magnesium alloy. Unfortunately, much less effort has been expended developing wrought-stock alloys for high temperature applications. Extrudability and high temperature performance of wrought material becomes an important factor in an effort to develop new wrought alloys and processing technologies.
This paper shows some results obtained from creep testing and studies of in-creep texture evolution, for several wrought magnesium alloys developed for use in elevated-temperature applications.

Introduction

Despite the enormous efforts expended to develop high-strength, high-temperature magnesium alloys, the most typical applications of magnesium in the automotive industry are still limited to a few select applications such as instrument panels, steering wheels, and valve covers. Limited success was attained in the use of magnesium in powertrain applications such as transmission cases and engine blocks [1-9]. These applications experience service conditions within the temperature range of 150-200°C under 50-70 MPa of tensile and compressive loads. In addition, metallurgical stability, fatigue resistance, corrosion resistance, and castability requirements need to be met. More than a decade of research and development has resulted in a number of creep-resistant magnesium alloys that are potential candidates for elevated-temperature automotive applications. These alloys are mostly based on rare-earth and alkaline-earth element additions to magnesium. A number of alloys are based on additions of Si, Sr, and Ca.

Although the majority of the material used in the automotive industry is in the form of castings, the use of wrought products in automotive applications is on the rise. Extruded sections provide opportunities for the mass-efficient design of structural and interior automobile components.

One of the main criteria for acceptable material performance in high temperature automotive applications is its resistance to creep. However, low resistance to creep deformation at elevated temperatures has been one of the main restricting factors in applying magnesium alloys.

Depending on the manufacturing route and resulting grain/crystallographic matrix, wrought magnesium alloys can exhibit quite different creep behavior, compared to similar cast alloys. Several creep-resistant wrought alloys have been studied in [10, 11]. It was shown that magnesium exhibits different creep properties under tension and compression. These studies also presented data gathered from the analysis of crystallographic texture and creep-induced residual stress –factors that would most certainly affect service properties of the material. The targeted alloying groups were magnesium-aluminum-rare earth, magnesium-aluminum-strontium, magnesium-aluminum-calcium, and magnesium-zinc-rare earth. The chemical compositions of the analyzed samples correspond to the following alloy designations: AE42, AE33, AX30, AZX310, AJ32, EZ33, and ZE10. The seven alloys targeted in this study were produced by the magnesium division of Timminco Corporation, now Advanced Magnesium International (AMI). The material was cast using a unique controlled-cooling static casting process, followed by hot extrusion. All alloys exhibited exceptionally good castability, as well as formability in the extrusion process.

The current analysis is part of a continuing effort to develop low-cost, wrought magnesium alloys with improved castability and formability, suitable for high temperature applications, and to add to the understanding of material behavior during high-temperature creep.

High-Temperature Creep Testing

The wrought material received from AMI was subsequently subjected to a tensile creep test at 150 and 175°C and then to a 200-hour compressive creep test at 150°C under a load of 50 MPa. All the samples were creep-tested along the extrusion direction. The following charts (Figure 1) illustrate some selected results of these tests.

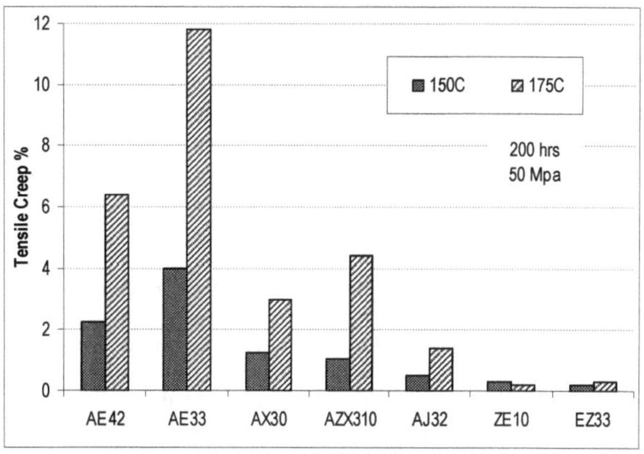

Figure 1. Resistance to creep for selected alloys at 150 and 175°C

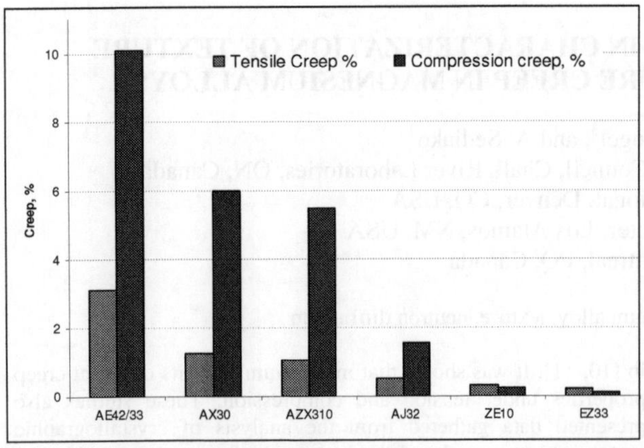

Figure 2 Tension-compression asymmetry in resistance to creep, at 150°C

Data shown in Figure 1 illustrate the comparison between results obtained in the tensile tests at 150 and 175°C, which included both primary and secondary creep. As expected, the 25°C temperature increase lead to significantly reduced resistance to creep, typically a reduction by a factor of three for all alloys, except for ZE10 and EZ33. Evidently, the manufacturing route applied in this study (permanent mould casting + hot extrusion) resulted in superior high temperature creep properties for the Mg-Zn-RE alloying system.

This finding was also confirmed in the compressive creep test, performed at 150°C. It is known that HCP crystallographic systems typically exhibit greatly reduced resistance to creep in compression. This can also be observed in Figure 2, which shows the tension/compression asymmetry in creep resistance for the alloys studied. Figure 2 indicates that for identical applied loads and test durations, resistance to creep was reduced by a factor of 3.5 to around 4 for all of alloys, with the exception, again, of ZE10 and EZ33. It can be concluded from these observations that, compared to the other samples studied, the applied manufacturing route favourably affected the ability of the Mg-Zn-RE alloying system to resist high-temperature creep, both in tension and in compression. In addition, almost no tension-compression asymmetry was observed for the ZE-group samples (in all tests, the resulting creep was within 0.25 to about 0.3% for the ZE10 samples and within 0.1 to about 0.2 for EZ33).
Figures 1 and 2 also show that in all tests the AE-group samples exhibited inferior creep properties, compared to the other three alloying systems. This was partially explained in [11] by the presence of β-phase $Mg_{17}Al_{12}$, along the grain boundaries in the magnesium matrix, which significantly reduced creep resistance at elevated temperatures. A reference was also made to the possible effect of initial texture (i.e., texture prior to creep).

Texture Evolution during Creep Deformation

It was suggested in [11] that the strength of the initial extrusion-type crystallographic texture may be another factor affecting the material resistance to creep. It can be assumed that, considering the amount of creep deformation obtained in this study (up to 10~12% strain), there could be noticeable texture evolution or modification during creep testing. These two assumptions were verified by texture analysis performed in the HIPPO (High-Pressure-Preferred-Orientation) time-of-flight spectrometer at Los Alamos Neutron Science Center. The time-of-flight method employed by HIPPO for texture analysis is well described in earlier study [12].

Similar studies have earlier been performed at the Canadian Neutron Beam Centre at Chalk River, ON, on pure magnesium and Mg-1.5wt.%Mn samples [13, 14]. It was confirmed that the tested material underwent significant texture modification during creep testing at 150°C and under a load of 50MPa. A typical "rolling" texture could be observed on the crept specimens. Important conclusions were reached regarding creep deformation mechanisms for the studied materials.

The pictures presented in Figures 3 and 4 show the results of texture calculation using the E-WIMF algorithm of the MAUD texture-analysis software [12], based on neutron measurements for several selected alloys.

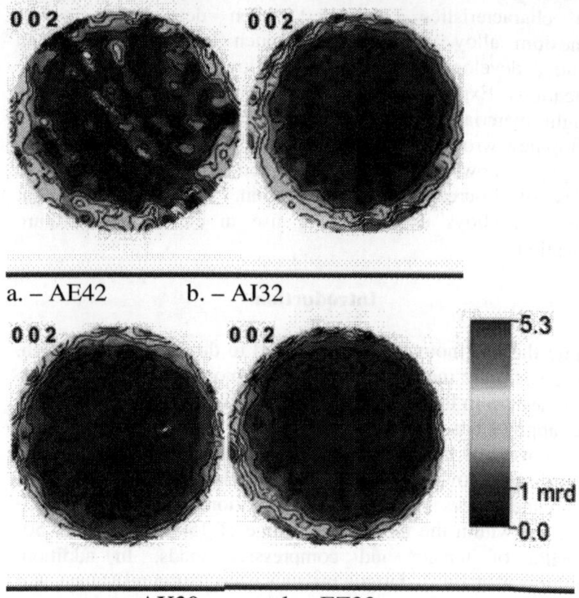

Figure 3 As-extruded texture for selected alloys, (0002) reflection

As expected, Figure 3 indicates that the selected alloys exhibited typical magnesium extrusion texture, with the (0002) poles aligned preferentially normal to the extrusion axis. The basal pole figures have similar strength for the basal texture, ranging from 4.5 m.r.d. (i.e. multiple of random distribution) for AE42 to 5.4 m.r.d. for EZ33.

Figure 4, however, shows somewhat different intensities, for the prismatic $\{10\bar{1}0\}$ orientation, for the initial texture, and for modified textures as a result of creep deformation. All samples in Figure 4 were identified in the following manner:

- As-extruded (i.e. not creep tested) – all *a.* samples;
- After the compression-creep test, 150°C – all *b.* samples
- After the tensile-creep test, 175°C – all *c.* samples

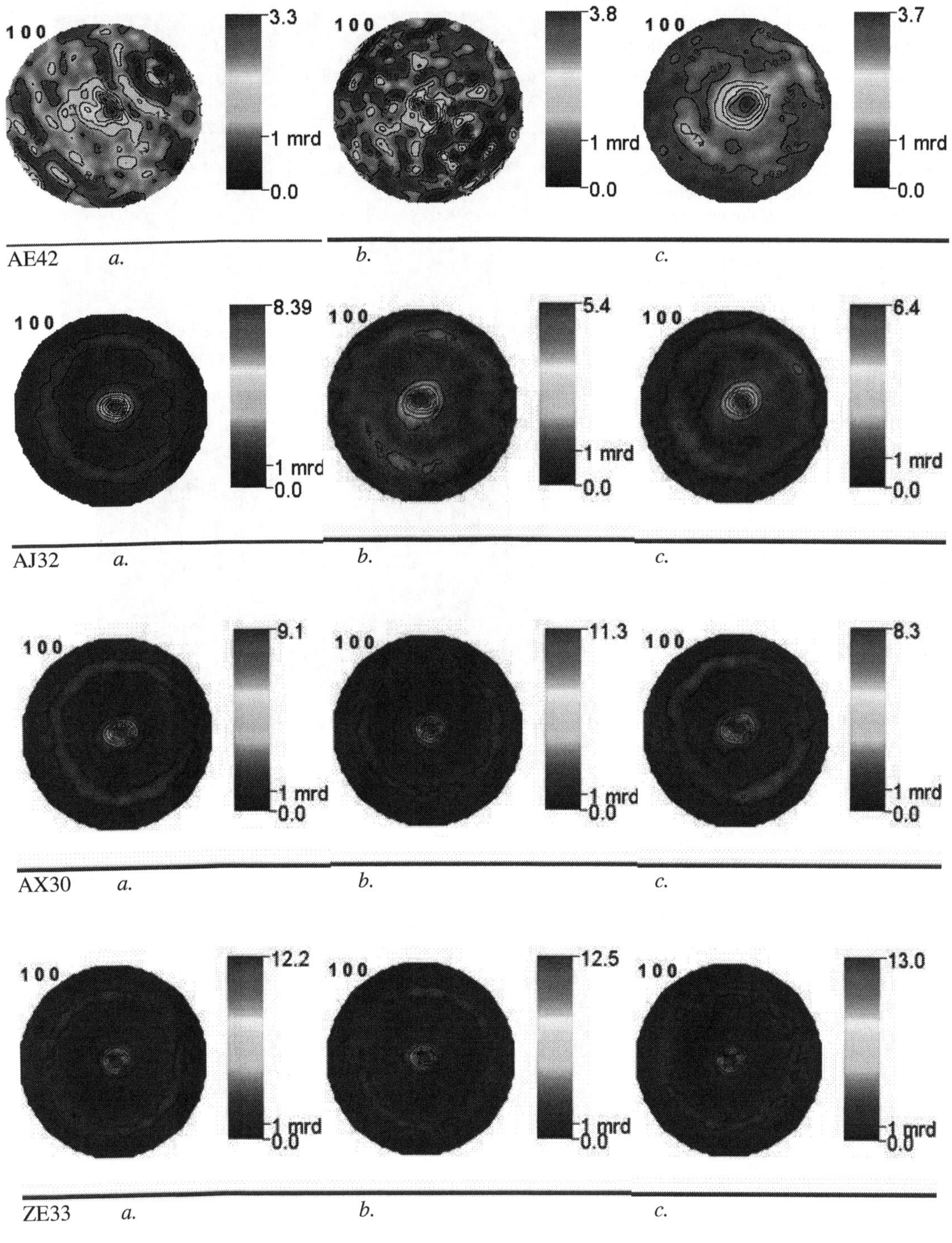

Figure 4 Texture evolution in creep tests for selected alloys, $\{10\bar{1}0\}$ reflection

Figure 4 shows that the strength of the initial as-extruded texture for the studied samples can vary for the prismatic reflection $\{10\bar{1}0\}$ pole figures, ranging from about 3 to about 12 m.r.d. Interestingly, the alloys that had relatively weak extrusion texture typically did not fare well in the creep testing (see Figure 1), an observation made in earlier studies [10, 11]. Fugure 4 also shows obvious texture modification on the AE42 specimen as a result of creep testing, particularly for the tensile creep specimen (see Figure 4, AE42, *c.*). Although the strength of the texture remains relatively the same, the grains exhibit further extrusion-type reorientation as a result of tensile creep deformation.

The other three alloys shown in Figure 4 exhibited much less texture evolution under creep, which is consistent with the observation that these alloys deformed much less than AE42. It can be suggested that the presence of the $Mg_{17}Al_{12}$ β-phase only partially explains the inferior performance of AE, AJ, and AX alloys compared to the EZ-group material.

It would be of interest to characterize in more details the phases present in the best-performing EZ33 alloy.

EZ33 – Second Phase Analysis

Unlike the other alloys, the EZ-group alloys clearly show the presence of a second phase in the diffraction pattern (see Figure 5). Along with easily identifiable peaks of α-magnesium, Figure 5 also depicts several [circled] peaks of another phase. This was further analyzed with application of SEM.

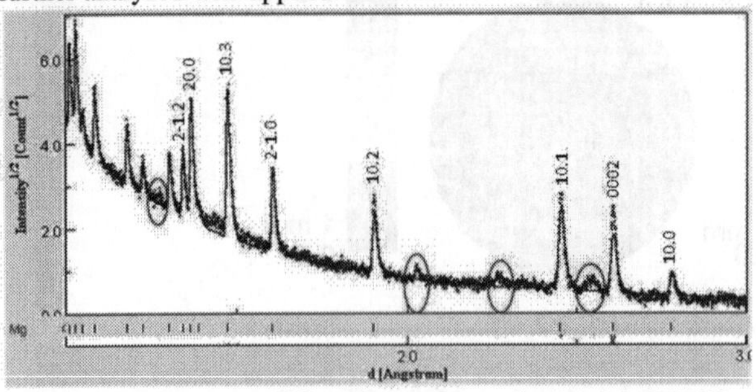

Figure 5 Neutron diffraction pattern obtained for EZ33 specimen

Figure 6*a* shows that a second phase is formed along the grain boundaries. Higher magnification (see Figure 6*b*) revealed that the second phase mostly consists of two different intermetallics, with most of the phase volume taken by magnesium-rich (up to about 85at.%Mg) complex $Mg_xZn_xZr_xRE_x$ intermetallic. The second (minority-phase) intermetallic that is formed inside this phase is mostly-binary Mn-Zr ($ZrMn_2$) inclusions, as seen in Figure 6*b* and Table 1. Diffraction peaks, however, associated with the $ZrMn_2$ crystallographic unit cell are different than the ones present in Figure 5, and are likely hidden in the background "noise", which could be expected considering that the volume fraction of this phase is minimal.

On the other hand, a significant volume fraction of the complex Mg-Zn-Zr-RE phase gives rise to the diffraction peaks detected in the neutron diffraction pattern in Figure 5.

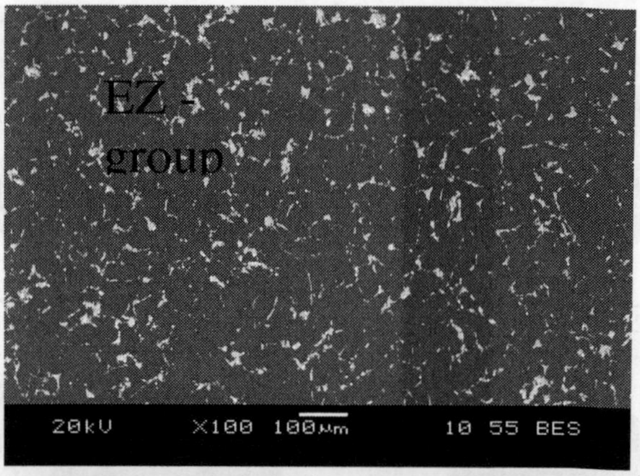

a. 100X

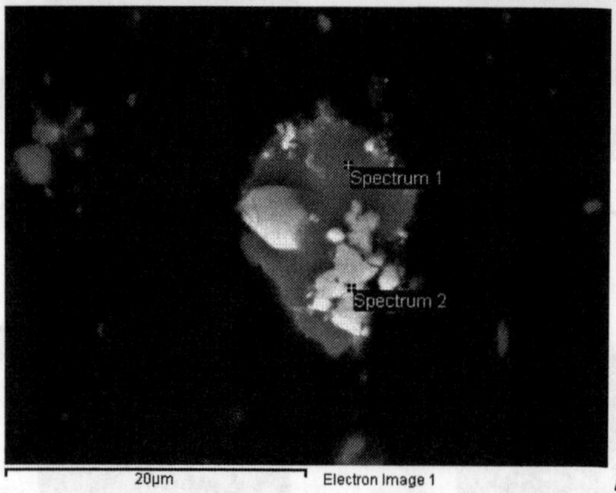

b. 1000X

Figure 6 SEM studies of EZ33 specimen

Table 1
X-ray phase analysis, EZ33, spectrum 1 and 2 (Figure 6*b*)

EZ33 Specimen 1 Element	Wt. %	At. %	EZ33 Specimen 2 Element	Wt. %	At. %
Mg K	48.2	76	Mg K	1.9	5.4
Mn K	12.3	8.6	Mn K	39.2	49.0
Zn K	10.2	6.0	Zn K	-	-
Zr K	9.0	3.8	Zr L	52.8	39.7
La L	5.5	1.5	La L	-	-
Ce L	11.2	3.2	Ce L	-	-
Nd L	2.1	0.6	Nd L	-	-

Conclusions

1. Crystallographic texture studies confirmed that there is a notable difference in the strength of extrusion-type texture for the extruded samples of studied alloys, which represent four alloying systems: Mg-Al-RE, Mg-Al-Ca, Mg-Al-Sr, and Mg-Zn-RE. Two alloys representing the last system developed the strongest texture in the extrusion process (12.3 mrd for the $\{10\bar{1}0\}$ reflection). These alloys also performed the best during high-temperature creep testing, at temperatures of 150 and 175°C, both under compressive and

tensile loading. In contrast, two alloys from the AE group developed the weakest texture in the extrusion process (3.3 mrd for the $\{10\bar{1}0\}$ reflection). These alloys also exhibited the lowest resistance to creep during creep testing.

2. The neutron diffraction spectrum for all selected alloys, except for the ZE-group, revealed the typical one-phase (α-magnesium) spectrum. On the other hand, the diffraction pattern for the ZE alloys exhibited second-phase peaks arising from the presence of complex Mg-Zn-Zr-RE intermetallic. Another minority phase was revealed in the material during the SAE analysis, namely $ZrMn_2$, that was forming inside the larger second-phase intermetallic. Due to its limited volume fraction, the $ZrMn_2$ phase is not detected in the neutron diffraction patterns.

3. The effect of the presence of complex Mg-Zn-Zr-RE intermetallic on the actual mechanism of creep deformation in wrought Mg-Zn-RE(Zr) alloys remains to be studied.

References

1 M. O. Pekguleryuz, A. A. Kaya, "Magnesium Diecasting Alloys for High Temperature Applications", Magnesium Technology 2004, Edited by Alan A. Luo, TMS (The Minerals, Metals & Materials Society), 2004, pp. 281-287.

2 A. A. Luo, Mg Alloys 2003, Mat. Sci.. Forum, Trans Tech, Switzerland, vols. 419-422, 2003, pp. 57-65.

3 M. Pekguleryuz, P. Labelle, US Pat. No. 6,322,644, Nov 27th, 2001 and PCT WO 01/44529

4 M. Pekguleryuz, P. Labelle, D. Argo, M. Dierks, T. Sparks, T. Waltematte, World Magnesium Conference, IMA, Brussels, May 2001, pp. 27-33.

5 M. Pekguleryuz, P. Labelle, E.Baril, D. Argo, SAE Paper 2003-01-190, Detroit, March 2003.

6 S. Koike, K. Wasizu, S. Tanaka, T. Baba, K. Kikawa, SAE Paper 2000-01-1117, Detroit, March, 2000.

7 E.Aghion, B.Bronfin, H.Friedrich, S.Schumann, F. von Buch, 2003 Magnesium Technology, TMS, San Diego, March 2003, pp. 177-182.

8 M. Lefebvre, M. Pekguleryuz, P. Labelle, US Pat. No 6, 342,180, Jan 29, 2002 and PCT WO 02/099147

9 A. Luo, M. Balough, B. R. Powell, SAE Paper 2001-01-0423, Detroit, Michigan, 2001.

10 S. Shook and D. Sediako, "Analysis of Stress Evolution in High Temperature Creep Testing of Creep-Resistant Magnesium Alloys", 66th Annual World Magnesium Conference, San Francisco, California, 2009, pp. 43-51.

11 D.Sediako, S. Shook, "Application of Neutron Diffraction in In-Situ Studies of Stress Evolution in High Temperature Creep Testing of Creep-Resistant Magnesium Alloys", The Minerals, Metals & Materials Society, TMS, San Francisco, CA, February 2009, Magnesium Technology 2009, pp. 255-259.

12 H.-R.Wenk, L. Lutterottia, and S. Vogel, "Texture analysis with the new HIPPO TOF diffractometer", Nuclear Instruments and Methods in Physics Research Section A: Accelerators, Spectrometers, Detectors and Associated Equipment, Volume 515, Issue 3, 11 December 2003, pp. 575-588.

13 M. Çelikin, F. Zarandi, D. Sediako, M.O. Pekguleryuz, "Compressive Creep Behavior of Cast Magnesium under Stresses above the Yield Strength and the Resultant Texture Evolution", Canadian Metallurgical Quarterly, Vol. 48, No.4, 2009, pp. 419-432.

14 M. Çelikin, D. Sediako, and M. Pekguleryuz, "Effect of Compressive Creep-Deformation on the Crystallographic Texture of - Pure Mg and Mg-Mn Casting Alloys", The Minerals, Metals & Materials Society, TMS, Seattle, WA, February 2010, Magnesium Technology 2010, pp. 249-251.

Magnesium Technology 2011
Edited by: Wim H. Sillekens, Sean R. Agnew, Neale R. Neelameggham, and Suveen N. Mathaudhu
TMS (The Minerals, Metals & Materials Society), 2011

IMPROVED PROCESSING OF Mg-Zn-Y ALLOYS CONTAINING QUASICRYSTAL PHASE FOR ISOTROPIC HIGH STRENGTH AND DUCTILITY

Alok Singh[1], Y. Osawa[1], H. Somekawa[1], T. Mukai[1]

[1]Lightweight Alloys Group, Structural Metals Center, National Institute for Materials Science, 1-2-1 Sengen, Tsukuba 305-0047, Japan

Keywords: magnesium alloys, quasicrystal, extrusion, tensile strength, compressive strength, ageing

Abstract

The stable quasicrystal phase in Mg-Zn-Y alloys has been proved to be beneficial for strength and ductility and has also been shown to impart high fracture toughness. However, the strength of these alloys has been limited to 300MPa with an average grain size of about 1 μm. We show here a simple procedure in which a $Mg_{93}Zn_6Y$ alloy was chilled cast and then extruded. Yield strengths of up to 400 MPa in tension and compression, accompanied by ductility of ~14% were obtained with grain of size of about 1 μm. Compression to tension yield anisotropy ratios were in the narrow range of 0.95 to 1.03. These alloys also showed ageing response with two peaks. The effect of precipitation ageing on the mechanical properties has been studied.

Introduction

Various approaches are being taken to improve strength and ductility of magnesium alloys. These include alloying to form new precipitates as well as mechanical deformation to produce finer microstructure to activate non-basal slip and inhibit twinning. An icosahedral quasicrystalline phase (i-phase) occurs in Mg-Zn-RE alloys in which the rare earth element RE is Y, Gd, Tb, Dy, Ho or Er, which is now known to be a thermodynamically stable Mg_3Zn_6RE phase [1, 2]. It also exists in direct equilibrium with α-Mg phase [3, 4]. Since this phase has special properties such as high hardness and low surface energy [5], it becomes a natural choice as a strengthening phase in magnesium alloys. However, this phase forms interdendritically; therefore it must be finely dispersed in the alloy by means such as hot rolling or extrusion [6, 7, 8, 9, 10, 11, 12]. These alloys have been shown to have a good balance of strength and ductility, have very high fracture toughness [13], good fatigue strength [14], weak texture [15] and excellent formability [16].

By direct extrusion and hot rolling, the grain size has been limited to 1μm, with tensile and compressive yield strength (YS) of about 300MPa [11, 12]. To raise this limit, powder metallurgy was performed, which produced a grain size of about half a micron and tensile YS at the level of 400MPa [17]. It has been shown that the strength levels follow Hall-Petch relationship indicating grain boundary as the main strengthening mechanism [12, 17]. In this study, we show that higher strengths can be obtained by improved processing for better distribution of the i-phase. Precipitation in the matrix and its effect on tensile and compressive YS is shown.

Experimental Procedure

An alloy of composition Mg-6at%Zn-1at%Y was made by melting in an electric furnace under a cover gas, and then cast into a chill cast mold. The alloy was extruded at an extrusion ratio 25:1 at temperatures 390°, 300° and 250°C. Ageing was performed in an oil bath at 150°C. Microstructure was observed by transmission electron microscopy (TEM) using a JEOL 2000FX microscope. Samples for TEM were thinned by ion milling. Tensile and compression tests were performed on an Instron machine at a strain rate of 10^{-3}. Tensile specimens had a diameter 3mm and gauge length 15 mm. Compression specimens were 4mm in diameter and 8mm in height. Mechanical test data were averaged from three tests. To determine the average grain size, the average linear intercept length measured was multiplied by a factor of 1.74 to account for stereographic effects [18]. Vickers microhardness tester was used for measuring hardness of the samples.

Results and Discussion

Extruded microstructure

The chill cast alloy had a dendritic structure with dendritic arms about 20 μm thick. Fine grain size was obtained after extrusion. Figure 1 shows grain structures after extrusion at 300°C and 250°C. The grain size are

Figure 1: Bright field electron micrographs showing grain structures and i-phase distribution after extrusion at (a) 300°C and (b) 250°C. (c) Details from (b) show nano-sized precipitates in the matrix.

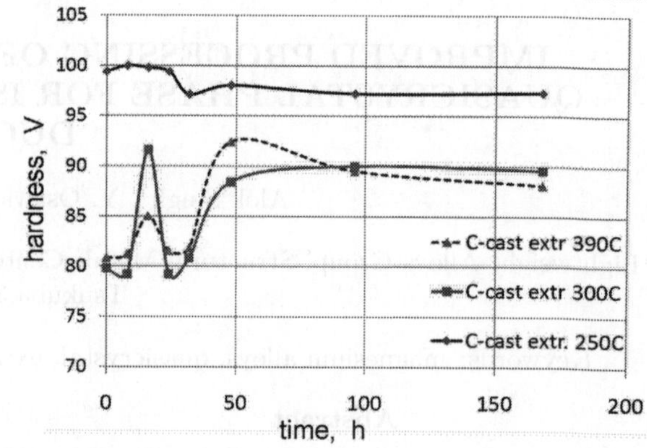

Figure 2: Vickers hardness curves on ageing for the alloys extruded at 390°C, 300°C and 250°C.

about a micron. Comparable size of i-phase particles (dark contrast) were distributed along the extrusion direction. The grain size were measured to be 5.14±0.87, 1.13±0.15 and 0.98±0.16 μm after extrusion at 390°, 300° and 250°C, respectively. Very fine nanometer-size precipitation was observed in the matrix.

Ageing response of the extruded alloys

Since extrusion temperatures of 300°C and higher were also used, some ageing effect was expected in the alloys. Therefore ageing response was measured, shown in Fig. 2. For extrusions at 390° and 300°C, a sharp peak is observed at about 17h. This was confirmed to be due to the precipitation of rod-like β_1' precipitates (Fig. 3). After this peak, the hardness rises again to peak at ~48h, beyond which the hardness remains nearly at the peak level, with very little decrease. The origin this second peak is yet to be ascertained.

The initial hardness of the sample extruded at 250°C was higher than the peak hardness of the other two alloys (Fig. 2). There were no sharp peaks, but instead a small plateau corresponding to the first peak in the other two alloys, and a small rise corresponding to the second peak. The final hardness was lower than the initial. Since the extrusion temperature was lower than 300°C, it is expected that precipitation has occurred during the extrusion. In such cases the precipitate morphology is round, instead of rod-shaped.

Fig. 3 shows precipitates in the alloy extruded at 390°C aged at 150°C for 17h and 48h. Rod-like β_1' precipitates are clearly observed in (a), aligned along the c axis of the matrix. Dark rounded particles are of the i-phase. Attached to these are precipitates of monoclinic Mg_4Zn_7 phase. It is observed in (b) that some of these rod-like precipitates persists after longer ageing for 48h. Fig. 3(c) shows both rounded and rod-like precipitates in the alloy extruded at 250°C and aged for 48h. The rounded precipita

Mechanical Properties

The mechanical properties of the as-extruded alloys are shown as tensile and compression stress-strain (SS) curves in Fig. 4. After extrusion at 390°C , the tensile yield stress (TYS) was 268.3±2.0 MPa and compression yield stress (CYS) was 252.9±0.7 MPa. For extrusion at 300°C, the TYS was 340.5±3 and CYS was 342.3±9.0 MPa. Thus the CYS was comparable or even larger than the TYS. This is similar to the extruded alloy $Mg_{93}Zn_6Ho$ of a comparable grain size [11]. In case of the alloy extruded at 250°C, TYS was 386.3±2.8 MPa and the CYS was 354.6±8.0 MPa. In this case, the TYS is larger than the CYS again. This may be due to the precipitation condition of the 250°C extruded alloys, in comparison to the 300°C extruded alloy, as mentioned above. It is noted that after yielding in compression, there is a plateau in the SS curve. This plateau becomes longer with finer grain size, as reported earlier [19]. Elongation to failure in compression are in the range 14-16% strain.

The tensile properties show a decrease in strain hardening as the grain size decreased. This may be understood as less number of dislocations are generated in a small grain. What is noteworthy is that the strain to failure (elongation or the ductility) increased with decreasing grain size. The elongation for the alloys extruded at 390°, 300°and 250°C were 12.88±1.03, 14.98±0.46 and 16.00±0.30%, respectively.

Effect of Ageing on Strength and Ductility

The extruded alloys were aged at 150°C for 48h and mechanically tested again. There was a slight increase in the TYS, but a decrease in the CYS in case of the finer grained alloys. The TYS for the aged 390°, 300° and 250°C extruded alloys were 288.7±1.6, 346.4±7.7 and 399.9±2.3 MPa, respectively. The corresponding CYT were 258.1±1.1, 326.2±0.8 and 377.6±9.5 MPa. The yield asymetry ratio YAS (=CYS/TYS) can decrease after ageing because the β_1' precipitates are easily sheared [20]. Particularly in the case of extrusion at 300°C, the

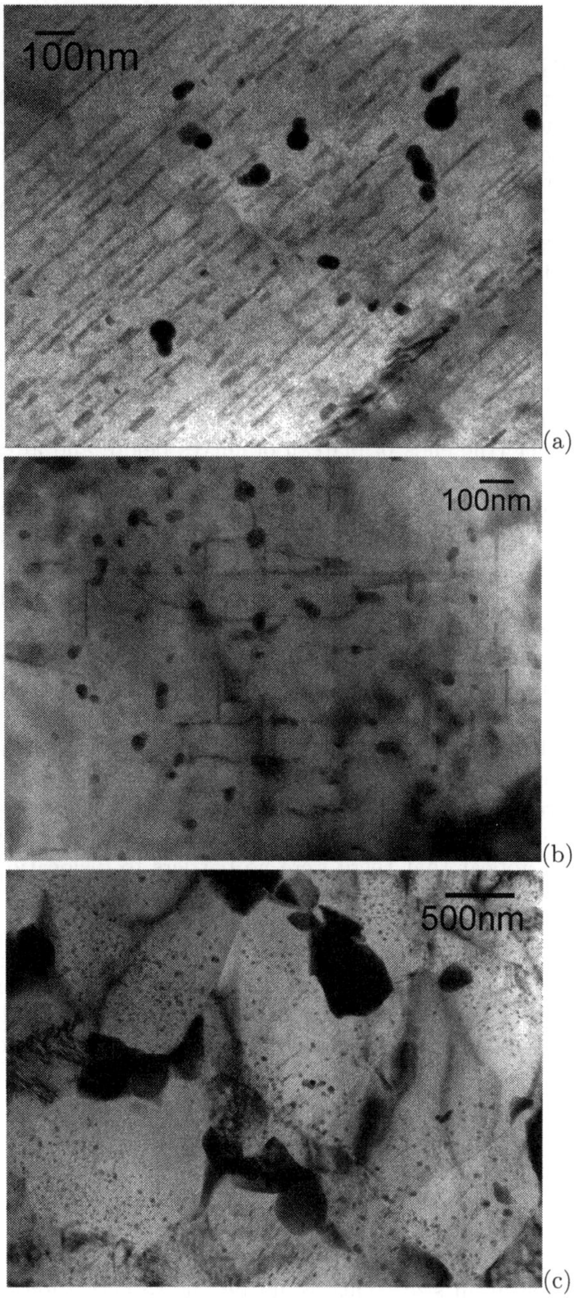

Figure 3: Precipitation in the matrix of alloy extruded at 390°C after ageing at 150°C for (a) 17h and (b) 48h, and (c) in alloy extruded at 250°C after ageing for 48h. Round dark particles in (a) and (b) are i-phase.

YAS ratio decreased from 1.1 before ageing to 0.94 after ageing.

Fig. 5 compares the tensile and compressive SS curves of as-extruded with the aged alloys for extrusions of 390° and 250°C. In case of the tensile curves of the alloy extruded at 390°C, the as-extruded sample has higher strain hardening and lower elongation, as compared to the aged sample. After ageing the TYS is higher, as well as the elongation is higher at nearly 18%. Precipitates would slow the slip, resulting in lower strain hardening and delaying pile-ups at grain boundaries, resulting in larger elongation. The compressive SS curves of the same samples (Fig. 5(b)) shows similar curves before and after ageing; however, after ageing the yield point is more pronounced, and the plateau after the yielding is longer (the ends of plateau of curves before and after ageing are marked with arrows). The plateau marks the region of predominant twinning. The plateau region ends when, following twinning, considerable slip occurs. The extended plateau region indicates that slip is impeded by precipitation. The rod-like precipitates which are bent or sheared by twins do not have significant effect on reduction of twinning [20]. Precipitation in Mg-Zn-RE alloys is reported to reduce the YAS ratio [21].

A slight increase in the TYS is observed after ageing in case of alloy extruded at 250°C too, Fig. 5(c), but there is no strain hardening after yielding in case of the aged sample, leading to an early failure (elongation 10.5±1.3%). It would appear that there were no barriers to dislocation motion after ageing. In case of compression (Fig. 5(d)), the SS curves before and after the ageing are similar, except that the CYS after the ageing is lower. The width of the plateau after yielding is (about 0.5% in strain) smaller than before ageing, indicating that slip is more active after ageing. Both in the case of extrusion of 300° and 250°C the CYS fell after ageing.

Effect of microstructure on ductility

Tensile elongations in as-extruded condition and after ageing are plotted against reciprocal of grain size in Fig. 6. In the as-extruded condition, the ductility rises with decreasing grain size. This would be due to activation of non-basal slip near grain boundaries to maintain compatibility stress across the grain boundaries [22]. Texture would also play a role. Finer grains after wrought processing are accompanied by a weaker texture, which is associated with better ductility [23]. In coarse grained alloys, in absence of precipitates to impeded dislocation

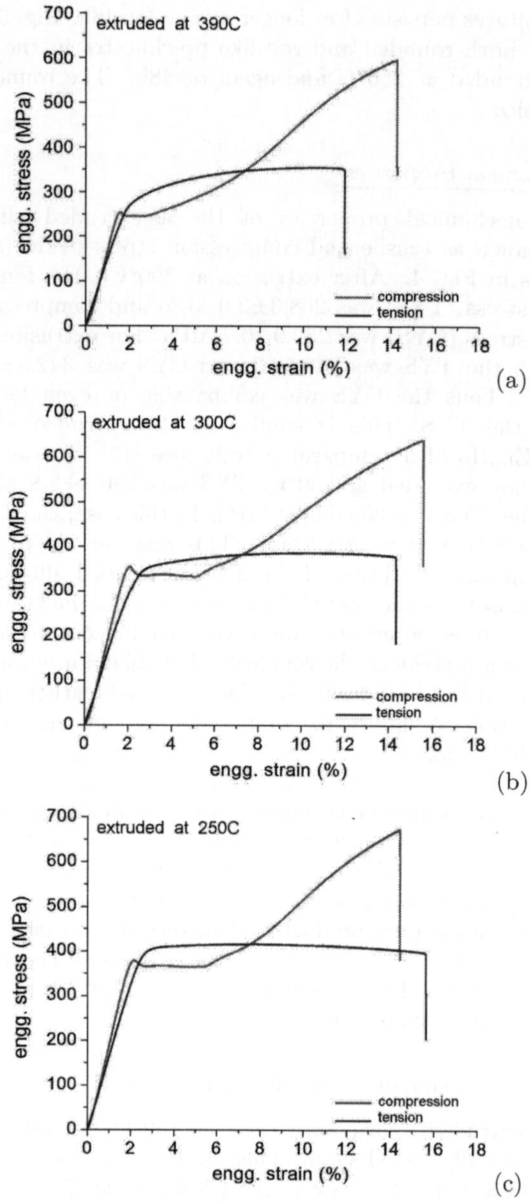

Figure 4: Stress-strain curves in tension and compression after extrusion at (a) 390°C, (b) 300°C and (c) 250°C.

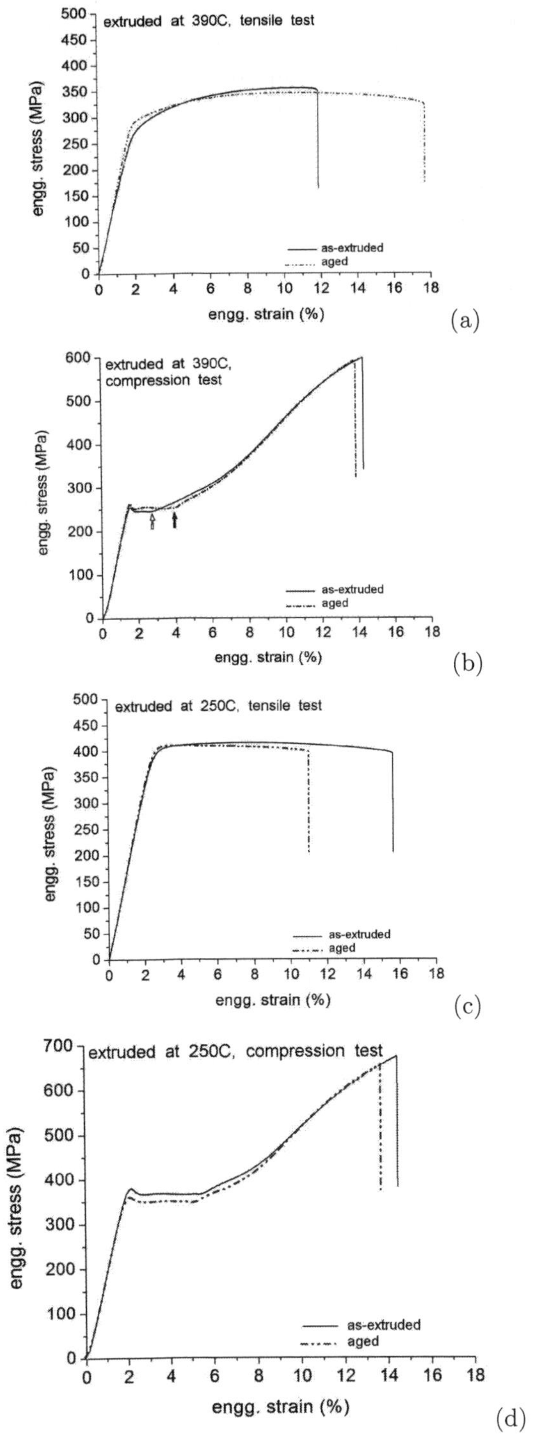

Figure 5: Comparison of stress-strain curves in as-extruded and aged (150°C for 48h) for extrusion at 390°C in tension (a) and compression (b), and for extrusion at 250°C in tension (c) and compression (d).

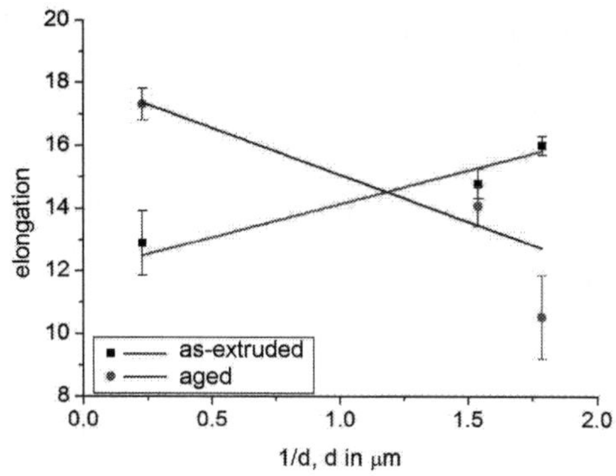

Figure 6: Tensile elongation to failure of alloys in as-extruded and aged condition plotted against reciprocal of grain size.

motion, pile-ups will occur easily, resulting in crack formation. Twinning also occurs easily in larger grained alloys which, coupled with dislocation pile ups, would lead to fracture. After ageing, the precipitates impede the motion of dislocations, preventing and early fracture. However, the lowering of ductility in fine grained alloys is not immediately clear. Long precipitates in fine grains may themselves become a source of brittleness.

Conclusions

A simple modification of processing of a quasicrystalline i-phase containing Mg-Zn-Y alloy resulted in excellent mechanical properties. The alloy was chill cast, and then wrought processes by direct extrusion. Very high yield strengths of about 400 MPa, both in tension and compression, were obtained with average grain size of about a micron. This strength was accompanied by reasonable ductility between 12 to 18%. On ageing at 150°C, two peaks were observed at 17h and 48h. Beyond 48h, the hardness showed very little decrease. After ageing for 48h, there was a slight increase in the tensile yield strength, but a slight decrease in the compressive yield strength in case of the fine grain size. The ductility increased in case of larger average grain size (~5μm) but decreased in case of fine grained (~1μm) alloys.

References

[1] Z.P. Luo, S. Zhang, Y. Tang and D. Zhao, "Quasicrystals in as-cast Mg-Zn-RE alloys", *Scripta metallurgica* 28 (1993) 1513-1518.

[2] A.P. Tsai, A. Niikura, A. Inoue and T. Masumoto, "Stable Zn-Mg-rare-earth face-centered icosahedral alloys with pentagonal dodecahedral solidification morphology", *Philosophical Magazine Letters* 69 (1994) 351-355.

[3] A.P. Tsai, Y. Murakami, A. Niikura, "The Mg-Zn-Y phase diagram involving quasicrystals", *Philosophical Magazine Letters* 80 (2000) 1043-1054.

[4] N. Lebrun, A. Stamou, C. Baetzner and J. Robinson, "Magnesium-Yttrium-Zinc" in Ternary Phase Diagrams (eds. G. Petzow & E. Effenberg), VCH, (1988) 702-710.

[5] D.J. Sordelet and J.-M. Dubois (eds.) "Quasicrystals: Perspectives and Potential Applications", *Materials Research Society Bulletin* 22 (1997) 34-.

[6] D.H. Bae, S.H. Kim, D.H. Kim and W.T. Kim, "Deformation behavior of Mg-Zn-Y alloys reinforced by icosahdral quasicrystalline particles", *Acta Materialia* 50 (2002) 2243-2356.

[7] D.H. Bae, M.H. Lee, K.T. Kim, W.T. Kim, D.H. Kim, "Application of quasicrystalline particles as a strengthening phase in Mg-Zn-Y alloys", *J. Alloys Compounds* 342 (2002) 445-450..

[8] A. Singh, M. Nakamura, M. Watanabe, A. Kato, A.P. Tsai, "Quasicrystal strengthened Mg-Zn-Y alloys by extrusion", *Scripta Materialia* 49 (2003) 417-422.

[9] A. Singh, M. Watanabe, A. Kato, A.P. Tsai, "Microstructure and strength of quasicrystal containing extruded Mg-Zn-Y alloys for high temperature applications", *Materials Science and Engineering A* 385 (2004) 382-396.

[10] H. Taniuchi, H. Watanabe, H. Okumura, S. Kamado, Y. Kojima, Y. Kawamura, "Microstructures and tensile properties of Mg-Zn-Y alloys containing quasicrystals", *Materials Science Forum* 419-422 (2003) 255-260.

[11] A. Singh, H. Somekawa and T. Mukai, "Compressive Strength and Yield Asymmetry in Extruded Mg-Zn-Ho Alloys Containing Quasicrystal Phase", *Scripta Materialia* 56 (2007) 935-938.

[12] A. Muller, G. Garces, P. Perez and P. Adeva, "Grain refinement of Mg-Zn-Y alloy reinforced by an icosahedral quasicrystalline phase by severe hot rolling", *J. Alloys Compounds* 443 (2007) L1-L5.

[13] H. Somekawa, A. Singh and T. Mukai, "High fracture toughness of extruded Mg-Zn-Y alloy by synergetic effect of grain refinement and dispersion of quasi-crystalline phase", *Scripta Materialia* 56 (2007) 1091-1094.

[14] D.K. Xu, L. Liu, Y.B. Xu, E.H. Han, "The fatigue behavior of I-phase containing as-cast Mg-Zn-Y-Zr alloy", *Acta Materialia* 56 (2008) 985-994.

[15] J.Y. Lee, H.K. Lim, D.H. Kim, W.T. Kim and D.H. Kim, "Effect of icosahedral phase particles on the texture evolution in Mg-Zn-Y alloys", *Materials Science & Engineering A* 491 (2008) 349-355.

[16] H. Somekawa, A. Singh, T. Mukai,"Superplastic behavior in Mg-Zn-Y alloy with dispersed quasicrystal phase particle", *Advanced Engineering Materials* 11 (2009) 782-787.

[17] E. Mora, G. Garces, E. Onorbe, P. Perez and P. Adeva, "High-strength Mg-Zn-Y alloys produced by powder metallurgy", *Scripta Materialia* 60 (2009) 776-779.

[18] A. W. Thompson, "Calculation of true volume grain diameter", *Metallography* 5 (1972) 366-369.

[19] M.R. Barnett, Z. Keshavarz, A.G. Beer and D. Atwell, "Influence of grain size on the compressive deformation of wrought Mg-3Al-1Zn", *Acta Materialia* 52 (2004) 5093-5103.

[20] J.B. Clark, "Transmission Electron Microscopy study of age hardening in a Mg-5wt% alloy", *Acta Metallurgica* 13 (1965) 1281-1289.

[21] A. Singh, H. Somekawa, T. Mukai and A.P. Tsai, "Deformation structures and yield asymmetry in extruded Mg-Zn-Y alloys containing quasicrystal phase", *Magnesium Technology in the Global Age* (eds. M.O. Pekguleryuz and L.W. F. Mackenzie), Canadian Inst. Mining, Metals & Petroleum, Montreal, Canada, 2006, pp. 441-451.

[22] J. Koike, T. Kobayashi, T. Mukai, et al. "The activity of non-basal slip systems and dynamic recovery at room temperature in fine-grained AZ31B magnesium alloy", *Acta Materialia* 51 (2003) 2055-2065.

[23] T. Mukai, M. Yamanoi, H. Watanabe and K. Higashi, "Ductility enhancement in AZ31 magnesium alloy by controlling its grain structure", *Scripta Materia* 45 (2001) 89-94.

Magnesium Technology 2011
Edited by: Wim H. Sillekens, Sean R. Agnew, Neale R. Neelameggham, and Suveen N. Mathaudhu
TMS (The Minerals, Metals & Materials Society), 2011

Precipitation hardenable Mg-Ca-Al alloys

J. Jayaraj, C. L. Mendis, T. Ohkubo, K. Oh-ishi, K. Hono
National Institute for Materials Science, 1-2-1 Sengen, Tsukuba, 305-0047, Japan

Keywords: Magnesium alloys, Aging, G. P. zone

Abstract

The age-hardening responses and the corresponding microstructures of Mg–0.5Ca–*x*Al (*x* = 0, 0.1, 0.3, 0.5, 1 wt. %) alloys were investigated by hardness tests and transmission electron microscopy. For the optimum Al addition of 0.3 wt. % an enhanced age-hardening response with the highest peak hardness of HV=72 was achieved. TEM analyses confirmed that the improvement in the peak hardness is associated with the dense precipitation of ordered monolayer G.P. zones. Whereas, lower content of Al resulted in the formation of G.P. zones and Mg_2Ca and the excess addition of Al causes the formation of the G.P. zones and the grain boundary Al_2Ca phase.

Introduction

The precipitation hardening of Mg–Ca binary alloys is very low due to the coarse precipitation of the equilibrium Mg_2Ca phase [1]. However, the age-hardening response of the Mg–Ca alloys can be enhanced due to the precipitation of plate type G.P zones by microalloying of Zn, or both Zn and Nd [1,2]. At elevated temperatures, the alloy strengthened by such plate type G.P zones in a Mg–1Ca–1Zn–1Nd–0.6Zr alloy exhibits better properties in strength, ductility and creep resistance compared to AZ91 (Mg–9Al–1Zn) [2]. The formation of the plate like precipitates on the basal plane is related with the microalloying elements. The combinations of oversized element like Ca and RE and undersized element like Zn in Mg causes the formation of plate like internally ordered G.P zones on the basal plane. The age hardening response and good creep properties in both Mg–Ca–Zn and Mg–RE–Zn alloys are attributed to the formation of the plate like G.P zones [2-7].

In order to understand the mechanism of the formation of the internally ordered GP zones, it will be useful to see whether or not other undersized elements like Al exhibit the similar effect in the formation of G.P zones in Mg–Ca system. Like Zn in Mg–Ca–Zn system, Al and Zn have similar atomic sizes and the heat of mixing of Al in Mg is similar to that of Ca in Mg. Based on this concept, we selected Al as a microalloying element to the Mg–0.5Ca binary alloy and found that the similar G.P. zones are formed in the Mg–0.5Ca–*x*Al alloys [8]. This paper updates the results reported in reference [8] for this symposium.

Experimental procedures

Mg–0.5Ca–*x*Al (*x* = 0.1, 0.3, 0.5, 1, wt. %) alloys were prepared by induction melting high purity Mg, Al and Mg–13.3Ca (wt. %) master alloy in an Ar atmosphere. The compositions of the alloys are shown in Table 1 with both wt. % and at. %. Hereafter, all alloy compositions are indicated in wt. %. Samples sectioned from the ingots were solution-heat treated at 500 °C for 2 h and then quenched into water. Aging treatments were performed at 200 °C in an oil bath for different time intervals. The hardness values were measured by a Vickers hardness tester with a 500g load. Scanning electron microscopy (SEM) observation was performed using a Carl Zeiss Cross beam 1540 microscope. TEM specimens were prepared by twin jet electro-polishing as explained elsewhere [8]. Microstructure observations were carried out using a TECNAI G^2 20 TEM. High-angle annular dark field (HAADF) observations were carried out using a TECNAI G^2 F30 TEM.

Designations	at. %			wt. %		
	Mg	Ca	Al	Mg	Ca	Al
Mg–0.5Ca	Bal.	0.3	-	Bal.	0.5	-
Mg–0.5Ca–0.1Al	Bal.	0.3	0.1	Bal.	0.5	0.1
Mg–0.5Ca–0.3Al	Bal.	0.3	0.3	Bal.	0.5	0.3
Mg–0.5Ca–0.5Al	Bal.	0.3	0.5	Bal.	0.5	0.5
Mg–0.5Ca–1Al	Bal.	0.3	1.0	Bal.	0.5	1.0

Table 1 Alloy compositions in wt.% and at.%.

Result and discussion

Figure 1 shows the age hardening response of Mg–0.5Ca–*x*Al (*x* = 0.1, 0.3, 0.5, 1, wt. %) alloys at 200 °C. The hardness value of all these alloys at as-quenched condition was approximately 40 HV. The peak hardness of Mg–0.5Ca binary alloy was low at about 50 HV. However, the addition of Al to Mg–0.5Ca significantly improved the age-hardening responses. In particular, the peak hardness of the Mg–0.5Ca–0.3Al alloy reached to 72 HV in 2 h and the increment in hardness (ΔHV=32) was the highest among all the alloys. For the other alloys, the peak hardness reached about 60 HV. As aging progress the hardness decreased to a lower value due to the coarsening of the microstructure. However, the Mg–0.5Ca–0.3Al alloy shows better resistance against coarsening with the over-aged hardness value remained high at about 60 HV until 1000 h of aging times. The developed ternary Mg–0.5Ca–0.3Al alloy shows similar peak hardness compared to that of the reported quinary Mg–1Ca–1Zn–1Nd–0.6Zr alloy. This indicates that the micro-alloying of Al to Mg–Ca system is highly advantageous considering the lower cost of Al compared to Nd.

Figure 2a and 2b shows the backscattered electron SEM image of the solution treated Mg–0.5Ca–0.3Al and Mg–0.5Ca–1Al alloy respectively. No secondary phase can be seen in the solution treated Mg–0.5Ca–0.3Al alloy, indicating the formation of super saturated solid solution of Mg–(Ca–Al). In contrast, insoluble white bright phases were observed along the grain boundaries of the grey dark phase for the solution treated Mg–0.5Ca–1Al alloy. Since heavy elements (high atomic number; Ca = 20, Al =13) backscatter electrons stronger than the light elements (low atomic number; Mg = 12), the appeared brighter phase in Fig. 2b could be Ca–Al bearing phase. Microstructure investigations on the solution treated Mg–0.5Ca–0.1Al and Mg–0.5Ca–0.5Al alloys showed similar features to that of the Mg–0.5Ca–0.3Al and Mg–

0.5Ca–1Al alloys respectively (figures not shown).The insoluble phase formed could be identified as Al_2Ca was reported for the alloys containing Al higher than 0.3 wt.% [9].This result indicate that the highest concentration of solute atoms would exists for the solution treated Mg–0.5Ca–0.3Al alloy and suggests that the optimum Al content for the age hardening Mg–0.5Ca alloy is 0.3 wt.%.

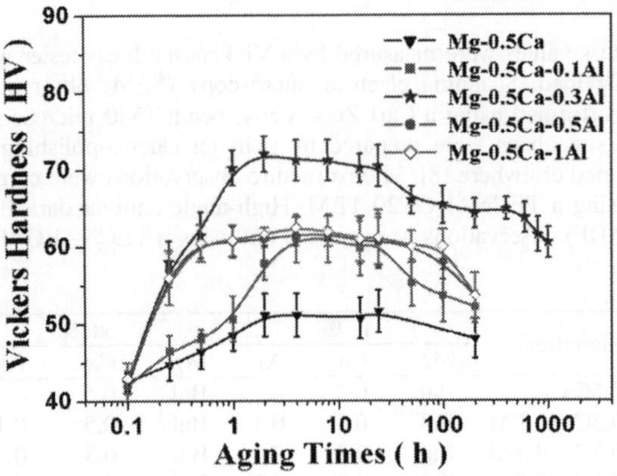

Figure 1 Age hardening response of Mg–0.5Ca–*x*Al (*x* = 0.1, 0.3, 0.5, 1, wt. %) alloys during isothermal aging at 200 °C. Reproduced from reference [8].

The microstructure causing the peak aging and over aging of the alloys with different Al content was characterized by TEM. Figure 3a and 3b show the TEM bright field images of the peak aged and over aged Mg–0.5Ca–0.1Al samples respectively. The images are taken with the incident beam along the $[10\bar{1}0]$ zone axis of the Mg matrix. The peak aged and over aged samples showing the precipitates with the strain contrast are expected to be Mg_2Ca phase as shown for the Mg–0.5Ca binary alloy [4]. The Mg_2Ca phase in the over aged is coarser than the peak aged sample. In addition to the Mg_2Ca precipitates, a speckled contrast implying the presence of extremely fine plate-like precipitates lying on the (0001) basal plane of this peak aged and over aged conditions of Mg–0.5Ca–0.1Al samples. Selected area electron diffraction (SAED) patterns of the peak aged and over aged Mg–0.5Ca–0.1Al in the $[10\bar{1}0]$ zone axis shows continuous streaks at the $1/3\{11\bar{2}0\}$, $2/3\{11\bar{2}0\}$ positions and parallel to [0001]. The

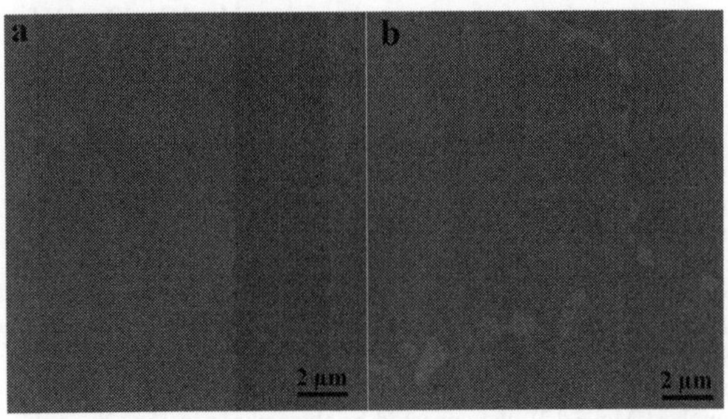

Figure 2 Backscattered electron SEM images of the alloys after solution treatment (a) Mg–0.5Ca–0.3Al (b) Mg–0.5Ca–1Al

SAED pattern in the $[11\bar{2}0]$ zone axis shows the streaks parallel to [0001]. Precipitates exhibiting such diffraction features have been reported in Mg–Ca–Zn (–Nd) [2-4], Mg–RE–Zn–Zr [5] and Mg–Gd–Zn [6] based alloys, which were denominated as ordered Guiner-Preston (G.P.) zones [4-6]. Figure 4a and 4b show the TEM bright field images of the peak aged and over aged Mg–0.5Ca–1Al samples respectively. The images are taken with the incident beam along the $[10\bar{1}0]$ zone axis of the Mg matrix. The speckled contrasts in the bright field images and the positions of streaks in the SAED pattern as explained above indicates the presence of G.P zones for the peak aged and over aged Mg–0.5Ca–1Al samples. Apart from the G.P zones, plate type precipitates with the length of 50 nm are present in the over aged condition. Those plate precipitates causing over aging formed on the basal planes have not been identified in this work. Figure 5a and 5b show the TEM bright field images of the peak aged and over aged Mg–0.5Ca–0.3Al samples respectively. The images are taken with the incident beam along the $[10\bar{1}0]$ zone axis of the Mg matrix. The speckled contrasts in the bright field images and the positions of streaks in the SAED pattern as explained for Fig.3 indicates the presence of G.P zones for the peak aged and over aged Mg–0.5Ca–0.3Al samples. The additional weak spots appeared in the SAED pattern are considered to be the reflections from the oxide layer that were formed during the TEM sample preparation. For the over aged sample, several disc shaped fine precipitates with different length scales of about 5-10 nm and 60 nm were observed. From our previous work, the micro-diffraction pattern analysis revealed that the coarse precipitate is the Al_2Ca phase with the C15 structure [8]. The crystallographic orientation relationship between the Al_2Ca phase and the Mg-matrix is consistent with that was reported by Suzuki et al. [10].

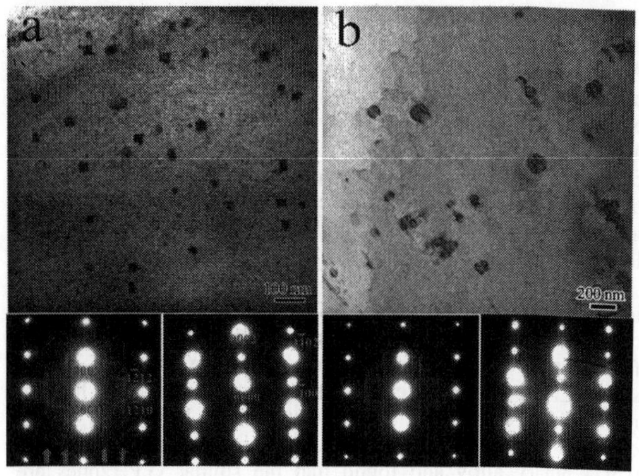

Figure 3 TEM bright field images of the Mg–0.5Ca–0.1Al aged alloy taken from the $[10\bar{1}0]$ and corresponding SAED pattern of the $[10\bar{1}0]$ and $[11\bar{2}0]$ directions. (a) Peak aged at 200 °C for 4h (b) Over aged at 200 °C for 200h.

Figure 6a and 6b show the HAADF image of the peak aged and over aged Mg–0.5Ca–0.3Al samples respectively. The image is taken with the incident beam along the $[10\bar{1}0]$ zone axis of the Mg matrix. For the peak aged condition, no precipitates were observed in the HAADF image. In the over aged condition the precipitates appear as bright lines along the basal plane for a fine (short) and coarse (long) precipitate. Since, the contrast of the HAADF image is related to the square of the atomic number [11], indicating that the precipitates are enriched with Ca and Al. It is clear from the image that the fine precipitate is a (0002) monolayer precipitates enriched with solute atoms, whereas the coarse precipitate is composed of several (0002) layers. Considering the speckled contrast in Fig. 5 and the streaks in the

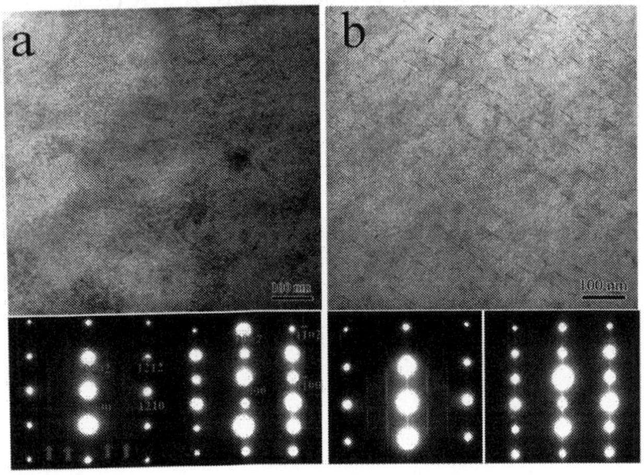

Figure 4 TEM bright field images of the Mg–0.5Ca–1Al aged alloy taken from the [10$\bar{1}$0] and corresponding SAED pattern of the [10$\bar{1}$0] and [11$\bar{2}$0] directions. (a) Peak aged at 200 °C for 4h (b) Over aged at 200 °C for 200h.

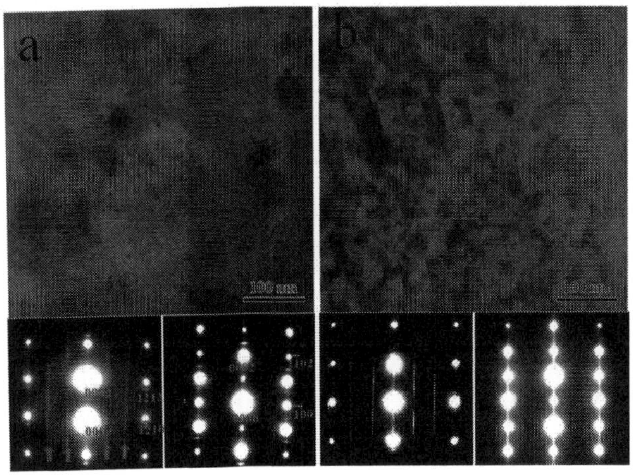

Figure 5 TEM bright field images of the Mg–0.5Ca–0.3Al aged alloy taken from the [10$\bar{1}$0] and corresponding SAED pattern of the [10$\bar{1}$0] and [11$\bar{2}$0] directions. (a) Peak aged at 200 °C for 2h (b) Over aged at 200 °C for 1000h. The figure is reproduced from [8].

inset SAED pattern, we conclude the fine precipitates are ordered monolayer G.P. zones as reported in the Mg–Ca–Zn and Mg–RE–Zn–Zr systems [4, 5].

Our earlier 3DAP investigations on the Mg–0.5Ca–0.3Al peak aged alloy shows that the solute atoms, Ca and Al are distributed essentially within a few atomic planes of $(0002)_{Mg}$ which revealed the presence of disc shaped G.P. zones [8]. The thickness of the G.P zone is within two atomic layers (~ 1 nm) and the disc diameter is approximately 3 nm. The average composition of the G.P. zone in the peak aged sample was estimated to be Mg–6Ca–7Al (at. %). The total solute concentration in the ordered monolayer G.P. zones of this peak aged Mg–Ca–Al alloy is about 13 at. %. It should be noted that these value are lower than the ~ 33 at. % of the solute concentration from the model proposed by

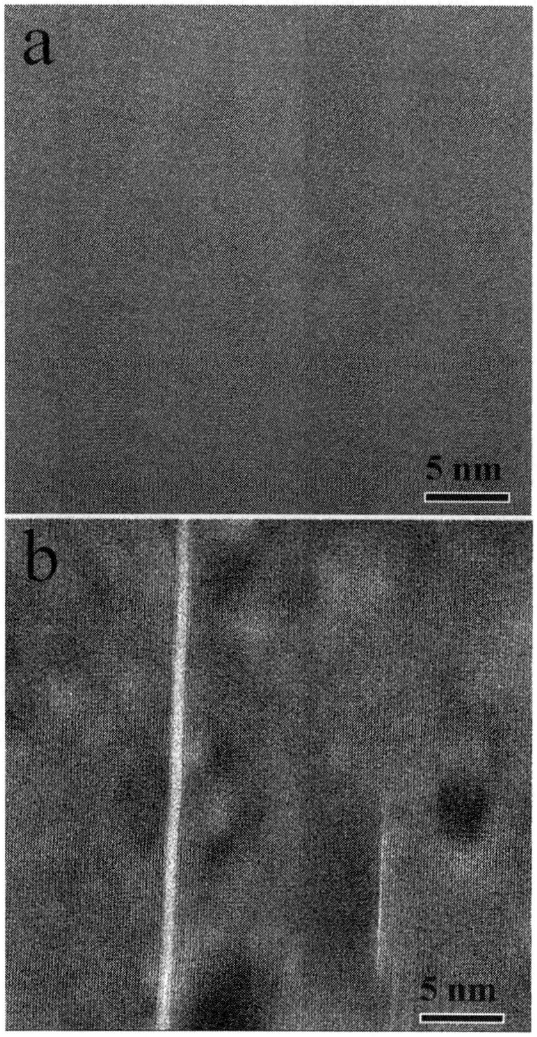

Figure 6 HAADF images of the Mg–0.5Ca–0.3Al aged alloy taken from the [10$\bar{1}$0] (a) Peak aged at 200 °C for 2h (b) Over aged at 200 °C for 1000h.

Ping et.al for the fully ordered G.P. zone formed on a single (0001) Mg plane [5].

In the studied Mg–Ca–Al system, for the optimum Al content, TEM analyses confirmed that the improvement in the peak hardness is associated with the dense precipitation of ordered monolayer G.P. zones. Whereas, lower content of Al resulted in the formation of G.P. zones and Mg_2Ca and the excess addition of Al causes the formation of G.P. zones and the grain boundary

Al_2Ca phase. It is interesting to compare the Mg–Ca–Al system with the other G.P zone forming system such as Mg–Ca–Zn [4] and Mg–Gd–Zn [6]. As that of Mg–Ca–Al, Mg–Ca–Zn system was strengthened by monolayer G.P. zones whereas the Mg–Gd–Zn forms tri-layered G.P. zones. The common features in these systems are that the solute atoms are oversized and undersized to Mg atoms. The atomic radius of Ca (1.97 Å) and Gd (1.78 Å) are larger than that of Mg (1.60 Å), while the atomic radius of Zn (1.48Å) and Al (1.43Å) are smaller than that of Mg atom. Another feature is that the mixing enthalpy between two solute elements is negatively large, i.e., there is a strong attractive interaction between the two solute elements. The mixing enthalpy values of Ca–Zn, Gd–Zn and Ca–Al are –22, –31 and –20 kJ/mol, respectively. These values are one order of magnitude larger than the other combinations, like Mg–Ca, Mg–Al, Mg–Zn and Mg–Gd. The difference in atomic size and large negative heat of mixing favors the formation of the G. P. zones enriched with Ca and Al atoms rather than Mg–Ca. Without microalloying or lower content of Al, the Mg_2Ca phase precipitates [1, 4]. For higher content of Al, the insoluble grain boundary Al_2Ca phase forms. Thus it could be speculated that high number density of G.P zones enriched with Ca and Al atoms is present in the alloy containing optimum Al content of 0.3 wt. %. Also, by comparing the atomic sizes, a large lattice misfit is expected for the Mg_2Ca precipitate, whereas the lattice misfit is lower for the G.P. zones composed of Ca–Al, resulting in coherency with the matrix. Thus due to the combinations of coherency, extremely fine size and high numerical density of the G.P. zones in the peak aged Mg–0.5Ca–0.3Al alloy the highest hardness value is achieved.

Conclusion

In summary, it was confirmed that the effect of addition of Al as an undersized element to Mg–Ca system promotes the formation of ordered monolayer G.P zones as similar to Mg–Ca–Zn. The composition for the highest age hardening response has been optimized to Mg–0.5Ca–0.3Al. The major strengthening precipitate is ordered monolayer G.P. zones and shows high resistance to coarsening. Since only a trace addition of inexpensive alloying elements contribute to the age hardening, this alloy may be suitable for wrought applications.

Acknowledgement

J. Jayaraj acknowledges a JSPS post-doctoral fellowship provided by the Japan Society for the Promotion of Science, Japan. This work was partially supported by CREST-JST and the Grant-in-Aid for Scientific Research (B) 21360348.

References

[1] J. F. Nie, B. C. Muddle, Scripta Mater. 37 (1997) 1475-1481

[2] X. Gao, S. M. Zhu, B. C. Muddle, J. F. Nie, Scripta Mater. 53 (2005) 1321-1326

[3] J. C. Oh, T. Ohkubo, T. Mukai, K. Hono, Scripta Mater. 53 (2005) 675-679

[4] K. Oh-ishi, R. Watanabe, C. L. Mendis, K. Hono, Mater. Sci. Eng. A 526 (2009) 177-184

[5] D. H. Ping, K. Hono, J. F. Nie, Scripta Mater. 48 (2003) 1017-1022

[6] J. F. Nie, K. Oh-ishi, X. Gao, K. Hono, Acta Mater. 56 (2008) 6061-6076

[7] J. F. Nie, X. Gao, S.M. Zhu, Scripta Mater. 53 (2005) 1049-1053

[8] J. Jayaraj, C. L. Mendis, T. Ohkubo, K. Oh-ishi, K. Hono, Scripta Mater. 63 (2010) 831-834

[9] E. F. Emley, Principles of Magnesium technology, Pergamon Press, (Oxford, New York) 1966, p.248,

[10] A. Suzuki, N. D. Saddock, J. R. Terbush, B. R. Powell, J. W. Jones, T. M. Pollock, Metall. Mater. Trans. A 39A (2008) 696-702

[11] D. E. Jesson, S. J. Pennycook, Proc. R. Soc. A 449 (1995) 273-293

Magnesium Technology 2011
Edited by: Wim H. Sillekens, Sean R. Agnew, Neale R. Neelameggham, and Suveen N. Mathaudhu
TMS (The Minerals, Metals & Materials Society), 2011

MICROSTRUCTURE, PHASE EVOLUTION AND PRECIPITATION STRENGTHENING OF Mg-3.1Nd-0.45Zr-0.25Zn ALLOY

G. Atiya, M. Bamberger, A. Katsman

Department of Materials Engineering, Technion, Haifa 32000, Israel

Keywords: Mg-Nd-Zn-Zr alloys, precipitation sequence, crystallographic structure, orientation relationship, hardening

Abstract

The microstructure of the Mg-3.1Nd-0.45Zr-0.25Zn (wt.%) alloy has been investigated after solution treatment at 540^0C for 24hr, followed by isothermal aging at 175^0C up to 32 days. Various electron microscopy techniques, like EDX mapping and TEM with SAED, have been used to characterize the phase composition and orientation relationships between different phases. After solution treatment, the BCT $(Mg_{1-x}Zn_x)_{12}Nd$ phase present in the as-cast alloy dissolved, and small tetragonal Zn_2Zr_3 rod-like particles precipitated in the α-Mg grain interiors. Zn_2Zr_3 particles are elongated along [001] direction, have an orientation relationship with the Mg matrix $[\text{-}2110](0001)_{Mg} \parallel [001](110)_{Zn2Zr3}$ and appear to be stable in the Mg matrix. Precipitation during isothermal aging involves the formation of metastable phases $\beta''(Mg_3Nd)_{HCP}$ (DO_{19} structure) and $\beta'(Mg_3Nd)_{FCC}$ (DO_3 structure). The β'' precipitates formed during the first 8 days of aging have a plate shape and are fully coherent with the Mg matrix, with the orientation relationship $[\text{-}2110](0001)_{Mg} \parallel [\text{-}2110](0001)_{\beta''}$ and $[\text{-}1100]_{Mg} \parallel [\text{-}1100]_{\beta''}$. The Zn_2Zr_3 rods serve as additional nucleation sites for precipitation during aging. The heterogeneous nucleation occurs in two ways: precipitates nucleate on the basal planes and on the side planes of the Zn_2Zr_3 rods. After 8 days of aging, these precipitates were identified as β'. During the 16÷32 days of aging, β'' precipitates in the grain interior transform into β' precipitates with an FCC structure. The β' precipitates are semi-coherent ith the Mg matrix and have the following orientation relationship: $[0001](2\text{-}1\text{-}10)_{Mg} \parallel [101](11\text{-}1)_{\beta'}$ and $[\text{-}1100]_{Mg} \parallel [\text{-}112]_{\beta'}$. In the late stage of aging, the β' precipitates transform into a stable incoherent β $(Mg,Zn)_{12}Nd$ phase. The growth, coarsening and phase transformations were followed by microhardness tests.

1. Introduction

Magnesium alloys containing rare earth (RE) elements have received extensive attention in recent years due to their excellent mechanical properties, especially high temperature creep resistance, light weight and good die-castability [1-5]. Mg alloys containing RE elements are age hardenable and the thin plate-like precipitates that are uniformly dispersed in the matrix are probably attributed to the mechanical properties and creep resistance. Small additions of Zn to Mg-Nd alloy systems would enhance the hardening ability. In addition, most Mg-RE alloys usually contain 0.4-0.6wt.% Zr as a grain refiner [1,6]. However, attempts to improve the hardness and the strength of these alloys are limited by lack of understanding of the structure, orientation and composition of precipitates and the influence of different alloying elements on the precipitation hardening. The research described in the present paper contributes to the investigation of the precipitation sequence in Mg-Nd based alloys in the presence of Zr and Zn. In addition, the crystallographic orientation of precipitates was studies in the grain interior and in the grain boundary regions.

2. Experimental

An alloy with a nominal composition of Mg-3.1Nd-0.45Zr-0.25Zn (wt.%) was prepared by the Dead Sea Magnesium Research Institute (MRI) . The samples were encapsulated in quartz tubes under 300mmHg Argon atmosphere and were solution treated (ST) for 24hr at 540°C followed by water quenching at 25°C. The treatment included heating to 225°C (heating rate of 10°C/hr), then heating to 540°C at a rate of 15°C/hr. Then the samples were aged in an oil bath at 175°C for up to 32 days. Phase analysis was carried out with X-ray diffractometer (XRD), the microstructure was examined using optical microscopy and FEI Scanning Electron Microscope (SEM), Quanta 200. EDS Analysis was conducted on TEM samples with a special specimen holder in order to increase the reliability and ensure that the photons are scattered from a specific analyzed region. Vickers Hardness testing was performed using 50gr load and holding time of 15sec. Precipitate microstructures were examined in FEI Tecnai G2 T20 S-Twin (TEM) and FEI Titan 80-300 FEG-S/TEM (HRTEM).

3. Results

3.1 As-cast Microstructure

The microstructure of the Mg-3.1Nd-0.45Zr-0.25Zn (wt. %) alloy in as-cast condition is shown in Figure 1. The microstructure consists of α-Mg matrix, eutectic compounds and Zr-rich particles. The average grain size is about 40μm. The eutectic compounds were analyzed by XRD analysis and TEM SAED pattern (Figure 2), and were identified as $Mg_{12}Nd$ with BCT structure (a=1.031nm c=0.593nm).

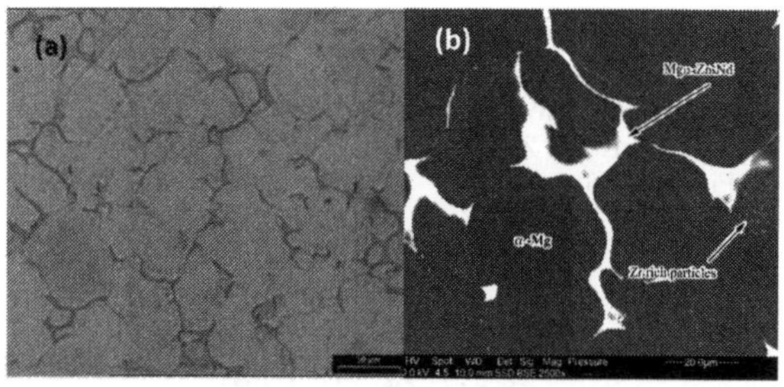

Figure 1. (a) Optical micrograph and (b) BSE SEM micrograph of Mg-3.1Nd-0.45Zr-0.25Zn showing as- cast microstructure.

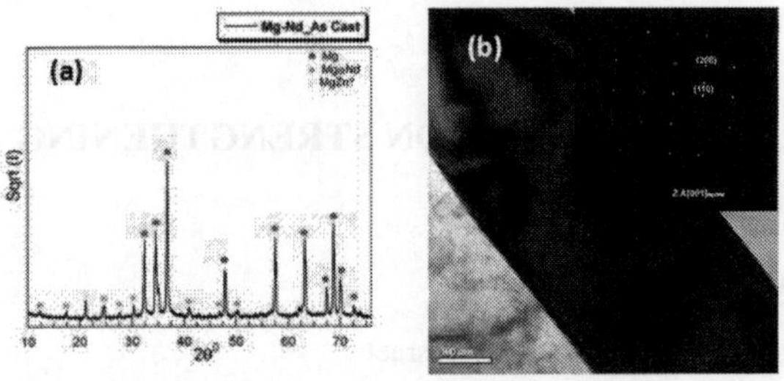

Figure 2. (a) XRD pattern of the Mg-3.1Nd-0.45Zr-0.25Zn as-cast alloy; (b) TEM BFI and corresponding SAED of the eutectic compound zone axis $[001]_{Mg_{12}Nd}$.

It was revealed by SEM EDS analysis performed on the TEM sample that Zn atoms substitute Mg atoms in the eutectic compound (~2.6at.%) as shown in Figure 3 and Table 1.

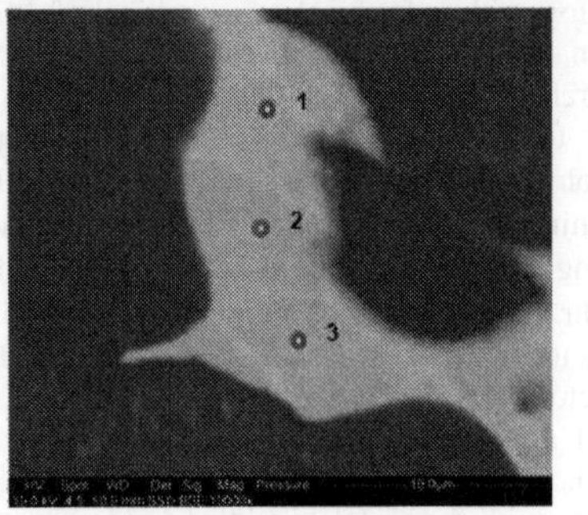

Figure 3. BSE SEM micrograph of the eutectic compound ir as-cast state.

Table 1. EDS element analysis conducted by SEM on the TEM sample

Spectrum	Mg(%at)	Zn(%at)	Nd(%at)
1	90.38	2.53	7.08
2	89.84	2.85	7.31
3	89.68	2.45	7.88
Mean	89.97	2.61	7.42
Std. deviation	0.37	0.21	0.41

3.2 ST Microstructures

As follows from XRD data (Figure 4), only α-Mg phase remains after solution treatment. However, detailed SEM and TEM investigations revealed residual eutectic compounds and Zn_2Zr_3 particles that are gathered in clusters (Figures 5, 6).

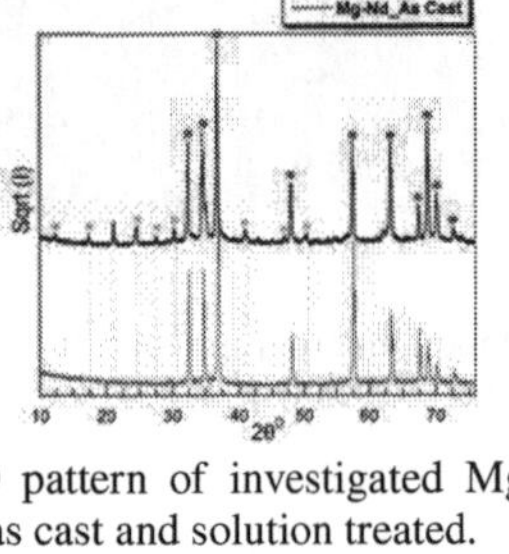

Figure 4. XRD pattern of investigated Mg-3.1Nd-0.45Zr-0.25Zn (wt.%) as cast and solution treated.

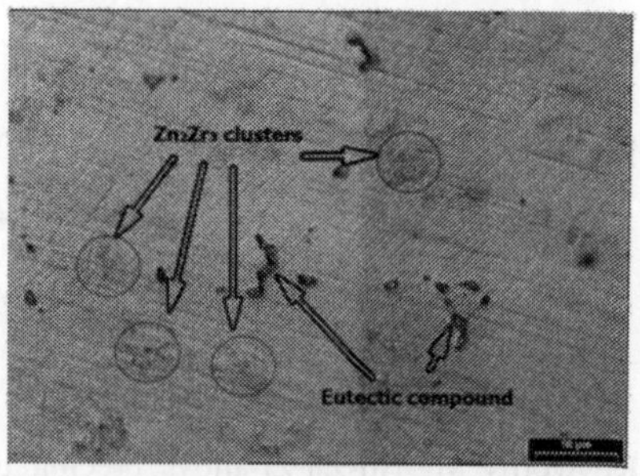

Figure 5. Solution treated microstructure showing Zn_2Zr_3 clusters and residual intermetallic phase at grain boundaries.

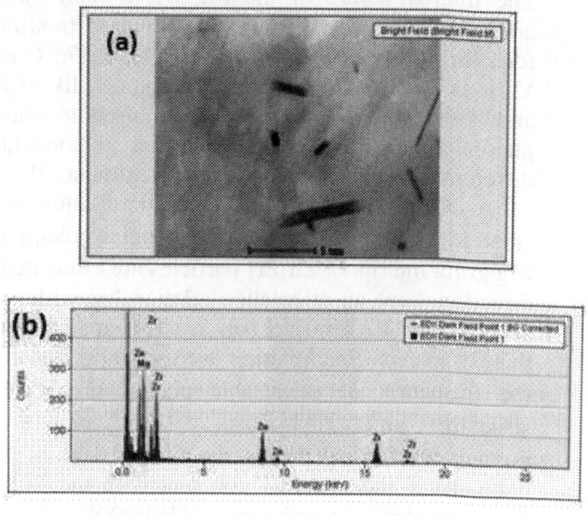

Figure 6. (a) STEM BF micrograph and (b) corresponding EDS spectrum of the Zn_2Zr_3 particle.

3.3 Precipitation Hardening During Aging

The aging of solution treated samples was performed at 175°C up to 32 days. The microhardness increases and reaches the maximum value of about 80HV after 8 days (Figure 7) and then decreases with the aging time.

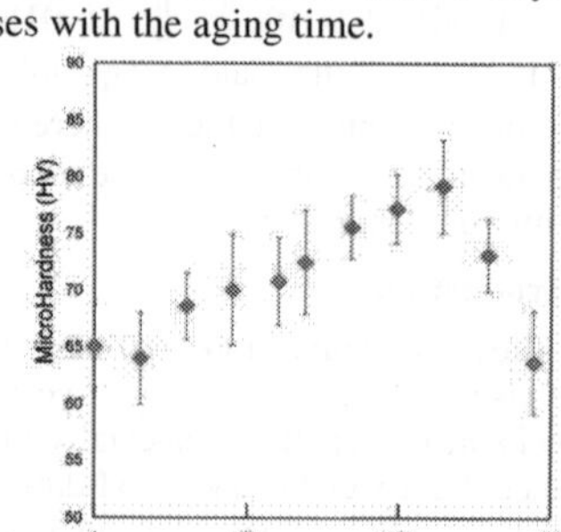

Figure 7. Microhardness as a function of time during isothermal aging at 175°C.

3.4 Phase Composition And Crystallographic Orientations of Precipitates After Aging

The peak aged (8 days aging) sample was investigated by TEM and HRTEM (Figures 8, 9). TEM micrograph and corresponding SAED from grain interior indicate the presence of metastable β''$(Mg_3Nd)_{HCP}$ (DO_{19} structure) precipitates. The precipitates have a plate shape and are fully

coherent with the Mg matrix, with the orientation relationship (OR) [-2110](0001)$_{Mg}$‖[-2110](0001)$_{\beta''}$ and [-1010](-12-10)$_{Mg}$‖[-1010](-12-10)$_{\beta''}$.
HRTEM image along [-1010]$_{Mg}$ zone axis (Figure 9) reveals β'' plate precipitate that is fully coherent with the Mg matrix.

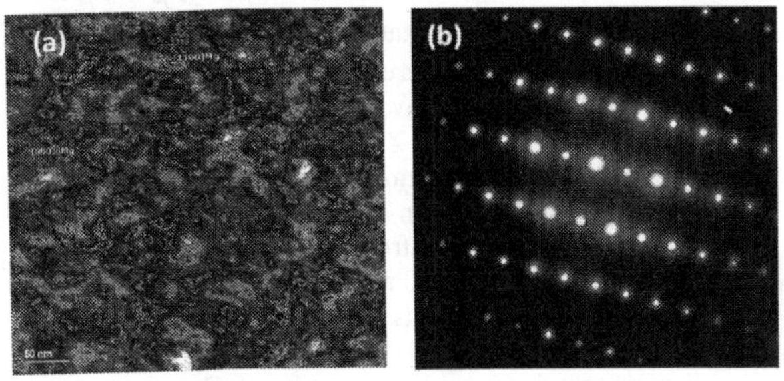

Figure 8. (a) BF TEM micrograph of β'' precipitates in the Mg matrix after 8 days of aging and (b) corresponding SAED [-2110](0001)$_{Mg}$‖[-2110](0001)$_{\beta''}$.

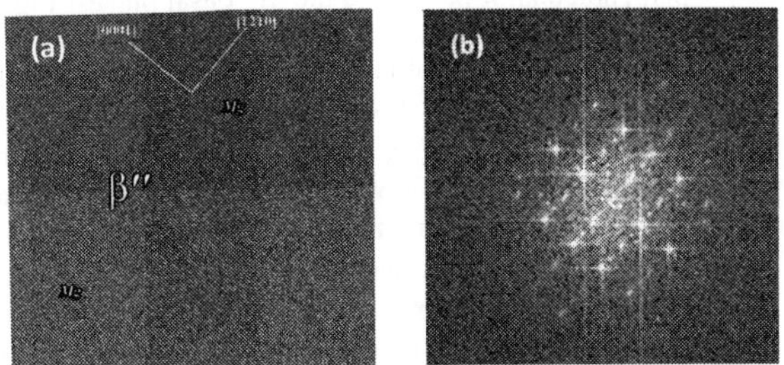

Figure 9. (a) HRTEM micrograph of β'' precipitates in the Mg matrix and (b) corresponding Fast Fourier Transform (FFT) [-1010](-12-10)$_{Mg}$‖[-1010](-12-10)$_{\beta''}$.

The Zn_2Zr_3 rods, distributed in the grain interior, serve as additional nucleation sites for precipitates. The tetragonal Zn_2Zr_3 particles (*P4₂/mnm*: a=0.768nm and c=0.699nm) which seem to have a near octagon cross-section (Figure 10), are elongated along [001] direction, have an orientation relationship with the Mg matrix [-2110](0001) $_{Mg}$‖[001](110) $_{Zn2Zr3}$ and appear to be stable in the Mg matrix.

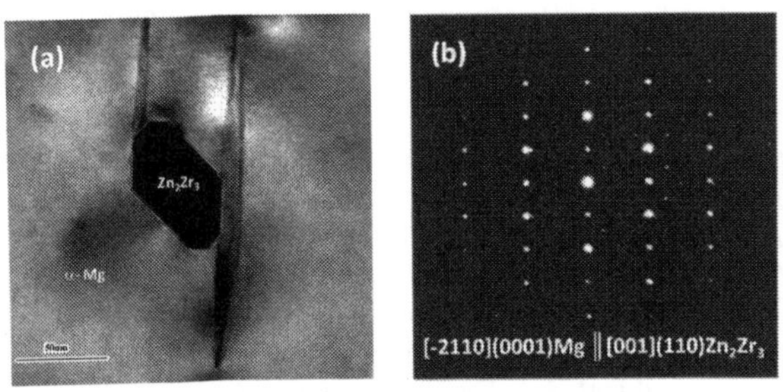

Figure 10. (a) BF TEM micrograph of Zn_2Zr_3 rod particle elongated to [001] direction; (b) corresponding SAED of Mg and Zn_2Zr_3 with the OR [-2110](0001)$_{Mg}$ ‖[001](110)$_{Zn2Zr3}$.

Nd-rich plate-like precipitates nucleate on the side and basal planes of the Zn_2Zr_3 as identified by EDS mounted on the TEM (Figures 11, 12). It is accompanied by the formation of precipitation free zones (PFZs) near Zn_2Zr_3 rod particles.

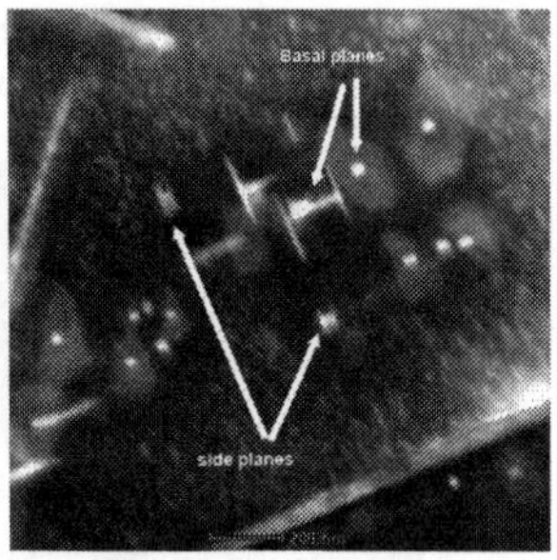

Figure 11. DF STEM micrograph of Zn_2Zr_3 particles with Nd-rich plate-like precipitates on their basal and side planes.

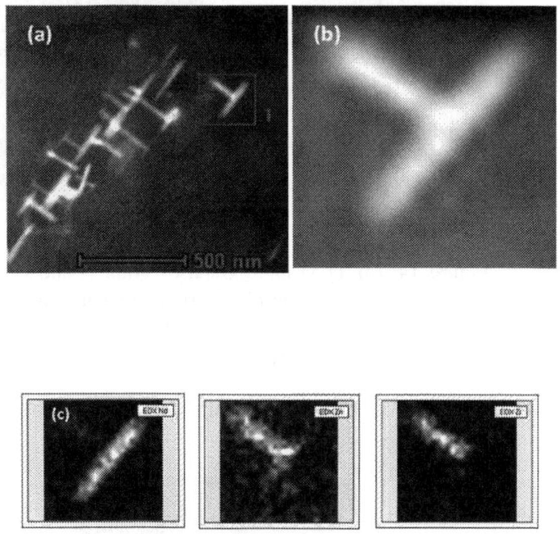

Figure 12. (a) DF STEM micrograph of Zn_2Zr_3 with Nd-rich plates on the basal and side planes; (b) EDS mapping area; (c) mapping of elements by EDS.

These plate-like precipitates were identified by SAED as β'$(Mg_3Nd)_{FCC}$ (DO_3 structure) with the following OR [0001](2-1-10)$_{Mg}$‖[101](11-1)$_{\beta'}$ (Figure 13).

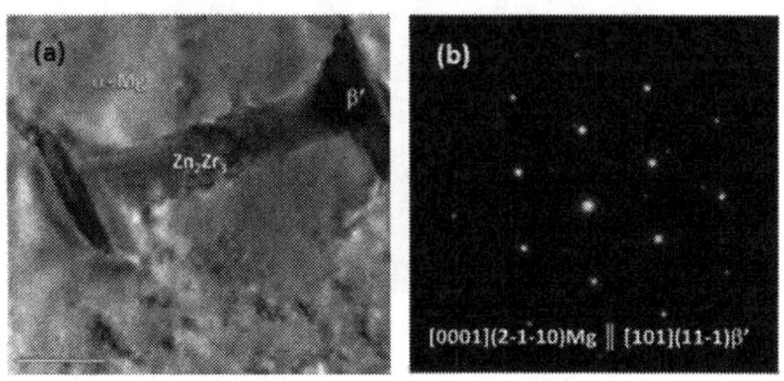

Figure 13. TEM micrograph and corresponding SAED of β' precipitate nucleated on Zn_2Zr_3 rod particle.

The β'-phase was also found in the grain boundary regions. Figure 14 provides TEM micrograph and corresponding SAED from the grain boundary region. A small number of large precipitates (few hundreds of nm in size) and a high density of fine β'-precipitates were observed.

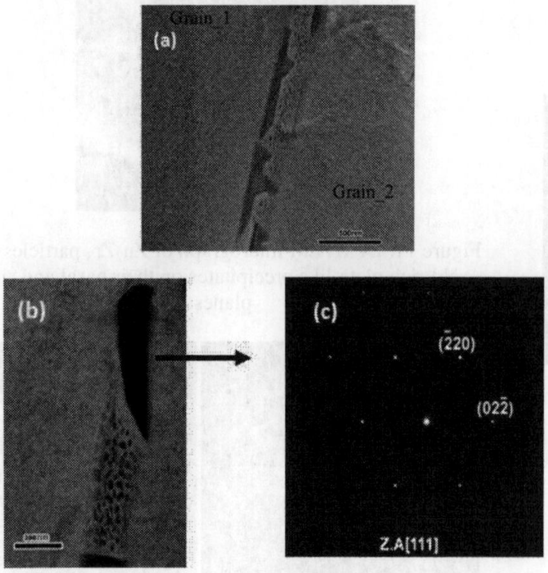

Figure 14. (a) TEM micrograph of grain boundary region; (b) BF TEM micrograph of β' precipitate in the grain boundary region; (c) corresponding SAED from large β'(Mg_3Nd)$_{FCC}$ ZA [111].

It was proved by HRTEM analysis (Figure 15) that the small precipitates (5-50nm in size) are β' as well.

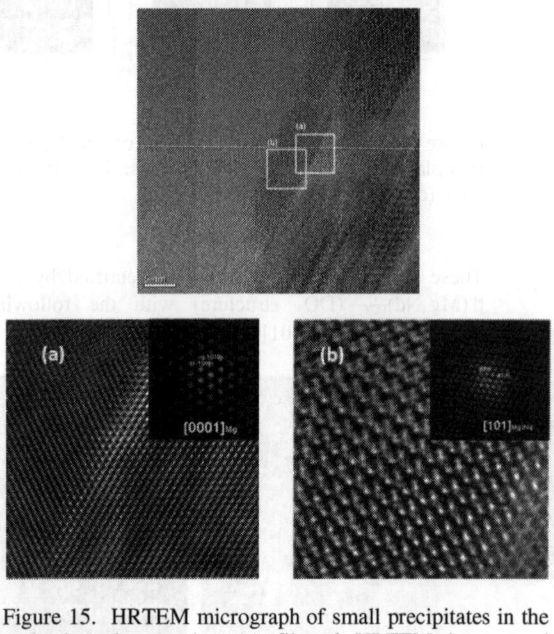

Figure 15. HRTEM micrograph of small precipitates in the grain boundary region (a) filtered HRTEM image and corresponding FFT of Mg matrix (b) filtered HRTEM image and corresponding FFT of β' precipitate

After 32 days of aging, it was found that β" phase transforms into β' in the grain interior and β' precipitates transform into β in the grain boundary regions.

4. Discussion

Precipitation during isothermal aging involves the formation of metastable phases β"(Mg_3Nd)$_{hcp}$ (DO_{19} structure) and β'(Mg_3Nd)$_{fcc}$ (DO_3 structure), and this is consistent with the precipitation sequence of [1,7]:

$$ssss(Mg) \rightarrow GP\ Zones \rightarrow \beta'' \rightarrow \beta' \rightarrow \beta .$$

Among these precipitates, the coherent β" and semicoherent β' are the primary strengthening phases while the β phase exists usually in an overaged condition and exhibits a lower strengthening effect. Zirconium-rich cores are often present in the center of the grains, when zirconium is used as a grain refiner. In addition, in an alloy containing Zn, Zn_2Zr_3 phase forms after solution treatment [8]. As found in the present research, the Zn_2Zr_3 rods are elongated along $[001]_{Zn2Zr3}$ direction and has OR with the Mg matrix $[-2110](0001)_{Mg} \| [001](110)_{Zn2Zr3}$ (Figure 10) in correspondence with the results of Gao et al. [9]. It means that Zn_2Zr_3 rods grow parallel to the basal planes of the Mg matrix. Therefore it probably has minor influence on the alloy hardness. However, they serve as additional nucleation sites for β" and β' plate-like precipitates, which may be perpendicular or parallel to the Mg basal planes (Figures. 11, 12). They may substantially influence basal and non-basal dislocation movement resulting in additional hardening. The β' plates adjacent to Zn_2Zr_3 are much bigger than β" plates distributed in the α-Mg matrix. It can be explained by their earlier nucleation followed by coarsening. The formation of PFZs around β'-precipitates confirms the coarsening process. At the peak aged condition, the β' precipitates are formed in the grain boundary regions, whereas β" are formed in the grain interior, as discussed above (Figures. 8, 14). Faster diffusion of Nd along grain boundaries may explain earlier transformation of β" into β' in the grain boundary region. High density of fine β' and a small number of large β' precipitates illustrate the coarsening process at the grain boundary area accompanied by formation of PFZs.

5. Conclusion

Eutectic phase $(Mg_{1-x}Zn_x)_{12}Nd$ with BCT crystal structure dissolves during ST, and small tetragonal Zn_2Zr_3 rod-like particles precipitate in the α-Mg matrix and near grain boundaries. Zn_2Zr_3 rod-like particles were found elongated along $[001]_{Zn2Zr3}$ direction and has the following orientation relationship: $[-2110](0001)_{Mg}//[001](110)_{Zn2Zr3}$. The Zn_2Zr_3 rods serve as additional nucleation sites for precipitation during aging. After 8 days of aging at 175°C, plate-like precipitates of a metastable β" (Mg_3Nd)$_{HCP}$ phase with DO_{19} structure formed in the grain interior. The β" precipitates are fully coherent with the Mg matrix with the OR $[-2110](0001)_{Mg} \| [-2110](0001)_{\beta''}$. Plate-like precipitates of β'(Mg_3Nd)$_{FCC}$ (DO_3 structure) phase were found in the grain boundary regions and on the side and basal planes of Zn_2Zr_3 rod particles. The β' precipitates are semi-coherent with the Mg matrix with the OR $[0001](2\text{-}1\text{-}10)_{MG} \| [101](11\text{-}1)_{\beta'}$. During the 16÷32 days of aging, β"- precipitates in the grain interior transform into the β' precipitates with an FCC structure. In the late stage of aging, the β'-precipitates transform into a stable incoherent β $(Mg,Zn)_{12}Nd$ phase.

References

1. Pike, T. J.; Noble, B. *Journal of the Less Common Metals* **1973,** 30, (1), 63-74.
2. Nie, J. F.; Muddle, B. C. *Acta Materialia* **2000,** 48, (8), 1691-1703.
3. Zheng, K. Y.; Dong, J.; Zeng, X. Q.; Ding, W. J. *materials Science and Engineering: A* **2008,** 489, (1-2), 44-54.
4. Penghuai, F.; Liming, P.; Haiyan, J.; Lan, M.; Chunquan, Z. *Materials Science and Engineering: A* **2008,** 496, (1-2), 177-188.
5. Apps, P. J.; Karimzadeh, H.; King, J. F.; Lorimer, G. W. *Scripta Materialia* **2003,** 48, 8), 1023-1028.
6. Wilson, R.; Bettles, C. J.; Muddle, B. C.; Nie, J. F. In *Precipitation hardening in Mg-3 wt%Nd(-Zn) casting alloys*, Materials Science Forum, 2003; 2003; pp 267-272.
7. Rokhlin, L. L., *Magnesium alloys containing rare earth metals*. 2003; Vol. 3.
8. Fu, P.; Peng, L.; Jiang, H.; Zhai, C.; Gao, X.; Nie, J. F. In *Zr-containing precipitates in Mg-3wt%Nd-0.2wt%Zn-0.4wt%Zr alloy during solutionTreatment at 540 °C*, Materials Science Forum, 2007; 2007; pp 97-100.
9. Gao, X.; Muddle, B. C.; Nie, J. F., Transmission electron microscopy of Zrâ€"Zn precipitate rods in magnesium alloys containing Zr and Zn. In Taylor & Francis: 2009; Vol. 89, pp 33 - 43.

Magnesium Technology 2011
Edited by: Wim H. Sillekens, Sean R. Agnew, Neale R. Neelameggham, and Suveen N. Mathaudhu
TMS (The Minerals, Metals & Materials Society), 2011

PRECIPITATION PROCESS IN Mg-Nd-Zn-Zr-Gd/Y ALLOY

J.H. Li[1,2*], G. Sha[2, 3], P. Schumacher[1, 4], S. P. Ringer[2, 3]

[1*] Chair of Casting Research, Department of Metallurgy, the University of Leoben, A-8700, Leoben, Austria (jiehua.li@mu-leoben.at)
[2] Australian Centre for Microscopy and Microanalysis, The University of Sydney, Madsen Building F09, Sydney, NSW 2006, Australia
[3] ARC Centre of Excellence for Design in Light Metals, The University of Sydney, NSW 2006, Australia
[4] Austrian Foundry Research Institute, Parkstrasse 21, Leoben, Styria, A-8700, Austria

Keywords: Mg-Nd-Zn-Zr-Gd/Y alloys; Age hardening; Precipitate; TEM; Atom probe tomography.

Abstract

Transmission electron microscopy (TEM) and atom probe tomography (APT) were employed to investigate the solute clusters and precipitates formed in different Mg-Nd-Zn-Zr-Gd/Y alloys aged at 200 °C up to 100 h. TEM characterizations confirmed the precipitation in these alloys involved the formation of precipitates such as β″, β′, β1 and β. Most precipitates habited on $(01\bar{1}0)_{\alpha\text{-Mg}}$ and a few thin plate (of 1-2 atomic layers in thickness) lying on $(0001)_{\alpha\text{-Mg}}$. APT analyses revealed that the precipitates were enriched with Nd, Zn and Gd. Moreover, Nd partitioned more strongly into the precipitates than Gd and Zn. In contrast, Y was less prone to partition into precipitates than other alloying elements in these alloys.

Introduction

Magnesium alloys have important applications in the automotive and aerospace industries because of their high specific strength for the weight reduction and better fuel economy [1]. However, the mechanical properties of conventional Mg alloys are often not suitable for high temperature applications. The addition of heavy rare-earth (HRE) elements, such as Nd, Gd, Y, Dy, Er, Sc, Tb and Sm etc, has been found to be effective to promote precipitation hardening and to improve the high temperature performance of Mg alloys [2-14]. Indeed, most high-strength Mg alloys such as WE54/43 and QE22 contain HRE elements. The level of alloying addition of the HRE elements is a critical concern in the alloy development and design of Mg alloys because of both material costs and the desire to have the alloy as light as possible. There is a strong interest in developing advanced Mg alloys with a low HRE addition.

The Mg-Nd-Zn based alloys [2, 3, 4], such as ZM-6 in China and ML10 in Russia, exhibit a strong age hardening response and have been used in various structural airframe components. Our previous researches [5, 6, 7] have revealed that the mechanical properties of Mg-Nd-Zn based alloy can be improved further through optimizing alloy composition and using proper heat-treatment, as well as the addition of Gd with a lower content. Specifically, an Mg-3.6Gd-2.8Nd-0.6Zn-0.4Zr (wt %, used through the paper, in case not specified otherwise) alloy has a yield strength and ultimate strength at 300 °C about 10% and 18% higher than those of an Mg-2.8Nd-0.6Zn-0.4Zr alloy, respectively. On the other hand, the addition of Y into magnesium alloys is also well-known to be one of the most effective ways to improve their mechanical properties at elevated temperatures. Mg-Y-Nd based alloys with higher Y contents, e.g. WE54 and WE43 [8], have always been regarded to be very important commercial Mg alloys. However, the addition of Y in high quantities is less attractive due to increasing alloy density and cost. Similar to the Gd addition into Mg-Nd-Zn based alloy, the only 0.2 % Y addition into Mg-2.8Nd-0.6Zn-0.4Zr based alloy has also been investigated in our previous researches [6, 7]. The Mg-2.8Nd-0.2Y-0.6Zn-0.4Zr alloy has also shown a significant improvement on tensile properties, especially the excellent tensile yield strength (TYS) at higher temperatures (200 °C -350 °C), compared with the Y-free Mg-2.8Nd-0.6Zn-0.4Zr based alloy. For example, the TYS (190 MPa) of the Y-containing alloy at room temperature is much higher than that of Y-free alloy (158 MPa) and the Gd-containing alloy with relatively higher Gd concentration 3.6 wt % (165 MPa). When tested at 200 °C, the TYS (183MPa) of the Y-containing alloy is also much higher than that of Y-free alloy (141MPa) and the Gd-containing alloy (143MPa).

The objective of this research is to characterize precipitate microstructure in order to understand their strengthening effect in these Mg-Nd-Zn-Zr-Gd/Y alloys. To date, there is a lack of information about the solute partitioning during precipitation in the alloys. In this contribution, transmission electron microscopy (TEM) and atom probe tomography (APT) were employed to characterize the solute clusters and precipitates in the Mg-Nd-Zn-Zr-Gd/Y alloys aged at 200 °C.

Experimental material and procedures

Two experimental alloys with compositions of Mg-3.6Gd-2.8Nd-0.6Zn-0.4Zr and Mg-2.8Nd-0.2Y-0.6Zn-0.4Zr were prepared with high purity Mg (99.9 %), Zn (99.9 %), Nd (99.9 %), Mg–28Gd, Mg–28Y and Mg–33Zr master alloys in an electric resistance furnace under the protection of an anti-oxidizing flux, and cast into a sand mould. The chemical compositions of the experimental alloys were determined by inductively coupled plasma atomic emission spectrum (ICP-AES) apparatus. Solution treatment were conducted at 525 °C for 18 h in a salt bath, then samples were quenched in cold water, and subsequently were aged in a oil bath at 200 °C. The Vickers hardness testing was undertaken on LECO Hardness Tester (LV700AT) with 1 kg load and 10 s dwelling time. Each data point reported in this paper represents an average of at least 10 measurements. The foil specimens for TEM were prepared by twin jet electro-polishing in a solution of 25 % $HClO_4$ and 75 % methanol at -40 °C with a voltage of 20 V and subsequently low energy beam ion thinning was used for surface cleanness. The TEM examinations were performed in a Philips CM12 operating at 120 kV and a high resolution TEM (JEOL-3000F) operating at 300 kV. The samples for atom probe analysis were prepared by micro-electro-polishing. Atom probe analyses were performed using an Imago LEAP™ 3000 operating at a specimen temperature of 20 K, 20 % pulse fraction and under ultrahigh vacuum conditions.

Results and discussions

Figure 1 shows the age hardening response for the three Mg alloys. The Gd-containing alloy and the Y-containing alloy demonstrate the higher hardening response than the base alloy. The hardness of the Gd-containing alloy increased quickly during the first 1.5 h at 200 °C and then reached plateau-like range before reaching peak hardness after 70 h. Further ageing led to over-ageing and a progressive decrease in hardness.

The age hardening response of the Y-containing alloy is higher than that of the base alloy, but lower than that of the Gd-containing alloy. This is consistent with the fact that the Y addition of 0.2 % in the Y-containing alloy is much lower than the Gd addition of 3.6 % in the Gd-containing alloy. It should be noted that the hardness of the Y-containing alloy also exhibits a strong age-hardening response during the first 1.5 h at 200 °C, with a hardness increase of 32 % from the initial hardness (58 HV) of the as-quenched sample. A similar age hardening response was also observed in other Mg alloys containing Nd [9, 10]. The significant ageing response of GN72 at 200 °C has been attributed to the rapid formation of β′, and possibly β″, precipitates during the early stages of ageing [10]. Our previous atom probe tomography data analysis [11] has indicated that the fast age-hardening response is directly correlated to the stronger partitioning of Nd in the Mg-Nd based alloy. Further ageing leads to a maximum hardness of 76 HV at about 14 h and then a progressive decrease in hardness.

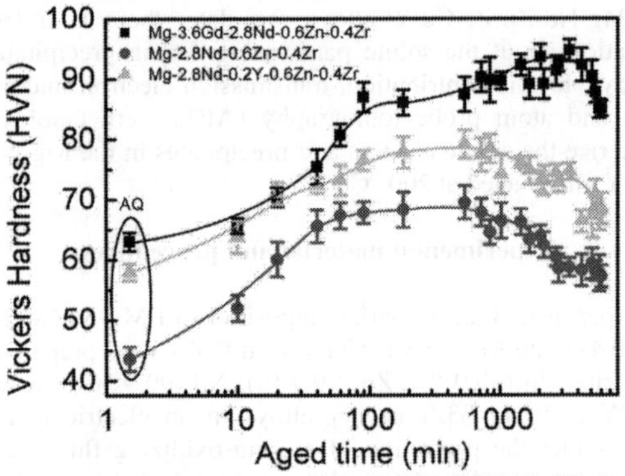

Fig. 1. Age hardening curve of Mg-2.8Nd-0.6Zn-0.4Zr (wt %) alloy aged at 200 °C. The results of the 0.2 % Y-containing and 3.6 % Gd-containing alloys are also shown.

Figure 2 is a series of representative $[01\bar{1}0]_{\alpha\text{-Mg}}$ bright field (BF) TEM images and the corresponding selected area diffraction patterns (SADP) of Mg-3.6Gd-2.8Nd-0.6Zn-0.4Zr alloy aged for 3 h, 14 h and 70 h at 200 °C. After ageing for 3 h, Fig. 2a reveals that the precipitates are in a high number density and are small in their sizes. The corresponding $[01\bar{1}0]_{\alpha\text{-Mg}}$ SADP provided in Fig. 2b indicates that the precipitates have a general habit plane or trace parallel to $\{2\bar{1}\bar{1}0\}_{\alpha\text{-Mg}}$ and the weak diffraction streaks at $½(2\bar{1}\bar{1}0)_{\alpha\text{-Mg}}$ and $½(2\bar{1}\bar{1}4)_{\alpha\text{-Mg}}$ (marked with a white arrow) are consistent with evidence for β″ precipitation, since this phase possesses a DO_{19} crystal structure with a hexagonal unit cell of a=b=0.64 nm and c=0.52 nm [8] and stoichiometry Mg_3X [13]. The extremely weak diffraction at $½(2\bar{1}\bar{1}2)_{\alpha\text{-Mg}}$, marked with a black arrow, suggests that a low number density of β′ precipitates co-exist with the more numerous β″ precipitates at this stage of ageing. The precipitate phase β′ is known to have a base-centred

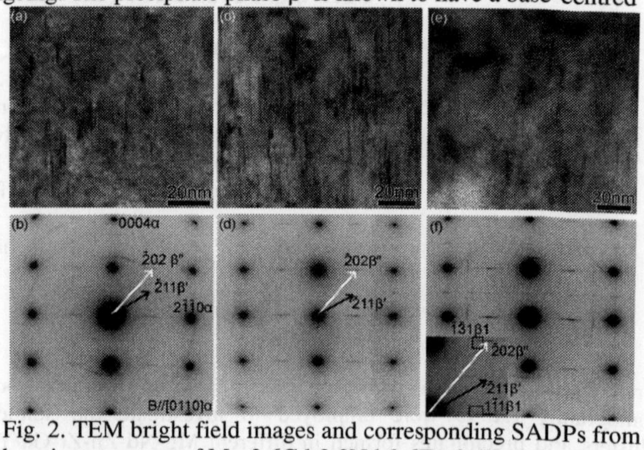

Fig. 2. TEM bright field images and corresponding SADPs from the microstructure of Mg-3.6Gd-2.8Nd-0.6Zn-0.4Zr (wt. %) alloy samples aged at 200 °C for various times: (a-b) 3 h, (c-d) 14 h and (e-f) 70 h. Incident electron beam direction is parallel to $[01\bar{1}0]_{\alpha}$.

orthorhombic unit cell a=0.640 nm, b=2.223 nm, c=0.521 nm [8] and stoichiometry Mg_5X [13]. After ageing for 14 h, stronger diffraction effects at $½(2\bar{1}\bar{1}2)_{\alpha\text{-Mg}}$ were evident in the $[01\bar{1}0]_{\alpha\text{-Mg}}$ SADP (marked with a black arrow in Fig. 2d), suggesting that β″ precipitates have transformed into or nucleated β′. After ageing for 70 h, a weak diffraction spot clearly adjacent to $½(2\bar{1}\bar{1}0)_{\alpha\text{-Mg}}$, as marked with a black box in Fig. 2f, indicates the presence of β_1 in the microstructure. The β_1 phase possesses a face-centred cubic unit cell with a=0.72 nm and stoichiometry of Mg_3X [8, 13]. The orientation relationships between the precipitates such as β″, β′, β_1 and the matrix are in agreement with those reported previously [8, 12, 13]. Thus, the precipitation sequence in Gd-containing alloy during ageing is supersaturated solid solution (SSSS) → β″ ($D0_{19}$) → β′ (bco) → β_1 (fcc).

A similar precipitation reaction was also observed in the Y-containing alloy. Fig. 3 shows a series of representative bright field (BF) TEM images and the corresponding SADP obtained from the Y-containing alloy samples after ageing at 200 °C at 14 h. After ageing at 200 °C for 14 h, the morphology of precipitates become easily distinguished in TEM images, as shown in Figs.3a and c. The $[0001]_{\alpha\text{-Mg}}$ and $[01\bar{1}0]_{\alpha\text{-Mg}}$ SADPs show some diffraction spots in the midway of those from the reflection of Mg matrix ($[01\bar{1}0]_{\alpha\text{-Mg}}$ or $[2\bar{1}\bar{1}0]_{\alpha\text{-Mg}}$), as shown in Figs.3b and d, respectively, suggesting that the β″ precipitates with $D0_{19}$ structure (a=b=0.64 nm and c=0.52 nm) still existed in α-Mg matrix after ageing for 14 h, as marked with a black solid arrow in Fig. 3d. The extremely weak diffractions at $½(2\bar{1}\bar{1}1)_{\alpha\text{-Mg}}$ in $[01\bar{1}0]_{\alpha\text{-Mg}}$ SADP, as marked with a white solid arrow in Fig. 3d, suggest that β′ precipitates co-existed with β″ precipitates at this stage of ageing. The precipitate phase β′ was also known to have a base-centred orthorhombic unit cell a = 0.640 nm, b = 2.223 nm, c= 0.521 nm [8], and has a lamella-like morphology with a longitudinal axis parallel to $[0001]_{\alpha\text{-Mg}}$. In contrast to the Fig. 2d, there is no clear diffraction spot adjacent to $½(2\bar{1}\bar{1}0)_{\alpha\text{-Mg}}$ in $[01\bar{1}0]_{\alpha\text{-Mg}}$ SADP, as marked with a white box in Fig. 3d. However, the HRTEM images and corresponding fast Fourier transform (FFT), as shown in Figs. 4b and c, indicates that the β_1 precipitate does exist in the microstructure. It should be noted that the FFT analysis was obtained from the areas only containing β_1 precipitate. As shown in Fig. 4a, the β_1 precipitates seam to be isolated.

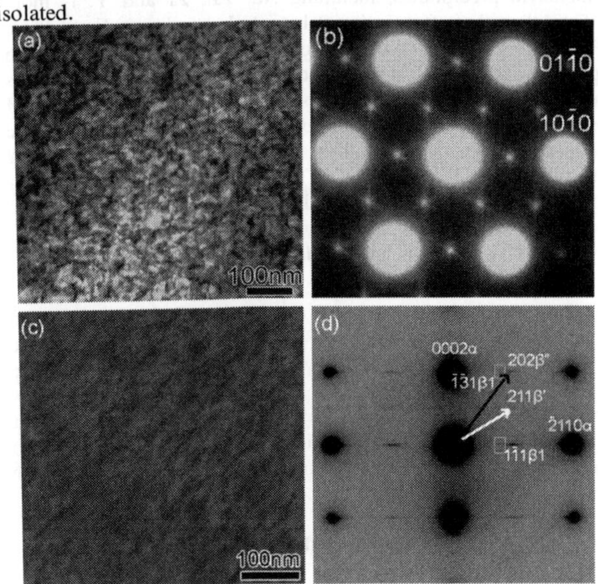

Fig. 3. TEM (bright field) images and SADPs of Mg-2.8Nd-0.2Y-0.6Zn-0.4Zr (wt %) alloy aged at 200 °C for 14 h. (a), (b) Incident electron beam direction is parallel to $[0001]_{\alpha\text{-Mg}}$; (c), (d) Incident electron beam direction is parallel to $[01\bar{1}0]_{\alpha\text{-Mg}}$.

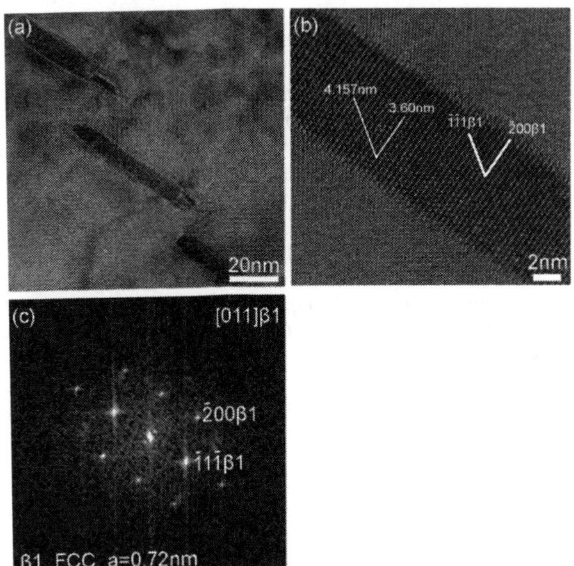

Fig. 4. Low magnification (a), high magnification (b) HRTEM image and FFT (c) of the precipitates in Mg-2.8Nd-0.2Y-0.6Zn-0.4Zr (wt %) alloy aged at 200 °C for 14 h. Incident electron beam direction is parallel to $[0001]_{\alpha\text{-Mg}}$.

No β' precipitates are associated with them at all. This is different from the report in [8, 12, 13] in which the β_1 phase is suggested to form via an invariant plane strain transformation of the magnesium lattice, with the specific morphology of the rhombic shape connecting with the β' phase. The β' phase present in contact with the smaller facets of the β_1 plates were suggested to be a consequence of shear strain accommodation. As suggested in [10], the nucleation of the β_1 phase may be independent of β' globules, although the β_1 phase is also connecting with β' phase in all of the alloys investigated in [10]. The formation of the β_1 precipitate may be correlated to a solute-rich region between the two β' phases caused by the nucleation of β' phases (Mg_5X) on the existing β'' plates along the length of β'' phase (Mg_3X). This solute-rich region provides an idea nucleation sites for the β_1 phase. The alloying elements partitioning into the precipitates can be further confirmed with our atom probe analysis.

The equilibrium β phase was not observed in both the Y-containing alloy and the Gd-containing alloy after ageing at 200 °C for 14 h and 70 h, respectively. We suppose that the equilibrium β phase, which possess a face-centred cubic unit cell with a=2.223 nm and stoichiometry of Mg_3X [8] eventually precipitates in this system, though it was not observed over the time-scale of our ageing experiments at 200 °C. Thus, the precipitation sequence in the Gd-containing alloy during ageing is, supersaturated solid solution (SSSS) → β'' ($D0_{19}$) → β' (bco) → β_1 (fcc)→ β (fcc). We propose that the β'' precipitates are primarily responsible for the age hardening effects at 200 °C observed in Fig.1 up to around 3 h ageing whereafter the effect of β' precipitates predominates.

Viewed from $[11\bar{2}0]_{\alpha\text{-Mg}}$, a few thin precipitates on the basal plane are also observed in Y-containing alloy after aged at 200 °C up to 14 h. As marked with white solid arrow in Figs. 5a and b, these precipitates are only about 2 atomic layers in thicknesses in Y-containing alloy. For clarity, the β series precipitates are also marked with a white box in Figs. 5a and b. With increasing ageing time, the length increases from about 10 nm (3 h) to about 20 nm (14 h), but the thickness does not change greatly. The very thin precipitates on the basal plane have been observed in [14] and designated as the γ series precipitates. The further investigations about the γ series precipitates in this alloy system are in progress.

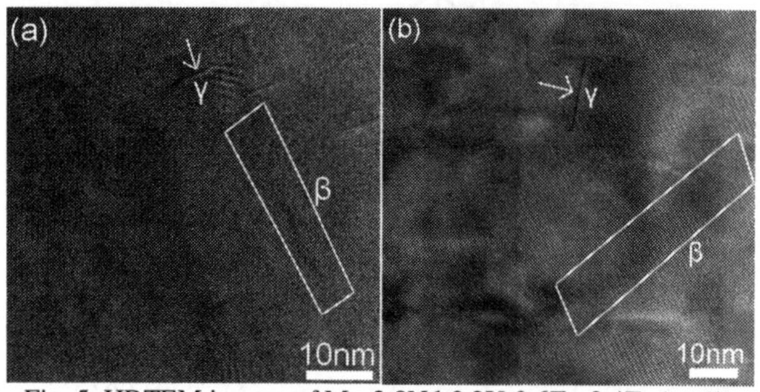

Fig. 5. HRTEM images of Mg-2.8Nd-0.2Y-0.6Zn-0.4Zr (wt %) alloy aged at 200 °C for 3 h and 14 h, respectively. (a) 3 h; (b) 14 h, Incident electron beam direction is parallel to $[11\bar{2}0]_{\alpha\text{-Mg}}$.

Some very thin precipitates on the basal plane are also observed in Gd-containing alloy after aged at 200 °C for 14 h, as shown in Figs. 6d, e and f. The γ precipitates are perpendicular to the β precipitates, and are rich in Nd, Gd and Zn. Figs. 6 a, b and c also provides a series of three-dimensional atom maps recorded from our atom probe experiments on specimens of the Mg-3.6Gd-2.8Nd-0.6Zn-0.4Zr alloy after ageing at 200 °C for (a) 3 h (b) 14 h and (c) 70 h. It is clear that all three of the main solute elements, Nd, Gd and Zn are directly imaged within the precipitates. By examining the precipitates from different directions, it is also clear that most precipitates are elongated with their longitudinal axis parallel to $[0001]_{\alpha\text{-Mg}}$ and this is in agreement with previous TEM observations for β'', β' and β_1. From the $[0001]_{\alpha\text{-Mg}}$ view direction, it is clear that the precipitate size increases with ageing time. It is also noteworthy that the number density of precipitates appears to

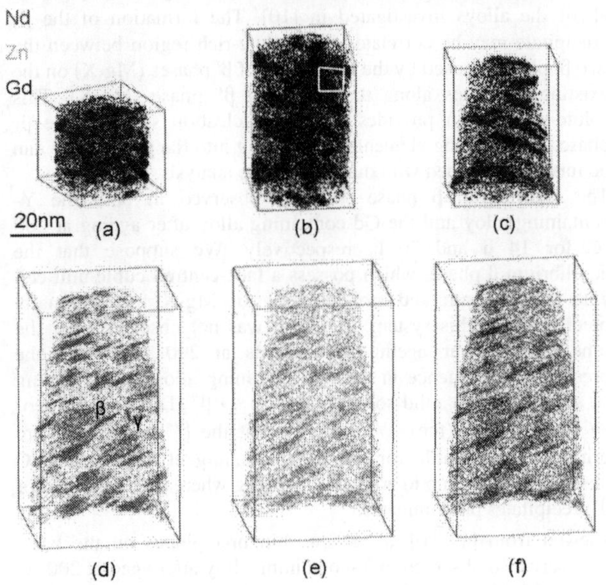

Fig. 6. Combined atom maps of Nd (red), Gd (blue) and Zn (green) obtained from Mg-3.6Gd-2.8Nd-0.6Zn-0.4Zr alloy samples aged at 200 °C for (a) 3 h, (b)14 h and (c) 70 h, in a view direction close to the $[0001]_{\alpha\text{-Mg}}$ zone axis, and (d), (e) and (f) are high resolution atom maps of a small region in the 14 h sample (marked with white box in Fig.6(b)).

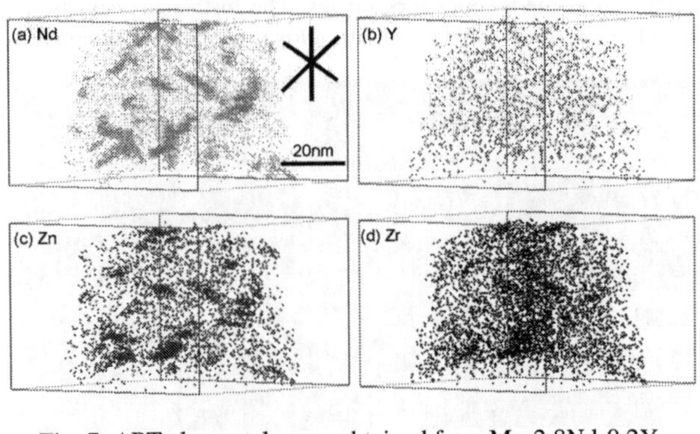

Fig. 7. APT elemental maps obtained from Mg-2.8Nd-0.2Y-0.6Zn-0.4Zr (wt %) alloy aged at 200 °C for 14 h. (a) Nd (green); (b) Y (blue); (c) Zn (red); (d) Zr (purple)

be highest in the sample aged 3 h and that this progressively decreases when compared to the precipitate number density after ageing for 14 and 70 h. In fact, our analysis of total number density (not shown here) reveals that there are less precipitates by a factor of 10 in the material aged 70 h at 200 °C compared to the material aged 3 h and yet the hardness is 10% higher in the 70 h material. Consistent with other reports [8, 13], this suggests that the β' and β_1 are more potent hardening obstacles than the β'' phase.

In contrast, the Y solute element shows less prone to partition into precipitates than other alloying elements in Y-containing alloy. After ageing at 200 °C for 14 h, the Y solute element is still distributed in the α-Mg matrix, rather than partition into these precipitates, as shown in Fig. 7, although the distribution of solute elements in precipitates, including Nd, Zn, Zr and Y, is more uniform than that after ageing 3 h (not shown here). From these APT element maps, especially the element maps of Nd shown in Fig. 7a, it can be easily found that the length of the precipitates is about 20nm (14 h). This is consistent with our previous TEM observation (Fig. 3). More importantly, the distribution of the precipitates is nearly along three different directions, as marked with black solid line in Fig. 7a. This is also consistent with our previous TEM observation viewed from $[0001]_{\alpha\text{-Mg}}$ (Fig. 3a).

Conclusions

The precipitation microstructure evolution and the precipitation sequence in the Mg-Nd-Zn-Zr-Gd/Y alloys involve in the formation of phase as β'', β', β_1 and the γ precipitates. These precipitates are enriched with Nd, Zn and Gd. Y appears to be less prone to partition into precipitates than any other alloying element in these alloys. A combined addition of Gd and Y into Mg-Nd-Zn alloy can be expected to be a promising way to further improve the age hardening response.

Acknowledgements

The authors are grateful for scientific and technical input and support from the Australian Microscopy & Microanalysis Research Facility (AMMRF) node at the University of Sydney. Jiehua Li also wishes to thank the China Scholarship Council for the financial support to his study in the University of Sydney.

References

[1] B.L.Mordike, *"Creep-resistant Magnesium Alloys,"* Mater. Sci. Eng. A ., 324 (1-2) (2002), 103-112.

[2] T.J. Pike, B Noble, *"The Formation and Structure of Precipitates in a Dilute Magnesium-Neodymium Alloy,"* J. Less Common Metals., 30 (1) (1973), 63-74.

[3] P.A. Nuttall, T.J. Pike, and B Noble, *"Metallography of dilute Mg-Nd-Zn alloys,"* Metallography., 13 (1980), 3-20.

[4] L.R. Gill, G.W. Lorimer, P. Lyon, *"The Effect of Zinc and Gadolinium on the Precipitation Sequence and Quench Sensitivity of Four Mg-Nd-Gd alloys,"* Adv. Eng. Mater., 9 (9) (2007), 784.

[5] J.H. Li, W.Q. Jie, G.Y. Yang, *"Effect of Gadolinium on the Aging Hardening Behavior, Solidification Microstructure and Mechanical Properties of Mg-Nd-Zn-Zr Cast Magnesium Alloys,"* Trans. Nonferrous Met. Soc. China., 18 (2008), s27-32.

[6] J.H. Li, W.Q. Jie, G.Y. Yang, *"Effect of Gadolinium on Microstructures and Mechanical Properties of Mg -Nd-Zn-Zr Cast Magnesium Alloys,"* Rare Metal Materials and Engineering, 37 (2008), 1751-1755.

[7] J.H. Li, W.Q. Jie, G.Y. Yang, *"Influences of Alloying Element Zn on the Microstructure and Mechanical Properties of GW series Magnesium Alloys,"* Acta Metallurgica Sinca. 10 (43) (2007), 1077-1081.

[8] J.F. Nie, and B.C. Muddle, *"Characterisation of Strengthening Precipitate Phases in a Mg–Y–Nd Alloy"*, Acta Mater., 48 (8) (2000), 1691-1703.

[9] K.Y. Zheng, et al., *"Precipitation and its Effect on the Mechanical Properties of a cast Mg–Gd–Nd–Zr Alloy,"*

Mater. Sci. Eng. A., 489 (1-2) (2008), 44-54.

[10] P.J. Apps, et al., *"Precipitation Reactions in Magnesium-Rare Earth Alloys Containing Yttrium, Gadolinium or Dysprosium,"* Scripta Mater., 48 (8) (2003), 1023-1028.

[11] G Sha, et al., "Atom Probe Tomography Characterization of Early-stage Precipitates in an Mg-Gd-Nd-Zn Alloy," *8th International Conference on Magnesium Alloys and Their Application.*, ed. K.U. Kainer, (Wiley-VCH, 2009), 40-45.

[12] T Honma, et al., *"Chemistry of Nanoscale Precipitates in Mg–2.1Gd–0.6Y–0.2Zr (at.%) Alloy Investigated by the Atom Probe Technique,"* Mater. Sci. Eng. A. 395 (1-2) (2005), 301-306.

[13] T Honma, et al., *"Effect of Zn Additions on the Age-Hardening of Mg–2.0Gd–1.2Y–0.2Zr Alloys,"* Acta Mater., 55 (12) (2007), 4137-4150.

[14] J.F. Nie et al., *"Solute Segregation and Precipitation in a Creep-resistant Mg–Gd–Zn Alloy,"* Acta Mater., 56 (20) (2008), 6061-6076.

Magnesium Technology 2011
Edited by: Wim H. Sillekens, Sean R. Agnew, Neale R. Neelameggham, and Suveen N. Mathaudhu
TMS (The Minerals, Metals & Materials Society), 2011

MECHANICAL PROPERTIES AND MICROSTRUCTURES OF TWIN ROLL CAST Mg-2.4Zn-0.1Ag-0.1Ca-0.16Zr ALLOY

C.L. Mendis[1] J. H. Bae[2] N.J. Kim[2] and K. Hono[1]

[1]National Institute for Materials Science, Tsukuba, Japan

[2]Pohang University of Science and Technology (POSTECH), Pohang, Korea

Keywords: Twin roll casting, Precipitation hardening, Microstructure

Abstract

In our previous study, we reported that the additions of 0.1 at.% Ag and 0.1 at.% Ca to Mg-2.4Zn-0.16Zr (at.%) alloy enhanced the age hardening response, and extruded alloy showed tensile yield strength of 325 MPa with the T6 heat treatment. Considering its excellent age hardenability, we attempted to develop high strength sheets from the alloy by twin-roll casting (TRC). TRC sheet of 2 mm in thickness were hot rolled to ~1.2 mm. The TRC Mg-2.4Zn-0.1Ag-0.1Ca-0.16Zr alloy sheet showed tensile yield strength of ~ 320MPa and an elongation to failure of 17% after T6 heat treatment. EBSD study indicated the average grain size is ~18±2.5μm and the grains have a weak basal texture. TEM, showed a uniform distribution of ~5 nm diameter $MgZn_2$ phase. The high yield strength was attributed to the dispersion of rod-like precipitates.

Introduction

Magnesium alloys receive continued attention as a structural material from the transport and personal electronics sectors as weight reduction becomes more important [1]. Magnesium cast products have been recently used in automotive settings as part of engine blocks and other components in the power train. However, use of wrought magnesium alloys in structural components such as space frames or body panels has not been commercially realized at present. This is mainly due to the lack of formability and lower final strength of magnesium alloys as compared with aluminium alloys. Hot rolled magnesium alloys have shown appreciable strength but they have a strong basal texture which severely affects the deformability [2-3]. Twin roll casting (TRC) has been considered to be a viable option to developing magnesium sheets but with significantly lower basal texture [4-5]. It has also been reported that the microstructure features such as grain size is significantly refined due to the relatively fast solidification rates associated with the TRC process [5].
Recently, we reported that the trace addition of 0.1at% Ag and 0.1at% Ca to Mg-2.4Zn-0.2Zr (at%) alloy significantly enhanced the age hardening response of the base alloy by the refinement of rod like $MgZn_2$ precipitates [6]. The extruded and heat treated Mg-2.4Zn-0.1Ag-0.1Ca-0.16Zr (at%) alloy showed a yield strength of 325 MPa, ultimate tensile strength of 360 MPa after a T6 heat treatment [7]. The high yield strength was attributed to the dispersion of fine rod-like $MgZn_2$ precipitates that formed during ageing at 160°C, and the high ultimate tensile strength was attributed to the fine grain structure of about ~500 nm that formed by the dynamic crystallization during the extrusion process. The number of applications that require sheet products is expected to be larger than that of the extruded products. The extruded and heat treated Mg-2.4Zn-0.1Ag-0.1Ca-0.2Zr alloy has a tensile yield strength of 325 MPa, ultimate tensile strength of 360MPa and ductility of ~14%. This is comparable to that of many automotive aluminium alloys. Therefore, the development of a high strength magnesium alloy with appreciable ductility sheet while retaining the yield strength will expand the number of possible applications. In this contribution we report the mechanical properties and microstructure of twin roll cast and rolled Mg-2.4Zn-0.1Ag-0.1Ca-0.2Zr sheet.

Experimental procedure

An alloy with a nominal composition of Mg-2.4Zn-0.1Ag-0.1Ca-0.1Zr (at%) (Mg-6.3Zn-0.16Ca-0.5Ag-0.05Zr (wt%)) was prepared by induction melting in a steel crucible in an argon atmosphere. The alloy was then re-melted under a CO_2 and SF_6 mixture and transferred into a pre heated tundish held at 680-700°C. The roll gap in the twin roll caster was set at 2 mm and a roll speed of 4 m/min was used to cast sheets of ~2 mm in thickness. The cast alloy was homogenized at 350°C for 48 h and then hot rolled at 350°C with a total reduction of ~50% to the final thickness of 1.2 mm. The rolled samples were solution heat treated for 0.5 h at 400°C quenched into cold water and aged at 160°C. The hardness response of the alloys was measured using Vickers hardness testing with a load of 0.5 kg with an average of 10 measurements. Tensile properties of TRC and rolled and heat treated samples were measured using flat tensile specimens with gauge length of 12.5 mm, gauge thickness of 1 mm and gauge width of 5 mm at a strain rate of 6.4×10^{-4} s^{-1}.
The microstructures of the alloy were characterized with optical microscopy, scanning electron microscopy and transmission electron microscopy. TEM observations were conducted using a FEI Tecnai F30 microscope operating at 300kV. TEM specimens were prepared using twin-jet electo-polishing at -50°C, with a polishing voltage of 90V and a current of ~0.8mA in a solution consisting of 15.9 g of LiCl 33.48 g $Mg(ClO_3)_2$ 300 ml of 2-butoxy-ethanol in 1500 ml methanol. The electron back scattered diffraction (EBSD) was used to evaluate the micro-texture observed for TRC and rolled alloys.

Results and discussions

<u>Microstructures of TRC processed alloy</u>
Optical micrographs of twin roll cast (TRC) and SEM and TEM micrographs of TRC and rolled sheets are shown in Figure 1. The TRC sheet consisted of cellular microstructure with cell size of approximately ~ 20±5 μm, Fig 1 (a and b). The microstructure parallel to the rolling plane showed cellular structure that is typical of cast microstructures, Fig 1 (a) while the microstructure perpendicular to the rolling plane, Fig 1 (b) showed that cellular structure was elongated along the rolling plane illustrating that some of the microstructure is deformed during the twin roll casting. The TRC microstructures also showed a semi-continuous network of intermetallic particles. Following hot rolling at 350°C, the amount of intermetallic particles observed in the

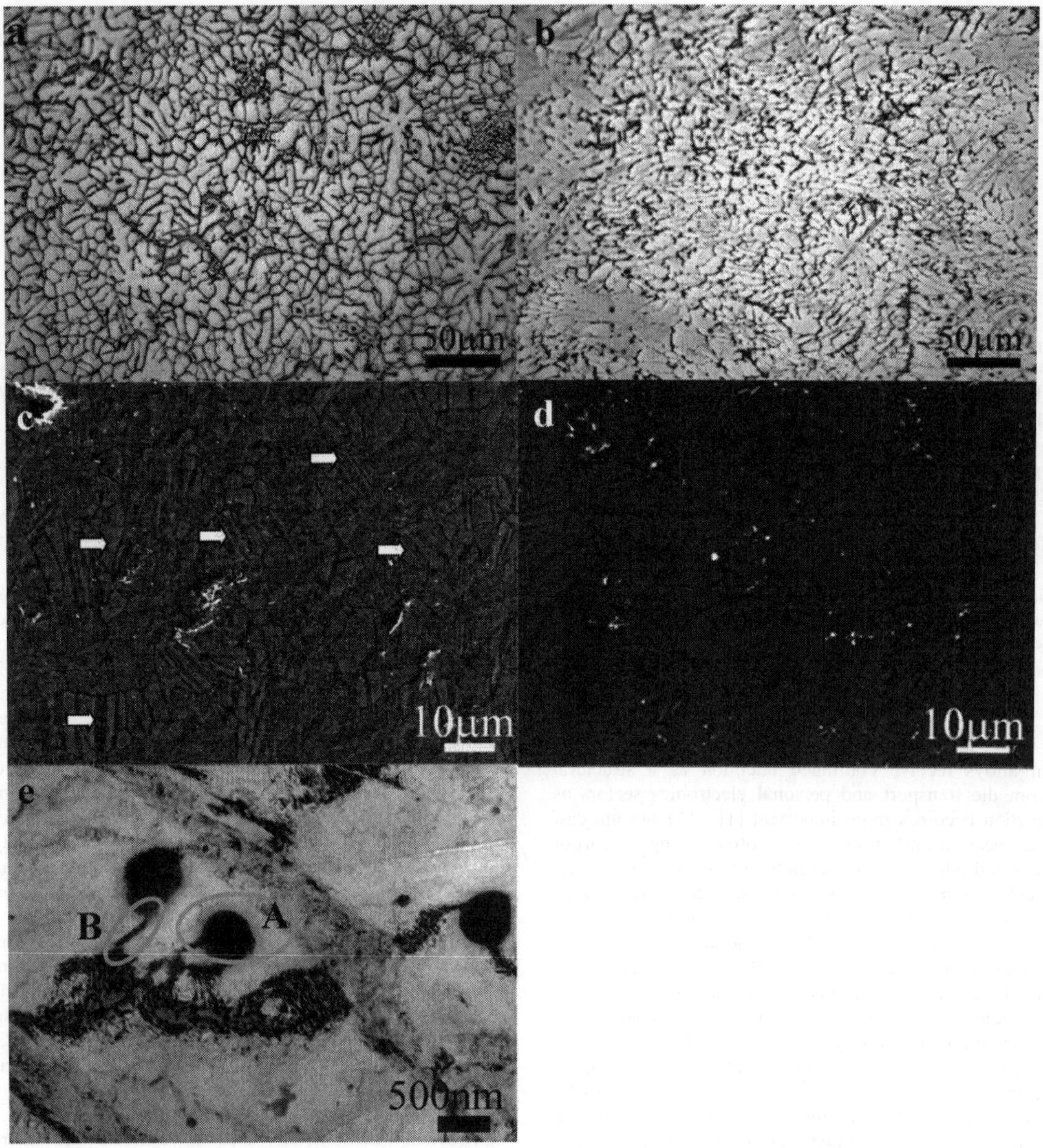

Figure 1 The Mg-2.4Zn-0.1Ag-0.1Ca-0.1Zr alloy (a, b) optical microstructure of TRC alloy; (c, d) SEM microstructure of TRC and rolled alloy and (e) TEM image of the TRC and rolled alloy showing precipitates observed, where A and B represent the two types of precipitates observed in the microstructure . (a and c) are parallel to the rolling plane while (b and d) are perpendicular to the rolling plane. The arrows in (c) indicate twinning observed in the microstructure.

microstructure decreased significantly, Fig 1 (c and d). The grain size remained approximately 18±5 μm, Fig 1 (c,d) following rolling. There is a large number of twins in the rolled samples and this can be clearly observed on the microstructure parallel to the rolling plane, Fig 1 (c). This showed that the rolling reductions did not result in extensive recrystallization and grain growth during hot rolling. This can be contrasted with the hot extrusion results reported previously where a fine recrystallized grain size without significantly twining activity [7].

The TRC and hot rolled samples were examined with TEM and showed that there is multiple twinning within grains. The TEM results for the TRC and rolled alloy, Fig 1 (e). . Some coarse precipitates were also observed in the rolled microstructure, marked A and B. The electron diffraction patterns recorded from the coarse precipitate particles will be reported elsewhere. The spherical shaped particles, marked as A, were characterized as Mg_7Zn_3 orthorhombic phase (space group Immm a=1.403 nm, b=1.048 nm and c=1.449 nm) [8]. The lath-like particles designated B are $MgZn_2$ phase, (hexagonal with space group

P63/mmc a=0.521 nm and c=8.78 nm [9]). However, unlike with the as-extruded Mg-2.4Zn-0.1Ag-0.1Ca-0.16Zr alloy [7, 11], there are no fine rod-like precipitates of Mg(Zn,Zr)observed in the microstructure of the TRC and rolled alloy.

The EBSD orientation map recorded from the TRC and rolled alloy in the rolling plane is illustrated in Figure 2. The orientation maps collected from the rolling plane showed that majority of the grains were oriented with basal plane parallel to the rolling plane, Fig 2 (a). However there are some grains with their orientation away from the basal orientation observed with the orientation map recorded parallel to the rolling plane. There are some grains with defined twins within the grains and these can be clearly seen with the orientation map. The inverse pole figure, Fig 2 (b) recorded for the TRC and rolled alloy showed that grains are not oriented with basal plane parallel to the rolling plane but ~ 10° from the rolling plane.

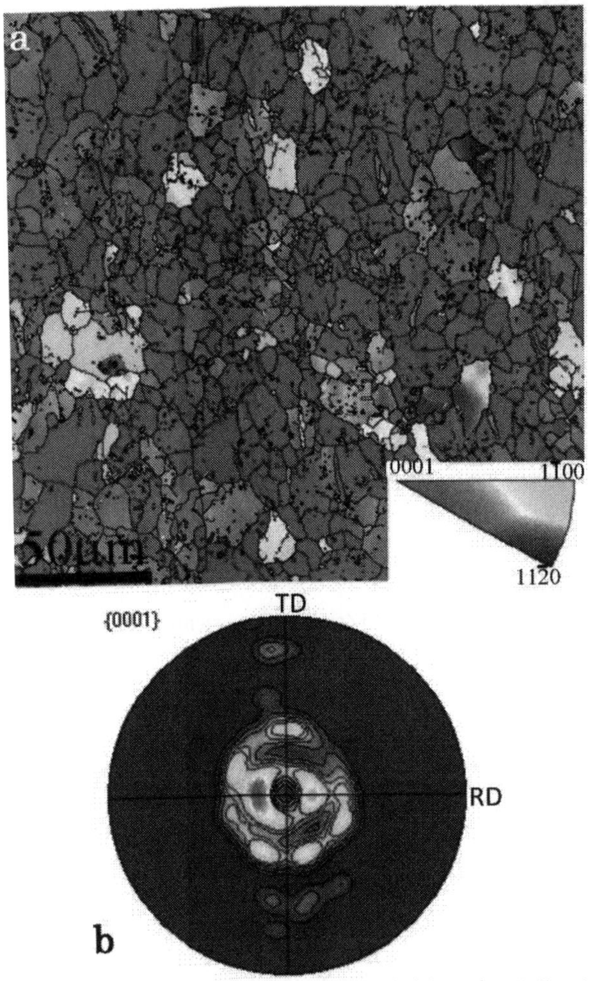

Figure 2 (a) EBSD orientation map recorded from the TRC and rolled alloy parallel to the rolling plane. And (b) the pole figure recorded from same EBSD orientation map showing the orientation of the $[0001]_{Mg}$ with respect to rolling plane.

Age hardening response and precipitate microstructures

The age hardening response of the TRC and hot rolled samples was measured following the solution heat treatment, Figure 3 (a). The hardness increases from as quenched hardness of ~62VH to a peak hardness of 96VH after ageing for 24 h at 160°C. The maximum hardness observed in the TRC alloy was similar to that observed for the cast and extruded alloy. However the time to reach peak hardness is shorter for the TRC alloy at 24h as compared with that for the extruded alloy at 48h [7].

The microstructure of the peak aged alloy samples were examined by transmission electron microscopy, Fig 3. It showed that the microstructure consisted two size of particles when observed with the electron beam parallel to the $[0001]_{Mg}$ showed that there is two different types of contrast from the precipitate particles, Fig 3 (b). The first is black in and was designated as diamond shaped precipitates and had a diameter of approximately 5±1.5 nm. The second is light grey, designated plate-like precipitates had a diameter of ~15±2.5 nm,. The microstructure observed was tilted approximately 40° from the $[0001]_{Mg}$ zone to show the microstructure perpendicular to the $[0001]_{Mg}$, Fig 3 (c). The black contrast precipitates now show elongated microstructure parallel to the $[0001]_{Mg}$ direction, thus these precipitates were deemed to be the rod-like precipitates of $MgZn_2$ phase. The diffraction patterns recorded from the black precipitates with $[0001]_{Mg}$ zone axis can be indexed according to $MgZn_2$ with the β_1' orientation relationship with magnesium as described in the previous investigations Fig 3 (d). The precipitates with grey contrast does not covert into rod-like microstructure but show more plate or lath-like structure when tilted away from the $[0001]_{Mg}$ zone axis . The electron diffraction patterns from these precipitates are indexed according to the $MgZn_2$ with the β_2' orientation relationship with magnesium, Fig 3 (e). The diameter of the rod-like precipitates observed in the TRC alloy could be compared with that for the extruded and heat treated Mg-2.4Zn-0.1Ag-0.1Ca-0.2Zr alloy [7] and the cast alloy [6, 7] and the number density of these precipitates (not reported here) is also similar regardless of the processing route.

Mechanical properties of TRC alloys

The tensile stress-strain curves of the twin roll cast and rolled alloys are shown in Fig. 4 (a). The TRC and rolled alloy showed a tensile yield strength (0.2%YS) of approximately 177 MPa with an ultimate tensile strength (UTS) of 285 MPa with a tensile strain to failure of approximately 28%. The tensile properties were lower than that for the as-extruded alloy [7]. The reasons for the lower yield strength is attributed to the significant differences observed in the microstructures; i.e., the extruded alloy contained much finer grain size compared to the TRC alloy [10] and a uniform distribution of fine scale precipitate distribution within the grains [7,10]. The fine scale precipitates found in the extruded alloy was not observed for the TRC and rolled alloy. However, the TRC and rolled alloy showed very high strength after solution treating at 400°C and ageing at 160°C for 24 h; i.e., yield strength of 320 MPa and the UTS of ~342 MPa with an elongation to failure of ~17%. The tensile properties reported for the TRC and rolled alloy is comparable with the extruded and heat treated alloy [7]. In both these alloys the major contribution to strengthening is the rod-like precipitate.

The tensile properties of Mg-2.4Zn-0.1Ag-0.1Ca-0.1Zr alloy is compared to ingot cast and rolled alloys [11] and other experimental TRC alloys as shown in Figure 4 (b). The tensile properties of the ingot cast and rolled Mg-Zn base alloys (ZM21) are comparable with the TRC Mg-2.4Zn-0.1Ag-0.1Ca-0.1Zr, but

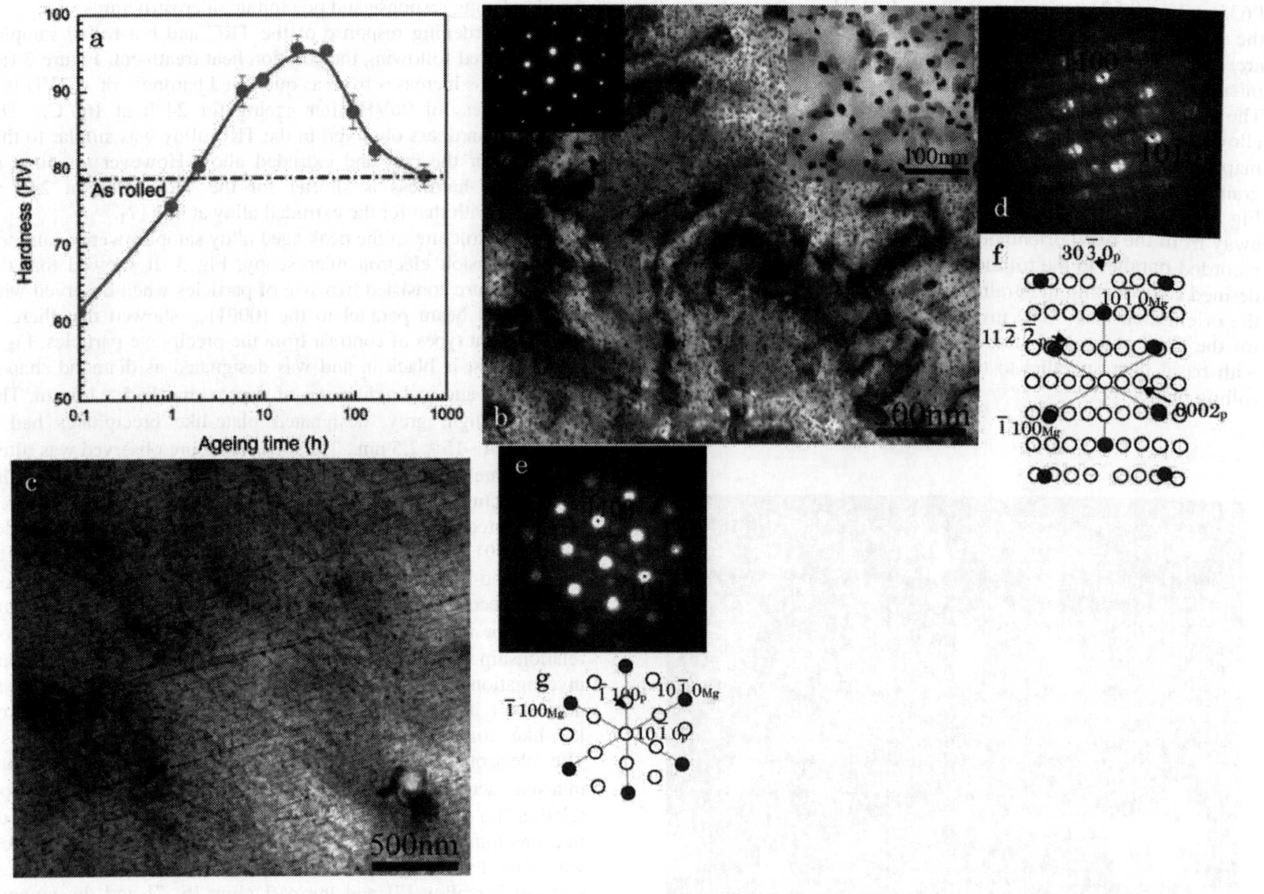

Figure 3 (a) the age hardening response of TRC Mg-2.4Zn-0.1Ag-0.1Ca-0.1Zr alloy at 160°C (b,c) TEM micrograph of the alloy peak aged at 160°C (b) electron beam is parallel to [0001]Mg (c) microstructure tilted 40° from [0001]Mg. electron micro-diffraction patterns from (d) diamond-shaped and (e) plate-like particles and (f, g) the schematic representation of diffraction patterns in (c and d) respectively.

with significantly lower elongation to failure. However the ZM21 and Mg-Th based HK31 alloys show significantly lower tensile yield strengths compared to the heat treated Mg-2.4Zn-0.1Ag-0.1Ca-0.1Zr TRC alloy. The Mg-1.7Gd-0.3Y (at%) alloy [12] has comparable yield strength but with far smaller elongations to failure. The yield strength of the as-TRC and rolled Mg-2.4Zn-0.1Ag-0.1Ca-0.1Zr alloy is comparable to that of the ZM61 alloy which has a similar composition. The ZMA611 and ZMA613 which contain a higher concentration of solute have higher yield strengths. Following ageing at 160°C, the TRC Mg-2.4Zn-0.1Ag-0.1Ca-0.1Zr alloy showed higher yield strength than the TRC processed and heat treated ZM61 and ZMA611 alloys subjected to a duplex ageing treatment. The mechanical properties of the TRC and heat treated Mg-2.4Zn-0.1Ag-0.1Ca-0.1Zr alloy is comparable to the tensile properties of heat treated 6009 series Al-Mg-Si based aluminium alloys both in yield strength and % elongation to failure [12]. In addition the formability of the TRC and rolled Mg-2.4Zn-0.1Ag-0.1Ca-0.1Zr alloy was found to be superior to any other reported Mg alloy sheets, which will be reported elsewhere.

Conclusions

Mg-2.4Zn-0.1Ag-0.1Ca-0.1Zr alloy show significant potential as a twin roll cast and rolled alloy for sheet applications. The grain size of the TRC and rolled alloy was approximately 18±5 μm. A tensile yield strength of 177MPa for the as twin roll cast and rolled sheet was substantially enhanced to 320 MPa following ageing at 160°C. Fine scale distribution of $MgZn_2$ precipitates phase with a rod-like morphology observed contributes to the increased strengthening after heat treatment.

Acknowledgments

This work was in part supported by the Ministry of Education, Science, Sports and Culture, Grant-in-Aid for Scientific Research (B), 21360348, 2009, and by World Premier International Research Center for Materials Nanoarchitectonics (MANA). This work was also supported in partially by the he Seoul Research and Business Development Program (10555) and the World Premier Materials (WPM) Program funded by the Ministry of Knowledge Economy through Research Institute of Advanced Materials..

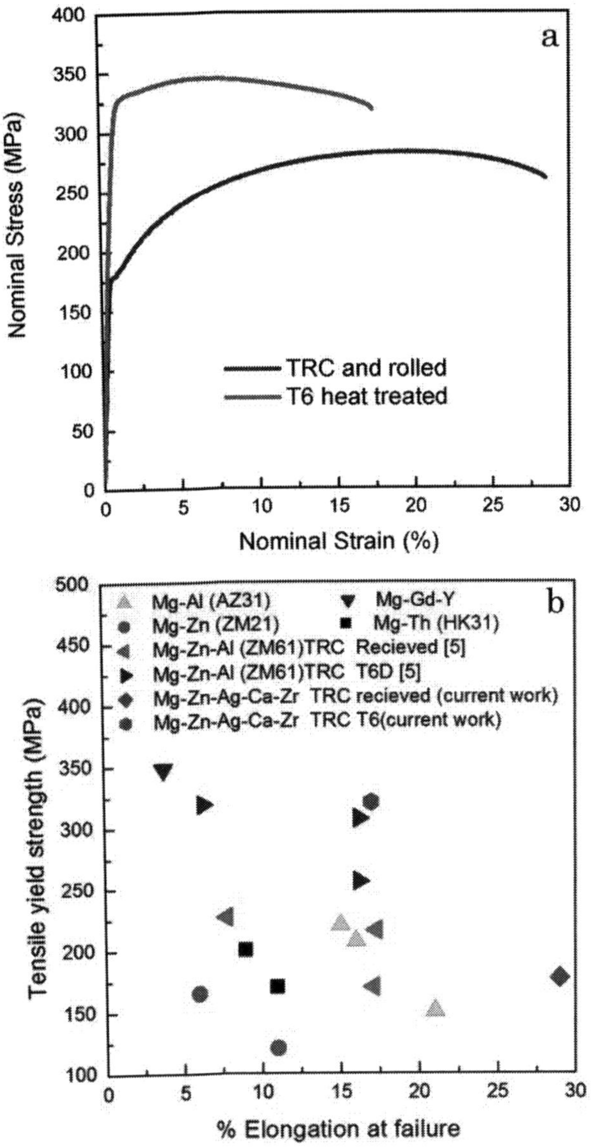

Figure 4 (a) Tensile stress strain curves for the twin roll cast (TRC) and rolled and heat treated Mg-2.4Zn-0.1Ag-0.1Ca-0.1Zr alloy and (b) comparison with the literature data for TRC and rolled or ingot cast and rolled magnesium alloys unless otherwise referred to data from [11 and 12]

References

[1] Magnesium vision 2020: A North American Automotive strategic vision for magnesium www.uscar.org (2006) U.S. Council for Automotive Research (USCAR)

[2] A. Jain, O Duygulu, D.W. Brown, C.N. Tomé, S.R. Agnew "Grain size effects on the tensile properties and deformation mechanisms of magnesium alloy AZ31B sheet" Mater. Sci. Eng. A 486 (2008) 545-555

[3] D.H. Kang, D.W. Kim, S. Kim, G.T. Bae, K.H. Kim, N.J. Kim "Relationship between stretch formability and work hardening capacity of twin-roll cast Mg alloys at room temperature" Scripta Mater. 61 (2009) 768-771

[4] K.H. Kim, B. C. Suh, J.H. Bae, M.S. Shim, S. Kim, N.J. Kim, "Microstructure and teturre evolution of Mg alloys during twin-roll casting and subsequent rolling" Scripta Mater. 63 (2010) 716-720

[5] S.S. Park, G.T. Bae, D.H. Kang I.H. Jung, K.S. Shin, N.J. Kim "Microstructure and tensile properties of twin-rol cast Mg-Zn-Mn-Al alloys" Scripta Mater 57 (2007) 793-796.

[6] C.L. Mendis, K. Oh-ishi, K. Hono "Enhanced age hardening in a Mg-2.4at%Zn alloy by trace additions of Ag and Ca" Scripta Mater. 57 (2007) 485-488.

[7] C.L. Mendis, K. Oh-ishi, Y. Kawamura, T. Honma, S. Kamado, K. Hono "Precipitation hardenable Mg-2.4Zn-0.1Ag-0.1Ca-0.16Zr (at%) wrought magnesium alloy" Acta Mater 57 (2009) 749-760

[8] X. Gao and J.F. Nie "Structure and thermal stability of primary intermetallic particles in a Mg-Zn casting alloy" Scripta Mater. 57 (2007) 655-658.

[9] P. Villars and L.D. Calvert, Pearson's Handbook of Crystallographic Data for Intermetallic Phases. 2nd ed. Materials Park Ohio: ASM International 1991.

[10] K. Oh-ishi, C. L. Mendis, T. Honma, S. Kamado, T. Ohkubo, and K. Hono "Bimodally grained microstructure development during hot extrusion of Mg-2.4Zn-0.1Ag-0.1Ca-0.16Zr (at%) alloy" Acta Mater 57 (2009) 5593-5604

[11] M. M. Avedesian, H. Baker, ASM Specialty Handbook Magnesium and Magnesium Alloys ASM International Materials Park (1999)

[12] I.J. Polmear, Light metals, 4th Edition, Butterworth Heinemann, Oxford, 2006.

Magnesium Technology 2011
Edited by: Wim H. Sillekens, Sean R. Agnew, Neale R. Neelameggham, and Suveen N. Mathaudhu
TMS (The Minerals, Metals & Materials Society), 2011

THE SOLIDIFICATION MICROSTRUCTURE AND PRECIPITATION INVESTIGATION OF MAGNESIUM-RICH ALLOYS CONTAINING Zn AND Ce

Chuan Zhang[1], Alan A. Luo[2], Y., Austin Chang[1]
[1]Department of Materials Science and Engineering, University of Wisconsin-Madison, 1509 University Avenue, Madison, WI 53706, USA
[2]Chemical Sciences and Materials Systems Lab, General Motors Global Research & Development, 30500 Mound Road, Warren, MI 48090-9055, USA

Keywords: Magnesium alloys, Mg-Zn-Ce, Phase diagram, Precipitation, Solidification

Abstract

Mg-Zn-Ce system has been identified as the alloy system for high-ductility wrought magnesium alloy development for automotive applications. The solidification microstructure and precipitation of Mg-Zn-Ce alloys were investigated to understand the phase equilibria and strengthening phases in this alloy system. The characterized microstructures of four directionally solidified Mg-Zn-Ce alloys agree well with the calculated solidification paths using computational thermodynamics coupled with the Scheil model. The precipitation of Mg_7Zn_3 intermetallic phase upon heat treatment was characterized using high-resolution transmission electron microcopy (HRTEM), a potential strengthening mechanism which will be explored in future research.

Introduction

Magnesium alloys with their weight saving advantages have been gradually accepted in automotive applications for vehicle lightweighting and environmental protection [1-4]. However, the relatively poorer mechanical properties of magnesium alloys compared to steel and aluminum alloys have limited their applications in critical structural subsystems. Recently, significant efforts have been made to develop high-strength and high-ductility magnesium alloys for automotive structural applications. It is well known that Zn is a major alloying element in magnesium alloys and is often used to improve the mechanical properties by solid solution strengthening and age-hardening. Zinc also helps overcome the harmful corrosive effect of iron and nickel impurities in the alloys [5]. However, the binary Mg-Zn alloys have some issues such as brittleness and coarse grains [6]. Therefore, binary Mg-Zn alloys are always modified by further alloying elements, such as Zr and/or rare earth (RE), in commercial alloys. RE elements have been used in magnesium alloys for many years and usually in the form of misch-metal (MM). Their strong influence on the creep resistance of Mg alloys is due to the strengthening of the α(Mg) matrix by solid solution and/or precipitation of RE-containing phases. There has been some reported work on the precipitation, morphology, structure and thermal stability of RE-containing intermetallic phases [7-9]. For better understanding the microstructure of Mg-Zn-RE alloys and subsequently enhancing their properties, the phase equilibria and thermodynamic properties of key sub-systems in the Mg-rich region, such as Mg-Zn-Ce, are of fundamental importance. Recently, the thermodynamic description of the Mg-Zn-Ce system at the Mg-rich region was reported [10].

In this paper, the solidification paths of four Mg-Zn-Ce ternary alloys were calculated using the thermodynamic description obtained by Chiu et al. [10] and Pandat [11]. Directional solidification technique was then employed to experimentally investigate the effect of Zn and/or Ce additions on the solidified microstructure of Mg-Zn-Ce alloys and validate the thermodynamic calculations. Solution and aging treatments were also carried out on one Mg-Zn-Ce alloy, and transmission electron microscopy (TEM) work was performed on the heat treated alloy to investigate the precipitating phases.

Experimental Procedure

Commercially pure magnesium and zinc, and Mg-26%Ce master alloy ingots were used to make the Mg-Zn-Ce (ZE) alloy compositions shown in Table I. The ZE alloys were prepared and melted in a 200 lb steel crucible under SF_6/CO_2 protection. The cast alloy was then cut into small pieces of about 50 g and was used for directional solidification experiment [12-14]. The axial temperature gradient was maintained at 4°C/mm in the furnace by three independently controlled heating coils. The sample was directionally solidified as cylindrical rods in a steel tube, by moving the tube downward in the furnace at 30 μm/s for about 100 mm in length. To study the precipitation microstructure, alloy samples were sealed in a quartz tube and back-filled with high-purity argon (>99.999%) atmosphere for solution treatment at 450°C for 24 h, followed by quenching into hot water. The samples were subsequently aged at 250°C for about 200 h. The X-ray diffraction (XRD), scanning electron microscopy (SEM) and High Resolution Transmission electron microscopy (HRTEM) were used for characterizations of the directionally solidified and heat-treated samples.

Table I. Alloy composition and Scheil calculated fraction of solid phase.

Alloy #	Alloy Designation	Scheil calculated solid phase fractions			
		α(Mg) %	$Ce(Mg,Zn)_{12}$ %	Mg_7Zn_3 %	MgZn %
1	Mg-2Zn-0.2Ce	97.98	0.4	1.5	0.05
2	Mg-6Zn-0.2Ce	93.60	0.47	5.7	0.2
3	Mg-2Zn-1Ce	97.19	1.79	0.98	0.3
4	Mg-7Zn-1Ce	91.48	2.28	6.02	0.2

Results and Discussion

Thermodynamic Calculations

Figure 1(a) shows the calculated liquidus projection of the Mg-Zn-Ce system in the Mg-rich corner with the compositions of Zn and Ce varying from 0 to 50%. All the alloy compositions in this paper are given in wt.% unless noted otherwise. The solidification paths of four Mg-Zn-Ce alloys were also shown in the same diagram, and they were calculated using the Scheil model, which is based on the assumptions of complete mixing in the liquid but no diffusion in the solid. As shown in this diagram, black lines are the monovariant liquidus lines and the arrows on the monovariant liquidus lines indicate the directions of decreasing temperature. The monovariant liquidus line at the Mg-rich corner is in equilibrium with the α(Mg) and $Ce(Mg,Zn)_{12}$. It should be emphasized that the $Ce(Mg,Zn)_{12}$ is a Ce-Mg binary phase with a large solid solubility of Zn extending into the Mg-Zn-Ce ternary system [10]. As can be seen from Figure 1(a), the solidification path of Alloy 1 is as follows:

L → L + α(Mg) (643°C) → L + α(Mg) + $Ce(Mg,Zn)_{12}$ (552°C) → L + α(Mg) + T_2 (II_2, 420°C) → L + T_2 + Mg_7Zn_3 (341°C) → T_2 + Mg_7Zn_3 + MgZn

The temperatures were calculated based on the thermodynamic description of Chiu et al. [10]. In order to discuss the last few solidification stages of Alloy 1 in Figure 1(a), Figure 1(b) is needed since the II_1 is in a very low Ce concentration region. Figure 1(b) shows an enlarged ordinate displaying the compositions of Ce from 0 to 0.2% and Zn from 0 to 60%. The four-phase invariant, II_2, at constant pressure is referred to as a type II invariant or reaction according to Rhines [15]. The subscript 2 indicates it is the second highest type II reaction in terms of temperature, following the notation of Chang et al. [16-18]. The sequences of phase formation of the other three alloys are identical to that of Alloy 1. The primary solidification phase of these alloys is α(Mg). After the composition of the liquid reaches the α(Mg) + $Ce(Mg,Zn)_{12}$ monovariant liquidus line, the α(Mg) and $Ce(Mg,Zn)_{12}$ form simultaneously. The T_2 and Mg_7Zn_3 come out subsequently during solidification and the last liquid to solidify at ternary eutectic invariant reaction: L → T_2 + Mg_7Zn_3 + MgZn. It is clear from Figure 1 that the phases formed in the interdendritic regions of the solidified Mg-Zn-Ce alloys (Zn < 10% and/or Ce < 5%) are identical, while the fraction of each phase may vary with different Zn and/or Ce concentrations.

In order to investigate the effect of various additions of Zn or Ce on the microstructure of solidified Mg-Zn-Ce alloys, the fraction of solid vs. temperature diagram of the selected four Mg-Zn-Ce Alloys was superimposed and shown in Figure 2. It indicates that the Zn addition decreases the alloy liquidus temperature. For example, the liquidus temperature of Alloy 1 with 2%Zn and 0.2%Ce is 634°C, while that of Alloy 2 containing the same amount of Ce (0.2%) but higher Zn concentration (6%) decreases to 631°C. And the same trend was also found on Alloys 3 (Mg-2Zn-1Ce) and 4 (Mg-7Zn-1Ce). Figure 2 also shows that with the addition of more Ce to the Mg-Zn-Ce alloys, the primary α(Mg) fraction decrease and more interdendritic phases formed during the solidification. Based on the Scheil calculations in this study, the primary α(Mg) fraction of Alloy 1

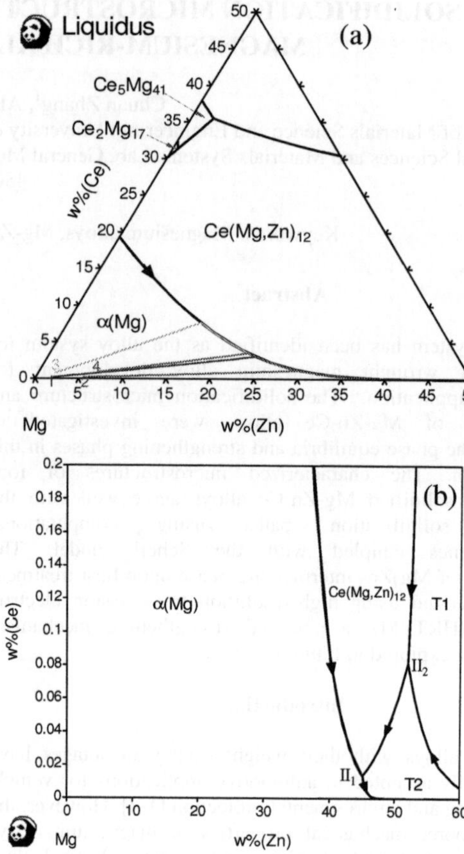

Figure 1. (a) Calculated liquidus projection of the Mg-Zn-Ce system at the Mg-rich region with the solidification path of four selected alloys using Scheil model; (b) Enlarged diagram – alloy 1: Mg-2Zn-0.2Ce; alloy 2: Mg-6Zn-0.2Ce; alloy 3: Mg-2Zn-1Ce; alloy 4: Mg-7Zn-1Ce (wt.%).

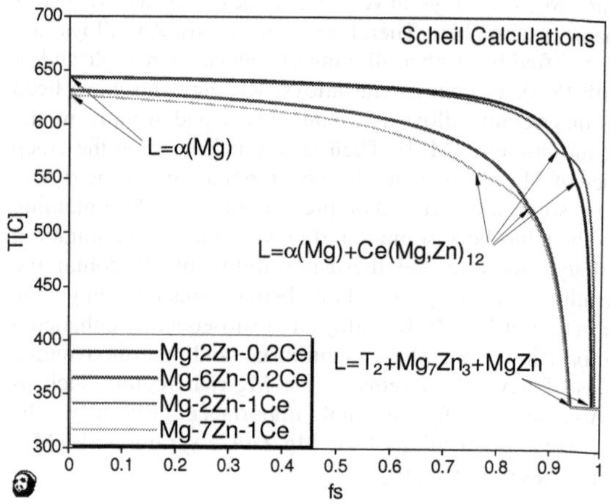

Figure 2. Fraction of solid vs. temperature diagram of four Mg-Zn-Ce alloys using Scheil model.

(Mg-2Zn-0.2Ce) is around 95%. However, the primary α(Mg) fraction of Alloy 3 (Mg-2Zn-1Ce) with higher Ce concentration is about 4% less than that of alloy 1. The Scheil calculations of the fractions of all interdendritic phases are summarized in Table I.

Solidification Microstructure

Figure 3 shows the SEM backscattered electron (BSE) images of the directionally solidified microstructure of the Mg-Zn-Ce alloys in the longitudinal direction in the steady state region. As shown in Figure 3(a), the directionally solidified Alloy 1 (Mg-2Zn-0.2Ce) shows the existence of one dominant phase with dark contrast and some regions containing devoiced eutectic-like microstructure with bright contrast. The XRD and SEM energy dispersive spectroscopic (EDS) phase identification results show that the dark matrix is α(Mg) and the bright phase is Mg_7Zn_3, which is in good agreement with the thermodynamic calculation results in Figures 1 and 2. It is not surprising that the $Ce(Mg,Zn)_{12}$ phase was not found in the solidified Alloy 1 due to its small volume fraction (<0.5%). The microstructure of Alloy 2 shown in Figure 3(b) is similar to that of Alloy 1. Two phases α(Mg) and Mg_7Zn_3 were identified. However, the fraction of Mg_7Zn_3 in the interdendritic region is more than that of Alloy 1, which is due to its higher Zn concentration and was predicted based on our Sheil calculations shown in Figure 2. The microstructure of Alloys 3 (Mg-2Zn-1Ce) and 4 (Mg-7Zn-1Ce) are shown in Figure 3 (c) and (d), respectively. For Alloy 3 shown in Figure 3(c), the XRD phase identification indicates that the dark α(Mg) coexisting with bright contrast interdendritic phases $Ce(Mg,Zn)_{12}$ and Mg_7Zn_3. However, it is very difficult to distinguish these two bight contract phases by SEM due to their similar morphology and contract within the interdendritic regions. The TEM selected area diffraction (SAD) was then carried out to confirm the existence of these two phases. The results will be reported in a future publication due to the limited space available in this paper. As shown in Figure 3(d), the microstructure of Alloy 4 (Mg-7Zn-1Ce) is similar to that of Alloy 3 (Mg-2Zn-1Ce), but the volume fraction of interdendritic phases is more due to its higher Zn concentration than that of Alloy 3. Therefore, the directional solidification results are in good agreement with the thermodynamic calculations using the Scheil model.

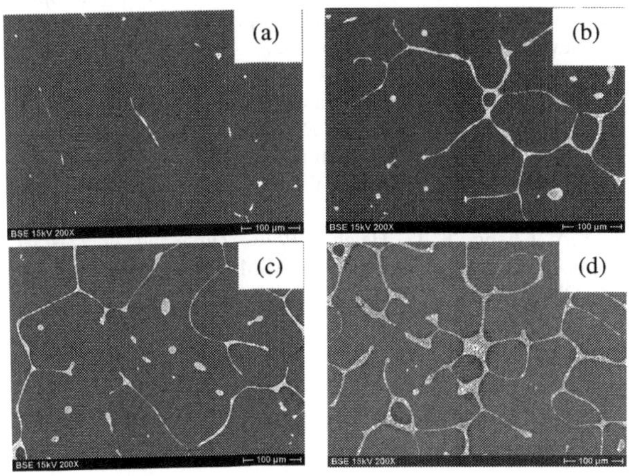

Figure 3. SEM/BSE images show the microstructure of directionally solidified Mg-Zn-Ce alloys (a) alloy 1: Mg-2Zn-0.2Ce; (b) alloy 2: Mg-6Zn-0.2Ce; (c) alloy 3: Mg-2Zn-1Ce; (d) alloy 4: Mg-7Zn-1Ce (wt.%).

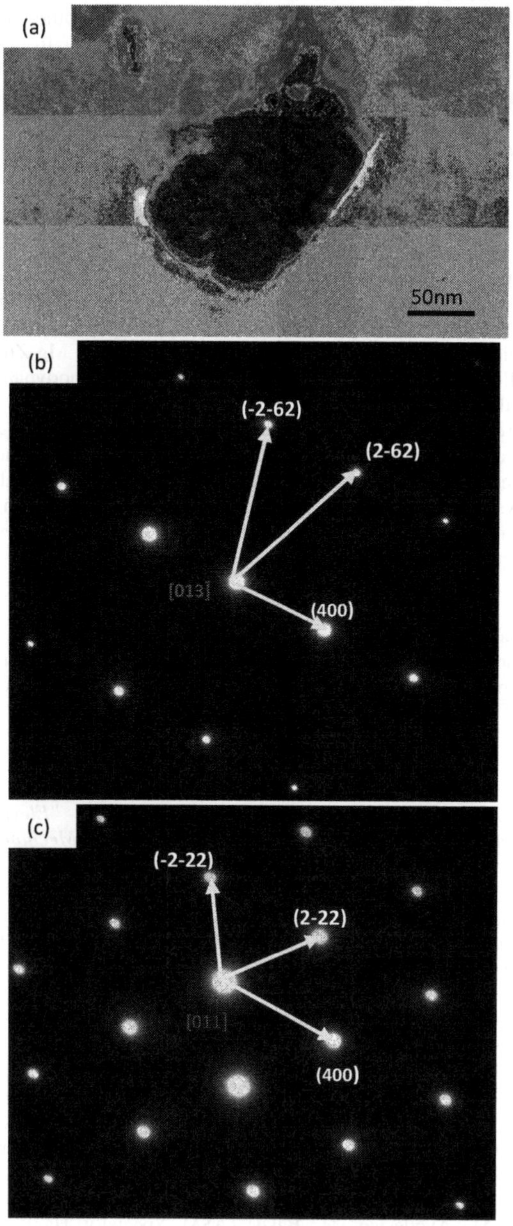

Figure 4. The TEM BF image and SAD of the Mg_7Zn_3 precipitate with α(Mg) matrix of heat treated Alloy 4 (Mg-7Zn-1Ce).

Precipitation Microstructure

Alloy 4 (Mg-7Zn-1Ce) was selected for the precipitation microstructure study due to its higher Zn and Ce concentrations. Figure 4 shows the HRTEM bright-field (BF) image and SAD of Alloy 4 (Mg-7Zn-1Ce) after solution and ageing treatments. The precipitating phase in the α(Mg) matrix was identified to be Mg_7Zn_3 by the diffraction patterns. As in many other alloy systems, the strengthening of the alloys is not only from the formation of thermodynamically stable intermetallics in the grain boundaries, but also the precipitates in the α(Mg) matrix after proper heat treatment processing. The precipitation of Mg_7Zn_3 phase in Mg-Zn-Ce alloys should be explored to improve the strength of the alloys.

Conclusions

1. The calculated solidification paths of four Mg-rich Mg-Zn-Ce alloys are in good agreement with the directional solidification microstructures of these alloys. The as-cast microstructures of four directionally solidified Mg-Zn-Ce alloys agree well with the computational thermodynamics calculations using the Scheil model.
2. Preliminary precipitation investigation found the Mg_7Zn_3 precipitation in the α(Mg) matrix after solution and age heat-treatments, indicating potential age-hardening for this alloy system, which should be explored in future research to improve the strength of the high-ductility Mg-Zn-Ce alloys.

Acknowledgements

The authors acknowledge the support of GM Global R&D, the financial support from NSF DMR No. 1005762 and Wisconsin Distinguished Professorship. Dr. Anil Sachdev of GM Global R&D is gratefully acknowledged for technical discussions.

References

1. E. Baril, P. Labelle, M.O. Pekguleryuz, "*Elevated Temperature Mg-Al-Sr: Creep Resistance, Mechanical Properties, and Microstructure*", JOM, (2003), pp. 34-39.

2. M.O. Pekguleryuz, A.A. Kaya. *In: K.U. Kainer, editor. Magnesium Alloys and their Applications.* Germany, Wiley-VCH Verlag GmbH, (2003), p. 74.

3. A.A. Luo, *"Recent magnesium alloy development for elevated temperature applications"*, Int. Mater. Rev., 49 (2004), pp. 13-29.

4. Y. Nakamura, A. Watanabe, K. Ohori, *"Effect of Ca, Sr Additions on Properties of Mg-Al Based Alloys"*, Mater Trans., 47(4) (2006), pp. 1031-1039.

5. M.M. Avedesian, H. Baker, ASM Specialty Handbook, American Society for Metals, Materials Park, OH, 1999.

6. C.J. Boehler, *"The tensile and creep behavior of Mg-Zn alloys with and without Y and Zr as ternary elements"*, J Mater Sci., 42 (2007), pp. 3675–3684.

7. I.J. Polmear, *"Recent developments in light alloys"*, Mater. Trans. JIM, 37 (1) (1996) 12.

8. L.Y. Wei, G.L. Dunlop, H. Westengen, *"Development of microstructure in cast Mg-Al-rare earth alloys"*, Mater. Sci. Technol., 12 (9) (1996) 741.

9. R. Ferro, A. Saccone, G. Borzone, *"Rare earth metals in light alloys"*, J. Rare Earths, 15 (1) (1997), pp. 45-61.

10. C.N. Chiu, J. Gröbner, A. Kozlov, R. Schmid-Fetzer, *"Experimental study and thermodynamic assessment of ternary Mg–Zn–Ce phase relations focused on Mg-rich alloys"* Intermetallics, 18(4) (2010), pp. 399-405.

11. Pandat 8.0–Phase Diagram Calculation Software for Multi-component Systems, Computherm LLC, 437 S. Yellowstone Dr., Suite217, Madison, WI, 53719, USA, 2008.

12. C. Zhang, D. Ma, K. Wu, H. Cao, J. Zhu, G. Cao, S. Kou, Y.A. Chang, in Magnesium Technology 2006 (Editors: Luo A.A., Neelameggham N.R. and Beals R.S.), TMS (The Minerals, Metals and Materials Society), 2006, p. 45.

13. C. Zhang, D. Ma, K.-S. Wu, H.-B. Cao, G.-P. Cao, S. Kou, Y. A. Chang, X-Y. Yan, *"Microstructure and microsegregation in directionally solidified Mg–4Al alloy"*, Intermetallics, 15 (2007), pp. 1395-1400.

14. C. Zhang, H.B. Cao, V. Firouzdor, S. Kou, Y.A. Chang, *"Microstructure investigations of directionally solidified Mg-rich alloys containing Al, Ca and Sn"*, Intermetallics, 18(8) (2010), pp. 1597-1602.

15. F.N. Rhines, Phase Diagrams in Metallurgy, McGraw-Hill Book Company, NY, 1956.

16. Y.A. Chang, J.P. Neumann, A. Mikula, D. Goldberg, *"Phase Diagrams and Thermodynamic Properties of Ternary Copper-Metal Systems"*, The International Copper Research Association, Inc., New York, N.Y. (1979), pp. 702.

17. Y.A. Chang, J.P. Neumann, U.V. Choudary, *"Phase Diagrams and Thermodynamic Properties of Ternary Copper-Sulfur-Metal Systems"*, The International Copper Research Association, Inc., New York, N.Y. (1979), pp. 200.

18. Y.A. Chang, K.-C. Hsieh, *"Phase Diagrams of Ternary Copper-Oxygen-Metal Systems"*, ASM International, Metals Park, Ohio, 1989.

Magnesium Technology 2011

Deformation Mechanisms I

Session Chairs:

Carlos H. Caceres
(University of Queensland, Australia)
Eric A. Nyberg
(Pacific Northwest National Laboratory, USA)

Magnesium Technology 2011
Edited by: Wim H. Sillekens, Sean R. Agnew, Neale R. Neelameggham, and Suveen N. Mathaudhu
TMS (The Minerals, Metals & Materials Society), 2011

CRYSTAL PLASTICITY ANALYSIS ON COMPRESSIVE LOADING OF MAGNESIUM WITH SUPPRESSION OF TWINNING

Tsuyoshi Mayama[1], Tetsuya Ohashi[2], Kenji Higashida[3], Yoshihito Kawamura[1]
[1]Kumamoto University, 2-39-1, Kurokami, Kumamoto, 860-8555, JAPAN
[2]Kitami Institute of Technology, 165 Koen-cho, Kitami, Hokkaido, 090-8507, JAPAN
[3]Kyushu University, 744 Motooka, Fukuoka, 819-0395, JAPAN

Keywords: Finite element analysis, Crystal plasticity, Compressive loading, Heterogeneous deformation, Twinning

Abstract

The compressive loading behavior of single crystals and bicrystals of magnesium without consideration of deformation twinning has been investigated by crystal plasticity finite element analysis with the aim of fundamental understanding of kink band formation in magnesium alloys with long period stacking ordered structure (LPSO) phase. The basal plane of the single crystal model is set to be parallel to the compressive direction. The result of the compressive loading analysis of single crystals indicates the significant influence of suppression of twinning on the activation of nonbasal slip systems and stress-strain behavior. The compressive analysis of symmetric bicrystal is also performed to clarify the influence of the angle between basal plane and the loading axis. The influence of the introduction of grain boundary and the slight change of crystal orientation is discussed in terms of activated deformation modes.

Introduction

Magnesium alloys with long period stacking ordered (LPSO) structure have attracted great attention. The wrought magnesium alloys with LPSO phase show the excellent mechanical properties [1-3]. It is considered that one of the strengthening mechanisms of these alloys is the formation of kink bands in the LPSO phase. However, the kink band formation mechanism and the influence on the subsequent deformation have not been clarified in detail. Therefore, fundamental research on kink band formation in magnesium is required to gain insight into the strengthening mechanism of magnesium alloys. Fig. 1 shows a TEM image of typical kink bands observed in extruded magnesium alloy $Mg_{96}Zn_2Y_2$.

The previous research on FCC crystals by using crystal plasticity finite element analysis indicates that kink bands perpendicular to the primary slip direction are formed in compatible bicrystals with tilted angle grain boundary under tensile loading [4]. By using the similar numerical method, this study investigates compressive behavior in HCP crystal with suppression of twinning.

Figure 1. TEM image of kink band in extruded $Mg_{96}Zn_2Y_2$

Crystal plasticity finite element method

A crystal plasticity finite element code developed for face centered cubic (FCC) metals [5, 6] is modified for magnesium. In this study, the words "slip system" and "twin system" are used to indicate the equivalent slip/twin system (slip/twin system family), collectively. The word "deformation mode" is used to indicate the specific one out of equivalent slip/twin system.

Constitutive equations

The critical resolved shear stress $\theta^{(n)}$ on deformation mode n is given by the Schmid law as

$$\theta^{(n)} = \mathbf{P}^{(n)} : \boldsymbol{\sigma} \tag{1}$$

where $\boldsymbol{\sigma}$ denotes the stress tensor. $\mathbf{P}^{(n)}$ is the Schmid tensor of deformation mode n defined by

$$\mathbf{P}^{(n)} = \frac{1}{2}\left(\mathbf{v}^{(n)} \otimes \mathbf{b}^{(n)} + \mathbf{b}^{(n)} \otimes \mathbf{v}^{(n)}\right) \tag{2}$$

where $\mathbf{v}^{(n)}$ and $\mathbf{b}^{(n)}$ are the slip/twin plane normal and the slip/twin direction, vector respectively. The increment of the critical resolved shear stress is written as the following equation:

$$\dot{\theta}^{(n)} = \sum h^{(nm)} \dot{\gamma}^{(m)} \tag{3}$$

where $h^{(nm)}$ and $\dot{\gamma}^{(m)}$ indicate the strain hardening coefficient and the plastic shear strain rate on deformation mode m.
Assuming that the deformation is small and lattice rotation is ignored, the constitutive equation for slip and twin deformation is given by

$$\dot{\varepsilon} = \mathbf{S}^{e} : \dot{\boldsymbol{\sigma}} + \sum\sum \left\{h^{(nm)}\right\}^{-1} \mathbf{P}^{(n)} : \left(\mathbf{P}^{(m)} : \dot{\boldsymbol{\sigma}}\right) \tag{4}$$

where $\mathbf{S}^{e}$ denotes the elastic compliance, and the summation is made over the active deformation modes. We solve the deformation with a rate-independent plasticity formulation.

The following Bailey-Hirsch type model of the critical resolved shear stress is used,

$$\theta^{(n)} = \theta_0 + \sum \Omega^{(nm)} a \mu \tilde{b}^{(a)} \sqrt{\rho_S^{(m)}} \tag{5}$$

where θ_0, a, μ, $\tilde{b}^{(a)}$ and $\rho_S^{(m)}$ denote the lattice friction term, a numerical factor on the order of 0.1, the elastic shear modulus, the magnitude of the Burgers vector for **<a>** dislocation and the statistically stored dislocation (SSD) density that accumulates on deformation mode m, respectively. Although Burgers vector for **<a>** dislocation and **<a+c>** dislocation are different, in this study, for simplicity, the magnitude of Burgers vector of **<a>** dislocation is used in Eq.(5) for all slip/twin systems. By substituting the derivation of Eq.(5) into Eq.(3), the strain hardening coefficient $h^{(nm)}$ in Eq.(3) is specified. The increment in the SSD density on deformation mode n is given as follows:

$$\dot{\rho}_S^{(n)} = \frac{c\dot{\gamma}^{(n)}}{\tilde{b}^{(n)} L^{(n)}} \tag{6}$$

Here, $\tilde{b}^{(n)}$ is the magnitude of Burgers vector for deformation mode n. For magnesium, the magnitude of Burgers vector $\tilde{b}^{(a)}$ is used for slip system with slip direction <a> and the magnitude of Burgers vector $\tilde{b}^{(a+c)}$ is used for slip system with slip direction <a+c>. Burgers vector $\tilde{b}^{(a+c)}$ is calculated by

$$\tilde{b}^{(a+c)} = \tilde{b}^{(a)} \sqrt{1+(c/a)^2} \tag{7}$$

where c/a is axial ratio for HCP crystal structure.
In Eq.(6), c is a numerical coefficient on the order of 1 and $L^{(n)}$ is the mean free path of dislocations on deformation mode n defined as

$$L^{(n)} = \frac{c^*}{\sqrt{\sum_m \omega^{(nm)} \left(\rho_S^{(m)} + \left\| \rho_G^{(m)} \right\| \right)}} \tag{8}$$

where c^* and $\omega^{(nm)}$ are a material constant of the order of 10-100 and the weight matrix, respectively. $\left\| \rho_G^{(m)} \right\|$ denotes the norm of geometrically necessary dislocation (GND) density on deformation mode m defined as

$$\left\| \rho_G^{(m)} \right\| = \sqrt{\left(\rho_{G,edge}^{(m)} \right)^2 + \left(\rho_{G,screw}^{(m)} \right)^2} \tag{9}$$

where the edge and screw components are defined by the strain gradient:

$$\rho_{G,edge}^{(m)} = -\frac{1}{\tilde{b}^{(m)}} \frac{\partial \gamma^{(m)}}{\partial \xi} , \qquad \rho_{G,screw}^{(m)} = \frac{1}{\tilde{b}^{(m)}} \frac{\partial \gamma^{(m)}}{\partial \zeta} . \tag{10}$$

Here, ξ and ζ denote directions parallel and perpendicular to the slip direction, respectively.

Slip systems

Because of the lower symmetry in materials with an HCP structure compared to materials with a cubic crystal structure, slip systems for magnesium are classified into different slip system families. The operative slip systems in magnesium have been previously investigated [7-11]. Basal slip system $(0001)\langle 11\bar{2}0 \rangle$, prismatic slip system $\{10\bar{1}0\}\langle 11\bar{2}0 \rangle$ and first order pyramidal (pyramidal-1) slip system $\{10\bar{1}1\}\langle 11\bar{2}0 \rangle$ have long been known. Although Yoshinaga and Horiuchi [12] suggested that magnesium single crystal showed no activation of second order pyramidal (pyramidal-2) slip system $\{11\bar{2}2\}\langle 11\bar{2}3 \rangle$, Obara et al. [13] reported the activation of this pyramidal-2 slip system based on slip line analysis and dislocation observation. Recent crystal plasticity studies also incorporate the pyramidal-2 slip system to compensate deformation in the c-axis direction and the results of calculations show quantitatively good agreements with experimental results. Therefore, in the present study, basal, prismatic, pyramidal-1 and pyramidal-2 slip systems in Fig.2 are considered. Table 1 shows the slip plane, the shear direction and the number of deformation modes.

Deformation twinning

Besides slip deformation, another important deformation mechanism for HCP metals is deformation twinning. In magnesium, the contribution of c-axis tensile twinning system $\{10\bar{1}2\}\langle 10\bar{1}1 \rangle$ as shown in Fig.1(e) to macroscopic deformation is significantly larger than those by the other twinning systems observed in the experimental studies [12, 14, 15]. Therefore, only the c-axis tensile twinning system is considered in this study because the focus of this study is on the relationships between heterogeneous deformation and activated deformation modes in magnesium. The twin plane, shear direction and the number of deformation mode of twin system $\{10\bar{1}2\}\langle 10\bar{1}1 \rangle$ are shown in Table 2. In this study, the twinning deformation is incorporated into crystal plasticity analysis in a similar manner to slip deformation [16]. That is, it is assumed that the twinning plane normal vector and the shear direction vector by twinning are equivalent to the slip plane normal $\mathbf{v}$ and the slip direction vector $\mathbf{b}$, respectively. However, the following two points are considered only for twinning system.

1. In order to simulate the polarity of twinning, shear slip rate is set to be zero when the compressive loading in c-axis.
2. Shear strain by twinning is limited to the maximum value determined by

$$\gamma^{max} = \frac{\sqrt{3}}{c/a} - \frac{c/a}{\sqrt{3}} . \tag{11}$$

When the accumulated shear strain reaches the maximum value in Eq.(10), the shear strain rate for twinning is set to be zero.

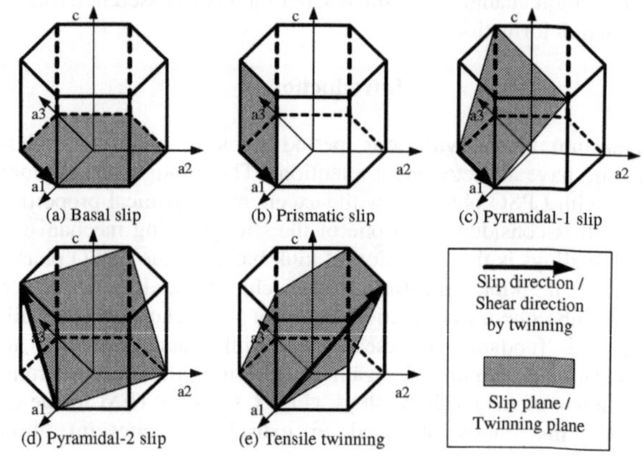

Figure 2. Schematic diagram of slip/twin systems in magnesium

Table 1. Deformation mechanisms in magnesium

Slip/twin system	Slip/twin plane	Shear direction	Number of deformation modes
Basal slip	(0001)	$\langle 11\bar{2}0 \rangle$	3
Prismatic slip	$\{10\bar{1}0\}$	$\langle 11\bar{2}0 \rangle$	3
Pyramidal-1 slip	$\{10\bar{1}1\}$	$\langle 11\bar{2}0 \rangle$	6
Pyramidal-2 slip	$\{11\bar{2}2\}$	$\langle 11\bar{2}3 \rangle$	6
Tensile twinning	$\{10\bar{1}2\}$	$\langle 10\bar{1}1 \rangle$	6

Anisotropic elasticity for HCP crystal

The anisotropy of elastic deformation of HCP single crystal is taken into account in the present method. The elastic compliances used in this study are shown in Eq. (12).

$$\begin{pmatrix} s_{11} & s_{12} & s_{13} & 0 & 0 & 0 \\ s_{12} & s_{11} & s_{13} & 0 & 0 & 0 \\ s_{13} & s_{13} & s_{33} & 0 & 0 & 0 \\ 0 & 0 & 0 & 2(s_{11}-s_{12}) & 0 & 0 \\ 0 & 0 & 0 & 0 & s_{55} & 0 \\ 0 & 0 & 0 & 0 & 0 & s_{55} \end{pmatrix} \quad (12)$$

here $2(s_{11}-s_{12})$ and s_{55} show shear components of elastic compliance on basal plane and plane perpendicular to basal plane, respectively.

Material parameters for pure magnesium single crystal

Identification of material parameters used in the constitutive equations is very difficult for magnesium. One of the reasons of this difficulty is that the deformation mechanisms and the critical resolved shear stresses (CRSSs) for magnesium have not yet been fully clarified experimentally. The difference of CRSSs between basal slip and nonbasal slip is so significant that deformations by single slip of nonbasal slip hardly occur. Furthermore, the details of strain hardening behavior for each deformation mode and the interaction between dislocations have not been clarified so far. In this study, CRSSs for basal and nonbasal slip systems are set to be 0.5 and 40 MPa based on the experimental reports for pure magnesium single crystal in the literatures [12, 17-21],. To suppress the activation of twinning, CRSS for tensile twinning system is set to be large enough value. The material parameters *a*, *c*, *c** used in this study are typical values from the previous studies for FCC metals [5, 6]. Assuming initially uniform dislocation density, the initial dislocation density 10^9 -/m^2 is used for all deformation modes. For the interaction matrix $\Omega^{(nm)}$ in Eq.(5), $\Omega^{(nm)}=1$ for the diagonal components and $\Omega^{(nm)}=1.01$ for the off-diagonal components are assumed. For the weight matrix $\omega^{(nm)}$ in Eq.(8), $\omega^{(nm)}=0$ when *n*=*m* or deformation mode *n* and deformation mode *m* are the same slip/twin system family. For all other components $\omega^{(nm)}=1$ is assumed. Lattice friction stresses for each slip/twin system are calculated by Eq.(5) using the CRSSs and initial SSD density. Anisotropic elastic constants [22] and the axial ratio [23] for magnesium are used.

Analysis models

Although the deformation modes for HCP materials are conventionally described by Millar-Bravais system as shown in Fig.3(a), an orthonormal system shown in Fig.3(b) is used for convenience of numerical implementation. In the present definition of orthonormal system, three axes $\mathbf{e}_1$, $\mathbf{e}_2$ and $\mathbf{e}_3$ are corresponding to $[10\bar{1}0]$-, $[\bar{1}2\bar{1}0]$- and $[0001]$- directions, respectively. This orthonormal system is similar to the one used in Staroselsky and Anand [16] except that they defined three axes $\mathbf{e}_1$, $\mathbf{e}_2$ and $\mathbf{e}_3$ as $[\bar{1}2\bar{1}0]$-, $[\bar{1}010]$- and $[0001]$- directions, respectively. Crystal orientations for each grain are specified by Euler angles κ, θ and ϕ. In the present definition of Euler angles, first, a crystal lattice whose $[10\bar{1}0]$-, $[\bar{1}2\bar{1}0]$- and $[0001]$-directions in crystal coordinate system are corresponding to x-, y- and z-directions in sample coordinate system, is rotated by ϕ around $[0001]$-axis. Then, the lattice is rotated by θ around $[10\bar{1}0]$- axis. Finally, the lattice is rotated by κ around $[0001]$-axis.
In the literature [24, 25], the HCP material whose basal plane is nearly parallel to the loading direction, deforms with kink band formation during compression. Therefore, in this study, the analysis models shown in Fig. 4 are used to investigate the compressive deformation behavior in HCP crystal with suppression of twinning. The compressive directions of the single crystal models $\langle 11\bar{2}0\rangle$ SC and $\langle 10\bar{1}0\rangle$ SC in Fig. 4(a) and (b) are $\langle 11\bar{2}0\rangle$- and $\langle 10\bar{1}0\rangle$- directions, respectively. Euler angles (κ,θ,ϕ) for $\langle 11\bar{2}0\rangle$ SC and $\langle 10\bar{1}0\rangle$ SC are $(60°,0°,0°)$ and $(30°,0°,0°)$. Fig. 4(c) shows the bicrystal model in which the angle between basal plane and compressive direction is α. In this study, four angles are chosen. Euler angles and angles α for bicrystal models are shown in Table 2.

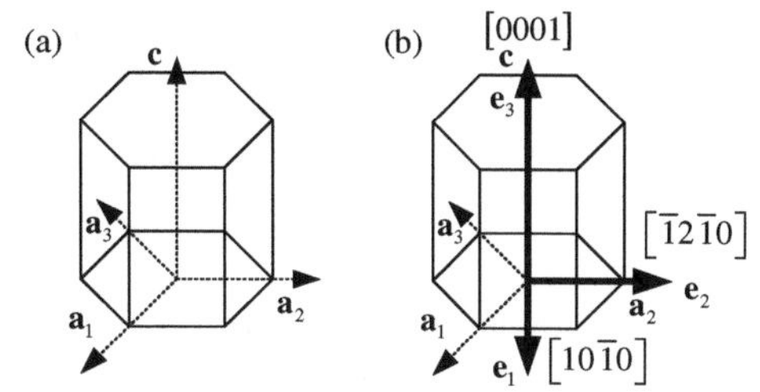

Figure 3. Coordinate system for HCP crystal, (a) Miller-Bravis coordinate, (b) Orthonormal coordinate

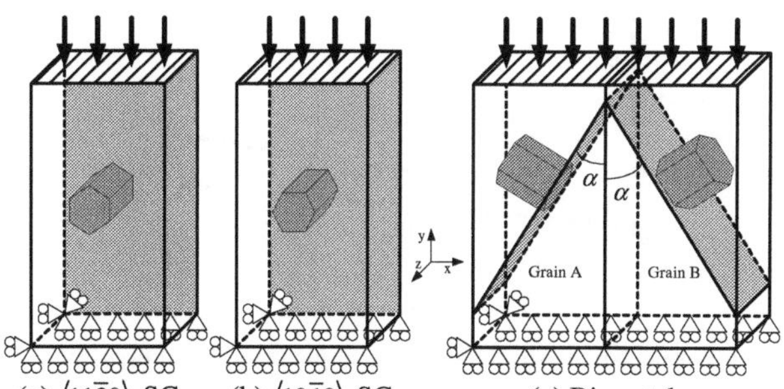

Figure 4. Schematic diagrams of analysis models

Table 2. Euler angles and angle α for bicrystal models

Model name	Euler angles (κ,θ,ϕ)	Angle α
BC5	$(330°,270°,85°)$ for grain A $(330°,90°,95°)$ for grain B	5°
BC15	$(330°,270°,75°)$ for grain A $(330°,90°,105°)$ for grain B	15°
BC30	$(330°,270°,60°)$ for grain A $(330°,90°,120°)$ for grain B	30°
BC44	$(330°,270°,46°)$ for grain A $(330°,90°,134°)$ for grain B	44°

Results and Discussion

Single crystal compression

In Fig. 5, solid lines show the compressive stress-strain curves of $\langle 11\bar{2}0\rangle$ SC and $\langle 10\bar{1}0\rangle$ SC models. The results with consideration of twinning using CRSS of 3MPa for tensile twinning system are also shown in Fig. 5 by broken lines. Depending on the crystal orientation and the consideration of twinning, different stress-strain curves are obtained. Table 3 shows activated deformation modes during compression. The models without consideration of twinning deform by multiple slip of nonbasal slip system. Activated slip systems in $\langle 11\bar{2}0\rangle$ SC and $\langle 10\bar{1}0\rangle$ SC are pyramidal-2 and prismatic slip system, respectively. This difference in activated slip systems results in the different stress-strain curves shown in Fig. 5. In the case with consideration of twinning, single crystals are deformed by a twinning system. The number of activated twin systems is different in $\langle 11\bar{2}0\rangle$ SC and $\langle 10\bar{1}0\rangle$ SC. Owing to the number of activated twin systems, stress-strain responses are different as shown in Fig. 5. In all cases, the basal slip system is not activated because the Schmid factor for basal slip system is zero. Additionally deformation of these single crystals is almost uniform. These results suggest that any triggers for non-uniform deformation are necessary to activate the basal slip system.

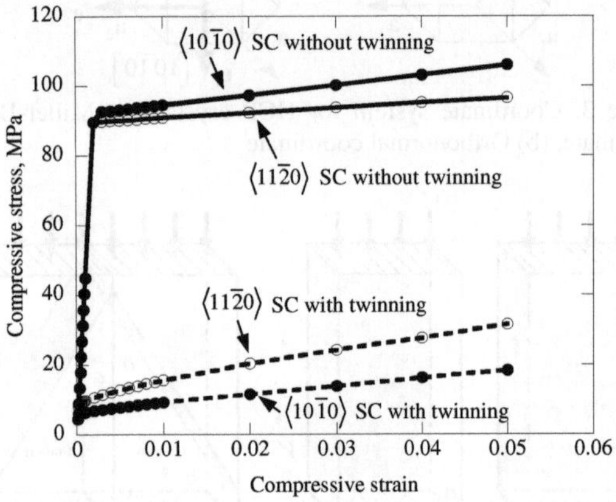

Figure 5. Stress-strain curves in single crystal compression

Table 3. Activated deformation modes in single crystal compression

Model	Activated deformation modes
$\langle 11\bar{2}0\rangle$ SC without twinning	$(11\bar{2}2)[11\bar{2}3]$, $(\bar{1}\bar{1}22)[11\bar{2}3]$
$\langle 10\bar{1}0\rangle$ SC without twinning	$(1\bar{1}00)[11\bar{2}0]$, $(10\bar{1}0)[\bar{1}2\bar{1}0]$
$\langle 11\bar{2}0\rangle$ SC with twinning	$(10\bar{1}2)[\bar{1}011]$, $(01\bar{1}2)[0\bar{1}11]$ $(\bar{1}012)[10\bar{1}1]$, $(0\bar{1}12)[01\bar{1}1]$
$\langle 10\bar{1}0\rangle$ SC with twinning	$(01\bar{1}2)[0\bar{1}11]$, $(0\bar{1}12)[01\bar{1}1]$

<u>Bicrystal compression</u>

Fig. 6 shows the calculated stress-strain curves of compression of bicrystals. With decreasing angle α the flow stress increases. Table 4 shows activated deformation modes in each bicrystal model. In BC5 and BC15 models, nonbasal slip systems are activated as well as the activation of basal slip systems. The increase in the flow stress with decreasing angle α shown in Fig.6 is considered a consequence of more significant hardening by larger accumulated slip of basal slip system with lower Schmid factor. Comparing BC5 with $\langle 11\bar{2}0\rangle$ SC, it is suggested that the introduction of grain boundary and the slight change of crystal orientation result in the activation of basal slip systems. The activation of the basal slip systems probably result in the lower flow stress in BC5 as compared with $\langle 11\bar{2}0\rangle$ SC.

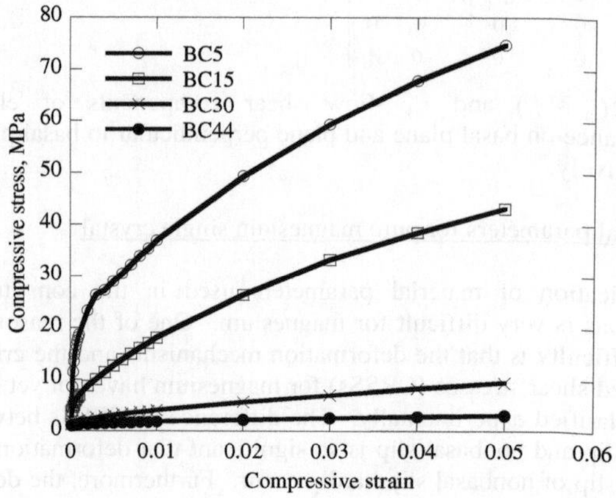

Figure 6. Stress-strain curves in bicrystal compression

Table 4. Activated deformation modes in bicrystal compression

Model	Activated deformation modes
BC5	$(0001)[11\bar{2}0]$, $(0001)[\bar{1}2\bar{1}0]$, $(0001)[\bar{2}110]$ $(10\bar{1}0)[\bar{1}2\bar{1}0]$, $(01\bar{1}0)[\bar{2}110]$, $(\bar{1}\bar{1}22)[11\bar{2}3]$
BC15	$(0001)[11\bar{2}0]$, $(0001)[\bar{1}2\bar{1}0]$, $(0001)[\bar{2}110]$ $(10\bar{1}0)[\bar{1}2\bar{1}0]$, $(01\bar{1}0)[\bar{2}110]$, $(\bar{1}\bar{1}22)[11\bar{2}3]$
BC30	$(0001)[11\bar{2}0]$, $(0001)[\bar{1}2\bar{1}0]$, $(0001)[\bar{2}110]$
BC44	$(0001)[11\bar{2}0]$, $(0001)[\bar{1}2\bar{1}0]$, $(0001)[\bar{2}110]$

Conclusions

In this study, compressive loading behavior of HCP crystal with suppression of twinning was investigated using a crystal plasticity finite element method. From the calculated results of single crystals and bicrystals the following conclusions were obtained.

1. Single crystals compressed in $\langle 11\bar{2}0\rangle$ – and $\langle 10\bar{1}0\rangle$ – directions deformed uniformly by the multiple slip of nonbasal slip system although the similar analysis with consideration of twinning resulted in the activation of twinning system.
2. In the results of compression analyzes of bicrystal models which had symmetric grain boundaries, the stress-strain behavior and activated deformation modes depended on the angle between basal plane and compressive direction.
3. Comparison between $\langle 11\bar{2}0\rangle$ SC and BC5 suggested that the introduction of grain boundary and the slight change of crystal orientation lead to the activation of basal slip systems. The lower flow stress in BC5 seems to be caused by the basal slip activation.

Comparison of the present results with experimental ones was one of the most interesting points, while there were some difficulties in preparation of LPSO phase single crystals and direct comparison was not made in this paper. Further development of experimental and numerical technique is required in the future research.

Acknowledgements

The work of T.M. was partially supported by a Grant-in-Aid for Scientific Research from the Ministry of Education, Culture, Sports, Science and Technology of Japan (KAKENHI: No.21760082) and the Kumamoto Prefecture Collaboration of Regional Entities for the Advancement of Technological Excellence from Japan Science and Technology Agency.

References

1. Y. Kawamura, K. Hayashi, A. Inoue and T. Masumoto, "Rapidly solidified powder metallurgy $Mg_{97}Zn_1Y_2$ alloys with excellent tensile yield strength above 600MPa" *Materials Transactions* 42 (2001) 1172-1176.
2. S.Yoshimoto, M. Yamasaki and Y. Kawamura, "Microstructure and mechanical properties of extruded Mg-Zn-Y alloys with 14H long period ordered structure" *Materials Transactions* 47 (2006) 959-965.
3. Y.Kawamura and M.Yamasaki, "Formation and mechanical properties of $Mg_{97}Zn_1RE_2$ alloys with long-period stacking ordered structure" *Materials Transactions* 48 (2007) 2986-2992.
4. R.Kondou and T.Ohashi, "Grain boundary accumulation of geometrically necessary dislocations and asymmetric deformations in compatible bicrystals with tilted angle grain boundary under tensile loading" *JSME International Journal A* 49 (2006), 581-588.
5. T.Ohashi, "Three dimensional structures of the geometrically necessary dislocations in matrix-inclusion systems under uniaxial tensile loading" *International Journal of Plasticity* 20 (2004), 1093-1109.
6. T.Ohashi, "Crystal plasticity analysis of dislocation emission from micro voids" *International Journal of Plasticity* 21 (2005), 2071-2088.
7. R.E. Read-Hill and W.D. Robertson, "Deformation of magnesium single crystals by nonbasal slip" *Journal of Metals Transactions of AIME* (1957) 496-502.
8. R.E. Read-Hill and W.D. Robertson, "Pyramidal slip in magnesium" *Transactions of the Metallurgical Society of AIME* 212 (1958) 256-259.
9. W.F. Sheely and R.R. Nash, "Mechanical properties of magnesium monocrystals" Transactions of Metallurgical Society of AIME 218 (1960) 416-423.
10. P. Ward Flynn, J. Mote and J.E. Dorn, "On the thermally activated mechanism of prismatic slip in magnesium single crystals" *Transactions of Metallurgical Society of AIME* 221 (1961) 1148-1154.
11. P.B. Hirsch and J.S. Lally, "The deformation of magnesium single crystals" *Philosophical Magazine* 12 (1965) 595-648.
12. H. Yoshinaga and R. Horiuchi, "Deformation mechanisms in magnesium single crystals compressed in the direction parallel to hexagonal axis" *Transactions of JIM* 4 (1963) 1-8.
13. T. Obara, H. Yoshinaga, S. Morozumi, "{11-22}<-1-123> slip system in magnesium" *Acta Metallurgica* 21 (1973) 845-853.
14. E.W. Kelly and W.F. Hosford, "Plane-strain compression of magnesium and magnesium alloy crystals" *Transactions of Metallurgical Society of AIME* 242 (1968) 5-13.
15. M.R. Barnett, "Twinning and the ductility of magnesium alloys part II. Contraction twins" *Materials Science and Engineering A* 464 (2007) 8-16.
16. A. Staroselsky and L. Anand, "A constitutive model for hcp materials deforming by slip and twinning: Application to magnesium alloy AZ31B" *International Journal of Plasticity* 19 (2003) 1843-1864.
17. E.C. Burke and W.R. Hibbard, "Plastic deformation of magnesium single crystals" Journal of Metals Transactions of AIME (1952) 295-303.
18. H. Yoshinaga and R. Horiuchi, "On the flow stress of alpha solid solution Mg-Li alloy single crystals" Transactions of JIM 4 (1963) 134-141.
19. H. Yoshinaga and R. Horiuchi, "On the nonbasal slip in magnesium crystals" Transactions of JIM 5 (1963) 14-21.
20. A. Akhtar and E. Teghtsoonian, "Solid solution strengthening of magnesium single crystal-I alloying behaviour in basal slip" Acta Metallurgica 17 (1969) 1339-1349.
21. A. Akhtar and E. Teghtsoonian, "Solid solution strengthening of magnesium single crystal-II the effect of solute on the ease of prismatic slip" Acta Metallurgica 17 (1969) 1351-1356.
22. G. Simmons and H. Wang, *Single crystal elastic constants and calculated aggregate properties, second ed.* (The MIT Press, 1971)
23. M.H. Yoo, "Slip, twinning and fracture in hexagonal close-packed metals" *Metallurgical Transactions A* 12 (1981) 409-418.
24. E. Orowan, "A type of plastic deformation new in metals" Nature 149 (1942) 643-644.
25. J.B. Hess and C.S. Barrett, "Structure and nature of kink bands in zinc" Metals Transactions 185 (1949) 599-606.

Magnesium Technology 2011
Edited by: Wim H. Sillekens, Sean R. Agnew, Neale R. Neelameggham, and Suveen N. Mathaudhu
TMS (The Minerals, Metals & Materials Society), 2011

CRYSTAL PLASTICITY MODELING OF PURE MAGNESIUM CONSIDERING VOLUME FRACTION OF DEFORMATION TWINNING

Yuichi Tadano[1]
[1]Saga University; 1 Honjo-machi; Saga 840-8502, Japan

Keywords: Crystal Plasticity, Pure Magnesium, Deformation Twinning, HCP metal.

Abstract

In this study, a novel crystal plasticity model for pure magnesium involving the deformation twinning is presented. The deformation twinning is an important deformation mechanism of magnesium and other HCP metals. The deformation twinning has two important issues: first, the large rotation of crystal lattice caused by twinning occurs. Second, in the crystalline scale, the twinned and untwinned regions may simultaneously exist in a grain. Therefore, a crystal plasticity analysis of magnesium should introduce both of them, and the present framework takes these two key features into account. To represent the second issue, the volume fraction of deformation twinning is considered. This paper provides a framework of crystal plasticity model involving the effect of tensile twinning, and a numerical example is conducted to evaluate the evolution of volume fraction of twinned region. It is shown that the present scheme can describe the mixed state of twinned and untwinned regions. The obtained results suggest that the twinned and untwinned regions simultaneously exist even under the large deformation and the volume fraction of twinned region should be considered.

Introduction

Magnesium has drawn much attention as a structural material, especially in the transportation industry. Magnesium, with a density of 1.74×10^3 kg/m^3, is the lightest of the practical metals, and has excellent specific rigidity and specific strength in comparison with other industrial materials. Therefore, it is expected to be used as the next-generation lightweight material to reduce the weight of products [1]. On the other hand, it has poor formability because of two principal reasons: (1) strong anisotropy in the crystalline scale induced by a hexagonal close-packed (HCP) structure, and (2) direction-dependent c-axis deformation twinning [2]. Understanding the deformation mechanism in the crystalline scale and its effect on the macroscopic scale are important in improving the formability of magnesium.

The mechanical properties of polycrystalline metals are strongly affected by their microstructure such as crystalline aggregates. Most industrial metals have a face-centered cubic (FCC), body-centered cubic (BCC), or HCP structure as the crystal lattice, and pure magnesium exists in HCP form at room temperature. In HCP metals, basal slip $(0001)<11\bar{2}0>$ is generally the most easily activated slip system, because the critical resolved stress is much lower than those of other slip systems. Other slip systems that might be active in plastic deformation of magnesium have been reported: the prismatic slip system $\{10\bar{1}0\}$ $<11\bar{2}0>$ and pyramidal slip systems $\{10\bar{1}1\}$ $<11\bar{2}0>$ and $\{11\bar{2}2\}$ $<11\bar{2}3>$. Another important deformation mechanism of magnesium is twinning due to c-axis tension ($\{10\bar{1}2\}$ $<10\bar{1}1>$ twinning) [2]. In contrast with FCC metals, whose slip systems have identical mechanical properties, each slip system in HCP metals shows significantly different behaviors, causing strong anisotropy in most HCP metals.

Recently, multiscale simulations of crystalline scale structures have been actively studied; one model that can describe the microscale behaviors of metals is the crystal plasticity model. Studies of HCP materials using the crystal plasticity approach were actively reported in the last decade. Dawson and Marin showed the framework of polycrystalline plasticity of HCP metals [3], and many reports on investigation of HCP with the polycrystal plasticity approach followed [4-14]. These studies provided many fruitful results and productive discussions to clarify the essential deformation mechanism of HCP metals.

This study focuses on polycrystalline pure magnesium. The deformation twinning is an important deformation mechanism of magnesium. The deformation twinning has two important issues: first, the large rotation of crystal lattice caused by twinning occurs. Second, in the crystalline scale, twinned and untwinned regions may simultaneously exist in a grain. Therefore, a crystal plasticity analysis of magnesium should introduce both of them. To represent the second issue, the volume fraction of deformation twinning is considered, and the material behavior of a grain is described as mixed state of twinned and untwinned regions. This paper provides a framework of crystal plasticity modeling, and a numerical example is conducted to evaluate the evolution of volume fraction of twinned region.

Formulation

Crystal Plasticity Formulation

This study adopts the crystal plasticity model formulated by Peirce et al. [15] and Asaro and Needleman [16]. The velocity gradient $\mathbf{L}$ is assumed to be decomposed into nonplastic and plastic parts:

$$\mathbf{L} = \mathbf{L}^{*} + \mathbf{L}^{p} \tag{1}$$

The plastic part of velocity gradient $\mathbf{L}^p$ is defined by the sum of the slip deformation of all slip systems as

$$\mathbf{L}^{p} = \sum_{\alpha=1}^{N} \dot{\gamma}^{(\alpha)} \left(\mathbf{s}^{(\alpha)} \otimes \mathbf{m}^{(\alpha)} \right) \tag{2}$$

Here $\mathbf{m}^{(\alpha)}$ and $\mathbf{s}^{(\alpha)}$ are the slip plane normal and slip direction vectors, respectively, $\dot{\gamma}^{(\alpha)}$ is the slip rate, and N is the number of slip systems. The superscript (α) denotes a specific slip system. The plastic deformation rate tensor $\mathbf{D}^p$ and plastic spin tensor $\mathbf{W}^p$ are given by the symmetric and skew-symmetric parts of $\mathbf{L}^p$ as

$$\mathbf{D}^p = \left(\mathbf{L}^p + \mathbf{L}^{pT}\right)/2 = \sum_{\alpha=1}^{N} \dot{\gamma}^{(\alpha)} \left[\left(\mathbf{s}^{(\alpha)} \otimes \mathbf{m}^{(\alpha)} + \mathbf{m}^{(\alpha)} \otimes \mathbf{s}^{(\alpha)}\right)/2\right] \equiv \sum_{\alpha=1}^{N} \dot{\gamma}^{(\alpha)} \mathbf{p}^{(\alpha)} \tag{3}$$

$$\mathbf{W}^p = \left(\mathbf{L}^p - \mathbf{L}^{pT}\right)/2 = \sum_{\alpha=1}^{N} \dot{\gamma}^{(\alpha)} \left[\left(\mathbf{s}^{(\alpha)} \otimes \mathbf{m}^{(\alpha)} - \mathbf{m}^{(\alpha)} \otimes \mathbf{s}^{(\alpha)}\right)/2\right] \equiv \sum_{\alpha=1}^{N} \dot{\gamma}^{(\alpha)} \mathbf{w}^{(\alpha)} \tag{4}$$

The nonplastic deformation rate tensor $\mathbf{D}^*$ and plastic spin tensor $\mathbf{W}^*$ are given by

$$\mathbf{D}^* = \left(\mathbf{L}^* + \mathbf{L}^{*T}\right)/2 \tag{5}$$

$$\mathbf{W}^* = \left(\mathbf{L}^* - \mathbf{L}^{*T}\right)/2 \tag{6}$$

The elastic constitutive law is

$$\overset{\circ}{\boldsymbol{\sigma}}{}^* = \dot{\boldsymbol{\sigma}} - \mathbf{W}^*\boldsymbol{\sigma} + \boldsymbol{\sigma}\mathbf{W}^* = \mathbf{C} : \mathbf{D}^* \tag{7}$$

Here $\mathbf{C}$ is the fourth-order elastic stiffness tensor, and $\overset{\circ}{\mathbf{A}}$ denotes the Jaumann derivative of $\mathbf{A}$. From Eqs. (5) and (7), and the additive decompositions of the deformation velocity and the spin, $\mathbf{D} = \mathbf{D}^* + \mathbf{D}^p$ and $\mathbf{W} = \mathbf{W}^* + \mathbf{W}^p$ respectively, the following relationship is obtained:

$$\overset{\circ}{\boldsymbol{\sigma}} = \mathbf{C} : \mathbf{D} - \sum_{\alpha=1}^{N} \dot{\gamma}^{(\alpha)} \left[\mathbf{C} : \mathbf{p}^{(\alpha)} + \mathbf{w}^{(\alpha)}\boldsymbol{\sigma} - \boldsymbol{\sigma}\mathbf{w}^{(\alpha)}\right] \tag{8}$$

The changes in the direction of $\mathbf{m}^{(\alpha)}$ and $\mathbf{s}^{(\alpha)}$ caused by the rotation of the crystal lattice are calculated by the following equations:

$$\dot{\mathbf{m}}^{(\alpha)} = \mathbf{W}^* \cdot \mathbf{m}^{(\alpha)} \tag{9}$$

$$\dot{\mathbf{s}}^{(\alpha)} = \mathbf{W}^* \cdot \mathbf{s}^{(\alpha)} \tag{10}$$

The evolution equation of the slip rate $\dot{\gamma}^{(\alpha)}$ is assumed as the rate dependent form

$$\dot{\gamma}^{(\alpha)} = \dot{\gamma}_0 \,\mathrm{sgn}\left(\tau^{(\alpha)}\right) \left|\frac{\tau^{(\alpha)}}{g^{(\alpha)}}\right|^{1/m} \tag{11}$$

Here $\tau^{(\alpha)}$ is the resolved shear stress obtained by $\tau^{(\alpha)} = \mathbf{s} \cdot \boldsymbol{\sigma}\mathbf{m}$. $\mathrm{sgn}(x) = 1$ if $x \geq 0$ and $\mathrm{sgn}(x) = -1$ if $x < 0$; m and $\dot{\gamma}_0$ are the slip rate sensitivity parameter and the reference strain rate, respectively, and $g^{(\alpha)}$ is the reference slip resistance, which generally depends on both the slip system and the loading history; the evolution equation of $g^{(\alpha)}$ is given as

$$\dot{g}^{(\alpha)} = \sum_{\beta} h_{\alpha\beta} \dot{\gamma}^{(\beta)} \tag{12}$$

Here the matrix $h_{\alpha\beta}$ is the interaction between slip systems, given as

$$h_{\alpha\beta} = \begin{cases} h(\gamma_a) & (\alpha = \beta) \\ q_{\alpha\beta} h(\gamma_a) & (\alpha \neq \beta) \end{cases} \tag{13}$$

$$\gamma_a = \sum_{\alpha} \int_0^t \left|\dot{\gamma}^{(\alpha)}\right| d\tau \tag{14}$$

$q_{\alpha\beta}$ and $h(\gamma_a)$ are the matrices describing the latent hardening and the characteristic hardening function, respectively, and their concrete forms are described in the following section. In addition, the rate tangent modulus method [15] is used for numerical integration of the constitutive equation.

Modeling of Slip Deformation

This study focused on pure magnesium at room temperature, whose crystal lattice is an HCP structure. The slip systems considered in this study are illustrated in Figure 1 and Table I. The pyramidal slip system $\{10\bar{1}1\}$ $<11\bar{2}0>$ is also an active

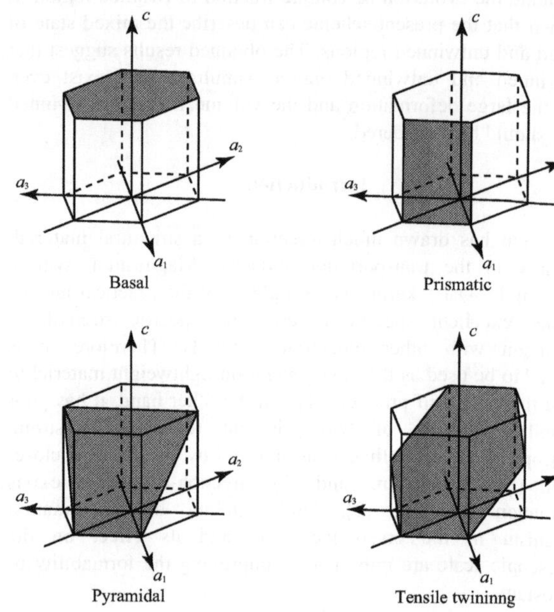

Figure 1. Slip and twining systems in the present analysis.

Table I. Slip and twin systems of HCP crystal.

	Number of Slip Systems	Slip Plane	Slip Direction
Basal	3	(0001)	$<11\bar{2}0>$
Prismatic	3	$\{10\bar{1}0\}$	$<11\bar{2}0>$
Pyramidal	6	$\{11\bar{2}2\}$	$<11\overline{23}>$
Tensile Twin	6	$\{10\bar{1}2\}$	$<10\bar{1}1>$

system in pure magnesium. However, it is neglected in the present study because it can be represented by a superposition of the basal (0001) $<11\bar{2}0>$ system and prismatic system $\{10\bar{1}0\}$ $<11\bar{2}0>$ [7]. Therefore, 12 slip systems and 6 twinning systems are considered here.

On the basis of the experimental results of Kelly and Hosford [17,18], Graff et al. established the following strain hardening laws for pure magnesium [7]:
Linear hardening for a basal system

$$h(\gamma_a) = h_0 \tag{15}$$

Voce hardening for prismatic and pyramidal systems

$$h(\gamma_a) = h_0\left(1 - \frac{\tau_0}{\tau_\infty}\right)\exp\left(-\frac{h_0\gamma_a}{\tau_\infty}\right) \tag{16}$$

In this framework, the deformation twinning is taken into account as a slip-like deformation, which can occur only in the tension direction, while general slip can occur in both directions of slip orientation. No reorientation after the saturation of twinning is considered. Therefore, this study takes the lattice rotation caused by the deformation twinning into account as the following manner.

Modeling of Deformation Twinning

If the deformation twinning occurs on a specified twinning plane, the crystal lattice in the twinned region takes the mirrored configuration of the original crystal lattice. The geometrical relation between twinned and untwined regions is expressed with an orthogonal tensor $\mathbf{T}^{\text{twin}}$.

$$\tilde{\mathbf{m}}^{(\alpha)} = \mathbf{T}^{\text{twin}} \cdot \mathbf{m}^{(\alpha)} \tag{17}$$
$$\tilde{\mathbf{s}}^{(\alpha)} = \mathbf{T}^{\text{twin}} \cdot \mathbf{s}^{(\alpha)} \tag{18}$$
$$\mathbf{T}^{\text{twin}} \equiv \mathbf{I} - 2\mathbf{m}^{\text{twin}} \otimes \mathbf{m}^{\text{twin}} \tag{19}$$

Here $\mathbf{m}^{\text{twin}}$ is the vector normal to the twinning plane. $\mathbf{m}^{(\alpha)}$ and $\mathbf{s}^{(\alpha)}$ are the normal to the slip plane and the slip direction vectors in the untwinned region, and $\tilde{\mathbf{m}}^{(\alpha)}$ and $\tilde{\mathbf{s}}^{(\alpha)}$ are those in the twinned region. The norm of each vector is unit. $\mathbf{I}$ denotes the second order unit tensor.

In this study, it is assumed that twinned and untwinned regions simultaneously exist in a material point, and f denotes the volume fraction of the twinned region. After the classical Taylor assumption, the twinned and untwinned region in a material point considered to be subjected a same strain.

$$\mathbf{L}^{\text{parent}} = \mathbf{L}^{\text{twin}} = \mathbf{L}^0 \tag{20}$$

The superscripts parent and twin indicate the untwinned and twined regions, respectively, and $\mathbf{L}^0$ is the macroscopic velocity gradient tensor. In this framework, the twinned and untwinned regions have different stress; therefore the macroscopic stress is defined as the volume average with the volume fraction f

$$\bar{\boldsymbol{\sigma}} = (1-f)\boldsymbol{\sigma}^{\text{parent}} + f\boldsymbol{\sigma}^{\text{twin}} \tag{21}$$

In the same way, the macroscopic constitutive tensor $\bar{\mathbf{C}}$ is give as

$$\bar{\mathbf{C}} = (1-f)\mathbf{C}^{\text{parent}} + f\mathbf{C}^{\text{twin}} \tag{22}$$

The proposed model presents the continuous transition of material properties with respect to deformation twinning.

Numerical results

Analysis Condition

The material parameters for the hardening laws and the components of the latent hardening matrix $q_{\alpha\beta}$ in Eq. (13) are shown in Tables II and III. Graff et al. determined these parameters by fitting simulation results to the channel-die compression tests of single crystal and polycrystal magnesium specimens [7]. In this study, most of parameters are identical as Graff's one; however, τ_0 for the basal slip system is modified to represent the material behaviors after deformation twinning more precisely. Young's modulus, Poisson's ratio, and m and $\dot{\gamma}_0$ in Eq. (11) are set to 45 [GPa], 0.3 [-], 0.02 [-], and 0.001 [s], respectively.

Table II. Material parameters for strain hardening law.

	Basal	Prismatic	Pyramidal
τ_0 / MPa	1	20	40
τ_∞ / MPa	---	150	260
h_0 / MPa	100	7500	7500

Table III. Components of the latent hardening matrix.

α	β			
	Basal	Prismatic	Pyramidal	Twin
Basal	0.2	0.5	0.5	0.5
Prismatic	0.2	0.2	0.2	0.5
Pyramidal	1.0	1.0	0.2	0.25
Twin	1.0	1.0	0.2	0.25

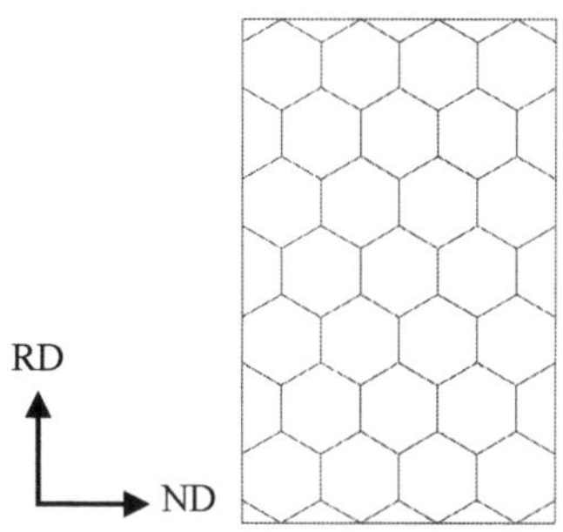

Figure 2. Initial configuration.

The analysis is conducted under the plane strain condition, i.e., $L_{33} = L_{23} = L_{32} = L_{31} = L_{13} = 0$, and the evolution of the volume fraction of twinned region is evaluated. The analysis model of polycrystal is shown in Figure 2. This model is divided by triangular finite elements and the number of finite elements is about 70,000. Note that the three dimensional configuration of crystal lattice is taken into account although the analysis condition is the plane strain. The top and bottom edges are the shear free condition and a compression of 10% nominal strain along RD direction is subjected. Each crystal grain has different initial orientation. A texture similar as the rolled material (rolled texture) and a texture with randomly distributed orientation (random texture) are considered as the initial crystal orientation.

Evolution of Twinned Region

Figure 3 shows the distribution of twinned region at 5% and 10% nominal strain. In case of the rolled texture, each crystal grain has the orientation whose Schmidt factor of a twinning system takes the highest value; therefore, the deformation twinning occurs in several grains even in the case of 5% nominal strain. Almost all material points in some grains are twinned, while only a part of material points are twinned in some grains. When the nominal strain is 10%, the twinned region expands; however, some grains remain to be not twinned. In case of the random textures, a number of twinned grains is fewer than the rolled texture because several grains have orientation hard to occur the deformation twinning. If a specimen is a single crystal, it is expected that all material points are twinned at the same time. However, the present results show that twinned and untwinned regions may exist in the same grain. This tendency is caused by the

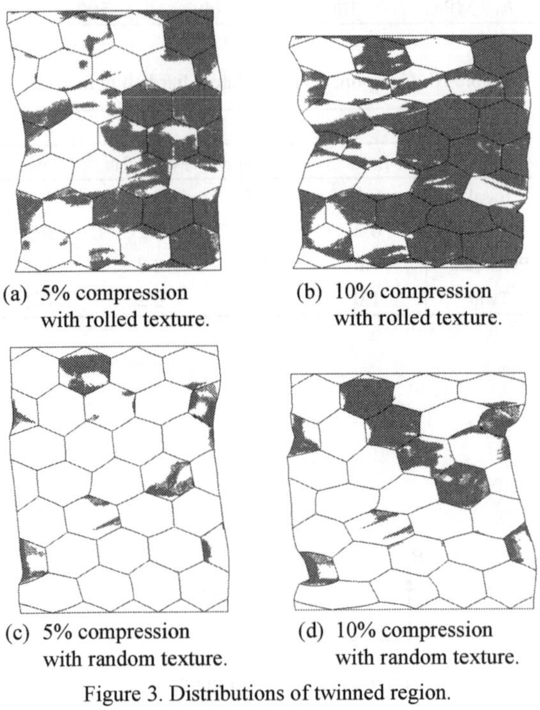

(a) 5% compression with rolled texture.
(b) 10% compression with rolled texture.
(c) 5% compression with random texture.
(d) 10% compression with random texture.

Figure 3. Distributions of twinned region.

inhomogeneous deformation due to the constraint between grains in the polycrystalline specimen. Note that the obtained results qualitatively agree with the experimental observations of polycrystalline magnesium [19, 20].

The evolution of volume fraction of twinned region is indicated in Figure 4. In case of the rolled texture, the deformation twinning occurs at about 2% nominal strain. In more than half of grains, the volume fraction reaches 100%, i.e., all material points are twinned. On the other hand, the volume fraction converges less than 100% value in some grains. In case of the random texture, the onset of

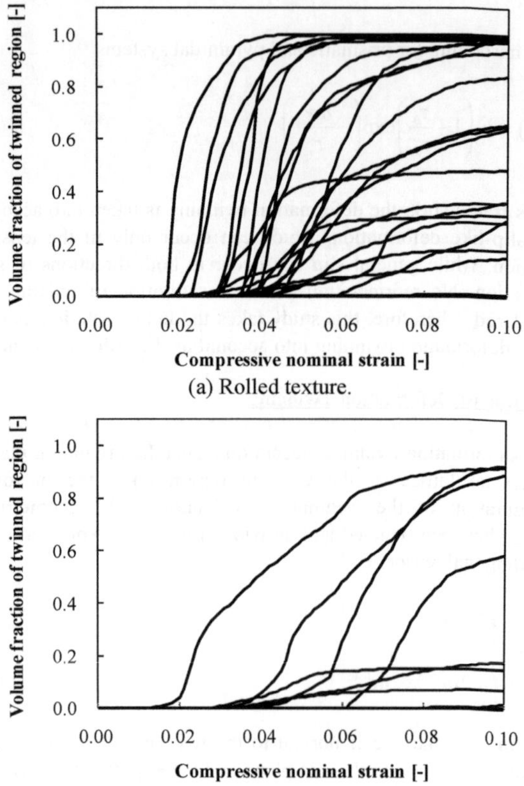

(a) Rolled texture.

(b) Random texture.

Figure 4. Evolution of volume fraction of twinned region.

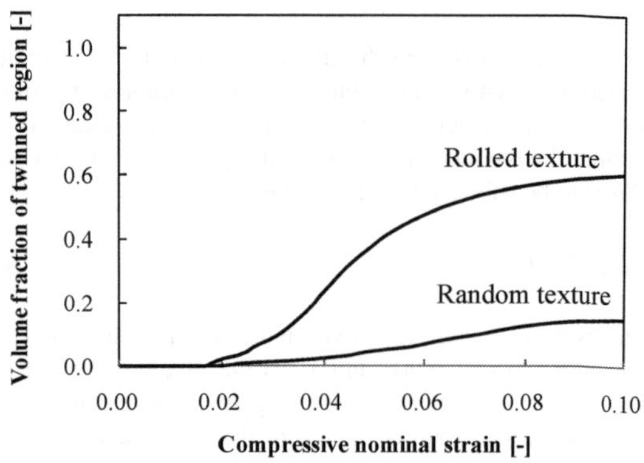

Figure 5. Evolution of volume fraction of twinned region with respect to whole volume of specimen.

deformation twinning delayed than the rolled texture, and the number of twinned grains is fewer; however, the qualitative tendency is similar as the rolled texture. Finally, the volume fraction of twinned region with respect to the whole volume of specimen is shown in Figure 5. The onset of twinning arises at about 2% nominal strain in both cases. At 10% nominal strain, the volume fraction reaches 60% in case of the rolled texture while the fraction is about 15% in case of the random texture. The present results suggest that the twinned and untwinned regions simultaneously exist even under the large deformation, and the volume fraction of twinned region should be considered to develop a constitutive model of polycrystalline pure magnesium.

Concluding Remarks

In this study, a novel crystal plasticity model for pure magnesium involving the deformation twinning is presented. The volume fraction of deformation twinning is considered, and a material behavior of a grain is described as mixed state of twinned and untwinned regions. A numerical example is conducted to evaluate the evolution of volume fraction of twinned region. The obtained results suggest that the twinned and untwinned regions simultaneously exist even under the large deformation, and the volume fraction of twinned region should be considered.

References

1. H.E. Friedrich and B.L. Mordike, *Magnesium technology, metallurgy, design data, applications*, Springer, Berlin, 2006.
2. F.H. Hosford, *Mechanical behavior of materials*, Cambridge University Press, New York, 2005.
3. P.R. Dawson and E.B. Marin, "Computational mechanics for metal deformation processes using polycrystal plasticity," *Adv. Appl. Mech.*, 34 (1998), 77-169.
4. A. Staroselsky and L. Anand, "A constitutive model for hcp materials deforming by slip and twinning: application to magnesium alloy AZ31B," *Int. J. Plasticity*, 19 (2003), 1843-1864.
5. S.R. Agnew et al., "Texture evolution of five wrought magnesium alloys during route A equal channel angular extrusion: Experiments and simulations," *Acta Mater.*, 53 (2005), 3135-3146.
6. S.R. Agnew and O. Duygulu, "Plastic anisotropy and the role of non-basal slip in magnesium alloy AZ31B," *Int. J. Plasticity*, 21 (2005), 1161-1193.

7 S. Graff, W. Brocks, and D. Steglich, "Yielding of magnesium: From single crystal to polycrystalline aggregates," *Int. J. Plasticity*, 23 (2007), 1957-1978.

8. B. Beausir, L.S. Tóth, and K.W. Neale, "Ideal orientations and persistence characteristics of hexagonal close packed crystals in simple shear," *Acta Mater.*, 55 (2007), 2695-2705.
9. B. Beausir et al., "Analysis of texture evolution in magnesium during equal channel angular extrusion," *Acta Mater.*, 56 (2008), 200-214.
10. A. Jain et al., "Grain size effects on the tensile properties and deformation mechanisms of a magnesium alloy, AZ31B, sheet," *Mater. Sci. Eng.*, A 486 (2008), 545-555.
11. G. Proust et al., "Modeling the effect of twinning and detwinning during strain-path changes of magnesium alloy AZ31," *Int. J. Plasticity*, 25 (2009), 861-880.
12. C.J. Neil and S.R. Agnew, "Crystal plasticity-based forming limit prediction for non-cubic metals: Application to Mg alloy AZ31B," *Int. J. Plasticity*, 25 (2009), 379-398.
13. Y. Tadano et al., "A polycrystalline analysis of hexagonal metal based on the homogenized method," *Key Eng. Mat.*, 340-341 (2007), 1049–1054.
14. Y. Tadano, "Polycrystalline behavior analysis of pure magnesium by the homogenization method," *Int. J. Mech. Sci.*, 52 (2010), 257-265.
15. D. Peirce, R.J. Asaro, and A. Needleman, "Material rate dependence and localized deformation in crystalline solids," *Acta Metall.*, 31 (1983), 1951-1976.
16. R.J. Asaro and A. Needleman, "Texture development and strain hardening in rate dependent polycrystals," *Acta Metall.*, 33 (1985), 923-953.
17. E.W. Kelley and W.F. Hosford, "Plane-strain compression of magnesium and magnesium alloy crystals," *Trans. Metall. Soc. AIME*, 242 (1968), 5-13.
18. E.W. Kelley and W.F. Hosford, "The deformation characteristics of textured magnesium," *Trans. Metall. Soc. AIME*, 242 (1968), 654-661.
19. L. Jiang et al., "Twinning and texture development in two Mg alloys subjected to loading along three different strain paths," *Acta Mat.*, 55 (2007), 3899-3910.
20. L. Jiang et al., "Influence of $\{10\bar{1}2\}$ extension twinning on the flow behavior of AZ31 Mg alloy," *Mater. Sci. Eng. A*, 445-446 (2007), 302-309.

Magnesium Technology 2011
Edited by: Wim H. Sillekens, Sean R. Agnew, Neale R. Neelameggham, and Suveen N. Mathaudhu
TMS (The Minerals, Metals & Materials Society), 2011

Nucleation Mechanism for Shuffling Dominated Twinning in Magnesium

Sungho Kim[1], H. El Kadiri[2], and M. F. Horstemeyer[1]
[1]Center for Advanced Vehicular Systems, Mississippi State University, Mississippi State, MS 39762
[2]Department of Mechanical Engineering, Mississippi State University, Mississippi State, MS 39762

Keywords: Magnesium, Twinning, Nucleation, Shuffling

Abstract

We observed nucleation of {10-12} twinning under tensile loading in magnesium rectangular rod system using atomistic molecular dynamic simulation. The rod axis is normal to basal plane of Mg crystal. The tensile deformation in c-axis nucleates {10-12} twinning starting at the corner of square of cross section of the rod. The twin boundary is spherical at the beginning and become a linear boundary in {10-12} planes as time goes by. The twinning and shuffling processes are described. The nucleation mechanism of the shuffling dominated twinning is explained.

Introduction

Twinning together with dislocation slip is a fundamental plastic deformation method. Many twinning nucleation mechanisms are proposed. Homogeneous twinning nucleation model in highly perfect crystals was first introduced by Orowan[5]. If the applied shear stress on twinning plane resolved along twinning direction reach theoretical strength of the material, homogeneous twinning occurs. Paxton et al.[4] considered the homogeneous twinning nucleation in this way against twinning- and anti-twinning shear stresses for five b.c.c. transition metals and for f.c.c. using density functional theory calculations. Thermal fluctuations are supposed to overcome the free energy barrier to the formation of a small twinned region.

The model of a lenticular twin consists of a series of loops of twinning dislocation. Each loop is in a lattice plane parallel to twinning plane. Loop diameter increases as the central plane of the lens is approached. The twinning dislocations in the loops are glissile and the embryonic twin could extend very rapidly in all directions contained within twinning planes. A possible mechanism for the thickening of a twin in the direction normal to twinning plane is the thermally-activated formation of a closed loop of twinning dislocation. The rate at which new loops form is big if the burger vector of twinning dislocation is small.

Defect-assisted twinning nucleation mechanisms are based on the dissociation of some dislocation which produce stacking fault for twin nucleus. The pole mechanism is one of defect-assisted twinning nucleation mechanisms.

Even though many theories of twin nucleation and dislocation loop in {10-12} twin planes are reported, none have reported a naturally occuring {10-12} tensile twin nucleation without artificial creation of twin structure in molecular dynamic simulations. The purpose of this article is to report the {10-12} twin nucleation, the nucleation mechanism, and structural analysis of twin in Mg crystal simulation.

Simulation setup

Figure 1 show the system configuration in the simulation. The system size is 9.33 nm by 5.51 nm by 5.09 nm. Total number of atom is 11520. The embed atom method magnesium potential of Sun et. al.[2] is used. No periodic boundary conditions are used in x, y, and z directions. The atoms in 1 nm of left and right ends of crystal are fixed after incremental strains are applied. The rest of atoms are free to move after strain applications. The constant NVE thermodynamics are applied where N represents the total number of atoms in system, V represents the volume of the system, and E represents the energy of system. Initial temperature are set to 100 K by assigning Gaussian random velocity on atoms. For 5 pico seconds equilibrium condition of system arrives. The ramping velocities are applied to all atoms correstponding to the strain rate of 1G/s.

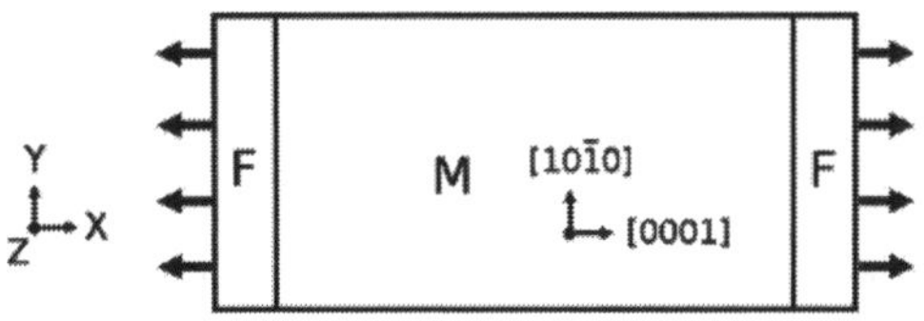

Figure 1. The schematic of the simulation configuration. Tensile strains are applied on Mg crystal in [0001] direction. No periodic boundary conditions are applied in all x, y, and z directions. The letter F indicates that atoms in the region are fixed and M indicates that atoms are mobile in the region in molecular dynamic steps after strains are applied.

Simulation results

The defect surface of quad sphere shape nucleates from the middle edge and propagates away from nucleation point. The defect surface leaves behind a HCP crystal structure which has different crystal oriention from parents. Figure 2 (a) shows three dimensional nucleation of twin. The colors represent the different structures classified by Ackland method[2]. Yellow represent h.c.p. structure and green f.c.c. The twin structure has the same h.c.p. as the parent and has different crystal orentation. The twin boundary is distinguished by different colors. The nucleated twin h.c.p. structure grows spherically from the nucleation point of a edge in the middle of simulation box.

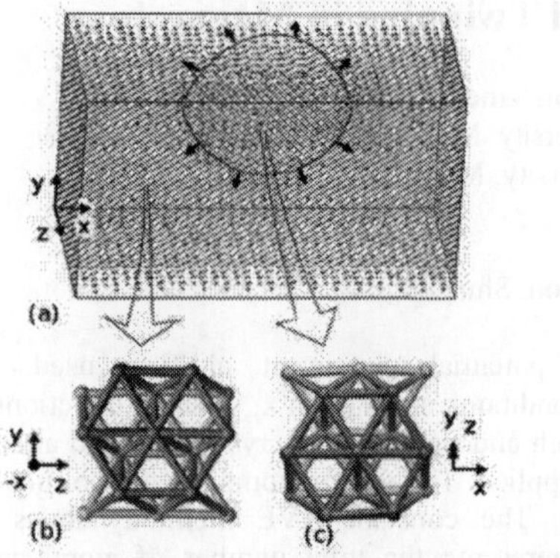

Figure 2. (Color online) (a) Twin nucleation and spherical growth of twin region from the middle edge. The colors of atoms are assigned by Ackland method[3]. The parent and twin structures are h.c.p. structures which represented yellow colors. (b) The structures of parent is magnified and viewed from x direction. The basal planes of parent is normal to x axis. (c) Twin region is magnified at specific view points to show the basal planes of twin. The colors in (b) and (c) represent depth.

Figure 3 (a) shows two layers of parent basal planes including twin nucleation point colored by Ackland method. The twin nucleates from top right conner and propergates spherically toward bottom left conner. As the twin region grows, two prismatic planes in parent region are aligned and form one basal plane in twin region. The angle between parent basal plane and twin basal plane look like right angle. The angle between parent and twin basal planes in {10-12} twin are 87 degrees according to theory. The twin boundary in –z direction is faster in speed and wider in width than –y direction.

Figure 3 (b) one layer colored by x positions of atoms. The red color and + sign mean the atoms are close, and the blue color and – sign mean the atoms are far. The basal plane of parent become the corrogated prismatic plane of twin.

Figure 3. (Color online) (a) Two layers of parent basal planes including twin nucleation point colored by Ackland method[3]. (b) One layers of the same view point colored by coordinate x positions of atoms.

Figure 4 (a) shows a few layers of (1-210) planes slicing the forefront of simulation box. The twinning boundary toward –z direction is quite thicker than others. Figure 4 (b) shows a few layers of (10-10) planes slicing the upper part of simulation box. The three dimensional shape of twin boundary is a ellipsoid and the growing speed and thickness of twin boundary are different.

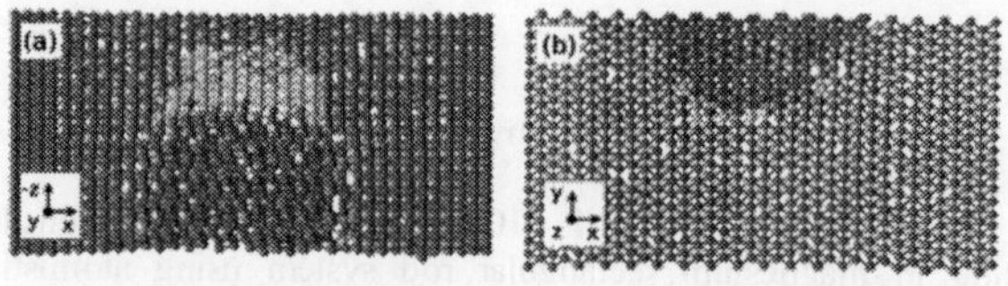

Figure 4. A few layers of top planes (a) and front planes (b) colored by Ackland method.

Figure 5 shows a few layers of (11-20) planes sliced to show the plane of shear for {10-12} twinning colored by Ackland method. In order to get rid of the interaction between surface and twin boundary more biger system is used for Figure 5. The twinning boundary shape in the plane of shear forms a retangle starting from a circle. The retangle consists of four twinning planes. The black solid line indicates the basal planes of parent and twin. The angle between parent and twin basal planes is consistant with the theoretic expectation.

Figure 5. A few layers of (11-20) planes slicing the twinning structure are colored by Ackland method[3].

Discussion

Our early twin structure just after nucleation is very different from common expectation that a certain number of atomic layers of twin structure nucleate at a time by twinning dislocations[6]. The twin boundary shape is spherical. The twin nucleation point is always at the edge of retangular rod. The thermal Brownian motion of atoms at edge creates twin nucleation point because the atoms at edge have more freedom to move away from the crystal lattice position and form nucleation points. The spherical shape of the twin boundary surface doesn't continue because the growing speeds of the twin boundaries are not same. Each part of the twin boundary grows with different velocity which depends on boundary structures and growing directions. As the twin region grows it might form the experimentally observed lenticular shape of twin.

Our overall simulation results suggest a surprising idea that the driving force to {10-12} twinning is tensile stress rather than shear stress. The {10-12} twinning forms by shearing and shuffling. The burger vector of twinning dislocation is extremely small and the shuffling is dominent. The shuffling lower the energy of system which is increased by applied tensile load. When the shear stress is applied to a crystal, dislocations are nucleated to accomadate loading stress instead of whole crystal shear deformation. In h.c.p {10-12} twinning, the shearing load is not the main contribution to twinning nucleation but the tensile load is. The shear-dominent twinning nucleates and grows by resolved shearing load. The shuffling-dominent twinning nucleates and grows by the resolved loading stress which induce shuffling to lower the system energy, for example, tensile loading stress in the current system. More evidences will be investigated in future.

Conclusion

We found a {10-12} twinning nucleation in rectangular rod system of Mg crystal under tensile stress in molecular dynamic simulation. The twinning nucleates from the edge which have more freedom of atom motion by thermal activation. The nucleated twinning grows spherically at early stage. The growing speeds of each part of spherical twin boundary are different and depend on boundary structures and growing directions.

Reference

[1] M. I. Baskes, "Application of the Embedded-Atom Method to Covalent Materials: A Semiempirical Potential for Silicon," *PRL*, 59, 2666, 1987.

[2] D. Y. Sun, J. W. Liu, X. G. Gong, and Zhi-Feng Liu, "Empirical potential for the interaction between molecular hydrogen and graphite," *PRB*, 75, 075424, 2007.

[3] G. J. Ackland and A. P. Jones, "Applications of local crystal structure measures in experiment and simulation," *PRB*, 73, 054104, 2006.

[4] A. T. Paxton, P. Gumbsch and M. Methfessel, Phil. Mag. 63A, 267, 1991.

[5] E.Orowan, Dislocations in Metals (edited by M. Cohen) p. 116. Amer. Inst. Min. (metall.) Engrs, New York (1964).

[6] J. Wang, J.P. Hirth, C.N. Tome, "(-1012) Twinning nucleation mechanisms in hexagonal-close-packed crystals", Acta Materialia 57, 5521, 2009.

Magnesium Technology 2011
Edited by: Wim H. Sillekens, Sean R. Agnew, Neale R. Neelameggham, and Suveen N. Mathaudhu
TMS (The Minerals, Metals & Materials Society), 2011

ON THE IMPACT OF SECOND PHASE PARTICLES ON TWINNING IN MAGNESIUM ALLOYS

M.R. Barnett[1], N. Stanford[1], J. Geng[2], J. Robson[3]
ARC Centre of Excellence for Design in Light Metals
[1]Centre for Material and Fibre Innovation, ITRI, Deakin University, Pigdons Rd, Geelong 3217, Australia
[2]Department of Materials Engineering, Monash University, Clayton 3800, Australia
[3]Manchester Materials Science Centre, Grosvenor St, Manchester M17HS, UK

Keywords: Magnesium, twinning, precipitation, TEM

Abstract

Deformation twinning is an important deformation mode in magnesium alloys. Despite this, little is known on the extent to which the stress for twinning can be altered by a dispersion of second phase particles. The current paper presents a series of findings on the role of differently shaped particles on the characteristics of the twins that form. It is shown that coherent rod shaped particles in Mg-Zn alloys have little obvious effect locally on the twin boundaries but that the twin number density is increased by their presence. Plate particles in a Mg-Al-Zn alloy cause obvious perturbations to the twin interface. Loops of twinning dislocations left around the particles eventually collapse into the particle interface, a phenomenon that is evidently facilitated by stress concentration on leading twin dislocation and stress relaxation in the adjacent material.

Introduction

The $\{10\bar{1}2\}$ twin plays an important role in the mechanical response of magnesium and its alloys. In some cases it dominates it. Two important examples are the compression of wrought plate in its plane and the compression of extruded material along its extrusion direction. In these cases the first few percent of strain can sometimes be attributed entirely to the $\{10\bar{1}2\}$ twin [1]. The practical consequence of these facts is a stress-strain curve in compression that differs in both the yield strength and strain hardening response to that seen in tension. It is not uncommon for the yield stress of an extrusion tested in compression to fall below that found in a tension test by a factor of 2.

When twinning dominates the plastic response, the sensitivity to other processing or "service" variables differs to that seen when slip dominates. For example, it is known that the stresses required for twinning are less sensitive to temperature and strain rate than those seen for slip. Consequently, the yielding stresses in extrusions tested in compression along the extrusion direction show relatively little change with temperature and strain rate [2-4].

The grain size also has a different impact upon yielding when twinning is prevalent. In their review of the literature, Meyers et al. [5] found that the Hall-Petch slope – i.e. the sensitivity of yield stress to grain size – is invariably higher when twinning dominates. In magnesium alloys, it has been found that the Hall-Petch slope for extrusions tested in compression falls approximately twice that seen in tension (i.e. ~10 $MPa.mm^{1/2}$ compared to ~5 $MPa.mm^{1/2}$) [6]. This obviously has important consequences for the strength of compression members and for the tension-compression yield asymmetry.

We turn now to the strengthening effect of fine dispersed precipitates, the subject of this short communication. For slip the basic principles are well known (e.g. [7]). At low particle volume fractions, small inter-particle distances, which force passing dislocations to bow to smaller radii, give higher strengths. Non-shearable particles facilitate higher work hardening rates – due primarily to geometrically necessary Orowan loops left in the vicinity of particles. At higher volume fractions, the back-stress generated by non-relaxed elastic stresses becomes important and kinematic work hardening is enhanced. For twinning, the effects are less well understood.

First, one must determine how the separate processes of twin nucleation (formation) and growth respond to the presence of particles. Here, we restrict our interest to growth. Richman [8], in his survey of twinning phenomena in bcc metals, examined the interaction between twins and $FeBe_2$ particles in an alloy of Iron and 25 at% Be. Upon encountering a particle, the twins "pinched off". The scenario painted by Richman following his microscopy work is sketched in Figure 1, in which his original diagram is closely followed. Twinning dislocations loop around the particle in a manner similar that followed by slip dislocations. Only here an inclined twin interface is built up around that particle. The particle and the material around it are not sheared in this case, though they are highly stressed. It is imagined that Orowan type precipitation hardening occurs in such an instance.

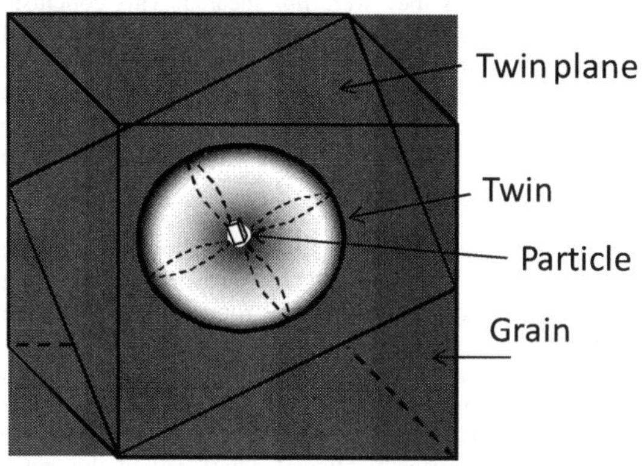

Figure 1. A twin centred on a particle producing a "donut" effect where loops of twinning dislocations build up around the particle. Diagram sketched from a schematic provided by Richman [8].

In his transmission electron microscope study on a magnesium-5% zinc alloy, Clark [9] observed the interaction between dislocations, twins and MgZn′ particles. The particle phase is a transition structure with a high coherency with the magnesium lattice. Basal dislocations were not able to shear the precipitates, but twins did. Sheared precipitates were seen to be "engulfed" by the twin. A schematic based on Figure 11 of Clark's paper is provided in Figure 2. This shearing is in marked contrast to the non-shearing particles observed by Richman. It is interesting that the same obstacle was not sheared by slip dislocations. Clark ascribed this to two effects: i) the relatively small shear associated with the twinning dislocations (Burgers vector $b_T = h\gamma$ where h is the step height, $2d$=0.38 nm in the current twin, and γ=0.13 is the twinning shear, thus b_T~0.05 nm) and ii) the supposed difficulty for a twin dislocation to bow around a particle due to its confinement in the twin interface.

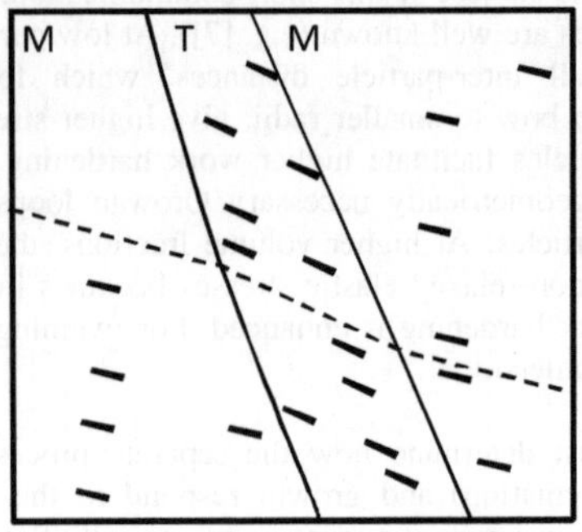

Figure 2. Particles of MgZn' sheared and thereby "engulfed" by the progress of a twin through a grain in Mg-5Zn [9]. Schematic based on Figure 11 of the paper.

Gharghouri et al. [10] studied the interaction between twins and precipitates in Mg-7.7Al, which forms plate shaped particles upon aging. In their study, twins were observed to be held up by particles, to engulf particles and in some instances to bypass particles. This last situation is shown schematically in Figure 3, which follows a published transmission electron microscope image. Important for the present consideration is the fact that Ghargouri et al. maintain that although the precipitates were engulfed by their twins they were *not* sheared. This conclusion appears to be based largely on their determination of the yield stress of the precipitate, which was found to be in the order of 1 GPa, approximately an order of magnitude higher than the yield stress of the matrix. The engulfed precipitates were seen to be inclined ~4° to their initial orientation in the direction of the twinning shear.

Two of the present authors have examined an alloy similar to that employed by Clark – Mg-5%Zn [9]. In that work, we concluded that the precipitates did not show any rotation, nor were they sheared by either twins or dislocations. However, precise measurements of rotation were hindered by foil warping and difficulties in obtaining sufficient contrast. Basal faults in the twin were used as fiducial markers but we acknowledge that this approach assumes a perfectly aligned c-axis rod in the matrix prior to twinning and this may not always be the case. Nevertheless, the study showed that twin nucleation was enhanced by precipitation but that twin growth was suppressed by it.

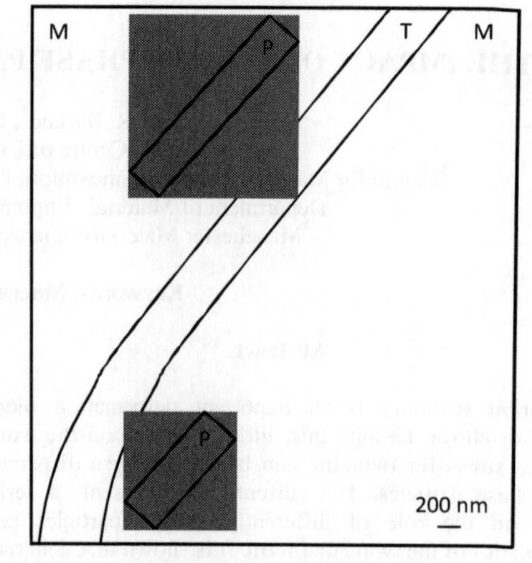

Figure 3. Sketch of a twin bypassing two precipitate particles in aged Mg-7.7Al as seen by Ghargouri et al. [10].

In the present study, we employ some additional microscopy to determine how twins interact with precipitates in magnesium alloys strengthened by MgZn' rods (Mg-5Zn) and $Mg_{17}Al_{12}$ plates (Mg-9Al-1Zn).

Methodology

Extruded magnesium round bar containing 5 wt% Zn (Z5) was employed in the study. A two step solution treatment was performed following extrusion, 2 hr at 330°C followed by 6 hr at 400°C. Aging was performed at 250°C, 200°C, 150°C and 110°C for 10 hr, 18 hr, 8 days and 32 days. After Clark [9] and Gao and Nie [11], the rod precipitates thus produced are a coherent transition phase referred to as MgZn′ or β_1' with their long axis parallel to the c-axis of the magnesium matrix. Compression specimens 8 mm in height and 5.4 mm in diameter were examined using Electron Backscatter Diffraction and transmission electron microscopy after deformation to a strain of 5%. The grain size was 30 µm and the texture was typical for a magnesium extrusion. More information is presented in reference [12].

Rolled magnesium alloy with 8.6 wt% Al and 0.8 wt% Zn with 0.24% Mn (AZ91) was also examined. Solution treatment was performed at 420°C for 24 hours. Aging was performed at 200°C for 2 hr, 30 hr and 20 days. The precipitates produced are plates of $Mg_{17}Al_{12}$ that lie on the basal plane. The grain size was 70 µm and the texture was typical for a magnesium plate although there were a few grains that had their c-axis tilted towards the transverse direction. Compression specimens were cut to enable compression to be performed along the rolling direction. Transmission electron microscopy was carried out after deformation to different strains. More information will appear in a future publication.

Specimens for EBSD were prepared using electropolishing in 5% nitric acid in ethanol at 20V for 30-45 s. The same solution was used for chemical cleaning for 5-10 s. For TEM sample preparation, a Gatan plasma ion polishing system was employed.

Results and Discussion

To gain insight into how twins interacted with precipitates at their encounter, precipitates in the vicinity of $\{10\bar{1}2\}$ twin interfaces were inspected using transmission electron microscopy. In all cases images were produced on the $<11\bar{2}0>$ zone axis. This is the "edge on" condition where the twin plane normal lies in the plane of the image, as does the twin shear direction $<\bar{1}011>$.

In Figure 4 two images taken near to a twin boundary in the Z5 alloy aged at 150°C are shown. In each case instances can be seen where the inclination of the precipitate long axis differs on either side of the twin boundary. As mentioned above, this was not always detected in this alloy. The effect appears similar to that seen by Clark [9] but due to the scarcity of the observations, further work is required to be certain. The sense of the shear provided by the twin, determined from inspection of the inclination of the twin boundaries to the matrix c-axis and noting the polarity of the twinning is such as to give extension along c. The slight rotation of the precipitates that can be seen near these twin boundaries is the right sense according to the twinning shear. Of considerable interest in Figure 4 is that there is no obvious sign of any "pinching off" or dislocation looping where the twins meet the precipitates. The twin boundaries show quite strong contrast in these images such that it is possible to detect ledges in the twin interface. These seem to be separated by longer regions of coherent twin interface on the lower boundary in Figure 4a compared to the upper boundary. At this lower interface some overlap of the flat portions of twin interface can be seen and this reflects the bowing of a twinning dislocation between the upper and lower surfaces of the foil. No such effects are seen to be concentrated at the particles. The precipitates appear to have been sheared with minor disruption to the twin interface.

As an aside, the height of the ledges in the twin boundary in Figure 4a is in the order of 2 nm. This equates to a segment of incoherent twin boundary comprised of 5 twinning dislocations.

In the case in Figure 4, the precipitates are rod shaped and are located well within the foil, whereas the twin boundary traverses the foil width. This may obscure features of the interaction. However, in the case of the plate shaped precipitates seen in AZ91, the plates are sufficiently large (295-825 nm) to traverse the whole foil thickness. An example of the twin-particle interaction seen in alloy AZ91 is shown in Figure 5. In this instance, the twin appears to have been held up at the particle in the lower right corner. It is also clear that twinning dislocations are looped around the precipitate to the right of centre in this image. Here the precipitate remains unsheared.

In Figure 5, the twin is thin. The thickening of the twins is not prevented in this material, however, and an image of a thicker twin is presented in Figure 6. In this case there is no obvious looping of twinning dislocations around the precipitates. However, some larger scale curvature concave to the twin can be seen on the right of the image. This requires the existence of loops of twinning dislocations. The twin interface is also considerably perturbed in the top left of the image.

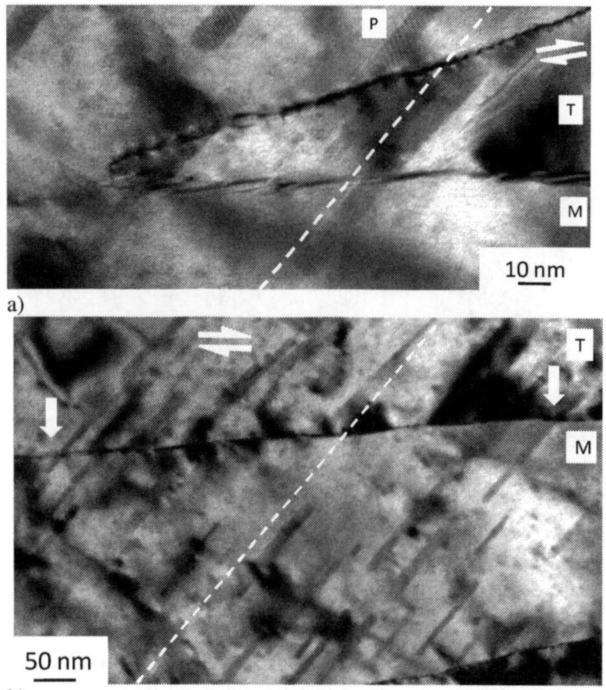

Figure 4. TEM images of $\{10\bar{1}2\}$ twin interfaces seen in alloy Z5 aged at 150°C and compressed along the extrusion direction to 5% strain. The twins shown share the $<11\bar{2}0>$ zone axis.

Figure 5. TEM image of a $\{10\bar{1}2\}$ twin seen in alloy AZ91 peak aged at 200°C and compressed along the extrusion direction to 8.5% strain.

Importantly, Figure 6 shows that the plate particles are strained considerably. Whether elastic or plastic, such a large strain requires considerable stresses. This consequently entails i) stress concentration, most likely in the form of pile-ups of twinning dislocations (seen as regions of twin boundary with an "elliptical" curvature) and ii) stress relaxation, most likely in the form of basal slip at the twin tip (possibly along with other emissary dislocations [13]) and basal slip inside the twin. Indeed, we have noted previously [12] that the hardening of accommodation slip is likely to comprise an important contribution to the hardening of twinning by particle addition.

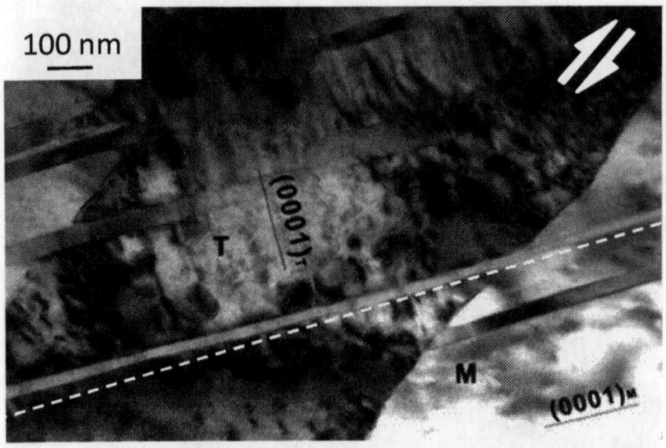

Figure 6. TEM image of a $\{10\bar{1}2\}$ twin seen in alloy AZ91 peak aged at 200°C and compressed along the extrusion direction to 8.5% strain.

A difficult observation to reconcile is that in Figure 6, the twinning dislocations appear to have collapsed into the matrix-precipitate interface. For this to occur at the interface of a strong (σ_y~1 GPa) rigid (E~80 GPa) particle is not easy to understand. Shear along the particle interface will go some way towards accommodating the incompatability strains. Such slip is shown in Figure 7. Considerable dislocation activity can be seen emanating from the tip of the particle and extending into the twin interior. Indeed, Gharghouri et al. [10] noted high "accommodating" slip activity on the perimeter of equivalent particles in their alloy.

Figure 6 also shows the process of collapse of loops of twinning dislocations into the particle interface. On the right of the particle, the "collapse" has occurred to a further extent than on the left. In the case of the latter, the twinning dislocations have to glide "past" each other and through the stable vertical array configuration so as to meet the particle interface. These dislocations are not subject to the same pile-up stress magnification that drives the dislocations into the other face of the particle. A similar asymmetry is seen in the loops in Figure 5.

It now remains to observe the interaction between twins and precipitates at a lower magnification. EBSD analysis (see [14]) was employed to observe twin growth and formation in alloy Z5 as a function of aging condition. The results are plotted in Figure 8 as a function of the yield stress. As hardening due to aging progresses, the twin number density rises. This can be attributed to increased resistance to twin growth over nucleation. The higher stresses required for twin propagation enable more twins to form. At the same time the twin volume fraction drops. With increasing aging up to peak hardness, the volume fraction continues to drop. So too does the number density. However, the number density does not drop below that seen in the unaged condition, even during overaging. Although overaging does not alter the twin number density much, it lifts the volume fraction considerably. The effect of this exceeds what one would expect from a stress based argument alone; that is, the overaged datum falls outside of the relationship between stress and volume fraction seen during aging. This may be due to simultaneous effects of particles on slip. The extent of twinning depends on its competition with slip.

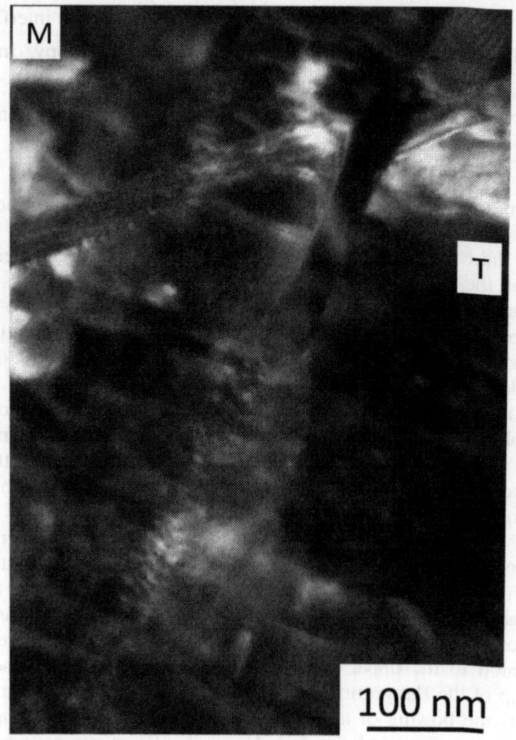

Figure 7. TEM image of a $\{10\bar{1}2\}$ twin seen in alloy AZ91 peak aged at 200°C and compressed along the extrusion direction to 8.5% strain.

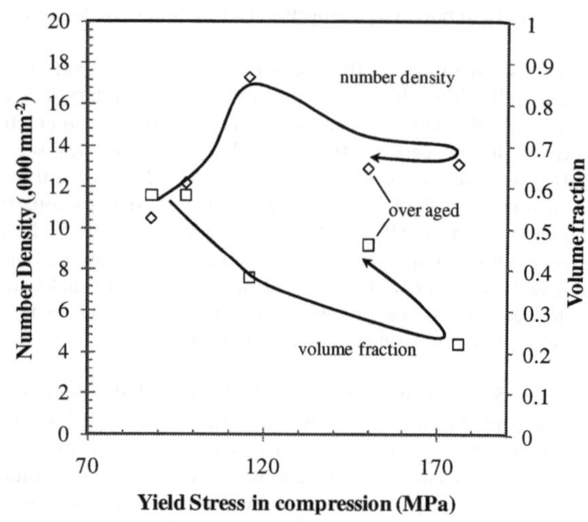

Figure 8. Twin volume fraction and number density determined using EBSD for magnesium alloy Z5 extruded and tested in compression to a strain of 5% after aging to different levels of yield stress.

Conclusions

1. Rod shaped precipitates in magnesium alloy Z5 seem to shear during twinning although more accurate measurements of the effect are required.
2. No obvious Orowan type looping is seen where the rods intersect a twin interface.
3. Orowan type looping is seen where twins meet plate shaped precipitates in alloy AZ91.
4. These loops collapse into the particle interface, driven by pile-up stress magnification and eased by slip relaxation into the twin.
5. In Z5 aging initially promotes twin nucleation but consistently depresses the twin volume fraction.

Acknowledgements

Dr C.H.J Davies, Dr J-F Nie, Dr Y. Chun are acknowledged for the support in the ARC CoE project within which this work was carried out.

References

1. Brown, D.W., S.R. Agnew, M.A.M. Bourke, T.M. Holden, M. Vogel, and C. Tome, *Internal strain and texture evolution during deformation twinning in magnesium.* Materials Science and Engineering A, 2005. 399: p. 1-12.

2. Beer, A.G. and M.R. Barnett, *Influence of Initial Microstructure on the Hot Working Flow Stress of Mg–3A1–1Zn.* Materials Science and Engineering A, 2006. 423 (Stuctural Materials: Properties, Microstructures and Processing; Elsevier SA, Switzerland): p. 292-299.

3. Barnett, M.R., *A Taylor Based Description of the Proof Stress of Magnesium AZ31 during Hot Working.* Metallurgical and Materials Transactions A, 2003. 34A: p. 1799-1806.

4. Klimanek, P. and A. Pötzsch, *Microstructure Evolution under Compressive Plastic Deformation of Magnesium at Different Temperatures and Strain Rates.* Materials Science and Engineering, 2002. A324 (1-2 Special Issue SI): p. 145-150.

5. Meyers, M.A., O. Vohringer, and V.A. Lubarda, *The onset of twinning in metals: a constitutive description.* Acta Materialia, 2001. 49 (19): p. 4025-4039.

6 Barnett, M.R., Z. Keshavarz, A.G. Beer, and D. Atwell, *Influence of grain size on the compressive deformation of wrought Mg–3Al–1Zn.* Acta Materialia, 2004. 52 (17): p. 5093-5103.

7. Martin, J.W., *Precipitation hardening.* 1968, Oxford: Pergamon Press.

8. Richman, R.H. *The diversity of twinning in body centred cubic structures.* in *TMS-AIME Conf., Deformation Twinning.* 1964. Gainesville, Florida: American Institute of Mining, Metallurgical and Petroleum Engineers, INC., Printed in Great Britain.

9. Clark, J.B., *Transmission electron microscopy study of age hardening in a Mg-5 wt.% Zn alloy.* Acta Metallurgica, 1965. 13 (12): p. 1281-1289.

10. Gharghouri, M.A., G.C. Weatherly, and J.D. Embury, *The Interaction of Twins and Precipitates in a Mg-7.7at.%Al Alloy.* Philosophical Magazine A, 1998. 78 (5): p. 1137-1149.

11. Gao, X. and J.F. Nie, *Characterization of strengthening precipitate phases in a Mg-Zn alloy.* Scripta Materialia, 2007. 56 (8): p. 645-648.

12. Stanford, N. and M.R. Barnett, *Effect of particles on the formation of deformation twins in a magnesium-based alloy.* Materials Science and Engineering: A, 2009. 516 (1-2): p. 226-234.

13. Agnew, S.R., J.A. Horton, and M.H. Yoo, *Transmission Electron Microscopy Investigation of <c +a> Dislocations in Mg and Solid Solution Mg-Li Alloys.* Metallurgical and Materials Transactions A, 2002. 33A: p. 851-858.

14. Robson, J.D., N. Stanford, and M.R. Barnett, *Effect of particles in promoting twin nucleation in a Mg-5 wt.% Zn alloy.* Scripta Materialia. 63 (8): p. 823-826.

Magnesium Technology 2011
Edited by: Wim H. Sillekens, Sean R. Agnew, Neale R. Neelameggham, and Suveen N. Mathaudhu
TMS (The Minerals, Metals & Materials Society), 2011

Influence of Crystallographic Orientation on Twin Nucleation in Single Crystal Magnesium

C.D. Barrett[1], M.A. Tschopp[1], H. El Kadiri[1], B. Li[1]

[1]Center for Advanced Vehicular Systems
Mississippi State University, Starkville, MS USA

Keywords: Magnesium, Twin Nucleation, Molecular Dynamics, Single crystal

Abstract

Experimental plasticity on single crystals has found substantial non-Schmid effects in both twinning and non-basal slip in pure magnesium. The deviation from Schmid's law has been attributed to the strong sensitivity of both twinning and slip to small lattice heterogeneities [1] and the effect of pre-slip and non-planar dislocation dissociation [2]. However, most molecular dynamics simulations use heterogeneities so the effect of slip on twin nucleation and vice-versa has been shrouded. This has motivated us to investigate the influence of crystal loading orientation on *homogeneous* slip and twin nucleation using molecular dynamics. These simulations allowed us to appreciate the propensity and nature of twin nucleation when pre-existing defects are absent. Analyses of deformation mechanisms and stress-strain responses shows that homogeneous dislocation nucleation on the basal slip system is correlated with the highest Schmid resolved shear stress, while homogeneous nucleation of tensile twins did not always correlate with the highest Schmid resolved shear stress.

Introduction

Twinning readily occurs in double lattice structures such as hexagonal-close-packed (HCP) structures primarily because the close-packed directions contained within the basal planes cannot accommodate c-axis deformation. Other slip modes having a Burgers' vector with a non-zero component along the c-axis, such as pyramidal $\langle c+a \rangle$ slip are at least five times harder than basal slip in polycrystals. Although twinning is believed to be athermal, the critical stress to activate prismatic and pyramidal slip decreases with increasing temperature, thereby minimizing the contribution of twinning at higher temperatures unless the strain rate is increased. Twinning sharply reorients the parent crystal, transmutes parent dislocations [3], and introduces substantial interfaces, thereby causing strong crystallographic anisotropy and a rapidly increasing hardening rate behavior, particularly when twinning is able to consume the entire parent grain in sharply textured materials. As high-strength wrought magnesium alloys are systemically textured, twinning is highly sought to be prohibited or mitigated so to promote forming at cost-effective rates without risking distortion of components. In part, these have motivated renewed efforts towards a fundamental understanding of the mechanisms underlying twin nucleation in HCP lattices.

While some insight into twin nucleation behavior may be able to be obtained via in situ high resolution transmission electron microscopy (TEM) experiments, these experiments are often very difficult to perform and are not without complications. The use of molecular dynamics (MD) simulations to probe plasticity in HCP structures has received much interest in the last decade. The primary objective of these simulations has been to elucidate the mechanisms of twin nucleation as well as some interest dedicated to non-basal slip activities. In these simulations, the twin nucleation problem tends to be a highly-debated subject. The influence of the interatomic potentials for HCP metals on twinning is important in the results of MD simulations. Also, the use of constrained free surfaces introduces a planar defect into the system that influences twinning in many MD simulations. Planar defects, such as grain boundaries, and line defects, such as dislocations, play an important role in twin nucleation. Twins systemically nucleate at low angle grain boundaries in polycrystals [4]. Furthermore, recent experiments have shown that even under profuse twinning conditions, twins nucleate only after 1.3% plastic strain. These studies show that prior slip plays a key role in twin nucleation and have actually substantiated the theories developed earlier by Mendelson [2] on the role of prior slip in twin nucleation. This slip-twin interaction may have also an important influence on the non-Schmid behavior observed when single crystals have been experimentally tested. For instance, Kelley and Hosford [1] showed a surprising behavior of non-basal slip and twins nucleating in samples designed to favor only one single deformation mode based on assumptions from the Schmid law classically used in single lattice structures such as FCC and BCC. This had led to considerably conflicting values of threshold stresses of non-basal slip and twinning systems.

So, how can the influence of resolved stress components *and* heterogeneities be addressed using molecular dynamics simulations? Molecular dynamics simulations that use pre-existing defects in a crystal lattice make it difficult to distinguish what is the effect of resolved stress components and what is the effect of the introduced heterogeneities. However, previous simulations on homogeneous and heterogeneous dislocation nucleation in FCC crystals may provide the framework for answering this question. Tschopp, Spearot, and McDowell [5] approached this problem in FCC Cu by: 1) analyzing the influence of resolved stresses on *homogeneous* dislocation nucleation in single crystal Cu, 2) formulating a phenomenological nucleation that accounted for resolved stress components, 3) analyzing the influence of added defects on *heterogeneous* dislocation nucleation, and 4) inserting a damage-like parameter to capture the resulting influence due to lattice hetereogeneities. While twin and dislocation nucleation in HCP metals is considerably different from dislocation nucleation in FCC crystals, the methodology can be applied to HCP metals. The key first step in this methodology is understanding how resolved stresses impact homogeneous nucleation in single crystals and developing a fundamental phenomenological model that can capture this. In this paper, we have used MD simulations to explore *homogeneous* dislocation and twin nucleation in single crystal Mg under uniaxial loading conditions as a function of crystallographic orientation. The crystallographic orientation of loading will impact the resolved shear stresses on various dislocation and twin systems within Mg. This will enable us to explore the relationships between the actual mechanisms, multiple potential slip/twin systems, and the corresponding Schmid and non-Schmid resolved shear stresses to try to develop better phenomenological models for twin nucleation as a function of crystallographic orientation.

Simulation Methodology

A parallel molecular dynamics code (LAMMPS, [6]) that incorporates domain decomposition is used to deform the single crystal atomistic models. A similar methodology to that previously used to examine homogeneous dislocation nucleation in single crystal FCC crystals is used here [7; 8; 9]. First, the configuration is equilibrated using MD in the isobaric-isothermal (NPT) ensemble at a pressure of 0 bar and a temperature of 100 K for 10 ps. Next, the configuration is deformed in uniaxial tension at a constant strain rate of 10^9 s^{-1} with a stress-free condition

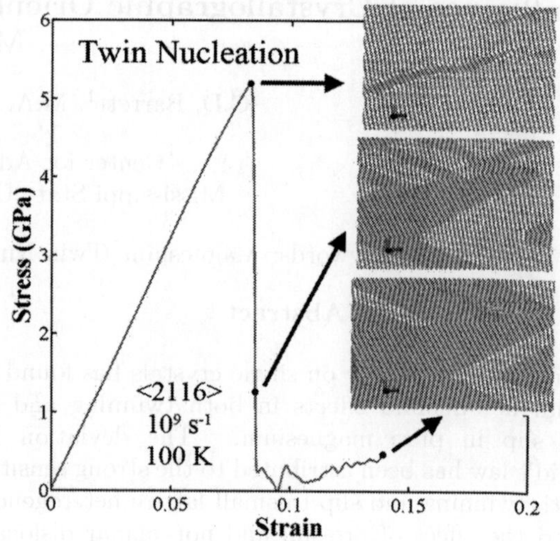

Figure 1: An example stress-strain curve showing twin nucleation at the maximum tensile stress.

for the other two boundaries. For mechanical properties, the system stress is calculated using the virial definition. The stress required for twin or dislocation nucleation is defined as the maximum uniaxial tensile stress. Figure 1 shows a stress-strain curve for the ⟨2116⟩ loading orientation at a strain rate of 10^9 s^{-1} and a temperature of 100 K. As shown in Fig. 1, visualization of selected tensile axis orientations along the appropriate planes showed that twins/dislocation are nucleated at a displacement very close to the maximum tensile stress for all single crystal models. In most cases, twins appeared to nucleate slightly before the maximum tensile stress is reached (<1.0% below the maximum tensile stress). However, in light of the difficulty of visually ascertaining exactly when the twin nucleates (i.e., how many spatially clustered, distorted atoms on the twinning plane constitute a single twin embryo), the maximum tensile stress provides an accurate indication of the stress required to homogeneously nucleate a twin in each single crystal model.

The Sun et al. [10] embedded-atom method potential for Mg is employed in this study. The Sun et al. potential was fit to crystal, liquid, and melting properties of Mg and dislocation core structures compare well compared to ab initio and experimental values [11]. Furthermore, Yasi and colleagues [11] found that the Sun et al. potential better captured the splitting distance of dissociated edge and screw partial dislocations on the basal plane than the Liu et al. [12] potential when compared to ab initio results. Additionally, the Peierls stresses for basal (0.3 and 3.6 MPa for edge and screw) and prismatic slip (13 and 44 MPa for edge and screw) are in agreement with experiments. For all these reasons, the aforementioned potential is deemed sufficient to characterize homogeneous twin nucleation in single crystal Mg.

A 3D periodic computational cell is used to investigate homogeneous twin nucleation in single crystal Mg under uniaxial tension at a constant strain rate. Cell dimensions are chosen to properly enforce the 3D periodic boundary conditions for each orientation, with a minimum length of 20 nm in all directions. This length is chosen to minimize both the effect of periodic boundaries on twin nucleation and the number of atoms in the system for computational efficiency. The non-integer nature of the c/a ratio for the interatomic potential complicates the generation of a 3D periodic cell in HCP metals. Although not explicitly shown in this work, simulations investigating the effect of the periodic length validate that the stress required for twin nucleation is essentially unaffected by further increases in the size of the simulation cell. This minimum periodic length results in cells with sizes on the order of 10^5-10^6 atoms. Therefore, the minimum length of 20 nm is deemed sufficient to avoid significant effects of periodic boundaries on the 3D dislocation nucleation dynamics.

Figure 2 shows the tensile axis for the 13 crystallographic orientations examined in this work, within the basic stereographic triangle with $\langle 0001\rangle$, $\langle 10\bar{1}0\rangle$, and $\langle 1\bar{2}10\rangle$ vertices. Each single crystal model is deformed under uniaxial tension at a temperature of 100 K. The expected twinning system is the $\{10\bar{1}2\}$ $\langle 10\bar{1}\bar{1}\rangle$ twin system for all tensile axis orientations in the $\langle 0001\rangle$-$\langle 10\bar{1}0\rangle$-$\langle 1\bar{2}10\rangle$ triangle.

The computational methodology used to investigate twin nucleation in this work differs from MD simulations for HCP metals described in previous literature. First, we have chosen to use a larger number of orientations to explore the anisotropy in twin nucleation throughout the stereographic triangle. Second, the simulation boundary conditions are different than previous simulations. Many previous simulations have used a fixed boundary to apply tension or simple shear, thereby admitting heterogeneous twin nucleation at the free surface and fixed regions. The present simulations incorporate a 3D periodic computational cell with no free surfaces or fixed regions and twin nucleation occurs homogeneously as a function of the local stress state. The current uniaxial deformation simulations of twin nucleation take into account the resolved shear stress, the resolved normal stress, and natural couplings that exist between resolved stresses on the twinning and slip planes.

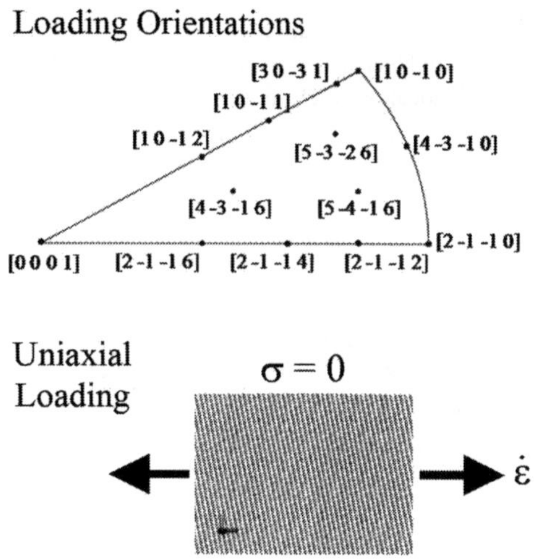

Figure 2: Uniaxial loading orientations and conditions for single crystal deformation simulations.

Simulation Results

The stress-strain curves for the 13 crystallographic loading orientations are shown in Figure 3. As in Fig. 2, the single crystals are loaded in tension and the stress increases until a peak stress, at which either a dislocation, twin, or void/crack nucleates within the single crystal. For each loading orientation, the deformation mechanism was characterized by examining the relevant slip/twin planes and directions to ascertain what defect corresponded to the peak nucleation stress. The stress-strain curves are colored according to deformation type. In many loading orientations, the stress required to nucleate dislocations tended to be lower than that for twins. In the curve with the highest nucleation stress, neither dislocations or twins were observed to nucleate. It should be noted that these stresses are the uniaxial tensile stresses required for the nucleation events; the resolved stresses are lower.

Of the 13 crystallographic orientations studied, five orientations nucleated twins and seven orientations deformed via slip. The final orientation was loaded in the basal $\langle 0001\rangle$ direction and showed no evidence of twinning or slip. Interestingly, this orientation has the highest likelihood of twinning under tensile stress, yet no twinning is observed. Again, recall that the single crystal models used here are absent of point defects, pre-existing dislocations, or free surfaces. Perhaps the lack of twin nucleation

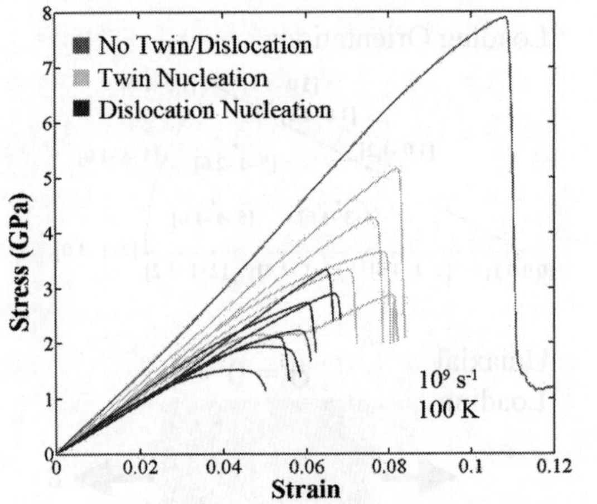

Figure 3: Stress-strain curves for all single crystal models.

suggests that other point, line, or planar defects need to be present in the lattice in order to spur the twin nucleation process. Additionally, analysis of resolved stress components may shed light on this phenomenon.

The crystallographic loading orientations which nucleated dislocations by slip were all characterized as basal slip, as would be expected from the Schmid resolved shear stress. The twins were easily identified, but further classification of twinning plane and directions is still ongoing work. It appears that a number of twin modes were observed in these simulations.

The influence of temperature (100 K, 200 K, 300 K, 400 K) and strain rate (10^8-10^{10} s^{-1}) was examined for a few loading orientations. In the loading orientations examined, the temperature range examined affected the nucleation stress and strain, but the mechanism tended to be unaffected. Higher temperature simulations at 400 K showed a lower elastic modulus, nucleation stress, and strain at nucleation. Interestingly, the rate of propagation of the twinned structure appeared more rapid in simulations run at higher temperatures. Simulations at strain rates from 10^8 s^{-1} to 10^{10} s^{-1} showed very little variation in results. These conditions nucleated the same deformation mechanism, at near the same stress, with the nucleating stress decreasing slightly as strain rate decreases. The choice of a strain rate of 10^9 s^{-1} appears valid. Interestingly, in many of the loading orientations which nucleated twins, one variant was observed to nucleate with multiple twin variants observed thereafter as the strain was increased. In only one loading orientation, the $\langle 10\bar{1}0\rangle$ orientation, were multiple twin variants nucleated at the peak stress.

The resolved stress components may shed light on why each loading orientation nucleated either twins or dislocations. Figure 4 shows the resolved shear stresses calculated from the nucleation stress and the Schmid factor for various twinning/slip systems. Here, we calculated the resolved shear stresses for $\{10\bar{1}2\}$ $\langle 10\bar{1}1\rangle$ twins, $\{0001\}$ $\langle 11\bar{2}0\rangle$ basal dislocations, $\{10\bar{1}0\}$ $\langle 11\bar{2}0\rangle$ prismatic dislocations, $\{10\bar{1}1\}$ $\langle 11\bar{2}0\rangle$ pyramidal dislocations, and $\{11\bar{2}2\}$ $\langle 11\bar{2}3\rangle$ pyramidal $\langle c{+}a\rangle$ dislocations. The resolved stress for all systems are plotted along with the observed twin/dislocation system (circled in black). The observed resolved shear stress varies with loading orientation, but in nearly all the loading orientations the observed mechanism had the highest Schmid factor and, therefore, the highest resolved shear stress. All of the loading orientations characterized as nucleating dislocations on the basal or prismatic slip systems predicted that that slip system had the highest resolved shear stress. The $\langle 10\bar{1}0\rangle$ loading orientation nucleated twins at a very low resolved shear stress, but the '2' indicated on the plot refers to the simultaneous nucleation of two twin variants. In this case, the low Schmid factor may be accommodated by the ability to simultaneously nucleate on two twin systems. Additionally, while the observed twinning system did not always have the highest resolved shear stress, the prismatic and pyramidal slip systems are known to have higher critical

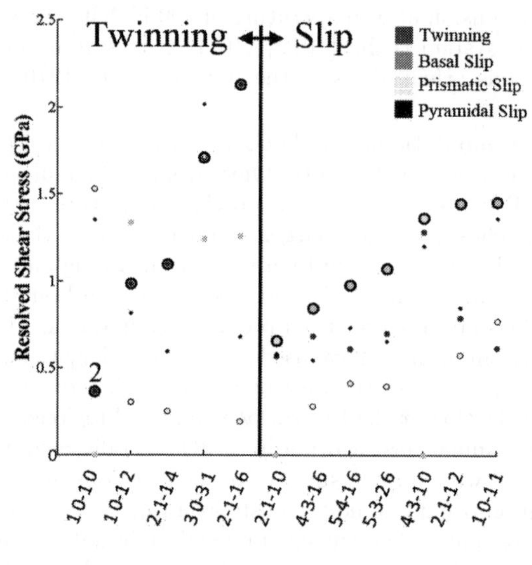

Figure 4: Resolved shear stresses on the twinning and slip systems for different loading orientations and the observed twin/dislocation system (circled in black).

resolved shear stresses, possibly explaining the nucleation of twins for these orientations. The $\langle 10\bar{1}2 \rangle$ loading orientation appeared to nucleate twins even though the basal slip mechanism has a higher resolved shear stress. It is not known why this occurs; future work will investigate the potential influence of non-Schmid resolved stresses on nucleation as well as the role of the interatomic potential. It is clear that only calculating resolved stresses via the Schmid factor may not be able to completely capture the complicated slip-twin nucleation event. Furthermore, examining in greater detail the atomic positions and crystallography corresponding to the twin nucleation event itself may shed light on whether dislocation nucleation plays a key initial role in the nucleation of twins on these systems.

Summary

The objective of this research is to elucidate the mechanisms of dislocation/twin nucleation in single crystals and grain boundaries for Mg. Here, we concentrate on single crystal calculations, where we used molecular dynamics simulations to study the influence of crystallographic orientation on dislocation/twin nucleation under uniaxial loading. Multiple crystallographic loading orientations from around the stereographic triangle were deformed under uniaxial tension to examine the role of resolved stresses on twin/dislocation nucleation. The results of this work detailed the influence of crystallographic loading orientation in single crystals on dislocation and twin nucleation mechanisms in magnesium. Here, we have shown that the Schmid resolved shear stress adequately predicts basal or prismatic slip in all orientations which were observed to nucleate dislocations on the basal and prismatic slip planes. However, in orientations where twins nucleated, the Schmid resolved stresses may not alone be able to completely capture the complicated slip-twin nucleation event.

Acknowledgments

This work was performed at the Center for Advanced Vehicular Systems (CAVS) at Mississippi State University. The authors would like to acknowledge funding from the Department of Energy.

References

[1] E.W. Kelley and W.F. Hosford, "Plane-strain compression of magnesium and magnesium alloy crystals," *Transactions of the metallurgical society of AIME*, 242 (1968) 5.

[2] S. Mendelson, "Dislocation dissociations in hcp metals," *Journal of Applied Physics*, 41 (1970) 1893.

[3] H. El Kadiri and A.L. Oppedal, "A crystal plasticity theory for latent hardening by glide twinning through dislocation transmutation and twin accommodation effects," *Journal of the Mechanics and Physics of Solids*, 58 (2010) 613.

[4] J. Wang, I.J. Beyerlein, C.N. Tome, "An atomistic and probabilistic perspective on twin nucleation in Mg," *Scripta Materialia*, 63 (2010), 741-746.

[5] M.A. Tschopp, D.E. Spearot, D.L. McDowell, "Influence of Grain Boundary Structure on Dislocation Nucleation in FCC Metals," *Dislocations in Solids, Volume 14*, ed. J.P. Hirth (Oxford, UK: Elsevier, 2008), 43-139.

[6] S. Plimpton, "Fast parallel algorithms for short-range molecular dynamics," *Journal of Computational Physics*, 117 (1995) 1.

[7] D.E. Spearot, M.A. Tschopp, K.I. Jacob, and D.L. McDowell, "Tensile strength of $\langle 100 \rangle$ and $\langle 110 \rangle$ tilt bicrystal copper interfaces," *Acta Materialia*, 55 (2007) 705.

[8] M.A. Tschopp, D.E. Spearot, and D.L. McDowell, "Atomistic simulations of homogeneous dislocation nucleation in single crystal copper," *Modelling and Simulation in Materials Science and Engineering*, 15 (2007) 693.

[9] M.A. Tschopp and D.L. McDowell, "Influence of single crystal orientation on homogeneous dislocation nucleation under uniaxial loading," *Journal of Mechanics and Physics of Solids*, 56 (2008) 1806.

[10] D.Y. Sun et al., "Crystal-melt interfacial free energies in hcp metals: A molecular dynamics study of Mg," *Physical Review B*, 73 (2006), 024116.

[11] J.A. Yasi et al., "Basal and prism dislocation cores in magnesium: comparison of first-principles and embedded atom methods predictions," *Modelling and Simulation in Materials Science and Engineering*, 17 (2009), 055012.

[12] X.Y. Liu et al., "EAM potential for magnesium from quantum mechanical forces," *Modelling and Simulation in Materials Science and Engineering*, 4 (1996), 293.

Magnesium Technology 2011
Edited by: Wim H. Sillekens, Sean R. Agnew, Neale R. Neelameggham, and Suveen N. Mathaudhu
TMS (The Minerals, Metals & Materials Society), 2011

TWINNING MULTIPLICITY IN AN AM30 MAGNESIUM ALLOY UNDER UNIAXIAL COMPRESSION

Q. Ma[1], H. El Kadiri[1,2], A.L. Oppedal[1], J.C. Baird[1], M.F. Horstemeyer[1,2]
[1] Center for Advanced Vehicular Systems, Mississippi State University, Mississippi State, MS 39762, USA
[2] Department of Mechanical Engineering, Mississippi State University, Mississippi State, MS 39762, USA

Keywords: Texture; Magnesium; twinning; plasticity; EBSD

Abstract

Twinning under compression along two perpendicular directions of an AM30 alloy with a $\{10\bar{1}0\}$ fiber texture was investigated. The primary extension $\{10\bar{1}2\}$ twinning occurred for both compression normal to the fiber and compression parallel to the fiber. These primary extension twins took forms of fully residual twins parallel to the fiber and the "stopped elastic" twins normal to the fiber. Both the $\{10\bar{1}1\}$-$\{10\bar{1}2\}$ and $\{10\bar{1}3\}$-$\{10\bar{1}2\}$ double twinning occurred in the matrix grains at the last stage of compression normal or parallel to the fiber, but as the "combined two shears" mode of the double twinning. Another time-sequence type of $\{10\bar{1}1\}$-$\{10\bar{1}2\}$ double twin also activated early normal to the fiber. The primary extension twinning and contraction twinning seem to obey the Schmid law according to the texture evolution.

Introduction

Twinning is common in deformed magnesium due to insufficient independent slip systems necessary for arbitrary shape changes. The $\{10\bar{1}1\}$, $\{10\bar{1}2\}$, $\{10\bar{1}3\}$, $\{30\bar{3}4\}$ twins and so-called $\{10\bar{1}1\}$-$\{10\bar{1}2\}$ and $\{10\bar{1}3\}$-$\{10\bar{1}2\}$ double twins were reported in deformed magnesium single crystal or polycrystalline aggregates [1-5]. The $\{10\bar{1}2\}$ and $\{10\bar{1}1\}$ twins, known as extension twins and contraction twins, respectively, are by far the most common twins observed in nearly all magnesium alloys [1-5]. The twinning which occurred depended upon chemistry, texture and stress state of the material [6-9]. Although twins' presence and their effects on magnesium's mechanical behavior were reported in both experiments and simulations in literature [1,2,4,10], these results were usually concluded based on the final texture and/or microstructure and even on the habit plane trace analysis of twins [1-5]. The above literature indicates some uncertainty on twinning activation because the final texture and trace analysis of habit planes are not enough to reveal twinning activity in detail. As such, the intrinsic behavior of various twinning systems along strain increase under deformation requires further study, especially with said twinning systems undergoing different loading directions at various intermediate strains. We aim to study various twinning systems' behaviors from initiation of yield to rupture compression along two perpendicular directions of a magnesium alloy based on the macrotexture and microtexture of the alloy. Twinning sequence and twins' morphology are the study points in this article.

Experimental Information

In this study, an extruded commercial magnesium AM30 billet (mass %, 2.54% Al, 0.40% Mn, Mg in balance) with a 178 mm diameter was selected as the experimental material. The simple uniaxial compression tests were conducted along the extrusion direction (ED) and the extrusion radial (defined as ERD) of the billet at strains -0.036, -0.06, -0.088 and rupture, respectively, at room temperature and strain rate $10^{-3}S^{-1}$. The third direction is defined as ETD. The ED and ERD cylindrical samples, with 6 mm in diameter and 6 mm in length, were cut by the electrical discharge machining (EDM) method from the AM30 billet along ED and ERD, respectively. For the results' consistency and stability, all the ED and ERD samples were cut in the same area which was near a mid-radius of the AM30 billet as shown in Figure 1a. The macrotexture and microtexture of the AM30 alloy at various strain levels were measured based on X-ray diffraction (XRD) and electron back-scattered diffraction (EBSD) techniques. Six pole figures $\{10\bar{1}0\}$, $\{0002\}$, $\{10\bar{1}1\}$, $\{10\bar{1}2\}$, $\{11\bar{2}0\}$ and $\{10\bar{1}3\}$ were characterized by a X-ray diffractometer Rigaku Smartlab using a reflection method, and background and defocusing effects were eliminated. The Orientation Distribution Functions (ODFs) were calculated based on the WIMV algorithm in the popLA package [11]. The complete pole figures and inverse pole figures were recalculated based on these ODFs and were plotted by the texture software MTEX [12]. The initial texture, based on XRD of the initial ED sample, is presented in Figure 1b. As shown in Figure 1b, this extruded AM30 alloy has a strong $\{10\bar{1}0\}$ fiber texture.

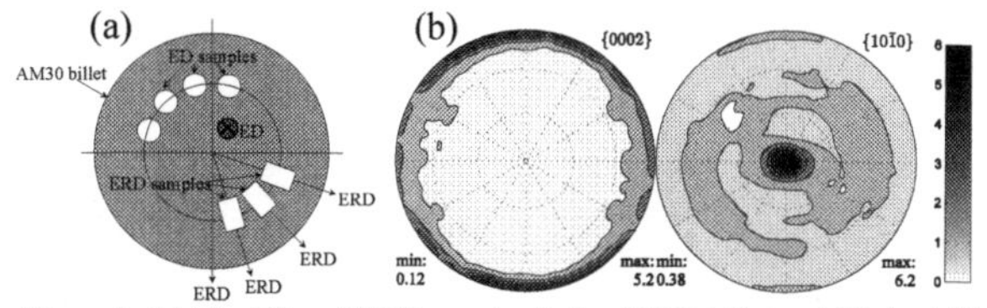

Figure 1: (a) The ED and ERD samples in the AM30 billet and (b) the initial texture of AM30 alloy based on XRD and the initial ED sample. The horizontal direction is the ETD and the ED is out of the paper.

Results and Discussion

In the case of compression parallel to the $\{10\bar{1}0\}$ fiber, EBSD inverse pole figure (IPF) maps were used to investigate twinning development. Figure 2 shows the EBSD inverse pole figure (IPF) maps of ED samples at strains 0, -0.036, and -0.088. As shown in Figure 2, profuse $\{10\bar{1}2\}\langle 10\bar{1}0\rangle$ twinning activated within both the elongate parent grains and small equaxied recrystallized grains at strain -0.036. These zonal shape twins crossing the matrix grain boundaries from end to end are the typical residual twins [11]. Figure 2 also shows that almost all large grains have been totally twinned by the primary $\{10\bar{1}2\}$ twinning at strain -0.088. The inverse pole figures of ED based on XRD are presented in Figure 3 at various strain levels. We selected 200 grains to represent initial texture of AM30 (Figure 1b). As shown in Figure 2, the primary extension twinning $\{10\bar{1}2\}\langle 10\bar{1}0\rangle$ activated first in the matrix, and then the contraction twinning $\{10\bar{1}1\}\langle 10\bar{1}2\rangle$ and/or the double twinning $\{10\bar{1}1\}$-$\{10\bar{1}2\}$ initiated in the twinned matrix grain in the rupture sample as shown in Figure 5. Accordingly, we assumed that the representative 200 grains evolved the $\{10\bar{1}2\}$ twinning first among 1: $(10\bar{1}2)[\bar{1}011]$, 2: $(1\bar{1}02)[\bar{1}101]$, 3: $(01\bar{1}2)[0\bar{1}11]$, 4: $(0\bar{1}12)[01\bar{1}1]$, 5: $(\bar{1}012)[10\bar{1}1]$, 6: $(\bar{1}102)[1\bar{1}01]$ according to the highest Schmid Factor (SF) of the twin variants. The second-order $\{10\bar{1}1\}$ twinning activated in the twinned grain as 1: $(10\bar{1}1)[10\bar{1}\bar{2}]$, 2: $(01\bar{1}1)[01\bar{1}\bar{2}]$, 3: $(\bar{1}101)[\bar{1}10\bar{2}]$, 4: $(\bar{1}011)[\bar{1}01\bar{2}]$, 5: $(0\bar{1}11)[0\bar{1}1\bar{2}]$, 6: $(1\bar{1}01)[1\bar{1}0\bar{2}]$ with the highest SF value of the six $\{10\bar{1}1\}$ twin variants. The predicted primary $\{10\bar{1}2\}$ twins and second-order $\{10\bar{1}1\}$ twins are presented in Figure 3f. The $\{10\bar{1}2\}$ twinning and $\{10\bar{1}1\}$ twinning SFs under compression distribution in inverse pole figure are presented in Figure 4.

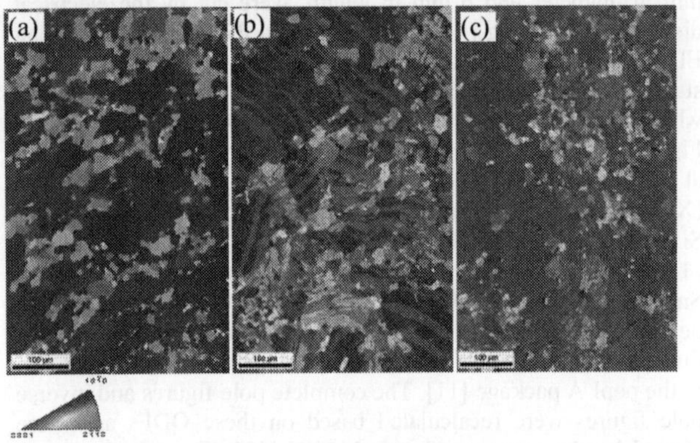

Figure 2: EBSD IPF maps of ED samples at strain (a) 0; (b) -0.036; (c) -0.088 showing $\{10\bar{1}2\}<10\bar{1}0>$ twin development. (The loading direction is out of paper. Inverse pole figure represents the ED direction.)

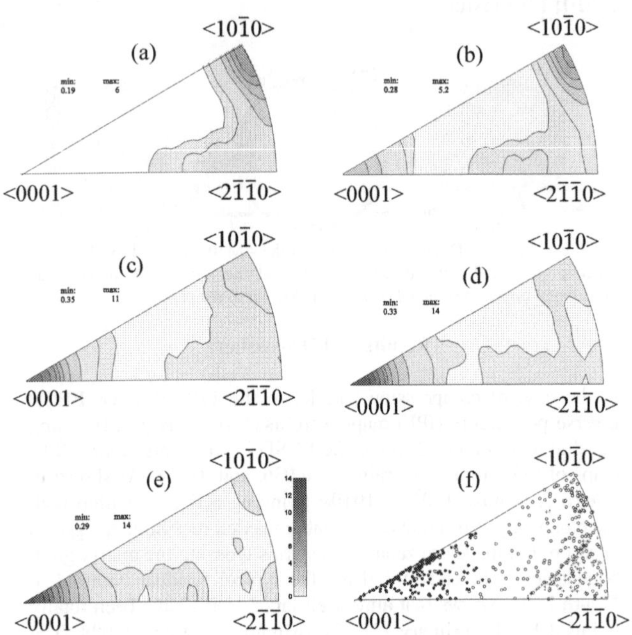

Figure 3: Inverse pole figures of AM30 alloy of ED at strain (a) 0; (b) -0.036; (c) -0.06; (d) -0.088; (e) rupture; (f) The primary $\{10\bar{1}2\}$ twins and the second-order $\{10\bar{1}1\}$ twins among the 200 representative grains compression along ED in inverse pole figure. The open circle symbols represent the initial 200 grains; the filled circle symbols represent the primary $\{10\bar{1}2\}$ twins; the cross symbols represent the secondary $\{10\bar{1}1\}$ twins.

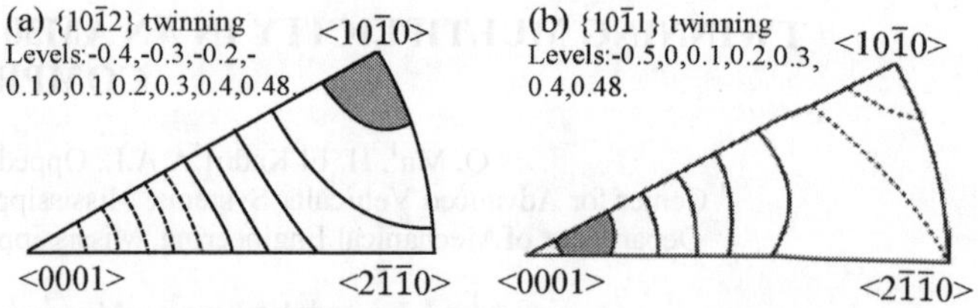

Figure 4: (a) The $\{10\bar{1}2\}$ twinning and (b) the $\{10\bar{1}1\}$ twinning Schmid Factor distribution under compression in inverse pole figure. (The shadow area notes SF >0.48; the dashed lines notes SF≤0.)

As shown in Figures 3a and 4a, the strong initial $<10\bar{1}0>$ fibre substantially favored the $\{10\bar{1}2\}$ twinning because it has a SF near the theoretical max value of 0.5. As a consequence, intensity around $<10\bar{1}0>$ quickly moved to <0001> area through primary $\{10\bar{1}2\}$ twinning (Figure 3f). As shown in Figure 4b, intensity around <0001> has also higher SF value for the $\{10\bar{1}1\}$ twinning; accordingly, the second-order $\{10\bar{1}1\}$ twinning would activate from strain about -0.088. Figures 3d and 3e illustrate that the third-order twinning could also initiated from strain -0.088, which was confirmed by the EBSD image quality map in Figure 5 below.

As for the $\{10\bar{1}1\}$ deformation twin, it is actually a degenerate compound twin with $K_1=\{10\bar{1}1\}$, $\eta_2=<30\bar{3}2>$, $K_2=\{10\bar{1}\bar{3}\}$, $\eta_1=<10\bar{1}\bar{2}>$ [13]. It has the plane of shear $\{1\bar{2}10\}$ that is a mirror plane, and the motif unit well lies in this plane of shear in this double lattice structure of magnesium. As such, simple shuffles of additional atoms movement all parallel to η_1 or normal to K_1 on shuffle mechanism and all atoms movement parallel to η_1 and η_2 on homogenous shear. As a consequence, the deformation $\{10\bar{1}1\}$ twinning with K_1, η_1, K_2, η_2 rational elements is favored by both a low twinning shear (0.137) and a simple shuffle mechanism even at a q=8 shuffle mode. However, the deformation $\{10\bar{1}1\}$ twins were observed rarely as single twins, but usually as double twins in matrix. Figure 5 shows some $\{10\bar{1}1\}$-$\{10\bar{1}2\}$ double twins and $\{10\bar{1}3\}$-$\{10\bar{1}2\}$ double twins in the ED rupture sample. However, these $\{10\bar{1}1\}$-$\{10\bar{1}2\}$ and $\{10\bar{1}3\}$-$\{10\bar{1}2\}$ double twins evolved in the matrix grains that have been totally twinned by the primary $\{10\bar{1}2\}$ twinning (Figure 2c). Specifically, third-order twinning occurred: $\{10\bar{1}2\}$-$\{10\bar{1}1\}$-$\{10\bar{1}2\}$ and $\{10\bar{1}2\}$-$\{10\bar{1}3\}$-$\{10\bar{1}2\}$ twinning activated in this case. The third-order twinning explains that the intensity around $<2\bar{1}\bar{1}0>$ in the inverse pole figure became weak in the rupture ED sample (Figure 3e).

In this study, the activated twinning system with $K_1=\{10\bar{1}3\}$, $K_2=\{10\bar{1}1\}$, $\eta_1=<30\bar{3}2>$, $\eta_2=<10\bar{1}2>$ is the reciprocal twinning of the $\{10\bar{1}1\}$ deformation twinning. The $\{10\bar{1}3\}$ twinning has the same plane of shear $\{1\bar{2}10\}$ and same twinning shear 0.137 with the $\{10\bar{1}1\}$ deformation twinning. Consequently, this $\{10\bar{1}3\}$ twinning was also favored due to the same reason mentioned above. However, the $\{10\bar{1}3\}$ deformation twinning is an extensional twinning [7,13] which could not activate in principle, in this case, under compression along *c*-axis of magnesium. The presence of $\{10\bar{1}3\}$-$\{10\bar{1}2\}$ twins in the ED rupture sample shown in Figure 5 implies a non-time-sequence activation of this double twinning. We found many $\{10\bar{1}3\}$-$\{10\bar{1}2\}$ and $\{10\bar{1}1\}$-$\{10\bar{1}2\}$ double twins in our other EBSD pictures of the ED rupture sample, existing together in the same twinned matrix grains similar to Figure 5.

This peculiar phenomenon is contradictory to the normal understanding of the double twinning mechanism that the secondary twin twins in the primary former twin. However, this special double twinning mechanism can be explained well by the Crocker's *Reciprocal Theorem* [14]. The Reciprocal theorem of double twinning mechanism of magnesium put forward by Crocker [14] in 1960s said: "If two double twinning mechanisms have first twinning shears which are either the same or reciprocal to each other and second twinning shears which are either the same or reciprocal to each other, then the resulting simple equivalent twinning modes are identical." According to the experimental observation (Figure 5) and Crocker's theory, the $\{10\bar{1}3\}$-$\{10\bar{1}2\}$ and $\{10\bar{1}1\}$-$\{10\bar{1}2\}$ double twins have the same simple double twinning mechanism: two shears plus one rotation in the plane of shear $\{1\bar{2}10\}$. In this case, the two shears must be combined together that means two shears evolving simultaneously because the primary $\{10\bar{1}3\}$ twin cannot exist independently under compression along *c*-axis of magnesium. As such, the $\{10\bar{1}3\}$-$\{10\bar{1}2\}$ and $\{10\bar{1}1\}$-$\{10\bar{1}2\}$ double twins should evolve in a process as: a small pre-existing or new nucleating $\{10\bar{1}3\}$ or $\{10\bar{1}1\}$ twin fault quickly retwins in the $\{10\bar{1}2\}$ twinning to the $\{10\bar{1}3\}$-$\{10\bar{1}2\}$ or $\{10\bar{1}1\}$-$\{10\bar{1}2\}$ double twin embryo, and this double twin embryo grows into the matrix grain. As such, we can call this double twinning mechanism as "combined two shears mechanism". This double twin system has a simple shear and an invariant plane strain similar with a simple normal twin system. The $\{10\bar{1}3\}$-$\{10\bar{1}2\}$ and $\{10\bar{1}1\}$-$\{10\bar{1}2\}$ double twin systems have the same shear of 0.258 with all four irrational twin elements [14]. The accommodation effect of this double twin system is not necessary in both the parent and the twin product [14], thus this mode is energetically favored. Accordingly, both $\{10\bar{1}3\}$-$\{10\bar{1}2\}$ and $\{10\bar{1}1\}$-$\{10\bar{1}2\}$ double twinning evolved as the combined two shears mode at the last stage under compression along ED, which may explain why single $\{10\bar{1}1\}$ or $\{10\bar{1}3\}$ primary twins were rarely observed in magnesium. In this study, the $\{10\bar{1}1\}$ or $\{10\bar{1}3\}$ twins in the tip of the double twins reported by Reed-Hill [1] were not observed in the ED rupture sample also validating this combined two shears mechanism. As shown in Figure 3, the activation of the primary $\{10\bar{1}2\}$ twinning seems to follow the Schmid law, but additional evidences are needed to validate the $\{10\bar{1}1\}$-$\{10\bar{1}2\}$ and $\{10\bar{1}3\}$-$\{10\bar{1}2\}$ twinning Schmid behavior.

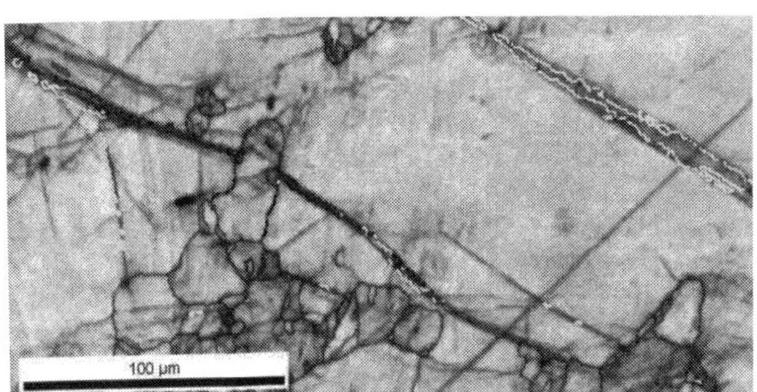

Figure 5: The $\{10\bar{1}1\}$-$\{10\bar{1}2\}$ double twins and $\{10\bar{1}3\}$-$\{10\bar{1}2\}$ double twins evolved in the same totally twinned matrix grain in the ED rupture sample. Fine black lines refer to grain boundaries >15°.

Axis angle pairs: 86±5°<$11\bar{2}0$>-----$\{10\bar{1}2\}$<$10\bar{1}0$>

38±5°<$11\bar{2}0$>-----$\{10\bar{1}1\}$-$\{10\bar{1}2\}$

22±5°<$11\bar{2}0$>-----$\{10\bar{1}3\}$-$\{10\bar{1}2\}$

With respect to compression normal to the $\{10\bar{1}0\}$ fiber, both inverse pole figures based on XRD and EBSD maps are used to investigate the twinning activation. Inverse pole figures of ERD at various strain levels based on XRD are presented in Figure 6. The primary $\{10\bar{1}1\}$ twins and the secondary $\{10\bar{1}2\}$ twins among the 200 representative grains under compression along ERD based on the highest SFs in inverse pole figure are presented in Figure 6f.

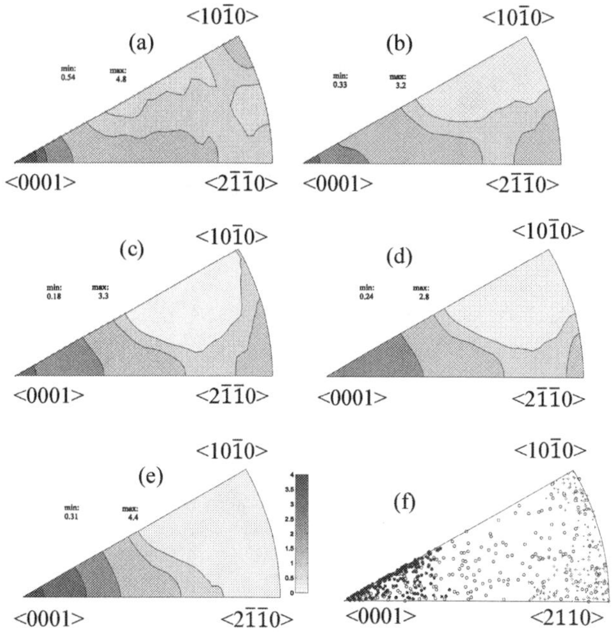

Figure 6: Inverse pole figures of AM30 alloy of ERD at strain (a) 0; (b) -0.036; (c) -0.06; (d) -0.088 and (e) rupture. (f) The primary $\{10\bar{1}1\}$ twins and the secondary $\{10\bar{1}2\}$ twins. The open circle symbols represent the initial 200 grains; the cross symbols represent the primary $\{10\bar{1}1\}$ twins; the filled circle symbols represent the secondary $\{10\bar{1}2\}$ twins.

As shown in Figure 6a and Figure 4, the primary $\{10\bar{1}2\}$ twinning and the primary $\{10\bar{1}1\}$ twinning will initiate simultaneously under compression along ERD. The intensity of <$10\bar{1}0$> weakened while <$2\bar{1}\bar{1}0$> intensity strengthened at same time along strain increase, as shown in Figure 6a, 6b and 6c. This measured result is consistent with Figure 6f, showing that the primary $\{10\bar{1}1\}$ twins will cause <$2\bar{1}\bar{1}0$> strength, while the primary $\{10\bar{1}2\}$ twins will result in <$10\bar{1}0$> weakness. Figures 6c, 6d and 6e show that the intensity of <$2\bar{1}\bar{1}0$> caused by primary $\{10\bar{1}1\}$ twinning annihilated again. Thus, the $\{10\bar{1}1\}$-$\{10\bar{1}2\}$ double twinning occurred starting from strain about -0.06 (Figure 6c), and a time-sequence of $\{10\bar{1}1\}$-$\{10\bar{1}2\}$ double twinning was clearly verified by <$2\bar{1}\bar{1}0$> intensity reduction from strain -0.06 to rupture. Moreover, this time-sequence of the $\{10\bar{1}1\}$-$\{10\bar{1}2\}$ twinning was confirmed further through EBSD analyses. As shown in Figure 7, a primary $\{10\bar{1}1\}$ twin was followed by a $\{10\bar{1}2\}$ twinning involved/contained within the $\{10\bar{1}1\}$ twin. Figure 7 substantially validated the sequence of the $\{10\bar{1}1\}$-$\{10\bar{1}2\}$ double twinning that the primary $\{10\bar{1}1\}$ contraction twinning activated first then the secondary extension $\{10\bar{1}2\}$ twinning evolved in the former contraction twin.

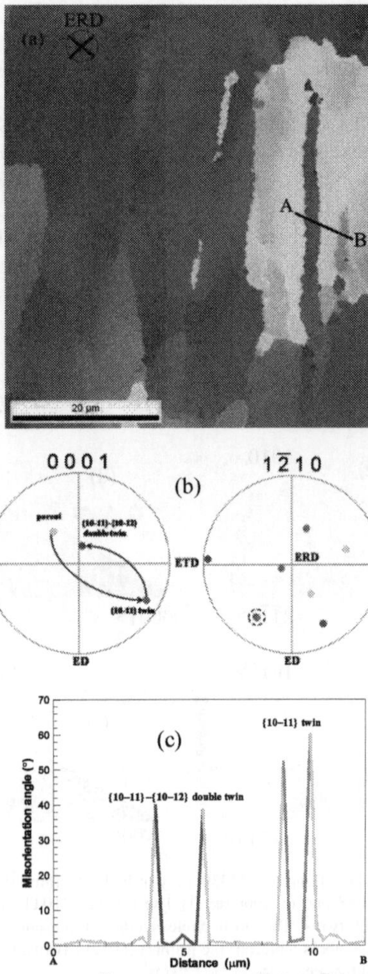

Figure 7: The primary $\{10\bar{1}1\}$ twin and the $\{10\bar{1}1\}$-$\{10\bar{1}2\}$ double twin in the ERD sample at strain -0.088 (a). The twinning sequence was proved by (b) the common rotation axis $<1\bar{2}10>$ and (c) the characteristic misorientation ~56° for $\{10\bar{1}1\}$ twinning and ~38° for $\{10\bar{1}1\}$-$\{10\bar{1}2\}$ twinning. The colors used here correspond to the colors of the IPF map in (a). The single $\{10\bar{1}1\}$ twin in (a) verified the time-sequence of $\{10\bar{1}1\}$-$\{10\bar{1}2\}$ double twinning in the ERD samples.

Some of the grains would evolve the primary $\{10\bar{1}2\}$ twinning and possibly the secondary $\{10\bar{1}1\}$ twinning according to the Schmid factor criteria in the case of compression along ERD. However, only the primary $\{10\bar{1}2\}$ twins were observed in the ERD sample, as shown in Figure 8a. The possible $\{10\bar{1}2\}$-$\{10\bar{1}1\}$ double twins cannot be found in ERD EBSD maps. The Schmid behavior suggests that the threshold stress of the $\{10\bar{1}1\}$ twinning evolved in the $\{10\bar{1}2\}$ twin should be very high, as it was in the ED sample at the last compression stage, but the stress compression along ERD cannot reach this required stress value. As a consequence, no $\{10\bar{1}2\}$-$\{10\bar{1}1\}$ double twinning occurred.

As shown in Figure 8a, the lenticular shape primary $\{10\bar{1}2\}$ twins exhibiting a tapering tip embedded within a matrix grain are the typical "stopped elastic" twins [13]. These stopped elastic twins did not disappear after stress removal due to the accommodation effect by slip systems in the matrix. The different forms of the extension $\{10\bar{1}2\}$ twins in the ED and ERD samples could result from the matrix grain morphology, necessary accommodation effect of twinning, neighbor grains shape and size and orientations.

Similar to the ED rupture sample, the $\{10\bar{1}3\}$-$\{10\bar{1}2\}$ and $\{10\bar{1}1\}$-$\{10\bar{1}2\}$ double twinning also occurred as the combined two shears mode at the last compression stage of ERD (Figure 8b). As shown in Figure 8b, the $\{10\bar{1}3\}$-$\{10\bar{1}2\}$ and $\{10\bar{1}1\}$-$\{10\bar{1}2\}$ double twins coexisted in the same matrix grain. Another important phenomenon, as shown in Figure 8b, is that the double twins preferentially favor the dynamic recrystallization (DRX) of magnesium, even at room temperature. In general, the active twinning systems in the ERD samples included the primary $\{10\bar{1}1\}$ twinning and the primary $\{10\bar{1}2\}$ twinning, the secondary $\{10\bar{1}2\}$ twinning and the last $\{10\bar{1}3\}$-$\{10\bar{1}2\}$ double twinning. The primary twinnings also seem to obey the Schmid factor criteria according to the texture evolution in Figure 6.

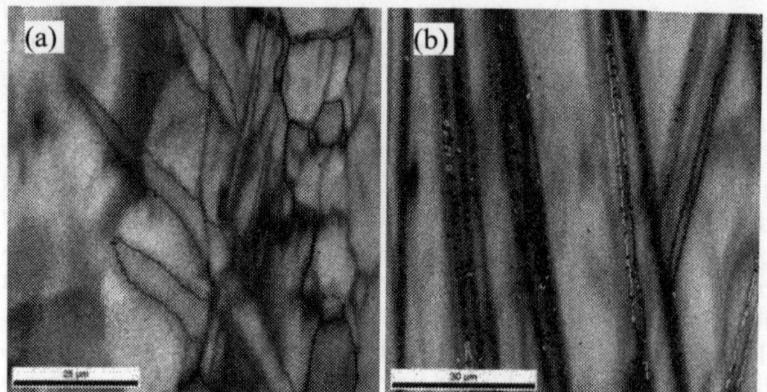

Figure 8: (a) The primary $\{10\bar{1}2\}$ twins in the ERD sample at strain -0.06 and (b) the $\{10\bar{1}1\}$-$\{10\bar{1}2\}$ and $\{10\bar{1}3\}$-$\{10\bar{1}2\}$ double twins in the ERD rupture sample. Fine black lines refer to grain boundaries >15°.

Axis angle pairs:

86±5°<11$\bar{2}$0>-----$\{10\bar{1}2\}$<10$\bar{1}$0>

38±5°<11$\bar{2}$0>-----$\{10\bar{1}1\}$-$\{10\bar{1}2\}$

22±5°<11$\bar{2}$0>-----$\{10\bar{1}3\}$-$\{10\bar{1}2\}$

Conclusions

In summary, an extruded AM30 alloy exhibited twinning multiplicity under compression along two perpendicular directions. The active twinning systems compression parallel to the $\{10\bar{1}0\}$ fiber included the first primary $\{10\bar{1}2\}$ twinning, the $\{10\bar{1}1\}$-$\{10\bar{1}2\}$ and $\{10\bar{1}3\}$-$\{10\bar{1}2\}$ double twinning evolving as the combined two shears mechanism at the last compression stage. For compression normal to the $\{10\bar{1}0\}$ fiber, the primary $\{10\bar{1}2\}$ twinning and the primary $\{10\bar{1}1\}$ twinning first activated early, followed by the secondary $\{10\bar{1}2\}$ twinning evolving within the primary $\{10\bar{1}1\}$ twin. The combined two shears mode of $\{10\bar{1}3\}$-$\{10\bar{1}2\}$ and $\{10\bar{1}1\}$-$\{10\bar{1}2\}$ double twinning activated also at last stage. The extension $\{10\bar{1}2\}$ twin took the form of full residual twin compression parallel to the fiber and the form of "stopped elastic" twin normal to the fiber. All the primary twinning systems seem to obey the Schmid factor criteria.

Acknowledges

The authors are grateful to the financial support from the Department of Energy, Contract No. DE-FC-26-06NT42755, and the Center for Advanced Vehicular Systems (CAVS) at Mississippi State University.

References

1. R.E. Reed-Hill, "A study of the $\{10\bar{1}1\}$ and $\{10\bar{1}3\}$ twinning modes in magnesium," *Transactions of the Metallurgical Society of AIME,* 218 (1960), 554-558.
2. E.W. Kelley and W.F. Hosford Jr, "Plane-strain compression of magnesium and magnesium alloy crystals," *Transactions of the Metallurgical Society of AIME,* 242 (1968), 5-13.
3. B.C. Wonsiewicz and W.A. Backofen, "Plasticity of magnesium crystals," *Transactions of the Metallurgical Society of AIME,* 239 (1967), 1422-1431.
4. R.E. Reed-Hill and W.D. Robertson, "The crystallographic characteristics of facture in magnesium single crystals," *Acta Metallurgica,* 5 (1957), 728-737.
5. S.L. Couling, J.F. Pashak, and L. Sturkey, "Unique deformation and aging characteristic of certain magnesium-based alloys," *Transactions of the American Society of Metals,* 51 (1958), 94-104.
6. P.G. Partridge, "The crystallography and deformation modes of hexagonal close-packed metals," *Metallurgical Reviews,* 12 (1967), 169-194.
7. M.H. Yoo, "Slip, twinning, and fracture in hexagonal close-packed metals," *Metallurgical and Materials Transactions,* A12 (1981), 409-418.
8. M.R. Barnett, "Twinning and the ductility of magnesium alloys part I: "tension" twins," *Materials Science and Engineering,* A464 (2007), 1-7.
9. M.R. Barnett, "Twinning and the ductility of magnesium alloys part II: "contraction" twins," *Materials Science and Engineering,* A464 (2007), 8-16.
10. A. Jain and S.R. Agnew, "Modeling the temperature dependent effect of twinning on the behavior of magnesium alloy AZ31B sheet," *Materials Science and Engineering,* A462 (2007), 29-36.
11. J.S. Kallend et al., "Operational texture analysis," *Materials Science and Engineering,* A132 (1991), 1-11.
12. R. Hielscher and H. Schaeben, "A novel pole figure inversion method: specification of the MTEX algorithm," *Journal of Applied Crystallography,* 41 (2008), 1024-1037.
13. J.W. Christian and S. Mahajan, "Deformation twinning," *Progress in Material Science,* 39 (1995), 1-157.
14. A.G. Crocker, "Double twinning," *Philosophical Magazine,* 7 (1962), 1901-1924.

Magnesium Technology 2011
Edited by: Wim H. Sillekens, Sean R. Agnew, Neale R. Neelameggham, and Suveen N. Mathaudhu
TMS (The Minerals, Metals & Materials Society), 2011

INHOMOGENEOUS DEFORMATION OF AZ31 MAGNESIUM SHEET IN UNIAXIAL TENSION

Jidong Kang[1], David S. Wilkinson[1], Raja K. Mishra[2]

[1]Department of Materials Science and Engineering, McMaster University,
1280 Main Street West, Hamilton, Ontario, L8S 4L7, Canada
[2]General Motors Research and Development Center,
30500 Mound Road, Warren, MI 48090-9055, USA

Keywords: magnesium alloy, plastic deformation, r-value, necking

Abstract

Inhomogeneous plastic deformation during uniaxial tensile test of AZ31 magnesium sheet has been studied using digital image correlation and electron backscatter diffraction techniques. It is shown that large strain gradients exist on the sheet surface parallel and perpendicular to the loading direction and very little deformation occurs in the thickness direction. The lack of thinning leads to abrupt fracture right after the formation of a premature but profound diffuse neck without transitioning to any localized neck. Such inhomogeneous deformation arises from the strong basal texture of the starting sheet and the resultant need for contraction and double twinning to accommodate strain. The strain distribution on the sheet surface evolves nonlinearly with strain, impacting the measured *r-value*.

Introduction

There is increasing interest in utilization of magnesium alloys in automotive applications due to their low density, superior specific tensile strength and rigidity compared to steel and aluminum alloys [1-2]. However, it is well recognized that the formability of wrought magnesium alloys at room temperature is inferior to that of steel and aluminum alloys. The poor formability of magnesium alloys is due to their hexagonal closed packed crystal structure and lack of many active slip systems at room temperature [1-2], requiring twinning to be activated.

The anisotropy of plasticity in Mg alloys affects yield asymmetry in tension and compression, the r-value (width to thickness strain ratio), and forming limit diagram, etc [3-8]. Most studies in the literature on the relationship between the mechanical properties and deformation mechanisms have relied on macroscopic stress-strain measurements and uniform strain assumptions in tensile gage length (common in studies of cubic metals such as steel and aluminum) to interpret microstructural observations in hexagonal Mg. For the first time, this study presents experimental data from simultaneous measurement of local strain and its evolution on the flat surface and the through thickness plane of the tensile sample using digital image correlation technique. Electron backscattered diffraction (EBSD) of the same surfaces is used to determine the nature of twinning accompanying the deformation. The dependence of the inhomogeneity of deformation on the deformation mechanism as well as the influence it has on the measurement of macroscopic r-values are discussed.

Experimental

A 2 mm thick AZ31 sheet material with nominal composition of 3 wt. % Aluminum and 1 wt. % Zinc was used in the present study. The sheets were annealed at 450 °C for 30 minutes in a vacuum furnace and air cooled to obtain a recrystallized starting microstructure. Uniaxial tensile tests were conducted at room temperate at a nominal strain rate of 6×10^{-4}/s. A commercially available optical strain measurement system, Aramis^{TD} [9], based on digital image correlation, was used to measure the surface strain during uniaxial tensile tests on both the sample surface and through-thickness plane simultaneously. The images were taken at a rate of 1 image per second.

A tapered sample was designed to investigate twinning during the tensile deformation. The test was stopped when maximum stress was reached in the tapered cross section. The sample was then cut into two halves at the center - the left part was mounted on the surface plane while the right part was mounted on the through thickness plane. The mounted samples were mechanically polished with 0.05 micron colloidal silica. The polished sample was first used for electron backscatter diffraction (EBSD) to determine the types of twins developed during the tensile tests and then etched using picric acid solution for optical microscopy to reveal twins in different regions.

For comparison, a strip cast 2mm thick AA5754 aluminum alloy sheet was tested for surface strain measurement on both surfaces in the present study.

Results and Discussion

The engineering stress – engineering strain curve for ASTM size sample of annealed AZ31 sheet pulled along the rolling direction (RD) is shown in Fig. 1. The data shows that there is an early onset of diffuse necking in this material followed by prolonged post necking deformation.

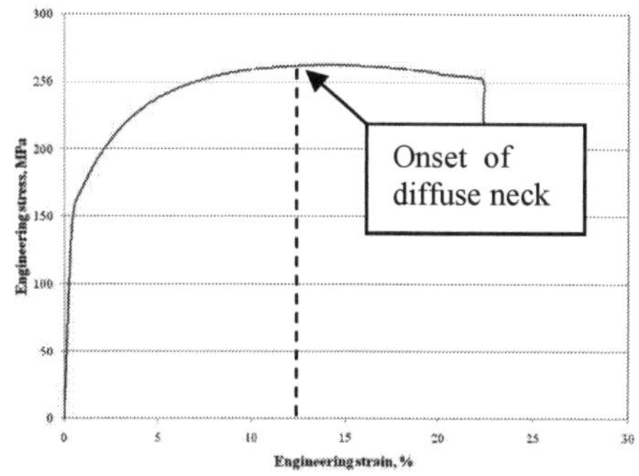

Fig. 1 Engineering stress – engineering strain curve of an AZ31 sheet material.

An examination of the DIC strain patterns on the surface during post diffuse necking stage of the deformation (Fig. 2 (a) (iv)-(vi)) shows that shear bands do not develop across the width of the sample. Fracture occurs right after a premature but profound diffuse necking without transitioning into localized necking.

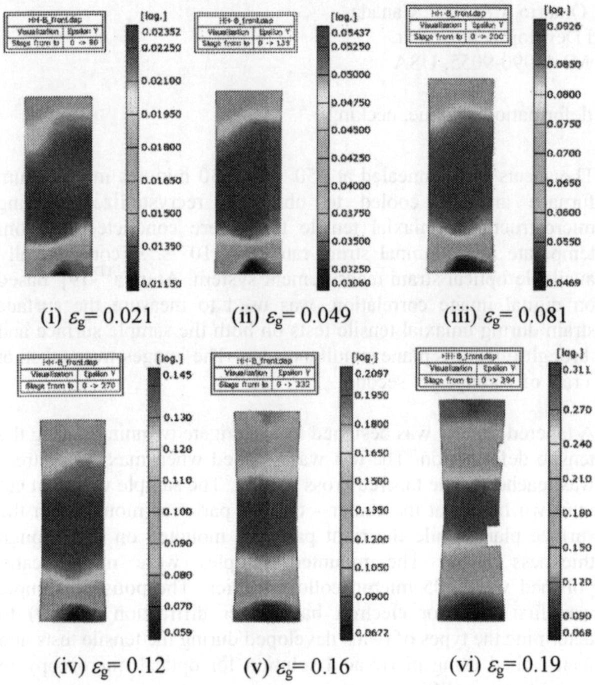

(a) Longitudinal strain

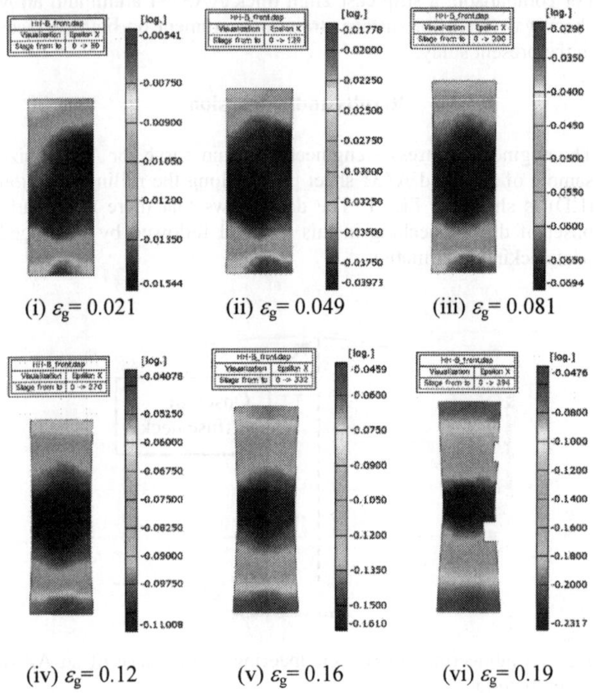

(b) width strain

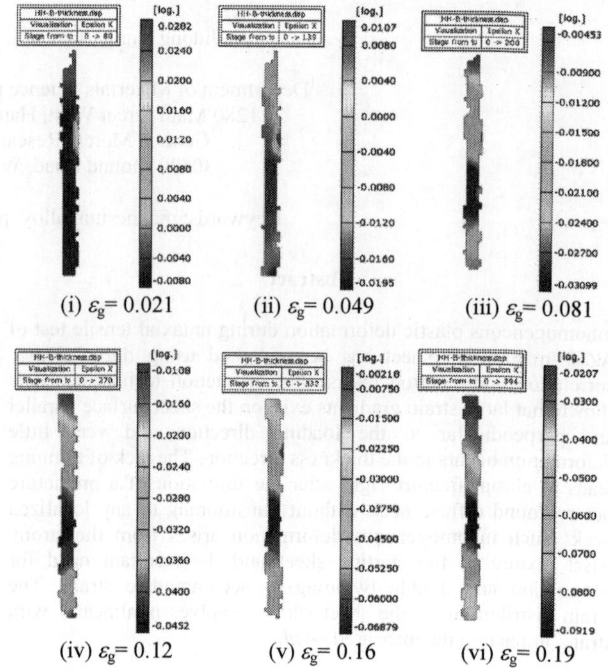

(c) thickness strain

Fig. 2 DIC maps showing the evolution of (a) longitudinal strain; (b) width strain and (c) thickness strain in a uniaixal tension sample of AZ31. ε_g represents global strain over a gage length of 25mm.

Data in Fig. 2 shows that both the longitudinal and width strains are non-uniformly distributed within the gage length. The nonlinearity of strain distribution is more evident in the line scan of longitudinal and width strains shown in Fig. 3. Both longitudinal and width strain at a given cross section keep evolving with global strain. The magnitude of the local width strain is nearly comparable to that of the longitudinal strain while the corresponding strain value in the thickness direction is very small. In other words, the necking process occurs in plane strain involving a narrowing process accompanied by minimal thinning.

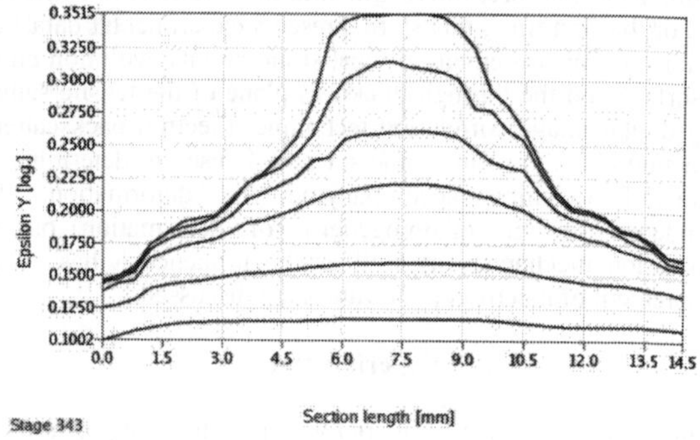

(a)

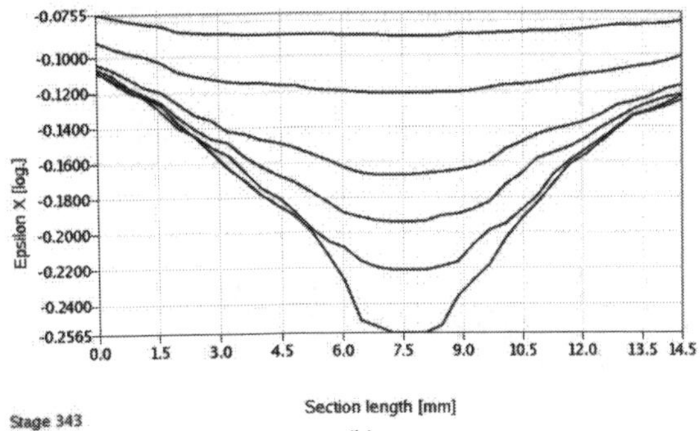

(b)

Fig. 3 Line scan of (a) longitudinal strain and (b) width strain at the center of the sample surface during "post diffuse neck" deformation in AZ31.

This data is to be contrasted with the evolution of local strain during uniaxial tensile testing of a strip cast AA5754 aluminum alloy sheet sample shown in Fig. 4. Strain within a narrow region increases with increasing global strain while strain outside this narrow region remains unchanged. The narrow region corresponds to the experimentally observed localized necking.

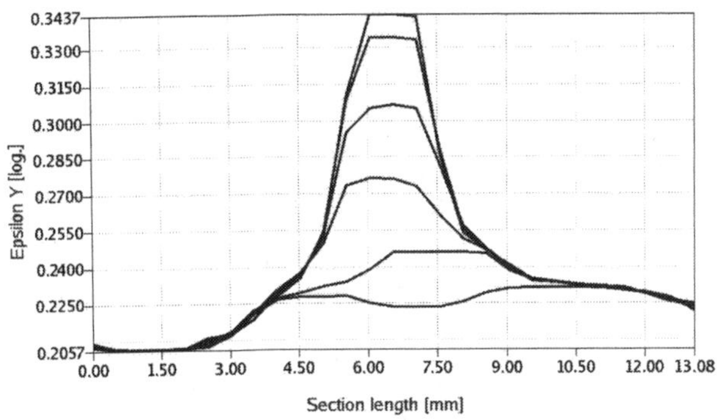

Fig. 4 Line scan of longitudinal strain during deformation in the "post diffuse neck" stage in AA5754 aluminum sheet. The strain gets localized when the shear band forms.

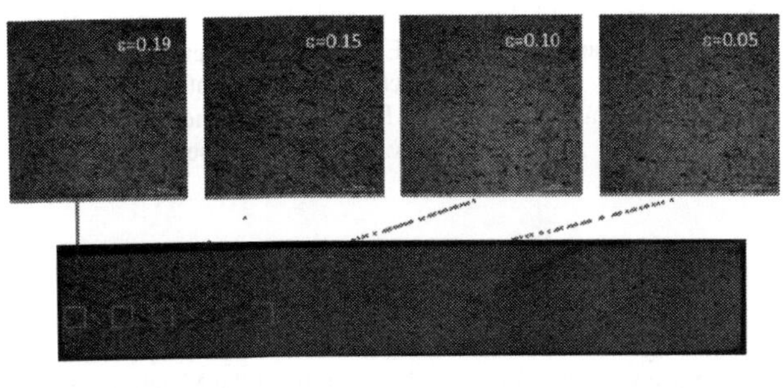

Fig. 5 Twinning development in a tapered AZ31 tensile sample.

Fig. 5 shows optical micrographs taken from the surface of a tapered sample deformed up to the necking strain in the tapered region. Images corresponding to regions with strains of 0.05, 0.10, 0.15 and 0.19 as measured by DIC show that even the region corresponding to a low strain value of 0.05 contains twins. The density of twins increases with increasing strain. Fig. 5 shows that twins tend to cluster by an auto catalytic process to form shear bands above a strain of 0.10. The shear bands get denser with increasing strain.

Electron back scatter diffraction (EBSD) data from these locations reveals that the location corresponding to the strain of 0.05 contains a number of grains deformed by extension twinning. Contraction twins are observed in areas corresponding to a strain of 0.10 or higher. The latter coincides with the onset of diffuse necking in the material. Contraction twins must form to accommodate the deformation in the thickness direction (which is along the c-axis). The low thickness strain measured by DIC is consistent with the observation of contraction twins which are difficult to form due to the high critical stress values needed for their formation, their low volume fraction and their clustering in closely spaced shear bands in the diffuse neck.

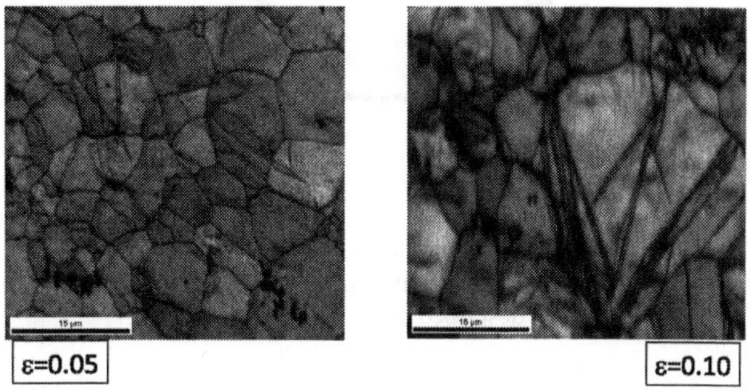

Fig. 6 Image quality maps from EBSD data showing extension twins with red boundaries. The dark straight lines are contraction and double twins (Jiang et al [10]).

The influence of the inhomogeneous deformation on mechanical properties of the material is revealed from the measurement of the anisotropy index, the so-called r-value, which is commonly defined as [11]

$$r = \frac{\varepsilon_w}{\varepsilon_t} = -\frac{\varepsilon_w}{\varepsilon_l + \varepsilon_w} \quad (1)$$

where $\varepsilon_l, \varepsilon_w, \varepsilon_t$ are longitudinal, width and thickness strains, respectively.

Thickness strain is difficult to measure using an extensometer during standard tensile testing. Two extensometers are usually used over a certain gage length in the longitudinal and width directions and incompressibility criterion along with the assumption of uniform strain distribution over the gage length is used to determine the r-value [11].

The current experimental method enables the independent measurement of all three strain components - the longitudinal, width and thickness strains at each point from which local r-value can be determined. Plots of the r-value evolution with global strain based on strain measurements using i) longitudinal and width strains (commonly used); ii) longitudinal and thickness strains at different locations; and iii) longitudinal and width strains measured at a given point are shown in Fig. 7. "Gage length, Y, X" in Fig. 7 represents r-values calculated using strain measurements on the sample surface within the gage length in loading (Y) and width directions (X), thereafter referred to as Case 1; "Gage length, X, t(center)" and "Gage length, X, t(quarter)" represent r-values calculated using measured width strain and thickness strain in the center or quarter length in the sample, thereafter referred to as Case 2; "Point on surface, center, Y, X" and "Point on surface, quarter, Y, X" represent r-values calculated using strain measurements at selected points on sample surface in the center (within the necked area) or quarter length, thereafter referred to as Case 3.

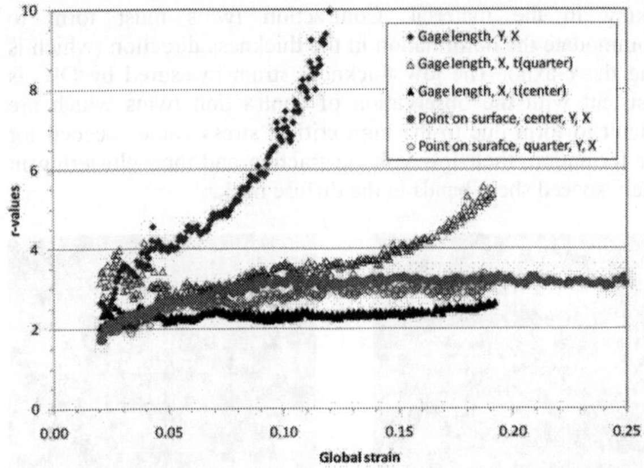

Fig. 7 Evolution of r-value with global strains in AZ31 based on different algorithms described in the text.

The r-values for Cases 1, 2 and 3 are quite different. In Case 1, r-value increases with increasing global strain, similar to what has been reported in the literature [7]. The variation of r-values with global strain for Case 2 is quite different. Since thickness strain is not uniformly distributed along the length direction of the sample, r-value varies with location of thickness strain. r-value data using strain values measured at "points" (Case 3) shows that r-value does not evolve above a strain of 0.10 regardless of the locations of the selected points, i.e. within or outside the necked area.

As shown in Fig. 2, the assumption of uniform strain in the gage section is not accurate, especially above 10% global strain. Since strain at a given point is calculated based on facet size in digital image correlation measurements, i.e. an actual spacing of 0.3mm in the present case, strain over this spacing can be regarded as macroscopically uniform. Therefore, r-value measurements based on "point" in Case 3, are consistent with the common definition of r-value. On the other hand, r-value data for Case 1 needs to be used with caution in view of the nonlinearity of strain distribution within the gage length in the longitudinal and width directions shown in Figs. 2 and 3.

The measured r-values using Case 1 and Case 3 criteria in AA5754 sheet material (Fig. 8) show no difference. Even with the existence of moving Portevin-Le Chatelier (PLC) bands in the Al-Mg alloy (AA5754 has 3 wt% Mg), the r-values based on strain measurements over the gage length tend to be constant beyond 10% strain, similar to the data in the literature [12-13].

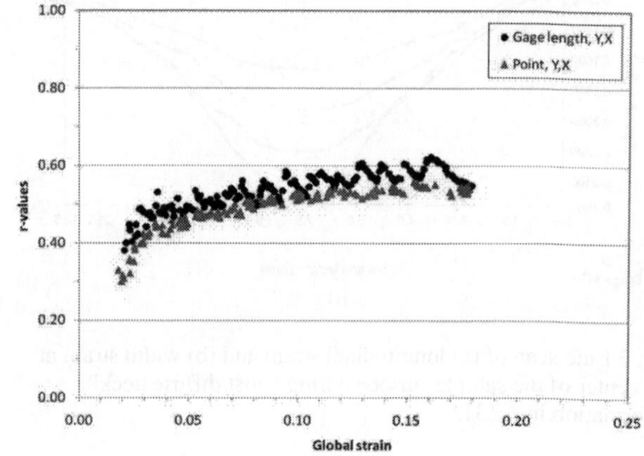

Fig. 8 Evolution of r-value with global strain in an AA5754 sheet material.

Summary

The inhomogeneous plastic deformation during uniaxial tensile test of an AZ31 magnesium sheet has been investigated using digital image correlation and electron backscatter diffraction.

(1) Very little deformation occurs in the thickness direction but large strain gradients exist on the sheet surface parallel and perpendicular to the loading direction during tensile deformation.
(2) The lack of thinning leads to the abrupt fracture right after a premature but profound diffuse necking without transitioning to localized necking.
(3) The observed inhomogeneous deformation arises from the strong basal texture and the necessity for the formation of contraction and double twins.
(4) *r-value* measurement in Mg alloys is shown to be sensitive to the method of measurement of the width and thickness strain. In view of the nonlinear evolution of strain distribution on the sheet surface, strain measured at "points" rather than "gage length" should be used to report r-values in Mg alloys.

Acknowledgements

The authors would like to acknowledge the support of the Resource for the Innovation of Engineering Materials program at the CANMET - Materials Technology Laboratory of Natural Resources Canada, and of Dr. Elhachmi Essadiqi in particular. Robert Kubic Jr. of GM R&D Center helped with EBSD measurements.

References

[1] Reed-Hill RE, Robertson WD. Acta Metall., 5 (1957) 728.
[2] Partridge PG. Metall Rev., 12(1967) 169.
[3] Wonsiewicz BC, Backofen WA. Trans of the Metall Society of AIME, 239 (1967) 1422.
[4] Kelly EW, Hosford WF. Trans of the Metall Society of AIME, 242 (1968) 5.

[5] Barnett MR, Keshavarz Z, Beer AG, Atwell D, Acta Mater. 52 (2004) 5093.
[6] Koike J. Metall and Mat Trans A, 36A (2005) 1689.
[7] Bohlen J, Nurnberg MR, Senn JW, Letzig D, Agnew SR. Acta Mater 55 (2007) 2101.
[8] Levesque J, Inal K, Neale KW, Mishra RK. Int J of Plasticity 26 (2010) 65.
[9] Aramis user manual, v.6, GOM mbH, Braunschweig, Germany, 2007.
[10] Jiang L. Scripta Mat, 54 (2006) 771.
[11] Hosford WF, Caddell RM, Metal Froming – Mechanics and Metallurgy (3rd edition), Cambridge University Press, 2007, p.207.
[12] Tong W. J Mech Phys Solids, 46 (1998) 2087.
[13] Kang J, Wilkinson DS, Embury JD, Jain M. SAE Transactions: J Mater and Manufacturing, 114 (2005) 156.

Magnesium Technology 2011
Edited by: Wim H. Sillekens, Sean R. Agnew, Neale R. Neelameggham, and Suveen N. Mathaudhu
TMS (The Minerals, Metals & Materials Society), 2011

Limitation of current hardening models in predicting anisotropy by twinning in hcp metals: Application to a rod-textured AM30 magnesium alloy

A. L. Oppedal[1], H. El Kadiri[1], C. N. Tomé[2], J. C. Baird[1], S. C. Vogel[2], M. F. Horstemeyer[1]

[1]Mississippi State University, Mississippi State, MS 39762, USA
[2]Los Alamos National Laboratory, Los Alamos, NM 87545, USA

Keywords: magnesium, deformation twinning, twin accommodation effects, crystal plasticity

Abstract

When a strongly textured hexagonal close packed (HCP) metal is loaded under an orientation causing profuse twinning or detwinning, the stress-strain curve is sigmoidal in shape and inflects at some threshold. Authors have largely attributed the dramatic stress increase in the lower-bound vicinity of the inflection point to a combined effect of a Hall-Petch mechanism correlated to grain refinement by twinning, and twinning-induced reorientation requiring activation of hard slip modes. We experimentally and numerically demonstrate that these two mechanisms alone are unable to reproduce the stress-strain behaviors obtained under intermediate loading orientations correlated to in-between profuse twinning and nominal twinning. We argue based on adopting various mechanistic approaches in hardening model correlations from the literature. We used both a physics dislocation based model and a phenomenological Voce hardening model. The HCP material is exemplified by an extruded AM30 magnesium alloy with a $\langle 10\bar{1}0\rangle$-fiber parallel to the extrusion direction.

Introduction

In this work, by using a $\langle 10\bar{1}1\rangle$ fiber texture and adopting the same correlations approaches used in the literature, we derive an important discrepancy in the prediction of the stress-strain behaviors measured under loading orientations with intermediate volume fractions of twins. Only the stress-strain behaviors under orientations completely prohibiting twinning and maximizing twinning could be acceptably reproduced. In between, the models do not provide enough precision to capture the hardening trend. We demonstrate that the reason lies behind a misunderstanding of the nature of twin-slip interactions.

A first misunderstanding comes from the Hall-Petch effect. In fact, on one hand we show that the measured twin volume fraction evolution is incompatible with the evolution of the Hall-Petch effect needed to capture the increasing hardening rate in Regime II of the stress-strain behavior. The dislocation based model is more sensitive to this effect, since no good correlations could even be achieved for the two classical limiting loading orientations. In fact, in contrast to Zr, the curves from profuse twinning and nominal twinning do not saturate at the same level in Mg, and this could not be captured by any simulation with Hall-Petch terms. On the other hand, the Hall-Petch effect accounted for by a sort of slip-twin latent hardening in the Voce type hardening model was not able to inflect the simulated curves with loading at an obtuse angle from ED.

A second misunderstanding comes from disregarding the dislocation transmutation effects upon twin incorporation [1]. The Voce model is unable to capture this effect, but the dislocation model [2] is. A pragmatic factor distinctly increasing the stored dislocations in the twin over that of the parent had an encouraging effect on modeling anisotropy in HCP metals using the transmutation effects.

We use an extruded, rod-textured, AM30 magnesium alloy obtained by relatively fast extrusion with high ratio at high temperature. Neutron diffraction, x-ray diffraction (XRD) and electron backscatter diffraction (EBSD) analyses were used to measure the deformation texture and evolution of volume fractions of twins. The VPSC code is used for the polycrystalline plasticity simulations with the dislocation-based model and an extended Voce hardening model.

Materials and Testing

The material is an AM30 Magnesium alloy received from Timminco in the form of a 200 mm diameter billet extruded from a 450 mm diameter as-cast ingot. The extrusion operated at 570 °C and at a 0.3 m/s extrusion speed.

Samples were prepared for testing along four different loading paths (Figure 1). Conventionally, previous investigators chose loading paths to maximize and minimize the effect of twinning on the mechanical be-

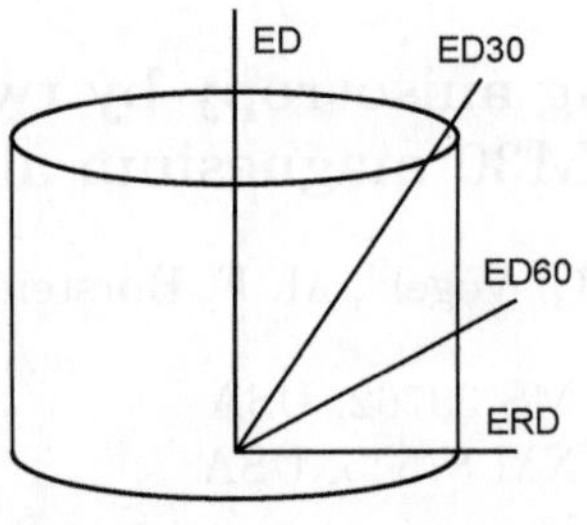

Figure 1: Schematic of sample orientations.

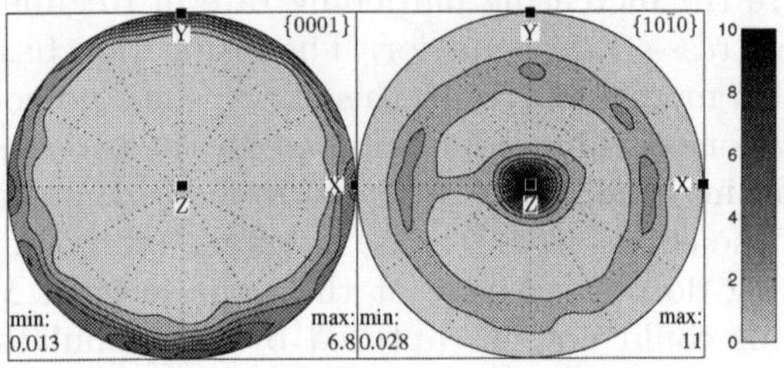

Figure 2: $\{0001\}$ and $\{10\bar{1}0\}$ pole figures obtained by neutron diffraction on an untested specimen. Z corresponds to extrusion direction; X and Y correspond to the extrusion tangent and extrusion radial directions, respectively. A pronounced ED $\parallel$ $\{10\bar{1}0\}$ fiber from the extrusion process is evident.

havior. In this study, additional intermediate loading paths were desired, and were obtained by extracting samples at intermediate orientations. Samples were prepared allowing loading along the Extrusion Direction (ED), 30 degrees from the extrusion direction (ED30), 60 degrees from the extrusion direction (ED60), and perpendicular to the extrusion direction, i.e. extrusion radial direction (ERD).

The macrotexture information in this study was collected by neutron diffraction using using the facilities of LANSCE (Los Alamos Neutron Science) at Los Alamos National Laboratory. Details of the LANSCE facility can be found in papers by Von Dreele [3] and Wenk et al. [4]. Texture analyses and pole figure plots were done using popLA [5] and pole8 [6] software from Los Alamos National Laboratory, and MTEX [7].

The mechanical tests corresponded to uniaxial compression at room temperature and at a rate of 0.001 s^{-1}. The intermediate loading directions (30ED and 60ED) were performed in an effort to test the hardening models and correlation approaches used in the literature to fit the entire yield surface domain corresponding to various levels of twin activity.

The specimens all yielded at approximately 100 MPa, but the strain hardening behavior was markedly different. This difference is not monotonic upon angle deviation from ED toward ERD. The ED curve is pronouncedly inflecting as all grains twinned. The ERD shows an apparent permanent decreasing hardening rate characteristic of slip preponderance. The difference between ED and ERD is in contrast with the difference between TTC and IPC typically encountered in compression of c-axis fiber textures. In fact, for c-axis fiber textures, the TTC loading orientation, equivalent here to ERD, is harder than IPC, because of the necessity of harder slip mode to accommodate deformation along the c-axis. Here, upon ERD, basal slip is the predominant deformation mode with a nominal contribution of twinning. The deformation texture in Figure 2 substantiates twinning of grains having their c-axis around the normal to the ERD (highest Schmid factors).

The ED sample, with profuse twinning, shows a higher saturation stress over all other curves. The 30ED sample broke before a perceptible completion of the saturation behavior. However, the trend indicates an overreaching saturation higher than 60ED curve. The higher saturation stress upon twinning than upon slip was recently reported for basal textured pure magnesium Oppedal et al. [8]. The 30ED and 60ED stress-strain curves (Figures 3b and 4b) shows not a quite intermediate response between ED and ERD curves. The 60ED starts higher and tends to saturate lower than 30ED. However, ERD seems to violate this tendency including the saturation regime. The overlapping between curves from all loading orientations is a signature of a complex twin-slip interaction in this rod-textured magnesium alloy.

The polycrystal plasticity model and hardening rules

This work uses and compares two different hardening rules and twinning behaviors within the VPSC code. The VPSC code is based on the Visco-Plastic Self-Consistent model to compute stress in grains and can incorporate hardening rules and texture information to simulate the stress-strain behavior and deformation texture evolution together with twinning. The code models polycrystalline aggregates using single crystal constitutive rules. The VPSC scheme, developed by Molinari et al. [9] and implemented by Lebensohn and Tomé [10], estimates the stress, σ, via micromechanics applied to crystals. The hardening response is based on updating the critical resolved shear stress τ^{a} following the constitutive power law that provides the shear strain rate $\dot{\gamma}^{a}$ in each slip mode α based on the Schmid factor m^α and a constant $\dot{\gamma}_0$ reflecting the macroscopic strain rate:

$$\dot{\gamma}^\alpha = \dot{\gamma}_0 \left| \frac{m^\alpha \overline{\overline{\otimes}} \sigma}{\tau^\alpha} \right|^n \text{sign}\left(m^\alpha \overline{\overline{\otimes}} \sigma \right) \tag{1}$$

The first hardening rule used in our simulations is the extended Voce model, described in the original work by Lebensohn and Tomé [10]. Agnew et al. [11] used the model in describing the behavior of Magnesium. In the extended Voce model, hardening in individual grains is described by:

$$\hat{\tau}_s = \tau_0^s + (\tau_1^s + \theta_1^s \Gamma) \left\{ 1 - \exp\left(-\frac{\theta_0^s}{\tau_1^s} \right) \right\} \tag{2}$$

where the evolution of the CRSS for each slip system, s, are represented by phenomenological parameters reflecting a curve fitting approach: τ_0^s, τ_1^s, θ_0^s, and θ_1^s. Γ is the accumulated shear strain in each grain.

Latent hardening effects between slip, and slip and twin systems, are incorporated by:

$$\delta \tau^s = \frac{d\hat{\tau}^s}{d\Gamma} \sum_{s'} h^{ss'} \delta \gamma^{s'} \tag{3}$$

where $h^{ss'}$ is a matrix of latent hardening parameters. Parameters for the Voce model simulations in this work are listed in Table 1.

The dislocation based model uses dislocation densities on different active modes to capture hardening. This model was fully described by Beyerlein and Tomé [2] where the authors used clock-rolled Zirconium, another HCP metal.

The hardening rule is a temperature and strain rate dependent formulation that assumes the effects of a dislocation forest, dislocation debris (or substructure), a Hall-Petch effect on slip propagation based on dislocation mean free path within the grain, and the effects of slip and Hall-Petch on twin propagation. If a denotes a slip mode having s as a slip system, and β denotes a twin mode having t as a twin system, both of these effects are given by the following:

$$\tau_c^\alpha = \tau_0^\alpha + \tau_{\text{forest}}^\alpha + \tau_{\text{sub}} + \tau_{\text{HP}}^s \tag{4}$$

$$\tau_c^t = \tau_0^\beta + \tau_{\text{HP}}^t + \tau_{\text{slip}}^\beta \tag{5}$$

The critical resolved shear stresses for the slip resistance equation take the following forms for each of the forest, debris, and Hall-Petch effects, respectively:

$$\tau_{\text{forest}}^\alpha = b^\alpha \chi \mu \sqrt{\rho^\alpha} \tag{6}$$

$$\tau_{\text{sub}}^\alpha = k_{\text{sub}} \mu b^\alpha \sqrt{\rho_{\text{sub}}} \log\left(\frac{1}{b^\alpha \sqrt{\rho_{\text{sub}}}} \right) \tag{7}$$

$$\tau_{\text{HP}}^s = \mu \text{HP}^\alpha \sqrt{\frac{b^\alpha}{d_g}} \quad \text{for } s \in \alpha \text{ without twins} \tag{8}$$

Here, b is the Burgers vector, χ is the dislocation interaction coefficient smaller that unity, μ is the shear modulus, and d is the spacing between twin lamellae.

Dislocation density evolution is modeled by:

$$\begin{aligned} \frac{\partial \rho^\alpha}{\partial \gamma^\alpha} &= \frac{\partial \rho_{\text{generation}}}{\partial \gamma^\alpha} - \frac{\partial \rho_{\text{removal}}}{\partial \gamma^\alpha} \\ &= \text{TSF} k_1^\alpha \sqrt{\rho^\alpha} - k_2^\alpha(\dot{\varepsilon}, T) \rho^\alpha \end{aligned} \tag{9}$$

The parameter k_1^α accounts for the dislocation generation, and the strain rate and temperature dependent $k_2^\alpha(\dot{\varepsilon}, T)$ accounts for dislocation removal. The TSF, or Twin Storage Factor, is a parameter first described by Capolungo et al. [12] for correlations to single crystal Magnesium work by Kelley and Hosford [13]. Oppedal et al. [8] performed experiments and numerical simulations using polycrystalline pure magnesium, providing evidence that a dislocation transmutation scheme described in El Kadiri and Oppedal [1] was responsible for the rapid strain hardening observed in HCP metals loaded to induce profuse twinning.

$$\frac{k_2^\alpha(\dot{\varepsilon}, T)}{k_1^\alpha} = \frac{\chi b^\alpha}{g^\alpha} \left(1 - \frac{kT}{D^\alpha (b^\alpha)^3} \log\left(\frac{\dot{\varepsilon}}{\dot{\varepsilon}_0} \right) \right) \tag{10}$$

Here D is termed as the drag stress, and g is the activation energy in an Arrhenius type rule for the strain rate dependence. The two critical resolved shear stresses for resistance to twin propagation relate to the Hall-Petch effect and slip and take the following forms, respectively:

$$\tau_{HP}^t = \frac{HP^\beta}{\sqrt{d_g}} \text{ with no twins or when } t \in \beta$$
$$\text{and } t \text{ is the predominant twin system (PTS)} \tag{11}$$

$$\tau_{\text{HP}}^t = \frac{HP^{\beta\beta'}}{\sqrt{d_{\text{mfp}}^s}} \text{ when } t \in \beta$$
$$\text{and } t \text{ is not the PTS, with PTS } \in \beta' \tag{12}$$

$$\tau_{\text{slip}}^\beta = \mu \sum_\alpha C^{\beta\alpha} b^\beta b^\alpha \rho^\alpha \tag{13}$$

In these equations, the Hall Petch (HP) terms are used to simulate the hardening response caused by grain segmentation as the twin grows. However, the experimental evidence for this Hall Petch type effect, especially for twin boundaries is not substantiated. In theses simulations, the HP terms are set to zero.

In addition slip could affect twinning through the following equation:

$$\tau^{\beta}_{\text{slip}} = \mu \sum_{\alpha} C^{\beta\alpha} b^{\beta} b^{\alpha} \rho^{\alpha} \tag{14}$$

The $C^{\beta\alpha}$ term represents the interaction between twin and slip and may be negative or positive depending on whether slip interferes with or contributes to twin growth. In simulation work presented here, we did not consider these two barriers for twin propagation for reasons that will be developed further in this paper.

The dislocation based model parameters with best fits are listed in Table 2.

Results of Simulations

In general, the several parameters in each hardening model were fit to the experimental stress-strain behavior of the ED (extrusion direction) and ERD (extrusion radial direction). The fit to the 30ED and 60ED experimental data, and comparison of the strain-hardening rate vs. stress plots were only compared after the initial fitting. Additionally, comparison of the predicted texture and experimentally measured texture at several intermediate strain levels was performed and in good agreement. Results of the texture comparison are not presented in this paper but can be found in [14].

The initial simulated values of CRSS are in agreement with those found in the literature for wrought magnesium alloys. In fact, our simulated initial strengths for basal: prismatic: pyramidal: tensile twinning were 23: 95: 111: 67 MPa, or as ratios to the easy basal slip, 1: 4.1: 4.8: 2.9. Agnew et al. [11] investigating an AZ31 alloy with c-axis fiber found initial strength ratios reported to the basal slip of pyramidal and tensile twinning to be 3 and 0.5. Proust et al. [15] investigating AZ31B and using a "composite grain" variant of the twin model reported values of ratios corresponding to prismatic: pyramidal: tensile twin to be 2.9: 5.2: 2.65.

The fit with the Voce hardening rule to the ED and ERD behaviors is satisfactory. The initial yields for the four loading orientations were all satisfactorily predicted. The hardening rate and overall response were satisfactorily captured for both ED and ERD. However, 30ED and 60ED behaviors were substantially mispredicted. 30ED prediction was the furthest from the test data since the simulated hardening rate in Regime II was of opposite sign to the measured one. It is clear that the current formulation of the Voce model does not incorporate yet a realistic understanding of twin-slip interaction. In fact, an only variation of the Schmid Factor for active deformation modes was sufficient to cause a substantial discrepancy in the prediction of anisotropy. This is atypical to the nature of the crystal plasticity framework. Therefore, it is imprudent to correlate hardening models based only on the two limiting cases of twinning contribution to deformation; intermediate contributions have to be tested as well.

Oppedal et al. [8] used the dislocation based rule to model the behavior of pure Magnesium with c-axis fiber. The corresponding initial slip resistances were found to be scalable to the ones simulated for AM30 in this study, although with a $\langle 10\bar{1}0\rangle \parallel$ ED fiber. For each of basal, prismatic, and pyramidal modes, the values producing good correlations for AM30 were 23: 95: 111 MPa (or 1: 4.1: 4.8) compared to 11: 30: 50 MPa (or 1: 3: 4.5) for pure Magnesium. For tensile and compression twins in AM30, τ_{crit} and τ_{prop} were 67 and 25 MPa, and 275 and 275 MPa, respectively, compared to 15 and 10 MPa, and 185 and 185 MPa, respectively for pure Magnesium. As a ratio to basal slip, tensile twinning had a higher initial resistance in AM30 (2.91) than in pure Mg (1.4). This reflects the higher sensitivity to the alloying elements of twinning compared to slip.

In the case of the dislocation density model, reference must be made to the strain hardening rate versus stress plots. Figure 5 shows the result of varying the twin storage factor (TSF). It was not possible to correlate simultaneously ED and ERD stress strain behaviors unless a TSF of three was used. The effect of the TSF on the stress strain behavior is illustrated in Figure 5. Not only under ED was the saturation stress substantially underestimated with TSF=1, but under ERD as well. This is due to nominal tensile twinning under ERD not possible to prohibit under any other loading. This is in variance to the in the c-axis fiber where compression along the fiber completely prohibits tensile twinning.

In contrast to the Voce model, the dislocation density based model with the TSF was able to reproduce the mechanical responses under 30ED and 60ED when the fit was done based on ED and ERD behaviors. The robustness of this model is offered by the TSF strategy that hypothesized that the twin hardens more importantly than parent. This hypothesis is of striking consistency with the progressive decrease in the severity of the hardening slope when the loading direction is progressively being deviated from ED toward ERD, i.e., with the progressive decrease of twinning contribution to the overall deformation.

Conclusions

The simulations are consistent with the idea of an increased hardening in the twin during profuse twinning caused by the dislocation transmutation idea put forward by El Kadiri and Oppedal [1].

In the simulations using the dislocation density hardening model, it was not possible to achieve a good model correlation to the experimental data without using a twin storage factor of three. This also gave the desirable effect of correlating fairly well the strain hardening rate versus stress behavior as well. With a twin storage factor of three, the simulations could reach the proper saturation stress levels that the experimental test data showed. The simulations also predicted well the experimentally observed stress-strain and texture behavior without utilizing a Hall-Petch type effect that has been attributed to the grain refinement by twinning. Also, the Hall Petch effect for magnesium twin boundaries has been experimentally observed to be quite low [16].

The use of the VPSC simulations with the dislocation density based hardening rule allowed a coupling between slip, twinning, texture, and stress state behavior analysis that has not fully explained by the use of empirical latent hardening parameters such as used in the Voce hardening models. Furthermore, the quantification of the slip activity, twin activity, texture evolution, and hardening as a function of strain allows for macroscale internal state variable.

Acknowledgments

The authors would like to recognize the Center for Advanced Vehicular Systems (CAVS) at Mississippi State University for supporting this work. Also, the authors gratefully acknowledge funding support through the DOE Southern Regional Center for Lightweight Innovative Design award DE-FC26-06NT42755 to carry out a portion of this research work. The authors acknowledge Alan Luo (General Motors Company), and Joy Hines Forsmark and John Allison (Ford Motor Company) for their leadership and encouragement of the larger USAMP/DOE Integrated Computational Materials Engineering for Magnesium Program. Part of this work was also supported by the U.S. Army Tank Automotive Research, Development and Engineering Center (TARDEC) under Contract No. W56HZV-04-2-0003.

References

[1] H. El Kadiri, A. L. Oppedal, J. Mech. Phys. Solids 58 (2010) 613–624.

[2] I. J. Beyerlein, C. N. Tomé, Int. J. Plast. 24 (2008) 867–895.

[3] R. B. Von Dreele, J. Appl. Crystallogr. 30 (1997) 517–525.

[4] H. R. Wenk, L. Lutterotti, S. Vogel, Nucl. Instrum. Meth. A 515 (2003) 575–588.

[5] U. F. Kocks, J. S. Kallend, H. R. Wenk, A. D. Rollett, S. I. Wright, popLA Preferred Orientation Package – LA-CC-89-18, Los Alamos National Laboratory, 1995.

[6] C. N. Tomé, Program Pole (version 8b), Los Alamos National Laboratory, 2008.

[7] R. Hielscher, H. Schaeben, J. Appl. Crystallogr. 41 (2008) 1024–1037.

[8] A. L. Oppedal, H. El Kadiri, C. N. Tomé, G. C. Kaschner, S. C. Vogel, J. C. Baird, M. F. Horstemeyer, 2010. Submitted.

[9] A. Molinari, G. R. Canova, S. Ahzi, Acta Metall. 35 (1987) 2983–2994.

[10] R. A. Lebensohn, C. N. Tomé, Acta Metall. Mater. 41 (1993) 2611–2624.

[11] S. R. Agnew, M. H. Yoo, C. N. Tomé, Acta Mater. 49 (2001) 4277–4289.

[12] L. Capolungo, I. J. Beyerlein, C. N. Tomé, Scr. Mater. 60 (2009) 32–35.

[13] E. W. Kelley, W. F. Hosford, Jr., Trans. Met. Soc. AIME 242 (1968) 5–13.

[14] A. L. Oppedal, H. El Kadiri, Q. Ma, M. F. Horstemeyer, 2010. In preparation.

[15] G. Proust, C. N. Tomé, A. Jain, S. R. Agnew, Int. J. Plast. 25 (2009) 861–880.

[16] C. H. Cáceres, P. Lukáč, A. Blake, Philos. Mag. 88 (2008) 991–1003.

Table 1: Voce parameters used in simulations.

Mode	τ_0	τ_1	θ_0	θ_1	$h^{ss'}$ Prism.	$h^{ss'}$ Basal	$h^{ss'}$ Pyr.	$h^{ss'}$ T. Twin	$h^{ss'}$ C. Twin
Prismatic	95	7	462	0	1	1	1	1	0
Basal	23	51	683	0	1	1	1	1	0
Pyramidal	111	914	3661	0	1	1	1	1	0
Tensile Twin	67	0	0	0	1.4	1.4	1.4	1	0
Compression Twin	315	0	0	0	0	1	1	0	1

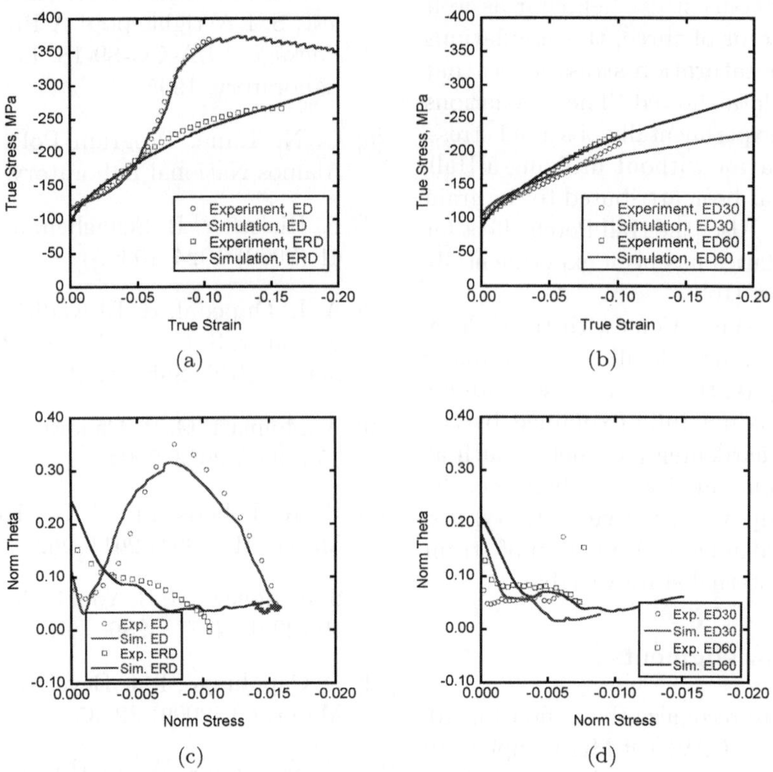

Figure 3: Comparison between experimental results and VPSC simulations using the Voce hardening rule. Stress-strain curves of the (a) extrusion (ED) and extrusion radial direction (ERD) and (b) 30 degrees (ED30) and 60 degrees (ED60) from the extrusion direction, and strain hardening rate versus stress curves for the (c) extrusion and extrusion radial direction and (d) 30 degrees and 60 degrees from the extrusion direction. In the strain hardening rate versus stress plots, stress values were normalized by $\sigma_{\text{normalized}} = (\sigma - \sigma_y)/\mu$ where σ_y is yield stress and μ is shear modulus.

Table 2: Dislocation density hardening rule parameters.

	Prism. ($\alpha = 1$)	Basal ($\alpha = 2$)	Pyr. ($\alpha = 3$)	T. Twin ($\beta = 1$)	C. Twin ($\beta = 2$)
k_1 $1/m$	1.00E+09	0.60E+09	2.00E+09		
$\dot{\epsilon}_0$ s^{-1}	1.00E+07	1.00E+07	1.00E+07		
g^α	0.0035	0.0035	0.003		
D_0^α MPa	3.40E+03	10.0E+03	0.08E+03		
τ_0^α MPa	95	23	111		
χ	0.9	0.9	0.9		
τ_{crit}^β MPa				67	275
τ_{prop}^β MPa				25	275
Twin Storage Factor				1, 3	1, 3

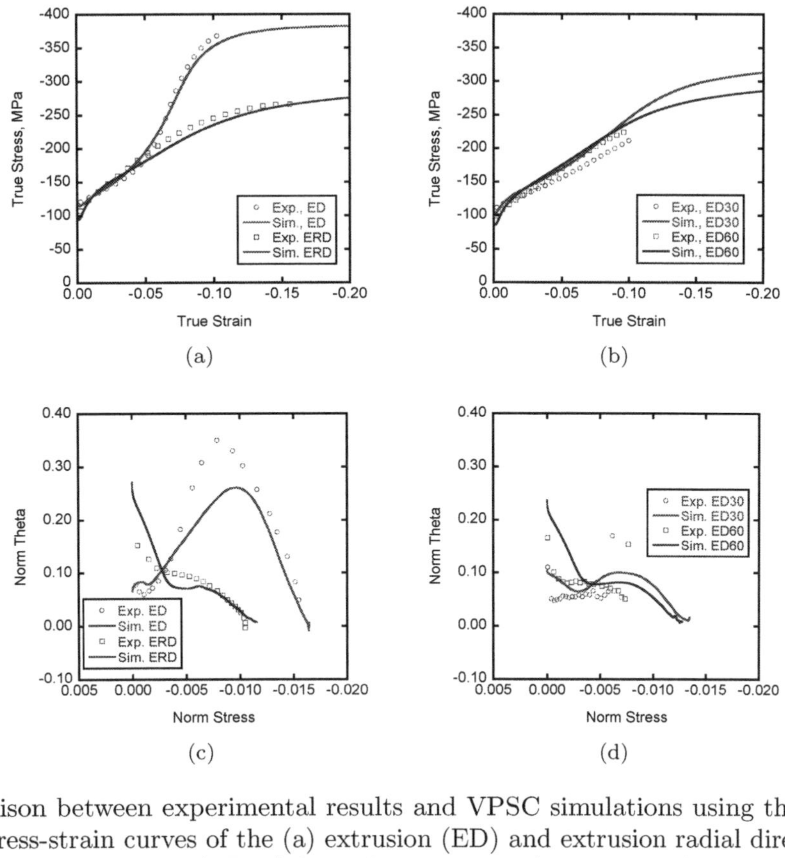

Figure 4: Comparison between experimental results and VPSC simulations using the dislocation density hardening rule. Stress-strain curves of the (a) extrusion (ED) and extrusion radial direction (ERD) and (b) 30 degrees (ED30) and 60 degrees (ED60) from the extrusion direction, and strain hardening rate versus stress curves for the (c) extrusion and extrusion radial direction and (d) 30 degrees and 60 degrees from the extrusion direction. The simulations used the parameters in Table 2 with twin storage factor equal to 3. In the strain hardening rate versus stress plots, stress values were normalized by $\sigma_{\text{normalized}} = (\sigma - \sigma_y)/\mu$ where σ_y is yield stress and μ is shear modulus.

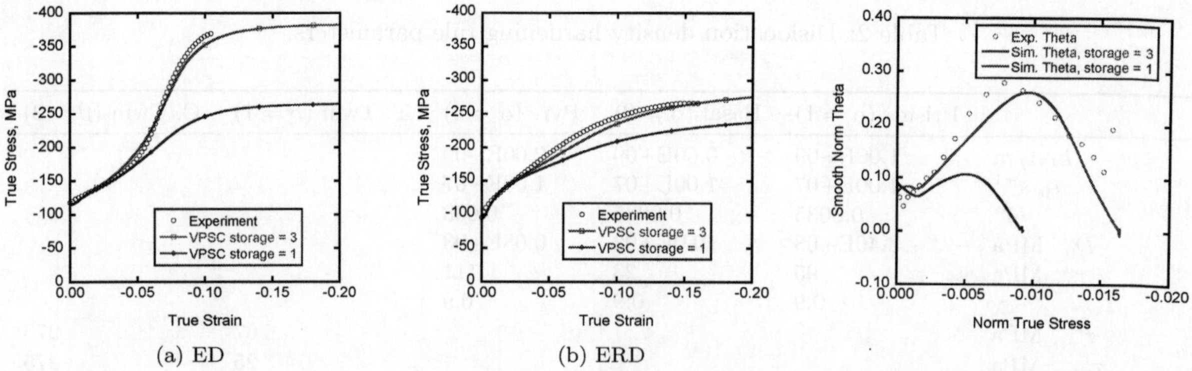

Figure 5: Results of VPSC simulations using the dislocation density hardening rule and experimental results, true stress versus true strain for two different twin storage factors. The simulation and experimental results are from uniaxial compression in a) the Extrusion Direction (ED) and b) Extrusion Radial Direction (ERD). The strain hardening rate versus stress is plotted in c), for two different values of twin storage factor.

Magnesium Technology 2011
Edited by: Wim H. Sillekens, Sean R. Agnew, Neale R. Neelameggham, and Suveen N. Mathaudhu
TMS (The Minerals, Metals & Materials Society), 2011

DEFORMATION BEHAVIOR OF MG FROM MICROMECHANICS TO ENGINEERING APPLICATIONS

E. Lilleodden, J. Mosler, M. Homayonifar, M. Nebebe, G. Kim, and N. Huber*
Institute of Materials Research, Materials Mechanics,
Helmholtz-Zentrum Geesthacht, Max-Planck-Str. 1, 21502 Geesthacht, Germany

Keywords: Mg, microcompression, twinning, crystal plasticity, forming limit

Abstract

We have investigated the influence of the underlying microstucture of Mg on its macroscopic deformation behavior. Using microcompression experiments on Mg single crystals, we have identified the orientation dependent deformation slip and twinning systems. These experiments have aided the development of an energy based crystal plasticity model for the twinning in polycrystals. This model has been applied to representative volume elements to identify the flow surface for a macroscopic model, which is used to predict the forming limit diagram. Validation of the results has been realized through Nakazima tests. Additionally an outlook is given on the prediction of the forming behavior of laser beam welded Mg sheets and its dependence on welding direction.

Introduction

Mg, the lightest of all structural metals, shows great promise for reducing weight in structural automotive applications, greatly improving fuel consumption and CO2 emissions. The development of new Mg alloys and joining technologies is an active area of research. In order to optimize alloy design and welding process parameters, the influence of alloying constituents and heterogeneity of microstructure on the mechanisms of deformation across the laser welded interface must be understood.

In order to develop a continuum model of such complex structural components, both a fundamental understanding of their constituent mechanical behavior at the microstructural level is needed, along with a continuum description of the mechanical behavior at the component level. While both ends of this broad spectrum necessitate in-depth investigations, the bridging of these scales is a critical challenge. In this paper, we present an overview of the scientific efforts across multiple length scales directed towards a comprehensive understanding of the mechanical behavior of Mg, with an outlook as to the applicability of such an approach to laser welded sheets.

Micromechanical characterization

Due to the potential involvement of both slip and twinning mechanisms during the deformation of Mg, the identification of the relevant deformation mechanisms under an arbitrary stress state cannot be known *a priori*. This is in strong contrast to face center cubic materials which show slip limited to the {111} planes. While the basal slip system is without debate the most easily activated system in Mg, it is less clear under what conditions prismatic slip and pyramidal slip systems are activated, or what the critical dependencies are for the activation of twinning. In order to identify slip and twinning activity in magnesium, the recently exploited method of microcompression testing [4,5] has been applied to single crystals of four model orientations: <0 0 0 1>, < 1 1 -2 0 >, < 1 0 -1 0 > and <2 -1 -1 2>. This same approach can be easily extended to studying the effect of alloying content and strain rate, which is the subject of a current investigation. The microcompression method is described elsewhere in detail [6].

Microcompression columns were fabricated from the bulk single crystals using focused ion beam machining, resulting in columns with a nominal diameter of 5μm and a height to diameter aspect ratio of approximately 3:1. Compression experiments were conducted with a Nanoindenter XP (Agilent) outfitted with a flat ended conical indenter with a 15 micron diameter circular punch. Experiments were run to varying maximum strain using a nominally constant strain rate of 0.001 /s. After compression, the columns were imaged in the SEM to assess the deformation morphology, such as slip steps. Cross sections were fabricated through columns with the FIB for conducting electron back-scattered diffraction (EBSD) analysis of orientation for the investigation of twinning. The FIB was also used to fabricate thin lamella for transmission electron microscopy (TEM) analyses of slip activity.

The engineering stress-strain curve of representative columns from each orientation is given in Figure 1. The associated deformed columns are shown in the SEM micrographs given in Figure 2. Unsurprisingly, the deformation behavior is strongly governed by the initial crystallographic orientation, as shown in the stress-strain response as well as the deformation morphology.

In the case of <0001> compression, relatively high stresses are achieved, with considerable hardening. Post-compression EBSD and TEM analyses have shown that no twinning occurs [6] and that pyramidal slip with a < c + a > Burgers vector is the dominant mechanism of deformation. The high hardening observed can be easily rationalized by the activation of 6 equivalent slip systems, as dictated by the symmetry of this orientation.

In the case of compression along the <2 -1 -1 2> axis, the so-called 45° orientation, basal slip is the governing deformation mechanism, as expected; a large slip band is observed along the 45° plane. It is important to note that the slip band occurs in the middle of the column, rather than at a stress concentration at the contact or base. The associated stress-strain response shows a large strain burst at about 25 MPa, the lowest stress of all orientations. Of course this is still a relatively high stress for Mg

* Corresponding author: norbert.huber@hzg.de

single crystal oriented for easy slip on the basal plane. This high stress is due to the widely observed size effect in microcompression testing, where the flow stress is shown to increase with decreasing diameter [4-8]. While the basis for size effects in microcompression testing has been attributed to various factors, such as the depletion of dislocation sources with decreasing size [7] or a size dependent dislocation source length [8], it is beyond the scope of this paper to investigate the basis of the observed size effect. While the presence of the size effect complicates the use of microcompression testing for the validation of continuum models, in the context of this paper the microcompression data are used to establish the relevant modes of deformation and relative stress-strain responses as a function of orientation; quantification of the resolved shear stresses is not our objective here.

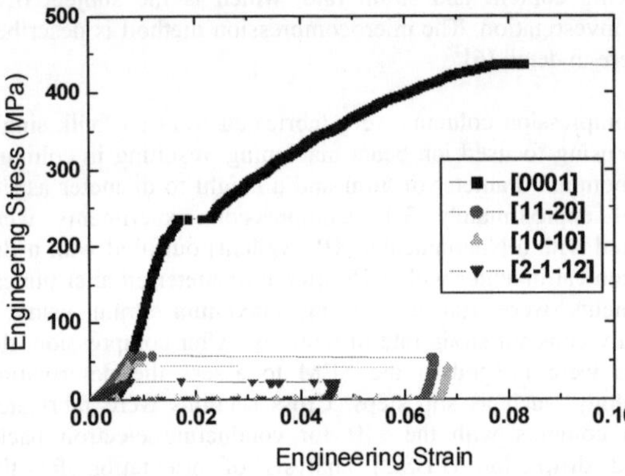

Fig. 1 Engineering stress-strain responses from representative columns of the 4 tested orientations, with nominal diameter of 5 μm.

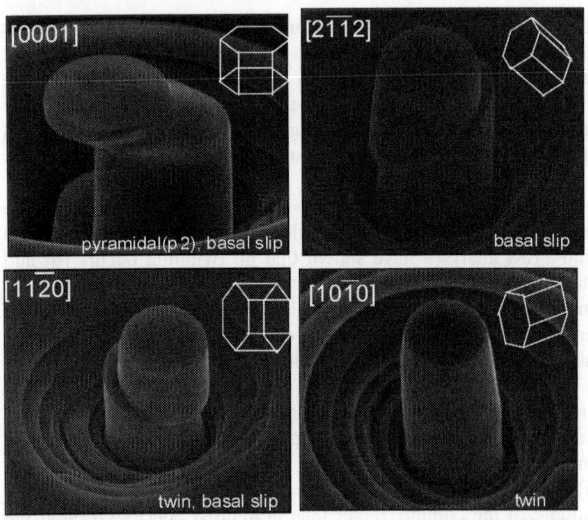

Fig. 2 SEM images of 5 μm diameter columns in Mg single crystals after compression.

The other two compression orientations: <11-20> and <10-10>, show intermediate stress levels relative to the <0001> and 45° orientations. Furthermore, they both show negligible hardening relative to the <0001> orientation, and a similar massive strain burst relative to the 45° orientation. However, deformation twinning is the primary mode of deformation for these two orientations, whereas no twinning was observed in the <0001> and 45° orientations. Compression along the <11-20> or <10-10> direction results in a tensile strain along the <c> axis which serves as the driving force for twinning. The massive strain burst shown for the <11-20> and <10-10> orientations has been shown to be due to a destabilization of the propagating twin [9], rather than a massive slip such as is observed in the 45° orientation. In case of <11-20> compression, the twin formed is well oriented for easy basal slip, which leads to massive shear on the basal system within the twin. This contributes to the strain burst observed in the stress-strain response in fig. 1, and typically leads to a larger burst than that observed in the <10-10> orientation. However, the strain burst is initiated prior to massive basal slip, as discussed in greater detail elsewhere [9].

Such a qualification of the deformation mechanisms in pure single crystal Mg as a function of orientation, as has been achieved here with microcompression testing, establishes which mechanisms are needed to be incorporated within a continuum framework. This can simplify both the development of a physically-based model, as well as allow higher efficiency in carrying it out. While at present, the values of critical resolved shear stresses for the individual slip and twinning mechanisms can not be obtained from the microcompression method, the relative stresses provide a starting point for a more accurate picture of deformation in Mg. The relevant slip and twinning systems identified from microcompression tests for a given stress state are used to develop a more physically based model of twinning. Ongoing work using dislocation dynamics simulations is being used to address the size effects in order to establish quantitative inputs for crystal plasticity modeling.

Micromechanical energy-based twinning model

While experimental characterization is an important element in understanding the mechanical behavior of materials, computational simulations are essential to the quantitative interpretation of results. This is particularly true for the aforementioned testing methods at the micron scale. For example, it is difficult if not impossible to observe the internal structure of a deforming body. Since the goal is the realistic prediction of mechanical behavior at the macro-scale, atomistic models are not promising. However, purely phenomenological macroscopic models are not suitable either, since the experimental observations cannot be directly included in this case. For this reason, the framework of crystal plasticity theory has been chosen. It naturally accounts for the different orientations of dislocations, but it is still numerically comparably efficient. In line with 0, twinning has been modeled as phase decomposition. For instance, highlighting the initial untwinned phase by the subscript (ini) and the twinned domain by (Tw), the total Helmholtz energy Ψ can be computed as the volume average, i.e.,

$$\Psi = (1 - \xi)\Psi_{\text{ini}} + \xi\,\Psi_{\text{Tw}} + \Psi^{\text{mix}}(\xi). \tag{1}$$

Here, $\xi \in [0;1]$ and Ψ^{mix} denote the relative twinning volume ratio and a mixture energy, respectively. Assuming that energy minimization is the underlying physical principle, the twinning volume ratio follows from minimizing Helmholtz (1). Since this function is highly non-linear and non-convex, its minimization cannot be computed in a straightforward manner. For this reason, only the limiting cases (completely untwinned or completely twinned crystal) have been considered. Since without additional modifications of the constitutive model, this would lead to a discontinuous stress response, the concept of twinning *pseudo dislocations* has been adopted, cf. [10]. Within this concept, the deformation associated with twinning is decomposed into a reorientation of the crystal lattice and a simple shear mode. The latter is modeled in an analogous fashion as dislocations.

The approach briefly discussed here can be generalized such that dissipative effects resulting from dislocations and twins are consistently included. Conceptually, the Helmholtz energy has to be replaced in this case by an incrementally defined potential denoted as I^{inc}. Further details are omitted. They can be found in [10]. The resulting optimization problem reads

$$I^{\mathrm{inc}} = \min[I^{\mathrm{inc}}_{\mathrm{ini}}; I^{\mathrm{inc}}_{\mathrm{Tw}}]. \qquad (2)$$

This variational approach has been implemented into a finite element code. Subsequently, it has been applied to the analysis of texture evolution in a Mg polycrystal. The prescribed strain history corresponds to that of a rolling process. According to Fig. 3, the initial isotropic texture evolves and leads finally to a material with pronounced anisotropy. The computed results agree well with those previously reported in [11].

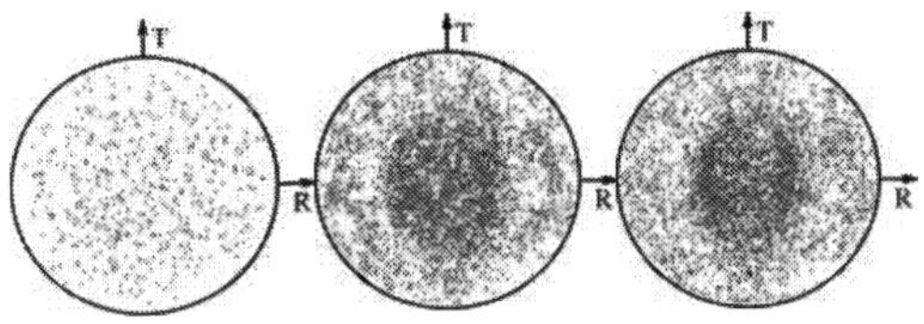

Fig. 3 Simulation of a rolling process by using the energy-based twinning model: Pole figure showing the evolution of the basal texture for different strain amplitudes. From left: ε = 0.00, ε = -0.68, ε = -1.20.

Macroscopic phenomenological plasticity model

Although the novel crystal plasticity model is computationally efficient and therefore can be used for the analysis of texture evolution in polycrystals, it is still numerically too complex for large scale engineering problems. For this reason, an additional model is introduced. It falls into the range of macroscopic finite strain plasticity theory. Its yield function was originally proposed in [12]. In tensor notation it can be rewritten as

$$\phi = \mathrm{J}_2^{3/2}\,(\mathbb{H}_1 : \boldsymbol{\Sigma}) + \mathrm{J}_3\,(\mathbb{H}_2 : \boldsymbol{\Sigma}) + \mathrm{h} \leq 0. \qquad (3)$$

Here, J_2, J_3 are the second and the third invariant of the second order tensors $\mathbb{H}_i : \boldsymbol{\Sigma}$ and $\boldsymbol{\Sigma}$, is the stress tensor (of Mandel type), h denotes the current yield strength (isotropic hardening) and the fourth order tensors $\mathbb{H}_i$ define the material's anisotropy. In contrast to [12], they may evolve (distortional hardening) and thus, the shape of the yield surface may change during deformation, cf. [13]. Following computational homogenization techniques (see e.g. [14]), the material parameters of the macroscopic model briefly presented here are computed by considering representative volume elements (RVE) with periodic boundary conditions. More precisely, using the aforementioned novel crystal plasticity model within such RVEs allows the computation of macroscopic yield functions which can be subsequently used for calibrating the material parameters of the macroscopic model. For that purpose, a standard least-squares approach is used, i.e., the difference between the macroscopic yield function, which depends on the unknown material parameters, and the yield function computed from the RVE is minimized. Accordingly, these yield functions are associated with a rolled sheet showing a pronounced texture (see Fig. 3). The resulting model has been implemented into a finite element code. Results predicted by this model compared to experimental measurements on rolled sheets with thickness 1.3mm are shown in Fig. 4. Accordingly, the model captures the strain distribution in good agreement with the Nakazima forming limit test. This test has been re-computed for different sheets with different shapes. For defining the ultimate load, the localization criterion proposed in [15] was implemented. By way of contrast, the forming limit curve associated with the conducted experiments has been determined by synchronizing the ultimate load with the respective principal strains. The results are summarized in Fig. 5. As evident, the numerically computed forming limit diagram matches its experimentally obtained counterpart very well.

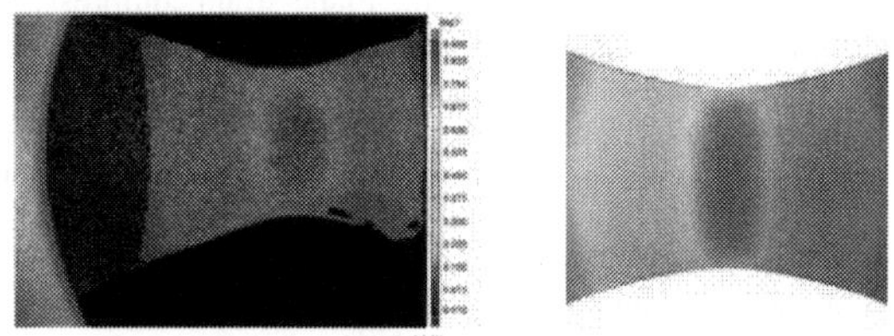

Fig. 4 Nakazima forming limit test for AZ31 with thickness 1.3 mm: contour plots of the maximum in-plane principal strain. Left: experiment; right: numerical results as predicted by the phenomenological macroscopic constitutive model, see Eq. (3).

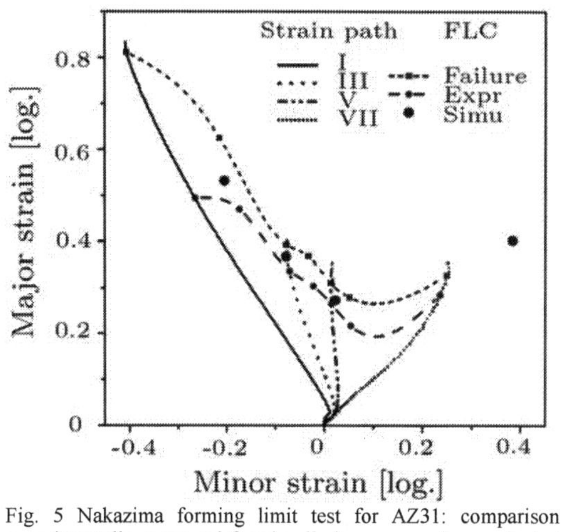

Fig. 5 Nakazima forming limit test for AZ31: comparison between experimentally measured forming limit curves and numerical simulations based on by the phenomenological macroscopic constitutive model, see Eq. (3).

Concluding remarks

We have shown that the deformation at the microstructural level can be integrated into a continuum framework for the simulation of sheet forming behavior of Mg alloys. The presented approach relies on the overlap between experiment and simulation at the micro-scale. Microcompression experiments provide the appropriate deformation mechanisms as constitutive inputs in crystal plasticity simulations. Through homogenization with representative volume elements the texture for a polycrystalline rolled material has been predicted. The resulting macroscopic yield functions, including their evolution, have been used to extend the existing macroscopic plasticity model of Cazacu and Barlat. Validation was achieved through experiment and simulation of Nakazima forming limit test for AZ31. The approach presented is currently being applied to laser-welded AZ31 sheets, a material system with a strong heterogeneity in microstructure and property distribution.

Acknowledgements

The authors want to thank J. Bohlen and D. Letzig, Institute of Materials Research, Magnesium Innovation Centre, for their experimental support with the Nakazima forming limit tests and the valuable discussions.

References

1. E.W. Kelley and W.F. Hosford, "Plane-strain compression of magnesium and magnesium alloy crystals", *Transactions of the Metalurgical Society,* 242 (1968) 5-13.
2. T.Obara, H.Yoshinga, and S.Morozumi, "{11-22}<-1-123> slip system in magnesium", *Acta Materialia,* 21 (1973) 845-53.
3. A. Couret and D. Caillard, "An *in situ* study of prismatic glide in magnesium - I. The rate controlling mechanism", *Acta Materialia,* 33 (1985) 1447-54.
4. M. D. Uchic, D. M. Dimiduk, J. N. Florando, and W. D. Nix, "Sample dimensions influence strength and crystal plasticity", *Science,* 305 (2004) 986-9.
5. M. D. Uchic and D. M. Dimiduk, "A methodology to investigate size scale effects in crystalline plasticity using uniaxial compression testing", *Materials Science and Engineering A,* 400–401 (2005) 268-78.
6. E.T. Lilleodden, "Microcompression study of Mg (0001) single crystal", *Scripta Materialia,* 62 (2010) 532-5.
7. J. R. Greer and W. D. Nix, "Nanoscale gold pillars strengthened through dislocation starvation", *Physical Review B,* 73 (2006) 245410.
8. C.V. Volkert and E.T. Lilleodden, "Size effects in the deformation of sub-micron Au columns", *Philosophical Magazine,* 86 (2006) 5567-79.
9. G. Kim, S. Yi, Y. Huang, and E. Lilleodden, "Twining and slip activity in magnesium <11-20> single crystal", *Mechanical Behavior at Small Scales — Experiments and Modeling,* ed. J. Lou, E. Lilleodden, B. Boyce, L. Lu, P.M. Derlet, D. Weygand, J. Li, M.D. Uchic, E. Le Bourhis (Warrendale, PA: Materials Research Society, 2010) 1224-FF05-03.
10. M. Homayonifar and J. Mosler, "On the coupling of plastic slip and deformation-induced twinning in magnesium: A variationally consistent approach based on energy minimization", *International Journal of Plasticity,* 2010, IN PRESS. Preprint download: http://www.hzg.de/institute/materials_research/structure/materials_mechanics/simulation/staff/004616/index_0004616.html.en
11. S.R. Agnew, M.H. Yoo, and C.N. Tome, "Application of texture simulation to understanding mechanical behavior of Mg and solid solution alloys containing Li or Y", *Acta Materialia,* 49 (2001) 4277-4289.
12. O. Cazacu and F. Barlat, "A criterion for description of anisotropy and yield differential effects in pressure-insensitive metals", *International Journal of Plasticity,* 20 (2004) 2027-2045.
13. M. Nebebe, L. Stutz, J. Bohlen, D. Steglich, D. Letzig, and J. Mosler, "Exprimental measurement and constitutive model for forming process of magnesium alloy sheet", *Magnesium-8th international conference on magnesium alloys and their application,* ed. K.U. Kainer, (Weinheim, Germany: Wiley VCH, 2009) 764–770.
14. C. Miehe, "Strain-driven homogenization of inelastic microstructures and composites based on an incremental variational formulation", *International Journal for Numerical Methods in Engineering,* 55 (2002) 1285-1322.
15. Z. Marciniak and K. Kuczynski, "Limit strains in the processes of stretch-forming sheet metal", *Journal of Mechanical Science,* 9 (1967) 609–620.

Magnesium Technology 2011
Edited by: Wim H. Sillekens, Sean R. Agnew, Neale R. Neelameggham, and Suveen N. Mathaudhu
TMS (The Minerals, Metals & Materials Society), 2011

EFFECT OF SUBSTITUTED ALUMINUM IN MAGNESIUM TENSION TWIN

K.N. Solanki[1], A. Moitra[1], M. Bhatia[1]
[1]Center for Advanced Vehicular Systems, Mississippi State University, Mississippi State, MS, USA

Keywords: Magnesium, Aluminum, Twin Nucleation, Molecular Dynamics, Single crystal

Abstract

Atomistic simulations are performed in order to study the Aluminum substitution effect on Magnesium twinning mechanism. Multiple twin boundaries are found in pure Mg single crystal under tensile loading condition along $(\bar{1}012)$ crystallographic direction. However, no twinning has occurred under compression loading. Al substitution has been done for 2, 5, 7, and 10% doping. For 2 and 5% Al substitution, number of twins increase when the system is monitored under tensile loading. On the other hand, for 7 and 10% Al doping under tensile loading, no twin has been found. We found that dislocation-twin and dislocation-dislocation interaction are majorly responsible for this behavior and it is important that which one is prevalent.

Introduction

Alloyed magnesium is the lightest metal with specific strength in practical use today. There are many practical application products such as vehicular industries. However its wide-spread use is greatly hampered by its anisotropic plastic response which limits its formability at room temperature [1]. This comes partly from the hexagonal-closed packed crystal structure, where slip in the basal plane is nearly one-hundred times easier than in perpendicular planes. Slip system of Mg is composed of three slips, which are parallel to basal, prismatic and pyramidal slips. Since the (critically resolved shear stress) CRSS of pyramidal slip is much larger than that of the other slips, it is difficult to activate pyramidal slip at room temperature. Although several slip systems have been reported, the slip is commonly in the closed packing basal direction, and does not produce strains parallel to the c axis. It is known that hcp Mg shows much poor ductility compared with fcc Al due to its crystal anisotropy. Magnesium also twins (a deformation mechanism involving extended stacking changes) at fairly low strains. However, because the wrought magnesium alloy is expensive and is poor in formability, the reduction of manufacturing costs and the improvement of formability are required.

Mechanistic explanation of the plastic behavior of material is based upon the dislocation and twin nucleation. This plastic behavior of the material influences on the formability of the material. Formability of elements and their alloys are affected by strain hardening rate, strain rate sensitivity, degree of anisotropy, stress-state at fracture etc. It is well known that traditional magnesium sheet alloys exhibit poor formability via conventional forming operations at ambient temperatures. However, it is also known that at high temperatures and moderate strain rates, magnesium alloys can be extremely ductile and readily formable. Traditional explanations for these phenomena relate to the behavior of the hexagonal close packed (hcp) magnesium single crystals. The plastic properties of magnesium single crystals were studied in detail during the 1950s and 60s with extensive application of slip trace analyses and constrained deformation experiments [2, 3]. A major conclusion of these studies was that magnesium crystals only possess two independent easy slip systems within the (0001) or basal plane. Thus, it appeared that magnesium, with its hcp crystal structure fails to satisfy the Taylor criterion [4] requiring five independent easy slip systems for homogeneous, generalized ductility of a polycrystalline aggregate. Other slip mechanisms observed [5, 6, 7, 8] during the early studies include the slips in prismatic and pyramidal planes. With reference to the Taylor criterion, it is noted that prismatic slip would offer two independent slip modes beyond the original two of basal slip for a total of four (still short of the required five for uniform plasticity). Pyramidal slip would offer four independent slip modes on its own (the same as basal and prismatic together). Although many investigators [9] performed c-axis compression experiments on single crystals designed to activate dislocations with non-basal Burgers vectors, no evidence of their activity was observed through slip trace analysis. Instead, some twinning mechanism was observed to barely precede fracture [8, 9]. Only after transmission electron microscopy (TEM) was applied to the problem during the early 1970s [10], non-basal dislocations were confirmed to be active within magnesium. In addition to dislocation slip, magnesium exhibits a strong propensity for mechanical twinning, especially the $\{10\bar{1}2\}\langle 10\bar{1}\bar{1}\rangle$ tension twin [2]. Twinning may help a material to satisfy the Taylor criterion [11], however, it must be recognized that twinning is a polar mechanism (only allowing simple shear in one direction, rather than both forward and backward directions like dislocation slip). Furthermore, the amount of strain that twinning can accommodate is directly proportional to the volume fraction of crystal which has twinned. Thus there is a finite amount of strain that may be accommodated by twinning, even under ideal circumstances. Deformation twinning results in a sudden reorientation of the volume fraction of the crystal that is being twinned rather than the gradual reorientation that characterizes slip. Finally, although twinning mechanisms offer more options for plastic accommodation, they generally require an accommodation mechanism of their own in order to avoid large internal stresses leading to failure. Wang et al. recently studied [12] in detail the mechanism of $(10\bar{1}2)$ twin nucleation, however the postulated mechanisms for twin nucleation in models have not been validated so far.

Molecular dynamics (MD) simulations have been successful and powerful tool in revealing the deformation mechanisms in materials. Studies related to the deformation process speculate the major role of deformation twinning for magnesium. New alloy development involves new solute additions and deformation to affect mechanical properties. In present paper, we observe the Al substitution effect on the Mg twinning for four different amount of Al substitution.

Method

For the purpose of present study, single crystal magnesium samples were created of $20 \times 20 \times 5.5$ nm^3 dimension. The samples were first relaxed using an NVT ensemble for 10 nS, and then they were relaxed independently in three directions with NPT ensemble for another 10 nS. Atomistic simulations were carried out using an MD time step of 1 fS. A periodic boundary condition was adopted in all directions. A tensile deformation along the Y $(10\bar{1}2)$ direction is performed at a strain rate of 10^8 S^{-1}, whereas deformation in X and Z directions that are on the $(10\bar{1}2)$ plane, are carried out by maintaining a zero pressure by Parrinello-Rahman method [13]. Throughout our simulations the temperature was maintained at 300 K. At a particular interval of time we collect the snapshots of the atomic structure to analyze the microstructure.

It has been already found [14] that the interstitial Al atoms in Mg crystal raise the internal energy prohibitively large via density functional theory calculations. However it is easier for the Al atoms to substitute the Mg atoms, and occupy their positions. Accordingly, we randomly doped different amount of Al atoms in Mg single crystal by substitution. The size of the simulation cell, strain rate, and temperature were kept same as for the case of single crystal Mg simulations and Mg with Al substitution. The molecular dynamics simulations in this work employed the embedded atom method (EAM) potential of Sun [15]. This potential has a better agreement on vacancy formation energy than the Liu's [16] potential, and apart from that, the potential is developed by fitting also the vacancy migration energy. It is important to note here that a recent study [17] while investigating the effect of precipitate in aluminum alloys, proposed that thermally activated vacancy movement plays a central role in grain growth mechanism.

Results and Discussion

Stress strain curves for tensile loading with several % of Al substitution are shown in Figure 1. For pure Mg sample, two distinctive peak stress can be observed, where the first peak represent the dislocation nucleation, the second peak correspond to the twin nucleation. A substitutional atom solution relieves hydrostatic stresses and interacts with nearby dislocations. Dislocations surrounded by atmospheres can produce plastic flow in two ways: at low stress dislocations cannot escape from the atmosphere, and solute atoms migrate with the dislocations, however at large stress those dislocations can tear apart and become highly mobile and able to produce rapid flow under small stress. Interstitial Al atoms locally distort the Mg lattice and produce a strain field locally. For 2% Al substitution, the dislocation nucleation peak is increased, however the twin nucleation peak is decreased compared to the pure Mg sample. With increased Al substitution (more than 2%) these two peaks converged to one peak, such that it is indistinguishable to identify the dislocation or twin nucleation from the stress-strain curve. Applied stress generates the dislocations (and hence the yielding) easily for the samples where Al exists more as substitutional atoms.

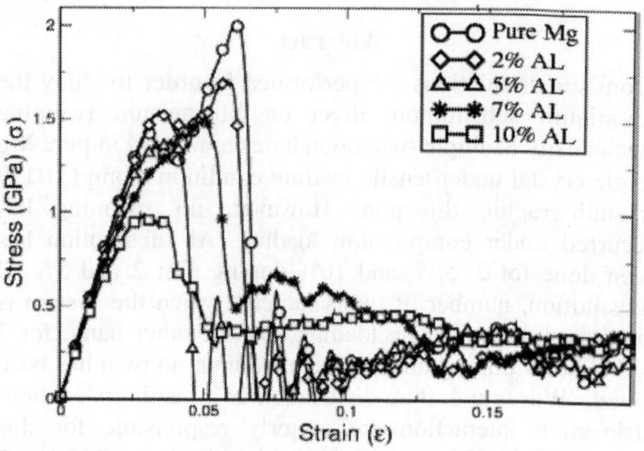

Figure 1: Stress strain curves for the pure Mg, and Mg with several % of Al substituted, under tensile loading.

We found that under tensile loading a twinning occurs, however even at quite larger strain no twinning occurs for the same direction if compressive load is provided. That confirms the fact that twinning is a polar mechanism, unlike dislocations that occurs for applied load in both directions. Figure 2 shows the structures of the sample with pure Mg for compressive loading in a) and tensile loading in b). We found that the twin formation occurs at 6.8 % of strain for pure Mg in tension. Also, we found that the initial twin volume grows with strain and secondary twins nucleate within the primary twin.

Figure 3 shows the microstructure of the samples with different amount of Al substitution. We found that in Figure 3a at 2% of Al substitution number of twins are four at 6.8% of strain. At Figure 3b with 5% Al substitution number of twins are four at 5.8% strain. In Figures 3c and 3d at 7% and 10% Al substitution, respectively we found no twinning occurs even at large strain. It has been also found that strain at which twin nucleates, decreases with increasing amount of Al doping up to 5% of substitution.

Figure 4 shows the evolution of twin-volume with strain for our models. It can clearly be seen that the twin nucleation occurs at higher strain for pure Mg model compared to the models where up to 5% Al is substituted. We can also find that the twin volume increases faster for the cases where Al is substituted compared to the pure Mg case. In our calculations, we have found that the angular rotation of twinned volume have increased up to 87° that in a very good agreement with the previous results. Since for 7 and 10% Al substitution, we do not find any twinning formation, the figure does not contain any data for that higher % s of Al substitution.

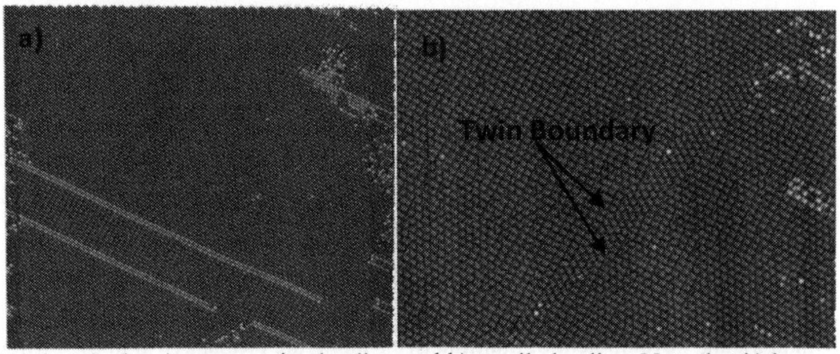

Figure 2: Sample at 6.8% of strain for a) compressive loading and b) tensile loading. Note that b) has a twin formation, while a) does not.

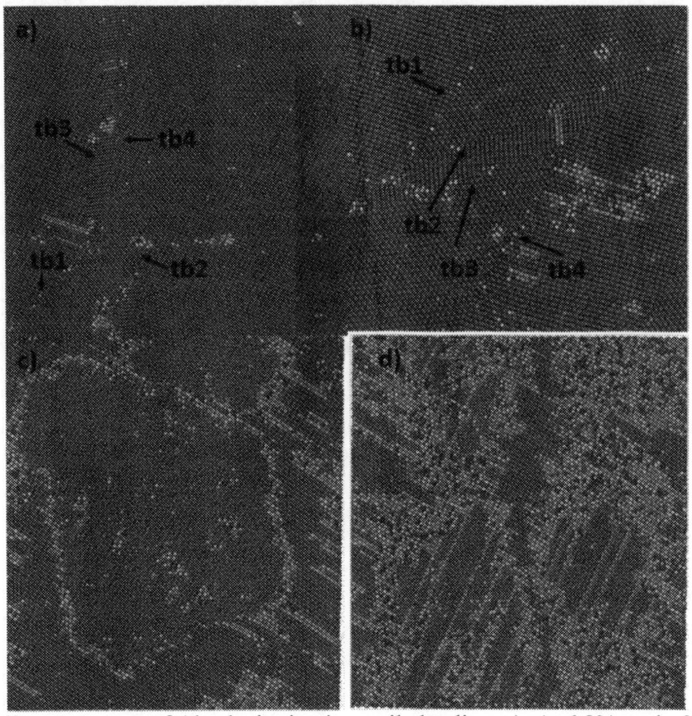

Figure 3: Mg samples with different amount of Al substitution in tensile loading. a): At 6.8% strain for 2% Al substitution, b): At 5.8% strain with 5% Al substitution, c) At 7% strain with 7% Al, and d) At 10% strain with 10% Al substitution. Note that you may find 2 more twins in b), but due to periodic boundary condition, they represent the same twins. Also note that c) and d) has no twins.

We further analyze the microstructure and revealed an interesting interaction of twin boundary with dislocation loop. Figure 5 shows a twin emission and the tip of twin is marked as 'Tip'. Figures 5a and 5b are for a pure Mg tensile deformation at 4.6% and at 9% strain. Figure 5b shows that at larger strain how the number of twins increases and how the secondary twin forms from the primary twin boundary. We have observed that secondary twin always forms from the primary twin boundary plane. Twin boundaries are marked by numbers in Figure 5b.

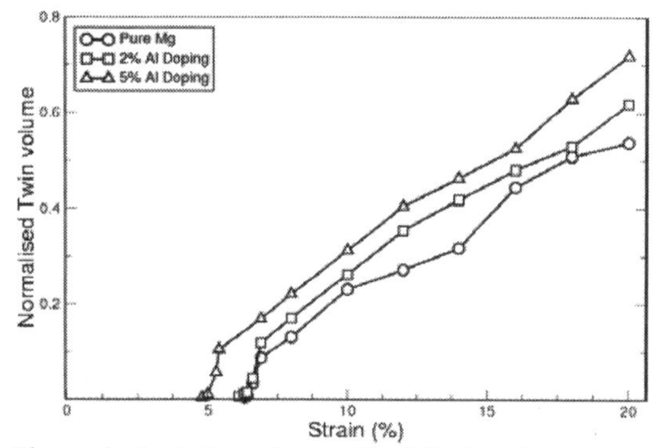

Figure 4: Evolution of normalized Twin volume.

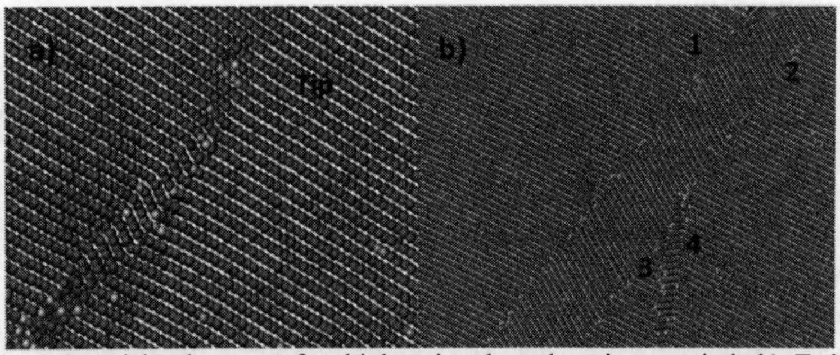

Figure 5: Emission of a twin at a), and development of multiple twins along the primary twin in b). Twin tip is marked by tip.

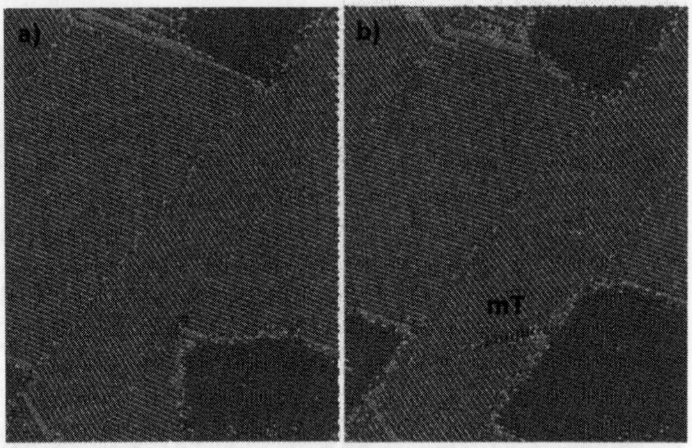

Figure 6: Interaction of dislocation with twin. In a) a dislocation loop emission toward the twin boundary is shown. In b) dislocation growth along the twin boundary is shown. Note a secondary twin formation in b) marked by mT.

In Figures 6a and 6b, we show that the twin boundary interaction with a dislocation loop. It is found that the twin boundary is very rigid and the dislocation loop cannot pass through the twin boundary. In Figure 6a, we show the initial emission of dislocation towards an existing twin boundary at 5% strain. Figure 6b shows the same dislocation-twin interaction at 9.6% strain. We found that even at large strain the dislocation loop grows along the twin boundary plane. This behavior of dislocation-twin interaction certainly due to the fact that twin boundary is very rigid, such that a dislocation cannot penetrate through the twin boundary; instead the dislocation grows along the twin boundary plane. In Figure 6b, however a micro-twin is observed inside the primary twin, marked as mT in the figure. At a later strain (not shown here) this micro-twin also grows. Thus, the interaction compels us in believing the fact that twin- dislocation interaction gives rise to more twins. When there are substituted Al atoms in the Mg system, dislocations occur at less stress state (See Figure 1); these dislocations enhance the twin-dislocation interaction that finally produces more twins. However, when the amount of Al substitution reaches to 7% atomic percentage, enormous number of dislocations are generated. These dislocations then interact mutually. As such, the twin-dislocation interaction seizes and dislocation-dislocation interaction becomes prevalent. When these dislocation-dislocation interactions become predominant, twin formation becomes sluggish. The interaction between two dislocations or a dislocation with a twin is inherently a dynamic nature of interaction of two defected lines. The atoms inside those defected-lines although possess a symmetrical nature of crystal structure (hcp for this case). These atoms being intelligent being at a given time knowing all the forces that causes them to move, try to dissipate the stress in such a fashion that the interaction energy of those two defected-lines minimized. Typically for two dislocation loops interaction, the atoms inside them are trying to reorient themselves to minimize the interaction energy, and hence no effort is being taken to penetrate the twin boundary. Naturally, when the dislocation-dislocation interaction becomes prevalent than the dislocations-twin interaction (which is 7% or higher for the present study), new twin formation does not take place.

In order to increase the formability of Mg we need to get the homogeneous and generalized ductile behavior for Mg. Since the slip systems are limited, and Mg fails to satisfy Taylor criterion, twinning is the alternative to increase the formability of Mg. We found in the present paper that how the evolution of twinning changes with the Al substitution. It should be kept in mind that the present calculation deals a tensile loading condition only for a particular orientation of single crystal Mg. In reality, there are several grains oriented in different directions with multiple grain boundary and different amount of multi-dimensional defects in it. Nevertheless, the nature of the dislocation-dislocation interaction or the dislocation-twin interactions discussed in the paper remains the same for polycrystalline aggregate. Perhaps it could be a nice future study to see the dislocation-twin interactions for a polycrystalline Mg sample with low angular misorientation.

Conclusion

Atomistic simulations are performed in order to study the Aluminum substitution effect on Magnesium twinning mechanism. Deformation twinning are found in pure Mg under tensile loading condition, however no twinning has occurred under compressive loading. It has been found that for 2 and 5% of Al substitution, twinning occurs at low strain compared to pure Mg case, but for 7 and 10% Al substitution, no twinning has been found. The interaction of dislocations with other dislocations and twin are held responsible for this behavior particularly for this orientation. For small (2 and 5%) amount of Al substitution the dislocation interacts with twin and their interaction gives rise to more twins, however for 7% or more Al substitution, dislocation-dislocation interaction becomes prevalent and no twins occur.

Acknowledgment

This material is based upon work supported by the Department of Energy and the National Energy Technology Laboratory under Award Number DE-FC26-02OR22910 and the Office of Naval Research under contract No. N00014-09-1-0661. This report was prepared as an account of work sponsored by an agency of the United States Government. Neither the United States Government nor any agency thereof, nor any of their employees, makes any warranty, express or implied, or assumes any legal liability or responsibility for the accuracy, completeness, or usefulness of any information, apparatus, product, or process disclosed, or represents that its use would not infringe privately owned rights. Reference herein to any specific commercial product, process, or service by trade name, trademark, manufacturer, or otherwise does not necessarily constitute or imply its endorsement, recommendation, or favoring by the United States Government or any agency thereof. The views and opinions of authors expressed herein do not necessarily state or reflect those of the United States Government or any agency thereof. Such support does not constitute an endorsement by the Department of Energy of the work or the views expressed herein. This work was in part supported by the Center for Advanced Vehicular Systems (CAVS) at the Mississippi State University. Computer time allocation has been provided by the High Performance Computing Collaboratory (HPC2) at Mississippi State University.

References

[1] M. Avedesian, H. Baker, ASM Specialty Handbook (1999).
[2] C. Roberts (1964).
[3] B.Wonsiewicz, W. Backofen, Trans. TMS-AIME 239 (1967) 1422–1431.
[4] G. Taylor, Journal of Inst. Metal. 62 (1938) 307–338.
[5] P. Ward Flynn, J. Mote, J. Dorn, TMS-AIME 221 (1961) 1148–1154.
[6] R. Quimby, J. Mote, J. Dorn, Trans. ASM 55 (1962) 149–157.
[7] R. Reed-Hill, W. Robertson, Trans. ASM 220 (1957) 496–502.
[8] R. Reed-Hill, W. Robertson, Trans. ASM 221 (1958) 256–259.
[9] F. Yoshinaga, R. Horiuchi, Trans. JIM 4 (1963) 1–8.
[10] T. Obara, H. Yoshinga, S. Morozumi, Acta Metall. 21 (1973) 845–853.
[11] U. Kocks, D. Westlake, Trans. ASM 239 (1967) 1107–1109.
[12] J. Wang, J. P. Hirth, C. N. Tom´e, Acta Materialia 57 (2009) 5521–5530.
[13] M. Parrinello, A. Rhaman, J. Appl. Phys. 52 (1982) 7182.
[14] B. Jelinek, J. Houze, S. Kim,M. F. Horstemeyer, M. I. Baskes, S.-G. Kim, Phys. Rev. B 75 (2007) 054106.
[15] D. Y. Sun, M. I. Mendelev, C. A. Becker, K. Kudin, T. Haxhimali, M. Asta, J. J. Hoyt, A. Karma, D. J. Srolovitz, Phys. Rev. B 73 (2006) 024116.
[16] X.-Y. Liu, J. B. Adams, F. Ercolessi, J. A. Moriarty, Modelling Simul. Mater. Sci. Eng. 4 (1996) 293–303.
[17] P. S. De, J. Q. Su, R. S. Mishra, Scripta Materialia 64, 1 (2010) 57-60.

Magnesium Technology 2011

Deformation Mechanisms II; Formability and Forming

Session Chairs:

Paul E. Krajewski
(General Motors, USA)
Wim H. Sillekens
(TNO Science and Industry, Netherlands)

Magnesium Technology 2011
Edited by: Wim H. Sillekens, Sean R. Agnew, Neale R. Neelameggham, and Suveen N. Mathaudhu
TMS (The Minerals, Metals & Materials Society), 2011

INFLUENCE OF SOLUTE CERIUM ON THE DEFORMATION BEHAVIOR OF A Mg-0.5 wt.% Ce ALLOY

L. Jiang[1], J.J. Jonas[1] and R. Mishra[2]
[1]Department of Materials Engineering, McGill University, Montreal, QC, H3A 2B2, Canada
[2]General Motors, Research & Development Center, Warren, MI 48090, USA

Keywords: Extruded Mg alloy bar, Dynamic strain aging, Solute cerium, Heat treatment

Abstract

The effect of solute Ce was investigated on the deformation behavior of a Mg-Ce binary alloy at elevated temperature. Compression and tension tests were performed on samples taken from extruded bars of Mg-0.5 wt.% Ce. Dynamic strain aging (DSA) was observed to occur during both tension and compression testing. Under the latter conditions, the effect of DSA on increasing the flow stress was greater than that of extension twinning. The occurrence of DSA also led to a decrease in the ductility. After heat treatment, the detrimental effect of the DSA taking place during compression was reduced significantly to an extent that depended on the cooling rate employed. However, these heat treatments did not lead to appreciable improvements in the tensile ductility.

Introduction

A number of studies have pointed out that rare earth alloying additions can significantly weaken texture sharpness in Mg alloys and thereby improve formability [1-6]. Stanford et al. [7] have shown that RE elements display stronger solute/dislocation interactions than non-RE elements. In the case of solid solution alloys such as the Mg-Y [3, 8] and Mg-Gd [9] alloys, texture weakening is observed even in the absence of second phase particles. Addition of the soluble elements Y and Gd at levels of 0.05 to 0.1 at.% also approximately doubles the extrusion force and refines the grain size by 75% [9]. This indicates that the interaction between RE solutes and dislocations and boundaries is very strong in magnesium-based alloys. It has also been observed that Ce is an effective texture modifier, probably because of its large atomic radius [9] and the effect on the Ce/Mg and Mg/Mg atomic binding states [10]. The observation of dynamic strain aging (DSA) in Mg-Ce [11] and other RE-containing alloys [7, 12-14] provides additional evidence that RE elements interact with dislocations.

In extrusion, the appropriated levels of RE alloying additions not only weaken the texture but also lead to the formation of a special component (referred to as the RE texture component): <11-21> parallel to the extrusion direction [15, 16]. Some researchers have found that the nucleation of recrystallization at shear bands plays a role in the formation of the RE texture component and texture weakening [4, 17, 18]. In steel rolling, it is well known that the formation of shear bands is favored by increasing the amount of solute carbon and rolling at temperatures where dynamic strain aging (DSA) occurs, leading to flow localization. The question therefore arises whether Ce in solution can play a similar role in Mg-based materials. Studies concerning DSA in magnesium alloys have mainly been conducted by tension testing [7, 12-14]. However, serrated flow can also occur during compression [19] and torsion [20] testing. In the current work, both uniaxial tension and compression were employed to characterize the DSA effects. The effect of different cooling rates on the Ce solute content was also investigated.

Experimental

The Mg-0.5 wt.% Ce bars were provided by the GM R&D Center, Warren, MI. The billets were preheated to 400ºC for 2h and extruded at 300ºC and a billet speed of 10mm/s. An extrusion of 25:1 was employed. The extrudate was water quenched right after exit from the die. The tension and compression samples were machined from the as-extruded rods with their axes aligned along the extrusion direction. The tensile specimens had gage sections 30mm long and 6mm in diameter, while the compression samples had heights of 11.6mm and diameters of 7.6mm. Various heat treatments (Table I) were carried out on the as-extruded samples in an argon atmosphere and the resulting Ce solute concentrations were measured. The mechanical tests were conducted on a servo-controlled MTS machine at 200ºC. Four different strain rates were investigated: 10^{-3}, 10^{-2}, 10^{-1} and $1s^{-1}$. The samples were held for 10 min at temperature prior to testing. In order to minimize oxidation, argon gas was passed through the quartz tube enclosing the specimen and grips.

The crystallographic texture was measured by means of EBSD techniques. The Ce solute content was measured using a JEOL JXA-8900L electron probe microanalyzer (EPMA). At least 20 spots were analyzed in each case so as to provide reliable data. In preparation for the EBSD measurements, specimens were polished using standard metallographic methods followed by immersion in a solution of 10 ml HNO3, 30 ml acetic acid, 40 ml H2O and 120ml ethanol for 2s. A similar polishing procedure was employed for the EPMA samples, except that no colloidal silica was used to avoid contamination.

Table I. The Ce solute contents of the specimens after different heat treatments.

	As extruded	H1	H2	H3	H4	H5
Ce solute content (wt.%)	0.146	0.087	0.095	0.077	0.083	0.160

H1: Heated at 400ºC for 4hrs, then placed on wood and cooled in air (~50ºC/min)
H2: Heated at 450ºC for 4hrs, then placed on a steel block and cooled in air (~75ºC/min)
H3: Heated at 450ºC for 4hrs, then cooled at 3ºC/min

H4: Heated at 450°C for 4hrs, then cooled in the furnace (~ 25 °C/min when the temperature dropped from 450 to 250C and ~10 °C/min for temperatures from 250°C to 200°C)
H5: Heated at 450°C for 4hrs, then quenched in water to room temperature (~2.7x10^5°C/min) the vertical direction on the micrographs.

Results and Discussion

A typical optical microstructure from a longitudinal section of an as-extruded profile is illustrated in Figure 1. The microstructure is fully recrystallized and the average grain size is 10µm. The as-extruded texture is quite weak, with the maximum intensity peak located at <11-21> parallel to the extrusion direction (i.e. at the RE component). This is consistent with the results reported by Stanford [9].

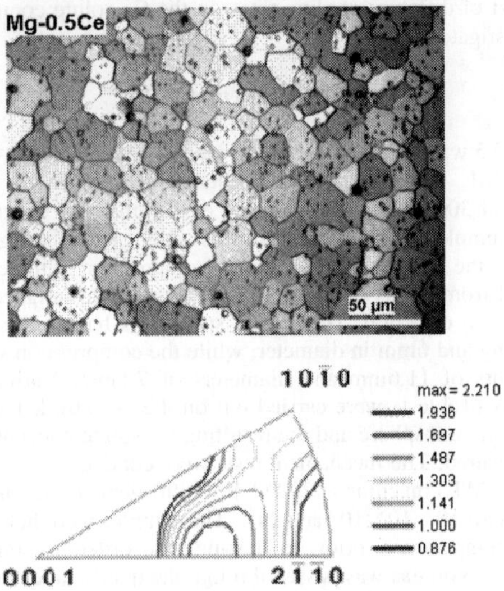

Figure 1. Microstructure and texture (presented in inverse pole figure form) of the as-extruded Mg-0.5 wt.% Ce alloy bar. The inverse pole figure refers to the extrusion direction, which is vertical in the micrograph.

The compressive true stress-strain curves determined at selected strain rates are presented in Figure 2(a). All the flow curves were sigmoidal and displayed yielding at around 70-80 MPa; this is a sign that extension twinning is taking place. Up to a strain of -0.18, the work hardening behavior at the highest strain rate (1/s) is similar to that at the lower strain rate of 0.1/s. However, on further compression, the 0.1/s flow curve rises *above* the 1/s flow curve. It is of interest that the peak stress at the highest strain rate (1/s) is the lowest among the strain rates selected. Similar flow characteristics were observed in the 0.01/s and 0.001/s tests. The increase in strain rate produced a *decrease* in flow stress. This is indicative of a negative macroscopic strain rate sensitivity (SRS).

The corresponding tensile stress-strain curves determined under the same conditions are illustrated in Figure 2(b). When the strain rate was decreased from 0.1/s to 0.001/s, the flow stress increased and the curve displayed serrations that are symptomatic of dynamic strain aging.

Serrated flow curves (also referred to as the PLC effect) are the most commonly observed macroscopic manifestations of the DSA process. There are other anomalies associated with DSA, such as negative strain rate sensitivity (SRS) [21]. In the present experiments, serrated curves were not observed in the compression tests because of the very short gage length. Nevertheless, the negative SRS that was observed indicated that DSA was taking place.

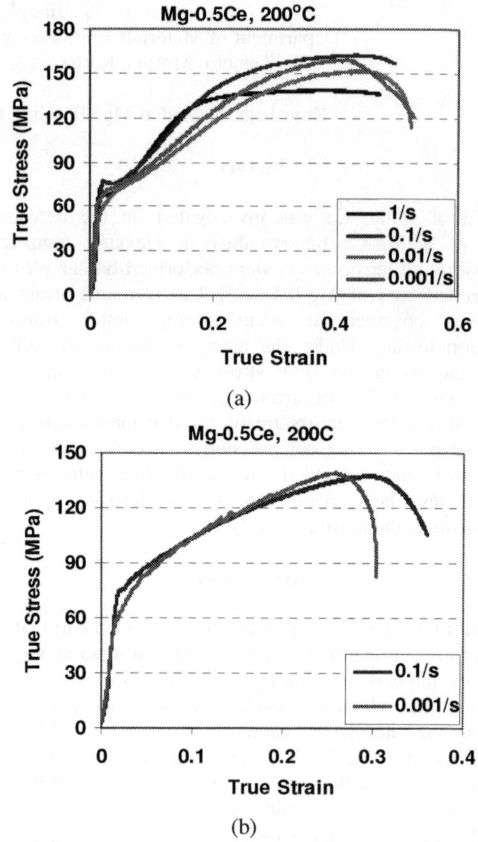

Figure 2. True stress-true strain curves for the (a) compression and (b) tension tests conducted at 200°C at selected strain rates.

During the compression tests, the DSA effects were superimposed on those of twinning and glide. Nevertheless, the higher compression flow stress at 0.001/s than at 0.01/s indicates that the effect of DSA on the flow stress is greater at the lower strain rate. This can be attributed to the solute drag associated with Ce addition being greater at the lower strain rate at 200°C because there is a better match between the Ce diffusivity and the dislocation velocity at that strain rate. The specimens exhibited negative rate sensitivity over the entire strain rate change. Thus DSA can be considered to take place in the strain rate range from 0.001/s to 0.1/s. The occurrence of DSA also led to a decrease in the tensile ductility.

The Ce solute contents of the as-extruded sample and the samples after the different heat treatments were displayed in Table I. As a reference, the equilibrium solubility of Ce in solid magnesium is given in Table II. The fastest cooling rate (about 2.7x10^5°C/min) employed in treatment H5 led to the highest Ce level (0.16 wt.%), while treatment H3 with the slowest cooling rate (3°C/min) led to the lowest Ce level (0.076wt.%, only half of that in the H5 sample). These observations reveal the importance of cooling rate after extrusion on the solute concentration in Ce-modified alloys.

They show that rapid cooling leads to a supersaturation of Ce at lower temperatures and conversely that minimization of the solute Ce content requires annealing followed by slow cooling.

Table II. The solubility of Ce in solid magnesium at different temperatures [22].

	590 ºC	500ºC	400 ºC	300 ºC	200 ºC
wt. %	0.74	0.26	0.08	0.06	0.04
at. %	0.13	0.045	0.014	0.010	0.007

The optical microstructures after the various heat treatments are displayed in Figure 3. Compared to the as-extruded material (10μm), the average grain sizes of the heat treated materials have increased into the range 15-25μm. The H3 specimen had the largest average grain size, ~25μm, although its texture was similar to that of the as-extruded sample.

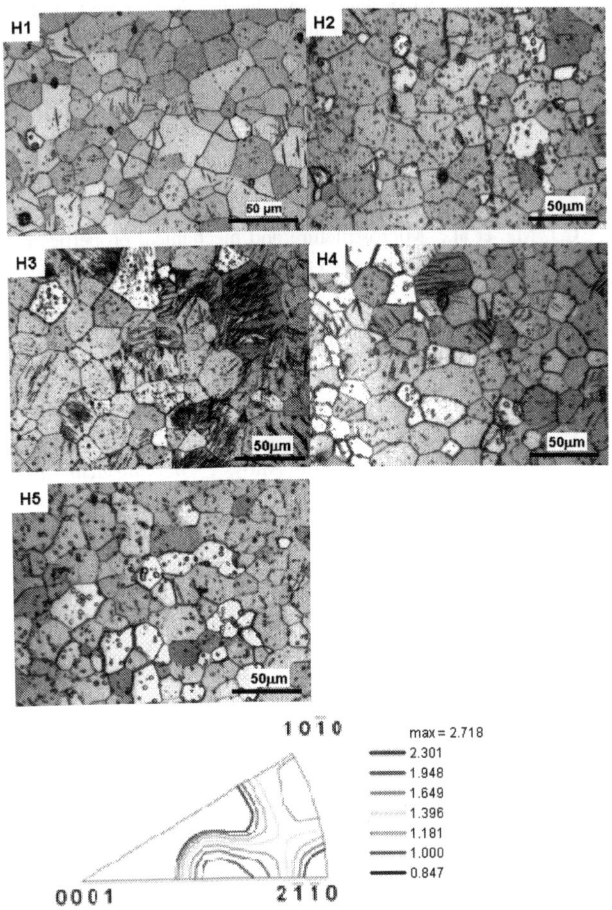

Figure 3. Optical microstructures of the Mg-0.5 wt.% Ce alloy samples after the various heat treatments. The texture of the H3 sample is presented in the inverse pole figure of the extrusion direction, which is vertical in the micrographs.

As shown in Figure 2, a strain rate of 0.001/s led to the lowest strain-to-fracture during compression and to serrated flow curves during tension. It was therefore concluded that the interaction between Ce atoms and dislocations was most pronounced at a strain rate of 0.001/s. To determine whether decreasing the Ce solute content can reduce the detrimental effects of DSA, compression testing was carried out at 200°C/0.001s^{-1} on the samples that had undergone the various heat treatments. The stress-strain curves obtained are presented in Figure 4. Here it can be seen that all the heat-treated specimens had improved ductilities, compared to the as-received material.

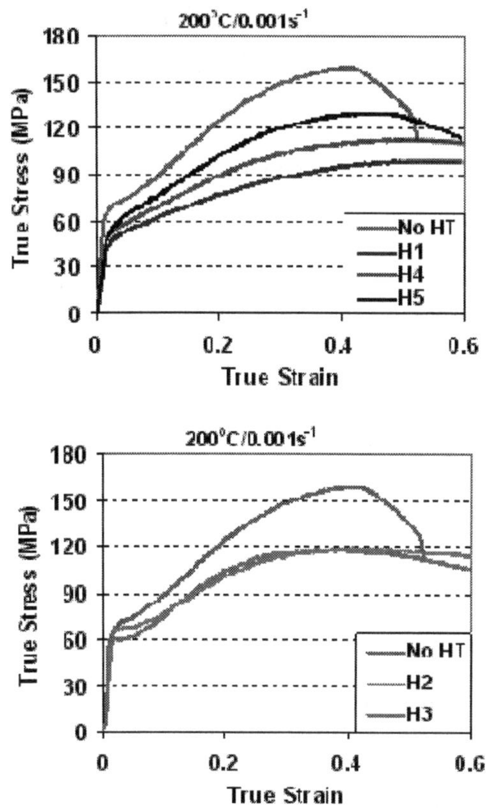

Figure 4. Compression stress-strain curves for the specimens subjected to the different heat treatments.

During heat treatment, the dislocations generated during the extrusion process were annihilated, which reduced the extent of Ce solute/dislocation interaction. Therefore, the H5 specimen did not fracture shortly after reaching peak stress as did the as-extruded sample, although it had the highest Ce solute level (0.16wt.%). Nevertheless, after the peak stress, the load-carrying ability was not maintained over appreciable strains due to the high Ce solute concentration. By contrast, although the H3 specimen had the lowest Ce solute content (0.077wt.%), its flow curve displayed a more marked sigmoidal shape and the lowest strain-to-peak. This is probably because the H3 specimen had a relatively large grain size. In addition, due to the slowest cooling rate, the precipitates in this material were the coarsest and their pinning effect was reduced. The detrimental effect due to DSA appears to have been reduced appreciably in the H1 and H4 specimens during compression. This can also be seen in Figure 5 (a). When the strain rate was increased from 0.001/s to 0.01/s, no negative SRS was observed in the H1 and H3 specimens.

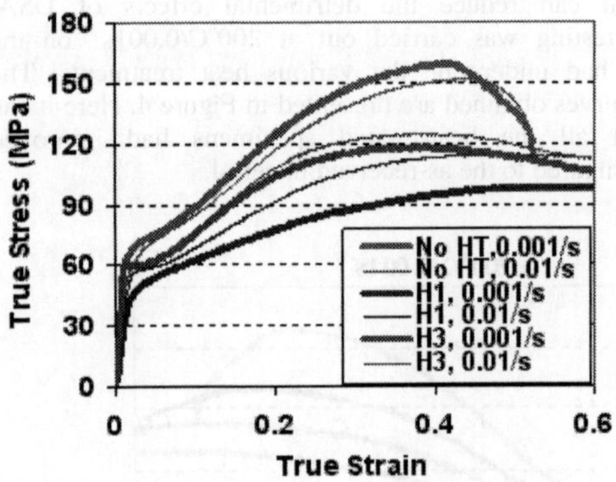

(a)

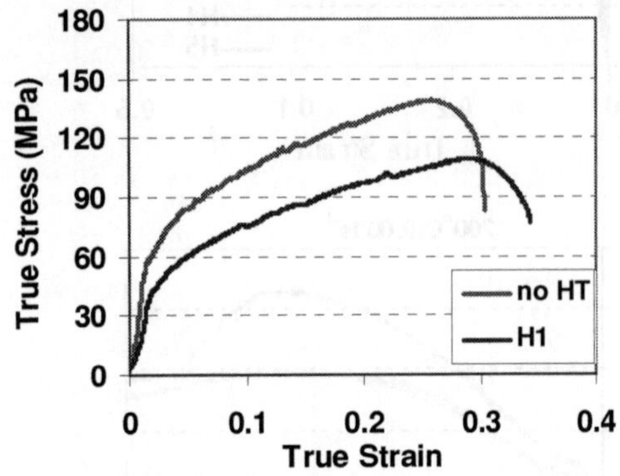

(b)

Figure 5. Stress-strain curves comparison for the specimens undergone different heat treatments during (a) compression at strain rates of 0.01/s and 0.001/s; (b) tension at a strain rate of 0.001/s. All the tests were conducted at 200ºC.

However, the H1 specimen still exhibited a serrated flow curve during tension, as displayed in Figure 5(b). This is consistent with the previous observations of Zhu and Nie [13], where their WE54 alloy (Mg-Y-Nd system) exhibited serrated flow even when tested in the peak-aged condition (solution treated at 525°C for 8h and aged at 250°C for 16h). The absence of negative SRS in the compressive H1 samples indicates that annealing reduced the level of solute Ce in those samples and therefore eliminated the DSA effect. This is probably also the reason why the detrimental effect of DSA was essentially absent in the H1 compression sample but not in the tension test.

Conclusions

The effect of solute cerium on the deformation behavior of the Mg-0.5Ce wt.% alloy was investigated at 200ºC by means of uniaxial tension and compression tests. At 200°C, DSA took place in the strain rate range from 0.001/s to 0.1/s and was most pronounced at 0.001/s. The occurrence of DSA leads to a decrease in the ductility in both compression and tension. The heat treatment results show that cooling at 25ºC/min-50ºC/min can significantly reduce the detrimental effect of DSA in compression but not as markedly in tension tests.

Acknowledgments

This research was sponsored by General Motors of Canada and the Natural Sciences and Engineering Research Council of Canada. Discussions with Prof. Matthew Barnett of Deakin University, Australia are gratefully acknowledged.

References

1. R.K. Mishra, et al., "Influence of cerium on the texture and ductility of magnesium extrusions", *Scripta Materialia*, 59 (2008) 562-565.

2. J. Bohlen, et al., "The texture and anisotropy of magnesium-zinc-rare earth alloy sheets", *Acta Materialia*, 55 (2007) 2101-2112.

3. Y. Chino, K. Sassa and M. Mabuchi, "Texture and stretch formability of a rolled Mg-Zn alloy containing dilute content of Y", *Materials Sci. and Eng.: A*, 513-514 (2009) 394-400.

4. J.W. Senn and S.R. Agnew. "Texture randomization of magnesium alloys containing rare earth elements", in the Proceeding of *Magnesium Technology 2008*, TMS, 2008), 153-158.

5. C.J. Ma, et al., "Microstructure and mechanical properties of extruded ZK60 magnesium alloy containing rare earth", *Materials Science and Technology*, 20 (2004) 1661-1665.

6. J. Gröbner and R. Schmid-Fetzer, "Thermodynamic modeling of the Mg-Ce-Gd-Y system", *Scripta Materialia*, 63 (2010) 674-679.

7. N. Stanford, et al., "Effect of Al and Gd Solutes on the Strain Rate Sensitivity of Magnesium Alloys", *Metallurgical and Materials Transactions A*, 41 (2010) 734-743.

8. B.L. Wu, et al., "Ductility enhancement of extruded magnesium via yttrium addition", *Materials Science and Engineering: A*, 527 (2010) 4334-4340.

9. N. Stanford, "Micro-alloying Mg with Y, Ce, Gd and La for texture modification--A comparative study", *Materials Science and Engineering: A*, 527 (2010) 2669-2677.

10. Y.Chino, M. Kado and M. Mabuchi, "Compressive deformation behavior at room temperature–773 K in Mg–0.2 mass% (0.035at.%)Ce alloy", *Acta Mater.*, 56 (2008) 387-394.

11. L. Jiang, et al., "Effect of cerium on the deformation behavior of two Mg-Ce alloys". in the proceeding of *Magnesium Technology 2010*. TMS, 2010), 343-346.

12. S.L. Couling, "Yield points in a dilute magnesium-thorium alloy", *Acta Metallurgica*, 7 (1959) 133-134.

13. L. Gao, R.S. Chen and E.H. Han, "Characterization of dynamic strain ageing in Mg-3.11wt.%Gd alloy", in the proceeding of *Magnesium Technology 2009*. TMS, 2009), 269-272.

14. S.M. Zhu and J.F. Nie, "Serrated flow and tensile properties of a Mg-Y-Nd alloy", *Scripta Materialia*, 50 (2004) 51-55.

15. N. Stanford, "Effect of composition on the texture and deformation behaviour of wrought Mg alloys", *Scripta Materialia*, 58 (2008) 179-182.

16. N. Stanford, A. Beer, C. Davies, M.R. Barnett, "Effect of microalloying with rare-earth elements on the texture of extruded magnesium-based alloys", *Scripta Materialia*, 59 (2008) 772-775.

17. L.W.F. Mackenzie and M.O. Pekguleryuz, "The recrystallization and texture of magnesium-zinc-cerium alloys", *Scripta Materialia*, 59 (2008) 665-668.

18. N. Stanford and M.R. Barnett, "The origin of "rare earth" texture development in extruded Mg-based alloys and its effect on tensile ductility", *Mater. Sci. Eng. A*, 496 (2008) 399-408.

19. A.O. Humphreys, et al., "Warm rolling behaviour of low carbon steels", *Mater. Sci. Tech.*, 19 (2003) 709-714.

20. N. Christodoulou, et al., "Analysis of steady-state thermal creep of Zr-2.5Nb pressure tube material", Metallurgical and Materials Transactions A, 33 (2002) 1103-1115.

21. J.M. Robinson and M.P. Shaw, "Microstructural and mechanical influences on dynamic strain aging phenomena", *International Materials Reviews*, 39 (1994) 113-122.

22. L. Rokhlin, "Dependence of the rare earth metal solubility in solid magnesium on its atomic number", *Journal of Phase Equilibria*, 19 (1998) 142-145.

Magnesium Technology 2011
Edited by: Wim H. Sillekens, Sean R. Agnew, Neale R. Neelameggham, and Suveen N. Mathaudhu
TMS (The Minerals, Metals & Materials Society), 2011

TEXTURE WEAKENING EFFECT OF Y IN Mg-Zn-Y SYSTEM

S. A. Farzadfar[1], M. Sanjari[1], I.-H. Jung[1], E. Essadiqi[2] and S. Yue[1]

[1]McGill University, Department of Materials Engineering, Montreal, QC, Canada H3A 2B2
[2]CANMET- Materials Technology Laboratory, Ottawa, Ontario, Canada K1A 0G1

Keywords: Magnesium, Yttrium, Compression test, Thermodynamics, Texture

Abstract

The CALPHAD (Calculation of Phase Diagram) method was successfully used in this study to select the alloys from Mg-Zn-Y system, aimed at determining the mechanism(s) of texture weakening in Y-containing Mg alloys: PSN (particle-stimulated nucleation of recrystallization) and/or solute effects. The selected alloys are Mg-6Zn-1.2Y, Mg-5Zn-2Y, Mg-2.9Y and Mg-2.9Zn (in wt%). After heat treatment, the ternary alloys contain (nearly) the same amount of ternary intermetallics in equilibrium with α-Mg at 350 °C, and the microstructure of the binary alloys is composed of α-Mg solid solution at 350 °C containing the same solute amount as that in the ternary alloys. The hot deformation and post-deformation annealing of these alloys at a constant temperature (i.e., 350 °C) showed that the texture weakening happens during growth of recrystallized grains in deformed and annealed samples where Y element is in α-Mg solid solution.

Introduction

The use of wrought Mg alloys, and especially flat-rolled sheets, in industrial applications is still limited mainly due to their low formability [1]. This poor formability is a direct consequence of the development of the basal texture upon deformation, resulting in plastic anisotropy and tension-compression asymmetry [2-4]. However, it has recently been found that the addition of rare earth (RE) elements, such as yttrium (Y), significantly weakens the deformation texture [3-4, 5, 6, 7], e.g., after multi-pass rolling, or the annealed texture [6, 8, 9]. To date, the mechanism(s) of texture weakening, which are most likely connected with the presence (or absence) of intermetallics and/or solutes, are not well understood. There is practically no systematic work on RE-containing Mg alloys with controlled levels of solutes and precipitates during deformation and annealing that allows for separating the possible mechanisms of texture weakening. The objective of this study is therefore to perform systematic alloy selection by CALPHAD (Calculation of Phase Diagram) method and FactSage™ software [10] to clarify the mechanisms involved in texture weakening after hot deformation of the selected alloys.

Alloy Selection

The Mg-Zn-Y system was chosen as the alloy system in this study firstly because of the reported texture weakening observed in this system [2, 5, 8, 11], and secondly because of the large solubility of Zn or Y in Mg as compared to other elements such as Mn, La, Ce or Nd [12, 13]. There is also the possibility of generating two-phase microstructures (α-Mg + ternary intermetallic) with different types of intermetallics. Since equilibrium state at deformation temperature is desired, the temperature of deformation and therefore that of thermodynamic calculations must first be determined. Based upon the typical temperature range in industrial processes such as rolling, i.e., 300 °C to 450 °C [14], the temperature of 350 °C was chosen in this study. Two categories of alloys with the following microstructures at 350 °C were selected: (a) two-phase microstructure Mg-Zn-Y alloys with the same amount of intermetallic phase in each alloy, in equilibrium with α-Mg, and (b) single-phase α-Mg binary alloys (Mg-Zn and Mg-Y) containing the same solute amount as those in group (a). The result of hot deformation of the alloys in group (a) and (b) will reveal the effect of intermetallics and Y solute, respectively, on the texture evolution.

To determine the alloys group (a), the variation of the amount of intermetallics and composition of α-Mg as a function of Zn and Y content at 350 °C were calculated by FactSage™ software using the FTlite database [10]. Based on cost and density considerations, Zn amounts lower than 6 wt% and Y amounts lower than 2 wt% were considered for thermodynamic calculations. It was found that only two alloy compositions generate the same amount of different intermetallics in equilibrium with α-Mg that contains the same solute amount: Mg-5Zn-2Y and Mg-6Zn-1.2Y, in which the intermetallic phase in equilibrium with α-Mg at 350 °C is W ($Mg_3Y_2Zn_3$) and I (Mg_3YZn_6), respectively. The α-Mg in the mentioned alloys contains only Zn solute in equal amounts.

Experimental Procedure

High purity elements 99.9% Mg, 99.99% Zn, and 99.9% Y (all concentrations are given in wt% unless otherwise mentioned) were used to make the alloys by induction melting under the protective gas of $SF_6 + CO_2$. The melt was held at 740 °C for 30 min and stirred at regular intervals. The molten alloys were cast into rods 15 mm in diameter and 140 mm in height in a Cu mould pre-heated at 250 °C. Inductively coupled plasma atomic emission spectroscopy (ICP-AES) was used to determine the composition of alloys, which was close to the nominal one (− 0.1 wt% of maximum difference). The major impurity levels were 0.004 wt% for Si and Mn, and 0.03 wt% for Al and Fe. The as-cast alloys were initially heat treated at a temperature just below the liquid formation in heating of cast material, and then annealed at 350 °C which was chosen as the deformation temperature. The temperature of initial heat treatment was determined by simultaneous thermal analysis TG-DSC (Thermogravimetry-Differential Scanning Calorimetry) for melting of as-cast alloys heated at 10 K/min under Ar in pure iron crucible. All heat treatments were carried out in an air-circulating furnace with ±1 °C precision, followed by quenching the alloys into room-temperature water. The time required for heat treatment was obtained by measuring the variation of room-temperature Vickers macro hardness and the composition of Mg solid solution with time (2 h to 24 h). The latter was investigated by EPMA (electron probe micro-analysis in a JEOL JXA-8900L SEM) using 10 kV and MgO, $Y_3A_{15}O_{12}$, and ZnS standards. Intermetallics were identified by EPMA (forty to fifty analyzed points) and bulk

X-ray diffraction (XRD) data. The latter was collected at room temperature with a Bruker D8 Advance diffractometer on a rotating and oscillating stage using a Cu *Kα* source. Compositional contrast in backscattered electron (BSE) imaging was utilized to identify the form and distribution of intermetallics in a field emission gun SEM (FE-SEM Hitachi S-4700).

Compression tests on cylindrical samples (6 mm in diameter and 9 mm in height) were carried out under a steady argon flow at 350 °C (in a radiant furnace) and true strain rate of 1 s^{-1}. The samples were quenched in room-temperature water within 1 s after unloading at different strains (half of peak strain, peak strain, and strain of 1) or after annealing of deformed material. The post-deformation annealing was carried out immediately after unloading in the radiant furnace at 350 °C for 1 min to 30 min. The bulk texture was evaluated by XRD using a Bruker D8 diffractometer with a Co *Kα* source. The orientation distribution function (ODF) was constructed from the incomplete pole figures of {10.0}, {00.2}, and {10.1}.

For microstructural examination, the samples were ground with 1200 grit SiC paper, and polished with 3 μm and 1 μm diamond suspension followed by 0.04 μm colloidal silica (Struers OP-S) polishing. The samples were then etched with acetic picral (10 ml acetic acid, 10 ml H_2O, 4.2 g picric acid, 70 ml ethanol) for 3 to 5 s. The grain sizes were measured using the linear intercept method based on ASTM E 112 standard, and the volume fractions of recrystallized material were determined using the point counting method from optical micrographs according to ASTM E 562 standard [15]. The deformed and recrystallized grains were differentiated based on their size and shape. The standard deviations are given in parenthesis, which refer to the last digit, e.g., 5.1(8) indicates 5.1 ± 0.8.

Results and Discussion

Alloy Design

BSE images show the presence of eutectic structure in as-cast alloys (Fig. 1).

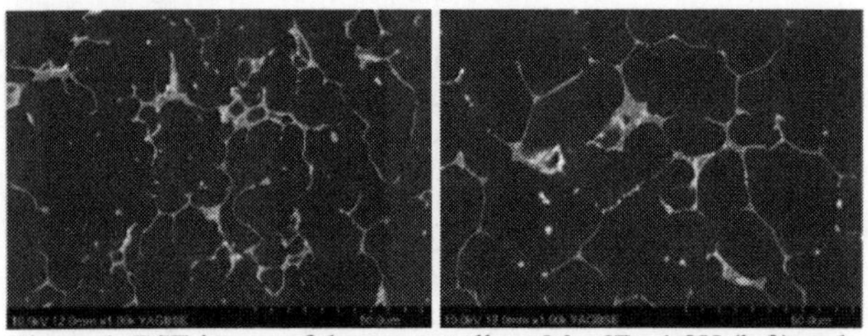

Figure 1. BSE image of the as-cast alloys Mg-6Zn-1.2Y (left) and Mg-5Zn-2Y (right).

Based on the results of XRD (Fig. 2) and EPMA (Zn/Y ratio was used to identify the intermetallics), the constituent phases in as-cast Mg-6Zn-1.2Y alloy are *α*-Mg and I phase (Zn/Y: 5.1(8) at%). The as-cast Mg-5Zn-2Y alloy is composed of *α*-Mg, I (Zn/Y: 5.4(8) at%) and W (Zn/Y: 1.5(1) at%) phases (Fig. 3).

The onset of melting, detected by simultaneous TG-DSC, occurs at 450 °C in both Mg-6Zn-1.2Y and Mg-5Zn-2Y as-cast alloys. The initial heat treatment of ternary alloys was therefore carried out at 430 °C just below this temperature.

Based on the results of XRD (Fig. 2) and EPMA, the I phase in as-cast Mg-6Zn-1.2Y alloy transforms completely to the W phase (Zn/Y: 1.7(1) at%) at 430 °C after the shortest time of 2 h, and the W phase remains in the microstructure after 24 h. The I phase forms again upon annealing of the heat treated alloy (430 °C-8 h) at 350 °C for 24 h. This indicates that either the kinetics of W to I phase transformation is slow or the equilibrium intermetallics at 350 °C are I and W phases. To verify this, the Mg-6Zn-1.2Y as-cast alloy was heat treated at 350 °C from 2 h to 24 h. The W phase forms again after 2 h and remains in the microstructure along with the I phase after 24 h. However, at 330 °C, no W phase was detected by XRD and EPMA after heat treatment for 2 h to 24 h.

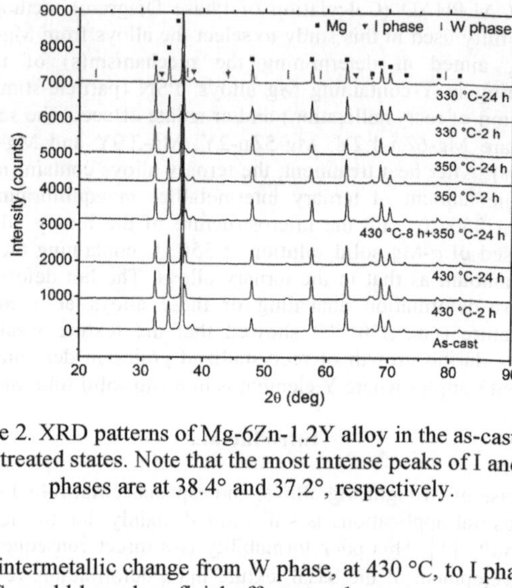

Figure 2. XRD patterns of Mg-6Zn-1.2Y alloy in the as-cast and heat treated states. Note that the most intense peaks of I and W phases are at 38.4° and 37.2°, respectively.

The intermetallic change from W phase, at 430 °C, to I phase, at 330 °C, could have beneficial effects on the texture evolution of deformed material. This alloy was therefore heat treated at 330 °C and 430 °C to obtain two distinct microstructures: *α*-Mg + I phase obtained at 330 °C (denoted by Mg-6Zn-1.2Y(I)), and *α*-Mg + W phase obtained at 430 °C (denoted by Mg-6Zn-1.2Y(W)).

As can be seen in the XRD patterns of Mg-5Zn-2Y alloy (Fig. 3), the I phase is not detected after heat treatment at 430 °C. Instead, the W phase is found at all tested times.

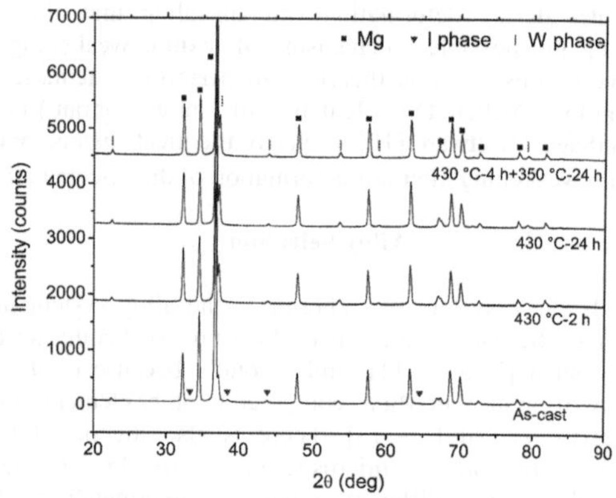

Figure 3. XRD patterns of Mg-5Zn-2Y alloy in the as-cast and heat treated states. Note that the most intense peaks of I and W phases are at 38.4° and 37.2°, respectively.

The minimum time of heat treatment, which was determined based on the onset of steady state regime in Vickers macro-hardness values and the amounts of Zn and Y solutes, was determined to be 8 h for Mg-6Zn-1.2Y alloy (at both temperatures 330 °C and 430 °C), and 4 h for Mg-5Zn-2Y alloy at 430 °C. The annealing of the heat treated Mg-5Zn-2Y alloy (430 °C-4 h) at 350 °C from 2 h to 24 h did not give rise to any change in the composition of *α*-Mg solid solution, macro hardness and microstructure. This alloy was therefore heat treated only at 430 °C for 4 h.

Since the intensity of X-ray peaks corresponding to intermetallics are low (due to their little amount), the volume percents of I and W phases were calculated by balancing the measured Zn amount, and are given in Table I. The density of I phase was taken from [16], and that of W phase was calculated from the crystal structure obtained by Rietveld refinement in [17]. The two binary alloys were selected based on the amount of Zn solute in Mg-6Zn-1.2Y (330 °C-8 h) and Mg-5Zn-2Y (430 °C-4 h) alloys, i.e, 2.9 wt%. The same procedure was undertaken for Mg-2.9Zn (1.1 at% Zn) and Mg-2.9Y (0.8 at% Y) alloys to find the temperature and time of heat treatment resulting in a fully solid solution microstructure at 350 °C. These binary alloys were therefore thermo-mechanically processed in the single-phase solid solution region, according to phase diagram data [12,13].

The heat treatment procedure and the characteristics of the four alloys explained above are summarized in Table I. It should be noted that the intermetallics in ternary alloys are in the form of interdendritic lamellar eutectic.

Table I. Heat treatment conditions and characteristics of the selected alloys. The standard deviations are given in parenthesis, which refer to the last digit.

Alloy (nominal composition in wt%)		**Mg-6Zn-1.2Y**		**Mg-5Zn-2Y**	**Mg-2.9Y**	**Mg-2.9Zn**
Heat treatment (T °C-t h)		330-8	430-8	430-4	430-24	430-8
Present phase(s)		*α*-Mg + I	*α*-Mg + W	*α*-Mg + W	*α*-Mg	*α*-Mg
Calculated amount of intermetallic*	**vol%**	1.68	1.24	2.01	–	–
	wt%	4.57	3.09	4.95	–	–
Predicted amount of intermetallic † (wt%)		7.32	I (6.94) + W (0.1)	6.85	–	–
Grain size (μm)		50(5)	51(7)	52(4)	124(12)	120(8)
Solute amount (wt%)	**Zn**	2.8(3)	4.7(2)	2.9(2)	–	2.8(1)
	Y	0.25(6)	0.10(6)	0.11(4)	2.8(2)	–

* Obtained by balancing the measured Zn amount
† Obtained by FactSage™ using FTlite database

Except for the Mg-6Zn-1.2Y alloy heat treated at 430 °C, the nature of the intermetallics predicted by thermodynamic calculations and obtained experimentally are similar. However, the amount of intermetallics obtained from experimental data appears to be lower than that predicted by thermodynamic calculations (Table 1).

Shao et al. [18] performed a recent assessment of the Mg-Zn-Y system, and generated a new database. Their database [18] was inputted in FactSage in this study to compare with the results obtained by the FTlite database. The result of thermodynamic calculations of the equilibrium phases for the alloys selected in this study revealed similar results from the two databases.

<u>Hot Compression and Post-deformation Annealing</u>

Since major texture changes occurred at the strain of 1, the following section includes the results of microstructure and texture study at this strain.

The typical basal texture develops in all alloys with nearly the same maximum intensity of basal pole figures (Fig. 4) at the strain of 1. It should be noted that the maximum error value in the maximum intensity of basal pole figures is ± 10%.

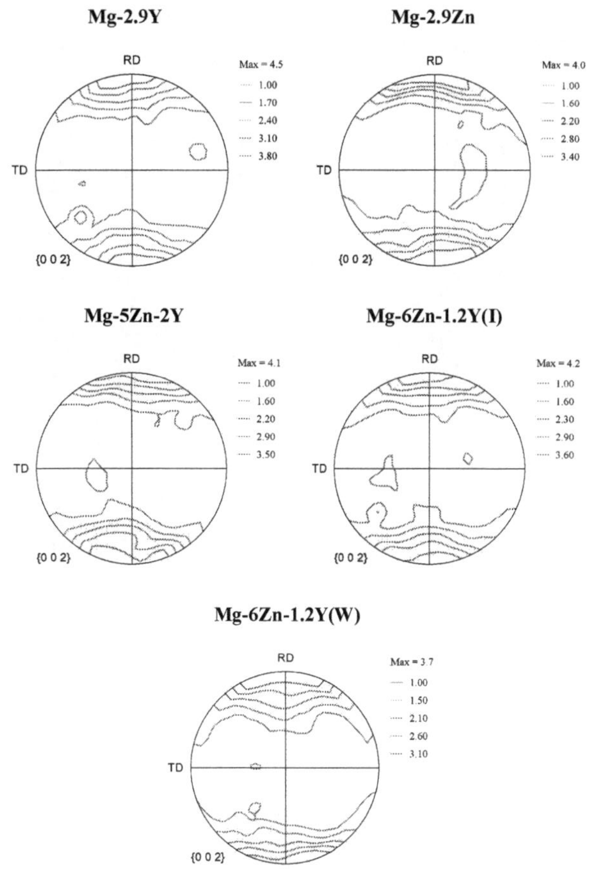

Figure 4. X-ray (00.2) basal pole figures after compression to the strain of 1. RD is the compression direction.

The noticeable anomaly in the microstructure of Mg-2.9Y alloy is the retardation of dynamic recrystallization (Fig. 5). While a nearly fully-recrystallized microstructure develops in all other alloys at strain of 1, the amount of recrystallized material in Mg-2.9Y is only 29(1) vol%.

The intermetallics in ternary alloys were observed to preserve their eutectic structure at peak strain, and deformed mostly into stringers at the strain of 1 (Fig. 6).

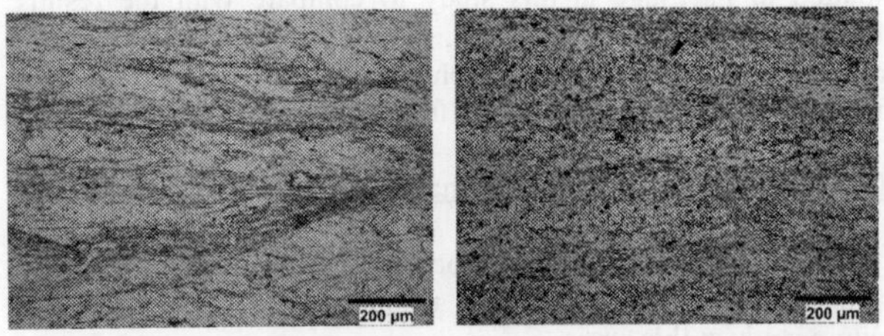

Figure 5. Optical micrographs showing the microstructures of Mg-2.9Y alloy immediately after compression to the strain of 1 (left) followed by 30-min annealing (right). The compression direction is parallel to the vertical direction on the micrographs.

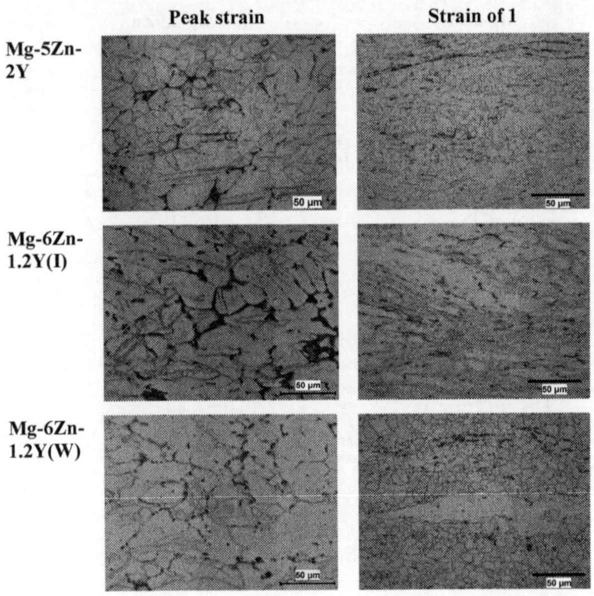

Figure 6. Optical micrographs showing the change in the form of intermetallics from eutectic structures at peak strain to stringers at the strain of 1.

Upon subsequent annealing of the compressed samples for 30 min, the maximum intensity of basal pole figures remains nearly unchanged, except in Mg-2.9Y alloy (Fig. 7). The texture evolution of Mg-2.9Y alloy with post-deformation annealing time is shown in Fig. 8 by inverse pole figures, illustrating the texture weakening in this alloy after sufficient annealing time (i.e., 30 min). It could therefore be concluded that the presence of the intermetallics investigated in this study does not have any beneficial effect in terms of texture weakening under the conditions investigated in this study.

The variation in volume percent (V_{rec}) and size (D_{RX}) of recrystallized grains with annealing time after strain of 1 in Mg-2.9Y alloy are given in Table II. Despite the large increase in V_{rec} after 1-min annealing (from 29(4) to 81(1) vol% in 1 min), the deformation texture remains nearly unchanged. However, after 30-min annealing, the texture weakening is accompanied by little rise in V_{rec} (from 81(1) to 94(1) vol%) and a large increase in D_{RX} (from 3.2(1) to 6.4(2) μm) with respect to 1-min annealing.

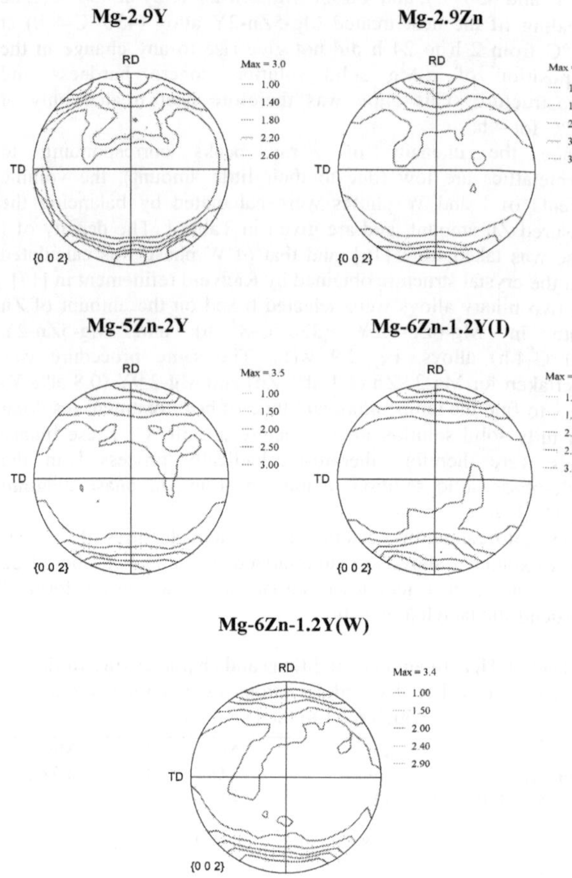

Figure 7. X-ray (00.2) basal pole figures of the alloys compressed to the strain of 1 followed by 30-min annealing. RD is the compression direction.

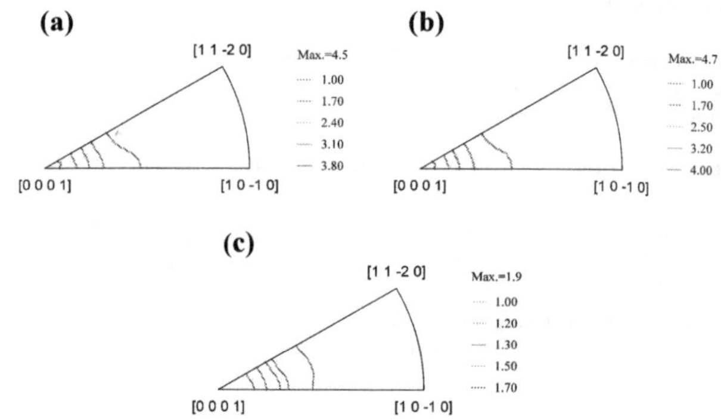

Figure 8. Result of X-ray texture measurement shown as IPF's referring to the compression direction in Mg-2.9Y alloy compressed to the strain of 1(a) and subsequently annealed (350 °C) for 1 min (b) and 30 min (c).

Table II. Volume percent (V_{rec}) and size (D_{RX}) of recrystallized grains with annealing time after strain of 1 in Mg-2.9Y alloy.

	No minutes	1 min	30 min
V_{rec}	29(4)	81(1)	94(1)
D_{RX}	2.0(2)	3.2(1)	6.4(2)

It could be concluded that the presence of Y in α-Mg solid solution can lead to texture weakening during the growth of recrystallized grains. It is hypothesized that a form of an orientation-dependent solute drag that could alter the mobility of specific high-angle boundaries could be responsible for the observed texture weakening in the Mg-2.9Y alloy.

Summary

The FactSage™ thermodynamic package was successfully used to select the alloys from Mg-Zn-Y system, aimed at determining the mechanism(s) of texture weakening in Y-containing Mg alloys.

The selected alloys are Mg-6Zn-1.2Y, Mg-5Zn-2Y, Mg-2.9Y and Mg-2.9Zn. After heat treatment, the ternary alloys contain (nearly) the same volume fraction of ternary intermetallics (I and W phases, respectively) in equilibrium with α-Mg at 350 °C, and the microstructure of the binary alloys is composed of α-Mg solid solution at 350 °C containing the same solute amount as that of Zn solute in the ternary alloys.

The results of hot deformation and post-deformation annealing of these alloys at a constant temperature (i.e., 350 °C) clarified the effect of Y in texture weakening of Mg. The obtained results show that the presence of Y in α-Mg solid solution can lead to texture weakening during the growth of recrystallized grains in deformed and annealed specimens.

Acknowledgements

The work described in this paper was supported in part by funding from the NSERC (Natural Sciences and Engineering Research Council of Canada) Magnesium Strategic Research Network. The author appreciates the financial support by the Werner Graupe International Fellowship from the Faculty of Engineering at McGill University. The assistance of Pierre Vermette with casting of the alloys is also gratefully acknowledged.

References

1. H.E. Friedrich, and B.L. Mordike, *Magnesium Technology, Metallurgy, Design Data, Applications* (Springer, 2006).

2. J. Bohlen, M.R. Nurnberg, J.W. Senn, D. Letzig, and S.R. Agnew, "The Texture and Anisotropy of Magnesium–Zinc–Rare earth Alloy Sheets," *Acta Materialia*, 55 (2007), 2101–2112.

3. J. Bohlen, J. Swiostek, D. Letzig, and K. Kainer, "Influence of Alloying Additions on the Microstructure Development of Extruded Mg-Mn Alloys," *TMS Magnesium Technology*, (2009), 225–230.

4. N. Stanford, and M. Barnett, "Effect of Composition on the Texture and Deformation Behavior of Wrought Mg Alloys," *Scripta Materialia*, 58 (2008), 179–182.

5. A. Hänzi, T. Ebeling, R. Bormann, and P. Uggowitzer, "New Microalloyed Magnesium with Exceptional Mechanical Performance," *TMS Magnesium Technology*, (2009), 521–526.

6. K. Hantzsche, J. Bohlen, J. Wendt, K.U. Kainer, S.B. Yia, and D. Letzig, "Effect of Rare Earth Additions on Microstructure and Texture Development of Magnesium Alloy Sheets," *Scripta Materialia*, 63 (2010), 725–730.

7. N. Stanford, "Micro-alloying Mg with Y, Ce, Gd and La for Texture Modification—A Comparative Study," *Materials Science and Engineering A*, 527 (2010), 2669–2677.

8. J. Wendt, K. Kainer, G. Arruebarrena, K. Hantzsche, J. Bohlen, and D. Letzig, "On the Microstructure and Texture Development of Magnesium Alloy ZEK100 During Rolling," *TMS Magnesium Technology*, (2009), 289–293.

9. Jeremy W. Senn, and Sean R. Agnew, "Texture Randomization of Magnesium Alloys Containing Rare Earth Elements," *TMS Magnesium Technology*, (2008), 153–158.

10. C.W. Bale, E. Bélisle, P. Chartrand, S.A. Decterov, G. Eriksson, K. Hack, I.-H. Jung, Y.-B. Kang, J. Melançon, A.D. Pelton, C. Robelin, and S. Petersen, "FactSage Thermochemical Software and Databases — Recent Developments", *CALPHAD*, 33 (2009), 295–311.

11. L.W.F. Mackenzie, and M.O. Pekguleryuz, "The Recrystallization and Texture of Magnesium-Zinc-Cerium Alloys," *Scripta Materialia*, 59 (2008), 665–668.

12. A.A. Nayeb-Hashemi, and J.B. Clarc, *Binary Alloy Phase Diagrams* (Second Edition, Ed. T.B. Massalski, ASM International, Materials Park, Ohio, 1990).

13. L.L. Rokhlin, *Magnesium Alloys Containing Rare Earth Metals* (London: Taylor & Francis, 2003).

14. R.W. Cahn, P. Haasen, and E.J. Kramer, *Materials Science and Technology* (Vol. 8, VCH, 1996), 151.

15. *ASTM Book of Standards*, Volume 03.01, Metals Test Methods and Analytical Procedures: Metals - Mechanical Testing; Elevated and Low-Temperature Tests; Metallography, 2009.

16. S. Bruhne, E. Uhrig, C. Gross, W. Assmus, A.S. Masadeh, and S.J.L. Billinge, "The Local Atomic Quasicrystal Structure of the Icosahedral $Mg_{25}Y_{11}Zn_{64}$ Alloy," *J Phys: Condens Matter*, 17 (2005), 1561–1572.

17. M. Sahlberg, and Y Andersson, "Hydrogen Absorption in Mg-Y-Zn Ternary Compounds," *J Alloys Compd*, 446–447 (2007), 134-137.

18. G. Shao, V Varsani, and Z. Fan, "Thermodynamic Modeling of the Y-Zn and Mg-Zn-Y system," *CALPHAD*, 30 (2006), 286-295.

Magnesium Technology 2011
Edited by: Wim H. Sillekens, Sean R. Agnew, Neale R. Neelameggham, and Suveen N. Mathaudhu
TMS (The Minerals, Metals & Materials Society), 2011

IN-SITU SCANNING ELECTRON MICROSCOPY COMPARISON OF MICROSTRUCTURE AND DEFORMATION BETWEEN WE43-F AND WE43-T5 MAGNESIUM ALLOYS

Tomoko Sano[1], Jian Yu[1], Bruce Davis[2], Richard DeLorme[2], Kyu Cho[1]
1. US. Army Research Laboratory, Materials and Manufacturing Science Division, Aberdeen Proving Ground, MD 21005 USA
2. Magnesium Elektron North America, 1001 College St. Madison, IL 62060 USA

Keywords: In-situ, magnesium, microstructure, deformation

Abstract

In-situ tensile testing in the scanning electron microscope was used to investigate the quasi-static deformation behavior and fracture mechanism of WE43 magnesium alloys. The in-situ tensile experiments were conducted at room temperature at a constant crosshead speed of 0.5 mm / min. One set of samples was a rolled and quenched F temper alloy and the other set was an artificially aged T5 temper alloy. The objective of this research was to determine the effect of tempering on precipitates chemistries, microstructure, and mechanical properties. The sample orientation is known to affect the tensile properties. Hence tensile specimens with different sample orientation were tested. The crystallographic orientations were characterized by electron backscattered diffraction. Strong textures were observed with rolling plane crystals indicating a basal plane orientation.

Introduction

Magnesium Elektron's Elektron™ WE43, developed initially as a sand casting alloy, was found to have good mechanical properties, creep and corrosion resistance, and good strength retention after exposure to elevated temperatures[1]. Elektron™ WE43 has the nominal wt% composition of Mg-(3.7 – 4.3)Y- (2.0-2.5)Nd-(0.4-2.4) Heavy Rare Earth elements-(0.4)Zr. Because of its castability, and creep and corrosion resistances, it is well known that there is interest by the aerospace and automotive industries. Due to Mg's low density[2], the Army is also interested in Mg alloys for numerous lightweight structural and engineered material applications.

There have been some recent research in the area of casting and process control of WE43, using computer simulations to determine the best casting parameters and optimizing the processsing[3]. This paper explores further into the processing, but at the mesoscale for optimization of the microstructure. The objective of this research is to understand the strengthening effects based on the homogenization and the T5 tempering process. Homogenization, which is a heat treatment at a temperature ranging from 200 to 600 °C for several hours, affects the precipitate formation. Some precipitates have been observed to affect hardness[4] and improve mechanical properties[5, 6] by Y additions[7] and by dispersion strengthening[8]. The objective is to characterize the precipitates chemistries, microstructure, and mechanical properties of the as-rolled WE43 –F and homogenized and artificially aged WE43-T5 tempered alloys. In this paper, current results identifying the effects of T5 heat treatments on the quasistatic tensile response of Elektron™ WE43 is discussed.

Experimental Procedures

Elektron™ WE43 -F and WE43 -T5 tensile specimens with a gauge length of 15 mm, width of 3 mm, and thickness of 1 mm (see Fig. 1) were milled in the longitudinal direction and transverse directions from rolled plates. The samples were mechanically polished starting with 600 grit SiC disc coated with wax, then with glycol based diamond solution of incrementally decreasing particle size on polishing cloth until reaching the 0.25 micron particle size. Further polishing with 0.04 micron colloidal silica on final polishing cloth was performed for samples to be examined by the scanning electron microscope (SEM), energy dispersive spectroscopy (EDS), and electron backscattered diffraction (EBSD).

The FEI Nova NanoSEM 600 SEM was used to characterize the microstructure before and after the tensile tests. EDS (EDAX Genesis) was used to determine the precipitate chemistry for both F and T5 samples. EBSD characterization of the crystallographic orientation texturing of the F and T5 samples were conducted at 20 kV accelerating voltage, spot size of 5, and at 70° tilt with the EDAX/TSL EBSD system in the SEM. EBSD patterns were collected from surfaces perpendicular to the rolling direction, normal direction, and the transverse direction for both WE43 samples. The collected data was minimally "cleaned" with the TSL OIM Analysis 5 "cleanup" program to correct incorrectly indexed points based on neighboring point's orientation correlation.

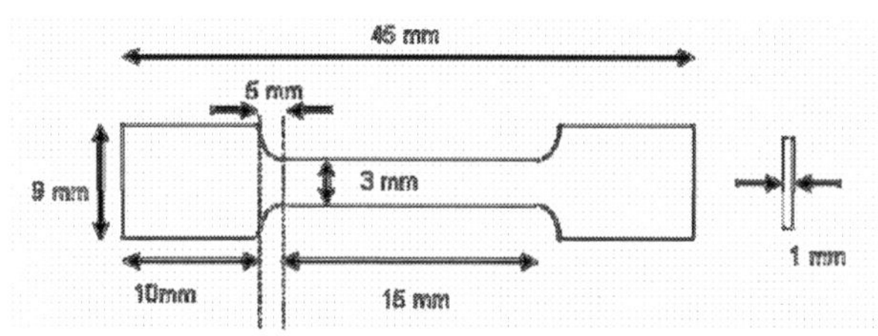

Figure 1. Tensile specimen geometry

The Ernest Fullam in-situ tension and compression stage for the SEM (see Fig. 2) coupled with the ADMET's MTEST Quattro™ interface and application program was used for the tension experiments. Strain rates on the order of 10^{-4} s^{-1} were applied for all samples. To determine a more accurate Young's modulus, digital image correction (DIC) technique was applied. During the tensile test, in the elastic regime, a series of SEM images of the microstructure were captured and corresponding live loads and positions were recorded. The image size was 1024 × 943 pixels; the field of view was about 4 × 4 mm^2. Then, the TIFF images were imported into GOM mbH's ARAMIS, a photogrammetric software, for strain analysis. The analysis

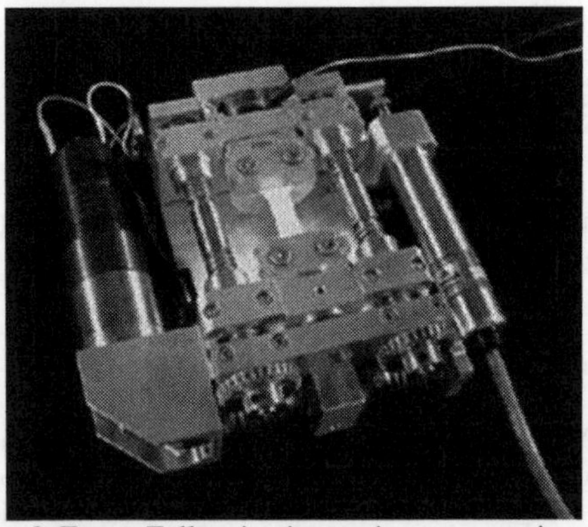

Figure 2. Ernest Fullam in-situ tension-compression stage

procedure for ARAMIS was detailed in earlier report[9]. Here, we used the 2-dimensional analysis procedure, in which only one camera view was required. The optimal facet size is 51 × 51 pixels and the optimal facet step overlap is 25 pixels. A signal filter was applied to remove noise (2 × 7 filter runs; replacing the facet point value with the median value among the 49 neighboring facet points and itself). In addition to the local strain contour plots, three virtual digital strain gauges were placed on the specimen surface to obtain the average bulk strain. The average bulk strain and load values were used to plot more accurate stress strain curves to calculate the Young's modulus.

Results

When examined under the SEM, the microstructures of both the F and T5 tempers revealed obvious precipitation, as shown in Fig 3. Precipitates are known to contribute to the strengthening of WE43 alloys[4, 10]. The finer precipitates are most likely attributed to the metastable β", β', intermediate β_1, and stable β phases[4, 10, 11]. Analysis of those and other precipitates are currently being conducted with a transmission electron microscope for this research. The larger rectangular precipitates and the oval shaped precipitates are believed to be $Mg_{24}Y_5$, and $Mg_{41}Nd_5$, respectively[12, 13]. EDS analysis of these larger precipitates showed that the oval precipitates were Nd-rich and the rectangular precipitates were Y-rich (see Fig. 4 and 5).

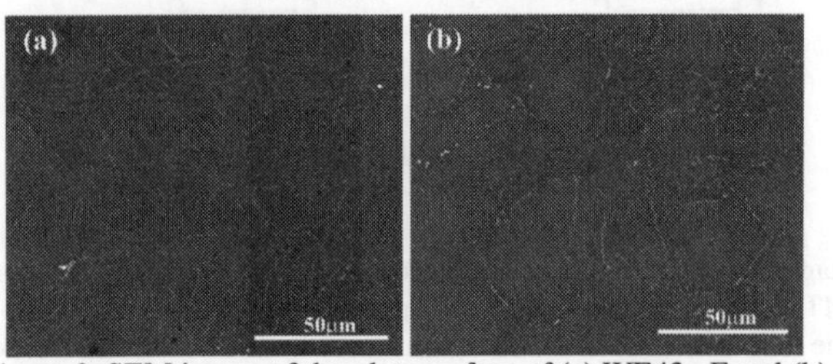

Figure 3. SEM image of the plate surface of (a) WE43 –F and (b) WE43 –T5.

From the in-situ micro tensile experiments, load over elongation data were obtained and engineering stress – strain curves were plotted, as shown in figure 6. DIC technique was used to collect data and plot the stress – strain curves in the elastic regime. By comparison of the moduli calculated from the stress – strain curves from the ADMET data to those calculated from DIC data, it was determined that the DIC data was more accurate along

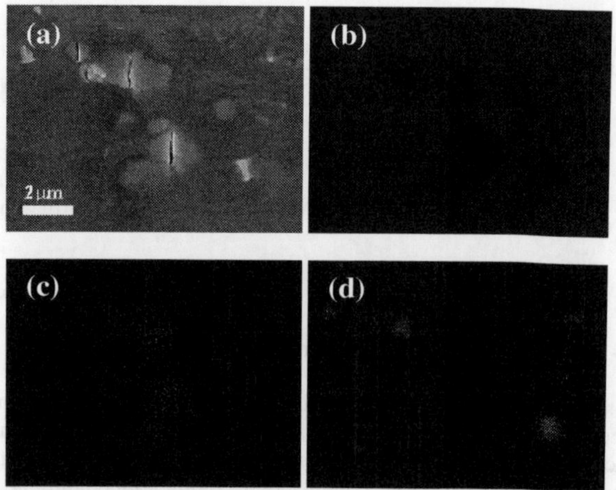

Figure 4. (a) Secondary electron image of the mapped area of WE43 -F. (b) Blue points indicating the presence of Mg, (c) Nd in purple and (d) Y in yellow.

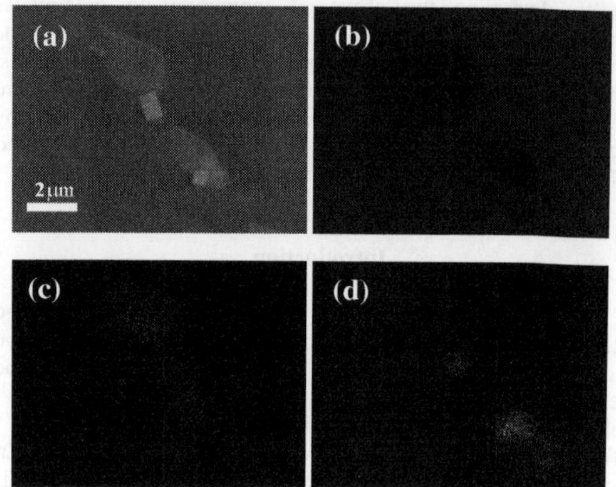

Figure 5. (a) Secondary electron image of the mapped area of WE43 –T5. (b) Blue points indicating the presence of Mg, (c) Nd in purple and (d) Y in yellow.

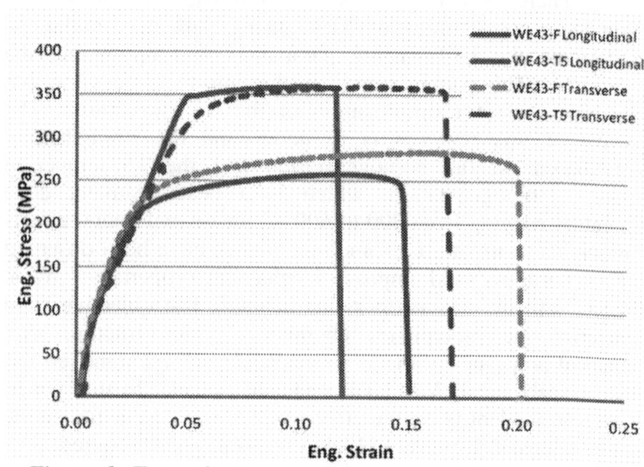

Figure 6. Example stress – strain curves of the –F and –T5 samples in the longitudinal and transverse directions.

the elastic regime. The average ultimate tensile stresses from the ADMET data, elongations, and moduli from the DIC data are tabulated in Table I. Post mortem analyses of the fracture surfaces were conducted. Though both F and T5 samples in both longitudinal and transverse directions failed by microvoid coalescence, the T5 sample showed the influence of less ductility (see Fig 7). In addition, there is evidence of a more brittle rupture mechanism in play for the T5 samples.

TABLE I. UTS, Elongation, and Young's Modulus for the F and T5 samples

Sample	UTS (MPa)	% Elongation	Young's Modulus (GPa)
WE43–F Longitudinal	273.2	20.5	25.0
WE43–F Transverse	283.2	20.0	34.5
WE43–T5 Longitudinal	331.7	10.5	25.8
WE43–T5 Transverse	358.6	13.3	34.1

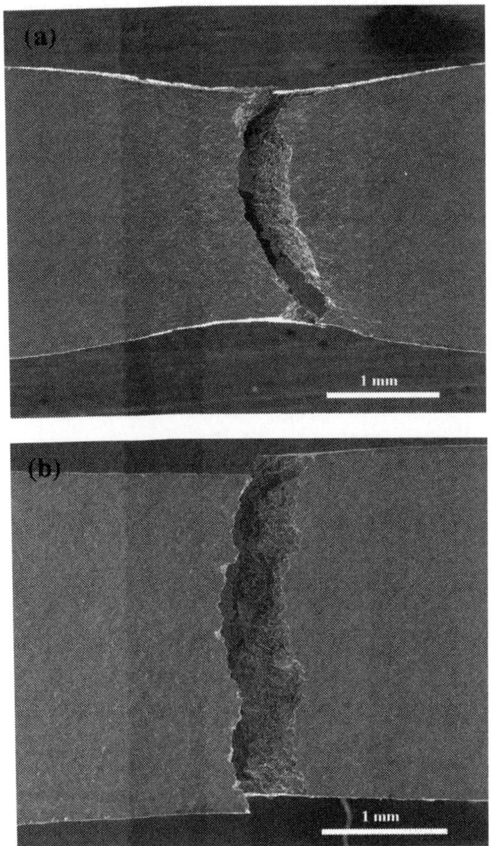

Figure 7. Failure region of (a) an F sample and (b) T5 sample.

Texturing of the surfaces was observed for both –F and –T5 samples from the collected EBSD patterns. The basal plane orientation was the most observed orientation in the surface parallel to the plate, while the plane orientations in the surfaces perpendicular to the rolling direction and transverse direction tended to be prismatic planes. This is in agreement with observations reported by Senn and Agnew[14]. Figure 8 shows a schematic of the orientations in the corresponding surfaces of the T5 plate. The schematic's surfaces are made up of one set of EBSD inverse pole figure maps from EBSD patterns obtained from each of the plate surfaces. The F tempered samples also showed similar texturing of the surfaces in relation to the plate's rolling and transverse directions. To better compare the differences between the F and T5, or the effect of the homogenization and aging on the orientation distribution, inverse pole figures of the surface orientations were plotted in Fig. 9.

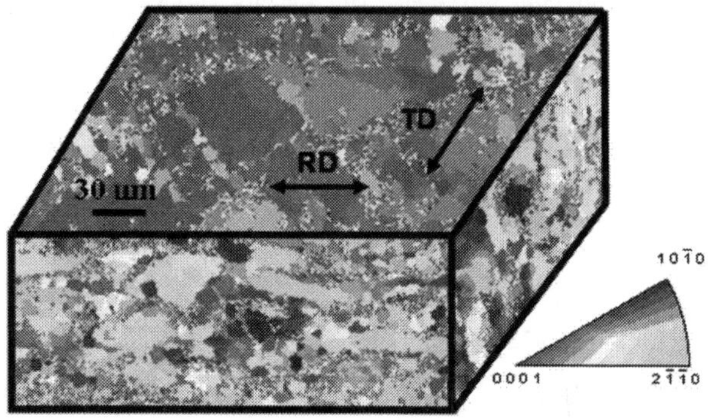

Figure 8. Schematic of the orientation texture of WE43-T5 in relation to the plate directions (not to scale).

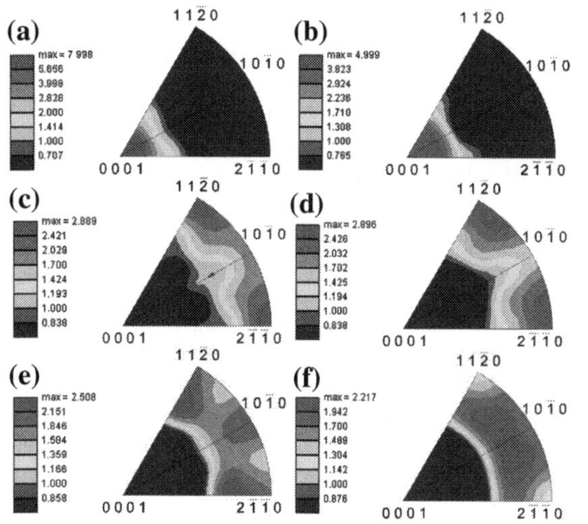

Figure 9. Inverse pole figures with texture index of (a) sample F surface parallel to the plane of the plate, (b) sample T5 surface parallel to the plane of the plate, (c) sample F surface perpendicular to the long transverse direction, (d) sample T5 surface perpendicular to the long transverse direction, (e) sample F surface perpendicular to the rolling direction, and (f) sample T5 surface perpendicular to the rolling direction.

Surface parallel to the plate had a higher amount of texturing than the surfaces perpendicular to the rolling and transverse directions. Although this is based on a limited dataset, Fig. 9 shows the homogenization and aging processes changes the orientation distribution, and decreases the texture.

Summary

The microstructure and tensile properties of Elektron™ WE43 –F and –T5 samples were characterized by microscopy techniques and in-situ tensile tests. Crystallographic orientation distribution results showed texturing of the surfaces, with the surface parallel to the plate surface having a basal plane orientation, and surfaces perpendicular to the rolling and transverse directions having a prismatic plane orientation. The in-situ tensile experiment results indicated a decrease in ductility but increase in the strength of the T5 alloys in both directions. The fracture surfaces suggest a change in the failure mechanism after the homogenization and artificial aging in the T5 alloys. More microstructural characterization and experiments are needed to determine the full effect of the homogenization and T5 tempering on the microstructure and the mechanics for future property optimization.

References

1. Magnesium Elektron, Elektron WE43 Wrought Alloy Datasheet: 478, www.magnesium-elektron.com

2. K. Cho, T. Sano, K. Doherty, C. Yen, G. Gazonas, J. Montgomery, and P. Moy, "Magnesium Technology and Manufacturing for Ultra Lightweight Armored Ground Vehicles," *Proceedings from the Army Science Conference* (2008).

3. M. Turski, J. F. Grandfield, T. Wilks, B. Davis, R. DeLorme, K. Cho, "Computer Modeling of DC Casting Magnesium Alloy WE43 Rolling Slabs," *Magnesium Technology 2010*, ed. S. R. Agnew, N. R. Neelameggham, E. A. Nyberg, and W. H. Sillekens (2010).

4. P. Mengucci, G. Barucca, G. Riontina, D. Lussana, M. Massazza, R. Ferragut, and E. Hassan Aly, "Structure Evolution of a WE43 Mg Alloy Submitted to Different Thermal Treatments," *Mater. Sci. Eng. A* 479 (2008) 37-44.

5. O. A. Lambri, W. Riehemann, L. M. Salvatierra, and J. A. Garcia, "Effects of Precipitation Processes on Damping and Elastic Modulus of WE 43 Magnesium Alloy," *Mater. Sci. Eng. A* 373 (2004) 146-157.

6. D. Lussana, M. Massazza, and G. Riontino, "A DSC Study of Precipitation Hardening in a WE43 Mg Alloy," *J. Thermal Analysis and Calorimetry* 92 (2008) 1, 223-225.

7. Z. Zhao, Q. Chen, Y. Wnag, and D. Shu, "Microstructures and Mechanical Properties of AZ91D Alloys with Y Addition," *Mater Sci. Eng. A* 515 (2009) 152-161.

8. J. F. Nie, "Effects of Precipitate Shape and Orientation on Dispersion Strengthening in Magnesium Alloys," *Scripta Mater.* 48 (2003) 1009-1015.

9. ARL report number ARL-TR-5212

10. J. Nie and B. C. Muddle, "Precipitation in Magnesium Alloy WE54 During Isothermal Ageing at 250 ºC," *Scripta Mater.* 40 (1999) 1089-1094.

11. J. G. Wang, L. M. Hsiung, T. G. Nieh, and M. Mabuchi, "Creep of a Heat Treated Mg-4Y-3RE Alloy," *Mat. Sci. Eng. A* 315 (2001) 81-88.

12. F. Penghuai, P. Liming, J. Haiyan, Z. Zhenyan, Z. Chunquan, "Fracture Behavior and Mechanical Properties of Mg-4Y-2Nd-1Gd-0.4Zr (wt%) Alloy," *Mater Sci. Eng. A* 486 (2008) 572-579.

13. F. Hnilica, V. Janík, B. Smola, I. Stulíková, and V. Očenášek, "Creep Behaviour of the Creep Resistant MgY3Nd2Zn1Mn1 Alloy," *Mater Sci. and Eng. A* 489 (2008) 93-98.

14. J. W. Senn and S. R. Agnew, "Texture Randomization of Magnesium Alloys Containing Rare Earth Elements," *Magnesium Technology 2008*, ed. M. O. Pekguleryuz, N. R. Neelameggham, R. S. Beals, and E. A. Nyberg (2008).

Magnesium Technology 2011
Edited by: Wim H. Sillekens, Sean R. Agnew, Neale R. Neelameggham, and Suveen N. Mathaudhu
TMS (The Minerals, Metals & Materials Society), 2011

A molecular dynamics study of fracture behavior in magnesium single crystal

Tian Tang, Sungho Kim, Mark F Horstemeyer, and Paul Wang

Center for Advanced Vehicular System, Mississippi State University, Box 5405, Mississippi State, MS 39762, USA

Keywords: molecular dynamics simulation, crack propagation; magnesium single crystal; temperature effects; strain rate

Abstract

The analysis of crack growth in magnesium crystals was performed using molecular dynamics simulation with Embedded Atom Method (EAM) potentials. Four specimens with increasing sizes were used to investigate the influences of material length scale on crack growth of magnesium single crystals. Furthermore, the effects of temperature, loading strain rate, and the size of the initial crack were also verified. The specimens were subjected to uniaxial tension strain up to the total strain level of 0.2 with a constant strain rate of $10^9 s^{-1}$ except in the studies of strain rate effects and the uniaxial stress strain curve was monitored. The simulation results show that the specimen size, loading strain rate, temperature, and the size of initial crack strongly influence the yield strength at which the twin nucleated and subsequently the crack grew. The initial slope of the uniaxial stress strain curve is independent of the loading strain rate and temperature. Moreover, high temperatures induce increased atomic mobility, and thereby atom reorganization, which, in turn, releases the stress at the crack tip

Introduction

The study of materials' fracture behavior has been the topic of intensive research in the physical and engineering sciences since last century. Numerous achievement has well established the continuum-based theoretical framework of fracture mechanics at the macro scale. Griffith first proposed a continuum fracture model for predicting the onset of crack growth in a brittle material based on energetic and thermodynamics considerations [1, 2] and the earlier work of Inglis [3]. After his seminal contribution, fracture mechanics theories were developed to account for the nonlinear behavior of materials such as plasticity and viscoplasticity at the crack tip [2].

Meanwhile, the fracture mechanism in nanocrystalline materials is not clearly understood. To reveal the mechanism of fracture at the nanoscale requires discerning the evolution process of the crack tip under the application of increased stress intensity, which is strongly affected by the local atomic structure, lattice orientation, interatomic potentials, and generation of defects (e.g., dislocations and twins). With the rapid development of computer technique as impetus, molecular dynamics (MD) has become an increasingly important tool for studying the fracture behavior at atomic scale. MD simulations predict the motion of atoms in an atomistic system through Newton's equations of motion. Combined with the initial and boundary conditions, these equations can be solved using numerical integration method to obtain the trajectories of atoms such as the position and velocity as a function of time.

During the past several decades, MD has been widely employed to investigate fracture problems. These studies include the fast brittle fracture in a material [4, 5, 6, 7], instability of dynamic fracture in three-dimensional fracture [8, 9], brittle-to-ductile transition (BDT) in crack growth [10, 11], stress-induced phase transformations and grain nucleation at crack tip [12], influences of grain boundary [13, 14], and large-scale MD simulation of three-dimensional fracture [15]. Moreover, Rafii-Tabar *et al.* [16] used MD simulations to study the effects of nanoscale inhomogeneities near the crack tip on the crack propagation in fcc metallic plates. Recently, Xu and Deng [17] performed MD simulations of ductile crack growth in an aluminum single crystal and provided insight to the stress field around the crack tip and its evolution during crack growth. They concluded that the ductile crack growth in aluminum single crystal was achieved via void nucleation, growth, and coalescence ahead of the crack tip. However, few studies have been published on the fracture behavior of magnesium.

In this study, we performed MD simulation of crack propagation in magnesium crystals using Embedded Atom Method (EAM) potential. The goal of the present work was to elucidate the influences of specimen size, temperature, strain rate, and initial crack length on the fracture behavior of magnesium single crystals. The uniaxial stress strain responses were monitored during uniaxial tension. The simulation results will help provide a better understanding of magnesium's fracture behavior at the nanoscale.

Simulation Method

The computation method used in this study is Embedded Atom Method (EAM) potential developed by Sun *et al.* [18]. This form of EAM consists of two contributions to the total potential energy, E, of the entire system composed of N atoms. The functional form of the total embedded energy can be expressed as

$$E = \sum_i G_i \left(\sum_{j \neq i} \rho_j^a \left(r_{ij}\right) \right) + \frac{1}{2} \sum_{ij} U^{ij} \left(r_{ij}\right) \quad (1)$$

where G_i is the embedding energy as a function of the local electron density, ρ_j^a is the speherically averaged atomic electron density, U^{ij} is the pair potential, and r_{ij} is the distance between atom i and j. Many examples have demenstrated that EAM can be an accurate representation of inter-atomic forces in a metallic lattice. In molecular dynamics, the energy is employed to determine the forces acting on each atom. At each atom the dipole force tensor, β_{ij}, is given by

$$\beta_{km}^i = \frac{1}{\Omega^i} \sum_{j(\neq i)}^{N} f_k^i \left(r^{ij}\right) r_m^{ij} \quad (2)$$

where f_k is the force vector between atoms, r_m is the displacement vector between atoms i and j, N is the number of neareast neighbour atoms, and Ω^i is the atomic volume. If stress could be defined at an atom, then β_{ij} would be the stress tensor at that point. Since stress is defined at a continuum point, the stress tensor can be determined as a volume average over the block of material,

$$\sigma_{mk} = \frac{1}{N^*} \sum_{i}^{N^*} \beta_{mk}^i \quad (3)$$

in which the stress tensor is defined in terms of the total number of atoms, N^*, in the block of material.

Specimens for atomistic studies

The size effects on crack growth were studied using four specimens of increasing size as shown in Fig. 1.

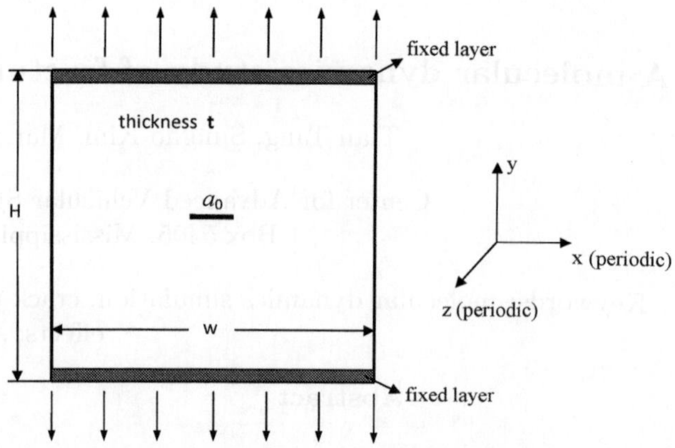

Figure 1: Model geometry of magnesium single crystal containing a center initial crack.

Table 1: Geometrical dimensions and the resulting number of atoms of the specimens of increasing size.

sp #	w (nm)	H (nm)	t (nm)	# of atoms
1	6.0	9.0	2.21	4,884
2	12.0	18.0	2.21	20,100
3	24.0	36.0	2.21	82,740
4	40.0	60.0	2.21	229,900

The sizes of the four specimens were generated by increasing the width and the height of the specimen but maintaining the thickness, t, as a constant. The ratio of height to width of in each of the specimens is kept constant as $H/w = 1.5$. A center crack introduced in each specimen by removing the atoms from the perfect crystals and maintaining the ratio of the initial crack length to the width of the specimen to $a_o/w = 0.1$. The resulting number of atoms varied from 4884 to 229,900. Table 1 presents the geometrical dimensions and the resulting number of atoms in the four specimens of increasing size. The edges of the specimens were aligned with the global coordinate system in which the x axis represents $[12\bar{1}0]$ direction, the y axis represents $[0001]$ direction and the z axis represents $[101\bar{0}]$ direction.

Simulation process

For both single crystal and bicrystal specimens, the top and bottom boundaries are free surfaces. About 1 nm deep atomic surface layers at the top and bottom boundaries were fixed for applying Mode I cyclic loading. For single crystal specimens, the periodic

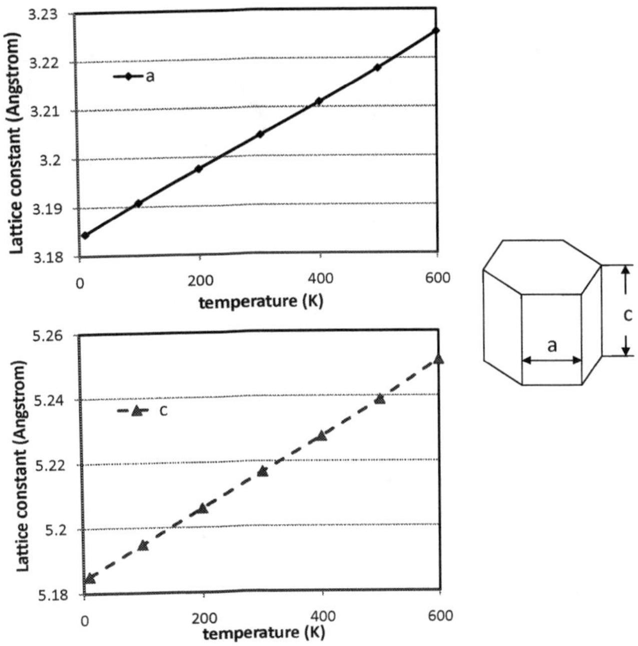

Figure 2: Variation of lattice constants of magnesium crystal with temperature.

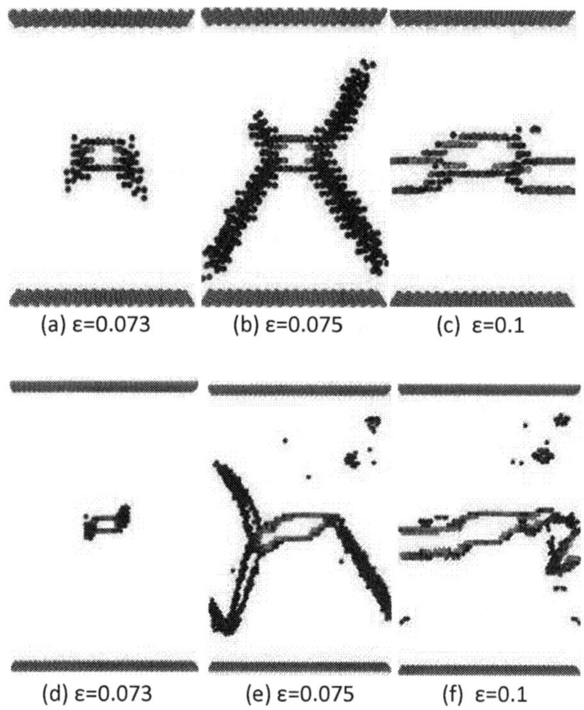

Figure 3: Crack growth and twin pattern of different specimen size at various applied strain: (a), (b), and (c) are the contour plot of the smallest size specimen-1 with 4884 atoms; (d), (e), and (f) are the contour plot of specimen-2 with 20,100 atoms.

boundary conditions were assigned in the $x-$ and $z-$ directions. In all simulations, the specimens were equilibrated at 100 K (except in the studies of the effects of temperature) by running 2000 timesteps before applying uniaxial tension loading. The uniaxial tension strain loading was applied along the y axis up to the total strain of $\epsilon_y = 20\%$. For the sake of eliminating the stress oscillation resulting from the sudden loading employed on the top and bottom boundary, the loading was applied such that the velocity was linearly distributed along the $y-$ direction from the bottom to the top. The velocities were maximum positive value and minimum negative value at the top and bottom boundary, respectively. The uniaxial tension loading was applied at a constant strain rate of $10^9 \mathrm{s}^{-1}$ except in the studies of strain rate effects. All simulations were performed at a constant temperature of 100 K except in the studies of the effects of temperature.

Lattice constants

To study the temperature effects on fatigue crack propagation, we calculated the lattice constants of magnesium at various temperatures. In this calculation, we created a three dimensional model without defects. Periodic boundary conditions were assigned to all of three directions so that the thermal expansion of the model was isotropic. This model was firstly equilibrated at constant temperature and zero pressure. The temperature was controlled by rescaling the velocities of atoms while the pressure of the model was set to zero by using a Berendsen barostat. Finally, constant NPT integration was performed upon the group of atoms to update positions and velocities at each time step. P is pressure and T is temperature. This created a system trajectory consistent with the isothermal-isobaric ensemble. The variation of lattice constants c and a are plotted in Figure 2.

Simulation results and discussions

This section presents the MD simulation results for verifying the effects of the specimen size, temperature, strain rate, and grain boundary on the crack growth in magnesium crystals. The centrosymmetry parameter defined by Kelchner *et al.* [19] was used to highlight the defects including void surface and twin.

The expression of this parameter is given by

$$P = \sum_i |R_i + R_{i+6}|^2 \quad (4)$$

where R_i and R_{i+6} are the vectors or bonds corresponding to the six pairs out of twelve nearest neighbors in the lattice. The six pairs are chosen to minimize P. The P values for the atoms in fcc and hcp structures, dislocations, twinning boundaries and crack surfaces are all different so that extensive information about crack growth and many defect behaviors can be obtained.

Size effects

The material length scale has apparent influences on the strength, toughness, ductility, and the stress strain response of materials [20, 21, 22, 23]. In this study, the investigation of size effects on the crack growth was performed using the four specimens of increasing size as specified in Fig. 1 and Table 1. The simulation results of different specimen sizes are shown in Fig. 3 and 4. The uniaxial stress strain curves of different specimen sizes obtained from MD simulation are plotted in Fig. 5. One common feature of all specimen sizes is that the twin nucleated at the crack tips due to the stress concentration just after the yield strength is reached as shown in Fig. 3(a), Fig. 3(d), Fig. 4(g), and Fig. 4(j). This means that the yield strength corresponds to the nucleation of twin bands. The stress resistance dropped down abruptly after yielding. As the external loading continues increasing, more twin bands are initiated and grow from the crack tips. Meanwhile, the crack starts growing just after the yield is reached. The paths of crack extension are significantly influenced by the twin bands being nucleated and grown from the crack tips. Finally, the stress resistance decreased to almost zero when the crack reached the specimen edges.

From Fig. 3, 4, and 5, we can observe the significant size effects. Different specimen sizes demonstrate different twin patterns and crack growth paths. At larger specimen sizes as shown in Fig. 4, the right crack propagates along the twin band with mixed Mode I+Mode II. It is observed that the crack propagated rapidly along the twin band with a manner of brittle fracture. The initial yield strength decreases with increasing specimen size. At the smallest specimen size, the yield strength is about 11.2 GPa while it decreased to about 6.3 GPa at the largest specimen size. On the other hand, the size effect de-

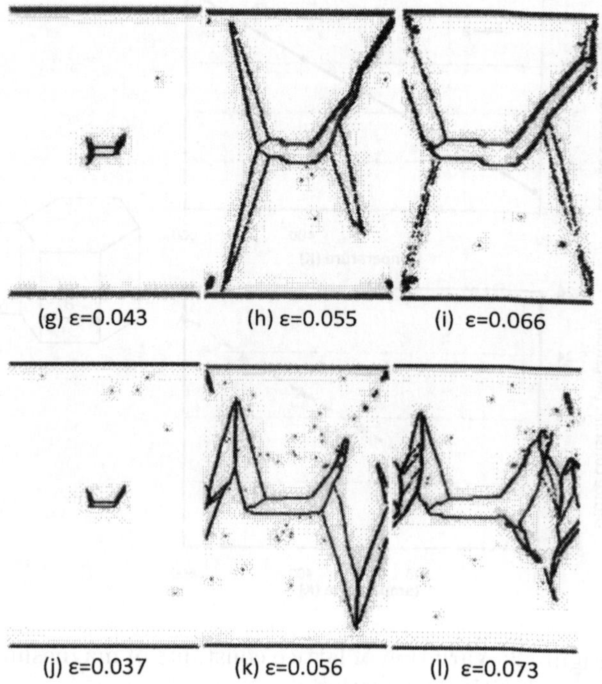

Figure 4: Crack growth and twin pattern of different specimen size at various applied strain: (g), (h), and (i) are the contour plot of specimen-3 with 82,740 atoms; and (j), (k), and (l) are the contour plot of specimen-4 with 229,900 atoms.

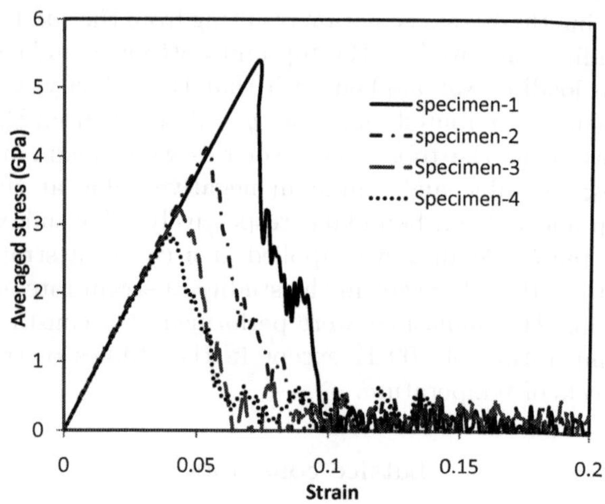

Figure 5: Influences of specimen size on the averaged stress strain response of magnesium single crystal specimen containing a center initial crack.

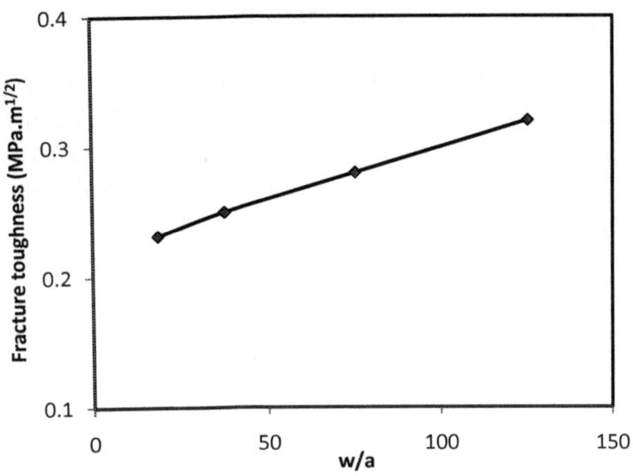

Figure 6: Variation of fracture toughness with the normalized specimen size.

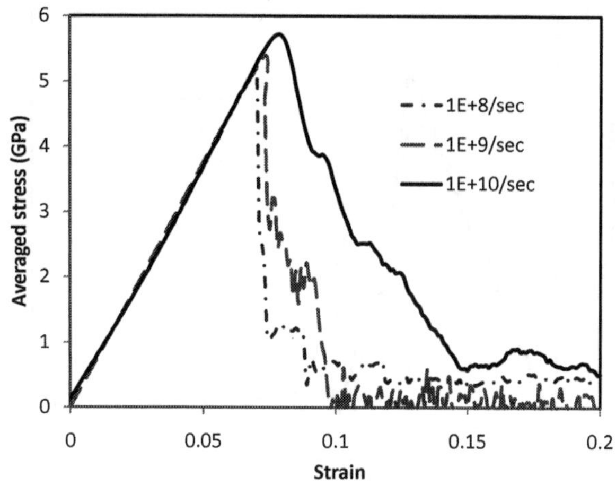

Figure 7: Effects of strain rate on the averaged stress strain response of magnesium single crystal specimen containing a center initial crack.

creases with increasing specimen size. The initial slope of the uniaxial stress strain curve of the smallest specimen is little lower than those of larger specimens. As the material length increasing, the initial slopes of the uniaxial stress strain curves are independent of the specimen size. Since the crack propagated rapidly and unstably after the initial yielding strength is reached, the values of the stress intensity factors at the yielding strength, K_{IC}, of different specimen sizes are nanoscale fracture toughness and can be expressed as

$$K_{IC} = Y\sigma\sqrt{\pi a} \tag{5}$$

where σ is the volume averaged stress in the $y-$ direction; a is the initial crack length; the value of geometrical factor Y is about 1.016. For simplicity, Y is assumed to be 1. Fig. 6 presents the variation of nanoscale fracture toughness with the normalized specimen size, the ratio of the length w of the specimens to the lattice constant a of magnesium. We can observe that the fracture toughnesses at nanoscale increase with increasing specimen size even though the values of peak stress points decrease with increasing specimen size. This effect results from the initial crack length increases with increasing specimen size at the same time.

Strain rate effects

For the sake of reducing computational cost, the investigation of strain rate effects was performed on the smallest specimen. Fig. 7 shows the stress-strain responses obtained from three strain rate levels, namely, 10^8, 10^9, and 10^{10}s^{-1}. One apparent finding is that the initial slopes of the uniaxial stress strain curves were independent of the strain rate. After the yielding strength is reached, the stress response decreases rapidly. The yield strength increases with increasing strain rates. On the other hand, the yield strength is getting insensitive to the strain rates with decreasing strain rate. The results are qualitatively similar to those obtained from Potirniche *et al.*'s studying on the strain rate effects on void growth and coalescence [23].

Temperature effects

In the previous studies, the simulations were carried out at a constant temperature of 100 K. It is well known that temperature has significant influences on the material properties. To study the effects of temperatures, we constructed the atomistic models with different lattice constants obtained from various temperatures, namely, 10 K, 100 K, 300 K, and 500 K. The lattice constants of various temperatures are shown in Fig. 2. The size of atomistic models of different temperatures is the same as $w(12nm) \times H(18nm) \times t(2.21nm)$. Fig. 8 shows the uniaxial stress strain curves of various temperatures. In general, the stress resistance dropped down rapidly after the yield strength is reached at all temperatures. It is seen that the initial slopes of the uniaxial stress strain curves are independent of temperature. The yield strength decrease with increasing temperature due to the degradation of materials at higher temperature. This phenomenon implies that the critical stress of twin nucleation decreases with increasing

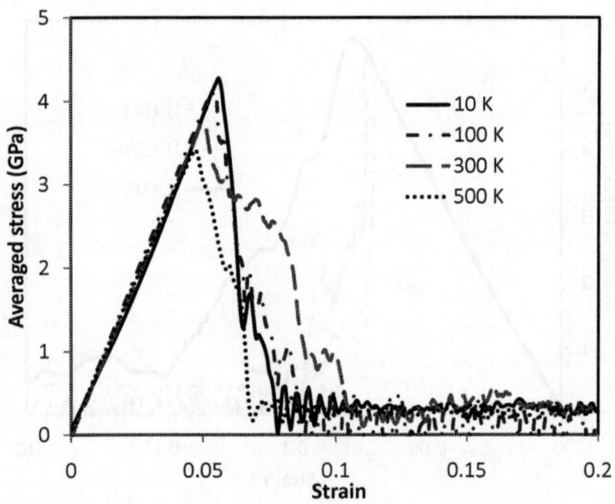

Figure 8: Temperature effects on the averaged stress strain response of magnesium single crystal specimen containing a center initial crack.

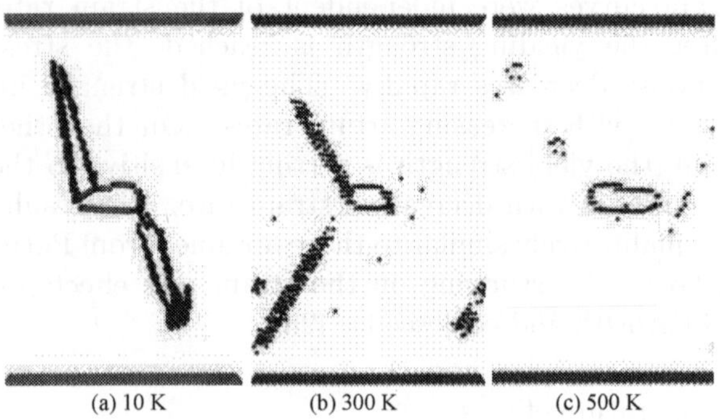

Figure 9: Simulation of crack growth in magnesium single crystal at various temperatures when the applied tension strain is $\epsilon_y = 0.06$.

temperature. Fig. 9 shows the contour plot of crack growth at different temperatures. The twinning at the crack tip becomes weak with increasing temperature and eventually disappears at 500 K. At high temperature 500 K, the stress concentration at the crack tip was significantly alleviated by the reorganization of atoms due to the high mobility of atoms at high temperature. This effect causes the twinning vanishing and the crack tip being blunted. This phenomenon is similar to the observation of Latapie and Farkas [12]. We can deduce that a brittle-to-ductile fracture behavior is induced by increasing temperature.

Conclusions

We have performed the MD simulations of crack growth in magnesium crystals using EAM potential at nanoscale. Single crystal specimens were used to study the influences of specimen size, strain rate, temperature, and initial crack length. The investigation of the effects of grain boundary was carried out using three bicrystal models with different misorientations. According to the simulation results, the following conclusions were obtained:

1. The significant size effects were clearly demonstrated by uniaxial averaged stress strain responses. The yield strength decreases with increasing specimen size. The initial slope of the average stress strain curve is independent of the specimen size except the smallest specimen size. Meanwhile, it is also found out that the size effects decrease with increasing specimen size.

2. Furthermore, the yield strength decreases with increasing strain rate, and temperature. On the other hand, the initial slopes of average stress strain curves are independent of strain rate and temperature. The twinning becomes weak and the material becomes ductile at high temperature so that the resistance of crack growth increase with increasing temperature. This effect results from the stress at the crack tip being released by the reorganization of atoms due to the high mobility of atoms at high temperature.

3. In all loading conditions and specimen sizes, the decrease of averaged stress strain response is mainly due to the nucleation of twinning at the crack tip followed by the crack starting propagation.

Acknowledgement

The authors acknowledged the Center for Advanced Vehicular Systems at Mississippi State University and the Department of Energy for supporting this research.

References

[1] A. A. Griffith. The phenomena of rupture and flow in solids. *Philosophical Transactions*, 221A:163–198, 1920.

[2] T. L. Anderson. *Fracture Mechanics Fundamentals and Applications.* Boca Raton, FL: CRC Press. 3rd edition., 2005.

[3] C. E. Inglis. Stress in a plate due to the presence of cracks and sharp corners. *Transactions of the Institute of Naval Architects*, 55:219, 1913.

[4] F. F. Abraham, D. Brodeck, W. E. Rudge, and X. Xu. A molecular-dynamics investigation of rapid fracture mechanics. *Journal of the Mechanics and Physics of Solids*, 45:1595–1619, 1997.

[5] J. G. Swadener, M. I. Baskes, and M. Nastasi. Molecular dynamics simulation of brittle fracture in silicon. *Physical Review Letter*, 89:085503, 2002.

[6] J. A. Hauch, D. Holland, M. P. Marder, and H. L. Swinney. Dynamic fracture in single crystal silicon. *Physical Review Letter*, 82:3823–3826, 1999.

[7] M. L. Falk. Molecular-dynamics study of ductile and brittle fracture in model noncrystalline solids. *Physical Review B*, 60:7062, 1999.

[8] F. F. Abraham, D. Schneider, R. A. Rafey, and W. E. Rudge. Instability dynamics in three-dimensional fracture: An atomistic simulation. *Journal of the Mechanics and Physics of Solids*, 45:1461–1471, 1997.

[9] F. F. Abraham. Unstable crack growth is predictable. *Journal of the Mechanics and Physics of Solids*, 53:1071–1078, 2005.

[10] Y. Guo, C. Wang, and D. Zhao. Atomistic simulation of crack cleavage and blunting in bcc-fe. *Materials Science and Engineering*, 349:29, 2003.

[11] F. Abraham and J. Q. Broughton. Large-scale simulations of brittle and ductile failure in fcc crystals. *Computational Materials Science*, 10:1–9, 1998.

[12] A. Latapie and D. Farkas. Molecular dynamics simulations of stress-induced phase transformations and grain nucleation at crack tips in fe. *Modelling and Simulation in Materials Science and Engineering*, 11:745, 2003.

[13] D. Farkas, H. V. Swygenhoven, and P. M. Derlet. Intergranular fracture in nanocrystalline metals. *Physical Review B*, 66:060101–1, 2002.

[14] A. Luque, J. Aldazabal, J. M. Martinez-Esnaola, and J. G. Sevillano. Molecular dynamics simulation of crack tip blunting in opposing directions along a symmetrical tilt grain boundary of copper bicrystal. *Fatigue Fracture of Engineering Materials and Structure*, 30:1008–1015, 2007.

[15] S. J. Zhou, P. S. Lomdahl, A. F. Voter, and B. L. Holian. Three-dimensional fracture via large-scale molecular dynamics. *Engineering Fracture Mechanics*, 61:173–187, 1998.

[16] H. Rafii-Tabar, H. M. Shodja, M. Darabi, and A. Dahi. Molecular dynamics simulation of crack propagation in fcc materials containing clusters of impurities. *Mechanics of Materials*, 38:243, 2006.

[17] S. W. Xu and X. M. Deng. Nanoscale void nucleation and growth and crack tip stress evolution ahead of a growing crack in a single crystal. *Nanotechnology*, 19:115705, 2008.

[18] D. Y. Sun, M. I. Mendelev, C. A. Becker, K. Kudin, T. Haxhimali, M. Asta, and J. J. Hoyt. Crystal-melt interfacial free energies in hcp metals: A molecular dynamics study of mg. *Physical Review B*, 73:024116, 2006.

[19] C.L. Kelchner, S.J. Plimpton, and J.C. Hamilton. Dislocation nucleation and defect structure during surface indentation. *Physical Review B*, 58:11085–11088, 1998.

[20] A. Carpinteri. Size effects on strength, toughness, and ductility. *Journal of Engineering Mechanics*, 115:1375, 1989.

[21] M. F. Horstemeyer, M. I. Baskes, and S. J. Plimpton. Length scale and time scale effects on the plastic flow of fcc metals. *Acta Materialia*, 49:4363–4374, 2001.

[22] K. J. Zhao, C. Q. Chen, Y. P. Shen, and T. J. Lu. Molecular dynamics study on the nano-void growth in face-centered cubic single crystal copper. *Computational Materials Science*, 46:749, 2009.

[23] G. P. Potirniche, M. F. Horstemeyer, G. J. Wagner, and P. M. Gullett. A molecular dyanmics study of void growth and coalescence in single crystal nickel. *International Journal of Plasticity*, 22:257–278, 2006.

Magnesium Technology 2011
Edited by: Wim H. Sillekens, Sean R. Agnew, Neale R. Neelameggham, and Suveen N. Mathaudhu
TMS (The Minerals, Metals & Materials Society), 2011

MICROSTRUCTURAL RELATIONSHIP IN THE DAMAGE EVOLUTION PROCESS OF AN AZ61 MAGNESIUM ALLOY

M. Lugo[1], J.B. Jordon[2], M. F. Horstemeyer[1,3], M. A. Tschopp[1]
[1] Center for Advanced Vehicular Systems (CAVS), Mississippi State University, Mississippi State, MS 39762, USA
[2] Department of Mechanical Engineering, The University of Alabama, Tuscaloosa, Al 35487, USA
[3] Department of Mechanical Engineering, Mississippi State University, Mississippi State, MS 39762, USA

Key Words: Damage evolution, microstructure, magnesium, intermetallics.

Abstract

The damage evolution process of magnesium AZ61 alloy under monotonic tensile loading conditions is investigated. Specimens that have been subjected to interrupted tensile loading were examined under optical microscopy to quantify the number density of cracked intermetallic particles as a function of applied strain. Digital image analysis of the optical images was employed to automatically quantify damage by separating cracked from non-cracked particles. Lastly, an internal state variable damage model was shown to adequately capture the experimentally-observed damage of intermetallic particles in the magnesium AZ61 alloy.

Introduction

A marked interest has been placed on magnesium alloys in recent years by both researchers and industry because of the attractive mechanical properties that these alloys have regarding design applications. The automotive industry sees magnesium alloys as a solution for reducing weight and cost [1]. Magnesium alloys are candidates to replace aluminum and steel alloys [2], and plastics also [1].

Despite that magnesium alloys have been incorporated in various industrial components, their mechanical behavior is still not well known. A complete characterization of magnesium alloys is available for only a few alloys. Some studies [2-8] have reported the tensile and compression properties and the microstructure changes under various temperature conditions for magnesium alloys. AZ31 alloy is a commercial alloy that has been investigated more frequently than other magnesium alloys. Although much interest in mechanical properties of magnesium alloys has been shown by researchers, damage models that incorporate the microstructure deterioration are scarce.

In magnesium alloys, twinning plays a primarily role in damage during plastic deformation, and numerous studies are found in the literature [3-4, 9-10]. However, studies regarding the role of second phases and intermetallics on damage evolution of magnesium alloys are lacking. As in other types of metals, cracking and/or debonding of particles could play a significant role in damage evolution in magnesium alloys. Currently it is not known how the cracking activity influences the damage process in these alloys. An understanding of the sources of damage in this alloy will provide insight on the plasticity and fracture in the AZ61 magnesium alloy.

The objective of this paper is to evaluate the microstructural damage evolution for the AZ61 magnesium alloy. A stereological analysis of the material microstructure was conducted on interrupted monotonic tensile loading tests to determine the damage evolution via microstructural statistics.

Damage of Magnesium Alloy

The failure process in ductile materials is associated with local failure of second phase particles, e.g., inclusions, intermetallic particles, and precipitates [11]. In aluminum alloys, damage is produced by cracking of intermetallic inclusions/particles, growth of voids at cracked particles and void coalescence [12]. However, magnesium alloys show limited ductility [3]. Additionally, slip dislocation and twinning have been identified as the main sources of plastic deformation in magnesium alloys [5, 9, 13]. Nonetheless, intermetallics and second phases have been observed in the magnesium alloys [2, 5, 14-19].

Damage evolution process leading to fracture is typically divided into three stages: void nucleation, growth, and coalescence [20]. Horstemeyer et al. [21] showed that these components develop simultaneously and have developed a model for damage evolution that includes void nucleation, growth, and coalescence implemented within the finite element method. Using this model, they showed that void nucleation is highly influenced by the stress state of the material. The model framework is represented in Figure 1 by a fictitious material with increasing nucleation density and void growth as a function of increasing strain over domain R.

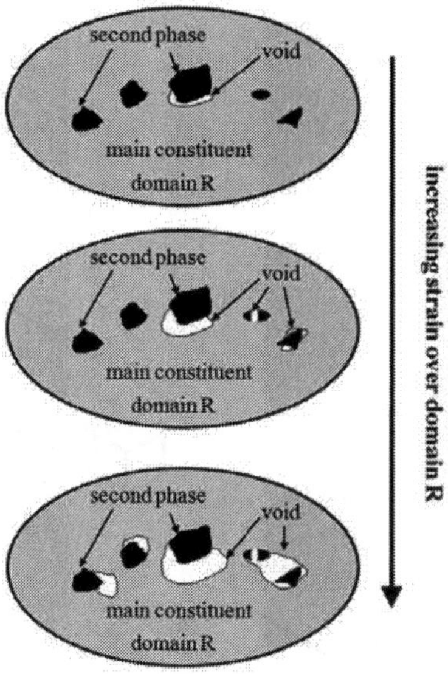

Figure 1. Conceptual Framework showing a fictitious material with increasing nucleation density and void growth.

Damage evolution can be quantified by measuring the evolution of microstructural features with metrics such as the number density, number fraction, and volume fraction [1-2]. Horstemeyer and Gokhale [22] proposed a model capable of capturing damage in ductile materials. This model was defined in terms of void nucleation and void growth. The void nucleation parameters in this model are length scale *d*, and particle volume fraction, *f*. They developed the following equation for the number voids per area.

$$\eta(t) = C \cdot \exp\left(\frac{\varepsilon(t) d^{1/2}}{K_{IC} f^{1/3}} \left\{ a\left[\frac{4}{27} - \frac{J_3^2}{J_2^3}\right] + b\frac{J_3}{J_2^{3/2}} + c\left\|\frac{I_1}{\sqrt{J_2}}\right\| \right\}\right) \quad (1)$$

where *C*, *a*, *b* and *c* are material constants determined of tension, compression and torsion tests. Further details of this damage model can be found in [22].

Test and Procedures

The material employed in this research was an extruded AZ61 magnesium alloy. The material was provided in the form of an automotive crash rail profile with a thickness of approximately 3 mm.

Tension Test

A flat tensile specimen in conformity with ASTM standard E8 was prepared with a 50.8 mm gage length, 12.7 mm width, and 3mm in thickness. A program of interrupted tensile tests was conducted at strain levels of 1, 2, 4, 6, and 8 percent. The tests were conducted using an INSTRON electromechanical load frame under strain control at a strain rate of 0.001/s at ambient temperature and relative humidity. An extensometer was used to measure and control strain. At each interrupted strain level, the specimens were removed and sectioned for metallographic analysis.

Metallographic Analysis

Figure 2 shows the (L-S) planes that were cut from the gage section for metallographic analysis. The metallographic specimens contained a vertical plane parallel to the applied loading direction (extruded direction). The (L-S) planes provide a clear picture of cracked particles, while the planes perpendicular to the applied loading do not show these features very readily [23].

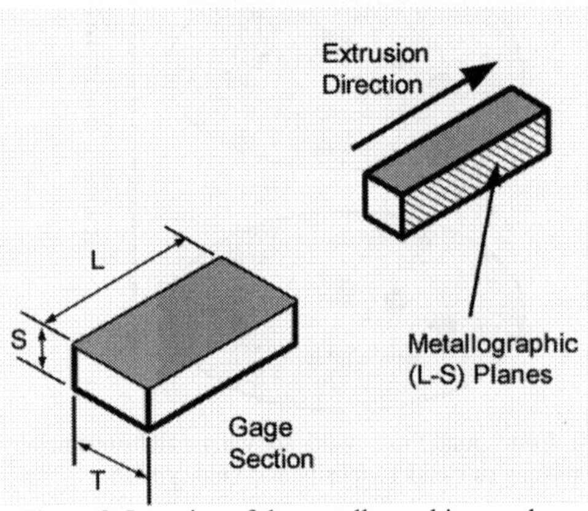

Figure 2. Location of the metallographic samples.

The metallographic samples were polished to obtain a highly reflective surface with no scratches. Further information on preparing magnesium specimens can be found in [24]. The samples were then imaged in an Axiovert Zeiss optical microscope with magnification of 500X in the unetched condition. Manual methods for identification are tedious, susceptible to error, and time-consuming. Hence, automating microstructural feature characterization via digital image analysis is essential for quantifying the evolution of microstructure. Digital image processing was used to measure the evolution of features within the microstructure images as a function of deformation. A program was developed to perform the digital image analysis using the Image Processing Toolkit in MATLAB. The code performed the following main tasks: a) segmented all particles based on a threshold parameter, b) distinguished the light-colored particles from the dark particles, c) separated the cracked gray particles from the non-cracked gray particles, and d) computed the microstructure statistics.

Results

Tension Tests

A typical true stress-true strain curve for magnesium alloy AZ61 is shown in Figure 3. The following average mechanical properties were determined from the tensile tests: yield strength= 177 MPa; ultimate tensile strength= 285 MPa; elongation at failure =10%.

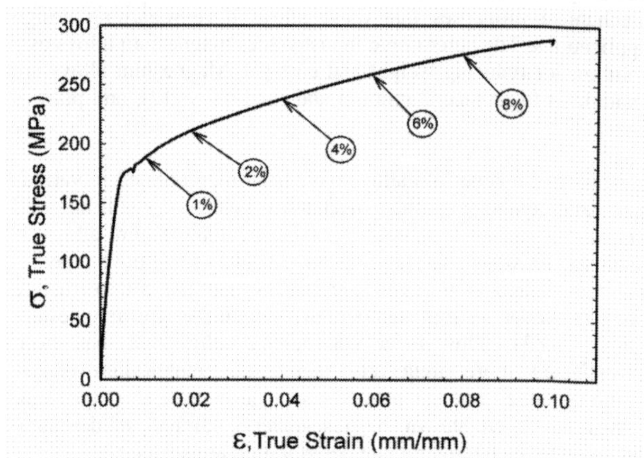

Figure 3. True stress-true strain curve for magnesium alloy AZ61.

Microstructure of AZ61

Various types of particles were observed in the metallographic study. Figure 3 shows the intermetallics that subsist in the alloy in an image taken with a 1000X of magnification at 8% of strain. Three shapes of particles are distinguished, round particles (A), slender particles (B), irregular particles (C), and small dark particles (D). Cracks were observed in all the particles except in dark particles.

Stereology Results

The microscope used in this study was equipped with an automatic stage capable of creating montages of many adjacent fields. Since the distribution of intermetallic particles is no uniform and the particle volume fraction is low in this alloy, a reasonable sample size must be taken. Therefore, montages composed of 25 adjacent fields arranged in a 5 x 5 grid were defined. In addition, we observed several samples at each strain level to increase the reliability of our analysis. Using the equation developed by DeHoff [25] and setting (the relative accuracy) %RA to 10%, it was determined that three samples were required for a statistically accurate analysis. However, in this metallographic analysis five samples were analyzed.

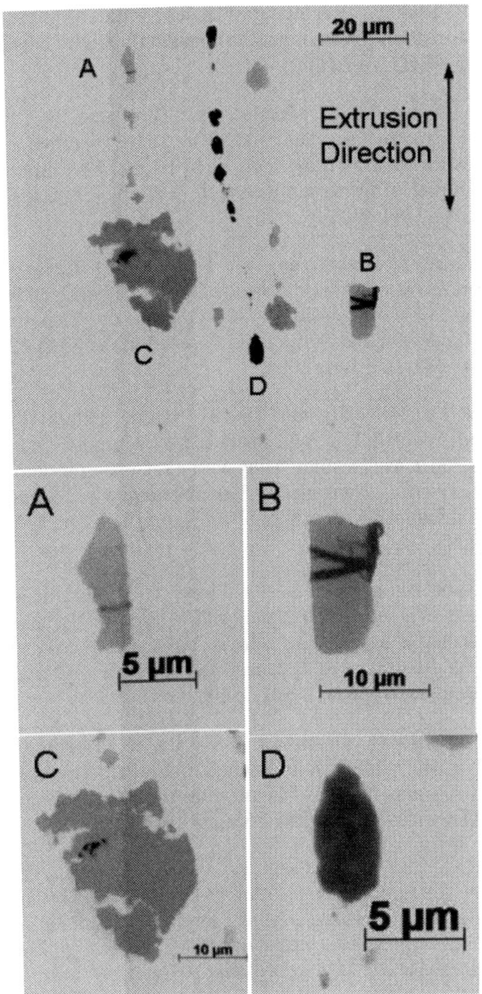

Figure 4. Particles observed at 1000X in an AZ61 magnesium alloy.

The evolution of microstructure in a material can be described by the stereological parameters. For instance; the area fraction and the number of cracked particles (number density) were measured and their evolution is shown in Figures 5 and 6. The number density is defined as the number of cracked particles divided by the total area of the field observed. The number density was determined based on the gray-colored particles only, since cracks in dark particles were difficult to observe using optical images. Figures 5 and 6 show the evolution of number density and number fraction, respectively, and their corresponding standard deviation as a function of strain.

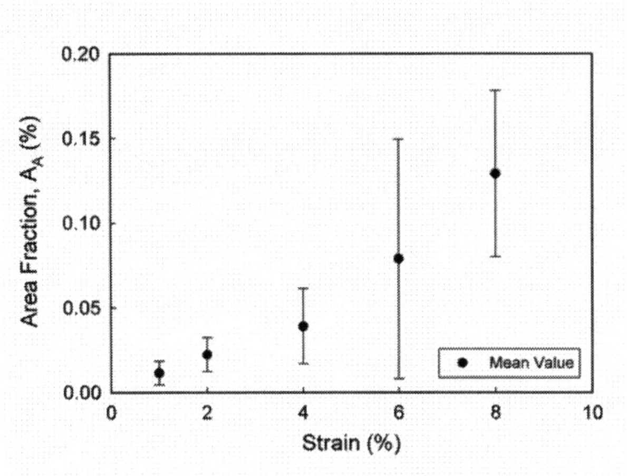

Figure 5. Area fraction for AZ61 magnesium alloy.

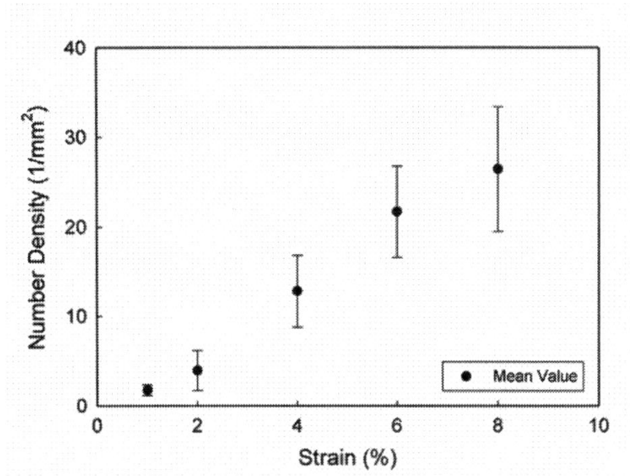

Figure 6. Number density for AZ61 magnesium alloy.

Discussion

Intermetallics in the AZ61 Magnesium Alloy

Hort et al. [14] have investigated the intermetallics that exist in magnesium alloys. They found that intermetallics and inclusions can adopt shapes like blocky, rosette-like, Chinese script, lamellar and needle-like among others. Other researchers have also identified these types of particles [2, 5, 14-19]. Some of these shapes were observed in this alloy, as shown in Figure 4.

In the AZ series alloys, these intermetallics are based on $Mg_{17}Al_{12}$, MnAl, $MnAl_{14}$, and $MnAl_{16}$ compounds [15]. Tartaglia and Grebetz [16] in their studies of intermetallic and inclusions in magnesium alloys found three types of particles. They found oxide/intermetallic, and Mn-Al, and Mg-Al phases. The oxide/intermetallic that they found had diverse morphologies ranging from small particles to a larger complex shapes. These oxide/intermetallics are dark colored.

Regarding the color and shape of particles observed in Figure 4, three type of intermetallics can be distinguished in the AZ61 alloy: oxide/intermetallic (D), Mn-Al, and Mg-Al phases (A, B, and C). Chemical analysis was conducted to determine the composition of these particles. This analysis showed that the gray-colored particles were based on Mn-Al and Mg-Al phases, and the dark particles were oxide/intermetallics. Since the interest here was on stereological parameters, we deemed it sufficient to quantify distinguishable cracked particles. Therefore, the particles were classified into two groups (gray-colored and dark particles), and the gray-colored particles were used to quantify cracking as a function of strain while the dark particles were neglected. In addition, the area fraction of dark particles was found to be 0.038 %.

Damage Evolution

Figure 7 shows the number of cracked particles per volume against applied tensile strain. Damage in this type of alloy is due mainly to slip and twinning as mentioned previously. However, Figure 7 shows a strong correlation between progression of cracking particles and damage. The constants used in the void nucleation model (Equation 1) are: $a=0$, $b=55000$, $c=50000$, $C=1.9$, $f=0.004$, $d=2.4$ μm, and $K_{IC}=20$ MPa·m$^{1/2}$. Constants C, a, b, and c were determined from the interrupted tension tests using a best fit method. A good correlation between the predicted nucleation model and experimental results was observed. The nucleation model predicts an exponential increase in the number density with increasing strain. For tensile loading, the number density estimated with a 2D technique is approximately equal to the number using a 3D methodology [26].

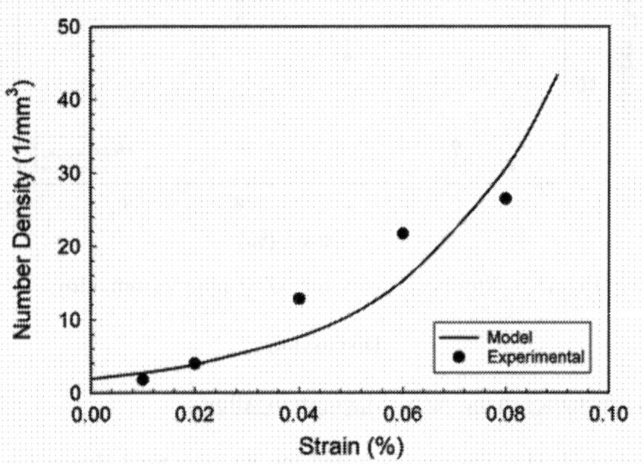

Figure 7. Number density against strain in AZ61 magnesium alloy.

Conclusions

The damage evolution process of magnesium AZ61 alloy under monotonic tensile loading conditions has been investigated. Using stereological methods and digital image analysis, a quantitative microstructural analysis was conducted. With regards to the microstructural properties relating to damage, the following conclusions can be made:

a) Under tensile loading conditions the area fraction of cracked particles composed of (Mn-Al) and (Mg-Al) varies in an exponential pattern as function of strain.
b) The number density (1/mm^2) of cracked particle follows a near linear pattern.
c) Three types of particles were identified: Oxide/intermetallic and Mn-Al, and Mg-Al phases.
d) Cracks were seen on the (Mn-Al) and (Mg-Al) particles.
e) The evolution of damage was modeled using the Horstemeyer and Gokhale nucleation model and good agreement were seen between experimental results and the model.

References

[1] T. Ruden, "Replacing Plastics with Magnesium for Structural Automotive Components", Light Metals Age, pp. 36-41, April 2005.

[2] S. Müeller, K. Müeller, H. Tao, and W. Reimers, "Microstructure and Mechanical Properties of the Extruded Mg-Alloys AZ31, AZ61, AZ80", International Journal of Materials Research, Volume 97, Issue 10, pp. 1384-1391, 2006.

[3] Y. Lu, M. Gharghouri, and F. Taheri, "Effect of Texture on Acoustic Emission Produced by Slip and Twinning in AZ31B Magnesium Alloy-Part I: Experimental", Nondestructive Testing and Evaluation, Volume 23, No. 2, pp. 141-161, June 2008.

[4] Y. Lu, M. Gharghouri, and F. Taheri, "Effect of Texture on Acoustic Emission Produced by Slip and Twinning in AZ31B Magnesium Alloy-Part II: Clustering and Neural Network Analysis", Nondestructive Testing and Evaluation, Volume 23, No. 3, pp. 211-228, September 2008.

[5] A. N. Chamos, Sp. G. Pantelakis, G.N. Haidemenopolud, and Kamoutsi, "Tensile and Fatigue Behaviour of Wrought Magnesium alloys AZ31 and AZ61", *Fatigue and Fracture of Engineering Materials and Structures*, Volume 31, pp. 812-821, 2008.

[6] T. Sasaki, H. Somekawa, A. Takara, Y. Nishikawa, and K. Higashi, "Influence of Second Phase Particles on Fracture Toughness in AZ31 Magnesium Alloys", Key Engineering Materials, Volume 261-263-, pp. 369-374, 2004.

[7] H. T. Zhou, X. Q. Zeng, L. L. Liu, J. Dong, Q. D. Wang, W. J. Ding, and Y. P. Zhu, "Microstructural Evolution of AZ61 Magnesium Alloy During Hot Deformation", *Materials Science and Technology, Institute of Materials, Minerals and Mining*, Volume 20, pp. 1397-1402, 2004.

[8] H. Yang, S. Yin, C. Huang, Z. Zhang, S. Wu, and S. Li, "EBSD Study on Deformation Twining in AZ31 Magnesium Alloy During Quasi-in-Situ Compression", *Advanced Engineering materials*, Volume 10, No. 10, pp. 955-960.

[9] M. R. Barnett, "Twinning and the Ductility of Magnesium Alloys Part I: Tension Twins", *Materials Science and Engineering A*, Volume 464, pp. 1-7, 2007.

[10] M. R. Barnett, "Twinning and the Ductility of Magnesium Alloys Part II: Contraction Twins", *Materials Science and Engineering A*, Volume 464, pp. 8-16, 2007.

[11] Y. Yeh, and W. Liu, "The Cracking Mechanism of Silicon Particles in an A537 Aluminum Alloy", *Metallurgical and*

Materials Transactions A, Volume 27A, Issue 11, pp. 3558-3568, 1996.

[12] A. Balasundaram, A. M. Gokhale, S. Graham, and M. F. Horstemeyer, "Three-dimensional Particle Cracking Damage Development in an Al-Mg-base Wrought Alloy", *Materials Science and Engineering A*, Volume 355, pp. 368-383, 2003J.

[13] I. J. Beyerlein, L. Capolungo, P. E. Marshall, R. J. McCabe, and C. N. Tome, "Statistical Analyses of Deformation Twinning in Magnesium", *Philosophical Magazine*, Volume 90, No. 16, pp. 2161-2190, 2010.

[14] N. Hort, Y. Huang, and K. U. Kainer, "Intermetallics in Magnesium Alloys", *Advanced Engineering Materials*, Volume 8, No. 4, 2006.

[15] ASM Speciality Hand book, "Magnesium and Magnesium Alloys", The Materials Information Society, 1999.

[16] J. M. Tartaglia, and J. C. Grebetz "Observations of Intermetallic Particle and Inclusion Distributions in Magnesium Alloys", *Magnesium Technology 2000, The Minerals, Metals and Materials Science Society*, pp 113-121, 2000.

[17] M. Marya, L. G. Hector, R. Verma, and W. Tong, "Microstructural Effects of AZ31 Magnesium Alloy on its Tensile Deformation and Failure Behaviors", *Materials Science and Engineering A*, Volume 418, pp. 341-356, 2006.

[18] R. Zeng, E. Han, W. Ke, W. Dietzel, K. U. Kainer, and A. Atrens, "Influence of the Microstructure on Tensile Properties and Fatigue Crack Growth in Extruded Magnesium Alloy AM60", *International Journal of Fatigue*, Volume 32, pp. 411-419, 2010.

[19] A. Srinivasan, J. Swaminathan, M. K. Gunjan, U. T. S. Pillai, and B. C. Pai, "Effect of Intermetallic Phases on the Creep Behavior of AZ61 Magnesium Alloy", *Materials Science and Engineering A*, Volume 527, pp. 1395-1403, 2010.

[20] W. M. Garrison and N. R. Moody, "Ductile Fracture", *The Journal of Physics and Chemistry of Solids*, Volume 48, No. 11, pp. 1035-1074, 1987.

[21] M. F. Horstemeyer, J. Lathrop, A.M. Gokhale, M. Dighe, "Modeling Stress State Dependent Damage Evolution in a Cast Al-Si-Mg Aluminum Alloy", *Theoretical and Applied Fracture Mechanics*, Volume 33, pp. 31-47, 2000.

[22] M. F. Horstemeyer, and A. M. Gokhale, "A Void-crack Nucleation Model for Ductile Metals", *International Journal of Solids and Structures,* Volume 36, pp. 5029-5055, 1999.

[23] J. Harris, "Particle Cracking Damage Evolution in 7075 Wrought Aluminum Alloy under Monotonic and Cyclic Loading Conditions", Master Thesis, Georgia Institute of Technology, 2005.

[24] G. V. Voort, "Using Microstructural Analaysis to Solve Practical problems", Tech-Notes, Published by Buehler, Volume 4, Issue 2.

[25] ASM International, Practical Guide to Image Analysis, 2000.

[26] H. Agarwal, A. M. Gokhale, M.F. Horstemeyer, and S. Graham, "Quantitative Characterization of Three-dimensional Damage as a Function of Compressive Strains in Al 6061", *Proceedings of the TMS Annual meeting, Aluminum Automotive and Joining Symposia*, Aluminum 2001, pp. 43-50, 2001.

Magnesium Technology 2011
Edited by: Wim H. Sillekens, Sean R. Agnew, Neale R. Neelameggham, and Suveen N. Mathaudhu
TMS (The Minerals, Metals & Materials Society), 2011

FORMABILITY ENHANCEMENT IN HOT EXTRUDED MAGNESIUM ALLOYS

Raja K. Mishra[1], Anil K Gupta[2], Rajiv Sikand[3], Anil K. Sachdev[1], Li Jin[4]
[1] General Motors Global R&D Center, Warren, MI 48090, USA
[2] Advanced Materials and Processes Research Institute, Bhopal, India
[3] National Physical Laboratory, Dr. K.S. Krishnan Marg, New Delhi, India
[4] Shanghai Jiao Tong University, Shanghai, China

Keywords: Magnesium alloy, Hot extrusion, Texture, Microstructure

Abstract

The effects of cerium (Ce), aluminum (Al) and manganese (Mn) additions on the microstructure and mechanical properties of hot extruded magnesium alloys have been investigated. Seven different compositions of Mg alloys - Pure Mg, Mg-0.2Ce, Mg-0.5Ce, AM30, AM50, Mg-3Al-0.2Ce and Mg-5Al-0.2Ce were hot extruded under optimized process conditions. Minor addition of Ce to Mg was found to enhance its ductility from <10% to >30% and 3 - 5% Al addition resulted in ~40% and 50% increase in strength respectively, compared to pure Mg. A combination of 5% Al and 0.2% Ce addition resulted in concurrent improvement in strength and ductility as a result of concurrent grain size refinement, texture randomization and solute strengthening.

Introduction

Magnesium (Mg) is emerging as a potential material for automotive applications, primarily because of its light-weight and cost competitiveness. When alloyed, Mg has higher strength-to-weight ratio compared to other structural metals and alloys. Mg alloys possess good damping characteristics and machinability. However, limited ductility and poor room temperature formability are major obstacles for wider use of wrought Mg. Higher temperature is required to allow activation of prismatic {1010}<11 2 0> and pyramidal {1011}<11 2 0> slip systems, in addition to basal (0001)<11 2 0> slip systems [1-2] to enable Mg alloys to be formed to any shape. The demand for lightweight, energy efficient materials provides a strong impetus to develop magnesium alloys with high strength and improved ductility, formable at room temperature.

Ductility in Mg can be increased by several methods. Chapman et al [3] have used grain refinement to improve ductility and strength. Menzen et al [4] in 1940 showed that alloying of Mg alloys with Ca, Th, Mn and certain rare earth (RE) elements improved cold rolling behavior at ambient temperature [5]. More recently, Mohri et al [6] have reported a combination of high strength and high ductility in hot extruded Mg-Y-RE (rare earth) alloy. Mukai et al [7] have found that grain refinement in WE43 Mg alloy results in enhanced ductility even at a dynamic strain rate of $\sim 2\times10^{3}$ s^{-1}. Barnett et al [8] have studied cold rolling behavior of pure Mg, AZ31 and Mg-0.2% Ce alloys and reported that, unlike pure Mg and AZ31 alloys, Mg-0.2% Ce alloy could be cold rolled to 90% reduction without failure. This enhanced cold rollability has been ascribed to modification of texture sharpness and shear banding. Ma et al [9] have reported grain refinement in extruded ZK60 alloy where higher RE content leads to increased tensile strength but reduced ductility. Mukai et al [10] have reported grain size refinement and texture modification in AZ31 by ECAP method with enhanced room temperature ductility.

There still continues to be a lack of data on the effect of processing parameters on rare earth alloyed Mg produced by single-pass extrusions. In the first part of this paper, the results of a systematic study of the effect of small Ce addition on the strength and ductility of Mg extruded at different temperatures and at different extrusion ratios are presented. The second part of the paper presents the effect of individual and combined addition of Ce and Al on hot extruded rods under optimized extrusion conditions to identify an alloy with combined high strength and high ductility. Microstructure and texture studies to establish the underlying mechanisms for the enhancement of strength and ductility are reported.

Experimental

The chemical composition of the billets extruded in this work is shown in Table 1. Hot extrusion experiments were conducted using a 500-ton Wellman Enefco vertical hydraulic press from 75 mm x 200 mm size billets. The effects of extrusion ratio (ER) and billet temperature on microstructure and properties of the extruded rods were investigated and optimized for Mg-0.2Ce billets for a constant extrusion speed of 10 mm (billet)/s, which is within the acceptable range for magnesium extrusions [11]. Extrusion ratios (ER) of 9:1, 25:1 and 36:1 and billet temperatures of 350°C, 400°C and 450°C were used. The effect of alloy chemistry was investigated employing the optimized processing parameters listed in Table 2 on the rest of the billets.

Table 1. Composition of billets.

Material	Composition (wt%)				
	Ce	Al	Mn	Zn	Mg
(a) Pure Mg	--	--	--	--	~100
(b) Mg-0.2Ce	0.20	0.009	0.027	<0.001	99.69
(c) Mg-0.5Ce	0.60	0.011	0.029	<0.001	99.63
(d) AM30	<0.001	2.45	0.42	<0.001	96.89
(e) AM50	<0.001	4.64	0.32	<0.001	94.74
(f) Mg-3Al-.2 Ce	0.17	2.440	0.150	0.003	97.23
(g) Mg-5Al-.2 Ce	0.18	4.490	0.220	0.001	95.10

The metallographic samples for porosity, grain size and second phase studies were prepared by etching polished samples of longitudinal and transverse sections of extruded rods using a solution containing 20 ml acetic acid - 6 gm picric acid - 20 ml H_2O - 50 ml ethanol. Microstructural evaluation as well as fractography was carried out using a Nikon Optical Microscope interfaced with a Leco Image Analyzer and a Leo 1450 scanning electron microscope (SEM) operating at 20 kV.

Table 2. Optimized hot extrusion processing parameters.

Container dia (mm)	75
Billet temperature (°C)	400
Soaking time (hrs)	2
Extrusion ratio (ER)	25:1
Pressing speed (mm/s)	10
Lubricant	Boron Nitride

Vickers Microhardness was measured using a Future Tech Microhardness Tester employing a load of 300g. Dumbbell shaped tensile specimens with 25 mm gauge length and 6.25 mm gauge diameter were used for tensile testing at room temperature at a strain rate of 0.66×10^{-3}/s using a Universal Testing Machine. Only the ultimate tensile strength (UTS) and % elongation were measured to assess and optimize strength and ductility values. Electron backscatter diffraction (EBSD) analysis of the samples were done using TSL camera and software in conjunction with a Leo 1455 SEM operating at an accelerating voltage of 20kV and a camera length of 18mm.

Results

The average volume fraction of pores in the cast Mg-0.2Ce billet is measured to be 2.48%, which drops to a level of 0.54% for the rods extruded at 450^{0}C at ER of 9:1. At ERs of 25:1 and 36:1 the porosity content is found to be negligible (~0.02%). Optical micrographs of Mg-0.2Ce rods extruded at 450^{0}C in Figure 1 show that the grains are nearly equiaxed in the transverse section for all three ERs. Few elongated bands of fine grains are present in the longitudinal section. The fraction of fine-grained bands is higher at higher ER. Figure 2 shows the EBSD grain maps and texture maps from the transverse section. It is clear that the texture of the extruded rod depends on the extrusion ratio. The rod extruded at ER of 25:1 has the c-axes of the grains at ~45^0 to the extrusion axis while that at 9:1 and 36:1 have their c-axes nearly normal to the extrusion axis [12]. Average grain size and mechanical properties are shown in Figure 3. The grain size in Figure 3 progressively increases with ER in both the transverse and longitudinal directions. Maximum ductility of ~28% with an optimum strength of 171 MPa is obtained at ER=25:1. Extrusion ratio of 25:1 is chosen as the optimum ER for the rest of the study from this data.

Figure 4 shows optical micrographs of Mg-0.2Ce alloy, extruded at billet temperatures of 350, 400 and 450°C at an extrusion ratio of 25:1. Both longitudinal and transverse directions exhibit equiaxed grains at all extrusion temperatures and the grain size increases with temperature. Figure 5 shows the mechanical properties (UTS and % elongation) of extruded Mg-0.2Ce alloy at different billet temperatures. The average grain size in the longitudinal direction at the three temperatures of 350, 400 and 450°C is 26, 34, and 47 µm, respectively, consistent with data in the literature [11, 13]. It is observed that while the UTS does not change with extrusion temperature, an optimum combination of strength and ductility is obtained at a billet temperature of 400°C. This temperature is considered to be optimum for subsequent extrusion experiments.

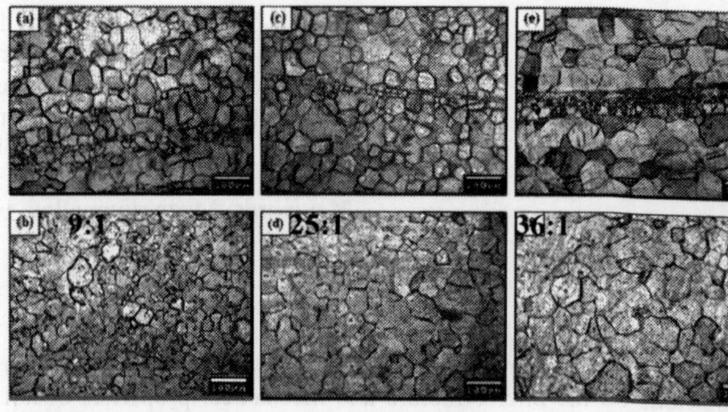

Figure 1. Optical micrographs from longitudinal (Top row, ED horozontal) and transverse (bottom, ED normal to image plane) sections of Mg-0.2Ce rods extruded at 450^{0}C.

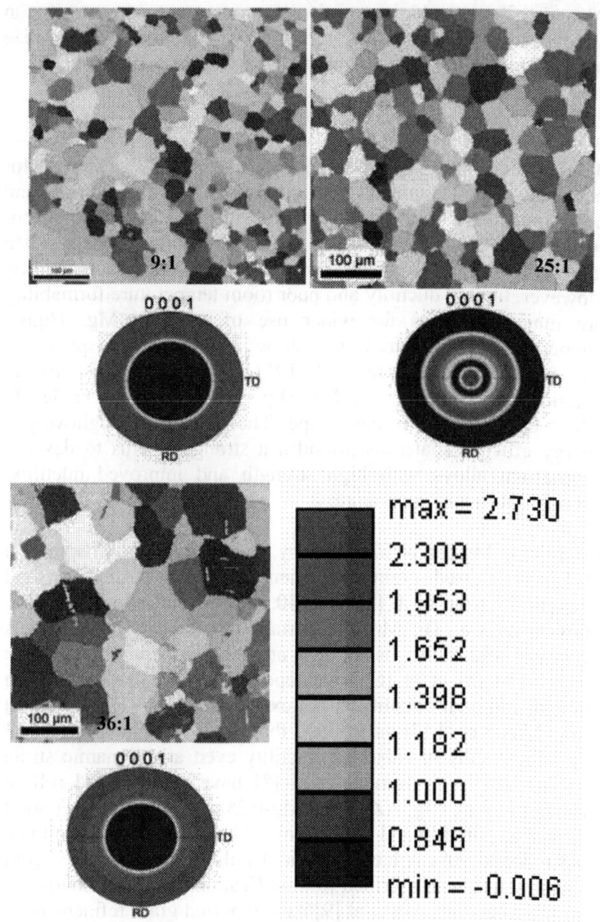

Figure 2. EBSD inverse pole figure maps and corresponding (0001) axial (symmetry) texture plots from transverse sections of the rods.

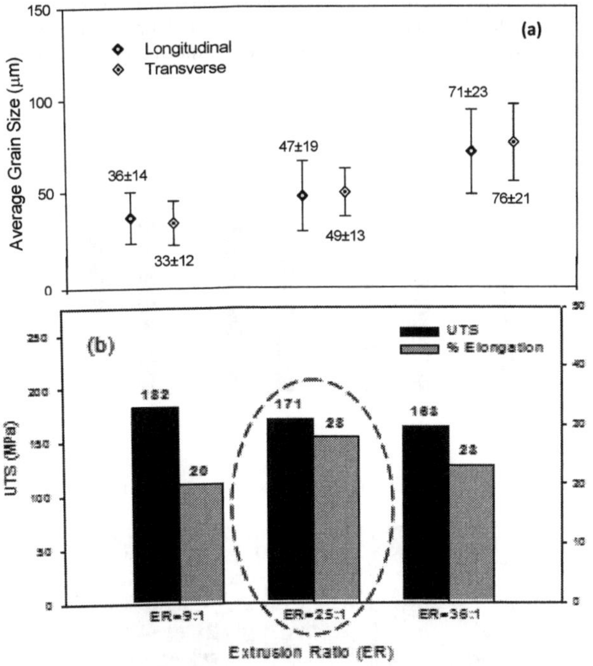

Figure 3. Effect of extrusion ratio (ER) on (a) average grain size, (b) tensile properties, of Mg-0.2Ce alloy [billet temperature: 450°C, pressing speed: 10 mm/s].

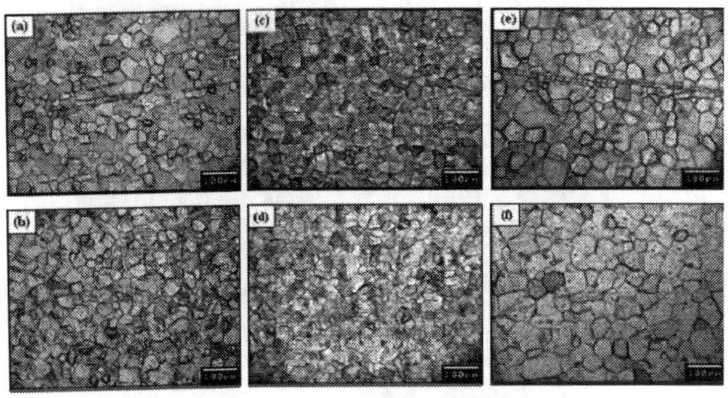

Figure 4. Microstructure of extruded Mg-0.2Ce alloy showing the effect of billet temperature (350 let, 400 center and 450 right) in longitudinal (top) and transverse (bottom) direction (ER = 25;1).

Figure 6 shows optical micrographs from longitudinal section of rods of different Mg alloys listed in Table 1. The average grain size and Vickers microhardness of each of the alloy is also shown in the inserted table. The maximum refinement of microstructure is obtained for Mg alloyed with Al and 0.2% Ce. Microhardness value increases marginally with Ce addition but considerably with Al addition. SEM micrographs taken in back scattered mode and their corresponding EDS patterns show the microstructure of pure Mg to be single-phase α-Mg. Presence of Mg-Ce intermetallics in Mg-Ce alloys and Mg-Ce-Al or Mg-Al-Mn intermetallics in Mg-Ce-Al and AM series alloys are seen. The fraction and size of intermetallics increased with alloying content. The EBSD texture data from the longitudinal sections of the rods for Mg, Mg-0.2Ce, Mg-0.5Ce and Mg-3Al-0.2Ce, shown in Figure 7, confirm that the c-axes in Mg-0.2Ce and Mg-0.5Ce are oriented at an angle to the extrusion axis while those in pure Mg and Mg-Al-Ce are nearly normal to the extrusion axis.

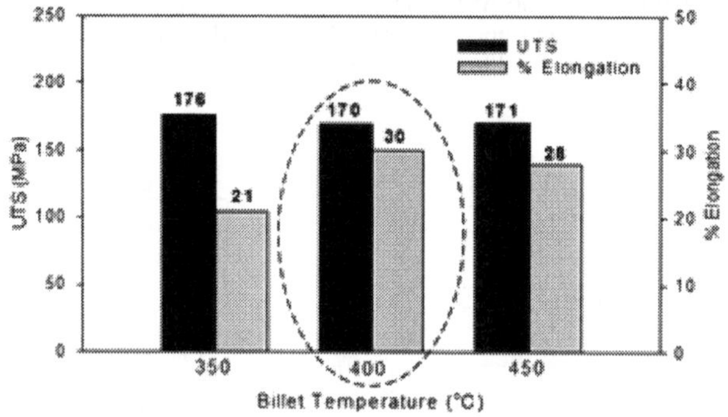

Figure 5. Effect of billet temperature on tensile properties of Mg-0.2Ce alloy [ER=25:1, pressing speed: 10 mm/s].

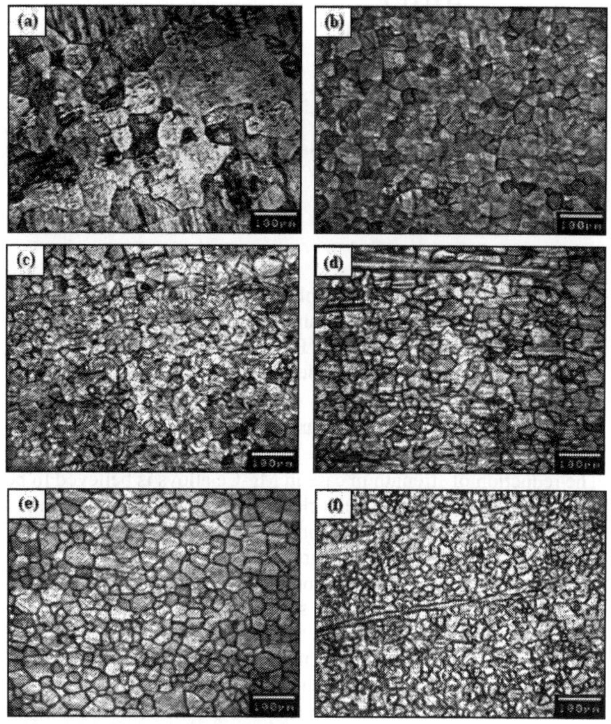

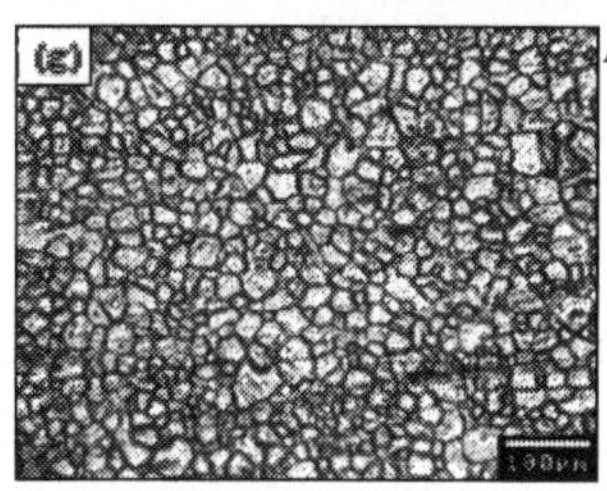

Alloy	Grain Size (mm)	Microhardness HV300
a	61	32
b	34	35
c	25	39
d	30	47
e	31	51
f	27	50
g	26	54

Figure 6. Optical metallographs from seven extrusions with their microhardness and grain sizes in the insert table. Alloy compositions in Table 1.

Figure 8 shows the UTS and elongation data for all seven extrusions. The data can be grouped into four groups. The maximum strength of 286 MPa was observed for Mg-5Al-0.2Ce alloy. It can be concluded that pure Mg has low strength and low ductility, Mg-0.2Ce and Mg-0.5Ce have high ductility but poor strength; AM30 and AM50 fall under the category of low ductility with high strength and Mg-3Al-0.2Ce and Mg-5Al-0.2Ce alloys are high strength and high ductility alloys.

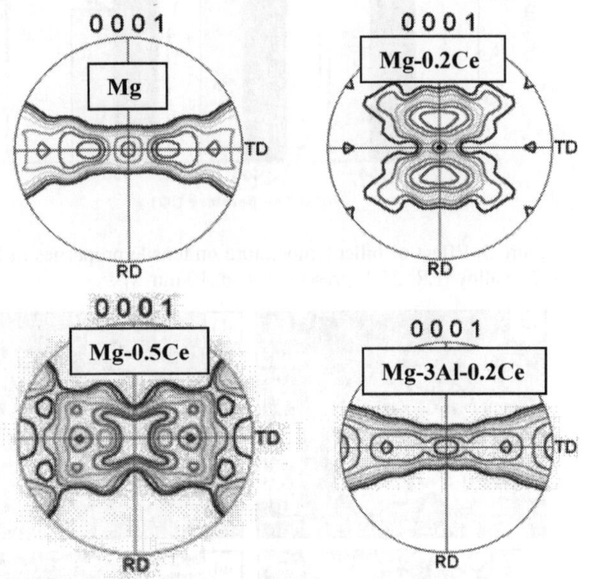

Figure 7. EBSD (0001) texture plots from longitudinal sections of pure Mg, Mg-.2Ce, Mg-.5Ce and Mg-3Al-.2Ce extrusions. RD parallel to extrusion direction. Pole figures are calculated with orthotropic symmetry for this orientation of the sample

Discussion

The reduction of strength in case in Mg-Ce alloys is believed to be due to "texture softening" seen clearly in Figure 7. Ce addition to Mg modifies the texture to favor basal slip (the Schmid factor of the grains oriented at 45 degrees to extrusion axis is high), enhancing ductility (% elongation). The increase in strength with Al addition is from solute strengthening and through the formation of dispersions of Al-Ce-Mn intermetallics [14]. Equiaxed grains in samples up to ER 25:1 form due to dynamic recrystallization. Bands of extremely fine grains observed at ER 36:1 form mainly due to non-uniform dynamic recrystallization. The increase in average grain size with ER is opposite of previously reported results [15], a result of grain growth due to high extrusion temperature employed in the present investigation (450°C) and the additional adiabatic heating for higher ERs. Addition of Al and Mn in AM30 and AM50 alloys led to substantial grain refinement while only minor addition of Ce (0.2%) resulted in considerable grain refinement. The grain size further decreased with increasing Ce to 0.5%. Minimum grain size of 25±9 μm was observed in Mg-0.5Ce and nearly comparable grain sizes were observed with alloying additions of Al and Ce. In all these cases, substantial number of intermetallic particles are present. The addition of Ce and Al as alloying elements seems to have provided additional nucleation sites for dynamic recrystallization and stronger pinning sites for grain boundaries to prevent grain growth. Increasing Al content from 3% to 5% and keeping Ce addition to 0.2% didn't have much effect on the grain size, most likely because all the Ce was tied up with 3 % Al already and thus the number of these intermetallics did not increase as a result.

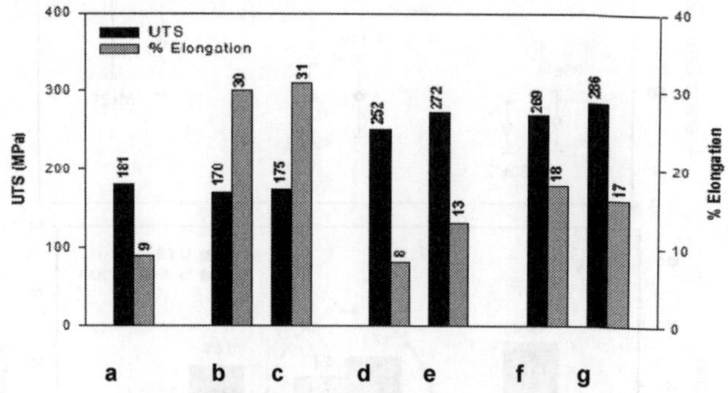

Figure 8. UTS and Elongation values of extruded rods

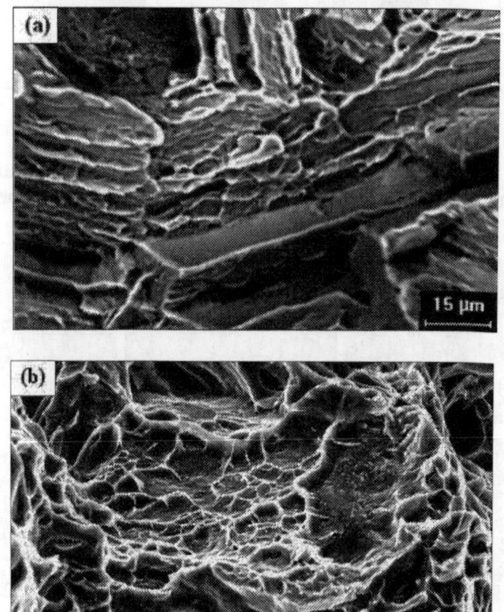

Figure 9. SEM images of fractured tensile bars from optimized (a) pure Mg, (b) Mg-0.5Ce and (c) Mg-3Al-0.2Ce rods.

The ductility of Mg-0.2Ce alloy is ~30%, which, in absolute terms, amounts to an enhancement of ~ 233% over that of pure Mg with a marginal decrease in the UTS value. It is believed that texture sharpness and severity of shear banding are responsible for the enhanced ductility of hot extruded Mg alloys with minor addition of Ce [12]. Though Al serves as solid solution strengthening element as observed in the case of AM30 and AM50 alloys, strengthening effect is more prominent in Mg-Al-Ce alloys due to the combined effect of solid solution effect, formation of complex Al-Ce-Mn intermetallics as well as the accompanying changes in grain size and texture.

The data suggests that addition of Ce to Mg-Al alloys does not contribute to ductility enhancement because Ce is tied up by Al-Ce intermetallics and is not available to alter deformation and recrystallization behavior that ultimately contributes to texture modification. In the absence of Al, there is a finite amount of Ce still dissolved in Mg that alters the deformation and recrystallization behavior. In a separate paper in this Proceeding [16], the influence of dynamic strain ageing in Mg-Ce alloys at high temperature and specific strain rates has been shown to influence the RE texture formation and this is believed to be behind the presence of RE texture in the sample extruded at 25:1 extrusion ratio. Since dynamic strain ageing occurs in a specific temperature/strain rate window, the observations strongly suggest this to be the governing physical mechanism for the texture modification in extruded Mg-Ce alloys.

Substantial improvement in ductility without much change in strength has been obtained with minor Ce addition to Mg. Mg-Al-Ce alloys have high strength coupled with high ductility. The fractured tensile specimens of Mg and Mg-0.5Ce were examined under SEM to correlate the fracture behavior with tensile properties [12]. SEM micrograph of the fractured surfaces of Pure Mg in Figure 9 shows quasi-cleavage fracture mode, while Mg-0.5Ce and Mg-Al-Ce alloy shows nearly spherical dimples from ductile failure. This fracture behavior contributes to improved ductility of Mg-0.5Ce and Mg-Al-Ce alloy.

Conclusions

Minor addition of cerium (0.2% Ce) to magnesium resulted in texture softening and grain refinement under specific extrusion conditions. Increasing Ce content from 0.2% to 0.5% did not improve ductility further. SEM and EDAX investigations revealed the presence of uniform distribution of fine intermetallic particles. Fractured samples of Mg-0.5Ce showed near spherical dimples. The enhanced ductility in extruded Mg-Ce alloys is attributed to less texture sharpness and grain size reduction with accompanying changes in the deformation mechanisms. Addition of 3% and 5% Al to Mg (AM30 and AM50) resulted in strength improvement without significant improvement in ductility. Addition of Al in combination with 0.2% Ce enhanced both strength as well as ductility. Extruded Mg-5Al-0.2Ce rods had the best strength and ductility combinations.

Acknowledgements

Funding from General Motors R&D Center is gratefully acknowledged. The authors are thankful to Mr. R. Khanna and Mr. R Shyam for assistance in the extrusion, Mr. V. Jain for optical metallography and Mr. R. Kubic for EBSD data collection.

References

[1] ASM Specialty Handbook. Magnesium and Magnesium alloys. Materials Park, Ohio: ASM International, (2000) 2.

[2] Yukihiro Oishi, Nozomu Kawabe, Akihito Hoshima, Youich Okazaki and Akira Kishimoto, SEI Technical Review, 56 (2003) 54 .

[3] J.A. Chapman and D.V. Wilson, J. Inst. Met, 91 (1962) 39.

[4] P. Menzen, in: A. Beck (Ed.) The Technology of Magnesium and its Alloys, F.A. Hughes and Co. Ltd., London, (1940) 34.

[5] S.L. Couling, J.F. Pashak, L. Strukey, Trans. ASM, 51 (1959) 94 .

[6] T. Mohri, M. Mabuchi, N. Saito and M. Nakamura, Mater. Sci. Eng., A 257 (1998) 287.

[7] T. Mukai, T. Mohri, M. Mabuchi, M. Nakamura, K. Ishikawa and K. Higashi, Scripta Mater, 39 (1998) 1249.

[8] M.R. Bernett, M.D. Nave and C.J. Bettles, Mater. Sci. Eng., A 386 (2004) 205.

[9] Chunjiang Ma, Manping Liu, Guohua Wu, Wenjiang Ding and Yanping Zhu, Mater. Sci. Eng., A349 (2003) 207.

[10] Toshiji Mukai, Masashi Yamanoi, Hiroyuki Watanabe, and Kenji Higashi, Scripta Materialia, 45 (2001) 89.

[11] Matthew R. Barnett, Dale Atwell, Chris Davies and Roman Schmidt, Proceedings of 2nd International Light Metals Technology Conference 2005.

[12] Raja K. Mishra, Anil K. Gupta, P. Rama Rao, Anil K. Sachdev, Arun Kumar and Alan Luo, Scripta Materialia, 59, (2008) 562.

[13] Mihriban O. Pekguleryuz, Magnesium Technology 2004, Alan A. Luo ed., TMS (The Minerals, Metals & Materials Society), (2004)

[14] K. Kubota, M. Mabuchi and K. Higashi, Journal of Materials Science, 34 (1999) 2255.

[15] Y. Uematsu, K. Tokaji, M. Kamakura, K. Uchida, H. Shibata and N. Bekku, Mater. Sci. Eng., A434 (2006) 131.

[16] L. Jiang, J. Jonas and R. Mishra, "Influence of Cerium Solute on the Deformation Behavior of an Mg-0.5wt% Ce Alloy", in this Proceeding.

Magnesium Technology 2011
Edited by: Wim H. Sillekens, Sean R. Agnew, Neale R. Neelameggham, and Suveen N. Mathaudhu
TMS (The Minerals, Metals & Materials Society), 2011

Deformation and Evolution of Microstructure and Texture during High Speed Heavy Rolling of AZ31 Magnesium Alloy Sheet

Tetsuo Sakai[1], Akinori Hashimoto[2], Go Hamada[2], Hiroshi Utsunomiya[2]□

[1]Center for Advanced Science and Innovation, Osaka University; Suita 565-0871, Japan
[2]Division of Materials and Manufacturing Science, Graduate School of Engineering, Osaka University; Suita 565-0871, Japan

Keywords: AZ31, Heavy rolling, High speed rolling, Rolling, Recrystallization, Texture evolution

Abstract

An AZ31 magnesium alloy sheet was rolled to 71% by one pass operation at 473K at a rolling speed of 500m/min. During rolling, the mill was suddenly stopped and the sheet was withdrawn from the gap of work rolls. The evolution of microstructure and texture of the AZ31 magnesium alloy sheet during rolling deformation was revealed by observing microstructure and texture at the plane perpendicular to the transverse direction from the entry to the exit of the zone of deformation of the withdrawn sheet. Grains with their orientation other than basal texture preferentially deform at the initial stage of rolling deformation. Dynamically recrystallized grains are observed in the deformation zone. The double peak texture developed during recrystallization.

Introduction

Magnesium alloy is the lightest metal among practical alloys and is helpful for weight saving of transportation vehicles and electronic equipments. In addition, it has excellent properties such as high-specific strength, high rigidity, good recycling potential, high damping capacity and high shielding efficiency of electromagnetic wave. Because of these properties, magnesium alloy is expected to be used widely.

Magnesium sheets are usually produced by rolling from cast slabs. In cold rolling, applicable reduction in thickness is less than 10% due to the low workability at room temperature. Therefore the thickness is mostly reduced in hot or warm rolling. In the process, multi-pass operation with small reduction accompanied by intermediate annealing is employed to suppress edge cracks or fracture of the material and to maintain the workability [1, 2]. Rolls are often heated to minimize the temperature drop. Because of these all the procedures, fabrication of magnesium alloy sheets is less productive and magnesium alloy sheets are more expensive than cast magnesium products as well as other metal sheets. If the productivity is improved, they should be used in quantity, especially in automotive industries.

It is necessary to provide sheets at a low price to apply magnesium alloy widely as structural materials. Rolling is the most appropriate process to mass-produce sheets at a low cost. However, as mentioned before, the efficiency of production is low and cost is high in rolling of magnesium alloy, due to its low ductility below 473K coming from its hexagonal crystal structure and inactiveness of non-basal slip systems.

The present authors succeeded in single pass large draught rolling of various magnesium alloy sheets below 473K without heating rolls by rolling at the speed higher than 1000m/min. In high speed rolling, heat transfer from a sheet to rolls was prevented due to the very short duration where the sheet and rolls were in contact, so the temperature of the sheet efficiently rose by plastic working and deformability even at cold or warm rolling region became comparable to hot rolling. Furthermore, this process was performed under high strain rate condition or high Zener-Hollomon parameter condition, so the sheet obtained by high speed rolling had fine grained microstructure of which the mean grain size was 2-3μm, and had good mechanical properties too [3-5]. For these advantages, the high speed rolling is a promising process to produce high-quality rolled magnesium alloy sheets at a low cost.

However, the mechanism that makes one pass large draught rolling possible and the evolution of microstructure and texture during high speed rolling are not fully understood. In this study, we tried to observe the evolution of microstructure and texture of the sheet during rolling by interrupting rolling and freezing the microstructure. We performed experimental high speed rolling of AZ31B magnesium alloy. During rolling, the mill was suddenly stopped and the sheet was withdrawn from the work rolls immediately. The evolution of microstructure and texture of the AZ31 magnesium alloy sheet during rolling deformation was revealed by observing microstructure and texture at the plane perpendicular to the transverse direction from the entry to the exit of the zone of deformation of the withdrawn sheet.

Experimental Procedures

A two-high laboratory rolling mill with φ530mm rolls shown in Fig.1 was used. The rolling speed can be changed from 200m/min to 2600m/min. Detailed specification can be found elsewhere [6, 7], where the authors studied microstructure and texture evolution of steels and other metals systematically.

Commercial 3.0mm thick AZ31B (Mg-3.2%Al-0.77%Zn-0.34%Mn) extruded sheets produced by Sankyo Material, Inc. were received. 30mm wide 300mm long specimens were cut from the sheets and subjected to the rolling experiment. The initial microstructure was covered by equiaxed grains of which mean

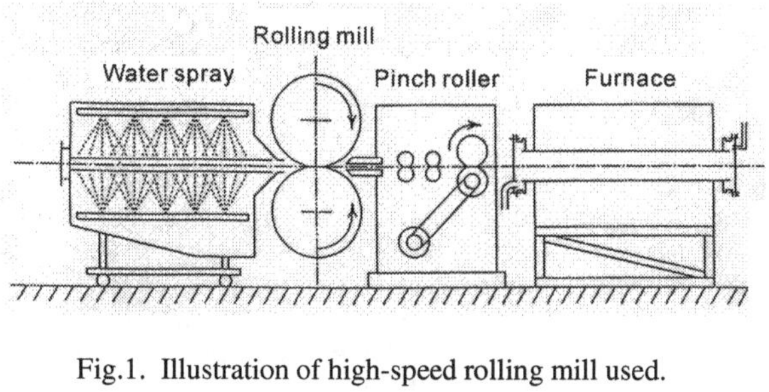

Fig.1. Illustration of high-speed rolling mill used.

grain size was 18μm. The grain size was inhomogeneously distributed ranging from 10 to 50μm.

Prior to the rolling, a specimen was held for 15 min at 473K in an electronic tube furnace, then supplied to the mill through the pinch roller. The rolling conducted in single-pass operation with reduction in thickness of 71%. The peripheral speed of the rolls was 500m/min. The estimated strain rate during 70% rolling is 4.9 $\times$ 10^2 s^{-1}. The rolls were neither heated nor lubricated. The rotating rolls were suddenly stopped during rolling and specimen was quenched immediately by cold (room temperature) rolls that are in contact with specimen. The rolled part of specimen was quenched by the water spray closely attached to the exit of the mill.

Microstructures on longitudinal section (TD plane) were observed by an optical microscope. The EBSP measurement by FE-SEM was performed also on TD plane.

Results

Partly Rolled Sheet

The shape of TD plane of the rolled and interrupted (partly rolled) sheet is shown in Fig.2. The length of contact arc is longer than calculated value (23.6mm) because of roll flattening. The areas at which microstructure and texture are observed on TD plane are indicated by small numbered squares. Reductions at several positions calculated from the thickness at that point are also given in Fig.2.

Change of Microstructure along Rolling Direction

Figure 3 shows the microstructure of the partly rolled sheet. At the middle stage of rolling deformation (a), a lot of deformation bands inclined about 45° to the rolling direction are introduced and initial crystal grains are divided to small grains. At the final stage of rolling deformation where reduction in thickness comes to 71%, the deformation microstructure prevails while fine recrystallized grains appear. Just after rolling, 12mm from the roll exit, recrystallization is completed and microstructure with fine equiaxed grains develops.

Change of Crystal Orientation along Rolling Direction

Figure 4 shows the change of IPF map of partly rolled sheet drawn by TSL-OIM software along rolling direction. The IPF maps shown here are drawn from the data (Kikuchi line) obtained from grains free from deformation microstructures. Deformation microstructures are indicated in black. These IPF maps are closely related to Fig.3. At the middle stage of rolling (a), initial grains are divided by deformation bands. Almost all the grains are oriented their 0001 plane parallel to the sheet plane in the initial stage. At the final stage of the rolling (b), initial grains are not observed and deformation microstructure prevails. A large number of fine recrystallized grains appear among deformation microstructure. These recrystallized grains nucleated during rolling because they are observed in the part of sheet just between the rolls. It suggests that dynamic recrystallization occurs during high speed large draught rolling. At the exit of the roll gap (c), the fraction recrystallized increases from (b), while deformation microstructure remains. Similar to the final stage of deformation, almost all the grains align their 0001 plane parallel to the sheet plane. After rolling, 12mm from the exit (d), recrystallization is completed. From these OIMs, for the formation of completely recrystallized equiaxed microstructure in the sheet heavily rolled

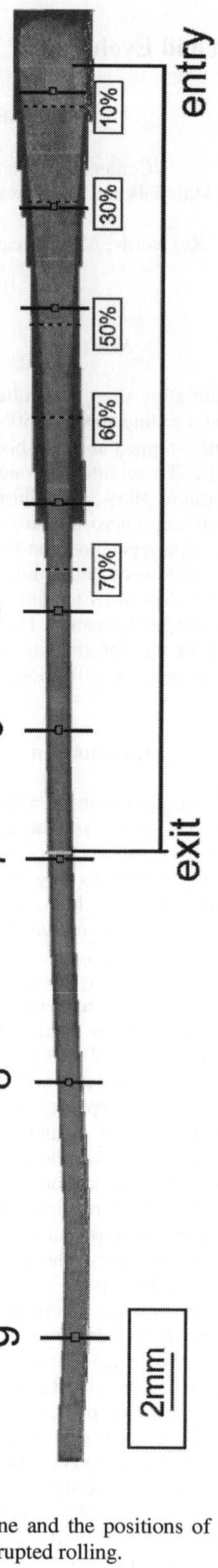

Fig.2. Shape of TD plane and the positions of observation of the sheet by interrupted rolling.

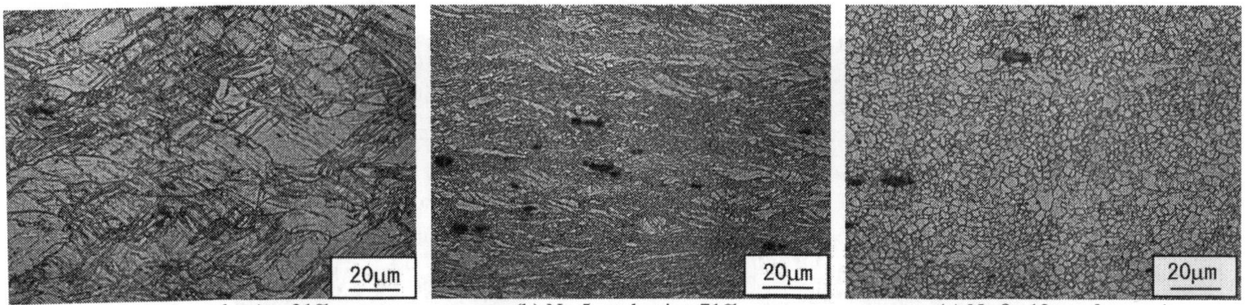

(a) No.2, reduction 31% (b) No.5, reduction 71% (c) No.8, 12mm from exit

Fig.3. Change of microstructure with the progress of rolling deformation.

by high speed rolling, contribution of static recrystallization cannot be disregarded.

After rolling, grains with orientations other than 0001 parallel to the rolling plane (colored in green or blue) that are not observed in the sheet at the roll gap appear here and there. The growth of grains other than basal orientation occurs during static recrystallization after rolling.

The change of pole figures drawn from the OIM data with the progress of rolling is shown in Fig.5. At the middle stage (a), 0001 peak shifts from ND to TD direction, which is the orientation resulting from extrusion texture originally formed in the specimen. At the final stage (b), 0001 peak splits to double peaks on RD axis originating from recrystallized grains. With the progress of recrystallization, the intensity of 0001 peak and the interval between two peaks increase. However, pole figures (b) and (c) only show the orientation of recrystallized grains. The

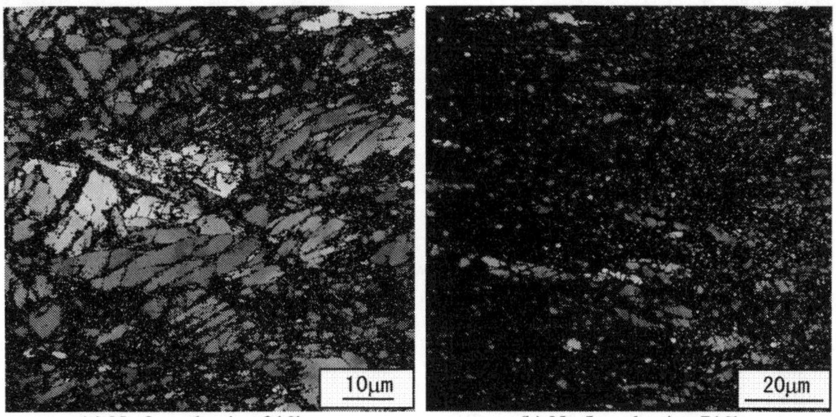

(a) No.2, reduction 31% (b) No.5, reduction 71%

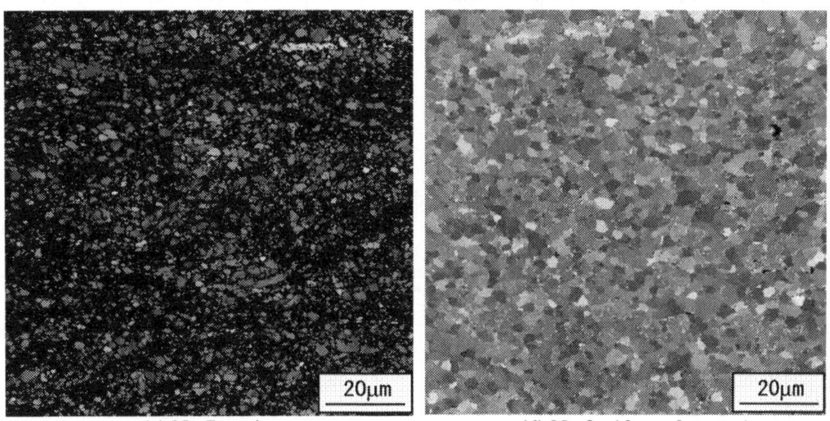

(c) No.7, exit (d) No.8, 12mm from exit

Fig.4. Change of IPF map with the progress of rolling deformation.

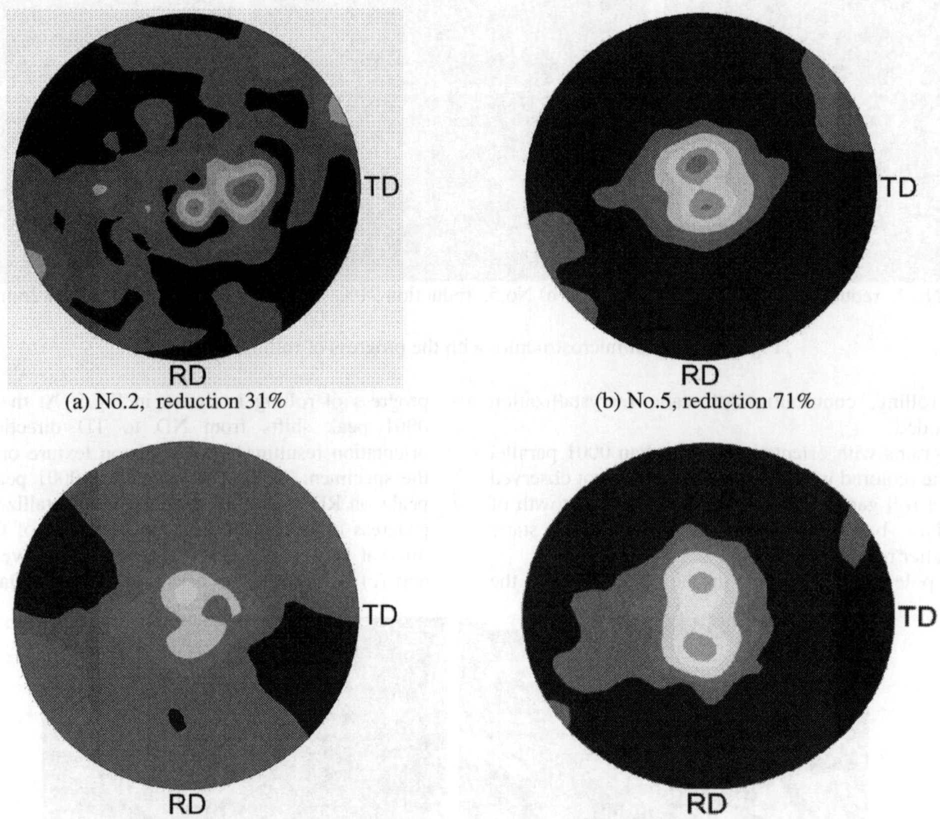

(a) No.2, reduction 31% (b) No.5, reduction 71%

(c) No.7, exit (d) No.8, 12mm from exit

Fig.5. Change of 0001 pole figure with the progress of rolling deformation.

orientation of deformed grains (colored black in IPFs) could not be obtained in the present study. The pole figures of heavily rolled sheets drawn by X-ray diffraction indicates that the orientation of deformed grains shows single 0001 peak elongated towards RD direction at ND position [4]. The texture with two 0001 peaks on RD axis, so called double peak texture, is the typical texture formed in magnesium alloy sheets recrystallized by high speed rolling to high reduction in one pass operation. These pole figures indicate that the static recrystallization just after rolling plays an important role in the formation of the texture of heavily rolled and recrystallized magnesium alloy sheets.

Conclusions

The evolution of microstructure and texture of the sheet during high speed large draught rolling was observed by sudden interruption of rolling operation followed by freezing of the microstructure. The results are summarized as follows.

1. The dynamically recrystallized grains are observed in the sheet deformed in the roll gap.
2. The static recrystallization proceeds in very short time after completion of rolling. The growth of grains with orientation other than near basal texture occurs during static recrystallization.
3. The recrystallized grains show the double peak texture with two 0001 peaks on RD axis and it develops after rolling.

References

1. C.S. Roberts, *Magnesium and Its Alloys*, (NJ, USA: John Wiley and Sons, Inc., 1960), 171-177.
2. E.F. Emley, *Principles of Magnesium Technology*, (Oxford, UK: Pergamon Press Ltd., 1966), 550-556.
3. T.Sakai, H.Utsunomiya, S.Minamiguchi, H.Koh: Magnesium Technology in the Global Age (2006), 205-215
4. S.Minamiguchi, H.Koh, H.Utsunomiya, T.Sakai: Magnesium Technology in the Global Age (2006), 217-227
5. Y. Watanabe, T. Sakai, H. Utsunomiya: Steel Research International Special Edition, Vol.1, 712-716 (2008)
6. K. Kato, Y. Saito, T. Sakai. "Investigation of Recovery and Recrystallization during Hot Rolling of Stainless Steels with High Speed Laboratory Mill", Transactions ISIJ, Vol. 24, 1984, 1050-1054.
7. T. Sakai, Y. Saito, K. Hirano, K. Kato, "Deformation and Recrystallization Behavior of Low Carbon Steel in High Speed Hot Rolling", Transactions ISIJ, Vol. 28, 1988, 1028-1035.

Magnesium Technology 2011
Edited by: Wim H. Sillekens, Sean R. Agnew, Neale R. Neelameggham, and Suveen N. Mathaudhu
TMS (The Minerals, Metals & Materials Society), 2011

FORMABILITY OF MAGNESIUM SHEET ZE10 AND AZ31 WITH RESPECT TO INITIAL TEXTURE

Lennart Stutz[1], Jan Bohlen[1], Dietmar Letzig[1], Karl Ulrich Kainer[1]
[1]GKSS-Forschungszentrum Geesthacht GmbH; Max-Planck-Straße 1; Geesthacht, D-21502, Germany

Keywords: Formability, magnesium sheet, forming limit, texture, dynamic recrystallisation

Abstract

The commercial application of magnesium alloy sheets is hindered by their low formability and therefore technological and economic constraints. Tailoring the texture has been identified as playing a major role in enhancing the formability of magnesium sheets. In this study, the formability of magnesium sheet ZE10 and AZ31 was investigated. While the texture of AZ31 is of unfavourable basal type, ZE10 shows a significantly different texture with the basal planes being randomly distributed. More basal planes are oriented favourably for basal slip, promising higher formability and thus, lower process temperature. Formability is assessed by means of forming limit diagrams as a representation of material response to various strain paths. Nakajima tests were carried out from room temperature to 250°C. Local strain data is correlated with the microstructural evolution. This study reveals a significant influence of texture on formability with ZE10 sheet showing superior formability compared to AZ31.

Introduction

Applying magnesium sheets in lightweight design bears a significant weight saving potential [1, 2]. Substitution of conventional materials, such as aluminium and steels, can help to reduce vehicle weight and thus, greenhouse gas emission and fuel consumption. Conventional sheets of alloys such as AZ31 show limited formability at low temperature which poses a serious obstacle to commercial applications. Distinct basal textures are formed during the rolling process and hamper the ability to strain hardening which is a prerequisite for improved formability [3]. For deformation in the thickness direction, the majority of basal planes are oriented unfavourably. To realise strain in thickness direction and thus in <c> direction of the majority of grains, twinning and <c+a> pyramidal slip are other options. While the former can only accommodate a low degree of plastic deformation, the latter is thermally not activated below 225°C in pure magnesium. Thus, forming components from magnesium sheets with typical basal type texture requires elevated process temperatures which results in high production costs.

Sheets such as ZE10 or ZW41 with weaker texture were reported to show generally superior formability [4-8] or competitive formability at lower process temperature. This study assesses the formability of two magnesium sheets with significantly varied texture, AZ31 representing a conventional sheet of typical basal type texture and ZE10 with a much more weak texture.

Material and Experimental Procedures

The forming behaviour of magnesium sheets (thickness 1.5mm) of two different alloys was investigated. Conventional AZ31-B (Mg-3Al-1Zn-Mn) sheet and sheet of alloy ZE10 (Mg-1Zn-Ce-La-Zr) were received in annealed condition. The initial microstructure shows a recrystallised microstructure consisting of equiaxed grains with an average grain size of 10 - 11 µm for both cases, see figures 1 and 2. It is notable that ZE10 shows a higher contrast between different grains compared to AZ31. This implies that the orientation of grains differs more in ZE10 than in AZ31.

Figure 1: Microstructure of AZ31 sheet

Figure 2: Microstructure of ZE10 sheet

Pole figures of the sheets were measured by a Panalytical X-ray diffractometer setup using Cu K_α radiation (40kV, 40mA). The resulting texture is presented as a recalculated (0002) pole figure from the orientation distribution function in figures 3 and 4. The measured distribution of basal planes differs significantly. AZ31 shows a typical texture of basal type where most grains are aligned parallel to the sheet plane. The angular spread of the orientation distribution is slightly broader to the rolling direction than to the transverse direction. This type of texture is typical for hot rolled magnesium sheets.

In contrary, the basal poles of ZE10 sheet are distributed more randomly. Furthermore, the maximum intensity is found at around 45° towards the transversal direction as a spread double peak. The overall distribution of basal planes is spread towards the transversal direction. Such a texture implies that more basal planes are oriented favourably for basal slip when loaded in sheet plane, promising higher formability.

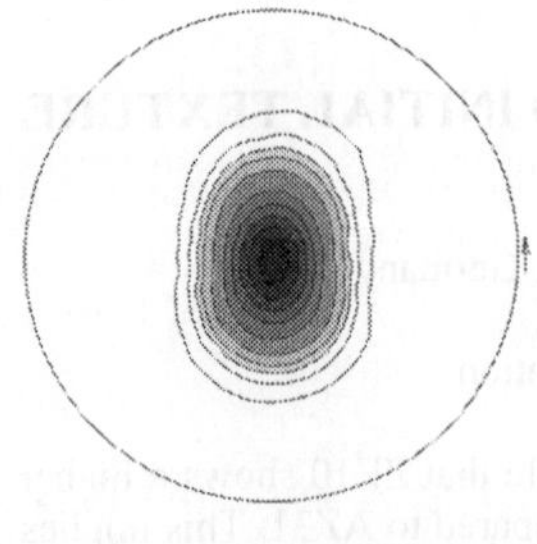

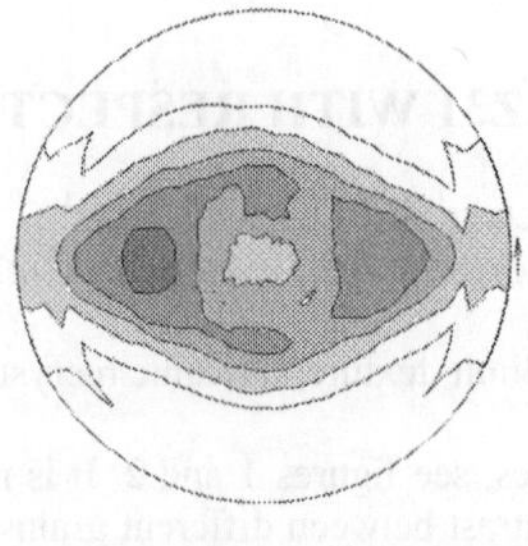

Figure 3: 0002 pole figure of AZ31 sheet, I_{max}= 14.6 m.r.d.

Figure 4: 0002 pole figure of ZE10 sheet, I_{max}= 3.6 m.r.d.

Nakajima type tests were conducted, using an Erichsen 145-60 sheet metal testing machine. A hemispherical punch deformed the sheets at a speed of 1mm/s until fracture. Punch displacement and punch force were recorded digitally at a rate of 5Hz. The blank holder force was 400kN. Oiled PTFE foil of 0.05mm thickness was used as lubricant. Hasek type specimens [9], manufactured by water jet cutting, covered various strain paths from biaxial stretching over plane strain to deep drawing, refer to figure 5. All specimens were cut along the rolling direction. For each geometry and testing temperature, three specimens were tested. Tests were conducted at room temperature, 150°C, 200°C and 250°C respectively. All involved components of the testing device were heated to isothermal condition before testing. The use of an optical deformation measuring system ARAMIS allowed in situ strain measurements. A statistical pattern was applied on the sheet surface and high resolution pictures were recorded digitally during the tests at a rate of 12Hz. From the calculated strain data, forming limit curves according to ISO 12004 as well as strain path plots, were derived in the forming limit diagram (FLD). The FLD, as introduced by Keeler [10] is a powerful tool to assess the formability of sheets. The major strain as highest local strain in sheet plane is plotted versus the perpendicular local minor strain. Figure 5 shows a schematic FLD, specific strain paths are highlighted. At the onset of necking, severe localisation of strain, the so called forming limit curve is plotted.

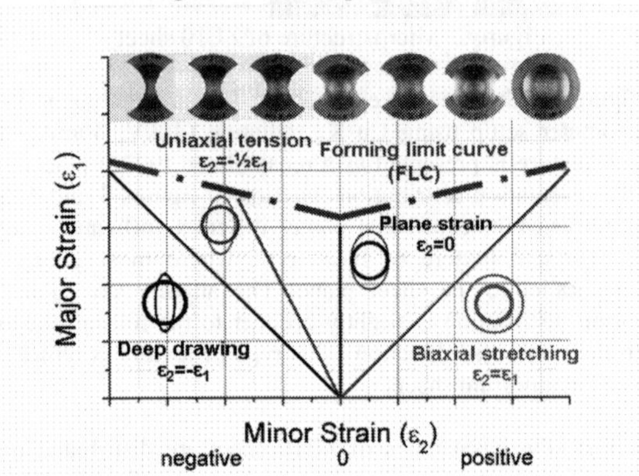

Figure 5: Schematic forming limit diagram. Hasek type specimens cover various strain paths.

Micrographs of the microstructure of deformed sheets were taken, using standard preparation techniques based on picric acid etching [11]. Samples were cut from the deformed specimen close to the crack location. The micro sections were grinded with SiC sandpaper and polished with SiO_2 suspension. Observation direction for all micrographs was the transversal direction (TD).

Results

The mechanical response of the sheets is plotted as punch force – punch displacement curves in figure 6 for AZ31 and 7 for ZE10. For both sheets, specimens are shown that represent the outermost edges of the FLD. While the specimens marked with squares are deformed by biaxial stretching and thus, show a positive minor strain, the specimens marked with triangles represent specimens with negative minor strain. As a general tendency the specimens with positive minor strain require higher punch forces compared to their counterparts with negative minor strain. This can easily be explained by the larger area of contact and thus, amount of deformed material. In the case of AZ31, specimens tested at room temperature and specimens with negative minor strain tested at 150°C show an abrupt drop in punch force upon fracture of the specimens. The specimen tested at room temperature fracture at comparable low forces. All other specimens show a curve that can be separated in four segments: after an initial increase in punch force, a linear relationship between punch force and punch displacement is observed. The slope of this linear relationship decreases with increasing temperature and is generally higher for specimens with positive minor strain (biaxial stretching). Following up, a plateau force is reached and maintained until a rather sudden drop in punch force. The plateau is associated with diffuse necking of the specimen while the latter drop in force is due to crack initiation. It is notable that while the maximum punch displacement, the limiting dome height, increases with temperature in the case of biaxial stretching, the maximum limiting dome height is reached at 200°C for specimens with negative minor strain followed by a decrease of maximum punch displacement at 250°C.

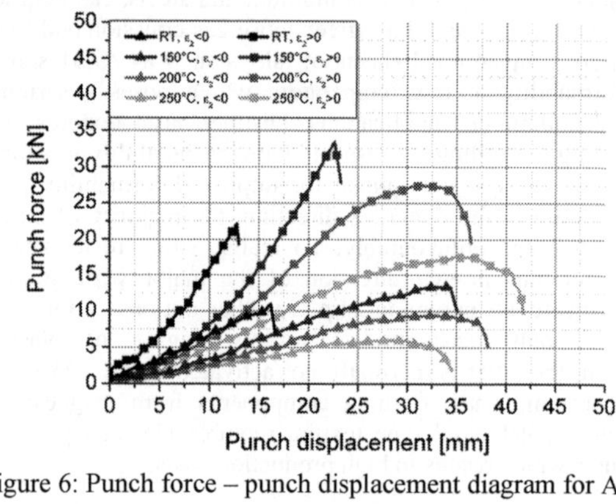

Figure 6: Punch force – punch displacement diagram for AZ31

In comparison, ZE10 shows simpler trends. Only specimen tested at room temperature show a brittle behaviour. With increasing temperature, the necessary punch force decreases and maximum punch displacement – the maximum dome height – increases. For biaxially stretched specimens, only slight differences in maximum dome height are achieved with increasing temperature from 150°C to 250°C. The limiting dome height of specimens with negative minor strain is always below their biaxially stretched counterparts. It is noteworthy that the maximum punch forces of biaxially stretched ZE10 are generally much higher than the punch forces of AZ31 specimens at the same temperature.

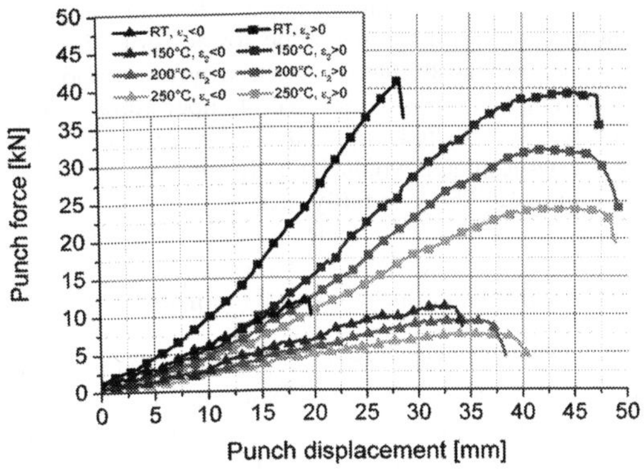

Figure 7: Punch force – punch displacement diagram for ZE10

By evaluation of the strain measurements as obtained by the ARAMIS system, the strain response of the material in terms of major strain vs. minor strain can be plotted in the FLD. Figures 8 and 9 show the strain response of both sheets until fracture of the specimens. For all tested specimens, there is an approximately linear relationship between minor and major strain. Deviations from this linearity are found at high values of major strain, close to fracture of the specimens.

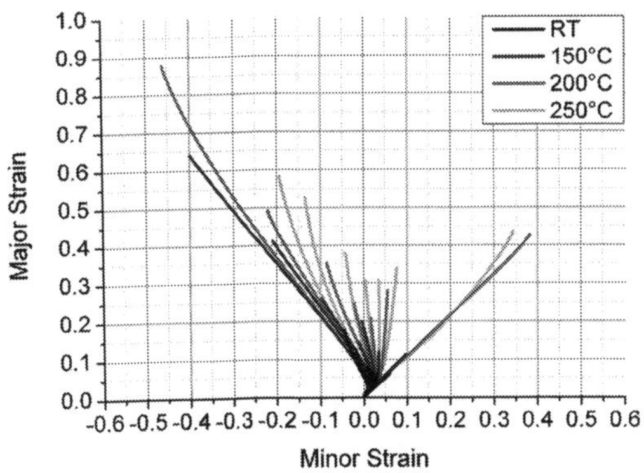

Figure 8: Strain paths of ZE10 sheet, testing temperature RT to 250°C

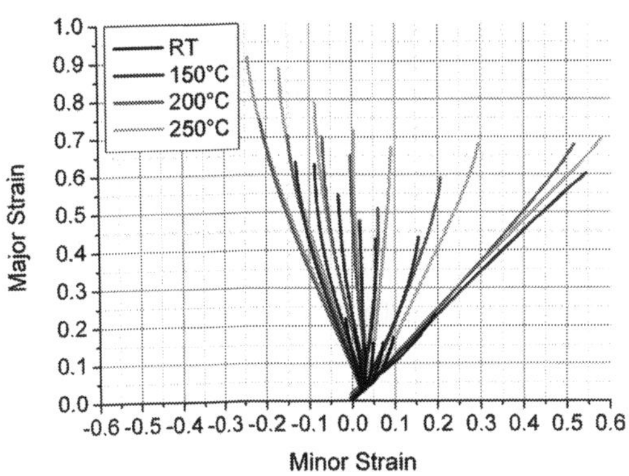

Figure 9: Strain paths of ZE10 sheet, testing temperature RT to 250°C

As a matter of tendency it is found that achieved values of major strain are higher for ZE10 than for AZ31. However, especially on the left hand side of the FLD, the slope of the measured strain paths is generally steeper for ZE10 than for AZ31. While for ZE10 it is found that an increase in temperature results in higher values of maximum major strain, AZ31 shows an exception from the rule at the outermost left side of the FLD. As already reported for the maximum dome height, maximum formability is reached at 200°C here and the strain path of specimens, tested at 250°C, does not even reach the maximum major strain as measured at 150°C.

Forming limit curves of the sheets according to ISO 12004 are shown in figures 10 and 11. The forming limit curve approximates the onset of necking of the material, beyond, strain localisation takes place until fracture. For AZ31, room temperature formability turns out to be poor. An increase of testing temperature to 150°C leads to a significant increase in formability under strain paths of negative minor strain, but gain in the domain of positive minor strain is low. Significant increase in stretch formability is found when increasing the test temperature to 200°C and beyond. The maximum major strain is reached with specimen tested at 200°C under negative minor strain. This result corresponds with the already observed findings in punch force – punch displacement curves and strain paths. The minimum major strain is found for specimen with minor strain close to zero, regardless of the temperature. This is a typical finding for sheet metals, the corresponding major strain value is commonly denoted FLD_0. For AZ31, FLD_0 increases with temperature from 0.10 at room temperature to 0.27 at 250°C.

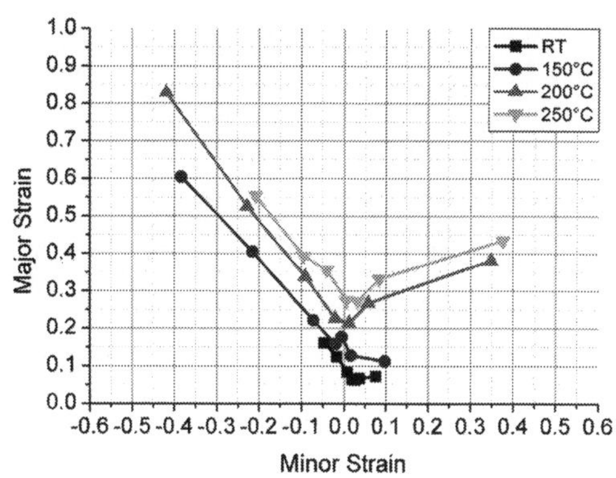

Figure 10: Forming limit curves of AZ31 sheet, testing temperature RT to 250°C

The forming limit curves of ZE10 sheet from room temperature to 250°C, as presented in figure 11, show generally higher values of major strain compared with AZ31. At room temperature formability is already higher than for AZ31 at 150°C under biaxial stretching. FLD_0 depends strongly on temperature, showing an increase from 0.14 to 0.53 from RT to 250°C. ZE10 generally outperforms AZ31 in terms of formability. At 150°C, competitive formability can be achieved, the degree of deformation under plane strain condition and biaxial stretching is unmatched by AZ31 even at 250°C. Figure 11 confirms that formability under positive major strain does not increase significantly from 150°C to 250°C, for negative minor strain and plane strain condition a certain gain in formability is notable.

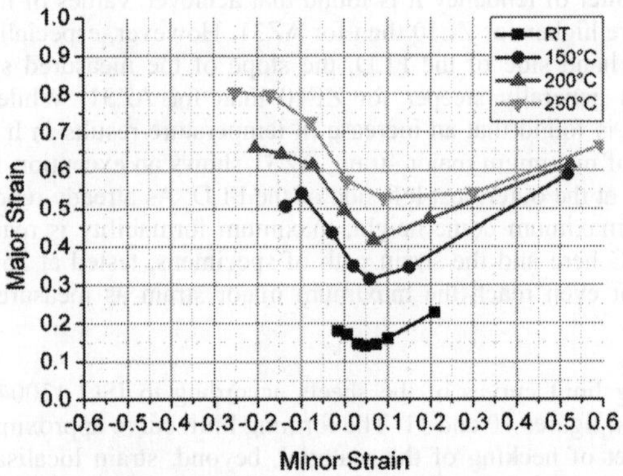

Figure 11: Forming limit curves of ZE10 sheet, testing temperature RT to 250°C

The microstructure of tested specimens is shown in figures 12 and 13. The micrographs show the microstructure close to the crack location. Figure 12a shows clearly that twins are formed during deformation of an AZ31 specimen with negative minor strain at room temperature. Twinning can contribute to the overall deformation of hcp material. Dynamic recrystallisation takes place at elevated temperatures. For the same strain path, figure 12b shows a specimen, deformed at 200°C. A duplex structure of fine, recrystallised grains at the former grain boundaries of large, unrecrystallised grains is found. Figure 12c shows a similar structure, however, the fine grains at the former grain boundaries are larger. Under biaxial stretching, fewer twins are found at room temperature, figure 12d. At elevated temperatures, dynamic recrystallisation is initiated, at 200°C the microstructure is partially recrystallised and voids have formed in the material, figure 12e. Figure 12f shows a fairly recrystallised microstructure of a specimen, tested at 250°C under a strain path with positive minor strain (biaxial stretching). The microstructure consists of small grains, much smaller than the initial grain size.

ZE10 shows a different behaviour. At room temperature, fewer twins are formed as shown in figure 13a. At elevated temperatures, no evidence of dynamic recrystallisation can be found under a strain path with negative minor strain, see figures 13b and 13c. Instead, stretched unrecrystallised grains are found, forming a pan cake like microstructure. Under positive minor strain, severely deformed microstructurse can be observed in the vicinity of the crack location, see figures 13d-f. Twins obviously play no role for the deformation under this strain path at elevated temperatures. Only at room temperature, very few twins can be observed in the microstructure, see figure 13d. With increasing temperature, the degree of elongation of the grains increases. At testing temperature of 250°C, individual grains are hard to identify, the microstructure consists of very flat, stretched grains.

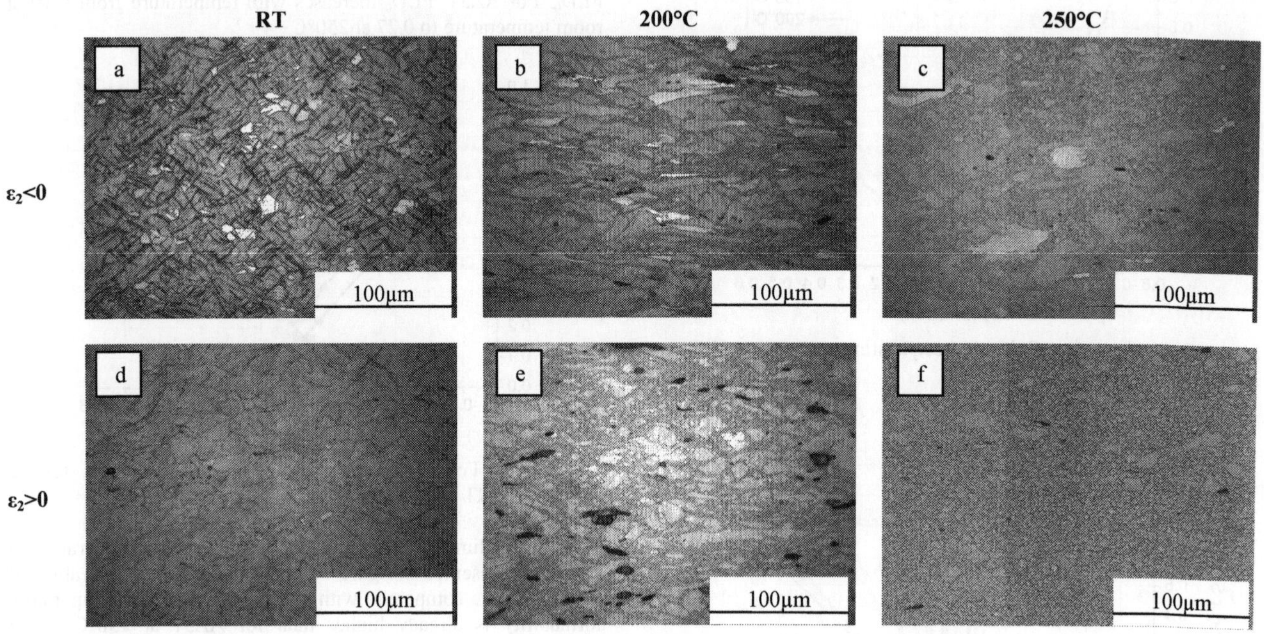

Figure 12: Micrographs of AZ31 sheet, tested at RT, 200°C and 250°C, negative minor strain (a-c) and positive minor strain (d-f)

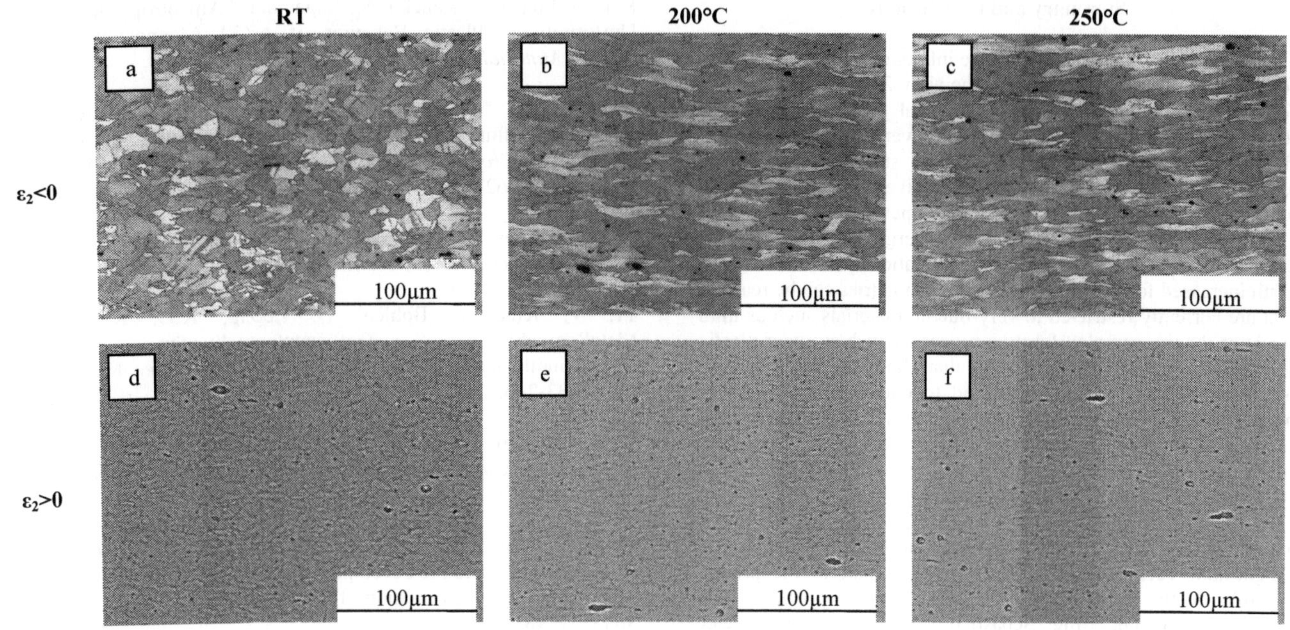

Figure 13: Micrographs of ZE10 sheet, tested at RT, 200°C and 250°C, negative minor strain (a-c) and positive minor strain (d-f)

Discussion

The punch displacement - punch force curves of the both materials provide a first impression on the superior formability of ZE10 compared to conventional AZ31 sheet. It is notable that the limiting dome height and the achieved maximum major strain in the forming limit curves are in good, yet qualitative agreement. High limiting dome height equals good formability in the FLD as reported by Dreyer et al. [4]. The significant differences in strain response of the two sheets are mainly determined by their difference in texture. It is well known that a less pronounced texture can result in an overall higher formability of magnesium sheets [4-8]. The commonly found basal type texture of AZ31 basically does not favour basal slip as a result of the orientation even if a low CRSS favours it in comparison to other slip mechanisms. The deformation via this main slip system is hampered for all strain paths. In case of uni-axial tension tests this was confirmed by Agnew [12]. Strain in thickness direction and thus, perpendicular to the majority of basal planes, is difficult to realise at low temperatures. Twinning can only accommodate low degrees of plastic deformation in <c> direction, formability of AZ31 at room temperature turns out to be poor. At elevated temperatures of 200°C and above, activation of pyramidal <c+a> slip can compensate for the unfavourable orientation of basal planes, with increasing temperature, formability of AZ31 rises significantly especially for strain paths with negative minor strain. Under these strain paths, increasing deformation can be achieved up to 200°C. Interestingly, a further increase of temperature does not lead to a further increase in deformation but to a decrease. This indicates that a consideration of the activity of slip modes does not cover all aspects that are involved. ZE10 shows generally higher formability at any testing condition. The texture is much weaker and many basal planes are oriented favourably for basal slip. <a> slip can be activated much more easily and can realise higher deformation compared to AZ31 under any strain path. In summary, it can be considered that differences in the activation of deformation mechanisms are related to the texture of the material and that the main reason for the higher formability of ZE10 compared to AZ31 is the weaker texture. This is also concluded by Bohlen et al. [6] where the main features of uni-axial tensile tests were modelled based on the different textures rather than based on significant differences in critical stresses required for respective deformation mechanisms. Figures 12a-c reveal one specific feature which changes in the case of AZ31 if the temperature increases. Starting from a deformed microstructure at room temperature there is a considerable formation of necklace type structures of small recrystallised grains. It is noteworthy that this formation, which is observed after fracture of the samples, is visible up to tests at 200°C whereas higher temperatures lead to considerable grain growth. This behaviour was reported before [13, 14] and can explain highly localised deformation in the fine recrystallised grains between the stretched, unrecrystallised grains. At 250°C, the grains highly ductile zones are not observed anymore due to grain growth. In case of ZE10 the finding is different where no hint of ongoing recrystallisation is observed in this sheet. It is described that rare earth containing magnesium alloys generally show hampered dynamic recrystallisation [15, 16]. Furthermore, the texture of this sheet itself leads to changes in the activity of deformation mechanisms including a decrease of <c+a> pyramidal slip activity. This also results in the delay of dynamic recrystallisation [5]. It is noteworthy that ZE10 shows superior formability especially under biaxial and plane strain conditions. The majority of failures in deep drawing processes happen under plane strain conditions. It should be pointed out that ZE10 shows superior formability under plane strain conditions at 150°C which is not surpassed by AZ31 even at 250°C.

Summary and Conclusions

The present study proves the massive influence of texture on sheet formability of magnesium alloys. The alloy ZE10 shows superior formability, compared to the conventional alloy AZ31. This is illustrated by forming limit curves that cover various strain paths from deep drawing processes to biaxial stretching. At testing temperatures as low as 150°C, competitive formability can be achieved. Lowering the process temperatures in forming procedures such as deep drawing saves energy and cost as well as enables the use of less sophisticated lubrication systems and tools. With increased formability, component geometries can be realised that are currently restricted to very ductile materials such as mild steels or require complex forming processes such as super plastic forming. This study clearly proves that changing the texture of magnesium sheets can pave to road to commercial success of magnesium sheets by making forming processes less expensive or offering bigger process windows.

Acknowledgement

The material used in this work is supplied by Salzgitter Magnesium Technology GmbH (SZMT) as part of the project "Mobile with magnesium". The authors appreciate this contribution by Dipl. Ing. Sebastien Wolff and Dr.-Ing. Peter Juchmann from SZMT and the financial support from the WING programme of the German Ministry of Education and Research (BMBF) under contract no. 03X3012H.

References

1. B.L. Mordike, T. Ebert, "Magnesium Properties – applications – potential", *Materials Science and Engineering A*, 302 (2001), 37-45.

2. H. Friedrich, S. Schumann, "Research for a "new age of magnesium" in the automotive industry", *Journal of Materials Processing Technology*, 117 (2001), 276-281.

3. W.F. Hosford, R.M. Caddell, *Metal forming: Mechanics and Metallurgy*, (Upper Saddle River, NJ: PTR Prentice Hall, 1993) 68–79.

4. C.E. Dreyer, W.V. Chiu, R.H. Wagoner, S.R. Agnew, "Formability of a more randomly textured magnesium alloy sheet: Application of an improved warm sheet formability test", *Journal of Materials Processing Technology*, 210 (2010), 37-47.

5. S. Yi, J. Bohlen, F. Heinemann, D. Letzig, "Mechanical anisotropy and deep drawing behaviour of AZ31 and ZE10 magnesium alloy sheets", *Acta Materialia*, 58 (2010), 592-605.

6. J. Bohlen, M.R. Nürnberg, J.W. Senn, D. Letzig, S.R. Agnew, "The texture and anisotropy of magnesium–zinc–rare earth alloy sheets", *Acta Materialia*, 55 (2007), 2101–2112.

7. Y. Chino, K. Sassa, M. Mabuchi, "Texture and stretch formability of a rolled Mg–Zn alloy containing dilute content of Y", *Materials Science and Engineering A*, 513-514 (2009) 394-400.

8. E. Yukutake, J. Kaneko, M. Sugamata, "Anisotropy and Non-Uniformity in Plastic Behavior of AZ31 Magnesium Alloy Plates", *Materials Transactions*, 44 (2003), 452-457.

9. V. Hasek, " Untersuchung und theoretische Beschreibung wichtiger Einflußgrößen auf das Grenzformänderungsschaubild", *Blech Rohre Profile*, 25 (10) (1978), 493-499, 213-220, 285-292, 493-499, 613-627.

10. S.P. Keeler, "Plastic instability and fracture in sheet stretched over rigid punches", *ASM Transaction*, 56 (1964), 25-48.

11. V. Kree, J. Bohlen, D. Letzig, K.U. Kainer, " Metallographische Gefügeuntersuchungen von Magnesiumlegierungen", *Praktische Metallographie*, 41 (2004), no. 5: 233-246.

12. S.R. Agnew, "Plastic Anisotropy of Magnesium Alloy AZ31B Sheet", *Magnesium Technology 2002*, (Warrendale, PA: TMS, 2002), 169-174.

13. S.E. Ion, F.J. Humphreys, S.H. White, "Dynamic recrystallisation and the development of microstructure during the high temperature deformation of magnesium", *Acta Metallurgica*, 30 (10) (1982), 1909-1919.

14. J.A. del Valle, M.T. Pérez-Prado, O.A. Ruano, "Texture evolution during large-strain hot rolling of the Mg AZ61 alloy", *Materials Science and Engineering A*, 355 (2003), 68-78.

15. K. Hantzsche, J. Bohlen, J. Wendt, K.U. Kainer, S.B. Yi, D. Letzig, "Effect of rare earth additions on microstructure and texture development of magnesium alloy sheets", *Scripta Materialia*, 63 (2010), 725–730.

16. J. Bohlen, S. Yi, D. Letzig, K.U. Kainer, "Effect of rare earth elements on the microstructure and texture development in magnesium–manganese alloys during extrusion", *Materials Science and Engineering A*, 527 (2010) 7092–7098.

HOT WORKABILITY OF ALLOY WE43 EXAMINED USING HOT TORSION TESTING

F. John Polesak III[1], Bruce Davis[2], Rick DeLorme[2], Sean R. Agnew[1]

[1]Department of Materials Science and Engineering, University of Virginia, Charlottesville, Virginia, USA 22904-4745
[2]Magnesium Elektron North America, Madison, Illinois, USA 62060

Keywords: WE43, hot torsion, dynamic strain aging, dynamic recrystallization, homogenization

Abstract

While rare earth additions can impart a variety of property improvements in magnesium alloys, they can also limit the processing parameter window inside which the alloys may be successfully wrought in commercial production routes such as extrusion and rolling. In the present work, hot torsion testing is used to explore the temperature and rate sensitivity of the flow stress of the alloy WE43 in order to establish the boundaries of this processing window. Two mechanisms appear to be important from a fundamental perspective: first, dynamic recrystallization appears to be a prerequisite for significant plastic flow; second, serrated flow and negative strain rate sensitivity are observed at lower deformation temperatures (T ≈ 150-250°C). The Sellars-Tegart model and simple power law, both of which have been successfully applied to describe the temperature/strain-rate/flow stress relationship in AZ31 and many other Mg alloys, are explored in the context of WE43.

Introduction

Magnesium is of interest for myriad applications due to its low density and high specific strength, as compared to competing materials such as aluminum, steel, and polymers. One of the main areas that traditional magnesium alloys lack is high temperature (above 100°C) strength [1]. Magnesium alloy WE43, which contains magnesium, yttrium, neodymium, zirconium, and other rare earth elements, retains its strength at temperatures up to 300°C, and remains stable for long term exposure up to 250°C [2]. In addition, WE43 has good corrosion and ignition resistance [2, 3].

The objective of the current study is to determine the deformation processing conditions (temperature and strain rate) for alloy WE43. Hot torsion testing data were used to compare the flow behavior of samples with different heat treatments. WE43 was examined in the as-cast and vendor-prescribed solutionized conditions. Optical microscopy, scanning electron microscopy (SEM), and energy dispersive spectroscopy (EDS) were employed to qualitatively compare the microstructure and composition of WE43 in the different conditions. Previous studies have reported a wide variety of microstructures at different conditions (as-cast and several heat treatments) including various second phases such as a grain boundary eutectic, Mg-RE precipitates, nanoscale cuboids, and zirconium-rich particles [4, 5, 6, 7, 8]. The presence of these and others phases and their effect on the mechanical behavior was investigated.

The flow behavior of magnesium alloys is often described by a conventional power relation, the Sellars-Tegart equation, or similar equations involving the strain-rate, temperature, and activation energy [7, 8, 9, 10, 11, 12, 13]. It will be shown that a power law can describe the relationship between the flow stress and the Zener-Holloman parameter (i.e. the temperature-compensated strain rate) in the high temperature (T ≥ 350°C) regime.

Experimental Procedures

Hot torsion tests were carried out on a custom Instron torsion machine at equivalent strain rates of 10^{-3}, 10^{-2}, 0.1, 1.0, and 5.0 s^{-1}. The motivation for torsion testing is that the samples do not undergo significant shape change during tests to high strains. Localizations such as necking and barreling that occur during tension and compression, respectively, are avoided in torsion, making it an appealing method of deformation. Another attractive feature of torsion is that a constant true strain rate is imposed by twisting one end of the sample at a constant angular velocity relative to the other end [14]. The solid cylindrical test specimens were machined on a lathe, and had a gage length of 22.225 mm, and a radius of 3.175 mm (Fig. 1). All strains and strain rates reported in this paper are the values at the surface of the gage region. Tests were performed at temperatures of 150, 250, 350, 375, 400, 425, and 450°C. An infrared furnace was used to heat the samples and hold them at temperature for 10 minutes prior to testing. At the higher rates of 0.1, 1.0 and 5.0 s^{-1}, the effect of adiabatic heating was taken into account and calculated to obtain an estimate of the actual testing temperature, which affects the subsequent analyses.

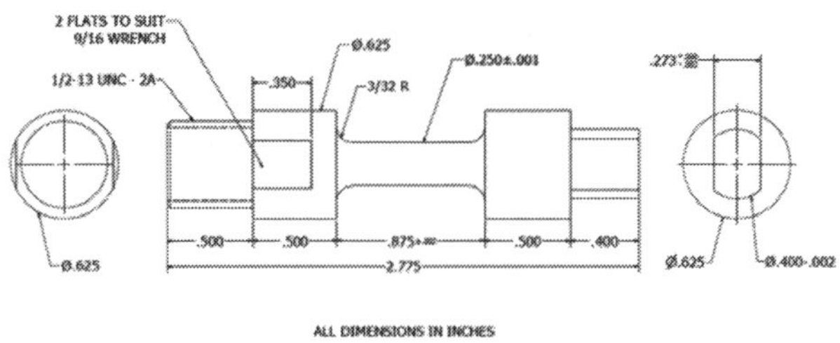

Figure 1. Torsion sample geometry and dimensions, courtesy of Professor J.J. Jonas, McGill University, Montreal, QC, Canada.

The effective stress and strain were calculated according to the von Mises yield criterion from the acquired torque-twist data using the following relationships:

$$\sigma = (\sqrt{3}\,M)/(2\pi r^3)\,(3 + m + n) \tag{1}$$

$$\varepsilon = 2\pi N r/(\sqrt{3}\,l) \tag{2}$$

where M is the measured torque, r is the sample radius, N is the number of revolutions, l is the gage length of the sample, m is the strain-rate sensitivity coefficient at constant strain, and n is the strain hardening coefficient at constant strain rate [12, 14, 15]. The value of m can obtained from log-torque-log-twist rate plots, and in this paper is initially taken to be 0.132 from Gao et al. [15]. The value of n can be obtained from log-log torque-twist plots, but is initially taken as zero in this study, as is commonly done for hot torsion testing [14].

Some readers may question the use of the von Mises isotropic flow rule, but, as detailed below, the material is tested in a randomly textured, i.e. initially isotropic condition. A deformation texture could obviously develop during the tests. However, as shown by Ball and Prangnell for related alloy, WE54, the texture is expected to evolve rather slowly relative to other Mg alloys during hot deformation [16]. Further, Jain and Agnew showed that twinning-induced strength asymmetry diminishes at about 200°C, even in relatively strongly textured alloy AZ31 [17].

The torsion samples were received in the as-cast condition. Torsion specimens and metallographic samples were solutionized at 525°C for 8 hours, followed by a hot water quench. The metallographic samples were then cold-mounted for optical and scanning electron microscopy. Standard metallographic grinding and polishing was performed with SiC grinding paper and an oil-based diamond suspension, respectively. After polishing to ¼-µm, samples were polished with 0.02-µm colloidal silica. Care was taken after each step to fully cleanse the sample with ethanol. Select samples were etched with an acetal-nitric and/or an acetal-picric solution for optical microscopy.

SEM and EDS were performed on a JEOL 840 with a LaB_6 filament. EDS was performed on the polished samples at an accelerating voltage of 15.0 kV, a working distance of 15mm and a probe current necessary to keep the dead time at ~30%. Different areas of the material were analyzed using EDS, including the bulk matrix, grain boundaries, and second phases. The nominal composition of alloy WE43B is shown in Table I, along with the characteristic x-ray energies of interest. It can be seen that some of the M_α energies of the rare earth elements are very close to the K_α energy of magnesium. Because of this, only a qualitative comparison of the various spectra was used to assess compositional variation throughout the samples.

Table I. Nominal composition of WE43B and the corresponding characteristic x-ray energies (keV) [2].

	wt%	K_α	L_α	M_α
Yttrium	3.7-4.3	14.933	1.922	-
Neodymium	2.0-2.5	37.185	5.230	0.978
Ytterbium	0-2.4	52.035	7.414	1.521
Erbium	0-2.4	48.818	6.948	1.406
Dysprosium	0-2.4	45.728	6.495	1.293
Gadolinium	0-2.4	42.761	6.059	1.185
Zirconium	0.4 min	15.746	2.042	-
Magnesium	Balance	1.254	-	-

Results and Discussion

Comparison of as-cast and homogenized microstructure

Scanning electron and optical microscopy of the as-cast and solutionized material revealed much about their microstructure and composition. Optical microscopy of the as-cast material etched with acetal-picric acid showed extensive coring around zirconium particles and eutectic phase at the grain boundaries (Fig. 2(a)). Using a mean linear intercept method, the average grain size of the as-cast and solutionized material was determined to be 37.5 µm and 81 µm respectively. Figure 2(b) shows the eutectic around the grain boundaries, and some second phase particles inside the as-cast grains. EDS analysis indicated that the intermetallic phase in the eutectic structure is primarily composed of Mg, Nd, Y, and small amounts of other rare earth elements, such as Gd. The SEM image also shows Y-rich cuboids lying around the edges of the eutectic (encircled within the inset in Fig. 2(c)). Figure 3(a) shows the EDS spectrum from one of the Y-rich cuboidal phases observed in the solutionized material.

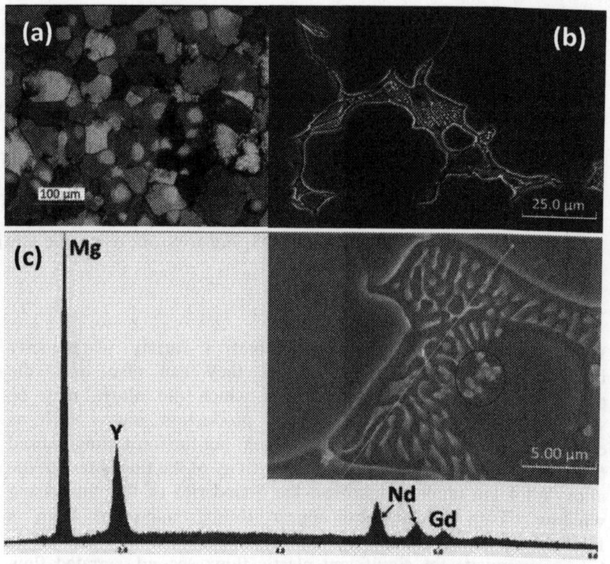

Figure 2. (a) Optical micrograph of contrast etched as-cast WE43. (b) SEM image of the same material, showing grain boundary eutectic and other 2nd phase particles. (c) EDS spectrum of the eutectic phase, showing a Mg-Y-rare earth composition. Y-rich cuboids around the eutectic are circled in the inset image.

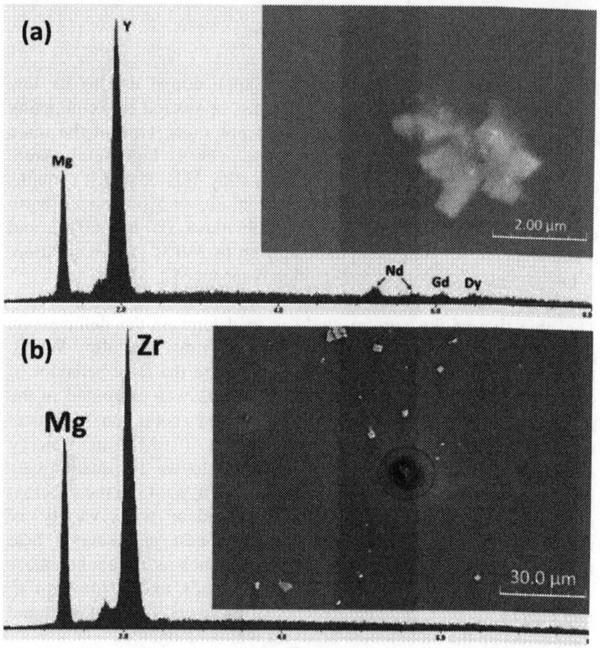

Figure 3. EDS spectra from second phase particles in solutionized WE43 revealed the chemistry of (a) the Y-rich cuboidal precipitates and (b) Zr-rich particles.

In summary, SEM-EDS revealed a eutectic interdendritic region, yttrium-rich cuboidal precipitates around the eutectic, and zirconium-rich particles throughout the as-cast material (Figs. 2 & 3). Only the latter two phases are present in the solutionized material (Fig. 3), i.e. the treatment is effective for dissolution of the eutectic second phase, but not all phases present in the casting. Different homogenization heat treatments will be required if a complete solutionization treatment is desired.

Flow Curves

Effective stress-strain curves obtained at three different temperatures and strain rates are presented in Figure 5, and provide a comparison of the as-cast and solutionized conditions. At the rates examined, the alloy did not exhibit sufficient ductility for hot deformation processing below 350°C, so only those conditions are shown in this first analysis. As there is no reduction in cross-sectional area during torsion testing, the peaks in these torsional flow curves are viewed as evidence of dynamic recrystallization (DRX) [18, 19]. Optical micrographs of deformed samples offered further confirmation of the presence of DRX. The average grain size increased in both the as-cast and solutionized samples. In the solutionized sample shown in Figure 4, new, smaller grains can be observed to have nucleated, mainly at the grain boundaries. The figure also shows evidence of significant precipitation within the grains and along the grain boundaries. These particles may serve to pin the original grain boundaries and impede growth of the new grains which nucleate during DRX. It is concluded that DRX is a pre-requisite for hot workability in this alloy to avoid premature failure by cracking [20].

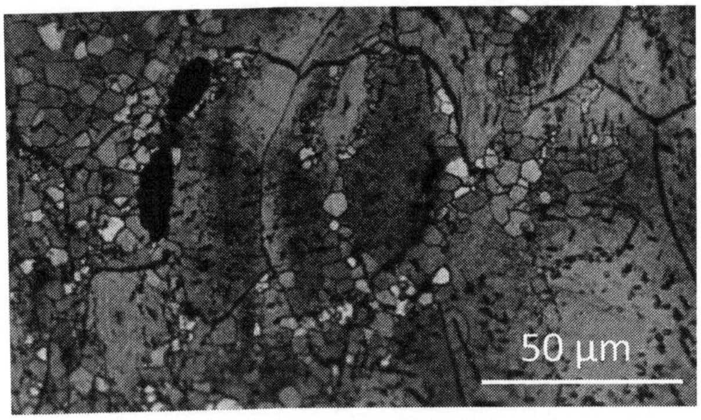

Figure 4. Optical micrograph of a WE43 torsion sample tested at 400°C and 1.0 s^{-1}. The micrograph, taken near the periphery of the sample, shows the presence of dynamic recrystallization.

As seen in Figure 5, the solutionized material was stronger than the as-cast under most conditions. This suggests that having the rare earth elements in solution (and/or dynamically precipitating) is a more effective high temperature strengthener than having the alloy additions bound up in an interdendritic eutectic second phase. The failure of the as-cast samples is thought to be caused by void nucleation at the large, brittle interdendritic second phase particles. The similar behavior of the alloy in the as-cast and solutionized conditions at the higher testing temperatures suggests that the working temperatures may have been sufficiently high that a degree of homogenization occurred during testing in the as-cast material. There was no obvious effect of the larger initial grain size of the solutionized material on the strength or ductility.

Adiabatic heating and Dynamic Strain Aging

Figure 6(a) shows the flow stress at an effective strain of 0.1 plotted as a function of strain rate. This type of graph is used to determine the positive stress exponent of an assumed power law relationship between strain rate and flow stress [10]. However, these data show a different trend. At the lowest temperature and strain rates, there was either rate insensitivity or negative rate sensitivity. At the higher rates (>0.1 s^{-1}), negative rate sensitivity was exhibited at almost all temperatures examined. In the former case, the low rate sensitivity was attributed to dynamic strain aging. In the latter, adiabatic deformation heating was likely the cause.

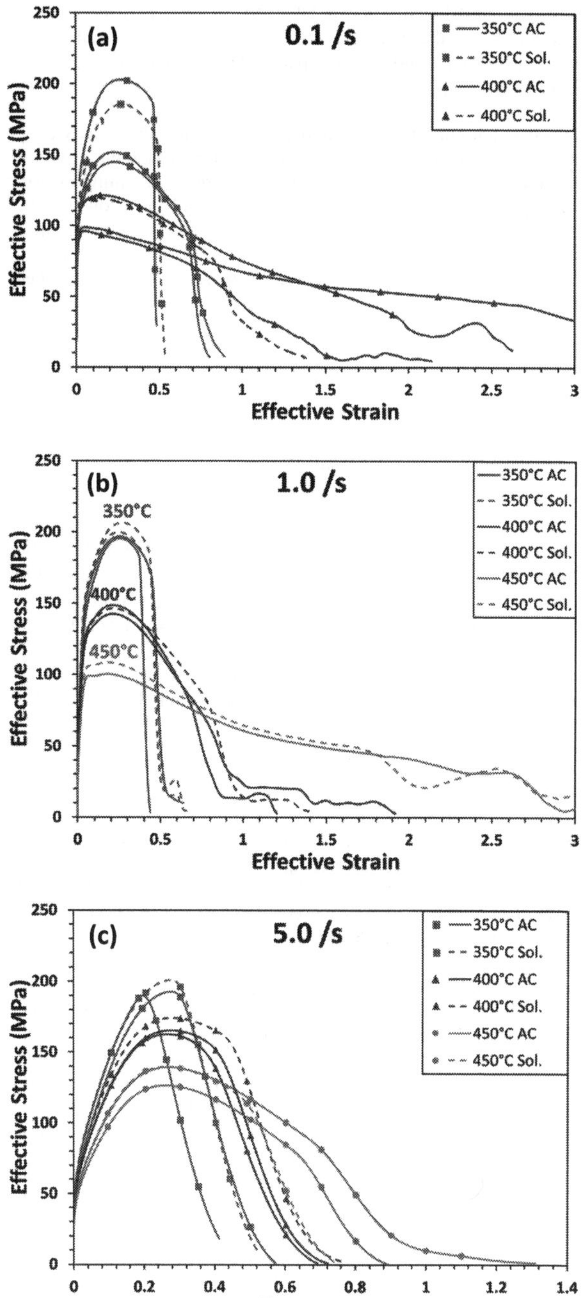

Figure 5. (a- c) Flow curves for as-cast and solutionized WE43 at three different temperatures and strain rates. Multiple curves with the same line style indicate duplicate tests at the same condition.

During torsion testing, work is put into the sample and, under certain conditions, can result in a significant temperature increase [14]. To determine the rate of temperature increase at a strain of 10%, the following equation was used:

$$\dot{T} = (\eta \dot{\varepsilon} \sigma_{0.1})/(\rho c_p) \quad (3)$$

where η is an efficiency factor taken to be 0.95, $\dot{\varepsilon}$ is the nominal strain rate, $\sigma_{0.1}$ is the stress at a strain of 0.1, ρ is the material density taken as 1.84 g/cm^3, and c_p is the specific heat taken to be 0.966 J/kg/K.

While it is admitted that heat transfer can occur radially as well as axially, only heat conduction along the axis and through the grips was considered. Samples were held at temperature for ten minutes prior to testing to minimize any radial temperature gradient. However, the temperature gradient from the sample to the grips that extend out of the furnace into the air may be significant. The dissipation of heat to the grips was calculated using a Fourier series solution to the heat equation:

$$u(x,t) = \sum_{n}^{4} c_n \sin(n\pi x/L) \exp(-n^2\pi^2\alpha t/L^2) \quad (4)$$

$$c_n = \frac{2}{L}\int_0^L \Pi(\xi)\sin(n\pi\xi/L)d\xi \quad (5)$$

where L is the gage length of the sample, x is the position along the gage length, t is time, and $\alpha = \kappa/(\rho c_p)$, where κ is the thermal conductivity, and ρ and c_p have the same meanings as in Equation 3. $\Pi(\xi)$ is the initial condition of the sample, and was taken to be a Heaviside step function with amplitude equal to the total adiabatic temperature gain calculated from Equation 3. It was determined that at rates of 0.01 s^{-1} and lower, the heat could dissipate faster than it was produced, so the tests can be assumed to be isothermal. At strain rates of 1.0 and 5.0 s^{-1}, the heat is produced in a much shorter time than is required to dissipate any appreciable amount, and thus these tests are assumed to be adiabatic. Tests performed at 0.1 s^{-1} were found to be neither isothermal, nor completely adiabatic. In other words, heat was produced and dissipated at comparable rates. To estimate the temperature rise during these tests, Equations 3 and 4 were integrated incrementally by dividing the time required to reach an effective strain of 0.1 into ten equal steps, and the heat addition and dissipation were estimated for each step.

The data presented in Figure 6(b) has been adjusted to account for this adiabatic heating effect. The rate sensitivities at higher temperatures all appear positive, after making this correction. However, a strength anomaly, where the flow stress remains constant or increases with increasing temperature or decreasing strain rate, still occurs within the lower temperature regime (150-250°C) at which some of the as-cast samples were tested (see left hand side of Fig. 6(b)). This anomaly is associated with dynamic strain aging (DSA) [21, 22, 23, 24]. The final piece of evidence for DSA is presented in Figure 7, a magnified plot of flow curves that clearly exhibit the serrated flow known as the Portevin-Le Châtelier (PLC) effect.

It has been shown that the PLC effect is caused by dynamic strain aging, which is the result of the interaction between solute atoms and dislocations, where the dislocations are repeatedly pinned by solute atmospheres and then able to break away before being pinned again [22, 23, 25]. It is often associated with a loss of uniform ductility [25]. Several different studies of Mg-RE alloys have shown the temperature and strain rate regime for DSA and PLC to be roughly 150-350°C and on the order of 5.5×10^{-5}- 6.0×10^{-3}, respectively [22, 26, 27, 28]. Surprisingly, these alloys do not appear to suffer a loss in ductility. It has been speculated that this is due to the relatively high homologous temperature at which the effect is observed, and the fact that dynamic recrystallization may be occurring simultaneously [21].

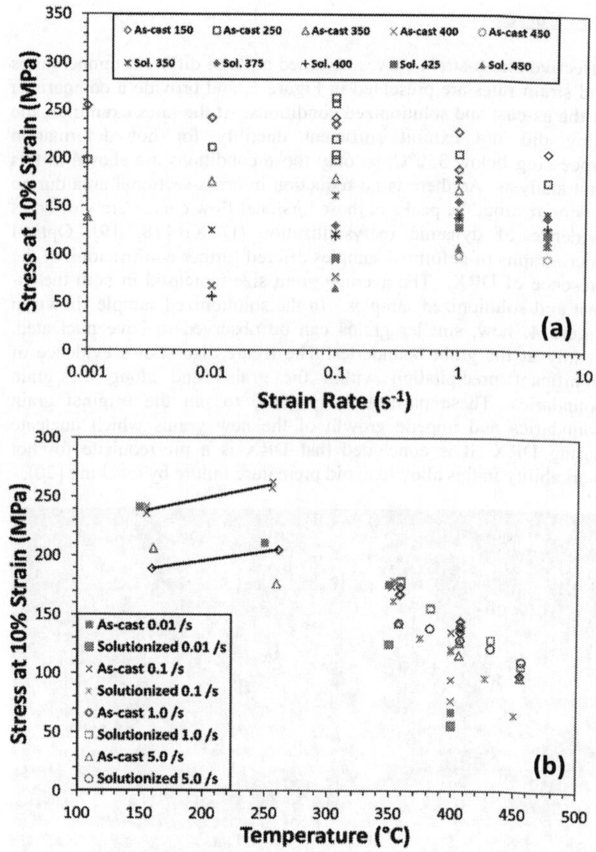

Figure 6. (a) Strain rate vs. flow stress at a strain of 0.1, showing certain regimes of negative strain-rate sensitivity. (b) Temperature vs. flow stress, showing rates that have a positive correlation.

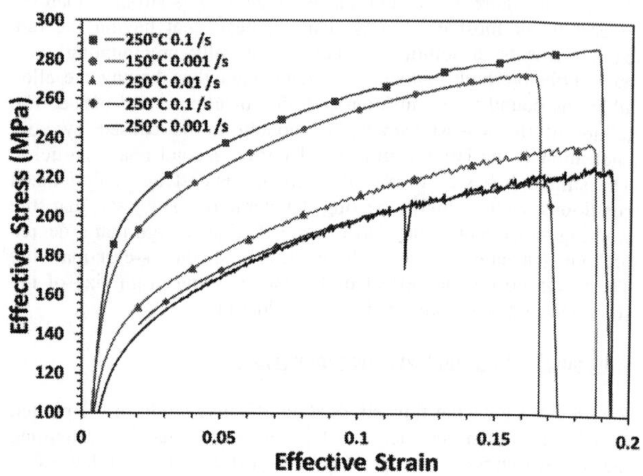

Figure 7. Serrated flow curves showing evidence of the PLC effect in as-cast material.

<u>Constitutive modeling</u>

The temperature-compensated strain rate, or the Zener-Holloman parameter, Z, is defined as,

$$Z = \dot{\varepsilon} \exp(Q/RT) \quad (6)$$

where $\dot{\varepsilon}$ is the strain rate, Q is the activation energy, R is the gas constant, and T is the absolute temperature. Using a single value of Q = 297 kJ/mol, as reported by Gao et al. for similar conditions [15], the low Z data did collapse into a single linear band (Fig. 7). Thus, at low stresses and values of Z, there is a power law relationship between Z and stress (i.e. linear in the log-log plot, Fig. 8).

$$Z = A\sigma^{n^*} \quad (7)$$

where A is a constant, σ is the flow stress, n^* is the stress exponent (or inverse strain rate sensitivity.) In this region, the power law fit, typical of high temperature creep [15, 29], is achieved with values of $A = 10^{8.9}$ and $n^* = 6.8$. These values agree well with those reported in the literature, even those obtained by other modes of deformation, such as tension [30].

It is very convenient that the Zener-Holloman parameter and power law fit work well over the range of temperatures and strain rates relevant to deformation processing at temperatures greater than 350°C. At lower temperatures, the material does not exhibit enough ductility for processing at feasible rates. Furthermore, the alloy exhibits different, sometimes negative, strain rate sensitivities at these lower temperature conditions. While the data from the low temperature tests can be collapsed into the same band in Figure 7 if lower activation energies are used for the 150 and 250°C tests, the fit would still not capture the observed negative strain rate sensitivity. Similarly, the Sellars-Tegart equation, which has successfully described the behavior of other Mg alloys over a wide range of temperatures and strain rates, cannot capture the behavior of low temperature tests for WE43. While the Sellars-Tegart equation permits the stress dependence to become much stronger (i.e., exponential) at high stress levels, typical of power law breakdown [10, 11], it does not have any provision to describe the observed negative strain rate sensitivity.

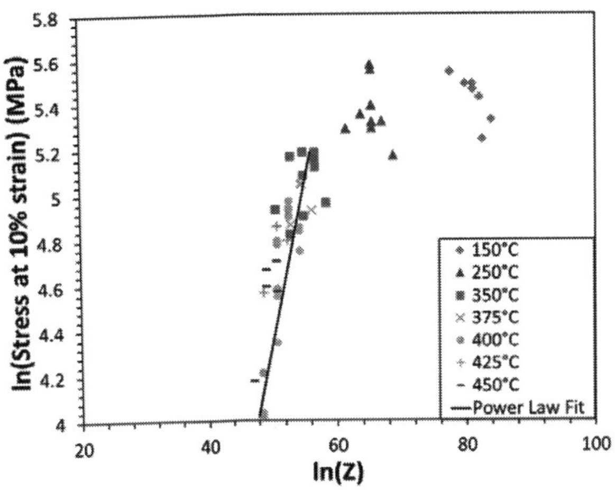

Figure 8. All torsion data with a simple power law fit that reasonably captures the low Z behavior. Data points labeled according to the nominal testing temperature (though the data are adjusted for adiabatic heating.)

Finally, the initial assumption that the strain rate sensitivity, m = 0.132, and strain hardening coefficient, n = 0, are valid for some, but not all test conditions. The strain rate sensitivity (inverse of the stress exponent), n^*, calculated for low Z tests is ~0.147, resulting in negligible error when calculating the flow stress. The assumption that the strain hardening coefficient is zero is only valid at the highest temperature and lowest rate tests. At all other conditions, there is obvious strain hardening occurring, meaning that the value of n needs to be calculated and included in subsequent analyses.

Conclusions

(1) Scanning electron microscopy showed the presence of grain boundary eutectic in the as-cast material, which dissolved during conventional solutionization treatment.

(2) Yttrium-rich cuboidal precipitates and zirconium-rich particles present in the as-cast alloy were not dissolved by the conventional solutionization treatment.

(3) During hot torsion testing, adiabatic heating occurred in magnesium alloy WE43 at high strain rates, namely 1.0 and 5.0 s^{-1}.

(4) If the adiabatic heating is taken into account, the plastic flow of WE43 can be modeled using a simple power law relationship, $n^* \sim 7$, with a single apparent activation energy, $Q \sim 297$ kJ/mol, for the high temperature (350≤T≤450°C) conditions presently examined which appear relevant to deformation processing.

(5) To achieve necessary ductility during processing, deformation must be at a temperature of at least 350°C.

(6) At the lower temperatures, the behavior could not be modeled by the power law or even the more complex hyperbolic sine function of the Sellars-Tegart equation. A model that accounts for dynamic strain aging must be used to capture the behavior of as-cast or solutionized alloy WE43 at low strain rates ($\leq 10^{-1}$ s^{-1}) and low temperatures (150-250°C), where the Portevin-Le Châtelier effect, i.e. serrated flow, was observed.

Acknowledgements

This research was sponsored by the Army Research Laboratory and was accomplished under Cooperative Agreement No. W911NF-07-2-0073. The views and conclusions contained in this document are those of the authors and should not be interpreted as representing the official policies, either expressed or implied, of the Army Research Laboratory or the U.S. Government. The U.S. Government is authorized to reproduce and distribute reprints for Government purposes notwithstanding any copyright notation hereon. The authors would also like to gratefully acknowledge Mr. Pritom Das for performing many of the torsion tests.

References

1. Turski, M., J. F. Grandfield, et al. (2010). "Computer Modeling of DC casting magnesium alloy WE43 rolling slabs." *Magnesium Technology 2010*: 6.

2. Magnesium Elektron Data Sheet 467. http://www.magnesium-elektron.com/data/downloads/DS467WE43.pdf.

3. Ravi Kumar, N.V. et al. “Effect of alloying elements on the ignition resistance of magnesium alloys,” *Scripta Materialia*, 49 (2003), 225-231.

4. Riontino, G., M. Massazza, et al. (2008). "A novel thermal treatment on a Mg-4.2Y-2.3Nd-0.6Zr (WE43) alloy." *Materials Science & Engineering A* **494**: 4.

5. Sanchez, C., G. Nussbaum, et al. (1996). "Elevated temperature behaviour of rapidly solidified magnesium alloys containing rare earths." *Materials Science & Engineering A* **A221**: 10.

6. Mengucci, P., G. Barucca, et al. (2008). "Structure evolution of a WE43 Mg alloy submitted to different thermal treatments." *Materials Science & Engineering A* **479**: 8.

7. Pekguleryuz, M. O. and A. A. Kaya (2003). "Creep resistant magnesium alloys for powertrain applications." *Advanced Engineering Materials* **5**(12): 13.

8. Pekguleryuz, M. and M. Celikin (2010). "Creep resistance in magnesium alloys." *International Materials Review* **55**(4): 21.

9. McQueen, H. J. and M. Sauerborn (2005). "Hot workability and extrusion modelling of magnesium alloys." *Z. Metallkd.* **96**: 7.

10. Takuda, H., H. Fujimoto, et al. (1998). "Modelling on flow stress of Mg-Al-Zn alloys at elevated temperatures." *Journal of Materials Processing Technology* **80-81**: 4.

11. Wang, M., S. Wang, et al. (2010). "Dynamic modeling of twin roll casting AZ41 magnesium alloy during hot compression processing." *Transactions of Nonferrous Metals Society of China* **20**: 6.

12. El Mehtedi, M., S. Spigarelli, et al. (2009). "Comparative study of the high-temperature behaviour of Mg-Al and Mg-Zn wrought alloys." *International Journal of Materials Research* **100**(3): 5.

13. Crossland, I. G. and R. B. Jones (1972). "Dislocation creep in magnesium." *Metal Science Journal* **6**: 5.

14. Semiatin, S. L. and J. J. Jonas (2003). Torsion testing to assess bulk workability. *ASM International: Handbook of Workability and Process Design*. G. E. Dieter, H. A. Kuhn and S. L. Semiatin. **14A**: 33.

15. Gao, J., Q. Wang, et al. (2008). "Microstructure and kinetics of hot deformation WE43 magnesium alloy." *Rare Metals* **27**(4): 5.

16. Ball, E. A. and P. B. Prangnell (1994). "Tensile-compressive yield assymmetries in high strength wrough magnesium alloys." *Scripta Metallurgica et Materialia* **31**(2): 6.

17. Jain, A. and S. R. Agnew (2007). "Modeling the temperature dependent effect of twinning on the behavior of magnesium alloy AZ31B sheet." *Materials Science & Engineering A* **462**: 8.

18. Yang, Z., Q. Guo, et al. (2008). "Plastic deformation and dynamic recrystallization behaviors of Mg-5Gd-4Y-0.5Zn-0.5Zr alloy." *Materials Science & Engineering A* **485**: 5.

19. Al-Samman, T. and G. Gottstein (2008). "Dynamic recrystallization during high temperature deformation of magnesium." *Materials Science & Engineering A* **490**: 10.

20. McQueen, H. J. and C. A. C. Imbert (2004). "Dynamic recrystallization: plasticity enhancing structural development." *Journal of Alloys and Compounds* **378**: 9.

21. Stanford, N., I. Sabirov, et al. (2009). "Effect of Al and Gd solutes on the strain rate sensitivity of magnesium alloys." *Metallurgical and Materials Transactions A* **41A**: 10.

22. Trojanova, Z., P. Lukac, et al. (2005). "Dynamic strain ageing during stress relaxation in selected magnesium alloys containing rare earth elements." *Advanced Engineering Materials* **7**(11): 6.

23. Kubin, L. P. and Y. Estrin (1990). "Evolution of dislocation densities and the critical conditions for the Portevin-Le Chatelier effect." *Acta Metallurgica* **38**(5): 12.

24. Kok, S., A. J. Beaudoin, et al. (2002). "A finite element model for the Portevin-Le Chatelier effect based on polycrystal plasticity." *Modelling and Simulation in Materials Science and Engineering* **10**: 19.

25. Lebyodkin, M., Y. Brechet, et al. (1996). "Statistical behavior and strain localization patterns in the Portevin-Le Chatelier effect." *Acta Materialia* **44**(11): 11.

26. Wang, C. Y., X. N. Zhang, et al. (2007). "Serrated flow in cast ZE43 alloy." *Journal of Materials Science* **42**: 3.

27. Wang, Z. F., W. Jia, et al. (2007). "Study on the deformation behavior of Mg-3.6% Er magnesium alloy." *Journal of Rare Earths* **25**: 5.

28. Zhu, S. M. and J. F. Nie (2004). "Serrated flow and tensile properties of a Mg-Y-Nd alloy." *Scripta Materialia* **50**: 5.

29. Polesak III, F. J., C. E. Dreyer, et al. (2009). "Blind study of the effect of processing history on the constitutive behaviour of alloy AZ31B." *Magnesium Technology* **2009**: 6.

30. Wang, J. G., L. M. Hsiung, et al. (2001). "Creep of a heat treated Mg-4Y-3RE alloy." *Materials Science & Engineering A* **315**: 8.

Magnesium Technology 2011
Edited by: Wim H. Sillekens, Sean R. Agnew, Neale R. Neelameggham, and Suveen N. Mathaudhu
TMS (The Minerals, Metals & Materials Society), 2011

ENHANCEMENT OF SUPERPLASTIC FORMING LIMIT OF MAGNESIUM SHEETS BY COUNTER-PRESSURIZING

Wonkyu Bang[1], Hyun-Seok Lee[1], Hyung-Lae Kim[2], Young-Won Chang[2]
[1]RIST; San 32, Hyoja-dong, Nam-gu; Pohang, 790-330, South Korea
[2]POSTECH; San 31, Hyoja-dong, Nam-gu; Pohang, 790-784, South Korea

Keywords: Magnesium, Superplasticity, Cavitation, Hydrostatic Pressure

Abstract

As often reported for various metallic materials, fine-grained wrought magnesium can exhibit extensive superplasticity at elevated temperature, which makes superplastic forming (SPF) of magnesium a promising process for manifacturing complex-shaped, lightweight thin-walled structural components. The superplastic tensile deformation of magnesium is commonly accompanied by substantial cavitation in its failure stage, which is also typical to quasi-single phase aluminum alloys.

In this regard, a series of bulge test have been conducted with independent inflating/counter-pressure control. By evaluating the superplastic forming limit along Limit Dome Height (LDH), considerable improvement was reproduced by counter-pressurized conditions. Constitutive modeling and microstructural analysis imply that this is mainly achieved by the retardation of nucleation and growth of cavities under hydrostatic stress.

Introduction

Magnesium alloys are considered promising structural materials due to their lightweight, high specific strength, excellent machinability and superior damping capacity [1,2]. Wrought magnesium products are drawing more interest recently along the development of continuous casting and coil rolling integrated process, which gives a significant cost-competitive edge [3].

As for the high temperature deformation of wrought magnesium, a considerable degree of superplasticity has been reported [4,5]. Additionally, cavity formation followed by intergranular fracture have also been identified in previous studies [6,7]. Since those phenomena are also common in the superplastic forming (SPF) of aluminum alloys, the superimposition of hydrostatic pressure has been introduced in order to retard the nucleation and growth of cavities, and enhance the elongation accordingly [8-11].

In the present study, superplastic formability of strip cast and warm rolled commercial AZ31B sheets has been investigated. Simple spherical bulge tests with counter-pressure variation have been conducted to identify quantitative relationship to the forming limits.

Experimental

Equipments and Materials

In order to conduct superplastic formability tests, a hot bulge tester was fitted into a 600 tonf hydraulic servo press as illustrated in fig. 1. The die set was enclosed into split furnace and connected with gas inlet/outlet whose pressures are controlled independently. Strip cast and warm rolled commercial AZ31B sheets have been selected as test materials. Samples have thickness of 1.5 mm and as-received grain size was measured to be 8.3 μm (fig. 2).

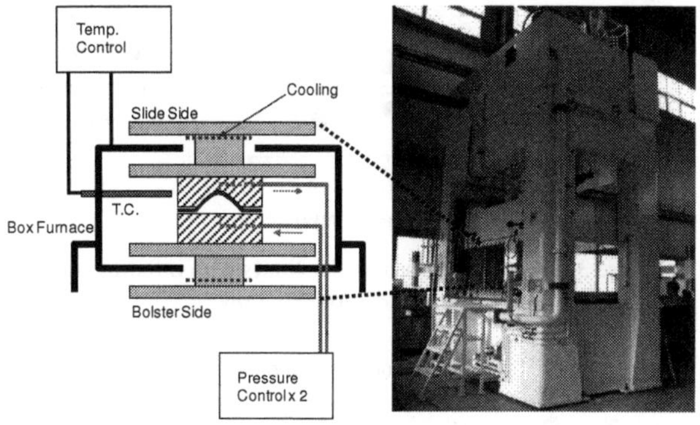

Figure 1. Schematic configuration of the hot bulge tester.

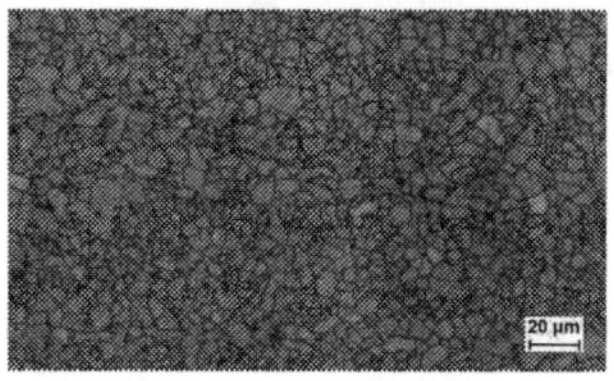

Figure 2. As-received microstructure of magnesium sheet.

Experiment Conditions

As aforementioned, simple spherical bulge tests have been conducted with various pressurizing configuration, as noted in table I.

Table I. Bulge test process window.

Parameters	Values
Bulge Geometry	Circular w/Dia = 100mm
Temperature	400°C
Inflating Pressure	0.1 ~ 0.5 MPa
Counter-Pressure	0 ~ 3 MPa

Results

Superplastic formability limits have been evaluated by recording limit dome height (LDH) up to fracture, which considered to be the most straightforward measure.

As shown in fig. 3, LDH increases with decreasing inflating pressure and increasing counter-pressure. It should be noted that counter-pressurizing does not exhibit considerable enhancement when inflating pressure is 0.4 MPa and higher.

The evolution of cavities has been investigated by measuring cavity volume fraction in the top region of the dome using image analysis of optical microscopy (fig. 4). As the amount of inflation (i.e., strain) was fixed at 50 mm, the cavitation was significantly decreased with increasing counter-pressure, which verifies the retardation of cavity nucleation and growth under high hydrostatic pressure.

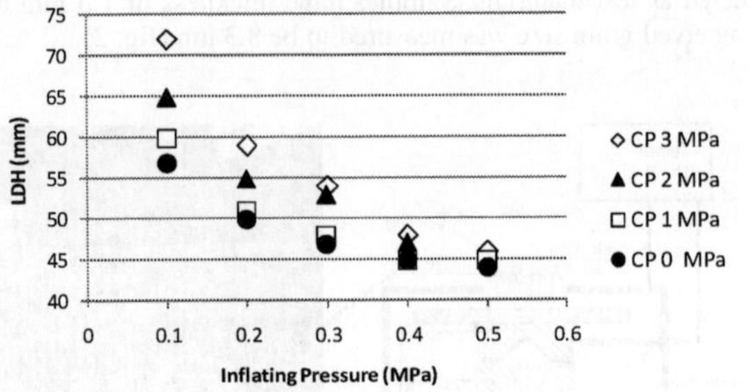

Figure 3. Superplastic formability limits with inflating and counter-pressure variation.

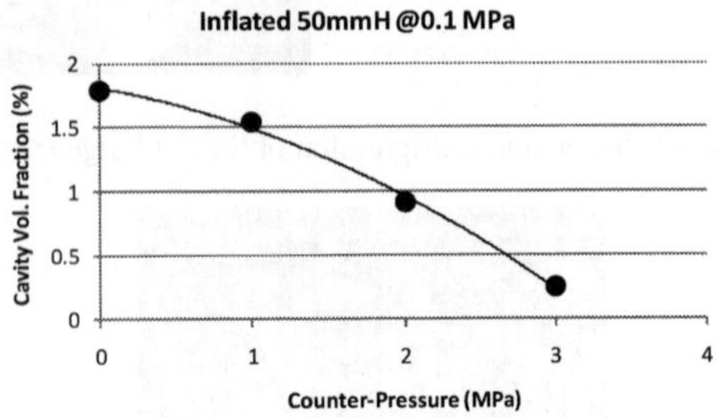

Figure 4. Cavity volume fraction as a function of counter-pressure.

From previous studies, it has been reported that the cavity growth during the superplastic deformation is predominantly controlled by plastic strain (ε) and the governing equation of the cavity growth rate (φ) can be formulated as [12-15]:

$$\varphi = \varphi_0 \exp(R \cdot \varepsilon) \,,$$

where φ_0 is the initial volume fraction of the cavities.

The exponential constant R is a function of constitutive parameters and stress states with following formulation;

$$R = 3/2\left\{\frac{m+1}{m}\right\}\sinh\left[2\left(\frac{2-m}{2+m}\right)\{Q/3 - P/\sigma\}\right],$$

where m is the strain rate sensitivity of the material, P is the superimposed pressure, and Q is a geometric factor which equals 1, 2 and 1.73 for uniaxial stress, equibiaxial stress and plane strain, respectively.

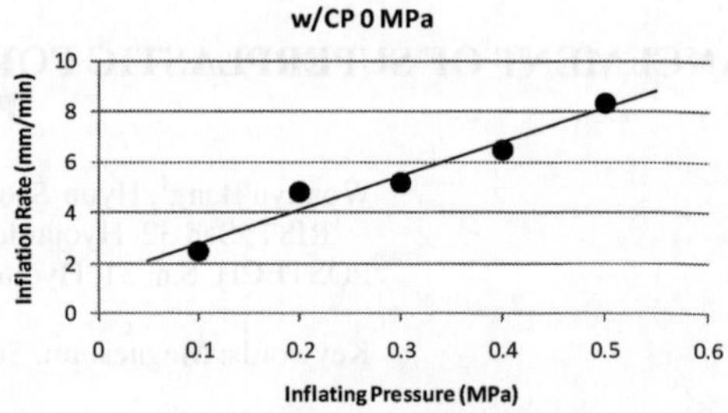

Figure 5. Inflation rate as a function of inflating pressure.

Since the inflation rate, which obviously consistent with the strain rate, is proportional to inflating pressure as shown in fig. 5, the test results under high inflating pressure corresponds to the high strain rate deformation regime with relatively lower strain rate sensitivity.

In this regard, the cavity growth rate exponent R is expected to have larger value at higher inflating pressure due to smaller m, which gives analogy for the insensitiveness to counter-pressurizing in that region.

Summary and Concluding Remarks

In order to control the cavitation, counter-pressurizing was implemented on the superplastic forming of commercial AZ31B sheets and promising results are obtained as follows.

1. Hydrostatic compressive stress imposed by counter-pressure effectively retarded cavitation during the bulge tests, which resulted in enhanced forming limits.

2. In the region of high inflating pressure and speed, the effect of counter-pressurizing was diminished, which considered to be attributable to low strain rate sensitivity.

Acknowledgements

Financial support from POSCO is greatly acknowledged. The authors also thank to Profs. N.J. Kim (POSTECH) and K.S. Shin (Seoul National Univ.) for their helpful advice and encouragement.

References

1. E. Aghion, B. Bronfin, D. Eliezer, *Journal of Materials Processing Technology* 117 (2001), 381-385.
2. A.A. Luo, *Journal of Metals* 54 (2002), 42-48.
3. I.H. Jung, W. Bang, I. J. Kim, H. J. Sung, W.J Park, D. Choo, S. Ahn, "Mg coil production via strip casting and coil rolling technologies", *Magnesium Technology 2007*, TMS Annual Meeting (2007), 85-88.
4. W.J. Kim, S.W. Chung, C.S. Chung, D. Kum, *Acta Materialia* 49 (2001), 3337-3345.
5. M. Mabuchi, K. Ameyama, H. Iwasaki, K. Higashi, *Acta Materialia* 47 (1999), 2047-2057.
6. K. Zhang, D. Yin, G. Wang, W. Han, *Journal of Materials Science and Technology* 21 (2005), 510-514.
7. J.J. Blandin, *Materials Science Forum* 551-552 (2007), 211-217.
8. J. Pilling, N. Ridley, *Acta Metallurgica* 34 (1986), 669-679.

9. C.C. Bampton, R. Raj, *Acta Metallurgica* 30 (1982) 2043-2053.
10. A.H. Chokshi, A.K. Mukherjee, *Materials Science and Engineering* 171A (1993), 47-54.
11. H.S. Yang, H.K. Ahmed, W.T. Roberts, *Materials Science and Engineering* 122A (1989), 193-203.
12. M. Zaki, *Metallurgical and Materials Transactions* 29A (1998), 2555-2561.
13. H.L. Xing, Z.R. Wang, *Journal of Materials Processing Technology* 75 (1998), 87-93.
14. P. Hu, Y. Cao, X. Liu, *Journal of Materials Processing Technology* 74 (1998), 36-43.
15. M.A. Nazzal, M.K. Khraisheh, *Journal of Materials Engineering and Performance* 16 (2007), 201-207.

Magnesium Technology 2011
Edited by: Wim H. Sillekens, Sean R. Agnew, Neale R. Neelameggham, and Suveen N. Mathaudhu
TMS (The Minerals, Metals & Materials Society), 2011

MICROSTRUCTURE EVOLUTION DURING ROLLER HEMMING OF AZ31B MAGNESIUM SHEET

Amanda Levinson[1], Raja K Mishra[2], John Carsley[2], Roger D Doherty[1], Surya R Kalidindi[1]

[1]Drexel University; 3141 Chestnut Street; Philadelphia, PA, 19104, USA
[2]General Motors R&D Center; 30500 Mound Road; Warren, MI, 48092, USA

Keywords: AZ31, Roller Hemming, EBSD, Twinning, Shear Bands

Abstract

The evolution of microstructure and texture during multi-pass roller hemming (flanging, pre-hemming and flat hemming) of commercial grade AZ31B-O sheet has been studied using electron backscatter diffraction. The pre-hemming operations were performed with and without local heating using a laser source. It was observed that samples pre-hemmed at room temperature could not be flat hemmed even after applying heat; whereas flat hemming was possible in the sample pre-hemmed with laser heating. The major difference between these samples was the formation of contraction/double twins on the outer radii of the bend in the sample pre-hemmed at room temperature. It is believed that such twins contributed directly or indirectly to the fracture of this sample upon the third pass by leading to the formation of shear bands and/or by significantly hardening the material and not allowing for further deformation.

Introduction

With a lower density and higher specific strength than both aluminum and steel, magnesium has the potential to help the auto industry realize the goal of vehicle weight reduction [1]. Thus far, cast magnesium alloys have found many automotive applications, but the limited room temperature formability of Mg alloys, owing to their hexagonal close-packed (hcp) structure, has been a major hurdle for implementing applications with wrought materials.

The inability of Mg alloys to accommodate large plastic strains at room temperature can be overcome by processing at elevated temperatures. At temperatures as low as 423K Mg alloys undergo dynamic recrystallization (DRX) [2]. Both continuous and discontinuous DRX mechanisms can be activated in Mg alloys at different temperature and strain rate regimes [3]. Temperature, strain rate, initial texture, and the formation of double twins have been shown to influence DRX [4-6].

If magnesium is to be used in automotive outer body panels, it must withstand numerous processing operations, including hemming. After forming to meet shape and design requirements, the inner and outer panels are joined in a hemming operation. This requires the edges of the outer body panel to be bent around the edge of an inner panel. Conventional hemming procedures are conducted at room temperature, but typical Mg sheet alloys cannot be bent to such extreme deformation at room temperature without fracturing.

Previous research has shown that hemming of Mg sheets is possible at elevated temperature [7]. Both press-and-die (Fig. 1(a)) and roller hemming (Fig. 1(b)) have been performed with Mg sheet. Press hemming is a plane strain bending operation where hem steels bend the flange approximately 90° in a two-step process. Roller hemming is a different process in which a roller is driven around the edge of an outer panel at various angles to gradually bend the flange in multiple passes. Roller hemming, unlike press hemming, is not a plane strain bending action but includes a shear component of strain that distributes strain through a different path [8]. Press hemming of AZ31B sheet has been successfully done at a temperature of ~270°C [7]; while AZ31B sheet roller hemmed in four steps using a laser source to locally heat the sample ahead of the roller (Fig. 1) has been performed at ~170°C [9].

This study presents the microstructural evolution accompanying the successive laser-roller passes used to hem AZ31. Electron backscatter diffraction (EBSD) has been used to examine the microstructural changes in the hemmed zone following successive roller passes with and without laser heating. The effects of heat on the deformation mechanisms during this process are discussed.

Experimental Methods

The samples used in this study were prepared from 1.3 mm thick AZ31B-O sheet as follows : 1) flanged in an oven at 300°C in a wiping motion with a flange steel to an open angle of ~98°; 2) flanged as in (1) then roller-hemmed with a laser-roller speed of 50mm/s in four passes to open angles of 60, 30, 0 degrees and a final pass to flatten the hem; and 3) flanged as in (1) then roller hemmed *without* the laser until the second pass (an open angle of 30°). Samples for the EBSD study were prepared by sectioning the flanged and hemmed samples through the thickness of the bend to examine the mircrostructure. Heat treatment of some of the hemmed samples was done at 375°C in a salt bath. EBSD data from the polished samples were collected using a Leo 1450 scanning electron microscope fitted with a Hikari high speed EBSD camera and TSL data collection system.

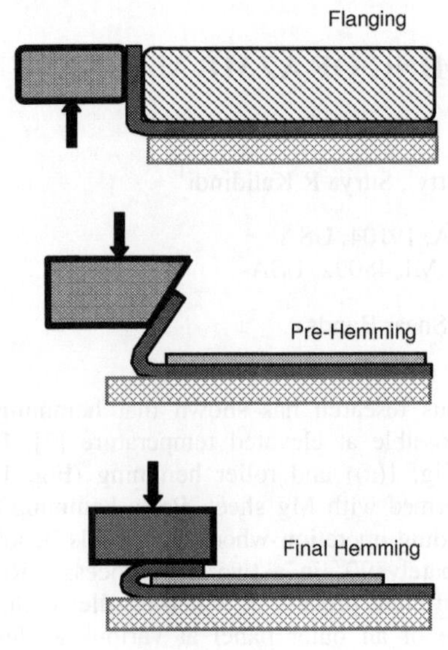

(a)

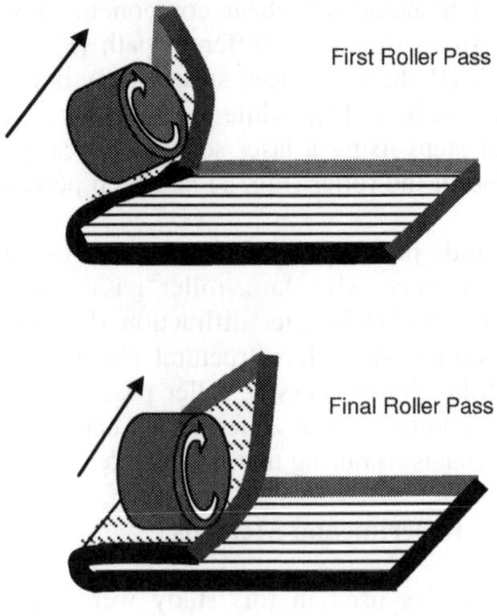

(b)

Figure 1. Schematic of, (a) press and (b) roller hemmning experimental setup. In laser roller hemming, a laser is placed ahead of the roller.

Results

Flanged Sheet

Figure 2 shows the microstructure of the flanged sample. A banded grain structure was observed at this stage that persisted in each sample after subsequent roller hemming steps. The three bands corresponded to the top tensile zone, bottom compressive zone and the central "neutral" zone of the bent sheet. In regions near the outer (OD) and inner (ID) radii of the bend, nearly defect-free grains, larger than that of the initial material were observed suggesting that these areas have recrystallized. The texture through the thickness of the flanged sheet remains constant, similar to the texture of the sheet before flanging, having a c-axis fiber texture parallel the sheet normal (ND) in all three bands. As will be discussed later, this data suggests that extension twinning did not play any role in altering the texture in the ID (compression band) at 300°C.

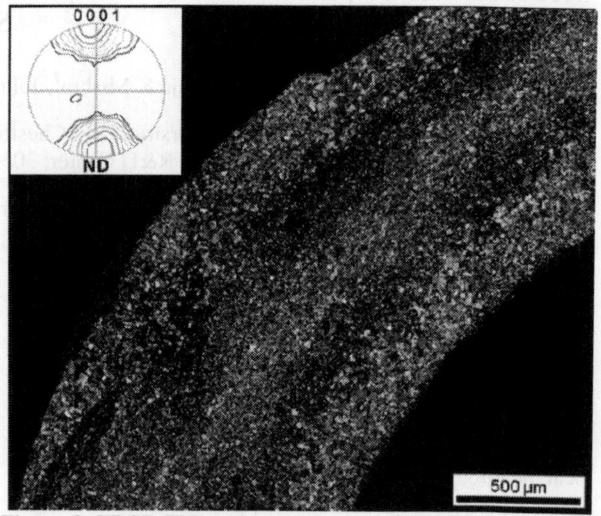

Figure 2. EBSD IPF with respect to the direction normal to the scan after flanging to an open angle of ~98°. The sheet extends vertically on the left of the image and the flange extends horizontally to the right of the image. The pole figure shows the texture through the thickness of the sheet.

Multi-Pass Laser-Roller Hemmed Sheet

Figure 3 shows the inverse pole figure EBSD maps with respect to the direction normal to the plane of the EBSD scan that illustrate the microstructural evolution during each of the four passes of laser-roller hemming at a roller speed of 50mm/s. The microstructure after each pass exhibited an asymmetrical, banded structure. The grain size along the OD and ID was found to increase until after the second pass. A smaller grain size along the OD and ID developed after the third pass, and it continued to decrease until the fourth and final pass. In this set of samples it was difficult to define an average grain size due to grain size gradients present through the thickness of the sheet as well as around the bend.

The textures for the OD, neutral band and ID along the axis in which the neutral band was the thinnest were measured separately. This location is illustrated on Figure 3(b) and was chosen to maintain consistency of location (and possibly the stress state) in each sample. The OD and neutral bands have a texture similar to the initial microstructure with the basal poles parallel with ND. The ID band has a basal texture component that is rotated ~90° from the initial orientation. These texture differences are due to the formation of extension twins in the compressed inner bend that rotated the c-axes by 86° from the ND to the sheet plane, normal to the bend axis.

a)

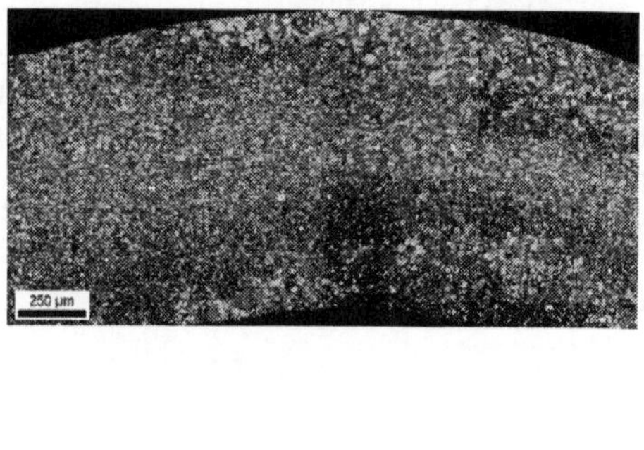

b)

c)

d)

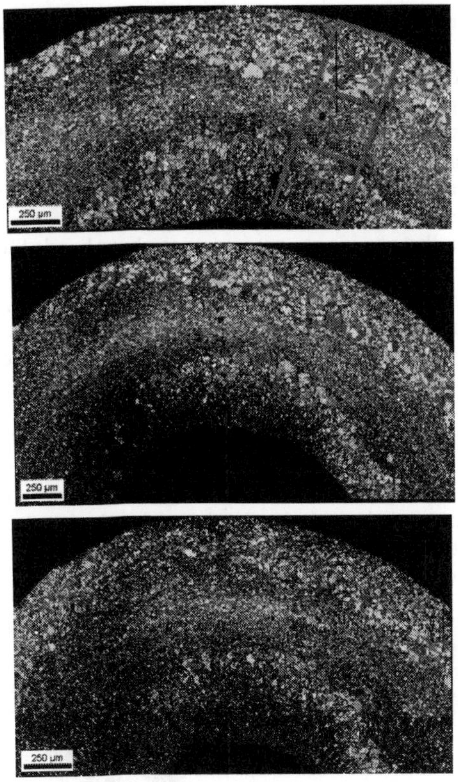

Figure 3. EBSD IPF with respect to the direction normal to the scan of the cross-section after each successive laser-roller hemming pass to open angles of (a) 60°, (b) 30°, (c) 0° and (d) final flat condition. The hem is oriented so the sheet is on the right in all scans. The boxes in (b) show the location of the section chosen to measure texture.

Roller Hemmed Sheet Without Laser Heating

Figure 4(a-b) compares a section of the microstructures of the sample hemmed *without* laser heating to that hemmed with laser heating after pass 2. The sample *without* laser heating was bent through the second pass successfully (to an open angle of 30°); however, further bending of this sample with or without the laser was unsuccessful, resulting in splitting of the hem.

After pass 2, both samples deformed with and without the laser, exhibited a larger grain size along the OD and ID compared to that of the flanged sheet. The textures of the samples hemmed with and without heat were similar to that of the flanged sheet along the OD and neutral axis. Unlike the flanged sample, the textures in the ID band in the rolled samples were rotated by ~90°. The sample hemmed *without* laser heating exhibited a stronger texture component associated with tensile twinning compared to the laser rolled sample.

The major difference between the two samples is the amount and distribution of contraction twinning along the OD. In Figure 4(b), multiple contraction and double twins were observed in the sample hemmed *without* the laser. Bands of low image quality were also present in this sample along the OD and seem to be distributed in a manner suggestive of insipient shear band formation. By comparison, the sample hemmed with a laser exhibited no contraction or double twinning or any evidence of shear band formation in Figure 4(a).

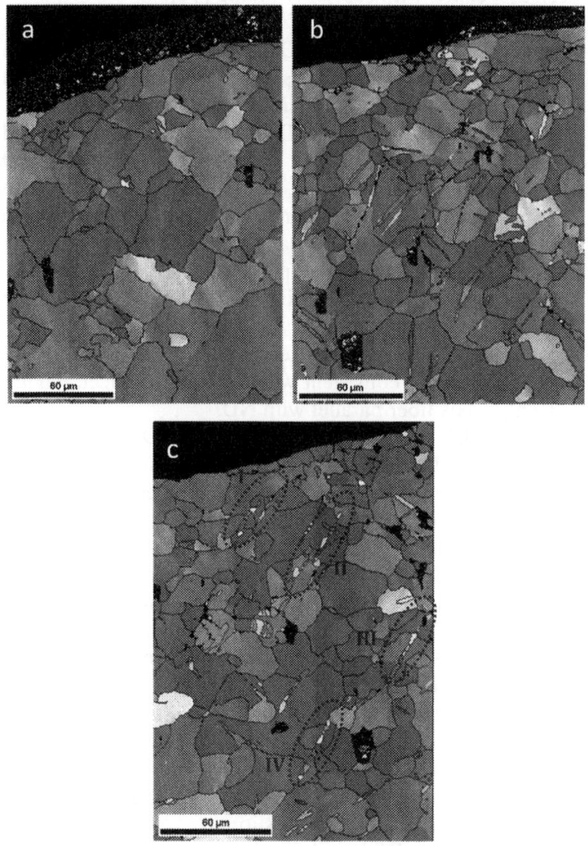

Figure 4. EBSD IPF map with respect to the normal direction of the sheet for the OD of samples hemmed to the second pass with an open angle of 30° under the following conditions; a) with a laser b) *without* a laser, and c) sample (b) followed by a 375°C heat treatment for 1 second in a salt bath. Four areas where recrystallized contraction or double twins have propagated through at least 2 grains in (c) are indicated on areas I-IV.

Discussion

Microstructure and Texture Evolution

High temperature flanging was a critical step during the hemming process as it provided the starting microstructure for the subsequent hemming steps. The banded microstructure that was developed by flanging, with larger grains in the OD and ID regions, was a product of recrystallization. The texture data suggests that during deformation at 300°C, non-basal deformation must have been accommodated by <c+a> type dislocations and very few tensile twins which rotate the c-axis by ~86° must have formed. These few tensile twins are retained in the recrystallized microstructure [10]. Recrystallization and grain growth in the deformed ID and OD bands during and after deformation contribute to a larger grain size whereas no recrystallization and grain growth has occurred in the neutral band which was exposed to 300°C. This banded structure remained throughout the rest of hemming and likely influenced the extent of different deformation mechanisms and texture evolution, as larger grains have been associated with ease of twinning [11].

At the outer edge of the flange and hem, a tensile strain is present along the OD band causing a contraction along the c-axis in most grains. In this orientation, Mg can only accommodate strain by either <c+a> pyramidal slip or contraction/double twinning. Contraction twins are thin and hard to detect using EBSD [12] due to their size and increased dislocation density. Although some twins were observed along the OD band, especially in the room temperature roller hemmed samples, either by measuring the misorientation relationships or by poor image quality bands, they did not comprise a large enough volume fraction of the material along the OD to significantly alter the texture. Therefore, the texture along the OD in all samples remained consistent with the starting sheet, having a strong c-axis fiber parallel with ND.

The compressive strain in the ID band can be accommodated by stretching the grains in this band normal to the sheet normal (ND). At low temperatures, where <c+a> slip is difficult, Mg alloys typically accommodate this strain state through tensile twinning. The texture observations from the ID bands in hemmed samples suggest that the grains in this band deformed by tensile twinning aligning the c-axis parallel to the sheet plane. This texture component became more pronounced after subsequent roller passes, indicating a larger volume fraction of twinned material. The stronger texture component in the ID band observed in samples rolled without the laser indicates more of the deformation in the laser heated sample may have been accommodated by tensile twinning than non-basal slip.

Failure in Sample Without Laser Heating

From the results presented above, it is clear that the heat from the laser affects the deformation behavior of AZ31B during hemming. Without the heat applied by the laser, the defect density after hemming to 30° in 2 passes is high enough to inhibit any further deformation from occurring in subsequent hemming passes. The fact that this sample could not be hemmed even after applying a laser in the third pass suggests that the stored defect density in this sample could not be recovered with the addition of laser heating.

One definitive difference in the microstructures is that contraction and double twins were observed more frequently in the large grains of the OD band in the sample hemmed *without* heat. The lack of these twins in the sample hemmed with heat indicates that the additional heat from the laser activated non-basal <c+a> slip to a greater extent; thus retarding the activation of contraction/double twinning. The addition of heat also must alter the dislocation mobility and dislocation interactions and affect stored dislocation density, allowing the material to accommodate further deformation in subsequent passes.

Although the exact failure mechanism of the sample hemmed *without* heat is unknown; the contraction/double twins are thought to play a key role in failure initiation. The twins may directly cause crack initiation and/or be responsible for abnormally high strain hardening rates and subsequent failure upon further straining. By reorienting the lattice either 56° or 38° for contraction and double twinning respectfully, basal slip can be easily activated inside the twins, resulting in the accumulation of a large amount of shear in these twins. In experiments on single crystal magnesium strained in tension parallel to the basal planes, Hartt and Reed-Hill found that double twins can exhibit strains approaching 1000 % leading directly to fracture along the twin boundaries [13]. It also has been suggested that these twins are precursors to shear bands [10]; which if formed in this material, could promote a larger scale failure mechanism. Figure 5(a) shows the OD of the sample hemmed *without* a laser with all poorly indexed points partitioned out. These poorly indexed bands were oriented at an angle, reminiscent of shear bands, along the OD. By examining some of these bands in more detail it becomes clear in Figure 5(b) that they are composed of contraction or double twins that have created a network through multiple grains. The networking of contraction twins could support their evolution into shear bands that do not anneal-out upon transient laser heating in the third pass and lead to fracture.

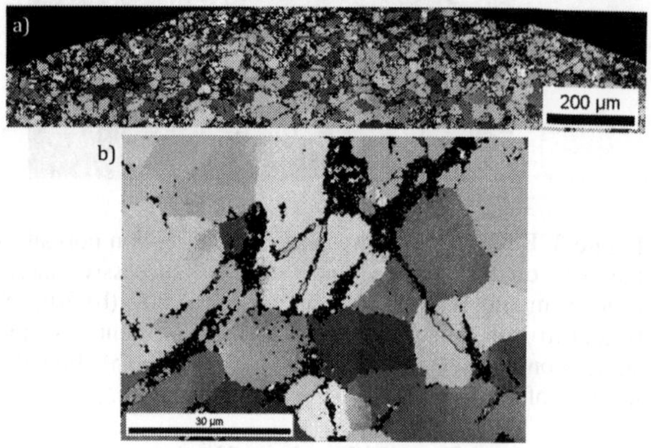

Figure 5. (a) OD of a sheet hemmed to the 2nd pass, an open angle of 30°, *without* the laser. (b) Higher resolution EBSD scan of red box in (a) illustrating that the bands of low quality, which may be shear bands (or precursors to them) are partially composed of double twins that have started to network with twins in neighboring grains.

To explore the potential effect of additional heat on the sample that was hemmed *without* a laser, the sample was heat treated at 375°C (the highest temperature that the surface of the sheet experienced under the laser) for a very short period of time (approximately 1 second) using a salt bath, then quenched in water to approximate the fast heat dissipation between the laser and the roller, the sheet acting as a heat sink. A comparison of the microstructure in the OD region before and after 375°C anneal is shown in figure 4(b-c). It can be seen that some of the contraction/double twins recrystallized from this short heat treatment. Areas in the heat treated sample, Figure 4(c), where recrystallized contraction/double twins extend through at least two grains are indicated by areas I-IV. Recrystallization by bulging did not occur to sweep these small grain zones that would have created a microstructure similar to that in the laser rolled sample after pass 2, leaving areas of very soft material, akin to insipient shear band formation, surrounded by a hard orientation matrix. Such a microstructure can promote plastic instability leading to shear banding and ultimate failure.

In addition to the differences in the occurrence of contraction/double twins, Fig. 4 shows that many grains retain high intra-grain misorientations. The grains in the laser rolled sample and the annealed sample seem to have retained a high degree of dislocation density. Without quantifying the dislocation density in these samples it is not possible to conclude if the sample rolled without laser has a higher amount of stored work caused by the different deformation modes activated at room temperature, including contraction/double twins that lead to higher work hardening [14] which ultimately led to early failure in pass 3.

References

1. B.L. Mordike, and T. Ebert, "Magnesium: Properties -- applications – potential," *Materials Science and Engineering A*, 302 (1) (2001), 37-45.

2. A. Galiyev, R. Kaibyshev, and G. Gottstein, "Correlation of plastic deformation and dynamic recrystallization in magnesium alloy ZK60," *Acta Materialia*, 49 (7) (2001), 1199-1207.

3. J.C. Tan and M. J. Tan, "Dynamic Continuous Recrystallization Characteristics in Two Stage Deformation of Mg-3Al-1Zn Alloy Sheet," *Materials Science and Engineering A*, 339 (1-2) (2003), 124-32.

4. Al-Samman, T. and G. Gottstein, "Dynamic recrystallization during high temperature deformation of magnesium," *Materials Science and Engineering: A*, 490 (1-2) (2008), 411-420.

5. J.A. del Valle and O.A. Ruano, "Influence of texture on dynamic recrystallization and deformation mechanisms in rolled or ECAPed AZ31 magnesium alloy," *Materials Science and Engineering: A*, 487 (1-2) (2008) 473-480.

6. S.W. Xu, S. Kamado, N. Matsumoto, T. Honma, Y. Kojima, "Recrystallization mechanism of as-cast AZ91 magnesium alloy during hot compressive deformation," *Materials Science and Engineering: A*, 527 (1-2) (2009), 52-60.

7. J. Carsley and S. Kim, "Warm Hemming of Magnesium Sheet," *Journal of Materials Engineering and Performance*, 16 (3) (2007), 331-338.

8. J.E. Carsley, "Microstructural evolution during bending: Conventional Vs. Roller hemming," *Trends in Materials and Manufacturing Technologies for Transportation Industries*, TMS (2005) 169-174.

9. J.E. Carsley, "Warm Bending Magnesium Sheet for Automotive Closure Panels," MS&T 2009, Pittsburgh, PA, 25-29 October 2009

10. X. Li, P. Yang, L.-N. Wang, L. Meng, and F. Cui, "Orientational Analysis of Static Recrystallization at Compression Twins in a Magnesium Alloy AZ31," *Materials Science and Engineering: A*, 517 (1-2) (2009), 160-69.

11. M. Barnett, Z. Keshavarz, A. Beer, and D. Atwell, "Influence of Grain Size on the Compressive Deformation of Wrought Mg-3Al-1Zn," *Acta Materialia*, 52 (17) (2004), 5093-103.

12. M.R. Barnett, "Twinning and the ductility of magnesium alloys: Part II. "Contraction" twins," *Materials Science and Engineering: A*, 464 (1-2) (2007), 2007, 8-16.

13. W.H. Hartt and R. E. Reed-Hill, "Internal Deformation and Fracture of Second-order {10-11}-{10-12} Twins in Magnesium," *Transactions of the Metallurgical Society of AIME*, 242 (1968), 1127-133.

14. M. Knezevic, A. Levinson, R. Harris, R.K. Misrha, R.D. Doherty, S.R. Kalidindi. "Deformation twinning in AZ31: Influence on strain hardening and texture evolution" *Acta Mat*erialia, 58 (2010), 6230-6242.

Magnesium Technology 2011
Edited by: Wim H. Sillekens, Sean R. Agnew, Neale R. Neelameggham, and Suveen N. Mathaudhu
TMS (The Minerals, Metals & Materials Society), 2011

The Warm Forming Performance of Mg Sheet Materials

P. E. Krajewski[1], P.A. Friedman[2] and J. Singh[3]

[1]General Motors Company LLC, Warren, MI
[2]Ford Motor Company, Dearborn, MI
[3] Chrysler Group LLC, Auburn Hills, MI

Keywords: Magnesium, Warm Forming, Sheet, Formability, AZ31

Abstract

The warm forming performance of five different magnesium sheet materials was evaluated using a heated pan die. Forming maps were generated to determine the optimum temperature and die conditions for successful forming. All five materials exhibited excellent formability above 325°C, with a robust forming window. Below 300°C, significant differences in the materials were observed. Some materials could be formed into a pan at temperatures as low as 175°C while others could not be formed below 300°C. The differences in forming performance were investigated considering material properties, tribology, and microstructure. These trials included both continuously cast and direct chill cast materials, and demonstrated that the continuously cast materials can be successfully warm formed.

Procedure

AZ31B-O magnesium sheet, 1 mm gage, was donated to the USCAR AMD602 warm forming project by Magnesium Elektron North America (MENA), POSCO, Magnesium Flachprodukte GmbH, Yinguang Magnesium Group Co., and Australian Commonwealth Scientific and Research Organization (CSIRO). These sheets were sheared into 500 mm x 550 mm blanks and then lubricated with a boron nitride (BN) containing synthetic lubricant from Fuchs. The blanks were then formed into a 300 mm x 400 mm x 70 mm deep rectangular pan under a variety of temperature and binder pressure conditions using the hydraulic press and die shown in Figure 1. The dimensions of the pan were selected to represent a shape that could be stamped at room temperature using steel, but not with aluminum or magnesium. The die was designed to have excellent thermal control to ensure a uniform temperature during forming. The goal was develop a forming map for each material and compare their relative range of forming, with special focus on comparing continuously cast and direct chill cast materials.

Figure 1 – Hydraulic press and insulated die used to stamp magnesium sheet.

Results/Discussion

Fully formed pans were successfully made without splitting or excessive wrinkling using all the provided materials; however, the range of temperature and die conditions which enabled successful forming were different depending on the material. All five materials showed excellent formability above 325°C. Examples of five pans, one formed from each material are shown in Figure 2. The pans were formed to the full depth of 70 mm without any fractures or local necking. A robust forming window over a wide range of binder pressures was observed. At temperatures below 300°C, forming results were varied; however four of the five materials were able to successfully form parts at temperatures as low as 200°C, including one of the continuously cast materials.

Summary

The present study demonstrated the magnesium sheet can be successfully used in a warm stamping process; including the lower cost, continuously cast sheet materials. Successful forming was observed over a range from 175°C to 350°C with the most robust forming found at 350°C.

Acknowledgements

The stamping trials were performed by Dennis Cedar at Troy Tooling Technologies. The work was funded by USCAR. This material is based upon work supported by the Department of Energy National Energy Technology Laboratory under Award Number DE-FC26-02OR22910 This report was prepared as an account of work sponsored by an agency of the United States Government. Neither the United States Government nor any agency thereof, nor any of their employees, makes any warranty, express or implied, or assumes any legal liability or responsibility for the accuracy, completeness, or usefulness of any information, apparatus, product, or process disclosed, or represents that its use would not infringe privately owned rights. Reference herein to any specific commercial product, process, or service by trade name, trademark, manufacturer, or otherwise does not necessarily constitute or imply its endorsement, recommendation, or favoring by the United States Government or any agency thereof. The views and opinions of authors expressed herein do not necessarily state or reflect those of the United States Government or any agency thereof.

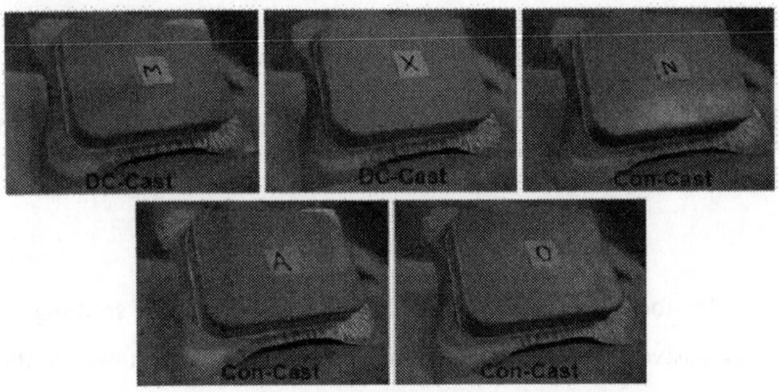

Figure 2 – Warm formed pans made from five different AZ31B-O magnesium sheet materials at 350C. The labels M, X, N, A, O indicate the supplier of the material (not shared publicly).

Magnesium Technology 2011

New Applications (Biomedical and Other)

Session Chairs:

Norbert Hort
(Helmholtz-Zentrum Geesthacht, Germany)
Michele V. Manuel
(University of Florida, USA)

Magnesium Technology 2011
Edited by: Wim H. Sillekens, Sean R. Agnew, Neale R. Neelameggham, and Suveen N. Mathaudhu
TMS (The Minerals, Metals & Materials Society), 2011

CURRENT RESEARCH ACTIVITIES OF BIOMEDICAL MAGNESIUM ALLOYS IN CHINA

Yufeng Zheng
Peking University, College of Engineering, Department of Advanced Materials and Nanotechnology; No.5 Yi-He-Yuan Road, Hai-Dian District, 100871 Beijing, China

Keywords: biomedical magnesium alloys, biodegradable implants

Abstract

The potential biomedical application of magnesium alloys as bioabsorbable / biodegradable implants in the human body has been extensively studied worldwide, and becomes an emerging and promising application direction.

The author served as the chairman in two recent conferences relating to biomedical magnesium alloys: (1) the *12th Annual Conference on Biomaterials* in December 2009, organized by the Chinese Society of Biomedical Engineering, and (2), a symposium on Biodegradable Metallic Materials in May 2010 at Peking University. Based on the papers presented in these conferences (about 50 in each), the frontier research activities of biomedical magnesium alloys in China for orthopedic and cardiovascular implants will be systematically summarized and comprehensively reviewed in this paper [1]. The research highlights the alloying system design, novel structure, degradation rate control, and surface modification methods; results will be demonstrated via plenty of in vitro and in vivo study data.

Reference

[1] Y.F. Zheng, "Current research activities of biomedical magnesium alloys in China", *JOM* 63/4 (2011), *to be published.*

Magnesium Technology 2011
Edited by: Wim H. Sillekens, Sean R. Agnew, Neale R. Neelameggham, and Suveen N. Mathaudhu
TMS (The Minerals, Metals & Materials Society), 2011

DESIGN CONSIDERATIONS FOR DEVELOPING BIODEGRADABLE MAGNESIUM IMPLANTS

Harpreet S. Brar[1], Benjamin G. Keselowsky[2], Malisa Sarntinoranont[3], Michele V. Manuel[1]

[1]University of Florida, Department of Materials Science and Engineering; Gainesville, FL 32611, USA
[2]University of Florida, Department of Biomedical Engineering; Gainesville, FL 32611, USA
[3]University of Florida, Department of Mechanical and Aerospace Engineering; Gainesville, FL 32611, USA

Keywords: Bioabsorbable, Design, Magnesium Alloy

Abstract

The integration of biodegradable and bioabsorbable magnesium implants into the human body is a complex undertaking that faces major challenges. The complexity arises from the fact that biomaterials must meet both engineering and physiological requirements to ensure the desired properties. Historically, efforts have been focused on the behavior of commercial magnesium alloys in biological environments and their resultant effect on cell-mediated processes. Developing causal relationships between alloy chemistry and microstructure, and its effect on cellular behavior can be a difficult and time intensive process. A systems design approach driven by thermodynamics has the power to provide significant contributions in developing the next generation of magnesium alloy implants with controlled degradability, biocompatibility, and optimized mechanical properties, at reduced time and cost. This approach couples experimental research with theory and mechanistic modeling for the accelerated development of materials. The aim of this article is to enumerate this strategy, design considerations and hurdles for developing new magnesium alloys for use as biodegradable implant materials [1].

Reference

[1] H.S. Brar, B.G. Keselowsky, M. Sarntinoranont, M.V. Manuel, "Design Considerations for Developing Biodegradable Magnesium Implants", *JOM* 63/4 (2011), *to be published.*

Magnesium Technology 2011
Edited by: Wim H. Sillekens, Sean R. Agnew, Neale R. Neelameggham, and Suveen N. Mathaudhu
TMS (The Minerals, Metals & Materials Society), 2011

COATING SYSTEMS FOR MAGNESIUM-BASED BIOMATERIALS - STATE OF THE ART

J. Waterman [1], M.P. Staiger [1]
[1] Department of Mechanical Engineering, University of Canterbury, Christchurch, New Zealand

Keywords: magnesium, biodegradable implants, coating, surface treatment, corrosion

Abstract

Magnesium and its alloys have the potential to be used for biodegradable orthopedic implants. However, the corrosion rate in physiological conditions is too high for most applications. For this reason, surface modification to slow the corrosion rate is of great interest. Such modifications must remain biologically compatible as well as protective in corrosive environments. What follows is a brief review of recent research in inorganic coatings and surface modifications to create coatings for magnesium-based biomaterials.

Introduction

Magnesium (Mg) and magnesium alloys have the potential to be useful in creating better orthopedic implants [1]. Magnesium alloys can offer the strength and toughness required for load bearing implants where ceramics and polymers fall short. Other metals currently used for implants, such as stainless steels and titanium alloys, have elastic moduli that are much higher than natural bone, leading to unwanted stress shielding. The elastic modulus of magnesium and many magnesium alloys are much closer to bone [2]. Also, a second surgery is required to remove current metallic implants. Magnesium shows promise as the material for biodegradable implants that degrade in the body without requiring removal. It is found in abundance in the body, and the degradation by-products have been shown to be non toxic [3]. In addition, magnesium may actually serve to stimulate new bone growth [1]. With all of these favorable properties magnesium looks very promising for this application. However, there are some challenges to be overcome. Magnesium and its alloys in general have low corrosion resistance, which is important for metal implants given the very aggressive environment in the physiological system [4]. Toxic degradation products and loss of mechanical properties are the major concerns. The low corrosion resistance of magnesium leads to loss of mechanical properties too quickly. It also leads to rapid hydrogen gas evolution within the body. For these reasons, pure unaltered magnesium metal is not an ideal implant material. Furthermore, it is desirable to optimize the biological response to these implants to maximize recovery. The biological response to the surface of the implant is important for compatibility of the implant with surrounding tissue [5]. Since corrosion and biocompatibility are ultimately surface phenomena, surface modification through treatments or coating systems are an obvious route to optimizing implant properties.

A good coating or surface treatment will be one that will control corrosion of the implant, maintaining mechanical integrity for the duration the implant is required. To be effective on many types of implants, it will be ideal for the coating to cover complex surfaces completely to ensure corrosion does not occur too quickly. Also, the coating must have good adhesion to the metal substrate, and have acceptable wear resistance to protect the implant during the insertion operation and throughout the load cycles of the implant life. It is also of great concern that the implant be non-toxic and fully degradable itself after the duration required. Additionally, good cellular response and attachment is desirable allow the implant to become fully integrated with the biological system. A suitable coating that meets these criteria will allow biodegradable magnesium implants to become a feasible alternative to current metallic orthopedic implants. What follows is a discussion of a few approaches taken to overcome the problems facing magnesium implants.

<u>The Corrosion of Magnesium</u>

In order to determine how best to control the corrosion of magnesium, it is important to understand the mechanism of corrosion, especially in physiological environments. In general, magnesium metal corrodes in aqueous environments to form magnesium hydroxide and hydrogen gas [6]. The overall corrosion reaction of magnesium is:

$$Mg + 2H_2O \rightarrow Mg(OH)_2 + H_2$$

Corrosion will in general form a surface coating of $Mg(OH)_2$ and/or MgO [6]. This forms a passivation layer that will generally slow the corrosion rate. The passive layer protects until the oxide layer is penetrated, exposing unoxidized metal to solution. When this happens, the corrosion rate accelerates locally. However, the corrosion degradation products passivate the surface again, slowing further corrosion [7].

When Magnesium is put into contact with other metals, galvanic corrosion will occur. This can be due to either external metals or internal secondary or impurity phases. Metals that have a low hydrogen overvoltage can cause severe galvanic corrosion. Examples are Ni, Fe, and Cu. Metals with lower hydrogen overvoltage like Al, Zn, Cd, and Sn are not as bad [6].

Physiological environments are typically very aggressive environments for metals. Corrosion tests carried out in simulated body fluids (SBF), which are fluids that contain many of the ions commonly found in the body, give some indication of how a metal will perform in an actual environment. In simulated body fluid, the rapid corrosion of magnesium can mainly be attributed to the presence of chloride ions. The Cl^- is absorbed into the $Mg(OH)_2$ surface and $MgCl_2$ is formed. This compound is quite soluble, and thus increases the rate of corrosion by destroying the passivation layer of magnesium hydroxide [8]. The effect of the Cl^- can be seen when corrosion is compared to a sample in a similar solution with a minimal amount of Cl^-. Pitting and surface cracking is the major mechanism of corrosion in SBF due to Cl^- ions [9, 10].

Alloying the magnesium with appropriate elements to overcome problems of corrosion and biocompatibility has been the focus of much research, and will be important to perfecting magnesium implants [2, 11, 12]. Alloying elements that create passivation layers that are more stable than $Mg(OH)_2$ can slow down corrosion, but often alloys with multiple phases exhibit microgalvanic corrosion [6].

<u>Biocompatibility</u>

Increasing the biocompatibility of the implant is also important. For the implant to have the maximum effect and a speedy recovery time, the manner in which the cells and tissue react to the implant is critical. [13] Toxicity is crucial for the purposes of a biodegradable implant. If the implant is to degrade in vivo, then the inclusion of toxic elements or compounds must be eliminated from the implants. This condition eliminates many of the common techniques used for corrosion resistance of magnesium in other applications. For example, heavy metals and chromate conversion coatings that have been used for magnesium corrosion resistance [14] should not be used. In addition to toxicity, cellular response while the implant is in place is of great importance. The biocompatibility of titanium implants has been investigated fairly extensively. A magnesium implant will also need to exhibit good biocompatibility if it is to be effective.

Coatings

A simple but effective method to reduce corrosion is to coat the metal so as to provide a barrier between the metal and the environment. Corrosion resistant coatings are commonly used for metals in many applications. In this case, we want the coating to keep the corrosive ions in the physiological system (especially Cl^-) away from the magnesium until sufficient time has passed and the bone is healed. Once this has occurred, it is desirable that the coating yield to the environment and degrade along with the magnesium leaving no harmful traces.

Anodization

A common practice is to use anodization to form a protective corrosion resistant layer on metals. Anodization uses and electrical current to form a thick, dense passivation layer that is more protective than the natural layer that develops. Magnesium oxide layers can be formed by anodizing Mg which slow short term corrosion rate in SBFs like Hanks solution [15, 16]. Furthermore, qualities of the coating like density can be optimized by controlling the voltage profiles during anodization [17]. The corrosion protection of the oxide layer can be increased by performing the anodization in a silicate solution, creating Mg_2SiO_4 on the surface as well as MgO [18]. However, magnesium oxide layers will convert to magnesium hydroxide in aqueous solutions, and magnesium hydroxide is soluble in chloride solutions such as body fluid [19]. Xue, for example, found anodization greatly increased polarization resistance in a NaCl solution, but after 2 hours the resistance decreased by a factor of 2 [18]. Because these films are not stable in biological solutions, anodization alone is unlikely to produce the protective coating required for many bioapplications. Instead, anodization can be used as a pretreatment to another coating system. For example, anodized layers can be used to control the amount and rate of calcium phosphate compounds precipitated on the surface in SBFs [20].

Metal coatings

Metal coatings have been used to prevent degradation of magnesium. Pure magnesium coating on a magnesium alloy particularly susceptible to corrosion has been show to decrease the corrosion. If alloying elements increase the corrosion potential, then a high purity deposition coating of pure magnesium on the surface will slow the corrosion [21]. Coatings of other metals may be achieved as well. Physical vapour deposition coating of aluminum has been successfully applied to magnesium AZ31 alloy [22]. The coating did corrode in a NaCl solution, however, and aluminum is not the best choice for biocompatibility. Still, a metal coating for corrosion resistance is a viable option for protection, provided the coating metal has a low toxicity. Zirconium [23] and zinc [24] are possible candidates in this respect. The unfortunate downside to metallic coatings on magnesium is galvanic corrosion when any defect in the coating is present. If the coating metal is more noble than the substrate, the substrate will corrode preferentially [9]. A gap in the coating will lead to sever galvanic corrosion of the substrate underneath, and loss of mechanical properties will follow. Since magnesium has a lower corrosion potential than all other engineering metals the applications of metallic coatings is limited.

Calcium Phosphates

One of the most biocompatible coating options for orthopedics is calcium phosphate coatings. Coatings of apatite have been researched extensively for use in biomedical applications [25, 26]. Several biologically important types of apatite are hydroxyapatite (HA), octocalcium phosphate (OCP), tricalcium phosphate (TCP), dicalcium phosphate dihydrate (DCPD) and amorphous calcium phosphates, often containing other secondary ion substitutions. The mineral component of bone itself is an apatite, but also contains other ions such as carbonate and phosphate groups. The mineral portion of bone is built up on a collagen matrix. The similarity of some of these apatites to bone minerals, such as stoichiometric HA, gives them good biocompatibility. Calcium phosphate coatings have been used to increase the integration of the implant to the bone [27, 28]. HA coated titanium implants have been found to increase cell proliferation and bone formation [29]. Because calcium phosphate compounds can be fairly insoluble in physiological conditions and are very biocompatible they are a perfect candidate for a protective coating for magnesium implants. To provide adequate corrosion protection, the coating must be complete and adherent to the substrate. The quality of the coating created is dependent on the process and process parameters used for formation.

Plasma Spray

The most popular commercial procedure to attach calcium phosphate coatings to a metallic implants is the plasma spray method. This entails using a jet of neutral gas, inserting the material or precursors for the material to be coated into the jet and plasmatizing by some means, for example a DC arc. The plasma spray is deposited onto the substrate where the coating forms. The process makes precise control of the coating thickness possible, as well as the composition by control of the feed powders [30]. Plasma sprayed hydroxyapatite has been used to coat implants to increase biocompatibility of implants [31, 32]. However, the high temperatures required for this process means care must be taken to avoid the presence of unwanted phases, as well as decomposition of the coating and/or substrate [30]. For Mg and its biocompatible alloys, the temperatures reached with plasma spray will be great enough to melt or change the substrate, making this technology difficult to apply to Mg. Plasma spray is also limited by the line of sight to the substrate, making complex shapes and porous structures difficult to coat uniformly. Other techniques to apply calcium phosphate coatings to metallic substrates have been attempted to overcome problems related to plasma spray such as poor integrity and adhesion, low crystallinity, and mechanical failure of the coating [33]. Some of these methods might be more appropriate for magnesium based materials.

Chemical Vapor Deposition

Coatings can be created by chemical reactions of gaseous chemicals near a heated substrate. This technique is known as chemical vapor deposition. Coating with this method allows the production of multilayer and composite coatings as well as complex shapes without line of sight. [34] For example, CVD has been used as an alternative to plasma spray to create a stable, crystalline, bioactive hydroxyapatite coating on 316 L stainless steels [35]. Most CVD processes are fairly high temperature, often requiring the substrate to be stable at temperatures above 600 deg C. However, there are lower temperature processes being explored to limit temperatures to around 180 deg C [14].

Pulsed Laser Deposition (PLD)

PLD uses a laser to vaporize a target and allow the vapor to condense on the surface of the substrate. This method allows for greater control of the crystallinity and composition and thickness of the coating [26]. Pulsed laser deposition has been studied on other implant materials, notably titanium [33, 36-38], however their protectiveness in corrosive solutions on magnesium is not widely reported.

Ion Beam Assisted Deposition (IBAD)

IBAD can produce coatings with good adhesion and allows precise control of the coating chemistry, including the Ca:P ratio. Yang et al reported using IBAD to coat AZ31 [39]. Calcium phosphate coatings created were heat treated transform into HA. Annealing improved the mechanical properties. Coatings were tested in a 3% NaCl solution for corrosion properties. Coated samples were more protective, although pitting did occur through cracks in the coating.

Solution Coatings

Solution chemistry methods for coating metals with calcium phosphates have a number of advantages. Simple and low cost setup, the ability to coat complex and porous materials, and the ability to use low temperatures make solution chemistry synthesis of calcium phosphates attractive for Mg substrates. The oxidation of Mg creates a local pH rise which promotes calcium phosphate deposition in solutions containing calcium and phosphate ions [40]. This can be leveraged to easily create coatings in simple solutions.

However, this method also has its drawbacks, as magnesium is highly reactive in aqueous environments it tends to corrode during the coating process. Furthermore, for calcium phosphate compounds, substitutions by Mg^{2+} ion in the crystal lattice of compounds like Hydroxyapatite are known to promote defects, limit crystallization [41], and decrease the stability of the created compound [42]. Hiromoto and Yamamoto reported HA coatings created on Mg and alloys in single step solution treatments [43]. Coating solution ionic concentration and pH affect the coating deposited. They reported reduction in corrosion current density of 10^3 to 10^4 times lower in 3.5% NaCl than uncoated Mg using potentiodynamic polarization (PDP). Hu et al. reported creation of a DCPD coating in solution on AZ91 alloy by titrating K_2HPO_4 into a $Ca(NO^3)^2$ solution [44]. The DCPD coating was transformed into HA over time in SBF, and the corrosion resistance of the coating increased to 4210 ohms from 331 on uncoated mg. By PDP tests, corrosion current density dropped from 70 to 2.6 $\mu A/cm^2$. Tomozawa reported solution chemistry techniques to form HA on pure Mg in solution [45]. Findings included increasing temperature to 333K and above increased HA formation and also $Mg(OH)_2$ formation. However, by increasing Ca concentration HA formation could be increased without affecting $Mg(OH)_2$ formation, which may be undesirable to have underneath the coating, due to its high solubility. Xu et al. reported in vivo studies using calcium phosphate coatings prepared by solution methods on Mg Mn Zn alloys [46]. The surface properties of the coated samples were much more friendly to cell growth and overall osteoconductivity than uncoated samples. Because these solution coatings can avoid using any toxic elements, the biocompatibility is good. However the issue remains creating a coating that is dense and adherent enough to remain crack free and fully protective in solution for the required amount of time.

Electrodeposition

Calcium phosphate coating formation can be assisted by the application of external potentials and currents. These processes are collectively referred to as electrodeposition. The setup for these methods is inexpensive and relatively simple, and the process can be done at low temperatures. The processing parameters can be easily controlled to optimize the coating created.

Electrochemical Assisted Deposition (ECAD)

ECAD uses the reduction of water in an aqueous solution to promote the precipitation of calcium phosphates on the surface of a metallic substrate. Reduction of H_2O generates H_2 gas and leaves behind OH^- at the cathode. This leads to a local rise in pH at the surface of the substrate. An increase in pH decreases the solubility of calcium phosphates in solution, leading to precipitation at the surface.

The ECAD process for calcium phosphate can be controlled using a number of methods [26]. A constant potential can be held between the working electrode (the surface to be coated) and the counter electrode, typically made of an inert material such as platinum or graphite. Constant potential between the working and counter electrodes means the potential between the solution and the electrode is not directly controlled. Potential and current are therefore related to aspects of the coatings such as cell geometry, solution composition, counter electrode material, etc.

This method has been used to form coatings of HA on AZ91D and have been shown to reduce corrosion currents by the electrochemical methods PDP and electrochemical impedance spectroscopy (EIS) [47].

Alternately, the coating process can be performed potentiostatically, where the working electrode is held at a constant voltage compared with a reference electrode. The reference electrode is placed near the working substrate in order to maintain a constant potential difference between the solution and the coating. This is useful for keeping the potential at a desired level to cause reduction of H_2O, without rising to levels that can reduce other actors the solution. In this setup, the current will decrease as the substrate becomes coated and the exposed area of the electrode drops. Lower OH production at the cathode can lead to a pH drop near the working electrode, and therefore a drop in deposition rate. DCPD coatings have been created on Mg alloys using the potentiostatic method, and while coatings do reduce corrosion rates, total coverage from thick, dense coatings that completely protect the substrates remain an issue [48].

To keep the hydroxide ion production constant, constant current, or galvanostatic methods have been used to coat magnesium with this method. The standard three electrode cell is used, but the controls are set to keep current applied between the working and counter electrodes constant. H_2O is the only molecule undergoing oxidation and reduction in solution, then the rate of OH^- production near the surface remains constant, keeping the pH profile roughly equal during the process. The voltage can spike during the process, especially after the substrate is partially coated. Song found that galvanostatic method of coating to form a calcium phosphate coating that was protective in SBFs, dropping the measured corrosion current significantly over a 48 hour test [47]. Wen used the galvanostatic method to coat AZ31 with HA. PDP results show protective effects. Corrosion potential (E_{corr}) was increased and corrosion current density (I_{corr}) decreased. Post treatment in an alkali solution can result in greater stability of the coating, resulting in a lower rate of mass loss over 30 days [49]. Calcium phosphate and chitosan composite coatings have also been reported on Mg alloys by Wu et al. [50]. By performing the deposition in a solution containing a HA suspension as well as chitosan, composite coatings could be formed during the deposition process.

Finally, the voltage profile can be controlled to whatever is allowed by the control systems used to monitor a 3 cell system. Other voltage profiles used in this process include pulsed profiles. Ion diffusion in the coating solution can limit the rate of coating, needing longer than the current can source. Additionally, the reduction of water at the cathode produces hydrogen gas when the voltage is high. The net result of these factors can be loose, porous coatings [51]. Pulse duration can be modified to change properties of the coatings, including crystal size, with longer durations leading to larger crystals [42], others To limit this pulsed current deposition on MgZnCa alloys has been studied, with the results being increased E_{corr} by PDP and decreased I_{corr} [51].

One downside of ECAD methods is the lack of complete dense adhesive coatings [48]. and others. Hydrogen evolution at the surface of the metallic substrate creates gas bubbles that block chemical formation of the ceramic at the interface, resulting in volcano-like interfaces as reported by Kumar et al. [52]. Unfortunately the evolution of hydrogen gas is unavoidable for these type of ECAD coatings in aqueous solutions, and is indeed necessary to raise the pH at the surface and drive the coating process.

For electrodeposition processes that are not based around pH solubility, it is possible to avoid the reduction of water by performing the reactions in no aqueous solutions. Due to the need for an eclectically conductive fluid, ionic liquids present themselves as a alternative plating medium. Bakkar and Neubert have reported successful plating of Mg substrates with metallic Zn to increase corrosion resistance [24]. Ionic liquid composition, applied current density, and substrate alloy composition was found to affect the coating created.

Surface Treatments

Another method for increasing the corrosion resistance of a magnesium alloy is to modify the surface structure of the metal itself. Magnesium alloys undergo microgalvanic corrosion when multiple phases exist in an alloy, one more cathodic than another [9]. This can be detrimental to the corrosion resistance, as it accelerates local corrosion at the anodic phase. Furthermore, there exists differences in energy between the grains and grain boundaries in magnesium and magnesium alloys. Galvanic cells therefore can form between grains and grain boundaries. For pure Mg, grain boundaries are more cathodic than grains, leading to undercutting of grains near the boundaries [19]. Local corrosion can cause a loss of mechanical properties at rates faster than predicted by bulk corrosion rate measurements. For load bearing implant design, controlling and eliminating localized corrosion is desirable.

A method for removing these effects is to modify the surface to homogenize it. If it is possible make the surface completely amorphous, it will eliminate the formation of galvanic cells between grains and boundaries.

Bulk Metallic Glasses

Making the matrix a completely amorphous bulk metallic glass is the obvious approach to completely remove corrosion difference due to crystal structure in the metal. Casting metallic glasses requires specific favorable alloy compositions and extremely high cooling rates to freeze the liquid metal without the formation of crystalline grains [53]. This makes it difficult to create fully amorphous structures of dimensions larger than a few millimeters. Alloy compositions can be optimized to form BMGs. Recently MgZnCa alloys at compositions with good glass forming ability have been of interest for biomedical and other applications [54-58]. As these ternary alloys do not contain toxic elements, enabling them to be used in degradable implants.

Amorphous MgZnCa alloys have been tested in vivo to show reduced hydrogen evolution [58]. Amorphous alloys above 28 at. % Zn showed particularly good passivation properties, due to zinc oxide layer. However, the BMG samples created were sheets only 0.5mm thick, so again, dimensional constraints due to the formation of BMGs may limit the use of fully glassy materials for larger implants. Gu et al. reported similar results for 2mm thick MgZnCa BMG samples, including reduced corrosion rate and increased response in cell culture tests [54].

An alternative to forming bulk metallic glasses for the substrate is to make the surface amorphous on a crystalline material using a surface treatment technique. Ion implantation provides a possible route of creating a modified surface like this. Accelerating ions to high velocities and implanting them into the surface of the substrate can cause a collision cascade that destroys the long range order of crystals in the metal, leaving a glassy surface. Chatterjee has demonstrated such formation in aluminum substrates by ion implantation [59]. Glassy surfaces can also be formed after ion beam mixing a coated surface layer with the substrate [60].

Ion Implantation

Ion implantation offers a method of surface modification to increase corrosion resistance. Advantages of ion implantation include modification of the existing substrate surface, often creating a gradual transition between the modified surface and the bulk of the material. This generally tends to make strong, adherent treatments that do not have the problems of adhesion, thermal stresses, and crackings that separate secondary coating phases tend to have. Plasma immersion ion implantation (PIII or PI^3) of Al, Zr, and Ti has been used to create corrosion resistance on AZ91. The mechanism is introducing the elements near the surface increases the density of the corresponding oxide during corrosion, resulting in a more protective passivation layer [61]. However, for biodegradable implants, additional elements may not be desirable if they are linked to toxicity like Al, or do not degrade such as Ti. However, using a thin implanted layer at the surface as opposed to complete alloying can decrease the amount of toxic ions required. Of course, toxic ions should be avoided if possible, limiting the species that may be used. Wan et al. used Zn ions due to their biocompatibility, but found Zn ion implantation in MgCa alloys increased corrosion rate rather than decreasing it [62]. Oxygen ion implantation has also been attempted but with little success against chloride solutions [63]. Nitrogen ion implantation has been used to improve corrosion resistance of magnesium as well. Nakatsugawa et al. reported N ion implantation reducing the corrosion rate of AZ91D to 15% of the untreated metal in 5% NaCl [64]. Similarly, Tian et al. used PIII to improve the corrosion resistance of AZ31B [65]. With ion implantation, the implant energy and dosage are critical to maximizing the implant performance.

Conclusion

Effective surface treatment of biodegradable implants will be able to slow corrosion for the period of time necessary to retain the mechanical strength required of the implant. The corrosion rate must be low enough that hydrogen evolution and corrosion products do not cause problems for the surrounding tissue. To do this, the coating must be adherent and consistent on the substrate. The coating system must protect the substrate from corrosion due to chloride ion attack and galvanic corrosion between phases. Defects such as cracks and voids in the coating should be avoided to prevent localized corrosion compromising the mechanical integrity of the device. Furthermore, the coating must be fully biocompatible and biodegradable. Toxic elements should not be incorporated into the coating system. When coatings can be created that fully provide the corrosion protection necessary, biodegradable orthopedic implants will become possible.

References

1. Staiger, M.P., et al., *Magnesium and its alloys as orthopedic biomaterials: A review.* Biomaterials, 2006. **27**(9): p. 1728-1734.
2. Witte, F., et al., *Degradable biomaterials based on magnesium corrosion.* Current Opinion in Solid State and Materials Science. **12**(5-6): p. 63-72.
3. Li, L., J. Gao, and Y. Wang, *Evaluation of cyto-toxicity and corrosion behavior of alkali-heat-treated magnesium in simulated body fluid.* Surface and Coatings Technology, 2004. **185**(1): p. 92-98.
4. Park, J.B. and J.D. Bronzino, eds. *Biomaterials Principles and Applications.* 2003, CRC Press: Boca Raton. 250.
5. Williams, D.F., *On the mechanisms of biocompatibility.* Biomaterials, 2008. **29**(20): p. 2941-2953.
6. Song, G.L. and A. Atrens, *Corrosion Mechanisms of Magnesium Alloys.* Advanced Engineering Materials, 1999.
7. Hänzi, A.C., et al., *On the biodegradation performance of an Mg-Y-RE alloy with various surface conditions in simulated body fluid.* Acta Biomaterialia, 2009. **5**(1): p. 162-171.
8. Wang, Y., et al., *Corrosion process of pure magnesium in simulated body fluid.* Materials Letters, 2008. **62**(14): p. 2181-2184.
9. Song, G. and A. Atrens, *Understanding Magnesium Corrosion - A Framework for Improved Alloy Performance.* Advanced Engineering Materials, 2003. **5**(12): p. 837-858.
10. Xin, Y., et al., *Influence of aggressive ions on the degradation behavior of biomedical magnesium alloy in physiological environment.* Acta Biomaterialia, 2008. **4**(6): p. 2008-2015.
11. Kirkland, N.T., et al., *A survey of bio-corrosion rates of magnesium alloys.* Corrosion Science, 2010. **52**(2): p. 287-291.
12. Gu, X., et al., *In vitro corrosion and biocompatibility of binary magnesium alloys.* Biomaterials, 2009. **30**(4): p. 484-498.
13. Ramaswamy, Y., C. Wu, and H. Zreiqat. *Orthopedic coating materials: considerations and applications.* 2009.
14. Gray, J.E. and B. Luan, *Protective coatings on magnesium and its alloys -- a critical review.* Journal of Alloys and Compounds, 2002. **336**(1-2): p. 88-113.
15. Lei, T., et al., *Enhanced corrosion protection of MgO coatings on magnesium alloy deposited by an anodic electrodeposition process.* Corrosion Science. **52**(10): p. 3504-3508.
16. Zhao, L., et al., *Growth characteristics and corrosion resistance of micro-arc oxidation coating on pure magnesium for biomedical applications.* Corrosion Science, 2010. **52**(7): p. 2228-2234.
17. Liu, X., et al., *Effect of alternating voltage treatment on the corrosion resistance of pure magnesium.* Corrosion Science, 2009. **51**(8): p. 1772-1779.
18. Xue, D., et al., *Corrosion protection of biodegradable magnesium implants using anodization.* Materials Science and Engineering: C, 2010. **In Press, Accepted Manuscript**.
19. Friedrich, H.E. and B.L. Mordike, *Magnesium technology : metallurgy, design data, applications.* 2006, Berlin ; New York: Springer. xxii, 677 p.
20. Hiromoto, S., et al., *Precipitation control of calcium phosphate on pure magnesium by anodization.* Corrosion Science, 2008. **50**(10): p. 2906-2913.
21. Yamamoto, A., et al., *Improvement of corrosion resistance of magnesium alloys by vapor deposition.* Scripta Materialia, 2001. **44**(7): p. 1039-1042.
22. Wu, G., X. Zeng, and G. Yuan, *Growth and corrosion of aluminum PVD-coating on AZ31 magnesium alloy.* Materials Letters, 2008. **62**(28): p. 4325-4327.
23. Xin, Y., et al., *Corrosion behavior of ZrN/Zr coated biomedical AZ91 magnesium alloy.* Surface and Coatings Technology, 2009. **203**(17-18): p. 2554-2557.
24. Bakkar, A. and V. Neubert, *Electrodeposition onto magnesium in air and water stable ionic liquids: From corrosion to successful plating.* Electrochemistry Communications, 2007. **9**(9): p. 2428-2435.
25. Barrere, F., *Biomimetic Calcium Phosphate Coatings: Physicochemistry and Biological Activity.* 2002, University of Twente: Enschede.
26. León, B. and J.A. Jansen, *Thin Calcium Phosphate Coatings for Medical Implants.* 2008: Springer.
27. Barrere, F. and C.M.v.d. Valk, *Osteogenecity of octacalcium phosphate coatings applied on porous metal implants.* Journal of Biomedical Materials Research Part A, 2003. **66A**(4): p. 779-788.
28. Barrere, F., et al., *Osteointegration of biomimetic apatite coating applied onto dense and porous metal implants in femurs of goats.* Journal of Biomedical Materials Research Part B: Applied Biomaterials, 2003. **67B**(1): p. 655-665.
29. Zhu, L., et al., *Biomimetic coating of compound titania and hydroxyapatite on titanium.* Journal of Biomedical Materials Research Part A, 2007. **83A**(4): p. 1165-1175.
30. Kou, M., T. Toda, and O. Fukumasa, *Production of fine hydroxyapatite films using the well-controlled thermal plasma.* Surface and Coatings Technology, 2008. **202**(22-23): p. 5753-5756.
31. Chen, D., et al., *Apatite formation on alkaline-treated dense TiO2 coatings deposited using the solution precursor plasma spray process.* Acta Biomaterialia, 2008. **4**(3): p. 553-559.
32. Gross, K.A. and C.C. Berndt, *Thermal processing of hydroxyapatite for coating production.* Journal of Biomedical Materials Research, 1998. **39**(4): p. 580-587.
33. Koch, C.F., et al., *Pulsed laser deposition of hydroxyapatite thin films.* Materials Science and Engineering: C, 2007. **27**(3): p. 484-494.
34. Choy, K.L., *Chemical vapour deposition of coatings.* Progress in Materials Science, 2003. **48**(2): p. 57-170.

35. Trommer, R.M., L.A. Santos, and C.P. Bergmann, *Alternative technique for hydroxyapatite coatings.* Surface and Coatings Technology, 2007. **201**(24): p. 9587-9593.
36. Mróz, W., et al., *Characterization of calcium phosphate coatings doped with Mg, deposited by pulsed laser deposition technique using ArF excimer laser.* Micron, 2009. **40**(1): p. 140-142.
37. Sygnatowicz, M. and A. Tiwari, *Controlled synthesis of hydroxyapatite-based coatings for biomedical application.* Materials Science and Engineering: C, 2008. **29**(3): p. 1071-1076.
38. Zeng, H. and W.R. Lacefield, *The study of surface transformation of pulsed laser deposited hydroxyapatite coatings.* Journal of Biomedical Materials Research, 2000. **50**(2): p. 239-247.
39. Yang, J.X., et al., *Modification of degradation behavior of magnesium alloy by IBAD coating of calcium phosphate.* Surface and Coatings Technology, 2008. **202**(22-23): p. 5733-5736.
40. Gray-Munro, J.E. and M. Strong, *The mechanism of deposition of calcium phosphate coatings from solution onto magnesium alloy AZ31.* Journal of Biomedical Materials Research Part A, 2009. **90A**(2): p. 339-350.
41. Bigi, A., et al., *Magnesium influence on hydroxyapatite crystallization.* Journal of Inorganic Biochemistry, 1993. **49**(1): p. 69-78.
42. Lin, S., R.Z. LeGeros, and J.P. LeGeros, *Adherent octacalciumphosphate coating on titanium alloy using modulated electrochemical deposition method.* Journal of Biomedical Materials Research, 2003. **66A**(4): p. 819-828.
43. Hiromoto, S. and A. Yamamoto, *High corrosion resistance of magnesium coated with hydroxyapatite directly synthesized in an aqueous solution.* Electrochimica Acta, 2009. **54**(27): p. 7085-7093.
44. Hu, J., et al., *Microstructure evolution and corrosion mechanism of dicalcium phosphate dihydrate coating on magnesium alloy in simulated body fluid.* Materials Chemistry and Physics, 2010. **119**(1-2): p. 294-298.
45. Tomozawa, M., S. Hiromoto, and Y. Harada, *Microstructure of hydroxyapatite-coated magnesium prepared in aqueous solution.* Surface and Coatings Technology. **204**(20): p. 3243-3247.
46. Xu, L., et al., *In vitro and in vivo evaluation of the surface bioactivity of a calcium phosphate coated magnesium alloy.* Biomaterials, 2009. **30**(8): p. 1512-1523.
47. Song, Y.W., D.Y. Shan, and E.H. Han, *Electrodeposition of hydroxyapatite coating on AZ91D magnesium alloy for biomaterial application.* Materials Letters, 2008. **62**(17-18): p. 3276-3279.
48. Chun-Yan, Z., et al., *Comparison of calcium phosphate coatings on Mg-Al and Mg-Ca alloys and their corrosion behavior in Hank's solution.* Surface and Coatings Technology, 2008. **204**(21-22): p. 3636-3640.
49. Wen, C., et al., *Characterization and degradation behavior of AZ31 alloy surface modified by bone-like hydroxyapatite for implant applications.* Applied Surface Science, 2009. **255**(13-14): p. 6433-6438.
50. Wu, C., et al., *Fabrication of calcium phosphate/chitosan coatings on AZ91D magnesium alloy with a novel method.* Surface and Coatings Technology, 2010. **204**(20): p. 3336-3347.
51. Wang, H.X., et al., *In vitro degradation and mechanical integrity of Mg-Zn-Ca alloy coated with Ca-deficient hydroxyapatite by the pulse electrodeposition process.* Acta Biomaterialia, 2010. **6**(5): p. 1743-1748.
52. Kumar, M., H. Dasarathy, and C. Riley, *Electrodeposition of brushite coatings and their transformation to hydroxyapatite in aqueous solutions.* Journal of Biomedical Materials Research, 1999. **45**(4): p. 302-310.
53. Greer, A.L., *Metallic glasses...on the threshold.* Materials Today, 2009. **12**(1-2): p. 14-22.
54. Gu, X., et al., *Corrosion of, and cellular responses to Mg-Zn-Ca bulk metallic glasses.* Biomaterials, 2010. **31**(6): p. 1093 - 1103.
55. Li, Q.-F., et al., *Microstructure and mechanical properties of bulk Mg-Zn-Ca amorphous alloys and amorphous matrix composites.* Materials Science and Engineering: A, 2008. **487**(1-2): p. 301-308.
56. Senkov, O.N. and J.M. Scott, *Formation and thermal stability of Ca-Mg-Zn and Ca-Mg-Zn-Cu bulk metallic glasses.* Materials Letters, 2004. **58**(7-8): p. 1375-1378.
57. Senkov, O.N. and J.M. Scott, *Glass forming ability and thermal stability of ternary Ca-Mg-Zn bulk metallic glasses.* Journal of Non-Crystalline Solids, 2005. **351**(37-39): p. 3087-3094.
58. Zberg, B., P.J. Uggowitzer, and J.F. Loffler, *MgZnCa glasses without clinically observable hydrogen evolution for biodegradable implants.* Nat Mater, 2009. **8**(11): p. 887-891.
59. Chatterjee, P. and A.K. Batabyal, *Metallic glass formation by CH4+ ion implantation into Al, Fe and Ni thin films.* Journal of Non-Crystalline Solids, 1990. **124**(2-3): p. 131-138.
60. Liu, B.X., W.S. Lai, and Q. Zhang, *Irradiation induced amorphization in metallic multilayers and calculation of glass-forming ability from atomistic potential in the binary metal systems.* Materials Science and Engineering: R: Reports, 2000. **29**(1-2): p. 1-48.
61. Liu, C., et al., *Corrosion behavior of AZ91 magnesium alloy treated by plasma immersion ion implantation and deposition in artificial physiological fluids.* Thin Solid Films, 2007. **516**(2-4): p. 422-427.
62. Wan, Y.Z., et al., *Influence of zinc ion implantation on surface nanomechanical performance and corrosion resistance of biomedical magnesium-calcium alloys.* Applied Surface Science, 2008. **254**(17): p. 5514-5516.
63. Wan, G.J., et al., *Corrosion properties of oxygen plasma immersion ion implantation treated magnesium.* Surface and Coatings Technology, 2007. **201**(19-20): p. 8267-8272.
64. Nakatsugawa, I., R. Martin, and E.J. Knystautas, *Improving Corrosion Resistance of AZ91D Magnesium Alloy by Nitrogen Ion Implantation.* Corrosion, 1996. **52**(12): p. 921-926.
65. Tian, X.B., et al., *Corrosion resistance improvement of magnesium alloy using nitrogen plasma ion implantation.* Surface and Coatings Technology, 2005. **198**(1-3): p. 454-458.

Magnesium Technology 2011
Edited by: Wim H. Sillekens, Sean R. Agnew, Neale R. Neelameggham, and Suveen N. Mathaudhu
TMS (The Minerals, Metals & Materials Society), 2011

CORROSION, SURFACE MODIFICATION, AND BIOCOMPATIBILITY OF MG AND MG ALLOYS

Sannakaisa Virtanen[1], Ben Fabry[2]
[1]University of Erlangen-Nuremberg, Institute for Surface Science and Corrosion, Department of Materials Science (WW-4), Martensstrasse 7, 91058 Erlangen, Germany
[2]University of Erlangen-Nuremberg, Center for Medical Physics and Technology, Department of Physics, Henkestraße 91, 91052 Erlangen, Germany

Keywords: Magnesium, corrosion, biodegradation, biocompatibility

Abstract

This manuscript summarizes recent studies on the corrosion behavior of Mg and Mg alloys in simulated biological environments, as well as interactions between corroding Mg (alloy) surfaces and cells. The influence of different types of simulated body solutions on the corrosion behavior of Mg is discussed. The effects of different types of chemical surface treatments on cell adhesion and spreading is presented. Moreover, possible routes to further optimize the corrosion and the biological performance of Mg alloys by surface modification are discussed.

Introduction

Mg-based materials corrode in aqueous biological environments and are of growing interest for use as biodegradable implants (see e.g. reviews [1,2]). For a successful and biologically safe application, a thorough understanding of the corrosion behavior in (simulated) physiological environments is required. In spite of the increasing number of publications in this field, many open questions still exist. In particular, detailed mechanisms of the effects of proteins and cells on Mg corrosion are unknown. Concerning the effects of proteins on the dissolution rate of Mg alloys, contradicting effects have been reported. For the Mg alloy AZ31, albumin addition has been found to show relatively little effect on the electrochemical behavior, but for pure Mg and the alloy LAE442 albumin addition led to strongly accelerated anodic dissolution [3]. On the other hand, corrosion inhibition by albumin addition has been reported for the AZ91 alloy in SBF [4], as well as for Mg-Ca alloy in water and in NaCl solutions [5]. These seemingly contradicting findings could be due to different experimental approaches, but may also indicate that albumin effects on corrosion are alloy-dependent.

Moreover, the effect of Mg dissolution on the biological environment needs to be characterized, as the corrosion reaction is coupled with H_2 gas liberation and surface alkalization. An increasing number of *in vitro* cell culture studies on corroding Mg alloy surfaces have been recently reported (see e.g., [6-11]. However, as different alloys and different cell lines have been used in these studies, it is not easy to compare the results from different studies or to draw general conclusions on the critical Mg alloy corrosion induced effects on the biological performance.

The aim of our studies is to further elucidate the interactions between corroding Mg (alloy) surfaces and cells, by an interdisciplinary approach of electrochemistry, surface modification and analysis, and cell culture testing. Of special interest is the development of novel chemical and biological surface functionalization methods, for tailoring the corrosion rate and biocompatibility.

Experimental

Experiments were carried out on cp-Mg (99.9%, Chempur) as well as on commercial WE43 and AZ91 Mg alloys. Corrosion behavior of the materials was studied in simulated body fluids (SBF 5[12]) and in cell culture medium [Dulbecco's Modified Eagle's cell culture medium (DMEM) with addition of 10% fetal bovine serum (FBS)].

Corrosion rates and modes were studied by Mg ion release measurements (atomic absorption spectroscopy), H_2 gas collection, pH measurements of the medium, electrochemical impedance spectroscopy, and various surface characterization methods.

Different types of surface treatments, including simple chemical passivation in 1 M NaOH or soaking in simulated body fluids, were explored regarding their ability to optimize the corrosion behavior and the biocompatibility. Moreover, functionalization of Mg surface by protein layers was explored. Surface coating with bovine serum albumin (BSA) was achieved by silanization of the surface with 3-aminopropyltriethoxysilane (APTES) and using an ascorbic acid (vitamin C) linker for attachment of BSA. For comparison, magnesium discs were directly incubated in aqueous BSA solution for 24 h at room temperature under gentle stirring of the solution. Details to the experimental procedures for preparation of the protein layers can be found in Ref. [13].

For the cell culture experiments, Mg samples were sterilized under UV irradiation with a wavelength of 260 nm. Cells from a human cervical cancer cell line (HeLa) were cultured in an incubator (5% CO_2, 37°C, 95% relative humidity) for 24 h in 24-well plates containing the differently treated Mg samples. DMEM with addition of 10% FBS and 100 U/ml Penicillin/Streptomycin was used as a cell culture medium. After 24 h in the incubator, the cells were fixed in 2% paraformaldehyde and stained with Alexa red phalloidin to visualize the actin cytoskeleton of the cells, and with Hoechst 33342 to visualize the cell nucleus. Fluorescence microscopy of the stained cells was carried out with a Leica DMI 6000B microscope.

Results and Discussion

Time-dependent corrosion behavior

Corrosion behavior of cp-Mg as well as Mg alloys was studied in SBF solutions as well as in cell culture medium. In agreement with findings from other laboratories [14,15], significantly lower corrosion rates were observed in cell culture medium than in SBF. The good corrosion resistance in cell culture medium has been brought into context of surface passivation by Mg-carbonates

[14,15]; however, detailed mechanisms on the reactions taking place on Mg surface in SBF vs. cell culture medium still need to be elucidated.

In previous studies, different types of simulated physiological solutions have been explored regarding their effect on the corrosion behavior of Mg alloys with an emphasis on the biodegradation behavior. A critical summary of the influence of the experimental parameters, including the electrolyte used, on the corrosion behavior of Mg alloys has been recently presented in [16]. A large number of investigations on Mg corrosion has been carried out in non-buffered NaCl solutions, however, these data have only limited value for understanding the corrosion behavior of Mg alloys in biological environments because body fluids are buffered. For Mg alloys, solution buffering is more critical than for many other metals, as a very strong alkaline pH shift takes place in the vicinity of corroding Mg surfaces. In non-buffered solutions, the pH can reach values >10-11 (starting from a neutral solution). In this case, the pH value is approaching the region of passivity of Mg [17]. Even though the presence of chloride ions in the environment hinders the formation of a stable passive surface, the high pH value suppresses the active dissolution rate. In simulated body fluids, buffered to pH 7.4, less strong pH effects occur. Therefore, different type of surface reactions take place on Mg in buffered or in non-buffered solutions.

Of course, not only buffering is important for the type of surface reactions taking place, but also the exact solution composition, as already discussed above in the case of SBF solutions versus cell culture medium. The nature of the corrosion layers changes from crystalline $Mg(OH)_2$ to an amorphous hydrated, carbonated (Mg,Ca)-phosphate in SBF [18]. The corrosion product layers from Mg and Mg alloys in SBF solutions are not highly protective, and active dissolution of Mg takes place even though the corrosion rate is gradually decreasing with time, with increasing coverage of the surface by the corrosion product layer [18,19].

Bovine serum albumin (BSA) addition showed a surprisingly complex time-dependent influence on the corrosion rate: in the first 6 h of immersion, a higher polarization resistance value was measured in presence of BSA in the solution, but during longer immersion times the polarization resistance again decreased, and for longer exposure times remained even lower than the polarization resistance in SBF in absence of albumin (Fig. 1). The initial increase of the polarization resistance could indicate adsorption of BSA on the surface, which partially blocks the dissolution reaction. The non-steady time-dependence suggests that the albumin adsorption layer changes with time, which is not surprising considering that the Mg alloy surface dissolves with a relatively fast rate during the experiment. This could for instance lead to removal of albumin from the surface. However, it is not clear why in the longer-term experiment the R_p-values remain significantly lower in the presence of albumin in the solution than in albumin-free solution. Clearly, a detailed surface characterization after different exposure times is required to understand the effects of albumin on the electrochemical behavior of Mg.

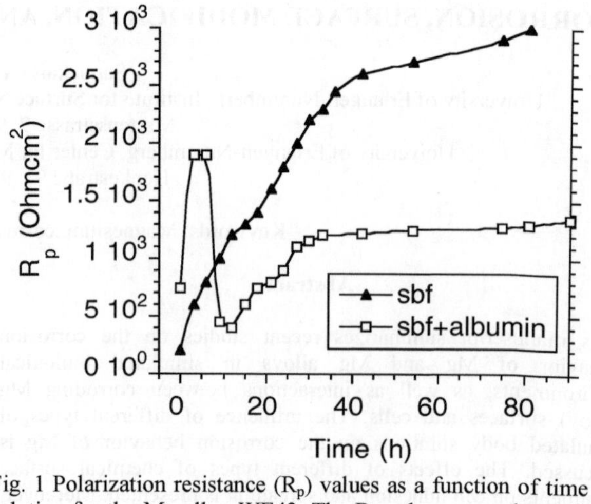

Fig. 1 Polarization resistance (R_p) values as a function of time in solution for the Mg alloy WE43. The R_p-values were determined from electrochemical impedance spectroscopy measurements in sbf or in sbf + 40 g/l bovine serum albumin (BSA) at 37°C.

<u>Cell culturing on corroding Mg surfaces</u>

For conventional cell culture experiments with direct adhesion of the cells on the sample surface, the surface of cp-Mg, without any surface modification was far too reactive for cell adhesion and survival. Even though strong retardation of the corrosion rate takes place for longer immersion times of Mg in cell culture medium, the initial corrosion rate was very high. Immediate hydrogen gas liberation and alkalinization of the culture medium could be observed. Cytotoxicity testing indicated that high Mg-concentrations - corresponding to the dissolution of the samples in the cell culture medium - did not reduce cell survival [20]. Rather, the single most important cause for reduced cell adhesion and survival was the increased pH in the cell culture medium due to Mg corrosion [21].

A viable strategy to achieve cell adhesion on corroding Mg (alloy) surfaces under *in vitro* cell culture conditions is to reduce the initial surface reactivity [22]. For instance, Figure 2 shows the effect of simple Mg surface passivation by soaking in 1 M NaOH. On the NaOH passivated surface, good spreading of the cells can be seen (2a). This in contrast to the polished Mg sample, in this case only very few non-spread cells were seen on the surface (2b). Although the $Mg(OH)_2$ passive film formed on Mg in NaOH is not stable in cell culture medium, it provides sufficient short-term protection to enable initial cell adhesion.

In general, all surface treatments that reduced the initial reactivity of the surface also increased the cell survival rate in our studies, but distinct differences were observed between different types of surface treatments as they altered not only surface reactivity but also chemistry, roughness, and wettability. For example, Mg samples incubated in SBF showed a biphasic behavior. During incubation in SBF, the Mg surface was covered by an amorphous hydrated, carbonated (Mg,Ca)-phosphate layer. Monitoring the cell behavior on such a surface as a function of time showed that initially strong cell spreading took place, but over time cell death occurred. These findings indicate that the surface chemistry and roughness of coatings formed by soaking in SBF are initially beneficial for cell adhesion and spreading, but after longer times the high reactivity of the surface induces cell death, presumably due to the local surface pH increase as these coatings are porous and hence not highly protective. The best coating, until now, in view of an efficient reduction of the corrosion rate and a good biocompatibility was the layer formed by incubating the Mg sample in cell culture medium [20].

Electrochemical experiments (open-circuit potential measurements and impedance spectroscopy EIS) of Mg surfaces were carried out in the absence and presence of a cell adhesion layer on the surface [21]. The electrochemical response correlated with the spreading of the cell layer on the surface. The formation of a cell layer on the Mg alloy surface significantly reduced the corrosion rate, as indicated by a reduced pH increase with time for Mg samples covered with a cell layer.

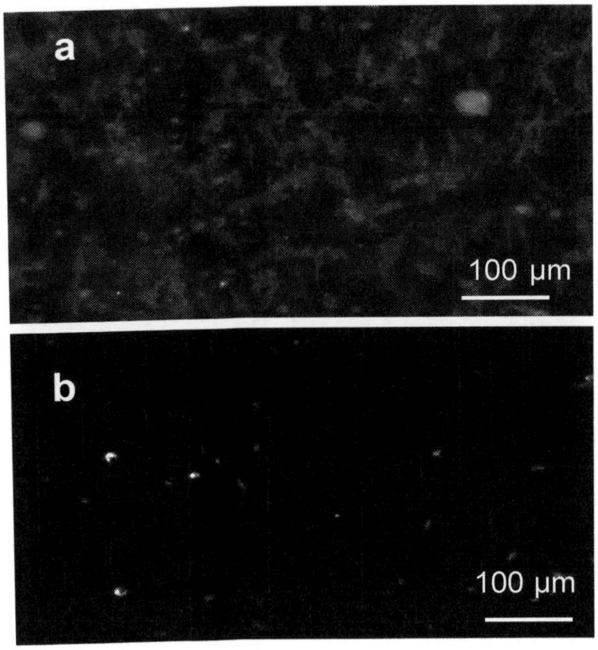

Fig. 2 Actin staining of HeLa cells on cp-Mg surfaces after 24 h culture. a: Mg surface passivated for 24 h in 1 M NaOH prior to cell culture testing; b: polished Mg surface.

Functionalization of Mg surface by proteins

As shown in Fig. 1, protein adsorption on Mg surfaces can have a significant effect on the corrosion behavior. We further studied protein adsorption on Mg surfaces, and also explored the possibility to use protein adsorption layers as a pre-treatment for tailoring the corrosion behavior and biocompatibility. First trials to simply coat Mg by directly soaking in BSA containing aqueous solutions were only partially successful; even though BSA adsorption took place, this was accompanied by corrosion (and H_2 liberation), and hence no homogeneous albumin layer could be prepared on the surface. To achieve a homogeneous protein coating on Mg surfaces, we therefore exploited silane coupling chemistry of a OH^- -terminated Mg surface. To enhance the OH^- termination, pre-passivation in NaOH is feasible, but the method works also for samples simply passivated in air (native oxide layers formed in air after surface polishing). The success of albumin-coating was verified by XPS (X-ray photoelectron spectroscopy) and ToF-SIMS (Time-of-flight Secondary Ion Mass Spectroscopy) measurements [13].

One advantage of such selective protein adsorption is to prevent non-selective protein adsorption from body fluids, and hence to condition the surface to the desired cell reaction. In our work, we are also interested in the effect of such protein coatings on the corrosion behavior. All experiments until now indicate that homogeneous protein coatings can be highly efficient in decreasing the corrosion rate. An example is shown in Fig. 3, illustrating the cumulative H_2 gas liberation as a function of time for albumin-coated Mg, in comparison to ground Mg surface. An efficient protection of the Mg surface by the albumin layer is observed, and the protective effect remains clearly detectable for a period of 5 days at least (longer-term experiments are running, results not shown). The exact amount of H_2 liberated from protein-covered samples varies, depending on the details of the surface pre-treatments. But it is also evident from Fig. 3 that no complete protection of the surface can be achieved, i.e., slow dissolution still takes place, which is desired for the material to be biodegradable.

This approach is currently being further explored on different proteins and different Mg alloys. Moreover, optimization of the protein coatings by a variation of surface pre-treatments, linker molecules, and soaking time is being carried out. Especially in the case of Mg alloys, specific pre-treatments of the heterogeneous multiphase alloy surface may be required, to ensure a homogeneous reactivity towards the linker molecules in the first coating step. For the optimized (homogeneous and well-structured) protein coatings, *in vitro* cell culture studies with suitable cell lines will be carried out in order to study the biological performance of the layers.

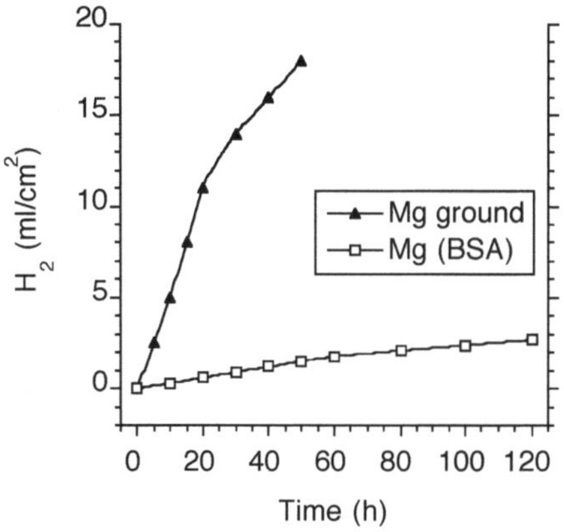

Fig.3 H_2 gas volume as a function of time for ground Mg sample and an albumin-coated Mg sample exposed to SBF.

References

1. M.P. Staiger, A.M. Pietak, J. Huadmai, G. Dias, Magnesium and its alloys as orthopedic biomaterials: a review, *Biomaterials* 27 (2006)1728-1734
2. F. Witte, N. Hort, C. Vogt, S. Cohen, K.U. Kainer, R. Willumeit, F. Feyerabend, Degradable biomaterials based on magnesium corrosion, *Current Opinion in Solid State and Materials Science* 12 (2008) 68-72.
3. W.-F. Mueller, M.F. L. de Mele, M. L. Nascimento, M. Zeddier, Degradation of magnesium and its alloys: Dependence on the composition of the synthetic biological media, *J. Biomed. Mater. Res.* 90A (2009) 487-495.
4. C. Liu, Y. Xin, X. Tian, P.K. Chu, Degradation susceptibility of surgical magnesium alloy in biological fluid containing albumin, *J. Mater. Res.* 22 (2007) 1806-1814.
5. C.L. Liu, Y.J. Wang, R.C. Zheng, X.M. Zhang, W.J. Huang, P.K. Chu, In vitro corrosion degradation behaviour of Mg-Ca alloy in the presence of albumin. *Corrosion Science* 52 (2010) 3341-3347.
6. L. Li, J. Gao, Y. Wang, Evaluation of cyto-toxicity and corrosion behavior of alkali-heat-treated magnesium in simulated body fluid. *Surface & Coating Technology* 185 (2004) 92-98.
7. Y. Zhang, H.-R. Tao, Y.-H. He, G.-Y. Zou, Y. Jian, S.-X. Zhang, B.-L. Zhang, J.-N. Li, C.-L. Zhao, X.-N. Zhang, Cytotoxicity and hemolytic properties of biodegradable Mg-Zn alloy, *Journal of Clinical Rehabilitative Tissue Engineering* 12 (2008) 8162 – 8166.
8. X. Gu, Y. Zheng, Y. Cheng, S. Zhong, T. Xi, In vitro corrosion and biocompatibility of binary magnesium alloys, *Biomaterials* 30 (2009) 484-498.
9. Y.H. Yun, Z. Dong, D. Yang, M.J. Schulz, V.N. Shanov, S. Yarmalenko, Z. Xu, P. Kumta, C. Sfeir, Biodegradable Mg corrosion and osteoblast cell culture studies, *Materials Science and Engineering C* 29 (2009) 1814-1821.
10. E. Zhang, D. Yin, L. Xu, L. Yang, K. Yang, Microstructure, mechanical and corrosion properties and biocompatibility of Mg-Zn-Mn alloys for biomedical application, *Materials Science and Enigineering C* 29 (2009) 987-993.
11. S. Zhang, J. Li, Y. Song, C. Zhao, X. Zhang, C. Xie, Y. Zhang, H. Tao, Y. He, Y. Jiang, Y. Bian, In vitro degradation, hemolysis and MC3T3-E1 cell adhesion of biodegradable Mg-Zn alloy, *Materials Science and Engineering C* 29 (2009) 1907-1912.
12. L. Müller, F.A. Müller, Preparation of SBF with different HCO_3^- content and its influence on the composition of biomimetic apatites. *Acta Biomaterialia* 2 (2006) 181-189.
13. M. Killian, V. Wagener, P. Schmuki, S. Virtanen: Covalent functionalization of metallic magnesium with protein layers, *Langmuir* 26 (2010) 12044-12048.
14. X.N. Gu, Y.F. Zheng, L.J. Chen, Influence of artificial biological fluid composition on the biocorrosion of potential orthopedic Mg-Ca, AZ31, AZ91 alloys, *Biomedical Materials* 4 (2009) 1-8.
15. A. Yamamoto, S. Hiromoto, Effect of inorganic salts, amino acids and proteins on the degradation of pure magnesium in vitro, *Materials Science and Engineering C* 29 (2009) 1559-1568.
16. W.-D. Mueller, M. Lucia Nascimento, M. Fernández Lorenzo de Mele, Critical discussion of the results from different corrosion studies of Mg and Mg alloys for biomaterial applications, *Acta Biomaterialia* 6 (2010) 1749–1755.
17. M. Pourbaix (1974) *Atlas of Electrochemical Equilibria in Aqueous Solutions, 2nd ed.* NACE, Houston.
18. R. Rettig, S. Virtanen, Composition of corrosion layers on a magnesium rare-earth alloy in simulated body fluids, *J. Biomed. Mater. Res. 88A* (2009) 359-369.
19. R. Rettig, S. Virtanen, Time-dependent electrochemical characterization of the corrosion of the magnesium alloy WE43 in simulated body fluids, *J. Biomedical Materials Research A* 85A (2008) 167-175.
20. S. Keim, J.G. Brunner, B. Fabry, S. Virtanen, Control of magnesium corrosion and biocompatibility with biomimetic coatings, *J. Biomedical Materials Research B* (in press, 2010).
21. F. Seuss, S. Keim, M. Can Turhan, B. Fabry, S. Virtanen (to be submitted)
22. C. Lorenz, J. Brunner, P. Kollmannsberger, L. Jaafar, B. Fabry, S. Virtanen, Acta Biomaterialia 5 (2009) 2783-2789.

Magnesium Technology 2011
Edited by: Wim H. Sillekens, Sean R. Agnew, Neale R. Neelameggham, and Suveen N. Mathaudhu
TMS (The Minerals, Metals & Materials Society), 2011

MAGNESIUM ALLOYS FOR BIOABSORBABLE STENTS: A FEASIBILITY ASSESSMENT

Charles Z. Deng[1], Rajesh Radhakrishnan[1], Steve R. Larsen[1], Dennis A. Boismer[1], Jon S. Stinson[1], Adrienne K. Hotchkiss[1], Eric M. Petersen[1], Jan Weber[2], and Torsten Scheuermann[3]

[1]Boston Scientific Inc., Maple Grove, Minnesota, USA
[2]Boston Scientific, Maastricht, The Netherlands
[3]Boston Scientific Technology Center GmbH, Munich, Germany

Keywords: magnesium, stent, bioabsorbable

Abstract

Today, stent designs consist of permanent metal alloy scaffolds which hold arteries open after percutaneous coronary intervention (PCI) to maintain arterial blood flow. Bioabsorbable stents are being investigated as an alternate for permanent stents, that disintegrate and dissolve in the body. In this article, we profile magnesium (Mg) alloy as a candidate for bioabsorbable stent material, and discuss aspects of its properties and challenges. Experimental data are generated in effort to draw correlations between in vivo vessel absorption and in vitro degradation, and to provide an overview of alloy mechanical properties, stent designs, and electrochemical behaviors. Preclinical porcine coronary model test results exhibit early on-set and rapid corrosion presenting a challenge to researchers to establish material design concepts that balance degradation time, duration for need of scaffolding, and healing.

Introduction

Cardiovascular disease is the leading cause of death in the world. Millions of people undergo PCI every year to treat the many varieties of vascular disease. Most often, the therapies utilized include the placement of a permanent stent implant to keep arteries open after the procedure. Over the past several years, many researchers in industry and academia have been exploring the concept of an absorbable stent that goes away over time [1-6].

Potential benefits of absorbable stents include cessation or elimination of antiplatelet therapies that complicate the necessity for the treatment of other medical conditions the patients require - adding risk to the overall health. Other benefits are possible including restoration of vessel properties such as vasomotion which is the ability for the vessel to expand and contract as the body attempts to modulate blood flow between times of rest and exercise. Additionally, absorbable stents would likely create the ability to more efficiently treat a previously stented site, or to enable the option of a bypass graft without the hindrance of an existing implant. Absorbable stents may also be used successfully as a way to deliver anti-restenosis drugs such as Everolimus or Paclitaxel.

Metal stents made of magnesium alloy have been studied for several years in animals [7-10] and humans [11-13] with reportedly encouraging results in terms of strength and degradation time. However there are many complexities associated with engineering the stent that range from its basic strength, biocompatibility, healing response, and rate of absorption. Of fundamental concern are the many interactive biological, chemical, and metallurgical mechanisms that dictate how these metal stents dissolve in the body and the yet to be discovered pathways of how the body responds to the material and its corrosion induced by-products.

Scope of this study was the evaluation of different magnesium alloys and coatings to gain knowledge about their corrosion rates and modes in animals and its predictability through bench tests.

Assessments

<u>Design of absorbable stent</u>

Magnesium alloys have generated a significant increase in interest from researchers in the search for biodegradable implant materials [14-30]. The design criteria for a magnesium stent challenge conventional permanent stent implant thinking due to its relatively low modulus, low ductility, and high corrosion rate. The stent must function mechanically for basic design performance such as radial strength and deliverability, and must compete in the marketplace not just with current stents available, but with next generation stents in the next 5-10 years. Starting a priori from a thin design (less than 100 micrometer strut thickness) in combination with a minimum corrosion period of at least 3-6 months so as to allow encapsulation in the neointima while providing sufficient radial force against elastic recoil, leads to the realization that many commercially available magnesium alloys are not compatible with stent design criteria, forces one to consider new alloys and a protective coating to be used in the design. The protective coating itself should be thin, pinhole free and robust, not only to avoid adding overall stent thickness, but also to withstand deformations occurring during assembly, delivery, and expansion of the stent. Figure 1 demonstrates the technical feasibility of such a coating.

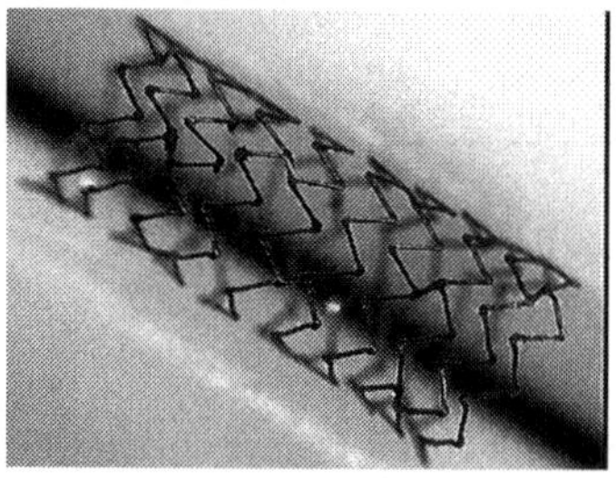

Figure 1. A thin coated magnesium alloy stent immersed in simulated body fluid for 4 weeks at 37°C showed little corrosion and remained intact, while uncoated stents only remained intact for a few days.

Additionally, the stent must be biocompatible, and the luminal surface should permit rapid endothelialization and should be compatible with anti-restenotic drugs eluting from the stent. A too rapid corrosion process in later phases can result in a deposition of large amounts of magnesium phosphate as shown in the left of Figure 2. The magnesium phosphate may in fact simply replace the initial magnesium volume, beating the overall biodegradability. The right image in Figure 2 is a slow corroding MZX, a non-commercial alloy covered by magnesium hydroxide coating and tissue.

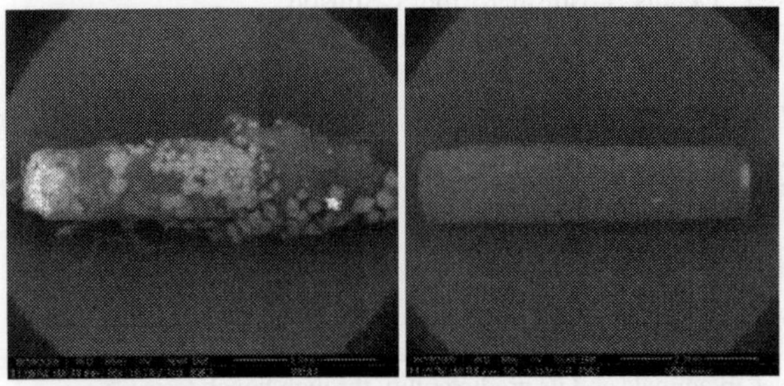

Figure 2. Explanted bare magnesium rods (left–WE43, right–MZX, a non commercial alloy) from mouse subcutaneous implant in 1- month.

Mechanical properties

Tensile test data was generated for each of the magnesium alloys evaluated utilizing samples of annealed stent-sized tubing. Tensile tests were performed using parameters in general accordance with ASTM E8, "Standard Test Methods for Tensile Testing of Metallic Materials". Mandrels were inserted into the tube at the sample ends to aid gripping during the test which was performed on a Sintech 1 G load frame, with a grip separation of 76.2 mm and elongation to fracture was measured over 50.8 mm. Test speeds were ranged from 0.254-5.08 mm/min. A 50.8 mm gage length extensometer was used for measuring strain.

Table I presents average tensile test data for the annealed magnesium alloy tubing. The materials evaluated have a relatively narrow range in strength and ductility at levels which represent a challenge for stent design and performance. No significant strain rate effect was observed for any of the materials over the range of rates tested.

Table I. Tensile test data of annealed Mg alloy and 316L stainless steel tubing.

Material	Elastic Modulus (GPa)	0.2% Yield Strength (MPa)	Tensile Strength (MPa)	Elongation (%)
316L SS	193	275	595	50
WE43B	43	169	254	18
ZK31	45	223	287	13
AZ80	43	163	255	7
MZX*	43-46	165-203	265-308	8-23

* MZX is a series of non commercial alloy which contains magnesium, zinc, and other elements.

Electrochemical properties

A survey of magnesium alloy bio-corrosion rates was published in reference [31]. Static simple immersion test shows the effect of time on the degradation rate. Bare MZX magnesium alloy demonstrated very low corrosion resistance in simulated body fluid (SBF) solution as shown in Figure 3. Surface oxidizing treatment increased corrosion resistance by nearly a factor of 6. However, it still corrodes too fast from an absorbable stent point of view. Biodegradable polymer coated magnesium alloy showed a reasonable corrosion rate. It seems that a thin layer of ceramic coating could be protecting the corrosion in a period of time as an example in Figure 1.

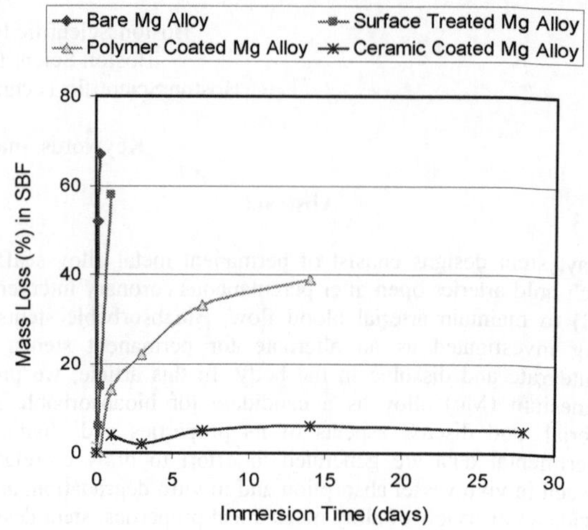

Figure 3. Mass loss of bare and surface coated magnesium alloy stents in SBF solution at 37 °C.

Potentiodynamic polarization curve gives an idea about corrosion behavior and a rapid estimation of initial corrosion rate. The polarization testing was performed in general accordance with ASTM F2129 "Standard Test Method for Conducting Cyclic Potentiodynamic Polarization Measurements to Determine the Corrosion Susceptibility of Small Implant Devices". One aim of electrochemical testing is to identify the similarity of body fluids in order to understand the correlation between in vitro and in vivo magnesium degradation. Potentiodynamic polarization of bare magnesium stents in different media are exhibited in Figure 4.

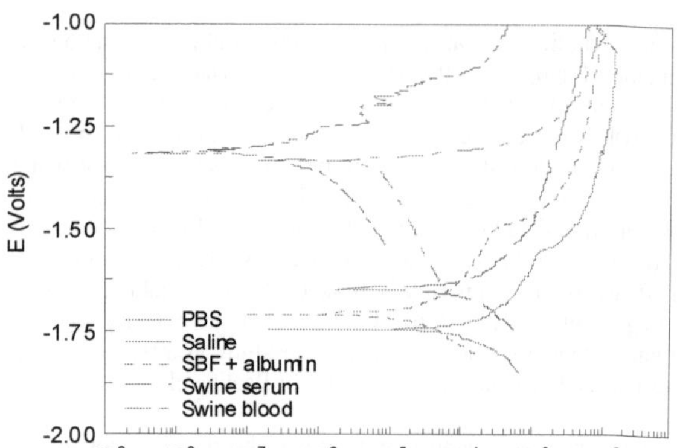

Figure 4. Potentiodynamic polarization measurements of magnesium alloy stents in different media with SCE reference electrode in Saline and PBS, and Ag/AgCl reference electrode for the other media.

Surface passivation was observed in both phosphate buffered saline (PBS) and SBF solution due to the formation of a phosphate layer. The rest potentials were higher in swine fluids than in simulated body fluids. This indicates that in vitro degradation tests in SBF are significantly different from the real degradation condition in swine fluids and perhaps in vivo. A much lower degradation rate is expected for in vivo tests than in vitro tests.

Pre-clinical results analysis

The stent implantation investigations conform to the "Guide for the Care and Use of Laboratory Animals published by the US National Institutes of Health" (NIH Publication 85–23, revised 1996). Both bare magnesium (MZX) bioabsorbable stent and platinum chromium bare metal stents (BMS) (12-mm length, 3.0-mm diameter) were implanted in the left anterior descending (LAD), left circumflex (LCX), right coronary (RCA), internal thoracic (ITA), renal and iliac arteries of female crossbred swine (Genetiporc, LLC, Alexandria, Minn). The stents were implanted at a targeted ratio of stent to artery diameter of 1.1:1.0 in the coronary arteries and 1.3:1.0 in the peripheral arteries as assessed by quantitative coronary angiography. Animals were euthanized after implant durations of 30 and 120 days.

For histology, randomly selected explanted coronary arteries were dehydrated; embedded in plastic; imaged using micro computed tomography (micro-CT); sectioned at proximal, mid, and distal stented locations and proximal and distal to the stented segment, and stained with hematoxylin and eosin and elastic trichrome stains. In selected cases, von Kossa staining was performed to confirm the presence of calcium. Morphological analysis of the in-stent sections (proximal, mid, distal) by light microscopy was performed with ordinal grading assessments of the following parameters: mural thrombus, endothelialization, strut tissue coverage, para-strut leukocytes, disruption of the internal elastic lamina (IEL), disruption of the external elastic lamina (EEL), and medial smooth muscle cell (SMC) loss.

At 30 days, neointimal thickness and percent stenosis were higher in all arteries implanted with MZX when compared to the single BMS device. Additional observations included an irregular distribution/distortion/cramping of struts in the neointima, irregularity of neointima thickness, and arterial wall calcifications in all arteries implanted with MZX devices. These findings were not observed in the control BMS stent (Figure 5). At 120 days, near complete absorption of the MZX was observed; any remaining fragments were well incorporated and covered by neointima. These fragments were surrounded either by cells (macrophages, monocytes or multinucleated giant cells) or by matrix with some calcifications (Figure 6).

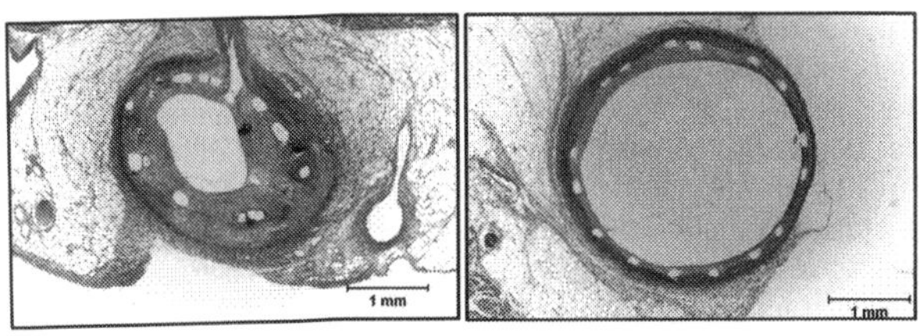

Figure 5. 30-day follow-up. Left - Histopathology of MZX vessel. Right - Histopathology of BMS vessel.

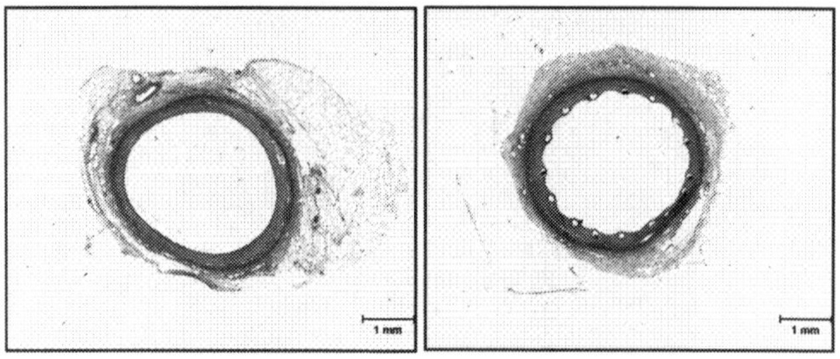

Figure 6. 120-day follow-up. Left - Histopathology of MZX vessel. Right - Histopathology of BMS vessel.

Magnesium stents demonstrated limited parastrut inflammation, absence of thrombi, and complete endothelial cell coverage similar to the bare metal control at 30 days. Lumen surface irregularities, late strut malaposition/staggering, mineralization, and reduced EEL area were observed in almost all magnesium stents in all vessel types at 30 days. These observations indicate the stents are exhibiting weakened radial strength by 30 days. All above mentioned changes may be related to the fragility of stents due to fast degradation. The discontinuity of struts can cause arterial wall damage and consequent dystrophic calcification. The fact that distortion of the stent struts coexists with mineralization suggests that it is not a result of the stents "handling" during explantation at necropsy but, most likely, the result of degradation of MZX which was quite prominent at 30 days. Arteries stented with magnesium stents have been reported in the literatures to have smaller diameters at follow-up; reduced diameters been attributed to negative remodeling [8] or recoil due to loss in radial strength [10].

Analysis of explanted stents

Conventional analytical methods can be used to characterize the condition of the magnesium stent implant after it has been harvested from the animal. It is of interest to examine the transformation of the stent metal to its by-products during degradation and to see how those products are distributed within the arterial wall. A method used to qualitatively characterize the progression of degradation is to make light microscopy examinations of stented arteries in cross-sections. Explanted stented arteries are cast in a translucent epoxy resin. Transverse sections are cut from in-stent regions (e.g., proximal, middle, distal). The wafer sections are cast in translucent metallography epoxy resin within molds to make sample configurations that can be ground and polished to prepare a metallographically smooth surface. The prepared planes are imaged with an optical stereozoom microscope and optical metallography microscope.

This technique allows for observations to be made with regard to the condition of stent struts within the artery wall and the distribution of metal degradation products. The cross-sections are first examined at relatively low magnification with an optical stereozoom microscope to see the entire stent strut ring and artery wall (Figure 7). An optical metallograph is used to examine individual struts in more details (Figures 8).

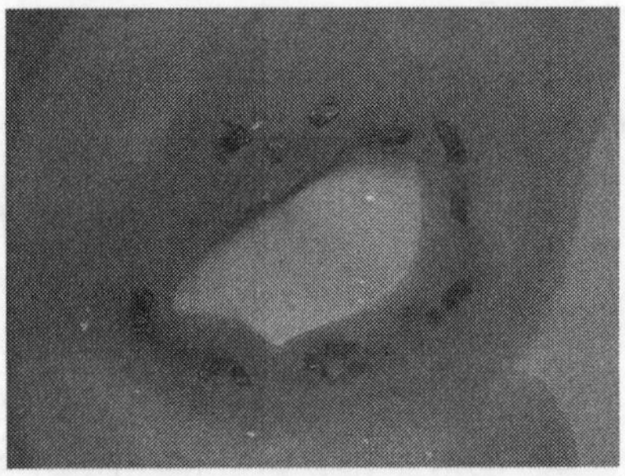

Figure 7. Optical stereozoom digital image of transverse cross section through a magnesium-stented artery. At 30 days time point, the stent struts are still present and there are degradation products adjacent to the struts.

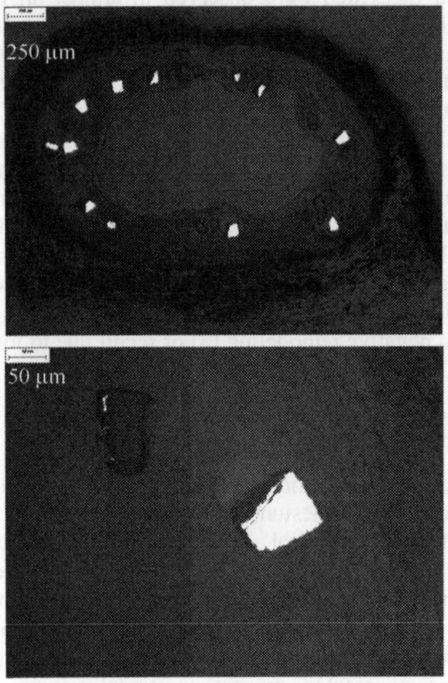

Figure 8. Optical metallograph image of a transverse cross-section through a magnesium-stented artery. The stent lumen and artery wall are visible in the top image. The remaining metal struts (white) and degradation products (grey around metal struts) can be seen in a higher magnification in the bottom. The strut next to it appears to be nearly completely degraded (dark grey).

The prepared cross-sections can then be imaged in the scanning electron microscope (SEM) in order to further resolve degraded strut features (Figure 9). The SEM image grey-scale contrast is related to the atomic number of the elements present in the features in the strut and tissue, so the degradation product can be more readily distinguished from the strut metal than with the optical microscopes. Energy dispersive spectroscopy analysis (EDS) can be used to qualitatively identify the elements that are present in the strut and degradation products. The EDS results can be mapped on the SEM image to show the distribution of elements.

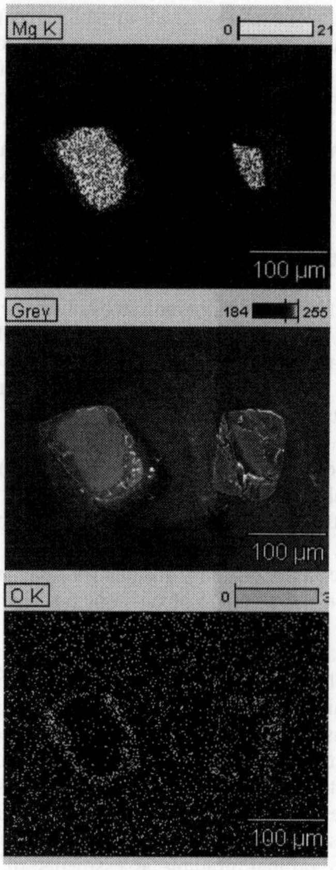

Figure 9. SEM-EDS elemental maps for magnesium (Mg, top) and oxygen (O, bottom) from two struts and their degradation products in a transverse cross-section through a magnesium-stented artery. The grey image (middle) is the SEM image that corresponds to the metal (light grey) and degradation product (dark grey).

Correlation in vivo and in vitro

The aim of a correlation is to predict in vivo outcomes through in vitro testing. Instead of running various costly animal studies, in vitro degradation studies can be used to preselect concepts ideally in a more timely manner. The experimental work included a comparison of measured mass loss values on explants (vessel containing implanted stent) with in vitro mass loss results from corrosion tests on stents. A non-destructive test procedure was developed using epoxy resin embedded vessel explants as test material and x-ray micro computed tomography (micro-CT) as analyzing method. The mass loss can be derived from the known magnesium alloy density and the micro-CT measured volume of the non-corroded stent metal. The deviation of the method depends from various parameters but mostly from pixel resolution, signal-to-noise ratio, bit depth of reconstructed image slices, strut position to beam line and polychromatic x-ray spectrum related artifacts (beam hardening).

Qualitative micro-CT assessment of magnesium stents, explanted from an animal internal thoracic artery (ITA) after 30 days, demonstrates advanced material corrosion which leads to low stent radial strength. The magnesium stent illustrated in the top of Figure 10 has numerous discontinuous struts which have led to the stent radially collapsing from vascular motion. Qualitative micro-CT assessment also shows magnesium stents do not corrode uniformly but rather focally, with corrosion generally initiating in high-strain regions. Quantitative mass loss measurements can be made by imaging stents of known mass (Figure 10 bottom) along with stents implanted during an in vivo study. The purpose of imaging stents of known mass is to determine an appropriate micro-CT threshold factor signifying the difference between magnesium and epoxy resin/tissue.

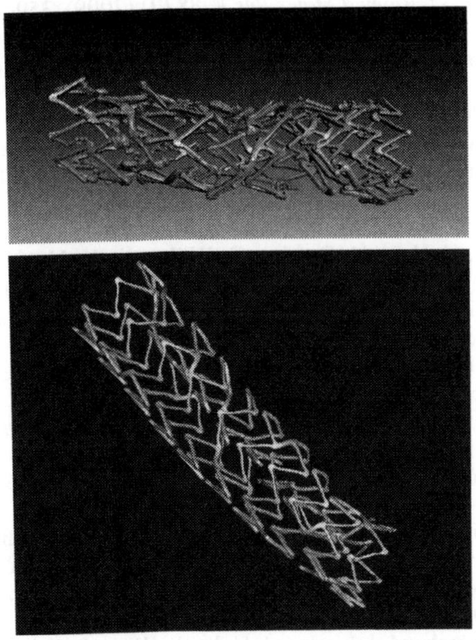

Figure 10. A magnesium alloy stent explanted from ITA after 30 days (top). A known mass stent implanted in epoxy resin (bottom).

Since the in vitro mass loss test is accelerated compared to in vivo mass loss, a more complex approach has to be used to determine corrosion as shown in Figure 11. Relative mass loss determined through in vitro test is fitted into a curve over different time points. The relative mass loss determined through measurement on explants is then found in this curve and related to a time point on the in vitro time axis. This is done for every time point from explants that creates a correlation curve between in vitro and in vivo time points as shown in the upper right part of Figure 11. Detailed correlation experiments are ongoing and will be published elsewhere.

Conclusion

Magnesium alloys have reduced mechanical strength relative to conventional metal stent materials, but the amount of strength required and stent designs to maintain stent diameters over a certain time frame is relatively unknown. Magnesium alloy stents provide a suitable corrosion rate in the physiologic environment with surface coating. Magnesium is an essential mineral for human metabolism which is a plus for bioabsorbable stent material.

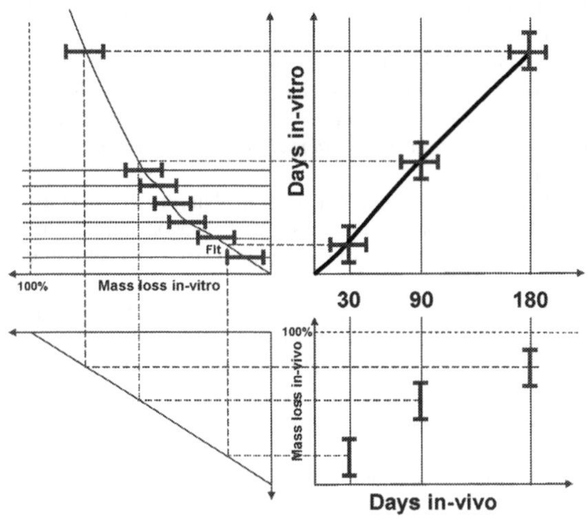

Figure 11: In-vitro / in-vivo correlation.

Intravascular ultrasound analysis at the termination of the 120-day animal provided no evidence of remaining stent material in the treated area of the vessels; histology confirmed nearly complete absorption within 120 days of implantation. At both 30 and 120 days post implantation, vascular response of coronary and peripheral arteries demonstrated absence of thrombi, complete strut coverage and minimal parastrut inflammation in all arteries implanted with either magnesium alloy or BMS devices. There was no evidence of myocardial abnormalities of any kind at 30 and 120 days. Evidence of negative remodeling at 30 days and positive at 120 days was noted by angiography and intervascular ultrasound (IVUS) at termination.

Both in vitro and preclinical in vivo data suggest that delayed and more uniform corrosion of the stent material is required. The vessel requires strength for a longer period of time than 30 days to allow for the completion of vessel healing. Therefore, new alloys and corrosion protection schemes such as coatings may be beneficial in generating a more complete understanding of the design requirements of magnesium alloy stents. In vitro experiments inherently exhibit a direct correlation to the real case, whereas a direct simulation of the in vivo case seems feasible.

References

1. J. E. Moore, J. S. Soares, and K. R. Rajagopal, "Biodegradable Stents: Biomechanical Modeling Challenges and Opportunities", *Cardiovascular Engineering and Technology*, 1 (1) (2010), 52-65.
2. R. Jabara, L. Pendyala, J. Chen, and N. Chronos, "Bioabsorbable Stents: The Future Is Near", *Cardiac Interventions Today*, June/July (2009), 50-53.
3. P. Erne, M. Schier, and T. J. Resink, "The Road to Bioabsorbable Stents: Reaching Clinical Reality?", *Cardiovas Inttervent Radiol*, 29 (2006), 11-16.
4. G. Mani, M. D. Feldman, D. Patel, and C. M. Agrawal, "Coronary Stents: A Materials Perspective", *Biomaterials*, 28 (2007) 1689-1710
5. H. Hermawan, M. Moravej, D. Dube, M. Fiset, and D. Mantovani, "Degradation Behaviour of Metallic Biomaterials

for Degradable Stents", *Advanced Materials Research*, 15-17 (2007), 113-118.
6. C. Di Mario, H. Griffiths, O. Goktekin, N. Peeters, J. Verbist, M. Bosiers, K. Deloose, B. Heublein, R. Rohde, V. Kases, C. Ilsley, and R. Erbel, "Drug-Eluting Bioabsorbable Magnesium Stent", *J, Interventional Cardiology*, 17 (6) (2004), 391-395.
7. B. Heublein, R. Rohde, V. Kaese, M. Niemeyer, W. Hartung, and A. Haverich, "Biocorrosion of Magnesium Alloy: A New Principle in Cardiovascular Implant Technology?" *Heart*, 89 (2003), 651-656.
8. M. Maeng, L. O. Jensen, E. Falk, H. R. Andersen, and L. Thuesen, "Negative Vascular Remodelling after Implanttation of Bioabsorbable Magnesium Alloy Stents in Porcine Coronary Arteries: A Randomised Comparison with Bare-metal and Sirolimus-eluting Stents", *Heart*, 95 (2009), 241-246.
9. R. Waksman, R. Pakala, P. K. Kuchulakanti, R. Baffour, D. Hellinga, R. Seabron, F. O. Tio, E. Wittchow, S. Hartwig, C. Harder, R. Rohde, B. Heublein, A. Andreae, K. Waldmann and A. Haverich, "Safety and Efficacy of Bioabsorbable Magnesium Alloy Stents in Porcine Coronary Arteries", *Catheterization and Cardiovascular Intervention*, 68 (2006), 607-617.
10. T. L. Pinto Slottow, R. Pakala, T. Okabe, D. Hellinga, R. J. Lovec, F. O. Tio, A. B. Bui, and R. Waksman, "Optical Coherence Tomography and Intravascular Ultrasound Imaging of Bioabsorbable Magnesium Stent Degradation in Porcine Coronary Arteries", *Cardiovascular Revascularization Medicine*, 9 (2008), 248-254.
11. P. Peeters, M. Boisers, J. Vernist, K. Deloose, and B. Heublein, "Preliminary Results after Application of Absorbable Metal Stents in Patients with Critical Limb Ischemia", *J Endovasc Ther*, 12 (2005), 1-5.
12. R. Erbel, C. Di Mario, J. Bartunek, J. Bonnier, B. De Bruyne, F. Eberli, P. Erne, M. Haude, B. Heublein, M. Horrigan, C. Ilsley, D. Bose, J. Koolen, T. F. Luscher, N. Weissman, and R. Waksman, "Temporary Scaffolding of Coronary Arteries with Bioabsorbable Magnesium Stent: A Prospective, Non-randomised Multicentre Trial", *The Lancet* 369 (2007), 1869-1875.
13. R. Waksman, R. Erbel, C. Di Mario, J. Bartunek, B. de Bruyne, F. R. Eberli, P. Erne, M. Haude, M. Horrigan, C. Ilsley, D. Bose, H. Bonnier, J. Koolen, T. F. Luscher, and N. J. Weissman, "Early- and Long-Term Intravascular Ultrasound Angiographic Findings After Bioabsorbable Magnesium Stent Implantation in Human Coronary Arteries", *JACC Cardiovascular Interventions* 2 (4) (2009), 312-320.
14. X. N. Gu and Y. F. Zheng, "A Review on Magnesium Alloys as Biodegradable Materials", *Front. Mater. Sci. China*, 4 (2) (2010), 111-115.
15. X. Gu and Y. Zheng, "In Vitro Corrosion and Biocompatibility of Binary Magnesium Alloys", *Biomaterials*, 30 (2009), 484-498.
16. B. Zberg, P. J. Uggowitzer, and J. F. Löffler, "MgZnCa GlassesWithout Clinically Observable Hydrogen Evolution for Biodegradable Implants", *Nature Materials*, 8 (11) (2009), 887-891.
17. A. C. Hanzi, F. H. Dalla Torre, A. S. Sologubenko, P. Gunde, R. Schmid-Fetzer, M. Kuehlein, J. F. Löffler, and P. J. Uggowitzer, "Design Stategy for Microalloyed Ultra-ductile Magnesium Alloys", *Philosophical Magazine Letters*, 89 (6) (2009), 377-390.
18. A. C. Hanzi, P. Gunde, M. Schinhammer, and P. J. Uggowitzer, "On the Biodegradation Performance of an Mg-Y-Re Alloy with Various Surface Conditions in Simulated Body Fluid", *Acta Biomaterialia*, 5 (2009), 162-171.
19. J. G. Brumer, R. Hahn, J. Kunze, and S. Virtanen, „Porosity Tailored Growth of Black Anodic Layers on Magnesium in an Organic Electrolyte", *J. Electrochem. Soc.*, 156 (2) (2009), C62-C66.
20. R. Rettig and S. Virtanen, "Composition of Corrosion Layers on a Magnesium Rare-earth Alloy in Simulated Body Fluids", *J. Biomed Mater Res A*, 88 (2) (2009), 359-369.
21. R. Rettig and S. Virtanen, "Time-dependent Electrochemical Characterization of the Corrosion of a Magnesium Rare-earth Alloy in Simulated Body Fluids", *J. Biomed Mater Res A*, 85 (1) (2008), 167-175.
22. F. Witte, "The History of Biodegradable Magnesium Implants: A Review", *Acta Biomaterialia*, 6 (2010), 1680-1692.
23. N. Hort, Y. Huang, D. Fechner, M. Stormer, C. Blawert, F. Witte, C. Vogt, H. Drucker, R. Willumeit, K. U. Kainer, and F. Feyerabend, "Magnesium Alloy as Implant Materials – Principles of Property Design for Mg-Re Alloys", *Acta Biomaterialia*, 6 (2010), 1714-1725.
24. C. Janning, E. Willbold, C. Vogt, J. Nellesen, A. Meyer-Lindenberg, H. Windhagen, F. Thorey, and F. Witte, "Magnesium Hydroxide Temporarily Enhancing Osteoblast Activity and Decreasing the Osteoclast Number in Peri-implant Bone Remodelling", *Acta Biomaterialia*, 6 (2010), 1861-1868.
25. F. Witte, J. Fischer, J. Nellesen, C. Vogt., T. Donath, and F. Beckmann, "In vivo Corrosion and Corrosion Protection of Magnesium Alloy LAE442", *Acta Biomaterialia*, 6 (5) (2010), 1792-1799.
26. F. Feyerabend, J. Fischer, J. Holtz, F. Witte, R. Willumeit, H. Drucker, C. Vogt, and N. Hort, "Evolution of Short-term Effects of Rare Earth and Other Elements Used in Magnesium Alloys on Primary Cells and Cell Lines", *Acta Biomaterialia*, 6 (5) (2010), 1834-1842.
27. F. Witte, J. Fischer, J. Nellesen, H-A. Crostack, V. Kaese, A. Pisch, F. Beckmann, and H. Windhagen, "In Vitro and In Vivo Corrosion Measurements of Magnesium Alloys", *Biomaterials*, 27 (2006), 1013-1018.
28. W. D. Muller, M. L. Nascimento, M. Zeddies, M. Corsico, L. M.Gassa, and M. A. F. L. de Mele, "Magnesium and Its Alloys as Degradable Biometerials, Corrosion Studies Using Potentiodynamic and EIS Electrochemical Techniques", *Materials Research*, 10 (1) (2007), 5-10.
29. Y. Yun, Z. Dong, N. Lee, Y. Liu, D. Xue, X. Guo, J. Kuhlmann, A. Doepke, H. B. Haslsall, W. Heinemann, S. Sundaramurthy, M. J. Schulz, Z. Yin, V. Shanov, D. Hurd, P. Nagy, W. Li, and C. Fox, "Revolutionizing Biodegradable Metals", *Materials Today*, 12 (10) (2009), 22-32.
30. M. B. Kannan and R. K. S. Raman, "In Vitro Degradation and Mechanical Integrity of Calcium-Containing Magnesium Alloys in Modified-Simulated Body Fluid", *Biomaterials*, 29 (2008), 2306-2314.
31. N. T. Kirkland, J. Lespagnol, N. Birbilis, and M. P. Staiger, "A Survey of Bio-corrosion Rates of Magnesium Alloys", *Corrosion Science*, 52 (2010), 287-291.

Magnesium Technology 2011
Edited by: Wim H. Sillekens, Sean R. Agnew, Neale R. Neelameggham, and Suveen N. Mathaudhu
TMS (The Minerals, Metals & Materials Society), 2011

PROCESSING ASPECTS OF MAGNESIUM ALLOY STENT TUBE

R.J. Werkhoven, W.H. Sillekens, J.B.J.M. van Lieshout
TNO Science and Industry; PO Box 6235; 5600 HE Eindhoven, Netherlands

Keywords: Biomedical Implants, Stents, Equal Channel Angular Pressing, Extrusion, Drawing

Abstract

Biomedical applications are an emerging field of interest for magnesium technology, envisioning biodegradable implants that resorb in the human body after having cured a particular medical condition (such as artery clogging or bone fractures). This challenges research in a sense that the materials to be used need to dissolve *in vivo* in a controlled fashion without leaving harmful remainders and while maintaining sufficient strength and other (mechanical) attributes as long as necessary.

To comply to the requirements, magnesium alloys as well as their processing routes into implants need to be tailored. While new alloy compositions are receiving ample attention, the paper at hand addresses the processing issue. The application of choice is the (cardio)-vascular stent. Different steps in manufacturing magnesium AZ-alloy stent tube are considered, including equal channel angular pressing, extrusion and subsequent drawing operations. Results show that the processing route has an important influence on the microstructure of the finished stent tube and hence on its functional performance.

Introduction

Life expectancy has increased considerably over the last fifty years due to the many advances in healthcare. In line with this, the market for implantable medical devices is booming due to upward trends in medical conditions and patient activity, changing patient treatment approaches and technical advances. For instance, the demand for cardiac implants quadrupled, for orthopaedic implants doubled and for all other implants tripled between 1997 and 2007 [1].

Current (metallic) biomaterials for these implants are essentially neutral *in vivo*. Such bio-passive implants remain either in the body (with the risk of loosening, fracturing and tissue inflammation) or are explanted after healing (implying additional surgical risk, patient discomfort and costs). Degradable materials that are under development – notably polymers – partially meet these drawbacks as they do need less repeated invasive surgery, since once absorbed they leave behind only the healed natural tissue. Further trends are towards surface modification of biomaterials to better control blood and tissue compatibility, implants and devices that are also vehicles for drug delivery, and the development of new classes of biomaterials for tissue engineering.

As for magnesium, its biocompatibility including its *in vivo* resorption (corrosion in the electrolytic environment of the body), supposed non-toxicity to the human body, as well as its functional role in the physiological system render it an interesting candidate for degradable implants. Its mechanical properties – notably strength and modulus of elasticity – as well as its density are quite similar to natural bone material and hence of interest for hard-tissue engineering applications.

Pioneering work in the mid-20th century indicated a significant reduction in healing time and acceleration in mineralization of bone fractures, while no toxic effects were observed [2]. The formation of hydrogen in the corrosion process, however, posed a problem and the interest shifted in favor of stainless steel and titanium. It was not before the late 1970's that corrosion resistance was substantially improved by the use of specific alloying elements and high-purity alloys, so that hydrogen formation was suppressed as well and implanted styles remained *in vivo* for months. Most recently the use of magnesium for medical applications has seen a growing research interest, also because of its perceived osteo-conductive activity (favoring bone apposition). The current state-of-the-art for biomedical implants and the envisaged role for magnesium are outlined in **Figure 1**.

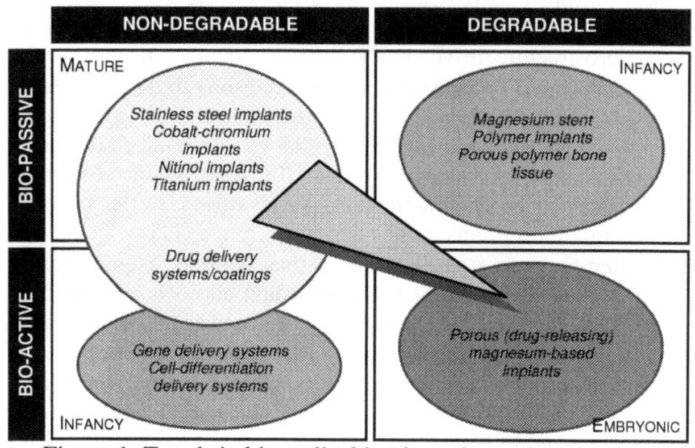

Figure 1. Trends in biomedical implants (represented by the arrow) and the position of magnesium.

As one of the targeted applications, the development of magnesium (cardio)-vascular stents is currently being undertaken with initial successes reported [3–5]. Stents are implants that are surgically placed in clogged arteries to restore adequate blood flow. A stent essentially consists of a thin-walled mesh tube with typical overall dimensions being 3 mm in diameter and 20 mm in length. Wall thickness and strut size can be as low as 100–150 μm. A stent is implanted by shrinking it on a balloon catheter and subsequently expanding it upon proper positioning in the artery. Thus, the stent material must comply to such mechanical requirements as strength and ductility in addition to biological demands, as indicated in **Table I**. Apart from these, there are supplementary requirements for stents that are not listed but also relevant, such as on geometrical accuracy, manufacturability and fatigue resistance.

While in the early investigations the commercial-grade alloy WE43 was used, more recently a variety of other alloy systems has been proposed. This also includes alloy modifications with calcium, aluminum or lithium as the primary alloying element [6–9]. Main objective of these efforts is to slow down resorption (or corrosion) rate while maintaining or improving mechanical properties. Although not all biocompatibility effects of the alloying elements are cleared and certification remains a concern, magnesium alloy stents (with an indicative weight of 5 mg) are among the least mass-critical as compared to other implants.

Table I. Primary requirements for magnesium alloy stent tube.

ASPECT	DESCRIPTION
Resorption	Scaffolding integrity 3–6 months Full dissolution within 1–2 years
Biocompatibility	Non-toxic, no inflammatory tissue response No harmful release and/or residence of particles
Mechanical properties	Tensile yield stress TYS>200 MPa, ultimate tensile strength UTS>300 MPa, tensile elongation >15–18% Elastic recoil <4% (in conjunction with stent design)

Stents usually are manufactured in a series of extrusion, drawing, (laser) cutting and (electro-chemical) polishing operations. For magnesium alloy stent tube, the strategy is to realize a possibly fine microstructure, driven by the next considerations.

- Small grains enhance yield stress according to the Hall-Petch relationship with a sensitivity which is typically greater for metals like magnesium that yield by mechanical twinning rather than by dislocation glide [10]. Also, grain refinement reduces (tension-compression) yield anisotropy to become marginal for grains in the micrometer range [11].
- Literature consistently suggests that as grain size decreases, the corrosion resistance of magnesium alloys is improved in neutral and alkaline electrolytes [12], such as blood.
- A coarse microstructure will induce inhomogeneous mechanical and other behavior across the thin walls and struts of the stent as well as add to the risk of large-particle release and/or residence.

From the perspective of process design, it is thus important to understand how the processing parameters affect the microstructural development along the manufacturing chain.

The paper at hand addresses some of the main processing aspects of magnesium alloy stent tube. More in particular, it is concerned with the extrusion and drawing operations. In a further elaboration, equal channel angular pressing is added as a means to refine the microstructure prior to extrusion.

Processing Outline and Adopted Approach

Prior reported efforts on manufacturing magnesium alloy stent tube are based on the route known from conventional metal stents. This consists of (1) hot working of cast feedstock to obtain a solid rod, (2) deep drilling of a hole in the rod to obtain a tube, and (3) multiple cold drawing and intermediate annealing of the tube to reduce diameter and wall thickness [3]. The mesh is subsequently made by laser cutting and the cutting edges are smoothened by electro-chemical polishing. An alternative route departs from this in that the hot working process is conducted by extrusion with a mandrel and tubular billets (enabling the direct manufacturing of seamless tube), and that the drawing is done at elevated temperature (reducing the number of drawing steps) [6].

The processes for manufacturing stent tube that were adopted in the current investigations can be amplified as follows.

Extrusion

In hot extrusion, heated material is pushed through a die, thereby reducing its cross-section to obtain solid or hollow profiles. Although the shape alterations that can be imposed in a single processing step are high, the geometrical tolerances of the obtained extrusions are limited and not high enough to reach the required accuracy of a finished stent tube. Thus hot extrusion is used to provide an intermediary shape. Extrusion is known to impose heavy plastic deformation within the warm-working regime and as such induces dynamic recrystallization and other microstructural effects to the material (e.g., [13]).

The extrusion trials have been conducted using a laboratory extrusion press (make Loire/ACB, 50 t capacity, 1-inch container). In this case the direct extrusion process is conducted with a porthole die, enabling the manufacturing of tube using solid billets. An earlier study has shown that the longitudinal weld seams that are implied by using this kind of tooling are uncritical in terms of ductility for alloys that do readily recrystallize during warm working [14]. Billet size is Ø25×100 mm; tube size is Ø9.5×2 mm, implicating an extrusion ratio (or area-reduction ratio) of R_E=10.4. Billet temperature and extrusion exit speed can be varied between extrusions.

Drawing

In drawing, bar or tube is pulled through a tapered die, thereby reducing the cross-section to obtain a product with the same basic shape. When conducted at room temperature (cold drawing), dimensional accuracy of the product is high, noting that a tolerance in concentricity of the initial tube of for instance ±10% propagates equally into the finished tube. The deformation that can be imposed in a single drawing step is limited so that a substantial reduction in tube diameter and wall thickness can only be achieved by repetitive drawing with progressively smaller tooling. Thus it is used to provide the final shape of the stent tube. Cold drawing is known to induce distortion of the grains and strain hardening of the material. Intermediary heat treatment (annealing) restores ductility but may also provoke microstructural effects like recrystallization and grain growth.

The drawing trials have been conducted by an industrial manufacturer. The implemented variant uses a cylindrical mandrel which passes through the die along with the tube for control of the bore dimension (so-called "moving-mandrel drawing"). In a sequence of drawing steps, the extruded tubes are reduced to Ø1.6×0.13 mm, which is an appropriate geometry for any following *in vivo* (animal) trials to test functionality including bio-compatibility and resorption behavior. Total drawing ratio (or area-reduction ratio) for this sequence is R_D=78.

Equal Channel Angular Pressing

In equal channel angular pressing (ECAP), heated material is repeatedly pushed through an angular die without altering the cross-section to obtain a product with basically the same shape and size. Due to the severe plastic deformation, microstructures can be refined which has been demonstrated to impart favorable mechanical properties to a variety of materials including magnesium alloys (e.g., [15]). Against this background, it is considered here as a sidetrack of development to potentially improve the quality of the feedstock prior to extrusion.

The ECAP trials have been conducted using the laboratory press that was also used for hot extrusion. In this case a die is installed which redirects the material flow from the container in lateral direction (90° intersection angle). Channel size is Ø25 mm.

Materials and Testing

For the investigations, aluminum-based magnesium alloys were used according to the specifications given in **Table II**. AZ80 is a commercial-grade baseline alloy, while AZM and AZNd are two experimental alloys that include manganese and neodymium as modifiers, respectively. These modifiers are added with the purpose of improving biodegradation properties (notably pitting corrosion resistance) as well as mechanical performance, which was experimentally confirmed [8]. The latter is suggested to be associated with the formation of the Al_8Mn_5 phase in AZM and the $Al_{11}Nd_3$ phase in AZNd. In view of the composition limits for AZ80 (7.8–9.2 wt% Al, 0.20–0.80 wt% Zn, 0.12–0.50 wt% Mn [16]), the baseline alloy turned out to be overdosed with manganese. Apart from the listed elements, traces of copper, iron and nickel (that notably affect corrosion behavior) remain below the common high-purity limits at the parts-per-million level.

Table II. Chemical compositions of the employed AZ alloys (in wt %) – spectrometric analysis.

ALLOY	Al	Zn	Mn	Nd	Mg
AZ80	7.8	0.44	0.93		bal.
AZM	8.1	0.53	0.45		bal.
AZNd	8.0	0.53	0.24	0.55	bal.

AZ80 was available from a commercial supplier in extruded and tempered (T5) condition, while AZM and AZNd were cast as ingots of about 7 kg each in a laboratory setting.

Analysis of the materials during the sequential processing stages was primarily done by means of light optical microscopy (LOM). Microstructures were further characterized with (linear intercept) grain-size measurements. By repeated measurements, including various directions, sections and samples, overall reproducibility is expressed in terms of the standard deviation (typically 10–15 % of the average value) which will be used further on to represent the variation in results.

Results and Discussion

To visualize the intermediate stages in the manufacturing of magnesium alloy stent tube, **Figure 2** shows some cut-out samples of the feedstock/billet, extruded tube, extruded and drawn tube, and finished stent (from left to right).

From the previous section it follows that the quality of the stent tube in terms of its geometrical tolerances, mechanical and corrosion properties will depend (through its microstructure) on the sequence of forming operations and their interaction. Two illustrations of these (inter)relations will be outlined below. The first relates to the working order of hot extrusion followed by cold drawing, the second to ECAP followed by hot extrusion.

Figure 2. Magnesium alloy stent: from feedstock to finished product.

Extrusion followed by Drawing

Main process parameters during hot extrusion are the degree of deformation (represented by extrusion ratio and die design), the temperature (which in turn depends on starting or billet temperature, degree of deformation and ram speed) and the rate at which the material cools down after processing.

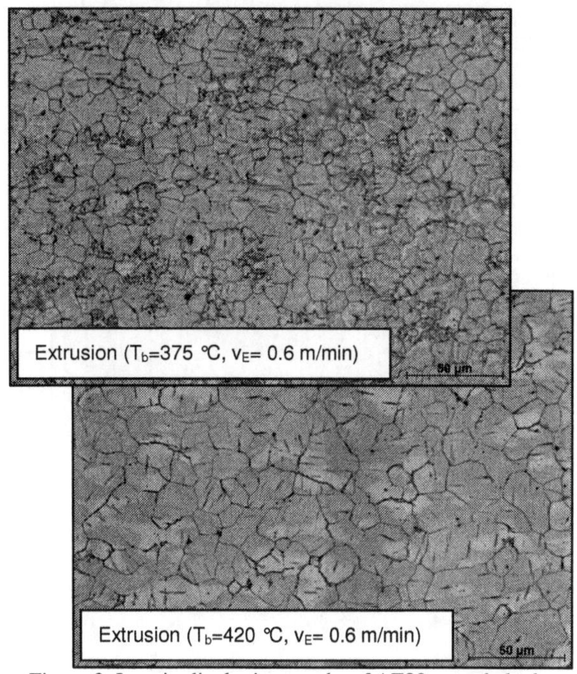

Figure 3. Longitudinal micrographs of AZ80 extruded tube (Ø9.5×2.0 mm) for different billet temperatures.

Figure 3 demonstrates the influence of the billet temperature T_b on the microstructure of the obtained tube for the alloy AZ80. Samples were taken from the last part (the so-called tail) of the extrusion. Extrusion exit speed v_E was kept low in order to avoid hot shortness (surface cracking of the extrusion); further, the tubes were cooled by forced air upon leaving the die. The LOM images reveal equi-axed, recrystallized microstructures with average grain sizes of 10.7±0.7 µm and 15.2±1.8 µm for the extrusion at T_b=375 °C and T_b=420 °C, respectively. This follows the general notion that extrusion temperature should be kept as low as possibly permitted by the press capacity to reach a fine microstructure (e.g., [17]).

Figure 4 shows microstructures of extruded tubes for the three alloys, again by taking samples from the tail of the extrusion. Extrusion settings were identical, including the forced air cooling. The LOM images reveal equi-axed, recrystallized microstructures. For AZ80, precipitates show at the grain boundaries, while for AZM and AZNd (fragmented) intermetallic compounds appear rather as stringers oriented in the extrusion direction. Average grain size is typically in the 10 µm range, but it should be noted that AZ80 on the one and AZM and AZNd on the other hand cannot be directly compared as the former started from a pre-extruded billet (average grain size ~13 µm) and the latter from as-cast billets (200–300 µm).

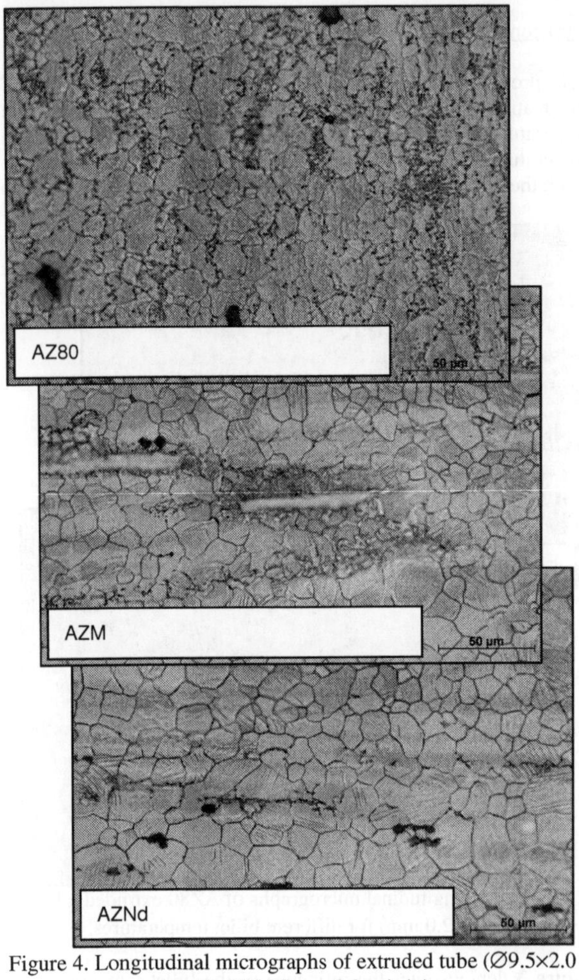

Figure 4. Longitudinal micrographs of extruded tube (Ø9.5×2.0 mm; T_b=375 °C, v_E=0.3 m/min).

Main process parameters during cold drawing are the degree of deformation (represented by drawing ratio and die designs), the drawing and annealing sequence (partitioning into individual drawing and annealing steps), and the annealing parameters (temperature and time).

Figure 5 shows microstructures of finished drawn tubes for the three alloys (starting with extruded tubes as depicted in Figure 4). Intermediate annealing temperature followed an initial recommendation of 385 °C for AZ80 [16], but was fine-tuned somewhat in conjunction with the drawing operations. Further a heat treatment was carried out at 340 °C after the final drawing step. The LOM images reveal equi-axed, re-crystallized microstructures with average grain sizes typically in the 30 µm range. Similar microstructures appear after each intermediate annealing step, supporting the view that static recrystallization is the underlying mechanism.

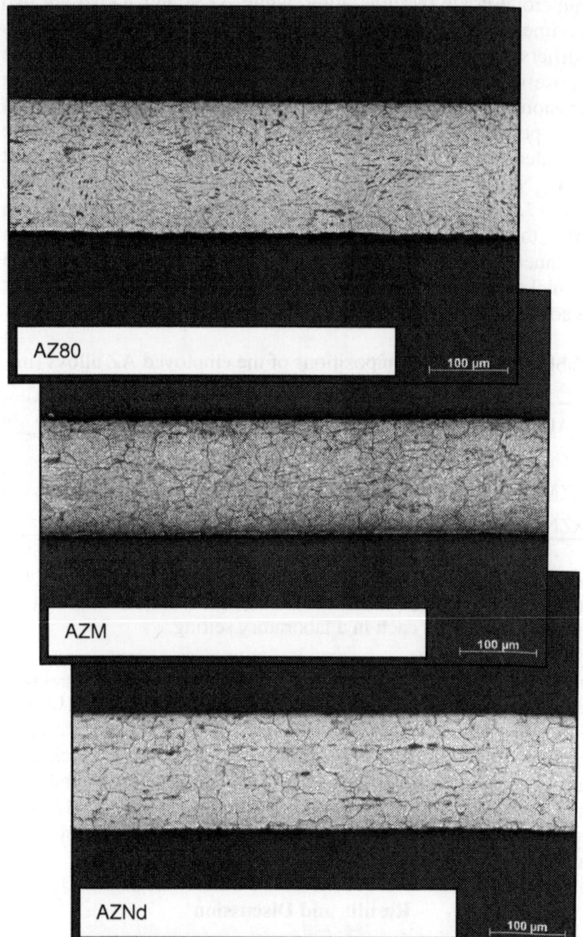

Figure 5. Longitudinal micrographs of extruded and drawn tube (Ø1.6×0.13 mm).

Figure 6 summarizes the development of the average grain size for the three alloys during the successive manufacturing steps. Extrusion refines the microstructure for all alloys to a similar extent. Notably, an earlier investigation with these alloys in which Ø5 mm bars were extruded with the same set-up and under comparable conditions yielded finer as-extruded microstructures (average grain size typically 5 µm) [8]. Extrusion of hollow sections through a porthole die generally requires more mechanical work than extrusion of solid sections and as a consequence entails more generated heat and associated higher temperatures, so that grain growth is likely to have a larger effect in these cases. Further, subsequent drawing at room temperature (RT) with intermediate heat treatment (HT) coarsens the microstructure, where the effect appears to be marginally more pronounced for AZ80 and AZM than for AZNd.

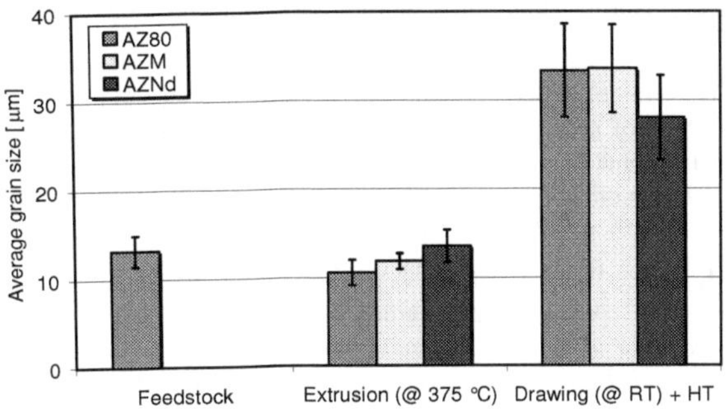

Figure 6. Microstructural evolution during the extrusion-drawing sequence.

Equal Channel Angular Pressing followed by Extrusion

Main process parameters during ECAP are the degree of deformation (represented by die design and number of passes) and the temperature (which in turn depends on starting or billet temperature and ram speed).

Figure 7 shows microstructures for the three subsequent stages of the concerned manufacturing sequence for AZ80. In ECAP the temperature of the billet and chamber were set at T_b=275 °C, with the billets being reheated between the passes. As for billet orientation, samples were reversed and turned 90° between individual passes. ECAP effectively refines the microstructure to an average grain size of about 3 µm after three passes which does not further decrease within the explored range of up to nine passes. After extrusion, however, average grain size has increased to the 10 µm range. Microstructures remain essentially equi-axed at the successive stages, suggesting that dynamic recrystallization is the predominant effect that determines the microstructural evolution.

Figure 8 summarizes the development of the average grain size for the three alloys during the successive manufacturing steps. After ECAPing (which involved nine passes), the average grain size ranges between 3 and 5 µm, with some variation between the alloys. As a result of the following extrusion process, average grain size increases for all alloys to a similar extent and comparable to those obtained without prior ECAPing (already presented in Figure 6).

Table III lists the microstructural results for some series of extrusion trials without and with prior ECAPing. From this it appears that the influence of ECAP on the average grain size of the semi-finished tube is marginal if not insignificant. Within the explored settings, billet temperature in the extrusion process is the most effective parameter for controlling eventual grain size.

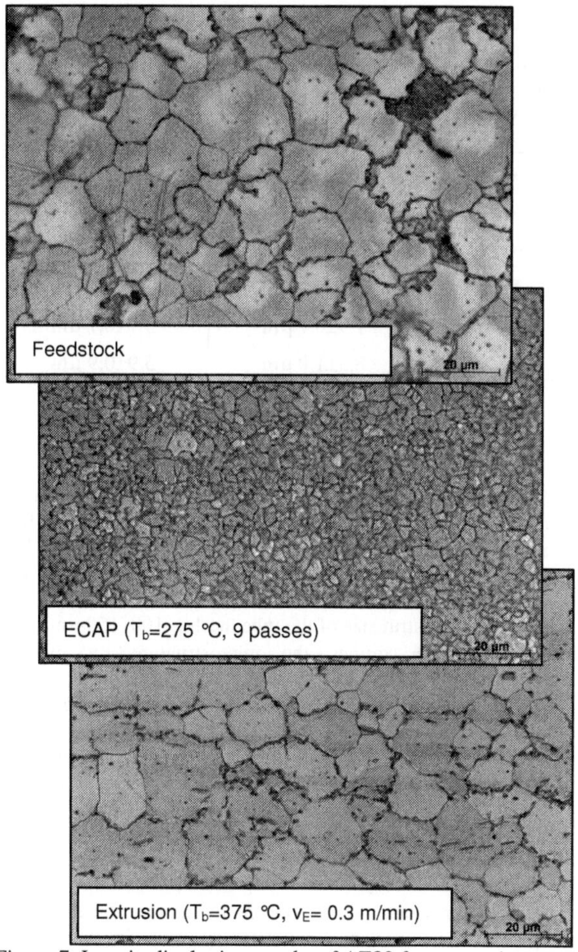

Figure 7. Longitudinal micrographs of AZ80 for successive stages of the ECAP-extrusion sequence.

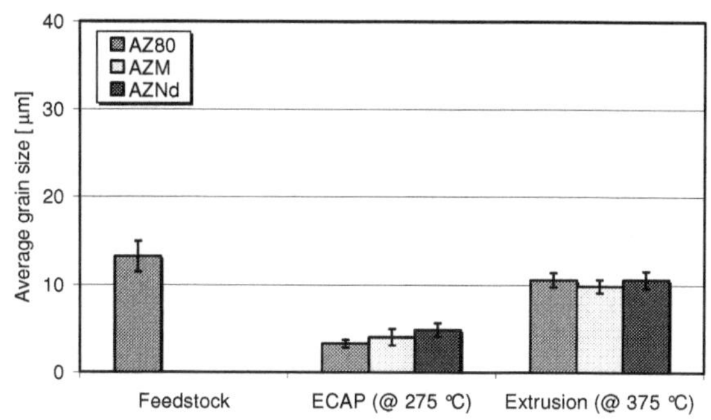

Figure 8. Microstructural evolution during the ECAP-extrusion sequence.

Table III. Average grain size of semi-finished tube (Ø9.5×2.0 mm) in dependence on the processing route.

T_b	ALLOY	EXTRUSION (@ T_b)	ECAP (@ 275 °C) + EXTRUSION (@ T_b)
375 °C	AZ80	10.5±1.4 µm	10.6±1.4 µm
	AZM	11.8±0.9 µm	9.8±1.0 µm
	AZNd	13.5±1.8 µm	10.6±1.3 µm
325 °C	AZ80	6.5±0.4 µm	7.7±0.8 µm
	AZM	7.9±0.7 µm	5.7±0.7 µm
	AZNd	8.3±1.0 µm	5.9±0.9 µm

Conclusions and Outlook

Results of the reported work can be recapitulated as follows.

- Stent tube can be manufactured from the baseline alloy AZ80 as well as from the experimental alloys AZM and AZNd by a sequence of extrusion and drawing operations. The microstructure of the obtained stent tube (average grain size ~30 µm) is still on the coarse side in view of the wall thickness and strut size of the tube mesh (~100–150 µm).
- Hot extrusion refines the microstructure by thermo-mechanically induced dynamic recrystallization, while subsequent cold drawing with intermediate annealing results in grain coarsening due to static recrystallization.
- ECAPing of the material prior to the extrusion-drawing sequence effectively refines the microstructure to an average grain size of 3–5 µm. The fine grains, however, are not maintained during subsequent extrusion, so that its effectiveness to improve feedstock quality is limited.

Based on the previous, possible means to advance the efficiency and quality of the current manufacturing route for magnesium alloy stent tube are to better attune the extrusion and the drawing process (in order to optimize microstructural development versus geometrical accuracy), and to better attune the order of the individual drawing operations in conjunction with intermediate annealing (in order to reduce grain growth during this stage of manufacturing). On a higher level of innovation, cold drawing may be replaced by a combined warm-cold drawing process with the objectives of enhancing microstructural control and simultaneously improving manufacturing efficiency while maintaining geometrical tolerances. On a fundamental level, textural development could be monitored "through process" with such methods as electron back-scatter diffraction in order to enhance the basic understanding of microstructural development versus (directionality of) mechanical properties.

References

1. D. Apelian, "Looking beyond the last 50 years: The future of materials science and engineering", *JOM*, 59 (2) (2007), 65–73.

2. M.P. Staiger, A.M. Pietak, J. Huadmai, G. Dias; "Magnesium and its alloys as orthopedic biomaterials: A review", *Biomaterials*, 27 (2006), 1728–1734.

3. B. Gerold, H. Müller, "Konzept für biologisch abbaubare Implantate aus Magnesium", *4. Ranshofener Leichtmetalltage 2006 – Vom Werkstoff zum Bauteilsystem*, LKR-Verlag, Ranshofen Austria (2006), 173–183.

4. H. Müller, "Biodegradation von Magnesiumlegierungen – Ein neues Konzept für temporäre Gefäßimplantate", *5. Ranshofener Leichtmetalltage 2008 – Zukunft // Leichtmetalle von der Forschung in die Anwendung*, LKR-Verlag, Ranshofen Austria (2008), 111–123.

5. C.Z. Deng, R. Radhakrishnan, S.R. Larsen, D.A. Boismer, J.S. Stinson, A.K. Hotchkiss, E.M. Petersen, J. Weber, T. Scheuermann, "Magnesium alloys for bioabsorbable stents: A feasibility assessment" – *this proceedings volume.*

6. Th. Hassel, F.-W. Bach, A.N. Golovko, "Production and properties of small tubes made from MgCa0,8 for application as stent in biomedical science", *Proceedings of the 7th International Conference on Magnesium Alloys and Their Applications*, Wiley-VCH, Weinheim (2006), 432–437.

7. M. Erinc, W.H. Sillekens, "An assessment on modified AZ80 alloys for prospect biodegradable CV stent applications", *Proceedings of the 8th International Conference on Magnesium Alloys and Their Applications*, Wiley-VCH, Weinheim (2010), 1201–1207.

8. M. Erinc, X. Zhang, W.H. Sillekens, "Modified AZ80 magnesium alloys for biomedical applications", TMS, *Magnesium Technology 2010*, (2010), 641–646.

9. M.A. Leeflang, J. Zhou, J. Duszczyk, "Deformability and extrusion behaviour of magnesium-lithium binary alloys for bio-medical applications", *Proceedings of the 8th International Conference on Magnesium Alloys and Their Applications*, Wiley-VCH, Weinheim (2010), 1182–1188.

10. M.R. Barnett, "A rationale for the strong dependence of mechanical twinning on grain size", *Scripta Materialia*, 59 (2008), 696–698.

11. J. Bohlen, P. Dobron, E.M. Garcia, F. Chmelík, P. Lukáč, D. Letzig, K.-U. Kainer, "The effect of grain size on the deformation behaviour of magnesium alloys investigated by the acoustic emission technique", *Advanced Engineering Materials*, 8 (5) (2006), 422–427.

12. K.D. Ralston, N. Birbilis, "Effect of grain size on corrosion: A review", *Corrosion*, 66 (7) (2010), 075005/1–13.

13. S. Müller, K. Müller, W. Reimers, M. Rosumek, "Microstructure development of extruded Mg alloys", TMS, *Magnesium Technology 2005*, (2005), 165–170.

14. W.H. Sillekens, D.C.W. van der Linden, A.J. den Bakker, "Weld-seam quality of hollow magnesium alloy extrusions", TMS, *Magnesium Technology 2008*, (2008), 189–194.

15. Y. Yan, G. Sun, B. Wei, Y. Hu, "Influence of thermoplastic ECAP on mechanical properties and microstructures of an AZ61A magnesium alloy", TMS, *Magnesium Technology 2004*, (2004), 97–100.

16. M.M. Avedesian, H. Baker, eds., *Magnesium and Magnesium Alloys – ASM Specialty Handbook* (Materials Park, OH: ASM International, 1999), 260–261.

17. Y. Uematsu, K. Tokaji, M. Kamakura, K. Uchida, H. Shibata, N. Bekku, "Effect of extrusion conditions on grain refinement and fatigue behaviour in magnesium alloys", *Materials Science and Engineering A*, 434 (2006), 131–140.

Magnesium Technology 2011
Edited by: Wim H. Sillekens, Sean R. Agnew, Neale R. Neelameggham, and Suveen N. Mathaudhu
TMS (The Minerals, Metals & Materials Society), 2011

BALLISTIC ANALYSIS OF NEW MILITARY GRADE MAGNESIUM ALLOYS FOR ARMOR APPLICATIONS

Tyrone L. Jones[1], Katsuyoshi Kondoh[2]
[1]U.S. Army Research Laboratory; APG, MD 21005-5066, USA
[2]Joining and Welding Research Institute (JWRI), Osaka University; 11-1 Mihogaoka; Ibaragi, Osaka, 567-0047, Japan

Keywords: Magnesium Alloys, Ballistic Performance, Armor, Rapid Solidification Process, Spinning Water Atomization Process (SWAP)

Abstract

Since 2006, the U.S. Army has been evaluating magnesium (Mg) alloys for ballistic structural applications. While Mg-alloys have been used in military structural applications since WWII, very little research has been done to improve its mediocre ballistic performance. The Army's need for ultra-lightweight armor systems has led to research and development of high strength, high ductility Mg-alloys. The U.S. Army Research Laboratory contracted through International Technology Center-Pacific Contract Number FA-5209-09-P-0158 with the Joining and Welding Research Instituteof Osaka University to develop the next generation of high strength, high ductility Mg-alloys using a novel Spinning Water Atomization Process for rapid solidification. New alloys AMX602 and ZAXE1711 in extruded bar form were characterized for microstructure, mechanical, and ballistic response. Significant increases in ballistic performance were evident when compared to the baseline alloy AZ31B.

Introduction

The U. S. Army Research Laboratory's (ARL) ballistic assessment of magnesium (Mg) alloys over the last 5 years has led to an increased understanding of the material's failure mechanisms and relationship between Mg alloy strength and ductility requirements for lightweight armor applications [*1*]. While Mg alloys have been used for military structural applications since WWII, very little research has been done to improve its mediocre ballistic performance [2]. The highest strength commercial Mg alloy available in plate form, AZ31B, has proven to be a very good substitute armor material for AA5083 against armor-piercing projectiles on an equal weight basis [*3*]. It is an adequate substitute armor material against fragment simulating projectiles (FSPs) within an areal density range that is FSP dependent [*4*]. The ballistic data generated by ARL was used to develop the first set of Mg alloy acceptance standards, MIL-DTL-323333 (MR), titled, "Armor Plate, Magnesium Alloy, AZ31B, Applique" [*5*]. Ultimate tensile strength (UTS), tensile yield strength (YTS), ductility, and grain size are all key performance parameters in determining the ballistic performance of metals. The bulk material properties of AZ31B are shown in Table I. Figure 1 correlates impact energy absorption (J) vs. Mg armor alloy AZ31B grain size (μm) [*6*].

In 2009, ARL collaborated with the Joining and Welding Research Institute (JWRI) of Osaka University under contract through International Technology Center-Pacific to develop and evaluate high-strength, high-ductility Mg alloy plate for structural applications. Initial evaluation of conventionally rolled AZ31B plate versus powder formed AZ31B plate showed that grain refinement is not the only parameter that is needed to significantly improve the viability of Mg armor alloy plate [*7*].

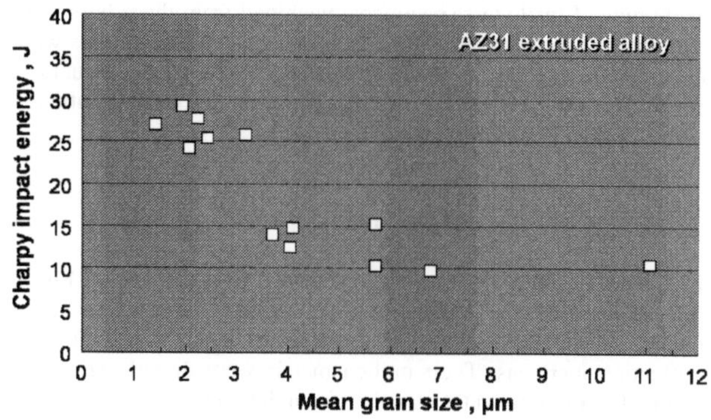

Figure 1. AZ31B Grain Size versus Impact Energy Absorption.

New fundamental Mg alloying is needed to increase the impact energy and thus the performance of Mg alloy plates. The result showed a research opportunity to make Mg alloys a viable armor material that could compete with current aluminum armor alloy solutions [*4*]. Based on our preliminary material and ballistic analysis, the ARL/JWRI program set goals to develop Mg alloys with the mechanical properties shown in Table I.

Table I. Mechanical Properties/Goals of Mg Alloys

	Ultimate Tensile Strength (MPA)	Tensile Yield Strength (MPA)	Elongation to Failure (%)
AZ31B [*7*]	245	150	7
New Mg Alloy	400	350	20

Clearly, there were two potential paths forward towards achieving these set goals:

1. Create new chemical compositions to develop high strength, high ductility Mg alloys.
2. Improve grain refinement through novel processing techniques to produce high-strength, high-ductility Mg alloys.

As a result, ARL and JWRI collaboratively developed two new experimental Mg alloys, AMX602 and ZAXE1711, using the advanced metallurgical powder process. The initial material development and ballistic evaluation of Mg alloy AMX602 and Mg alloy ZAXE1711 are discussed in the next sections.

Material Development

AMX602 (Mg-6Al-0.5Mn-2Ca/mass%) and ZAXE1711 (Mg-1Zn-7Al-1Ca-1La/mass%) Mg alloy powders produced by the Spinning Water Atomization Process (SWAP) were used as raw input materials [*8, 9*]. The coarse Mg alloy powders were 1-5mm length. It was previously verified that the coarse Mg powders of these lengths were non-combustible. The α-Mg grain size of the raw powders was less than 0.5 μm. The powder compaction and hot extrusion were applied to these raw powders to fabricate the extruded bars. The bar had a cross-section of 24.5 mm x 40 mm x 1000 mm. Tensile test specimens machined from these bars were evaluated at room temperature. The material microstructures were observed using an optical microscope. The microstructural evaluation of ZAXE1711 is withheld from discussion until the patent application is processed.

Experimental Evaluation of Raw materials

In SWAP powder preparation, schematically illustrated in Fig. 2 (a), non-combustive AMX602 magnesium alloy ingots were melted at 1053 K in the ceramic crucible covered with the protection inert gas. Their molten metals were directly streamed inside the spinning water chamber from the crucible nozzle. Table II shows chemical compositions of AMX602 alloy powders prepared by SWAP. The calcium is necessary because it promotes the non-combustive properties of the magnesium alloys. The impurity content of Fe and Cu is s controlled to less than 0.005% because they are corrosive elements in magnesium alloys. As shown in Figure 2 (b), a length of the coarse AMX602 powders prepared by SWAP is 1 ~ 4 mm, and they have an irregular shape. The cast ingot with the same composition was also prepared as a reference input material.

Powder Consolidation

The powder was consolidated at room temperature using a 2000kN hydraulic press machine to fabricate the green compact. The green compact had a relative density of 85% and 42 mm diameter. The columnar compact and cast ingot were heated at 573 ~ 673K for 180 seconds in an argon gas atmosphere, then immediately consolidated into full density material by hot extrusion. An extrusion ratio of 37 and an extrusion speed 1 m/s were used in this study.

Table II. Chemical Compositions of Non-Combustive Magnesium Alloy Powders.

Al	Zn	Mn	Fe	Si	Cu	Ca	Mg
6.01	0.007	0.26	0.002	0.038	0.004	2.09	Bal.

Microstructural Analysis

Figure 3 indicates optical microstructures of AMX602 cast ingot (a), as-received fine powders less than 0.5 mm (b) and coarse powders over 1 mm (c) prepared by SWAP. The cast ingot material consists of coarse α-Mg grains of 60 ~ 150 μm diameter, and some intermetallic compounds are observed at their grain boundaries.

(a)

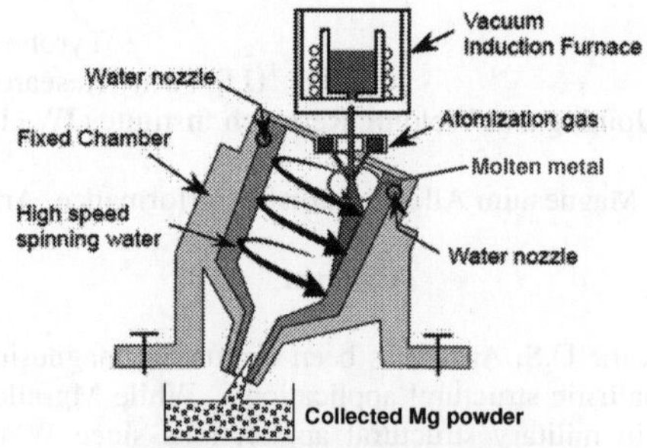

(b)

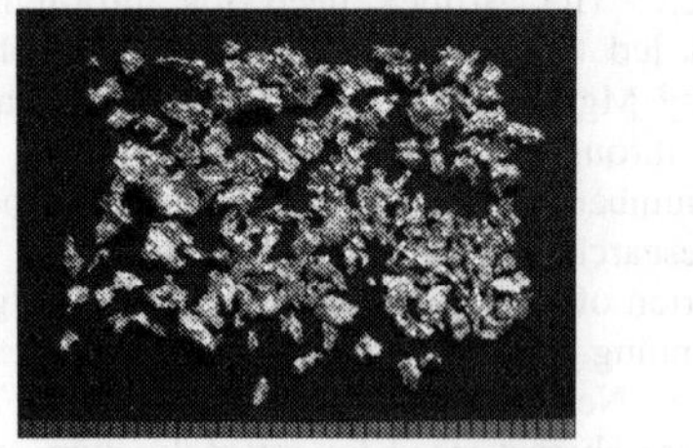

Figure 2. (a) Schematic Illustration of SWAP Equipment to Produce Rapidly Solidified Mg Alloy Powders; (b) Morphology of Coarse Magnesium Alloy Powder Prepared by SWAP.

As shown in Figure 3(b), the fine powders revealed small dendrite structures that were formed during the rapid solidification of molten Mg alloy droplets after atomization. A mean Dendrite Arm Spacing (DAS) of fine powders Figure 3(b) was 0.97 μm. The coarse powders shown in Figure 3(c) indicate large grains of 4 ~ 8 μm diameter that were caused by a slower solidification rate during atomization. The mean grain size was 6.7 μm. The DAS value was estimated using the following Equation 1 [*9*], the DAS value suggests the estimated solidification rate (R) of the fine AMX602 powders:

$$\lambda = 35.5R^{-0.31} \qquad (1)$$

where,

λ = DAS (μm),

R = Solidification rate (K/s)

Hence the estimated solidification rate of fine SWAP Mg alloy powders is 1.1×10^5 K/s. The estimated solidification rate of coarse SWAP Mg alloy powders is 6.8×10^3 K/s. The mean particle size of fine and is 0.36 mm. The mean particle size of coarse powders is 1.87 mm. Theses particle sizes are significantly larger than other alloy powders that with 100 ~ 200 μm via the conventional atomization process. This indicates that SWAP Mg alloy powders are very safe in handling. Additionally the SWAP powders showed very fine microstructures due to the high solidification rate of the SWAP. On the other hand, the cast AMX602 alloy had a very small solidification rate of 0.6 K/s calculated by Equation 1. This cast solidification rate was equivalent to the solidification rate for conventional cast Mg alloys [*10*]. Additionally, the high rapid solidification rate makes inter-metallic compounds nearly undetectable at the grain boundaries.

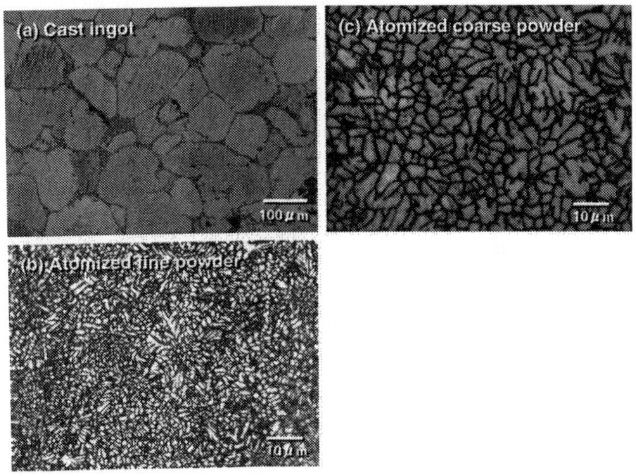

Figure 3. Optical Microstructures of Input Raw Materials: (a) As-Cast Ingot Billet; (b) As-Received Fine SWAP Powder; (c) Coarse Powder.

Figure 4 shows differential thermal analysis (DTA) profiles of the atomized coarse powders and cast ingot. The break on the slope at 723 K in both profiles are caused by changing the heating rate from 10 to 5 K/min. The large endothermic peaks were detected at 798 ~ 800 K at both rates, and correspond to a melting point of Al_2Ca intermetallics [8]. Based on these results, the extrusion temperature of the atomized powders and cast ingot should be controlled less than the melting point of Al_2Ca intermetallics at 798 K. In this study, the pre-heating temperature of the green compact and cast ingot material is set at 573 ~ 673 K.

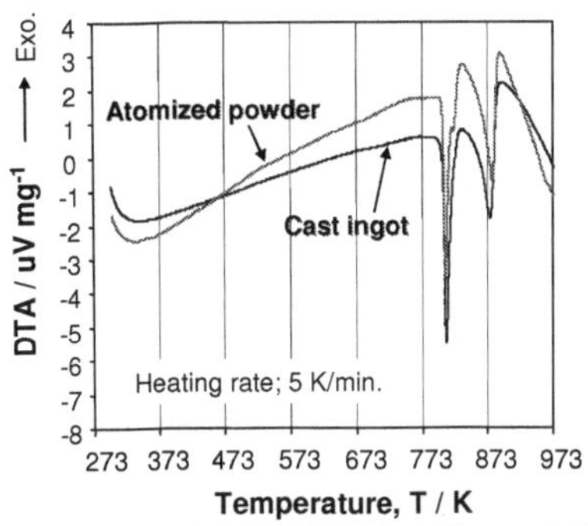

Fig. 4. DTA Profiles of AMX602 Cast Ingot Billet and As-Received Coarse SWAP Powder.

The non-combustive AMX602 magnesium alloys were fabricated by extruding the green compacts of rapidly solidified coarse powders with 1~4 mm via SWAP. They showed extremely fine α-Mg grains of 0.3~1.1μm in diameter via dynamic recrystallization. Fine Al_2Ca compounds with a particles size of 100~300 nm were precipitated during hot extrusion, and uniformly distributed in the matrix. Compared to the AMX602 extruded alloys using the cast ingot billets, TS and YS of the powder extruded materials showed the significant increment of 30~45%. The optimization of the pre-heating temperature before hot extrusion was effective to form very fine recrystallizedα-Mg grains and intermetallic dispersoids. For example, a good balance of 422 MPa TS and 14.2% elongation was obtained when employing the pre-heating temperature at 623 K.

Figure 5 shows x-ray diffraction patterns of the input raw material and its extruded alloys at 573 K and 673 K. In the case of SWAP powders shown in (a), magnesium oxide (MgO) peak is detected at $2\theta = 43°$, which is caused by oxidation of molten Mg alloy droplets during water atomization. MgO films covering the powder surfaces are thermally stable. It is, however, easy to break them by applying a very low pressure in consolidation because of their porous structures [*11*]. No Al_2Ca intermetallic peak is shown in the profile of the raw powder, but it is detected in the extruded alloys. This means that Al and Ca are solid-soluted in the matrix of the as-received raw powder by rapid solidification. However, Al_2Ca compounds are precipitated due to the high elevated temperature during hot extrusion. The peak intensity of Al_2Ca gradually increases with increase in the extrusion temperature. That is, the amount of precipitated Al_2Ca intermetallics increases by a larger thermal history in pre-heating. On the other hand, as shown in (b), the cast ingot material before extrusion originally contains coarse Al_2Ca intermetallics crystallized during casting [*12*]. There is no remarkable difference of the profile of the wrought alloys extruded at between 573K and 673K.

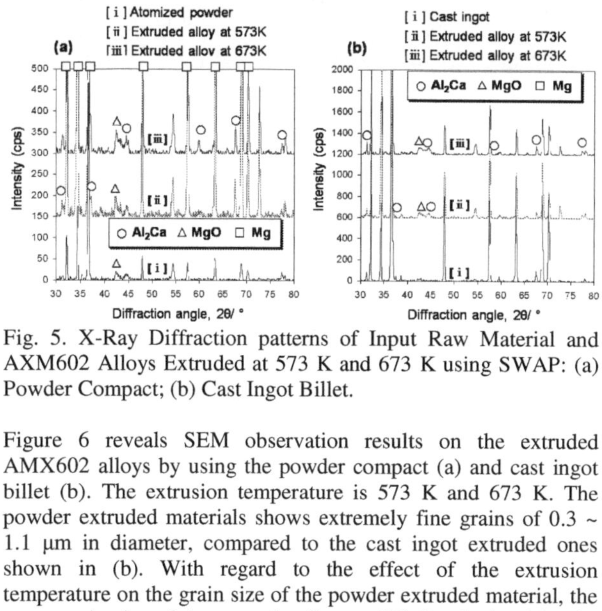

Fig. 5. X-Ray Diffraction patterns of Input Raw Material and AXM602 Alloys Extruded at 573 K and 673 K using SWAP: (a) Powder Compact; (b) Cast Ingot Billet.

Figure 6 reveals SEM observation results on the extruded AMX602 alloys by using the powder compact (a) and cast ingot billet (b). The extrusion temperature is 573 K and 673 K. The powder extruded materials shows extremely fine grains of 0.3 ~ 1.1 μm in diameter, compared to the cast ingot extruded ones shown in (b). With regard to the effect of the extrusion temperature on the grain size of the powder extruded material, the mean grain size of the wrought alloy at 573 K calculated by the image scanning soft ware is 0.45 μm, and that of the extruded alloy at 673 K is 0.8 μm. The larger thermal history during pre-heating in the latter at 673 K causes a small grain growth after dynamic recrystallization in hot extrusion [*13*]. The amount of precipitated Al_2Ca intermetallic compounds, which correspond to white fine particles of 100 ~ 300 nm distributed in the matrix, of the extruded alloy at 673 K is larger than that of the wrought alloy at 573 K. On the other hand, Fig. 6 (b-1) indicates that the AMX602 wrought alloy extruded at 573 K using the cast ingot billet consists of fine α -Mg grains of 1 ~ 3 μm via dynamic recrystallization. Some of large grains of 5 ~ 10 μm are observed in the matrix. The mean grain size of the extruded materials at 573

K is 1.96 μm. However, when employing the higher pre-heating temperature at 673 K, a lot of coarse grains over 10 μm exist in the matrix as shown in (b-2) and a few fine grains less than 3 μm are observed. The mean grain size is 3.29 μm. Compared to the microstructure shown in (b-1), the grain growth and coarsening of (b-2) certainly occurred during extrusion at higher temperature of 673 K after dynamic recrystallization. Both wrought alloys using cast ingots contain coarse intermetallic compounds with irregular morphologies. Compared to AMX602 cast ingot material, some small compounds are observed because severe plastic deformation during hot extrusion caused fragmentation of coarse and brittle intermetallics distributed in the matrix.

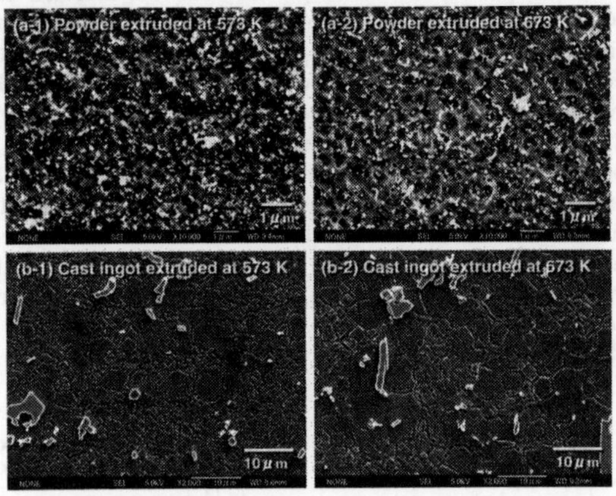

Figure 6. Scanning electron microscope observation on AMX602 alloys extruded 573 K and 673 K using SWAP powder compact (a) and cast ingot billet (b).

Figure 7 shows a micro hardness dependence on the grain size of the extruded AMX602 alloys fabricated by using SWAP powder compacts and cast ingot billets. Micro-Vicker's hardness of each material shown in Figure 7 is as follows; 113 HV (a-1), 94.3 HV (a-2), 77.0 HV (b-1), and 69.9 HV (b-2). The hardness of wrought alloys using atomized powder compacts is larger than that using the cast ingot. This is due to the extremely fine grains and very small intermetallic compounds of the former materials as mentioned in Figure 7. It reveals that the micro hardness is proportional to $d^{-0.5}$ (d: α -Mg grain size), and Hall-Petch relationship is shown in these data. Figure 9 indicates a dependence of tensile strength (TS) and yield stress (YS) of the extruded AMX602 alloys on the pre-heating temperature. TS and YS of the wrought alloys using the cast ingot billets is 298 ~ 311 MPa and 204 ~ 251 MPa, respectively. It is reported that the content of Ca included AM60 alloys is effective for not only their non-combustive but also mechanical properties because coarse intermetallics of Al_2Ca with irregular shapes cause the decrease of the tensile properties [*8*, *13*]. Therefore, when using the cast ingot billet, the maximum content of Ca is 2 mass%. On the other hand, the alloys using the SWAP powder compacts show an extremely high strength of 391 ~ 452 MPa TS and 358 ~ 428 MPa YS. Concerning their dependence on the temperature, both materials reveal the decrease of TS and YS with increase in the pre-heating temperature due to the grain growth and coarsening as shown in Figure 5. In particular, the decrement of the strength of SWAP powder extruded alloys is larger than that of cast ingot extruded materials. The fine microstructures via dynamic recrystallization of the former are very sensitive to the temperature [*14*], and the grain growth by solid-diffusion easily occurs when a higher pre-heating temperature is applied.

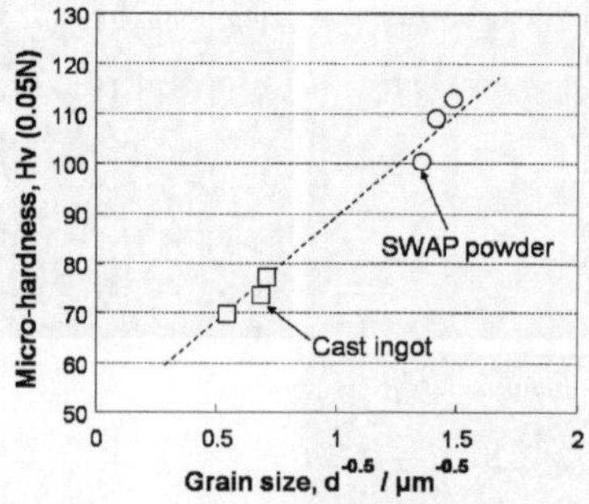

Figure 7. Dependence of micro-Vicker's Hardness on Grain Size of Extruded AMX602 Alloys.

As shown in Figure 8, the elongation increases with increasing the temperature. The SWAP powder extruded alloy at 623 K shows a good balance of 422 MPa TS and 14.2% elongation. As shown in Figure 9, fine dimple fractured patterns, which mean typical fractures inside α-Mg grains, are observed, and no fracture at the primary particle boundaries observed in the by SEM image. The fragmentation of intermetallic compounds at the fractured surface is not observed. Therefore, this material shows a high strength and good ductility. On the other hand, Figure 9 indicates the fractured surface of the cast ingot extruded alloy at 623 K. It also shows dimple fractured patterns. However, the coarse brittle intermetallic compounds include some cracks marked with white arrows, and they correspond to the initiation and propagation of the fracture.

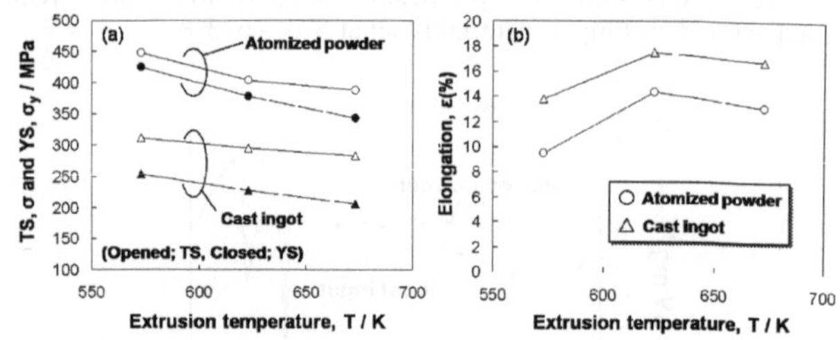

Figure 8. Tensile Strength, Yield Stress and Elongation Dependence on Pre-Heating Temperature.

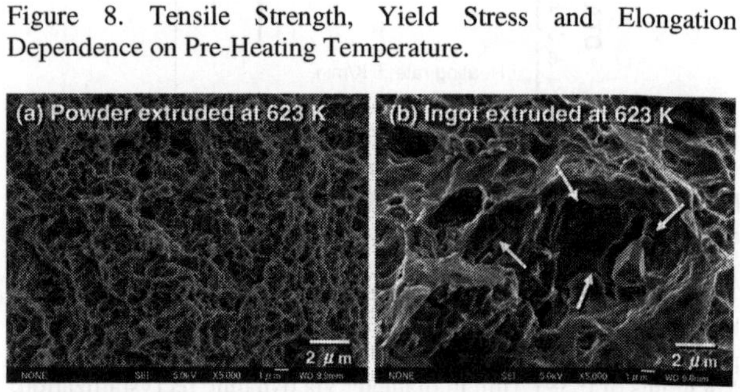

Fig. 9. SEM Observation on Fractured Surface of Tensile Test Specimen of Extruded at 623 K using Powder Compact (a) and Cast Ingot (b).

32333), while Mg alloy ZAXE1711 showed to a 37% increase in the V_{50} ballistic limit compared to Mg armor alloy AZ31B.

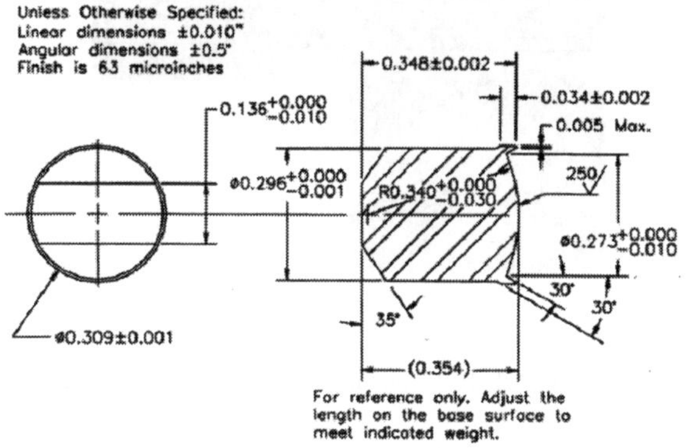

ALL UNITS ARE IN INCHES.

Figure 10. The .30-cal. FSP Schematic Diagram.

Mechanical Test Results

The mechanical properties of the Mg alloy AMX602 and Mg alloy ZAXE1711 samples are shown in Table III and Table IV. Mg alloy AMX602 and ZAXE1711 bars of three different tempers were produced for ballistic evaluation. Measurements at 0° were taken at the center and 0.25" width of the specimen. The temper designations (i.e. AMX602-temper) for each Mg alloy extrusion temperature were as follows:

For AMX602:	For ZAXE1711:
1: 350°C	B: 350°C
2: 300°C	C: 250°C
3: 250°C	D: 200°C

Table III. Mechanical Properties for Mg Alloy AMX602

	0° - center			0° - 1/4 width			45°			90°		
	σ	σy	El	σ	σy	El	σ	σy	El	σ	σy	El
	MPa	MPa	%	MPa	MPa	%	MPa	MPa	%	MPa	MPa	%
Rod 1	355	301	21.0	359	306	19.2	319	234	18	330	244	16.4
Rod 2	357	314	20.8	363	302	18.0	323	236	22.4	307	237	8.5
Rod 3	358	311	17.8	360	312	19.1	324	261	15.1	304	244	7.6

Table IV. Mechanical Properties for Mg Alloy ZAXE1711

	0° - center			0° - 1/4 width			45°			90°		
	σ	σy	El	σ	σy	El	σ	σy	El	σ	σy	El
	MPa	MPa	%	MPa	MPa	%	MPa	MPa	%	MPa	MPa	%
Rod B	377	268	17.4	380	276	17.4	356	245	15.8	363	248	13.0
Rod C	380	269	18.2	384	280	18.4	361	245	18.7	369	246	18.2
Rod D	388	274	18.7	391	286	18.3	362	238	22.5	374	236	19.4

Note the difference between the mechanical properties of Mg alloy AMX602 and Mg alloy ZAXE1711 at the selected tempers are very small. This means extrusion temperature will not be a significant factor on the mechanical properties of these alloys strength when fabricating the scale-up specimens.

Ballistic Evaluation

Based on the aforementioned MIL-DTL-32333 required baseline thicknesses, the Mg alloy bars were evaluated using the .30-cal. FSP. The test projectile schematic diagram, weights, and hardness specifications are shown in Figure 10 and Table V. The hardness of each Mg alloy test bar was measured using the 500-kg Brinell scale. The nominal hardness of each Mg alloy test bar is shown in Table VI.

Table V. Projectile Weight and Hardness Requirements

FSP Type	Weight (gr)	Rockwell C Hardness
.30 Cal	44.0 ± 0.5	30 ± 1

All V_{50} ballistic limits (velocity at which the projectile is expected to perforate the armor 50% of the time) were calculated following MIL-STD-662F [*15*] and are shown in Table VI. Mg alloy AMX602 showed to a 33% increase in the V_{50} ballistic limit compared to Mg armor alloy AZ31B (attained from MIL-DTL-32333), while Mg alloy ZAXE1711 showed to a 37% increase in the V_{50} ballistic limit compared to Mg armor alloy AZ31B.

The thin plates were held horizontally in a test fixture by C-clamps on the ends of the bars. As evident in figures 11 & 12, the AMX602 showed confined impact. For the instances the projectile perforated the Mg alloy AMX602 bar and created an exit hole, the spall was typically localized within a 31.75-mm diameter. Although the edge of the spall after ballistic impact reached the edge of the AMX602 and ZAXE1711 bars, the V_{50} data significantly exceeded Mg alloy AZ31B V_{50} data at the same areal weight while keeping its structural integrity.

Table VI. Mg Alloy V_{50} Ballistic Velocities

Metal Alloys	Thickness	*Hardness	Ballistic Limit	Ballistic Limit
	mm	HBN	m/s	ft/s
AZ31B	25.400	61	833	2733
AMX602-1	25.190/25.235	80	1061	3480
AMX602-2	25.171/25.210/25.197	80/80/83	1092	3570
AMX602-3	25.171/25.178	80	1105	3624
ZAXE1711-B	25.248	80	1111	3646

(a) Front

(b) Back

Figure 13. Post-ballistic pictures of Mg alloy AMX602 bars.

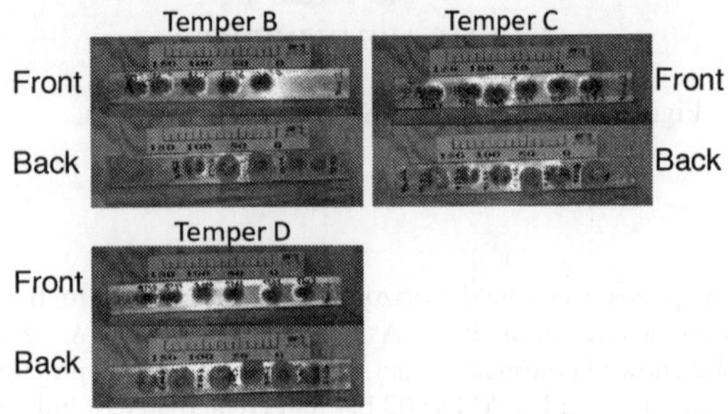

Figure 14. Post-ballistic pictures of Mg alloy ZAXE1711 bars.

Conclusions

The non-combustive AMX602 magnesium alloys were fabricated by extruding the green compacts of rapidly solidified coarse powders with 1~4 mm via SWAP. They showed extremely fine α-Mg grains of 0.3~1.1μm in diameter via dynamic recrystallization. Fine Al_2Ca compounds with a particle size of 100 ~ 300 nm were precipitated during hot extrusion, and uniformly distributed in the matrix. Compared to the AMX602 extruded alloys using the cast ingot billets, TS and YS of the powder extruded materials showed the significant increment of 30~45%. The optimization of the pre-heating temperature before hot extrusion was effective to form very fine recrystallized α-Mg grains and intermetallic dispersoids. For example, a good balance of 422 MPa TS and 14.2% elongation was obtained when employing the pre-heating temperature at 623 K.

New Mg alloy AMX602 and Mg alloy ZAXE1711 bars showed superior ballistic performance as an armor alloy when compared to baseline Mg armor alloy AZ31B plate. Through advanced powder metallurgy processing and chemical alloying, superior mechanical properties were achieved. Initial ballistic performance results of Mg alloy AMX602 showed up to a 33% higher ballistic limit compared to the baseline Mg armor alloy AZ31B. Initial ballistic performance results of Mg alloy ZAXE1711 showed up to a 37% higher ballistic limit compared to the baseline Mg armor alloy AZ31B. Future development will include the scaling up of Mg alloy AMX602 bars to plate for further analysis and full scale structural applications. It is expected that the ballistic limit and thus the overall ballistic performance of the scaled size will increase because of the reduction of the edge effects during an impact. Pending successful results of the large scale Mg alloy AMX602 development, Mg alloy ZAXE1711 will follow the same large scale development plan.

References

1. Jones, T.; DeLorme, R.; Burkins, M.; Gooch, W. *Ballistic Evaluation of Magnesium Alloy AZ31B*; ARL-TR-4077; U.S. Army Research Laboratory: Aberdeen Proving Ground, MD, April 2007.

2. Mathaudhu, S.; Nyberg, E.; *Magnesium Alloys in Army Applications: Past, Current and Future Solutions*; Proceedings of the 2010 Minerals, Metals and Materials (TMS) Annual Symposium, Seattle, WA, February 2010.

3. Jones, T.; DeLorme, R.; *Development of a Ballistic Specification for Magnesium Alloy AZ31B*; ARL-TR-4664; U.S. Army Research Laboratory: Aberdeen Proving Ground, MD; December 2008.

4. Jones, T.; DeLorme, R.; *A Comparison of the Ballistic Performance Between Rolled Plate in AZ31B-H24 Magnesium and 5083-H131 Aluminum*; 24th International Symposium on Ballistics, New Orleans, LA; September 2008.

5. MIL-DTL-32333 (MR); Armor Plate, Magnesium Alloy, AZ31B, Applique; 2009.

6. Liao, J.; Hotta, M.; Kaneko, K.; Kondoh, K.; *Enhanced Impact Toughness of Magnesium Alloy by Grain Refinement*; Scripta Materialia 2009, 61; 208–211.

7. Jones, T.; Kondoh, K.; *Initial Evaluation of Advanced Powder Metallurgy Magnesium Alloys for Dynamic Applications*; ARL-TR-4828; U.S. Army Research Laboratory, Aberdeen Proving Ground, MD; May 2009.

8. Sakamoto, M.; Akiyama, S.; Hagio, T.; Ogi, K.; *Control of Oxidation Surface Film and Suppression of Ignition of Molten Mg-Ca Alloy by Ca Addition*; Journal of Japan Foundry Engineering Society; 69 (1997) 227-233.

9. Nishida, S.; Motomura, I.; *Estimation of Heat Transfer Coefficient and Temperature Transition on Melt Drag Process of AZ31 Magnesium Alloy by Heat Transfer and Solidification Analysis*; Journal of Japan Institute of Light Metals; 58 (2008) 439-442.

10. Wei, L. Y.; Dunlop, G. L.; *The Solidification Behavior of Mg-Al-rare Earth Alloys;* Journal of Alloys and Compounds; 232 (1996) 264-268.

11. Šplíchal, Karel; Jurkech, Ludvík; *Comparison of Oxidation of Cast and Sintered Magnesium Materials in CO2*; Journal of Nuclear Materials; 48 (1973) 277-286.

12. Anyanwu, I.A.; Gokan, Y; Suzuki, A.; Kamado, S; Kojima, Y.; Takeda, S.; Ishida, T.; *Effect of Substituting Cerium-Rich mischmetal with Lanthanum on High Temperature Properties of Die-Cast Mg-Zn-Al-Ca-RE alloys*; Materials Science and Engineering A; 380 (2004) 93-99.

13. Chandrasekaran, M.; and John, Y.M.S.; *Effect of Materials and Temperature on the Forward Extrusion of Magnesium Alloys*; Materials Science and Engineering A; 381 (2004) 308-319.

14. Zhang, J.H.; Liu, Hai Feng; Sun, W.; Lu, H.Y.; Tang, D.X.; Meng, Jian; *Influence of Structure and Ionic Radius on Solubility Limit in the Mg–Re Systems*, Materials Science Forum; Vol.561-565 (2007) 143-146; doi10.4028/www.scientific.net/MSF.561-565.143; accessed October 2010.

15. MIL-STD-662F; V_{50} Ballistic Test For Armor; 1997.

Magnesium Technology 2011
Edited by: Wim H. Sillekens, Sean R. Agnew, Neale R. Neelameggham, and Suveen N. Mathaudhu
TMS (The Minerals, Metals & Materials Society), 2011

$MG_{17}AL_{12}$ INTERMETALLIC PREPARED BY BULK MECHANICAL ALLOYING

Kenji Sakuragi, Masashi Sato, Takamitsu Honjo, Toshiro Kuji
Department of Applied Chemistry, Tokai University; 4-1-1 Kita-Kaname, Hiratsuka, 259-1292, Japan

Keywords: Magnesium-Aluminum alloy, Mechanical alloying, Nano-crystalline materials, Hydrogen storage materials

Abstract

In a last decade, nano-crystallized or amorphous materials are widely investigated. Such materials provide other directions for solving several issues in the field of materials science. Ball milling processes are general tools for preparing the nano-crystallized or amorphous material. The samples, however, show the fine powder-form and the most of those are very sensitive against circumstance around. The bulk-form samples are desirable. The present work proposes the newly developed bulk mechanical alloying (BMA) technique as an alternative method for the ball milling. Successful solid-state reaction and the formation of nano-scale crystalline of $Mg_{17}Al_{17}$ phase will be demonstrated.

Introduction

A ball-milling technique is available as a powerful tool for the preparing of nano-crystalline or amorphous materials. In the case of hydrogen storage materials, for instance, the nano-crystallizing or amorphorization modifies thermodynamic and / or kinetic properties upon hydrogen absorption-desorption processes [1-4]. However, the ball-milling process provides the specimen as fine powder-form, which leads the very sensitive surface for the contaminations from the atmospheres around. The yield of a final sample by ball-milling process is, in addition, significantly low as compared with an amount of starting material.

Bulk Mechanical Alloying (BMA) process has been proposed as an alternative technique for the traditionally applied ball milling method [5, 6]. The loaded materials travel through the different diameters of a die cavity cell [7], and the plastic deformation takes place on the cycle of compaction-withdraw-extrusion-withdraw-ejection as shown in Figure 1. By repeating this cycle, finally, the samples can be received as nano-crystalline or amorphous materials. Previous reports on the synthesis of promising candidates of hydrogen storage media, nano-crystalline Pd [3], Mg_2Ni [4, 8] and $Mg_{17}Al_{12}$ [9], by the BMA demonstrated the possibility of BMA as a candidate on the novel solid-state reaction technique. Interest in BMA as desirable mechanical alloying process has emerged due to the received samples that show bulk-form. During BMA process, the sample contact with only the inner wall of cell and upper / lower punches, and the loss of the sample is efficiently low.

The present work shows more detailed on the material processes of BMA from the synthesis of $Mg_{17}Al_{12}$ intermetallic.

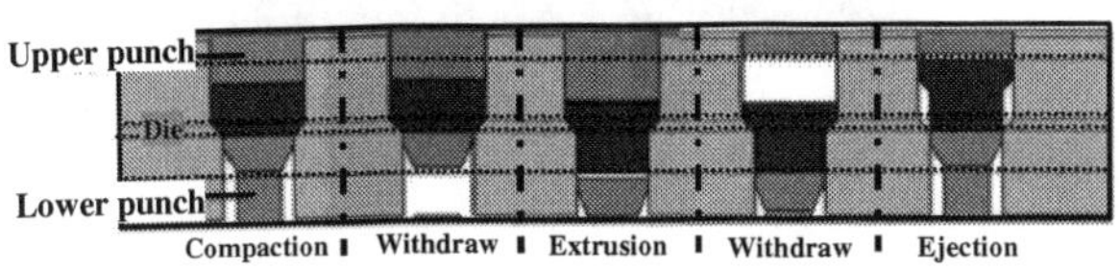

Figure 1. Schematic diagram of typical process for BMA.

Experimental

$Mg_{17}Al_{12}$ intermetallic was prepared by BMA technique. BMA set-up was supplied from KYB Co., Ltd, Japan. Actual set-up of BMA is shown in Figure 2. A stoichiometric mixture of Mg and Al powders with an amount of 5 g was loaded into a die cavity with 20 mm diameter of the die-set as a starting material. The sizes of Mg or Al powder were less than 180 μm or 300 μm, respectively. The purity grade of those was higher than 99.9 %. One compaction-withdraw-extrusion-withdraw-ejection cycle was controlled at the pressing 100 kN for 8 s (see Figure 1). The cycle was repeated for 100, 500, 1000 and 2000 cycles in air with water-cooling system at 298 K.

The formation of $Mg_{17}Al_{12}$ intermetallic phase was characterized by powder X-ray diffraction (XRD). Wavelength dispersive X-ray spectrometry (WDX) and transmission electron microscope (TEM) measurements were made for the confirmation of homogeneous distributions and microstructure of the alloy.

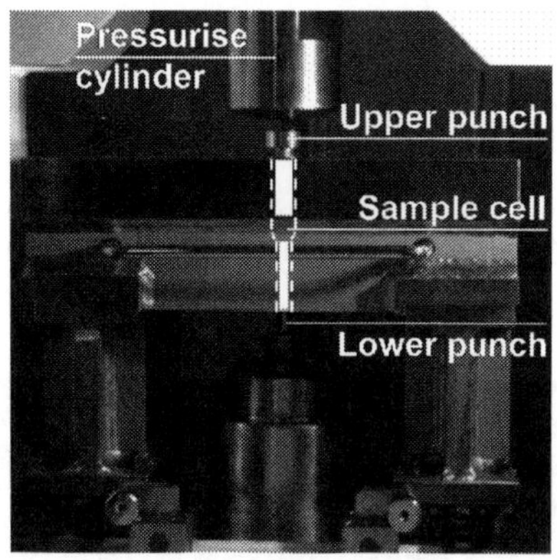

Figure 2. A photograph of BMA set-up.

Results and Discussion

Figure 3 shows a photo image of Mg-Al alloy prepared by BMA process. The sample shows bulk-form, and the sample yielded almost 100 % (5 g) as compared with an amount of starting materials. These are obviously advantages of BMA technique since further handling care for contaminations is not necessary. With increasing repeating cycles, the XRD profile shows broaden peaks, and small contributions from $Mg_{17}Al_{12}$ intermetallic phase appeared at 1000 cycles as shown in Figure 4. Finally, the single phase of the intermetallic $Mg_{17}Al_{12}$ can be seen at 2000 cycles (approximately 4.4 h). In previous work, 2 g of $Mg_{17}Al_{12}$ intermetallics (rate of yield was 80-90 %) was received by a high energy ball-milling for 5 h [10]. BMA can be emphasized as a pronouncedly effective mechanical alloying technique.

Figure 3. Photo image of Mg-Al alloy after BMA process. The sample shows bulk-form.

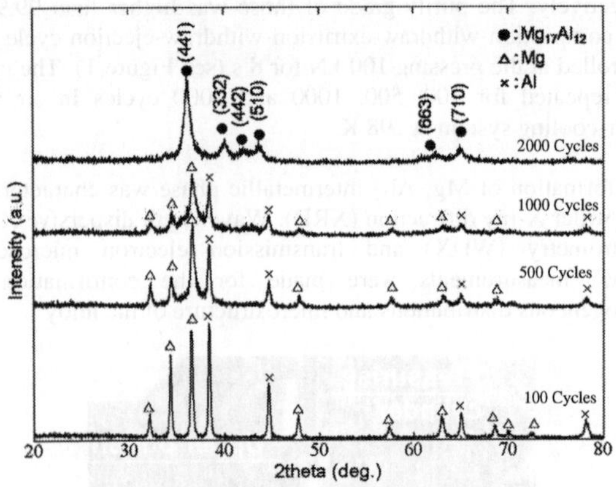

Figure 4. Collected XRD profiles of Mg-Al alloy with cyclic BMA showing (x) Al, (Δ) Mg and (•) $Mg_{17}Al_{12}$. Miller indexes of $Mg_{17}Al_{12}$ are shown together.

Figure 5. WDX images of Mg-Al alloy with cyclic BMA. The distributions of Mg and Al in alloy can be seen.

The changes in the elemental distributions of Mg / Al during cyclic BMA were monitored by WDX (see Figure 5). The proceeding of the alloying is clearly seen. The increases of cyclic BMA provide homogeneous distribution of Mg and Al, and the formation of intermetallic $Mg_{17}Al_{12}$ phase completely takes place at 2000 cycles. This is well agreed with the data from XRD profiles (see Figure 4). It should be noted as one of advantages in BMA that no significant impurities were found by WDX profile. In ball-milling process, the contamination from vial / balls or atmosphere is a common problem. On the process of BMA, the contamination from cell may be less likely, because the sample contacts with only inner wall of cell during the extrusion-compression cycles.

TEM observation was made for the sample at 2000 cycles as shown in Figure 6. The electron scattering rings are also presented. The crystalline shows around 20-30 nm scales. This concludes that BMA drives nano-crystallines upon mechanical alloying processes.

Conclusion

In this study, nano-crystalline $Mg_{17}Al_{12}$ intermetallic phase was successfully driven by newly developed bulk mechanical alloying (BMA) technique. The alloying time is rather short compared with conventional ball-milling, and the sample yields effectively. For the further possibility as attractive alternatives of ball-milling type technique, the controlling of atmosphere may be required. This is under development for the materials that is very sensitive against oxygen or water, such as Li-based alloy [7].

Acknowledgement

This work partly received financial support from the New Energy and Industrial Technology Development Organization (NEDO), Japan. We thank Mr. Takashi Kobayashi and Ms. Mayumi Watanabe from KYB Co., Ltd. for the kind discussion in this work.

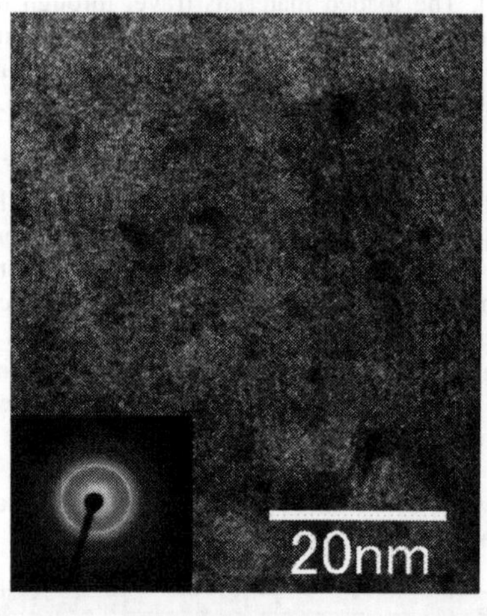

Figure 6. TEM image of $Mg_{17}Al_{12}$ intermetallic microstructure at 2000 BMA cycle. The electron scattering rings are presented together.

References

1. T. Haraki et al., “Properties of hydrogen absorption by nano-structured FeTi alloys,” Int. J. Mat. Res. (formerly Z. Metallkd.) 99 (2008), 507-512.
2. J. -C. Crivello, T. Nobuki and T. Kuji, “Improvement of Mg-Al alloys for hydrogen storage applications,” Int. J. Hydrogen Energy, 34 (2009) 1937-1943.
3. T. Kuji et al., “Hydrogen Absorption of Nanocrystalline Palladium,” J. Alloys Compd, 330–332 (2002), 718–722.
4. T.Kuji, H. Nakano and T. Aizawa, “Hydride Formation and Electrochemical Properties of $Mg_{2-x}Ni(X=2.0\text{-}1.5)$ Alloys Prepared by Bulk Mechanical Alloying,” J. Alloys Compd, 330–336 (2002), 590–596.
5. T. Aizawa, K. Tatsuzawa and J. Kihara, “Mechano-metallurgical Proceedings for Direct Fabrication of Solid Non-Equilibrium Phase Materials,” J. Faculty of Engineering University of Tokyo XLII, (1993), 261–279.
6. T. Aizawa, J. Kihara and D.J. Benson, “Nontraditional Mechanical Alloying by the Controlled Plastic Deformation, Flow and Fracture Processes,” Mater. Trans. JIM, 36 (1995), 138–149.
7. T. Honjo and T. Kuji, “LiAl alloy prepared by Bulk Mechanical Alloying,” (Proceeding at the Light Metals Division, TMS 2011 Annual Meeting and Exhibition, San Diego, 27 February - 3 March) accepted.
8. T. Aizawa, T. Kuji and H. Nakano, “Synthesis of Mg_2Ni Alloy by Bulk Mechanical Alloying,” J. Alloys Compd, 291 (1999), 248–253.
9. H. Yabe and T. Kuji, “Mechanically driven nanostructured magnesium-aluminium alloy,” J. Metastable Nanocryst. Mater, 24/25 (2006), 173–176.
10. J. -C. Crivello, T. Nobuki and T. Kuji, “Limits of the Mg-Al γ-phase range by ball-milling,” Intermetallics, 15 (2007) 1432-1437.

Magnesium Technology 2011
Edited by: Wim H. Sillekens, Sean R. Agnew, Neale R. Neelameggham, and Suveen N. Mathaudhu
TMS (The Minerals, Metals & Materials Society), 2011

CORROSION BEHAVIOUR OF MG ALLOYS IN VARIOUS BASIC MEDIA: APPLICATION OF WASTE ENCAPSULATION OF FUEL DECANNING FROM UNGG NUCLEAR REACTOR

David Lambertin, Fabien Frizon, Adrien Blachere, Florence Bart
CEA Marcoule, DTCD/SPDE/L2ED, BP 27171, 30207 Bagnols Sur Ceze, France

Keywords: Magnesium-zirconium alloy, encapsulation, electrochemistry

Abstract

The dismantling of UNGG nuclear reactor generates a large volume of fuel decanning. These materials are based on Mg-Zr alloy. The dismantling strategy could be to encapsulate these wastes into an ordinary Portland cement (OPC) or geopolymer (aluminosilicate material) in a form suitable for storage. Studies have been performed on Mg or Mg-Al alloy in basic media but no data are available on Mg-Zr behaviour. The influence of representative pore solution of both OPC and geopolymer with Mg-Zr alloy has been studied on corrosion behaviour. Electrochemical methods have been used to determine the corrosion densities at room temperature. Results show that the corrosion densities of Mg-Zr alloy in OPC solution is one order of magnitude more important than in a geopolymer solution environment and the effect of an inhibiting agent has been undertaken with Mg-Zr alloy. Evaluation of corrosion hydrogen production during the encapsulation of Mg-Zr alloy in both OPC and geopolymer has also been done.

Introduction

The dismantling of UNGG nuclear reactor generates a large volume of fuel decanning. These materials are based on Mg-Zr alloy. The dismantling strategy could be to encapsulate these wastes into a mineral binder, namely an ordinary Portland cement (OPC) or geopolymer (aluminosilicate material) in a form suitable for storage.

Since 1990, scrap metal from the MAGNOX reactor in Great Britain based on Mg-Al alloy were packed with a mixture of Portland cement and blast furnace ash or fly ash [1, 2, 3]. The strategy of this formulation is to use a minimum of water for hydration of cement and a high pH (around 12.5) [2]. Hydrogen gas generation has been observed with the formation of magnesium hydroxide on the surface of magnesium metal, causing damage to the package [4]. An alternative to the use of Portland cement has been proposed with the use of mineral geopolymer [1, 5]. Encapsulations of Magnox waste containing magnesium metal were made with organic polymers [6] to limit the amount of water to avoid a hydrogen gas.

In this work, we have studied the influence of representative solution of OPC and geopolymer with Mg-Zr alloy on corrosion behaviour. Electrochemical methods have been used to determine the corrosion densities at room temperature. Evaluation of encapsulation of Mg-Zr alloy in OPC and geopolymer has been done in term of corrosion hydrogen production.

Experimental procedure

The material used in the Mg-Zr corrosion tests was inactive fuel decanning plates from UNGG reactor whose composition was 0.49 % Zr and Mg balance. The samples were cut in order to obtain a rectangular shape, before being pickled in 1 M H_2SO_4 solutionfor 20 seconds and rinsed with distilled water.

For electrochemical measurements, the composition of representative solution of ordinary Portland cement (OPC) was NaOH 10g/L, KOH 3 g/L, $Ca(OH)_2$ 1 g/L,and the composition of geopolymer was NaOH 370 g/L, SiO_2 333 g/L.

A classical three electrodes cell was used for the electrochemical investigations. A platinum counter electrode and a saturated calomel reference electrode (SCE) were used. The working electrode is made of Mg-Zr pieces partially recovered of epoxy resin for a well defined surface. The experiments were performed with a Princeton Applied Research Potentiostat VersaSTAT model. For potentiodynamic polarization, the scan rate was equal to 1 mV/mn.

The encapsulation of Mg-Zr alloy in OPC and geopolymer was completed and the hydrogen evolution has been measured in order to obtain the corrosion rates. The composition of the OPC and geopolymer used for corrosion ratemeasurements is reported in Table 1. Two binders were used:

- Geopolymer (Metakaolin of Pieri Premix MK, SiO_2 Tixosil of Degussa)
- Ordinary Portland Cement(CEM I 52.5 Lafarge Le Teil

Table 1. Composition of the various mineral binders for the encapsulation of Mg-Zr alloy

	Matrix composition
Geopolymer	Metakaolin: 87.8 g SiO_2: 20 g NaOH: 22 g H_2O: 60 g
Ordinary Portland Cement	CEM I: 80 g H_2O: 32g

Measurements of hydrogen evolution are initiated by placing Mg-Zr samples encapsulated in binders in a fixed volume container.

The hydrogen analysis is undertaken by gas chromatography and the time dependant results are used to determine corrosion rates.

Corrosion rates measurement

The simplest and most fundamental measurement of the corrosion rate is the metal weight loss rate, ΔW (mg/cm^2/d). This can be converted to an average corrosion rate (mm/y) using [7]:

$$P_W = 3.65\Delta W/\rho \quad \text{Eq.1}$$

where ρ is the metal density (g/cm^3).
For Mg alloys, ρ is 1.74 g/cm^3, and Eq. (1) becomes:

$$P_W = 2.10\Delta W \quad \text{Eq.2}$$

In the overall corrosion reaction of pure Mg, one molecule of hydrogen is evolved for each atom of corroded Mg. One mole (*i.e.* 24.31 g) of Mg metal corrodes for each mole of hydrogen gas produced. Therefore, the hydrogen evolution rate, V_H (ml/cm^2/d), is related to the metallic weight loss rate, ΔWm (mg/cm^2/d), using [8]:

$$\Delta W = 1.085V_H \quad \text{Eq.3}$$

So the corresponding corrosion rate, P_H, is equal to:

$$P_H = 2{,}279V_H \quad \text{Eq.4}$$

For Mg corrosion, there is excellent agreement [8] between the corrosion rate measured by the weight loss rate and that evaluated from the hydrogen evolution rate. In the Tafel extrapolation method for measuring the Mg corrosion rate, the corrosion current density, Icorr (mA/cm^2) is estimated by Tafel extrapolation of the cathodic branch of the polarisation curve, and Icorr is related to the average corrosion rate using [9]:

$$Pi = 22.85\ \text{Icorr} \quad \text{Eq.5}$$

Results and discussion

Electrochemical measurements

The corrosion properties of Mg-Zr alloy in representative solution of OPC (with sodium fluoride addition) and geopolymer have been evaluated by open circuit and potentiodynamic polarization. Figure 1 shows the open circuit potential measurements (Ecorr) for Mg-Zr alloy in representative solution of OPC with addition of sodium fluoride ([NaF] =1.25 M and [NaF] =2.5 M). A previous study [10] has pointed out the effect of fluoride on corrosion rate decreasing in alkaline solutions with pure Mg. Consequently, the addition of sodium fluoride in representative solution of OPC has been performed to evaluate the performance of sodium fluoride inhibiting effect on Mg-Zr alloy. We can see in Figure 1 that the open circuit potential (Ecorr) trends to a more anodic value with greater fluorideconcentrations.. Gulbrandsen et al. [10] have pointed out that the open circuit values give trend to corrosion current density Icorr; the corrosion current density decreases when the open circuit potential is more anodic. So we can notice that in our basic media at a high fluoride concentration (2.5 M), the corrosion current density will be less important than at a medium fluoride concentration (1.25 M). In order to determine the corrosion current density, potentiodynamic polarization measurements have been carried out in OPC solution with fluoride addition and in activated geopolymer solution for 24 hours (Figure 2). Corrosion current densities have been evaluated with the cathodic branch provided a linear Tafel region [11], The evaluated Icorr value is included in Table 2. We can see that the corrosion current density of Mg-Zr alloy decreases in OPC solution with fluoride addition. In activated geopolymer solution, the corrosion current density of Mg-Zr is one order of magnitude less important than in OPC solution, and addition of fluoride in OPC is necessary to obtain equivalent performance to an activated geopolymer solution. By using Eq.5, corrosion rates at 24 hours have been calculated and reported in Table 2. Ratios P_{OPC-i}/Psol-i show that corrosion rates obtained with Tafel extrapolation of Mg-Zr alloy in OPC solution with fluoride at 1.25 M and in activated geopolymer solution are in the same order of magnitude. Corrosion rate value of Mg-Zr alloy in OPC solution with fluoride at 2.5 M evidenced high fluoride concentration for inhibiting efficiency.

Table 2. Icorr, Ecorr and Pi for Mg-Zr alloy at 24 hours evaluated from polarization curves, open circuit and Pi using Eq.5

	Icorr (mA/cm^2)	Ecorr (V/SCE)	$P_{i\text{-}24h}$ (mm/y)	$P_{OPC\text{-}I}$ / $P_{sol\text{-}i}$
OPC solution	5.5 10^{-4}	-1.51	1.25 10^{-2}	1
OPC solution +[NaF]=1.25 M	6.7 10^{-5}	-1.15	1.53 10^{-3}	8.16
OPC solution +[NaF]=2.5 M	3.1 10^{-5}	-0.71	7.08 10^{-4}	17
Activated geopolymer solution	5.9 10^{-5}	-1.28	1.34 10^{-3}	9.3

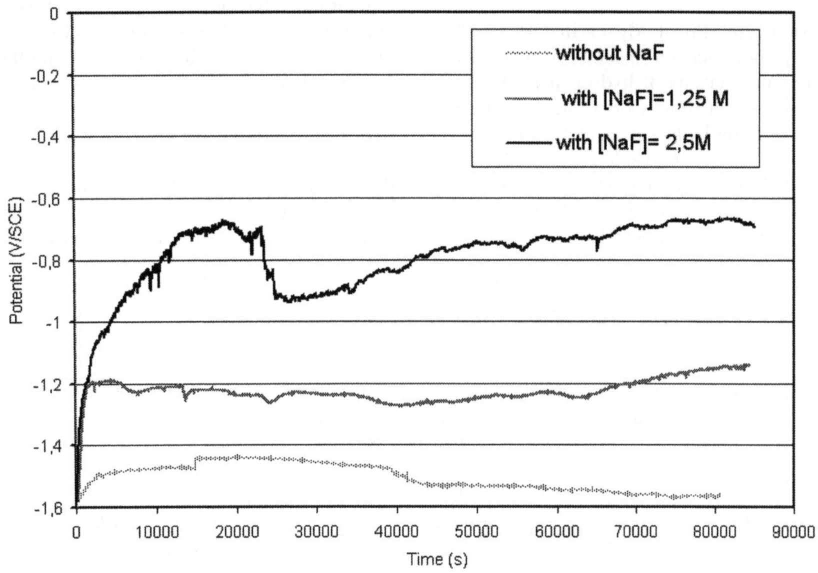

Figure 1.Open circuit evolution of Mg-Zr in OPC solution with addition of fluoride as a function of time

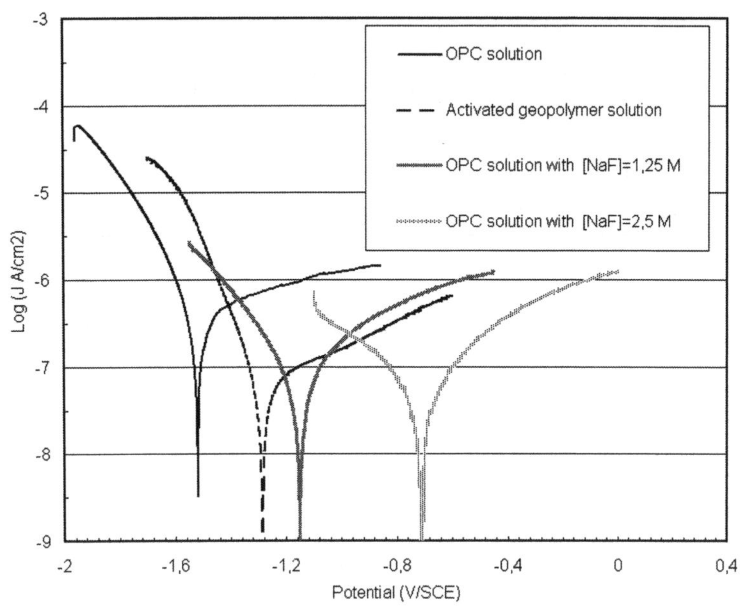

Figure 2. Potentiodynamic curves of Mg-Zr alloy in OPC solution with fluoride addition and in geopolymer solution at 24 hours

Encapsulation tests of Mg-Zr alloy in OPC and in geopolymer

In order to estimate corrosion rates P_H of Mg-Zr in OPC with addition of fluoride and in geopolymer, encapsulation of Mg-Zr pieces have been undertaken and analysis of hydrogen evolution have been performed as a function of time. Corrosion rates have been deduced by using Eq.4 and the hydrogen evolution rate V_H ($ml/cm^2/d$). The results are reported in Table 3. In order to compare results obtained by Tafel extrapolation and hydrogen evolution at 1 day, $P_{H\text{-}24h}$ / $P_{i\text{-}24h}$ ratios have been calculated and reported in Table 4. Table 5 represents calculated $P_{OPC\text{-}H}/P_{binders\text{-}H}$ ratios from Table 3 results.

Table 5 showed that at 1 day, corrosion rate P_H of Mg-Zr in OPC is higher than with fluoride at 1.25 M and in geopolymer. Ratios $P_{OPC\text{-}H}/P_{binders\text{-}H}$ at 8 and 29 days pointed out that corrosion rates obtained with hydrogen evolution of Mg-Zr alloy in OPC with fluoride at 1.25 M and in geopolymer are in the same order of magnitude. The ratio $P_{H\text{-}24h}$ / $P_{i\text{-}24h}$ showed that determination of corrosion rates obtained by Tafel extrapolation and hydrogen evolution rate are in the same order in OPC, but with fluoride and in geopolymer a deviation is observed. We can see also that corrosion rates ratio obtained by Tafel extrapolation and with hydrogen evolution are what??

To compare various corrosion environments, corrosion rates determined by Tafel extrapolation (P_i) give a good evaluation of Mg-Zr behaviour, however corrosion rates P_H are always larger than P_i due to the NDE effect evidenced by Song *et al.* [9].

Table 3.Measurements of hydrogen evolution rate V_H and corrosion rates P_H of Mg-Zr alloy encapsulated in OPC with fluoride addition and in geopolymer at room temperature

	1 day		**8 days**		**29 days**	
	V_H ($ml/cm^2/d$)	**$P_{H\text{-}1d}$ (mm/y)**	**V_H ($ml/cm^2/d$)**	**$P_{H\text{-}8d}$ (mm/y)**	**V_H ($ml/cm^2/d$)**	**$P_{H\text{-}29d}$ (mm/y)**
OPC	$6{,}7010^{-3}$	0,015	$3{,}30\ 10^{-3}$	$7{,}52\ 10^{-3}$	$1{,}13\ 10^{-3}$	$2{,}99\ 10^{-3}$
OPC with [NaF]=1.25 M	$2{,}82\ 10^{-3}$	$6{,}42\ 10^{-3}$	$1{,}73\ 10^{-3}$	$3{,}95\ 10^{-3}$	$5{,}62\ 10^{-4}$	$1{,}28\ 10^{-3}$
OPC with [NaF]=2.5 M	ND	ND	$3{,}28\ 10^{-4}$	$7{,}46\ 10^{-4}$	$2{,}58\ 10^{-4}$	$5{,}90\ 10^{-4}$
geopolymer	$2{,}2\ 10^{-3}$	$5\ 10^{-3}$	$1{,}42\ 10^{-3}$	$3{,}25\ 10^{-3}$	$5{,}70\ 10^{-4}$	$1{,}29\ 10^{-3}$

Table 4. Ratio $P_{H\text{-}24h}/P_{i\text{-}24h}$ from the data of table 2 and 3

	$P_{H\text{-}24h}$ / $P_{i\text{-}24h}$
OPC	1.2
OPC with [NaF]=1.25 M	4.2
OPC with [NaF]=2.5 M	ND
geopolymer	3.73

Table 5.Ratio $P_{OPC\text{-}H}/P_{binders\text{-}H}$ from the data of table 3 at 1, 8 and 29 days

	1 days	8 days	29 days
OPC	1	1	1
OPC with [NaF]=1.25 M	2.33	1.9	2.33
OPC with [NaF]=2.5 M	ND	10	5
geopolymer	3	2.3	2.3

Conclusions

The objectives of our study were to evaluate Mg-Zr alloy behaviour in OPC solution and activation geopolymer solution by electrochemical methods and to compare with hydrogen evolution of Mg-Zr in OPC and geopolymer binders. We have evidenced that the corrosion rate P_i of Mg-Zr alloy in OPC solution is one order of magnitude more important than in geopolymer solution environment. Evaluation of encapsulation of Mg-Zr alloy in OPC and geopolymer has been done by determination of corrosion rates using hydrogen evolution, so we have evidenced that order of corrosion rates P_H can be summarized as follows:

$P_{OPC\text{-}H} > P_{OPC+[NaF]=1.25M\text{-}H} \approx P_{GEO\text{-}H} > P_{OPC+[NaF]=2.5M\text{-}H}$

To compare various corrosion environments, corrosion rate order determined by Tafel extrapolation (P_i) gives a good evaluation of Mg-Zr behaviour.

References

[1] J. Morris, Proceedings of the 12th International Conference on Environmental Remediation and Radioactive Waste Management ICEM2009 October 11-15, (2009), Liverpool, UK.

[2] G.A. Fairhall, J.D. Palmer, Cement and Concrete Research, Volume 22, Issues 2-3, Pages 201-514 (1992).

[3] L.M. Spasova, M.I. Ojovan, Journal of Nuclear Materials 375 (2008) 347–358.

[4] A. Setiadi, N.B. Milestone, J. Hill, M. Hayes, Adv. Appl. Ceram. 105 (4), (2006) 191-196.

[5] A. Zosin, Atomic Energy, 85, 510-514, (1998).

[6] G. Turner, Proceedings of the 15th International Symposium on the Packaging and Transportation of Radioactive Materials (PATRAM 2007), (2007), Miami, Florida.

[7] D.A. Jones, Principles and Prevention of corrosion, Prentice-Hall, Englewood Cliffs, NJ, 1992.

[8] G. Song, A. Atrens, Advanced Engineering Materials 5 (2003) 837-858.

[9] G. Song, Advanced Engineering Materials 7 (2005) 563-586.

[10] E. Gulbrandsen, J. Tafto, A. Olsen, Corrosion Science, Vol 34, N°9, (1993), 1423-1440.

[11] X. Hallopeau, Ph D. University of Paris XII, 1996.

Magnesium Technology 2011

Advanced Materials and Processing

Session Chairs:

Karl U. Kainer
(Helmholtz-Zentrum Geesthacht, Germany)
Venkata Nagasekhar Anumalasetty
(Carpenter Technology Corporation, USA)

Magnesium Technology 2011
Edited by: Wim H. Sillekens, Sean R. Agnew, Neale R. Neelameggham, and Suveen N. Mathaudhu
TMS (The Minerals, Metals & Materials Society), 2011

CHARACTERIZATION OF HOT EXTRUDED Mg/SiC NANOCOMPOSITES FABRICATED BY CASTING

Sunya Nimityongskul[1], Noé Alba-Baena[1], Hongseok Choi[1], Milton Jones[2], Tom Wood[3], Mahi Sahoo[4], Roderic Lakes[5], Sindo Kou[6], and Xiaochun Li[1*]

1Department of Mechanical Engineering, University of Wisconsin-Madison, WI53706
2. Materials Science Program, University of Wisconsin-Madison, WI53706
3. GS Engineering, Inc (GSE), Houghton MI 49931
4. CANMET Materials Technology Laboratory, Ottawa, Ontario, Canada K1A 0G1
5. Department of Engineering Physics, University of Wisconsin-Madison, WI 53706
6. Department of Materials Science and Engineering, University of Wisconsin-Madison, WI53706
*Corresponding author, Ph (608) 262-6142, email: xcli@engr.wisc.edu

Keywords: Nanocomposites, Extrusion, Ultrasonic dispersion

Abstract

Mg-1%SiC nanocomposites were fabricated using an ultrasonic-cavitation based casting method, resulting in the dispersion of the reinforcing SiC nanoparticles to form Mg-metal matrix nanocomposite (Mg-MMNC) billets. The MMNC billets were then processed using hot extrusion at 350°C. Micrographic observations illustrate a significant grain size reduction and the presence of micro-bands that align the SiC nanoparticles parallel to the direction of extrusion for Mg-MMNCs. Observations from the cross-section at 90^0 of the extrusion direction show uniform nanoparticles dispersion contrasting previous observations. Results from the extruded Mg-MMNCs tensile testing at different temperatures (25^0C, 125^0C and 177^0C) reveal an increase of the yield strength, ultimate tensile strength, and ductility values as compared to the un-reinforced and extruded Mg-alloy; such increase was also observed from the microhardness testing results where an increase from 19 to 34% was measured.

Introduction

Magnesium alloys are becoming ever more prevalent in electronics, automotive and aerospace industries as energy conservation and performance demands increase. Mg alloys are one-third lighter than the equal volume of aluminum alloys, and this fact is one of the contributing factors that makes these alloys so desirable [1, 2]. Production with Mg alloys has been mostly in the field of pressure die casting because of its high productivity and dimensional accuracy [3], and extrusion is very useful for its technical and economic advantages in the production of structure components. Magnesium alloys offer high specific strength, superior damping capability, excellent machinability and good electromagnetic shielding characteristics [3]. However, Mg alloys exhibit poor mechanical performance at elevated temperatures [1, 2], their applications are usually limited to temperatures below 120 °C; in consequence, the improvement in the high-temperature mechanical properties of magnesium alloys will expand their industrial applications [4].

In general, the use of thermally stable ceramic reinforcements to create Mg based metal matrix composites (MMC) facilitates the retention of enhanced mechanical properties at elevated temperatures [4]. The desirable characteristics of MMCs include enhancements in stiffness, strength, creep resistance, and wear resistance [5]; however, MMCs normally provide lower ductility than the original matrix [4]. Cao et al. [6, 7] have been reported that nanocomposites can increase the ductility of as-cast Magnesium based metal matrix nanocomposites (MMNCs). Furthermore, by using silicon carbide (SiC) nanoparticles, Cao et al [6-13] (among others) have reported enhancement in yield strength and tensile strength without the loss of ductility by reinforcing the matrix with small volumes (<2%) of ceramic nanoparticles. However, the production of magnesium-based metal matrix nanocomposites (MMNCs) is extremely challenging for conventional casting and mechanical stirring methods. Such methods have not been successful in fabricating MMNCs due (among others) to the high specific surface area of nanoparticles and poor wettability between nanoparticles and molten metal [8]. On the other hand, the use of a technique that combined solidification processes with ultrasonic cavitation based dispersion of nanoparticles in metal melts achieved promising results for a uniform dispersion and distribution of reinforcing nanoparticles in Mg MMNCs [6-13]. During this process, ultrasonic cavitations generate micro "hot spots" with temperatures of approximately 5000°C, pressures exceeding 1000 atm, and heating rates exceeding 10^{10} K/s [14]. During MMNCs processing once nanoparticles are added to the melt, any air bubbles trapped around nanoparticle agglomerates serve as nucleation sites for cavitation generating the desired nanoparticle dispersion [9].

This paper shows results in implementing the ultrasonic cavitation based casting process to fabricate pure-Mg and Mg-1%SiC nanocomposite billets for hot extrusion. After extrusion, the mechanical properties of the MMNC rods were tensile tested at 25°C, 125°C, and 177°C. Finally, to observe the dispersion of nanoparticles and grain sizes in extruded MMNCs, optical micrographs and SEM images were obtained using samples from the tensile specimens.

Methodology

Figure 1 illustrates the experimental setup schematic of ultrasonic cavitation based solidification processing of SiC nanoparticles (SiCnp) in an Mg matrix. The experimental system was comprised of resistance heating furnace for melting the magnesium alloy, nanoparticle feeding mechanism, protective gas system, and an ultrasonic processing unit. The crucible used for melting and ultrasonic processing was made of mild steel with an inside diameter of 114 mm and a height of 127 mm. A Permendur power ultrasonic probe made of niobium C-103 alloy was used to generate a 17.5 kHz and maximum 4.0kW power output (Advanced Sonics, LLC, Oxford, CT) for melt processing. The niobium C-103 probe is 31.12 mm in diameter and 223.5 mm in length. The ultrasonic probe was dipped into the melt about 13mm. A thin walled niobium cage (31.8 mm upper diameter, 88.9 mm base diameter; 76.2 mm high, 254μm wall thickness) in a shape of truncated cone was used to hold nanoparticles inside the melt pool during the ultrasonic processing. The niobium cage has a total of 55 holes 7.94 mm in diameter to allow the circulation of the melt and nanoparticles.

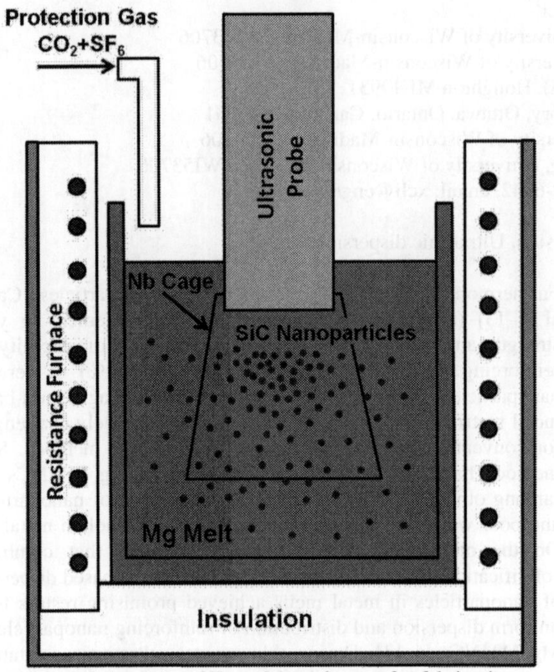

Figure 1 Schematic of Experimental setup for fabricating nanocomposites

800g of Mg was melted in the steel crucible while being protected by CO_2 and 0.75% SF_6. Once the melt temperature of 700°C was attained the niobium cage containing 1 wt. % β-SiC nanoparticles was submerged in the melt beneath the ultrasonic probe. The average size of the SiC nanoparticles used was 50nm. The melt was then ultrasonically processed at 3.5kW power level for 15 minutes for the samples with and without SiC nanoparticles. The ultrasonic probe and niobium cage were then removed from the melt and the melt was elevated to a pouring temperature of 725°C. The melt was cast into a steel permanent mold purged with CO_2 + 0.75% SF_6 and preheated to 350°C, which was designed and fabricated to produce a cast billet of 63.5 mm in diameter and 102mm in length. The casting was allowed to cool for 30 minutes before the mold was opened and the billet was removed. A graphite pouring cup was used to guide the melt into the mold, and also mounts a Pyrotech SIVEX ceramic foam filter (55mm x 55mm x 12mm, 20 pores/in).

The cast billets were preheated for two hours prior to extrusion, and were extruded at 350°C to obtain rods of 12.7 mm in diameter. The rods were extruded at a 25:1 ratio, with a ram speed of 10 mm/s and a load of 500 tons. The extruded rods of 12.7 mm (0.5 inches) in diameter were then machined into tensile bars of approximately 105 mm long with a 35 mm gauge length and 6.0 mm diameter. Tensile specimens were tested at 25°C, 125°C, and 177°C. The tensile specimens were tested in an Instru-Met TTC 90 in (228.6 cm) 4,535.92 Kg load frame using an Epsilon 3542-0100-050-HT2 extensometer with a 25.4 mm gage length clamped to each specimen. The furnace and tensile grips were preheated to the test temperature. The test specimens were then loaded into the tensile grips and the furnace was re-heated to testing temperature. The specimens were held at the testing temperature for 10 minutes before testing to ensure a uniform temperature distribution. The cross head velocity was set to 1.27 mm/min and the test was run until the specimen failure. The specimens were removed from the furnace and test fixture within 30 seconds and allowed to air cool to room temperature. The microstructures of the tested samples were studied by optical microscopy and scanning electron microscopy (SEM). Average grain size was measured using the linear intercept method from optical micrographs of the samples taken at room temperature. Samples were cut, mounted, and manually ground and polished from the grip sections of the tensile bars after tensile testing and samples were etched with 5% acetic acid in 95% water for 30 seconds. SEM was conducted in a LEO 1530 machine. Microhardness tests were conducted with a Buehler Micromet 2003 microhardness tester (load 500 gf, load time 30s). All microhardness tests were conducted at room temperature from specimens that had undergone tensile testing.

Results

Mechanical properties

The average yield strength, tensile strength, and ductility for the Mg and Mg-1%SiCnp tensile specimens conducted at various temperatures are shown in Table I. At room temperature and with the addition of 1%SiCnp to the Mg, there are significant enhancements in yield strength by 49% (from 93 to 133 MPa), tensile strength increases from 198 to 224 MPa (+13%), and ductility increase ~37% (from 5.9 to 8.1%). Below recrystallization temperature (150°C [15, 16]) for Mg and at 125°C there are moderate enhancements in yield strength (by 18%, 65MPa for pure Mg and 77 MPa for Mg-1%SiCnp) and tensile strength (by 15%), however a significant increase in ductility (by 85% measuring 13.8 for pure Mg and 25.4% for the Mg-1%SiC composite) with the addition of 1%SiC. Also shown in Table I, at 177°C (after the recrystallization temperature of Mg, 150°C) there is almost no enhancement in yield strength (by only 3%, 55MPa for pure Mg and 57 MPa for Mg-1%SiCnp), a moderate enhancement in tensile strength and 11% only (from 88 to 97 MPa), also a significant increase in ductility was averaged (91%) with the addition of 1%SiC (20.9% and 39.9% respectively).

Table I. Tested properties of hot-extruded pure Mg and Mg-1%SiC at 25°C, 125°C, and 177°C.

Testing Temperature		Yield Strength, MPa	Tensile Strength, MPa	Ductility, %	Microhardness, HV	Avg. Grain Size, μm
Room temp. (25°C)	Pure Mg	93±2.9	198±2.4	5.9±1.3	29.2± 1.1	33± 11
	Mg-1%SiCnp	133±0.7	224±1.3	8.1±1.1	39.0± 2.1	22 ± 4
	Variation	*49%*	*13%*	*37%*	*34%*	---
125°C	Pure Mg	65±2.9	118±1.4	13.8±1.0	30.8 ± 1.4	36± 27
	Mg-1%SiCnp	77±0.7	136±1.3	25.4±1.1	39.3± 0.8	25± 12
	Variation	*18%*	*15%*	*85%*	*28%*	---
177°C	Pure Mg	55±2.4	88±2.0	20.9±0.7	31.6± 1.5	30± 7
	Mg-1%SiCnp	57±3.2	97±1.8	40±1.8	37.7± 0.4	28 ± 7
	Variation	*3%*	*11%*	*91%*	*19%*	---

Table I shows the average microhardness of the specimens tested at room temperature (25°C), 125°C, and 177°C). For comparison, microhardness of 125°C and 177°C specimens were taken after mechanical testing and conducted at room temperature. Samples from room temperature testing show an average microhardness of Mg and Mg-1%SiC was averaged 29.2 and 39.0 HV, respectively. The microhardness enhancements of extruded Mg-1%SiCnp compared to the tested Mg at 25°C, 125°C, and 177°C are 34%, 28%, and 19%, respectively. The performance increases in microhardness at all temperatures are largely attributed to the reinforcement that SiC nanoparticles to the Mg matrix.

Characterization

Last columns in Table I show average grain size longitudinal to the direction of extrusion in the specimens tested at 25°C, 125°C, and 177°C. The average grain size longitudinal to the direction of extrusion for extruded Mg and extruded Mg-1%SiCnp at all temperatures is 33 μm and 25 μm, respectively. Beside the calculated averages, a large variation was found (see Table I). There is not significant difference in grain refinement between the Mg and Mg-1%SiCnp due to dynamic recrystallization (DRX) during hot extrusion at 350°C. As described by Buschow et al. [17] and reported by Liu et al. [18] grains begin to recrystallize at initial grain boundaries in hot extrusion growing in layers until the old grains are consumed. DRX is thought to initiate at original (prior to extrusion) grain boundaries due to an accumulation of dislocations at the grain boundaries during the hot extrusion [19]. Optical micrographs in the longitudinal (extruded) direction of extruded Mg and extruded Mg-1%SiCnp tested at 25°C, 125°C, and 177°C, are shown in Figure 2. Elongated nanoceramic agglomerations are shown in Figures 2b, 2d, and 2f where the SiC nanoparticles are distributed in the extruded Mg matrix.

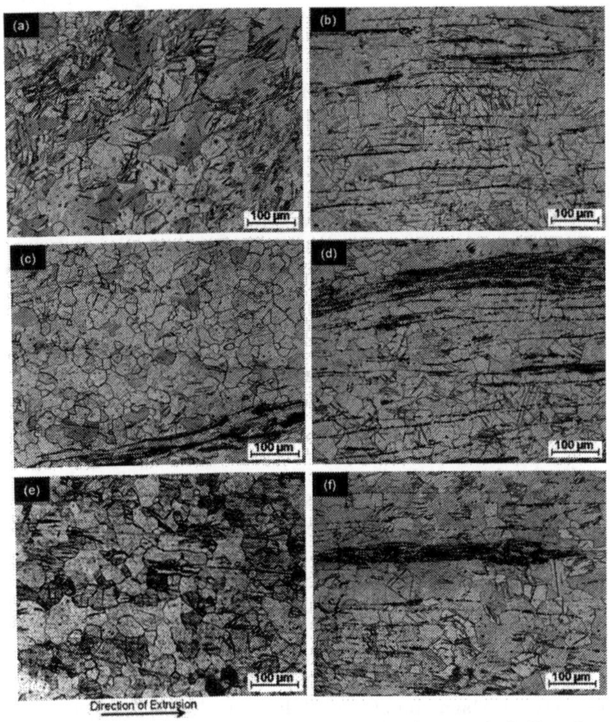

Figure 2. Optical micrographs in longitudinal (extruded) direction of extruded specimens: a) Mg tested at 25°C, b) Mg-1%SiC tested at 25°C, c) Mg tested at 125°C, d) Mg-1%SiC tested at 125°C, e) Mg tested at 177°C, f) Mg-1%SiC tested at 177°C.

The SiC nanoparticle-bands in Figures 2b, 2d, and 2f are not the only regions that SiCnp are located in; however, larger concentrations and clusters of SiC nanoparticles form these bands. The most distinguishing characteristic between the extruded Mg samples (Fig. 2a, c, e) and the extruded Mg-1%SiC (Fig.2b, d, f) are the dark SiC bands running parallel to the axis of extrusion. Such observation are similar to the described by Wang et al. [19] that observed similar phenomena in pure Mg with 40 μm sized SiC particles that were extruded at a ratio of 13:1. Streaks of SiC micro particles were observed parallel to the direction of extrusion; however, there were no clusters of SiC micro particles and the particles were bonded well to the matrix [8].

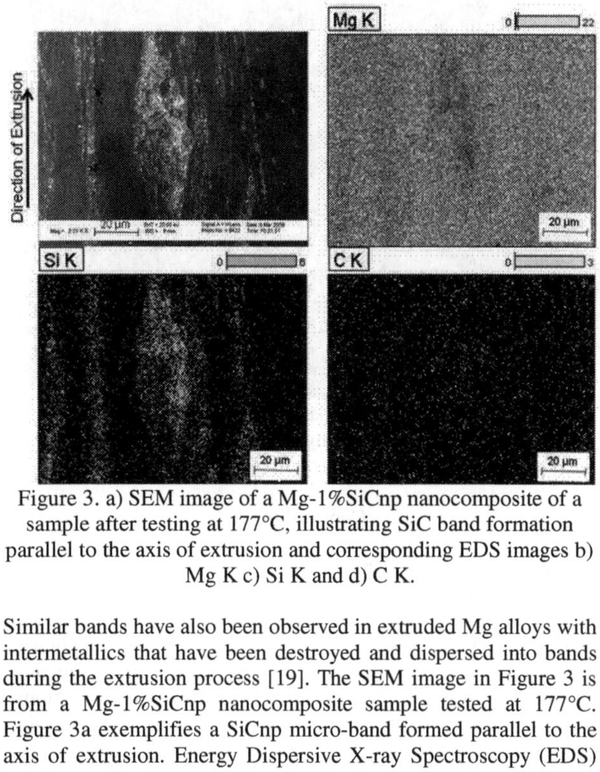

Figure 3. a) SEM image of a Mg-1%SiCnp nanocomposite of a sample after testing at 177°C, illustrating SiC band formation parallel to the axis of extrusion and corresponding EDS images b) Mg K c) Si K and d) C K.

Similar bands have also been observed in extruded Mg alloys with intermetallics that have been destroyed and dispersed into bands during the extrusion process [19]. The SEM image in Figure 3 is from a Mg-1%SiCnp nanocomposite sample tested at 177°C. Figure 3a exemplifies a SiCnp micro-band formed parallel to the axis of extrusion. Energy Dispersive X-ray Spectroscopy (EDS) images confirm that the micro bands observed in Figures 2b, 2d, and 2f are in fact elongated SiC nanoparticle agglomerations. EDS spectral imaging in Figure 3c shows the SiC particle bands in an Mg matrix in the longitudinal direction, while in contrast Mg spectral image (Figure 3b) shows its complement spectra.

Besides of the recrystallization process, the addition of 1%SiC nanoparticles enhanced yield strength, tensile strength, and ductility. At the tested temperatures below and higher recrystallization temperature the Mg-1%SiCnp nanocomposite preserves its uniform distribution, as seen in the SEM image in Figure 4 also showing the SiC nanoparticle agglomerations contouring the extrusion flow. The observed nanoparticle dispersion and mechanical properties' variation indicates a thermally stable bond between the reinforcing nanoparticles throughout the matrix. The observed increase in ductility with the addition of 1%SiC nanoparticles is in contrast to the described results shown in Cao et al, where the ductility decreases with the addition of micron-sized particles and fibers to Mg matrix [7]. Mg/SiCnp nanocomposite results are in concordance to the characterization reported by Cao et al. Cao et al [7] also described the characterization of the Mg/SiC interface behavior where is shown no intermediate phases at the interface of Mg and SiC, and found that SiC nanoparticles were well bonded to the matrix. Same authors also report that nanoparticles aid in maintaining the grain structure in the extruded Mg-1%SiC by pinning grain boundaries at elevated temperatures thus resulting in higher ductility which explain in part the effect of the SiC nanoparticles in the Mg matrix. Explaining why (partially at least) even though clusters of nanoparticles exist in the Mg-1%SiCnp samples and why the tensile tests revealed an increase in ductility at all tested temperatures.

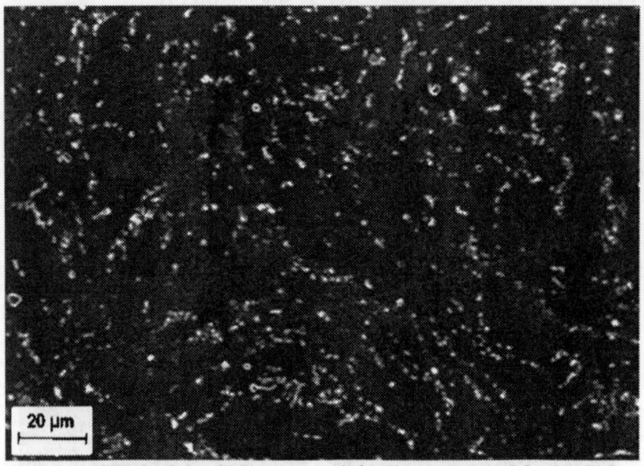

Figure 4. SEM of the SiC nanoparticle dispersion in Mg-1%SiCnp after tested at 177°C, illustrating the achieved dispersion of SiC agglomerations in transverse (extruded) direction

Conclusions

The incorporation of SiC nanoparticles by using ultrasonic cavitation based dispersion processing help to achieve a uniform distribution of reinforcing SiC nanoparticles in the Mg matrices. As an effect of the extrusion process, MMNC billets contained micro bands of SiC nanoparticles oriented parallel to the direction of extrusion evidencing the SiC nanoparticles dispersion. Along the cross section or transverse to the direction of extrusion the achieved dispersion of SiC agglomerations shows to be homogeneous. Micrographs show no significant grain refinement between the Mg and Mg-1%SiC MMNC samples after hot extrusion. In contrast to previous reports using SiC micron-sized particles or fibers, samples from Mg-1%SiCnp consistently exhibit larger ductility and enhancement of other mechanical properties. Yield strength, tensile strength, ductility and microhardness were improved also as compared to the extruded Mg samples at various testing temperatures.

Acknowledgements

This work was supported by National Science Foundation and American Foundry Society. The authors thank US Magnesium for its donation of Mg ingots.

References

1. H. Friedrich and B. Mordike *Technology of magnesium and magnesium alloys in Magnesium Technology - Metallurgy, Design Data, Applications* (Springer–Verlag, 2006), 219-430.

2. F. Czerwinski, *Magnesium Injection Molding* (Springer, 2007), 1-76.

3. Y. Chen et al., "Effects of extrusion ratio on the microstructure and mechanical properties of AZ31 Mg alloy," *Journal of Materials Processing Technology*, 182 (2007), 281–285.

4. H. Ye and X. Liu "Review of recent studies in magnesium matrix composites," Journal of Material Science, 39 (2004), 6153-6171.

5. T. Clyne and P. Withers, *An introduction to Metal Matrix Composites.* (Cambridge University Press, 1993) 454-457.

6. G. Cao, H. Konishi and X. Li, "Mechanical properties and microstructure of Mg/SiC nanocomposites fabricated by ultrasonic cavitation based nanomanufacturing," *Journal of Manufacturing Science and Engineering,* 130 (2008), 31105-1 to 31105-6.

7. G. Cao et al., "Mg-6Zn/1.5%SiC nanocomposites fabricated by ultrasonic cavitation-based solidification processing," *Journal of Materials Science* 43 (2008), 5521–5526.

8. J. Lan, Y. Yang and X. Li (2004) Microstructure and microhardness of SiC nanoparticles reinforced magnesium composite fabricated by ultrasonic method," *Materials Science and Engineering A*, 386 (1-2) (2004), 284-290.

9. G. Cao, J. Kobliska, H. Konishi, and X. Li "Tensile Properties and Microstructure of SiC Nanoparticles Reinforced Mg-4Zn Alloy Fabricated by Ultrasonic Cavitation Based Solidification Processing," *Metallurgical and Materials Transactions A*, 39 (4) (2008), 880-886.

10. X. Li, et al., "Ultrasonic Cavitation Based Solidification Processing of Bulk Mg Matrix Nanocomposite," *Transactions of the American Foundry Society,* 115 (2007), 747-752.

11. G. Cao, H. Konishi, and X. Li, "Mechanical properties and microstructure of SiC-reinforced Mg-(2,4) Al-1Si nanocomposites fabricated by ultrasonic cavitation based solidification processing," *Materials Science & Engineering A,* 486 (1-2) (2008), 357-362.

12. Y. Yang, J. Lan, X. Li "Study on Bulk Aluminum Matrix Nano-composite Fabricated by Ultrasonic Dispersion of Nano-sized SiC Particles in Molten Aluminum Alloy," *Materials Science and Engineering A,* 380 (2004), 378-383.

13. J. Lan, Y. Yang, and X. Li, "Microstructure and microhardness of SiC nanoparticles reinforced composite fabricated by ultrasonic method", *Materials Science and Engineering A,* 386 (1-2) (2004), 284-290.

14. K. Suslick et al., "Acoustic cavitation and its consequences," *Philosophical Transactions of the Royal Society A,*357 (1999), 335-353.

15. J. B. Chalmers, *Physical metallurgy,* (John Wiley & sons Inc, New York, 1962)

16. J. R. Davis *Metals Handbook,* (ASM International, New York. 1998)

17. K. Buschow et al., *Encyclopedia of Materials: Science and Technology* (Elsevier, 2008), 2375-2381.

18. K. Liu, et al "Microstructures and mechanical properties of extruded Mg–8Gd–0.4Zr alloys containing Zn," *Materials Science and Engineering A,* 505 (2009), 13-19.

19. R. Wang, et al., "Microstructure and interface structure of SiC-reinforced Mg metal matrix composite" in *Proceedings of the 2nd Israeli International Conference on Magnesium Science and Technology 2000* E Aghion and D Eliezer (eds.). *Magnesium 2000,* (Israel, Magnesium Research Institute, 2000), 229-234.

Magnesium Technology 2011
Edited by: Wim H. Sillekens, Sean R. Agnew, Neale R. Neelameggham, and Suveen N. Mathaudhu
TMS (The Minerals, Metals & Materials Society), 2011

EFFECTS OF SILICON CARBIDE NANOPARTICLES ON MECHANICAL PROPERTIES AND MICROSTRUCTURE OF AS-CAST Mg-12wt.%Al-0.2wt.%Mn NANOCOMPOSITES

Hongseok Choi; Hiromi Konishi; Xiaochun Li
Mechanical Engineering Department, University of Wisconsin-Madison, Madison, WI 53706, USA

Keywords: Mg-Al alloy, Nanocomposite, Grain refining, Intermetallic phase, Ductility

Abstract

Microstructure and tensile properties of as-cast Mg-12Al-0.2Mn alloys with SiC nanoparticles were studied. SiC nanoparticles were dispersed into Mg-12Al-0.2Mn melts through an ultrasonic based nanoparticle-dispersion method. The content of SiC nanoparticles varied from 0 to 2 wt.%. The microstructural analysis with optical and scanning electron microscopy (SEM) showed that the massive brittle intermetallic phase (β-$Mg_{17}Al_{12}$) as well as α-Mg grain was significantly refined, enhancing both strengths and ductility of as-cast Mg-12Al-0.2Mn alloys. Transmission electron microscopy (TEM) showed that there were SiC nanoparticles in both α-Mg and β-$Mg_{17}Al_{12}$ phases of the Mg-12Al-0.2Mn nanocomposites. It was found that there was a partial reaction between Mg/Al and SiC nanoparticles, producing Mg_2Si intermetallic phases. X-ray diffraction (XRD) analysis confirmed also that the content of Mg_2Si phases increased with increasing SiC content, limiting a further ductility enhancement for Mg-12Al-0.2Mn-SiC nanocomposite.

Introduction

Magnesium alloys are of great potential as structural materials. The demand for magnesium is continuously increasing due to the effort by automotive industry to use lighter metals to reduce fuel consumption and achieve lower emission levels. Moreover, magnesium alloys are also utilized widely in electric and aerospace industries, etc. The most applicable alloys are those based on Mg-Al, such as AZ91, because of their excellent castability, corrosion resistance, and strength at room temperature [1-4]. Furthermore, AZ91 is the Mg-Al alloys with the highest Al content for commercial applications today. It is well-known that the strength of the Mg-Al alloys is improved as the Al content increases. However, the ductility decreases significantly with increased Al content, which limits applications of the alloys.

The high strength of alloys can also be attained from grain refinement because grain size is one of the most important factors determining the mechanical properties of Mg alloys according to Hall-Petch equation [5]. There are a variety of different methods to achieve refined grains in alloys: by increasing the cooling rate, adding heterogeneous nuclei during solidification processing, etc [6]. Although much research has been reported in achieving fine grains in various Mg alloys [7-10], reliable and castable Mg-Al alloys with higher Al content still remain to be further grain refined and developed so that their potential higher mechanical strength can be utilized for commercial applications.

Recently, there have been increasing studies about a new class of nanostructured materials, Mg metal matrix nanocomposites (Mg MMNCs), consisting of Mg alloys reinforced with ceramic nanoparticles, due to their potential to strengthen the Mg matrix while maintaining or even enhancing ductility [11-14]. Ceramic materials, e.g. silicon carbide (SiC) and aluminum oxide (Al_2O_3), have been widely used as reinforcement particles in aluminum alloys due to their relatively good thermal and chemical stability as compared to other types of reinforcements [15, 16]. However, most processing techniques are not capable of producing bulk nanocomposites of complex shapes. Solidification processing, e.g. casting, is a desirable processing method. However, it is extremely difficult to uniformly disperse nanoparticles in metal melts due to their high specific surface areas, which easily induce agglomeration and clustering. A novel technique, ultrasonic-cavitation based solidification processing, was recently developed, and it was found that nanoparticles had a significant effect on the properties and microstructure of alloys [12-14].

The purpose of this study is to investigate the effect of ceramic nanoparticles on microstructure of Mg alloys with higher Al content in order to produce as-cast high performance Mg-Al alloys with much improved ductility. The mechanical properties of Mg-Al nanocomposites will also be tested and the mechanisms of properties change will be discussed based on the microstructure analysis.

Experimental Procedure

Commercial pure magnesium (99.9% purity), pure aluminum (99.8% purity), and aluminum-manganese master alloy (25wt.% Mn) were used to synthesize Mg-Al alloys. The Mg alloy with higher Al content (Mg-12wt.%Al-0.2wt.%Mn alloys in this study) were chosen as the matrix materials. β-SiC nanoparticles (97.5% purity; Si<0.15, Cl<0.15, C<0.75, O<1.25wt.%) with an average size of 50 nm (Nanostructured & Amorphous Materials, Inc., USA) were used for the nanocomposites of Mg-12Al-0.2Mn alloy through an innovative ultrasonic-cavitation based dispersion method developed earlier [12].

The processing procedure for casting nanocomposites includes melting the alloy, creating ultrasonic cavitation and streaming in the molten metal by an ultrasonic processing system, addition of preheated nanoparticles (150 °C for one hour), and casting the nanocomposite melt into a permanent mold. Figure 1 shows the schematics of experimental set-up for the ultrasonic-cavitation based dispersion processing of SiC nanoparticles for Mg-12Al-0.2Mn nanocomposites. The system consists of a resistance heating furnace to melt the alloys, a nanoparticle feeding system, gas protection system and an ultrasonic processing system. The ultrasonic processing system consists of an ultrasonic probe, booster and transducer. A niobium ultrasonic probe with a diameter of 12.7 mm and a length of 92 mm was attached to a booster (Sonicator 3000, Misonix, USA), which was mounted in a transducer working under a frequency of 20 kHz and a maximum power output of 600 W.

About 160 g of magnesium alloy was melted in a mild steel crucible under a protective CO_2+1%SF_6 gas. The tip of a niobium ultrasonic probe was inserted into the melt about 12.7 mm in depth when the melt temperature dropped to an ultrasonic processing temperature at about 700 °C. Ultrasonic vibrations with a peak-to-peak amplitude of 60 μm were used, and then nanoparticles were added into the melt during ultrasonic processing.

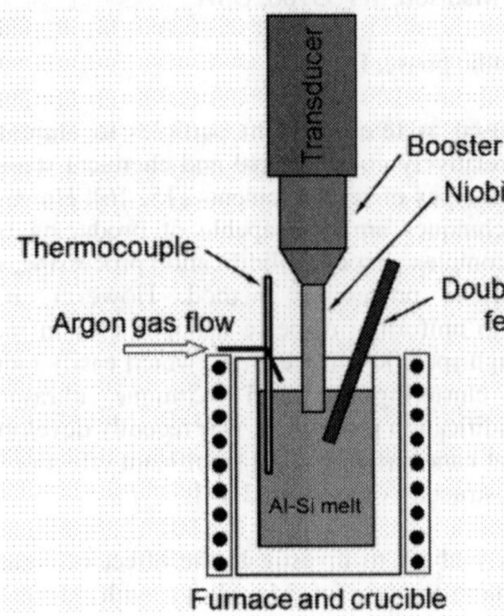

Figure 1. Schemetic of experimental set-up for ultrasonic processing.

Due to a low specific density of ceramic nanoparticles and their poor wettability with metal melt, the ceramic nanoparticles tend to float to the melt surface at first before ultrasonic cavitation helps to mix them into the melts. Therefore, an effective feeding technique (double-capsulate feeding method) was developed to improve feeding efficiency. In the double-capsulate feeding method, ceramic nanoparticles were first wrapped with a thin aluminum foil (Alloy 1000) with a thickness of 0.0127 mm, and then the aluminum foil was rolled up into a rod shape with a diameter of about 6 mm (making the first capsule). The aluminum foil rod was wrapped again with another thin aluminum foil (Alloy 1100) with a dimension of 355.6 mm × 152.4 mm and a thickness of 0.0254 mm (making the second capsule). The second aluminum foil would make the nanoparticles discharge into the melt gradually, resulting from the graduate melting of the thicker wall of the capsule. The aluminum foil capsule with a total weight of 4.6 g containing ceramic nanoparticles was then put into the melt during ultrasonic processing. One Al foil capsule was slowly fed into the melt and ultrasonically processed within the melt. The melt was ultrasonically processed for 10 minutes for each Al foil capsule, bringing a total processing time of 30 minutes. After the ultrasonic processing for 30 minutes, the ultrasonic probe was lifted out of the melt and the temperature of the melt was increased to 740 °C for pouring. The melt was cast into a low carbon steel permanent mold preheated to 400 °C. Two standard flat tensile specimens with a gage dimension of 6.35 mm × 6.35 mm and a gage length of 38.1 mm were obtained after each casting. For comparison reasons, samples were also made without nanoparticle additions (but still with aluminum foils and capsule of the same amount as in the nanocomposites).

Mechanical properties of samples were determined using a tensile testing machine (SINTECH 10/GL, MTS, USA) with a crosshead speed of 5.08 mm/minute. For accurate measurement of yield strength, an extensometer was used for up to 0.2% strain at which point the test was paused, and the extensometer was removed. Final elongation was measured manually after putting the two test bar pieces back together according to ASTM B 557-06. The microstructural characterization was conducted with polished samples of pure alloy and its nanocomposites. They were lightly etched using a 2% natal etchant. Optic microscope and Scanning Electron Microscopy (SEM) with Energy Dispersive X-ray spectroscopy (EDS) were used for detailed metallographic analysis. Transmission Electron Microscopy (TEM) samples were prepared using ion milling. The TEM analysis of samples was conducted using a FEI TITAN-80-200 microscope, operating at an accelerating voltage of 200 kV. X-ray diffraction (XRD) measurements were done using a Scintag Pad V Diffractometer with CuKα radiation. The alloy samples were polished with a 1 μm diamond paste. Scan parameters used were 2θ angle 10-80 degrees, step size 0.02 degrees and dwell time of 12 seconds.

Experimental Results and Discussion

The mechanical properties of as-cast pure Mg-12Al-0.2Mn alloy and nanocomposites with two different contents of β-SiC (0.9 wt.% and 1.4 wt.%) are shown in Figure 2. The results show increases in strength (ultimate tensile and yield) and ductility of the Mg-12Al-0.2Mn alloy with the addition of β-SiC nanoparticles. For Mg-12Al-0.2Mn-0.9SiC nanocomposite, the ultimate tensile strength and yield strength are increased by 14% and 8% respectively, compared with the pure alloy. The ductility is improved from 0.97% to 1.65% (increased by 71%). This implies that β-SiC nanoparticle is effective for the enhancement of mechanical properties (especially ductility) of Mg-Al alloys with higher contents of Al. The increase in ductility with the addition of nanoparticles is in contrast to the ductility decrease that is usually seen with the addition of micro particles or fibers to metal matrix [17]. As the content of β-SiC nanoparticles increases to 1.4 wt.%, the yield strength is increased not much, only about 2% higher than that of Mg-12Al-0.2Mn-0.9SiC nanocomposite. Moreover, the ultimate tensile strength and ductility are even lower by 5% and 40% than those of Mg-12Al-0.2Mn-0.9SiC nanocomposite respectively. However, it still shows significantly increased ultimate tensile strength and ductility than the pure alloy (increased by 9% and 22% respectively).

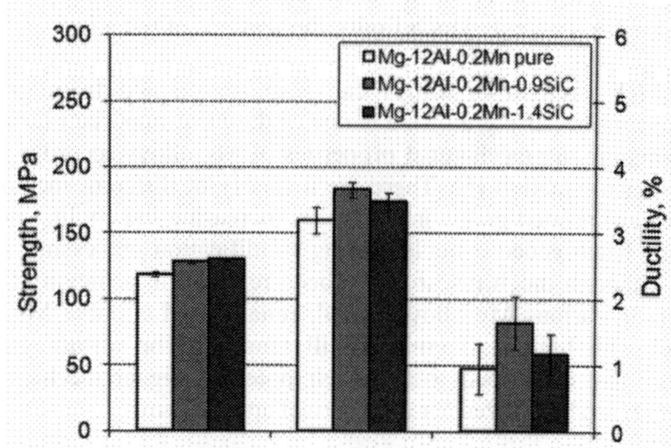

Figure 2. Mechanical properties of Mg-12Al-0.2Mn alloy and its nanocomposites.

The optical micrographs of the pure Mg-12Al-0.2Mn alloys and their nanocomposites samples are shown in Fig. 3a-c. The pure

Mg-12Al-0.2Mn alloy exhibited massive intermetallic phases (β-$Mg_{17}Al_{12}$) surrounding the primary magnesium (α-Mg) dendrites. The β-$Mg_{17}Al_{12}$ phases are mainly distributed along the boundaries of primary α-Mg grains, and have a morphology of long continuous network. Moreover, the formation of α-Mg islands of a few microns was observed inside the irregularly networked β-$Mg_{17}Al_{12}$ phases. A certain amount of lamella phases are also observed around the massive β-$Mg_{17}Al_{12}$ phases in the boundaries of primary α-Mg grains.

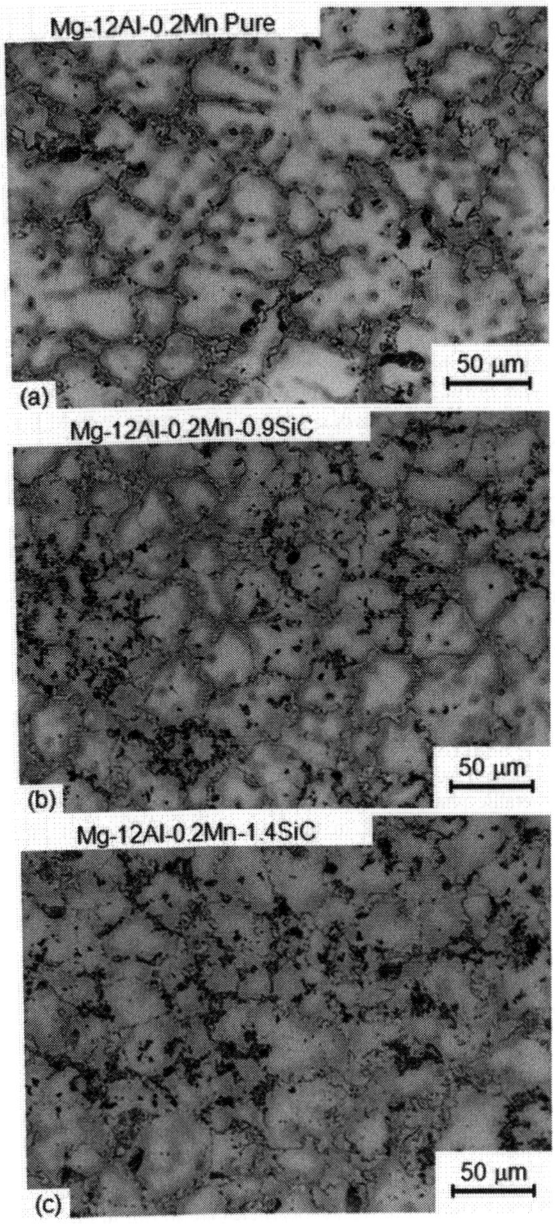

Figure 3. Optical micrographs of Mg-12Al-0.2Mn alloy modified with (a) 0% (b) 0.9% (c) 1.4% SiC nanoparticles.

Average grain size was measured using the linear intercept method from optical micrographs of the samples according to ASTM E 112-96. The grain size was 45.9 ± 7.1 μm, as shown in Fig. 4. However, figure 3b and c show that the Mg-12Al-0.2Mn nanocomposites with SiC nanoparticles have much finer globular primary α-Mg grains. With the addition of β-SiC nanoparticles, the morphology of α-Mg phases changes from large dendrite to small equiaxed globular structure. Figure 4 shows that the average size of primary α-Mg grains was noticeably reduced as compared with that of pure alloy (decreased by 52% in Mg-12Al-0.2Mn-1.4SiC nanocomposite). This is in full agreement with the previous studies [12-14], which reported that β-SiC nanoparticles would work as heterogeneous nucleation agents despite the large disregistry between Mg and β-SiC. It can be also seen from Fig. 3b and c that the β-$Mg_{17}Al_{12}$ phases in the nanocomposites are significantly modified form large continuous network to thinner and shorter structures, compared with that in the pure alloy. This might be attributed to the significant grain refinement, resulting from enhanced nucleation of α-Mg and β-$Mg_{17}Al_{12}$ phases caused by the dispersed β-SiC nanoparticles.

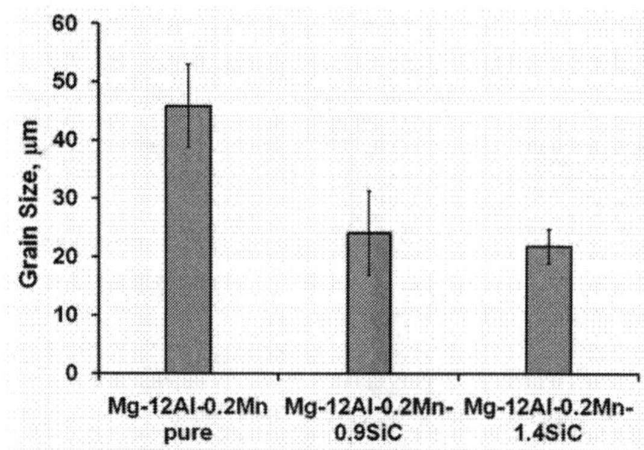

Figure 4. Average grain size of Mg-12Al-0.2Mn pure alloy and its SiC nanocomposites.

The modification of β-$Mg_{17}Al_{12}$ phases are more obviously shown in Fig. 5. Typical large continuous network of β-$Mg_{17}Al_{12}$ phases are observed in the pure Mg-12Al-0.2Mn alloy. As shown in Fig. 5a, the primary α-Mg phases of pure Mg-12Al-0.2Mn alloy have dendritic structures, and the β-$Mg_{17}Al_{12}$ phases are in the form of large continuous network with distributed small α-Mg islands. The β-$Mg_{17}Al_{12}$ phases are also formed in areas between the secondary dendrite arms, but the morphology mainly appears in a compact globular form. In the case of 0.9 wt. % SiC nanocomposites, the morphology of primary α-Mg shows typical globular equiaxed shape as shown in Fig. 5b. Some β-SiC nanoparticles micro-clusters are also observed inside the β-$Mg_{17}Al_{12}$ phases. The modification effect of β-SiC nanoparticles on the primary α-Mg is more obvious than the one on the β-$Mg_{17}Al_{12}$ phases. With higher amount of β-SiC nanoparticles (Fig. 5c), the β-$Mg_{17}Al_{12}$ phases are significantly refined. It can be found that the large long networks around primary α-Mg grains are modified to the thin and shortly discrete ones. It is noticed that the micro-clusters β-SiC nanoparticles are more easily observed in β-$Mg_{17}Al_{12}$ phases of Mg-12Al-0.2Mn-1.4SiC nanocomposite. Moreover, there is no significant change in the average grain size of α-Mg phases between two nanocomposites with different contents of β-SiC nanoparticles. These results indicate that higher content of β-SiC nanoparticles promote the modification of β-$Mg_{17}Al_{12}$ phases instead of the refinement of α-Mg phases. TEM micrographs of the α-Mg and β-$Mg_{17}Al_{12}$ phase areas of Mg-

12Al-0.2Mn-1.4SiC nanocomposite are shown in Fig. 6. Some β-SiC nanoparticles are located inside both α-Mg and β-$Mg_{17}Al_{12}$ phases, which is consistent with the observation in the optical micrographs in Fig. 5c. Therefore, the refinement of α-Mg phases might be attributed to the enhanced heterogeneous nucleation caused by the addition of β-SiC nanoparticles, and the modification of morphology of β-$Mg_{17}Al_{12}$ phases might be related to the addition of β-SiC nanoparticles, resulting in the geometrical constraint by increased grain boundary area as well as the restricted growth by neighboring β-SiC nanoparticles during the solidification process.

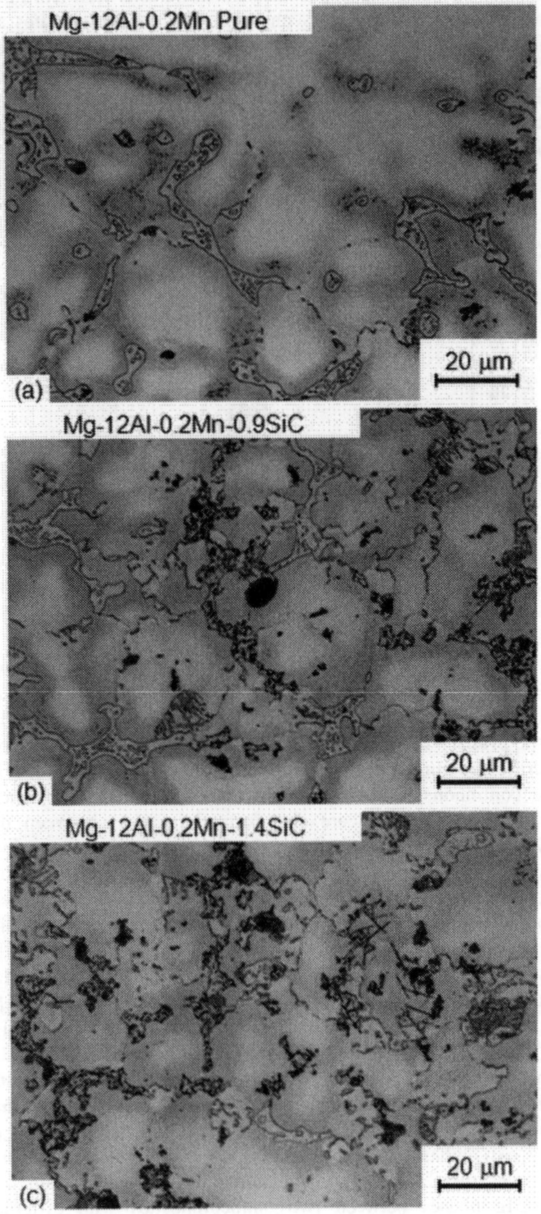

Figure 5. Optical micrographs of Mg-12Al-0.2Mn alloy modified with (a) 0% (b) 0.9% (c) 1.4% SiC nanoparticles.

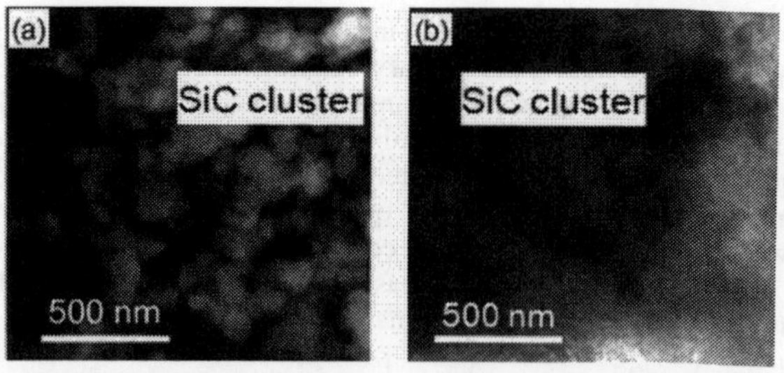

Figure 6. TEM photographs of (a) α-Mg and (b) β-SiC clusters in Mg-12Al-0.2Mn-1.4SiC nanocomposite.

The modified β-$Mg_{17}Al_{12}$ phases as well as the refined α-Mg grains would contribute to the enhancement of mechanical properties of the alloy. It is clearly shown in the experiment results that the beneficial effect of β-SiC nanoparticles on the mechanical properties of Mg-12Al-0.2Mn alloys is related to the refinement of primary α-Mg grains and modification of β-$Mg_{17}Al_{12}$ phases. The mechanical properties of cast materials typically depend on the grain size of primary and intermetallic phases. A fine microstructure means that the material contains a large amount of grain boundary area. It is well recognized in the literature that the cracks easily initiate at interfaces between the hard and brittle β-$Mg_{17}Al_{12}$ phases and α-Mg phases, and then propagate inside β-$Mg_{17}Al_{12}$ phases. The massive β-$Mg_{17}Al_{12}$ phases of body-centered cubic (b.c.c.) structure are incompatible with the primary α-Mg of hexagonal close-packed (h.c.p.) structure, which leads to the fragility of the interface of two phases. The results of the microstructural characterization show that the primary α-Mg and β-$Mg_{17}Al_{12}$ phases are refined and modified in nanocomposites. The considerable enhancement of ductility might be related to the combined effect of the significant refinement and enhanced uniformity of the primary α-Mg and β-$Mg_{17}Al_{12}$ phases achieved by dispersed β-SiC nanoparticles.

The deterioration in the enhancement of ultimate tensile strength and ductility of Mg-12Al-0.2Mn-1.4SiC is probably due to the increase of Si content caused by the reaction of SiC nanoparticles with alloy, resulting in the formation of hard, brittle Mg_2Si intermetallic phases. It was reported that the following reactions (1)-(2) can be observed in the microstructure of metal matrix composite of Mg-Al alloys with SiC particles [18-19].

$$4Al + 3SiC \rightarrow Al_4C_3 + 3Si \quad (1)$$
$$2Mg + Si \rightarrow Mg_2Si \quad (2)$$

It was also found that there is SiO_2 amorphous layer on the surface of some SiC particles [18]. Therefore, other reactions (3)-(4) can occur, supplying more free Si elements to form Mg_2Si phases.

$$2Mg + SiO_2 \rightarrow 2MgO + Si \quad (3)$$
$$Mg + 2Al + 2SiO_2 \rightarrow MgAl_2O_4 + 2Si \quad (4)$$

XRD patterns of pure Mg-12Al-0.2Mn alloy and its nanocomposite with β-SiC nanoparticles are shown in Fig. 7. For the pure alloy, the peaks from Mg and $Mg_{17}Al_{12}$ phases are clearly displayed. In the case of SiC nanocomposites, small peaks from Mg_2Si phase can be detected. In addition, the peak intensity of Mg_2Si phase in the nanocomposite with 0.9 wt.% SiC nanoparticles was obviously weaker than that in the 1.4 wt.% SiC nanocomposite.

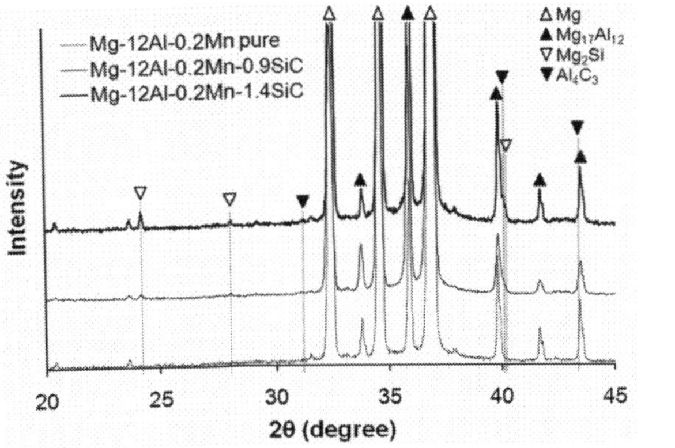

Figure 7. XRD patterns of Mg-12Al-0.2Mn pure alloy and its SiC nanocomposites.

As shown in Fig. 8, SEM photograph of Mg-12Al-0.2Mn-1.4SiC nanocomposite also shows the massive Mg_2Si phase at the grain boundary adjacent to β-$Mg_{17}Al_{12}$ intermetallic phases. It is also observed that SiC nanoparticles are located not only in the massive Mg_2Si but inside some $Mg_{17}Al_{12}$ phases. Furthermore, it is found that some α-Mg phases contain β-SiC nanoparticles.

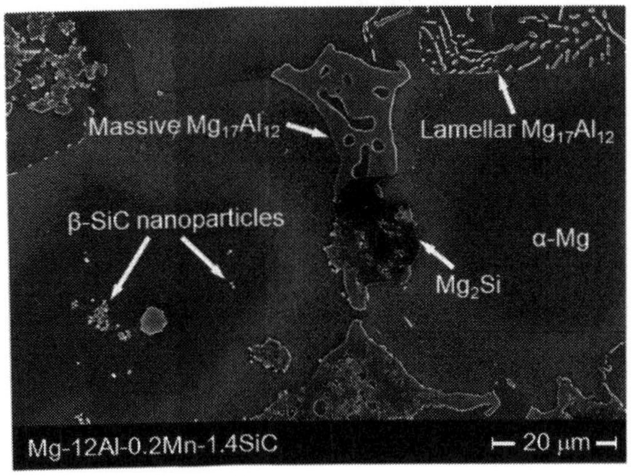

Figure 8. SEM photographs of Mg-12Al-0.2Mn-1.4SiC nanocomposite.

Conclusions

Simultaneous refinement of primary α-Mg and β-$Mg_{17}Al_{12}$ phases is achieved in Mg-12Al-0.2Mn alloy with the addition of β-SiC nanoparticles with a diameter of 50 nm. The primary α-Mg phases are refined from the dendrite to the globular equiaxed grain. The refining effect does not change significantly when β-SiC nanoparticle content is increased from 0.9 to 1.4 wt.%. The average grain size of primary α-Mg in Mg-12Al-0.2Mn-0.9SiC nanocomposite is about two times smaller than that of pure Mg-12Al-0.2Mn alloy. The massive β-$Mg_{17}Al_{12}$ phases are also significantly refined with the addition of β-SiC nanoparticles. The large continuous network phases are modified into discrete thin ones. The mechanical properties of Mg-12Al-0.2Mn alloy are improved with the addition of β-SiC nanoparticles. For Mg-12Al-0.2Mn-0.9SiC nanocomposites, the ultimate tensile strength increased by 14% and the ductility is improved by 71%, while the yield strength is increased by about 8%. The enhancement of yield strength in Mg-12Al-0.2Mn-1.4SiC nanocomposite was limited by the formation of hard brittle Mg_2Si phases by the reaction of nanoparticles with alloy.

Acknowledgments

This work is sponsored by National Institute of Standard and Technology through its Technology Innovation Program. In addition, the authors express their gratitude to US Magnesium LLC. for materials donation.

References

1. A. Rudajevova, M. Stanek, and P. Lukac, "Determination of thermal diffusivity and thermal conductivity of Mg-Al alloys," *Mater. Sic. Eng. A*, 341 (2003), 152-157.
2. M.X. Zhang annd P.M. Kelly, "Crystallography of Mg17Al12 precipitates in AZ91D alloy," *Scr. Mater.*, 48 (2003), 647-652.
3. G. Song, A.L. Bowles, and D.H. St. John, "Corrosion resistance of aged die cast magnesium alloy AZ91D," *Mater. Sci. Eng. A*, 366 (2004) 74-86.
4. A. Pardo, M.C. Merino, A.E. Coy, F. Viejo, R. Arrabal, and S. Feliu Jr., "Influence of microstructure and composition on the corrosion behaviour of Mg/Al alloys in chloride media," *Electrochin., Acta.*, 53 (2008) 7890-7902.
5. K. Kubota, M. Mabuchi, and K. Higashi, "Processing and mechanical properties of fine-grained magnesium alloys," *J.Mater. Sic.*, 34 (1999) 2255-2296.
6. F. Czerwinski, *Magnesium Injection Molding* (New York, NY: Springer, 2007), 19-41.
7. Y.C. Lee, A.K. Dahle, and D.H. St. John, "The role of solute in grain refinement of magnesium," *Metal. Mater. Trans. A*, 31 (2000) 2895-2905.
8. Y. Guangyin, S. Yangshan, and W. Zhen, "Effect of antimony on microstructure and mechanical properties of Mg-9Al based alloy," *J. Nonferrous Metals*, 9 (1999) 779-784.
9. Y. Lu, Q. Wang, X. Zeng, W. Ding, C. Zhai, and Y. Zhu, "Effects of rare earths on the microstructure, properties and fracture behavior of Mg–Al alloys," *Mater. Sci. Eng. A*, 278 (2000) 66-76.
10. S. Zhang, W. Li, K. Yu, and D. Tan, "The grain refinement processes of magnesium alloys," *Foundry*, 7 (2001) 373-375.
11. Z. Trojanova, P. Lukac, H. Ferkel, B.L. Mordike, and W. Riehemann, "Stability of Microstructure in Magnesium Reinforced by Nanoscaled Alumina Particles," *Mater. Sci. Eng. A*, 234–236 (1997) 798–801.
12. J. Lan, Y. Yang, and X. Li, "Microstructure and microhardness of SiC nanoparticles reinforced composite fabricated by ultrasonic method," *Mater. Sci. Eng. A*, 386 (2004) 284-290.
13. G. Cao, J. Kobliska, H. Konishi, and X. Li, "Tensile properties and microstructure of SiC nanoparticle-reinforced Mg-4Zn alloy fabricated by ultrasonic cavitation-based solidification processing," *Metal. Mater. Trans. A*, 39 (2008) 880-886.
14. M. DeCicco, H. Konishi, G. Cao, H. Choi, L. Turng, J. Perepezko, S. Kou, R. Lakes, and X. Li, "Strong, ductile magnesium-zinc nanocomposites," *Metal. Mater. Trans. A*, 40 (2009) 3038-3045.
15. P.S. Robi, B.C. Pai, K.G. Satyanarayana, S.G.K. Pillai, and P.P. Rao, "The Role of surface treatments and magnesium

additions on the dispersoid matrix interface in cast Al-Si-Mg-15 wt-percent SiCp composites," *Mater. Charact.*, 27 (1991) 11-18.

16. D.Y. Ying and D.L. Zhang, "Processing of Cu–Al_2O_3 metal matrix nanocomposite materials by using high energy ball milling," *Mater. Sic. Eng. A.*, 286 (2000) 152-156.

17. G. Gao, H. Choi, H. Konish, S. Kou, R. Lakes, and X. Li, "Mg-6Zn/1.5%SiC nanocomposites fabricated by ultrasonic cavitation-based solidification processing," *J. Mater. Sci.*, 43 (2008) 5521-5526.

18. M.C. Gui, J.M. Han, and P.Y. Li, "Microstructure and mechanical properties of Mg-Al9Zn/SiCp composite produced by vacuum stir casting process," *Mater. Sci. Technol.*, 20 (2004) 765-771.

19. M.A. Easton, A.Schiffl, J. Yao, and H. Kaufmann, "Grain refinement of Mg-Al(-Mn) alloys by SiC additions," *Scripta Mater.*, 55 (2006) 379-382.

20. M.C. Gui, D.B. Wang, J.J. Wu, G.J. Yuan, and C.G. Li, "Microstructure and mechanical properties of cast (Al–Si)/SiC_p composites produced by liquid and semisolid double stirring process," *Mater. Sci. Technol.*, 16 (2000) 556-563.

Magnesium Technology 2011
Edited by: Wim H. Sillekens, Sean R. Agnew, Neale R. Neelameggham, and Suveen N. Mathaudhu
TMS (The Minerals, Metals & Materials Society), 2011

THERMALLY-STABILIZED NANOCRYSTALLINE MAGNESIUM ALLOYS

Suveen Mathaudhu[1], Laszlo Kecskes[2], Kristopher Darling[2]
[1]U.S. Army Research Office, Materials Science Division, P.O. Box 12211, Research Triangle Park, NC 27709-2211, USA
[2]U.S. Army Research Laboratory, Materials and Manufacturing Sciences Division, Building 4600, Aberdeen Proving Ground, MD 21005-5069, USA

Keywords: Magnesium, Nanocrystalline, Strength, Thermal Stability

Abstract

Advanced nanocrystalline alloys have shown remarkable property improvements, particularly, order-of-magnitude strength increases, when compared to their coarse-grained counterparts. However, a major obstruction to the widespread application of such materials is the degradation of properties via rapid grain growth at even ambient temperatures. Conventional methods for circumvention of this problem at low temperatures have largely steered toward kinetically pinning the boundaries with disperoids, or through misorientation of grain boundaries, yet even these methods have limited utility at elevated temperatures needed for routine sintering and forming operations. In this work, we will present a synergistic approach to the development of thermally stable nanostructured Mg-alloys which incorporates elements of predictive modeling of suitable alloy systems, fabrication of nanostructured alloy powders by high energy ball milling and consolidation of the powders at elevated temperatures.

Introduction

Mg based alloys have many desirable mechanical properties, such as low density and high specific strength, compared to conventional structural materials such as steel and aluminum. Reports from the private sector have shown that weight reduction is the single most important factor in the reduction in fuel consumption [1]. Recently, as a consequence, the drive to use Mg based alloys has increased in commercial automotive applications, and thus, this driving force has lead to the extensive evaluation of magnesium alloys. Despite the desirable mechanical properties, current Mg alloys show limited strength. This is especially true at elevated temperatures. As a result, the application of Mg based alloys in vehicles has been heavily restricted, particularly high strength applications, currently reserved for alloyed steels [2-3].

While mechanisms such as solid solution strengthening and precipitate dispersion can be used to strengthen metallic alloys, it has been shown that much larger increases can be realized by refining the grain size to the nanoscale [4]. However, as the grain size decreases, the percent microstructure constituted by grain boundaries increases and can be in excess of 50%. Thus, grain boundaries in nanocrystalline materials can account for a large increase of the total free energy of the system. The reduction of this excess free energy, via the removal of grain boundary area, represents a large driving force for grain growth. It has been demonstrated that extensive grain growth at room temperature occurs in low melting point, nanocrystalline metals, such as Mg, which results in a dramatic loss of mechanical strength [4]. Thus, the thermal stability of nanocrystalline Mg alloys must be realized if these materials are to ever be adopted for extended high-temperature structural applications.

For this work, we present a synergistic approach to the development of thermally stable nanostructured Mg-alloys which incorporates elements of predictive modeling as a guiding hand for selecting suitable alloy systems for fabrication of thermally stabilized nanostructured Mg-alloy. Using a regular solution approximation derived elsewhere [6], an estimation of the effect of various solutes on decreasing grain boundary energy can be made. The model takes into consideration the elastic enthalpy change and the change in cohesive energy for both the lattice and grain boundary area for an estimation of the enthalpy change associated with grain boundary segregating solutes. Equation 1 is used to estimate the effect of stabilizing a nanocrystalline microstructure;

$$\Delta\gamma = \gamma^{*} - \gamma_{o} = \Gamma \Delta H_{seg} \tag{1}$$

where $\Delta\gamma$ is the reduction in grain boundary energy, γ^* is the effective grain boundary energy after solute segregation, γ_0 is the initial grain boundary energy for Mg, ΔH_{seg} is the change in the enthalpy of the system due to the segregation of solute to the grain boundaries, and Γ is the excess amount of segregated solute per unit grain boundary area. If the interaction of a solute with a Mg grain boundary has an effect of lowering the interfacial energy of the associated boundary, then a collection of solute at that grain boundary occurs. If the boundary energy can be lowered to zero, then the immediate effect is the existence of an equilibrium grain boundary area. Furthermore, if the effects of precipitation and can be kinetically hindered, then the system will reside in a deep metastable state. This approach has been shown to be successful in stabilizing nanocrystalline Fe-, Cu-, and Pd-alloys systems [6-9]. Herein we report how these principles, for the first time, have been applied to nanocrystalline Mg systems with the effect of drastically increasing the high temperature stability of the matrix. In the best case, we show that a Mg 4at%Fe alloy is capable of retaining an average grain size of 45 nm after annealing at 200°C for one hour (51% of the absolute melting temperature of pure Mg).

Experimental Methodology

Using Equation (1) and reference [6], four solutes were selected as candidates for stabilization of nanocrystalline Mg: Fe, V, Mn, and Y. Out of the four solutes, Y was suggested to have the least effect (25% reduction in grain boundary energy). It has been experimentally shown that the bulk or global solute content has an effect of stabilizing a high grain boundary area, and hence, a smaller grain size [7]. Therefore, eight magnesium-based materials were synthesized, which comprised of pure Mg plus 1 and 4 at% of each of the candidate solutes.

All alloying reactions were prepared in a high purity argon filled glove box (m-Braun, O_2 and humidity contents <0.1 ppm). Mg compounds were synthesized by charging a milling vessel with Mg (99.8% Alfa Aesar) and either Fe (99.99% Sigma-Aldrich), Mn (99.99% Sigma-Aldrich), V (99.5% Alfa Aesar) or Y(99.9% ESPI) using a 10:1 mass ratio of milling balls to powder. Hexane (3.5:10 surfactant to well volumetric ratio) was added to inhibit agglomeration of the metal powder during milling. The reactions were sealed under argon and ball milled for 20 hours using a proprietary high energy milling process at Wildcat Discovery Technologies (San Diego, CA).

The materials were recovered under an inert atmosphere and dried at room temperature under vacuum for approximately 15 hours to give ~ 1g of loose powder for each composition. Each of the eight materials was separated into seven batches. One of the batches was left without heat treatment, while the remaining six were heated to 100, 200, 300, 400, and 500°C for 1hr under vacuum.

The particle size of the 4 at% samples was assessed by scanning electron microscopy (JEOL). EDAX spectroscopy was used to confirm the absence of contaminants. X-ray diffraction patterns were obtained and the crystallite size was estimated by applying the Scherrer equation [10]. The (002) and (101) reflection peaks were used to measure the grains size. A Mg-ingot standard and Gaussian peak calibration were used to correct XRD peak broadening from instrumental errors. To verify the x-ray estimated grain size measurements, Mg with 4at% Fe and Y were investigated with transmission electron microscopy. Additionally, equal channel angular extrusion [11] was performed on all powders at 200°C as a means to consolidate the powder into a bulk form.

Results and Discussion

The SEM images in Figures 1 show that the as-milled particle morphologies for all four 4at% solute containing alloys is approximately 20μm and are somewhat spherical. While not shown here, it was observed that the 1at% alloys showed similar morphology and grain size.

Figure 1. SEM images of as-milled particle morphologies for (a) Mg4at%Fe, (b) Mg4at%Mn, (c) Mg4at%V, and (d) Mg 4at% Y showing ~20 μm average particle size.

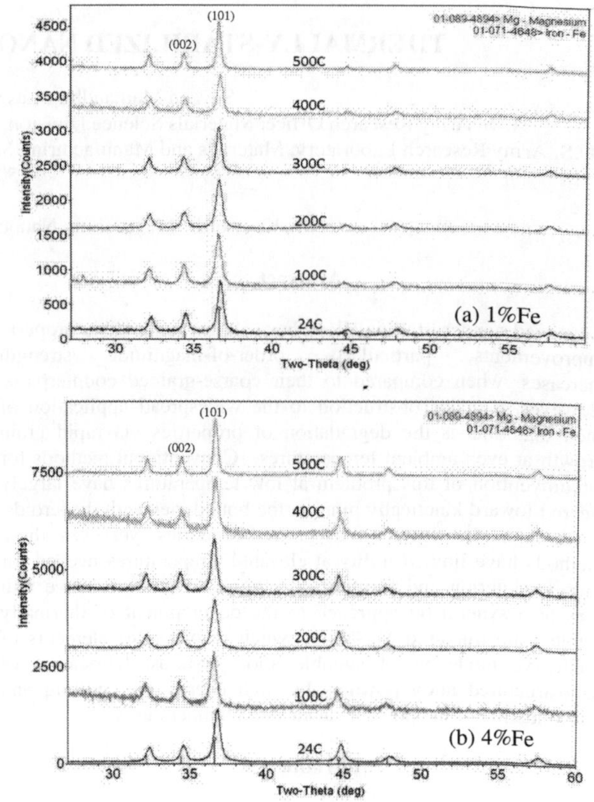

Figure 2. X-ray diffraction analysis patterns as a function of increasing annealing temperature for (a) Mg1at%Fe and (b) Mg4at%Fe.

Figure 2 shows the x-ray diffraction patterns for Mg1at%Fe and Mg4at%Fe annealed at various temperatures for one hour after the high-energy milling. The six other alloy systems (V, Mn, and Y in 1at% and 4at% concentrations) show very similar x-ray analysis trends, and are not shown for brevity. The broadening (as measured at half the maximum intensity of the peak) was observed to decrease with increasing annealing temperature for all alloys, suggesting that some grain growth has taken place. The appearance of secondary phases is not observed in the x-ray spectrum up to the highest temperatures tested (500°C), but this does not necessarily indicate that secondary phases are not present. Such phases might be in volume fractions too small to detect, or perhaps the phase itself is nanocrystalline, and therefore, cannot be distinguished above the background.

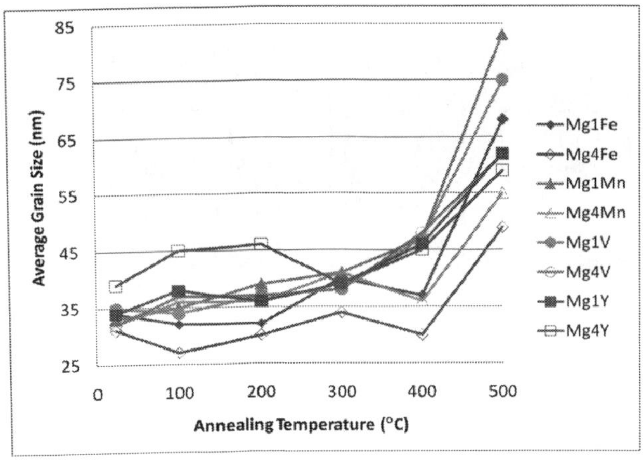

Figure 3. X-ray approximated average grain size as a function of annealing temperature for high-energy milled nanocrystaline Mg-alloys.

The average grain size approximations from the x-ray spectrum data for each alloy are given in Figure 3. The Fe-containing alloys are observed to have the greatest influence on preventing grain growth, with the 4at%Fe alloy retaining finer grain sizes than the 1at%Fe alloy. While Fe additions show the best thermal stability, Fe is known to cause severe galvanic corrosion when in contact with Mg. Metals that are higher in the electrochemical series displace metals which are lower in the sequence. This means that when connecting two metals, the metal with lowest potential will corrode. The same is true on the atomic scale, which plays importance in corrosion of solid solutions and grain boundary embrittlement. Thus, while Y was shown to induce less thermal stability, Mg and Y have the same potential (-2.37V) while Fe does not (-0.44V). Therefore, MgY nanocrystalline alloys should show superior corrosion resistance but have a larger average grain size.

TEM observation of the microstructure of Mg 4at% Fe and Mg 4at% Y, annealed at 200°C, are given in Figure 4. From these images, it can be seen that the average grain size is larger in the Y-containing alloy (~70 nm) than in the Fe-containing alloy (~45 nm) annealed at the same temperature for the same amount of time, which indicates that Fe is likely a better stabilizer.

Preliminary trial extrusions with equal channel angular extrusion (ECAE) for consolidation of milled powder resulted in complete bulk consolidate with no evidence of particle boundaries and no porosity. The samples were prepared by loading billets made of Nickel 100 alloy with the alloyed powder separated by thin metals disks. The powder was pre-compacted with a Ni steel rod, which was then welded shut. The can was then extruded at 160°C around a 90° die intersection angle for four passes via route A (no rotation about billet long axis during subsequent passes) [11]. The billet sectioned along the extrusion direction mid- plane is shown in Figure 5. ECAE resulted in full consolidation of the powder. Preliminary investigations on the effect of severe plastic deformation at high temperatures suggest that some grain growth has taken place. Future studies will be used to optimize extrusions parameters to limit excessive grain growth in these samples.

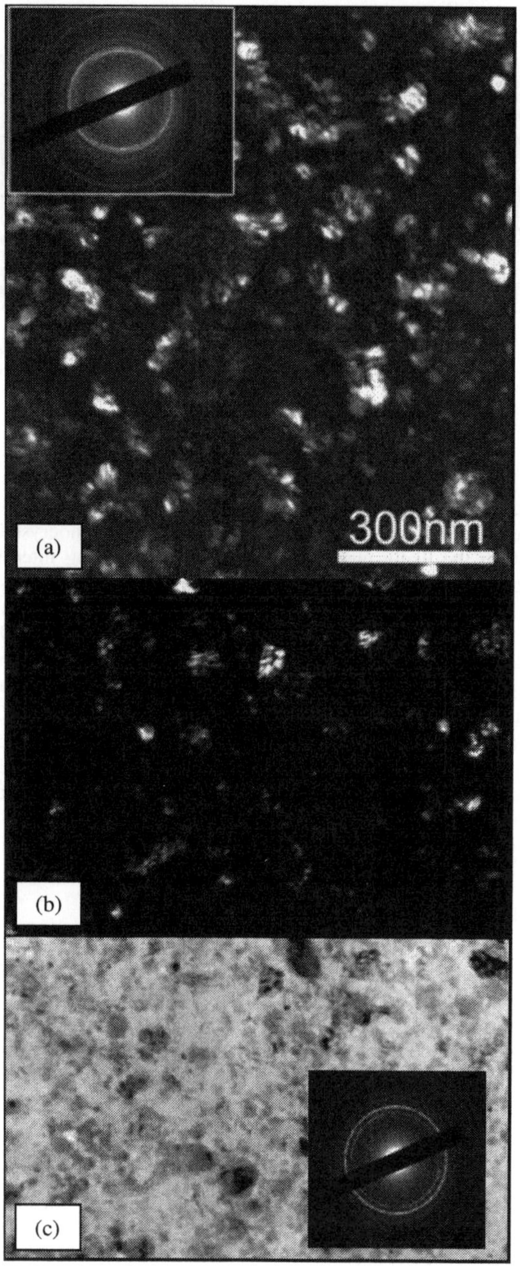

Figure 4. TEM images showing nanocrystalline grain size after 200°C for (a) Mg4at%Fe – dark field, (b) Mg 4at%Y – dark field and (c) Mg4at%Y – bright field.

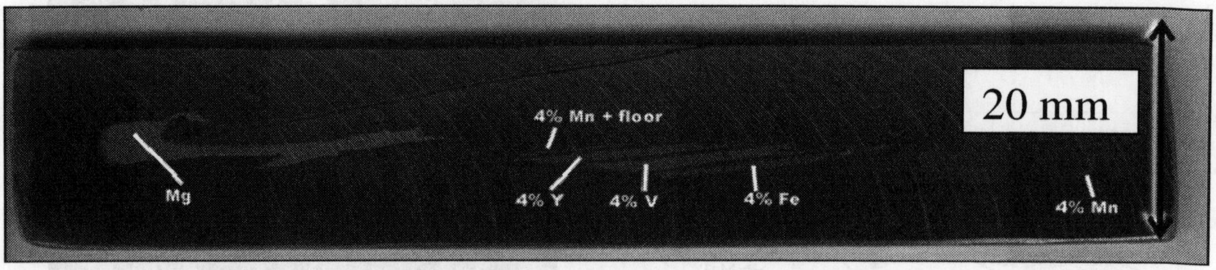

Figure 5. Cross-section of ECAE billet showing fully dense pure microcrystalline Mg, Mg4at% Fe, Mg 4at% Mn, Mg 4at% V and Mg 4at% Y alloys.

Summary

Select solutes concentrations of Fe, Mn, V and Y were successfully used to produce thermally stabilized nanocrystalline Mg with the likely mechanism of stabilization being interfacial energy reduction and/or solute drag. These results are consistent with findings and observations made in several other alloy systems [6-9], X-ray analysis indicates that increasing the bulk solute content increases the thermal stability. The work presented shows Fe to be an excellent solute for stabilization and retention of the nanocrystalline structures (~45 nm after 200°C/1h annealing). Preliminary studies show ECAE to be an effective method of producing bulk samples from the nanocrystalline powders. Future studies will investigate this stabilizing effect in more advanced alloys, such as the AZ or WE series of alloys, as it may be possible to improve and tailor engineer properties.

References

1. H. Dieringa and K.U. Kainer, "Magnesium – Der Zukunftswerkstoff für Die Automobilindustrie?," *Materialwissenschaft und Werkstofftechnik*, 38 (2007), 91-96.
2. G.S. Cole, "Issues that Influence Magnesium's use in the Automotive Sector," *Materials Science Forum*, 419 (2003), 43-50.
3. S. Schumann and H. Friedrich, "Current and Future Use of Magnesium in the Automobile Industry," *Materials Science Forum*, 419 (2003), 51-56.
4. M.A. Meyers, A. Mishra and D.J. Benson, "Mechanical Properties of Nanocrystalline Metals," *Progress in Materials Science*, 51 (2006), 427-556.
5. R. Birringer, "Nanocrystalline Materials," *Materials Science and Engineering, A*, A117 (1989), 33-43.
6. K.A. Darling, B.K. VanLeeuwen, C.C. Koch and R.O. Scattergood, "Thermal Stability of Nanocrystalline Fe-Zr Alloys," *Materials Science and Engineering, A*, 527 (2010) 3572-3580.
7. K.A. Darling, R.N. Chan, P.Z. Wong, J.E. Semones, R.O. Scattergood and C.C. Koch, "Grain-Size Stabilization in Nanocrystalline FeZr Alloys," *Scripta Materialia*, 59 (2008) 530-533.
8. B.K. VanLeeuwen, K.A. Darling, C.C. Koch, R.O. Scattergood, and B.G. Butler, "Thermal Stability of Nanocrystalline $Pd_{81}Zr_{19}$," *Acta Materialia*, 58 (2010) 4292-4297.
9. K.A. Darling, R.K. Guduru, C.L. Reynolds, Jr., V.M. Bhosle, R.N. Chan, R.O. Scattergood, C.C. Koch, J. Narayan and M.O. Aboelfotoh, "Thermal Stability, Mechanical and Electrical Properties of Nanocrystalline Cu_3Ge," *Intermetallics*, 16 (2008) 378-383.
10. H.P. Klug and L.E. Alexander, X-ray Diffraction Procedures for Polycrystalline and Amorphous Materials, Wiley, New York, 1974.
11. K.T. Hartwig, I. Karaman, M. Haouaoui and S.N. Mathaudhu, "Consolidation of Cu and Amorphous Zr-based Powders by Severe Plastic Deformation," *Metallic Materials with High Structural Efficiency- NATO Science Series II: Mathematics, Physics and Chemistry*, 146 (2004), 91-100.

Magnesium Technology 2011
Edited by: Wim H. Sillekens, Sean R. Agnew, Neale R. Neelameggham, and Suveen N. Mathaudhu
TMS (The Minerals, Metals & Materials Society), 2011

TiNi Reinforced Magnesium Composites by Powder Metallurgy

Ziya Esen
Industrial Engineering Department, Çankaya University, Öğretmenler Cad. No:14, Balgat/Ankara, 06530, TURKEY

Keywords: Magnesium, Composites, Powder Metallurgy, Mechanical Properties

Abstract

Rod shaped Mg-TiNi composite samples were manufactured by powder metallurgical route in which the samples were heated and deformed simultaneously using rotary hot swaging technique. Firstly, encapsulated argon filled copper tubes which contained compacts of pure magnesium and pre-alloyed TiNi alloy powder mixtures were deformed about 45% in two steps at 450°C. Pre/post annealing heat treatments were applied at 450°C for 20 mins between the stages of coaxial deformation to enhance the sintering degree and to homogenize the heavily deformed composite structures. Next, copper peeled and machined samples were compression tested under quasi-static conditions to investigate the mechanical properties, i.e. yield and peak strength, and ductility. Transmission and Scanning Electron Microscopy studies were carried out to examine the Mg-TiNi interface and fracture surfaces of the compression tested composites, respectively.

Introduction

Magnesium and magnesium alloys are preferred especially in weight critical applications such as automotive and aerospace industry where high specific strength is needed. An increase in the usage of magnesium and its alloys is anticipated since energy is saved as a result of reduced fuel consumption due to relatively low weight of magnesium and magnesium alloy based parts [1, 2]. Accordingly, aluminum and aluminum alloy parts used in various components of automobile body and its engine have been replaced by magnesium and magnesium alloys, i.e. AZ91, AM50/60. In addition, some of magnesium alloys such as WE43 are being considered as attractive lightweight materials to manufacture ultra lightweight armored ground vehicles.

Some of the critical material properties of parts designed for many structural applications in automotive and aerospace industry are specified as density, mechanical properties, i.e. strength and ductility, corrosion and wear resistance. For high temperature applications, on the other hand, the material must carry various types of loads without excessive deformation, which necessitates high creep resistance. For further improvement of mechanical properties of magnesium and its alloys, deformation grain refinement and alloying are used. However, the former is difficult because Mg has poor formability at room temperature due to its HCP crystal structure which is deformed only by basal slip of (0001)<11-20> [3]. With respect to alloying recently high strength and ultra high strength magnesium alloys are manufactured by adding rare earth elements such as Y and Re [4, 5]. However, aging heat treatments conducted for increasing the strength of alloyed magnesium [6] may sometimes degrade both creep and corrosion resistance of the alloy. For example β-$Mg_{17}Al_{12}$ phases, formed during aging of Mg-Al-Zn alloys are reported to reduce the creep resistance as they precipitate along the grain boundaries [7].

The magnesium based parts with higher yield strength and wear resistance compared to their counterparts produced by alloying can be obtained only by the production of magnesium composites. Magnesium composites are manufactured by solid state, i.e. powder metallurgy, or liquid state processing techniques by introducing ceramic or metallic reinforcements in fiber, whisker or particulate form. In liquid processing technique, some problems such as inhomogeneous distribution of reinforcements, formation of pores and inclusions, limited wettability of reinforcement by liquid matrix phase, formation of secondary brittle phases in the matrix/reinforcement interface may arise and degrade the mechanical properties of composites. Because of that, the use of powder metallurgy in composite production has become essential mainly due to relatively low temperatures used, feasibility of adjusting the composite's average grain size by controlling the prior powder particle size, homogenous distribution of reinforcement particles and possibility of utilizing relatively higher amounts of reinforcement, i.e. 50 vol.%.

Although low density ceramic reinforcements such as SiC, TiC, Al_2O_3, B_4C, Y_2O_3 and TiB_2 provide desired high yield strength, the ductility of magnesium composites decrease due to formation of brittle phases in the interface region or due to pores caused by poor wettability of reinforcements in liquid processing techniques. The observed decrease in ductility of such composites is attributed to the elastic moduli and coefficient of thermal expansion differences between matrix and ceramic reinforcements [8]. In addition, differences in the crystal structures of magnesium and ceramic reinforcement particles prevents load transfer to reinforcement particles and ceramic reinforcement fractures prior to matrix phase. Because of this, the metallic reinforcements in particulate or fiber form have been used to eliminate such disadvantageous effects of ceramic reinforcements. Metallic micrometer sized reinforcements such as copper and nickel [9, 10] have been utilized in manufacturing of magnesium composites. Moreover, some studies were also carried out to investigate the effect of carbon nana-tube addition on mechanical properties of magnesium composites [11]. However, either the required ductility level cannot be achieved or the specific weights of the composite parts increase due to high density of metallic reinforcements. A recent study carried out by Hassan and Gupta [1] has shown that notable ductility levels is achieved in addition to increased yield and tensile strength by using titanium as a reinforcement particle.

Magnesium matrix composites are produced using either conventional compaction and sintering technique or extrusion process [12, 13], which make use of powders or pre-sintered compacts of composites. Conventional techniques are not preferred since they are not effective in obtaining fully dense samples. As these techniques cannot break the oxide layers effectively available diffusion paths for sintering are limited. Because of that, simultaneous application of stress and temperature is an effective way in magnesium composite production by powder metallurgy. Rotary swaging with the characteristics of multidirectional forging and high-frequency

stroking, which results in rod-shaped components, seems to be an effective way of magnesium composite production [14].
This study investigates the powder metallurgy production of TiNi alloy powders added magnesium composites rods via rotary hot swaging. The influence of processing conditions on microstructure has been presented by considering particle boundaries and interfacial regions between matrix and TiNi particles. In addition, the effect of microstructure and TiNi content on mechanical properties of Mg-TiNi composites have also been discussed using compression stress-strain curves and the analysis at the fractured surfaces of samples.

Experimental

Materials

During the composite production, elemental magnesium and pre-alloyed TiNi powders were utilized. Irregular shaped magnesium powders (99.8 %purity) with an average powder particle size of 40 μm were supplied from Alfa-Aesar. Spherical pre-alloyed Ni-rich Ti-50.6 at. %Ni powders (99.9% purity) had an average particle size of 25 μm and they were supplied from Nanoval GmbH & Co. KG, respectively (Figure 1).

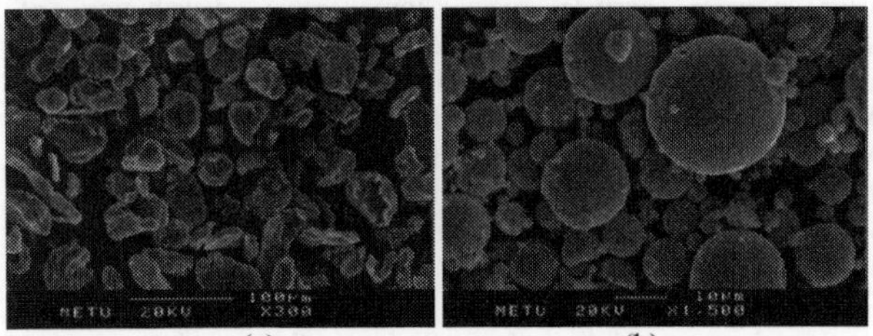

(a) (b)

Figure 1. SEM micrographs of utilized powders, (a) angular magnesium, (b) pre-alloyed spherical Ti-50.6 at. %Ni.

Sample Preparation

Initially, Mg-TiNi powder mixtures containing 5, 10 and 15 vol. % TiNi have been prepared by mixing the powders for ½ hour. Then, 10 grams of powder mixtures were pre-shaped in a double ended steel die at around 600 MPa using a hydraulic press to break the possible oxide layers present on magnesium powders. Prior to manufacturing by rotary hot swaging, magnesium and Mg-TiNi green compacts were filled in copper tubes which had one end closed by oxy-acetylene welding and had internal diameter of 12 mm and 1 mm wall thickness. In the next step of production, the green compacts filled tubes are vacuum and argon treated for five cycles to remove the air and/or moisture that might present in copper tubes. This step was finalized by enclosing the tubes when they contained a mixture of vacuum and argon atmosphere. After that, encapsulated copper tubes were heated up to 450°C temperature at which they were kept for 20 minutes prior to feeding them to swaging machine for coaxial deformation. In the deformation stage, about 45% deformation was applied to samples in two steps at a feeding temperature of 450°C. Annealing heat treatments were also conducted at 450°C for 20 minutes between the stages and at the end of the coaxial deformation to increase the sintering degree and to homogenize the heavily deformed structures. Eventually, magnesium and its composite rod samples with 6.5 mm diameter were obtained by peeling and machining copper tubes containing pure magnesium and Mg-TiNi composite samples after they cooled down to room temperature in furnace.

Characterization

Particle size characterization of as-received magnesium and TiNi pre-alloyed powders were done with a Malvern Mastersizer 2000 using wet method. The density and total porosities of manufactured rod samples were determined on the basis of Archimedes' principle with a Sartorius precision balance equipped with a density determination kit by dipping the samples in a xylol solution ($CH_3C_6H_4CH_3$). X-ray diffractogrammes of as-received powders and manufactured rod samples were taken using a Rigaku D/Max 2200/PC model X-Ray Diffractometer by continuous scanning at 40 kW between 30° to 90° 2θ angles. Microstructures of samples were examined by optical microscope and scanning electron microscope (SEM) Jeol JSM 6400 equipped with Noran System 6 X-ray microanalysis system, on the other hand, was used for both micro structural and fracture surface analysis. Commercial software (Clemex Vision, Professional Edition, version 3.5.020) was used for quantitative analysis to determine the average grain sizes of pure magnesium and magnesium composite rod samples. Manufactured samples were examined both in polished and etched conditions using 5% Nital solution as an etchant.
Further microscopic studies were carried with a Jeol JEM 2100F Transmission Electron Microscope to investigate especially magnesium matrix-TiNi particle interfaces present in Mg-TiNi composite samples. Thin foils of specimens were prepared by ion milling after their thicknesses were reduced using a dimpler. Both dark field (DF) and bright field (BF) techniques were used to investigate the microstructure. In addition, EDS line analysis was carried out to investigate compositional change across the magnesium matrix, TiNi and magnesium-TiNi interface.
Mechanical properties of manufactured samples were determined under compression loads using machined samples having height to diameter ratio of 1.5. The compression tests were conducted under quasi-static conditions with a Shimadzu ACS-J of 10 kN capacity universal tension-compression test machine at a cross-head speed of 0.5 mm/min.

Results and Discussion

Density and Porosity

Table 1 shows the densities and porosity contents of pure magnesium and Mg-TiNi composites produced by rotary hot swaging performed at 450°C.

Table I. Density and porosity content of rotary swaged specimens.

Material	Density (gr /cm³)	Porosity (%)
Magnesium	1.734±0.002	0.34
Mg-5 vol. %TiNi	1.97±0.002	0.4
Mg-10 vol. %TiNi	2.207±0.002	0.44
Mg-15 vol.% TiNi	2.44±0.001	0.6

As can be seen from Table I, the utilized manufacturing technique resulted in nearly dense materials. The porosity content of the manufactured materials are observed to increase as the content of TiNi reinforcement particles increase presumably due to an increase in interfacial area between matrix and TiNi particles. In addition, spherical pores with 1-5 μm diameters, which were observed rarely, were found both in the matrix of magnesium and its composites perpendicular to swaging direction. Probably, these micro pores formed due to air entrapped in the powders during cold compaction stage which combined together to form larger pores as deformation and sintering proceeded.

X-Ray Diffraction

X-Ray diffractogrammes of both as-received and processed Mg and pre-alloyed TiNi powders are shown in Figure 2. Completely same X-ray diffractogrammes have been obtained for both as-received magnesium powders and manufactured magnesium rod samples such that mainly Mg peaks have been detected in the profile. In addition, a slight change is observed in the relative intensities of (10-10) and (0002) peaks detected at 32.24° and 34.48°, respectively, in the X-Ray diffractogrammes of processed Mg powders or rotary swaged Mg samples. The higher intensity of (10-10) peak of rotary swaged magnesium was attributed to the texture formation which is usually characterized as fiber texture. Another shallow peak characterized as MgO was also detected both in as-received magnesium powders and in swaged samples which indicates presence of MgO on the starting Mg powder particles and/or a slight oxidation during processing.

As can be seen in Figure 2, as-received TiNi powders contain only one phase which is known as B2 ordered austenite phase and exhibits super elastic behavior. On the other hand, X-Ray diffractogramme of Mg-TiNi composites labeled as (d) in Figure 2 consisted of peaks of both magnesium and B2 austenite phase. Similar to magnesium samples MgO peak was also observed in addition to Mg and B2 phase peaks.

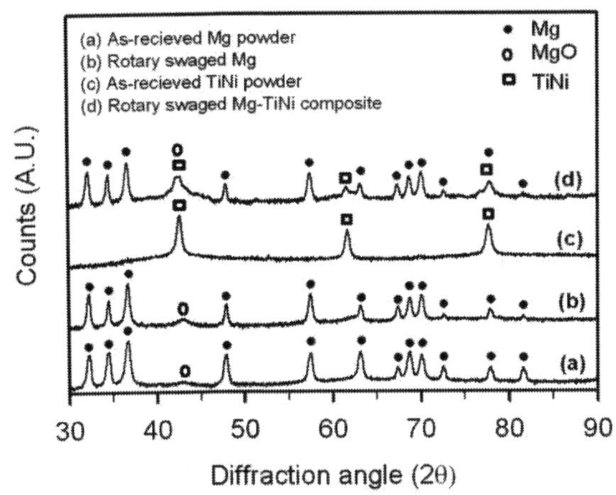

Figure 2. X-ray diffraction spectra of starting and processed Mg and TiNi pre-alloyed powders.

Microstructure

Microstructures of rotary swaged pure magnesium and magnesium composite samples are shown in Figure 3 and Figure 4. Nearly equi-axed particles, which are separated by partly continuous dark thin lines, were observed in the cross-sections of pure magnesium rods perpendicular to the rotary-swaging direction (Figure 3(a)). While the bright regions exhibited nearly 100% of magnesium as a result of EDS analysis, the dark lines which separate the starting powder particles revealed oxygen and magnesium contents about 8.19 wt. % and 91.81 wt. %, respectively, indicating the presence of MgO as previously found in X-Ray diffractogrammes. Magnesium starting powders generally contain oxide layers ranging between 1 and 2 nm [15]. EDS analysis taken from some starting powders exhibited oxygen content as high as 10 wt. %, while the oxygen content of the most of the powders couldn't be measured or observed to change around 1 wt. %. Because of that, the observed oxide layers some of which encountered inside the powder particles probably come from the starting powders and retain in the structure throughout the manufacturing stage. In addition to thin oxide layers observed in microstructure, at some regions of manufactured samples, especially at tripole junctions, nearly spherical small pores with 1-5 μm are detected. Similar to observed oxide layers, these pores appeared during the initial stage of production, namely cold pressing, as in irregular shaped ones and became spherical during rotary swaging at elevated temperatures as they cannot escape out of the structure.

The average grain sizes of pure magnesium and Mg-TiNi composite rods in the cross-sections perpendicular to rotary swaging direction were calculated as 45 μm, which is comparable to initial powder particle size. The oxide layers present on the surfaces of starting powder particles possibly hinder the grain growth and act as barriers to diffusion. The available sintering paths for this type of structures seem to be the broken oxide regions only, formed during powder consolidation stage, i.e. cold pressing and rotary hot swaging stage.

Micro structural examination of Mg-TiNi composites revealed a homogenous distribution of TiNi reinforcement particles throughout the magnesium matrix as shown in Figure 3(b). Similar to processed pure magnesium roads, MgO phase was also detected in Mg-TiNi composites as dark lines which separate the magnesium powder particles. In addition, as shown in Figure 4(a), a highly elongated structure was observed in the cross-sections of both magnesium and Mg-TiNi composite rods in directions parallel to the rotary swaging. As can be seen in Figure 4(a), the shape of the starting TiNi powders were preserved although the specimen matrix elongated in one direction upon hot swaging. The preservation of TiNi powder particle's shape was attributed to super elastic behavior of Ti-50.6 at. % Ni pre-alloyed powders as they contain only B2 austenite phase which is responsible for shape recovery. Apart from these, in the regions close to the surface of manufactured rods, each elongated structure were observed to contain nearly equi-axed grains whose dimensions are limited to the width of the elongated powder particles as shown in Figure 4(b). The equi-axed grains are believed to form as a result of dynamic recovery and recrystallization during rotary-hot swaging stage since the utilized deformation temperature during manufacturing was higher than the recrystallization temperature of the magnesium. Presence of equi-axed grain containing elongated structures, especially in the regions close to the surface, is the indication of non-homogenous deformation resulted by hot swaging operation.

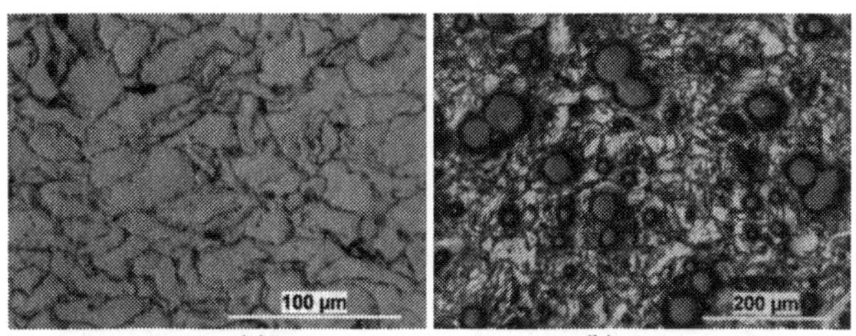

(a) (b)

Figure 3. Optical micrographs of rotary swaged specimens in the cross section perpendicular to rotary swaging direction, (a) pure magnesium, (b) Mg-10 vol. % TiNi composite

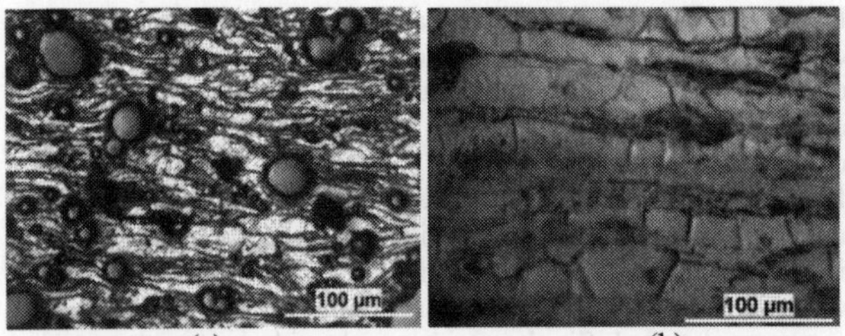

(a) (b)

Figure 4. Optical micrographs of Mg-10 vol. % TiNi composite in the cross section parallel to the rotary swaging direction, (a) Elongated magnesium particles and spherical TiNi particles, (b) Recrystallized grains in elongated structures.

In addition, micro structural examination conducted in the interface region of Mg matrix/TiNi particles revealed neither cavities nor reactions. A non-porous continuous interface and good interfacial bonding have formed between dark TiNi particles and bright Mg matrix as shown in Figures 5(a) and 5(b).

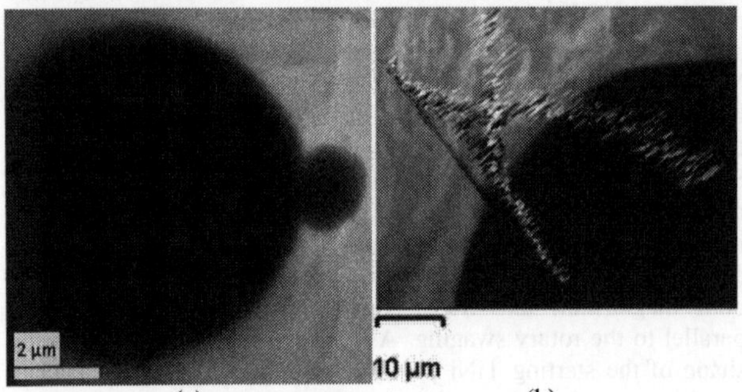

(a) (b)

Figure 5. (a) TEM micrograph showing non-porous interface between magnesium matrix and TiNi particles in Mg-10 vol.% TiNi composite, (b) EDS spectra across a TiNi particle

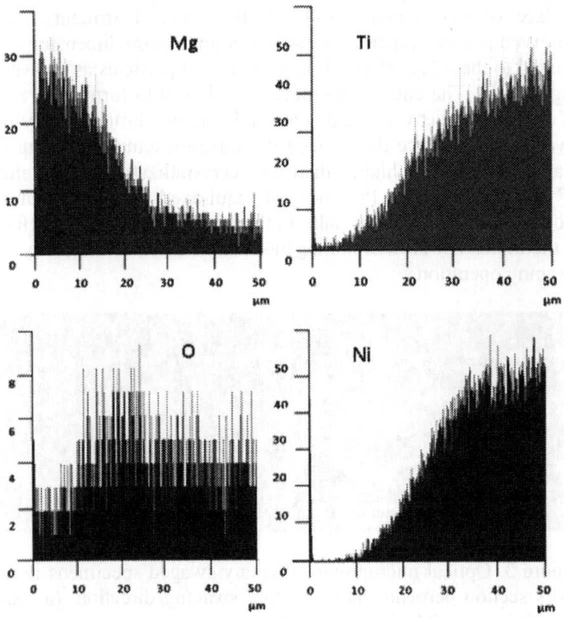

Figure 6. EDS line analysis of the elements across TiNi particle and Mg/particle interface in Mg-10 vol. % TiNi composite

Figure 5(b) and Figure 6 show EDS line spectra across a TiNi particle and magnesium matrix in Mg-10 vol. % TiNi composite. Although, both Ti and Ni peaks were observed only in TiNi particles, a slight amount of magnesium was detected in TiNi particles. On the other hand, oxygen is distributed both in magnesium matrix and TiNi particles, but with a clear enrichment in the Mg matrix-TiNi interface (Figure 6) which indicates the presence of oxides in the interfacial region. In this study, detected oxygen peaks were attributed to the oxygen mainly coming from MgO in addition to oxides formed on TiNi powder surfaces.

Compression Tests

Stress-strain diagrams of compression tested pure magnesium and Mg-TiNi composites are shown in Figure 7. Upon compression the samples under quasi-static conditions stress-strain curves similar to wrought alloys has been obtained with almost linear elastic behavior at small strains, followed by yield and strain hardening up to an ultimate stress. Subsequent to a peak stress fracture occurred after a small straining whose magnitude depends on the sample purity and content of the TiNi reinforcement particles.

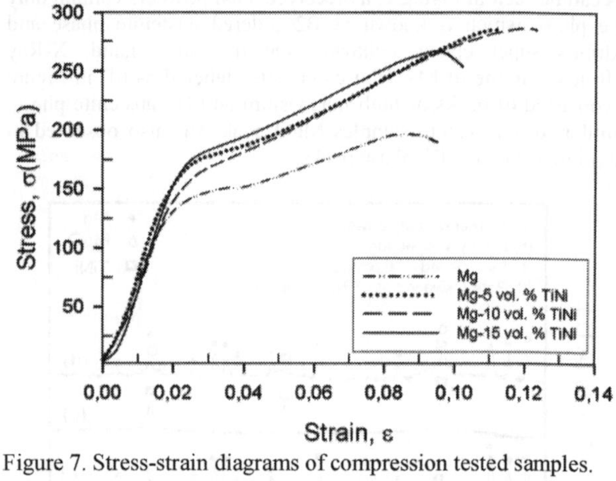

Figure 7. Stress-strain diagrams of compression tested samples.

The mechanical properties of magnesium and magnesium alloys, i.e. elastic modulus and yield strength, may be enhanced by grain refinement, alloying and subsequent heat treatment or by the composite production through the addition of stiffer particles to the matrix in solid or liquid states. In this study, only the effect of reinforcement particles on mechanical properties are discussed and the results of the compression behavior revealed a significant increment in compression yield and peak stresses of magnesium as a result of TiNi addition, which were around 130 and 210 MPa, respectively. The yield strength of the composites was observed to increase linearly with an increase in TiNi reinforcement particle content as shown in Figure 8 and about 25% increment in yield strength was detected in Mg-15 vol. % TiNi samples. Yield strength of an alloy depends on the obstacles, i.e. grain size of the alloy, and the size and the distance between precipitates, which render the motion of dislocations. TiNi particles in this study possibly serve as obstacles for dislocation motion and result in an increased $\sigma_{0.2}$ as their content in the matrix increase. In contrast to yield strength, the effect of TiNi particles on peak or maximum stress at which internal micro pores or cracks form was not linear. Although the peak stress of magnesium is enhanced by 35% as a result of adding around 5 or 10 vol. % of TiNi, presence of 15 vol.

% TiNi particles in matrix didn't exhibited an increment as high as the other composites show. As can be seen in Figure 8, the effect of reinforcement TiNi particle content on maximum stress diminishes when the reinforcement quantity reaches to 15 vol. %. The observed non-linear effect of TiNi particle content on peak stress of Mg-TiNi composites was attributed to the higher interfacial area created by increased TiNi reinforcement content, which may serve as crack initiation sites. Observed increment in peak stress of composites is generally attributed to two facts, i.e. grain refinement and load transfer from matrix to reinforcement. In addition to these two reasons it has been shown that MgO in magnesium may act as secondary reinforcement although it reduces the ductility of the alloy.

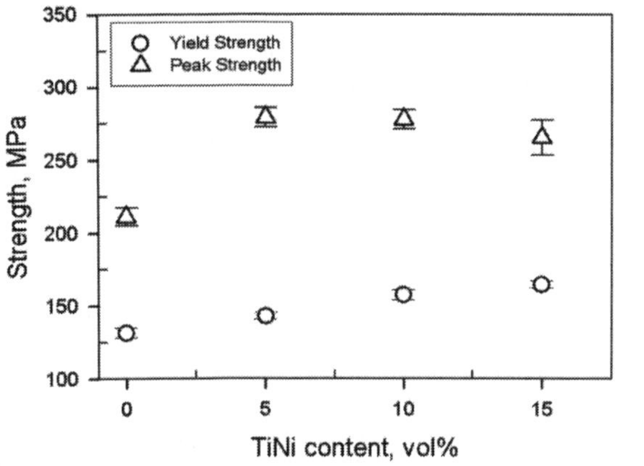

Figure 8. Yield and peak strength values of compression tested magnesium and Mg-TiNi composites.

In contrast to other magnesium composites containing various types of reinforcement particles, i.e. SiC, ductility of the manufactured Mg-TiNi composites were increased notably compared to pure magnesium samples. As shown in Figure 7, about 40% increment is observed in ductility level of pure magnesium when 10 vol. % TiNi is added. Similar effect was also seen in the study of Hassan and Gupta [1] conducted on Mg-Ti composites in which the increment in ductility level by the addition of Ti was attributed to diffusional dissolution of Ti in magnesium matrix. In the present study, the increase in ductility level is attributed to the load transfer from magnesium matrix to interface and TiNi particle. For Mg- 15 vol. % TiNi composite, on the other hand, elongation value at fracture point was decreased and composite's elongation value reach to that of pure magnesium probably due to increased number defects in matrix or interfacial area due to increased quantity of reinforcement particles.

Upon testing the samples under compression loads, fracturing occurred after small straining beyond their peak stress values. Occasionally fracturing took place at an angle 45° to the loading axis in some samples, while most of the samples failed in regions parallel to the loading directions which were also parallel to the rotary swaging direction (Figure 9 (a)). Failure took place in central weakest regions of rod samples since the severity of deformation in rotary swaged samples decreases as going from sample's surface to center. Examination of fracture surfaces of pure magnesium revealed brittle type intergranular and transgranular mode of fractures which followed the MgO present between the powder particles and the magnesium matrix, respectively, as shown in Figure 9(b). Application of loads beyond a critical value of composite materials results in fracture due to concentration of load in matrix, reinforcement particle or the interfacial area between matrix and reinforcement. As it can be seen in Figure 9(c) and 9(d), all the Mg-TiNi composites fractured by de-bonding of interface as a result of initiation of localized damage upon compression loading. During failure de-bonded TiNi particles also caused fracturing of the magnesium matrix (Figure 9(d)).

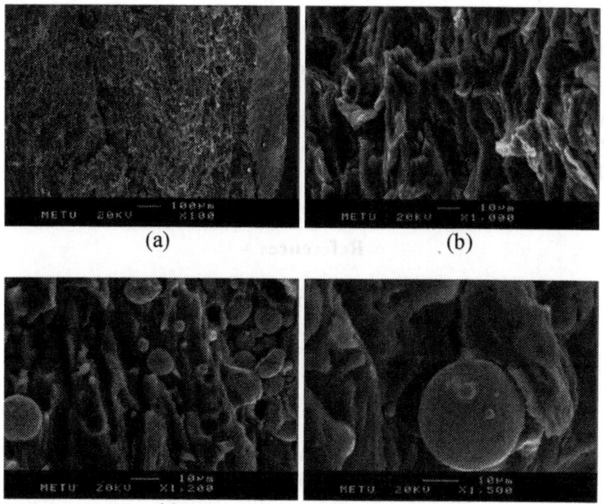

(a) (b)

(c) (d)

Figure 9. Fracture surfaces of samples, (a), (b) magnesium, (c), (d) Mg-10 vol. % TiNi.

Conclusions

Present study have shown the feasibility of pure magnesium and Mg-TiNi composite rods production by rotary hot swaging which uses cold pressed green compacts. Highly dense rod samples were manufactured and they were observed to contain very few amount of spherical micro pores formed due to entrapped air during cold compaction step. Recrystallized equi-axed grains containing elongated structures were produced due to combined effect of high deformation rate and high temperature. Recrystallized grains were observed especially in the surface regions since the rotary swaging deformation is more severe in the surfaces of samples. Although the grains of magnesium matrix elongated in the direction parallel to the rotary swaging direction, the shape of the spherical TiNi particles was preserved due to their super elastic behavior. EDS analysis has shown that the processed samples of pure magnesium and Mg-TiNi composites contain thin layers of MgO which is mainly due to starting powders and processing conditions. In addition, MgO was also detected in interfacial region of Mg-TiNi composites which may degrade the mechanical properties of composites. The addition of TiNi particles resulted in a linear increase in the yield strength of Mg-TiNi composites while the enhancement of maximum or peak strengths of composites was limited. As high as 280 MPa peak strength value was obtained in Mg-TiNi composites, which contain about 10% TiNi by volume. However, higher content of TiNi particles in magnesium resulted in a decrease both in peak stress and ductility of the samples presumably due to an increase in defective interface areas between TiNi particles and magnesium matrix. Fracture in magnesium and Mg-TiNi composite samples was observed to start in the central regions parallel to rotary swaging deformation probably due to non-homogenous nature of hot-swaging in surface and central regions of samples. In magnesium and Mg-TiNi composites both intergranular and transgranular modes of failure were observed and fracture cracks passed followed either the MgO present between prior powder particles and the grains of magnesium matrix. Fracture surfaces of Mg-TiNi composites were similar to those of pure magnesium samples except the circular cup regions formed due to de-bonding of TiNi particles upon compression loading.

Acknowledgments

Author would like to thank Dr. Özgür Duygulu, Dr. Kaan Pehlivanoğlu and Dr. Hasan Akyıldız for their help during the characterization studies.

References

1. S. F. Hassan, M. Gupta, ''Development of Ductile Magnesium Composite Materials Using Titanium as Reinforcement'', Journal of Alloys and Compounds, 345 (2002), 246-251.

2. H. Z. Ye, X. Y. Liu, ''Review of recent studies in magnesium matrix composites'', Journal of Materials Science, 39 (2004), 6153-6171.

3. Y. Oishi, N. Kawabe, A. Hoshima, Y. Okazaki, ''Development of High Strength Magnesium Alloy Wire'', SEI Technical Review, 56 (2003), 54-58.

4. M. Sagumata, S. Hanawa, J. Kaneko, ''Structure and Properties of Rapidly Solidified Mg-Y Based Alloys'', Mater. Sci. Eng. A, 226-228 (1997), 861-866.

5. T. Mohri, M. Mabuchi, N. Saito, M. Nakamura, ''Microstructure and Mechanical Properties of a Mg-4Y-3RE Alloy Processed by Thermo-Mechanical Treatment'', Mater Sci.Eng. A, 257 (1998) 287-294.

6. J. Kojima, ''Project of Platform Science and Technology for Advanced Magnesium Alloys'', Mater. Trans. JIM ,42 (2001), 1154-1159.

7. N. Balasubramani, U. T. S. Pillai, B. C. Pai, ''Optimization of Heat Treatment Parameters in ZA84 magnesium alloy'', Journal of Alloys and Compounds, 457 (2008), 118-123.

8. Z. R. Yang, S. Q. Wang, M. J. Gao, Y. T. Zhao, K. M. Chen, X.H. Cui, ''A New-Developed Magnesium Matrix Composite by Reactive Sintering '', Composites: Part A, 39 (2008), 1427-1432.

9. S.F. Hassan., M. Gupta, ''Development of a Novel Magnesium–Copper Based Composite with Improved Mechanical Properties'', Materials Research Bulletin, 37(2) (2002), 377-389.

10. A.Mortensen, "Mechanical and Physical Behaviors of Metals and Ceramic Compounds", (Riso National Laboratory, Roskilde, Denmark, 1988), 141.

11. C. S. Goh, J. Wei, L. C. Lee and M. Gupta, ''Development of Novel Carbon Nanotube Reinforced Magnesium Nanocomposites Using the Powder Metallurgy Technique'', Nanotechnology, 17 (2006), 7-12.

12. Mamoru Mabuchi, Kohei Kubota and Kenji Higashi, ''New Recycling Process by Extrusion for Machined Chips of AZ91 Magnesium and Mechanical Properties of Extruded Bars'', Materials Transactions, JIM, 36 (10) (1995), 1249-1254.

13. P. J. Burke and Georges J. Kipouros, ''Powder Metallurgy of Magnesium: Is it Feasible?'', Magnesium Technology 2010, TMS (The Minerals, Metals and Materials) Ed. S. R. Agnew, N. R. Neelameggham, E. A. Nyberg and W. H. Sillekens, pp. 115-120 ,Seattle, WA, USA,February 14-18, 2010.

14. Rong Li, Nie Zuo-ren, Zuo Tie-yong, ''Effect of Reduction of Diameter on Microstructure and Surface Roughness of Rotary Swaged Magnesium by FEA'', Transactions of Nonferrous Metals Society of China, 18 (2008), 263-268.

15. V. Fournier, P. Marcus and I. Olefjord, ''Oxidation of Magnesium'', Surface and Interface Analysis, 34 (2002), 494-497.

Magnesium Technology 2011
Edited by: Wim H. Sillekens, Sean R. Agnew, Neale R. Neelameggham, and Suveen N. Mathaudhu
TMS (The Minerals, Metals & Materials Society), 2011

NANOCRYSTALLINE Mg-MATRIX COMPOSITES WITH ULTRAHIGH DAMPING PROPERTIES

Babak Anasori, Shahram Amini, Volker Presser and Michel W. Barsoum
Department of Materials Science and Engineering, Drexel University, Philadelphia, PA 19104, USA

Keywords: Mg-matrix composite, Nano composites, Liquid melt infiltration, Powder Metallurgy.

Abstract

Recently, we reported on the processing of 50 vol.% Ti_2AlC-nanocrystalline magnesium, nc-Mg, matrix composites using a pressureless melt infiltration method. Herein we report on composites with up to 80 vol.% Mg. These composites are readily machinable, relatively stiff, strong and light, and exhibit ultrahigh damping. Increasing the nc-Mg volume fraction leads to lighter composites with higher damping characteristics at lower stresses (~30% of the mechanical energy is dissipated at 250 MPa). In some cases the Mg nanograins are also extraordinarily thermally stable which renders these composites good candidates for applications at temperatures higher than ambient. Due to the simple inexpensive melt infiltration technique used to fabricate these novel nanocomposites, it is possible to produce samples as large as ones made via normal powder metallurgy methods.

Introduction

The MAX Phases

The $M_{n+1}AX_n$ (MAX) phases are layered hexagonal solids, with two formula units per unit cell, in which near close-packed layers of M are interleaved with layers of pure A-group elements, with the X-atoms filling the octahedral sites between the M layers. These solids combine some of the best attributes of metals and ceramics. Like metals, they are electrically and thermally conductive, most readily machinable [1, 2], not susceptible to thermal shock, plastic at high temperatures, and exceptionally damage tolerant [1-6]. Like ceramics, some of them are elastically rigid (Young's mod. > 300 GPa), lightweight (≈ 4 gm/cc) and maintain their strengths to high temperatures. The ternaries Ti_3SiC_2 and Ti_2AlC are creep, fatigue and oxidation resistant [7-10], which is one reason they were chosen for this work. The Ti-based MAX phases are also more conductive, both electrically and thermally than Ti metal at all temperatures [6, 11].

Due to their layered nature and the fact that basal slip is operative at all temperatures, the MAX phases possess unique mechanical properties that are atypical for structural ceramics. The fracture toughness of Ti_3SiC_2 exhibits R-curve behavior and varies from 8-16 $MPam^{1/2}$ depending on crack length [12, 13]. The latter value is one of the highest ever reported for a single-phase, monolithic ceramic. Such good damage tolerance is attributed in part to kinking and the formation of the heavily deformed lamellar bridges in the crack wake that are reminiscent of those that form in wood and other layered composites. The MAX phases also have excellent low and high-cycle fatigue properties [14]. For example, fine-grained Ti_3SiC_2 samples can be compressively loaded to 700 MPa for 100 cycles with no apparent fatigue [15, 16]. The reason for this state of affairs is believed to be the formation of fully reversible, dislocation-based incipient kink bands, IKBs, discussed in the next section. Note that in the MAX phases, every basal plane is a potential delamination and/or slip plane.

Kinking Nonlinear Elasticity

More recently it has been shown that the MAX phases are a subset of a much greater class of solids termed kinking nonlinear elastic, KNE, for which the only requirement for membership is plastic anisotropy, in the sense that dislocations are confined to two dimensions [15, 17, 18]. Plastic anisotropy insures that the formation of, non-basal dislocations is prohibitively expensive. These solids then deform by the formation of kink bands, KBs. Prior to the formation of the KBs, however, it has been shown that these solids must first form incipient kink bands, IKBs. IKBs are fully reversible dislocation-based loops that are nucleated on the easy slip planes of stressed, plastically anisotropic solids. IKBs are comprised of coaxial, parallel dislocation loops, with $2\alpha >> 2\beta$ (Fig. 1a) [15]. Frank and Stroh [19] showed that by virtue of their shape, IKBs, can only exist when a load is applied; removal of the load results in their shrinkage and ultimate annihilation and restores the perfect lattice. In recent papers, it was shown that IKBs are the precursors of mobile dislocation walls, MDWs, or low angle boundaries that, in turn, result in regular kink boundaries, KBs [20-22].

Using a Griffith-like approach, Frank and Stroh [19] showed that at a critical threshold stress, σ_t:

$$\tau_c \approx \frac{\sigma_t}{M} \approx \sqrt{\frac{G^2 b}{2\alpha\gamma_c}} \tag{1}$$

a subcritical IKB becomes critical and would run to the ends of the two-dimensional crystal they modeled. In Eq. 1, M, G, b, and γ_c are, respectively, the Taylor factor, shear modulus, Burgers vector and the kinking angle, which is small and of the order of 3° or less. Based on our work to date it has been shown that 2α can be equated to grain dimension along [0001] [23]. Interestingly, this implies σ_t is proportional to 1/√grain size, i.e. follows a Hall-Petch like relationship.

As noted above, removal of the stress restores the perfect crystal. The driving force for the collapse of the IKBs stems from their shape and is contingent on their ends remaining attached [21]. If the ends are sundered, the IKBs devolve into MDWs, also known as, low angle grain boundaries, which ultimately give rise to KBs; the latter are irreversible.

The signature of IKBs is the formation of fully reversible stress-strain loops [15, 22, 23] (Fig. 1b), whose size and shape are a strong function of grain size. Before discussing the relationship between IKBs, microyielding and damping in KNE solids, it is important to briefly summarize the major elements of the IKB-based model, which as noted above, is based on the aforementioned work by Frank and Stroh paper [19]. Referring to Fig. 1b:

i) a threshold stress, σ_t, given by Eq. 1, is needed to nucleate an IKB;

ii) the energy dissipated per unit volume per cycle, W_d, is given by [21, 23, 24]:

$$W_d = \frac{4\pi(1-\nu)N_k\alpha^3}{G^2\gamma_c M^2}\frac{\Omega}{b}(\sigma^2 - \sigma_t^2) \quad (2)$$

where N_k is the number of IKBs per unit volume, and Ω is the energy dissipated by a dislocation line sweeping a unit area. It follows that Ω/b is the critical resolved shear stress, CRSS, of the IKB dislocations. According to Eq. 2 W_d, scales with σ^2, with a x-axis intercept at σ_t^2 as observed (see below).

iii) Non-linear strain, ε_{NL} (defined in Fig. 1b). Recasting Eq. 2 in terms of ε_{NL} yields [21, 23, 24]:

$$W_d = 3k_1\frac{\Omega}{b}\varepsilon_{IKB} \quad (3)$$

It follows that Ω/b should be proportional, if not equal, to the CRSS of an IKB dislocation loop. k_1 relates the IKB volumetric strain to the axial strain along the loading direction. k_1 and typically is of the order of 1 or 2 depending on texture [25]. It follows that W_d should scale linearly with ε_{NL} as, repeatedly, observed. The slopes of such curves are directly proportional to the CRSS of the IKB dislocations [21, 22, 25]. In other words, using the IKB-based model the CRSS of the IKB dislocations can be extracted from compression of polycrystalline samples. For hexagonal metals, the values obtained are in good agreement with values measured in single crystal work [26].

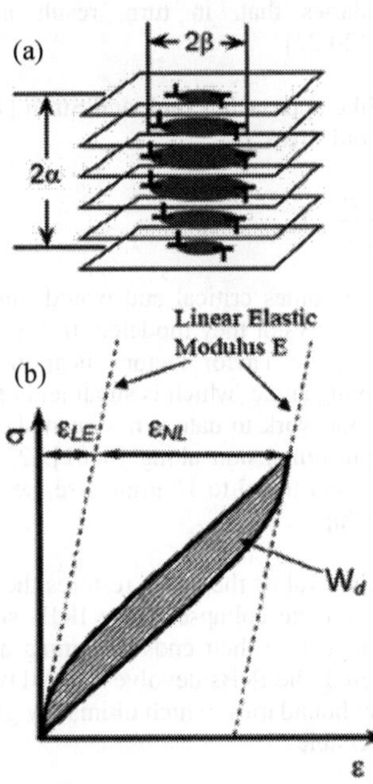

Figure 1. Schematic of (a) dislocation loops comprising an IKB, (b) typical stress-strain curve for a KNE solid and definitions of non-linear strain, ε_{NL} and energy dissipated per cycle, W_d.

With this insight we were not only able to explain our own results, but also the deformation of many diverse and seemingly unrelated solids such as graphite [27], the MAX phases [24, 28], sapphire [21], ZnO [22], $LiNbO_3$ [29], $LiTaO_3$ [30], mica [18, 31] and presumably other layered silicates and thus much of geology among many others.

Microyielding and Damping in Hexagonal Metals

The hexagonal close-packed metals, Ti, Mg, Co, Zr and Zn are of immense technological importance and have been studied extensively for the past 70 years. Despite this large amount of work, the nature of their damping and microyielding has remained unclear. Recently, we showed that when polycrystalline Co [23], Mg [20] or Ti [26], samples are loaded in simple compression fully reversible loops form above a threshold stress. The size and shape of these loops was found to be a strong function of grain size; large grained samples had significantly larger W_d values. The size of the loops, or W_d, was also a function of pre-strain. Increasing the deformation strain, reduces the domain or grain sizes, 2α, which increases σ_t which ultimately reduces W_d. Based on this IKB-based model it is possible to associate damping at stresses > σ_t in these solids with the to-and-fro motion of IKB dislocations. Results of this model, for example, explain why Mg and Zn are high damping metals, while Al is not.

Currently there is no direct evidence for the existence of IKBs. The circumstantial evidence for their existence, however, is compelling. Three dislocation-based possibilities for microyielding and damping exist: reversible dislocation pileups, reversible twinning and IKBs. For the former two, it can be shown that W_d should scale with σ, and not σ^2, as observed. The same physics also applies to solids in which twins form, such as Mg and Co, as to those that do not, such as the MAX phases. Lastly and most importantly, the CRSS values obtained from W_d vs. ε_{NL} plots are in good agreement with the same values measured from single crystal work [20, 22, 26].

Metal Matrix Composites Reinforced with MAX Phases

Based on the aforementioned ideas it was speculated that fabricating composites with two KNE solids should result in exceptionally high values of W_d. A natural choice was to combine Mg with commercially available MAX powders, such as Ti_3SiC_2 and Ti_2AlC. The results for Ti_2AlC-Mg composite were as unexpected as they were surprising in that the Mg matrix grains were at the nano-scale. The processing and microstructural characterization of 50 vol. % Ti_2AlC-nc Mg-matrix composites fabricated by pressureless spontaneous melt infiltration at 750 °C for 1 h can be found in Ref. [3]. X-ray diffraction and transmission electron microscopy, TEM, both confirmed that the Mg grain size was at nano scale [3].

The 50 vol.% Ti_2AlC/nc-Mg composites are readily machinable, stiff (effective E ≈ 70 GPa), strong, light (2.9 g/cm^3) and exhibited – at the time - record W_d values at σ levels of the order of ≈ 500 MPa. Since the IKB-based model predictions were well adhered to, it was concluded that the composite, like its individual components, was a KNE solid in which the nucleation and annihilation of IKBs was responsible for the high W_d values. The CRSS values extracted from the compression results were closer to those of Ti_2AlC than for Mg, indicating that it is the former than is responsible for the damping [32]. In other words, because it is at the nano-scale, the Mg does not form IKBs. The role of the Mg is to thus to impart strength to the composite and allow for high compressive stresses. In addition to their record damping – and presumably for the same reason, i.e. IKB formation – we have recently shown that the MAX/Mg composites are also quite fatigue resistant [33]. The aim of this work was to reduce the vol.

% of the reinforcement to 20 vol. % to obtain lighter samples with higher damping properties that would also be less expensive.

Experimental Details

In contrast to almost all the processes that used for the fabrication of nc metals, in general, and low melting point metals, like Mg, in particular, the method used here is quite simple and unique. The method has been used to fabricate Mg–50 vol. % Ti_2AlC by pressureless, spontaneous Mg melt-infiltration into a porous Ti_2AlC preforms [3]. In addition to this process, powder metallurgy was also used to make composites with higher volume fractions of Mg. The starting powders - Ti_2AlC (-325 mesh, 3-ONE-2, Voorhees, NJ) and Mg (-325 mesh, 99.8 % pure, Alfa Aesar, Ward Hill, MA) - were ball-milled for 12 h. Porous samples were fabricated from the powder mixtures in the form of rectangular bar (1.3 x 1.3 x 70 mm^3) by cold pressing at 50 MPa. Also for the sake of comparison, Mg-20 vol. % Ti_3SiC_2 composites were fabricated, using the same processing steps. In this case, the starting powders were Ti_3SiC_2 (- 325 mesh, 3-ONE-2, Voorhees, NJ), and the same Mg powder used above. The samples were placed in alumina, Al_2O_3, crucibles (AdValue Technology, Tucson, AZ) that were in turn covered with Al_2O_3 lids and placed in a graphite-heated vacuum-atmosphere hot press, (Series 3600, Centorr Vacuum Industries, Somerville, MA) heated at 10°C/min to 750°C, held at that temperature for 1 hour, after which the furnace was turned off and the samples were furnace cooled. The samples' microstructures were observed in a field emission scanning electron microscope, SEM, (Zeiss Supra 50VP, Germany). X-ray diffraction was carried out on a diffractometer (Model 500D, Siemens, Karlsruhe, Germany) and the spectra were collected using step scans of 0.01 in the range of 10°–80° 2 theta and a step time of 2 s. Scans were made with Cu Kα radiation (40 KV and 30 mA).

The room temperature compressive and cyclic uniaxial loading–unloading compression tests were measured using a hydraulic testing machine (MTS 810, Minneapolis, MN). The strains were measured by a capacitance extensometer (MTS, Minneapolis, MN) – attached to the samples – with a range of 1% strain. All the loading–unloading compression tests were performed in load-control mode at a loading rate of 54 MPa/s. The Vickers microhardness values, V_H, were measured using a microhardness indenter (LECO-M400, LECO Corp. St. Joseph, MI) at 10N.

Results and Discussion

Comparing the full width at half maximum, FWHM, of the various Mg peaks from the XRD diffractogram of the 20 vol. % Ti_2AlC composite with those of pure Mg powder ($d_{av} \approx 150$ μm), 50 vol. % Ti_2AlC and 50 vol. % Ti_3SiC_2, revealed that the Mg peaks broadening happened only when Ti_2AlC was used as the reinforcement (Fig. 2). Using the Scherrer formula [34] the particle size was estimated to be 90 ± 15 nm. Note that even by decreasing the volume fraction of Ti_2AlC to 20 vol. %, surprisingly, Mg grains are still at the nano scale.

The microstructure of a mounted, and polished, 20 vol. % Ti_2AlC sample (Fig. 3a) was quite homogeneous and apparently fully dense. The Ti_2AlC grain size was 20±10 μm. The fractured surface (Fig. 3b), however, showed the presence of nano Mg grains, some as fine as 35 nm, as well as sub-micron Mg single crystals (pointed to by arrows).

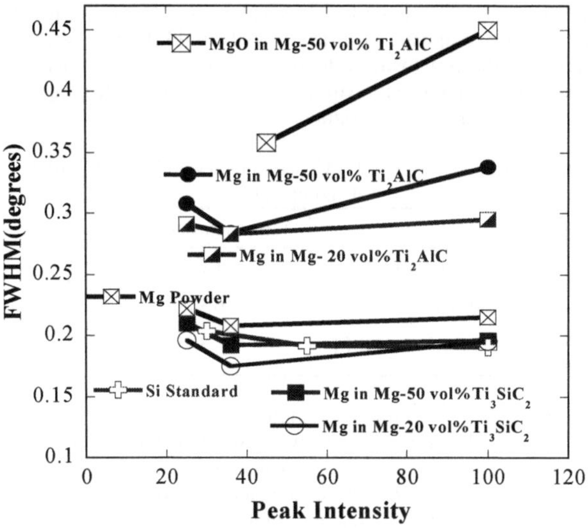

Figure 2. FWHM of Mg and MgO vs. peak intensity. The three highest intensity peaks in Mg and two in MgO are compared with those of a Si standard, pure as-received Mg powder peaks.

Figure 3. Secondary electron SEM images of, (a) polished surface of Mg-20 vol.% Ti_2AlC composite, (b) a fractured surface; arrows point to nano and sub-micron Mg grains.

Table I compares the microhardness values of the composites as a function of Ti_2AlC volume fractions and the hardness of a Mg- 30 vol. % SiC composite reported elsewhere [35]. The Vickers hardness of the composite with 20 vol. % Ti_2AlC is significantly higher than the composite with 30 vol. % of SiC confirming the nanoscopic nature of the Mg-grains.

Table I. Hardness of Mg-Ti_2AlC composites and Mg- 30 vol.% SiC composite

Material	Micro Hardness (GPa)
Mg- 50 vol.% Ti_2AlC	1.6±0.2
Mg- 20 vol.% Ti_2AlC	0.9±0.1
Mg- 30 vol.% SiC [35]	0.5

At 350±10 MPa, the ultimate compressive strength, UCS, of the 20 vol. % Ti_2AlC–Mg composite was lower than the 700±10 MPa of the 50 vol. % Ti_2AlC–Mg composites reported on earlier [3]. However, to confirm that the nanometer scale of the Mg-grains was mainly responsible for the high strengths observed we measured the UCS of Mg–50 vol.% Ti_3SiC_2 and Mg–20 vol.% Ti_3SiC_2 composites. In the latter two composites, as mentioned above, the Mg-grains were not in the nanometer scale and their UCSs, at 400±20 and 240±20 MPa, respectively, were lower than those of the Ti_2AlC–Mg composites with comparable volume fractions of reinforcement.

Typical compressive hysteretic stress–strain loops, at four different loads, up to 75% of the UCS, shifted horizontally for clarity, are shown in Fig. 4a. Based on these curves and the fact that W_d (Fig. 4b) and ε_{NL} both scale as σ^2 as predicted from IKB-based model are consistent with the fact that IKBs are responsible for both [15, 17, 18, 27, 36]. In other words, W_d and ε_{NL} are due to the formation and annihilation of IKBs. The value of W_d for the composite with 20 vol. % Ti_2AlC is ≈ 0.34 MJ/m^3 at 250 MPa. To the best of our knowledge, this value is the highest ever reported at 250 MPa. These materials can dissipate ≈ 30 % of the total mechanical energy applied during each cycle.

The effective Young's moduli – calculated from least squares fits of the entire data set and shown as diagonal lines bisecting the loops in Fig. 4a – range from 38 to 50 GPa. While there are many solids for which damping is higher (e.g., elastomers) we believe that the combination of W_d, compressive strengths and moduli values reported herein is unique and outside the Ashby envelope [37] (see intersection of dashed lines in Fig. 5). In the case of the composites the relatively softer Mg phase in between the Ti_2AlC grains allows the latter to kink. However, and in contradistinction to pores, that essentially achieve the same feat (for e.g. it was shown that a 10% porous Ti_2AlC sample had higher W_d values on an absolute scale than a fully dense one [28]) the presence of the nc-Mg allowed h much higher stress values to be reached before failure.

As noted above, W_d scales linearly with ε_{NL} with a slope that is equal to $3k_1$ Ω/b. Table II compares the calculated CRSSs of different materials tested. At 37 MPa, the Ω/b values obtained in the previous work [3], and for the 20 vol. % Ti_2AlC composites tested herein, are comparable to each other, and to those of bulk monolithic Ti_2AlC. This is important because it implies that most of the energy is dissipated in the Ti_2AlC phase. The decrease in Ω/b by increasing the volume fraction of Mg in the matrix can be

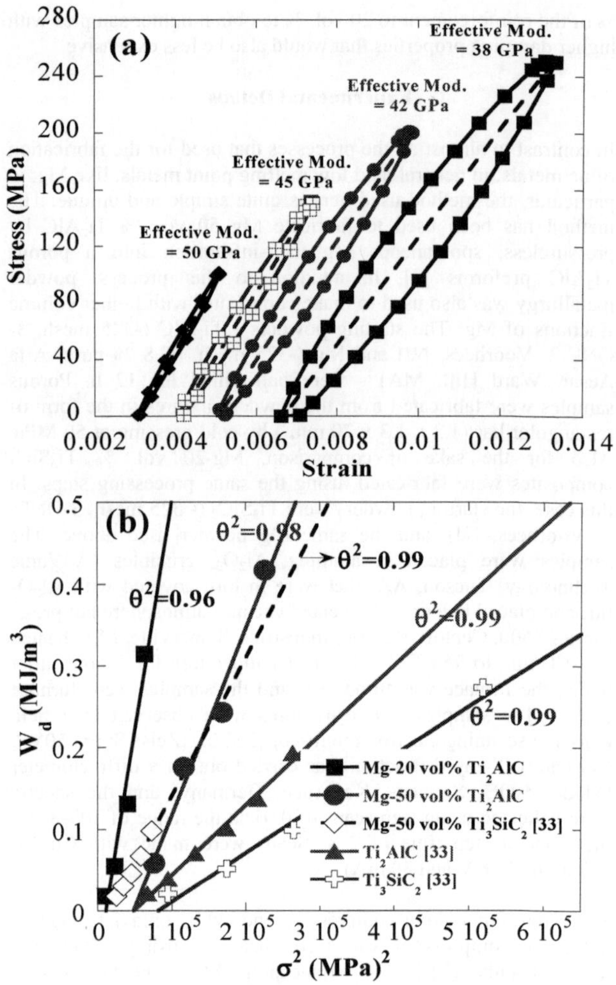

Figure 4. (a) Fully reversible hysteretic loops in a Mg-20 vol.% Ti_2AlC composite. The sample was compressed to ~ 75% of its failure stress; the loops are shifted horizontally for clarity; (b) Plot of W_d vs. σ^2 obtained from the uniaxial compression stress–strain curves.

Table II. Calculated Ω/b values from Eq. (3)

Material	$3k_1\Omega/b$	Ω/b (MPa)
Ti_2AlC [32]	228	38
Mg-50 vol% Ti_2AlC	233	37
Mg-20 vol% Ti_2AlC	200	33
Ti_3SiC_2 [32]	192	32
Mg-50 vol% Ti_3SiC_2 [32]	93	15.5
Mg [26]	24	4

related to the grain size of the Mg; by having larger Mg grains the possibility of forming IKBs in the latter increases. As noted above, the fact that a 10% porous Ti_2AlC sample dissipates more energy per unit volume per cycle on an absolute scale than its fully dense counterpart [28] essentially eliminates all mechanisms, such as dislocation pileups and/or twinning that scale directly with the volume of the material tested. It is, however, in agreement with an IKB-based model in which kinking – which is a form of plastic instability, or buckling - is more prone to happen in a less rigid or porous solid than a fully dense one.

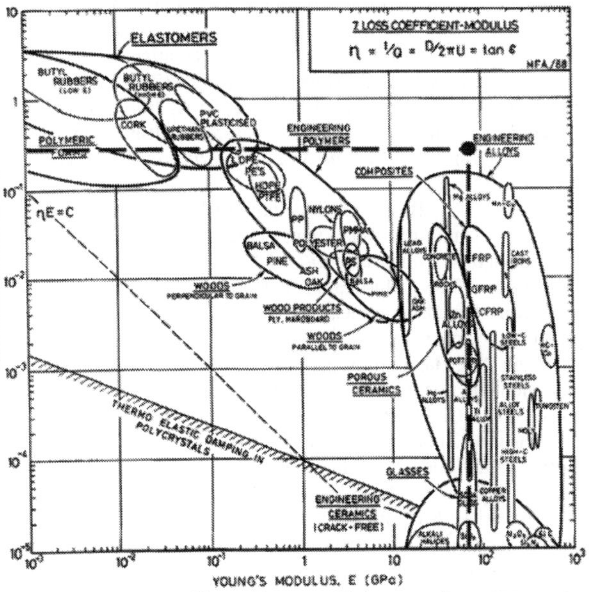

Figure 5. Ashby map of log of loss coefficients vs. log of Young's moduli [37]. The intersection of the two dotted lines is where some of our composites fall; clearly outside the envelope.

In contrast to the aforementioned case where the Ti_2AlC phase is responsible for most of the energy dissipated per cycle, the situation for the Mg-Ti_3SiC_2 composites is substantially different. At 16 MPa, the Ω/b value appears to be an average of that of Ti_3SiC_2 (32 MPa) and Mg (4 MPa). This suggests that in this case, both the Mg and Ti_3SiC_2 contribute to Ω/b, and hence W_d [32].

Accompanying the record strengths and damping is a nc-Mg matrix that is remarkably stable [38]. The microstructure is so stable that heating the composite three times to 700 °C – 50 °C over the melting point of Mg – not only resulted in the repeated melting of the Mg, but surprisingly and within the resolution of our differential scanning calorimeter did not lead to any coarsening (Fig. 6). The reduction in the Mg melting point due to the nano-grains in the 20 vol.% Ti_2AlC composite was ≈ 20 °C which is less than the 50 °C reported for the Mg- 50 vol.% Ti_2AlC [38]. This is most probably because the Mg nano-grains in the latter are smaller than the former. For the Mg-50 vol. % Ti_2AlC composites, XRD and neutron diffraction results suggested that a thin, amorphous and/or poorly crystallized rutile/anatase/magnesia layer separate the Mg nanograins and prevent them from coarsening [38]. That layer is presumably thin enough and thus mechanically robust enough to survive the melting and solidification stresses encountered during thermal cycling.

Summary and Concluding Remarks

Stiffness, in general, and high specific stiffness in particular are desirable qualities in solids. Typically, the price paid for high stiffness is lack of damping (Fig. 5). High specific stiffness exacts another price: difficulty in machinability, which can add considerable complexity, weight and cost to components and structures. From a design point of view it would be advantageous if multi-functionality could be engineered into alloys or composites, such that the same load-bearing material could also dampen vibrations or noise while remaining easily machinable and stiff. The Mg-Ti_2AlC composites fabricated here are readily machinable, with strengths that range from 350 to 700 MPa in compression, stiffness values that range from 40 to 70 GPa, a density of 2 to 2.9 Mg/m^3, that can also dissipate ≈ 30% of the mechanical energy during each cycle.

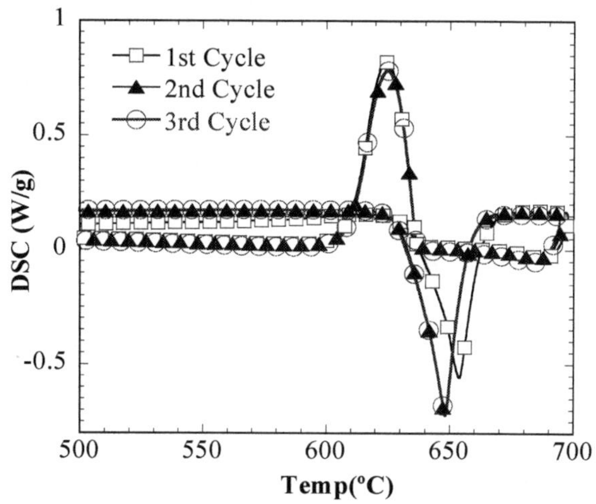

Figure 6. Three consecutive DSC cycles from 100°C to 700 °C of Mg- 20 vol. % Ti_2AlC composite.

More recently, much emphasis has been given to nano scaled solids for structural applications and while the advantages of nanostructured solids in some applications are clear, making the latter economically and on an industrial scale has been more of a challenge. The techniques used here - either simple spontaneous melt infiltration of Mg into porous ceramics preforms or powder metallurgy processes - are simple and inexpensive. Note that since the wetting and subsequent infiltration are spontaneous, there should be, in principle, no limits to the sizes or shapes of the samples, which in turn would allow for the production of large, net-, or near net-shape parts or components.

In addition to the simple fabrication techniques, the Mg nano grains, in contrast to other nano-grained solids, are extraordinary thermally stable which make them potentially good candidates for application at temperatures higher than ambient.

Acknowledgement

This work was supported by the Army Research Office (No. W911NF-07-1-0628).

References

1. Barsoum MW, Brodkin D, ElRaghy T. "Layered machinable ceramics for high temperature applications," *Scripta Materialia*, 36 (1997), 535-541.
2. Barsoum MW, ElRaghy T. "Synthesis and characterization of a remarkable ceramic: Ti3SiC2," *Journal of the American Ceramic Society*, 79 (1996), 1953-1956.

3. Amini S, Ni CY, Barsoum MW. "Processing, microstructural characterization and mechanical properties of a Ti2AlC/nanocrystalline Mg-matrix composite," *Composites Science and Technology*, 69 (2009), 414-420.
4. Barsoum MW, El-Raghy T. "Room-temperature ductile carbides," *Metallurgical and Materials Transactions a-Physical Metallurgy and Materials Science*, 30 (1999), 363-369.
5. Barsoum MW, El-Raghy T, Rawn CJ, Porter WD, Wang H, Payzant EA, Hubbard CR. "Thermal properties of Ti3SiC2," *Journal of Physics and Chemistry of Solids*, 60 (1999), 429-439.
6. Barsoum MW, Flemings MC, Kramer EJ, Mahajan S, Veyssiere P. "Physical properties of the MAX phases," *In: Encyclopedia of materials science and technology*, (2006).
7. Amini S, McGhie AR, Barsoum MW. "Isothermal Oxidation of Ti2SC in Air," *Journal of the Electrochemical Society*, 156 (2009), P101-P106.
8. Barsoum MW, Ho-Duc LH, Radovic M, El-Raghy T. "Long time oxidation study of Ti3SiC2,Ti3SiC2/SiC, and Ti3SiC2/TiC composites in air," *Journal of the Electrochemical Society*, 150 (2003), B166-B175.
9. Barsoum MW, ElRaghy T, Ogbuji LUJT. "Oxidation of Ti3SiC2 in air," *Journal of the Electrochemical Society*, 144 (1997), 2508-2516.
10. Low IM, Lee SK, Lawn BR, Barsoum MW. "Contact damage accumulation in Ti3SiC2," *Journal of the American Ceramic Society*, 81 (1998), 225-228.
11. Barsoum MW, Yoo HI, Polushina IK, Rud VY, Rud YV, El-Raghy T. "Electrical conductivity, thermopower, and hall effect of Ti_3AlC_2, Ti_4AlN_3, and Ti_3SiC_2," *Physical Review B*, 62 (2000), 10194-10198.
12. Chen D, Shirato K, Barsoum MW, El-Raghy T, Ritchie RO. "Cyclic fatigue-crack growth and fracture properties in Ti3SiC2 ceramics at elevated temperatures," *Journal of the American Ceramic Society*, 84 (2001), 2914-2920.
13. Gilbert CJ, Bloyer DR, Barsoum MW, El-Raghy T, Tomsia AP, Ritchie RO. "Fatigue-crack growth and fracture properties of coarse and fine-grained Ti3SiC2," *Scripta Materialia*, 42 (2000), 761-767.
14. Zhang H, Wang ZG, Zang QS, Zhang ZF, Sun ZM. "Cyclic fatigue crack propagation behavior of Ti3SiC2 synthesized by pulse discharge sintering (PDS) technique," *Scripta Materialia*, 49 (2003), 87-92.
15. Barsoum MW, Zhen T, Kalidindi SR, Radovic M, Murugaiah A. "Fully reversible, dislocation-based compressive deformation of Ti3SiC2 to 1GPa," *Nature Materials*, 2 (2003), 107-111.
16. Zhen T, Barsoum MW, Kalidindi SR. "Effects of temperature, strain rate and grain size on the compressive properties of Ti3SiC2," *Acta Materialia*, 53 (2005), 4163-4171.
17. Barsoum MW, Zhen T, Zhou A, Basu S, Kalidindi SR. "Microscale modeling of kinking nonlinear elastic solids," *Physical Review B*, 71 (2005), 134101.
18. Barsoum MW, Murugaiah A, Kalidindi SR, Zhen T. "Kinking nonlinear elastic solids, nanoindentations, and geology," *Physical Review Letters*, 92 (2004), -.
19. Frank FC, Stroh AN. "On the Theory of Kinking," *Proceedings of the Physical Society*, **65** (1952), 811-821
20. Zhou AG, Barsoum MW. "Kinking Nonlinear Elasticity and the Deformation of Magnesium," *Metallurgical and Materials Transactions a-Physical Metallurgy and Materials Science*, 40A (2009), 1741-1756.
21. Basu S, Barsoum MW, Kalidindi SR. "Sapphire: A kinking nonlinear elastic solid," *Journal of Applied Physics*, 99 (2006).
22. Basu S, Barsoum MW. "Deformation micromechanisms of ZnO single crystals as determined from spherical nanoindentation stress-strain curves," *Journal of Materials Research*, 22 (2007), 2470-2477.
23. Zhou AG, Brown D, Vogel S, Yeheskel O, Barsoum MW. "On the kinking nonlinear elastic deformation of cobalt," *Materials Science and Engineering A*, 527 (2010), 4664-4673.
24. Amini S, Zhou A, Gupta S, DeVillier A, Finkel P, Barsoum MW. "Synthesis and elastic and mechanical properties of Cr2GeC," *Journal of Materials Research*, 23 (2008), 2157-2165.
25. Reed-Hill RE, Dahlberg EP, Slippy WA, Jr., "Some anelastic effects in zirconium at room temperature resulting from prestrain at 77K," *Transactions of the Metallurgical Society of AIME*, 233 (1965), 1766-1771.
26. Zhou AG, Basu S, Barsoum MW. "Kinking Nonlinear Elasticity, Damping and Microyielding of Hexagonal Closed-Packed Metals," *Acta Materialia*, 59 (2008), 60-67.
27. Barsoum MW, Murugaiah A, Kalidindi SR, Zhen T, Gogotsi Y. "Kink bands, nonlinear elasticity and nanoindentations in graphite," *Carbon*, 42 (2004), 1435-1445.
28. Zhou AG, Barsoum MW, Basu S, Kalidindi SR, El-Raghy T. "Incipient and regular kink bands in fully dense and 10 vol.% porous Ti2AlC," *Acta Materialia*, 54 (2006), 1631-1639.
29. Basu S, Zhou AG, Barsoum MW. "Reversible dislocation motion under contact loading in LiNbO3 single crystal," *Journal of Materials Research*, 23 (2008), 1334-1338.
30. Anasori B, Sickafus KE, Usov IO, Barsoum MW. "Spherical Nanoindentation Study, and Effects of Ion Irradiation on the Deformation Micromechanisms of LiTaO3 Single Crystals," *Sub. for pub.*, (2010).
31. Basu S, Zhou A, Barsoum MW. "On spherical nanoindentations, kinking nonlinear elasticity of mica single crystals and their geological implications," *Journal of Structural Geology*, 31 (2009), 791-801.
32. Amini S, Barsoum MW. "On the effect of texture on the mechanical and damping properties of nanocrystalline Mg-matrix composites reinforced with MAX phases," *Materials Science and Engineering a-Structural Materials Properties Microstructure and Processing*, 527 (2010), 3707-3718.
33. Kontsos A, Hazeli K, Anasori B, Loutas T, Sotiriadis G, Kostopoulos V, Barsoum MW. Grain Size Effect on the Fatigue Response of Nanocrystalline Magnesium Composites Reinforced with MAX Phases. 9th HSTAM International Congress on Mechanics. Limassol, Cyprus, (2010).
34. Scherrer P. "Bestimmung der Größe und der inneren Struktur von Kolloidteilchen mittels Röntgenstrahlen," *Nachrichten von der Gesellschaft der Wissenschaften zu Göttingen*, 2 (1918), 98-100.
35. Saravanan RA, Surappa MK. "Fabrication and characterisation of pure magnesium-30 vol.% SiCP particle composite," *Materials Science and Engineering a-Structural Materials Properties Microstructure and Processing*, 276 (2000), 108-116.
36. Barsoum MW, Radovic M, Zhen T, Finkel P, Kalidindi SR. "Dynamic elastic hysteretic solids and dislocations," *Physical Review Letters*, 94 (2005), -.
37. Ashby MF. "On the Engineering Properties of Materials," *Acta Metallurgica*, 37 (1989), 1273-1293.
38. Amini S, Cordoba JM, Daemen L, McGhie AR, Ni CY, Hultman L, Oden M, Barsoum MW. "On the Stability of Mg Nanograins to Coarsening after Repeated Melting," *Nano Letters*, 9 (2009), 3082-3086.

Magnesium Technology 2011
Edited by: Wim H. Sillekens, Sean R. Agnew, Neale R. Neelameggham, and Suveen N. Mathaudhu
TMS (The Minerals, Metals & Materials Society), 2011

Effect of Fiber Reinforcement on Corrosion Resistance of Mg AM60 Alloy-based Composites in NaCl Solutions

Qiang Zhang, Henry Hu
Mechanical, Automotive & Materials Engineering
University of Windsor, 401 Sunset Ave.
Windsor, Ontario, Canada N9B 3P4
E-mail: qiangz@uwindsor.ca, huh@uwindsor.ca

ABSTRACT

There is great interest in developing low-cost, magnesium-based MMCs because of their high stiffness-to-weight ratio for aerospace and automotive applications. However, corrosion resistance of metal-matrix composites is often a concern for components to be used in harsh environment. In this study, the corrosion behavior of Al_2O_3 fibres reinforced magnesium AM60 composite, in aqueous solutions containing various concentrations of NaCl, was studied in comparison to that of matrix AM60 alloy by potentiodynamic polarization measurements. The microstructure of the composite and matrix alloy AM60 before and after corrosion testing was analyzed by optical microscopy, and scanning electron microscopy (SEM). The results show that the presence of Al_2O_3 fiber deteriorates the corrosion resistance of the matrix magnesium alloy AM60. The effect of Al_2O_3 fiber reinforcement and NaCl concentrations on the corrosion behavior of the composites are discussed. The corrosion mechanisms of the composite are proposed in light of metallographic observation on the formation of corrosion products.

INTRODUCTION

The need for high-performance and lightweight materials in automobile and aerospace industries has led to extensive research and development efforts generating metal matrix composites (MMCs) and cost-effective fabrication technologies. Magnesium-based composite materials allow constituent variation such that the properties of the materials can be tailored. Compared to the unreinforced monolithic metal, magnesium-based composites have been recognized to possess superior mechanical properties, such as high elastic modulus and strengths as well as enhanced wear resistance [1].

Extensive studies on mechanical properties and solidification behaviors of light metals and their composites have been carried out. Light metal-based composites are often fabricated by various manufacturing processes such as high pressure die casting, stir mixing, pressure infiltration and squeeze casting. Among the available techniques for manufacturing magnesium-based composites, preform-squeeze casting technique is considered to be an effective process. This is because it offers low cost, high efficiency, and uniform distribution of reinforcements to the fabrication of the composites, in which different volumes, especially relatively low volumes, of reinforcements are required [1-5]. Since magnesium alloys possess a very active behavior in moist environment, addition of an appropriate reinforcement into metal matrix which improves physical and mechanical properties often deteriorates the corrosion resistance of composites.

However, relatively scarce investigation about corrosion behavior of Al_2O_3 reinforced magnesium based composites, in particular under aggressive environments, is performed. In this study, the influence of Al_2O_3 fiber on the corrosion behavior of Al_2O_3/AM60 composites, obtained by preform-squeeze casting, was evaluated in aqueous solutions containing various concentrations of NaCl by electrochemical experimentation and microstructure analysis.

EXPERIMENTAL PROCEDURE

Composites preparation

Al_2O_3 short fibres were employed as the raw materials for preparation of reinforcements since they are relatively inexpensive and possess adequate properties. The matrix alloy was magnesium alloy AM60 with the chemistry composition (wt%) of 6.0Al-0.22Zn-0.4Mn-0.1Si-0.01Cu-0.004Fe-0.002Ni-Mg due to its wide usage in the automotive industry. The thermophysical properties of the Al_2O_3 fibre as well as the matrix alloy (AM60) are shown in Table 1 [7-9]. Composite specimens with 9%vol. Al_2O_3 fibres were prepared by a fibre preform casting process. The detailed information about the specific process can be found in references [10-12]. The unreinforced AM60 was also cast at the same condition. The rectangular samples were taken from the casting allowing sufficient surface area as required for the testing procedure.

Table 1 Thermoplysical properties of property of ceramic Al_2O_3 fibre [7-9]

Material	Al_2O_3 fibre	AM60
Elasticity modulus / GPa	200	35-44
Density /kg/m³	3400	1740
Diameter / μ m (average)	4.0	
Heat expansion coefficient /10^{-6}/K	6	45
Specific heat / J/(kg K)	1000	1250
Thermal conductivity / W/(m K)	5	85

Electrochemical experimentation

Electrochemical studies were carried out by using EC-LAB SP-150 electrochemical apparatus with corrosion analysis EC-lab software. A three-electrode cell was set up through assigning the samples as working electrode, Ag/AgCl/sat'd KCl electrode as a reference electrode and a Pt metal electrode as counter-electrode. At the beginning of experiments, samples were held in the salt solution allowing the open circuit potential to settle to a constant value. Potentiodynamic polarization scans were conducted at a rate of 10mv/s from -0.5v versus open circuit potential in a more noble direction up to 0.5v versus the reference electrode. Machined samples were ground with silicon carbide papers with various grades from 280 to 2500 grit, and then cleaned in acetone, rinsed with deionized water and dried prior to potentiodynamic polarization.

Microstructural Analysis

Specimens were sectioned, mounted, and polished from the centre of cast cylindrical specimens with a diameter of 100mm and a height of 20mm and prepared following the standard metallographic procedures. A Buehler (Lake Bluff, IL) optical image analyzer 2002 system was used to determine the primary characteristics of the specimens. The detailed features of the microstructure were also characterized at high magnification using a JSM-5800LV (Tokyo, Japan) scanning electron microscope (SEM) with a maximum resolution of 100 nm in a backscattered model/l um in X-ray diffraction mapping mode, and maximum useful magnification of 30,000. To maximize the composition reading of the energy dispersive spectroscopy (EDS) data an etchant was applied to the polished specimens for microscopic examination.

RESULTS AND DISCUSSION

Microstructure

The cast microstructure of unreinforced AM60 alloy and 9 vol.% Al_2O_3 fibre-reinforced AM60 composite is depicted in Figures 1 and 2. From Figure 1, the microstructure analysis of the unreinforced AM60 indicates the presence of eutectic along the grain boundaries, within which the β-$Mg_{17}Al_{12}$ is present. The precipitates are hard and brittle and have certain contribution to the hardness values. Figure 2 illustrates that the short fibres are distributed in a random and isotropic orientation in the matrix alloy without agglomeration and caves.

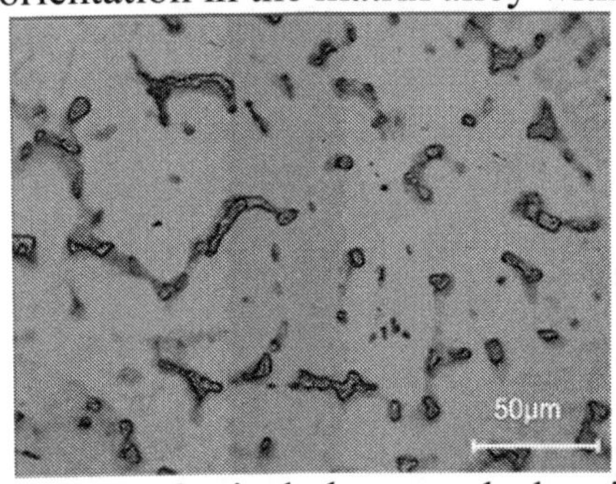

Figure 1 Optical photograph showing the microstructure of matrix AM60 alloy.

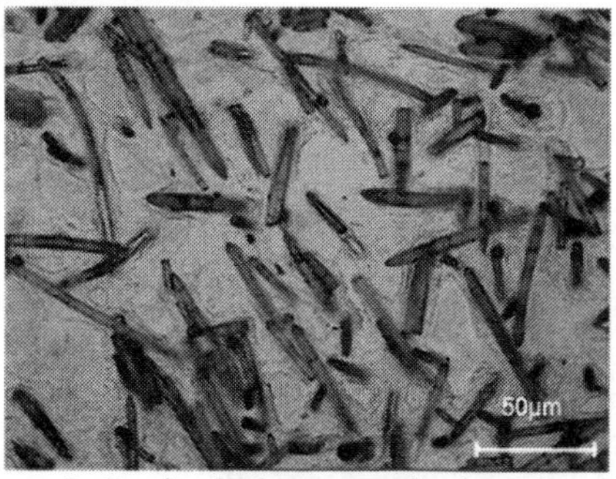

Figure 2 Optical photograph showing the microstructure of composite (9 vol% Fibres/ AM60).

Despite of the implication of a finer grain structure in the composite from Figures 1 and 2 resulting from the addition of fibres, it is difficult to distinguishthe difference in grain sizes between unreinforced AM60 alloy and Al_2O_3 fibre-reinforced AM60 composite. Thus, a common practice in the magnesium industry is to subject the as-cast specimens to a solution heat treatment (T4), which dissolves the β-$Mg_{17}Al_{12}$ phase and reveals the grain boundaries. Figure 3 distinctly reveals the grain boundaries of the unreinforced AM60 alloy and Al_2O_3 fibre-reinforced composites in the T4 condition. As can be seen, the grains in the matrix of the Fibre/AM60 composites are significantly refined compared with those of the AM60 alloy. Figure 4 presents the grain size measurements for both the unreinforced AM60 alloy and its composites. It is worth noting that the grain size of the AM60 alloy matrix is reduced to around 50μm in the composites due to the grain refinement effect of the fibres.

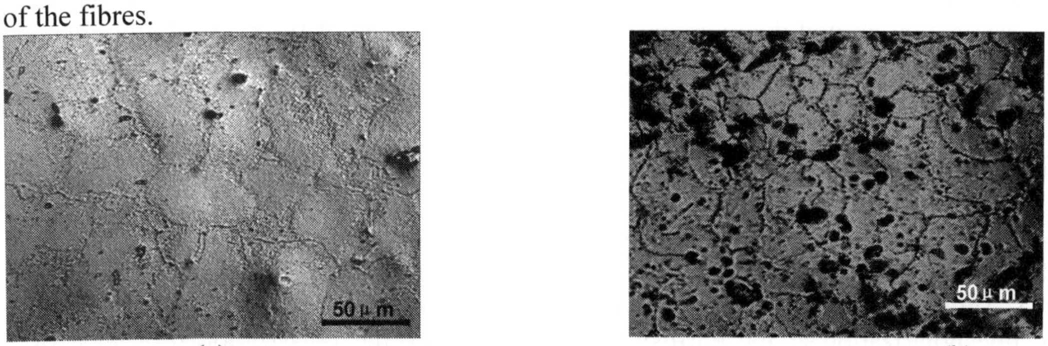

(a) (b)

Figure 3 Optical micrograph showing grain structure of (a) AM60 alloy and (b) Fibre/AM60 in T4 condition

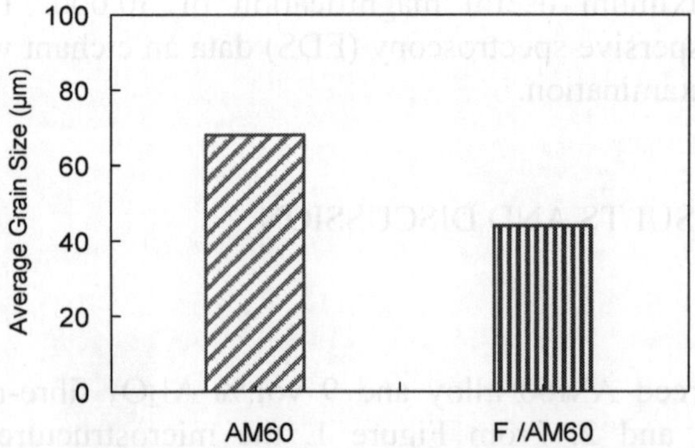

Figure 4 Measured grain size of the matrix alloy and F/AM60.

It is evidently observed from Figure 5 that the eutectic phases attach to the surface of the reinforcement. This observation indicates that the eutectic phases are able to wet Al_2O_3 fibres. As a result, they can heterogeneously nucleate on the substrate of Al_2O_3 fibres, and grows on the reinforcement. The heterogeneous nucleation of the eutectic phase on reinforcement may be attributed to the relationship of coherent interface between the eutectic phase and the reinforcement. .

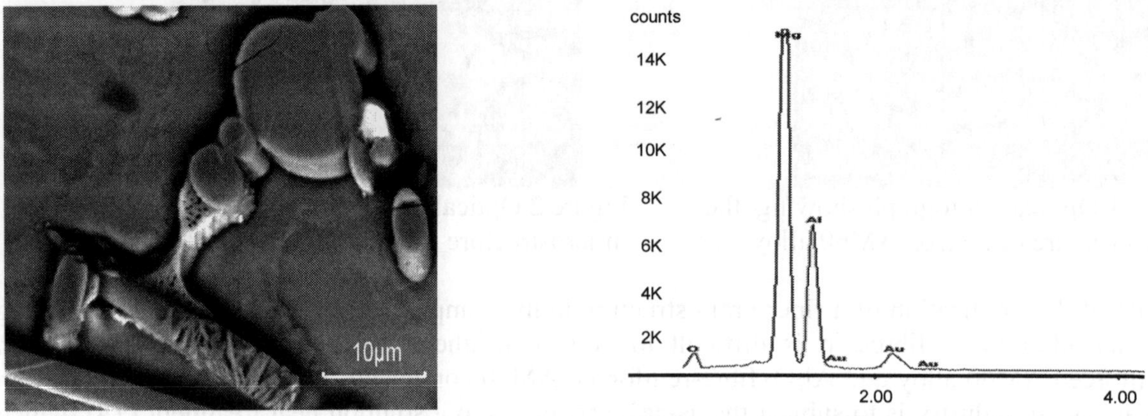

(a) (b)

Figure 5 Fibre serve as nucleation of eutectic phase, (a) Arrow A-fibre is heterogeneous nucleation substrate and (b) EDS result of Al-Mg eutectic phases in Fibre/AM60 composites

Comparing the microstructure of the composite with that of the alloy, it is obvious that, the role of alumina fibers in the microstructure development of the composite is to refine grain structure in the magnesium alloy matrix. Due to the presence of the eutectic phases adjacent to the reinforcement, on the other hand, Al content around the alumina fiber is higher than that of the original alloy matrix. This is because the bulk analysis shows the eutectic phase contains 30-36 wt. % Al compared with 6 wt.% Al in average in the matrix magnesium alloy.

Potentiodynamic Polarization

Potentiodynamic testing results yields the polarization curves shown in Figures 6 and 7 for the 1wt% and 3.5wt% NaCl solutions, respectively. The corrosion potentials, corrosion current density, and anodic/cathodic Tafel slopes (β_A and β_C) were derived from the test data. Based on the approximately line polarization at the corrosion potential (E_{corr}), the polarization resistance (R_p) values were determined from the relationship [13, 14]:

$$R = \frac{\beta_A \beta_C}{2.303\, i_{corr}(\beta_A + \beta_C)}$$

where i_{corr} is the corrosion current density. A summary of the results of the potentiodynamic corrosion tests is given Table 2.

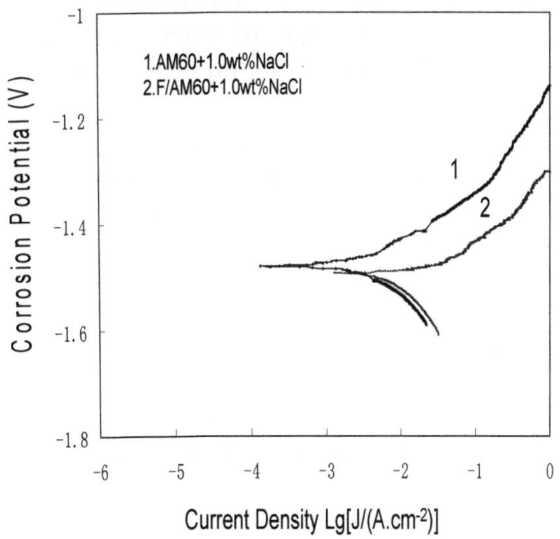

Figure 6 Potentiodynamic polarization curves for of the Al_2O_3 fiber/AM60 composites and its pure matrix alloy in 1.0wt% NaCl solution.

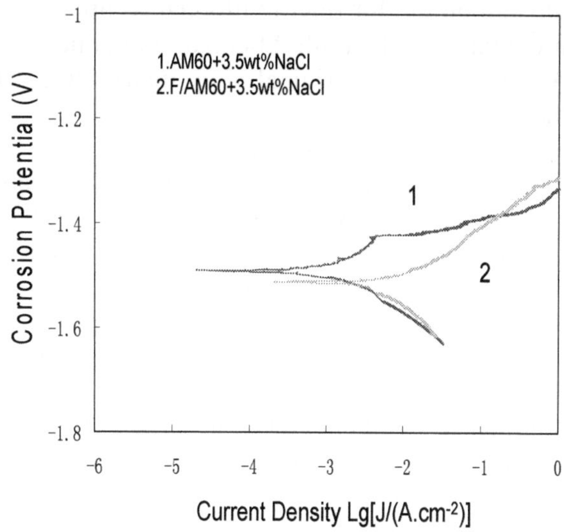

Figure 7 Potentiodynamic polarization curves for of the Al_2O_3 fiber/AM60 composites and its pure matrix alloy in 3.5wt% NaCl solution.

Table 2 Summary of polarization curve characteristics and calculated polarization resistance

	Materials type	$I_{pitting}$(μA/cm^2)	$E_{pitting}$(V)	I_{corr}(μA/cm^2)	E_{corr}(V)	R(kΩ.cm^2)
1%	AM60	14.9	-1.413	4.888	-1.4778	5.06
NaCl	Al_2O_3 fiber/AM60	N/A	N/A	7.028	-1.4959	1.28
3.5%	AM60	4.44	-1.423	1.038	-1.4912	12.9
NaCl	Al_2O_3 fiber/AM60	N/A	N/A	6.42	-1.5137	2.82

The main differences in corrosion behavior between the compositeand matrix alloy are illustrated in Figures 6 and 7. The composite as compared with matrix alloy, shifts the polarization curves to higher current densities, and consequently become more active than matrix alloy and more noticeable in the stronger salt solution. The polarization resistance (R_p) of composite is decreased by 5 times (Table 2) than that of matrix alloy. In both solutions, R_p is decreased by the addition of Al_2O_3 fibers and the values of composites are much lower than those of matrix alloy. In general, the parameter (E_{pit}-E_{corr}) is used to measure the extent of the passive region on the polarization curves. Given in Figures 6 and 7, all curves exhibit a different regime for the matrix alloy and composites. The results indicate the onset of pitting is not visible for the composites, which means that E_{pit} is very close to E_{corr}, compared with a well-defined pitting potential for matrix alloy. It is known that galvanic corrosion is the primary prospect when active magnesium is coupled with a relatively noble material [13]. But, no galvanic interaction between the alloy matrix and reinforcing fiber could take place since the Al_2O_3 fiber is an insulator. Hypothetically, it implies that the addition of Al_2O_3 fiber to the AM60 alloy should increase the corrosion resistance of the composite. However, the introduction of fiber reinforcements into the matrix alloy indeed generates new interfaces between the Mg matrix and Al_2O_3 fiber . The presence of Mg matrix/ Al_2O_3 fiber interfaces breaking the continuity of the Mg matrix and creating preferential locations for corrosion attack, and consequently decreases corrosion resistance of the composite. This type of corrosion was observed in SiC_p/ZC71 composites by Nunez-lopez [15].

CONCLUSIONS

- The addition of alumina fibers leads to a significant refinement in the grain size, which are 50 μm and 68 μm for fiber reinforced composites and matrix magnesium alloy, respectively.
- it containing The high Al-containing zones is present adjacent to the alumina fiber compared to Al content in the original alloy matrix since fibers serves as heterogeneous nucleation sites for the eutectic phase.
- Electrochemical testing results show that the presence of alumina fibers deteriorates the corrosion resistance of magnesium. The magnesium matrix alloy exhibits a higher corrosion resistance value (5.06 kΩ.cm^2 and 12.9kΩ.cm^2 in 1.0wt% and 3.5wt% NaCl solution, respectively) than the fiber reinforced composites (1.28 kΩ.cm^2 and 2.82kΩ.cm^2 in 1.0wt% and 3.5wt% NaCl solution, respectively).

ACKNOWLEDGEMENTS

The authors would like to take this opportunity to thank the Natural Sciences and Engineering Research Council of Canada and University of Windsor for financially supporting this work.

REFERENCES

[1] H. Ye, and X. Liu, Review of recent studies in magnesium matrix composites, Journal of materials science, (39), 2004, 6153-6171.

[2] J. Yao, W. Wang, and L. Fang, Morphology of the wear surface and sub-surface of net work ceramic reinforced aluminum matrix composites, Special Casting& Nonferrous alloy, (4), 2001, 5-9.

[3] M. Zhou, H. Hu, N. Li and J. Jo, Microstructure and tensile properties of squeeze cast magnesium alloy AM50, Journal of materials engineering and preformance, (14), 2005, 539-545.

[4] J. Lo, and R. Santos, Magnesium matrix composites for elevated temperature applications, 2007 SAE World Congress, SAE, Detroit, MI, 2007-01-1028.

[5] Q. Zhang, and H. Hu, Processing and Characterization of Al-based Hybrid Composites, EPD 2009, TMS, 2009, 769-775.

[6] H.Umehara, M.Takaya, Corrosion Resistance of Die Casting AZ91D Magnesium Alloys in the Atmosphere, Magnesium Alloys and Their Applications, Wiley-Vch, Germany, 2000, 506-513.

[7] D.R. Poirier, G.H.Geiger, Transport phenomena in Materials processing, TMS, USA, 1994.

[8] Magnesium and Magnesium Alloys, ASM Specialty Handbook, the Materials Information Society, USA, 1999.

[9] J. A. Dicarlo, High temperature structural fibers- status and needs, NASA .Technical memorandum 105174, May 1991.

[10] Qiang Zhang, Master thesis, University of Windsor, Ontario, Canada, 2009.

[11] Q. Zhang, and H. Hu, Development of Hybrid Magnesium-based Composites, Magnesium for Automotive Applications, 2010 SAE World Congress &Exposition, 2010, Detroit, USA.

[12] Q. Jing, Solidification microstructures in a short fiber reinforced alloy composite containing different fiber fractions, China Foundry. Vol.3, No.1, Feb 2006.

[13] M.G.Fontana, Corrosion Engineering, 3th ed., McGraw-Hill International Edition, 1996, 42.

[14] M.S.Phadke, Quality Engineering of Robust Design, AT&T Bell Laboratory, NY, USA, 1989.

[15] C.A.Nunez-lopez, The Corrosion Behaviour of Mg Alloy ZC71/SiCp Metal Matrix Composites, Corrosion Science, Vol.37, No.5, 1995, 689-708.

Magnesium Technology 2011
Edited by: Wim H. Sillekens, Sean R. Agnew, Neale R. Neelameggham, and Suveen N. Mathaudhu
TMS (The Minerals, Metals & Materials Society), 2011

THE PRODUCTION OF POWDER METALLURGY HOT EXTRUDED Mg-Al-Mn-Ca ALLOY WITH HIGH STRENGTH AND LIMITED ANISOTROPY

Ayman Elsayed[1], Junko Umeda[2], Katsuyoshi Kondoh[2]
[1]Graduate School of Engineering, Osaka University, 2-1 Yamadaoka, Suita, Osaka 565-0871, Japan
[2]Joining and Welding Research Institute, Osaka University, 11-1 Mihogaoka, Ibaraki, Osaka 567-0047, Japan

Keywords: Powder metallurgy, Hot extrusion, EBSD, Anisotropy.

Abstract

Rapidly solidified Mg-Al-Mn-Ca alloy produced by Spinning Water Atomization Process (SWAP) was hot extruded into rectangular bars, from which tensile and compression samples have been cut at 0°, 45° and 90° from the extrusion direction to study their anisotropy. Electron Back Scattered Diffraction (EBSD) has been used to investigate the texture evolution during the hot extrusion process. Both the Schmid factor and the intensity of the basal plane in the pole figure have been evaluated and correlated to the mechanical properties. Results have shown that the extruded rods exhibit high strength and limited anisotropy compared to many previously reported values for magnesium alloys. The reasons for that limited anisotropy were both the fine grained microstructure of the extruded material and the transverse component of the texture evolution.

Introduction

Concerns about power saving and environmental issues encouraged efforts to apply light weight materials, especially in automobiles. Magnesium is one of the most promising materials which show high strength to weight ratio. Wrought magnesium alloys are more likely to get increased markets due to their improved performance compared to cast alloys. However, applications of magnesium alloys are still limited due to their lack of formability at room temperature. This disadvantage arises from the features of the close packed hexagonal crystal structure which limit the deformation mechanisms at room temperature to basal slip [1]. Improvement of the mechanical properties of magnesium alloys could be obtained via various techniques, including, but not limited to, thermo-mechanical processing and chemical alloying [2, 3]. The Hall-Petch relationship, describing the trend of increasing the yield strength with decreasing of the grain size, was shown to prevail. The strengthening factor of Mg alloys in this relationship has shown higher values (0.2 to 0.34 MPa $\sqrt{m}$) compared to that of other materials [4]. This shows the strong effect of grain refinement on the properties of Mg alloys. The present authors have previously shown the considerable impact of rapid solidification powder metallurgy processing to improve the strength of Mg-Al-Mn-Ca alloy through the effects of grain refinement and the dispersion of tiny inter-metallic compound particles [5].

The anisotropy of the mechanical response of wrought Mg alloys caused by the strong texture has been shown as a barrier in the extension of their applications. Hence, the texture analysis became a very important tool for the evaluation of the mechanical response of Mg alloys [6]. The texture evolution during thermo-mechanical processing of Mg alloys has been investigated aiming at understanding their effects on properties of extruded bars [7-8]. It has been shown that the extrusion temperature has no effect on the texture formation [7]. Contrarily, it was also reported that it affects the amount of dynamic recrystallization [8]. However, most references agreed that the fiber texture was usually obtained after hot extrusion of Mg alloys in which the basal plane is aligned parallel to the extrusion direction. Texture evolution during hot deformation, like compression, has also been studied [9, 10], and the effects of both strain rate and processing temperature on the texture were shown through the Zener-Hollomon parameter.

Trials have been made to modify the texture formation in Mg alloys, and ultimately to improve their mechanical response, through the use of alloying elements, especially rare earth elements [11-14]. They have shown a remarkable modification of the texture of wrought Mg alloys owing to the reduced texture strength attained by the formation of randomized crystal orientation. That effect was shown to be mainly obtained through extensive grain refinement, solute atoms interactions, and nucleation of grains at shear bands rather than inside original grains [11-13]. In some other cases particle-stimulated nucleation of recrystallization has been reported to have a minor effect [14]. Alloying with Ca has been shown to reduce the texture strength due to the randomized nucleation of recrystallized grains near intermetallic compound particles [15]. Equal channel angular extrusion has also been reported to improve the ductility of Mg alloys in the longitudinal direction due to the inclined texture formation, but on the expense of reduced strength [16]. The effect of the texture on the mechanical properties of wrought Mg alloys has also been investigated and found to be mainly through favoring the deformation mechanism, namely slipping or twinning, based on the available texture pattern [17, 18]. The tensile and compression properties have shown to be of totally different nature for the same alloy due to the different possibilities of twinning activation [19, 20]. The grain size also has played an important role in determining whether the twinning mechanism becomes active or not, which if activated results in remarkable softening.

The aim of this work is to study the anisotropic behavior of hot extruded powder metallurgy Mg-Al-Mn-Ca alloy under both tension and compression loadings. That alloy has previously been shown to have good tensile properties due to the remarkable advantages of rapid solidification powder metallurgy processing of both grain and inter-metallic compound refinement and homogenous distribution [5].

Experimental work

Mg-6wt.%Al-0.26wt.%Mn-2.1wt.%Ca magnesium alloy powder produced using the spinning water atomization process (SWAP) was used. In SWAP process, gas atomization is combined with water atomization to produce very high cooling rates of about 10^6 K/s, which result in powders with fine microstructures. The details of the extensive study on the characteristics of the alloy SWAP powder can be found in the authors' previous reference [5]. For comparison, extruded cast material was also used to confirm the effect of rapid solidification of SWAP powder on the final properties of the extruded alloy. The cast alloy was produced using casting in a steel mold using the melting temperature of 1023 K. Microstructure analyses of the extruded bars were performed using Scanning Electron Microscope (SEM, JEOL: JSM-6500F) because of the fine grain size. X-ray Diffraction (XRD) analysis was used to investigate the phases existing in the alloy using an X-ray Diffractometer (Shimadzu: XRD 6100) over the range of 2θ of 20 to 80°.

As-received SWAP powders were consolidated using cold compaction at room temperature under the pressure of 600 MPa. Both consolidated powder billets and cast billets were then extruded at the temperature of 673 K. Extrusion was performed using a die that produces extrusion rods of 25 x 40 mm cross-section, which is equivalent to an extrusion ratio of 37. The preheating of billets was done just before extrusion using the heating rate of 1 K/s in Ar gas atmosphere. The billets were held at the extrusion temperature in the furnace for 60 min prior to extrusion to ensure homogenous temperature distribution.

The texture evolution in the extruded bars was investigated using the electron back-scattered diffraction (EBSD) analysis. The specimens for EBSD investigation were cut such that the observation plane is parallel to the extrusion direction. The specimen surface was prepared by grinding until 4000 grit emery paper, polishing using 0.25 µm diamond paste, electro-chemical etching using (37.5 vol.% H3Po4 + 62.5 vol.% ethanol) solution at 8 V for 60 s, and then by cleaning with methanol.

Both tensile and compression tests were performed on the extruded bars to evaluate their mechanical response at three different orientations, which have 0, 45 and 90° angles with the extrusion direction. Tests were performed using a universal testing machine (Shimadzu: Autograph AG-X 50 KN) at room temperature on three specimens for each material and processing condition, and the average of the three values was evaluated. The tensile specimens, having the diameter of 3 mm and the gage length of 12 mm, and compression specimens with 6 mm diameter and 15 mm length were evaluated using the strain rate of $5x10^{-4}$ /s.

Results and Discussion

The extrusion process of Mg-Al-Mn-Ca alloy in both its powder and cast forms resulted in the microstructures shown in Figure 1 observed on a plane normal to the extrusion direction. The extruded SWAP powder specimen shows a fine and homogenous microstructure with the average grain size of about 1.7 µm, as shown in Figure 1a, with grains shown by dotted ovals. The fine and homogenously dispersed inter-metallic compound particles in the structure of the extruded SWAP powder specimens, appearing as white dots shown by arrows, can also be seen. They exist both inside the grains and at the grain boundaries. It should be noted that the grain size, shown herein, is coarser than that previously reported by the authors for the same alloy SWAP powder extrusion in the case of smaller extrusion rods of 7 mm diameter (about 0.5 µm) [5]. This may be due to the grain growth that occurs in the case of bigger extruded bars because of longer heating times and slower cooling rates. In contrast, the microstructure of the extruded cast specimen, shown in Figure 1b, reveals coarser average grain size of about 7 µm with remarkable non-homogenity. However, that size of the extruded cast specimens of the Mg-Al-Mn-Ca alloy is still much finer than many other reported values of grain sizes of extruded Mg alloys [8, 12]. Bimodal microstructure with both coarse deformed and fine recrystallized grains can be observed. The inter-metallic compound particles, also shown by arrows reveal coarser size and lower dispersion compared with that of the extruded SWAP powder specimens. In both cases of extruded SWAP and cast billets, the inter-metallic compound of Al_2Ca can be detected after the dynamic recrystallization process accompanied by hot extrusion, as confirmed by the XRD result shown in Figure 2. However, the morphology of the compound particles varied significantly. In the case of the extruded SWAP powder specimens, the fine compound particles have been formed during the dynamic recrystallization process from the super saturated elements existing in the SWAP powder structure, as the very high cooling rate used in the fabrication of this powder did not permit the precipitation of these elements. On the other hand, the compound particles, originally existing in the structure of the cast billet of the investigated alloy, have either remained after extrusion or fractured during extrusion depending on their sizes and morphologies.

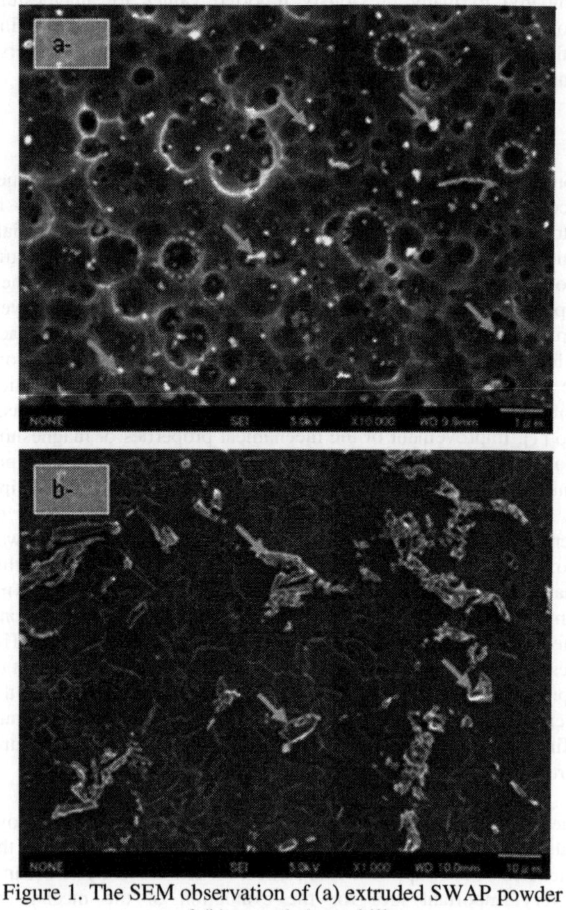

Figure 1. The SEM observation of (a) extruded SWAP powder and (b) extruded cast billet

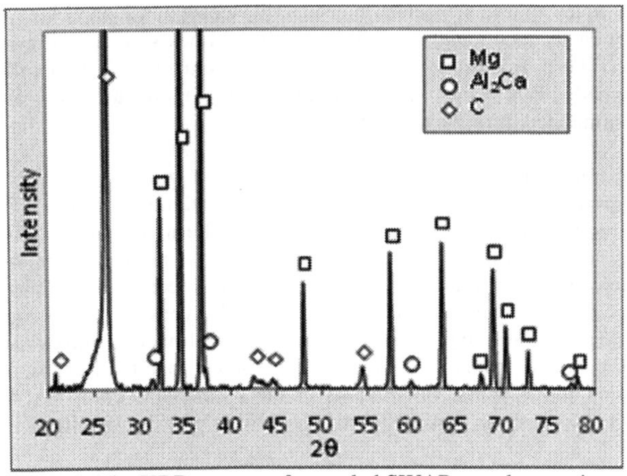

Figure 2. The XRD pattern of extruded SWAP powder specimen

The above shown microstructure of the extruded SWAP powder specimen has resulted in an improved mechanical response of this alloy compared to that of the extruded cast specimen, as shown in Figure 3, and in the authors' reference [5]. The tensile and compression test results at 0, 45 and 90° angles from the extrusion direction are shown in terms of the stress-strain plots for both extruded SWAP powder and cast billet. It can be shown that the extruded SWAP powder specimen possesses high values of yield and ultimate stresses while maintaining reasonable levels of ductility. It is evident that the strength of extruded SWAP powder is drastically increased compared to the extruded cast billet from the 50 % increase in tensile yield strength, and even 100 % increase in compressive yield strength. They are superior to previously reported values of the strength of Mg alloys via conventional routes. However, these results are lower than those presented earlier in the current authors' references for smaller extrusions of 7 mm extruded bars of the same alloy [5]. That is due to the grain growth that accompanied the larger size extrusions which is performed at longer times of homogenization, 60 min compared to 5 min for smaller size extrusions. Generally, both yield and ultimate strengths have decreased as the angle of the loading has increased from 0 to 45 and 90°. This behavior has occurred in both the tensile and compression cases. In contrast to the previously reported pattern of increasing the compression/tensile yield strengths' ratio with the angle, the investigated alloy SWAP powder has shown almost the same values of both tensile and compression yield strengths at all directions, with the ratio equals unity, as shown in Figures 3 a and b. However, the alloy has shown more strain hardening in the case of compression than that of tension. The elongation has also decreased simultaneously. However, the difference of both ultimate and yield strengths are much lower in the case of extruded SWAP powders than those for extruded cast billets. It could be shown from the figure that the extruded cast alloy shows more anisotropy than that of extruded SWAP powders in terms of wider difference in both yield strength and elongation at various directions, which is more noticeable for elongation levels, as shown in Figures 3 c and d.

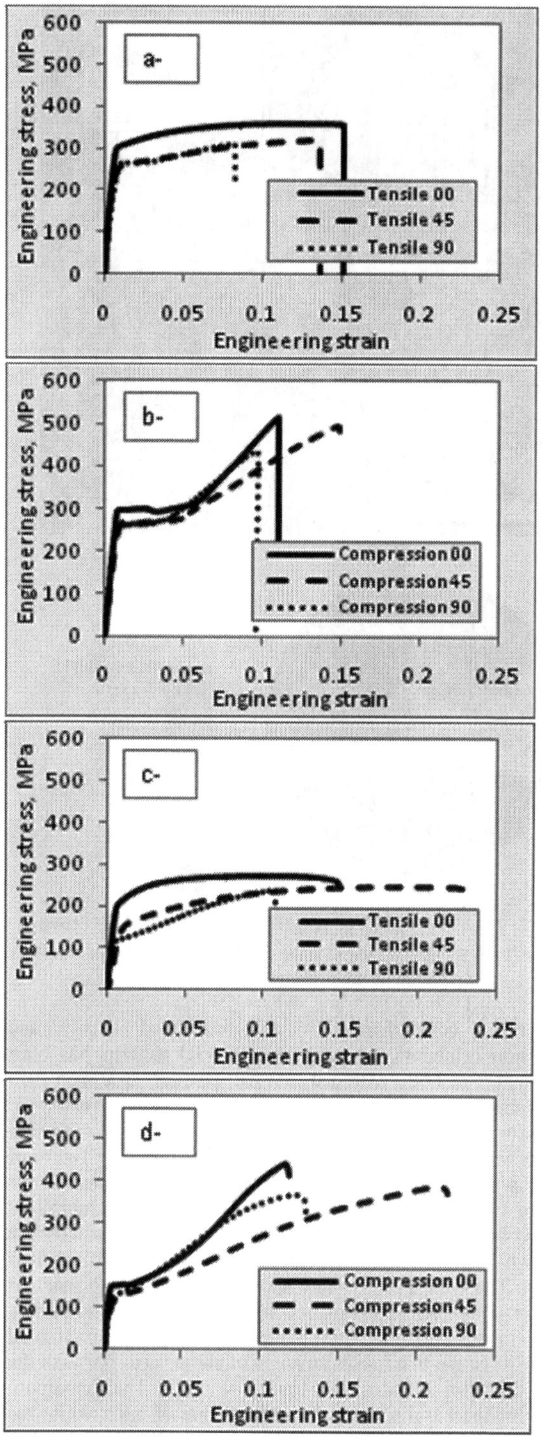

Figure 3. Stress-strain plots of tensile (a) and compression (b) of extruded SWAP powder, and tensile (c) and compression (b) of extruded cast billets

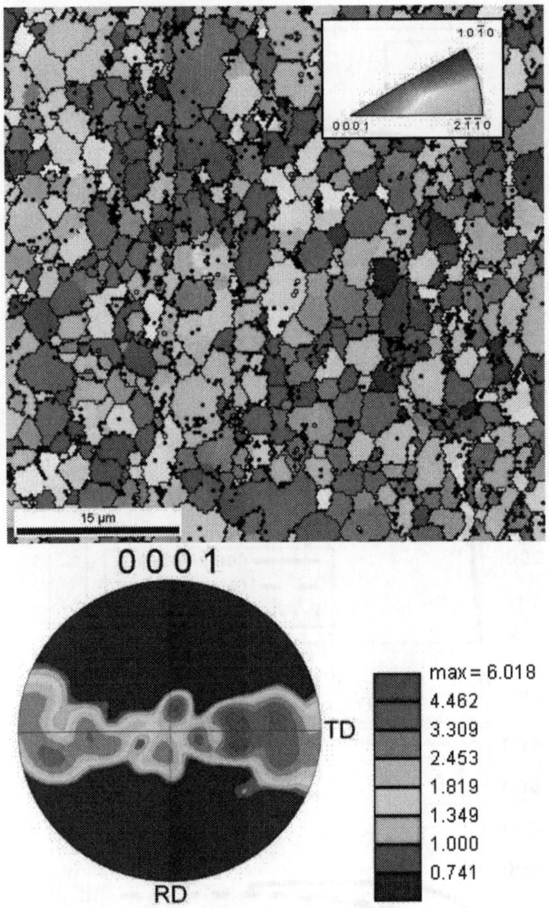

Figure 4. The EBSD analysis results of extruded SWAP powder specimen showing both the crystal orientation mapping and the basal plane orientation texture

In order to comprehend the aforementioned tensile and compression behaviors, the analysis of EBSD patterns has been investigated for both extruded SWAP powder specimen and extruded cast one, as shown in Figures 4 and 5. The inverse pole figure showing the crystal orientation mapping, with the extrusion direction vertically aligned, and the corresponding pole figure of the basal plane texture of the extruded SWAP powder specimen is shown in Figure 4. The homogenous distribution of grain sizes can easily be observed. The inter-metallic compound particles are not shown in this map, with the black dots representing their positions. The basal plane texture shows a mix of the normal and transverse components, which is consistent with the findings by other references [11]. The basal planes in most grains are aligned along the direction that makes an angle of about 60 to 80° with the normal direction towards the transverse side. The maximum intensity of the basal texture shows the value of 6.02, while the Schmid factor shows the average value of 0.18. On the other hand, the extruded cast specimen shows almost the same pattern of basal plane texture component, but with its maximum intensity increased to about 9.62, as shown in Figure 5. The Schmid factor has shown a very close value of 0.17 to that of the extruded SWAP powder specimen. However, the grain sizes have shown non-homogenous distribution, an observation that is consistent with that of SEM images. The observed area includes some examples of the fractured inter-metallic compound particles, as previously mentioned, in contrast with the ones shown in Figure 1b. These fractured particles appear as clusters of vertically aligned black dots in the inverse pole figure as they are accumulated during extrusion.

Separating both deformed and recrystallized grains in separate maps and pole figures provides deeper understanding of the effect of recrystallization during hot extrusion of the investigated alloy, as shown in Figure 6. Deformed grains show higher maximum intensity of the basal plane texture than that of recrystallized ones. However, as the volume fraction of both types of grains has similar values, the maximum intensity ultimately takes an intermediate value. It should also be noted that both types of grains show similar orientations of the basal plane compared to each other, and to that of the extruded SWAP powder specimen.

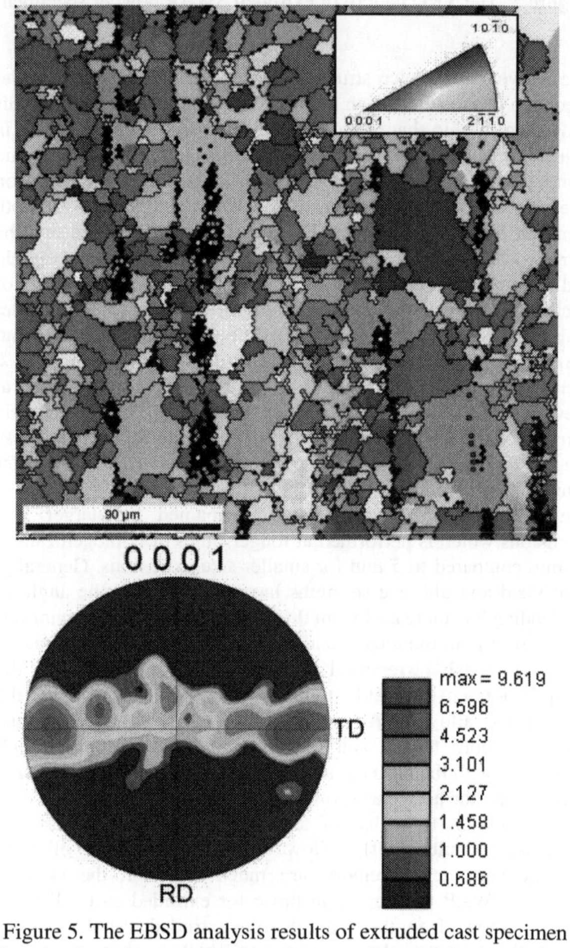

Figure 5. The EBSD analysis results of extruded cast specimen showing both the crystal orientation mapping and the basal plane orientation texture

It has been shown in previous reports that the Ca element has the effect of weakening of the texture of Mg alloys through the random nucleation of recrystallized grains due to the formation of inter-metallic compound particles [15]. This effect can also be observed in this study in terms of the decreased maximum intensity of the basal texture of the extruded SWAP powder specimen compared to that of extruded cast due to the fine distribution of intermetallic compounds, which did not occur in the case of extruded cast. Hence the decreased asymmetry of both tensile and compression results in various directions can be explained through the effect of randomized nucleation of recrystallized grains stimulated by compound particles. This claim may also be supported by the observation of the decreased tensile and compression strengths from 0 to 45°, and the similarity of both of 45 and 90° angle results, which can be correlated to orientation of the basal plane being randomized in both the normal and transverse directions. On the other hand, the tensile-compression isotropy shown in this alloy can be attributed to the remarkably refined grain sizes. It has been previously shown that the fine grain size results in the difficulty of the activation of the twinning mechanism, which if activated results in the remarkable softening during deformation in either tensile or compression directions based on the available texture [17, 18].

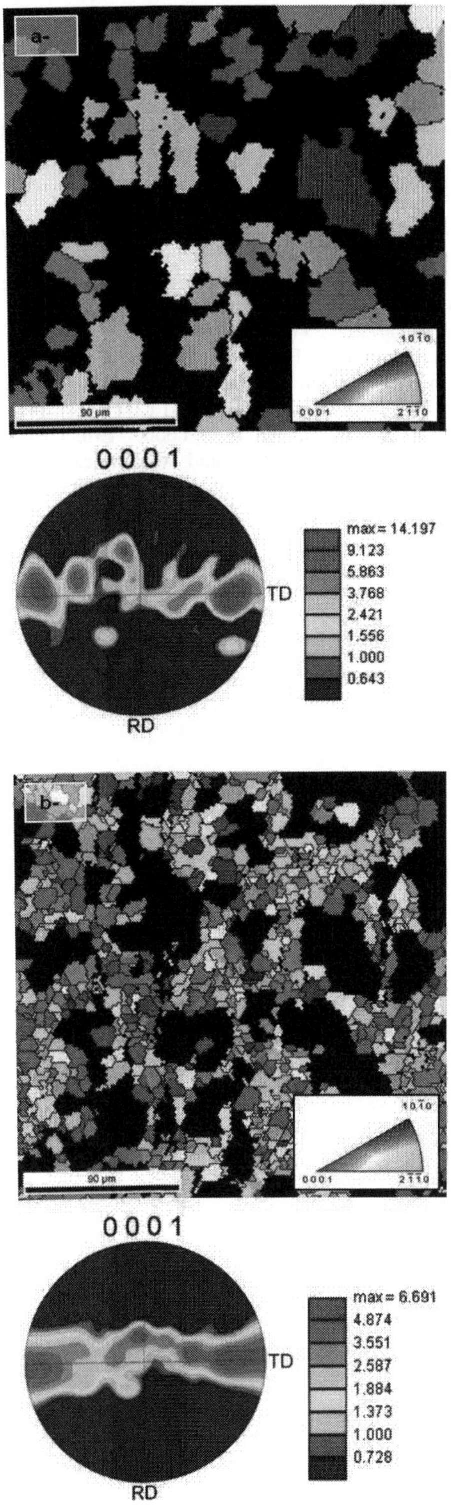

Figure 6. The EBSD results of extruded cast specimen showing (a) separated deformed grains and (b) separated recrystallized grains

Conclusion

Owing to the advantages of the powder metallurgy processing of Mg-Al-Mn-Ca alloy, the following conclusions can be drawn:

- The alloy has shown very promising tensile and compression results.
- Both the tensile and compression results have shown very low asymmetry in various loading angles with respect to the extrusion direction.
- Isotropic tension-compression behavior could be obtained at various loading directions.

References

1. Z. Yang et al., "Review on Research and Development of Magnesium Alloys," Acta Metallurgica Sinica, 21 (5) (2008), 313-328.
2. D.K. Xu et al., "Effect of Microstructure and Texture on The Mechanical Properties of As-Extruded Mg-Zn-Y-Zr Alloys," Materials Science and Engineering A, 443 (2007), 248-256.
3. Y. Yoshida et al., "Realization of High Strength and High Ductility for AZ61 Magnesium Alloy by Severe Warm Working," Science and Technology of Avvanced Materials, 6 (2005), 185-194.
4. P. Andersson, C.H. Caceres, and J. Koike, "Hall-Petch Parameters for Tension and Compression in Cast Mg," Materials Science Forum, 419-422 (2003), 123-128.
5. A. Elsayed et al., "Microstructure and Mechanical Properties of Hot Extruded Mg-Al-Mn-Ca Alloy Produced By Rapid Solidification Powder Metallurgy," Materials & Design, 31 (2010), 2444-2453.
6. W.N. Wang and J.C. Huang, "Texture Analysis in Hexagonal Materials," Materials Chemistry and Physics, 81 (2003), 11-26.
7. M. Shahzad and L. Wagner, "Influence of Extrusion Parameters on Microstructure and Texture Developments, and Their Effects on Mechanical Properties of Magnesium Alloy AZ80," Materials Science and Engineering A, 506 (2009), 141-147.

8. H. Ding et al., "Study of The Microstructure, Texture and Tensile Properties of As-Extruded AZ91 Magnesium Alloy," Journal of Alloys and Compounds, 456 (2008), 400-406.
9. E. Martin and J. Jonas, "Evolution of Microstructure and Microtexture During The Hot Deformation of Mg-3% Al," Acta Materialia, 58 (2010), 4253-4266.
10. A.S. Varzaneh, A. Hanzaki, and H. Beladi, "Dynamic Recrystallization in AZ31 Magnesium Alloy," Materials Science and Engineering A, 456 (2007), 52-57.
11. N. Stanford and M.R. Barnett, "The Origin of "Rare Earth" Texture Development in Extruded Mg-Based Alloy and Its Effect on Tensile Ductility," Materials Science and Engineering A, 496 (2008), 399-408.
12. B.L. Wu et al., "Ductility Enhancement of Extruded Magnesium Via Yttrium Addition," Materials Science and Engineering A, 527 (2010), 4334-4340.
13. N. Stanford, "Micro-Alloying Mg with Y, Ce, Gd and La For Texture Modification-A Comparative Study," Materials Science and Engineering A, 527 (2010), 2669-2677.
14. N. Stanford and M Barnett, "Effect of Composition on Texture and Deformation Behaviour of Wrought Mg Alloys," Scripta Materialia, 58 (2008), 179-182.
15. T. Laser et al., "The Influence of Calcium and Cerium Mischmetal on The Microstructural Evolution of Mg-3Al-1Zn During Extrusion and Resulting Mechanical Properties," Acta Materialia, 56 (2008), 2791-2798.
16. T. Mukai et al., "Ductility Enhancement in AZ31 Magnesium Alloy By Controlling Its Grain Structure," Scripta Materialia, 45 (2001), 89-94.
17. D.L. Yin et al., "On Tension-Compression Yield Asymmetry in An Extruded Mg-3Al-1Zn Alloy," Journal of Alloys and Compounds, 478 (2009), 789-795.
18. J. Bohlen et al., "Microstructure and Texture Development During Hydrostatic Extrusion of Magnesium Alloy AZ31," Scripta Materialia, 53 (2005), 259-264.
19. E.A. Ball and P.B. Prangnell, "Tensile-Compression Yield Asymmetries in High Strength Wrought Magnesium Alloy," Scripta Metallurgica et Materialia, 31 (2) (1994), 111-116.
20. S.S. Park, B.S. You, and D.J. Yoon, "Effect of Extrusion Conditions on The Texture and Mechanical Properties of Indirect-Extruded Mg-3Al-1Zn Alloy," Journal of Materials Processing Technology, 209 (2009), 5940-5943.

Magnesium Technology 2011
Edited by: Wim H. Sillekens, Sean R. Agnew, Neale R. Neelameggham, and Suveen N. Mathaudhu
TMS (The Minerals, Metals & Materials Society), 2011

THERMAL EFFECTS OF CALCIUM AND YTTRIUM ADDITIONS ON THE SINTERING OF MAGNESIUM POWDER

Paul Burke[1], Chloe Petit[2], Sonia Yakoubi[2] and Georges J. Kipouros[1]
[1]Dalhousie University, 1360 Barrington Street, PO Box 1000, Halifax, NS, Canada, B3J 2X4

[2]ICAM, 75 avenue de Grande-Bretagne, 31 300 Toulouse, France

Keywords: magnesium powders, magnesium powder metallurgy, sintering phenomena

Abstract

The low density and good mechanical properties make magnesium and its alloys attractive materials for use in automotive and aerospace applications. Powder metallurgy P/M can be used to alleviate the formability problem through near-net-shape processing, and also allows unique chemical compositions that can lead to new alloys with novel properties. However, the surface layer formed on the Mg powders during processing acts as a barrier to diffusion and sintering is problematic. The layer, characterized using focused ion beam milling and transmission electron microscopy (FIB-TMS), as well as x-ray photoelectron spectroscopy (XPS), contains oxides, hydroxides and carbonates of magnesium, formed by reactions with the atmosphere. To overcome this barrier, small additions were made of calcium and yttrium the oxides of which are thermodynamically more stable than magnesium oxide. The present work reports on the thermal effect of Ca and Y additions to magnesium powder during sintering, utilizing differential scanning calorimetry (DSC).

Introduction

In the past quarter of a century the increasing cost of energy and increased environmental awareness have lead to a global requirement for the reduction of automotive emissions. One strategy for the reduction of emissions is to reduce the gross weight of vehicles. As the weight is decreased, less fuel is consumed to propel the vehicle. The average vehicle in 1977 had a mass of 3665.5 lb, and in 2001 the average weight had been reduced to 3309.0 lb; an over 300 lb reduction [1]. Among the principal reasons for the weight reduction was the increased use of lightweight materials, especially aluminum. In the same 24 year period, the average amount of Al utilized per automobile climbed from 97 lb to 256.6 lb. The driving force behind the increased usage can be attributed to the intense research effort put forth to provide over 1600 aluminum alloys from multi-component systems and numerous production and fabrication improvements. More recently, aluminum products processed via powder metallurgy (P/M) routes have shown excellent properties and are increasing in utilization.

Magnesium, which has the lowest density of all structural metals, has not enjoyed the phenomenal increase in average mass per vehicle that aluminum has, but had an exceptional growth of 850%. Unfortunately, the actual amounts went from under 1 lb in 1977 to 8.5 lb in 2001. Aside for low density, magnesium has a number of other advantageous properties: Its stiffness to weight ratio is very high; one pound of magnesium is as stiff as 1.8 lb of aluminum and 2.1 lb of steel; the dimensional stability and damping capacity are high, it is easily machined and can be readily recycled [2]. For magnesium to penetrate the automotive industry as a structural material several requirements need to be met [3].

Magnesium has poor corrosion resistance due to heavy metal contamination [4] and lacks a catalogue of developed alloys [5]. The use of magnesium in under the hood components is also hindered by the low creep resistance of existing commercial alloys [2]. Also, magnesium has a hexagonal close packed (HCP) crystal structure, which leads to difficulty in forming, especially at room temperature [5].

Magnesium Powder Metallurgy

Powder metallurgy can be used to alleviate one of the largest problems with magnesium utilization, formability, with its inherent near-net-shape processing. Raw magnesium powders of a 75 μm typical size are blended with alloying elements and compacted in a die at high pressure. The "green" compacts are then sintered in a controlled atmosphere at a temperature lower than the melting point, but high enough to allow rapid diffusion between the powder particles. The method in which powders are produced allows unique chemical compositions that can lead to new alloys with novel properties [6, 7].

For magnesium P/M to compete on a level with aluminum P/M, simple, industrially relevant processing using existing P/M equipment is necessary. Because the economics of P/M require high volume, a robust process utilizing uni-axial compaction and mesh belt furnaces is needed to compete with established magnesium production methods like die casting.

The major issue in the development of the magnesium P/M is the availability of commercial magnesium powders. Unlike aluminum most commercial magnesium powders are produced by mechanical grinding. The low cost and less restrictive requirements for the main intended use of magnesium as a reactant make grinding attractive. For P/M applications the angular morphology of the powder gives good green strength because of mechanical interlocking, but the powder particles are typically covered by a thick surface layer. The layer is hypothesized to contain primarily oxides, but hydroxides, carbonates and hydrates are possible due to long exposure to atmospheric conditions during the grinding process. Recently magnesium powder was also produced commercially by centrifugal inert gas atomization by a small number of companies. The product has very little surface oxidation due to the inert conditions maintained during production and shipment, but the spherical morphology gives poor green strength. No fundamental studies of the sintering behaviour of magnesium powders have been completed to date, but preliminary studies have been reported [8, 9, 10, 11]. The goal of the present study is to determine if the surface layer is in fact the main obstacle to producing magnesium P/M parts without the addition of secondary hot working, by looking solely at solid state sintering of pure magnesium powders. A new surface characterization technique, focused ion beam/transmission electron microscopy (FIB/TEM), was applied to identify the nature of the surface layer [10] with very promising results.

A method for disrupting the surface layer is to add a small amount of a more reactive metal. During sintering, the additive reacts with the surface of the surrounding base powder, reducing the oxides of the base to pure metal and forming its own oxide. While the total amount of oxide is still the same, the cohesive layer that was the barrier to diffusion has been disrupted, providing pathways for unimpeded diffusion. This technique is used successfully in aluminum P/M [12], where a small amount of magnesium results in an increase in density and mechanical properties. In the case of magnesium P/M, calcium has shown promise as a surface layer reducing alloy addition [11]. Yttrium has also been identified as a possible reductant.

If the alloy being sintered has one or more lower melting point additions, a liquid phase forms during sintering. The liquid flows through the inter-granular space, and if the matrix is soluble in the liquid, the liquid acts as a diffusion bridge between adjacent powder particles. The presence of the liquid phase can make use of cracks or pores in the surface layer that may not have been located at a point contact. Because of this, there are many more areas for diffusion to proceed, and sintering is much more rapid. However, if the liquid is also soluble in the matrix, the liquid phase is termed transient, and will be absorbed into the matrix at some point and can no longer assist the sintering. Previous studies have shown that Ca [11] and Al [13] have the ability to form transient liquid phases during the sintering of Mg P/M, with Ca having the additional benefit of disrupting the surface layer of Mg. According to phase diagrams yttrium will also form a transient liquid during sintering, and enjoys the dual benefit of surface layer disruption. In the case of Ca and Y, transient behaviour is preferred because solubility in the Mg matrix will allow greater mobility and therefore interaction with the surface layer.

Traditionally sintering of P/M magnesium alloys has been completed in inert argon environments to provide the best protection from oxidation during the time at high temperature [8, 9, 10, 11]. However, it has been shown that in the case of aluminum P/M alloys [14], sintering in semi-inert nitrogen is beneficial because gases trapped in pores to react with Al to form nitrides, which then lowers the partial pressure of gases inside the pore. This reduced pressure allows the pore to be filled during sintering, and increases the final density. It was also found that the nitride formed on grains was more readily wet by the liquid phase allowing the liquid phase to travel more easily. If similarly beneficial nitrides are formed during the sintering of P/M Mg alloys, perhaps superior properties can be achieved by sintering in nitrogen.

The objective of this research is to improve the attractiveness of magnesium through non-traditional processing, by producing magnesium alloys via powder metallurgy utilizing the industrially dominant process of cold die compaction and controlled atmosphere sintering. The addition of calcium and yttrium to the base Mg powder has the potential to reduce oxides in the surface layer and create pathways for rapid diffusion, as well as assisting sintering with the formation of a liquid phase. Sintering under a nitrogen atmosphere may allow the formation of beneficial nitrides, resulting in improved densification. This work focuses on the initial steps of investigating the effects of Ca and Y additions, and a N_2 atmosphere by simulating sintering conditions and tracking thermal response utilizing a differential scanning calorimeter (DSC)

Experimental

Materials

Tangshan Weihao Magnesium Powders supplied the centrifugally atomized pure magnesium powder of 98.02 % active magnesium content, which had an average particle size of 38 μm. Pure calcium granules 99% were supplied by Acros Organics, and had an average particle size of 2mm. The granules where reduced to 100 μm through grinding a screening under inert conditions. Acros Organics was the manufacturer of the Y powder, which was -325 mesh size. All powders were characterized by laser Malvern size analyzer in water to determine the size distribution.

Methodology

Blending. Calcium and yttrium containing Mg alloy powders were mixed under a controlled atmosphere and sealed in an appropriate container. Blending was done in a Turbula mixer/shaker for 10 minutes.

Differential Scanning Calorimetery. Tests were completed using a SDT – Q600 DSC/TGA manufactured by TA instruments. All specimens were heated from room temperature to 600°C at a rate of 10°C/min under flowing nitrogen (100ml/min). Experiments were run with loose powder settled in the crucibles by light tapping.

Crucibles. Standard crucibles used for the DSC are alumina, and will be reduced by the highly reactive Mg, Ca and Y metals. Crucibles were gold coated by sputtering to minimize any interaction at temperatures of interest, and used for both the empty reference and sample containing crucibles.

Results and Discussion

There are several possible reactions between magnesium and atmospheric gases that may form compounds found in the surface layer covering Mg powders. Using FACTSage thermodynamic software, the plausible reactions are:

$$Mg + O_2 = MgO$$

$$Mg + CO_2 = MgCO_3$$

$$Mg + H_2O = Mg(OH)_2$$

Of those compounds formed, $MgCO_3$ and $Mg(OH)_2$have decomposition reactions that become spontaneous below the melting point of magnesium. FACTSage was utilized to estimate the decomposition temperatures:

$$Mg(OH)_2 = MgO + H_2O \qquad \sim 260°C$$

$$MgCO_3 = MgO + CO_2 \qquad \sim 450°C$$

Reduction to elemental Mg would be more beneficial, but is not possible below the melting temperature of Mg. However, the disruption to the surface layer caused by the decomposition may well expose areas of Mg underneath the layer, allowing for better diffusion between adjacent particles during sintering of P/M parts.

Pure Magnesium

The DSC trace of pure magnesium powder is shown in Figure 1. Area (A) shows a slight decrease in slope begging around 260°C, indication that an exothermic reaction is taking place. This co-insides with the FACTSage estimated decomposition of $Mg(OH)_2$. Area (B) shows a larger exothermic peak at ~450°C, the estimated decomposition temperature of $MgCO_3$. Although **area (C) is beyond the temperature of interest, it is prudent to note** that the large endothermic peak starting at ~540°C is likely the formation of Mg-Au compounds from the sample reacting with the gold crucible coating, and the inflection at 575°C corresponds to the eutectic liquid formation of Mg-Au.

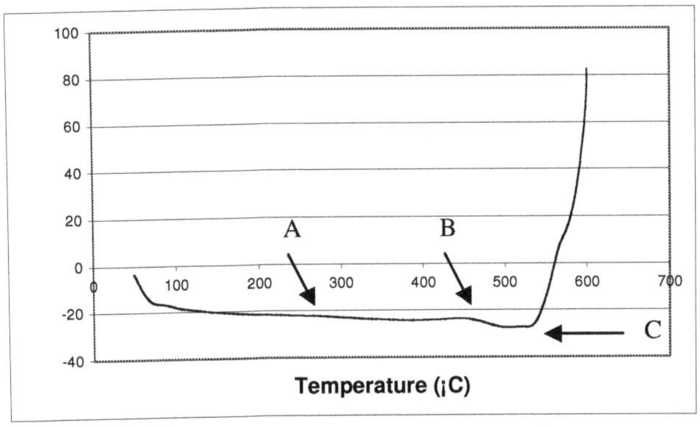

Figure 1 – DSC trace of pure magnesium powder heated from ambient to 600°C at 10°C/min. Area (A) represents hydroxide decomposition, area (B) carbonate decomposition. Area (C) is the formation of Mg-Au compounds followed by Mg-Au eutectic liquid formation.

Magnesium + 1wt% Calcium

Figure 2 shows the DSC trace of magnesium with a 1wt% addition of calcium. The presence of calcium, which is a more reactive metal than magnesium, disrupts magnesium surface layer through the following reactions:

$MgO + Ca = Mg + CaO$

$Mg(OH)_2 + Ca = Mg + Ca(OH)_2$

$MgCO_3 + Ca = Mg + CaCO_3$

In this case, not only do the reaction disrupt the cohesiveness of the surface layer, but the product is elemental Mg. The total amount of impurity does stay the same as now Ca is taking the place of Mg with the impurities. Since these reactions are spontaneous at any temperature, they are controlled by kinetics and are more difficult to estimate.

Area (A) at ~120°C indicates a slope decrease similar to that of the hydroxide decomposition for the pure Mg samples that may be the conversion of $Mg(OH)_2$ to $Ca(OH)_2$. The decomposition reaction for calcium hydroxide is possible below the melting temperature, but at relatively high temperature of 516°C. Area (B) indicates the possible Mg to Ca carbonate reaction at roughly 310°C. Area (C) corresponds to the lowest melting Mg-Ca eutectic at 445°C. Beyond that temperature, there is another Mg-Ca eutectic at 516°C and also the same Mg-Au reactions with the crucible are possible. The larger endothermic heat flow at 600°C compared to the pure Mg samples points to the combination of a ternary Mg, Ca and Au melting.

Magnesium + 2wt% Yttrium

Figure 3 shows the DSC trace of a mixture of pure magnesium with 2wt% yttrium. The $Mg\text{-}O_2\text{-}H_2O\text{-}CO_2\text{-}Y$ system is different from the Ca system because Y it is more reactive than Ca, and also it has only the ability to decompose the oxide of magnesium, but not the hydroxide or carbonate.

Examining Figure 3 shows little is happening up to 500°C. There are no exothermic peaks indicating hydroxide or carbonate decomposition, whether they stem from the effect of yttrium or are the basic reaction as shown in the pure magnesium case. It would seem that the presence of yttrium may be detrimental in this case, as it does not allow the surface layer disruption caused by the decomposition of magnesium hydroxide and carbonate to MgO. Beyond 500°C, there is a Mg-Y eutectic at 567°C, and the Mg-Au eutectic that contribute to the large endothermic peak in that area.

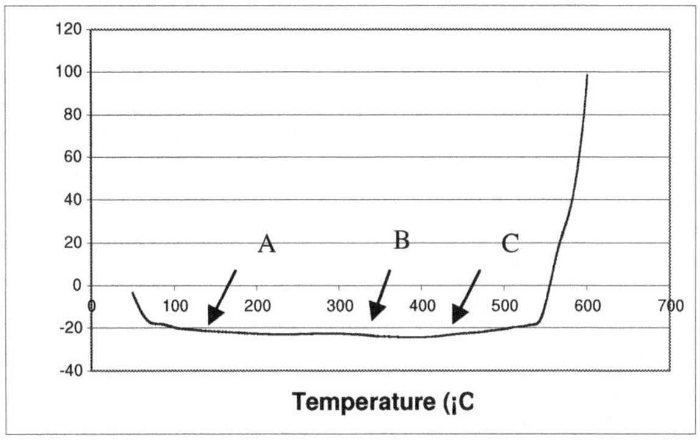

Figure 2 – DSC trace of magnesium with 1wt% calcium heated from ambient to 600°C at 10°C/min. Area (A) represents hydroxide decomposition, area (B) carbonate decomposition. Area (C) is the Mg-Ca eutectic liquid formation.

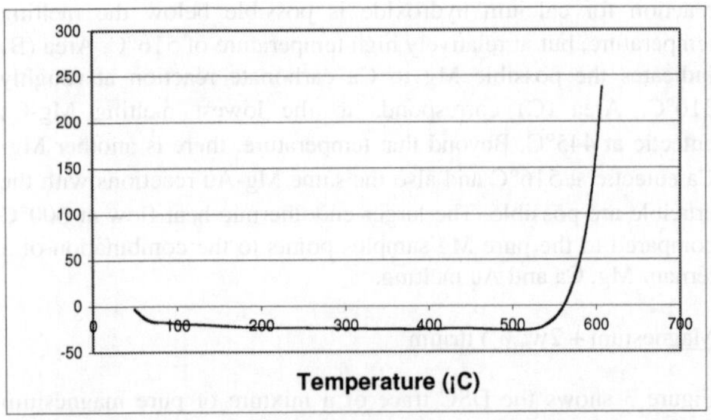

Figure 3 - An optical microscopy micrograph of the boundary between the eutectic liquid and the equi-axed regions. Calcium rich area at left.

Conclusions

In this work, differential scanning calorimetery was used to simulate sintering conditions to examine the thermal effects on the magnesium surface layer, with and without the sintering aids calcium and yttrium. Estimations using the FACTSage software suggested the presence of oxides, carbonates and hydroxides of magnesium. It was found that the process of heating the pure magnesium powder from room temperature to 600°C allowed for the spontaneous decomposition of the hydroxide and carbonate to oxide, which may disrupt the cohesiveness of the layer and allow better diffusion between particles.

Additions of calcium allow earlier and more complete surface layer disruption by forming calcium hydroxide and carbonate in the place of the original magnesium hydroxide and carbonate.

Yttrium additions, while having the greatest potential to reduce the oxide of magnesium, do not react with magnesium hydroxide or carbonate, and it appears to prohibit the decomposition of those compounds to MgO as well.

Acknowledgments

The authors wish to acknowledge the financial support of the Natural Sciences and Engineering Research Council (NSERC) of Canada, the Minerals Engineering Centre (MEC), and MATNET of Dalhousie University for the use of the characterization equipment.

References

1. "State of the North American P/M industry", International Journal of Powder Metallurgy, 37, 2001, 36.
2. M.M. Avedesian and H. Baker, Editors, "ASM specialty handbook, magnesium and magnesium alloys", ASM International, Materials Park, OH, 1999.
3. G.J. Kipouros, "Bringing Magnesium to Automobiles", Materials Solutions for Environmental Problems, The Metallurgical Society, The Canadian Institute of Mining, Metallurgy and Petroleum , Ed.H. Mostaghasi, Sudbury, Ontario, Canada, August 17-20, 1997, 265-267.
4. E. Ghali, W. Dietzel and K-U. Kainer, "General and Localized Corrosion of Magnesium Alloys: A Critical Review", Journal of Materials Engineering & Performance, 13, 2004, 7-23.
5. C.J. Bettles and M.A. Gibson, "Current wrought magnesium alloys: strengths and weaknesses", Journal of Metals, 57, 5, 2005, 46-49.
6. C.W. Hennessey, W.F. Caley, G.J. Kipouros and D.P. Bishop, "Development of Al-Si-Base P/M Alloys", International Journal of Powder Metallurgy, 41, 2005, 50-63.
7. G.J. Kipouros, W.F. Caley, and D.P. Bishop, "On the Advantages of Using Powder Metallurgy in New Light Metal Alloy Design", Metallurgical and Materials Transactions A, 37A, 12, 2006, 3429-3436.
8. P. Burke, D. Fancelli and G J. Kipouros, "Investigation of the Sintering Fundamentals of Magnesium Powders", Ed. M.O. Pekguleryuz, Light Metals in Transport Applications, COM 2007, Toronto, 2007, 183-195.
9. P. Burke, D. Fancelli. V. Laverdiere and G. J. Kipouros, "Sintering Fundamentals of Magnesium Powders", Canadian Metallurgical Quarterly, 48, (2), 2009, 123-132.
10. J. Li, G.J. Kipouros, P. Burke and C. Bibby, "Investigation of surface film Formed on Fine Mg Pacrticles", Microscopy & Microanalysis, VA, USA, July 26-30, 2009.
11. P. Burke and G.J. Kipouros, "Powder Metallurgy of Magnesium: Is it Feasible?", Ed. Sean R. Agnew, Magnesium Technology 2010, TMS 2010, Seattle, 2010, 115-120.
12. Lumley, R. N., Sercombe, T. B. and Schaffer, G. B. "Surface oxide and the role of magnesium during the sintering of aluminum". Metallurgical and Materials Transactions A, 30, 2, 1999, 457-463.
13. P. Burke, "Development of Magnesium Powder Metallurgy AZ31 Alloy Using Commercially Available Powders", M.A.Sc. Thesis, Dalhousie University, Halifax, NS, Canada, 2008.
14. G.B. Schaffer, B.J. Hall, S.J. Bonner, S.H. Huo, T.B. Sercombe, "The effect of the atmosphere and the role of pore filling on the sintering of aluminium", Acta Materialia, 54, 1, 2006, 131-138

Magnesium Technology 2011
Edited by: Wim H. Sillekens, Sean R. Agnew, Neale R. Neelameggham, and Suveen N. Mathaudhu
TMS (The Minerals, Metals & Materials Society), 2011

MICROSTRUCTURE AND MECHANICAL PROPERTIES OF SOLID STATE RECYCLED MG ALLOY CHIPS

Kunio Matsuzaki, Youichi Murakoshi and Toru Shimizu
National Institute of Advanced Industrial Science and Technology(AIST); 1-2 Namiki Tsukuba 305-8564, Japan

Keyword: Mg alloy chip, Recycling, Hot-extrusion, Forging, Formability

Abstract

Recycling of Mg alloy chips generated in machining processes such as turning and sawing through a melting process is difficult because the chips are very fine and can burn easily during heating. In this study, two machined Mg alloy chips were solid-state-recycled into a bar by hot pressing and hot extrusion, and the mechanical properties of the recycled chips were examined. The recycled AZ91 and AZX911 alloys showed a fine microstructure with a grain size of less than 10 μm. The compressed yield stresses at room temperature were 208 and 210 MPa for the recycled AZ91 and AZX911, respectively, which are higher than those for non-recycled samples. A backward extrusion test revealed that the recycled AZ91 and AZX911 alloys have good forgeability at temperatures above 573 K and slightly higher hardness than non-recycled samples. Therefore, solid-state-recycled Mg alloys have good formability for forging at elevated temperatures with good mechanical properties and have potential for use as forging material

Introduction

Mg alloy is the lightest among commercial structural materials and used in the housing of computers, cell phones, and consumer goods. Moreover, the application of Mg alloy to automotive parts reduces the weight of the car and helps improve fuel efficiency and CO_2 emission reduction. The use of Mg alloy in automotive markets has increased in the last ten years. The increase in Mg alloy components has also caused an increase in their waste; thus, the recycling of Mg alloys has become more important. The production of a kilogram of Mg from raw materials consumes 35 kWh. In contrast, a kilogram of refined recycled metal requires less than 3 kWh [1], implying that Mg alloy has excellent recyclability. Most recycling is carried out by a melting process. In general, the waste of Mg alloys is categorized into several classes; among them, clean scraps without impurities or oil, such as casting scraps, are easily recycled by melting. However, recycling of fine chips generated during machining by melting is difficult. Upon heating, Mg alloy chips are easily oxidized and burned even in a cover gas before melting. Increasing the Mg alloy parts causes an increase in Mg chips; thus, there is a demand for a suitable process for recycling of Mg chips. Studies have reported [2,3] that Mg alloy chips can be solid-state-recycled by undergoing hot extrusion into a bar, which then results in a fine structure with high strength. The solid-state recycling process is useful for enhancing the mechanical properties of Mg alloys. However, little has been reported on the formability of solid-state-recycled Mg alloy. This study was aimed at clarifying the microstructure and mechanical properties of solid-state-recycled Mg alloy chips and evaluating the formability of recycled Mg alloy for forging materials.

Experimental Procedure

In this study, AZ91 and AZX911 alloys were used; their chemical compositions are listed in Table 1. AZX911 alloy contains Ca, which improves creep resistance and flammability. The chips were produced by dry cutting non-recycled AZ91 and AZX911 bars prepared by casting and hot extrusion. The chips had a fibrous shape with a length of about 1 mm, width of 100 μm, and thickness of 20 μm, as shown in Fig.1. The chips were hot-pressed at a temperature of 573 K under a pressure of 350 MPa into a preform with a diameter of 70 mm and height of 50 mm; the preform was then hot-extruded at 623 K into a bar with a diameter of 18 mm. The higher the extrusion temperature, the lower is the extrusion force; however, extrusion above 673 K causes oxidation or partial melting of samples. Hot pressing and hot extrusion were carried out in air. The extrusion ratio was 1:15 because the consolidated chips with a low extrusion ratio below 1:10 showed no sound deformation during upsetting at elevated temperature. The bar prepared by the hot extrusion of chips is shown in Fig.2.

Table 1. Chemical composition of AZ91 and AZX911(wt%)

alloy	Al	Zn	Mn	Ca
AZ91	8.7	0.81	0.21	-
AZX911	8.8	0.84	0.17	1

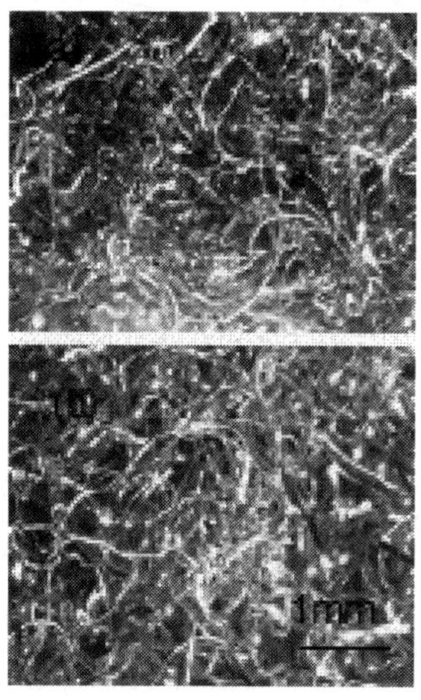

Figure 1 Mg alloy chips: (a) AZ91 and (b)AZX911

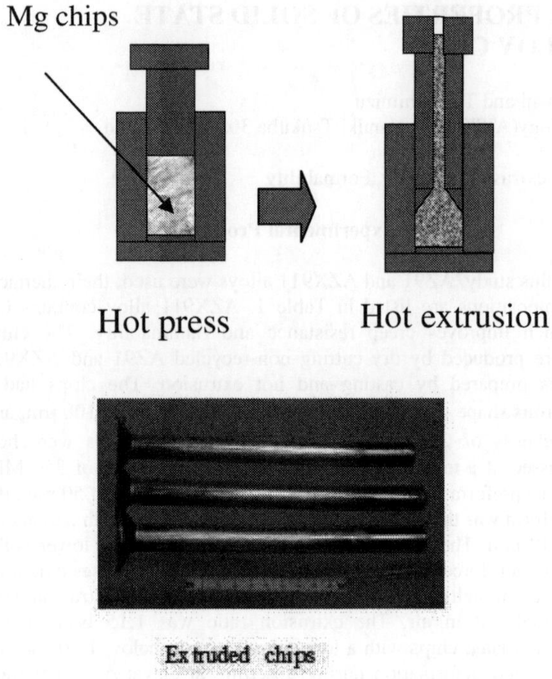

Figure 2 Consolidation process and extruded Mg alloy chips

Figure 3 Die set for the backward extrusion

The microstructure of the obtained bar was examined by optical microscopy and X-ray diffractometry. The mechanical properties at room temperature were examined by a compression test. The forgeability was evaluated by a backward extrusion test. The test piece was machined from extruded bars into a shape with a diameter of 16 mm and height of 10 mm. Figure 3 shows a die set for the backward extrusion test, installed on a crank press machine. The stroke number and maximum force of the crank press machine were 86 spm and 100 tf, respectively. The inner diameter of the die is 16 mm. Molybdenum disulfide was used as the lubricant. A test piece prepared by turning the hot-extruded bar was inserted into a die heated by rod-type heaters at the desired temperature. After maintaining these conditions for 10 min, the test piece was backward-extruded by a punch into a cup shape. The punch has a diameter of 13.4 mm. The extruded sample has an area reduction of 70%, which corresponds to an equivalent strain of about 2. Most of forged parts are subjected to deformation with an equivalent strain of 2, and therefore the area reduction of 70% is suitable for evaluating forgeability.

Results and discussion

Microstructure

Figure 4 shows the optical microstructure of AZ91 and AZX911 chips recycled by hot extrusion. Individual machined chips were not observed, and a fully dense structure was obtained for both samples. The density of AZ91 as measured by Archimedes' method was 1.81 g/cm^3, which is nearly the same as the theoretical value. The extruded AZ91 and AZX911 chips were composed of fine equiaxed grains. The average grain sizes were 10 and 7 μm for AZ91 and AZX911, respectively. The non-recycled AZ91 and AZX911 alloys showed grain sizes of 100 and 50 μm, respectively, as shown in Fig.5. A fine microstructure was obtained by solid-state recycling. This is owing to the dynamic recrystallization induced by hot extrusion. In the recycled samples, AZX911 showed a finer microstructure than AZ91. The addition of Ca was effective in preventing grain growth during recrystallization. This may be because Ca addition causes the formation of A_2Ca or an intermetallic compound containing Ca, which acts as a pin to inhibit grain growth.

Figure 6 shows the X-ray diffraction patterns taken from the cross-sectional area of the bar recycled by the hot extrusion of AZ91 and AZX911 chips. AZ91 alloy is mostly composed of hcp Mg. The peaks corresponding to $Mg_{17}Al_{12}$ were not observed,

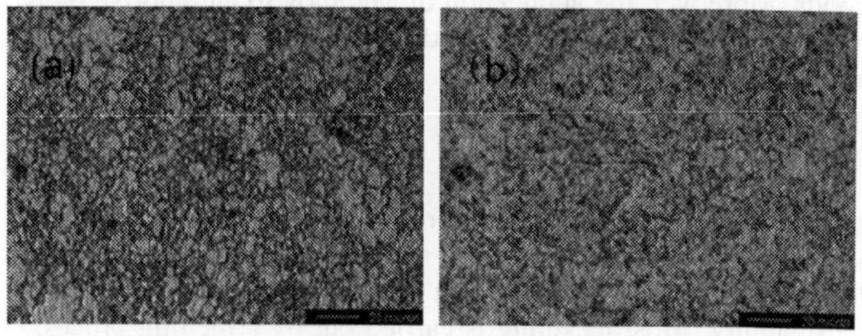

Figure 4 Optical microstructure of hot extruded AZ91 (a) and AZX911(b) alloy chips

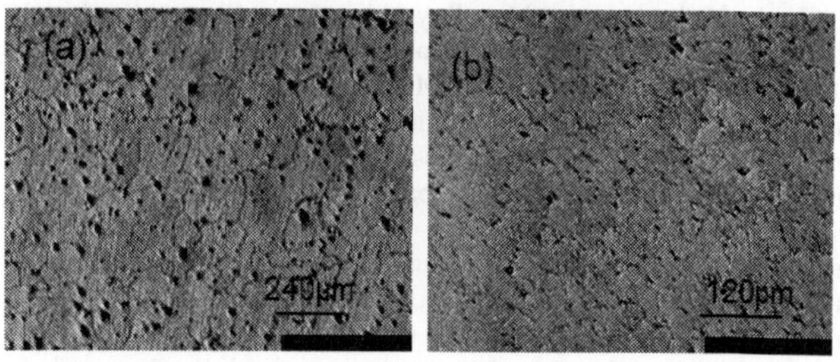

Figure 5 Optical microstructure of non-recycled AZ91(a) and AZX911 alloy(b)

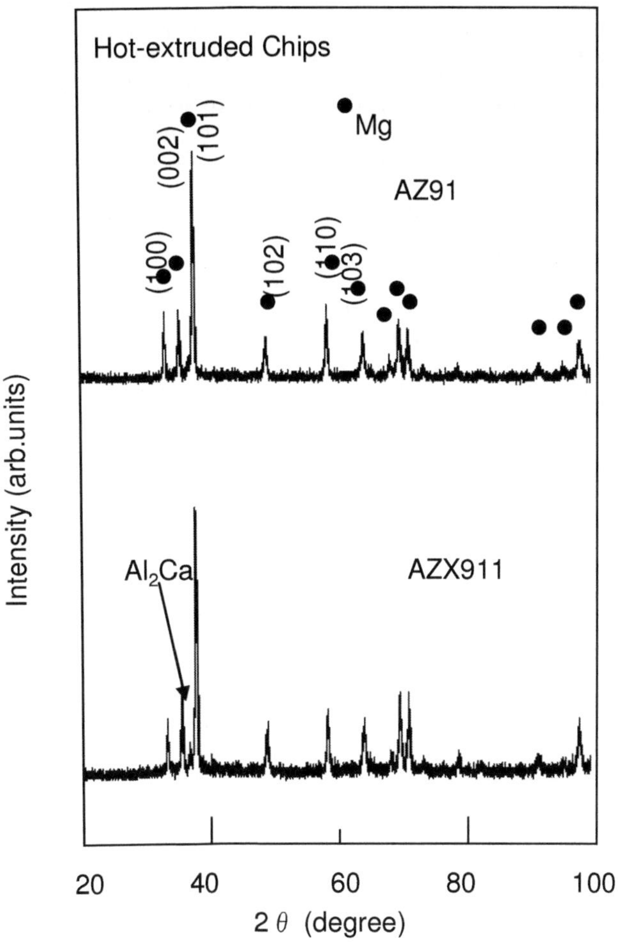

Figure 6 X-ray diffraction patterns of hot extruded AZ91 and AZX911 alloy chips

which might be because $Mg_{17}Al_{12}$ dissolved into Mg during the hot extrusion process. In the X-ray diffraction pattern of AZX911, peaks corresponding to Al_2Ca were observed along with those corresponding to hcp Mg. The X-ray diffraction patterns of both samples were similar to those of cast samples. This means that the samples recycled by hot extrusion of the chips have no texture. In general, a Mg wrought alloy produced by hot extrusion shows an oriented structure with a basal plane of hcp Mg that is parallel to the extrusion direction. The sample without texture should have a good formability.

Mechanical Properties

Figure 7 shows the compressive mechanical properties of solid-state-recycled AZ91 and AZX911 alloys at room temperature. For comparison, the compressive mechanical properties of non-recycled AZ91 and AZX911 are shown in Fig.8. The fracture stress (σ_f) and yield stress ($\sigma_{0.2}$) of recycled AZ91 alloy are 334 and 208 MPa, respectively. These values are higher than those of non-recycled AZ91 alloys. For AZX911, a similar tendency was observed. The recycled AZX911 also showed a higher strength, with σ_f of 327 MPa and $\sigma_{0.2}$ of 210 MPa, compared to a non-recycled sample. It is notable that Mg alloys produced by solid-state recycling have improved mechanical properties. This is owing to the grain refinement by dynamic recrystallization. The fracture strains were 0.47 and 0.38 for recycled AZ911 and AZX911, respectively, which are also higher than those for non-recycled AZ91 and AZX911. AZX911 showed a slightly lower fracture strain than AZ91, although AZX911 showed a finer microstructure than AZ91. This is caused by precipitation of intermetallic compounds such as Al_2Ca. However, the fracture strain of solid-state-recycled Mg alloy is higher than those of non-recycled materials, suggesting that ductility is also improved by solid-state recycling.

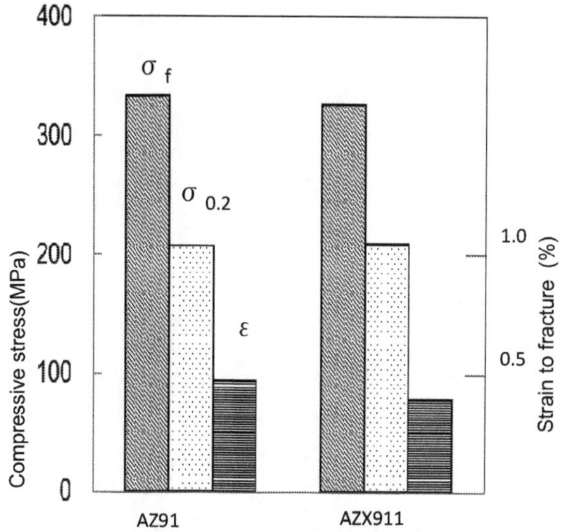

Figure 7 Compressive mechanical properties of solid state recycled AZ91 and AZX911 alloys

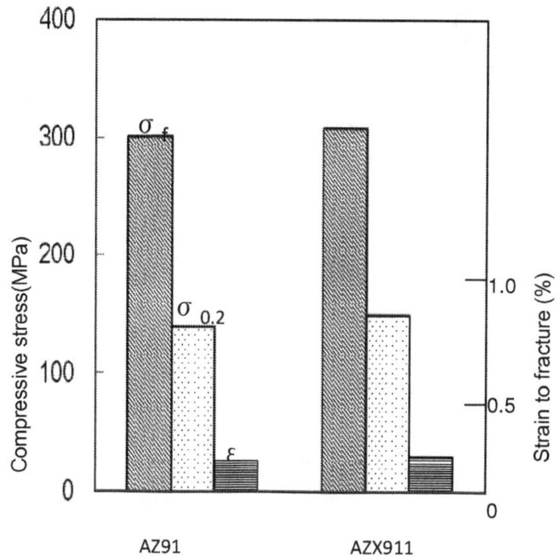

Figure 8 Compressive mechanical properties of solid state non- recycled AZ91 and AZX911 alloys

Forgeability

In order to evaluate the forgeability, the solid-state-recycled AZ91 and AZX911 alloys were backward-extruded into a cup shape at elevated temperatures. Figure 9 shows the outer surface of recycled and non-recycled AZ91 alloys after backward extrusion in the temperature range of 473–573 K. At 473 K, several cracks were observed on the surface of recycled AZ91, and the top of the cup was broken. The fracture surface was declined at 45° to the extruded direction, implying the occurrence of shear fracture. At 573 K, the recycled AZ91 showed no cracks on the surface, but a small fracture occurred on the top part. For the non-recycled AZ91, a similar tendency was observed. A further increase in test temperature improved the formability. However, a backward extrusion test at above 573 K caused partial melting owing to the generation of heat by working. Figure 10 shows the results of backward extrusion at 473–593 K for the recycled and non-recycled AZX911 alloys. For recycled AZX911, cracks were observed at 473 K and they decreased with increasing extrusion temperature. The cup extruded at 593 K shows no cracks on the surface. In contrast, non-recycled AZX911 showed cracks on the surface of the cup backward-extruded at 593 K. The forgeability of solid-state-recycled AZ91 and AZX911 was comparable to that of the non-recycled alloys. However, the solid-state-recycled Mg alloy has enough workability for forging at elevated temperatures. The generation of cracks may be caused by the temperature difference between the punch and the samples; this can be controlled by changing the forging condition. The punch geometry also influences the deformation behavior. An appropriate geometry of the punch should improve the forgeability. Furthermore, the deformation rate also influences the formability. In particular, Mg alloys show a strong strain rate dependence. The higher the strain rates, the lower is the elongation. In the present backward extrusion test, the punch speed was estimated to be 0.22 m/s, which indicates a high deformation rate. A lower deformation would make it possible to achieve a sound deformation.

Figure 11 shows an optical microstructure of the recycled AZ91 alloy after the backward extrusion test at 573 K. A grain flow is clearly found in the wall part and bend part. This metal flow introduced by plastic deformation helps enhance the mechanical properties, in which the structure is not obtained by casting. The backward-extruded AZ91 shows a fine microstructure with an average grain size of 5 μm, which is finer than the recycled bar. Further grain refinement occurs by dynamic recrystallization during backward extrusion. The backward-extruded AZX911 shows a similar fibrous and fine structure with an average grain size of less than 5 μm.

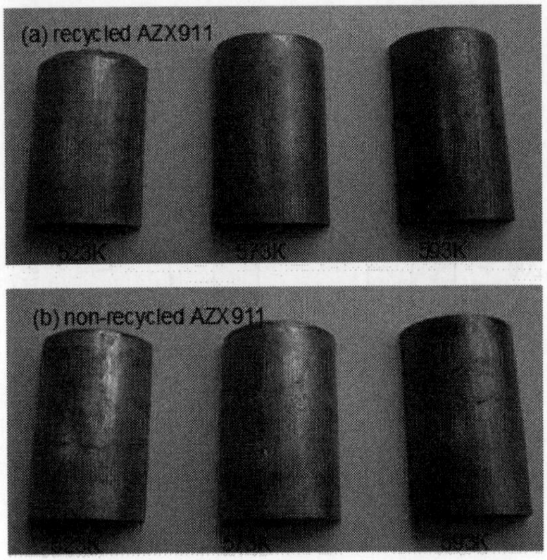

Figure 9 External appearance of recycled and non-recycled AZ91 alloy after backward extrusion at various temperatures

Figure 10 External appearance of recycled and non-recycled AZX911 alloy after backward extrusion at various temperatures

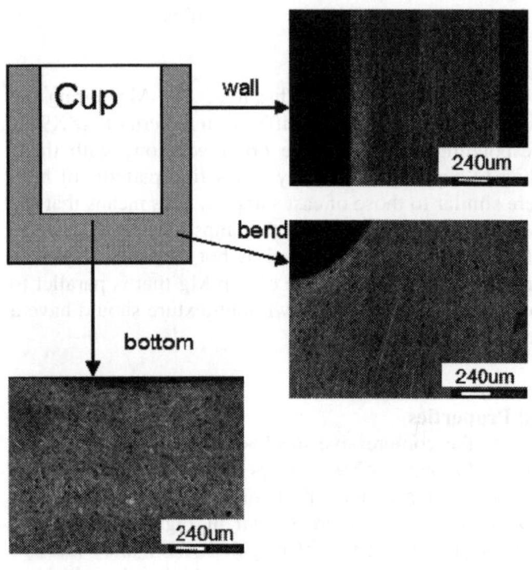

Figure 11 Optical microstructure of back ward extruded AZ91alloy at 573K prepared by hot extrusion of the chip

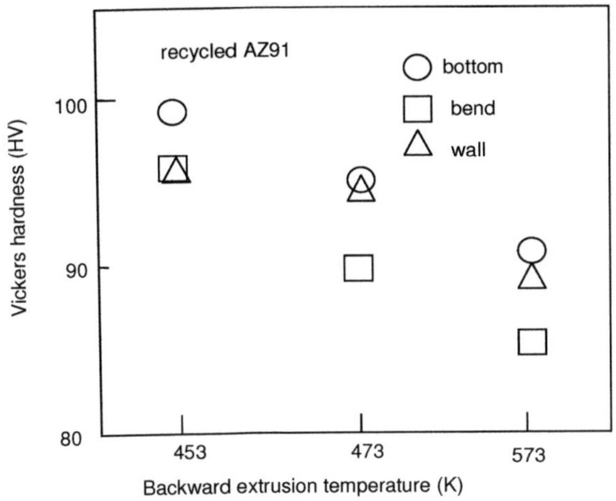

Figure 12 Change in the Vickers hardness of the recycled AZ91 alloy as a function of back ward extrusion temperature

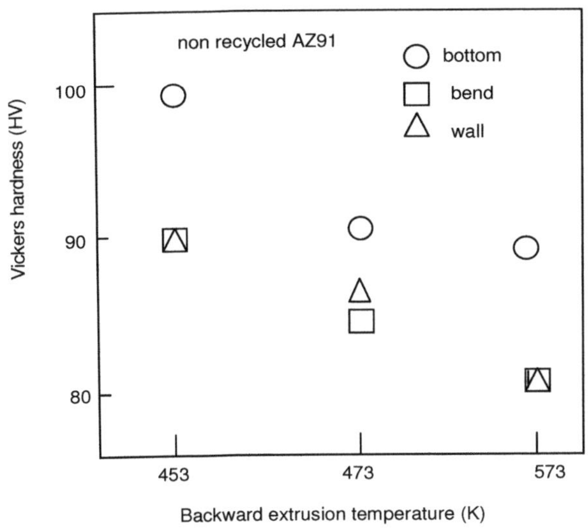

Figure 13 Change in the Vickers hardness of non-recycled AZ91 alloy as a function of back ward extrusion temperature

Figure 12 shows the change in Vickers hardness of the bottom, bend, and wall parts of the backward-extruded AZ91 alloy prepared by solid-state recycling as a function of backward extrusion temperature. Hardness was found to decrease with increasing extrusion temperature. The cup deformed at 573 K showed a hardness of 85–90 HV. The hardness of the AZ91 alloy prepared by backward extrusion of the non-recycled sample is shown in Fig.13. A similar change was observed, and the hardness of the recycled AZ91 was slightly higher than that of non-recycled AZ91. For AZX911, a similar tendency was observed. This suggests that the mechanical properties of the sample forged using the solid-state-recycled Mg alloy are superior to those of the sample forged using the non-recycled Mg alloy. Finally, the solid-state-recycled Mg alloy showed good forgeability, and the forged sample showed good mechanical properties. The AZ91 and AZX911 alloys used in this study are generally casting alloys with poor plastic formability. Thus, it is notable that the solid-state-recycled AZ91 and AZX911 alloys showed enough formability for forging, making it possible to use solid-state-recycled Mg alloy as forging materials. This is also caused by grain refinement by dynamic recrystallization. Furthermore, the recycling process investigated here was carried out at temperatures below 673 K without melting. This means that the energy consumption of solid-state recycling is less than that of conventional melting methods. In addition, the flux or cover gas used in the melting process is not necessary for solid-state recycling. Therefore, this is an environmentally friendly recycling process. In this study, dry processed Mg alloy chips were used. In most cases, machining of Mg alloys is carried out using machine oil, and thus, the Mg alloy chips are coated with oil. For solid-state recycling, deoiling treatment is needed because the coated oil prevents the connection of chips. We previously reported [4] that superheated steam is useful for cleaning oily chips; solid-state recycling with superheated steam treatment is under investigation.

Conclusion

Machined AZ91 and AZX911 alloy chips were solid-state-recycled into bars using hot extrusion. The obtained bars showed a fine microstructure with a grain size of less than 10 μm and higher strength than non-recycled materials. A backward extrusion test revealed that the solid-state-recycled AZ91 and AZX911 were forged into a cup without cracks at 573 and 593 K, respectively. The forged sample prepared by solid-state recycling showed a favorable grain flow and higher strength than the non-recycled samples. Therefore, we can conclude that the solid-state recycling process is useful for application to Mg alloy chips and that the materials recycled by this process can be used as a forging billet.

Acknowledgement

This work was supported by New Energy and Industrial and Technology Development Organization (NEDO) of Japan.

Reference

1.Riopelle,"The Recycling of Magnesium Makes Cents", Journal of Metals, October 1996,44-46.

2.M.Mabuchi,K.Kubota and K.Higashi, "New Reccycling Process by Extrusion for Machined Chips of AZ91 Magnesium and Mechanical Properties of extruded Bars", Mater.Trans, JIM, 36(1995), 1249-1254.

3.Y.Chono and M.Mabuchi,"Recycling of masgnesium Alloys Using Plastic Forming Process", Journal of the Japan Society for Technology of Plasticity, 44(2003)15-19.

4.Y.Murakoshi, K.Matsuzaki, M.Saigo, S.Koyanaka and M.Kimura, "Consolidation and Forging of AZ31 Chips Cleaned with Super Heat Steam", Magnesium ed.K.U.Kainer(Wiley-VCH,2009),417-422.

Magnesium Technology 2011

Corrosion and Coatings

Session Chairs:

Robert C. McCune
(Robert C. McCune and Associates LLC, USA)
Neale R. Neelameggham
(US Magnesium LLC, USA)

Magnesium Technology 2011
Edited by: Wim H. Sillekens, Sean R. Agnew, Neale R. Neelameggham, and Suveen N. Mathaudhu
TMS (The Minerals, Metals & Materials Society), 2011

SALT SPRAY CORROSION OF MECHANICAL JUNCTIONS OF MAGNESIUM CASTINGS

Sabrina Grassini, Paolo Matteis, Giorgio Scavino, Marco Rossetto, Donato Firrao
Politecnico di Torino – DISMIC department; 24 Corso Duca degli Abruzzi; Torino, It-10129, Italy

Keywords: AE44, galvanic corrosion, mechanical couplings

Abstract

The corrosion of cast, 3 mm thick, AE44 magnesium-alloy plates fastened to aluminum-alloy threaded counterparts, either constituting the screw or the nut, were tested in neutral salt spray for 48 hours, with or without interposed AA5051 spacers (washers). Steel nuts or screws were used, always insulating from corrosion the steel sides. Couplings between magnesium alloy plates and coated steel counterparts (screw heads) with interposed AA5051 washers were also tested, while insulating the nut side. Every 4 or 8 hours the test was halted and the samples were rinsed and photographed for manual image analysis. Then the plates were unmounted, slightly polished (highlighting the deep corrosion pits), and scanned for automatic image analysis. Different image analysis methods were compared. The least corrosion occurs, in couplings with aluminum alloy counterparts, when AA5051 washers are interposed; whereas the most effective coupling with steel counterparts is the one with nylon coated steel heads.

Introduction

A serious limitation to the application of magnesium alloys for the production of parts exposed to the environment is their susceptibility to corrosion, in particular when the alloy contains critical impurities, is coupled with other metals and is exposed to aggressive environments [1].

The corrosion of Mg and Mg alloys generally initiates as localized corrosion, while only in rare cases it is uniform and widespread. The poor corrosion behavior of Mg alloys derives from the following two main reasons.

Firstly, Mg and Mg alloys are generally susceptible to galvanic corrosion, which can be both internal and external. Low hydrogen overvoltage elements, such as Ni, Fe and Cu, that can be present as second phase and impurities in the alloy, or can be coupled with it, constitute efficient cathodes in contact with the magnesium anode, causing serious attacks. Instead, metals with high hydrogen overvoltage, like Al, Zn and Sn, can be less damaging [2]. The galvanic corrosion rate is increased by the conductivity of the medium, by a large potential difference and a small distance between the cathode and the anode, as well as by a large area ratio of cathode and anode [3].

Secondly, even if the quasi-passive hydroxide film can give a good protection in indoor and outdoor atmospheres, Mg alloys are characterized by a poor pitting resistance. Pitting corrosion occurs in the presence of chloride ions, both in neutral or alkaline salt solutions, e.g. seawater. The pitting corrosion resistance is influenced by the alloy's chemical composition and microstructure. In the case of Mg-Al alloys, such as AZ91 (which contains 9-10% of aluminum), pits are formed due to selective attacks along the primary intermetallic β-phase ($Mg_{17}Al_{12}$); the β-phase can act either as a micro-galvanic cathode increasing the corrosion rate, if present as small intergranular precipitates, or as an anodic barrier inhibiting the corrosion reaction, when it is more continuous, distributed homogeneously on the grain boundaries and present with high volume fraction [4].

Several technologies have been developed for improving the corrosion resistance of magnesium and its alloys, including coatings, surface treatments, and alloying.

Chromate conversion coatings are relatively simple and commonly used in the automotive industry and can lead to good corrosion performance. However, since the Cr^{6+} ion is toxic and carcinogenic, treatments with chromate compounds are undesirable for industrial safety control and environment protection. Recently, low pressure plasma treatments and PECVD (Plasma Enhanced Chemical Vapor Deposition) of organosilicon thin films, thanks to their versatility and their low environmental impact, have been proposed for corrosion prevention [5].

At the same time, alloying with Rare Earth (RE) elements, such as Lanthanum (La), Cerium (Ce), Yttrium (Y) and Neodymium (Nd), is increasingly used to enhance both the corrosion resistance and the mechanical properties. The addition of Y can markedly improve the corrosion performance of Mg-Al alloys, by forming intermetallic phases, both on the grain boundaries and on the grain bulk, able to improve the stability of the film forming on the surface of the β-phase, thus reducing the corrosion attack [6].

Finally, in automotive applications, Mg alloy parts are commonly assembled with either bare aluminum alloy counterparts or coated steel counterparts, with different design solutions being employed to reduce the likelihood of severe galvanic corrosion, including the usage of aluminum alloy spacers to avoid the direct contact between the Mg alloy and a counterpart with an higher hydrogen overvoltage, as well as the usage of non-conductive (i.e., polymeric) coatings on the counterpart.

The corrosion behavior of the AE44 magnesium alloy, developed by the Hydro company, mechanically joined with common automotive assembly counterparts, is examined here.

The general corrosion of the AE44 alloy has been previously studied in both salt water [4,9] and salt spray [9], and the galvanic corrosion of the same alloy coupled with either steel or aluminum alloys has been previously studied with polarization measurements [7] and free corrosion tests [8] in salt water. Therefore, this work aims to assess the effectiveness of different assembling methods, which could be applied in the automotive industry to limit the galvanic corrosion of this alloy in coupling with either aluminum alloy or coated steel counterparts, by subjecting several such assemblies to a salt spray environment.

Experimental Details

Samples Preparation

The examined AE44 Mg alloy [4] has the following nominal weight composition: 4% Al and 4% RE, with the following RE mixture: Ce > 50%, La 20 – 35%, Nd 10 – 20%, Pr 4 – 10%.

The Mg alloy was cast in the shape of 3 mm thick, 100 mm wide and 140 mm long plates, with 8.5 mm diam. trough holes, and was tested in the as-cast state, without any coating and without previous surface polishing.

Screws, or other male threaded parts, were mounted through the plates holes and fastened with nuts, or other female threaded parts, with or without washers (acting as spacers). All assembled items, except the plates themselves, were industrial components, or parts thereof. Either the screw side or the nut side of each assembly was exposed to the salt spray, and is described below; while the other side (made with aluminum washers and coated steel nuts or screws) was effectively sealed with a neutral silicone sealant (ISO 11600 standard).

Uncoated aluminum alloy spacers and uncoated aluminum alloy counterparts or coated steel counterparts were mounted as described in Table I and sketched in Figure 3 below (1st row).

Table I. Tested assembly types (or subtypes): tests number; metric thread and mounting torque; spacers and counterparts mounted on the Mg-alloy plates: material (steel and coating type, or aluminum alloy), thickness (Th.), inner and outer diameter (ID and OD).

type	tests	thread	torque	spacer (washer)				counterpart		
				alloy	Th.	ID	OD	alloy	coating	OD
-	#	mm	Nm	-	mm	mm	mm	-	-	mm
A	6	8	24	-	-	-	-	328.0	-	*
B	4	6	10	5051	2	6.5	22.6	328.0	-	18.6
C_1	2	18	24	†	1.6	18.5	23.4	2011	-	23.3
C_2	1	18	24	5051	2	18.5	23.4	2011	-	23.3
D_1	6	8	24	5051	2	8.5	25.5	steel	Zn-Al-Si 720 h‡	17.0
D_2	6	8	24	5051	2	8.5	25.5	steel	Zn-Al-Si 480 h‡	17.0
E	6	8	24	5051	2	8.5	24.2	steel	Zn, Nylon	16.2
F	6	8	24	5051	1	8.2	14.0§	steel	Zn	13.0

*Not circular †Unknown ‡Salt spray endurance §Contact area

The cast AA 328.0 counterparts (assembly types A and B) were cut from one automotive gear box with female threads; the wrought AA 2011 counterparts (type C) were oil plugs with a male thread (only in this case, the plate holes were enlarged to 18.5 mm diameter); and the steel counterparts were screw heads with different Zn-based coatings: either a plain Zn layer (type F), or a Zn inner layer and a polymeric outer layer (type E), or a Zn-Al-Si layer with 480 or 720 h nominal salt spray endurance (subtypes D_1 and D_2, respectively). Most spacers were plain washers, with the dimensions given in Table I; only the F assembly spacers were cup washers, 1 mm thick, with 8.2 mm hole diameter, 14 mm contact (plane) area outer diameter, 18 mm maximum outer diameter and 8 mm total axial height, with a large fillet radius between the contact area outer diameter and the maximum diameter. Most counterparts exhibited circular contact areas, with the diameters given in Table I, thus the path going from the counterpart edge to the Mg alloy plate, passing on the spacer, was constant along the same edge; only the A assemblies, which had no spacers, exhibited non-circular contact areas.

Salt Spray Tests

The 37 assemblies were mounted on 19 Mg-alloy plates, which were tested simultaneously on two polymeric racks placed in symmetric positions inside a 0.5 m^3 salt spray chamber, in accordance with the ASTM B117 standard. The plates were inclined of 30° in respect to the vertical. The two available mounting holes in each plate were vertically aligned in the rack, hence each assembly was either in a top (upstream) or bottom (downstream) position.

The 5 wt.% aqueous NaCl solution was prepared with deionized water and 97.7% pure NaCl (without Ni and Cu and with less than 0.1 wt.% NaI). The nozzle exit temperature was 48 °C and the chamber temperature was 35 ± 1.5 °C.

The test was halted after 4, 8, 16, 24, 32, 40 and 48 h of total salt spray exposure. After each stop, the chamber was rinsed with external air for 30 min to remove the salt spray, then it was opened and the samples were removed, rinsed with distilled water, dried with a compressed air jet, and photographed. Each plate was always tested in the same rack position. Since the samples were not subjected to a corrosive environment during the test stops, the overall salt spray test is deemed equivalent to a continuous test of equal total salt spray exposure time.

Image Analysis

All assemblies were photographed before the test start and after each test stop (after rinsing and drying), always with the same lighting, digital camera, camera settings, and angle of view. The assemblies were photographed from the direction perpendicular to the plate, except the A and B type assemblies, which exhibited large counterparts hiding the contact areas edges in such view, and were thus photographed from two mutually perpendicular directions forming angles of about 45° with the plate. 2288 x 1712 pixel, 24-bit RGB color, jpg compressed pictures, were recorded. The resolution of the perpendicular view pictures on the Mg-alloy plate plane was from 30 to 40 pixel/mm.

Then, the following quantitative analyses were performed on one assembly, mounted in a bottom position, for each type.

The pictures taken during the test stops were subjected to manual image analysis, by measuring the maximum linear dimension of each detectable corrosion pit in the plate region surrounding the contact area; a grand total of 1002 pits were measured.

Moreover, after the last salt spray test, the assemblies were unmounted and their Mg-alloy plates were first photographed in the same above-described manner (perpendicular view) and then polished with emery papers on a metallographic polishing machine (the plates were cut to reduce the area to be polished), in two successive steps, with 28 and 14 μm abrasive size, in order to remove both the salts and corrosion products and a surface layer about 50-100 μm thick. This included the layer affected by the original surface roughness and by most general corrosion effects, but not the bottom of the corrosion pits due to localized corrosion close to the mechanical junction (as well as of some pits due to general corrosion), which were deeper. Hence, the latter corrosion pits were clearly evidenced as opaque areas against the reflective metallic background.

Thereafter, the polished plates were scanned on a flatbed image scanner, and 8-bit grey, jpg compressed pictures were recorded with a resolution of 24 pixel/mm and processed with an automatic image analysis software (ImageJ, http://rsbweb.nih.gov/ij) to identify the corrosion pits occurring inside the 10 mm wide annulus surrounding the contact circle, the annulus inner diameter being equal to the washer contact area OD (given in Table I). In the A-assembly case, a 10 mm wide band surrounding the irregularly shaped contact area was similarly examined.

The corrosion pits area fraction was calculated both for the whole examined annulus or band and as a function of the distance from the contact area boundary; for the latter purpose, the 10 mm annulus or band was divided into 20 adjacent 0.5 mm wide narrower annuli or bands.

Results and Discussion

Manual Image Analysis

Nominally equal assemblies at equal test times exhibited similar corrosion effects, as evaluated qualitatively during the tests and on the ensuing photographic documentation.

The results of the quantitative manual image analysis performed on one assembly for each type are shown in Figure 1. Within 4 to 8 h from the start of the test, the Mg-alloy plate exhibits general corrosion effects and develops a surface layer of corrosion products and/or deposited salt, while some corrosion pits are detected close to the contact area edge and grow in time; only in the A assembly a continuous corrosion groove already occurs on the contact area edge after 4 h test duration. Thereafter, both the largest and the mean pit size measured after each test stop either decrease or exhibit large oscillations, because the corrosion pits are at times partially or totally hidden by the salts and/or corrosion products deposited on the surface during the test (and not removed by rinsing). For this reason, after 16 h test duration, the actual corrosion pit development can hardly be inferred from the external appearance of the samples.

For example, the corrosion groove found on the whole contact area edge in the examined A type assembly after 4 h (Figure 2a) was then partially hidden and detected as a set of separate smaller pits after 24 h (Figure 2b); even if the former corrosion pit was evident again after the slight surface polishing performed at the end of the 48 h salt spray test (Figure 3 below).

Neglecting this problem, on the basis of the manual image analysis, and particularly of the largest and mean pit size measurements, the corrosion behavior was worst in the A and D assemblies and best in the E and F ones. Finally, at the end of the test, the unmounted A assembly exhibited corrosion effects on the plate area which was covered by the counterpart during the test, as shown in Figure 3, which were not accounted in the image analysis (both manual and automatic). Similar effects were much less evident, but not completely absent, in other assemblies.

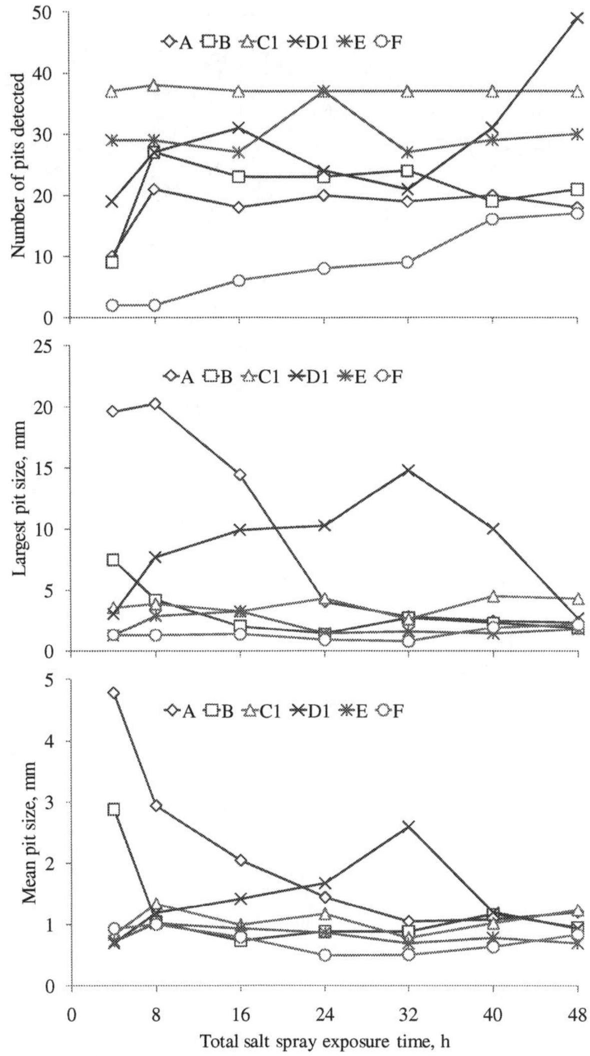

Figure 1. Manual image analysis. Number, largest size, and mean size, of the corrosion pits detected around the contact area after each test stop, for one assembly of each type (A to F).

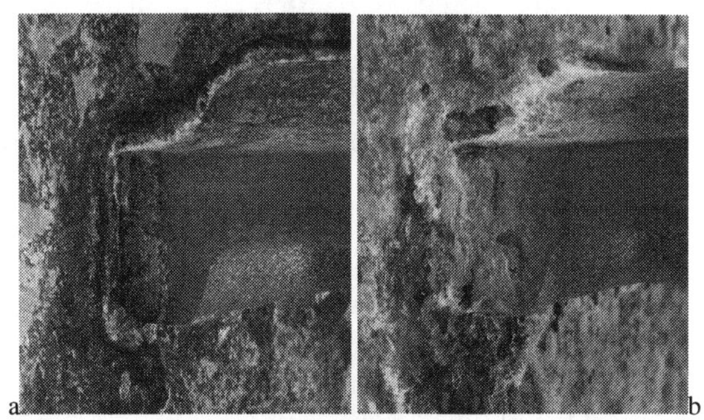

Figure 2. Continuous corrosion pit along the contact area in a type A assembly after 4 h salt spray exposure (a); same region partially covered by salts and/or corrosion products after 24 h (b).

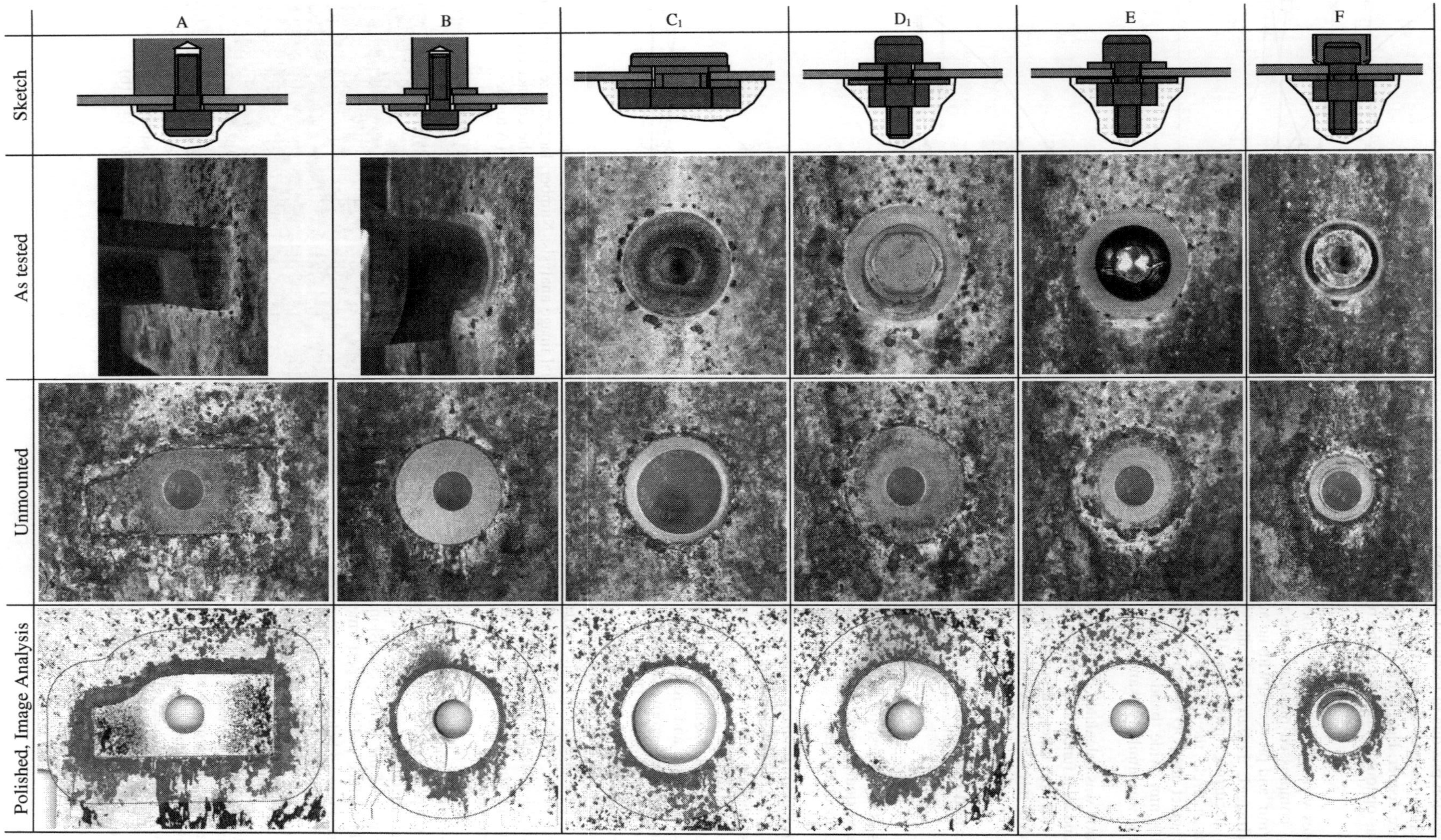

Figure 3. Photographic documentation and image analysis of 1 specimen for each assembly type (columns). Assembly sketches (row 1; Mg-alloy, Al-alloys, coated-steel and silicon and are shown in grey, blue, red and yellow, respectively). Digital-camera color pictures after 48 h salt bath exposure, rinsing and air drying (row 2) and after unmounting the counterpart and the spacer from the Mg-alloy plate (row 3); flatbed-scanner gray pictures after slight polishing of the Mg-alloy plate, highlighting the deep corrosion pits on the metallic background (row 4); and binary red/transparent pictures of the corrosion pits recognized by image analysis in a 10 mm wide region around the contact area (region bounded by red lines), superimposed on their corresponding gray pictures (row 4). Top-bottom pictures orientation as during the salt spray test. Pictures direction of view normal to the Mg-alloy plate and picture height on the plate plane 48 mm (magnification 0.77x) in all cases, except the A and B-type assemblies as-tested pictures (row 2, column 1 and 2).

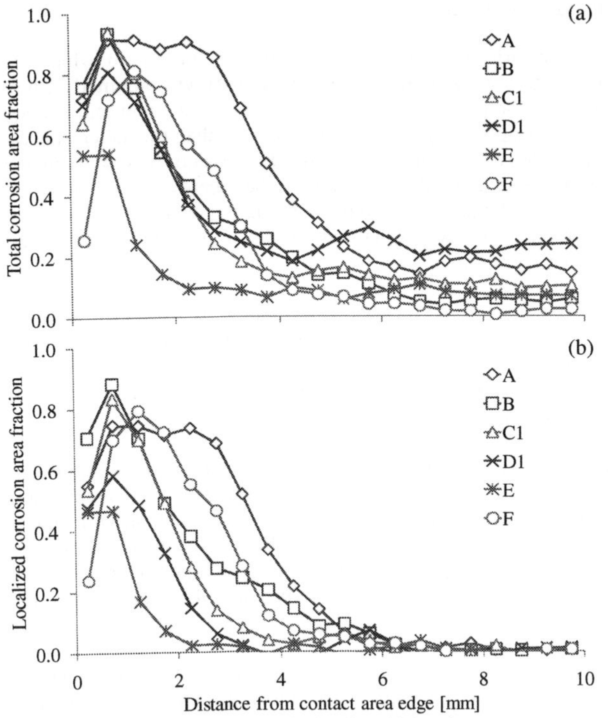

Figure 4. Automatic image analysis after 48 h salt spray exposure. Total corrosion area fraction (AF), at increasing distance from the mechanical junction (a), and AF due to localized corrosion only (b); 1 specimen for each assembly type (A to F).

Automatic Image Analysis

The results of the automatic image analysis, performed after the 48 h salt spray test, are given in Figures 3 and 4 and Table II.

In Figure 3, the polished plate surfaces are compared with the same regions photographed before polishing, both at the end of the test and after unmounting the counterparts and spacers, and with the corrosion pits recognized by image analysis in the 10 mm wide region around the contact area, showing that the polishing and automatic image analysis procedures were generally effective for the purpose of evidencing the deep corrosion pits. On the basis of the overall corrosion pits area fraction in the examined region, which ranges from 12 to 40 % (Table II), the six assembly types can be rated in the following order of apparent increasing corrosion sensitivity: E, F, B, C_1, D_1 and A. However, the latter results do not always discriminate the localized corrosion features related to the mechanical junction from the surrounding general corrosion ones.

This problem is addressed in Figure 4, where the corrosion pits area fraction is plotted as a function of the distance from the mechanical junction. The corrosion area fraction always exhibits an initial maximum due to localized corrosion, and then decreases to a lower value, due to general corrosion, which becomes almost constant (for each sample) at distances larger than 6 mm. Most often the maximum corrosion area fraction (Figure 4a and Table II) is detected in the 2nd nearer annulus or band, rather than in the 1st one, in respect to the junction edge; this is likely an artifact due to small deviations between the inner boundary of the 1st annulus or band and the actual contact area boundary, in turn due to geometrical irregularities of the latter boundary (this phenomenon is the most evident in the F assembly, in which the contact area boundary is the least clearly defined due to the lack of a sharp edge in the cup washer).

The general corrosion area fraction, defined as the mean corrosion area fraction in the 7 to 10 mm distance range from the junction, ranges from 2 to 22 % in different plates (Table II), likely due to the different sample positions in the salt spray chamber and/or to a more or less complete removal of the shallower general corrosion features during the polishing procedure. Thus, in Figure 4b the same data (already shown in Figure 4a) are plotted again after subtracting the general corrosion area fraction of each sample, in order to show only the contribution (to the total corrosion) due to the localized corrosion at the junction edge.

The latter plot, Figure 4b, evidences that the A, B, C_1 and F type assemblies all cause localized corrosion effect on more than 74% of the Mg-alloy surface near the junction edge, whereas the same measure is 58 and 47% in the D_1 and F assemblies, respectively (Table II). Moreover, the width of the region affected by localized corrosion, defined as the distance from the junction at which the localized corrosion area fraction eventually falls below half of its maximum value, is the largest in the A and F assemblies and the smallest in the E assembly (Table II).

Table II. Automatic image analysis after 48 h salt spray exposure. Mean and maximum total corrosion Area Fraction (AF) within 10 mm from the mechanical junction, AF due to general or localized corrosion only, and width of localized corrosion region; 1 specimen for each assembly type (A to F).

Corrosion type	Measure	Assembly type					
		A	B	C_1	D_1	E	F
Total	mean AF, %	40	22	23	31	12	17
	max. AF, %	91	93	94	81	54	81
General	mean AF, %	17	5	10	22	7	2
Localized	max. AF, %	74	88	83	58	47	79
	width, mm	3,5	2	2	2	1	3

Thus, by subtracting the contribution of the general corrosion, and by considering both the localized corrosion maximum area fraction and width of affected area (Figure 4b and Table II), the six assembly types can be rated in the following order of increasing corrosion sensitivity: E, D_1, C_1, B, F and A.

This latter order is different, and is deemed more correct, that the order which could be obtained both from the overall corrosion pits area fraction measured in the examined region after the salt spray test, and from the manual image analysis of the junctions photographed during the test stops, because the latter methods do not discriminate the localized effects, which are due to the mechanical junction, from the general effects, which may be different due to experimental details.

This correction is especially important in the examined F assembly type specimen. In fact, this was rated among the least affected by corrosion both on the basis of the manual image analysis of the assembled junctions and on the basis of the overall corrosion area fraction in the region examined by automatic image analysis. However, from Figures 3 and 4 it can be noted that the former observation was misleading because the localized corrosion occurring under the curved surface of the cup washer was almost completely hidden by the top edge of the same washer (Figure 3), whereas the latter measure was misleading because the specimen exhibited the least general corrosion contribution and the smallest contact area OD (hence the area closer to the junction was the smallest fraction of the total examined area).

Conclusions

Test Methods Effectiveness

The manual image analysis of the specimens exposed to the salt spray test for increasing durations was not reliable, because the actual extension of the localized corrosion effects was soon hindered by a layer of corrosion products and/or deposited salt, not removed by rinsing in water, and was hidden by the assembly itself in the case of the cup washer.

A slight polishing of the (unmounted) Mg-alloy plate, performed after the salt spray test, was effective to highlight the deep opaque corrosion pits against a reflective metallic background, allowing to perform quantitative automatic image analyses. Moreover, a plot of the corrosion pits area fraction against the distance from the edge of the contact area (between the Mg-alloy plate and the counterpart or spacer), up to 10 mm distance, obtained from the latter image analysis, allowed to effectively measure and compare the localized corrosion at the edge of the mechanical junction, notwithstanding differences among different specimens in the amount of general corrosion effects detected further away from the junction, likely due to differences in the salt spray test severity (due to different positions in the test chamber) and/or in the polishing procedure. On the contrary, the overall corrosion pits area fraction of the same 10 mm wide examined region is less significant, because the surrounding general corrosion introduces a variable and possibly large contribution to this measure.

Localized Corrosion Mechanisms

Localized corrosion close to the boundary of the contact area (between the Mg-alloy plate and the counterpart or spacer) was detected in all tested assemblies.

In most cases, the contact area boundary was defined by a sharp edge of the counterpart or spacer, and the corrosion occurred immediately outside, but not immediately inside, such edge, on the plate surface which was directly exposed to the salt spray, hence it is attributed to galvanic corrosion.

However, in the F assembly, in which a cup washer was used as a spacer (between the Mg-alloy plate and a coated steel counterpart), the corrosion occurred mainly in the interstice between the cup itself and the plate, and is attributed to a combination of galvanic corrosion and crevice corrosion.

Furthermore, some crevice corrosion was also detected, although not measured, in the A assembly, in which there was no spacer, on some Mg-alloy plate areas, which were not adjacent to the contact area edge and were completely covered by the cast aluminum alloy counterpart.

Spacers Effectiveness

The comparison of the B and A type assemblies, in which the same 328.0 cast aluminum alloy counterpart was mounted with or without interposed 5051 (AlMg2) wrought aluminum alloy spacers, respectively, allows to conclude that the latter spacers were effective to reduce the localized corrosion, both by reducing (from 3.5 to 2 mm) the width of the deeply corroded region adjacent to the junction, and by avoiding the occurrence of crevice corrosion (which was found only in the A assembly contact area, as mentioned above).

This conclusion is further corroborated by the fact that the localized corrosion observed in the other assemblies, which all mounted aluminum alloy spacers, albeit with different counterparts, was always less than that observed in the A assembly type.

Moreover, the comparison between the F assembly, exhibiting a cup-washer spacer, on one side, and the D and E assemblies, exhibiting plain-washer spacers, on the other side, all with coated steel counterparts (albeit with different coating) allows to conclude that the plain washer geometry is much more effective than the tested cup washer one, because the latter favors the crevice corrosion.

However, the crevice corrosion arises here because the employed cup washer exhibit a large fillet radius at the contact area edge, thus creating an interstice with the Mg-alloy plate; hence it is not excluded that cup washers with a sharp contact area edge may be more effective than plain washers, by imposing a longer path between a low hydrogen voltage counterpart and the magnesium alloy.

Assemblies with Aluminum Alloy Counterparts

Among the 3 tested assemblies with aluminum alloy counterparts, the A assembly, with a 328.0 cast aluminum alloy counterpart and without any spacer, clearly exhibits the worst corrosion behavior; whereas the corrosion behaviors of the B and C assemblies, with 5051 (AlMg2) aluminum alloy spacers and cast or wrought aluminum alloy counterparts, is almost equal; hence, the usage of such spacers is recommended.

Assemblies with Coated Steel Counterparts

Among the 3 tested assemblies with coated steel counterparts, all with bare aluminum alloy spacers, the E assembly, with a zinc and nylon double coating applied on the steel counterpart, clearly showed the best corrosion resistance, whereas the F assembly, with the cup-washer spacer, was the worst due to the above described crevice corrosion issue, and the D assembly with the Zn-Al-Si coating was intermediate. In particular, the corrosion behavior of the E assembly was the best of all examined assemblies (with both aluminum alloy and coated steel counterparts).

Acknowledgements

A. Regis and I. Orsingher, Magnesium Products of Italy (MPI) S.p.A, Verres, Aosta, Italy, for material procurement and useful discussion.

References

1. G.L. Song and A. Atrens, "Corrosion Mechanism of Magnesium Alloys", Advanced Engineering Materials, 1 (1999), 11-33.
2. A. Froats, T.K. Aune, D. Hawke, W. Usworth and J. Hillis, in Metals Handbook, 9th ed., Vol.13, ASM Int., Material Park, OH, USA, 1987, 740-754.
3. W.S. Loose, in: Corrosion and Protection of Magnesium, ASM Int., Materials Park, OH, USA, 1946, 173-260.
4. R.B. Alvarez, H.J. Martin, M.F. Horstemeyer, M.Q. Chandler and N. Williams, "Corrosion relationship as a function of time and surface roughness on a structural AE44 magnesium alloy", Corrosion Science, 52 (2010), 1635-1648.
5. E. Angelini, S. Grassini, F. Rosalbino, F. Fracassi and R. D'Agostino, "Electrochemical impedance spectroscopy evaluation of the corrosion behaviour of Mg alloy coated with PECVD organosilicon thin film", Progress in Organic Coatings, 46 (2003), 107-111.
6. T.J. Luo, Y.S. Yang, Y.J. Li and X.G. Dong, "Influence of rare earth Y on the corrosion behavior of as-cast AZ91 alloy", Electrochimica Acta, 54 (2009), 6433-6437.
7. K.B. Deshpande, "Validated numerical modelling of galvanic corrosion for couples: Magnesium alloy (AE44)–mild steel and AE44–aluminium alloy (AA6063) in brine solution", Corrosion Science, 52 (2010), 3514–3522.
8. K.B. Deshpande, "Experimental investigation of galvanic corrosion: comparison between SVET and immersion techniques", Corrosion Science, 52 (2010), 2819-2826.
9. H.J. Martin, M.F. Horstemeyer and P.T. Wang, "Comparison of corrosion pitting under immersion and salt-spray environments on an as-cast AE44 magnesium alloy", Corrosion Science, 52 (2010), 3624–3638.

Magnesium Technology 2011
Edited by: Wim H. Sillekens, Sean R. Agnew, Neale R. Neelameggham, and Suveen N. Mathaudhu
TMS (The Minerals, Metals & Materials Society), 2011

Comparing the Corrosion Effects of Two Environments on As-Cast and Extruded Magnesium Alloys

H.J. Martin[1], C. Walton[1], J. Danzy[1], A. Hicks[1], M.F. Horstemeyer[1], P.T. Wang[1]
[1]Center for Advanced Vehicular Systems (CAVS), Mississippi State University, Box 5405, Mississippi State, MS, USA 39762

Keywords: Magnesium Alloy, Pitting Corrosion, Intergranular Corrosion

Abstract

Magnesium is easily corroded in the presence of saltwater, limiting its use in the automotive industry. The magnesium microstructure greatly affects the corrosion rate, due to various additional elements. In the Center for Advanced Vehicular Systems at Mississippi State University, the effects of immersion and cyclical salt spray testing on various as-cast and extruded magnesium alloys is currently being examined. Previous work on an as-cast AE44 magnesium alloy has demonstrated that individual pit characteristics, such as pit depth, pit area, and pit volume, were deeper and larger following exposure to the immersion environment. However, the data elucidating the corrosion effects on individual pit characteristics has only been seen on as-cast magnesium containing rare earth elements, not on extruded magnesium alloys or zinc-containing magnesium alloys, both common magnesium forms. The research presented here will cover the effects of individual pit characteristics formed on various magnesium alloys due to the different environments.

Introduction

While magnesium is currently used in both the automotive and aerospace industries, its high corrosion rate means that it can only be used in areas that are unexposed to the environment [1-3]. Various elements, such as aluminum, zinc, manganese, and rare earth elements, have been added in an effort to improve the corrosion resistance of magnesium [2, 4-8].

Corrosion resistance has been shown to be highly affected by percentage of aluminum added [2]. The presence of aluminum, in the β-phase precipitate and appearing as $Mg_{17}Al_{12}$, can improve the corrosion resistance of magnesium when the β-phase is continuous [2, 9-11]. However, when the β-phase is small and unconnected, aluminum can lead to the creation of micro-galvanic cells, which reduces corrosion resistance [2, 9-11].

The presence of rare earth elements can also affect the corrosion properties and mechanical properties of magnesium [4-8]. When rare earth elements are present in the β-phase, the creep properties are improved, as well as the corrosion resistance [6-8, 12]. Corrosion resistance is improved due to shift of pitting corrosion, from along the magnesium grain – eutectic boundary to the interior of the magnesium grain [7, 13].

Besides the alloying elements, the presence of an as-cast skin can also affect the corrosion resistance of magnesium. It has been shown on AZ91 that the corrosion of the as-cast skin is 10-fold lower than the bulk AZ91, due to the presence of very small grains [9, 14]. However, extrusion removes the as-cast skin, resulting in a higher corrosion rate.

Pitting corrosion and general corrosion are also affected by the exposure conditions of magnesium. Salt spray testing and immersion testing are common testing environments, with ASTM standards that were developed separately [15-16]. These two ASTM standards, however, require different concentrations of salt, meaning a direct comparison between the results cannot be made [15-16]. In addition, field corrosion tests do not translate to ASTM standards, leading to an industrial development of cyclical tests [15-18]. Several industrial tests exist, such as Renault ECC1, Volkswagen PV1210, and General Motors GM9540P [18]. These industrial cyclical tests contain a pollution phase and a wet or dry phase in an effort to expose test alloys to environmental conditions that are associated with engine cradles, such as de-icing salt, mud, and condensation [18-19]. However, none of the industrial tests are the same, with varying amounts of NaCl, such as 1%, 5%, and 0.9%, respectively, varying pHs, such as a pH of 4, 6.5-7.2, and 6-9, respectively, and varying exposure times, such as 30 min/day, 4 hrs/day, and 4 x 30 min/day, respectively [18]. In addition, the General Motors tests include other chemicals, such as $CaCl_2$ and $NaHCO_3$ [18]. These differences mean that the results from the multitude of industrial tests cannot be compared [18-19]. In an effort to develop a cyclical test where the results could easily compare with the results from an immersion test, four cyclical test combinations were examined, which showed that a 3.5% NaCl solution which cycled through salt-spray, 100% humidity, and a drying phase proved to be the most corrosive [20]. The goal of this research, then, is to study various magnesium alloys in as-cast or extruded form in order to understand how individual pits grow in depth, surface area, and volume to determine how the various alloying elements affect individual pit characteristics based on environment.

Materials and Methods

Testing

Twelve AZ61 coupons and twelve AM30 coupons (2.54 cm x 2.54 cm x varying thicknesses) were cut from an extruded crash rail provided by Ford using a CNC Mill (Haas, Oxnard, CA). Twelve AZ31 coupons were cut from extruded sheets using a vertical band saw (MSC Industrial Supply Company, Columbus, MS). Twelve AE44 coupons (2.54 cm x 2.54 cm x varying thicknesses) were cut from an as-cast engine cradle provided by Meridian Technologies using a vertical band saw. The coupon surfaces were left untreated to test the corrosion effects on the extruded AZ31 and AZ61 magnesium alloys and on the as-cast AM60 and AE44 magnesium alloys.

Two different test environments were used in this study: salt spray testing and immersion. For salt spray testing, a Q-Fog CCT (Q-Panel Lab Products, Cleveland, OH) was used to cycle through three stages set at equal times, including a 3.5 wt.% NaCl spray at 35°C, 100% humidity using distilled water at 35°C, and a drying purge at 35°C. For immersion testing, an aquarium with an aeration unit was filled with 3.5 wt.% NaCl at room temperature. For both tests, the six coupons per test environment were hung at 20° to the horizontal, as recommended by ASTM B-117 [15]. The coupons were exposed to the test environment for 1 h, removed, rinsed with distilled water to remove excess salt, and dried. No chemicals were used to clean the surfaces following the corrosion experiments to ensure that the pits and surfaces were unchanged over time. Following the profilometer analysis, the coupons were then placed back into the test environment for an additional 3 h, an additional 8 h, an additional 24 h, and another 24 h. These times allowed for a longitudinal study to follow pit growth and surface changes over time, where $t_0 = 0$, $t_1 = 1$ h, $t_2 = 4$ h, $t_3 = 12$ h, $t_4 = 36$ h, and $t_5 = 60$ h. Between analyses and environmental exposures, the coupons were stored in a desiccator to ensure that no further surface reactions occurred.

Analysis

Following each time exposure, the coupons were analyzed using optical microscopy and laser profilometry. The coupons were weighed prior to testing and following each exposure on two different scales and an average was taken. Four thickness measurements were taken on each sample prior to and following the test. Because the coupons were cut from an engine cradle, the thicknesses of the coupons varied from side to side, meaning an average was taken per coupon based on the four measurements. Measurements for all figures were averaged from the data with error bars based on one standard deviation.

Laser profilometry was used to scan a 1 mm by 1 mm area on two coupons per environment following each test cycle (Talysurf CLI 2000, Taylor Hobson Precision Ltd, Leicester, England). The resulting 2-D and 3-D images were used to document the changes in the pit characteristics due to the different test environments over the six cycles (Talymap Universal, v. 3.18, Taylor Hobson Precision Ltd, Leicester, England). Data collected was averaged based on fourteen pits within the same 1 mm by 1 mm area, for a total of twenty-eight data points per environment per cycle. The software was used to calculate pit maximum depth, pit mean depth, pit surface area, and pit volume.

Results

Figures 1 and 2 show the average weight and thickness change, respectively, over the five exposure times for the immersion and salt spray surfaces on the various magnesium alloys being compared. As one can see, all surfaces follow similar logarithmic trends for weight change (Figure 1) and thickness change (Figure 2).

Figure 3 shows the maximum pit depth over the five exposure times for the immersion and salt spray surfaces on the various magnesium alloys being compared. As one can see, the salt spray surfaces followed second-order polynomial trends, while the immersion surfaces followed more linear trends. The as-cast AM60 surfaces showed the deepest pits as compared to the other surfaces, while the as-cast AE44 surfaces and extruded AZ61 surfaces showed the shallowest pit formation.

Figure 4 shows mean pit depth over the five exposure times for the immersion and salt spray surfaces on the various magnesium alloys being compared. As with maximum pit depth, most salt spray surfaces followed a second-order polynomial, while most immersion surfaces followed an almost linear trend. However, the trend on the as-cast AM60 surfaces switched, with the salt spray surface following an almost linear trend and the immersion

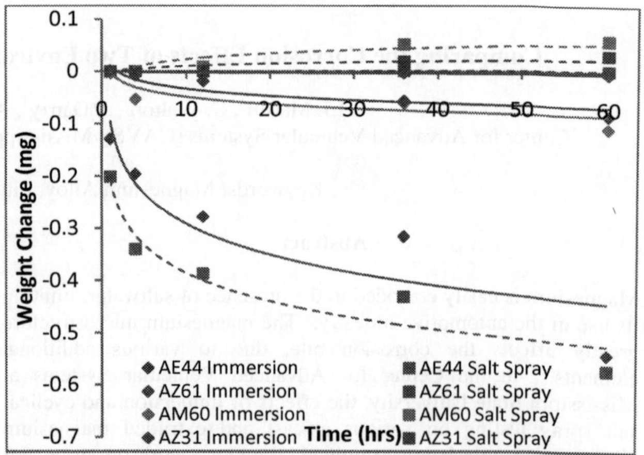

Figure 1: Average weight change of various magnesium alloys based on test environment over 60 h. Notice that all surfaces followed logarithmic trends.

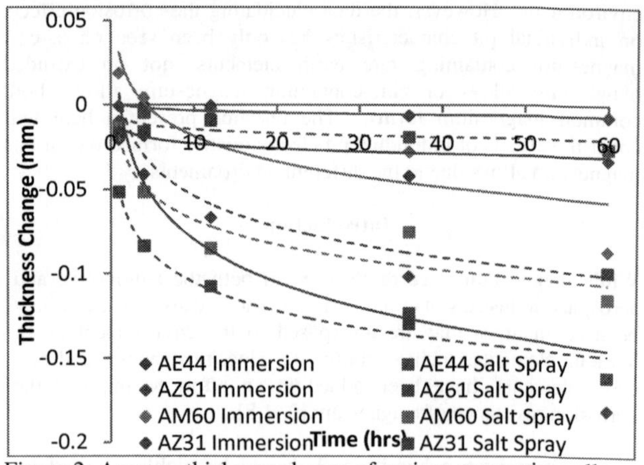

Figure 2: Average thickness change of various magnesium alloys based on test environment over 60 h. Notice that all surfaces followed logarithmic trends.

surface following a second-order polynomial trend. The as-cast AM60 surface still showed the deepest pits, though.

Figure 5 shows the changes in the pit surface area, which is the area calculated over the 3-D area covered by the pit using laser profilometry, over the five exposure times for the immersion and salt spray surfaces on the various magnesium alloys being compared. For all surfaces, the pit surface area followed second-order polynomial trends, with the largest pit surface area occurring on the as-cast AM60 surface and the smallest pit surface area occurring on the as-cast AE44 surface. Notice also that both AM60 surfaces were divided by 10 to bring the values for pit surface area within the range of the other values, in order to followed second-order polynomial trends, while the immersion surfaces appeared to follow more linear trends.

Figure 6 shows the changes in the pit volume over the five exposure times for the immersion and salt spray surfaces on the various magnesium alloys being compared, using laser profilometry. As with the pit surface area, all surfaces followed second-order polynomial trends, with the largest pit volume occurring on the as-cast AM60 surface and the smallest pit surface

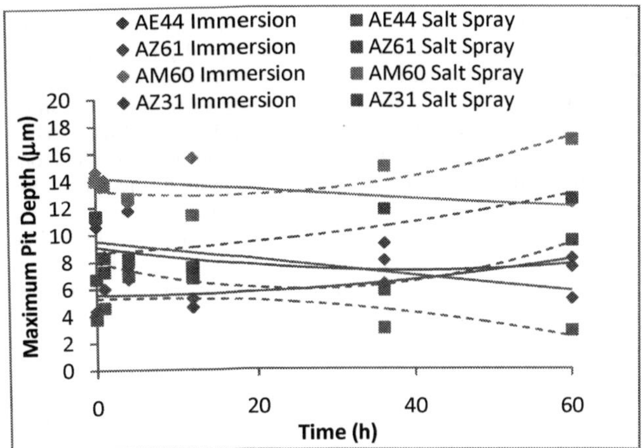

Figure 3: Maximum pit depth of various magnesium alloys based on test environment over 60 h. Notice that all salt spray surfaces exposure times for the immersion and salt spray surfaces on the various magnesium alloys being compared.

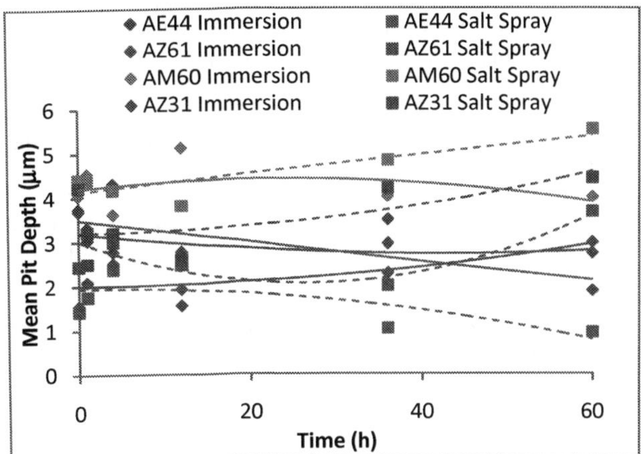

Figure 4: Mean pit depth of various magnesium alloys based on test environment over 60 h. Notice that the salt spray surfaces, except for AM60, followed second-order polynomial trends while the immersion surfaces followed mostly linear trends, again except for AM60.

area occurring on the as-cast AE44 surface. Notice also that both AM60 surfaces were divided by 10 to bring the values for pit volume within the range of the other values, in order to prevent a large y-axis that compressed the other three volume values.

Discussion

More weight loss is seen on the immersion surfaces as compared to the salt spray surfaces, expect with respect to the as-cast AE44 surfaces (Figure 1). Because the samples in the salt spray environment are not continuously exposed to water, the water could not react with the surface continuously, meaning that less weight was lost from the surface of the material. However, higher thickness loss was seen on the salt spray surfaces as compared to the immersion surfaces, likely due to changes around the outsides of the coupons exposed to the cyclical salt spray (Figure 2). Since water was not continuously removing debris and salt from the edges in the cyclical salt spray, the debris and salt remained on the

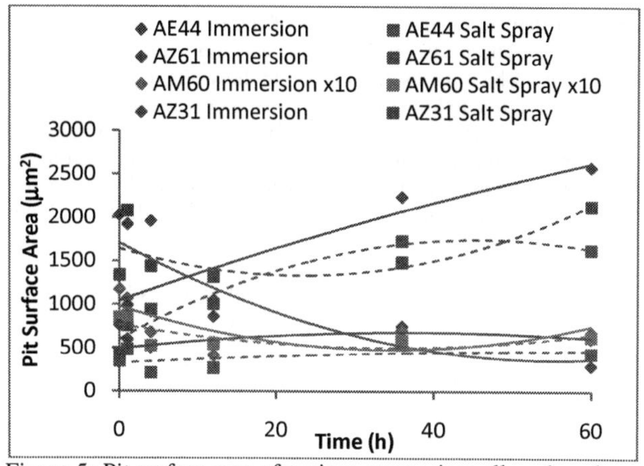

Figure 5: Pit surface area of various magnesium alloys based on test environment over 60 h. Notice that all surfaces followed second-order polynomial trends. Also notice that the as-cast AM60 surfaces had the largest pit surface area, which was divided by 10 to ensure all data could be seen. Notice also that the as-cast AE44 surfaces had the smallest pit surface area.

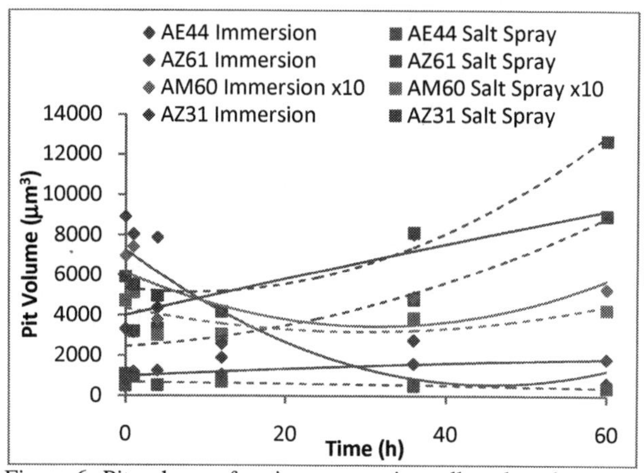

Figure 6: Pit volume of various magnesium alloys based on test environment over 60 h. Notice that all surfaces followed second-order polynomial trends. Also notice that the as-cast AM60 surfaces had the largest pit volume, which was divided by 10 to ensure all data could be seen. Notice also that the as-cast AE44 surfaces had the smallest pit volume.

edges, reacting with and degrading the magnesium. Since weight was measured with a scale, placement on the scale would not affect the weight. However, thickness was measured with calipers, so the placement of the calipers on the outside of each of the coupons would affect the thickness measurements, and the debris degrading the edges would negatively affect the thickness measurements.

Weight loss and thickness loss are not the only measure of corrosion, however. Pitting corrosion is highly detrimental but may not be detected by changes in weight and thickness. Instead, monitoring for pitting characteristics, such as pit depth, pit surface area, and pit volume allow one to determine how the pits grow over time. The maximum pit depth indicates how deep a pit is at its deepest spot, which is useful in determining the long term effect of that pit on the material or if the pit can cause a breech (Figure 3). The mean pit depth indicates, on average, how deep the pit is, which is useful in seeing how the pit is growing outwards relative to how the pit is growing downwards (Figure 4). The pit surface area is the 3-D area covered by the pit (Figure 5), while the pit volume is the volume of the material removed during the pitting process (Figure 6). Since pits are essentially cones, the 3-D area and volume incorporate the mean pit depth and the maximum pit depth in the calculation of surface area and volume, allowing one to determine how the growth of the pit relates to the depth of the pit. Table I shows the overall rankings based on the charts of the differences in pit characteristics.

Table I: Overall and Group Ranking of Pit Characteristics by Environment

Environment	Alloy	Depth		Sur. Area		Volume	
Immersion	AZ31	6	3	3	2	4	2
	AE44	7	4	8	4	7	4
	AZ61	5	2	6	3	6	3
	AM60	3	1	1	1	1	1
Salt Spray	AZ31	2	2	4	2	3	2
	AE44	8	4	7	4	8	4
	AZ61	4	3	5	3	5	3
	AM60	1	1	2	1	2	1

The initial number in each column is the ranking of each line on the graphs based on the final time. The second number in each column is the ranking of each line within the environment based on the final time. Notice that the trend changes for the depth, but stays the same for the pit surface area and pit volume. Also remember that AM60 was divided by 10 on the graphs to ensure that all data could be seen, which is why it is 1 and 2 in this table.

Four different factors affect the pit depth, pit surface area, and pit volume: general corrosion, the form of magnesium (as-cast versus extruded), the corrosive environment, and the type of magnesium alloy. When looking at the effects of general corrosion (Figures 1 and 2), one can see that some magnesium alloys, specifically both environments of AE44 and the immersion environment of AZ61, showing decreasing values in pit depth, while only AE44 shows a decreasing surface area and volume. The decrease, instead of increases, means that general corrosion is acting on the surface faster than the pits can grow. While the pits were initially able to begin corroding the AE44 material, over time, general corrosion stopped pit growth and instead corroded the entire surface equally.

The first factor that can affect the change in pit characteristics, general corrosion, only really affected the as-cast AE44. The other materials could experience differences in pit formation due to the form of corrosion, the environment, or the magnesium alloying elements. When one compares the pit depth based on the form of magnesium, the extruded AZ61 and AZ31 materials fall in the middle of the two as-cast magnesiums (Table I, Figures 3 and 4). In addition, when looking at the surface area and volume, the form of magnesium does not play a significant role in the growth of the pits, with the extruded magnesium alloys again occurring in the middle of the as-cast materials (Please remember that the AM60 material was divided by 10 in order to fit the lines on the graph and ensure all data could be seen) (Table I, Figures 5 and 6). By looking at the individual pit characteristics, averaged over 14 pits per environment per time, one can see that the form of magnesium plays very little role in the formation or growth of individual pits.

Because general corrosion only affected AE44 and the forming of the magnesium alloys did not affect individual pit formation and growth, one of the other two factors, the corrosive environment or the magnesium alloy, must account for the differences in pitting on the four magnesium alloys (Table I, Figures 3-6). When looking at the corrosive environments irregardless of alloy, one sees that, with the exception of AZ61, the pits in the immersion environment decreased throughout the experiment time (Table I, Figures 3 and 4). Because pitting corrosion is the initial mechanism of corrosion and general corrosion "catches" up to pitting corrosion, the continuous exposure of water to the magnesium surface allowed the magnesium surrounding the pits to degrade, resulting in less deep pits. However, because the pits in the salt spray environment were not continuously exposed to water, but the chloride ions could become trapped within the pits, the pits were able to grow in depth over time, with the exception of AE44. With respect to pit surface area, there were no overall trends when examining the different environments, as some materials increased throughout the experiment time (AZ31 immersion, AE44 salt spray), one decreased throughout the experiment time (AE44 immersion), and the remaining followed parabolic curves, positively (both AZ61 environments) or negatively (both AM60 environments, AZ31 salt spray) (Figure 5). The pit volume data showed the same lack of trends between the two environments (Figure 6).

The differences in environment can be attributed to the differences in pit characteristics, although the environment alone cannot be responsible for the differences in pit characteristics. If environment was solely responsible for the differences, then one would expect to see that all surfaces followed the same trends, regardless of alloying elements. Either the salt spray environment would produce the largest, deepest pits because of pit debris allowing the autocatalytic nature of pitting to proceed even during the drying phase [1] or the immersion environment would produce the largest, deepest pits because of the continuous presence of chloride ions. Instead, one sees that a difference between environments exists based on alloying elements, where the pit depth for AM60, AZ31, and AZ61 are higher on the salt spray surfaces and the pit depth for AE44 is higher on the immersion surfaces. In addition, the pit surface area is larger on the AZ31, AM60, and AE44 immersion surfaces, while AZ61 has a higher pit surface area on the salt spray surfaces. Lastly, the pit volume on AZ31 and AZ61 is higher on the salt spray surfaces, while the pit volume on AM60 and AE44 is higher on the immersion surfaces.

When it comes to alloying elements, though, one can see a major difference between the four different magnesium alloys (Table I). AM60 had the deepest pits, with the largest pit surface area and largest pit volume, while AE44 had the shallowest pits with the smallest pit surface area and smallest pit volume. The other two alloys, AZ31 and AZ61, showed differences in pit depth, pit surface area, and pit volume, but were within the differences experienced by AE44 and AM60. The differences in pit characteristics must then be related to the alloying elements, although they cannot be attributed to the percentage of aluminum. While it has been shown that up to 10% aluminum can increase corrosion resistance [2], if the percentage of aluminum alone was responsible for the differences in pitting corrosion, then one would expect that AZ31, with 3% aluminum, would be most heavily corroded, followed by AE44 (4% Al), with AZ61 (6% Al) and AM60 (6% Al) tied. As previously stated, though, that is not the case, as AM60 is the most heavily corroded and AE44 is the least corroded.

While the reason that AE44 is corroded less than the other three materials can be explained, the reason that AM60 is more heavily corroded, even in the as-cast state, at this point cannot be explained. Currently, other AM alloys compositions are being corroded to see if manganese has an influence on the corrosion or if the aluminum-manganese percentage is the cause. AE44 corroded less than either of the AZ alloys due to the way in which AE44 corrodes. It has been shown that on AE44, the pits form within the magnesium grains and not along the intergranular boundaries due to the addition of the rare earth elements [7, 13]. While the pit can begin growing, both in depth and in area/volume, once the pit has corroded the entire grain, there is no further way for that pit to grow. This means that, once the pit has corroded the entire grain, there is nowhere else for the pit to grow, resulting in a pit growth stoppage and preventing a breech through the material.

Conclusions

Four magnesium alloys in two forms, as-cast AE44, as-cast AM60, extruded AZ61, and extruded AZ31, were examined in two corrosive environments, immersion and salt spray. Bulk coupon characteristics, weight loss and thickness loss, as well as individual pitting characteristics, maximum pit depth, mean pit depth, pit surface area, and pit volume, were quantified over 60 hours. With respect to the individual pitting characteristics, the form of magnesium and the environment appeared to have minimal affect on the pit depth, pit surface area, or pit volume, indicating that once the pits began forming, the environment nor the form would affect their growth. However, alloying elements did affect pit growth. AM60 had deeper pits, with the largest pit surface areas and largest pit volumes, while AE44 had the shallowest pits, with the smallest pit surface areas and pit volumes. The extreme corrosion of AM60 cannot yet be explained, while the small corrosion of AE44 is attributed to the corrosion characteristics controlled by the addition of the rare earth elements. Overall, the most heavily corroded magnesium alloy, determined by combining general and pitting corrosion, was AM60, followed by AZ31, AZ61, and AE44, respectively.

References

[1] M.G. Fontana, Corrosion Principles, in: M.G. Fontana (Eds.), *Corrosion Engineering*, McGraw-Hill, Boston, 1986, pp. 12-38.

[2] G. Song, A. Atrens, "Understanding Magnesium Corrosion – A Framework for Improved Alloy Performance", *Advanced Engineering Materials* 5 (2003) 837-858.

[3] BA Shaw, Corrosion Resistance of Magnesium Alloys, in: L.J. Korb, ASM (Eds.), ASM Handbook, Vol. 13A: Corrosion, Ninth Ed., ASM International Handbook Committee, Metals Park, 2003, pg. 692.

[4] J.D. Majumdar, R. Galun, B. Mordike, I. Manna, "Effect of laser surface melting on corrosion and wear resistance of a commercial magnesium alloy", *Materials Science and Engineering A*, 361 (2003) 119-129.

[5] C. Blawert, E.D. Morales, W. Dietzel, K.U. Kainer, "Comparison of Corrosion Properties of Squeeze Cast and Thixocast MgZnRE Alloys", *Materials Science Forum*, 488-489 (2005) 697-700.

[6] W. Liu, F. Cao, L. Chang, Z. Zhang, J. Zhang, "Effect of rare earth element Ce and La on corrosion behavior of AM60 magnesium alloy", *Corrosion Science*, 51 (2009) 1334-1343.

[7] W. Liu, F. Cao, L. Zhong, L. Zheng, B. Jia, Z. Zhang, J. Zhang, "Influence of rare earth element Ce and La addition on corrosion behavior of AZ91 magnesium alloy", *Materials and Corrosion*, 60 (2009) 795-803.

[8] Y.L. Song, Y.H. Liu, S.R. Yu, X.Y. Zhu, S.H. Wang, "Effect of neodymium on microstructure and corrosion resistance of AZ91 magnesium alloy", *Journal of Materials Science*, 42 (2007) 4435-4440.

[9] G. Song, "Recent Progress in Corrosion and Protection of Magnesium Alloys", *Advanced Engineering Materials*, 7 (2005) 563-586.

[10] M.C. Zhao, M. Liu, G. Song, A. Atrens, "Influence of pH and chloride ion concentration on the corrosion of Mg alloy ZE41", *Corrosion Science*, 50 (2008) 1939-1953.

[11] G. Song, A. Atrens, X. Wu, B. Zhang, "Corrosion behavior of AZ21, AZ501, and AZ91 in sodium chloride", *Corrosion Science*, 40 (1998) 1769-1791.

[12] Y.L. Song, Y.H. Liu, S.H. Wang, S.R. Yu, X.Y. Zhu, "Effect of cerium addition on microstructure and corrosion resistance of die cast AZ91 magnesium alloy", *Materials and Corrosion*, 58 (2007) 189-192.

[13] N. Birbilis, M.A. Easton, A.D. Sudholz, S.M. Zhu, M.A. Gibson, "On the corrosion of binary magnesium-rare earch alloys", *Corrosion Science*, 51 (2009) 683-689.

[14] G. Song, A. Atrens, M. Dargusch, "Influence of microstructure on the corrosion of diecast AZ91D", *Corrosion Science*, 41 (1998) 249-273.

[15] ASTM B117 - 07a (2007) Standard Practice for Operating Salt Spray (Fog) Apparatus, Vol. 03.02, 2007.

[16] ASTM G31 - 72 (2004) Standard Practice for Laboratory Immersion Corrosion Testing of Metals, Vol. 03.02, 2004.

[17] K.R. Baldwin, C.J.E. Smith, "Accelerated corrosion tests for aerospace materials: current limitations and future trends", *Aircraft Engineering and Aerospace Technology*, 71 (1999) 239-44.

[18] N. LeBozec, N. Blandin, D. Thierry, "Accelerated corrosion tests in the automotive industry: A comparison of the performance towards cosmetic corrosion", *Materials and Corrosion* 59 (2008) 889-94.

[19] G. Song, D. St.John, C. Bettles, G. Dunlop, "The corrosion performace of magnesium alloy AM-SC1 in automotive engine block applications", *Journal of the Minerals, Metals, and Materials Society*, 57 (2005) 54-6.

[20] H.J. Martin, M.F. Horstemeyer, P.T. Wang, "Effects of Variations in Salt-Spray Conditions on the Corrosion Mechanisms of an AE44 Magnesium Alloy", *International Journal of Corrosion*, 2010 (2010) 1-10 doi: 10.1155/2010/602342.

[21] M.F. Horstemeyer, J. Lathrop, A.M. Gokhale, M. Dighe, "Modeling stress state dependent damage evolution in a cast Al-Si-Mg aluminum alloy", *Theoretical and Applied Fracture Mechanics*, 33 (2000) 31-47.

Magnesium Technology 2011
Edited by: Wim H. Sillekens, Sean R. Agnew, Neale R. Neelameggham, and Suveen N. Mathaudhu
TMS (The Minerals, Metals & Materials Society), 2011

Influence of Lanthanum concentration on the Corrosion Behaviour of Binary Mg-La Alloys

Rosario Silva Campos, Daniel Höche, Carsten Blawert, Karl Ulrich Kainer

GKSS Research Centre, Institute of Materials Research, Max-Planck-Straße 1, Geesthacht 21502, Germany

Keywords: Magnesium corrosion, Rare earth - Lanthanum, XPS, Auger

Abstract

Different contents of Lanthanum have been added to Magnesium and have been investigated on their influence on the microstructure and the corrosion properties. The microstructure was studied by optical microscopy. Corrosion performance was evaluated using potentiodynamic polarization measurements. Immersion tests were carried out using distilled water and 0.1 M sodium chloride solution. The corrosion products were investigated by X-ray induced photoelectron spectroscopy (XPS), Auger electron spectroscopy (AES) and X-ray diffraction (XRD) which lead to detailed information on phase formation. The oxide and hydroxide formation have been correlated to the chemical states and formed intermetallics, i.e by taking into account the XPS peak shift and peak splitting of the the Mg-2p state. Additionally, the results have been verified by means of AES on the Mg-KLL, O-KLL and La-MNN excitation and by XRD. Latter suggests the supplemental formation of a nanocrystalline phase.

Introduction

Magnesium alloys are known for their good properties, which are greatly attractive to the automotive and aerospace industries. Their use however is limited due to their poor corrosion properties [1]. There are two main reasons of this lack of corrosion resistance [2]. Firstly, there is internal galvanic corrosion caused by second phases or impurities [3]. Secondly, the quasi-passive hydroxide film on Mg is much less stable than the passive films which form on metals such as aluminium and stainless steels. This quasi-passivity provides only poor pitting resistance for Mg and Mg-alloys [4]. The early Mg-alloys suffered rapid attack in moist conditions due mainly to the presence of impurities, notably iron, nickel, and copper. These impurities or their compounds act as minute cathodes in a corroding medium. They create micro-cells with an anodic Mg matrix [5].

Previous studies [6-9] have showed that for Mg alloys, rare earths (RE) elements can enhance their corrosion resistance. For example addition of Ce as a typical RE element prompts the formation of an Al- enriched layer on the corrosion film, which contributes to the corrosion resistance of AZ91D alloy [10]. In case of Mg-Gd-Zr alloys the addition of La or Ce reduces the dendrite arm spacing of the as- cast alloy and improves the mechanical properties and age hardening response [11].

In this work several experiments were performed in order to study the corrosion behaviour of Mg-La alloys. The immersion test, potentiodynamic polarization and analytical techniques were combined to study the characteristics of the films formed on the surface of Mg-La specimens immersed in different corrosion environments.

Experimental Procedure

Casting

Several Mg-La alloys were prepared from their pure elements, amounts in weight percent were added (1, 5, 10 and 15 wt% La) and balance with high purity Mg. The alloys were casted in a resistance furnace at a starting temperature of 700°C under an Argon-SF6 atmosphere. After casting, chemical analysis was performed in order to ensure the composition homogeneity in the whole specimen.

Optical Characterization

The samples were grinded with SiC paper up to grit 4500; polished with SiO_2 and etched with a picric acid solution, ultrasonically cleaned using ethanol, and dried in hot air. After the pre-treatment the surfaces were investigated using a light microscope - PC system including metallographic software.

Immersion

Flat specimens with dimensions 20 x 15 x 4 mm^3 were cut using a diamond cutting disc, grinded with SiC paper up to grit 1200, ultrasonically cleaned using ethanol and dried in hot air. The coupons were immersed 10 days in deionised water. Subsequently these were washed with ethanol and dried in hot air. The samples were used for XPS analysis. Pure Mg samples were treated in the same way to compare the results.

Electrochemical Measurements

The coupons were cut from the cast ingot to the following dimensions 20 x 15 x 10mm^3, grinded with SiC paper up to grit 1200, ultrasonically cleaned using ethanol and dried in hot air. The electrochemical tests were conducted in an acrylic cell (330 ml electrolyte) using a Gill AC Potentiostat from ACM Instruments. The working electrode had ~0.5 cm^2 exposed area. A solution of 0.1 M NaCl was used as electrolyte, while reference and auxiliary electrodes were Ag/AgCl, and Pt grid respectively. The experiments were done at 25°C without deareation. Magnetic stirring was performed with a 4 cm length magnet at low speed. Experiments were carried out in triplicate. The sequence of measurements used was (a) 2 hours of open circuit potential followed by (b) potentiodynamic scanning starting at -200 mV relative to OCP and ending at 1500 mV and current limit was 0.5 mA/cm^2 with a scan rate of 0.5 mV/s. The corrosion density was determined from the current density at the intersection of cathodic slope with the vertical line through the OCP.

X-ray induced Photoelectron Spectroscopy (XPS)

XPS experiments were carried out on a Kratos Axis Ultra DLD attached with a 15 kV X-ray gun using monochromatic Al K_α radiation. The spot size was 700 x 300 microns and the pass energy 40 eV at the regions measurements. Due to physical limits the information depth is limited to approx. 5 nm. Additionally, argon ions (4 keV) have been used to etch the samples in order to obtain depth profiles. The used sputter rate was approx. 40 nm per minute for total sputter times of 250 min.

Auger Electron Spectroscopy (AES)
AES is an attached feature of the Kratos system. An electron gun having a spot size of about 100 nm and a beam current of 5 nA has been focussed on points of interests, which were analyzed at specific kinetic energies.

X-ray diffraction (XRD)
The measurements have been performed on a Panalytical system including a poly-capillary optic and an "Euler" circle using CuK_α radiation. The scans have been carried out at Bragg-Brentano geometry leading to an information depth of about 200 microns.

Results and Discussion

Microstructures of Mg alloys with different La compositions are shown in Fig. 1. In all cases two phases are present. Primary phase is α–Mg and second phase or eutectic corresponds to (α-Mg+ $Mg_{17}La_2$) (XRD data is shown in Fig. 7).At the highest compositions (Fig 1c and 1d), the α–Mg regions are smaller and are fully enveloped by the eutectic. The lamellar structure of the eutectic phase becomes also coarser as the La content increases.

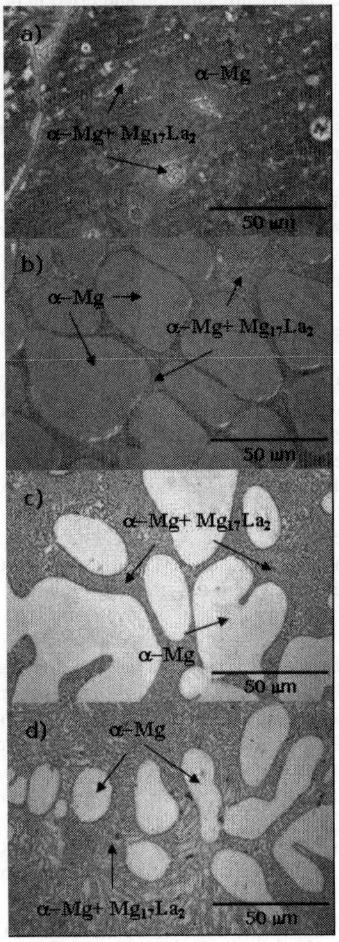

Figure 1. Microstructure of Mg-La alloys a) Mg1wt%La, b)Mg5wt%La, c) Mg10wt% La and d) Mg15wt% La

Electrochemical Tests
The corrosion potentials (Open circuit potential.) of Mg-La alloys were recorded for 2 hours. The potential evolution is shown in Fig. 2 and the potentials measured after 2 hours are reported in Table 1. Mg, Mg1La, Mg5La have similar potentials and are more active than the rest, while $Mg\text{-}La_{intermetallic}$ ($Mg_{17}La_2$), Mg10La and Mg15La alloys have more noble potentials than Mg but are more active compared to La. With increasing La content the potential shifts to more noble values which are consistent with the nobler potential of La compare with Mg.

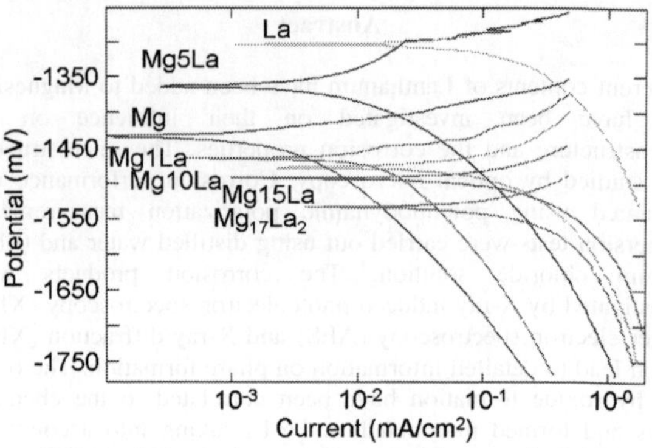

Figure 2. Potentiodynamic current-potential curves in 0.1 M NaCl at T_{amb} (pH=7 with stirring)

Table I. Electrochemical corrosion data for the alloys in 0.1M NaCl at T_{amb} (pH=7 with stirring)

Alloy	OCP (2h) $[mV]_{Ag/AgCl}$	E_{corr} $[mV]_{Ag/AgCl}$	I_{corr} $[\mu A/cm^2]$	Corrosion Rate [mm/year]
Mg1La	-1572 ± 2	-1464 ± 10	22.5 ± 1	0.51 ± 0.02
Mg5La	-1573 ± 2	-1364 ± 4	27.3 ± 20	0.62 ± 0.39
Mg10La	-1552 ± 10	-1477 ± 12	160 ± 32	3.67 ± 0.74
Mg15La	-1511 ± 1	-1492 ± 3	275 ± 56	3.70 ± 1.29
Mg_{pure}	-1576 ± 4	-1433 ± 11	4.6 ± 1.6	0.11 ± 0.04
$Mg_{17}La_2$	-1549 ± 2	-1533 ± 6	182 ± 44	4.15 ± 1.00
La_{pure}	-1330 ± 3	- 1303 ± 4	125 ± 8	3.06 ± 0.21

The current-voltage curves for the alloys are shown in Fig. 2 and the corrosion parameters are reported in Table 1. The microstructure changes due to La addition (Fig. 1) create variations on the corrosion resistance as shown in Fig. 3 and Table 1. Mg had the lowest corrosion rate value and additions up to 5 wt% La caused minor increase, while that with 10 wt% La or more causes a high increase on the corrosion rate. In other words, the eutectic phase (α-Mg + $Mg_{17}La_2$) gets more easily and rapidly corroded than the primary (α-Mg) phase. There are two main reasons to explain that behaviour. If the material is polarised, the potentials of Mg and $Mg_{17}La_2$ shift differently. The consequence is that in contrast to OCP the Mg matrix becomes more noble. In galvanic couple now the intermetallic is dissolving protecting the matrix. Furthermore the corrosion resistance of the intermetallic is low, thus with increasing La content in the alloy more of the eutectic forms with poor corrosion resistance which is additionally polarised anodically by the Mg matrix.

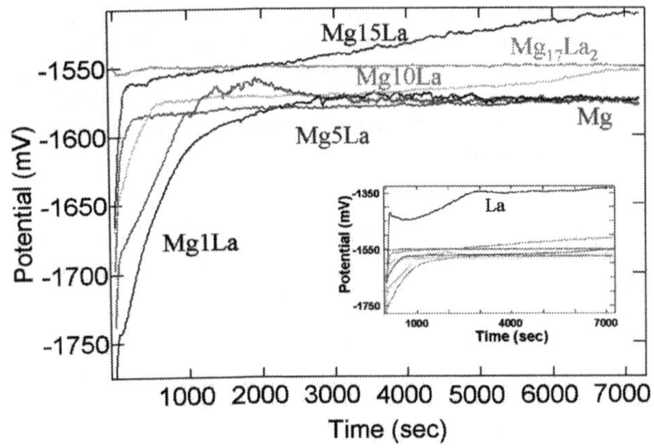

Figure 3. Open circuit potential vs time curves of the studied magnesium alloys in 0.1 M NaCl (pH=7) with stirring. Potentials are referred to Ag/AgCl electrode.

X-ray induced Photoelectron Spectroscopy (XPS)

The detailed analysis of corrosion processes needs the knowledge of the composition in the corroded material, especially of the resulting corrosion layers. Thus, XPS was applied to measure elemental depth profiles in an alternating process of sputtering and measuring. Fig. 4.represents the elemental distribution of the Mg-La alloys after immersion in distilled water for ten days.

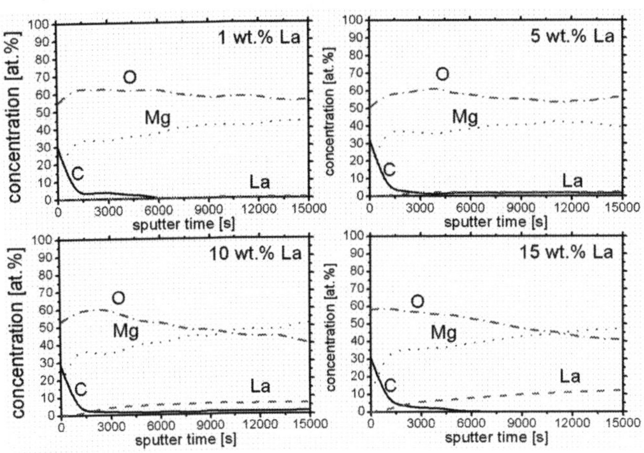

Figure 4. XPS depth profiles for Mg-La alloys measured by evaluating the peak areas of Mg 2p, O 1s, C 1s and La 3d states after different etch time steps.

First, it becomes visible that the carbon contamination is very similar for all samples. These C atoms are bonded in carbonates (mainly $MgCO_3$). As next it is obvious that the lanthanum content is very small close to the surface, which leads to the conclusion of a magnesium dominated process close to the surface. The Mg depth profiles (atomic content) are also very similar, but not the state of magnesium. This gets explained in the next paragraph. By means of the oxygen distribution it is possible to estimate the thickness of the corrosion layer. For higher La contents the layers are significantly thinner.

In order to get information on the chemical state inside of the corrosion layer, the Mg 2p state has been measured at a high resolution after sputtering. Fig. 5 shows the different atomic bonds of magnesium for varying lanthanum weight contents. It has to be mentioned that the analyzed area is 700x300 μm², which yields in an average determination of the phases across the surface. The four high resolution scans reveal the different oxidation states of Mg-La alloys after immersion. For the lowest lanthanum content the typical hydroxide formation [12] occurs at least to the crater depth of that measurement. Additionally, after drying MgO phase formation occurs. The situation changes for higher La contents. Magnesium hydroxide formation gets suppressed due to the formation of $La(OH)_3$ (see X-ray diffraction), small amounts of La_2O_3 and a phase known as La_2MgO_x [13-14].

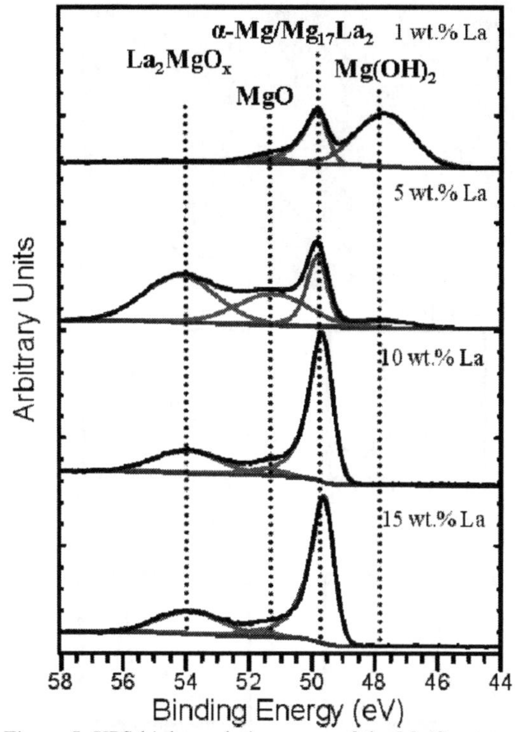

Figure 5. XPS high resolution scans of the Mg 2p state after completed etching. The corresponding chemical bond has been highlighted.

This corrosion behavior stays similar for higher La contents, but the layer thickness becomes thinner. As a result of the $Mg_{17}La_2$ intermettalic phase, La_2MgO_x formation seems to be the most common reaction product. It is interesting that this compound has catalytic properties especially with respect to CO_2 according to earlier studies in [13].

Auger Electron Spectroscopy (AES)

In order to verify the results detailed point analyses by means of auger spectroscopy was carried out for the 15wt.% lanthanum containing sample at the sputter crater. This allows studying the different atomic states at a very small localized area. In Fig. 6 some typical auger region scans of the Mg KLL, O KLL and the La MNNa excitation have been plotted, measured at the microstructure of the bulk and the intermettalic phase.

The corresponding Mg KLL transition at 1180 eV is related to the MgO phase [15] and the transitions in the O KLL auger spectra too [16]. Both spectra are convoluted by a weak signal related to the La_2MgO_x phase. There still not exists literature about auger spectra of this phase. Thus, it is very difficult to confirm this assumption, but the weak La MNNa excitation shows the existence of a small fraction of lanthanum at the measured point. According to the kinetic energies in [17] the La is in an oxidized state. Very interesting results arise from the studies at the intermetallic phase.

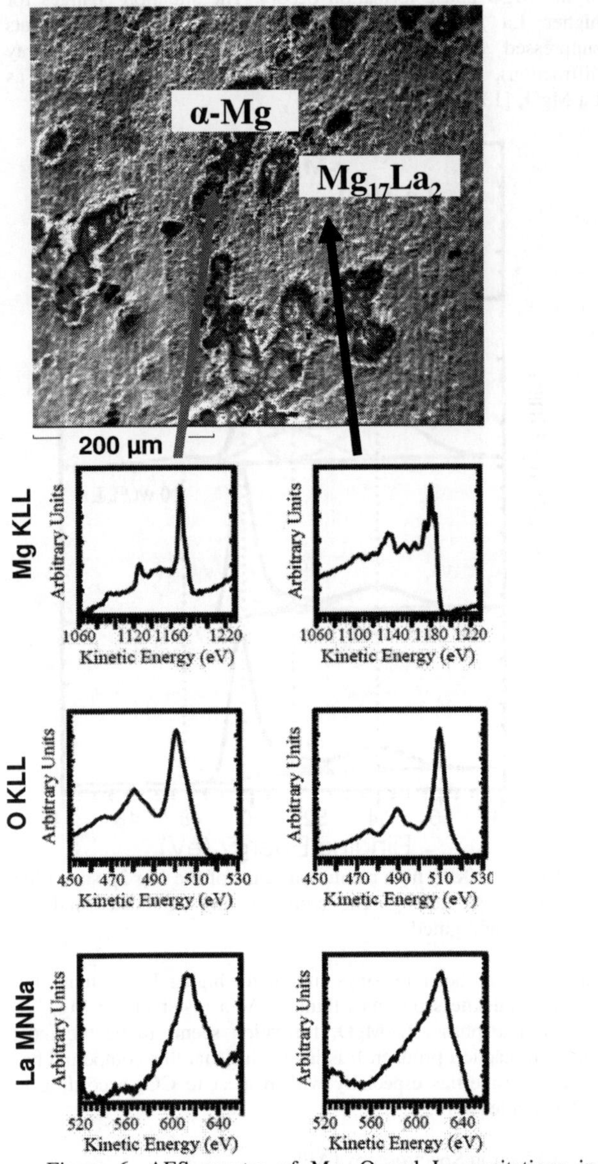

Figure 6. AES spectra of Mg, O and La excitations in different phases measured inside the sputter crater after completed etching for the Mg15La sample

The Mg KLL transition shows a strong peak splitting due to the interaction with the heavy lanthanum atoms at the $Mg_{17}La_2$ phase. At an energy of 1186 eV (the right shoulder) metallic magnesium has been detected [15]. Comparing the O KLL state to the bulk one another phase than MgO occurs. I can be assumed that the spectra is a convolution of La_2MgO_x, La_2O_3 and $La(OH)_3$ related signals. Further studies are required for verification. The same problem exists for the evaluation of the La MNNa transition. The right shoulder at 634 eV also corresponds to the metallic phase [18], but there are no information available (also in the literature) to separate the oxidized state of lanthanum by means of auger electron spectroscopy..

X-ray diffraction (XRD)

XRD is one of the most powerful tools for material analyses also in this case. The diffraction patterns in Fig. 7 offer a lot of additional information on the corrosion process. All theta -2theta scans have been plotted by means of a logarithmic scale in order to see the pattern occurring from compounds close to the surface (information depth of corrosion products of approximately 5 μm). Otherwise the bulk signal (x-ray penetration depth about 200 μm) becomes dominant.

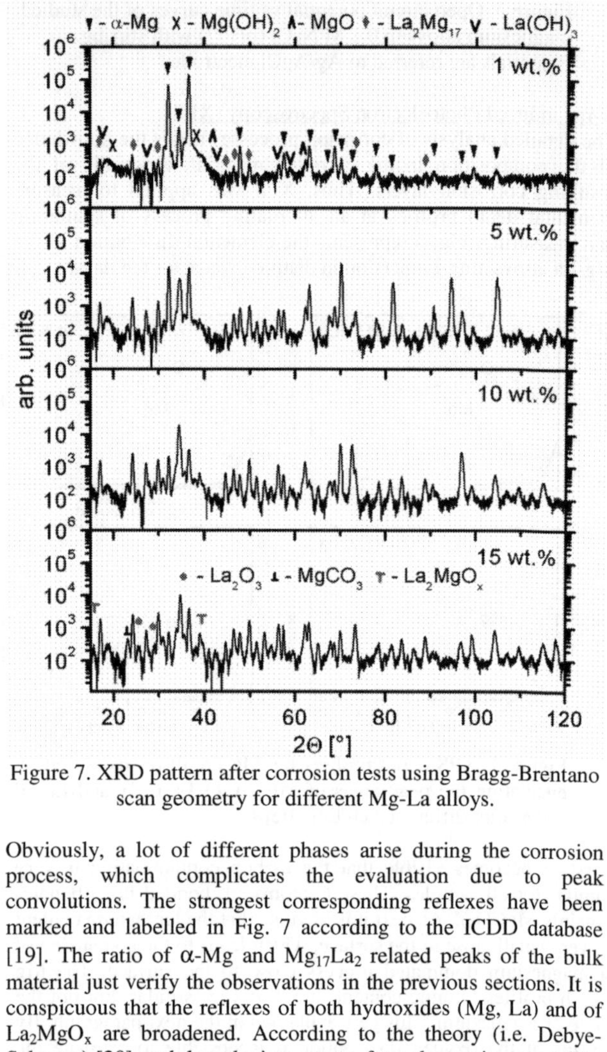

Figure 7. XRD pattern after corrosion tests using Bragg-Brentano scan geometry for different Mg-La alloys.

Obviously, a lot of different phases arise during the corrosion process, which complicates the evaluation due to peak convolutions. The strongest corresponding reflexes have been marked and labelled in Fig. 7 according to the ICDD database [19]. The ratio of α-Mg and $Mg_{17}La_2$ related peaks of the bulk material just verify the observations in the previous sections. It is conspicuous that the reflexes of both hydroxides (Mg, La) and of La_2MgO_x are broadened. According to the theory (i.e. Debye-Scherrer) [20] such broadening occurs for a decreasing crystallite size. In the case of La_2MgO_x a nanocrystalline microstructure would confirm the results in [13]. Besides, the typical oxides have been detected. The measurement of $MgCO_3$ is related to the strong carbon adsorption during the tests.

Conclusions

The presence of two phases: primary phase (α-Mg) and eutectic phase (α-Mg+ $Mg_{17}La_2$) will induce galvanic corrosion in Mg-La alloys. The corrosion rate depends strongly on the amount and distribution of these phases.

When the La content is higher than 5 wt% La, the alloys exhibit a potential even more noble than that of Mg, but the corrosion resistance decreases due to the increase of the eutectic phase.

For increasing lanthanum contents $Mg(OH)_2$ formation gets suppressed, but contrary $La(OH)_3$ formation becomes amplified. As a result the thickness of the corrosion layer is thinner. Besides the typical oxides are developing. It is likely that this has also a contribution to the decreasing corrosion resistance with increasing La content,

As next, a nanocrystalline La_2MgO_x phase with catalytic properties has been formed. The influence on the corrosion process is still unclear.

References

1. G. Makar, J. Kruger and A. Joshl, *The effect of alloying elements on the corrosion resistance of rapidly solidified magnesium alloys*, International Magnesium Association and the Non-ferrous Metals Committee, The Minerals, Metals and Materials Society, Phoenix, 1988.
2. G. L. Makar and J. Kruger, "Corrosion studies of rapidly solidified magnesium alloys," *J Electrochem Soc,* 137 (2) (1990), 414-421.
3. E. Emley, *Principles of magnesium technology*, London, Pergamon Press, New York, 1966.
4. G. Makar and J. Kruger, *Corrosion of magnesium*, vol. 38, Maney Publishing, 1993.
5. I. Polmear, *Magnesium alloys*, Sevenoaks, UK 1989.
6. T. Takenaka et al., "Improvement of corrosion resistance of magnesium metal by rare earth elements," *Electrochimica Acta,* 53 (1) (2007), 117-121.
7. J. H. Nordlien et al., "Morphology and structure of water-formed oxides on ternary mgal alloys," *J Electrochem Soc,* 144 (2) (1997), 461-466.
8. F. Rosalbino et al., "Effect of erbium addition on the corrosion behaviour of mg-al alloys," *Intermetallics,* 13 (1) (2005), 55-60.
9. G. Song and D. St. John, "The effect of zirconium grain refinement on the corrosion behaviour of magnesium-rare earth alloy mez," *Journal of Light Metals,* 2 (1) (2002), 1-16.
10. Y. Fan, G. Wu and C. Zhai, "Influence of cerium on the microstructure, mechanical properties and corrosion resistance of magnesium alloy," *Materials Science and Engineering: A,* 433 (1-2) (2006), 208-215.
11. Q. Peng et al., "The effect of La or Ce on ageing response and mechanical properties of cast mg-gd-zr alloys," *Materials Characterization,* 59 (4) (2008), 435-439.
12. S. Ardizzone et al., "Magnesium salts and oxide: An xps overview," *Appl Surf Sci,* 119 (3-4) (1997), 253-259.
13. A. Ivanova, "Structure, texture, and acid-base properties of alkaline earth oxides, rare earth oxides, and binary oxide systems," *Kinetics and Catalysis,* 46 (5) (2005), 620-633.
14. J. M. Fraile et al., "The basicity of mixed oxides and the influence of alkaline metals: The case of transesterification reactions," *Applied Catalysis A: General,* In Press, Corrected Proof (-) (2010).
15. C. D. Wagner and P. Biloen, "X-ray excited auger and photoelectron spectra of partially oxidized magnesium surfaces: The observation of abnormal chemical shifts," *Surface Science,* 35 ((1973), 82-95.
16. J. C. Fuggle et al., "High-resolution auger spectra of adsorbates," *Journal of Electron Spectroscopy and Related Phenomena,* 26 (2) (1982), 111-132.
17. W. H. Hocking et al., "Scanning auger microscopy study of lanthanum partitioning in sphene-based glass-ceramics," *Philosophical Magazine A,* 49 (5) (1984), 637-656.
18. L. Davis et al., *Handbook of auger electron spectroscopy,* Physical Electronics Industries Eden Prairie, MN, 1976.
19. "Powder diffraction file," *ICDD,* International Centre for Diffraction Data, Newtown, PA,, 2008.
20. T. Ungár, "Microstructural parameters from x-ray diffraction peak broadening," *Scripta Materialia,* 51 (8) (2004), 777-781.

Magnesium Technology 2011
Edited by: Wim H. Sillekens, Sean R. Agnew, Neale R. Neelameggham, and Suveen N. Mathaudhu
TMS (The Minerals, Metals & Materials Society), 2011

CRYOGENIC BURNISHING OF AZ31B Mg ALLOY FOR ENHANCED CORROSION RESISTANCE

Z. Pu[1], G.-L. Song[2*], S. Yang[1], O.W. Dillon, Jr.[1], D. A. Puleo[3], I.S. Jawahir[1]
[1]Department of Mechanical Engineering, Institute for Sustainable Manufacturing, University of Kentucky, Lexington, KY 40506, USA
[2]Chemical Sciences and Materials Systems Lab, GM Global Research and Development,
Mail Code: 480-106-212, 30500 Mound Road, Warren, MI 48090, USA
Tel: 248-807-4451, email: guangling.song@gm.com
[3]Center for Biomedical Engineering, Wenner-Gren Lab, University of Kentucky,
Lexington, KY 40506, USA

Keywords: cryogenic burnishing, magnesium alloys, grain refinement, corrosion

Abstract

Poor corrosion resistance is limiting applications of Mg alloys. However, the corrosion performance of an Mg alloy can be enhanced through modification of its microstructure. It has been reported in the literature that the microstructure, especially grain size of AZ31 Mg alloy, has a significant influence on its corrosion resistance. In this study, AZ31B discs were subjected to a novel mechanical processing method–cryogenic burnishing; the surface of AZ31B work piece was burnished with a custom tool under a liquid nitrogen spraying condition. The processing led to a more than 3 mm thick surface layer with remarkably changed microstructures formed on the disc surface. Significant grain refinement occurred within this surface layer due to dynamic recrystallization induced by severe plastic deformation and effective cooling by liquid nitrogen. Both electrochemical method and hydrogen evolution method indicate that the corrosion resistance of the burnished surface was significantly improved.

Introduction

Mg alloys are potential lightweight materials for automotive applications and the use of the alloys may remarkably improve vehicle fuel economy. However, the poor corrosion resistance of Mg alloys significantly limits their wide application. The corrosion performance of Mg AZ31 alloy is among the poorest when compared with other common cast Mg alloys, such as AZ91 or AM60.

Although various approaches, such as coating and alloying, have been extensively studied, the potential of grain refinement to improve the corrosion resistance of Mg AZ31 alloy has not been well investigated. Hot rolled AZ31 Mg alloy samples were reported to have increased corrosion resistance compared with squeeze cast samples, which was attributed to grain refinement from 450 μm to 20 μm [1]. Alvarez-Lopez et al. [2] reported that grain refinement from 25.7 μm to 4.5 μm after equal channel angular pressing (ECAP) led to better corrosion performance for Mg AZ31 alloy. Aung and Zhou [3] considered the grain boundary as physical corrosion barriers and claimed that smaller grain size improved the corrosion resistance. However, Song and Xu [4] reported that there was no evidence to support that changes in grain size and twin density by heat-treatment were principal causes of the reduced corrosion resistance in AZ31 Mg alloys.

While more efforts are needed to further investigate the relationship between grain size and corrosion of Mg alloys, there is a great demand for novel surface processing techniques which can lead to improved corrosion resistance. Most of the researchers use severe plastic deformation (SPD) processes, such as equal channel angular pressing (ECAP), friction stir processing and high pressure torsion, to introduce grain refinement in Mg alloys. However, these processes are often very slow and heating may be required, such as for ECAP, due to limited ductility of Mg alloys at room temperature. Also, only small samples with simple geometry have been processed. Burnishing is a widely used process in industry to reduce surface roughness, increase hardness and/or introduce compressive residual stresses. Although there have been many publications on the beneficial effects of burnishing on fatigue life and/or wear of various materials, the corrosion performance of burnished materials have rarely been reported. Recently, Al-Qawabeha and Al-Rawajfeh [5] reported that the weight loss of galvanized steel samples after roller burnishing was reduced to only 6.5% of the value in the initial material. This is a significant indication that burnishing may also lead to improved corrosion resistance of a material.

In this study, a novel SPD method based on burnishing is used to change the surface microstructure of the AZ31B Mg alloy in order to understand the influence of the novel surface processing technique on corrosion resistance.

Experimental Work

Work Material

The work material studied was the commercial AZ31B-O magnesium alloy. The work material was received in the form of a 3 mm thick sheet. Disc specimens having 130 mm diameter were cut from the sheet and subsequently subjected to burnishing.

Burnishing Experiments

The burnishing experiments were conducted on a Mazak Quick Turn-10 Turning Center equipped with an Air Products liquid nitrogen delivery system, which sprays liquid nitrogen to the processing zone for cooling. The experimental setup is shown in Figure 1.

The AZ31B Mg disc was fixed in the lathe chuck and rotated during processing. A roller made of high speed steel was pushed against the discs at certain feed rate. Different from the traditional burnishing method, the roller used here was fixed and does not rotate in order to introduce more severe plastic deformation. During processing, liquid nitrogen was sprayed to the processing zone as shown in Figure 1. The application of liquid nitrogen was to remarkably reduce the temperature during processing and suppress the growth of ultrafine/nano grains introduced by dynamic recrystallization (DRX).

The burnishing speed refers to the linear speed at the contact point between the fixed roller and the disc. It was 100 m/min. The feed rate was 0.01 mm/rev. The process was stopped when the final diameter reduced to 125 mm.

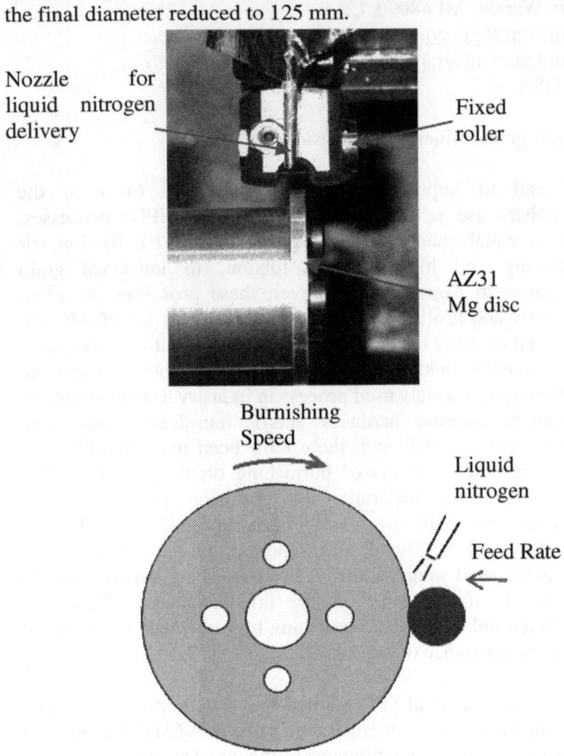

Figure 1 Burnishing setup with an Air Products liquid nitrogen delivery system

Grinding Treatment

To eliminate the possible influence of surface roughness on corrosion resistance [4], the unprocessed AZ31B Mg samples were ground successively using 4000 grit sand paper. Samples after grinding serve as the reference for corrosion resistance comparison.

Characterization Method

After burnishing, metallurgical samples were cut from the burnished discs. After cold mounting, grinding and polishing, acetic picric solution was used as an etchant to reveal the grain structure. A KEYENCE digital microscope VHX-600 was used to observe and record the microstructures of the burnished samples.

Surface roughness after grinding and burnishing were measured using a ZYGO New View 6000 measurement system which was based on white light interferometry.

The hardness of the samples was measured using a Hysitron TriboIndenter. The load used was 8 mN.

Electrochemical Measurement

A Solatron 1280 potentiostat system was used for polarization curve and AC impedance measurements. Only the processed surface was exposed to the testing solution and all the other surfaces are protected by a thick layer of MICCROSTOP lacquer. The exposed area was 1.5 cm^2. The testing solution was 5 wt. % NaCl. A platinum gauze was used as a counter electrode and a KCl-saturated Ag/AgCl electrode was used as a reference in the cell. During AC impedance measurements, the frequency ranged from 17,777 Hz to 0.1 Hz with 7 points/decade, and the amplitude of the sinusoidal potential signal was 5mV with respect to the OCP. Potentiodynamic polarization curve measurements were performed at a potential scanning rate of 0.1 mV/s from −0.3 V vs. OCP to −1.0 V vs. reference.

Hydrogen Evolution Measurement

In addition to electrochemical methods, hydrogen evolution method [6] was also used to compare the corrosion rates of samples after cryogenic burnishing and after grinding. The samples were mounted in epoxy resin and only the processed surface was exposed to 5 wt. % NaCl. The exposed area was 1.5 cm^2. Pipettes with 0.1 mL interval were used to collect the evolved hydrogen from the samples.

Results and Discussion

Microstructure

Figure 2 shows an overview of the microstructure after cryogenic burnishing. There is a clear interface between the processing-influenced zone and the bulk. This interface is also shown in Figure 3 under ×1000 magnification. The total thickness of the processing-influenced layer is 3.40 ± 0.01 mm.

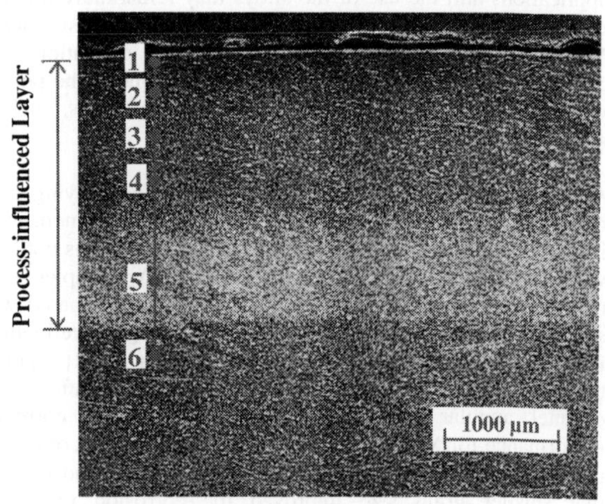

Figure 2 Microstructure after cryogenic burnishing (×30 magnification)

While no twinning can be seen in the initial material, there is a high density of deformation twins above the interface as shown in Figure 3. The location of twinning is near the bottom of the processing-influenced layer. Twinning gradually disappears as one moves from the bulk to the top surface. The deformation twinning indicates that the temperature near this interface is lower when compared with the top portion of the layer.

A clear evidence of dynamic recrystallization (DRX) is observed in Figure 4. The microstructures at different points in Figure 2 were shown at a ×5000 magnification using the VHX-600 digital microscope.

The image at Point 6 in Figure 4 represents the initial microstructure and Point 1 is the microstructure near the surface after cryogenic burnishing. It is clear that significant grain refinement occurred near the surface. As shown in Figure 5, the grain size after cryogenic burnishing is reduced to 1.03±0.26 µm from the initial grain size of 11.88±4.54 µm. Not only is the grain size reduced, but also the distribution of grain size becomes more uniform (smaller in scatter).

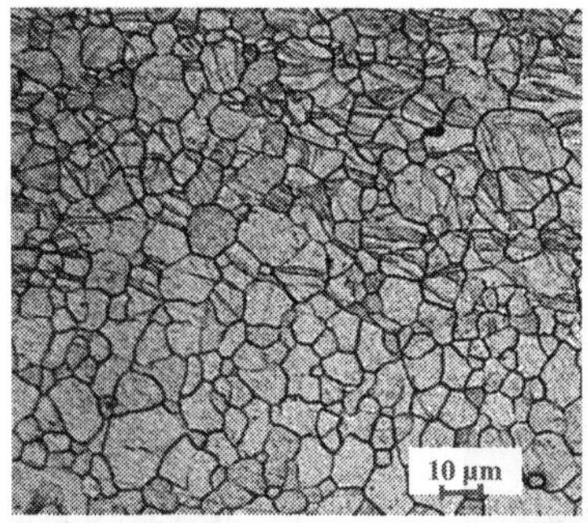

Figure 3 Interface between initial material and process-influenced layer after cryogenic burnishing (×1000 magnification)

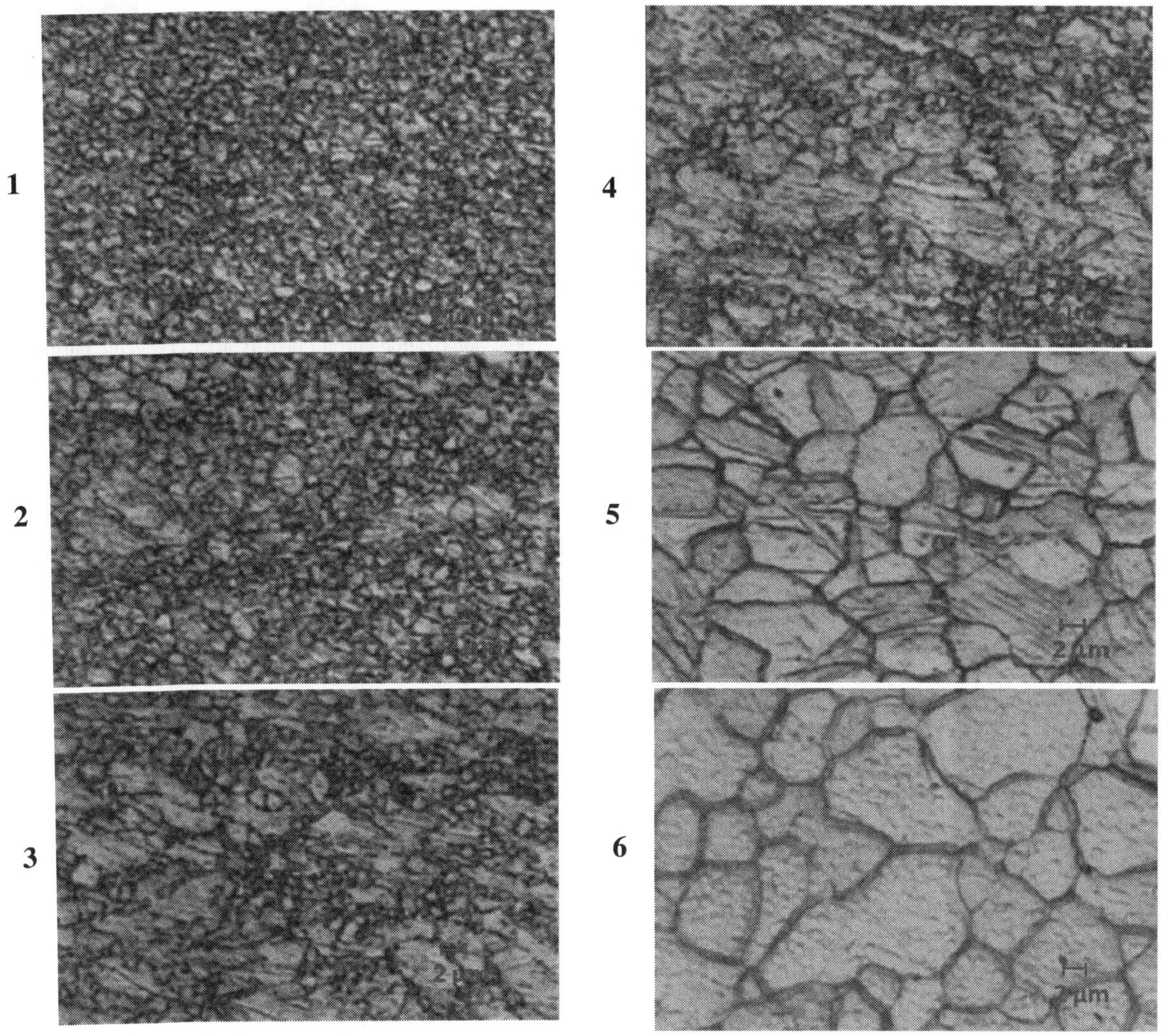

Figure 4 Microstructures at different depths after cryogenic burnishing (×5000 magnification)

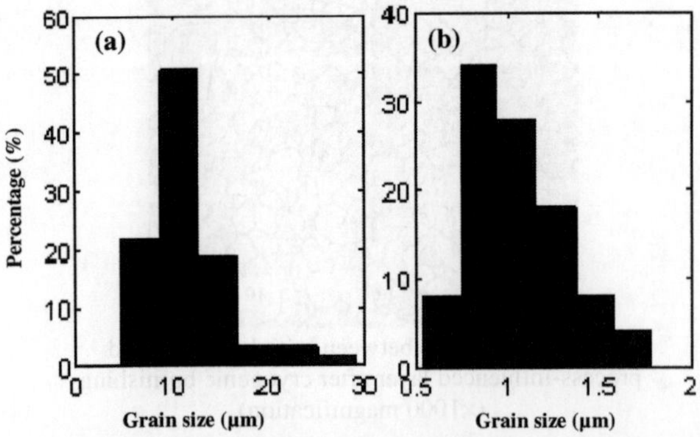

Figure 5 Distribution of grain size: (a) initial; (b) after cryogenic burnishing at Point 1.

From Point 2 to Point 4, there is a clear trend that the amount of ultrafined grains is decreasing. Fatemi-Varzaneh et al. [7] investiaged the effects of temperature, strain and strain rates on dynamic recrystallization of AZ31 Mg alloy in detail and reported that the amount of dynamically recrystallized grains increased with strain in a sigmoidal form. The strain induced by cryogenic burnishing should decrease from the surface to the bulk material where the material was not influenced by the process. This aggrees with the literature that the amount of dynamically recrystallized grains will decrease when the strain becomes smaller.

The microstructural features at Point 6 in Figure 4 further shows that deformation twins are dominant in the transition layer from the processing-influenced microstructure to the initial microstructure.

Hardness Measurement

As shown in Figure 6, the hardness far away from the surface, which is not influenced by the processing, is about 0.9 GPa. After cryogenic burnshing, the hardness near the surface reaches 1.35 GPa. The relationship between hardness and grain size in AZ31 Mg alloys has been frequently reported in literature [8]. The large increase in hardness agrees with the previous finding that signficant grain refinement occurs near the surface after cryogenic burnishing.

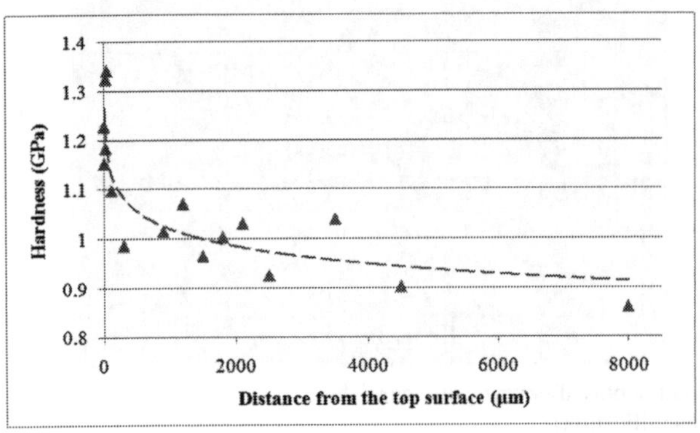

Figure 6 Hardness variation from the top surface to the bulk material

Surface Roughness

Figure 7 shows a comparison of surface roughness (Ra) between grinding and cryogenic burnishing. It shows that grinding creates slightly better surface roughness and should result in better corrosion resistance.

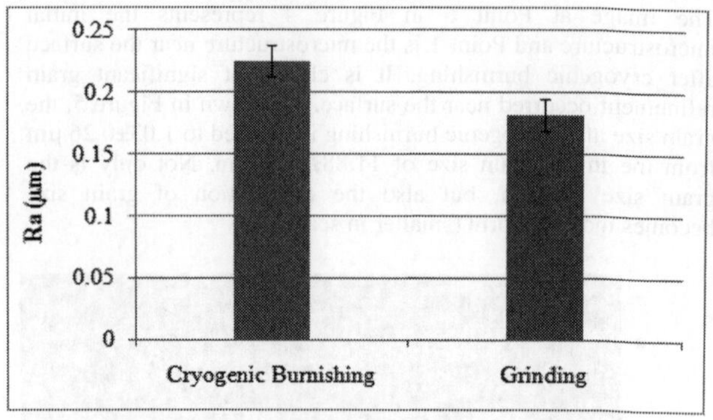

Figure 7 Comparison of surface roughness between cryogenic burnishing and grinding

Electrochemical Measurement

The polarization curves of samples after grinding and after cryogenic burnishing are presented in Figure 8. It shows that the cathodic polarization current density after cryogenic burnishing is smaller than the one after grinding, which suggests that cryogenic burnishing leads to improved corrosion resistance. However, there is a large shift in corrosion potential from -1.44 mV after grinding to -1.53 mV after cryogenic burnishing. Similar findings were reported by Balakrishnan et al. [9] where ultra-fine-grained (238 nm) Ti has a lower corrosion potential than coarse grained Ti (15.2 μm). While in general, metals with lower potential are prone to more corrosion; both the literature and the current study show the opposite trend. There is another possibility that quicker passivation of the surface layer may retard the corrosion process [10].

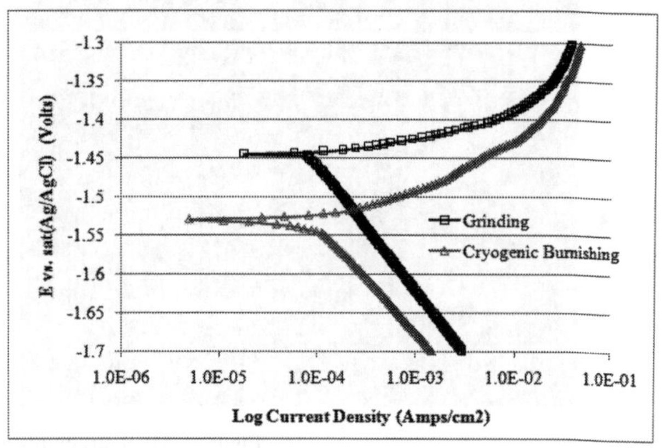

Figure 8 Polarization curves of AZ31B Mg samples after grinding and cryogenic burnishing in 5 wt. % NaCl

Figure 9 shows the Nyquist diagrams of AZ31B Mg samples after grinding and cryogenic burnishing in 5 wt.% NaCl. Both curves have a clear capacitive arc at the high frequency region. The diameter of this capacitive loop at the high frequency region is associated with the charge-transfer resistance. Makar and Kruger [11] shows that for magnesium alloys, larger diameter indicates better corrosion resistance. The diameter for the sample after cryogenic burnishing is remarkly larger than the one after grinding, which suggests the sample after cryogenic burnishing has better corrosion resistance than the ground sample. This finding agrees with the trend of cathodic polarization current densities as shown in Figure 8.

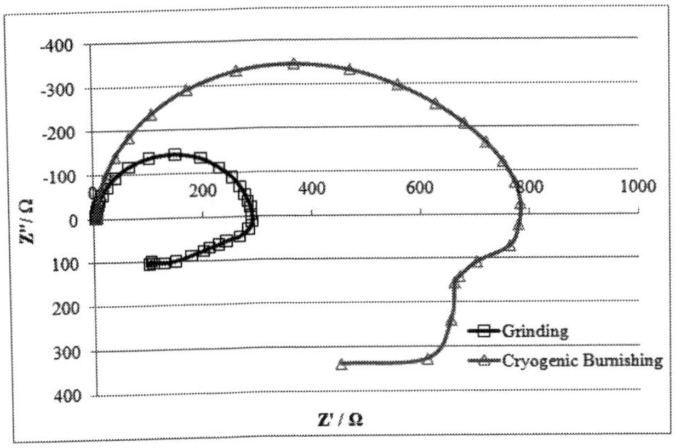

Figure 9 Nyquist diagrams of AZ31B Mg samples after grinding and cryogenic burnishing in 5 wt.% NaCl

Hydrogen Evolution Measurement

The hydrogen evolution of the samples in 5 wt. % NaCl after grinding and burnishing are presented in Figure 10. It shows that more hydrogen is generated from the ground samples. Also, the data scatter after grinding is larger than after cryogenic burnishing. Since cryogenic burnishing was carried out automatically on a CNC machine, it is expected that the process is more repeatable than grinding by hand. The finding from hydrogen evolution measurement further proves that the corrosion resistance of the AZ31B Mg alloy after cryogenic burnishing is improved compared with the one after grinding.

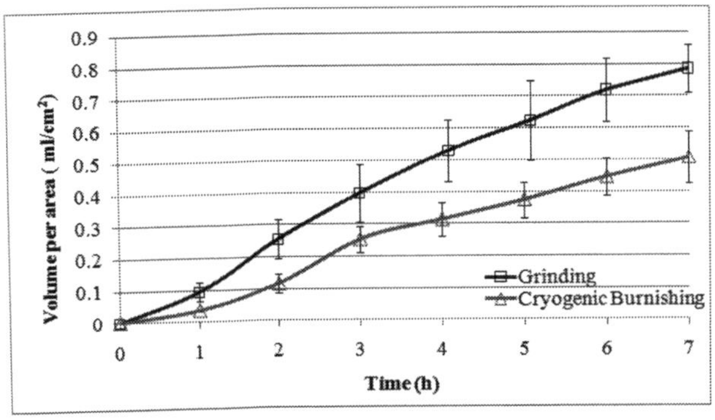

Figure 10 Hydrogen evolution of AZ31B Mg samples after grinding and cryogenic burnishing in 5 w.t.% NaCl

Conclusion

The present study shows that significant grain refinement as well as a large increase in hardness can be achieved in the surface layer of AZ31B Mg alloy after cryogenic burnishing. The microstructure of AZ31B at depths up to 3.4 mm away from the surface can be remarkably changed by cryogenic burnishing. The mechanism for grain refinement is dynamic recrystallization.

Both electrochemical method and hydrogen evolution method show that the corrosion resistance of AZ31B Mg alloy is improved after cryogenic burnishing. This agrees with other literatures that smaller grain size leads to better corrosion resistance of AZ31B Mg alloy.

In addition, it reveals a great opportunity to improve material performance through fabricating a grain refinement surface layer by cryogenic burnishing. Not only corrosion resistance, but other properties, such as fatigue and wear resistance may also be significantly enhanced if proper processing conditions are used.

Acknowledgement

The authors would like to thank Air Products and Chemicals for providing the ICEFLY® liquid nitrogen delivery system.

References

1. H. Wang, Y. Estrin, H.Fu, G.Song, Z. Zuberova, "The effect of pre-processing and grain structure on the bio-corrosion and fatigue resistance of magnesium alloy AZ31", *Advanced Engineering Materials*, 2007, 9: 967-972.

2. M. Alvarez-Lopez, María Dolores Pereda, J.A. del Valle, M. Fernandez-Lorenzo, M.C. Garcia-Alonso, O.A. Ruano and M.L. Escudero, "Corrosion behaviour of AZ31 magnesium alloy with different grain sizes in simulated biological fluids", *Acta Biomaterialia,* 2010, no.6:1763-1771.

3. Naing Naing Aung and Wei Zhou, "Effect of grain size and twins on corrosion behaviour of AZ31B magnesium alloy". *Corrosion Science*, 2010, no.52: 589-594.

4. Guang-Ling Song and ZhenQing Xu, "The surface, microstructure and corrosion of magnesium alloy AZ31 sheet", *Electrochimica Acta*, 2010, no. 55: 4148–4161.

5. Ubeidulla Al-Qawabeha and Aiman Eid Al-Rawajfeh, "Influence of roller burnishing on surface properties and corrosion resistance in steel", *Anti-Corrosion Methods and Materials*, 2009, no. 56: 261–265.

6. Song, G., Atrens, A., St John, D. H., "An hydrogen evolution method for the estimation of the corrosion rate of magnesium alloys", *Proceeding of Magnesium Technology 2001,* TMS Annual Meeting. New Orleans, LA. February 11-15, 2001.

7. S.M. Fatemi-Varzaneh, A. Zarei-Hanzaki, H. Beladi, "Dynamic recrystallization in AZ31 magnesium alloy", *Materials Science and Engineering A*,2007, no.456:52–57.

8. C.I. Chang, C.J. Lee and J.C. Huang, "Relationship between grain size and Zener–Holloman parameter during friction stir

processing in AZ31 Mg alloys". *Scripta Materialia*, 2004, no.51:pp.509-514.

9. A. Balakrishnan, B. C. Lee, T. N. Kim and B. B. Panigrahi, "Corrosion Behaviour of Ultra Fine Grained Titanium in Simulated Body Fluid for Implant Application", *Trends Biomater. Artif. Organs*, 2008, Vol 22(1), pp 0-0.

10. L. Krivián, "Meaning and measurement of corrosion potential", *British Corrosion Journal*, 1991, no.26:191-194.

11. G.L. Makar, K. Kruger, "Corrosion Studies of Rapidly Solidified Magnesium Alloys", *Journal of the Electrochemical Society*, 1990, no.137:414-421.

Magnesium Technology 2011
Edited by: Wim H. Sillekens, Sean R. Agnew, Neale R. Neelameggham, and Suveen N. Mathaudhu
TMS (The Minerals, Metals & Materials Society), 2011

Advanced Conversion Coatings for Magnesium alloys

Syam Nibhanupudi, Alp Manavbasi

Metalast International, Minden, NV

Keywords: conversion coating, chromium, hexavalent, trivalent, magnesium, corrosion.

Abstract

Magnesium and its alloys have excellent physical and mechanical properties due to their high strength-to-weight ratio and are ideal for various applications in automotive, aerospace and defense sectors. However, Mg alloys are also highly susceptible to corrosion under harsh environments. Owing to this carcinogenicity as well as environmental impact of hexavalent chromium fueled by stringent environmental regulations, an environmentally green alternative to the carcinogenic hexavalent chromium coatings on magnesium is due.

In this work, a novel trivalent chromium based conversion coating has been developed to improve the corrosion resistance and paint adhesion properties of Mg alloys. Coating performance characterization has been investigated via hydrogen evolution, weight loss measurement and electrochemical corrosion analysis techniques. Results have shown that the novel environmentally green trivalent chromium based coating on magnesium has indeed performed comparable to hexavalent chromium and thus establishing a viable alternative.

Introduction

Magnesium, the eighth most abundant metal on earth, has seen an increase in applications in a variety of sectors ranging from aerospace, military, defense, automobile to commercial mobile phones, sporting goods and handheld tools owing to its high strength-to-weight ratio, machinability, thermal conductivity and weldability [1-4]. Although magnesium alloys have the distinct advantage of high strength-to-weight ratio and high impact resistance, their extremely poor corrosion resistance in aggressive environments especially in assemblies with multiple galvanic couples, limits their usage because of the premature component degradation and replacement. [4-7].

Improving corrosion performance involves a variety of finishing processes including oil applications, wax coating, anodizing, electroplating, conversion coating, painting etc. Metal finishers are often faced with challenges to tailor a better combination of these processes to enhance the product corrosion performance considering various factors ranging from specific applications to environmental conditions. Even though development of new and high purity magnesium alloys and novel surface modification techniques such as ion implantation and laser annealing have mitigated the corrosion problem, magnesium alloys still suffer from performance issues especially under aggressive corrosive environments. High chemical reactivity, complex alloy microstructures, hazardous pretreatment procedures and recycling problems to an extent have discouraged many applications of magnesium [8-10].

Among the typical surface modification techniques, conversion coating, electroplating with alloys like Zn, Ni and Cr and anodizing processes such as DOW 17, HAE, and Tagnite have become widely popular for the surface treatment of magnesium alloys.

Majority of the magnesium alloys are either sand-cast or die-cast and hence are prone to microstructural inhomegeniety, surface imperfections such as porosity and impurities. Magnesium and its alloys are also prone to galvanic corrosion and alloy dissolution owing to lesser nobility of magnesium. Design of a process routine that results in an effective barrier to corrosion as well as wear would be ideal. Hexavalent chromium (Cr^{6+}) based conversion coatings have been used for a long time as a coating material on aluminum, magnesium and other metals in order to increase the corrosion protection, paint adhesion, and adhesive bonding characteristics. However, solutions containing Cr6+ are highly toxic and adversely affect the environment and human health.

A recent Under Secretary of Defense memo, dated April 2009, regarding the elimination and restriction of the Cr6+ from Department of Defense (DoD) weapon systems and platforms shows the importance of novel non-hexvalent technologies for the U.S. military departments. The U.S. and International market needs for a Trivalent Chromium Conversion Coating is predicated on three severe and stringent European Union (EU) Directives: Restriction of Hazardous Substances (RoHS), Waste Electrical & Electronic Equipment (WEEE), and End of Life Vehicle (ELV). Global manufacturing demands compliance even though these directives are implemented in the EU. EU member nation states have established regulations and enforcement of these Directives. In the U.S., heavy regulations from both EPA and OSHA are continuing to be issued and implemented. OSHA has mandated a Permissible Exposure Level (PEL) of 5 ppm of Hexavalent Chromium, an unrealistic level for the vast majority of metal finishers and manufacturers to attain. Compliance to the Directives and OSHA requirements are not an option, they are mandatory. In addition, the EPA Executive Order 12856 showed the need to reduce or eliminate the release of chromates during aircraft coating applications.

There is, therefore, a need for environmentally green conversion coatings that can provide high corrosion resistance and increase the adhesive bonding strength characteristics of the metal surface. Although there are other conversion coatings, which do not contain Cr6+, their corrosion performance and paint adhesion characteristics are not as effective as the Cr6+ based conversion coatings.

Our aim is to at develop an environmentally friendly conversion treatment for magnesium alloys. In this present study, trivalent chromium based environmentally conversion coating for AZ92A magnesium alloy was developed. Furthermore, the performance was gauged against hexavalent chromium coatings using hydrogen evolution test, weight loss measurement techniques and potentidynamic studies.

Experimental set up

Experimental process routine for the study was conducted in accordance with AMS -3171-C, [11]. Sand cast magnesium Alloy AZ92A discs (1.5 in. dia and 0.5 in. thickness) conforming to AMS 4434 [12], have been purchased from Metalmart Inc., with the following compositions.

Al	Zn	Si	Mn	Cu
9.0	2.0	0.3 max	0.10 min	0.25 max

Table 1 : Elemental compositions of AZ92A sand casting in percentages.

Conversion coating samples underwent the following identical process routine for neutral coating-performance comparisons.

Mechanical Grinding: Samples have been ground to 1200p using SiC wafers, rinsed in D.I water and dried.

Pretreatment Process: Ground samples are then wiped with acetone and rubbing alcohol. Samples are then subjected to the pretreatment routines for effective coating depositions as shown below.

Alkaline Cleaning: Sodium Hydroxide based cleaner has been chosen for alkaline cleaning of magnesium alloys for 5 minutes at 140F and a pH> 12.0. Magnesium is very resistant to corrosion by alkalis if the pH exceeds 10.5 which corresponds to the pH of a saturated $Mg(OH)_2$[13]; a $Mg(OH)_2$film is formed on the magnesium surface [3,13]. Dilute alkali solutions show negligible attack at temperatures up to the boiling point [14]. Consequently, a 10% caustic solution is commonly used for cleaning at temperatures up to the boiling point [15].

Surface activation via Acid Pickling: Chromic-Nitric-HF acid pickle in compositions recommended in the AMS-3171-C - specification [11], has been observed to be beneficial in contrast to phosphoric and sulfuric acid etches on AZ92A and hence has been adopted. A 2 minute immersion in an agitated mixture of Ammonium Bifluoride + Nitric acid has been observed to effectively remove the acid smut adhering to the surface in the trial studies.

Conversion Coating: Hexavalent chromium based conversion coatings were applied in accordance with the Type VIII treatment in the AMS-3171-C specification [11]. Iridite #15 was used for hexavalent chromium conversion coating of AZ92A.

Once conversion coated samples are stored in a dehumidifier chamber for a minimum of 24 hour curing period prior testing.

Performance characterization

Hydrogen Evolution and Weight loss measurement:

Hydrogen evolution studies have been conducted on bare AZ92A discs ground to 1200p grit and hexavalent chromium conversion coated samples for base line test criteria. Trivalent chromium conversion coated samples processed and cured were tested for amount of hydrogen evolved under 3.5% NaCl solution and was compared to the baseline criteria for rate of hydrogen evolution per surface area. Hot chromic acid 24 oz/gal at 190F, 8 minutes was used to remove any corrosive products on the alloy surface followed by a rinse in D.I. ,dried and weighed to evaluate the amount of weight loss per surface area due to salt immersion.

DC Polarization studies

Potentiodynamic measurements were performed on the conversion coated and bare AZ92A discs. The experiments were carried out in an aggressive 3.5% NaCl electrolyte. A three-electrode configuration was employed: conversion coated magnesium substrate as working electrode, a carbon rod as a counter electrode, and a SCE (Standard Calomel Electrode) as reference electrode. The amplitude of the perturbation was 10mV and the OCP (Open Circuit Potential) was chosen as bias potential. The examined frequency range was from 10mHz to 100kHz. Results obtained from these tests are used to determine the corrosion current, potential and corrosion rate.

Results and discussion

Hydrogen evolution experiment:

Magnesium dissolution in aqueous environments are often influenced by an electrochemical reaction with water that produces magnesium hydroxide and hydrogen gas. Under film-breakdown agents such as chlorides the stable oxide/hydroxide layers experience surface film break-down followed by the evolution of hydrogen. A stable coating negates this surface film limits this surface film break down vis-à-vis lesser hydrogen evolution. After conversion coated and cured, substrates are submerged in a bath of 3.5% NaCl solution to measure the amount of hydrogen released and the amount of weight loss per surface area. The volume of hydrogen displaced in the inverted burette is used to calculate the amount of hydrogen released per surface area.

Results obtained from the above measurements were used to evaluate the corrosion rate of the alloy (mm/y) with respect to their coating configurations using the following formulae.

Corrosion rate [mm/y] = 2.2785 (hydrogen evolution rate per Surface area [ml/cm^2/d]).

Corrosion rate [mm/y] = 2.10 (weight loss rate [mg/cm^2/d]);

Figure 1 and 2 details the results of the hydrogen evolution test conducted on various conversion coated and bare magnesium samples as well as the calculated corrosion rates. It is clear from these values that the novel trivalent coating enhanced the corrosion protection compared to the uncoated bare substrate and performed close to the hexavalent coating. Figure 3 details the calculated corrosion rates via weight loss measurement tests and it can be observed that in both cases Trivalent chromium conversion coating processed samples performed on par with hexavalent IRIDITE coating by offering comparably effective film breakdown protection.

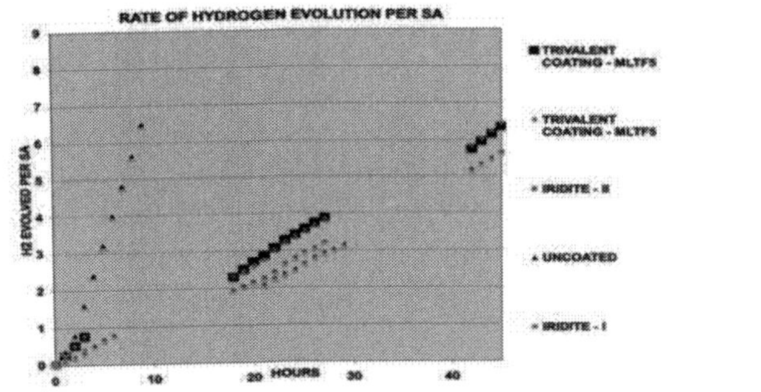

Figure 1: Rate of hydrogen evolution per surface area for various coated surfaces.

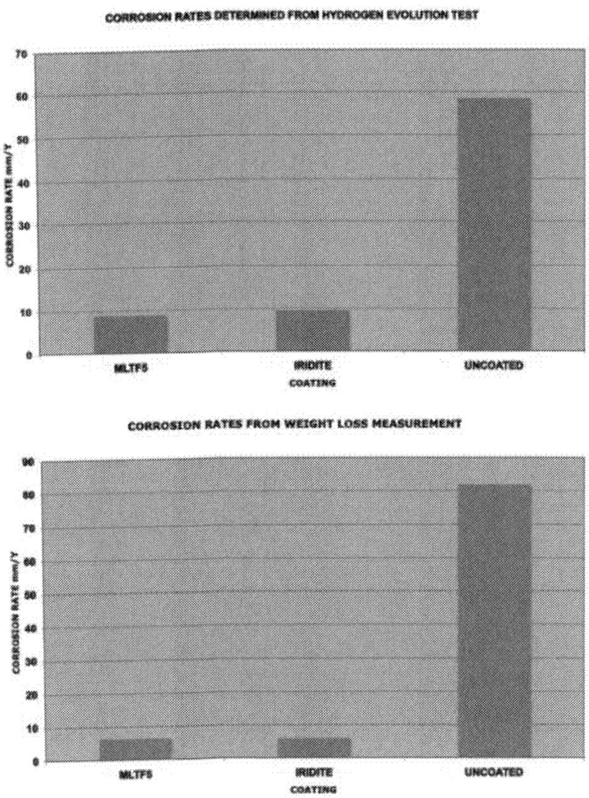

Figure 2 & 3: Corrosion rates calculated from the hydrogen evolution experiment and weight loss measurement tests on various coated surfaces.

DC Polarization studies:

DC polarization tests were conducted on bare as well as conversion coated substrates. Potentiodynamic studies revealed that the trivalent based conversion coating offered much nobler corrosion potential as shown in Figure 4 and Table 2 displays the corrosion current, corrosion potential and corrosion rates from Tafel plots generated using Gamry DC105 corrosion techniques software. It can be observed from these values that the corrosion rate for trivalent chromium based conversion coating on AZ92A has been reduced than for the hexavalent chromium based conversion coating as wells as the bare substrate.

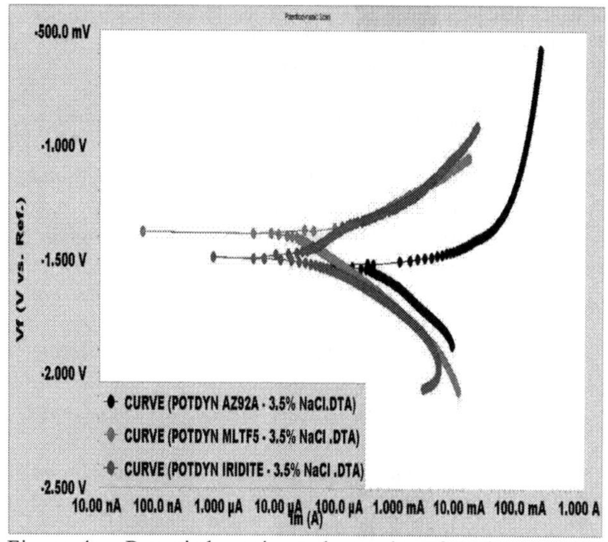

Figure 4: Potentiodynamic study results of various coating surfaces under 3.5% NaCl electrolyte.

Coating	Icorr-Amperes	Ecorr - Volts	Corrosion Rate (MPY)
Uncoated	4.48 mA	-1.53	6.82E+02
Iridite – Cr^{+6}	24.6 µA	-1.49	11.23
MLTF5 – Cr^{+3}	18.1 µA	-1.38	8.26

Table 2: Corrosion current, voltage and corrosion rates obtained from Potentiodynamic tests on AZ92A.

Conclusions

The novel trivalent chromium based conversion coating developed for magnesium alloys has in fact enhanced the corrosion resistance comparable to hexavalent chromium. Based on the results from various tests it can be concluded that;

- Rate of hydrogen evolution per surface area for magnesium substrates with trivalent chromium based conversion coatings is less than bare substrates and comparably same as hexavalent chromium coated substrates.
- Corrosion rates obtained from hydrogen evolution tests and weight loss measurement tests have shown that the trivalent chromium based coatings have indeed enhanced the corrosion resistance of the magnesium alloys.
- DC polarization studies have shown that the trivalent chromium based conversion coatings have offered significant amount of resistance when exposed to harsh chloride environment.

Acknowledgment

This work was performed as a part of the Phase I research for DoD - SBIR Grant proposal, F083-226-0271.

References

1. Natarajan.S, *Corrosion Prevention and Control*, v 51, n 4, p 142-163, December 2004
2. Avedesian.M.M, Baker.H, *Magnesium and magnesium alloys*, ASM International, 1999.
3. Emley.E.F, *Principles of Magnesium Technology*, Pergamon Press New York 1966.
4. G. Song.G, A. Atrens, *Advanced Engineering Materials* 5 (2003) 837–858.
5. G. Song, A. Atrens, *Advanced Engineering Materials* 1 (1999) 11–33.
6. G. Song, A. Atrens, *Advanced Engineering Materials* 9 (2007) 177-183.
7. G. Song, *International Conference on Magnesium*—Science Technology and Applications (2004) E01.
8. S.S. Cho, B.S. Chun, C.W. Won, S.D. Kim, B.S. Lee, H. Baek, *J. Mater. Sci.* 34 (1999) 4311.
9. Y. Li, J. Lin, F.C. Loh, K.L. Tan, *J. Mater. Sci.* 31 (1996) 4017.
10. S. Ono, N. Masuko, *Materials Science Forum* 419 (4) (2003) 897–902.
11. *"Magnesium Alloy, Proceses for Pretreatment and Prevention of Corrosion on"*, Society of Automotive Engineers, Warrendale PA.
12. "*Magnesium Alloy Castings, Sand 9.0Al – 2.0 Zn (AZ92-T6) Solution and Precipitation Heat Treated", specification*, Society of Automotive Engineers, Warrendale PA.
13. G. L. Makar, J. Kruger, *Int. Mater. Rev.* 1993, 38(3), 138.
14. W. S. Loose, *Corrosion and Protection of Magnesium*, ASM Int., Materials Park, OH 1946, pp. 173±260.
15. Froats, T. K. Aune, D. Hawke, W. Unsworth, J. Hillis, *Metals Handbook*, 9th ed., Vol. 13, ASM Int., Materials Park, OH 1987, pp. 740±754.

Magnesium Technology 2011
Edited by: Wim H. Sillekens, Sean R. Agnew, Neale R. Neelameggham, and Suveen N. Mathaudhu
TMS (The Minerals, Metals & Materials Society), 2011

DEVELOPMENT OF ZIRCONIUM-BASED CONVERSION COATINGS FOR THE PRETREATMENT OF AZ91D MAGNESIUM ALLOY PRIOR TO ELECTROCOATING

James Reck[1], Yar-Ming Wang[2], Hong-Hsiang (Harry) Kuo[2]
[1]United States Navy; 1240 Isaac Hull Ave. SE; Washington Navy Yard, DC, 20376, USA
[2]General Motors; GM R&D Center; Warren, MI, 48090, USA

Keywords: AZ91D, zirconia, conversion coating

Abstract
This work examines the use of hexafluorozirconic acid based solutions at concentrations from 0.025 M to 0.100 M and pH values of 2.0 to 4.0 for the creation of a zirconia-based conversion coating less than 1 micron thick to protect magnesium alloy AZ91D. Similar coatings have been found to give excellent protection for steel and aluminum alloys, but little research has been conducted on its application to magnesium. Work was performed to gain an understanding of the film formation mechanisms and related kinetics using x-ray photo-electron spectroscopy, scanning electron microscopy, and open circuit potential monitoring techniques. A design of experiments approach was taken to determine the effects of acid concentration, pH, and soak time on the corrosion properties both as-deposited and with an application of electrocoat. It was found that the application of the zirconia-based coating significantly increased corrosion resistance, and allowed for an acceptable e-coat application with excellent adherence.

Introduction
The use of Mg alloys for automotive applications has continued to increase in the industry, with hopes of increasing fuel economy by decreasing vehicle weight. Magnesium has a strength-to-weight ratio $^2/_3$ that of aluminum and ¼ that of iron while still exhibiting high thermal conductivity, high dimensional stability, good damping characteristics, and is easily recycled, making it one of the lightest structural metals available [1]. Magnesium, however, is currently limited in use due to a low resistance to corrosion. This issue can be addressed, to a limited extent, by alloying, but is more appropriately handled using a series of protective coatings.

The current automotive standard coating sequence for steel-based body structures involves a pre-treatment stage of phosphating prior to e-coat application. The phosphating process produces a conversion coating that assists in corrosion prevention by decreasing the active surface area of the metal available for corrosion reactions [2]. Some of the major issues with the phosphate process include sludge generation (which requires frequent desludging of process tanks in addition to expensive filters), the use of heated tanks requiring large amounts of energy, and a post phosphate sealing procedure to fill in the pores created during deposition [3]. Additionally, Mg dissolution in the traditional phosphate chemistries [4, 5] in conjunction with rising environmental restrictions and energy costs [3, 6, 7] are driving research efforts to find a cost efficient but equally protective coating process.

Several coating technologies have been researched to address the need for an Mg coating process that will provide cost effective corrosion protection. Popular methods include modifications of the phosphate process [4, 5, 8], Cr(III) conversion coatings [1], stannate conversion coatings [9], and cerium oxide conversion coatings [10, 11], but the emergence of ZrO_2-based nano-coatings have been of significant interest [2, 12-18]. These ZrO_2-based coatings have been well characterized on aluminum [7, 19-24] and steel [3, 7, 25-28], but significant work remains to be done on its use with magnesium. On aluminum and steel alloys, these coatings have been found to be very thin (e.g. <100 nm) [3, 6, 13, 24, 27], show acceptable corrosion protection up to 70 days of salt fog exposure [3, 20], and have been cited as providing "similar protection behavior as Zn and Fe phosphate coating" when combined with a layer of e-coat after deposition [3].

Experimental
Material and Sample Preparation. A single batch of die cast AZ91D test panels was sectioned into approximately 2"x4" coupons. The panels were polished using a series of 240, 320, 400, 600, and 1200 grit SiC sandpaper. All but the final sanding was performed using tap water for lubrication and dust removal.

The polished AZ91D plates were subjected to a pretreatment consisting of either an acid clean in 1wt% H_2SO_4 at 55°C followed by a 45°C DI water rinse (referred to as Ac or the acid clean) and/or an alkaline clean in 5wt% Na_2CO_3 followed by a DI water rinse (referred to as Alk or alkaline cleaning), all solutions at room temperature. The specimen was then immediately submerged in the prepared H_2ZrF_6 solution for the prescribed period of time.

If prescribed, the e-coat was applied with a prescribed voltage for a set time to attach the polymer chains to the nano-coated AZ91D plates. A post-deposition cure of 171°C for 25 minutes was used to reflow the polymers for a smoother, more adherent e-coat.

Solutions for the ZrO_2-based nano-coatings were prepared in 2 liter batches using a concentrated solution of 50 wt% hydrofluorozirconic acid (H_2ZrF_6) that was diluted to the desired concentration with DI water. All dilutions were specified by moles per liter (M) and prepared by weight. Titration to the desired pH was accomplished with concentrated NaOH solution.

E-coating was performed using a Darrah Electric 5-500 power supply which could supply a maximum of 500 V and 5 A, run under voltage control. The e-coat material used was a premixed solution of Dupont Cormax VI Gray.
Experimental Setup. Following initial coating studies, the general effectiveness of a ZrO_2-based nano-coating on Mg was examined using a design of experiments (DOE) regimen. The variables of interest were the concentration of H_2ZrF_6 (0.025, 0.050, and 0.100 M), solution pH (2,3, and 4), immersion time (30, 60, and 90 seconds), and surface pretreatment (alkaline or acid-alkaline). Given the large number of variables and the number of desired levels, a full factorial DOE (54 runs) was not feasible. So a Box-Behnken surface response, limited–sample DOE regimen was used (16 runs). The results of the DOE allowed for process optimization, resulting in a combination of factors that result in

the most desirable ZrO_2-based film.

Using the optimized ZrO_2-based nano-coating from the first DOE, a second DOE matrix was developed to optimize the deposition of the e-coat layer on top of the nano-coat. This regimen was designed as a full factorial matrix with voltage (150, 200, and 250 V) and time (1, 2, and 3 minutes). The results of the combined DOE results allowed the preparation of the best nano-coat plus e-coat bi-layer system on AZ91D possible within the design space studied for both DOEs for given the chosen optimization criteria.

Results and Discussion

The research was conducted in three stages: in-depth film characterization, nano-coat optimization, and e-coat optimization on the nano-coat.

In-Depth Film Characterization. Testing was performed to gain an understanding of the nano-coating's chemical, morphological, and growth rate characteristics on AZ91D. The chemical composition of two samples was studied via x-ray photoelectron spectroscopy (XPS), to determine if any chemical differences can account of the differences in corrosion behavior between the samples. SEM images on samples in both planar and cross-section orientation were also taken with a desire to detect morphological differences that may account for corrosion behavior, as well to assist in the development of a film growth model. A kinetics study was performed using three of the H_2ZrF_6 solutions to further assist with the same growth model. Lastly, the inductively coupled plasma (ICP) technique was performed on used H_2ZrF_6 and e-coat solutions to test for the presence of leached Mg.

XPS Characterization. XPS was performed on two of the trial runs with the H_2ZrF_6 solutions. These initial samples were at concentrations of 0.025 M and 0.100 M H_2ZrF_6 with as-diluted pH values of 1.83 and 1.67, respectively. The 0.025 M solution was used on an untreated (i.e., as-polished) AZ91D coupon, and the 0.100 M solution was tested on a two-stage acid-alkaline cleaned coupon. Both samples were analyzed with XPS for surface and bulk (~95nm sub-surface) chemistry (Table I). The surfaces of the films were very similar, showing a mix of Mg metal, $MgCO_3$, $Mg(OH)_2$, MgF_2, Al_2O_3, Al-hydroxides, ZrO_2, and ZrF_4.

Table I. Chemical composition of nano-coated specimens

		C	O	Mg	Al	Zr	F
Surface Chemistry (atomic %)	0.025 M No Pre-treat	11	50	18	2.1	2.1	14
	0.100 M Ac-Alk	9	50	20	2.6	1.3	16
Chemistry at ~95 nm (atomic %)	0.025 M No Pre-treat	4	48	23	2.1	16	8
	0.100 M Ac-Alk	2.7	42	27	2.7	6	20

The bulk of the films, however, did begin to show some of the differences between the two films. O-H bonds appear in the 0.025 M film that had received no pretreatment, that was not found in the other specimen. Likewise, the 0.100 M film with the acid-alkaline cleaning showed limited C-H-O bonding in the bulk that was not apparent in the 0.025 M film. Additional differences were found in the concentration of elements in the two films. The 0.025 M film exhibited a higher concentration of Zr and O in the bulk of the film by ~10 at% and ~6 at%, respectively, whereas the 0.100 M film was found to possess ~12 at% more F and ~4 at% more Mg than its counterpart. The bulk of the films did show some chemical similarities in the presence of both Mg and Al metal, MgO_x, MgF_2, AlO_x, ZrO_2, $Zr(OH)_4$, and ZrF_4. These results compare well with the literature, where many sources report finding similar coatings containing predominantly Zr, Mg, and Al oxides, hydroxides, hydrated oxyhydrides, and oxyfluorides [12, 16, 17].

Scanning Electron Microscopy (SEM). SEM analysis was performed on a selection of 5 samples chosen to allow the comparison of the morphological effects resulting from changes in H_2ZrF_6 concentration, solution pH, immersion time, and choice in surface pretreatment. Several common features were found in all of the films; including the presence of boulders ("massive" formations at random locations in the film), particulate matter surrounding said boulders (termed the transition zone), randomly dispersed particulate matter, and the presence of cracks in the base plane of the film.

It was observed that an increase in the concentration of H_2ZrF_6 in the coating solution decreased the size, frequency, and morphology of the boulder formations on the surface of the films. At 0.025 M H_2ZrF_6, the boulders appeared to possess a coral-like structure (Fig. 1), whereas the 0.100 M film produced boulders composed of rock-like material (Fig. 2). Additionally, the increased concentration resulted in a film with a less dense base composed of thin, cylindrical "worms" (Fig. 3) instead of the densely packed, pebble-like surface seen in the 0.025 M film (Fig. 4). Lastly, the amount of cracks present in the films was observed to decrease with the use of the higher concentration solution.

Increasing the pH also produced different coating morphologies of a more subtle nature. Raising the pH from 2.0 to 4.0 decreased the size and frequency of the boulders, as well as the size of the rocks composing the boulders. The crack severity (i.e. crack width) and number of cracks was also observed to decrease for an increase in pH. Lastly, the "worms" in the base were found to increase in diameter, and effectively increase the packing density of the base, at the elevated pH.

A markedly different coating morphology was found to occur when the immersion time was increased from 30 to 90 seconds. The base layer of "worms" was found to be significantly denser after the additional 60 seconds of immersion, as the diameter of the "worms" had greatly increased. The presence of crystallites growing from both the base and the boulders was also an additional feature of the longer immersion time (Fig. 5). Additional particulates, composed primarily of these crystallites, were observed to be randomly scattered throughout the surface of the film; appearing to grow through the base. The observed terminations of these crystallites suggests a monoclinic crystal structure, suggesting these crystals may be ZrO_2 or ZrF_4. As the energy required to form ZrO_2 (-1097 kJ/mol) is less than that for the formation of ZrF_4 (-1911 kJ/mol), it is presumed that the crystallites are ZrO_2 single crystals.

Another markedly different surface morphology was observed for a change in AZ91D surface pretreatment. An acid-alkaline two stage cleaning was observed to increase the number and size of the boulders on the film surface compared to a single stage alkaline cleaning. The two stage cleaning process also increased the number of cracks and the amount of particulate matter on the surface. Additionally, the single stage cleaning resulted in the formation of a denser base layer. The specimen subjected to the single stage cleaning also showed the initial presence of crystallite growth on the film surface.

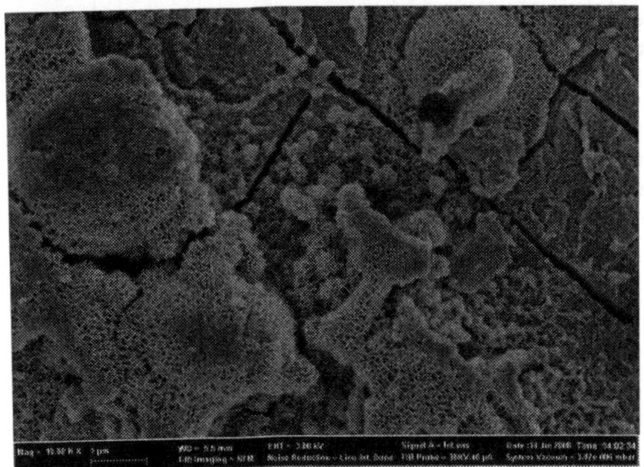

Figure 1. SEM micrograph of 0.025 M coating (10,000X)

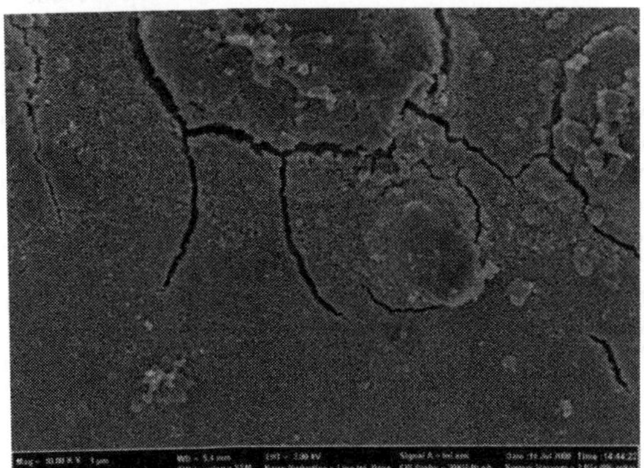

Figure 2. SEM micrograph of 0.100 M coating (10,000X)

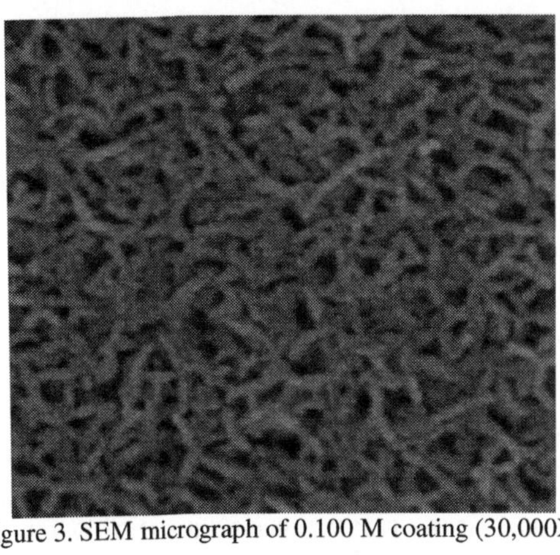

Figure 3. SEM micrograph of 0.100 M coating (30,000X)

Limited cross-section images have also been acquired (Fig. 6). These images show the presence of a two layer system on the surface of the AZ91D alloy, and also portray another difference due to the change in solution pH. Both layers were observed by electron diffraction spectrometry (EDS) to be composed of various amounts of Mg, Zr, and O, with limited amounts of F and Al present as well. Given the apparent "brightness" of the inner layer, it appears that the first layer to form is denser and contains a higher fraction of heavy elements (e.g. Zr). The outer layer, which was found to be thicker and rougher at the lower pH value, appears to contain varying concentrations of elements (chemical segregation) and the presence of multiple morphologies that may match with the topographic elements observed in the plan-view images.

Figure 4. SEM micrograph of 0.025 M coating (30,000X)

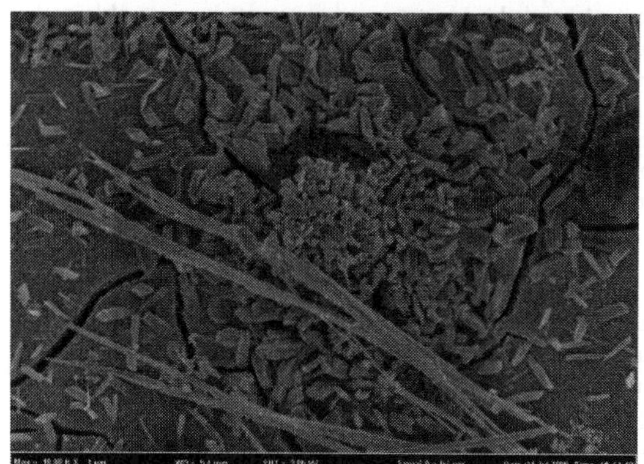

Figure 5. SEM micrograph of 90 second coating. (10,000X)

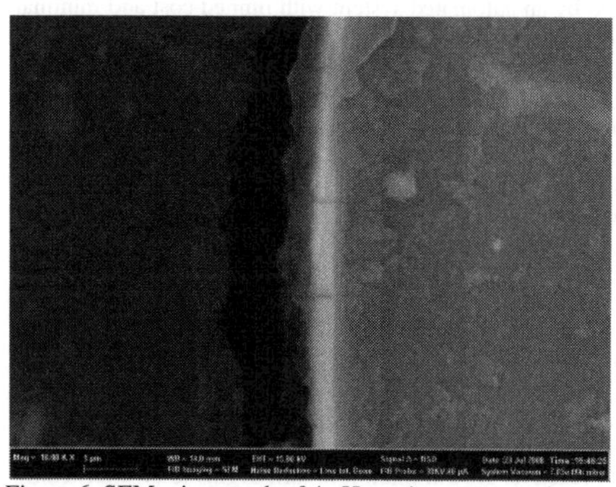

Figure 6. SEM micrograph of 4 pH coating cross-section.

Kinetics Study. Given the complex nature of the ZrO_2-based nano-coating, a better grasp of the growth mechanisms and film kinetics was desired. To accomplish this task, a standard OCP test was conducted using three different H_2ZrF_6 coating solutions: two samples that match the solutions used for the samples examined by XPS, and one sample that had an adjusted pH. In the case of the first two solutions, the AZ91D coupons were given pretreatments to match the specimens that were previously examined by XPS. The coupons tested in the 3rd solution were given an acid-alkaline two stage cleaning pretreatment. Three coupons were run for each condition for statistical purposes.

The average OCP response was calculated for the solutions from time 0 (initial immersion) to 600 seconds. First observations showed that the 0.100 M H_2ZrF_6 solutions resulted in a more noble response, indicating the presence of a more protective film being developed. The change in pH was also observed to slightly alter the OCP trace profile, but still held similar trends across the entire observation period.
Assuming that the OCP is an accurate indication of film growth, the response can be modeled using standard kinetics equations. The simplest form of the kinetics equation assumes a constant temperature:

$$V = A * \exp(k * t) \tag{1}$$

With V being the OCP (volts) at any time, t (sec), A as some initial coefficient (volts), and k is the kinetics constant (V/s). Using the average response curves for all three solutions, the coefficient, A, and the kinetics rate constant, k, were able to be calculated for the various regions that developed during the OCP measurements. It is interesting to note that the value of A was found to be roughly equal to the value of the OCP at the point in which the given region began. It was also found that a negative value for the rate constant showed periods of film growth (rising OCP) and those rate constants with positive (+) values matched with areas of film degradation (decreasing OCP).

Plotting the modeled kinetics data gathered from this analysis with the measured values shows the close correlation between the actual and the modeled values. The calculated R^2 of the models was found to be ~0.99 for all three cases, indicating that the OCP is likely measuring the film growth. This has far-reaching implications, as it may be possible to monitor film growth of all conversion coatings by following the OCP behavior, which is a simple, non-destructive test method that could be instantaneously monitored by an automated system with limited cost and minimal human interaction.

Growth Mechanisms – The OCP data from the kinetics study, coupled with SEM and XPS data, can also be used to gain a basic understanding of the film growth mechanisms. The OCP plots were observed to consist of 4 primary stages. The initial stage (Stage 0) occurs from the instant of solution contact with the Mg (time = 0) and continues for less than ½ second (Fig. 7). During this period of time the native oxide on the Mg coupon is attacked by the highly acidic coating solution [4, 13, 18, 29]. Once the native oxide has been dissolved, a magnesium-oxy-hydroxide, with the potential for a Mg-hydroxy-fluoride, deposits on the surface of the metal [9, 15, 17, 18, 27], and has been referred to as Stage I. This film continues to develop for approximately 2-4 seconds of total immersion time. As the new protective film begins to reach its growth limit, it in turn begins to be attacked by the coating solution (transition from Stage I to Stage II) [12, 15, 18, 30].

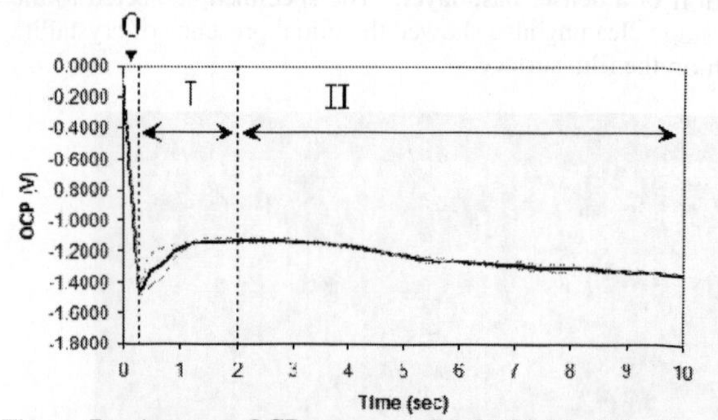

Figure 7. Average OCP response for 0.100 M, 2.1 pH coating from 0 to 10 seconds.

During this chemical attack, the film begins to accept higher concentrations of Zr, thus initiating the growth of various oxides (e.g. Al_2O_3, MgO, ZrO_2, etc.), hydroxides, and fluorides (Fig. 8) [13, 16, 19, 24, 31]. It should be noted that significant H_2 evolution was not observed until after a few seconds of immersion, further supporting the observed delay in the Zr deposition. These compounds continue to precipitate and grow on the surface of the Mg until the entire surface is fully covered (Stage II). The growth of these films continues in like fashion until the beginning of Stage III, after a total immersion time of approximately 100 seconds depending on the coating solution parameters. During the initial portions of Stage III the appearance of nano-crystallites on the surface of the film becomes apparent. Based on the SEM investigation, it appears that they initially form on the boulders, but with increased immersion time they begin to grow from or through the base of the coating as well. Based on the lack of change in the behavior of the OCP response past the onset of Stage III, it is believed that these crystals continue to grow and densify on the surface of the film with continued submersion [27].

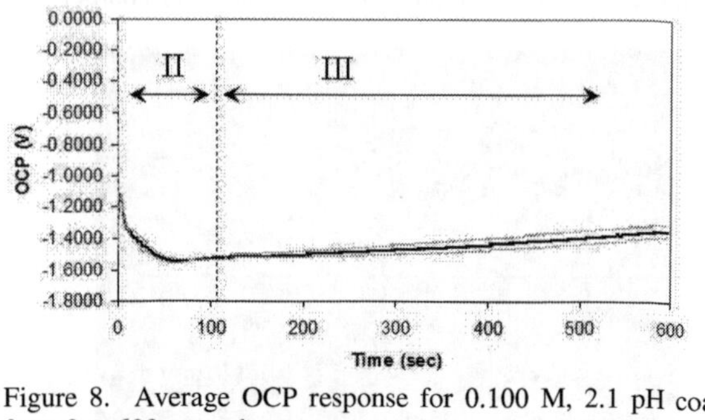

Figure 8. Average OCP response for 0.100 M, 2.1 pH coating form 0 to 600 seconds.

Fig. 9 shows the average OCP responses for all three tests. Based on these curves a few insights into the different film morphologies and growth mechanisms can be gained. By increasing the concentration of the H_2ZrF_6 solution, the OCP was found to significantly increase indicating the development of a more protective film. As magnesium fluoride is a very stable, insoluble compound, it is reasonable to assume that the addition of more F^- ions in the solution (as would be the case with increased H_2ZrF_6 concentration) would create a more dense protective film. This was confirmed by XPS with the increased F content found in the film deposited from the higher concentration solution. An increased H_2ZrF_6 concentration was found to decrease Stage I, increase Stage II, and flatten Stage III on the OCP curves, indicating, again, that the initial protective film (Stage I) was developed more rapidly due to the increased F^- content, and takes longer times to dissolve (Stage II) in order to initiate Stage III [16].

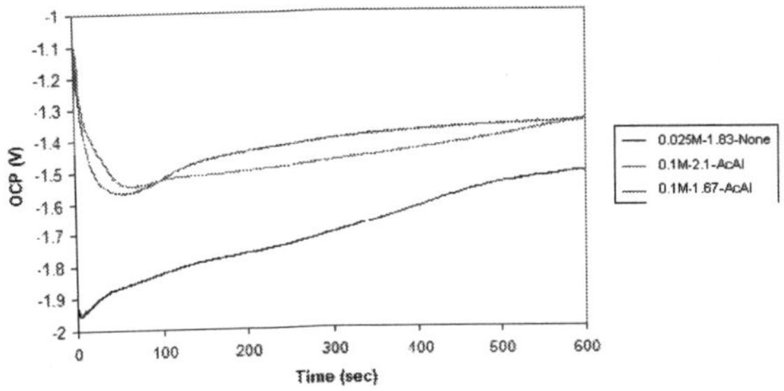

Figure 9. Average OCP response for three testing conditions.

An increase in pH was found to have a less significant impact on the film growth than was observed for a change in concentration. By increasing the pH, the only observable change was an increase in the Stage I response time, and an initial decrease in OCP values of Stage III which returned to similar values by the end of the test period, corresponding to trends observed in the literature [17]. The effects of pretreatment choice are compounded with the change in H_2ZrF_6 concentration in this series of tests, but based on the SEM observations appear to have the same general effects as a change in concentration. Namely, by choosing to use the two steps acid-alkaline cleaning treatment, there is a decrease in Stage I, increase in Stage II, and a flattening of Stage III, with an overall increase in OCP values.

Mg Dissolution Studies – The last characterization study performed on these films was used to determine how much, if any, Mg^{2+} ions were being leached into both the H_2ZrF_6 coating solution and the e-coat solution. For this portion of the investigation, the inductively coupled plasma (ICP) technique was used to measure the Mg^{2+} concentration in the H_2ZrF_6 solution after 10 runs and in the e-coat solution after 25 runs. It was found that after 10 runs in a 750 ml container, the H_2ZrF_6 solution had acquired a total Mg content of 50 ppm. Assuming equal amounts of Mg was leached into solution with each run, this averages to approximately 5 ppm per sample (each test area on the Mg coupon was approximately 2" wide by 3" long and $^1/_8$" thick). This result is promising for the use of Mg alloys with these solutions on an assembly line when compared to the 227 ppm of Mg in a phosphate solution after only one AZ91D run.

The same test was performed on an e-coat solution (1.75 liters) that was used to coat 25 samples of approximately the same size that had been coated with varying nano-coat compositions. For these 25 runs, 100 ppm of Mg was leached into the e-coat solution. Given that the e-coat solution was found to have 1.4 ppm Mg prior to any testing, the 25 coated samples added a total of approximately 98.6 ppm Mg, for an average of approximately 3.9 ppm Mg per run. It should be noted, however, that in both the H_2ZrF_6 and the e-coat solutions, there was no observable sludge or particulate buildup at the bottom of the container, or suspended in the solution. It is felt that continued research on these coatings, perhaps with the addition of a polymer in the H_2ZrF_6 solution [20, 22, 31], it may be possible to reduce the amount of Mg leached into the e-coat solution to an unobservable level.

Nano-Coat Optimization. Nano-coat optimization was conducted using the DOE discussed previously. Specimens were tested in the as-coated condition, as well as with a layer of e-coat deposited at 200 V for 3 min that had been cured at 171°C for 25 minutes. The OCP and EIS measurements of both the nano-coat single-layer, and nano-coat/e-coat bi-layer systems were acquired, the PD and coating weights gathered for the single layer nano-coat condition, and the surface roughness and adhesion values for just the bi-layer configuration.

Each response was placed into the DOE software and analyzed for statistical significance using a standard ANOVA procedure. All of the responses showed a significant statistical model could be fit to the data with varying degrees of fit (R^2). An important observation from the DOE results necessary for understanding this system is that all responses except for the e-coat adhesion show multiple parameter dependencies; further showing the complex nature of these coatings.

In addition to indicating the importance of various parameters and interactions, the more useful application of the DOE software is the calculation of a mathematical regression model to describe the measured data. These models can be used to optimize the deposition of the nano-coat to meet a series of chosen criteria. For the purposes of this project, a nano-coating was desired that minimized the corrosion current (I_{corr}) and the surface roughness (R_a); while maximizing the pore resistance (R_p), OCP_{ZrO2}, e-coat resistance (R_{el}), and e-coat adhesion (Table X). Using these criteria, a set of optimum conditions was derived by the software that would meet or exceed these demands. Using this optimized parameter set, another series of samples were made to validate the models.

Based on the results from the optimized condition, 9 of the 14 models were found to make predictions within 50% of the measured values; 3 of which were within 10% accuracy. It should be noted that given the complex nature of the variable interactions found in the ANOVA analysis, that is reasonable to assume that there are higher order interactions occurring (e.g. four-way interactions and/or cubic dependencies) that were not able to be tested due to the limited number of runs. To increase the fidelity and accuracy of the models further depositions and testing should be performed. Based on the available data, however, it can be stated that the nano-coating deposition is not a simple, linear process, but rather a very complex interaction with a multitude of competing forces.

E-coat Optimization. Using the optimized ZrO_2-based nano-coating as the base layer, a second DOE was developed (as discussed above) to optimize the addition of an e-coat layer for increased corrosion protection, and to present a surface ready for paint application. The responses were coating thickness, $OCP_{e\text{-}coat}$, EIS, surface roughness, and e-coat adhesion. As with the nano-coat DOE, all of the variables in this second series of experiments were able to be statistically modeled. In all but one of the cases, there was a complex parabolic interaction. The only case where this trend was not observed was for the OCP, which resulted in an R^2 of 0.246 indicating that the model does not accurately match the available data. Several of the other R^2

values are also questionable (i.e. C_{el}, R_{el}, and roughness), but are still within an acceptable range. The reason for these low R^2 values may be the presence of higher order variable dependencies (e.g. t^3 and/or V^3) that could not be analyzed with the selection of samples processed.

Optimization of the bi-layer system of e-coat on the nano-coat base layer was determined by minimizing the surface roughness to a value less than 45 nm and maximizing the OCP, R_{el}, and adhesion. The resulting optimized condition was found to be 150 V for 3 minutes, which was one of the runs tested for the DOE. Comparing the predicted values from the models back to the original run data showed that all predicted values were within 15% of the measured data. Given the sub-standard R^2 values associated with the models, this agreement between data and models is a good indication of the model's ability for optimization.

Summary

Investigations have been performed on the coating mechanisms of a ZrO_2-based nano-coating on Mg alloy AZ91D. A basic film growth model has been developed based on the morphologies observed in the SEM and the profiles gathered via the monitoring of OCP during deposition. The stages of the models are related to the chemistries found to be present in the films by XPS. Preliminary tests were also performed to determine that a small amount of Mg was being leached into the conversion coating solution (~5 ppm per samples) and the e-coat solution (~4 ppm per sample). Two series of DOE experiments were also conducted to optimize 1). the deposition of the ZrO_2-based nano-coating (at a concentration of H_2ZrF_6 of 0.100 M, solution pH of 2.10, and acid-alkaline two-stage pretreatment for 35 seconds), and 2). the e-coat that was deposited on top of the optimized ZrO_2-based coating (at 150 V for 3 minutes). The work has shown that the use of ZrO_2-based nano-coatings on Mg alloy AZ91D improves the corrosion resistance of the base metal, and that the use of e-coat on top of the nano-coating greatly improves the component's corrosion resistance. The final optimized sample (with optimum nano-coat and e-coat depositions) showed a smooth, adherent, and well protected magnesium component.

Acknowledgements

The authors would like to thank Sundaresan Avudaiappan for training James on the use of appropriate equipment for the necessary testing. Curt Wong for operating the SEM. Steve Gaarenstroom and Maria Militello for XPS and ED-XRF testing and analysis, and Nick Irish for the ICP work. Willie Dixon for sectioning the Mg plates, and for training on the appropriate sample preparation and polishing equipment.

References

1. Gray, J. and B. Luan, *Protective coatings on magnesium and its alloys—a critical review.* Journal of Alloys and Compounds, 2002. **336**(1-2): p. 88-113.
2. Ardelean, H., I. Frateur, and P. Marcus, *Corrosion protection of magnesium alloys by cerium, zirconium and niobium-based conversion coatings.* Corrosion Science, 2008.
3. Zhai, Y., et al., *A Replacement for Phosphate Conversion Coating Based on Hexafluorozirconic Acid.* 2008.
4. Zhou, W., et al., *Structure and formation mechanism of phosphate conversion coating on die-cast AZ91D magnesium alloy.* Corrosion Science, 2008. **50**(2): p. 329-337.
5. Gao, G. and M. Ricketts, *Evaluation of Protective Coatings on Magnesium for Phosphate Process Compatibility and Galvanic Corrosion Prevention.* 2002.
6. Giles, T., et al., *New Generation Conversion Coatings for the Automotive Industry.* 2007.
7. Talbert, R. (2007) *Pretreatments: The Next Generation.* **Volume,**
8. Zhao, M., et al., *Influence of surface pretreatment on the chromium-free conversion coating of magnesium alloy.* Materials Chemistry & Physics, 2007. **103**(2-3): p. 475-483.
9. Elsentriecy, H., K. Azumi, and H. Konno, *Effect of surface pretreatment by acid pickling on the density of stannate conversion coatings formed on AZ91 D magnesium alloy.* Surface & Coatings Technology, 2007. **202**(3): p. 532-537.
10. Maddela, S., et al., *Cerium-Based Conversion Coating on AZ91D Alloy-II.* 2008.
11. Maddela, S., et al., *Preliminary Studies of Cerium Based Conversion Coatings on AZ91D Alloy.* 2008.
12. Gray, J. and B. Luan, *Protective coatings on magnesium and its alloys□a critical review.* Journal of Alloys and Compounds, 2002. **336**(1-2): p. 88-113.
13. Knudsen, O.O. and A. Bjorgum, *Chromium free pretreatments - state of the art.* unknown.
14. Skar, J., L. Sivertsen, and J. Öster, *Chrome-free Conversion Coatings for Magnesium Die Castings□a Review.* ICEPAM-International Conference on Environmental friendly pre-treatment for Aluminum and other Metals, Oslo, Norway, 2004.
15. Walter, M., *Magpass-CoatSr as a Chrome-Free Pre-Treatment for Paint Layers and An Adhesive Primer for Subsequent Bonding.* 2004.
16. Verdier, S., et al., *Monochromatized x-ray photoelectron spectroscopy of the AM60 magnesium alloy surface after treatments in fluoride-based Ti and Zr solutions.* Surf. Interface Anal, 2005. **37**: p. 509-516.
17. Verdier, S., et al., *An electrochemical and SEM study of the mechanism of formation, morphology, and composition of titanium or zirconium fluoride-based coatings.* Surface & Coatings Technology, 2006. **200**(9): p. 2955-2964.
18. Verdier, S., et al., *The surface reactivity of a magnesium□aluminium alloy in acidic fluoride solutions studied by electrochemical techniques and XPS.* Applied Surface Science, 2004. **235**(4): p. 513-524.
19. Lunder, O., et al., *Formation and characterisation of Ti□Zr based conversion layers on AA6060 aluminium.* Surface & Coatings Technology, 2004. **184**(2-3): p. 278-290.
20. Mirabedini, S., et al., *Characterization and Corrosion Performance of Powder Coated Aluminium Alloy.* IRANIAN POLYMER JOURNAL, 2003. **12**: p. 261-270.
21. Nickerson, B.C. and E. Lipnickas, *Characterization of A Viable Non-Chromated Conversion Coating Aluminum and its Alloys by Electrochemical and Other Methods.* unknown.
22. Reghi, G.A., *Process for corrosion resisting treatments for aluminum surfaces.* 1992, US Patent 5,089,064.
23. Reghi, G.A., *IMPROVED CHROMIUM-FREE COMPOSITION AND PROCESS FOR CORROSION RESISTING TREATMENTS FOR ALUMINUM SURFACES.* 1994, EP Patent 0,555,383.
24. Schram, T., G. Goeminne, and H. Terryn, *Study of the composition of zirconium- based chromium free conversion layers on aluminum [JI Trans. Inst.* Metal Finishing, 1995. **73**: p. 91-95.

25. Puomi, P., et al., *Optimization of commercial zirconic acid based pretreatment on hot-dip galvanized and Galfan coated steel.* Surface & Coatings Technology, 1999. **115**(1): p. 79-86.
26. Romero Pareja, R., et al., *Corrosion behaviour of zirconia barrier coatings on galvanized steel.* Surface & Coatings Technology, 2006. **200**(22-23): p. 6606-6610.
27. Stromberg, C., et al., *Synthesis and characterisation of surface gradient thin conversion films on zinc coated steel.* Electrochimica Acta, 2006. **52**(3): p. 804-815.
28. Tepe, B. and B. Gunay, *Evaluation of pre-treatment processes for HRS (hot rolled steel) in powder coating.* Progress in Organic Coatings, 2007.
29. Ambat, R., N. Aung, and W. Zhou, *Evaluation of microstructural effects on corrosion behaviour of AZ91D magnesium alloy.* Corrosion Science, 2000. **42**(8): p. 1433-1455.
30. Say, W., C. Chen, and S. Hsieh, *Electrochemical characterization of non-chromate surface treatments on AZ80 magnesium.* Materials Characterization, 2008.
31. Nordlien, J., et al., *Formation of a zirconium-titanium based conversion layer on AA 6060 aluminium.* Surface & Coatings Technology, 2002. **153**(1): p. 72-78.

Magnesium Technology 2011
Edited by: Wim H. Sillekens, Sean R. Agnew, Neale R. Neelameggham, and Suveen N. Mathaudhu
TMS (The Minerals, Metals & Materials Society), 2011

USE OF AN AC/DC/AC ELECTROCHEMICAL TECHNIQUE TO ASSESS THE DURABILITY OF PROTECTION SYSTEMS FOR MAGNESIUM ALLOYS

Sen Song[1] , Robert C. McCune[2], Weidian Shen[1] and Yar-Ming Wang[3]
[1]Eastern Michigan University, Ypsilanti, MI, 48197, USA
[2]Robert C. McCune & Associates, LLC, West Bloomfield, MI, 48322, USA
[3]General Motors Co.,R&D Center, Warren, MI, 48090, USA

Keywords: magnesium, coatings, corrosion

Abstract

One task under the U.S. Automotive Materials Partnership (USAMP) "Magnesium Front End Research and Development" (MFERD) Project has been the evaluation of methodologies for the assessment of protective capability for a variety of proposed protection schemes for this hypothesized multi-material, articulated structure. Techniques which consider the entire protection system, including both pretreatments and topcoats are of interest. In recent years, an adaptation of the classical electrochemical impedance spectroscopy (EIS) approach using an intermediate cathodic DC polarization step (viz. AC/DC/AC) has been employed to accelerate breakdown of coating protection, specifically at the polymer-pretreatment interface. This work reports outcomes of studies to employ the AC/DC/AC approach for comparison of protective coatings to various magnesium alloys considered for front end structures. In at least one instance, the protective coating system breakdown could be attributed to the poorer intrinsic corrosion resistance of the sheet material (AZ31) relative to die-cast AM60B.

Introduction

The Magnesium Front End Research and Development Project (MFERD) has been the subject of several prior reports [1,2]. In brief, the project is an international collaboration between the U.S. Department of Energy, the U.S. Automotive Materials Partnership, Natural Resources – Canada, and the Ministry of Science and Technology of the People's Republic of China, with a goal of advancing both the knowledge base and manufacturing capabilities necessary to engineer and produce magnesium-intensive automotive body substructures, similar to the all-steel baseline structure shown in Figure 1, having markedly reduced weight but with equivalent mechanical strength and durability.

One task of the MFERD Project was aimed at corrosion prevention and surface treatment for magnesium components and assemblies such as that envisioned for the front end. An objective of this task was the evaluation of various coating systems as might be incorporated in the individual components and assembled structure. Figure 2 illustrates a "model" corrosion protection system for magnesium incorporating a "pretreatment" step which includes any metal cleaning and pickling or 'activation', followed by one of several possible processes to provide an anchoring layer for the topcoat (or adhesive) which is presumed to be a polymeric coating or "paint". Various pretreatment processes for magnesium have been considered over the years [3] including conversion coating (e.g. chromating), anodizing or other electrically-based processes used to develop primarily inorganic layers with substantial surface roughness to which the topcoat or adhesive can adhere. Considering the choices of base metal (die-casting, extrusion, formed sheet), pretreatment options, and possible topcoats or adhesives, the number of surface treating permutations is substantial. Prior work reported through TMS [4] has offered an approach to coating systems evaluation predicated on such metrics as scribe creepback and accumulation of surface corrosion deposits.

Figure 1. Baseline articulated automotive front-end substructure as implemented in steel

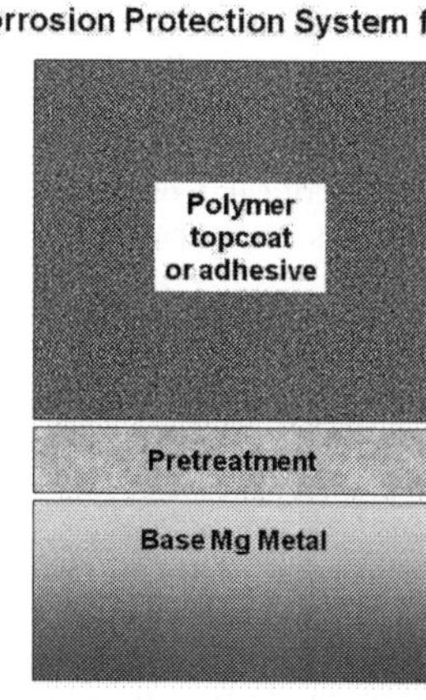

Figure 2. Model corrosion protection system proposed for magnesium components and front end subassembly.

Over the years, electrochemical approaches to the assessment of coatings durability have been advanced, and, in particular, the use of electrochemical impedance spectroscopy (EIS) and associated techniques. Unlike most DC electrochemistry techniques which produce substantial polarizations in excess of the unperturbed corrosion potential, EIS methods utilize only minor electrical polarizations over a wide range of frequencies, permitting an assessment of complex impedance and assignment of specific signatures to physical phenomena occurring in the "system" of the corroding workpiece and its environment. These are usually expressed in terms of electrical circuit analogs which can act as networks of capacitive and resistive elements. A simple primer on EIS methods may be found online [5].

While AC methods such as EIS offer a generally "non-destructive" probe of the state of various charge transfers occurring in coated metals in contact with an electrolyte, a more aggressive stress is required to accelerate the corrosion process for purposes of comparing the relative durability of different coating systems. Often, a cathodic reaction (i.e. electron consuming) at polymer-coated metal surfaces, leads to eventual breakdown of the attachment of the polymer to the metal permitting both undercutting of the protective polymer and concomitant corrosion of the metal. In the case of magnesium, the aqueous corrosion over a wide range of pH generates copious $Mg(OH)_2$, further aggravating polymer delamination, as well as hydrogen evolution as the principal cathode reaction, leading to increasing concentration of OH^- and further polymer delamination. Thus, DC cathodic polarization may be used to drive the polymer delamination process and accelerate the undercutting for painted magnesium. Before the delamination and corrosion product are visible, changes to the impedance spectrum can be observed, giving indications of such contributions as loss of polarization resistance at the metal surface, indicative of an increase in the corrosion process at that interface. This sequence of AC impedance measurement, followed by DC cathodic polarization and subsequent AC analysis has been referred to as "AC/DC/AC" measurement, and origins of the approach have been attributed to Hollaender, et al. [6].

The work reported here illustrates potential use of an AC/DC/AC method in the assessment of similar pretreatments and polymer coatings to differing magnesium alloys and inference of the influence of intrinsic corrosion differences of the materials on performance in these tests. Development of this, or comparable procedures, is seen as a critical tool for rapid assessment of corrosion protective coatings for magnesium.

Experimental Procedures

Materials and Surface Treatments. The magnesium materials used for this comparison were AM60B low-pressure die cast in the form of 100 x 150 x 3mm plates, and AZ31 sheet of thickness 2 mm, and comparable size. These plates were surface treated using a process line operated by MetoKote, Inc., (Lima, OH), and included alkaline cleaning, acid activation (etching), pretreatment with Henkel Alodine 5200® followed by a degassing (curing) cycle and application of an electrostatic epoxy powder topcoat (Protech KS-542-N49), with subsequent fusion. This sequence is nominally that employed commercially for current automotive front-end components [7].

Testing Procedures and Analysis. Corrosion evaluations of the two alloy forms were conducted using several techniques including:

a.) ASTM B-117 neutral salt-spray exposure with photography at 200,300,500,700 and 1000 hours, with a scribed panel,

b.) cathodic delamination according to ASTM D-1654 using 5% NaCl (per ASTM B-117) as the electrolyte, graphite counter electrode and polarization of -1.5 V vs. OCP, with photography at 500 and 1000 hours,

c.) conventional EIS using the same electrolyte and Gamry MultEchem 4 three-electrode cell, using 10 mV p-p AC voltage relative to OCP in the frequency range of 1 mHz to 100 kHz, at various times up to 35 days exposure,

d.) AC/DC/AC method (described below).

In each "cycle", the impedance of the electrochemical cell was first measured using the EIS measurement parameters described above. A cathodic DC voltage of – 4V vs. the open-circuit potential (OCP) was then applied to the sample or "working electrode" for 30 minutes to promote the cathodic reaction and presumed undercutting of the polymer film. The applied DC voltage was halted after this exposure, permitting the sample to relax to a "new" open circuit potential; this time period being two hours. The second cycle repeated the procedure, starting with the AC impedance measurement, to probe any changes of the tested sample after the previous cycle. All cycles were controlled by software using a laboratory PC. In this manner, the AC/DC/AC test generally accelerates the corrosion process. For this study, up to 30 cycles with a duration of 4 hours per cycle were typically applied. Figure 3 illustrates schematically the test cycle.

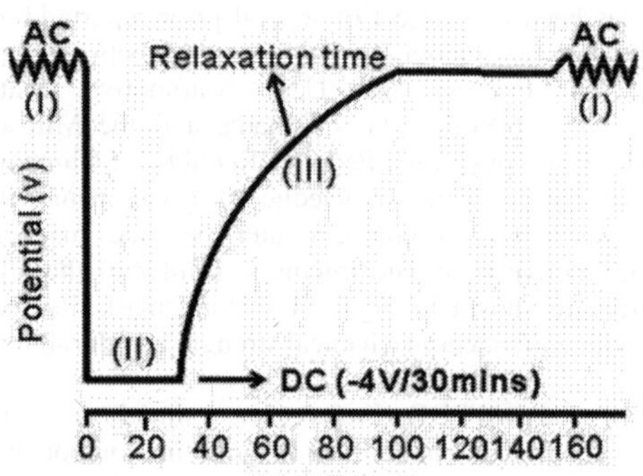

Figure 3. Schematization of AC/DC/AC polarization cycle.

The analysis of EIS spectra traditionally invokes the use of "equivalent" electrical circuit analogs such as that shown in Figure 4 for a simple coated metal in contact with a corrosive aqueous electrolyte. In the equivalent circuit suggested by Figure 4, R_s represents the resistance of the electrolyte or "solution" resistance; R_1 represents the resistance of coatings, including the pretreatment (e.g.,chemical conversion coating); C_1 represents the capacitance of the epoxy powder coating; R_2 represents the charge transfer resistance at the interface between the coatings and the metallic substrateor alternatively the "polarization resistance" or corrosion resistance of a metal electrode; C_2 represents the double-layer capacitance established at the interface. ZView® Electrochemical Analysis software (Scribner Associates, Inc., Southern Pines, NC), was used to record and analyze the measured data, *i.e.*, the measured impedance at the different frequencies, and calculate the values for the simulated elements in the equivalent circuit. Plots of impedance modulus |Z|, vs frequency (i.e. "Bode") plots were also instructive in observing changes occurring during coating exposure.

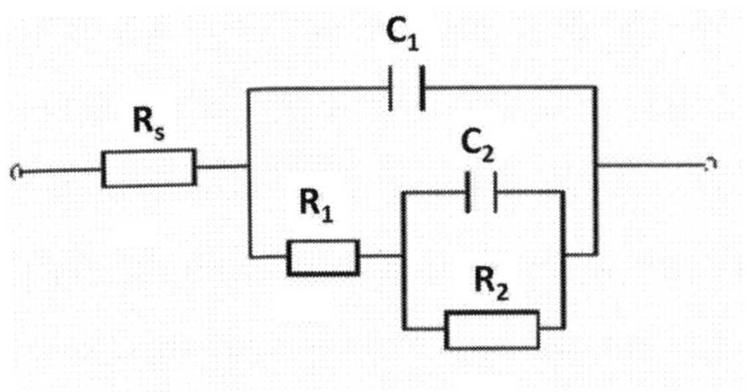

Figure 4. Equivalent circuit schematization for a corroding painted metal surface in an electrolyte.

Results and Discussion

Despite the similarity of the pretreatment and topcoating process for AM60B and AZ31 used here, the corrosion behaviors were markedly different by any of the methodologies employed; the AZ31 showing clearly inferior corrosion resistance to the AM60. This suggests that the "intrinsic" corrosion properties of the base metal, known to be poorer for AZ31 than AM60 are also manifest in the performance of the particular coating processes employed here. Because a variety of magnesium alloys are envisioned for the "front end" structure, any surface treatment and coating scheme must provide an optimized protection for all component materials, including ancillary materials of construction including other metal parts, fasteners, etc. A comparison of ASTM B-117 exposure behavior after 200 hours is shown in Figure 5. By this point, the AZ31 has already exhibited scribe undercutting or creep-back relative to the AM60 substrate.

Results of cathodic delamination studies are illustrated in Figure 6. For this testing, the AZ31 sheet exhibited pronounced paint layer debonding and corrosion at 500 hours of exposure, whereas the AM60B die cast material with comparable surface treatment showed virtually no debonding at 500 hours of exposure and beginning of noticeable debonding at 1000 hours.

Conventional EIS Bode magnitude plots for the initial and 35 day salt water exposure (5% NaCl according to ASTM B-117) of identically-treated AZ31 and AM60B are shown in Figure 7. The initial Bode plots are virtually indistinguishable suggesting that the aggregate metal and protection system are comparable. The 35 day exposure behavior does indeed, however, suggest that the AZ31 sheet material had undergone a greater loss of low-frequency impedance by about an order of magnitude. Assignment of specific circuit element impedances using the modeling software, however, did not suggest a likely candidate for this loss of impedance within the physical system.

Results of the AC/DC/AC approach as outlined are compared in Figure 8 for 27 cycles of operation as described (approximately 108 hours of exposure). While the initial Bode plots are indistinguishable (as with conventional EIS testing), the 27 cycle Bode plots now suggest a much greater loss of the initial impedance for the AZ31 sheet, by as much as three orders of magnitude. The Bode plot for the AZ31 now resembles that more of a poorly protected and corroding metal as opposed to a metal that is highly protected by the insulating polymer topcoat.

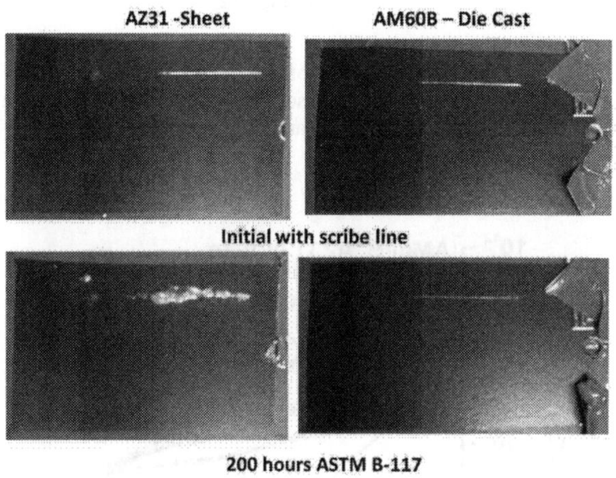

Figure 5. Compariosn of ASTM B-117 exposure behaviors for AZ31 sheet and AM60B die casting with comparable surface treatments, after 200 hours of exposure.

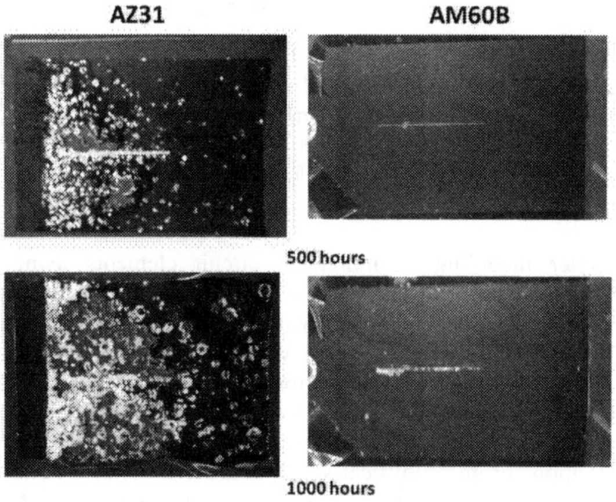

Figure 6. Comparison of cathodic debonding behavior at 500 hours and 1000 hours for AZ31 sheet and AM60B die cast magnesium with comparable surface treatments.

of non-equilibrium β-phase ($Mg_{17}Al_{12}$), exhibiting substantial corrosion resistance or having improved reaction with the conversion coating process used in this case.

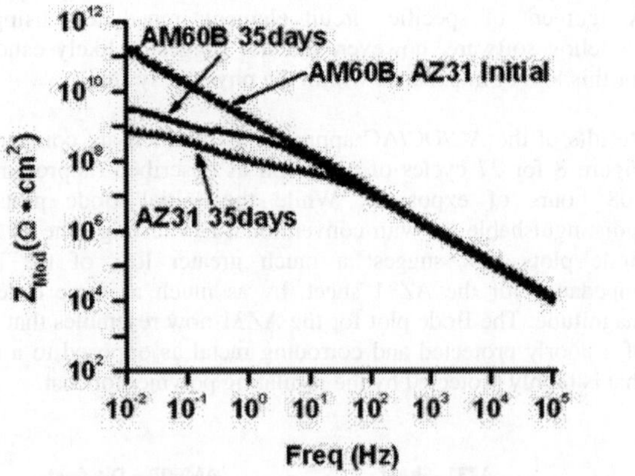

Figure 7. EIS Bode magnitude plots for AZ31 sheet and AM60B die casting with comparable surface treatments for initial measurements (indistinguishable) and after 35 day exposure to 5% NaCl solution.

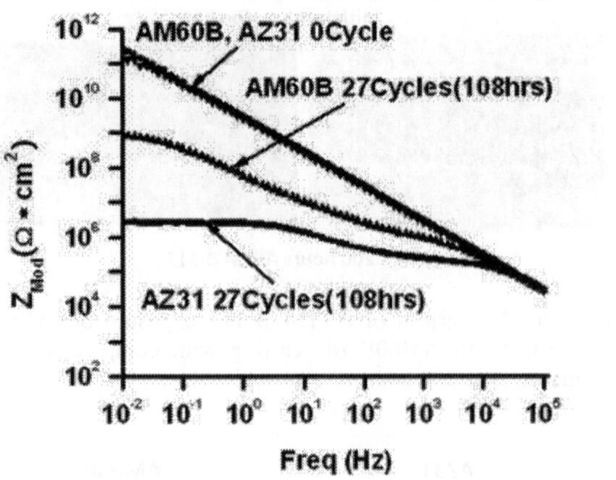

Figure 8. EIS magnitude Bode plots for AZ31 sheet and AM60B die casting with comparable surface treatments for the initial measurement and after 27 cycles of the AC/DC/AC protocol as described using 5% NaCl as the electrolyte.

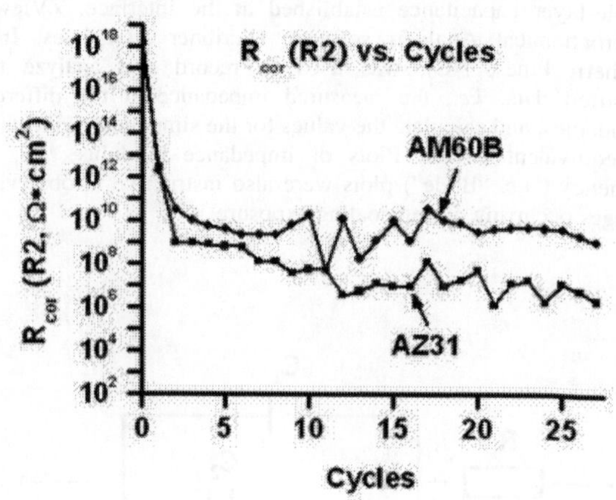

Figure 9. Plots of R_{cor} (polarization resistance) for AZ31 sheet and AM60B die casting having the same surface treatment as a function of AC/DC/AC cycle number. The situation is clearly worse for AZ31 at longer cycle times as the polymer layer and conversion coating are undercut.

Computer modeling of individual circuit elements using the AC/DC/AC protocol and presumed equivalent circuit now suggest the polarization or "corrosion" resistance at the corroding metal interface is appreciably lower for AZ31 than for AM60B. Plots of R_{cor} or "corrosion" or "polarization" resistance as a function of cycle are shown in Figure 9.In general, separate measurements of the 'intrinsic' corrosion resistance of the base metals support this observation. From a physical standpoint there is not a definitive explanation other than the poorer corrosion resistance of AZ31 in comparison to AM60, based primarily on differing aluminum content of the alloys, plus any artifacts of the processing (e.g. "skin" effects) which render the surface of AM60B either more highly enriched in aluminum or generation of appreciably levels

In summary, the AC/DC/AC approach offers an opportunity to induce an accentuation of differences in protective capabilities of comparable processes on differing substrates, and hopefully can be developed such as to distinguish overall protective capabilities of a protection system in a rapid and quantitative fashion. Correlation of performance in such accelerated testing with actual field performance remains, however, one of the great challenges for corrosion science as applied to difficult materials such as magnesium.

Acknowledgments

This material is based upon work supported by the Department of Energy National Energy Technology Laboratory under Award Number Nos.DE-FC05-95OR22363, DE-FC05-02OR22910, and DE-FC26-02OR22910. This report was prepared as an account of work sponsored by an agency of the United States Government. Neither the United States Government nor any agency thereof, nor any of their employees, makes any warranty, express or implied, or assumes any legal liability or responsibility for the accuracy, completeness, or usefulness of any information, apparatus, product, or process disclosed, or represents that its use would not infringe privately owned rights. Reference herein to any specific commercial product, process, or service by trade name, trademark, manufacturer, or otherwise does not necessarily constitute or imply its endorsement, recommendation, or favoring by the United States Government or any agency thereof. The views and opinions of authors expressed herein do not necessarily state or reflect those of the United States Government or any agency thereof. Such support does not constitute an endorsement by the Department of Energy of the work or the views expressed herein.

The authors thank the U.S. Automotive Materials Partnership (USAMP) and the U.S. Department of Energy for their support of this work.

References

1. A. A. Luo, E. A. Nyberg, K. Sadayappan, W. Shi, "Magnesium Front End Research and Development: A Canada-China-USA Collaboration," in Magnesium Technology 2008, eds. M. Pekguleryuz, N.R. Neelameggham, R.S. Beals and E.A. Nyberg, TMS (The Minerals, Metals and Materials Society), Warrendale, PA, 2008.

2. E.A. Nyberg, A.A. Luo, K. Sadayappan and W. Shi, "Magnesium for Future Autos," p. 35-37 in Advanced Materials and Processes, October 2008.

3. J. E. Gray and B. Luan, "Protective coatings on magnesium and its alloys – a critical review," Journal of Alloys and Compounds 336, 88-113 (2002).

4. P. J. Blanchard, D. J. Hill, G. T. Bretz and R. C. McCune, "Evaluation of Corrosion Protection Methods for Magnesium Alloys in Automotive Applications," in Magnesium Technology 2005, Neale R. Neelameggham, Howard I. Kaplan, and Bob R. Powell, eds.,TMS (The Minerals, Metals & Materials Society), Warrendale, PA, 2005.

5. Gamry, Inc. http://www.gamry.com/App_Notes/EIS_Primer/EIS_Primer.htm.

6. J. Hollaender, E. Ludwig, S. Hillebrand, " Assessing protective layers on metal packaging material by electrochemical impedance spectroscopy," Proceedings of the Fifth International Tinplate Conference, London, (1992) pp. 300-315.

7. J. S. Balzer, P. K. Dellock, M. H. Maj, G.S. Cole, D. Reed, T. Davis, T. Lawson and G. Simonds, "Structural Magnesium Front-End Support Assembly," Society of Automotive Engineers, Warrendale, PA, Paper 2003-01-0186 (2003).

Magnesium Technology 2011
Edited by: Wim H. Sillekens, Sean R. Agnew, Neale R. Neelameggham, and Suveen N. Mathaudhu
TMS (The Minerals, Metals & Materials Society), 2011

EFFECTS OF OXIDATION TIME ON MICRO-ARC OXIDIZED COATINGS OF MAGNESIUM ALLOY AZ91D IN ALUMINATE SOLUTION

Weiyi Mu, Zhengxian Li, Jihong Du, Ruixue Luo, Zhengping Xi
Northwest Institute for Nonferrous Metal Research, Xi'an 710016, China

Keywords: Magnesium alloy, Micro-arc oxidation, Coating, Corrosion resistance

Abstract

Micro-arc oxidation coatings were prepared on magnesium alloy AZ91D at different oxidation times in aluminate solution. The effects of the oxidation time on the microstructure, growth rate and corrosion resistance of the coatings were investigated. The results indicate that the coatings are uniform in thickness and mainly composed of $MgAl_2O_4$ and MgO. There were many residual discharging channels on the coatings surface. The coatings improved the corrosion resistance of magnesium alloy AZ91D considerably. With increased oxidation time, the crystalline substances content and thickness of the coatings increased, while the growth rate of the coatings decreased, and the resulting coatings surface had lower porosity and larger pore sizes. In addition, the corrosion resistance of the coatings on magnesium alloy AZ91D surfaces is superior to the magnesium alloy AZ91D substrate in the NaCl solution, and the effect is more remarkable with longer oxidation times.

Introduction

Magnesium (Mg) and its alloys have low density and high specific strength, so they are being increasingly applied to various industries like automotive and consumer electronics. Effective coatings will facilitate further application of magnesium and its alloys.

Micro-arc oxidization (MAO, which also named as plasma electrolytic oxidation or anodic spark oxidation), based on conversional anodic oxidization technology, is a new promising surface treatment method of forming functional coatings such as anti-corrosion coatings, anti-wear coatings or bioactive coatings on valve metals (such as Al, Mg and Ti) in a suitable electrolyte by increasing the anodic voltage to a high stage, usually accompanied by intensive gas evolution and sparking phenomenon at the anode surface[1-3]. Recently, MAO has successfully been used for producing protective coatings on Mg and its alloys to enhance the corrosion resistance [4–9].

The presence of $MgAl_2O_4$ in a composite coating on metal substrates is desirable for corrosion resistance [10]. Our previous work has reported that the $MgAl_2O_4$/MgO composite coatings on Mg and its alloys by MAO in an aqueous $NaAlO_2$-containing electrolyte [8, 9, 11]. The coating firmly bonds to the substrate, exhibits a high microhardness and has improved considerably the corrosion resistance of the substrate. The effects of $NaAlO_2$ concentration in the electrolyte and the applied voltage on the structure, phase composition, microhardness and corrosion resistance of coatings have been studied, respectively. On the other hand, the oxidation time has been proven to be a crucial factor in the microstructure and properties of MAO-formed coatings in silicate electrolyte [12]. In this work, it is mainly for above-mentioned reasons that the morphologies, phase components, and corrosion resistances of the MAO coatings prepared on Mg alloy AZ91D substrates at the different oxidation time in the $NaAlO_2$-containing electrolytes were investigated.

Experimental procedures

Samples (20 mm×15 mm×6 mm) of die-cast Mg alloy AZ91D (chemical composition in wt. %: Al 8.5-9.5, Zn 0.5-0.90, Mn 0.17-0.27, Mg balance.) were used as substrate material. The samples were ground with abrasive papers and ultrasonically cleaned with acetone and distilled water prior to MAO treatment. For MAO, a pulse power supply was employed, and the Mg alloy AZ91D sample was used as an anode while a stainless steel cylinder container with electrolyte inside was used as a cathode. The electrolyte is mainly composed of $NaAlO_2$ and some additional additives prepared with deionized water. During MAO treatment of each Mg alloy AZ91D, the oxidation time was fixed at 5, 20, 40, and 60 minutes, respectively. The applied voltage, pulse frequency and duty circle was fixed at 480 V, 100 Hz and 25%, respectively. The electrolyte bath was water-cooled and its temperature was maintained lower than 30 °C. After the MAO treatment, the samples were washed with deionized water and dried at room temperature.

The morphologies and phase components of the samples were investigated using scanning electron microscopy (SEM; JSM-6460, Japan) and X-ray diffraction (XRD; D/MAX-2200PC, Japan), respectively.

The corrosion resistance of Mg alloy AZ91D before and after the MAO treatment was tested through Zennium-type electrochemical corrosion testing system produced by Germany's ZAJNER Company. The samples with an exposed area of 1 cm^2 were immersed in a 3.5 wt. % NaCl solution prepared using analytical grade reagents in distilled water. After 10 minutes of initial delay, the scan was conducted with a constant rate 0.5 mV/s speed from -1.9 to -1.0 V. The reference electrode is a saturated calomel electrode (SCE), and the auxiliary electrode is a graphite electrode.

Results and discussion

Microstructure

The surface morphologies of the coatings formed at different oxidation time are shown in Figure 1. It can be seen in Figure 1 porous surface of the coating has been formed. There were many residual discharging channels on the coatings surface. As oxidation time increases, the resulting coatings surface has lower porosity, larger pore size and bigger melting particles. The coatings surface is composed of grains with different diameters,

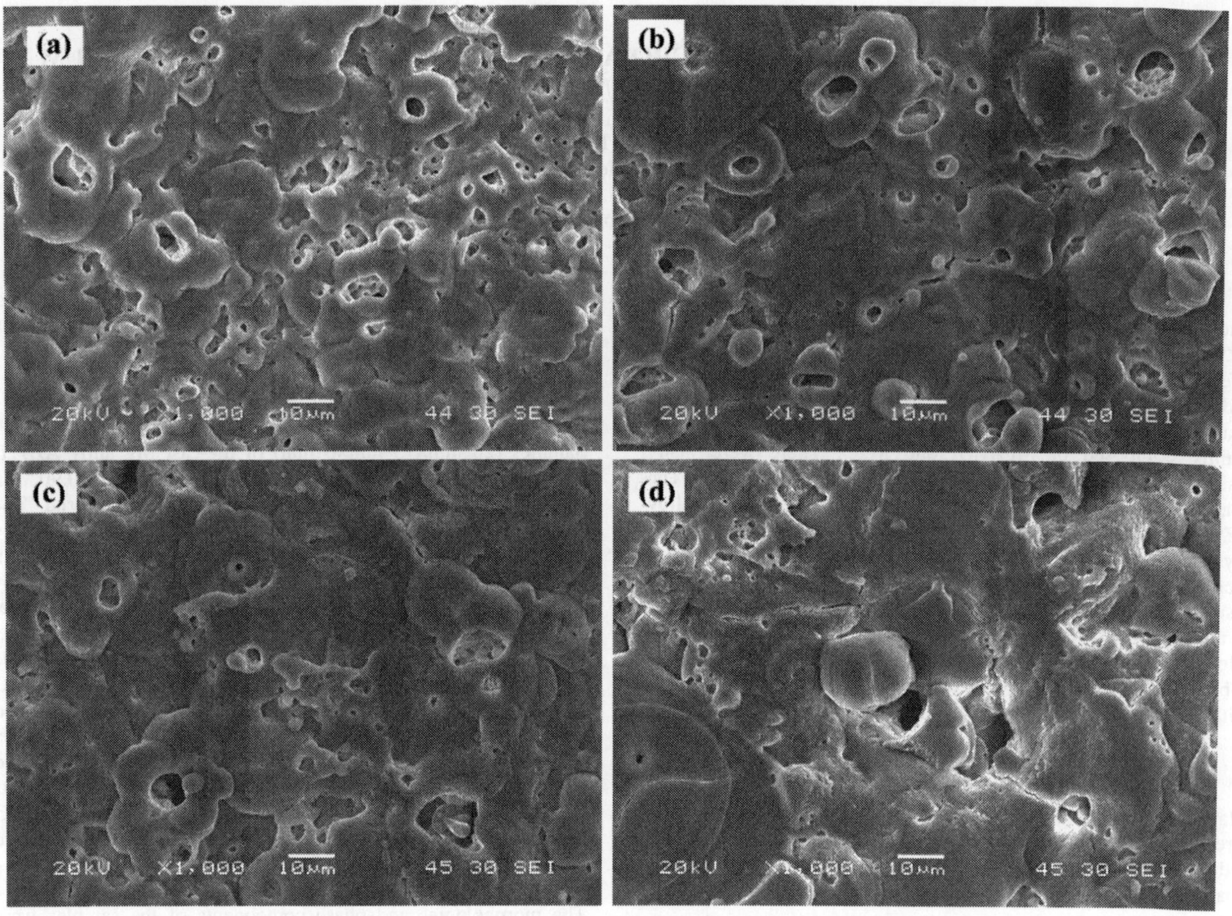

Figure 1. Surface morphologies of the coatings formed at (a) 5, (b) 20, (c) 40 and (d) 60 minutes.

which are melted and unevenly distributed on the coatings surface. The molten grains have integrated each other due to sintering effect during MAO process. The rough surface of coatings is due to uneven-melted solids. In addition, obvious extended micro cracks can be seen on the coatings surface. These micro cracks are caused by overlarge thermal stress in fast solidification of the melts when the coatings surface contact with the electrolytic solution [13].

Figure 2 shows the cross section of the coatings formed at different oxidation times. The thickness of the coating is approximately 21, 27, 30 and 32 µm, respectively. As oxidation time prolonged, the growth rate of MAO coatings was gradually decreased. The cross section of the coatings appears to be relatively uniform in thickness. However, even though they are not entirely transverse in the coatings, micro pores and micro flaws can be seen in Figure 2. It can also be seen in Figure 3 that the internal substrates are covered by a relatively intact protective layer, which are composed of relatively stable thermodynamically compounds. The preferential growth of coatings at a certain location is either due to the inherent electrochemical heterogeneity of basis Mg, or the concentration of current, or both, when sparking occurs. The preferential growth and the dielectric breakdown of coatings, together with the trapping and the evolution of oxygen and aqueous vapor should be responsible for the above mentioned micro pores and micro flaws. It is also noticed that there is no apparent discontinuity in the interface of the coatings and substrates. This indicates that the coatings can bond tightly to the substrates.

Figure 3 shows the XRD patterns of the coatings formed at different oxidation time. The coatings are mainly composed of $MgAl_2O_4$ and MgO. With increasing oxidation time, the $MgAl_2O_4$ content of the composite coatings increases, while no significant variation is observed in the MgO content. This suggests that the AlO_2^- anions in the electrolytes react with the substrate during the MAO process, and the oxidation time has great influence on the composition of the coatings formed on Mg alloy AZ91D. In addition, as observed from the experiment, the phase components of the MAO coatings differ from the amorphous coating on Mg alloy treated by the anodic oxidation. MAO can make amorphous phase into crystalline phase through high temperature and high pressure instant sintering. The method can form a dense layer to increase the hardness and corrosion resistance of Mg alloy [2]. The instantaneous temperature in the micro-spark zone can reach 10^3-10^4 K [3], thus the oxide products synthesized by plasma chemical interactions possess a stable chemical thermodynamic property.

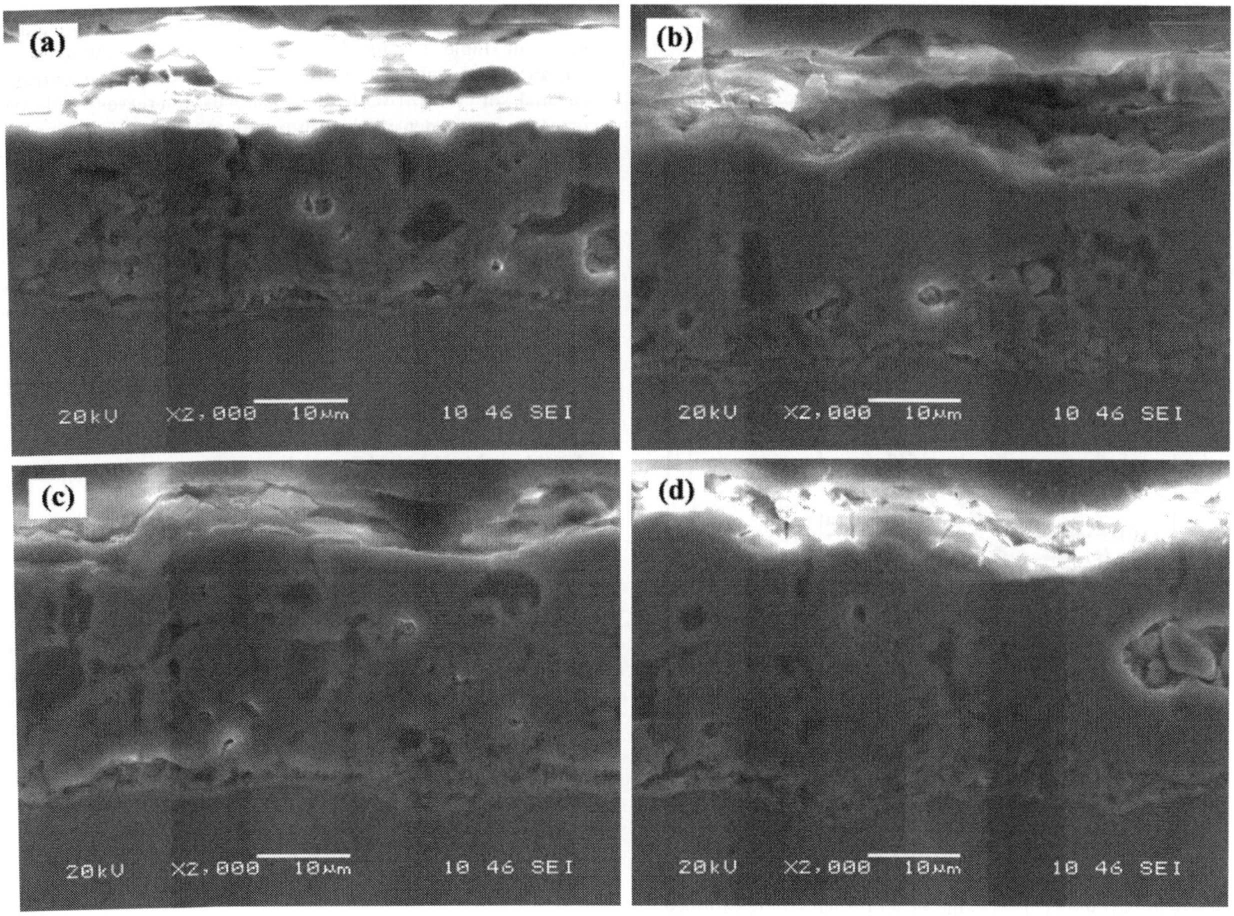

Figure 2. Cross-sectional morphologies of the coatings formed at (a) 5, (b) 20, (c) 40 and (d) 60 minutes.

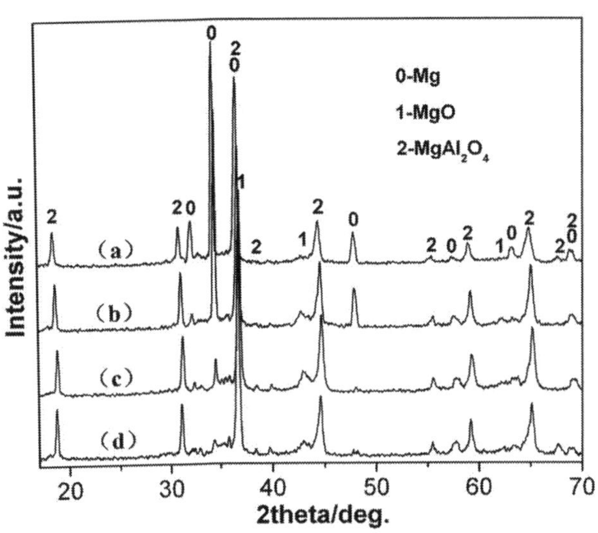

Figure 3. XRD patterns of the coatings formed at (a) 5, (b) 20, (c) 40 and (d) 60 minutes.

Above results demonstrate that a kind of $MgAl_2O_4$/MgO coating has been synthesized on Mg alloy AZ91D by the MAO. The presence of $MgAl_2O_4$ and MgO indicate that AlO_2^- and OH^- ions have incorporated into the coating intensively. A possible explanation for the formation of the coating is as follows.

During the MAO, due to the effect of the electric field, Mg^{2+} (Al^{3+})(reaction (1)) from the Mg alloy AZ91D substrate according to the reactions (2) and (3) combined with the OH^- and AlO_2^- from the $NaAlO_2$ solution to form $Mg(OH)_2$, $Al(OH)_3$ and $MgAl_2O_4$. Since the AlO_2^- is relatively unstable, it can partially interact with water in solution, forming $Al(OH)_4^-$ [14]. An extreme high thermal energy generated by sparks caused the dehydration of the hydroxides (reactions (4) and (5)) [15]. Notably, the local temperature inside the discharge channels can reach 3000~10,000 K in the MAO process [6]. Therefore, the phase transformations caused by the high local temperature. It is known that the only compound present in the phase diagram of the MgO-Al_2O_3 system is $MgAl_2O_4$ [16], so, when micro-arcs rise, the reaction (6) can be generated.

$$Mg \rightarrow Mg^{2+} + 2e^- \ (Al \rightarrow Al^{3+} + 3e^-) \tag{1}$$

$$Mg^{2+} + 2OH^- \rightarrow Mg(OH)_2 \ (Al^{3+} + 3OH^- \rightarrow Al(OH)_3) \quad (2)$$

$$Mg^{2+} + 2AlO_2^- \rightarrow MgAl_2O_4 \quad (3)$$

$$Mg(OH)_2 \rightarrow MgO + H_2O \ (2Al(OH)_3 \rightarrow Al_2O_3 + 3H_2O) \quad (4)$$

$$2Al(OH)_4^- \rightarrow Al_2O_3 + 2OH^- + 3H_2O \quad (5)$$

$$MgO + Al_2O_3 - MgAl_2O_4 \quad (6)$$

Corrosion resistance

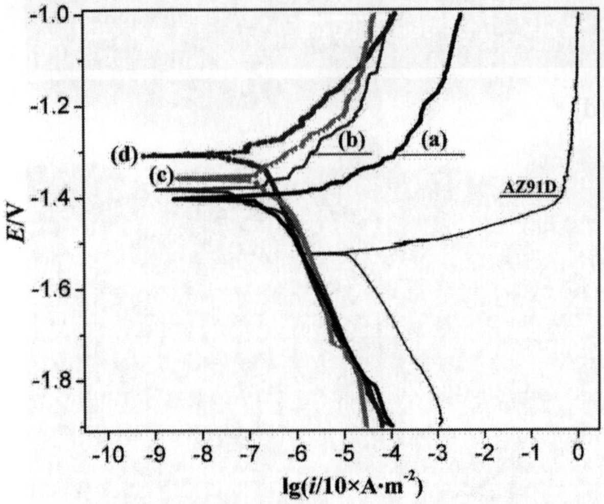

Figure 4. Polarization curves of the MAO-treated samples formed at (a) 5, (b) 20, (c) 40 and (d) 60 minutes and the Mg alloy AZ91D substrate in 3.5 wt. % NaCl solution.

Table.1 The result of potentiodynamic corrosion tests in 3.5 wt. % NaCl solution

Samples	E_{corr} (V)	i_{corr} ($\mu A/cm^2$)
AZ91D	–1.541	22.023
5 minutes	–1.425	1.748
20 minutes	–1.383	1.523
40 minutes	–1.352	1.202
60 minutes	–1.338	0.893

Figure 4 shows the potentiodynamic polarization curves of Mg alloy AZ91D and the samples treated at different oxidation time. It can be seen from the results that the anodic polarization curve of Mg alloy AZ91D enters the active dissolution zone when the potential reaches -1.524 V. The anodic current density of Mg alloy AZ91D electrode increases with the potential increasing in this zone. The anodic polarization curve of the each MAO-treated sample enters the active dissolution zone when the potential reaches -1.408, -1.387, -1.360 and -1.309 V, respectively, and the anodic current density of the each MAO-treated sample electrode increases with the potential increasing in this zone. The corrosion potential (E_{corr}) and the corrosion current density (I_{corr}) data are shown in Table 1. The data clearly show that the corrosion resistances are enhanced by the MAO process. The corrosion potential of the MAO-treated samples increases and their corrosion current density decreases with increasing oxidation time. The results indicate that the coating formed at longer oxidation time has a lower corrosion rate and better corrosion resistance. After the corrosion test, large corrosion craters can be clearly observed by naked eye on Mg alloy AZ91D surface, but there are no visible changes on the coatings surface. When formed at 60 minutes, the corrosion current density of coating decreases by two orders of magnitude compared with the Mg alloy AZ91D and its corrosion resistance is greatly improved. The above results illustrate that the coatings provide effective corrosion protection for the Mg alloy AZ91D substrate in solutions containing Cl^-. The effective corrosion protection provided by the $MgAl_2O_4$/MgO composite coatings can be attributed to intact microstructures and relatively stable chemical thermodynamic composition.

Conclusions

Through MAO process, $MgAl_2O_4$/MgO composite coatings with uniform thickness are directly prepared on Mg alloy AZ91D, and thus the surface property has been greatly improved. The coatings are mainly composed of $MgAl_2O_4$ and MgO. There were many residual discharging channels on the coatings surface. With the increase of the oxidation time, the crystalline substances content and the thickness of the coatings increase, while the growth rate of the coatings decrease, and the resulting coatings surface has lower porosity and larger pore size. When formed at 60 minutes, the coating is approximately 32 μm thick and has sufficient corrosion resistance. Such $MgAl_2O_4$/MgO composite coatings are expected to be an effective protection of Mg alloy substrates for its practical application.

References

1. Y. Ma et al., "Corrosion and Erosion Properties of Silicate and Phosphate Coatings on Magnesium,"Thin Solid Films, 469-470(2004), 472-477.

2. H.Y. Hsiao, W.T. Tsai, "Characterization of Anodic Films Formed on AZ91D Magnesium Alloy,"Surface and Coatings Technology, 190(2005), 299-308.

3. A.L. Yerokhin et al., "Plasma Electrolysis for Surface Engineering,"Surface and Coatings Technology, 122(1999), 73-93.

4. W. Y. Mu, Y. Han, "Characterization and Properties of the MgF_2/ZrO_2 Composite Coatings on Magnesium Prepared by Micro-arc Oxidation,"Surface and Coatings Technology, 202(2008), 4278-4284.

5. H.F. Guo, M.Z. An, "Growth of Ceramic Coatings on AZ91D Magnesium Alloys by Micro-arc Oxidation in Aluminate-fluoride Solutions and Evaluation of Corrosion Resistance,"Applied Surface Science, 246(2005), 229-238.

6. Y. Ma et al., "Systematic Study of the Electrolytic Plasma Oxidation Process on A Mg Alloy for Corrosion Protection,"Thin Solid Films, 494(2006), 296-301.

7. H. P. Duan, C.W. Yan, F.H.Wang, "Effect of Electrolyte Additives on Performance of Plasma Electrolytic Oxidation Films Formed on Magnesium Alloy AZ91D,"Electrochimica Acta, 52(2007), 3785-3793.

8. W.Y. Mu, Y. Han, "Study on Micro-Arc Oxidized Coatings on Magnesium in Three Different Electrolytes,"Rare Metal Materials and Engineering, 39(7) (2010), 1129-1134.

9. W.Y. Mu et al., "Effects of Applied Voltages on Micro-arc Oxidized Coatings of Magnesium Alloy Z91D in Aluminate Solution,"Advanced Materials Research, 2010, in press.

10. O. Khaselev, D. Weiss, J. Yahalom, "Structure and Composition of Anodic Films Formed on Binary Mg-Al Alloys in KOH-aluminate Solutions Under Continuous Sparking,"Corrosion Science, 43 (2001), 1295-1307.

11. W. Y. Mu, Y. Han, L.M. Zhang, "Effects of $NaAlO_2$ Concentration on Microstructure and Performance of Micro-Arc Oxidation Coatings Formed on Magnesium,"Journal of Xi'an Jiaotong University, 7(41) (2007), 847-851(in Chinese).

12. W.L. Lv et al., "Effects of Oxidation Time on Microstructures and Properties of Micro-arc Oxidation Coatings of AZ91D Magnesium Alloy,"The Chinese Journal of Nonferrous Metals, 8(19) (2009), 1385-1391(in Chinese).

13. W. B. Xue et al., "Growth Regularity of Ceramic Coatings Formed by Microarc Oxidation on Al-Cu-Mg Alloy,"Thin Solid Films, 372(2000), 114-117.

14. L. Rama, K.R.C Krishna, G. Sundararajan, "The Tribological Performance of Ultra-hard Ceramic Composite Coatings Obtained Through Microarc Oxidation,"Surface and Coatings Technology, 163-164(2003), 484-490.

15. A.L. Yerokhin, A. Leyland, A. Matthews, "Kinetic Aspects of Aluminium Titanate Layer Formation on Titanium Alloys by Plasma Electrolytic Oxidation,"Applied Surface Science, 200(2002), 172-184.

16. R. Sarkar, G. Banerjee, "Effect of Compositional Variation and Fineness on the Densification of $MgO\text{-}Al_2O_3$ Compacts,"Journal of the European Ceramic Society, 19(1999), 2893-2899.

Magnesium Technology 2011
Edited by: Wim H. Sillekens, Sean R. Agnew, Neale R. Neelameggham, and Suveen N. Mathaudhu
TMS (The Minerals, Metals & Materials Society), 2011

COMPOSITE COATINGS COMBINING PEO LAYER AND EPD LAYER ON MAGNESIUM ALLOY

Yongfeng Jiang, Yefeng Bao, Ke Yang
College of Mechanical and Electrical Engineering, Hohai University; Changzhou, Jiangsu, 213002, China

Keywords: Magnesium Alloy, Plasma Electrolytic Oxidation, Cathodic Electrophoretic Deposition, Corrosion Resistance, Adhesion

Abstract

Protective composite coatings were prepared combining plasma electrolytic oxidation (PEO) treatment and cathodic electrophoretic deposition on magnesium alloy AZ91D. The corrosion protection of composite coatings were evaluated using potentiodynamic polarization measurements in 3.5% NaCl solution, copper accelerated acetate salt spray (CASS) test and immersion test in acid solution. The adhesion of composite coatings was evaluated using cross-cut test and pull-off test. It is indicated that the corrosion resistance of magnesium alloy AZ91D with the composite coatings is improved obviously compared to it merely with PEO coating and it is also shown that pitting corrosion of PEO coating on magnesium alloy is decreased with EPD post-treatment. The adhesion of composite coatings could be up to 11.3 N/mm^2 in quantitative method due to the interlocking effect of organic layer in pores of PEO layer.

Introduction

Magnesium and its alloys are considered to be the most promising material in the field of structural materials for its advantageous properties including low density, high strength-to-weight ratio, good electromagnetic shielding and easy recycled. Therefore it is available in applications including automobile, aerospace components and computers [1-3]. Unfortunately, magnesium and its alloys are susceptible to galvanic corrosion, which greatly limits its use in harsh environments.

It is one of the effective ways to prepare coating on magnesium alloy to improve corrosion prevention. Thus a number of coating techniques appears, including electrochemical plating, conversion coating, gas-phase deposition, plasma electrolytic oxidation (PEO), cathodic electrophoretic deposition [4-6]. Among then, PEO is a new and effective way to magnesium alloy and the hard PEO coating containing outer layer and inner layer can be formed on magnesium alloy. Unfortunately, pores can be run through the PEO coating formed during continual and intense sparking discharges and gas bubbles on surface [7]. Furthermore, low Pilling-Bedworth (PB) ratio of magnesium oxide to metal substrate make PEO coating much loose. Therefore a number of posttreatment were developed to protect the magnesium alloy from corrosion ions more effectively. Duan Hongping [8] combined PEO and multi-immersion technique to improve corrosion resistance of magnesium alloy AZ91D. Zeng Liyun reported a composite coating produced by PEO and electroless plating on magnesium alloy AZ91D [9].

However, there are restrictions for previous technologies in severe environment. Thus it is necessary to develop new methods. At present EPD technology is one of best promising method to corrosion protection on the magnesium alloy. However, it is difficult to directly apply EPD to magnesium alloys due to the following reasons: (1) The magnesium alloys are severely corroded in aqueous electrolytes; (2) The loose MgO formed on surface of magnesium alloy rapidly inhibit successive electrodeposition and reduce the adhesion of the EPD coating; and (3) Magnesium alloys are easily dissolved in cathodic electrophoretic solutions below pH 7 [6]. A combination of the PEO and EPD to prepare composite coating on magnesium alloys is promising to improve corrosion prevention.

Therefore, a composite coating on magnesium alloy AZ91D prepared combing PEO and EPD has been developed in this paper. The surface and cross-section morphological of the composite coating on magnesium alloy AZ91D was investigated using SEM. The corrosion protection of composite coatings on magnesium alloy AZ91D was evaluated using potentiodynamic polarization, copper accelerated acetate salt spray and immersion test in acid solution. The adhesion of composite coatings was determined using cross-cut test and pull-off test.

Experimental

Rectangular samples in dimension of 50mm×50mm×1mm cut from die-cast magnesium alloy AZ91D (Al 8.5-9.5%, Zn 0.50-0.90%, Mn 0.17-0.27%, Mg balance) were used as working electrode and stainless steel were used as counter electrodes. The samples were polished with SiC paper (grade 400, 800, and 1200) and degreased with acetone followed by rinsing with distilled water before PEO process. Then PEO coating on magnesium alloy were prepared in 8 g/L Na_2SiO_4, 10 g/L KF and 2 g/L NaOH solution under pulsed DC mode, the current density was controlled constant at 0-10 A/dm^2.

Then the PEO samples were cleaned using ultrasonic cleaning machine. Organic coating was deposited at 100-200V voltage in 30-240 seconds on the PEO coating in cell as shown in Figure 1. the samples with PEO was used as the cathode, the stainless steel as the anode. Then the coated samples were rinsed with distilled water and baked in infrared oven at 160-170 °C for 20-30min.

The surface and cross-section morphologies of PEO coating and composite coatings were investigated in JSM-6360IA scanning electron microscopy (SEM). Thickness and roughness of PEO coating were evaluated using TR200 roughometer.

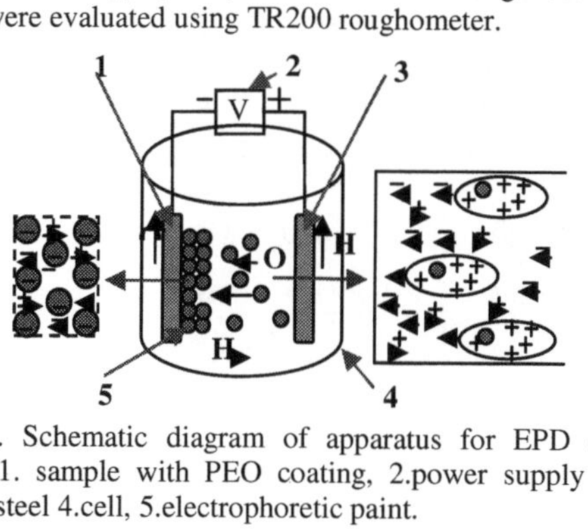

Figure 1. Schematic diagram of apparatus for EPD on PEO coating: 1. sample with PEO coating, 2.power supply unit, 3. stainless steel 4.cell, 5.electrophoretic paint.

The corrosion protection of PEO coating and composite coatings was investigated using PS-268A at scan rate of 1 mV/s. Copper accelerated acetate salt spray (CASS) was carried out up to 320 h on composite coatings. The corrosion solution, containing 50g/L NaCl, 0.26g/L $CuCl_2 \cdot H_2O$, was used, which pH was adjusted among 3.0-3.1 with glacial acetic acid. The test was operated in continuous spray conditions at temperature of 50 °C. Samples were inspected for every 24 h.

The corrosion rate of composite coatings with different parameters is evaluated using immersion test in acid solution. The test is operated at 24-26 °C in 0.1 mol/L H_2SO_4 solution. The weight loss of samples was evaluated to the FS-1006 electronic balance every 3-5h.

The adhesion of composite coatings was determined by qualitative method called cross-cut test according to GB/T 9286-1998 and quantitative method called pull-off test equipped with Elcometer F106 adhesive tester.

Results and Discussion

Microstructure

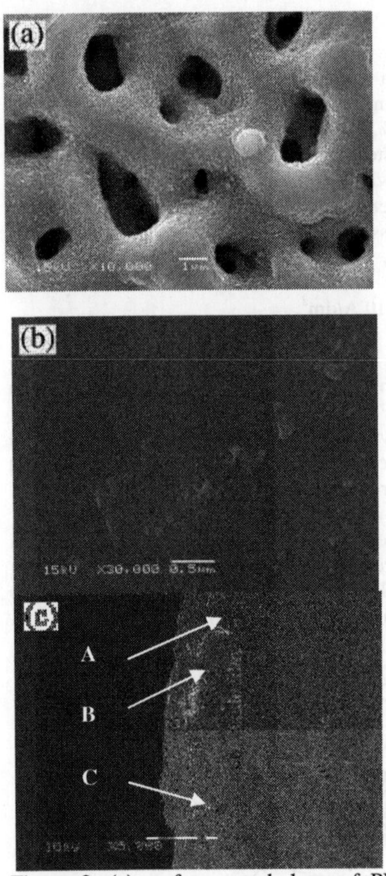

Figure 2. (a) surface morphology of PEO coating (b) surface morphology of composite coating (c) cross-section of composite coating.

Figure 2a shows surface of PEO coating. It is presented that micro-pores are randomly distributed on the PEO coating due to impact or tunneling ionization during PEO process. Figure 2b and Figure 2c shows the surface and cross-section morphologies of composite coatings. It is shown in Figure 2b that there are few micro-pores on composite coatings, whereas the formation of mechanical interlocking between PEO layer and EPD layer as in location of A, B and C is presented through the penetration of organic resin into PEO micropores as shown in Figure 2c.

Potentiodynamic Polarization

The corrosion protection of the PEO coating and the composite coatings are evaluated using potentiodynamic polarization in 3.5% NaCl solutions. Figure 3 presents the potentiodynamic polarization curves of the substrate, the PEO coating and composite coatings in different parameters. Table I is a summary of the potentiodynamic polarization parameters. It is indicated that the corrosion current(i_{corr}) of samples with PEO coating are very lower than that of magnesium alloy substrate, but the corrosion current (i_{corr}) of samples with PEO coating are higher than all that of composite coatings, the corrosion current of composite coatings was decreased by three orders of magnitude in comparison to magnesium alloy substrate and was decreased by two orders of magnitude in comparison to the sample with PEO coating, whereas the polarization resistance (R_p) of the composite coatings is higher than the sample with PEO coating by one order of magnitude. It is demonstrated that the corrosion resistance of magnesium alloy has been obviously improved with composite coating.

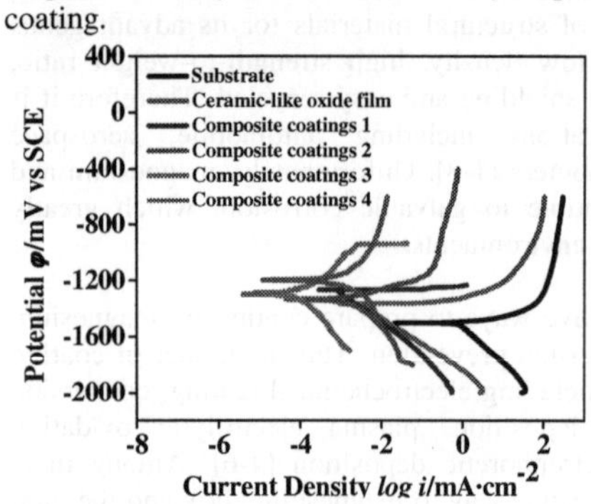

Figure 3. Potentiodynamic polarization curves of the substrate, the PEO coating and the composite coatings.

Table I. Results of the potentiodynamic corrosion tests in 3.5 % NaCl solution.

Samples	E_{corr}(V vs SCE)	i_{corr}(A/cm^2)	R_p(Ω)
Substrate	-1.501	3.0281×10^{-4}	7.1791×10^{1}
PEO coating	-1.367	1.3120×10^{-5}	1.6570×10^{3}
Composite coatings 1(EPD in100 V)	-1.267	8.8534×10^{-6}	2.4554×10^{3}
Composite coatings 2(EPD in130 V)	-1.334	1.5906×10^{-6}	1.3667×10^{4}
Composite coatings 3(EPD in160 V)	-1.298	2.5482×10^{-7}	8.5310×10^{4}
Composite coatings 4(EPD in200 V)	-1.198	8.7604×10^{-7}	2.4815×10^{4}

Copper Accelerated Acetate Salt Spray (CASS)

Copper accelerated acetate salt spray was carried out to investigate corrosion protection of PEO coating and composite coating. The surface morphology of PEO coating for 72 h and composite coatings for 320 h after copper accelerated acetate salt spray is shown in Figure 4. It is indicated that the noticeable corrosion pitting was presented on magnesium alloy with PEO coating after 72 h testing as show in Figure 4a. There is no noticeable corrosion on magnesium alloy with composite coatings with exception the edge regions until 320 h copper accelerated acetate salt spray testing (Figure 4b).

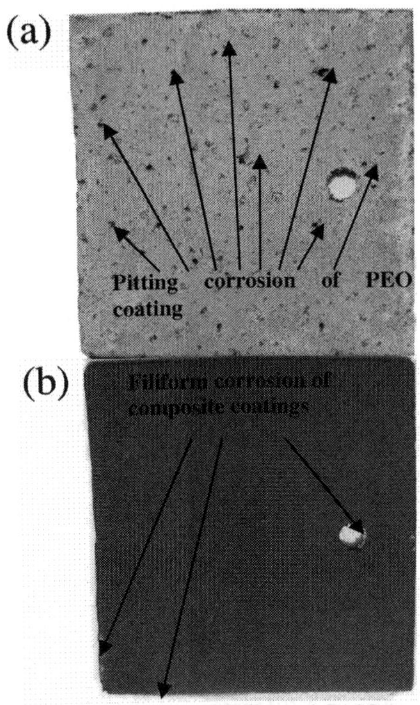

Figure 4. Surface morphology of CASS test for (a) PEO coating after 72 h; (b) composite coatings after 320h.

Weight Loss of Magnesium Alloy with Composite Coatings in Acid Corrosion

The immersion tests were performed in acid solution to evaluate corrosion rate of magnesium alloy with composite coatings. Sample with PEO coating is liberated gas in acid solutions in 25 s, whereas corroded completely in 80 s. The reaction is possible as follows:

$$Mg + H_2SO_4 = H_2 + MgSO_4 . \quad (1)$$

$$MgO + H_2SO_4 = MgSO_4 + H_2O . \quad (2)$$

The corrosion rate of magnesium alloy with composite coatings combining PEO layer about 17.2 μm in thicknesses for PEO 200s and EPD layer at 28.6 μm 42.3μ m and 48.6 μm in thicknesses for corresponding to EPD for 120V, 160V and 200V, respectively, is shown in Figure 5a. The corrosion rate of the sample with posttreatment EPD for 120V is lower than the sample with posttreatment EPD for 160V, it may be attributed to that water and sulfate radicals can permeate into the thinner PEO coating in a short time. However, the sample with posttreatment EPD for 200V with the high corrosion rate may be due to the high voltage in posttreatment EPD process.

The samples combing PEO time at 50s, 300s and 900s and EPD for 160V, respectively, have different corrosion rate. Sample 4 with PEO layer 21.6 μm in thickness and with EPD layer 40.7 μm in thickness has the higher corrosion rate than that of sample 5 with PEO layer 33.5 μm in thickness and with EPD layer 57.6 μm in thickness may because that the sample 4 has the thicker organic coating but also has the thinner PEO coating, the corrosion rate increased when the sulfate radical ions permeate into the PEO coating as shown in Figure 5b. However, sample 6 with PEO layer 40.6 μm in thickness and with EPD layer 71.3 μm in thickness has the high corrosion rate may due to the reason which is same as the sample 3. Therefore, good corrosion resistance of composite coatings was prepared combining PEO and EPD in the suitable parameters process. The corrosion rate of samples in 0.1 mol/L H_2SO_4 solutions with composite coatings in different parameters from 1 to 6 at 10h, 48h was also in Table II. It is indicated that the corrosion products mainly contain oxidation of substrate.

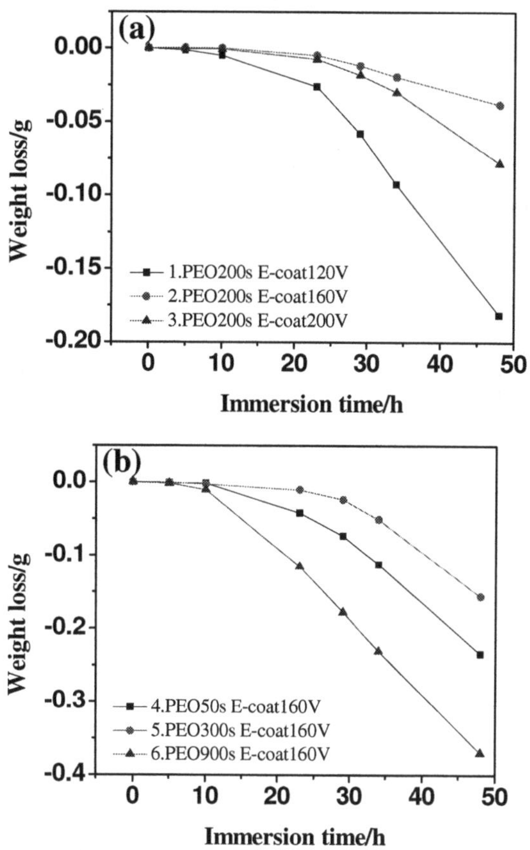

Figure 5. Dependence of weight loss of magnesium alloy with composite coatings on different parameters in acid solution: (a) PEO 200s, different EPD voltage; (b) different PEO time, EPD voltage 160V.

Table II. Corrosion rate of composite coatings in 0.1 mol/L H_2SO_4 solutions.

Samples	1	2	3	4	5	6
After 10h(mg/h)	0.0005	0.0003	0.0001	0.0002	0.0001	0.0010
After 48h(mg/h)	0.0038	0.0008	0.0016	0.0049	0.0032	0.0077

Adhesion of Composite Coatings

The formation of the mechanical interlocking between PEO layer and EPD layer occurs due to the organic layer permeate into the pores of PEO layer. Therefore, the adhesion of composite coatings is related to roughness of PEO coating resulting from the increase of the diameter of pores caused increase of roughness. The variation of roughness for PEO coating with the PEO time is presented as shown in Figure 6 (a). It was apparent that the roughness increased with prolong of the PEO treat time. Figure 6 (b) shows that adhesion of composite coatings was not increasing with increasing of roughness of PEO coating. The value of adhesion for composite coatings is up to 4 N/mm^2 when roughness of PEO coating is at 0.2245 μm in 3 specimens. However, wettability of electrophoresis paint on PEO coating decrease with increase of roughness of PEO coating, thus electrophoresis paint cannot completely run into the pores of PEO coating.

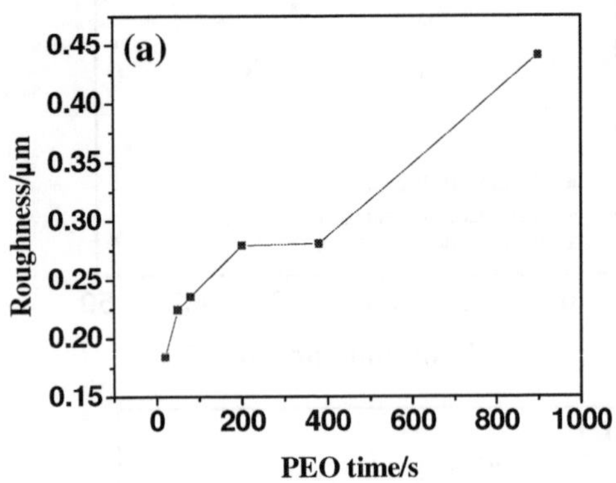

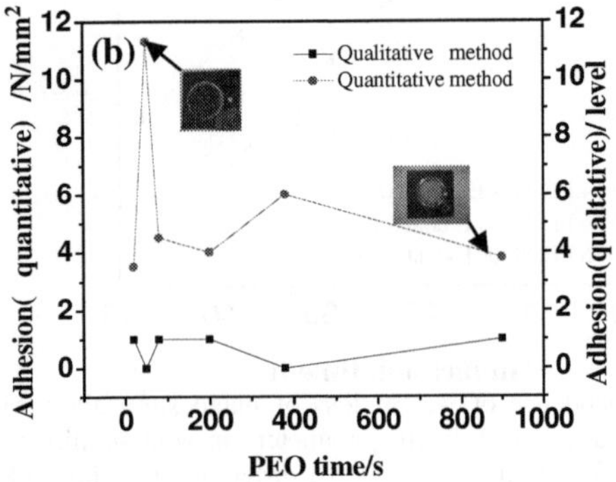

Figure 6. Adhesion (qualitative and quantitative method) variation of composite coatings with the roughness of PEO coating.

Conclusions

The composite coatings prepared combining PEO and EPD have good corrosion resistance in comparison to that with merely PEO coating on magnesium alloy AZ91D. The corrosion protection of composite coatings was superior to the PEO coating by three orders magnitudes. The principle of composite coatings was filiform corrosion rather than the pitting corrosion of PEO coating due to the copper accelerated acetate salt spray. The adhesion composite coatings is up to 4 N/mm^2 in quantitative method when roughness of PEO coating is 0.2245 μm because of good mechanical interlocking between organic layer and PEO layer.

Acknowledgements

The authors would like to thank the financial support of National Natural Science Foundation of China (No. 50901031) and the Fundamental Research Funds for the Central Universities (No. 2009B30314). This work has been also supported by the open research funds from Jiangsu Province Key Lab of Advanced Welding Technology in Jiangsu University of Science and Technology.

References

1. B.L. Mordike, T. Ebert, "Magnesium: Properties-applications-potential," *Materials Science and Engineering,* A302 (2001), 37-45.

2. Z. Yang, J.P. Li, J.X. Zhang, G.W. Lorimer and J. Robson, "Review on Research and Development of Magnesium Alloys," *Acta Metallurgica Sinica* (English Letters) 21(2008), 313-328.

3. John F.King, *Magnesium Technology* (Springer Berlin Heidelberg, chapter 6, 2006).

4. J.E. Gray and B. Luan, "Protective coatings on magnesium and its alloys — a critical review," *Journal of Alloys and Compounds Surface and Coating Technology,* 336(2002), 88-113.

5. Y.L. Song, Y.H. Liu, S.R. Yu, X.Y. Zhu, Q. Wang, "Plasma electrolytic oxidation coating on AZ91 magnesium alloy modified by neodymium and its corrosion resistance," *Applied Surface Science,* 254(2008), 3014-3020.

6. J. Zhang and C. Wu, "Corrosion protection behavior of AZ31 magnesium alloy with cathodic electrophoretic coating pretreated by silane," *Progress in Organic Coatings,* 66(2009),387-392.

7. J.A. Curran and T.W. Clyne, "Porosity in plasma electrolytic oxide coatings," *Acta Materialia,* 54(2006),1985-1993.

8. H. Duan, K. Du and C. Yan, "Growth process of plasma electrolytic oxidation films formed on magnesium alloy AZ91D in silicate solution," *Electrochimica Acta,* 52(2007), 5002-5009.

9. L. Zeng, S. Yang, W. Zhang, Y. Guo and C. Yan, "Preparation and characterization of a double-layer coating on magnesium alloy AZ91D," *Electrochimica Acta,* 55(2010), 3376-3383.

Magnesium Technology 2011

Poster Session

Session Chair:

Eric A. Nyberg
(Pacific Northwest National Laboratory, USA)

Magnesium Technology 2011
Edited by: Wim H. Sillekens, Sean R. Agnew, Neale R. Neelameggham, and Suveen N. Mathaudhu
TMS (The Minerals, Metals & Materials Society), 2011

GROWTH KINETICS OF γ-$Al_{12}Mg_{17}$ AND β-Al_3Mg_2 INTERMETALLIC PHASES IN Mg VS. Al DIFFUSION COUPLES

Sarah Brennan[1], Katrina Bermudez[1], Nagraj Kulkarni[2], Yongho Sohn[1]

[1]University of Central Florida, Advanced Materials Processing and Analysis Center
Department of Mechanical, Materials and Aerospace Engineering, Orlando, FL, 32816, USA
[2]Oak Ridge National Laboratory, Measurement Science & Systems Engineering Division
Oak Ridge, TN, 37831, USA

Keywords: Magnesium, Interdiffusion, Growth Kinetics, γ-$Al_{12}Mg_{17}$ and β-Al_3Mg_2

Abstract

Increasing use and development of lightweight Mg-alloys have led to the desire for more fundamental research in and understanding of Mg-based systems. As a strengthening component, Al is one of the most important and common alloying elements for Mg-alloys. In this study, solid-to-solid diffusion couple techniques were employed to examine the interdiffusion between pure Mg and Al. Diffusion anneals were carried out at 300°, 350°, and 400°C for 720, 360, and 240 hours, respectively. Optical and scanning electron microscopies (SEM) were employed to observe the formation of the intermetallics γ-$Al_{12}Mg_{17}$ and β-Al_3Mg_2, but not ε-phase. Concentration profiles were determined using X-ray energy dispersive spectroscopy (XEDS). The growth constants and activation energies were determined for each intermetallic phase.

Introduction

Benefits of lightweight alloys and composites in various applications including transportations and military have increased interest and research in Mg-alloys [1-5]. Accurate databases of diffusion kinetic parameters are of great value in designing and tailoring composition and microstructure of materials to achieve desired properties. Understanding of diffusion between Mg and Al is of great importance since Al is one of the primary alloying elements for Mg-alloys. In this study, we examined the interdiffusion between pure Mg and Al using solid-to-solid diffusion couples annealed at 300°C, 350°C and 400°C for 720, 360, and 240 hours, respectively. Development of phase constituents and concentration profiles were examined. In this paper, we report the thicknesses of intermetallic layers observed that were measured to determine the growth constants and activation energies. Concentration profiles from each diffusion couple have been analyzed to determine the composition- and temperature-dependent coefficients of interdiffusion, intrinsic diffusion (at the marker composition) and impurity diffusion (extrapolated), and will be published elsewhere.

Experimental Procedure

Polycrystalline Mg (99.9%) and Al (99.99%) from SCI Engineered Materials and Alfa Aesar, respectively, were sectioned into discs, 10 mm in diameter and 2 mm in thickness. The disc specimens were metallographically polished from 600 grit SiC paper down to 1 μm alumina using a non-oxidizing lubricant. The diffusion couples, Mg vs. Al, were then assembled with 2 mm-thick Al_2O_3 spacers in stainless steel jigs as shown in Figure 1.

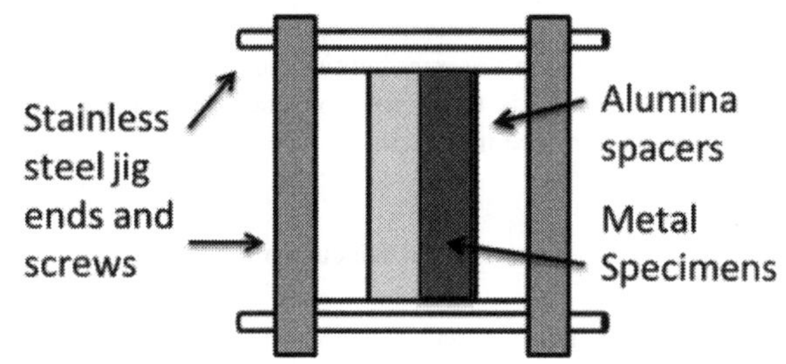

Figure 1: A schematic of Mg vs. Al diffusion couple assembly.

The jig assemblies were encapsulated individually in quartz tubes and evacuated to 10^{-6} Torr, then backfilled with ultra-high purity argon and hydrogen mixture. The quartz capsules were placed in a Paragon Bluebird furnace that was preheated to the annealing temperature. The temperature of the diffusion couples was monitored with an independent type K thermocouple. Anneals were carried out for three diffusion couples at 300°C, 350°C and 400°C for 720, 360, and 240 hours, respectively. After the anneal, the capsules were quickly removed from the furnace and quenched in water. The entire jig and diffusion couple assembly was mounted in epoxy and cross-sectioned using a Buehler IsoMet saw with a low-speed diamond wafering blade and a non-oxidizing lubricant. The cross-sectioned specimens were then cross-mounted for metallographic preparations, again using a non-oxidizing lubricant. Each diffusion couple was examined using optical microscopy first to check the quality of the diffusion bond, then using SEM (Hitachi S-3500N) equipped with XEDS to determine the concentration profiles of each couple. The presence of surface oxides of both the Mg and Al served as the marker in these diffusion couples. The location of the Kirkendall marker plane was identified by the presence of an oxygen

X-ray peak in addition to readily-visible plane of markers on the optical micrographs. Further studies carried out also employed electron probe microanalysis (EPMA) for the determination of concentration profiles. Results from EPMA and determination of composition- and temperature-dependent coefficients of interdiffusion, intrinsic diffusion (at the marker composition) and impurity diffusion (extrapolated), and will be published elsewhere.

Calculation of Growth Constant

To determine the diffusion growth constants for the intermediate phases that formed, a parabolic growth constant based on diffusion-controlled growth can be assumed, which is related to the thickness by:

$$Y = k\sqrt{t} \tag{1}$$

where Y is the thickness of the layer, k is the growth constant, and t is the time of anneal. The temperature dependence of the growth should follow Arrhenius relation expressed by:

$$k = k_o \exp\left[-\frac{Q}{RT}\right] \tag{2}$$

where k_o (m^2/s) is the pre-exponential factor, R (J/mol-K) is the gas constant, Q is the activation energy, and T is the annealing temperature in Kelvin.

Results and Discussion

Backscatter electron micrographs from the three diffusion couples are presented in Figure 2. The concentration profiles obtained by XEDS point-to-point counting are presented in Figure 3. The micrographs and concentration profiles both clearly reveal the presence of two intermetallic layers. These two intermetallic layers correspond to γ-$Al_{12}Mg_{17}$ (near the Mg) and β-Al_3Mg_2 (near the Al) phases based on the concentration profiles presented in Figure 3 and the phase diagram [6] shown in Figure 4. Thicknesses measured for each phase is presented in Table I for each diffusion couple. Also presented in backscatter electron micrographs is the presence of a marker plane, identified by x_m in β-Al_3Mg_2 phase near the β/Al(ss) interface. A large solubility range for the γ-$Al_{12}Mg_{17}$ phase was observed in all couples in accordance with the phase diagram.

Table I. Thickness measured from SEM and XEDS comparison.

Phase	γ-$Al_{12}Mg_{17}$	β-Al_3Mg_2
Couple	Thickness [std. dev] (μm)	
400°C 10 days	226.4 [2]	595.7 [3]
350°C 15 days	77.0 [6]	481.0 [5]
300°C 30 days	29.7 [2]	272.9 [2]

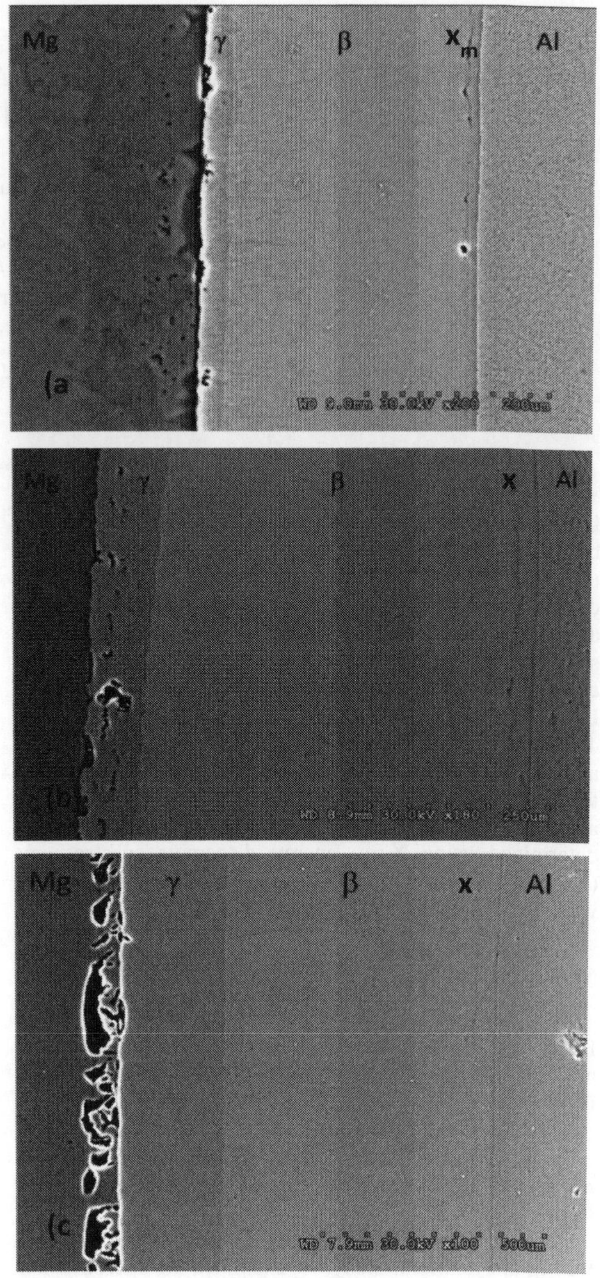

Figure 2: Backscatter electron micrographs from Mg vs. Al diffusion couples annealed at (a) 300°C for 720 hours, (b) 350°C for 360 hours, and (c) 400°C for 240 hours.

Clearly in this investigation, the ε-phase did not form in any of the diffusion couples. A diffusion couple study of the Mg-Al system by Brubaker and Liu [7] in the temperature range of 360° to 420°C only gave evidence of the existence of the ε-phase in diffusion couples investigated at 367° and 360°C. In contrast, an earlier investigation of the system in the range of 325° to 425°C by Funamizu and Watanabe [8] reported that the ε-phase did not appear in any of the diffusion couples, which is in agreement with our current study.

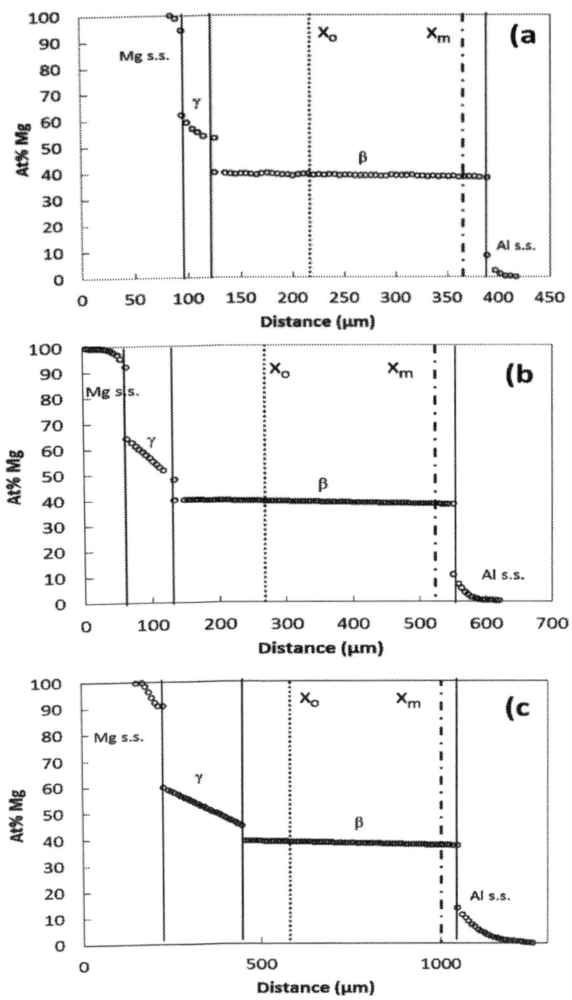

Figure 3. Concentration profiles of Mg and Al determined by XEDS from Mg vs. Al diffusion couples annealed at (a) 300°C for 720 hours, (b) 350°C for 360 hours, and (c) 400°C for 240 hours. X_o represents the Matano plane and X_m represents the marker plane.

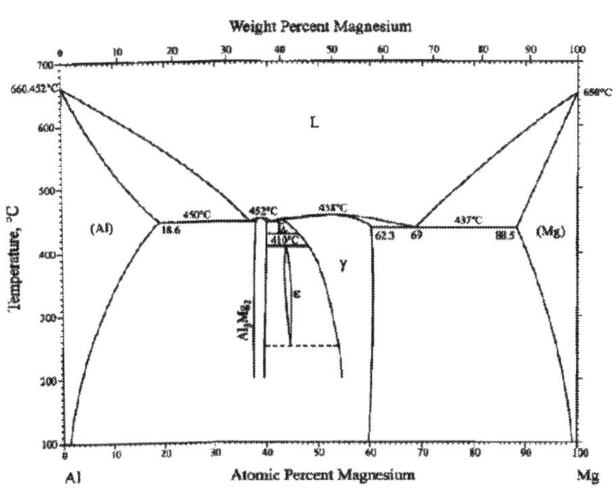

Figure 4: Al-Mg equilibrium phase diagram [6].

Table II. Activation energy and pre-exponential factor of the growth constants for γ-$Al_{12}Mg_{17}$ and β-Al_3Mg_2 phases.

Phase	γ-$Al_{12}Mg_{17}$	β-Al_3Mg_2
Pre-Exponential Factor, k_o (m^2/s)	0.36	$2.2x10^{-6}$
Activation Energy, Q (kJ/mol)	165.1	85.9
Q (kJ/mol) [7]	143.1	62.3

The growth constants, determined using Eq. (1), from the thickness measurements for the intermediate intermetallic phases, γ-$Al_{12}Mg_{17}$ and β-Al_3Mg_2 are plotted in an Arrhenius plot presented in Figure 5. From these plots, activation energy and pre-exponential factor for the growth constants were calculated as reported in Table II. The β phase with limited solubility range has the lower activation energy, and corresponds to the thicker layer observed in the diffusion couple. These results agree well with the growth constants reported by Brubaker and Liu [7]. Our activation energies for the γ and β phases are in good agreement with those reported by Funamizu and Watanabe [8].

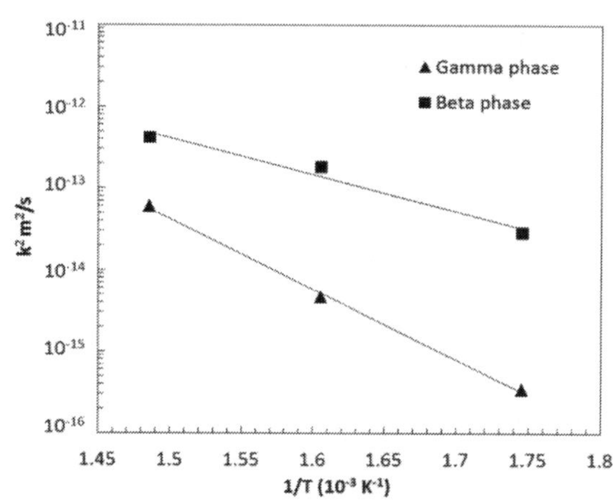

Figure 5. Temperature dependence of the growth constants in γ-$Al_{12}Mg_{17}$ and β-Al_3Mg_2 phases, determined from Mg vs. Al diffusion couples annealed at (a) 300°C for 720 hours, (b) 350°C for 360 hours, and (c) 400°C for 240 hours.

Conclusions

The solid-to-solid diffusion couple technique was employed to examine intermetallic phase formation and growth in the Mg-Al system from 300°C to 400°C. SEM and XEDS were used to identify the phases present as γ-$Al_{12}Mg_{17}$, and β-Al_3Mg_2. The ε phase did not appear in any of the diffusion couples studied. The pre-exponential factor and activation energy for the diffusional growth constant of each phase was calculated. The activation energy for β-Al_3Mg_2 phase growth (85.9 kJ/mole) was significantly lower than that for the γ-$Al_{12}Mg_{17}$ phase (165.0 kJ/mole).

Additional Note

Based on concentration profiles measured by XEDS and EPMA, temperature and composition-dependent coefficients of interdiffusion for each phase (e.g., Boltzmann-Matano), intrinsic diffusion (marker composition and Heumann analysis) in β-Al_3Mg_2 phase and impurity diffusion (extrapolated within dilute solid solutions of Mg and Al using Hall analysis) have been determined. Furthermore, corresponding activation energies and pre-exponential factors for these diffusion coefficients have been calculated. These results and relevant analyses will be published elsewhere.

References

[1] B.L. Mordike, T. Ebert, "Magnesium: Properties-applications-potential," *Materials Science and Engineering A*, 302, 2001, 37-45.

[2] A. Luo, "Magnesium: Current and Potential Automotive Applications," *Journal of Materials*, 54, 2002, 42-48.

[3] M.K. Kulekci, "Magnesium and its alloys applications in automotive industry," *International Journal of Advanced Manufacturing Technology*, 39, 2008, 851–865.

[4] R. Urbance, F. Field, R. Kirchain, R. Roth, J. Clark, "Market Model Simulation: The Impact of Increased Automotive Interest in Magnesium," *Journal of Materials*, 54, 2002, 25-33.

[5] A.K. Mondal, D. Fechner, S. Kumar, H. Dieringa, P. Maier, K.U. Kainer, "Interrupted creep behavior of Mg alloys developed for powertrain applications," *Materials Science and Engineering A*, 527, 2010, 2289-2296.

[6] H. Okamoto, "Al-Mg (Aluminum-Magnesium)," *Journal of Phase Equilibria*, 19, 1998, 598.

[7] C. Brubaker, Z.K. Liu, "Diffusion Couple Study of the Mg-Al System," *Magnesium Technology*, 2004, 229-234.

[8] Funamizu, K. Watanabe, "Interdiffusion in the Al-Mg System," *Transactions of the Japan Institute of Metals Supplement*, Vol. 13, 1972, 278-283.

Acknowledgement

This research was sponsored by the U.S. Department of Energy, Assistant Secretary for Energy Efficiency and Renewable Energy, Office of Vehicle Technologies, as part of the Lightweight Materials Program.

Magnesium Technology 2011
Edited by: Wim H. Sillekens, Sean R. Agnew, Neale R. Neelameggham, and Suveen N. Mathaudhu
TMS (The Minerals, Metals & Materials Society), 2011

DEVELOPMENT AND CHARACTERIZATION OF NEW AZ41 AND AZ51 MAGNESIUM ALLOYS

Md Ershadul Alam[1], Samson Han[2], Abdelmagid Salem Hamouda[1], Quy Bau Nguyen[2] and Manoj Gupta[2]
[1] Department of Mechanical and Industrial Engineering, College of Engineering, Qatar University, Doha, Qatar 2713
[2] Department of Mechanical Engineering, National University of Singapore, 9 Engineering Drive 1, Singapore 117576

Keywords: Magnesium alloy, AZ31, AZ41, AZ51, Mechanical properties.

Abstract

In the present study, new AZ41 and AZ51 magnesium alloys were successfully synthesized by adding 1 wt. % and 2 wt. % aluminum into AZ31B matrix, respectively, using an innovative disintegrated melt deposition technique followed by hot extrusion. Microstructural characterization studies revealed equiaxed grain morphology, reasonably uniform distribution of intermetallics in the matrix and minimal porosity. Physical properties characterization revealed that addition of Al reduced the coefficient of thermal expansion (CTE) of monolithic AZ31B. The presence of higher percentage of aluminum also assisted in improving overall mechanical properties including microhardness, modulus of elasticity, 0.2% yield strength, ultimate tensile strength, and work of fracture of AZ31. The results suggest that these newly developed AZ magnesium alloys have significant potential in diverse engineering applications when compared to AZ31 alloy.

Introduction

Increasing demand for the reduction of fuel consumption and environmental problems has led to intensive research effort into the design and development of light structural materials [1, 2]. Consequently, magnesium alloys are attracting much attention as light weight materials. Magnesium, with a density of 1.74 g/cm^3, is the lightest engineering metal available in the earth which is about two-thirds of the density of aluminum (2.70 g/cm^3) and one-quarter of that of iron (7.87 g/cm^3) [3]. Hence, its use is gaining significance in certain key engineering applications, particularly in automobile, aviation, electronics and consumer industries where weight is one of the most important criteria in the choice of materials [1,2, 4-7]. Besides low density, magnesium based materials exhibit good specific mechanical properties, machinability, castability, weldability, thermal stability, damping and resistance to electromagnetic radiation [8]. However, as magnesium has low ductility, modulus of elasticity and limited strength and creep resistance at elevated temperature, it cannot be extensively applied in structural applications [4-9]. Therefore, magnesium is rarely used in its pure form. Instead, it is usually alloyed with other elements like aluminum, zinc, silver and zirconium to improve its properties like corrosion resistance and ductility [1, 8-11]. Among these alloying constituents, the addition of aluminum seems to have the most favorable effect on magnesium, exhibiting much greater strength and hardness. Even though there has been extensive research and characterization on magnesium-aluminum alloys, it has mainly focused on specific compositions, like AZ31 (3 wt.% Al, 1 wt.% Zn, remaining Mg), AZ61 (6 wt.% Al, 1 wt.% Zn, remaining Mg) and AZ91 (9 wt.% Al, 1 wt.% Zn, remaining Mg) [8, 9-12]. Results of open literature search shows that only a few research groups have worked on developing and characterizing AZ41 sheets, mainly by twin-roll-cast technique [13-16], while Kim *et al.*[17] synthesized and characterized AZ51 alloys that contained pure Sn up to 9 wt%. No results are reported in which investigators have synthesized AZ41 and AZ51 alloys by incorporating 1 and 2 wt. % elemental Al, respectively, by superheating with AZ31B using DMD technique. Accordingly, in the present study, new AZ41 and AZ51 magnesium based alloys were synthesized and characterized their properties.

Characterization studies were carried out on these formulations to investigate their physical, microstructural and mechanical properties. These include density and porosity measurements, the coefficient of thermal expansion, grain morphology, x-ray diffraction analysis, optical and scanning electron microscopy, microhardness, tensile tests and fractography studies.

Experimental Procedures

Materials

In the present study, AZ31B magnesium alloy ingots (2.94% Al, 0.87% Zn, 0.57% Mn, 0.0027% Fe, 0.0112% Si, 0.0008% Cu, 0.0005% Ni and balance Mg, supplied by Tokyo Magnesium Company, Japan) were used as a matrix material and cut into small pieces so that they can be placed into the graphite crucible easily. The aluminum lumps of 99.5% purity (supplied by Alfa Aesar, USA) were also used for developing new AZ41 and AZ51 alloys.

Primary Processing

Synthesis of AZ31, AZ41 and AZ51 magnesium based alloys was carried out using disintegrated melt deposition (DMD) technique. Synthesis of the AZ41 and AZ51 alloys involved heating the AZ31B magnesium alloy pieces with the addition of 1 and 2 wt.% Al, respectively, to 750 ^{0}C in an inert Ar gas atmosphere in a graphite crucible using a resistance heating furnace. The crucible was equipped with an arrangement for bottom pouring. Upon reaching the superheat temperature, the molten slurry was stirred for 5 min at 450 rev./min using a twin blade (pitch 45^0) mild steel impeller to facilitate the incorporation and uniform distribution of reinforcement particulates in the metallic matrix. The impeller was coated with Zirtex 25 (86% ZrO_2, 8.8% Y_2O_3, 3.6% SiO_2, 1.2% K_2O and Na_2O, and 0.3% trace inorganic) to avoid iron contamination of the molten metal. The melt was then released through a 10-mm diameter orifice at the base of the crucible. The melt material was disintegrated by two jets of argon gas orientated normal to the melt stream. The argon gas flow rate was maintained at 25 L /min. The disintegrated alloy melt slurry was subsequently deposited onto a metallic substrate. Preform of 40-mm diameter was obtained following the deposition stage. The synthesis of monolithic AZ31 magnesium alloy was carried out using steps similar to those employed for the AZ41 and AZ51 alloys except that no Al lumps were added.

Secondary Processing

Pre-Extrusion: The 40 mm diameter ingots were machined down using a lathe machine to a diameter of 36 mm and cut into billets with heights of approximately 45 mm. They were then lathed in the direction perpendicular to the length of the ingot to ensure both ends are flat and perpendicular to the surface. The billets were then sprayed with colloidal graphite for lubrication purposes.

Extrusion: The billets were first soaked at 400 ^{0}C for 60 minutes in a constant temperature furnace before extrusion. Extrusion was performed on a 150 tonne hydraulic press using an extrusion ratio of 20.25:1, producing rods of 8 mm diameter.

Post-Extrusion: After extrusion, the extruded rods were machined to produce tensile specimens. Sections of approximately 10 to 15 mm in height were also cut by a low speed diamond blade and the surface graphite was then cleaned-off to use for various characterization studies.

Density and Porosity Measurements

Density (ρ) measurements were performed in accordance with Archimedes' principle on three randomly selected polished samples of AZ31, AZ41 and AZ51 alloys taken from the extruded rods. Distilled water was used as the immersion fluid. The samples were weighed using an A&D ER-182A electronic balance; with an accuracy of ±0.0001 g. Theoretical densities of materials were calculated assuming they are fully-dense. Rule-of-Mixture was used in all calculations. The porosity was calculated by using the theoretical and experimental densities.

Microstructural Characterization

Microstructural characterization studies were conducted on metallographically polished extruded samples to investigate morphological characteristics of grains, reinforcement distribution and interfacial integrity between the matrix and reinforcement. The etching solution (5 ml acetic acid, 6 g picric acid, 10 ml water and 100 ml ethanol) was applied for approximately 20 seconds using a swabbing technique and then washed under running water to reveal the grain boundaries [8, 18]. The sample was then analyzed using the Olympus BH2-UMA metallographic optical microscope equipped with Olympus DP-10 microscope digital camera. The grain boundaries were then traced out from the micrographs with the aid of the Adobe Photoshop program. Image analysis using the Scion system was carried out to determine the grain size of the materials.

The presence and distribution of the intermetallic phase was investigated using the JEOL JSM-5600LV Scanning Electron Microscope (SEM). Polished specimens were observed at 1000x magnification to reveal the intermetallic phase. Images were captured at an accelerating voltage of 15 kV and spot size of 20.

X-Ray Diffraction Studies

X-ray diffraction analysis was conducted using the automated Shimadzu LAB-X XRD-6000 x-ray diffractometer. Flat, ground and ultrasonically cleaned specimens of approximately 5 mm in height were exposed to CuK_α radiation (λ = 1.54056 Å) with a scanning speed of 2 ^{0}/min. The scanning range was 30^0 to 80^0 for all samples. A plot of intensity against 2θ (θ represents Bragg angle) was obtained, illustrating peaks at different Bragg angles. The Bragg angles corresponding to different peaks were noted; and the values of interplanar spacing (d-spacing) obtained from the computerized output were compared with the standard values from the International Centre for Diffraction Data's Powder Diffraction File (PDF).

Coefficient of Thermal Expansion

The coefficients of thermal expansion (CTE) of all the compositions were determined by measuring the displacement of the samples as a function of temperature in the temperature range of 50 ^{0}C- 400 ^{0}C using an automated SETARAM 92-16/18 thermo-mechanical analyzer.

Mechanical Characteristics

The mechanical properties carried out on all samples were investigated by conducting tensile and microhardness tests. All experiments were carried out at room temperature.

Microhardness Test: The microhardness tests were conducted on flat and metallographically polished specimens. The test was conducted using a Shimadzu HMV automatic digital microhardness tester with a Vickers indenter (square-based pyramidal-shaped diamond indenter with face angle of 136 ^{0}). An indenting load of 25 gf and a dwell time of 15 seconds were used. Testing was performed in accordance with ASTM test standard E384-08. Indentations were performed on the matrix and measurements were recorded in Vickers Hardness (HV).

Tensile Testing: The tensile properties of each sample were determined in accordance with ASTM test standard E8M-08. Round tension test specimens of 5 mm in diameter were machined from the 8 mm extruded rod. An average of five tensile specimens could be obtained from a single rod, and a minimum of four tensile tests were conducted for each sample. Tests were performed on the 810 Material Test System (MTS) with an extensometer of 25 mm gauge length and a crosshead speed of 0.254 mm/min was used. The raw stress-strain data recorded was extracted to be analyzed. Broken test specimens were labeled individually, carefully handled and packed into individual plastic bags to prevent damage of the fractured surface. A Microsoft Excel program was written to analyze the raw data extracted. Then, the modulus of elasticity (E), 0.2% offset yield strength (0.2% YS), ultimate tensile strength (UTS) and failure strain was computed via a series of Microsoft Excel functions. The work of fracture was also computed in the program by the use of the trapezium rule.

Fracture Behavior: The fracture surfaces of the broken tensile samples were analyzed to investigate the failure mechanisms that occurred during the tensile tests. Fractography studies were performed on the JEOL JSM-5600LV Scanning Electron Microscope. Images were captured at an accelerating voltage of 15 kV and a working distance of about 25 mm. Macromechanism of these fracture surfaces were also investigated.

Results

Macrostructure

The result of macrostructural characterization conducted on the as deposited monolithic AZ31B and alloyed materials did not reveal any presence of macropores, shrinkage, cracks or any other defects. Following extrusion, there was also no evidence of any macrostructural defects.

Density and Porosity Measurements

The results of the density measurements are shown in Table I. Results show that near dense monolithic AZ31B and alloyed AZ41 and AZ51 can be developed using the fabrication methodology adopted in the present study.

Table I. Results of Density and Porosity Measurements

Materials	Al (wt. %)	Density (g/cm^3)		Porosity (%)
		Theo.	Expt.	
AZ31	-	1.776	1.775	0.05
AZ41	1.0	1.786	1.785	0.05
AZ51	2.0	1.792	1.791	0.08

Microstructural Characteristics

Microstructural studies conducted on the extruded specimens showed near equiaxed grain morphology (see Table II and Figure 1). Average grain size of AZ31B samples decreased with the addition of Al. However, the variation of grain size among the samples, especially between the AZ41 and AZ51 alloys is statistically insignificant considering standard deviation.

Table II. Results of Grain Morphology

Materials	Grain Size (μm)	Aspect Ratio	Roundness[1]
AZ31	4.1 ± 1.8	1.5 ± 0.4	1.5 ± 0.4
AZ41	2.9 ± 1.3	1.6 ± 0.4	1.6 ± 0.6
AZ51	2.8 ± 1.3	1.7 ± 0.5	1.7 ± 0.7

1. Roundness is the shape of the grain expressed by the formula (perimeter)^2 / 4π (area)

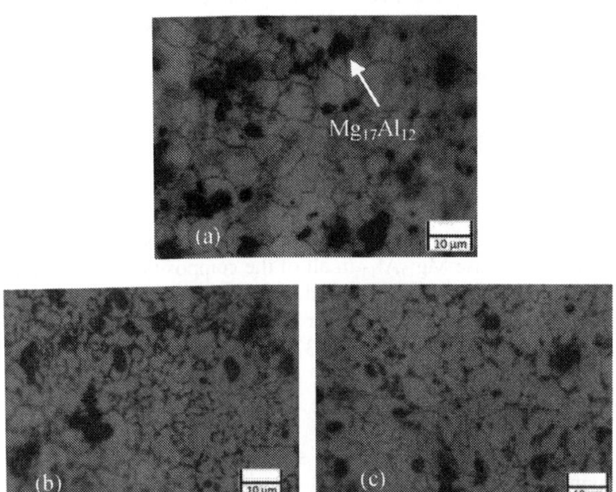

Figure 1. Representative optical micrographs showing the grain morphology (1000x) of: (a) AZ31B, (b) AZ41 and (c) AZ51 samples, respectively.

Microstructural characterization studies were also conducted on the extruded and polished samples to observe the shape and distribution of intermetallic phases in the matrix (see Figure 2).

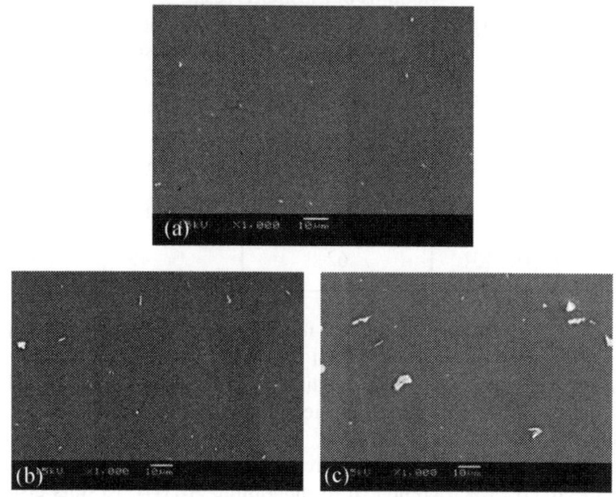

Figure 2. Representative SEM micrographs showing the shape and distribution of second phase in the case of: (a) AZ31B, (b) AZ41 and (c) AZ51 samples, respectively.

X-Ray Diffraction

X-ray diffraction (XRD) was carried out on all extruded samples (see Table III). The obtained lattice spacing's (d-spacing) and two-theta (2θ) value were compared with the standard values for magnesium and the Mg-Al intermetallic phases. Results showed the presence of intermetallic $Al_{12}Mg_{17}$ phase in the α-Mg phase.

Table III. Results of XRD and CTE Analysis

Materials	Mg	$Mg_{17}Al_{12}$	CTE (μm/mK)
AZ31	11[3]	1[1]	28.8
AZ41	11[3]	1[1]	27.1
AZ51	11[3]	2[1]	27.0

Coefficient of Thermal Expansion

Table III shows the results of coefficient of thermal expansion measurements obtained from AZ31, AZ41 and AZ51 samples. The results revealed a reduction in CTE of the AZ31B matrix with the addition of Al.

Mechanical Characteristics

Microhardness Test: Vickers microhardness tests were conducted on all extruded and polished samples (see Table IV). Results revealed that the addition of Al increased the microhardness value of AZ31 matrix. Around 29% microhardness value was increased when 2 wt. % aluminum was added into AZ31B to develop AZ51 magnesium alloy. Moreover, AZ51 magnesium alloy exhibited

37% higher hardness when compared to commercial AZ61A alloy (see Table IV) [3].

Table IV. Results of Room Temperature Microhardness and Modulus of Elasticity Tests

Materials	Microhardness (HV)	Modulus of Elasticity (GPa)
AZ31	66±2	34 ± 4
AZ41	77±3	65 ± 4
AZ51	85±3	69 ± 1
AZ61A [3]	62	45

Tensile Test: Results of ambient temperature tensile test on round tensile specimens revealed that the addition of Al in AZ31 magnesium increased the modulus of elasticity, 0.2% yield strength (YS) and ultimate tensile strength (UTS) (see Tables IV and V). In the case of AZ41 and AZ51 magnesium alloys, modulus of elasticity was increased around 91% and 103%, respectively, when compared to monolithic AZ31B. Results also showed that the AZ51 alloy exhibited 53% higher modulus of elasticity when compared to the commercial AZ61A magnesium alloy (see Table IV) [3]. However, failure strain (FS) dropped with the addition of Al while work of fracture (WoF) between AZ31 and AZ51 remained statistically unchanged.

Table V. Results of Room Temperature Tensile Properties

Matsrials	0.2 % YS (MPa)	UTS (MPa)	FS (%)	WoF (MJ/m^3)
AZ31	180 ± 3	273 ± 6	10.6 ± 1.3	27.9 ± 3.9
AZ41	218 ± 5	287 ± 6	8.2 ± 0.3	23.0 ± 1.4
AZ51	222 ± 4	302 ± 4	8.7 ± 0.4	27.3 ± 1.2
AZ61A [3]	228	310	16	-

Fracture Behavior: Macroscopic fracture characteristics of the broken tensile samples revealed that all samples exhibited shear-type ductile fracture mode. Microscopic features of fracture surface were also conducted on the tensile fractured surface using SEM and dimple-like features were observed in all the cases.

Discussion

Macrostructure

Synthesis of AZ31, AZ41 and AZ51 alloys were successfully accomplished by using DMD technique followed by hot extrusion. Observations made during DMD processing revealed: (a) minimal oxidation of melts, (b) undetectable reaction between graphite crucible and melts (AZ31, AZ41 and AZ51) and (c) absence of blowholes and macropores. The results suggest the appropriateness of processing parameters and methodology used in the present study. These observations are also consistent with the previous findings made on DMD processed magnesium based materials [1, 8, 10].

Density and Porosity Measurements

The experimental density obtained by the Archimedes' principle exhibited that the density of newly developed magnesium based materials increased negligibly with the addition of elemental Al into AZ31 matrix (see Table I). This can be attributed to the relatively higher dense Al (2.70 g/cm^3) addition into AZ31 matrix (1.78 g/cm^3) [3, 8]. Obtained results also revealed that the experimental values were relatively close to the theoretical values. This indicated that the experimental method used was accurate, and the experimental density of the fabricated materials could be calculated.

Porosity of the materials was calculated using the theoretical and the experimental densities obtained from each sample. The highest porosity obtained among the samples was 0.08% (see Table I), hence indicating that near dense materials was obtained and that the alloying constituent was successfully incorporated into the matrix. Therefore, as demonstrated by prior studies, the fabrication route of DMD followed by hot extrusion is capable of producing new AZ41 and AZ51 alloys with minimal porosity [8].

Microstructural Characteristics

The results of microstructural characterization revealed the presence of nearly equiaxed grains (see Table II and Figure 1) in AZ31, AZ41 and AZ51 samples. β-$Mg_{17}Al_{12}$ phase predominantly located at grain boundaries was observed in all samples (see Figure 1). Addition of elemental Al into AZ31 matrix decreased the average grain boundaries. This can be attributed to the pinning of grain boundaries by the increasing amount of second phases resulting in limited grain growth.

Optical and SEM micrographs of the etched samples revealed the presence and good distribution of the equilibrium intermetallic phase $Mg_{17}Al_{12}$ (see Figures 1 and 2). This could be attributed to the use of a layered arrangement for the raw materials during the solidification process and effective stirring parameters. The successful disintegration of the melt and the inert atmosphere caused by the argon gas jets were also the contributing factors. In the AZ41 system, the intermetallic particles appeared to have sharp edges. At these boundaries, the local stress concentration will be higher than that of the AZ31 system, which has blunt edges (see Figure 2). In contrast, these intermetallic phases became blunt with the addition of 2 wt% of Al into AZ31 matrix, thereby lowering the local stress concentration (see Figure 2).

X-Ray Diffraction

X-Ray diffraction results confirmed the presence of the intermetallic phase $Mg_{17}Al_{12}$ in all of the compositions (see Table III). For AZ51 samples, increased number of intermetallic peaks can be observed indicating higher amount of second phase were formed when compared to the AZ31 and AZ41 systems.

Coefficient of Thermal Expansion

The results of CTE measurement showed that the addition of Al alloying constituent into AZ31B matrix decreases the average CTE values of the materials (see Table III). The results suggested an appropriate integration of AZ31B alloy with elemental Al leading to low CTE values. Moreover, the lower CTE values of Al

(23.6 μm/mK) than Mg (25.2 μm/mK) also helped to lower the average CTE value of the newly developed compositions [3].

Mechanical Characteristics

Microhardness Test: The experimental results of microhardness measurement are shown in Table IV. AZ31 magnesium exhibited the lowest average hardness value and the average microhardness increased by about 17 % and 29% with the addition of 1 wt. % and 2 wt. % Al, respectively, into AZ31B. This can be attributed to the lower average grain size of these new compositions when compared to AZ31 which is associated with larger grain boundary area leading to higher hardness [19, 20]. The increase in hardness can be attributed to: (i) the presence of relatively harder elemental Al in the matrix, (ii) the reasonably uniform distribution of harder $Mg_{17}Al_{12}$ intermetallic in the matrix and (iii) higher constraint to the localized matrix deformation during indentation due to the higher presence of the intermetallic [21, 22].

Tensile Test: The elastic modulus measurement revealed that the addition of elemental Al into AZ31B system helped to increase the modulus of elasticity (see Table IV). The elastic modulus or stiffness of AZ51 samples was up to 103% higher than that of AZ31B samples. Increase in modulus was expected due to: (a) the addition of higher modulus of Al (69 GPa) into Mg (38 GPa) based compositions and (b) uniform distribution of second phase with good interfacial integrity [23, 24].

Room temperature tensile test results revealed that the addition of elemental Al increased the strength of monolithic AZ31 magnesium (~ 23% increment of 0.2% YS and ~11% increment of UTS in case of 2 wt. % Al addition) (see Table V and Figure 3). This can be attributed to the lower average grain size of AZ41 and AZ51 samples when compared to AZ31 samples (see Table II and Figure 1). Presence of additional Al modified the microstructural features, hence increasing strength. This follows a trend as the strength level of AZ61 and AZ91 is higher than that of AZ31 [3]. Failure strain decreased with Al addition into AZ31 magnesium. This can be attributed to the formation and presence of sharp edge intermetallic particles which might act as a stress concentration sites (see Figure 2). Work of fracture also dropped for AZ41 magnesium when compared to AZ31 magnesium. This can be attributed to the lower area under the stress-strain curve as the ductility of AZ41 is lower than AZ31. However, WoF of AZ51 samples remain unchanged when compared to AZ31B and increased when compared to AZ41 samples.

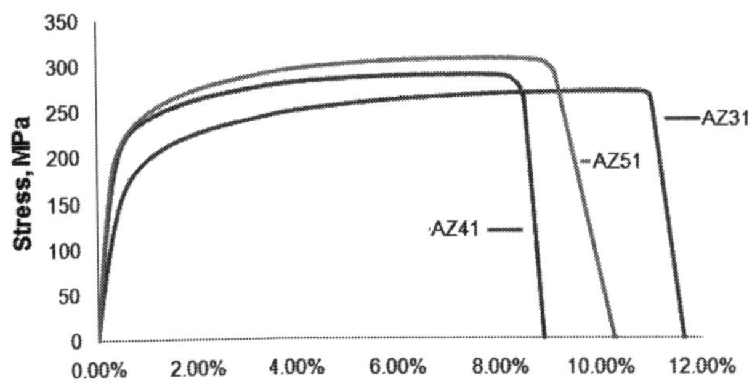

Figure 3. Engineering stress-strain diagram of AZ31, AZ41 and AZ51 samples.

Fracture Behavior: Macroscopic observations performed on the broken tensile samples revealed shear type fracture characteristics. Microscopic observations showed that dimple like features were predominately present in all compositions, indicating high plastic deformation (see Figure 4). However, more presence of microcracks were observed for new systems, hence are responsible for the decrease in failure strain of these composites.

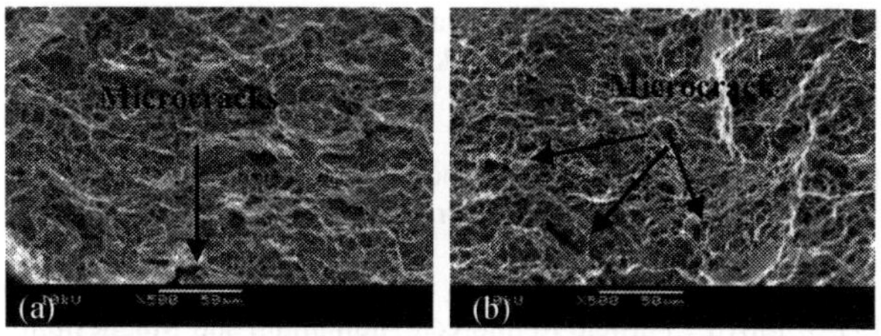

Figure 4. Representative SEM fractographs showing the microcracks and dimple like features for: (a) AZ31 and (b) AZ51 samples, respectively.

Conclusions

The following conclusions can be made from the present study:

1. Monolithic AZ31, AZ41 and AZ51 alloys can be successfully synthesized by using the disintegrated melt deposition technique followed by hot extrusion with minimal porosity.

2. The microstructure of all the AZ systems developed in this study consists of the magnesium phase and the intermetallic phase, $Mg_{17}Al_{12}$.

3. The coefficient of thermal expansion of AZ31 magnesium decreases with the addition of elemental aluminum.

4. Microhardness and modulus of elasticity values increase with the elemental Al addition as an alloying element to AZ31 magnesium.

5. A good combination of tensile properties (in terms of 0.2% YS, UTS and WoF) was observed for AZ41 and AZ51 samples when compared to AZ31B samples. These newly developed AZ magnesium based materials have significant potential in diverse engineering applications when compared to AZ31 alloy.

Acknowledgments

The authors gratefully acknowledge the support received for this research work ref: NPRP 08-424-2-171 from the Qatar National Research Fund (QNRF), Qatar.

References

[1] S. F. Hassan and M. Gupta, "Development of Nano-Y_2O_3 Containing Magnesium Nanocomposites Using Solidification Processing", Journal of Alloys and Compounds, 429 (1-2) (2007) 176-183.

[2] R. Fink, in: K. U. Kainer (Ed.), "Die-Casting Magnesium, Magnesium Alloys and Technologies", (Wiley-VCH Verlag GmbH and Co., Germany, 2003) 23-44.

[3] J. R. Davis, "Properties and Selection: Nonferrous Alloys and Special-Purpose Materials", Formerly Tenth Edition, (ASM Handbook, Volume 2, ASM, Metal Park, Ohio, 1993) 480-515, 1099-1100, 1118-1128,1199, 1132-1135.

[4] B. L. Mordike and K. U. Kainer, "Magnesium Alloys and Their Applications", (Werkstoff-Informationsgesellschaft, Frankfurt, Germany, 1998).

[5] B. L. Mordike and T. Ebert, "Magnesium: Properties-Applications-Potentials", Mater. Sci. Eng. A, 302 (2001) 137-45.

[6] D. J. Lloyd, "Particle Reinforced Aluminium and Magnesium Matrix Composites", Int. Mater. Rev., 39 (1994) 1-23.

[7] F. Moll and K. U. Kainer, "Magnesium Alloys and Technology", Kainer, K. U. (Ed), (Weinheim: Wiley-VCH, Germany, 2002) 197-217.

[8] Q. B. Nguyen and M. Gupta, "Increasing Significantly the Failure Strain and Work of Fracture of Solidification Processed AZ31B Using Nano-Al_2O_3 Particulates", J. Alloys Compd., 459 (2008) 244-250.

[9] W. Westengen, "Magnesium: Alloying", Encyclopedia of Materials: Science and Technology, (2008) 4739-4743.

[10] S. F. Hassan, K. F. Ho and M. Gupta, "Increasing Elastic Modulus, Strength and CTE of AZ91 by Reinforcing Pure Magnesium with Elemental Copper", Mater. Letters, 58 (2004) 2143-2146.

[11] K. Hirai, H. Somekawa, Y. Takigawa and K. Higashi, "Effects of Ca and Sr Addition on Mechanical Properties of a Cast AZ91 Magnesium Alloys at Room and Elevated Temperature", Mater. Sci. Eng. A, 403 (2005) 276-280.

[12] Presentations from Elektron 21 product launch CD provided by Magnesium Elektron, UK http://www.magnesium-elektron.com/site-map.asp (accessed on 26 September' 2010).

[13] X. Gong, H. Li, S. B. Kang, J. H. Cho and S. Li, "Microstructure and Mechanical Properties of Twin-Roll-Cast Mg-4.5Al-1.0Zn Sheets Produced by Differential Speed Rolling", Materials and Design, 31 (2010) 1581-1587.

[14] H. M. Chen, H. S. Yu, S. B. Kang and G. H. Min, "Optimization of Annealing Treatment Parameters in a Twin-Roll-Cast and Warm Rolled Mg-4.5Al-1.0Zn Alloy", Advanced Materials Research, 79-82 (2009) 2139-2142.

[15] S. Wang, M. Wang, R. Ma, Y. Wang and Y. Wang, Microstructure and Hot Compression Behavior of Twin-Roll-Casting AZ41M Magnesium Alloy", Rear Metals, 29 (2010) 396-400.

[16] K. Matsuzaki, K. Hatsukano, Y. Torisaka, K. Hanada, T. Shimizu and M. Kato, "Microstructure and Mechanical Properties of Mg Alloy Sheets Produced by Twin-Roll-Caster", Materials Science Forum, 539-543 (2007) 1747-1752.

[17] B. H. Kim, J. J. Jeon, K. C. Park, B. G. Park, Y. H. Park and I. M. Park, "Microstructural Characterization and Mechanical Properties of Mg-xSn-5Al-1Zn Alloys", International Journal of Cast Metals Research, 21 (2008) 186-192.

[18] A. Jager, P. Lukac, V. Gartnerova, J. Haloda, and M. Dopita, "Influence of Annealing on the Microstructure of Commercial Mg Alloy AZ31 after Mechanical Forming", Mater. Sci. Eng. A, 432 (2006) 20-25.

[19] G. E. Dieter, "Mechanical Metallurgy", (2nd Edition, McGraw-Hill, Inc., USA, 1976) 191-193.

[20] M. E. Alam and M. Gupta, "Tensile Behavior of Tin Sintered Using Microwave and Radiant Heating", International Conference on Mechanical Engineering (ICME' 07), December 29-31, 2007, Dhaka, Bangladesh.

[21] M. Paramsothy, N. Srikanth and M. Gupta, "Solidification Processed Mg/Al Bimetal Macrocomposite: Microstructure and Mechanical Properties", J. Alloys Compd., 461 (2008) 200-208.

[22] M. E. Alam, S. M. L. Nai and M. Gupta, "Development of High Strength Sn-Cu Solder Using Copper Particles at Nanolength Scale", Journal of Alloys and Compounds, 476 (2009) 199-206.

[23] M. Paramsothy, S. F. Hassan, N. Srikanth and M Gupta, "Enhancing the Performance of Magnesium Alloy AZ31 by Integration with Millimeter Length Scale Aluminium Based Cores", J. Comp. Mater., 44 (9) (2010) 1099-1117.

[24] S. F. Hassan and M. Gupta, "Development of High Strength Magnesium Based Composites using Elemental Nickel Particulates as Reinforcement", Journal of Mateials science, 37 (2002) 2467-2474.

Magnesium Technology 2011
Edited by: Wim H. Sillekens, Sean R. Agnew, Neale R. Neelameggham, and Suveen N. Mathaudhu
TMS (The Minerals, Metals & Materials Society), 2011

ENGINEERING A MORE EFFICIENT ZIRCONIUM GRAIN REFINER FOR MAGNESIUM

S. Viswanathan[1], P. Saha[1], D. Foley[2] and K. T. Hartwig[2]
[1]The University of Alabama, Tuscaloosa, AL 35487
[2]Texas A&M University, College Station, TX 77843

Keywords: Grain Refinement, Magnesium, ECAE, Solidification

Abstract

Zirconium is currently used as a grain refiner for high temperature magnesium alloys. However, currently around 1 wt% is used for grain refinement, and much of this is wasted as sludge due to settling, as zirconium has a density almost four times that of magnesium. Since a recent study suggested that only faceted particles were activated as nucleation sites during casting solidification, ECAE processing of a magnesium-15wt% zirconium grain refiner master alloy was attempted to try to increase the number of faceted particles. ECAE processing of the master alloy only resulted in a modest decrease in grain size, even though the percentage of faceted particles doubled as a result of ECAE processing, since only 1.6% of zirconium particles in the as-received master alloy were faceted and deemed to be likely nucleation sites. The results from this work provide insights for the future development of more efficient grain refiners.

Introduction

Grain refinement is an effective method for control of grain size and morphology during casting. The most widespread technique of grain refinement is the introduction of extraneous nucleating agents (grain refiners) to a liquid melt. The benefits from grain refinement include uniform microstructure, distribution of second phase and microporosity on a finer scale, better feeding, and improved mechanical properties. Finer grains also improve surface finish and impart superior machinability. [1] The addition of grain refiners has historically been made as master alloys (often called hardeners) in the form of shot, rod, waffle, or briquette.

Zirconium is well known as an excellent grain refiner for magnesium alloys that do not contain aluminum, since aluminum poisons the grain refining ability of zirconium. [2] However, zirconium is a heavy element with a density almost four times that of magnesium (ρ_{Mg} = 1740 kg/m^3, ρ_{Zr} = 6520 kg/m^3). As a result, the addition of zirconium into molten magnesium results in the loss of zirconium by settling, since during melting zirconium particles settle to the bottom of the crucible and are wasted as sludge. As zirconium is expensive, this represents a significant additional cost as well as wastage of material.

The magnesium rich region of the binary Mg-Zr phase diagram (see Figure 1) [3] shows that grain refinement of magnesium can occur in theory by a peritectic reaction at 654°C, where α-zirconium reacts with liquid and forms an α-magnesium phase. A recent modification [4] of the magnesium-zirconium phase diagram [5] suggests that the maximum solubility of zirconium is 0.5 wt% at 654°C. Currently, approximately 1 wt% zirconium is currently added to ensure grain refinement.

However, in a systematic study of the grain refinement of magnesium, Saha et al. [6] showed that the peritectic reaction was not required for grain refinement. In a further study involving SEM and TEM characterization of grain refined samples, Saha and Viswanathan [7] showed that only faceted particles are likely nucleation sites. Based on these results, this work attempts to reprocess an existing magnesium-15 wt% zirconium master alloy and improve its efficacy, and thereby make it more efficient, i.e. reengineer the grain refiner such that much less than the 1% that is currently used is sufficient for grain refinement.

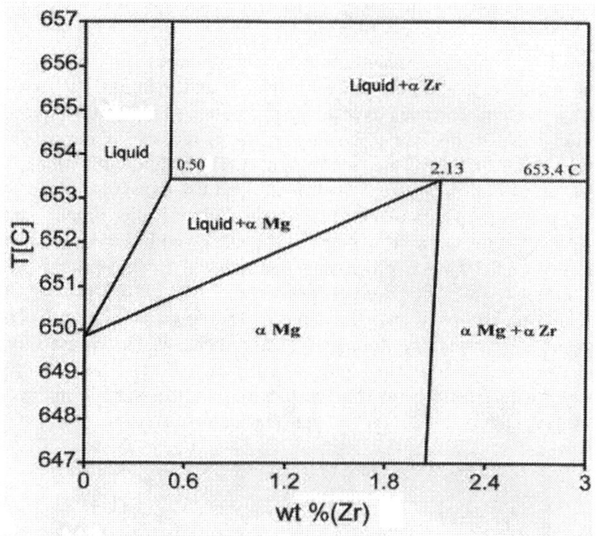

Figure 1. Relevant portion of magnesium-zirconium phase diagram [3] showing peritectic reaction.

Previous Work

Qian et al. [8] found zirconium-rich cores, the majority of which are pure zirconium particles, at the center of magnesium grains or at the grain edge. Electron microprobe analysis of these particles showed that the particle sizes were between 1 and 5 µm, suggesting that only particles within this size range acted as nuclei for the grains. Particles which were greater than 5 µm were not associated with zirconium-rich cores.

Qian et al. [9] also reported on the development of a new magnesium-zirconium master alloy (AM-cast) with zirconium particles smaller than 10 µm. They reported that the newly developed master alloy provided excellent grain refinement when added to pure magnesium. The excellent grain refining ability was attributed to the fine size and even distribution of the zirconium particles in the master alloy. Qian et al. [10] also demonstrated that the efficiency of a Zirmax® master alloy was improved by hot rolling the master alloy ingots into thin plates. The subsequent improvement in the grain refining behavior was attributed to improved particle size distribution in the rolled master alloy.

Approach

The goal of this program was to increase the grain refinement efficacy of the magnesium-15 wt% zirconium master alloy used in this work, by breaking up large particles and particle clusters and increasing the total number of faceted particles. Accordingly, the master alloy was subjected to severe plastic deformation at an elevated temperature by the Equal Channel Angular Extrusion (ECAE) process. Figure 2 shows an SEM micrograph of the master alloy containing large particles and particle clusters.

Rods 12 mm in diameter and 75 mm in length were machined from the magnesium-15 wt% zirconium master alloy ingot. One rod sample was encapsulated in a steel can (shown in Figure 3) and deformed by ECAE route 4A [11], i.e. 4 passes with no rotation between multiple passes, at 300°C.

The ECAE process spreads out grain boundaries (i.e., elongates the grains as if the material were rolled) and therefore disperses the zirconium particles over a larger volume. [11] For the square billet used in this work, the flow plane is the side plane as the billet exits the ECAE die (see Figure 4). The longitudinal plane is the top of the billet as it exits the die, and the transverse plane is the plane perpendicular to the long axis of the billet. The microstructure in the flow plane of the deformed billet determines how well the ECAE processing has worked for the processing conditions used. For further processing, the ECAE processed billet was sectioned into three pieces as shown in Figure 4. To ensure microstructural uniformity, the center portion was used for all studies. As shown in Figure 4, samples marked 1 and 2 were used for microstructural characterization, and the sample marked 3 was used for dissolution and grain refinement studies.

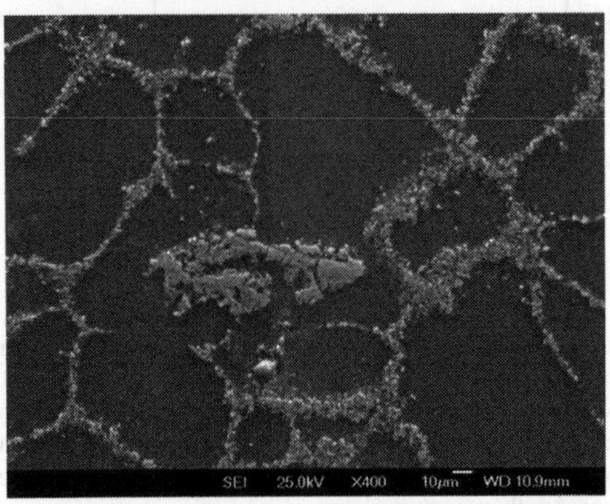

Figure 2. SEM micrograph of the magnesium-15 wt% zirconium master alloy showing large particles and particle clusters.

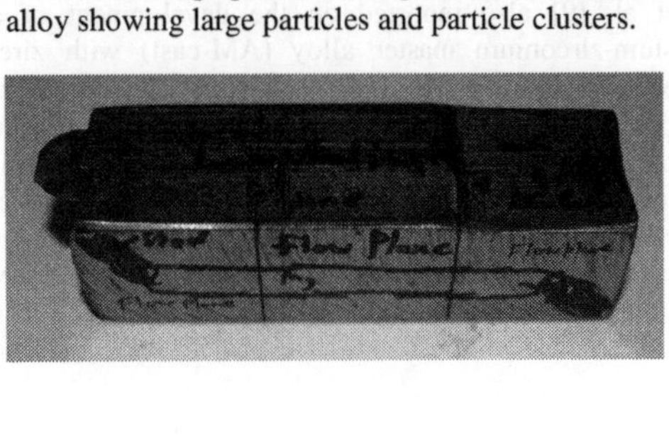

Figure 3. Photograph of magnesium-15 wt% zirconium ECAE processed rod encapsulated in a steel can.

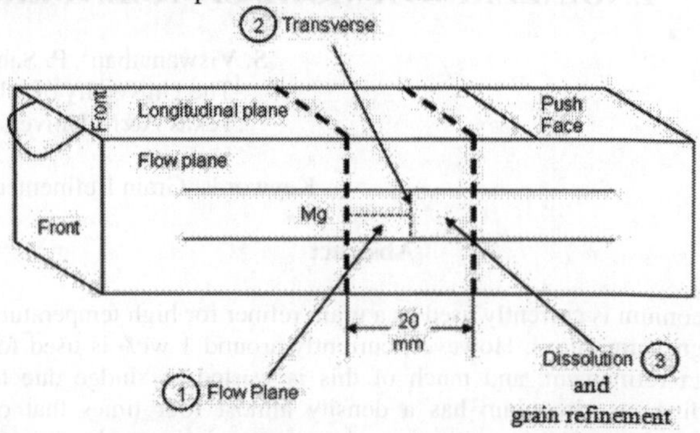

Figure 4. Schematic diagram of the ECAE processed billet showing the flow plane, the longitudinal plane, the push face, and samples (marked 1 to 3) sectioned for microstructural, dissolution, and grain refinement studies.

Characterization of ECAE Billet Microstructure

Figure 5 shows a comparison of the microstructure of the as-received master alloy and the ECAE processed billet. The microstructure of the as-received magnesium-15wt% zirconium master alloy (Figure 5a and Figure 2) contains large zirconium particles along the grain boundary. After ECAE processing, the grains become elongated and the particles are spread along grain boundaries (Figure 5b). The large zirconium particles and particle clusters which were originally at the grain boundary are sheared over a large volume and broken into small individual particles (Figure 5c). Figure 5c was obtained from the boxed region shown in Figure 5b. The micrographs in Figure 5 suggest that ECAE effectively broke up the zirconium particles and particle clusters.

Zirconium Particle Characterization

A sample of zirconium particles was prepared by dissolving the master alloy in acid, which dissolved away the magnesium but did not affect the zirconium. The purpose of the dissolution sample was to determine the particle size distribution, and count the number of faceted particles present in each sample. Samples were dissolved into 1% diluted hydrochloric acid (0.1 g/10 ml basis) and kept for 72 hrs to digest the sample completely. One ml of the solution was transferred into an EPPENDORF tube and centrifuged at 15000 rpm for 15 min for 3 cycles. After each cycle the liquid was decanted, and de-ionized water was added to the mixture, and the tube was sonicated for 10 min. This allowed magnesium chloride, which is highly soluble in water, to be washed away and for the zirconium particles to be collected at the bottom.

The final centrifuged sample was ultra-sonicated for 30 min and one drop of the solution was evenly distributed on a silicon wafer. The wafer was dried and kept under vacuum to prevent it from being contaminated. The silicon wafer with the zirconium particles was characterized by SEM at an accelerating voltage of 15 kV, working distance of 10 mm and spot size of 7 nm. Figure 6 shows SEM micrographs of zirconium particles from the as-received and ECAE processed master alloy. It is apparent that the zirconium particles in the as-received master alloy are irregular and that some large particles are present (Figure 6a). However, the majority of the particles in the ECAE processed sample appear to be regular and in the submicron to 2 µm range (Figure 6b). This is likely due to the shearing action of the ECAE process that tends to break up and distribute particles.

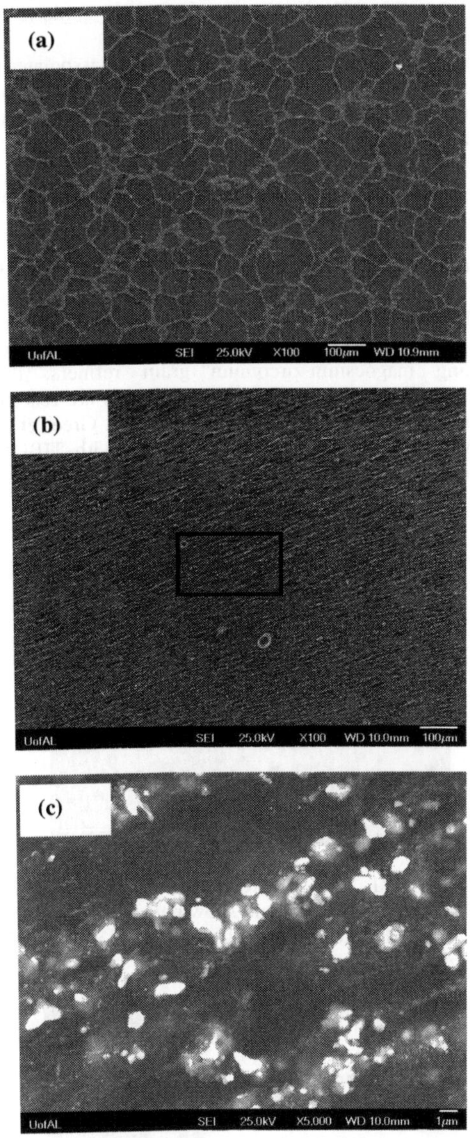

Figure 5. SEM micrographs of (a) as-received Mg-15 wt% Zr master alloy, (b) ECAE processed billet along the flow plane, and (c) enlarged view of a portion of the boxed region in (b) showing breakup and distribution of particles.

The zirconium particle size distributions in the as-received and ECAE processed master alloy samples were determined using an Accusizer model 770 particle size analyzer that uses the light obscuration/photozone method (Particle Technology Labs, Downers Grove, IL, USA). Figure 7 shows the particle size distributions of the as-received and the ECAE processed master alloys. It is apparent that ECAE processing increases the number of particles and narrows the size distribution. It is also apparent that the number of small particles (0.5 to 2 µm) increased in the ECAE processed alloy.

Since Saha and Viswanathan [7] indicated that only faceted particles are likely nucleation sites, the fraction of faceted particles in both the as-received and ECAE processed master alloy samples were estimated from SEM micrographs. Figure 8 shows a typical SEM micrograph showing both faceted and non-faceted particles. Faceted particles were outlined and quantified in terms of area using Nikon NIS-Elements image analysis software. It was found that 1.6% of particles present in the as-received master alloy were faceted, whereas 3.3% of particles in the ECAE processed sample were faceted. The increase in the number of faceted particles suggests that ECAE broke up particles and created more faceted particles.

Figure 6. SEM micrographs of zirconium particles obtained from dissolution samples for (a) as-received master alloy, and (b) ECAE processed sample.

Grain Refinement Experiments

Grain refinement experiments were conducted with the original and ECAE processed master alloys. A magnesium-2% zinc alloy was used for the grain refinement experiments. Pure magnesium (99.9%) and zinc were melted in a mild steel crucible using an electrical resistance furnace. The typical melt size was 150 g. A protective gas (0.5% SF6 + CO2) mixture was used to prevent magnesium from burning. Once the desired pouring temperature of 815ºC was reached, a small piece of the grain refiner master alloy was added and allowed to dissolve for one minute. The melt was then stirred vigorously for 15 s with a coated mild steel rod and poured into a "hockey puck" mold that was preheated to 350ºC. The mold was sectioned from 1018 steel pipe to form a ring mold 45 mm I.D., 63.5 mm O.D., and 12 mm height. A gray iron block, 100 x 100 x 25 mm, was used as the chill. The mold, the chill, and the stirring rod were coated with CONCOTE™

MAG 669 (Hill and Griffith) prior to preheating to 350°C. Figure 9 shows the hockey puck casting setup used for this work.

Metallographic specimens were sectioned from the center of the hockey puck samples using a Buehler Isomet® 1000 precision saw. Specimens were mounted in conductive epoxy using a hot compression mounting process (Buehler Simplimet® 1000). Mounted specimens were ground and polished using standard metallographic techniques. The final polished samples were etched with an acetic picral solution (5 mL acetic acid, 6 g picric acid, 10 mL water and 100 mL ethanol) for 5 s. Optical micrographs were captured with NIS-Elements image analysis software which was integrated with a NIKON EPIPHOT optical microscope. The grain size was calculated at 50x magnification using a standard grain size calculation method. [12, 13]

Since an important goal of this work was to demonstrate that the amount of zirconium could be significantly reduced from the current 1 wt% that is used, 0.1 wt% zirconium (0.67 wt% of master alloy) was added to the magnesium-2% zinc melt and hockey puck samples were poured. Figure 10 shows optical micrographs of the base magnesium-2% zinc alloy without grain refiner addition, and samples grain refined at the 0.1 wt% zirconium level with the as-received and ECAE processed master alloys.

It is clear from the micrographs in Figure 10 that the ECAE processed master alloy had a noticeable improvement on the overall grain refinement compared to the as-received master alloy. This improvement in the grain refinement behavior of the ECAE processed master alloy is attributed to the increased number of faceted particles in the ECAE processed master alloy. However, since ECAE processing only increased faceted particles from 1.6% to 3.3%, the decrease in grain size was modest (Note: particle density correlates with grain density rather than grain size).

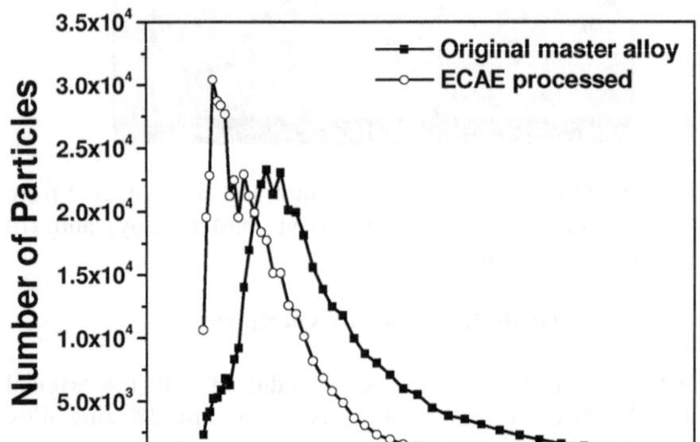

Figure 7. Zirconium particle size distributions in as-received and ECAE processed master alloy samples.

Analysis of Grain Refinement Efficacy

Although the result shown in Figure 10 is encouraging, it is clear that a significant improvement in grain refiner performance will require better up-stream processing of the grain master alloy that results in a higher fraction of faceted particles. To further highlight and explain the link between faceted particles and grain refinement efficacy, faceted particles in the magnesium-2% Zn-0.1% Zr grain refined (hockey puck) samples were also analyzed and counted using the same methodology as that used previously for the master alloys. The results are tabulated in Table 1.

In the data presented in Table 1, each grain is associated with a single nucleus; accordingly, efficiency is the ratio of the grain density to the particle density. Although ECAE processing increased the number of faceted particles, it also resulted in a significantly larger number of particles (i.e. a factor of seven). Accordingly, since there are many more particles, a far smaller percentage of zirconium particles in the ECAE processed master alloy resulted in grains compared to the as-received alloy: 0.17% for the ECAE processed alloy vs. 0.75% for the as-received alloy. However, clearly less than 1% of the zirconium particles in either master alloy are effective, suggesting that current grain refiners are very inefficient. This suggests that the currently used process for making magnesium-zirconium grain refiners, the salt replacement technique in which zirconium salts are reacted with magnesium, needs to be revisited. Interestingly, Greer et al. [14] in their work on aluminum grain refinement with TIBOR, also found that only 1% of TiB_2 particles were effective as nuclei.

Figure 8. Typical SEM micrograph of zirconium particles obtained from dissolution sample showing both faceted and non-faceted particles. Faceted particles were outlined and quantified in terms of area using image analysis software.

Figure 9. Hockey puck casting setup used for this work.

Role of Faceted Particles

The actual increase in the faceted particle density in samples grain refined with the ECAE processed alloy was quite small: 599 vs. 470/mm^3, resulting in only a modest decrease in grain size. (Note:

all faceted particles are not selected as nuclei since size is also a factor, as addressed in the next section.) However, the most significant figure in Table 1 is the faceted particle efficiency. While the total efficiency of zirconium particles is less than 1%, the faceted particle efficiency is 50% or higher, i.e. the ratio of grain density to faceted particle density is approximately half. This provides very strong evidence that only faceted particles likely result in grain refinement, i.e. serve as nuclei. These results further corroborate the hypothesis advanced by Saha and Viswanathan [7] that only faceted particles present in grain refiner master alloys are likely active nucleation centers.

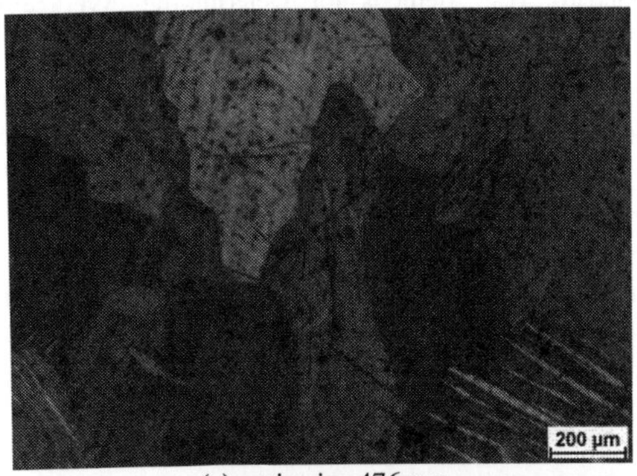

(a) grain size 476 µm

(b) grain size 131 µm

(c) grain size 110 µm

Figure 10. Optical micrographs of magnesium-2 wt% zinc samples, a) without grain refiner, b) with 0.1 wt% zirconium using as-received master alloy, and c) with 0.1 wt% zirconium using ECAE processed master alloy. The measured grain size is indicated below the micrographs. (magnification 50x)

Table 1. Analysis of grain refinement efficacy for Mg-2% Zn-0.1% Zr hockey puck samples.

Grain Refiner Used	As-Received	ECAE Processed
Grain size (µm)	131	110
Grain density (mm^{-3})	222	376
Zirconium Particle density (mm^{-3})	29553	214825
Zirconium Particle Efficiency (%)	0.75	0.17
Faceted Particle density (mm^{-3})	470	599
Faceted Particle Efficiency (%)	63	47

Engineering Efficient Grain Refiners

This work provides very strong evidence that grain refiners must be properly engineering to be efficient, and that current grain refiners are extremely inefficient. It is important to note, however, that faceting is not the only factor that determines grain refiner efficacy; another important factor is particle size. The size of the nucleant particle determines the undercooling needed for it to be activated, and the smaller the particle, the larger the undercooling that is required. (Note: the poor efficiency of TIBOR grain refiners is likely related to the size of the particles, since all TiB_2 particles appear to be faceted.) The available undercooling for nucleation is determined by the casting process and the alloy. Moreover, since a range of particle sizes and undercoolings are likely to be active during solidification [14] and measured undercoolings are typically 0 to 2ºC, [15] particle size distributions must be also be suitably tailored. Consequently, it is likely that any future effort to develop more efficient grain refiners must target the alloy or alloy family (at least cast vs. wrought) and the process. Finally, the number of suitable particles is also important, as grain density is dependent on nucleant particle density. Therefore, efficient and effective grain refiners must address particle density, particle size, particle size distribution, and particle morphology.

Conclusion

ECAE processing of the magnesium-15 wt% zirconium grain refiner master alloy used for this work provided only a modest improvement in grain refiner efficacy. This is due to the fact that only faceted zirconium particles appear to nucleate grains during solidification. Although ECAE processing doubled the number of faceted particles, the fact that only 1.6% of the particles in the as-received master alloy were faceted limited the extent of improvement that was possible by ECAE processing. This work suggests that future development of efficient and effective grain refiners must address particle density, particle size, particle size distribution, and particle morphology.

Acknowledgments

This work was supported by The University of Alabama and the High Integrity Magnesium Automotive Component project sponsored by the United States Automotive Materials Partnership. This material is based upon work supported by the Department of Energy National Energy Technology Laboratory under Award Number DE-FC26-02OR22910. This report was prepared as an account of work sponsored by an agency of the United States Government. Neither the United States Government nor any agency thereof, nor any of their employees, makes any warranty, express or implied, or assumes any legal liability or responsibility for the accuracy, completeness, or usefulness of any information, apparatus, product, or process disclosed, or represents that its use would not infringe privately owned rights. Reference herein to any specific commercial product, process, or service by trade name, trademark, manufacturer, or otherwise does not necessarily constitute or imply its endorsement, recommendation, or favoring by the United States Government or any agency thereof. The views and opinions of authors expressed herein do not necessarily state or reflect those of the United States Government or any agency thereof.

References

1. *ASM Handbook: Vol. 15 Casting* (ASM International, Materials Park, OH 44073, Dec 2008), pp. 255-262.

2. E.F.Emley, *Principles of Magnesium Technology*, (Pergamon Press, Oxford, United Kingdom, 1966), pp. 126-156, 254-260.

3. Pandat Software with PanMagnesium Database, Computherm LLC, Madison Wisconsin, USA.

4. H. Okamoto, "Mg-Zr (Magnesium-Zirconium)," *Journal of Phase Equilibria*, vol. 23, 2002, pp. 198-199.

5. M. Hämäläinen and K. Zeng, "Thermodynamic evaluation of the Mg-Zr system," *Calphad*, vol. 22, 1998, pp. 375-380.

6. Partha Saha, Katie Lolies, Srinath Viswanathan, Robert G. Batson, and Arun M. Gokhale, "A Systematic Study of the Grain Refinement of Magnesium by Zirconium," *Magnesium Technology 2010* (ISBN: 978-0-87339-746-9, The Minerals, Metals, and Materials Society, Warrendale, PA, 2010), p 425.

7. Saha, P. and Viswanathan, S. "Grain Refinement of Magnesium By Zirconium: Characterization and Analysis," *Magnesium Technology 2011*, (TMS, Warrendale, PA, 2011), submitted for publication.

8. Q. Ma, D. H. StJohn, and M. T. Frost, "Characteristic zirconium-rich coring structures in Mg-Zr alloys," *Scripta Materialia*, vol. 46, 2002, pp. 649-654.

9. M. Qian, D. H. StJohn, and M. T. Frost, "A new zirconium-rich master alloy for the grain refinement of magnesium alloys," *Magnesium: Proceedings of the 6th International Conference on Magnesium Alloys and Their Applications*, 2003, pp. 706-712.

10. M. Qian, D. H. StJohn, M. T. Frost, and R. M. Barnett, "Grain refinement of pure magnesium using rolled zirmax master alloy (Mg-33.3ZR)," *Magnesium Technology 2003*, (TMS, Warrendale, PA, 2003), pp. 215-220.

11. V. M. Segal, K. T. Hartwig, and R. E. Goforth, "In situ composites processed by simple shear," *Materials Science and Engineering A*, vol. 224, 1997, pp. 107-115.

12. B.R. Morris, A.M. Gokhale and G.F. Vander Voort, "Grain Size Estimation in Anisotropic Materials," *Metall and Mat. Trans*, 29A, 1998, 237-244.

13. Standard Test Methods for Determining Average Grain Size," ASTM E112-96 (2004).

14. A. L. Greer, A. M. Bunn, A. Tronche, P. V. Evans, and D. J. Bristow, "Modelling of inoculation of metallic melts: application to grain refinement of aluminium by Al-Ti-B," *Acta Materialia*, vol. 48, 2000, pp. 2823-2835.

15. Maxwell and A. Hellawell, "A simple model for grain refinement during solidification," *Acta Metallurgica*, vol. 23, 1975, pp. 229-237.

Magnesium Technology 2011
Edited by: Wim H. Sillekens, Sean R. Agnew, Neale R. Neelameggham, and Suveen N. Mathaudhu
TMS (The Minerals, Metals & Materials Society), 2011

MICROSTRUCTURE AND MECHANICAL PROPERTIES OF Mg-1.7Y-1.2Zn SHEET PROCESSED BY HOT ROLLING AND FRICTION STIR PROCESSING

V. Jain[1,2], J.Q. Su[2], R.S. Mishra[2], R. Verma[3], A. Javaid[4], M. Aljarrah[4], E. Essadiqi[4]
[1]Division of Engineering Materials, National Physical Laboratory, CSIR, New Delhi 110012, India
[2]Center for Friction Stir Processing and Department of Materials Science and Engineering, Missouri University of Science and Technology, Rolla, MO 65409, USA
[3]General Motors R&D Center, Warren, MI 48090, USA
[4]Materials Technology Laboratory-CANMET, Ottawa, Ontario K1A 0G1, Canada

Keywords: Friction Stir Processing, Magnesium-Rare Earth, Texture, Hot Rolling

Abstract

The use of lightweight structural materials is an integral part of mass reduction strategy in transportation applications. Magnesium based sheet products have gained significant interest in the automobile industry. Newer magnesium alloys such as Mg-Y-Zn have potential to develop sheet products with superior mechanical properties owing to improved precipitation hardening response. In the present work, rolled sheet of an Mg-1.7Y-1.2Zn alloy containing small amounts of Al and Ce was investigated. Microstructure and mechanical properties were examined in as-rolled, rolled+aged, friction stir processed (FSP) and FSP+aged conditions. Mechanical properties (YS, UTS and %El) of the sheet showed certain anisotropy in rolling and transverse directions, which was marginally reduced upon ageing. However, FSP led to a significant range of mechanical properties depending on the test direction. Ageing of the FSP sheet showed improvement in mechanical properties and reduced anisotropy in the two directions. The static recrystallization due to ageing caused reduced anisotropy in FSP treated sheet. The overall strength-ductility variation is discussed in terms of relative contributions of grain boundary strengthening, texture and precipitation strengthening.

Introduction

Magnesium alloys have gained significant interest among automobile manufacturers due to low density and high specific strength that translate to improved fuel economy and reduced gaseous emissions [1-3]. Despite significant potential of magnesium sheet for hot forming of complex automobile panels [4], its role in automotive applications is still limited due to poor ductility and strength. For example, AZ31 sheets produced by the conventional direct-chill (DC) process or even a more advanced twin roll casting (TRC) technique find limited applications primarily due to their low ductility, although specific strength of Mg alloys is high [5]. The highest yield strength achievable with the current commercial wrought Mg alloys is about 240 MPa and there is a requirement to develop improved wrought alloys and products having yield strength more than 300 MPa [6].

Grain refinement and precipitation hardening are two primary mechanisms for improving room temperature strength in Mg alloys, whereas, ductility could be improved by grain refinement as well as texture modification. Ductility improvement in fine grained Mg alloys is due to activation of prismatic and pyramidal slip systems at grain boundaries and limited tendency for twinning [7]. Extensive ductility in coarse grained Mg alloys is achievable by the addition of alloying elements, such as, Ca, Th, Mn or certain rare earth (RE) elements (*e.g.*, Ce), which was realized in 1940 [8]. Barnett *et al.* [9] have demonstrated cold rolling reduction of ~30, >90 and ~15% in pure Mg, Mg-0.2Ce and AZ31 alloys, respectively. Mishra *et al.* [10] have demonstrated extensive ductility in hot extruded Mg-0.2Ce alloy due to weakening of texture in the presence of cerium. Extrusions conducted at other than the optimized process parameters delivered low ductility in the alloy due to incomplete recrystallization [11]. Hantzsche *et al.* [12] have recently reported reduced planer and in-plane anisotropy in the mechanical behavior of ZEK100 Mg alloy (containing an RE-element) sheet compared to ZM21 alloy (having no RE-element), that were hot rolled at the temperature of 450°C up to a thickness of ~1.35 mm. The decreased anisotropy in the mechanical behavior of ZEK100 alloy was due to weakening of basal texture that was governed by different recrystallization kinetics compared to the ZM21 alloy. Sikand *et al.* [13] observed grain refinement and improved mechanical properties (ductility as well as strength) in the porthole extrusion of ~3 to 3.5 mm thick walled AM30 circular tubes compared to conical die extrusions. Porthole extruded 4 mm thick walled circular tubes of Mg alloys with <0.5% Ce addition showed improved ductility; but, found ineffective in Mg-3Al alloys due to higher affinity of Ce with Al [14]. The weakening of texture due to Ce addition in Mg is therefore offset by the presence of high Al content and requires attention during the design of newer alloy systems.

Severe plastic deformation (SPD) processes, such as, Equal Channel Angular Pressing (ECAP), High Pressure Torsion (HPT), Accumulative Roll Bonding (ARB), etc have opened up new opportunities for the development of ultra fine grained (UFG) lightweight materials [15]. Friction stir processing (FSP) is a relatively new process for developing UFG alloys with enhanced mechanical properties. A schematic diagram of the FSP is shown in Fig. 1. Important aspects of this process, such as, process parameters, process modeling, microstructure and resultant mechanical properties, etc, have been comprehensively addressed by Mishra and Ma [16]. The process uses a rotating tool comprising of a threaded pin and tool shoulder to apply severe plastic deformation and frictional heating that produces significantly refined grain size.

Grain boundary strengthening employing SPD has limited potential and newer Mg alloys having precipitation hardening response are required. Mg-Y-Zn is a promising alloy system that could deliver exceptional mechanical properties at ambient as well as elevated temperatures due to the presence of icosahedral (I) ternary phase [17]. Three different types of ternary phases in the Mg-Y-Zn alloy systems have been reported; which are, icosahedral quasicrystal Mg_3YZn_6 (I-phase), cubic $Mg_3Y_2Zn_3$ (W-phase) and 18-R long-period ordered (LPO) $Mg_{12}YZn$ (Z-phase) [18-19]. The formation of ternary phases depends on the Zn/Y ratio of the alloy composition. Lee *et al.* [19] have identified Zn/Y ratio as 5~7 for the presence of I-phase in as cast Mg-Y-Zn alloy systems. Room temperature strengthening in Mg-Y-Zn systems greatly depends on the fraction of quasicrystalline I-phase, since, the I-phase precipitates with a definite orientation relationship (OR) with the matrix and forms strong interface with it. The I-phase is resistant to coarsen up to 440°C and has low interfacial energy, thus impedes grain boundary sliding and hinders dislocation movement at elevated temperatures.

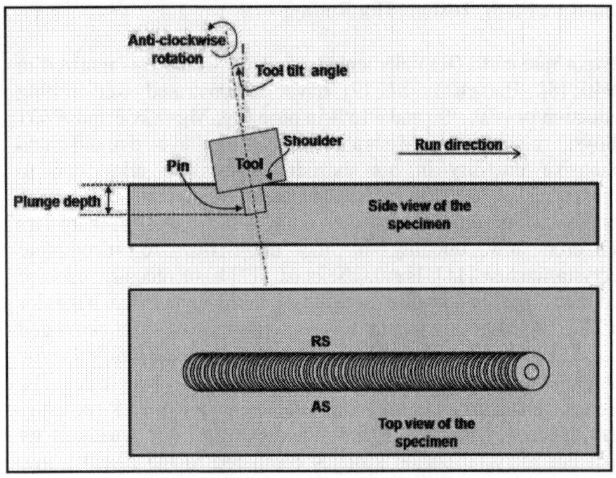

Fig. 1 Schematic diagram of the Friction Stir Processing (FSP). AS: Advancing Side; RS: Retreating Side

The present study was aimed to determine the effect of FSP on the room temperature mechanical properties of hot rolled sheets of an Mg-1.7Y-1.2Zn alloy. FSP was used as the technique to obtain refined microstructure. The observed mechanical properties have been discussed in terms of correlation with the texture anisotropy in the rolled and friction stir processed sheets, which were later aged under the chosen parameters.

Experimental Procedure

The composition of Mg-Y-Zn alloy was Y-1.7%, Zn-1.2%, Al-0.53%, Mn-0.27%, Ce-0.013%, Mg-balance. FactSage thermodynamic calculations predicted four equilibrium phases coexisting at 450°C, *i.e.*, $MgYZn_3$, Mg_2Y, Mg and β-Mn. It was predicted that at room temperature, two additional phases, *i.e.*, $Mg_{43}Y_4Zn_3$ and $Ce(Mg,Al)_{12}$ could precipitate. The alloy was cast as rectangular plate in size of 14×30×1.3 cm^3. The cast plates were hot rolled at 450°C to a final thickness of 1.9 mm in nine passes with a total reduction of ~85%. Reheating of the rolled material was carried out at 450°C after each pass. The rolled sheets after the final pass were annealed at 450°C for 10 minutes. FSP of rolled sheets was carried out with a simple cylindrical threaded tool. The tool had a pin of 4.5 mm dia and 1.5 mm height. Diameter of the tool shoulder was 11.5 mm. A single pass FSP was carried out employing tool rotation rate of 500 rpm and translation speed of 101.6 mm per minute. The tool tilt angle was kept as 2.5°. The depth of processing/plunge depth was chosen as ~1.5 mm and the direction of processing was kept same as rolling direction of the sheets. Ageing of as-rolled and FSP sheets was conducted at the temperature of 160°C for 96 hrs. The ageing parameters were selected based on optimized conditions observed for similar kinds of Mg alloys [6].

Dog bone tensile specimens of 1.5 mm gage length, 1.2 mm gage width and ~0.8 mm thickness were machined from the as-rolled, rolled-aged, FSP treated and FSP-aged sheets. Table I shows details of different types of samples used. In the case of FSP sheet, tensile samples were machined from centre line of the nugget region removing ~200 μm top layer by polishing. Tensile testing was carried on a computer controlled mini tensile testing machine employing a strain rate of 1×10^{-3} s^{-1}. Optical, stereo and scanning electron microscopy of the sheet material processed under different conditions were carried out to study the microstructure and fracture response of tensile specimens. For the microstructural investigations, samples were polished up to 0.5 μm diamond paste followed by etching in a solution containing 4.5 g picric acid, 10 ml acetic acid, 10 ml distilled water and 70 ml ethanol.

Table I. Sample description for mini tensile tests of Mg-1.7Y-1.2Zn sheet.

Sample ID	Material Condition	Test Direction
As-rolled_RD	As-rolled (450°C)	Rolling
As-rolled_TD	As-rolled (450°C)	Transverse
Aged_RD	Rolled and aged (160°C, 96h)	Rolling
Aged_TD	Rolled and aged (160°C, 96h)	Transverse
FSP_RD	Rolled and FSP*	Rolling
FSP_TD	Rolled and FSP*	Transverse
FSP_aged_RD	Rolled, FSP* and Aged	Rolling
FSP_aged_TD	Rolled, FSP* and Aged	Transverse

* FSP was conducted in the rolling direction

Results & Discussion

Microstructure

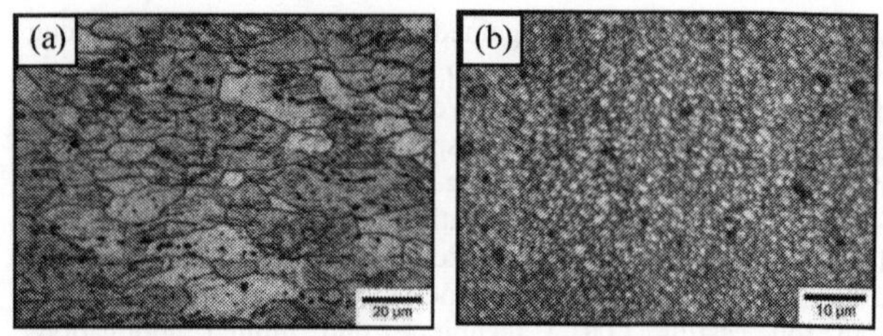

Fig. 2. Optical micrographs (longitudinal section) of Mg-1.7Y-1.2Zn alloy sheet, a) as rolled and b) after FSP. FSP processed micrograph was taken from nugget region showing substantial grain refinement

The average grain size of as-cast Mg-1.7Y-1.2Zn alloy was measured as ~730 μm. Figure 2 shows optical micrographs of the alloy sheet in the as-rolled and FSP conditions. Micrographs were taken in longitudinal section of the sheet. The average grain size of as-rolled sheet was determined as ~20 μm (Fig. 2a), which was refined close to 1 μm after FSP as observed in Fig. 2b. The microstructure of the as-rolled sheet showed a mix of coarse and fine grains with numerous coarse grains having twinned structure, suggesting that recrystallization was not fully complete after rolling. A few elongated bands of coarse grains were observed in the FSP sheet. Frictional heating and extensive plastic deformation during the FSP resulted in significant grain refinement due to dynamic recrystallization, which again appeared to have completed partially. No significant change in the grain size after aging was observed for both the as-rolled and FSP sheets; however, in the case of aged FSP sheet, bands of elongated coarse grains were slightly broken up due to the progress of static recrystallization during ageing.

Mechanical properties and textural effects upon FSP

Fig. 3 shows engineering stress-strain curves of the as-rolled and FSP treated sheet in rolling and transverse directions. The tensile properties, such as, yield strength (YS), ultimate tensile strength (UTS) and total elongation (%) determined from these curves are shown in Table II. For the as-rolled sheet, yield stress in rolling direction is 247 MPa, which is higher compared to 179 MPa in the transverse direction. The difference in UTS in the two directions is low and total elongation is similar (~25%). Several twins in the as-rolled microstructure were observed suggesting partial completion of dynamic recrystallization and significant deformation texture is expected due to the low Ce content (~0.013%) in the alloy. An inversion in the mechanical properties and considerable anisotropy is observed in the FSP treated sheet. Yield strength of 178 MPa in the rolling direction was measured, which is substantially lower than 306 MPa observed in the transverse direction. Correspondingly, UTS in the processing direction is very low, but, interestingly, even the elongation in this direction is poor (2.9%) compared to 24.5% in the transverse direction. FSP treated sheet therefore delivered high YS, high UTS and high elongation in transverse direction compared to the processing direction.

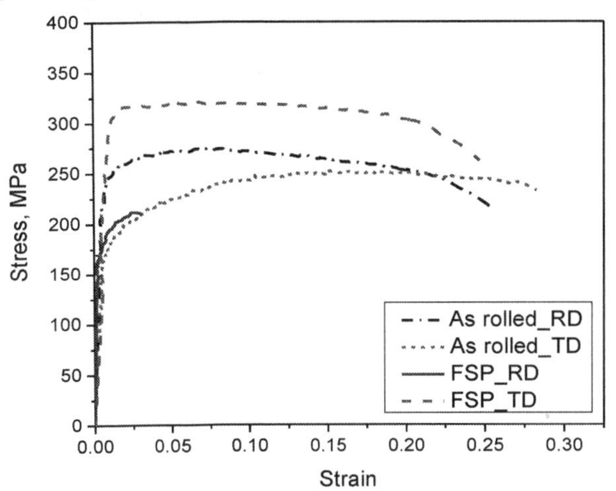

Fig. 3. Effect of FSP and test direction on the stress-strain curves of the rolled Mg-1.7Y-1.2Zn sheet

The observed mechanical behavior is due to strong textural effects produced during FSP. Earlier investigations have suggested that shear plastic flow during FSP resulted in a texture in which {0002} basal planes roughly surrounding the rotating pin surface within the nugget region leading to an "onion ring" structure [20,21]. The basal slip was presumed to be the dominant slip system within the nugget region for the shear plastic flow due to large difference in the critical resolved shear stress (CRSS) of basal and nonbasal slip that produced strong textural effects after FSP. A wide variation in texture was observed in regions within and away from the stir zone, *i.e.*, transition and base material regions, which correlated well with significant anisotropy in the mechanical behavior in different directions. Schuddin *et al.* [22] have reported alignment of {0002} basal planes with the tool shoulder surface in the upper stir zone. A similar texture response is anticipated in the present study. The low yield stress of 178 MPa observed in the processing direction is due to easier basal activity as a result of preferential alignment of basal planes. In transverse direction, a high yield stress of 306 MPa suggests predominant non-basal activity, since CRSS for non basal slip systems is about 2 to 6 times higher for polycrystalline Mg alloys [23]. Woo *et al.* [21] have reported early fracture in transverse direction for FSP AZ31B rolled plate due to plastic incompatibility created by the orientation difference between the stir zone and the transition zone. However, the observation of high yield strength coupled with high ductility in transverse direction in the present study was a result of sample extraction within the nugget region or stir zone. The large variation in yield stress in different directions in spite of similar grain size (~1 μm) reveals that Hall-Patch relation is significantly influenced by strong textural effects produced during FSP.

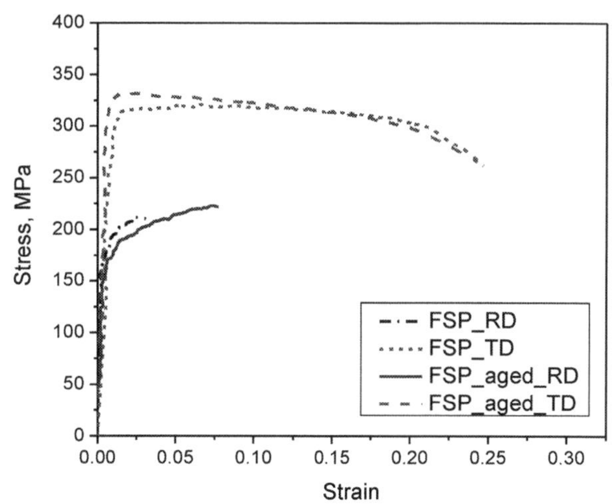

Fig. 4. Effect of ageing and test direction on the stress-strain curves of the FSP Mg-1.7Y-1.2Zn sheet

Table II. Mechanical properties of as-rolled, FSP and aged sheet.

Sample ID	Tensile Properties		
	Yield Strength (MPa)	UTS (MPa)	Total Elongation (%)
As-rolled_RD	247	274	25.1
As-rolled_TD	179	245	27.4
Aged_RD	240	266	20.4
Aged_TD	189	249	26.5
FSP_RD	178	212	2.9
FSP_TD	306	320	24.5
FSP_aged_RD	172	222	7.2
FSP_aged_TD	324	333	24.1

Effect of ageing on the mechanical behavior

Fig. 4 shows effect of ageing and test direction on the stress-strain behavior of FSP sheet. The compiled mechanical properties are shown in Table II, which suggest that ageing has marginal influence on the strength properties of as-rolled sheet; however, FSP treated and subsequently aged sheet showed improvement. Yield strength of 306 MPa in the transverse direction is increased to 324 MPa upon ageing of the FSP sheet; however, ductility is not changed (~25%). On the other hand, in the processing direction, yield strength is nearly unaffected, but, ductility is improved from 2.9% to 7.2% upon ageing. Fig. 5 is a graphical comparison showing effect of ageing and test direction on the UTS/YS ratio for the as-rolled and FSP sheets. In spite of high yield strength in the rolling direction (Table II), the UTS/YS ratio is low (~1.1) compared to that in the transverse direction (~1.35), both for as-rolled as well as aged sheets (Fig. 5a). Therefore, the as-rolled sheet possesses high strain hardening capacity in the transverse direction, which is nearly unaffected upon ageing. The total elongation for all four samples (*i.e.*, As-rolled_RD, As-rolled_TD, Aged_RD and Aged_TD) are comparable as observed from the figure. Mechanical properties after FSP and subsequent ageing treatment show large variation in the two test directions as shown in Fig. 5b. The UTS/YS ratio is little higher in the processing direction (~1.2) as compared to the transverse direction (~1.05) after FSP. Upon ageing, this ratio is nearly unchanged in the transverse direction, but, improved to ~1.3 in the processing direction). The total elongation values have a large variation in the two directions both for the FSP as well as FSP-aged sheet.

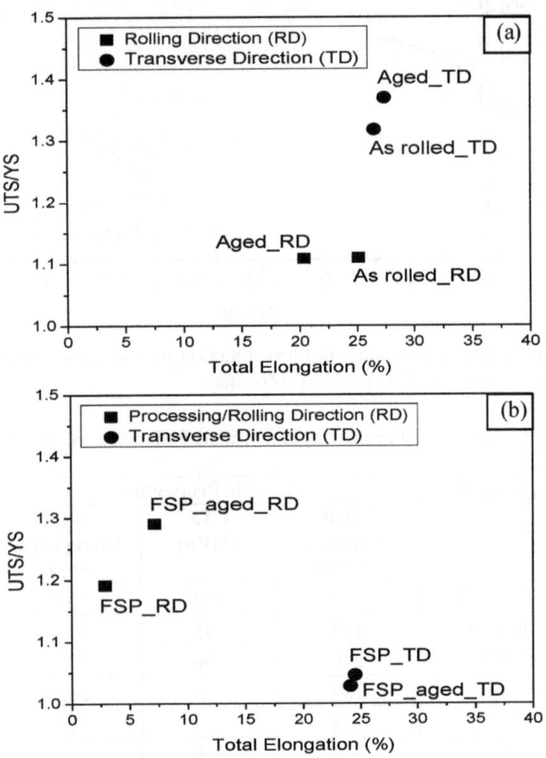

Fig. 5. Effect of ageing and test direction on UTS/YS ratio v/s total elongation for a) as rolled sheet and b) FSP sheet

FSP is carried out at strain rates of the order of 10^0 to 10^2 s^{-1} and up to strain of 40 depending on the process parameters. The process results in substantial grain refinement due to dynamic recrystallization, however, the final microstructure has deformation induced texture. FSP causes frictional heating and extensive mixing of material within the stir zone, hence, dissolution of existing phases takes place in the matrix. FSP leads to variation of texture in the stirred zone of Mg alloys but generally the grain size is relatively uniform. Ageing imposes two effects on microstructure in the stir zone, i) static recrystallization takes place that could lead to further refinement in the microstructure, at least in the elongated grained regions, since the low angle boundaries within these grains may change to high angle grain boundaries under the driving force of stored energy due to high dislocation density, and ii) precipitation of intermetallics phases occurs. Both of these effects favor improved yield strength in the aged material, whereas, grain refinement causes improvement in ductility. Extensive ductility observed in this study in the transverse direction for FSP and FSP-aged sheet is due to combination of microstructural refinement and texture that favors nonbasal slip. However, the reason for poor ductility in the processing direction is not clear. Detailed textural investigations are required to understand the anisotropic mechanical behavior observed in this work.

Stereo micrographs and correlation with the mechanical behavior

Fig. 6 shows stereo macrographs of fractured mini tensile specimens of as-rolled and FSP treated sheet tested in the rolling/processing and transverse directions. The right side-bottom portion of each tensile specimen (e.g., part D in Fig. 6a) was tilted 90° upright to observe the fracture response from the side plane. Thickness and width of the specimens were measured at necked region and compared with original dimensions so as to determine overall geometrical changes at fracture, such as, % reduction in width, thickness and area. The plane of initiation of fracture was identified in each case. The angle between the fracture plane and the tensile loading axis was also determined. The results are shown in Table III for different test specimens. Figs. 6a and b show that shear fracture in as-rolled specimens initiated from the top or bottom plane. The as-rolled specimen tested in rolling direction shows uniform reductions of ~16% in thickness as well as width accompanied with ~30% reduction in area. The shear fracture angle was found as 57°. Similar fracture response is obtained in transverse direction with slightly lower reductions in width, thickness as well as area. Shear fracture occurred at an angle of 52°. The lower values of geometrical changes in the transverse direction are due to higher strain hardening capability and lower post uniform elongation as observed in Fig. 3. For an isotropic ultrafine grained copper, Zhao *et al.* [24] reported nearly uniform reduction in thickness and width for tensile specimen geometry having width and thickness ratio (*W/T*) close to unity. Such geometry causes shear deformation both at the gage top or bottom plane as well as side planes resulting in homogenous development of shear bands. In the present study, *W/T* ratio was close to 1.5, which produced similar geometrical changes in width and thickness for the as-rolled tensile specimens, although, slightly lower values were observed in transverse direction compared to rolling direction, which was due to the presence of deformation texture after hot rolling and complete recrystallization could not occur.

Table III. Geometrical parameters of fractured tensile specimens of as rolled, FSP treated and aged sheet

Sample ID	Geometrical changes in fractured tensile specimen at necked region			Fracture initiation plane	Angle between loading axis and fracture plane (°)
	Reduction in width (%)	Reduction in thickness (%)	Area reduction (%)		
As rolled_RD	16.7	15.6	29.7	Top plane	57
As rolled_TD	13.6	10.5	22.3	Top Plane	52
Aged_RD	14.2	15.7	27.6	Top Plane	55
Aged_TD	13.4	14.2	29.1	Top Plane	53
FSP_RD	0.9	5.4	6.3	Top Plane	31
FSP_TD	25.9	11.1	34.1	Side Plane	58
FSP_aged_RD	3.3	7.3	10.4	Top Plane	30
FSP_aged_TD	30.2	8.3	35.9	Side Plane	60

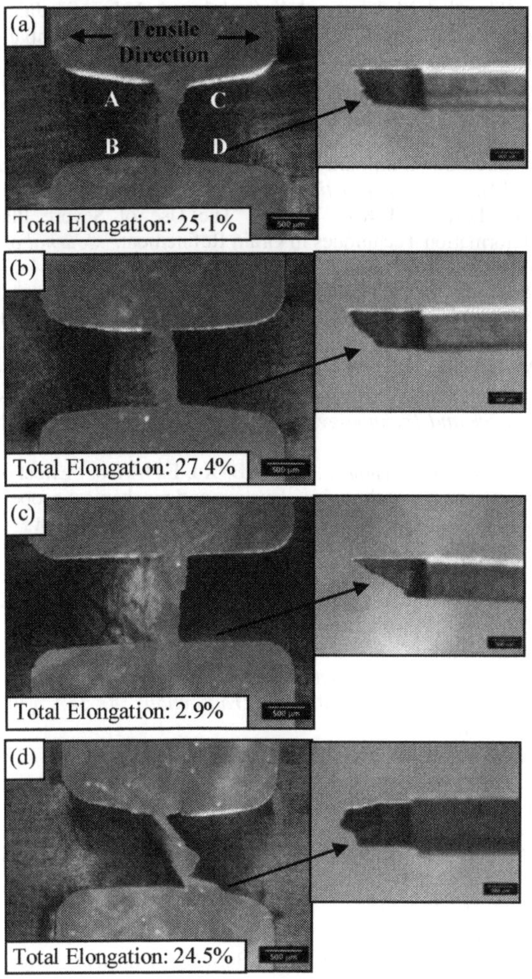

Fig. 6 Stereo macrographs of fractured tensile specimens as per sample ID shown in Table I, a) As rolled_RD, b) As rolled_TD, c) FSP_RD and d) FSP_TD. Macrographs on the right side show side plane of the fractures samples, which were taken by tilting right side bottom portion (e.g., part D shown in Fig. a) upright 90°. Fig. a, b and c show that fracture initiated from top plane, whereas, Fig. d shows that fracture initiated from side plane.

Figs. 6 c and d show that FSP treatment of rolled sheet greatly affected fracture response of tensile specimens tested in the processing and the transverse directions. In the processing direction, the FSP treated sheet showed fracture initiation from the top or bottom plane of tensile specimen with an acute shear failure angle of 31° (shown in Table III), whereas, in the transverse direction, failure is initiated from the side plane at a shear angle of 58°. It is interesting to observe that in the processing direction, reduction in width is mere less than 1%, whereas, thickness reduced by 5.4% accompanied with an area reduction of 6.3%. In the transverse direction, the geometrical changes in width and thickness are very high, but, reversed compared to the processing direction. The reduction in width and thickness is 25.9% and 11.1%, respectively, whereas, area reduction is 34.1%, which is substantially high. Textural anisotropy has played a significant role in the observed mechanical behavior. The results correlate well with the % elongation results shown in Table II. In the processing direction, the poor elongation of 2.9% suggests that shear bands could not develop well in the gage section, in spite of easy basal slip that caused yielding at a lower stress 178 MPa. In the transverse direction, the yield stress as well as total elongation both were very high and confirm with intense shear bands developed within the gage section, which is visible in Fig. 6d. The fracture localization occurred at side plane instead of top plane (see inset in Fig. 6d showing location of fracture initiation) due to higher reduction in the width, which led to the shear failure at an angle of 58°.

The effect of ageing of FSP treated sheet on the geometrical changes during tensile tests is shown in Table III. In the transverse direction, there is no appreciable change in the geometrical parameters of FSP-aged specimen compared to FSP alone and total reduction in area is similar in both cases, *i.e.*, ~35%, which is consistent with total elongation value of ~25% with or without ageing of the FSP sheet. In the processing direction, however, a noticeable change is observed upon ageing. Reduction in width and thickness of FSP_aged_RD specimen was measured as 3.3% and 7.3%, respectively, which is an improvement over the values of 0.9% and 5.4%, observed for the FSP_RD specimen. The total elongation of 7.2% (Table II) and area reduction of 10.4% (Table III), both are higher for the FSP_aged_RD condition compared to FSP_RD specimen that delivered respective values of 2.9% and 6.3%.

Conclusions

Hot rolled sheet of an Mg-1.7Y-1.2Zn alloy was processed by FSP and aged in as-rolled and FSP conditions. As-rolled sheet showed some anisotropy in the mechanical behavior in rolling and transverse directions that was marginally reduced after ageing. FSP resulted in substantial refinement in grain size, however, large difference in mechanical properties was observed in the two directions. The substantial anisotropic behavior was due to strong textural effects. The mechanical properties were found improved and anisotropy was reduced upon ageing. Small sized tool was employed for FSP in this study. Advances in FSP, such as, feasibility of using large dimension tools, multi-pass FSP and robotic controlled processing have potential scope in implementing FSP for the large-sized automotive body panels. Further optimization studies on the FSP and ageing parameters are necessary to achieve the said goal.

Acknowledgements

The first author, V. Jain, gratefully acknowledges J. William Fulbright Foreign Scholarship Board (FSB), the US Department of State's Bureau of Educational & Cultural Affairs (ECA), the Government of India, the Institute of International Education (IIE) and United States-India Educational Foundation (USIEF) for the award of *"Fulbright Nehru Doctoral & Professional Research Fellowship"* under the grant No. 1369/DPR/2010-2011 that enabled him to take up the present work at CFSP, Missouri S&T. Mr. Jain is a Graduate student at Deptt of Materials & Metallurgical Engg, IITK-Kanpur (India) and he is thankful to his PhD supervisors, Dr. Gouthama (Assoc. Professor, IITK-Kanpur) and Dr. Anil K. Gupta (Director, AMPRI, CSIR, Bhopal & Ex-Head Division of Engg. Materials, NPL, CSIR, New Delhi) for their constant encouragement and valuable suggestions during this work.

References

1. P. Krajewski, A. Sachdev, A.A. Luo, J. Carsley and J. Schroth, "Automotive aluminum and magnesium: Innovations and opportunities," *Light Metal Age*, 67 (5) (2009), 6-13.
2. A. I. Taub, P. E. Krajewski, A.A. Luo, and J. N. Owens, "The Evolution of Technology for Materials Processing over the Last 50 Years: The Automotive Example," *Journal of the Minerals, Metals and Materials Society*, 59 (2) (2007), 48-57.
3. C. Blawert, N. Hort and K.U. Kainer, "Automotive applications of magnesium and its alloys," *Transactions of Indian Institute of Metals*, 57 (4) 2004, 397-408.
4. R. Verma and J.T. Carter, "Quick Plastic Forming of a Decklid Inner Panel with Commercial AZ31 Magnesium Sheet," SAE 2006-01-0525, Society of Automotive Engineers, Inc., Warrendale, Pa.
5. E. Essadiqi, "Magnesium Sheet Technology Perspectives," *Journal of the Minerals, Metals and Materials Society*, 61 (8) (2009), 13.
6. K. Hono, C.L. Mendis, T.T. Sasaki and K. Oh-ishi, "Towards the development of heat-treatable high-strength wrought Mg alloys," *Scripta Materialia*, 63 (2010), 710–715.
7. J. Koike, T. Kobayashi, T. Mukai, H. Watanabe, M. Suzuki, K. Maruyama and K. Higashi, "The activity of non-basal slip systems and dynamic recovery at room temperature in fine-grained AZ31B magnesium alloys," *Acta Materialia*, 51 (2003), 2055-2065.
8. P. Menzen, in: A. Beck (Ed.), *The technology of magnesium and its alloys* (F.A Hughes and Co. Ltd., London, 1940), 34.
9. M.R. Barnett, M.D. Nave and C.J. Bettles, "Deformation microstructures and textures of some cold rolled Mg alloys," *Materials Science and Engineering A*, 386 (2004), 205–211.
10. R. K. Mishra, A. K. Gupta, P. Rama Rao, A. K. Sachdev, A. M. Kumar and A. A. Luo, "Influence of cerium on the texture and ductility of magnesium extrusions," *Scripta Materialia*, 59 (2008), 562–565.
11. A.K. Gupta, R. Sikand, V. Jain, A.K. Sachdev and A.M. Kumar, "Advanced Magnesium Extrusion Alloys" (Project Completion report, Project No. CNP040232, NPL-General Motors Internal Report, 2007).
12. K. Hantzsche, J. Wendt, K.U. Kainer, J. Bohlen, and D. Letzig, "Mg Sheet: the effect of Process Parameters and alloy Composition on texture and mechanical Properties," *Journal of the Minerals, Metals and Materials Society*, 61 (8) (2009), 38-42.
13. R. Sikand, A.M. Kumar, A.K. Sachdev, A.A. Luo, V. Jain and A.K. Gupta, "AM30 porthole die extrusions-A comparison with circular seamless extruded tubes," *Journal of Materials Processing Technology*, 209 (2009), 6010–6020.
14. A.A. Luo, W. Wu, R.K Mishra, L. Jin, A.K. Sachdev and W. Ding, "Microstructure and Mechanical Properties of Extruded Magnesium-Aluminum-Cerium Alloy Tubes," *Metallurgical and Materials Transactions A*, 41A (2010), 2662-2674.
15. T.C. Lowe and R.Z. Valiev, "The Use of Severe Plastic Deformation Techniques in Grain Refinement," *Journal of the Minerals, Metals and Materials Soc.*, 56 (10) (2004), 64-77.
16. R.S. Mishra and Z.Y. Ma, "Friction Stir Welding and Processing," *Materials Science and Engineering R*, 50 (2005) 1-78.
17. A. Singh, M. Watanabe, A. Kato and A.P. Tsai, "Strengthening in magnesium alloys by icosahedral phase," *Science and Technology of Advanced Materials*, 6 (2005) 895-901.
18. D.K. Xu, W.N. Tang, L. Liu, Y.B. Xu, E.H. Han, "Effect of Y concentration on the microstructure and mechanical properties of as-cast Mg-Zn-Y-Zr alloys," *Journal of Alloys and Compounds*, 432 (2007) 129-134.
19. J. Y. Lee, D.H. Kim, H.K. Lim, D.H. Kim, "Effects of Zn/Y ratio on the microstructure and mechanical properties of Mg-Zn-Y alloys," *Materials Letters*, 59 (2005) 3801-3805.
20. S.H.C. Park, Y. S. Sato and H. Kokawa, "Basal plane texture and flow pattern in friction stir-weld of a magnesium alloy," *Metallurgical and Materials Transactions A*, 34A (2003), 987-994.
21. W. Woo, H. Choo, D.W. Brown, P.K. Liaw and Z. Feng, "Texture variation and its influence on the tensile behavior of a friction-stir processed magnesium alloy," *Scripta Materialia*, 54 (2006) 1859-1864.
22. U.F.H.R. Suhuddin, S. Mironov, Y.S. Sato, H. Kokawa and C.W. Lee, "Grain structure evolution during friction-stir welding of AZ31 magnesium alloy," *Acta Materialia*, 17 (2009), 5406-5418.
23. W.B. Hutchinson, and M.R. Barnett, "Effective values of critical resolved shear stress for slip in polycrystalline magnesium and other hcp metals," *Scripta Materialia*, 63 (2010), 737–740.
24. Y.H. Zhao, Y.Z. Guo, Q. Wei, A.M. Dangelewicz, C. Xu, Y.T. Zhu, T.G. Langdon, Y.Z. Zhou and E.J. Lavernia, "Influence of specimen dimensions on the tensile behavior of ultrafine-grained Cu," *Scripta Materialia*, 59 (2008), 627–630.

Magnesium Technology 2011
Edited by: Wim H. Sillekens, Sean R. Agnew, Neale R. Neelameggham, and Suveen N. Mathaudhu
TMS (The Minerals, Metals & Materials Society), 2011

The microstructure and mechanical properties of cast Mg-5Sn based alloys

M. Keyvani, R. Mahmudi, G. Nayyeri
School of Metallurgical and Materials Engineering, University of Tehran, Tehran, Iran

Keywords: Mg-5Sn alloy; microstructure; creep resistance; mechanical properties.

Abstract

The effects of 1 wt.% Ce-rich misch-metal, 3 wt.% Bi, 0.4 wt.% Sb, and 2 wt.% Ca additions on the microstructure, mechanical properties, and creep resistance of a cast Mg-5Sn alloy were studied. Impression creep tests were carried out at 150 and 225 °C under constant stresses of 300 and 175 MPa. The results showed that creep resistance, hardness, and shear strength of the base alloy were significantly enhanced with the addition of ternary alloying elements. This was attributed to the concurrent formation of Mg_2Sn particles with the more thermally stable CaMgSn, Mg_3Sb_2, Mg_3Bi_2 and MgSnRE phases which strengthen both matrix and grain boundaries during deformation in the investigated system. Among all tested materials, the Mg-5Sn-2Ca alloy indicated the highest mechanical properties and creep resistance due to the refined microstructure as well as the type and volume fraction of the intermetallic phases present in this alloy.

Introduction

In recent years, research and development of magnesium alloys have been greatly promoted by the lightweight requirement in the automotive industry [1]. Despite their good combination of low density and high strength-to-density ratio, Mg alloys suffer from low mechanical properties at elevated temperatures [2]. Therefore, attempts have been made to develop new magnesium alloys having improved structural stability at the elevated temperatures. Among many possibilities, the Mg-Sn alloy system appears to be promising for high temperature applications. This is attributed to the formation of the thermally stable Mg_2Sn phase, which is mainly distributed along the grain boundaries in the as-cast condition [3,4].

The study of the microstructure, tensile properties, and creep resistance of cast Mg-Sn alloys by Liu et al. [5] has shown that the Mg-5Sn alloy exhibits the best combination of tensile properties at ambient temperature and a reasonable creep resistance at 150 °C. The compression [6] and impression [7] creep study of the Mg-5Sn alloy has indicated that the dispersive distribution of Mg_2Sn precipitates is the main cause of the enhanced creep resistance of the material. To further enhance the mechanical and creep properties of the Mg-Sn alloys, various alloying elements have been used. Nayyeri and Mahmudi reported that additions of Sb and Ca into the Mg-5Sn alloy significantly improved the creep resistance [8,9], hot hardness [10], and high temperature shear strength of the material [11]. Similar strength improvement has been observed after adding La [12], Di [13], and Bi [10] to the Mg-Sn alloys. The achieved enhancement in the strength and creep resistance has been attributed to the respective formation of the thermally stable Mg_3Sb_2, CaMgSn, MgSnRE, and Mg_3Bi_2 intermetallic particles, which strengthen both matrix and grain boundaries and oppose recovery and recrystallization processes during creep and deformation at elevated temperatures.

The aim of this study is to examine the effects of different alloying elements, such as RE, Bi, Sb, and Ca on the microstructure, creep properties, hardness, and shear strength of the Mg-5Sn alloy in the as-cast condition. The evaluation of strength is made by using the shear punch testing (SPT) technique, which has been attempted previously for evaluating the strength of some cast Mg alloys [14,15]. This is an efficient method that is capable of producing strength data which are well correlated with those found by the conventional tensile tests [16].

Experimental procedure

The five alloys designed for this investigation were based on Mg-5 wt.% Sn with 1%RE, 3%Bi, 0.4%Sb, and 2%Ca additions. High purity Mg, Sn, Bi, Sb, Ca, and a misch-metal composed of 50.6%Ce, 23.2%La, 20.3%Nd, and 5.9% Pr, were melted in an electrical furnace under a protective flux. The melts were held at 750 °C for 30 min and then poured into a steel die, preheated to 150 °C. To examine the microstructure of the materials, the specimens were studied by optical and scanning electron microscopy (SEM). For microstructural analysis, the polished samples were etched using a 5% nitric acid and 95% ethyl alcohol solution at room temperature. X-ray diffraction (XRD) analysis was carried out to identify the phases.

4-mm thick slices were tested in an impression tester. The details of impression testing arrangement are explained elsewhere [17] and will only be briefly described here. A SANTAM universal tensile testing machine equipped with a three-zone split furnace was used to perform constant-load impression tests. A flat ended cylindrical punch of 2 mm diameter was mounted in a holder which was positioned in the center of the vertical loading bar. The specimen was located on an anvil below the loading bar and the assembly of specimen and indenter was accommodated by the split furnace. Impression creep measurements were made at 150 and 225 °C and under respective punch stresses of 300 and 175 MPa. After application of the load, the impression depth was measured automatically as a function of time by the machine and the data were acquired by a computer.

The hardness of the alloys was measured in a Vickers hardness tester with an indenter load of 100 N and a loading time of 30 s. Each reading was an average of at least five separate measurements taken at random places on the surface of the specimens. The strength of the materials was assessed by the shear punch testing. The 1-mm thick slices were ground to a thickness of 0.6 mm, from which disks of 10 mm in diameter were punched for the SPT. A shear punch fixture with a 3.175 mm diameter flat cylindrical punch and 3.225 mm diameter receiving-hole was used for this experiment. Shear punch tests were performed at room temperature using a screw driven SANTAM material testing system with a load cell of 20 kN capacity and a

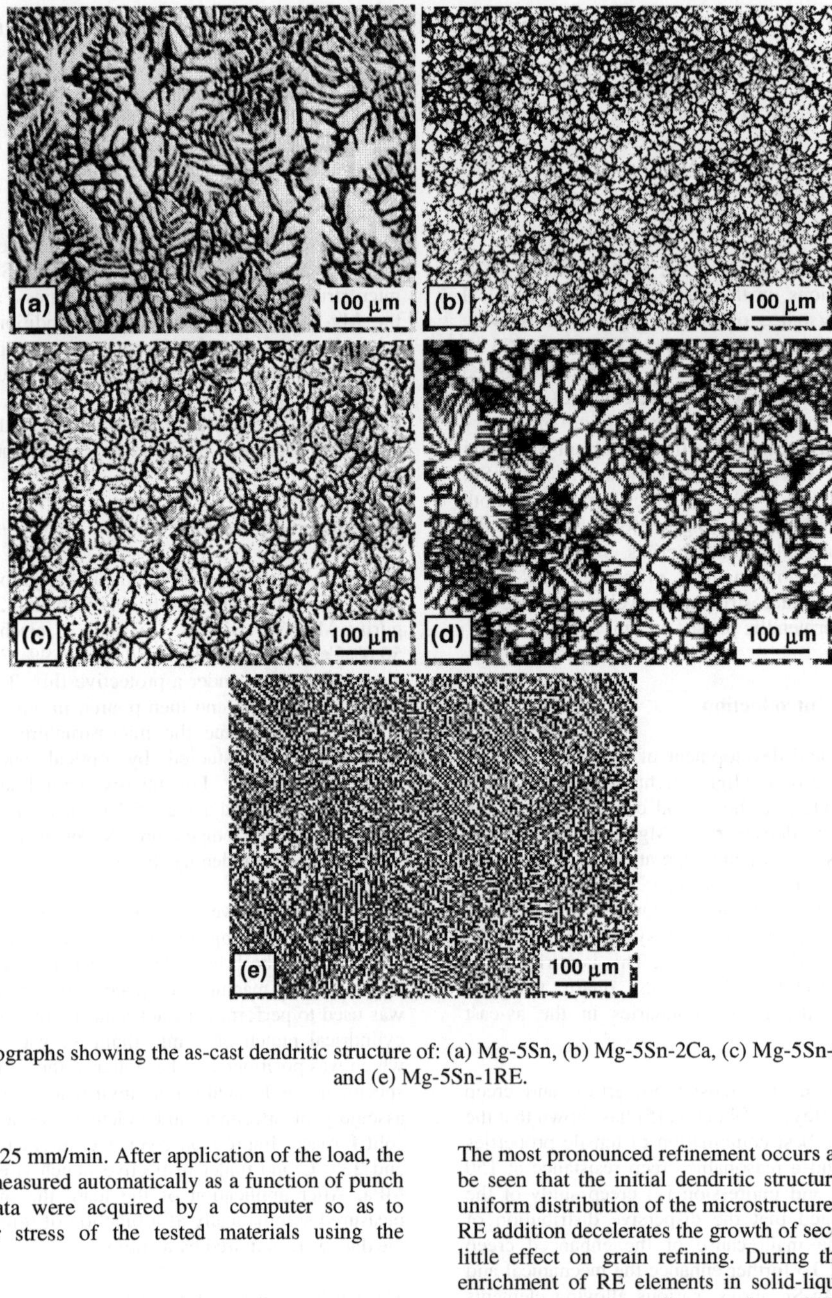

Fig. 1. Optical micrographs showing the as-cast dendritic structure of: (a) Mg-5Sn, (b) Mg-5Sn-2Ca, (c) Mg-5Sn-0.4Sb, (d) Mg-5Sn-3Bi, and (e) Mg-5Sn-1RE.

crosshead speed of 0.25 mm/min. After application of the load, the applied load P was measured automatically as a function of punch displacement; the data were acquired by a computer so as to determine the shear stress of the tested materials using the relationship:

$$\tau = \frac{P}{\pi d t} \tag{1}$$

where P is the punch load, t is the specimen thickness and d is the average of the punch and die diameters. Three different samples were tested for each condition and it was observed that the variation in the measured ultimate shear strength values was small.

Results and discussion

The optical micrographs of the investigated alloys are depicted in Fig. 1. It can be observed that after addition of Bi, Sb, and Ca the coarse dendritic structure of the Mg-5Sn alloy is generally refined. The most pronounced refinement occurs at 2 wt.% Ca where it can be seen that the initial dendritic structure is broken and a rather uniform distribution of the microstructure constituents is achieved. RE addition decelerates the growth of secondary dendrites, but has little effect on grain refining. During the solidification process, enrichment of RE elements in solid-liquid interface induces the constitution undercooling at the solidification interface front, decreasing the secondary dendrites' growth.

The XRD patterns exhibited in Fig. 2 indicate that α-Mg and Mg_2Sn are the only constituents of the Mg-5Sn alloy. In the Bi-, Sb-, and Ca-bearing materials, additional diffraction peaks emerged, corresponding respectively to the Mg_3Bi_2, Mg_3Sb_2, and CaMgSn intermetallic compounds. For the RE-containing material, some new peaks were observed near 24.5, 42, and 49.5°. However, these diffraction peaks could not be identified by the XRD database. It is worth noting that the intensity of the Mg_2Sn peaks decreases in the ternary alloys.

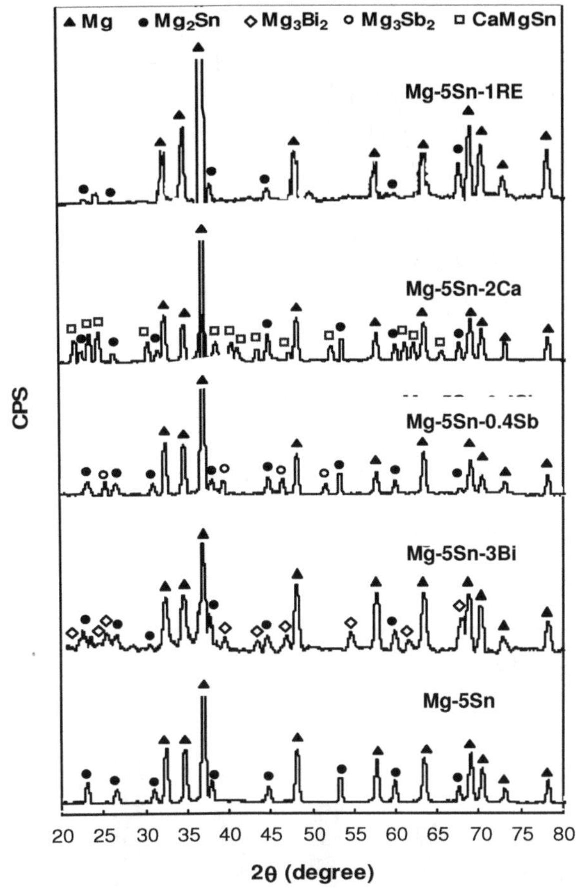

Fig. 2. XRD pattern of the investigated alloys.

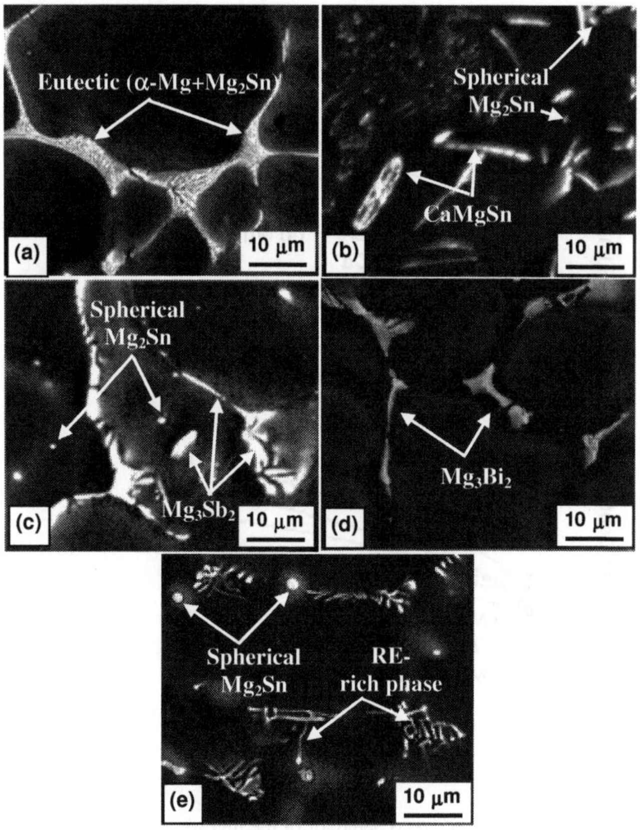

Fig. 3. SEM micrographs showing the distribution of the second phase particles in the α-Mg matrix for: (a) Mg-5Sn, (b) Mg-5Sn -2Ca, (c) Mg-5Sn-0.4Sb, (d) Mg-5Sn-3Bi, and (e) Mg-5Sn-1RE.

For a clearer view of the microstructure and phases present in the magnesium matrix, SEM images of all tested materials are shown in Fig. 3. The microstructure of the Mg-5Sn alloy is composed of the primary α-Mg matrix and the lamellar eutectic structure (α-Mg + Mg_2Sn) at the grain boundaries (Fig. 3a). In contrast to the microstructure of the base alloy, the α-Mg dendrites are refined and the intermetallic Mg_2Sn phase is present in the form of spherical particles in ternary alloys. In addition, a new rod-shaped phase can be detected mainly at the grain boundaries in Bi-, Sb-, and Ca-containing alloys. According to the XRD analysis, these intermetallic compounds should be Mg_3Bi_2, Mg_3Sb_2, and CaMgSn, respectively. These particles may have formed prior to the solidification of the Mg matrix, acting as powerful nucleation sites, the consequence of which is a refined microstructure [18]. In the Mg-5Sn-1RE alloy, however, some branch-like particles can be seen around the grain boundaries. These particles should be a RE-rich phase, the identification of which was not possible in the XRD pattern of the alloy.

Fig. 4 shows a comparison of impression creep curves under punching stresses of 300 MPa at 150 °C and 175 MPa at 225 °C. As can be seen, after a short primary creep stage almost all of the curves exhibit a relatively long steady-state region where depth increases linearly with time. Furthermore, it can be noticed from the level and slope of the curves that different materials demonstrate different creep behavior under both stress levels. According to this figure, third element addition into the Mg-5Sn alloy generally improves the creep resistance of the material, the effect which becomes more pronounced after Ca addition.

Fig. 5 exhibits the SPT data, presented as the variation of shear stress with normalized displacement obtained for all five alloys. It can be observed that, similar to conventional tensile stress-strain curves, after a linear elastic behavior the curves deviate from linearity before they reach a maximum point. The deviation point is taken as the shear yield stress (SYS) and the stress corresponding to the maximum point is referred to as the ultimate shear strength (USS). It is evident that all ternary alloys have higher USS values than the binary base alloy. In accordance with the creep results, the highest strength belongs to the Ca-containing alloy. To have a better comparison of the effect of alloying elements on the strength of the base material, the USS and SYS data are summarized in Fig. 6. The results indicate that addition of all alloying elements improves both shear and ultimate strength of the base Mg-5Sn alloy. Addition of Ca has the most pronounced effect, while RE addition exhibits the weakest effect. A similar trend is observed for the hardness of the alloys shown in Fig. 7.

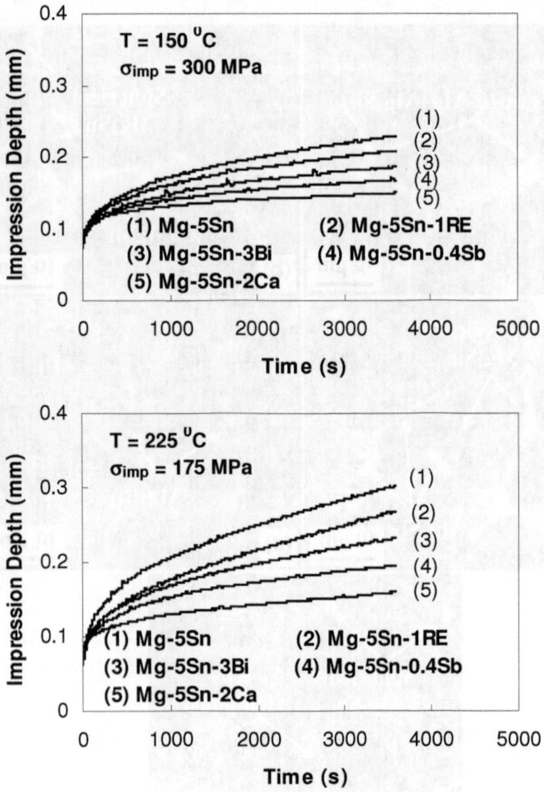

Fig. 4. Comparison of creep behavior of the investigated alloys at: a) $T = 150$ °C, $\sigma_{imp} = 300$ MPa, and b) $T = 225$ °C, $\sigma_{imp} = 175$ MPa.

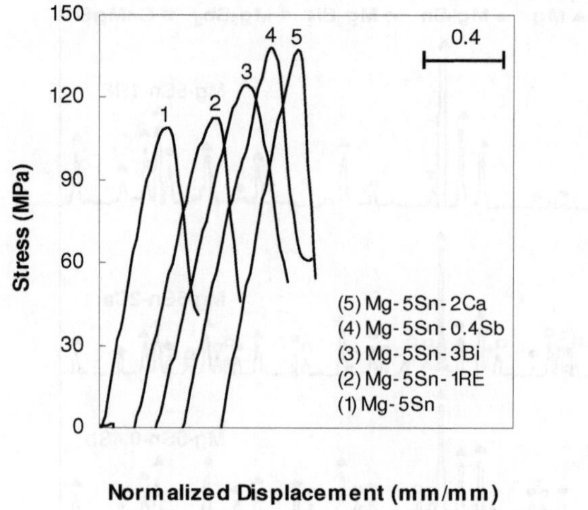

Fig. 5. Shear stress against normalized displacement for the tested materials.

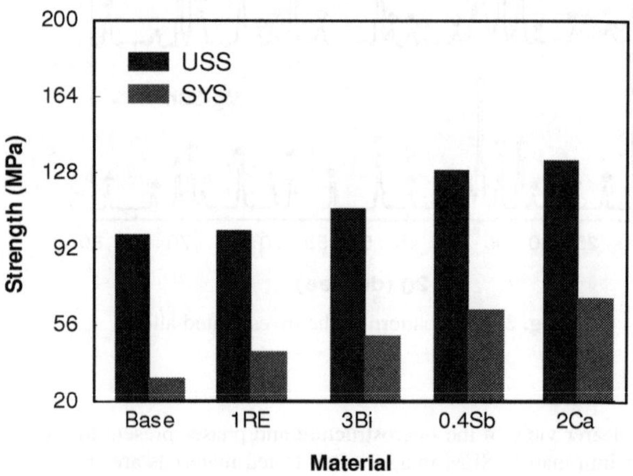

Fig. 6. Variation of USS and SYS values at room temperature for different materials in the as-cast conditions.

The highest hardness belongs to the alloys containing Ca, followed by Sb, Bi, and RE.

The enhancement in creep resistance, shear strength, and hardness of the ternary alloys respect to the Mg-5Sn base alloy can be ascribed to their microstructural features. Introduction of alloying elements into the as-cast Mg-5Sn alloy leads to the formation of thermally stable phases and changes the morphology of Mg_2Sn particles. The respective melting points of Mg_3Bi_2, Mg_3Sb_2, and CaMgSn intermetallic compounds are 823, 1228, and 1184 °C, which are all higher than 771 °C for the Mg_2Sn phase. In addition, as shown in Fig. 3, the second phase particles lie in the interior parts of the dendrites as well as at the grain boundaries. Therefore, they can improve the strength of the grains, by obstructing the dislocation annihilation and thus restricting the slip inside the grains. On the other hand, presence of thermally stable phases at the grain boundaries results in inhibiting the grain boundary migration or grain boundary sliding during high-temperature plastic deformation [13]. The superior creep resistance, hardness, and shear strength of the Mg-5Sn-2Ca alloy is attributed to its refined microstructure and the higher volume fraction of the more thermally stable CaMgSn particles.

4. Conclusions

The microstructural evolution, creep resistance, hardness and room temperature shear strength of Mg-5Sn with additions of RE, Bi, Sb, and Ca were investigated in the as-cast condition. Microstructural analysis indicated that the addition of these elements refined the dendritic microstructure, modified the morphology of Mg_2Sn phase and caused the formation of some rod-shaped precipitates of Mg_3Bi_2, Mg_3Sb_2, and CaMgSn and branch-like RE-rich particles. These particles are all thermally stable phases, which strengthen both grains and grain boundaries and oppose both recovery and recrystallization processes during deformation processes at room and elevated temperatures, decreasing the minimum creep rate and increasing the hardness and strength of the material. The highest creep resistance, hardness, and shear strength of the Mg-5Sn-2Ca alloy is related to

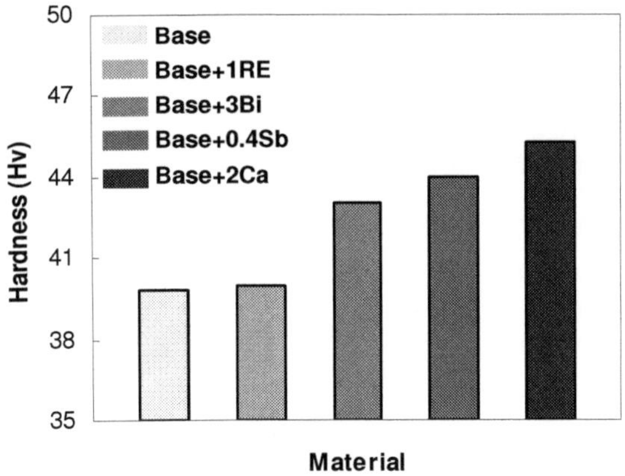

Fig. 7. Variation of hardness at room temperature for different materials in the as-cast conditions.

its refined microstructure and also to the higher volume fraction of the more thermally stable CaMgSn particles.

Reference

[1] B.L. Mordike, *J. Mater. Proc. Tech.*, 2001, 117, 391.

[2] B.L. Mordike, T. Ebert, Mater. Sci. Eng. A302 (2001) 37–45.

[3] N. Hort, Y. Huang, T.A. Leil, P. Maier, K.U. Kainer, Adv. Eng. Mater. 8 (2006) 359–364.

[4] G. Nayyeri, R. Mahmudi, Mater. Sci. Eng. A 527 (2010) 4613–4618.

[5] H. Liu, Y. Chen, Y. Tang, S. Wei, and G. Niu, *J. Alloys Compds.*, 2007, 440, 122.

[6] S. Wei, Y. Chen, Y. Tang, H. Liu, S. Xiao, G. Niu, X. Zhang, and Y. Zhao, *Mater. Sci. Eng. A,* 2008, 492, 20.

[7] G. Nayyeri, and R. Mahmudi, *Mater. Sci. Eng. A*, 2010, 527, 4613.

[8] G. Nayyeri, and R. Mahmudi, *Mater. Sci. Eng. A*, 2010, 527, 669.

[9] G. Nayyeri, and R. Mahmudi, *Mater. Sci. Eng. A*, 2010, 527, 2087.

[10] M. Keyvani, R. Mahmudi, and G. Nayyeri, *Mater. Sci. Eng. A,* doi: 10.1016/j.msea.2010.08.045.

[11] G. Nayyeri, and R. Mahmudi, *Mater. Sci. Eng. A*, 2010, 527, 5353.

[12] S. Wei, Y. Chen, Y. Tang, X. Zhang, M Liu, S. Xiao, and Y. Zhao, *Mater. Sci. Eng.*

[13] H. Liu, Y. Chen, Y. Tang, S. Wei, and G. Niu, *Mater. Sci. Eng. A*, 2007, 464, 124.

[14] R. Alizadeh, and R. Mahmudi, *Mater. Sci. Eng. A*, 2010, 527, 3975.

[15] R. Alizadeh, and R. Mahmudi, *Mater. Sci. Eng. A*, 2010, 527, 5312.

[16] R.K. Guduru, A.V. Nagasekhar, R.O. Scattergood, C.C. Koch, and K.L. Murty, *Metall. Mater. Trans.A,* 2006, 37, 1477.

[17] R. Mahmudi, A.R. Geranmayeh and A. Rezaee-Bazzaz, *Mater. Sci. Eng. A*, 2007, 448, 287.

[18] M. Bamberger, *Mater. Sci. Tech.*, 2001, 17, 15.

Magnesium Technology 2011
Edited by: Wim H. Sillekens, Sean R. Agnew, Neale R. Neelameggham, and Suveen N. Mathaudhu
TMS (The Minerals, Metals & Materials Society), 2011

EFFECT OF COOLING RATE AND CHEMICAL MODIFICATION ON THE TENSILE PROPERTIES OF Mg–5wt% Si ALLOY

Farshid Mirshahi, Mahmood Meratian, Mohsen Mohammadi Zahrani, Ehsan Mohammadi Zahrani

Department of Materials Engineering; Isfahan University of Technology (IUT); Isfahan, 84156-83111, Iran

Keywords: Mg–Si alloy, Cooling rate, Modification, Bismuth, Tensile properties.

Abstract

Hypereutectic Mg–Si alloys are a new class of light materials usable for aerospace and other advanced engineering applications. In this study, the effects of both cooling rate and bismuth modification on the microstructure and tensile properties of hypereutectic Mg–5wt% Si alloy were investigated. It was found that the addition of 0.5% Bi, altered the morphology of primary Mg_2Si particles from bulky to polygonal shape and reduced their mean size from more than 70 μm to about 30 μm. Also, the tensile strength and elongation of the modified alloy increased about 10% and 20%, respectively, which should be ascribed to the modification of Mg_2Si morphology and more uniform distribution of the primary particles. Moreover, an increase in tensile strength value with increase in cooling rate were observed which is attributed to finer microstructure of alloy in higher cooling rates. It was observed that Bi addition is significantly more effective in refining the morphology of primary Mg_2Si particles than applying faster cooling rates.

Introduction

In recent years, the growing demand for more fuel-efficient vehicles to reduce energy consumption and air pollution is a challenge for the automotive industry. Aluminum and magnesium based composites, reinforced with particulates of Mg_2Si have been recently introduced as a new group of particulate metal matrix composites that offer attractive advantages such as low density, good wear resistance and good castability [1-5]. Hypereutectic Mg–Si alloys can be considered as in-situ Mg matrix composites containing hard particles of Mg_2Si [1, 3, 4], since a maximum solid solubility of Si into Mg is only about 0.003 at.%, thus Si atoms react with Mg atoms and are precipitated as an intermetallic compound of Mg_2Si. Despite the fact that Mg_2Si intermetallic particles exhibit a high melting temperature, low density, high hardness, low thermal-expansion coefficient and equilibrium interface [4, 5], their undesirable morphology (size, shape and distribution) in the Mg matrix, during conventional casting, has led to the low ductility and strength observed in these alloys and has limited these alloys for use in high performance applications.

Various efforts have been made to modify the microstructure and improve the mechanical properties of the Mg–Si alloys. Other than expensive methods of rapid solidification and mechanical alloying [2, 5, 6]; high cooling rates and impurity modifications are well known techniques, which have been used so far for modifying the undesirable morphology of Mg_2Si in Al–Mg–Si and Mg–Si alloys. Many of these studies have been focused on the modification of Al–Mg–Si alloys [7, 8], while less works has been done on the modification of Mg_2Si morphology in Mg–Si alloys. So far, several modifying agents such as Ca, P, Sb and Sr, have been used to modify the coarse primary Mg_2Si in Al–Mg_2Si composites [7, 8]. It is noticeable that both P and Sb are in the VA group of the periodic table of the elements. The element Bi belongs to the VA group of the periodic table like P and Sb. Therefore, on the basis of similar physical and chemical properties between elements in the same group of the periodic table, Bi seems to be effective for morphology modification of Mg_2Si phase in Mg–Si alloys.

The objective of the present study is to improve the mechanical properties of a hypereutectic Mg–Si alloy through the modification of morphology of primary Mg_2Si particles by applying high cooling rates and Bi addition.

Experimental

Industrially pure magnesium ingot, with the composition is shown in Table I, and Si powder (99.4wt% purity) were used as starting materials to prepare Mg–5wt.% Si alloy, which corresponds to a composition of Mg–15wt% Mg_2Si in-situ composite. To study the effects of cooling rate on tensile properties, a special casting assembly shown in Figure 1 was used. It was designed in such way that the heat extracted only through the water-cooled system at the bottom, promoting vertical upward directional solidification. A steel mould was used having an internal diameter of 60 mm, height of 150 mm and wall thickness 2 mm. The lower part of the mould was closed with a water-cooling jacket made of stainless steel, with a wall thickness of 1 mm. An electric heater was used to preheat the peripheral side of the mould to about 700 °C, which was suitable for preventing lateral heat transfer and guaranteeing directional solidification from the bottom to the top of the mould.

Table I. Chemical Composition of Magnesium Ingot Used in the Present Work.

Elements	Content (wt.%)
Mn	0.144
Al	0.127
Zn	0.126
Si	0.045
Cu	0.005
Mg	Balance

The charge was melted in an electric resistance furnace using a steel crucible under argon protective atmosphere. After stirring, the melt was degassed using pure dry argon, injected into the melt by means of a steel pipe. Then, the molten alloy with temperature of about 800 °C was poured into the mould whose bottom plate was cooled down with the cooling jacket, and directionally solidified.

In order to evaluate the effects of Bi-modification on tensile properties, the alloy containing 0.5wt.% Bi was prepared. Pure elemental Bi was wrapped in aluminum foil and added to molten

Mg–5wt.% Si alloy. When the temperature reached about 800 °C, alloys with different compositions were poured into cylindrical moulds (60 mm in diameter and 80 mm in height) to prepare castings for metallographic studies and tensile testing.

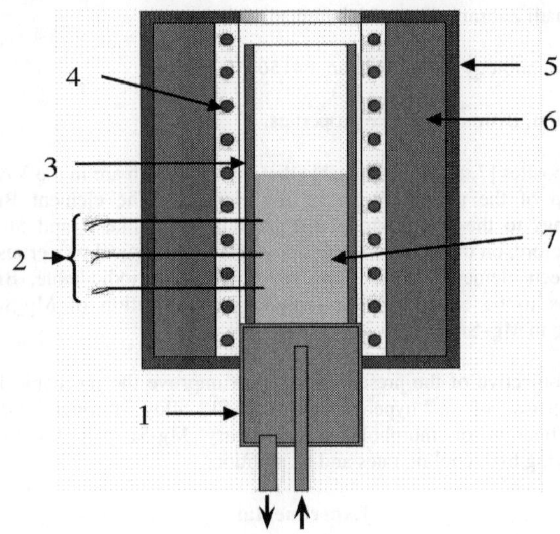

Figure 1. Schematic representation of the directional solidification apparatus: 1) cooling jacket, 2) thermocouples, 3) steel mould, 4) electric heater, 5) furnace cover, 6) heat insulator and 7) casting.

To evaluate the effect of cooling rate on tensile properties, tensile test bars were cut from the solidified ingots at different distances from the bottom of mould. Tensile test bars machined, according to ASTM-B577 standard for each condition. Tensile tests were carried out at room temperature with a constant cross-head speed of 1 mm/min. Specimens for microstructural characterization were sectioned from the same position as tensile test bars. Metallographic specimens were polished and etched with a solution of 5vol.% nitric acid + ethyl alcohol for 10-12 s at room temperature. Quantitative data on microstructure were determined using an optical microscope equipped with an image analysis system (Clemex Vision Pro. Ver.3.5.025). Also, scanning electron microscopy (SEM, Phillips XL30) and X-ray diffractometer (XRD, Philips Xpert) were used to investigate the microstructure characteristics of the specimens and phase analyses.

Results and Discussion

According to the Mg–Si binary phase diagram [9], the Mg–5wt.% Si alloy is a typical hypereutectic alloy and, therefore, the as-cast microstructure should be consisted of primary Mg_2Si and eutectic Mg_2Si + α-Mg phases. Figures 2a and b show a typical as-cast microstructure and XRD pattern of the alloy, respectively. XRD result reveals that the obtained alloy components are only Mg_2Si and Mg phases as expected. It should be mentioned that the bulky crystals are primary Mg_2Si, while the Chinese script type particles [4, 10] are eutectic Mg_2Si, as shown in Figure 2a. However, an interesting microstructural feature observed in Figure 2a is that all the primary Mg_2Si particles are surrounded by a layer of α-Mg halos, around which the fine eutectic structures of Mg_2Si + Mg are formed. The existence of α-Mg halos is related to non-equilibrium solidification of the alloy, which leads to the limited diffusion rate of Si in the liquid around the primary Mg_2Si particles and allows the formation of the Mg-rich phase. These non-equilibrium conditions, as described by Kurz and Fisher [11], could lead to a skewed coupled zone in the binary eutectic diagrams especially when one of the phases shows anisotropic growth characteristics such as Mg_2Si. As the temperature decreased to eutectic temperature, the fine eutectic structures of Mg_2Si + Mg were then formed around the α-Mg halos. This result is similar to some of the experimental findings in other hyper-eutectic systems, such as Al–Si [12] and Al–Mg–Si [13] alloys.

Effect of cooling rate

Figures 3a–e show the microstructures observed in the different sections of unmodified alloy. Quantitative image analysis was used to examine the microstructural changes in the alloy in different sections. The measured particle size and volume fraction of the primary Mg_2Si phase is plotted as a function of distance from bottom of mould in Figure 4a. As it is clear in Figures 3 and 4a, a noticeable decrease is seen in primary Mg_2Si particle size in lower section corresponding to faster cooling rates although there is no obvious change in the volume fraction of the primary Mg_2Si phase in different sections. Therefore, the number of primary Mg_2Si particles per unit area, increased with increasing cooling rate. Also, the eutectic structure in different sections remains mainly fibrous while it becomes finer with increasing cooling rate. The observations follow the general rule that faster cooling corresponding to lower sections results in higher undercoolings, which subsequently leads to faster nucleation and growth thus creating a finer microstructure.

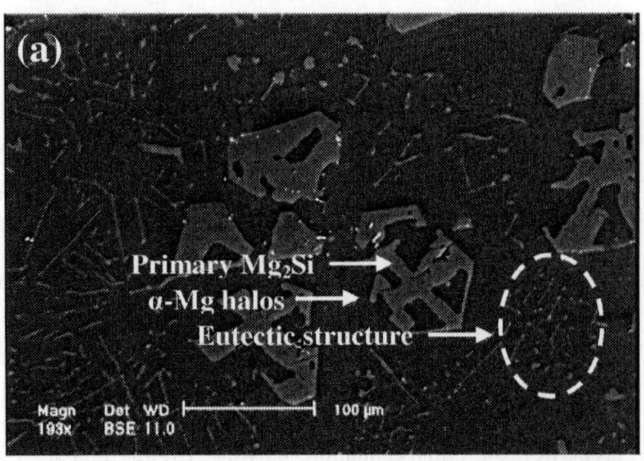

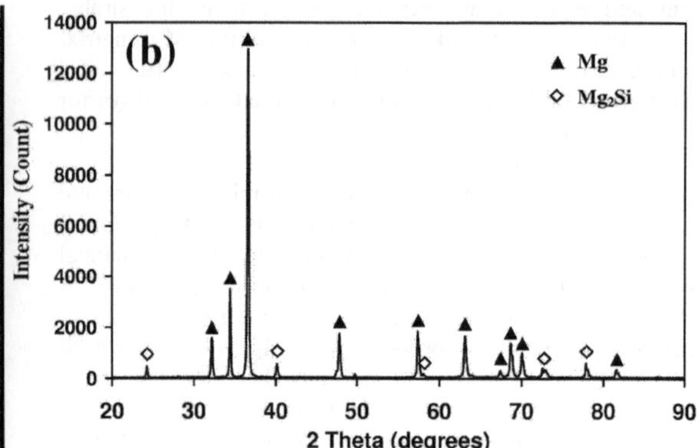

Figure 2. (a) Typical SEM micrograph and (b) XRD pattern of as-cast Mg–5wt.% Si alloy.

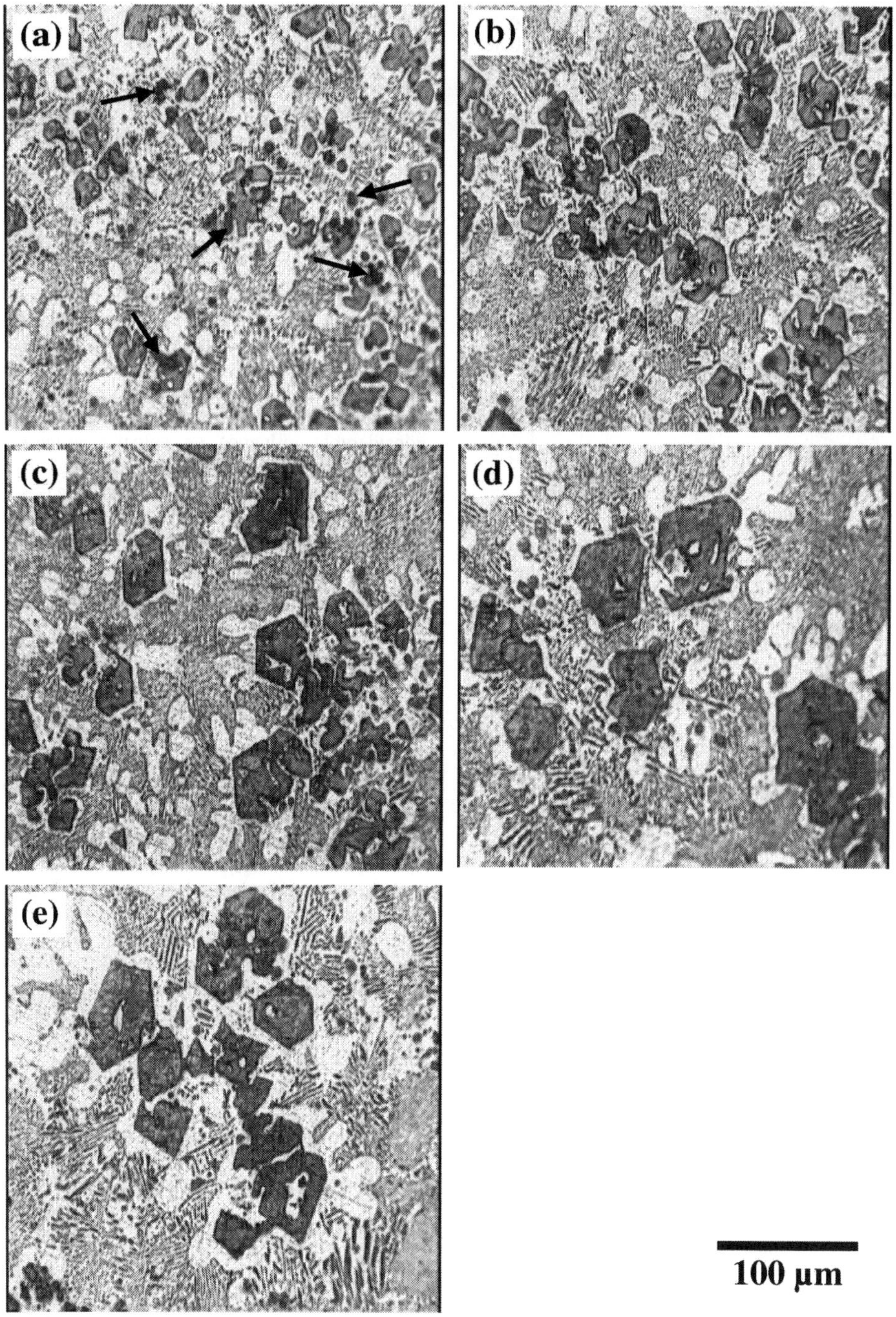

Figure 3. Effect of cooling rate on microstructure of Mg–5wt.% Si alloy without Bi modification. (a) 5 mm, (b) 15 mm, (c) 30 mm, (d) 50 mm and (e) 75 mm sections from bottom of mould. Arrows indicate the gas porosities.

Despite degassing treatment, the alloy still contains gas porosity, which is clearly seen in Figure 3a. Since Mg–5wt.% Si alloy enjoys a wide range of solidification, dispersed shrinkage porosity is more likely. In samples with major casting defects, fracture is initiated at the defects, and cracks tend to propagate from these defects, even bypassing Mg_2Si primary particles.

The ultimate tensile strength (UTS) values of the specimens are plotted against the distance from bottom of mould in Figure 4b. It is evident from the diagram that in general there is an increasing trend in the strength of the composite in lower sections corresponding to higher cooling rates. Thereby, this confirms expectations based on metallographic results. However, some scattered results are observed within 5mm section. The 5mm section result is an exception to the observed trend of improved UTS with increasing cooling rate. It seems reasonable to attribute the unexpected low tensile strength value for the 5mm section to presence of casting defects, which are exaggerated in this section (Figure 3a). The presence of casting defects implies that in addition to microstructural modification by applying high cooling rates, casting process improvement is also necessary to obtain higher tensile properties for this alloy. Also, bulky morphology of primary Mg_2Si particles and their clustering behavior, however, can result in stress concentration at sharp edges and corners of Mg_2Si and lead to anisotropic tensile properties. Therefore, in order to improve the tensile properties, a more isotropic microstructure is preferred.

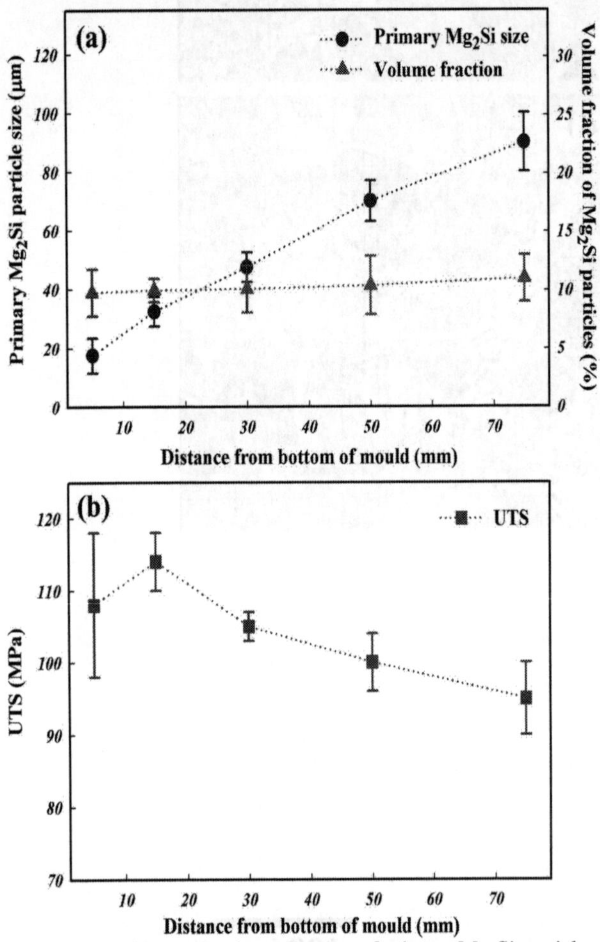

Figure 4. (a) Size and volume fraction of primary Mg_2Si particles and (b) UTS value vs. distance from bottom of the mould.

Effect of Bi addition

In Figures 5a and b, the general microstructures of unmodified and Bi-modified alloys are compared. As compared to the base alloy shown in Figure 5a, considerable microstructural refinement is obtained by the addition of Bi (Figure 5b). According to quantitative image analysis results, the mean size of primary Mg_2Si particles was about 70 μm in unmodified alloy. Modification by Bi reduced their size to about 30 μm, formed an even distribution of particles in the matrix and decreased reinforcement clustering (Figure 5b). Other than refinement of size and distribution, the morphology of the bulky primary Mg_2Si particles changed to a polygonal character after Bi addition. Comparing Figures 3 and 5b, it can be inferred that Bi addition is significantly more effective in refining the morphology of primary particles than the faster cooling rates applied in this study. Furthermore, it is interesting to note that addition of Bi has a considerable influence on eutectic structure and the eutectic Mg_2Si fibers exhibit modified morphology as a finer size (Figure 6). Also, some white point-like phases are observed in the eutectic microstructure of Bi-modified alloy (Figure 6b). The EDS analysis confirmed that the white point-like compound in modified alloy is a Bi-containing compound. According to Mg–Bi binary phase diagram [9], it may be presumed that the compound is Mg_3Bi_2.

Microstructural refinement could be attributed to a change in diagram due to high constitutional undercooling. There is also a hypothesis that a change in surface energy of Mg_2Si crystals caused by Bi atoms on its surface or entering its lattice may also be responsible for the change in growth morphology. Tani and Kido [14] reported that there is a high solubility limit of Bi in Mg_2Si and Bi atoms are primarily located at the Si sites in Mg_2Si lattice. Therefore, during solidification of the melt with Bi addition, some of Bi atoms should dissolve in the Mg_2Si lattice and thus change the surface energy of the Mg_2Si crystals by lattice distortion due to the larger atomic radius of Bi compare with Si (about 30%). This may poison the growth manner of primary Mg_2Si in preferred directions and significantly influence and suppress the growth of the undesirable morphology of primary Mg_2Si. Further observations of Figure 5b depict the volume fraction of α-Mg halos increased with addition of Bi but the eutectic structure area was appreciably reduced. The refining mechanism of eutectic Mg_2Si phase needs further study. But, one possible explanation is that the precipitation of Bi-containing compound around the surface of eutectic Mg_2Si rods results in restriction of growth in radial direction, leading to their finer size as shown in Figure 6b.

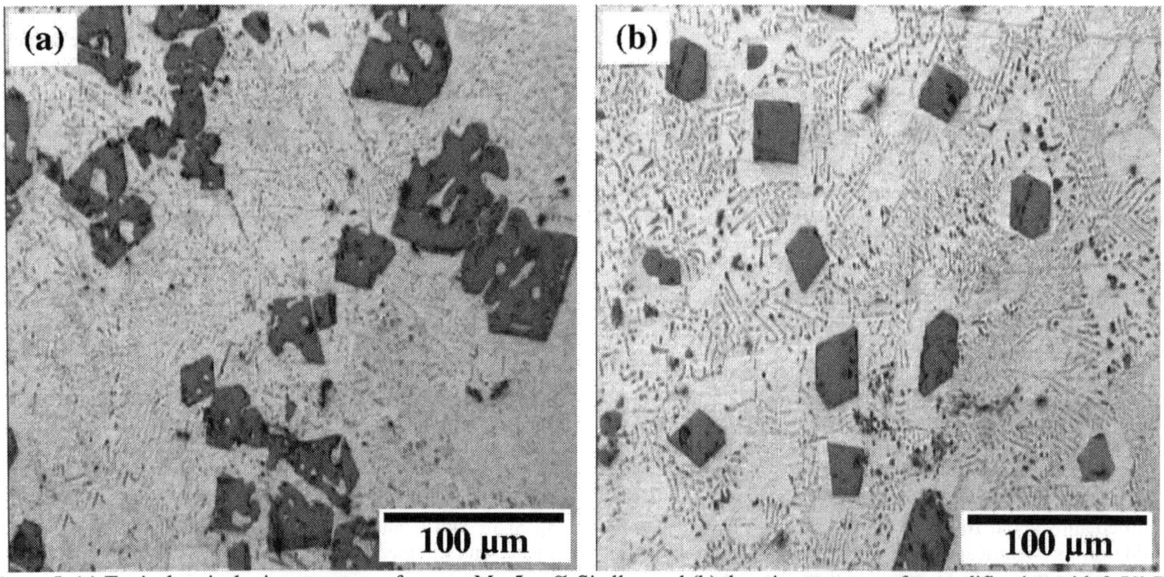

Figure 5. (a) Typical optical microstructure of as-cast Mg-5wt.% Si alloy and (b) the microstructure after modification with 0.5% Bi.

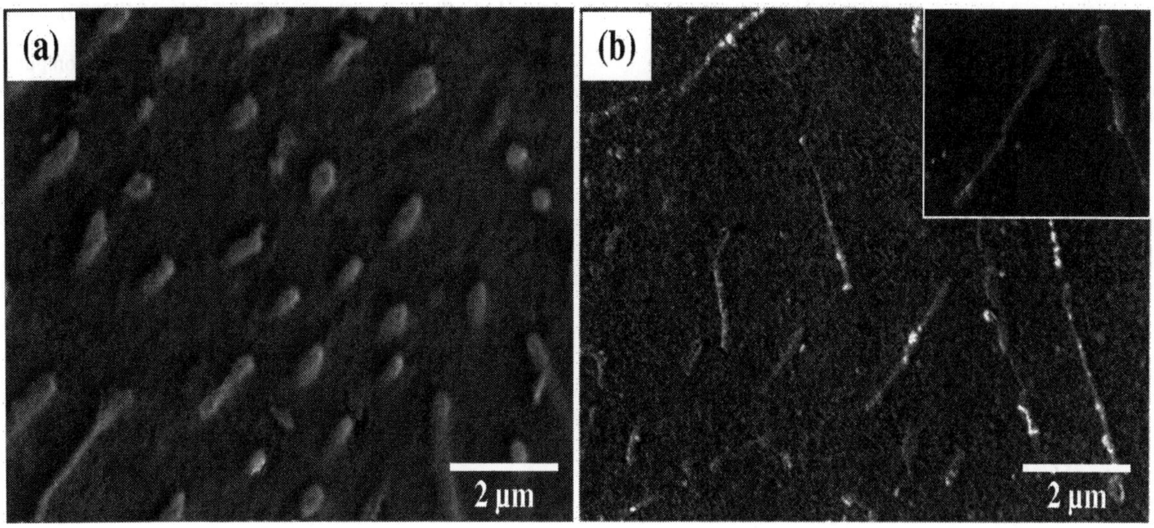

Figure 6. Typical SEM micrographs of eutectic Mg_2Si in the (a) unmodified and (b) Bi-modified alloys.

As mentioned above, the coarse Mg_2Si primary crystals in an unmodified alloy grow bulky, but the finer Mg_2Si particles in Bi-modified alloy do not show such morphology, which is considered an advantage since holes inside the particles could be potential sites for crack initiation. Therefore, Bi modification raises the tensile strength of the unmodified alloy about 10% (Figure 7). Some researchers emphasized the importance of homogeneous particle distributions to improve the ductility [15, 16, 17]. Adding Bi has clearly homogenized the structure, reducing the clustering behavior of primary Mg_2Si particles (Figure 5b) which would be expected to aid crack propagation. Large primary particles and their clusters seem to be the favored crack propagation path in unmodified alloy. Therefore, Bi addition increase the elongation of unmodified alloy about 20% as is observed in Figure 7.

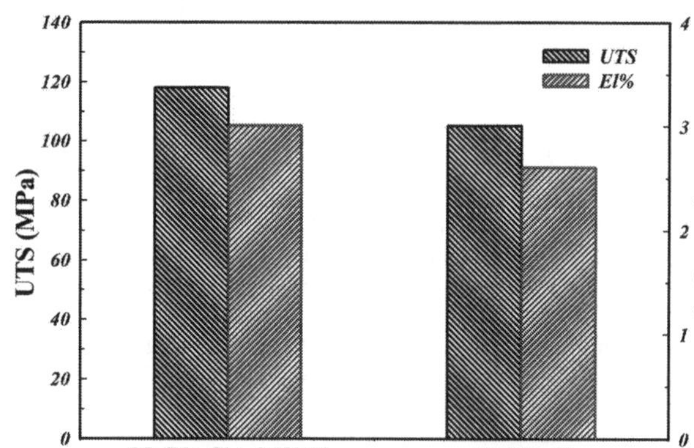

Figure 7. Comparison between tensile properties of unmodified and Bi-modified Mg–5wt% Si alloys.

Conclusions

The following conclusions may be drawn from the experiments conducted to study the influence of cooling rate and addition of Bi on the microstructure and tensile properties of Mg–5wt.% Si alloy.

1- The observed as-cast microstructure of Mg–5wt.% Si alloy consists of coarse primary Mg_2Si particles are surrounded by a layer of α-Mg halos, which is again surrounded by the eutectic Mg_2Si + Mg structures.

2- UTS values increased with increasing cooling rate. However, in the lowest cast section the UTS value showed an unexpected fall together with significant scatter in data attributed to casting defects.

3- Adding Bi refined the primary Mg_2Si morphology, reducing their mean size from more than 70 μm to about 30 μm. Furthermore, the bulky primary Mg_2Si particles exhibit modified morphology as a polygonal shape in Bi-modified alloy. In addition, the eutectic Mg_2Si also exhibits modified morphology as a fine fiber. Adding Bi also raised the UTS and elongation values of unmodified alloy about 10% and 20%, respectively.

4- Bi addition is significantly more effective in refining the morphology of primary Mg_2Si particles than the faster cooling rates applied in this study.

Acknowledgements

The authors are grateful for the financial supports from Isfahan University of Technology. The authors also greatly acknowledge Dr. M. Panjepour for valuable discussions in the course of this research.

References

[1] M. Mabuchi, and K. Higashi, "Strengthening Mechanisms of Mg–Si Alloys," Acta Materialia, 44 (1996), 4611–4618.

[2] L. Lu, K.K. Thong, and M. Gupta, "Mg-Based Composite Reinforced by Mg_2Si," Composites Science and Technology, 63 (2003) 627–632.

[3] Y. Pan, X. Liu, and H. Yang, "Microstructural Formation in a Hypereutectic Mg–Si Alloy," Materials Characterization, 55 (2005), 241–247.

[4] J.J. Kim et al., "Modification of Mg_2Si Morphology in Squeeze Cast Mg-Al-Zn-Si Alloys by Ca or P Addition," Scripta Materialia, 41 (1999), 333–340.

[5] H.Z. Ye, and X.Y. Liu, "Review of Recent Studies in Magnesium Matrix Composites," Journal of Materials Science, 39 (2004), 6153–6171.

[6] M. Mabuchi, K. Kubota, and K. Higashi, "High Strength and High Strain Rate Superplasticity in a Mg-Mg_2Si Composite," Scripta Metallurgica et Materialia, 33 (1995), 331–335.

[7] Q.D. Qin et al., "Effect of Modification and Aging Treatment on Mechanical Properties of Mg_2Si/Al Composite," Materials Science and Engineering A, 527 (2010), 2253–2257.

[8] L. Liao et al., "Influence of Sb on Damping Capacity and Mechanical Properties of Mg_2Si/Mg–9Al Composite Materials," Journal of Alloys and Compounds, 430 (2007), 292–296.

[9] A.A. Nayeb-Hashemi, and J.B. Clark, Phase Diagrams of Binary Magnesium Alloys (Metals Park, OH: American Society for Metals, 1988), 280.

[10] H.Y. Wang et al., "Modification of Mg_2Si in Mg–Si Alloys with K_2TiF_6, KBF_4 and KBF_4 + K_2TiF_6," Journal of Alloys and Compounds, 387 (2005), 105–108.

[11] W. Kurz, and D.J. Fisher, Fundamentals of Solidification (Switzerland, Trans Tech Publications, 1998), 12.

[12] Y.T. Pei, and J.Th.M. De Hosson, "Functionally Graded Materials Produced by Laser Cladding," Acta Materialia, 48 (2000), 2617–2624.

[13] J. Zhang et al., "Microstructural Development of Al–15wt.% Mg_2Si In Situ Composite with Mischmetal Addition," Materials Science and Engineering A, 281 (2000), 104–112.

[14] J.I. Tani, and H. Kido, "Thermoelectric Properties of Bi-Doped Mg_2Si Semiconductors," Physica B, 364 (2005), 218–224.

[15] J. Segurado, and J. LLorca, "A Computational Micromechanics Study of the Effect of Interface Decohesion on the Mechanical Behavior of Composites," Acta Materialia, 53 (2005), 4931–4942.

[16] J. Segurado et al., "A Numerical Investigation of the Effect of Particle Clustering on the Mechanical Properties of Composites," Acta Materialia, 51 (2003), 2355–2369.

[17] X.C. Tong, and A.K. Ghosh, "Fabrication of In Situ TiC Reinforced Aluminum Matrix Composites," Journal of Materials Science, 36 (2001), 4059–4069.

Magnesium Technology 2011
Edited by: Wim H. Sillekens, Sean R. Agnew, Neale R. Neelameggham, and Suveen N. Mathaudhu
TMS (The Minerals, Metals & Materials Society), 2011

ON PREDICTING THE CHANNEL DIE COMPRESSION BEHAVIOR OF HCP MAGNESIUM AM30 USING CRYSTAL PLASTICITY FEM

Q. Ma[1], E.B. Marin[1], A. Antonyraj[1], Y. Hammi[1], H. El Kadiri[1,2], P.T. Wang[1], M.F. Horstemeyer[1,2]
[1] Center for Advanced Vehicular Systems, Mississippi State University, Mississippi State, MS 39762, USA
[2] Department of Mechanical Engineering, Mississippi State University, Mississippi State, MS 39762, USA

Keywords: Magnesium, Texture, Plasticity, 3D microstructure, Modeling

Abstract

Deformation of polycrystalline aggregates of HCP crystals was investigated by employing crystal plasticity and finite element simulations. Results were validated using channel die compression tests. The three dimensional polycrystal, represented by Voronoi techniques, was assigned an initial distribution of crystallographic orientations determined from X-Ray Diffraction. The mechanical properties from the channel die compression tests were used to correlate the material parameters of the crystal plasticity model. Simulations predicted grain-to-grain interactions and the resulting texture evolution under channel die compression. The inhomogeneous deformation among grains was mapped by a stress and a strain distribution and grain orientation spread. The simulation results were compared with experimental observations of an HCP polycrystal subjected to channel die compression.

Introduction

Plastic straining of a polycrystal induces changes in the crystals orientation and the corresponding evolution of orientation distributions oftentimes called crystallographic texture. Texture governs the anisotropic mechanical properties of a polycrystal, e.g. yield surface and strain ratio (Lankford parameter) [1,2]. As texture evolution is intrinsically tied to the activation of slip systems and twinning systems, a number of deformation models incorporating texture, slip, and twinning have been developed in the last decades [3-12]. Two pioneering deformation models for polycrystals are the upper-bound and lower-bound theories, typically referred to as Taylor and Sachs models [3-4]. These models considered only strain compatibility and stress equilibrium, respectively, hence simplifying intricate interactions among grains within a polycrystal which resulted in a marked discrepancy between predicted and experimental textures [5]. Other models, called relaxed constraint models, are modifications of Taylor and Sachs models. For example, modified Taylor models [6,7] relaxed partly the strain compatibility condition; a modified Sachs model [8] considered reaction shear stresses between local grains and its neighborhood represented by the whole matrix; the Visco-Plastic Self-consistent (VPSC) model [9] satisfied the stress and strain equilibrium simultaneously by considering individual grains embedded in an effective medium, whose property was represented by the average of the polycrystalline aggregate. These modified models improved the texture prediction; however, most of these models did not consider the topological information of grains [5-12].

The crystal plasticity finite element model (CPFEM) combined the single crystal plasticity constitutive law with finite elements to compute the macroscale response of crystal aggregates [10,11]. In CPFEM, the equilibrium of forces (stresses) and the compatibility of displacements are implicitly satisfied by the finite element formulation. This technique can be used to model two dimensional (2D) or three dimensional (3D) microstructures [10-15] to investigate many aspects of the heterogeneous plastic response of a local grain within a polycrystal under various boundary conditions [14]. At present, 3D CPFEM deformation simulations attract more attention as 3D microstructure deformation resembles the realistic deformation of polycrystals. In this context, the use of a fine meshed regular polyhedron representing a grain with a particular orientation has been widely used to investigate intragranular orientation gradient and grain interaction effects during deformation. Many CPFEM studies have used brick-shape grains [16], rhombic dodecahedron grains [13,14,17], and tetradecahedron grains [15] to investigate grain interaction effects on texture evolution, grain subdivision and strain localization. However, as suggested by Mika and Dawson [18,19], crystal shape could also play a large role on texture scatter, and hence, the use of uniform and regular grain shapes can still produce a discrepancy between experimental textures and simulations.

This article investigates the plane strain deformation of a polycrystal aggregate using the CPFEM and 3D Voronoi grains. The focus is on revealing the effect of grain interaction on global texture evolution, intergranular heterogeneous plasticity, and intragranular orientation spread. The material studied is HCP AM30 magnesium alloy.

Crystal Plasticity Constitutive Framework

The constitutive framework adopted in this paper corresponds to the formulation presented in [20,21]. This formulation is an extension of a rigid viscoplastic model to account for elasticity effects, and incorporates a number of changes with respect to a previous formulation [22]. The model can be summarized as follows:

$$\mathbf{d} = \overset{\nabla}{\boldsymbol{\epsilon}^e} + \tilde{\mathbf{D}}^p, \quad \overset{\nabla}{\boldsymbol{\epsilon}^e} = \dot{\boldsymbol{\epsilon}}^e + \boldsymbol{\epsilon}^e\tilde{\Omega}^e - \tilde{\Omega}^e\boldsymbol{\epsilon}^e \tag{1}$$

$$\mathbf{w} = -skew(\dot{\boldsymbol{\epsilon}}^e\boldsymbol{\epsilon}^e) + \tilde{\Omega}^e + \tilde{\mathbf{W}}^p, \quad \tilde{\Omega}^e = \dot{\mathbf{R}}^e\mathbf{R}^{eT} \tag{2}$$

$$\tau = \tilde{\mathbb{C}}^e : \boldsymbol{\epsilon}^e \tag{3}$$

$$\tilde{\mathbf{D}}^p = \sum_{\alpha=1}^{N} \dot{\gamma}^\alpha sym(\tilde{\mathbf{S}}^\alpha \otimes \tilde{\mathbf{m}}^\alpha) \tag{4}$$

$$\tilde{\mathbf{W}}^p = \sum_{\alpha=1}^{N} \dot{\gamma}^\alpha skew(\tilde{\mathbf{S}}^\alpha \otimes \tilde{\mathbf{m}}^\alpha) \tag{5}$$

$$\dot{\gamma}^\alpha = \dot{\gamma}_0 \left(\frac{|\tau^\alpha|}{\kappa^\alpha}\right)^{1/m} sign(\tau^\alpha) \tag{6}$$

$$\tau^{\alpha} = \tau : sym(\tilde{\mathbf{S}}^{\alpha} \otimes \tilde{\mathbf{m}}^{\alpha}) \quad (7)$$

$$\dot{\kappa}^{\alpha} = h_0\left(\frac{\kappa_S - \kappa}{\kappa_S - \kappa_0}\right)\sum_{\alpha}|\dot{\gamma}^{\alpha}| \quad (8)$$

Where d is the deformation rate, $\overset{\nabla}{\epsilon^e}$ is the elastic deformation rate, $\tilde{\mathbf{D}}^p$ is the plastic deformation rate, w is the spin tensor and ϵ^e is the small elastic tensor. $\tilde{\Omega}^e$ is the elastic lattice spin tensor, $\tilde{\mathbf{W}}^p$ is the plastic spin tensor, $\mathbf{R}^e$ is the lattice rotation tensor, τ is the stress tensor, $\tilde{\mathbb{C}}^e$ is the fourth order anisotropic crystal elasticity tensor and $\tilde{\mathbf{S}}^{\alpha} \otimes \tilde{\mathbf{m}}^{\alpha}$ defines the Schmid tensor. $\dot{\gamma}^{\alpha}$ is the plastic shear rate on the α slip which follows a power law relationship shown in Equation 6, $\dot{\gamma}_0$ is the reference shear rate ($\dot{\gamma}_0$=0.001/S in this study), τ_α is the resolved shear stress on α slip, m(=0.05 in this case) is the strain rate sensitive exponent parameter and κ^α is the threshold stress of the α slip. h_0 is the initial hardness rate due to dislocation accumulation, κ_s and κ_0 are the saturation strength and the initial strength of the slip, respectively. Details of the formulations are given in the reference [20,21].

Modeling Framework

Magnesium has a hexagonal close-packed (HCP) structure whose deformation modes are different from those of aluminum. Typical deformation modes in magnesium are basal <a>-{0002}<11$\bar{2}$0> slip, prismatic <a>-{10$\bar{1}$0}<11$\bar{2}$0> slip, second pyramidal <c+a>-{11$\bar{2}$2}<11$\bar{2}$3> slip and extension twinning {10$\bar{1}$2}<10$\bar{1}$1>. In this study, a commercial extruded AM30 alloy (mass %, 2.54% Al, 0.40% Mn, Mg in balance) was selected as the HCP experimental material to conduct channel die compression at high temperature 200°C and at strain rate of $10^{-3}S^{-1}$ to strain 30%. At these loading conditions, twinning would not be profuse and, hence, only slip will be the predominant deformation mode. Figure 1 displays the three slip modes used in this study. The hardening parameters of the three slips modes were obtained using material point simulations to fit the experimental stress-strain curve recorded from the channel die compression test. The elasticity parameters used in this study are listed in Table I, while the slip system hardening parameters obtained from the calibration procedure are presented in Table II. Figure 2 compares the predicted and experimental stress-strain curves.

The texture of AM30 was measured by X-ray diffraction method (XRD). The recalculated pole figures were calculated based on the orientation distribution functions (ODFs) which was obtained using the measured six incomplete pole figures {10$\bar{1}$0}, {0002}, {10$\bar{1}$1}, {10$\bar{1}$2}, {11$\bar{2}$0} and {10$\bar{1}$3}. The initial and the channel die compressed texture of AM30 plotted by the texture software MTEX [23] are shown in Figure 4a and 4b, respectively.

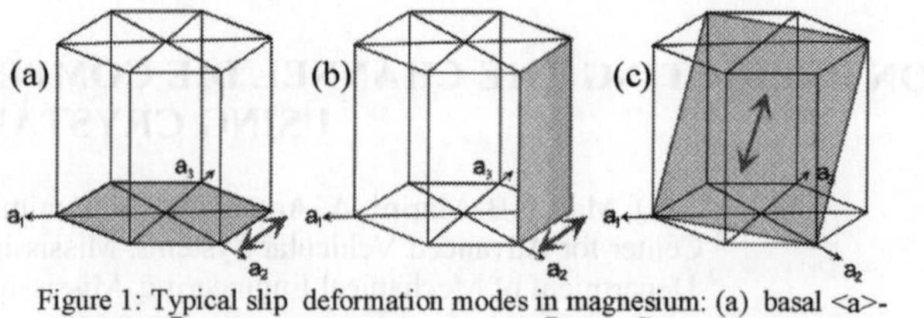

Figure 1: Typical slip deformation modes in magnesium: (a) basal <a>-{0002}<11$\bar{2}$0> slip, (b) prismatic <a>-{10$\bar{1}$0}<11$\bar{2}$0> slip, and (c) second pyramidal <c+a>-{11$\bar{2}$2}<11$\bar{2}$3> slip.

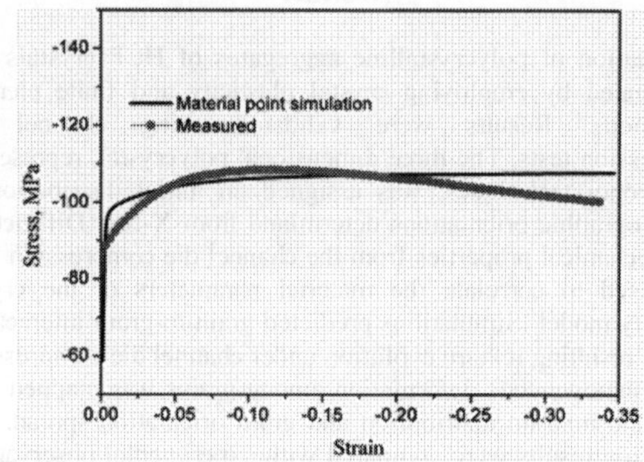

Figure 2: Correlation of the experimentally measured and the material point simulation stress-strain behavior of an AM30 magnesium alloy at 200°C.

Table I. Elasticity parameters for AM30 magnesium alloy.

C_{11}	C_{12}	C_{13}	C_{33}	C_{44}
59.52 GPa	25.6 GPa	21.4 GPa	61.74 GPa	16.47 GPa

Table II. Hardening parameters of the slip modes for AM30 magnesium alloy.

mode	h_0	κ_0	κ_s
Basal <a>	19.6 MPa	18 MPa	20 MPa
Prismatic <a>	37.5 MPa	30 MPa	34 MPa
Pyramidal 2nd <c+a>	18.8 MPa	44 MPa	46 MPa

The commercial finite element software ABAQUS 6.9 and a user subroutine UMAT incorporating the crystal plasticity constitutive theory and the magnesium AM30 materials parameters as listed in Table I and Table II were used to simulate texture evolution and mechanical response. The initial texture used in this CPFEM simulation was measured in the undeformed sample (Figure 4a) and was represented by 343 discrete orientations as shown in Figure 3c. The 3D polycrystal is represented by the 3D Voronoi grains created by the code Neper [24]. The Voronoi 3D grain is evidently more realistic than regular shaped grains used in other studies: brick grains [15,16], rhombic dodecahedron grains and other regular polyhedron grains [13-15,17]. In particular, 3D brick grains are too regular and too uniform in comparison to a realistic grain morphology. The element type used to discretize the Voronoi grains is the second order tetrahedral C3D10 element. As such, one element has four integration points, namely four orientations. There are totally 55092 elements in the Voronoi 3D microstructure. The undeformed polycrystal and the channel die compressed 3D polycrystal up to a strain of 30% are presented in Figures 3a and 3b.

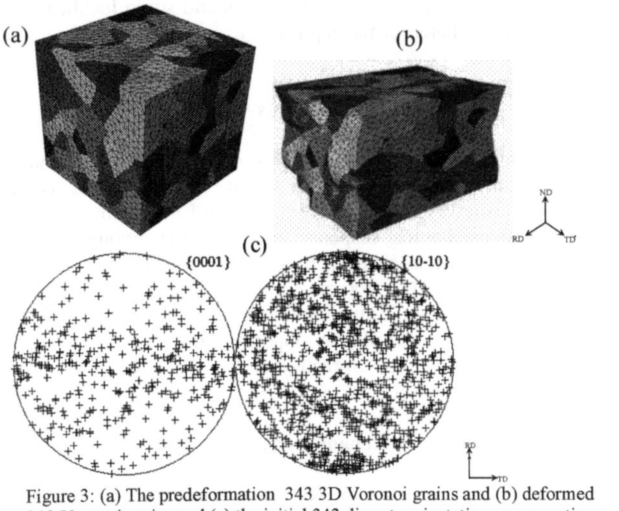

Figure 3: (a) The predeformation 343 3D Voronoi grains and (b) deformed 343 Voronoi grains and (c) the initial 343 discrete orientations representing the initial texture of AM30 magnesium alloy.

Plane strain compression simulations were performed on the 3D polycrystal to strain 30%. Deformation textures were plotted in the {0001} and {10$\bar{1}$0} pole figures using the texture software MTEX [23] with a half scatter width of 7.5 degree.

In order to study intragranular orientation spread in discrete grains, one special orientated grain was selected: grain number123 with a orientation {φ_1, Φ, φ_2}={190.62, 69.85, -162.48}.This grain has 331 elements with a total of 1324 orientations. Note that the analysis of local inhomogeneous deformations in a 3D Voronoi polycrystal has been scarcely treated in the literature [25,26].

In-grain misorientation, namely the orientation spread in one grain indicates the local inhomogeneous orientation distribution and the effect of grain interaction. In addition, the orientation spread of a grain can also characterize the stability property of the grain with a special orientation. The orientation spread $\Delta\theta_i$ in a grain g at various strain levels can be calculated by Equations 9-11. In the specimen coordinate system, one orientation g_i ={φ_1, Φ, φ_2} can be expressed into the rotation matrix given by Equation 9. There are 12 misorientations Δg_{ij} between two orientations g_i and $\bar{g}$, where g_i is the orientation of an integration point in the grain g and $\bar{g}$ is the mean orientation of all the orientations in the same grain. Δg_{ij} can be calculated by Equation 10 where M_j is the symmetry operator which depends on the symmetry of the HCP crystal system. Usually, the minimum rotation angle $\Delta\theta_i$ defines the misorientation angle or disorientation angle which is expressed as Equation 11.

$$g_i = \begin{pmatrix} \cos\varphi_1\cos\varphi_2 - \sin\varphi_1\sin\varphi_2\cos\Phi & \sin\varphi_1\cos\varphi_2 + \cos\varphi_1\sin\varphi_2\cos\Phi & \sin\varphi_2\sin\Phi \\ -\cos\varphi_1\sin\varphi_2 - \sin\varphi_1\cos\varphi_2\cos\Phi & -\sin\varphi_1\sin\varphi_2 + \cos\varphi_1\cos\varphi_2\cos\Phi & \cos\varphi_2\sin\Phi \\ \sin\varphi_1\sin\Phi & -\cos\varphi_1\sin\Phi & \cos\Phi \end{pmatrix} \quad (9)$$

$$\Delta g_{ij} = M_j \cdot g_i \cdot \bar{g}^{-1} = \begin{pmatrix} \Delta g_{11} & \Delta g_{12} & \Delta g_{13} \\ \Delta g_{21} & \Delta g_{22} & \Delta g_{23} \\ \Delta g_{31} & \Delta g_{32} & \Delta g_{33} \end{pmatrix}_{ij} \quad (10)$$

$$\Delta\theta_i = (\arccos(\frac{\Delta g_{11} + \Delta g_{22} + \Delta g_{33} - 1}{2})_{ij})\ \text{min}\,imum \quad (11)$$

Note that $\Delta\theta_i$ is the smallest value among the 12 misoientation angles.

Simulation Results and Discussion

The measured and CPFEM predicted texture are listed in Figure 4. Figure 5 shows the orientation distribution and its spread of grain number 123 as well as its stress and strain localization at strain 30%. The finite element simulation channel die compression stress-strain curve using crystal plasticity model and the measured stress-strain curve are presented in Figure 6.

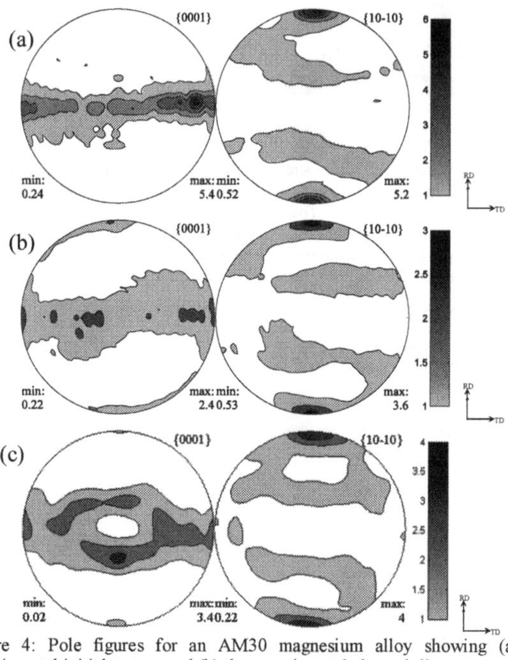

Figure 4: Pole figures for an AM30 magnesium alloy showing (a) the experimental initial texture and (b) the experimental channel die compression at a strain of 30% and (c) the finite element crystal plasticity simulation result at a strain of 30%.

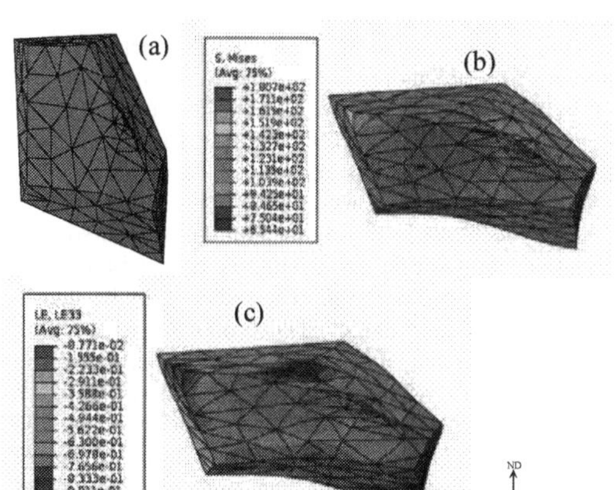

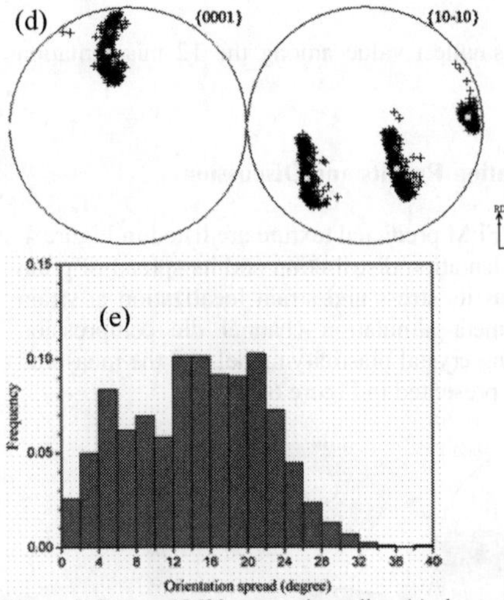

Figure 5: For an AM30 magnesium alloy (a) shows mesh of the initial grain number 123, (b) the local Mises stress and (c) local strain ε_{33} distribution of grain number 123 at a strain of 30%, (d) the orientation distribution of the grain number 123 at a strain of 30%. The cross symbols represent the orientations, the circle symbol represents the initial orientation of grain number 123 and the cube symbol represents the mean orientation among all the orientations, and (e) the orientation spread of the grain number 123 at a strain of 30%.

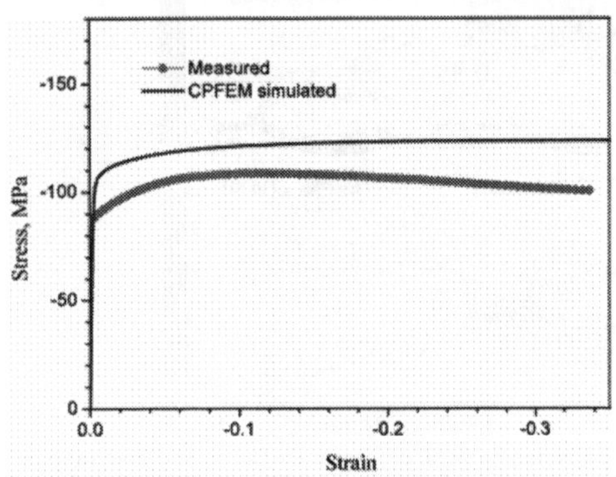

Figure 6: The finite element simulation stress-strain curve based on crystal plasticity model and the experimental stress-strain curve under channel die compression for AM30 magnesium alloy at 200°C.

CPFEM predicted textures of the 3D Voronoi microstructure that are consistent with experimental results as shown in Figure 4b and 4c. As shown by this figure, the measured and the predicted plane strain compression AM30 textures exhibit similar texture configuration. However, there is discrepancy that could stem from the fact that dynamic recrystallization, which typically happens in Mg alloys, is not accounted for in the simulations. Also, the CPFEM simulation stress-strain curve slightly overestimates the stress of the channel die compression. The reason may be due to the effect of grain interaction in the 3D Voronoi microstructure. Figure 5b and 5c present the Mises and ε_{33} strain distribution in grain number 123 at strain 30%. The inhomogeneous deformation of grain number 123 is clearly proved by the observed stress and strain heterogeneous distribution. In addition, as shown in Figure 5d, the grain orientation distribution in grain number 123 displays the different rotation tendency along RD direction. As such, the orientation spread of grain number 123 presents a big scatter with two peaks, as shown in Figure 5e. Accordingly, the effect of grain interaction among this 3D Voronoi polycrystal should play an important role on orientation spread, stress and strain localization and intergranular heterogeneous plasticity.

Conclusions

Plane strain compression simulations of a three dimensional microstructure of HCP magnesium AM30 were performed using an elastic-plastic crystal plasticity model and the finite element method. The mechanical response, global texture evolution and the intergranular heterogeneous plasticity of discrete grain were captured by the 3D Voronoi microstructure channel die compression simulation. Simulation results showed that the effect of grain interaction could play an important role on global texture evolution, the orientation spread and the local heterogeneity deformation of one grain.

Acknowledgements

The authors are grateful to the financial support from the Department of Energy, Contract No. DE-FC-26-06NT42755, and the Center for Advanced Vehicular Systems (CAVS) at Mississippi State University.

References

1. D. Raabe, "Yield surface simulation for partially recrystallized aluminum polycrystals on the basis of spatially discrete data," *Computational Materials Science,* 19 (2000), 13-26.
2. A. Fjeldly and H.J. Roven, "Observations and calculations on mechanical anisotropy and plastic flow of an AlZnMg extrusion," *Acta Materialia,* 44 (1996), 3497-3504.
3. G. Sachs, "Zur ableitung einer fleissbedingun," *Zeichschrift Verein Deutscher Ingenieur,* 72 (1928), 734-736.
4. GI. Taylor, "Plastic strain in metals," *Journal of the Institute of Metals,* 62 (1938), 307-324.
5. D. Raabe, Z. Zhao, and W. Mao, "On the dependence of in-grain subdivision and deformation texture of aluminum on grain interaction," *Acta Materialia,* 50 (2002), 4379-4394.
6. P. Van Houtte et al., "Deformation texture prediction: From the Taylor model to the advanced Lamel model," *International Journal of Plasticity,* 21 (2005), 589-624.
7. U.F. Kocks and H. Chandra, "Slip geometry in partially constrained deformation," *Acta Metallurgica,* 30 (1982), 695-709.
8. W. Mao and Y. Yu, "Effect of elastic reaction stress on plastic behaviors of grains in polycrystalline aggregate during tensile deformation,"*Materials Science and Engineering,* A367 (2004) 277-281.
9. R.A. Lebensohn and C.N. Tomé, "A self-consistent anisotropic approach for the simulation of plastic deformation and texture development of polycrystals: Application to zirconium alloys," *Acta Materialia,* 41 (1993), 2611-2644.
10. D. Peirce, R.J. Asaro, and A. Needleman, "Material rate dependence and localized deformation in crystalline solids," *Acta Metallurgica,* 31 (1983), 1951-1976.
11. D. Peirce, R.J. Asaro, and A. Needleman, "Analysis of nonuniform and localized deformation in ductile single crystals," *Acta Metallurgica,* 30 (1982), 1087-1119.
12. A. Alankar, Ioannis N. Mastorakos, and D.P. Field, "A

dislocation-density-based 3D crystal plasticity model for pure aluminum," *Acta Materialia,* 57 (2009), 5936-5946.
13. E.B. Marin and P.R. Dawson, "Elastoplastic finite element analyses of metal deformations using polycrystal constitutive models,"*Computer Methods in Applied Mechanics and Engineering,* 165 (1998), 23-41.
14. F. Roters et al., "Overview of constitutive laws, kinematics, homogenization and multiscale methods in crystal plasticity finite-element modeling: Theory, experiments, applications,"*Acta Materialia*, 58 (2010), 1152-1211.
15. Z. Zhao et al., "Influence of in-grain mesh resolution on the prediction of deformation textures in fcc polycrystals by crystal plasticity FEM," *Acta Materialia,* 55 (2007), 2361-2373.
16. G.B. Sarma and P.R. Dawson, "Effects of interactions among crystals on the inhomogeneous deformations of polycrystals ," *Acta Materialia,* 44 (1996), 1937-1953.
17. D.P. Mika and P.R. Dawson, "Effects of grain interaction on deformation in polycrystals ," *Materials Science and Engineering,* A257 (1998), 62-76.
18. P.R. Dawson, D.P. Mika, and N.R. Barton, "Finite element modeling of lattice misorientations in aluminum polycrystals," *Scripta Materialia,* 47 (2002), 713-717.
19. D.P. Mika and P.R. Dawson, "Polycrystal plasticity modeling of intracrystalline boundary textures," *Acta Materialia,* 47 (1999), 1355-1369.
20. E.B. Marin, "On the formulation of a crystal plasticity model" (Report SAND2006-4170, Sandia National Laboratories, CA, 2006).
21. S. Groh et al., "Multiscale modeling of the plasticity in an aluminum single crystal,"*International Journal of Plasticity,* 25(2009), 1456-1473.
22. E.B. Marin and P.R. Dawson, "On modeling the elasto-visoplastic response of metals using polycrystal plasticity," *Computer Methods in Applied Mechanics and Engineering,* 165 (1998), 1-21.
23. R. Hielscher and H. Schaeben, "A novel pole figure inversion method: specification of the MTEX algorithm," *Journal of Applied Crystallography,* 41 (2008), 1024-1037.
24. http://neper.sourceforge.net
25. M. Blicharski, R. Becker, and H. Hu, "Deformation texture of channel-die deformed aluminum bicrystals with *S* orientations," *Acta Metallurigica et Materialia,* 41 (1993), 2007-2016.
26. Y.L. Liu, H. Hu, N. Hansen, "Deformation and recrystallization of a channel die compressed aluminium bicrystal with (112)[111]/(123)[412] orientation," *Acta Metallurigica et Materialia,* 43 (1995), 2395-2405.

INVESTIGATION OF MICROHARDNESS AND MICROSTRUCTURE OF AZ31 ALLOY AFTER HIGH PRESSURE TORSION

Jitka Vrátná[1], Miloš Janeček[1], Josef Stráský[1], Hyoung Seop Kim[2], Eun-Yoo Yoon[2]
[1]Charles University, Department of Physics of Materials, Ke Karlovu 5, CZ-12116 Czech Republic
[2]Department of Materials Science and Engineering, POSTECH, Pohang, 790-784 Korea

Keywords: AZ31 alloy, High pressure torsion, Microhardness evolution

Abstract

Cast commercial magnesium alloy AZ31 was processed by high pressure torsion (HPT) at room temperature for 1, 3, 5 and 15 rotations (strain ranged from 1 to 7). Microstructure evolution with strain imposed by HPT was observed by light and electron microscopy. HPT was shown to be a very effective method of grain refinement. The initial coarse grain structure was refined by a factor of almost 200 already after one HPT turn ($\varepsilon \approx 4$). Mechanical properties were investigated by detailed 2-D microhardness measurements. HPT straining was found to introduce a radial inhomogeneity in the material which is manifested by a pronounced drop in the center and the maximum near the specimen periphery. With increasing strain due to HPT this inhomogeneity is continuously smeared out tending to saturate with increasing strain. Integrated 3-D meshes across the total surface of disks revealed the undulating character of microhardness variations. The strain imposed by HPT was shown to saturate with increasing number of HPT turns.

Introduction

Due to high specific strength, magnesium alloys are very attractive and promising materials for structural components in automotive and aerospace industries. However, the use of Mg alloys in more complex applications is limited because of problems associated with low ductility, poor corrosion and creep resistance. The limited ductility is a consequence of the lack of independent slip systems and the large difference in the values of the critical resolved shear stress in the potential slip systems. Moreover, the occurrence of strong deformation textures and stress anisotropy in magnesium alloys reduces significantly the variety of possible industrial applications [1].

The properties of magnesium alloys can be improved by refining the grain size to the submicrocrystalline (grain sizes in the range of 100-1000 nm) or even nanocrystalline level (grains sizes smaller than 100 nm) [2-3]. A variety of new techniques have been proposed for the production of the ultra-fine grain (UFG) structure in materials. All these techniques rely on the imposition of heavy straining and thus introduction very high density in the bulk solid material. Since these procedures introduce severe plastic deformation (SPD) into the material, it became convenient to describe all of these operations as SPD processing.

Several processes of SPD are now available but only three of them receiving the most attention at present time, in particular equal-channel angular pressing (ECAP) [4], accumulative roll-bonding (ARB) [5] and high-pressure torsion (HPT) [6].

Equal-channel angular pressing, first reported by Segal [7], became probably the most popular technique of SPD due to the combination of its effectiveness in producing UFG structure, the versatility and the scalability. Attractive results have been achieved using ECAP *e.g.* with aluminium alloys [8, 9], nevertheless the ECAP was less effective if applied to magnesium, namely dilute Mg alloys or pure Mg [10]. Recently, Horita [11] introduced a combined two-step processing route involving an initial extrusion step and subsequent processing by ECAP. This process designated by the acronym EX-ECAP, was used successfully to achieve UFG microstructures in many materials including magnesium alloys [12].

Although general principles of high pressure torsion were first proposed many years ago, it has become of general scientific interest only recently. It is only within approximately the last 5 years that numerous extensive reports documenting the processing and properties of materials fabricated by HPT have started appearing in the scientific literature showing that HPT is more effective technique of grain refinement than other SPD techniques. Extensive investigation of microstructure and mechanical properties of pure metals and solid solution alloys with FCC structure were conducted and reported in the scientific literature, e.g. in Cu [13-15], Ni [16] and Al [17-19]. Reports of successful application of HPT processing of metals with BCC structure can be found in the literature, e.g. on steels, Mo, Cr and W [20-21]. BCC metals were found to be more difficult to deform by HPT than FCC metals.

On the other hand, data about physical properties of UFG metals and alloys with hexagonal lattice are scarce. There are reports of HPT processing of titanium of commercial purity used for biomedical and dental applications [22] and zirconium for the use in surgical implants [23]. The reports describing the use of HPT with Mg alloys are almost lacking. Only recently Horita et al. reported significant grain refinement in Mg-9% Al alloy [24] and AZ61 [25].

This work is therefore motivated by this fact and its main objective is to extend this missing knowledge of the properties of UFG Mg based alloys processed by HPT. The objective of the paper is to provide a detail analysis of microstructure evolution in AZ31 alloy subjected to HPT straining and to correlate it with the observation of microstructure.

Experimental

Commercial AZ31 alloy, with a nominal composition of Mg-3%Al-1%Zn in the initial as cast condition was used in this investigation. Prior to high pressure torsion the alloy was homogenized at 390°C for 12 hours. After homogenization the disk specimens of the diameter of 20 mm and the thickness of 1 mm were cut from the billet. These specimens were processed by high pressure torsion at room temperature for 1, 3, 5 and 15 rotations by applying the hydrostatic pressure of 2.5 GPa. The experimental setup schematically illustrated at Fig. 1 comprises two anvils. The upper anvil is fixed with a load cell mounted on its top allowing measuring the hydrostatic pressure which is applied to the specimen during straining. The lower anvil with a sample placed in the groove is first lifted to its final position pressing the specimen to the symmetrical groove in the upper anvil. Once the operating hydrostatic pressure is reached, specimen compression is stopped and maintained for 10 seconds. Then the torsional straining stage starts by rotation of the lover anvil with the constant speed of 1 rpm while maintaining the preset hydrostatic pressure.

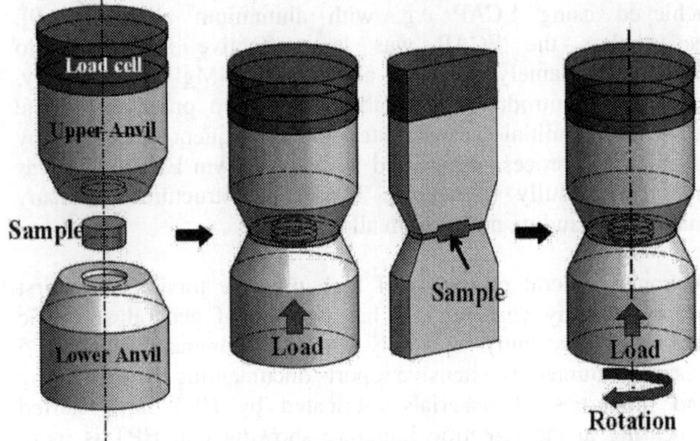

Figure 1. Schematic illustration of HPT device showing the compression and torsional stage.

Specimens for light microscopy and microhardness measurements were first mechanically grinded on watered abrasive papers, then polished with polishing suspension of grade 3 and 1 μm. Using this procedure, flat specimens with minimum surface scratches were obtained. Specimens for transmission electron microscopy were prepared by mechanical polishing followed by ion milling using PIPS ion mill at 4 kV. TEM observations were performed using Jeol 2000 FX electron microscope operated at 200 kV.

Vickers microhardness (100g load) was measured on the semi-automatic Wolpert tester allowing automatic indentation. The regular square network of indents with the step of 0.5mm was performed in one quarter of each specimen after HPT. In order to find the exact center of the specimen two additional lines of indents on the other half of the specimen were also done. In Fig. 2 the optical micrograph showing the layout of individual indents on the specimen is shown. Using this procedure the center of each specimen was found with the accuracy of +/- 0.25mm.

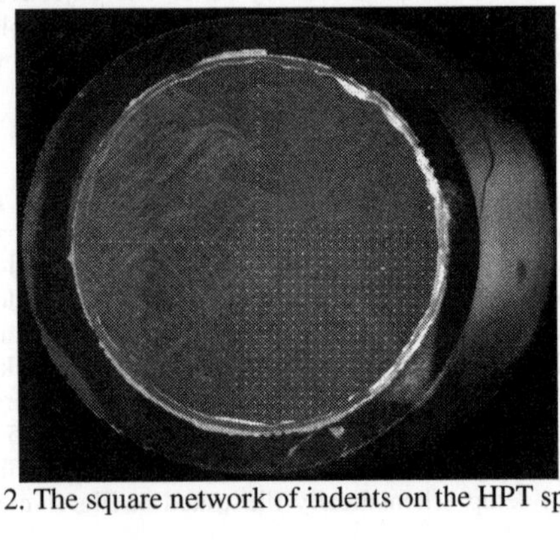

Figure 2. The square network of indents on the HPT specimen.

Results

The microstructure of AZ31 alloy in the initial condition (after homogenization treatment) is shown in Fig. 3. The microstructure consists of almost equiaxed grains with the average size of approximately 150 μm.

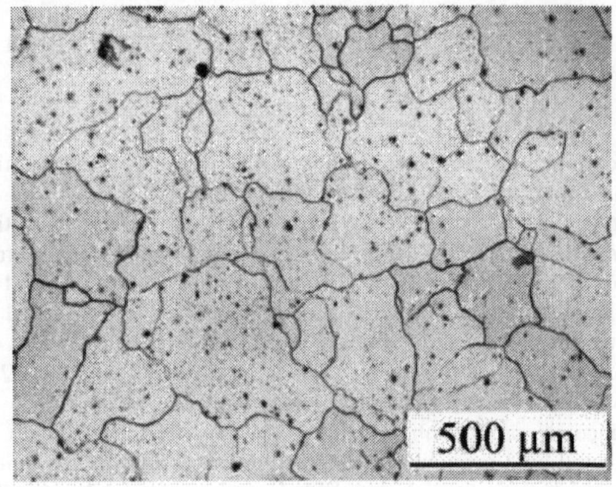

Figure 3. Microstructure of AZ31 alloy in the initial condition.

Microhardness profiles, i.e. the dependence of the microhardness on the distance r from the center of the disk were measured along two perpendicular lines passing through the center of the disc (see also Fig. 2). Fig. 4 displays the values of Vickers microhardness measured in linear traverses across the diameter of disks of AZ31 alloy after performing HPT at room temperature at different number of rotations (N=1, 3, 5 and 15). Each data point was obtained as the average of 4 Hv values obtained at the same distance from the center of the disk. The scatter of the data is in the range of approximately 10% (+/- 5 MPa). For clarity the error bars are not shown in Fig. 4. The base line of the compressed (N=0) specimen having the average value of 75 MPa is also shown in Fig. 4 to see the influence of compression and subsequent torsional straining on Hv values. Note that the average microhardness of the annealed specimen is 58 +/- 3 MPa (not shown in Fig. 4).

By detailed inspection of Fig. 4 it is seen that the initially homogeneous distribution of a compressed specimen changes. HPT staining introduces an inhomogeneity in the material which is manifested by a clear minimum in Hv values in the center of the specimen after one turn (N=1). In this specimen the microhardness increases with increasing distance from the center reaching its maxium of 110 MPa which is approximately 30 MPa higher than in the center (Hv ≈ 80 MPa). With increasing number of turns this difference gradually decreases by extension of the zone of maximum hardness from the periphery towards the center of the specimen. In the specimen after 5 turns (N=5) almost homogeneous distribution of microhardness was observed if the scatter of measured data is considered. In the specimen after 15 turns no differences in microhardness throughout the specimen were found. Note also that the maximum hardness does not change with increasing strain (number of turns). Within the experimental scatter the maximum value of approximately 112 MPa was observed in all specimens.

Microhardness results correspond well with our preliminary microstructure examinations by TEM. TEM proved significant grain refinement in specimens after HPT. One example of the microstructure is shown in Fig. 5. It displays the microstructure in the middle part of the specimen after one turn of HTP ($\varepsilon \approx 4$). It is seen that HPT resulted in strong grain refinement. New grains of the average size of approximately 800 nm are clearly seen in the micrograph. Most grains contain many dislocations. However, several recrystallized almost dislocation free grains are also seen in Fig. 5. The individual grains have high misorientation as confirmed by the contrast or by detail electron diffraction analysis.

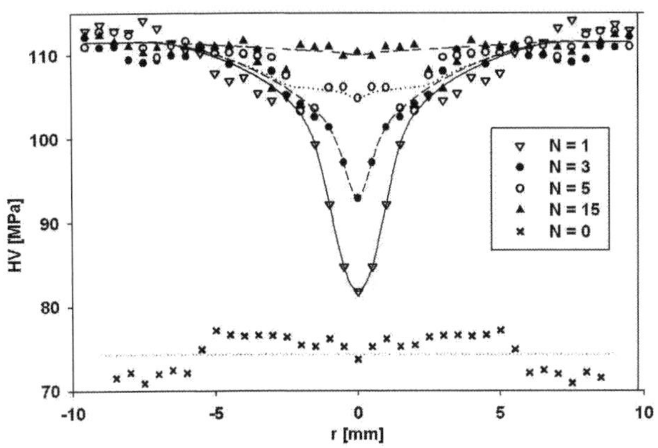

Figure 4. Microhardness distribution across the diameters of AZ31 disks subjected to a pressure of 2.5 GPa and up to 15 whole number of revolutions.

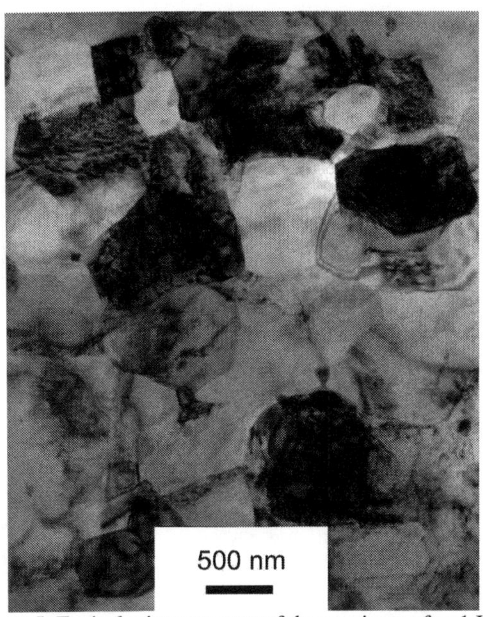

Figure 5. Typical microstructure of the specimen after 1 HPT rotation (middle section, $\varepsilon \approx 4$).

Most grain boundaries are very close to equilibrium. It is manifested by a typical thickness band structure. The detailed analysis of TEM observations in individual parts of specimens is still in progress. However, general conclusions obtained from the preliminary investigation may be summarized as follows: in the specimen after 1 turn significantly smaller grains were observed in the periphery of the disk than in the center. On the other hand, almost homogeneous distribution of grain sizes was observed in the specimen after 15 turns.

Changes of average Hv value in different parts of the disks are shown in Fig. 6 confirming the different ways towards the saturation in the center, in the middle section and at the periphery of the specimen.

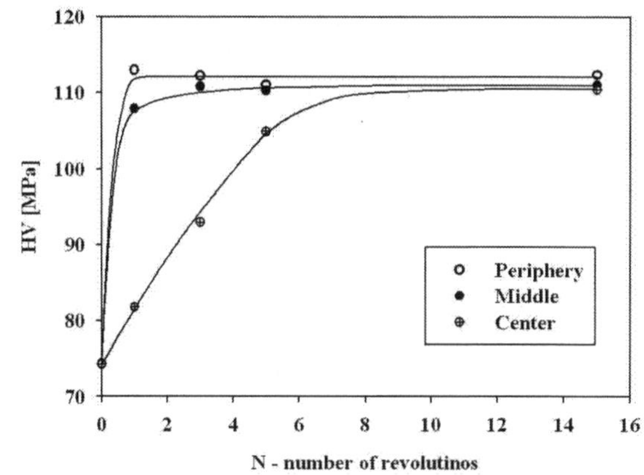

Figure 6. The evolution of average microhardness values in different parts of the disks due to HPT straining.

In order to obtain more complex image of microhardness evolution throughout individual specimens subjected to different numbers of HPT rotations three-dimensional meshes of Hv data were constructed using the following procedure. The microhardness data were measured in one quarter of each disk as described in the previous section (approximately 400-500 indents were made in each specimen). First, these data were depicted, minimal microhardness found and matched with the center of the sample. Due to the step of measurement the center of the disk was found with the accuracy of +/- 0.25 mm. Data for the whole specimen were constructed from measured data symmetrically with respect to the center (data at the same distance from the center that were measured twice were both used and the average value calculated). However, such "completed data" are not suitable for 3D-depicting. The following smoothing procedure was applied to remove each wrongly indented or evaluated datum.

Each value was recomputed using the original value and values of all close neighbors (4 edge neighbors and 4 corner neighbors). The distance of corner neighbors is √2 times higher than for edge neighbors, so the weight of corner neighbors is divided by √2 in the smoothing algorithm. Smoothing can be simply demonstrated by the following equation:

$$FV = c.OV + \frac{b}{4+2\sqrt{2}} \sum_{i=1}^{4} \left(EN_i + \frac{CN_i}{\sqrt{2}} \right) \quad (1)$$

where FV is the final value, OV is the original value, EN is the edge neighbor, CN is the corner neighbor; fraction denominator represents partial normalization and finally c and b are smoothing parameters. However b and c are not independent (because of total normalization, i.e. $b + c = 1$). Thus only e.g. c can be arbitrarily chosen. The choice $c = 1$ means that no smoothing occurs, whereas minimal meaningful value of c is 0.12. Otherwise the weight of each edge neighbor is higher than the weight of original values. In our case the value of $c = 0.3$ was used. It corresponds to the weight of 0.3, 0.1 and 0.07 for the original value, the edge neighbor and the corner neighbor, respectively.

Equation (1) holds only for interior points (a point that has all eight neighbors). Points that have incomplete number of neighbors are treated in the manner that nonexistent neighbors are ignored and weights of others are appropriately adjusted. Similarly, this procedure allows healing the missing data from the interior of the sample. The procedure extrapolates any missing value from values of the neighbors (if at least 5 of 8 possible neighbors exist). This procedure allows depicting 3D plots in a readable way even for partly damaged and/or missing data.

Unlike Fig. 4 which is based on taking measurements of the variation in the microhardness values following diameters across disks after HPT processing, Fig. 7 shows integrated data across the total surface of individual disks. In Fig. 7 three-dimensional meshes of microhardness for specimens subjected to different number of turns are displayed. These plots were obtained by symmetrical completion of measured data and by smoothing procedure described in the previous paragraph. The variations of microhardness with position within the disk are clearly displayed at these meshes. The pronounced drop of microhardness in the center of the specimen after one turn is clearly seen in Fig. 7a. In the specimen after 3 turns (Fig. 7b) the central drop is still visible even if its depth is much lower in comparison with the specimen after 1 turn. The tendency to saturation with increasing number of turns is demonstrated at the mesh for the specimen after 15 turns (see Fig. 7c) whose surface is almost flat with slight undulations only, confirming saturated Hv values within the whole specimen.

Discussion

Microhardness Behavior and the Development of Homogeneity in AZ31

The variations in the values of microhardness across the diameters of AZ31 disks processed by HPT are in full agreement with the well-known dichotomic behavior reported by other authors in materials with face centered cubic structure and low stacking fault energy, e.g austenitic steel [26], Cu [27] and Ni [28]: i) lower hardness values were reported in the centers of disks and higher values in peripheral regions and ii) when torsional straining is continued to a sufficiently high total strain these variations tend to saturate and almost homogeneous microhardness distribution throughout the diameter of the disk is observed. Inspecting Fig. 6 shows that the saturation is achieved after 5 HPT rotations. As reported by Somekawa et al. [29] AZ31 alloy has a stacking fault energy of 27.8 mJm^{-2}. Our measurements indicate that it is the stacking fault energy rather than the lattice structure which controls the microstructure evolution of the material which is macroscopically manifested by microhardness behavior. Contrary to these results, the centers of disks for pure aluminum exhibit higher values of Hv than the surroundings areas in early stages of torsional straining [30]. The authors explain this inverse behavior by exceptionally high rate of dynamic recovery in this material. Relatively low rate of dynamic recovery may be therefore expected in AZ31 alloy.

Detail inspection of 3D meshes in Fig. 7 reveals the undulating character of microhardness variations across HPT disks indicating the manner in which the deformation develops in AZ31 alloy. This problem was recently addressed by strain gradient plasticity modeling introduced by Estrin et al. [31]. The authors consider the material in the form of a composite material comprising cell walls and cell interiors. The dynamic recovery in the former occurs by climbing of dislocations whereas in the latter it is the cross slip of dislocations which controls the rate of dynamic recovery. The cross slip in AZ31 is very difficult due to high separation width between partial dislocations. For a low stacking fault energy material the model predicts the behavior observed in Fig. 4 with clear minimum in the center of the disk in early stages of straining tending to saturate with continuing deformation.

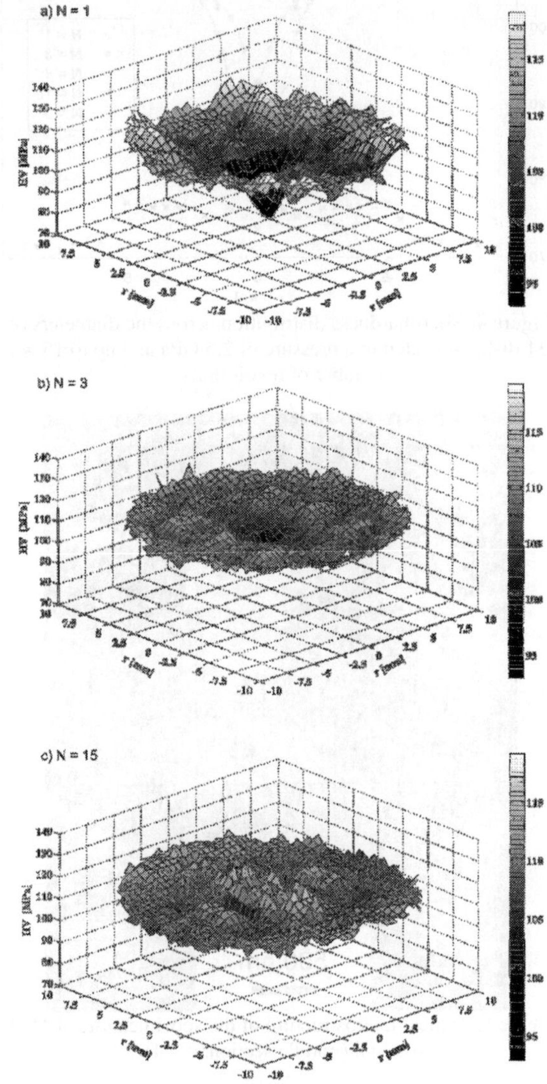

Figure 7. Three- dimensional meshes of microhardness as a function of the number of turns: a) N=1 turn, b) N=3 turns, c) N=15 turns.

Strain imposed by HPT

Due to the variability in strain across the radius of torsionally strained specimen there were many attempts how to express the strain imposed by HPT in terms of the radius of the disk and the number of HPT rotations.

Exhaustive discussion of the definition of strain imposed during HPT processing allowing calculation of different strain types (the true accumulated strain, shear strain, equivalent von Mises strain, the equivalent strain and the true strain) using different models can be found in the review by Zhilyaev [32].

The most precise quantity, taking directly into account the change of the specimen thickness during HPT processing, is the true strain ε defined as:

$$\varepsilon = \ln\left[1+\left(\frac{2\pi N.r}{h}\right)^2\right]^{\frac{1}{2}} + \ln\left(\frac{h_0}{h}\right), \qquad (2)$$

where N is number of revolutions, r is radius of the specimen, h_0 and h is the original and final thickness of the specimen, respectively. The first term of the equation (2) corresponds to the torsion of the specimen, whereas the second term refers to the compression strain due to thickness reduction.

The initial thickness h_0 of samples before HPT was 1 mm. The final thickness h of samples depends slightly on the number of HPT turns. Table I summarizes the values of final thickness of specimens after different number of HPT turns obtained as the average of thickness measurements at four specimens in the same condition. As the volume of the sample does not change significantly, the final thickness decrease was caused by outward flow of material due to a quasi-constrained setup of HPT anvils.

Table I. The dependence of the final thickness of HPT specimens on the number of turns.

N	h [mm]
0	1
1	0.8
3	0.76
5	0.75
15	0.74

The theoretical dependence of strain on the number of revolutions calculated according to eq. (2) is shown in Fig. 8. Radii under consideration are 10 mm (the periphery of the sample), 5 mm (middle section) and 0.25 mm (the center - it is the approximate value of the distance from sample center at which the closest microhardness value could have been measured). It is of particular interest that the strain saturates very soon (after 3 revolutions), especially for peripheral and middle regions. To illustrate this: another two revolutions (3 together) do not cause the strain of peripheral and middle section to rise for more than one third of the strain imposed after the first revolution. On the other hand, by second and third revolution the strain imposed in the center doubles. Similar relations hold between 3 and 15 revolutions. By straining the sample for additional 14 rotations (15 altogether) the imposed strain on the periphery and in the middle doubled only as compared to the strain after one rotation. Whereas by additional 14 rotations the imposed strain in the center tripled with respect to the strain after one rotation.

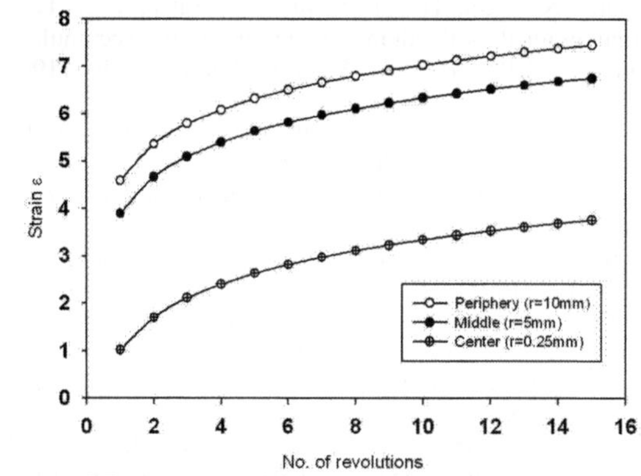

Figure 8 Dependence of the true strain on the number of HPT rotations in various parts of the specimen.

Conclusions

The influence of HPT straining on microhardness evolution and microstructure in cast magnesium alloy AZ31 was investigated. The following conclusions may be drawn from this work.

- HPT straining results in strong grain refinement.
- HPT straining introduces a radial inhomogeneity in the material with a pronounced minimum in the center and the maximum near specimen periphery.
- The inhomogeneity in Hv is continuously smeared out with increasing number of turns by continuous extension of the zone of maximum hardness towards the specimen center.
- The integrated Hv data across the whole surface of HPT specimens revealed undulating character of microhardness variations across HPT disks.
- The rate of dynamic recovery in AZ31 is very low due to its the low stacking fault energy.
- Preliminary observations of microstructure evolution correspond well with microhardness behavior.

Acknowledgements

The authors acknowledge funding through the research program MSM 0021620834 of Ministry of Education of the CR. Partial support by GACR under grant 106/09/0482 is also gratefully acknowledged. J. Vrátná acknowledges financial support by GAUK under the Grant 9594/2009 and by Grant SVV-261303.

References

1. S.R. Agnew, J.A. Horton, T.M. Lillo and D.W. Brown, "Enhanced Ductility in Strongly Textured Magnesium Produced by Equal Channel Angular (ECA) Processing", *Scripta Materialia*, 50 (2004), 377-381.

2. Y. Estrin, S.B.Yi, H.G. Brokmeier, Z. Zúberova, S.C. Yoon, H.S. Kim, R.J. Hellmig, "Microstructure, texture and mechanical properties of the magnesium alloy AZ31 processed by ECAP", *International Journal of Materials Research*, 99 (2008), 50-55.

3. R.B. Figueiredo, T.G. Langdon, "The characteristics of superplastic flow in a magnesium alloy processed by ECAP", *International Journal of Materials Research,* 100 (2009), 843-846.

4. R.Z. Valiev, T.G. Langdon, “Principles of equal-channel angular pressing as a processing tool for grain refinement”, *Progress in Materials Science,* 51 (2006), 881-981.
5. Y. Saito, N. Tsuji, H. Utsunomiya, T. Sakai, R.G. Hong, “Ultra-fine grained bulk aluminum produced by accumulative roll-bonding (ARB) process”, *Scripta Materialia,* 39 (1998), 1221-1227.
6. N.A. Smirnova, V.I. Levit, V.I. Pilyugin, R.I. Kuznetsov, L.S. Davydova, V.A. Sazonova, “Evolution of the structure of f.c.c. single crystal subjected to strong plastic deformation”, *Fizika Metallov i Metallovedeniya,* 61 (1986), 1170–1177.
7. V.M. Segal, “Materials processing by a simple shear“, *Mater. Sci. Eng.*, A 197 (1995), 157-164.
8. R.Z.Valiev, D.A Salimonenko, N.K. Tsenev, P.B. Berbon, T.G. Langdon, “Observations of high strain rate superplasticity in commercial aluminium alloys with ultrafine grain sizes“, *Scripta Materialia,* 37 (1997), 1945-1950.
9. Z. Horita, M. Furukawa, M. Nemoto, A.J. Barnes, T.G. Langdon, "Superplastic forming at high strain rates after severe plastic deformation", *Acta Materialia*, 48 (2000), 3633-3640.
10. A. Yamashita, Z. Horita, T.G. Langdon, “Improving the mechanical properties of magnesium and magnesium alloy through severe plastic deformation“, *Materials Science and Engineering*, A 300 (2001), 142-147.
11. Z. Horita, K. Matsubara, T.G. Langdon, “A two-step processing route for achieving a superplastic forming capability in dilute magnesium alloys“, *Scripta Materialia,* 47 (2002), 255-260.
12. K. Matsubara, Y. Miyahara, Z. Horita, T.G. Langdon, "Developing superplasticity in a magnesium alloy through a combination of extrusion and ECAP", *Acta Materialia*, 51 (2003), 3073-3084.
13. Z. Horita, T.G. Langdon, “Microstructures and microhardness of an aluminum alloy and pure copper after processing by high-pressure torsion”, *Materials Science and Engineering*, A 410–411 (2005), 422-425.
14. Z. Horita, D.J. Smith, M. Nemoto, R.Z. Valiev, T.G. Langdon, “Observations of grain boundary structure in submicrometer-grained Cu and Ni using high-resolution electron microscopy”, *Journal of Materials Research,* 13 (1998), 446-450.
15. Y.H. Zhao, Y.T. Zhu, X.Z. Liao, Z. Horita, T.G. Langdon, “Influence of stacking fault energy on the minimum grain size achieved in severe plastic deformation”, *Materials Science and Engineering, A* 463 (2007), 22-26.
16. A.P. Zhilyaev, G.V. Nurislamova, B.K. Kim, M.D. Barό, J.A. Szpunar, T.G. Langdon, “Experimental parameters influencing grain refinement and microstructural evolution during high-pressure torsion”, *Acta Materialia*, 51 (2003), 753-765.
17. Z. Horita, D.J. Smith, M. Furukawa, M. Nemoto, R.Z. Valiev, T.G. Langdon, “An investigation of grain boundaries in submicrometer-grain Al_Mg solid solution alloys using high-resolution electron microscopy”, *Journal of Materials Research,* 11 (1996), 1880-1890.
18. G, Sakai, Z, Horita, T,G. Langdon, “Grain refinement and superplasticity in an aluminum alloy processed by high-pressure torsion”, *Materials Science and Engineering*, A 393 (2005), 344-353.
19. S. Dobatkin, E.N. Bastarache, G. Sakai, T. Fujita, Z. Horita, T.G. Langdon, “Grain refinement and superplastic flow in an aluminum alloy processed by high-pressure torsion”, *Materials Science and Engineering*, A 408 (2005), 141-146.
20. Y.U. Ivanisenko, I. MacLaren, X. Sauvage, R.Z.Valiev, H.J. Fecht, “Shear-induced $\alpha \rightarrow \gamma$ transformation in nanoscale Fe–C composite”, *Acta Materialia,* 54 (2006), 1659-1669.
21. S.S.M. Tavares, D. Gunderov, V. Stolyarov, J.M. Neto, “Phase transformation induced by severe plastic deformation in the AISI 304L stainless steel”, *Materials Science and Engineering*, A 358 (2003), 32-36.
22. J. Petruželka, L. Dluhoš, D. Hrušak, J. Sochova, “Nanostructure titanium – new material for dental implants (in Czech)”, *Česká Stomatologická Ročenka,* 106 (2006), 72-77.
23. L. Saldana, A. Mendez-Vilas, L. Jiang, M. Multigner, J.L. Gonzalez-Carrasco, M.T. Perez-Prado et al. “In vitro biocompatibility of an ultrafine grained zirconium”, *Biomaterials,* 28 (2007), 4343-4354.
24. M. Kai, Z. Horita, T.G. Langdon, “Developing grain refinement and superplasticity in a magnesium alloy processed by high-pressure torsion”, *Materials Science and Engineering*, A 488 (2008), 117-124.
25. Y. Harai, M. Kai, K. Kaneko, Z. Horita, T.G. Langdon, “Microstructural and Mechanical Characteristics of an AZ61 Magnesium Alloy Processed by High-Pressure Torsion”, *Materials Transactions,* 49 (2008), 76-83.
26. A. Vorhauer, R. Pippan, "On the homogeneity of deformation by high pressure torsion," *Scripta Materialia*, 51 (2004), 921-925.
27. H. Jiang, Y.T. Zhu, D.P. Butt, I.V. Alexandrov, T.C. Lowe, “Microstructural evolution, microhardness and thermal stability of HPT - processed Cu”, *Materials Science and Engineering*, A 290 (2000) 128-138.
28. Z. Yang, U. Welzel, “Microstructure–microhardness relation of nanostructured Ni produced by high-pressure torsion”, *Materials Letters*, 59 (2005), 3406-3409.
29. H. Somekawa, K. Hirai, H. Watanabe, Y. Takigawa, K. Higashi, "Dislocation creep behavior in Mg–Al–Zn alloys“, *Materials Science and Engineering*, A 407 (2005), 53-61.
30. C. Xu, Z. Horita, T.G. Langdon, “The evolution of homogeneity in processing by high-pressure torsion”, *Acta Materialia*, 55 (2007), 203-212.
31. Y. Estrin, A. Molotnikov, C.H.J. Davies, R. Lapovok, "Strain gradient plasticity modelling of high-pressure torsion“, *Journal of the Mechanics and Physics of Solids*, 56 (2008), 1186-1202.
32. A.P. Zhilyaev, T.G. Langdon, “Using high/pressure torsion for metal processing: Fundamentals and applications”, *Progress in Materials Science,* 53 (2008), 893-979.

Magnesium Technology 2011
Edited by: Wim H. Sillekens, Sean R. Agnew, Neale R. Neelameggham, and Suveen N. Mathaudhu
TMS (The Minerals, Metals & Materials Society), 2011

PLASTIC DEFORMATION OF MAGNESIUM ALLOY SUBJECTED TO COMPRESSION-FIRST CYCLIC LOADING

Soo Yeol Lee[1,2], Michael A. Gharghouri[2], and John H. Root[2]

[1] Department of Materials Engineering, The University of British Columbia, Vancouver, B.C. V6T 1Z4, Canada
[2] Canadian Neutron Beam Centre, National Research Council Canada, Chalk River, ON K0J 1J0, Canada

Keywords: Magnesium, Plastic deformation, Twinning, Neutron diffraction

Abstract

In-situ neutron diffraction has been employed to study the deformation mechanisms in a precipitation-hardened and extruded Mg-8.5wt.% Al alloy subjected to compression followed by reverse tension. The starting texture is such that the basal poles of most grains are oriented normal to the extrusion axis and a small portion of grains are oriented with the basal pole parallel to the extrusion axis. Diffraction peak intensities for several grain orientations monitored in-situ during deformation show that deformation twinning plays an important role in the elastic-plastic transition and subsequent plastic deformation behavior. Significant non-linear behavior is observed during unloading after compression and appears to be due to detwinning. This effect is much stronger after compressive loading than after tensile loading.

Introduction

Magnesium and its alloy are currently the subject of many studies due to their potential use for lightweight structures in the automotive and aircraft industries, and for portable electronic devices [1,2]. The poor room-temperature formability of these alloys arises from the limited number of available slip systems [3,4]. The primary slip system in magnesium is <a> slip with a 1/3<11–20> Burgers vector on the close-packed (0002) basal plane. Non-basal <a> slip on the {10–10} prismatic and {10–11} pyramidal planes has also been observed. None of these slip modes provide deformation along the c-axis. Deformation twinning can provide deformation along the c-axis; {10–12}<10–11> tensile twinning is commonly observed at room temperature in favorably oriented grains, resulting in a strong tension-compression asymmetry in textured specimens due to the unidirectional nature of twinning [5-7]. Deformation twinning results in texture evolution because it gives rise to a large lattice reorientation relative to the parent grain [8-10]. Neutron diffraction has been successfully used, during deformation, to investigate bulk texture evolution due to twinning, as well as internal stress distributions among the different crystallographic orientations [11-14].

In this work, lattice strains and diffraction peak intensities for several grain orientations were measured in an extruded Mg-8.5wt.% Al alloys during deformation using neutron diffraction. The loading path consisted of compression followed by reverse tension.

Experimental details

The binary alloy was supplied by the Péchiney Research Centre, Voreppe, France. The material was solution treated and aged. The volume fraction of β-$Mg_{17}Al_{12}$ precipitates is ~10%. The precipitates are plate shaped, with an average thickness of 165nm, a length that varies from 6 to 20μm and a length to width ratio between 2 and 6. The grains are equiaxed with an average size of ~60 μm. Neutron diffraction experiments were conducted on the L3 spectrometer of the Canadian Neutron Beam Centre, Chalk River Laboratories, Canada. Data were acquired for the (10–10), (10–11) and (0002) diffraction peaks with the scattering vector parallel to and normal to the applied load. The experiments were conducted in-situ, which allowed us to observe lattice strain evolution and bulk texture development as a function of applied load. The material microstructure and neutron diffraction experiments are described in detail elsewhere [11].

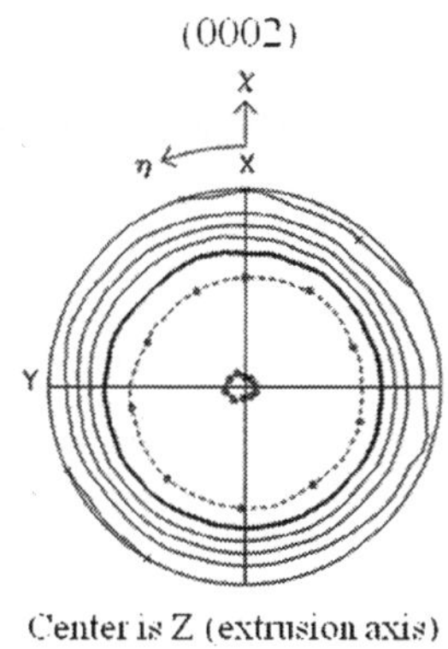

Figure 1. (0002) pole figure showing initial texture of the test material measured by neutron diffraction. The pole figure is contoured in multiples of random distribution (m.r.d) with the thick solid black line corresponding to 1 m.r.d. The contour levels above and below 1 m.r.d are shown by solid and dotted lines, respectively, in 0.5 m.r.d steps.

Results and Discussion

The (0002) basal pole figure for the as-received material is shown in Fig. 1. The center of the pole figure corresponds to the extrusion axis. The (0002) basal planes in most grains are parallel or close to the extrusion axis, with a correspondingly small fraction of grains with the basal planes perpendicular to the extrusion axis. Fig. 2 shows the macroscopic stress-strain response of the alloy subjected to compression followed by reverse tension. The symbols in the graph correspond to points in the loading history at which diffraction data were acquired. In compression up to –100 MPa, the alloy shows mainly elastic deformation, with some limited plasticity. Beyond –100 MPa, the elastic-plastic transition is well underway, though plasticity is not fully developed. During unloading after compression and reloading in tension, the stress-strain response is clearly non-linear, showing a typical s-shaped profile. The material undergoes general yielding in tension at ~+125 MPa. The unloading portion of the curve, after tensile loading, is again non-linear, but the effect is less significant than after compression.

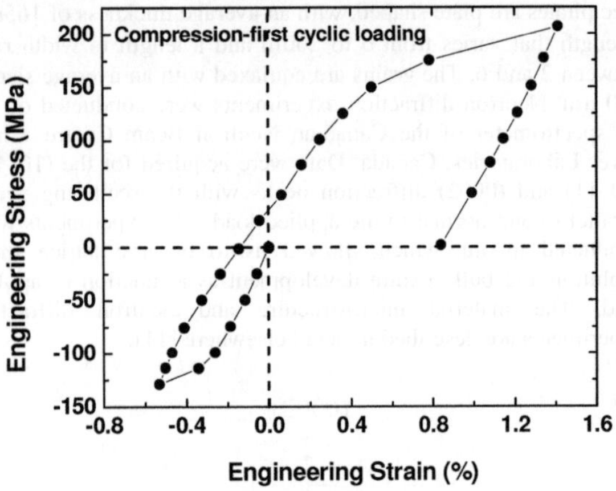

Figure 2. Macroscopic stress-strain response of the extruded Mg-8.5wt.%Al alloy. The symbols correspond to points in the loading history at which diffraction data were acquired.

The integrated intensity data acquired during loading for all measured reflections are shown in Fig. 3. For these measurements, the scattering vector was parallel to the loading direction. The data are plotted chronologically as functions of the applied load in order to show clearly how the measured values change during the test. The (0002), {10–11} and {10–10} orientations have the basal poles (c-axes) oriented at 0°, 61.9° and 90°, respectively, relative to the extrusion direction (ED), which is also the loading direction. Thus, the {10–10} and (0002) grains are favorably oriented for extension twinning in compression and tension, respectively. Neither family of grains is oriented favorably for <a>-slip on any plane. In contrast, the {10–11} grains are favorably oriented for basal <a> slip. The {411} reflection is from the β-$Mg_{17}Al_{12}$ precipitates.

During initial loading in compression, the (0002) intensity starts to increase beyond ~–100 MPa, while the {10–10} intensity decreases concurrently. This corresponds with the macroscopic yield point (~–100 MPa). The intensity of a diffraction peak changes when the amount of material oriented for Bragg diffraction changes. Since the volume sampled by the neutron beam is constant, this can only occur if there is a phase change, or if material is re-orientated. In the present case, no phase change occurs. Thus, a decrease in intensity of a given reflection means that the contributing grains have twinned – i.e. that some of the grain volume is no longer appropriately oriented for diffraction. Conversely, an increase in intensity means that a conjugate set of grains has undergone twinning, such that additional material contributes to the intensity. In the case of magnesium, {10–12} extension twinning results in an 86.6° reorientation of the crystal lattice, which would convert an {10–10} orientation into an (0002) orientation and vice versa.

As the load increases from –100 MPa to –130 MPa, the (0002) intensity increases sharply due to {10–12}<10–11> extension twinning in grains with the c-axis normal to the loading direction, such as the {10–10} grains. It is thus likely that this extension twinning is responsible for the observed plastic deformation behavior shown in Fig. 2. Since many grains in the starting material have the c-axis approximately normal to the loading direction, the lattice reorientation due to extension twinning will reorient the material into the (0002) orientation. The effect on the intensity of the (0002) reflection is thus very strong, resulting in a 3-fold increase even at very low macroscopic strain.

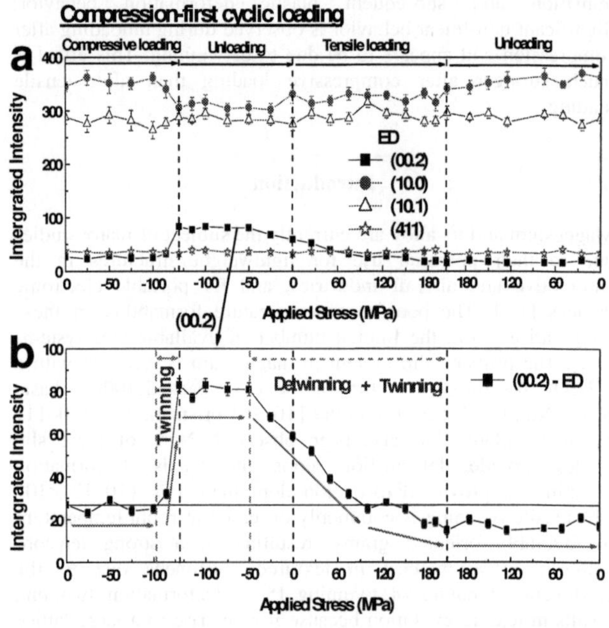

Figure 3. (a) Intensity variations as functions of applied load during the deformation shown in Figure 2; (b) Close-up of intensity variations for the (0002) reflection. Measured (*hkil*) plane normal is parallel to the applied loading direction, which is the same as the extrusion direction (ED).

When the sample is unloaded after compression, the intensity of the (0002) peak remains stable down to an applied stress of ~–50 MPa, and then decreases gradually. The intensity falls by about 40% during unloading to zero load, indicating that about 40% of the twinned volume has detwinned during unloading after compression. It is possible that the parent (0002) grains undergo compressive {10–11}<10–12> twinning during unloading after compression, but this is unlikely as this type of twinning is very uncommon, and is generally found to contribute little to lattice reorientation during deformation. This detwinning behavior is thought to contribute significantly to the non-linear behavior observed in Fig. 2.

During reverse loading in tension, detwinning continues until the (0002) intensity at the start of the test is fully recovered at about +100 MPa. As the applied load increases from 100 MPa to 205 MPa, the (0002) intensity continues to decrease, but at a lower rate. At this point, it appears that the (0002) minority grains undergo {10–12} extension twinning. During unloading after the tensile portion of the stress-strain curve, the (0002) peak intensity does not change, suggesting that the twinned material in the minority (0002) grains does not undergo significant detwinning.

Summary

Neutron diffraction has been used to study the plastic deformation behavior of a Mg-8.5wt.%Al alloy subjected to compression followed by reverse tension. It was found that the onset of extension twinning corresponds well with the macroscopic elastic-plastic transition. The non-linear behavior during unloading after compression was more significant than that after tension. Diffraction results show that about 40% of the twinned volume detwins during unloading after compression, but that no significant detwinning occurs during unloading after tension.

Acknowledgements

This work was supported by funding from the NSERC Magnesium Strategic Research Network. More information on the Network can be found at www.MagNET.ubc.ca.

References

1. M.M. Avedesian and H. Baker, Magnesium and Magnesium Alloys, ASM Specialty Handbook, ASM International, Materials Park, OH, 1999.
2. C. Wang, P. Han, L. Zhang, C. Zhang, X. Yan, B. Xu, J. All. Comp. 482 (2009) 540-543.
3. C.S. Roberts, Magnesium and its Alloys, John Wiley & Sons, Inc., New York, 1960.
4. O. Muránsky, D.G. Carr, M.R. Barnett, E.C. Oliver, P. Šittner, Mater. Sci. Eng. A496 (2008) 14-24
5. P.G. Partridge, Metall. Rev. 12 (1967) 169-194.
6. S.R. Agnew, C.N. Tomé, D.W. Brown, T.M. Holden, S.C. Vogel, Scripta mater. 48 (2003) 1003-1008.
7. J. Jain, W.J. Poole, C.W. Sinclair, M.A. Gharghouri, Scripta mater. 62 (2010) 301-304.
8. B. Clausen, C.N. Tomé, D.W. Brown, S.R. Agnew, Acta mater. 56 (2008) 2456-2468.
9. G. Proust, C.N. Tomé, A. Jain, S.R. Agnew, Int. J. Plasticity 25 (2009) 861-880.
10. O. Muránsky, M.R. Barnett, V. Luzin, S. Vogel, Mater. Sci. Eng. A 527 (2010) 1383-1394.
11. M.A. Gharghouri, G.C. Weatherly, J.D. Embury, J. Root, Phil. Mag. A 79 (1999) 1671-1695.
12. D.W. Brown, S.R. Agnew, M.A.M. Bourke, T.M. Holden, S.C. Vogel, C.N. Tome, Mater. Sci. Eng. A 399 (2005) 1-12.
13. L. Wu, S.R. Agnew, D.W. Brown, G.M. Stoica, B. Clausen, A. Jain, D.E. Fielden, P.K. Liaw, Acta mater. 56 (2008) 3699-3707.
14. O. Muránsky, M.R. Barnett, D.G. Carr, S.C. Vogel, E.C. Oliver, Acta mater. 58 (2010) 1503-1517.

Magnesium Technology 2011
Edited by: Wim H. Sillekens, Sean R. Agnew, Neale R. Neelameggham, and Suveen N. Mathaudhu
TMS (The Minerals, Metals & Materials Society), 2011

MICROSTRUCTURE EVOLUTION IN AZ61L DURING TTMP AND SUBSEQUENT ANNEALING TREATMENTS

T.D. Berman[1], W. Donlon[1], R. Decker[2], J. Huang[2], T.M. Pollock[3], J.W. Jones[1]

[1]Materials Science and Engineering, University of Michigan, 2300 Hayward St; Ann Arbor, MI, 48109, USA
[2]nanoMAG, LLC., 620 Technology Drive; Ann Arbor, MI, 48108, USA
[3]Materials Department, University of California, Santa Barbara; Santa Barbara, CA, 93106-5050, USA

Keywords: AZ61, Microstructures, Texture, Mechanical Properties, Thixomolding, Thermomechanical Processing

Abstract

Microstructure evolution is studied in Thixomolded® Thermomechanical Processed (TTMP) AZ61L sheet at various stages of processing. Transmission electron microscopy (TEM) is utilized to examine (1) grain refinement and recrystallization and (2) refinement and re-distribution of the β-$Mg_{17}Al_{12}$ phase in the as-Thixomolded, as-TTMP, and annealed conditions. Electron backscatter diffraction (EBSD) is used to study texture evolution through TTMP and annealing. The influence of microstructure produced by TTMP and annealing on the mechanical properties will be discussed.

Introduction

The high cost and poor formability of magnesium sheet has limited its commercial application [1]. If, through materials design and processing, formability could be enhanced, several markets, especially automotive, would benefit. The emerging consensus is that deformation modes required for high ductility materials are more easily activated in fine, micro- or nanoscale grains [2]. Several different processes to achieve fine scale microstructure have been developed, producing high strength, high ductility materials [3–10]. Thixomolding Thermal Mechanical Processing (TTMP) builds upon the fine grains, isotropy, and low porosity of Thixomolded Mg alloys containing eutectic phases. Intense thermomechanical processing is applied to further refine grain size and eutectic phases. Furthermore, additional thermal treatments can be applied to optimize strength, ductility and formability.

Experimental

Thixomolded and TTMP samples of AZ61L were obtained from Thixomat, Inc., composition shown in Table I. A rolling pass yielded a sheet thickness of 1.5 mm for the TTMP samples, a 50% reduction in the thickness of the as-Thixomolded plates. Annealing temperatures of 250 °C and 300 °C were chosen to explore effects of annealing on microstructure and mechanical properties.

Table I: Composition of received AZ61L in wt %

Al	Zn	Mn	Si	Fe	Mg
6.5	0.46	0.14	0.01	0.003	bal.

Dogbone tensile specimens with a gauge length of 31.75 mm and cross section of 7.94 mm, were machined from the sheets with the tensile axes parallel to the rolling direction. Room temperature tensile tests were performed with a displacement rate of 0.71 mm/min. An extensonometer was used to measure tensile elongation.

Samples for microscopy were removed from the grip ends of the tensile bars in the plane of the sheet. TEM specimens were prepared by electropolishing with a 8% perchloric acid in methanol electrolyte or dimpled and ion milled using a Gatan Precision Ion Polishing System (PIPS). TEM was conducted with a JEOL2000FX and a Phillips CM12 AEM systems. EBSD specimens were prepared by polishing with 1 μm diamond followed by ion polishing in a Gatan PIPS. EBSD examination was conducted on a JEOL 840A SEM system equipped with a HKL EBSD system. A step size of 0.5 μm was used.

Results

The as-Thixomolded material reveals equiaxed grains, regions of high dislocation density, and both agglomerated (indicated by the arrow) and isolated β particles (Fig. 1). In addition to agglomerated and isolated, equiaxed β grains, regions of eutectic β are also found in the as-Thixomolded condition. Fig. 2 highlights a region of eutectic β, likely composed of a series of small β grains along a boundary.

The TTMP-induced microstructure demonstrates significant grain refinement from the Thixomolded condition (Fig. 3). Highly elongated,deformed grain structures are seen (region A) in the TTMP condition. A collection of nanoscale grains, that may indicate local dynamic recrystallization occurred, are shown in region B. The elongated β structures and clumps of β particles

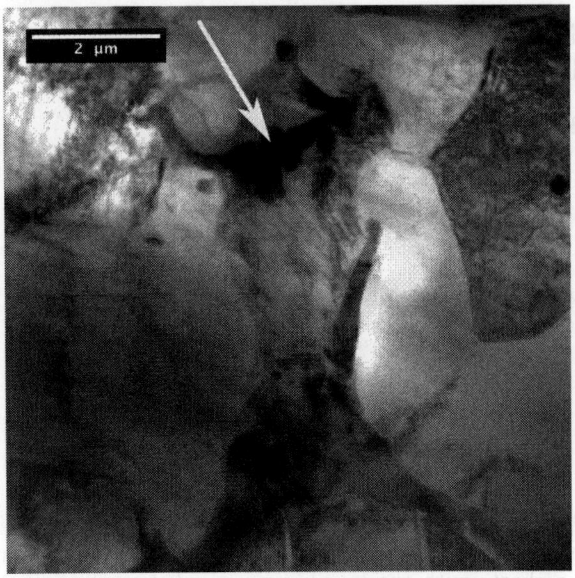

Figure 1: Microstructure of Thixomolded AZ61L. The arrow highlights a region of agglomerated β particles.

observed in the as-Thixomolded material are dissipated. β grains are heterogeneously distributed throughout the as-TMP microstructure.

Annealing treatments allow for further recrystallization. At 250 °C the material initially retains a fine grain microstructure. Grain coarsening becomes evident after 20 minutes at 250 °C (top row Fig. 4). Annealing at 300 °C initiates more rapid grain growth, evident even at the shortest annealing time (bottom row Fig. 4). Heterogeneous distribution of β grains is evident in Figs. 4(c-e). Nano-scale β particles in the TMP + annealed material were equiaxed and often exhibited a rounded morphology as can be seen in Fig. 5 and Fig. 6.

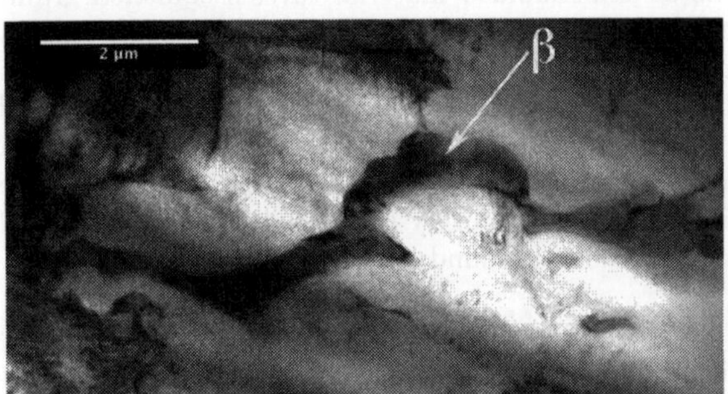

Figure 2: TEM micrograph showing the morphology of the eutectic β particles in the as-Thixomolded condition. Identification of phase was determined by EDAX measurements.

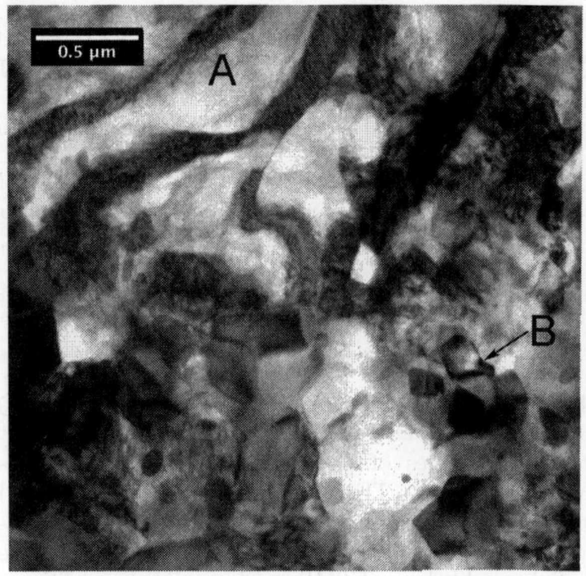

Figure 3: Elongated grains (A) and nanoscale α-Mg and β grains (B) are found in the as-TTMP condition.

β particles were frequently observed in the interior of grains and along α-Mg grain boundaries. Nano-scale recrystalized α-Mg grains tend to coincide with the β phase. Clusters of β particles and recrystallized α-Mg grains suggests that the β particles play a significant role in retarding grain growth.

A subset of the samples are analyzed by EBSD. As-TTMP and TTMP samples annealed for short times exhibit many (~75%) non-indexable EBSD diffraction patterns due to high deformation. The TMP samples annealed for longer durations yield fewer (~10%) non-indexable patterns. Band contrast images, with re-constructed grain boundaries (>5% misorientation), are shown in Fig. 7. Finer grains with a bimodal size distribution is evident for the TTMP + 300 °C 20 min anneal sample, demonstrating the effect of recrystallization. The maximum grain diameters for the as-Thixomolded and TTMP + 300 °C 20 min anneal are 36 μm and 14 μm respectively.

Pole figures were also obtained from the EBSD analysis. Nearly random texture patterns are observed in the as-Thixomolded and TTMP + 300 °C 20 min anneal sample (Fig. 8(a and d)). Strong (0001) texture patterns were observed for the as-TTMP sample(Fig. 8(b)) and the TTMP + 250 °C samples annealed for short times (Fig. 8(c)). In addition to the (0001) component, a weaker $\langle\bar{1}100\rangle$ texture was observed to follow the same trend parallel to the rolling direction.

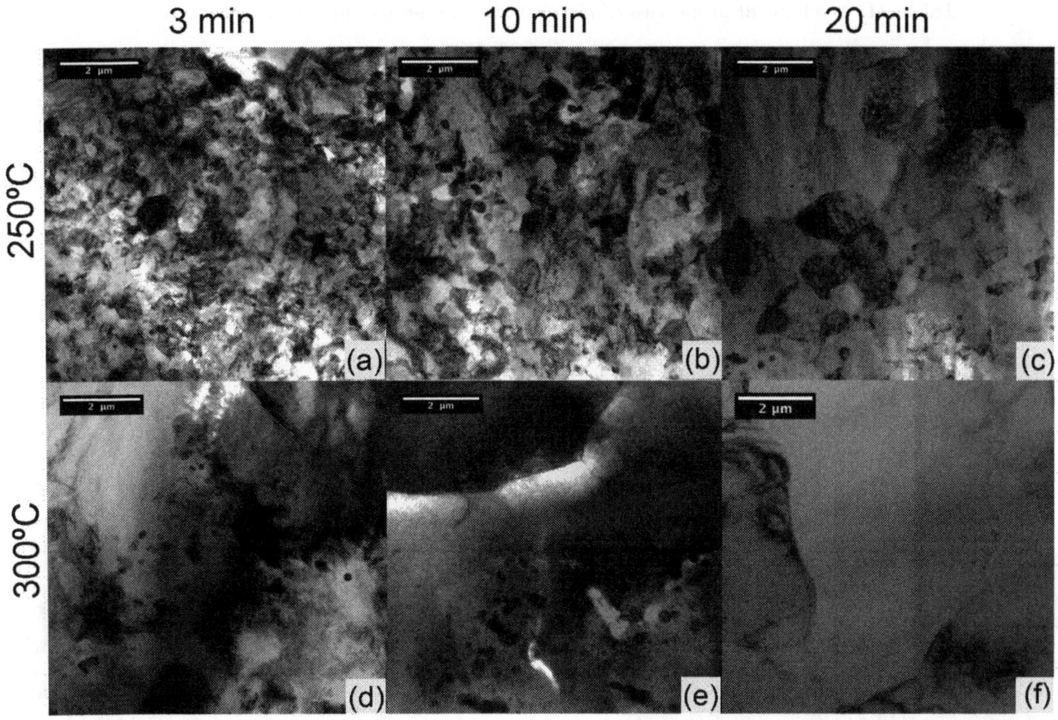

Figure 4: Microstructure evolution resulting from post TTMP annealing treatments (a) TTMP + 250 °C 3 min anneal, (b) TTMP + 250 °C 10 min anneal, (c) TTMP + 250 °C 20 min anneal, (d) TTMP + 300 °C 3 min anneal, (e) TTMP + 300 °C 10 min anneal, (f)TTMP + 300 °C 20 min anneal.

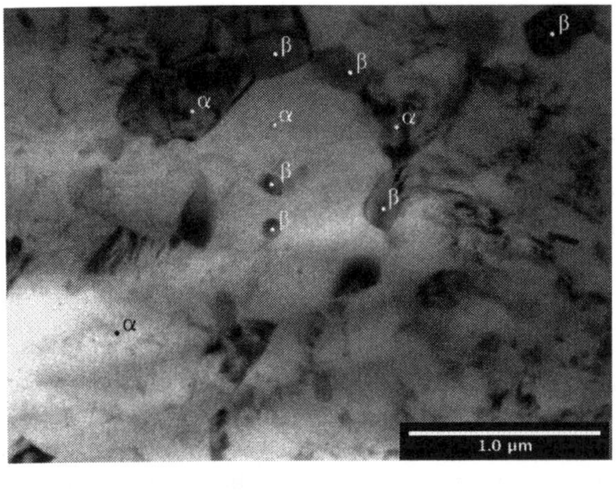

Figure 5: TEM micrograph of a sample annealed at 250 °C for 20 min with small α-Mg grains and β grains. Identification of the β phase was determined by EDAX measurements.

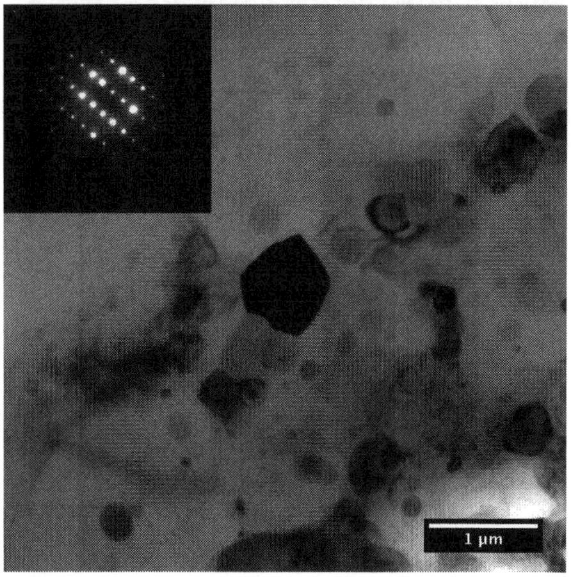

Figure 6: A larger β particle (dark) in a cluster of smaller, presumably β particles in the sample annealed at 300 °C for 20 min. SAED pattern from the $[\bar{1}13]$ zone axis of the β crystal.

Table II: Mechanical properties of the AZ61L samples examined in this study

Sample	YS (MPa)	UTS (MPa)	El (%)
as-Thixomolded	127.3 ± 14.7	191.5 ± 16.1	5.1 ± 0.4
as-TTMP	315.8 ± 6.8	370.7 ± 4.3	6.7 ± 3.9
TTMP + 250 °C 3 min Anneal	331.5 ± 6.4	372.3 ± 5.4	5.5 ± 2.4
TTMP + 250 °C 10 min Anneal	313.0 ± 17.2	365.3 ± 8.0	5.6 ± 3.4
TTMP + 250 °C 20 min Anneal	321.7 ± 3.1	361.3 ± 2.3	7.2 ± 1.4
TTMP + 300 °C 3 min Anneal	236.7 ± 4.1	315.3 ± 1.3	15.5 ± 2.1
TTMP + 300 °C 10 min Anneal	223.6 ± 1.2	308.5 ± 4.8	16.6 ± 3.0
TTMP + 300 °C 20 min Anneal	216.5 ± 5.7	307.6 ± 1.4	22.5 ± 1.5

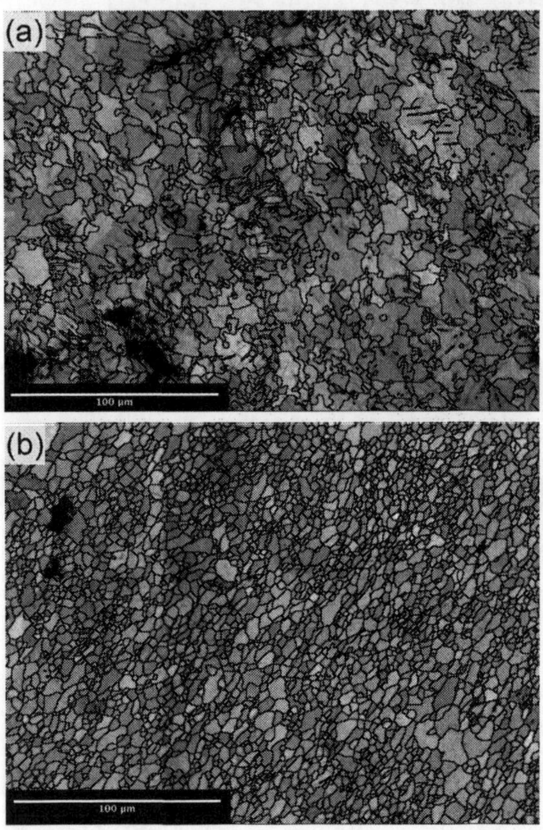

Figure 7: EBSD maps of (a) as-Thixomolded and (b) TTMP + 300 °C 20 min anneal

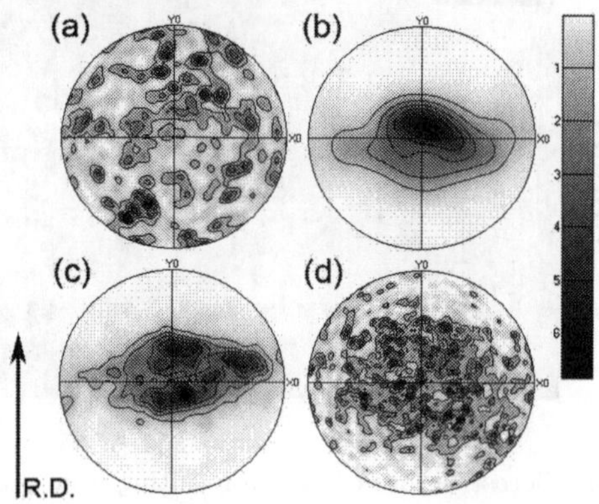

Figure 8: (0001) Pole figures of (a) as-Thixomolded, (b) as-TTMP, (c) TTMP + 250 °C 3 min anneal, and (d) TTMP + 300 °C 20 min anneal. Mapping was performed in the plane of the sheet with the rolling direction pointing up.

Discussion

It has previously been demonstrated that the effect of TMP on Thixomolded AZ61L is a marked increase in the yield strength and ultimate tensile strength of the material, without reduction of tensile ductility[3, 11, 12]. A summary of the mechanical characteristics of the material can be found in Table II. The TMP approximately doubled yield and tensile strength over that of the as-Thixomolded sample. Annealing at 250 °C had little affect on the tensile properties. As grains coarsen during the 300 °C annealing treatments, yield and tensile strength decreases . For the times and temperatures selected, annealing temperature had a more significant affect on tensile properties than annealing duration. An increase in ductility of samples annealed at 300 °C may result from a loss of deleterious texture as recrystallization occurs.

Low R value measurements taken in similarly processed AM60, indicate a decreased anisotropy from the as-TTMP condition, corresponding with the demonstrated loss of texture after the 300 °C 20 min anneal [11]. Texture evolution and R value measurements in TTMP AZ61L with anneal and/or aging treatments will be an area of future exploration. To our knowledge, non-textured magnesium sheet has not been seen in conventional rolled material.

Future investigations will further explore methods to (1) minimize texture after TTMP, (2) control and refine the distribution of β particles, and (3) optimize the heat-treatment to better control grain size, optimizing the mechanical properties.

Conclusions

TMP of Thixomolded AZ61L results in very fine grain sheet with a twofold increase in strength with no loss of ductility. Annealing at 250 °C had little affect on the mechanical properties from the as-TTMP condition and evidence of grain growth is not seen until 20 minutes. Grain coarsening and a drop in strength is seen after annealing the TTMP material at 300 °C. Annealing at 300 °C for 20 min was shown to lead to a nearly non-textured sheet with ductility several times greater than in the As-TTMP material. These results show the promise of optimizing the TTMP and post treatments to develop fine grain, non-textured magnesium sheet with high strength and formability.

Acknowledgement

This material is based upon work supported by the National Science Foundation under Grant No. 0847198.

References

1. S. Mathaudhu and E. Nyberg, "Magnesium Alloys in U.S. Military Applications: Past, Current and Future Solutions," *Magnesium Technology 2010*, eds. S. Agnew et al. (Minerals, Metals, and Materials Society, Warrendale, PA, 2010), 27–33.

2. J. Koike, "Enhanced Deformation Mechanisms by Anisotropic Plasticity in Polycrystalline Mg Alloys at Room Temperature," *Metall Mater Trans A*, 36 (7) (2005), 1689–1696.

3. R. Decker et al., "nanoMAG® High Strength/Density Mg Alloy Sheet," *Magnesium Technology 2009*, eds. E. Nyberg et al. (Minerals, Metals, and Materials Society, Warrendale, PA, 2009), 489–493.

4. J. Swiostek et al., "Hydrostatic Extrusion of Commercial Magnesium Alloys at 100 °C and its Influence on Grain Refinement and Mechanical Properties," *Mater Sci Eng A*, 424 (1-2) (2006), 223–229.

5. K. Matsubara et al., "Developing Superplasticity in a Magnesium Alloy Through a Combination of Extrusion and ECAP," *Acta Mater*, 51 (11) (2003), 3073–3084.

6. D. H. St. John et al., "Grain Refinement of Magnesium Alloys II . Grain Refinement of Magnesium Technical Status," *Metall Mater Trans A*, 36 (July) (2005), 1669–1679.

7. H. S. Di et al., "New Processing Technology of Twin Roll Strip Casting of AZ31B Magnesium Strip," *Mater Sci Forum*, 488-489 (2005), 615–618.

8. S. M. Zhang, Z. Fan, and Z. Zhen, "Direct Chill Rheocasting (DCRC) of AZ31 Mg Alloy," *Mater Sci Technol*, 22 (12) (2006), 1489–1498.

9. W.-J. Kim, G. E. Lee, and J. B. Lee, "Achieving Low Temperature Superplasticity from Ca-Containing Magnesium Alloy Sheets," *Adv Eng Mater*, 11 (7) (2009), 525–529.

10. G. Cao et al., "Study on Tensile Properties and Microstructure of Cast AZ91D/AlN Nanocomposites," *Mater Sci Eng A*, 494 (1-2) (2008), 127–131.

11. R. Decker et al., "Thixomolded and Thermomechanically Processed Fine-Grained Magnesium Alloys," *Mater Sci Forum*, 654-656 (2010), 574–579.

12. J. Huang et al., "On Mechanical Properties and Microstructure of TTMP Wrought Mg Alloys," *Magnesium Technology 2010*, eds. S. Agnew et al. (Minerals, Metals, and Materials Society, Warrendale, PA, 2010), 489–493.

Magnesium Technology 2011
Edited by: Wim H. Sillekens, Sean R. Agnew, Neale R. Neelameggham, and Suveen N. Mathaudhu
TMS (The Minerals, Metals & Materials Society), 2011

MODELING THE CORROSIVE EFFECTS OF VARIOUS MAGNESIUM ALLOYS EXPOSED TO TWO SALTWATER ENVIRONMENTS

H.J. Martin, C. Walton, J. Danzy, A. Hicks, M.F. Horstemeyer, P.T. Wang
Center for Advanced Vehicular Systems (CAVS), Mississippi State University, Box 5405, Mississippi State, MS 39762, USA

Keywords: Magnesium Alloy, Pitting Corrosion, Intergranular Corrosion, Modeling of Corrosion

Abstract

The use of magnesium within the automotive industry is limited by its corrosion rate in the presence of saltwater. By adding various elements, the magnesium microstructure and corrosion rate can be altered. In the Center for Advanced Vehicular Systems at Mississippi State University, a model is being developed to elucidate the total corrosion of magnesium alloys and is comprised of general corrosion and pitting corrosion, respectively, as shown below:

$$\phi = \phi_{gc} + \phi_{IC} \quad (1)$$

where pitting corrosion is based on the pit number density, pit surface area, and a nearest neighbor distance function, respectively, as shown below:

$$\phi_{IC} = \eta_{p} \nu_{p} C \quad (2)$$

The exposure environment resulted in differences in the amount of pit nucleation, in the size of the pits formed, and in the rate of coalescence. Time also affected the surface characteristics, as general corrosion began degrade the number and size of the pits. The research presented here will cover the model development, calibration, and validation.

Introduction

Magnesium has a high corrosion rate as compared to aluminum or steel, relegating its role in the automotive and aerospace industries in places that are unexposed to the environment [1-3]. In an effort to improve the corrosion resistance of magnesium, various elements are added, including aluminum, zinc, manganese, and rare earth elements [2, 4-8].

The addition of aluminum, up to 10%, has been shown to affect the corrosion resistance of magnesium [2]. Aluminum, which is present in the β-phase precipitate, appearing as $Mg_{17}Al_{12}$, leads to improved corrosion resistance when the β-phase is continuous and finely divided [2, 9-11]. However, the same β-phase precipitate can lead to the creation of micro-galvanic cells, thereby decreasing the corrosion resistance, when the β-phase is small and unconnected [2, 9-11].

In addition to the addition of aluminum, the presence of rare earth elements can affect the corrosion properties and mechanical properties of magnesium [4-8]. The formation of meta-stable rare earth element – containing phases along the grain boundaries improves the creep properties of magnesium and can also improve corrosion resistance due to trace amounts of the rare earth elements in the passive films formed during atmospheric exposure [6-8, 12]. The addition of the rare earth elements also shifts the location of pitting corrosion, from along the magnesium grain – eutectic boundary to the interior of the magnesium grain, resulting in unaffected rare earth intermetallic regions [7, 13].

The presence of alloying elements is not the only consideration when dealing with corrosion. The formation method, whether it is extrusion or casting, plays a significant role in the corrosion properties of magnesium alloys. Casting results in the formation of an as-cast "skin", with very small grains, which increases the corrosion resistance of magnesium, up to ten-fold higher than the bulk material [9, 14]. Since extrusion removes this as-cast skin, and does not allow the formation of another skin, extrusion can negatively affect the corrosion resistance of magnesium.

While the addition of elements to magnesium can increase the corrosion resistance of magnesium, currently, there are no available models to determine if the addition of certain elements will improve corrosion resistance. Models designed for stainless steel and aluminum attempt to quantify the causes of pitting, such as diffusion, energy, pH, and corrosion potential, but involve the use of a current to initiate the corrosion [15-22]. In addition, the only available model for magnesium requires a current to produce a polarization curve to predict galvanic corrosion, which is not applicable to this research [23]. The goal of this research, then, is to study various magnesium alloys in as-cast or extruded form in order to develop a model that accurately describes pit nucleation, pit growth, and pit coalescence.

Materials and Methods

<u>Testing</u>

Twelve AZ61 coupons (2.54 cm x 2.54 cm x varying thicknesses) were cut from an extruded crash rail provided by Ford using a CNC Mill (Haas, Oxnard, CA). Twelve AZ31 coupons (cm x cm x cm) were cut from extruded sheets using a vertical band saw (MSC Industrial Supply Company, Columbus, MS). Twelve AM60 coupons (2.54 cm x 2.54 cm x varying thicknesses) were cut from as-cast control arms using a vertical band saw. Twelve AE44 coupons (2.54 cm x 2.54 cm x varying thicknesses) were cut from an as-cast engine cradle provided by Meridian Technologies using a vertical band saw. The coupon surfaces were left untreated to test the corrosion effects on the extruded AZ61 and AZ31 magnesium alloys and on the as-cast AM60 and AE44 magnesium alloys.

Two different test environments were used in this study: salt spray testing and immersion. For salt spray testing, a Q-Fog CCT (Q-Panel Lab Products, Cleveland, OH) was used to cycle through three stages set at equal times, including a 3.5 wt.% NaCl spray at 35°C, 100% humidity using distilled water at 35°C, and a drying purge at 35°C. For immersion testing, an aquarium with an aeration unit was filled with 3.5 wt.% NaCl at room temperature. For both tests, the six coupons per test environment were hung at 20° to the horizontal, as recommended by ASTM B-117 [24]. The coupons were exposed to the test environment for 1 h, removed, rinsed with distilled water to remove excess salt, and dried. Following the profilometer analysis, the coupons were then placed back into the test environment for an additional 3 h, an additional 8 h, an additional 24 h, and another 24 h. These times allowed for a longitudinal study to follow pit growth and surface changes over time, where $t_0 = 0$, $t_1 = 1$ h, $t_2 = 4$ h, $t_3 = 12$ h, $t_4 = 36$ h, and $t_5 = 60$ h. Between analyses and environmental exposures, the coupons were stored in a desiccator to ensure that no further surface reactions occurred.

Analysis

Following each time exposure, the coupons were analyzed using optical microscopy and laser profilometry. The coupons were weighed prior to testing and following each exposure on two different scales and an average was taken. Four thickness measurements were taken on each sample prior to and following the test. Because the coupons were cut from an engine cradle, the thicknesses of the coupons varied from side to side, meaning an average was taken per coupon based on the four measurements. Measurements for all figures were averaged from the data with error bars based on one standard deviation.

Optical microscopy with an inverted light was used to take multiple images of the resulting surface at 5x magnification and 10x magnification (Axiovert 200M Mat, Carl Zeiss Imaging Solutions, Thornwood, NY). The 5x magnification images were combined and then analyzed using the ImageAnalyzer (v. 2.1-2) provided by Mississippi State University to determine the number of pits, the pit surface area, the nearest neighbor radius, and the intergranular corrosion area fraction necessary for the development of a corrosion model not detailed in this paper but previously outlined by Horstemeyer *et al.* [25]. The 10x magnification was used to pictorially show the changes over the six cycles. Laser profilometry was used to scan a 1 mm by 1 mm area on two coupons per environment following each test cycle (Talysurf CLI 2000, Taylor Hobson Precision Ltd, Leicester, England). The resulting 2-D and 3-D images were used to document the changes in the pit characteristics due to the different test environments over the six cycles (Talymap Universal, v. 3.18, Taylor Hobson Precision Ltd, Leicester, England). Data was collected based on fourteen pits within each 1 mm by 1 mm area, for a total of twenty-eight data points per environment per cycle.

Results

Figures 1 and 2 show the average weight and thickness loss, respectively, over the five exposure times for the immersion and salt spray surfaces on the various magnesium alloys being compared. As one can see, all surfaces follow similar logarithmic trends for weight loss (Figure 1) and thickness loss (Figure 2).

Figure 3 shows the pit number density over the five exposure times for the immersion and salt spray surfaces on the various magnesium alloys being compared. As one can see, all surfaces followed second-order polynomial trends. The AZ61 surfaces showed the highest amount of pit formation as compared to the other surfaces, while the as-cast AM60 surfaces showed the lowest amount of pit formation. In addition, all immersion surfaces had higher pit number densities as compared to the respective salt spray surface.

Figure 4 shows the changes in the pit area, which is the 2-D area covered by the pits as seen by micrographs for the immersion and salt spray surfaces on the various magnesium alloys being compared. As with pit number density, the most surfaces followed a second-order polynomial, with the highest pit area occurring on the AE44 immersion surface. Also notice that the AZ31 surface was divided by 10 so as not to compress the data on the the y-axis, making the three other surfaces indistinguishable. In addition, there was no decrease in pit area on the AZ61 salt spray surface or either AZ31 surfaces, which increased until the end of the experiment. The smallest pit area occurred on the AM60 as-cast surfaces.

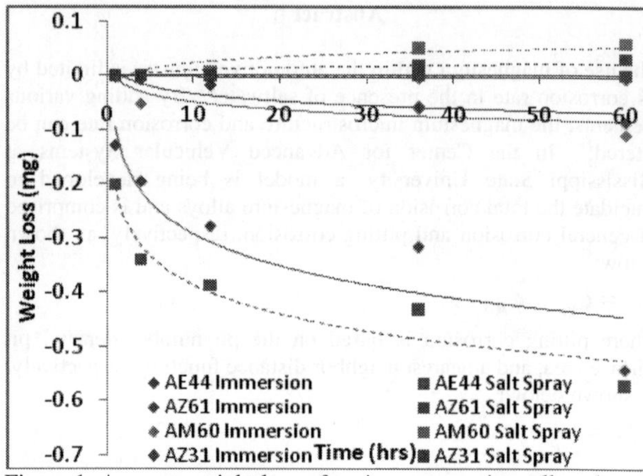

Figure 1: Average weight loss of various magnesium alloys based on test environment over 60 h. Notice that all surfaces followed logarithmic trends.

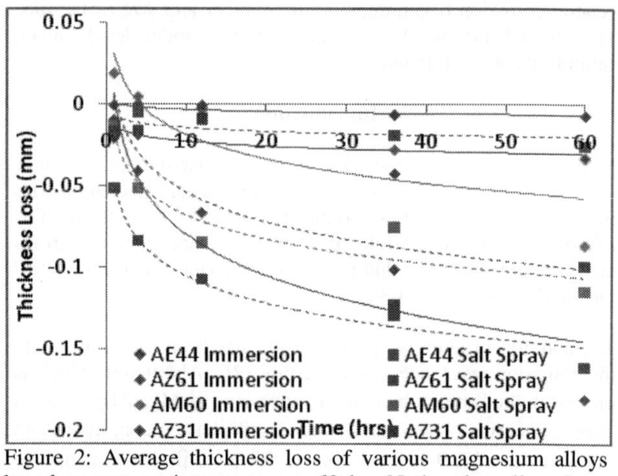

Figure 2: Average thickness loss of various magnesium alloys based on test environment over 60 h. Notice that all surfaces followed logarithmic trends.

Figure 5 shows the changes in the nearest neighbor distance, which is the distance between two pits, for the immersion and salt spray surfaces on the various magnesium alloys being compared. As with the pit number density and the pit area, the surfaces followed second-order polynomial trends, although in reverse of the pit number density and pit area. The extruded AZ61 surfaces showed the smallest nearest neighbor distance while the as-cast AM60 surfaces showed the largest nearest neighbor distance.

Figure 6 shows the intergranular corrosion area fraction (ICAF), which is the fraction of the surface that shows the corrosion that occurs in the β-phase precipitate phase of the alloy, for the immersion and salt spray surfaces on the various magnesium alloys being compared. All surfaces follow logarithmic trends, with the highest ICAF occurring on the as-cast AE44 surfaces.

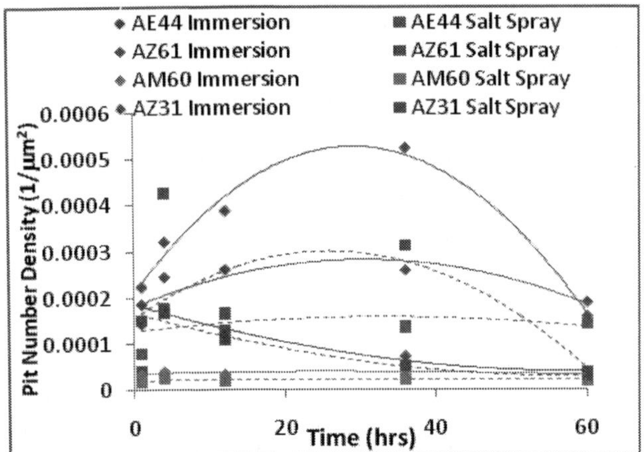

Figure 3: Pit number density of various magnesium alloys based on test environment over 60 h. Notice that all surfaces followed second-order polynomial trends. Also notice that all immersion surfaces had higher pit number densities as compared to the respective salt spray surface.

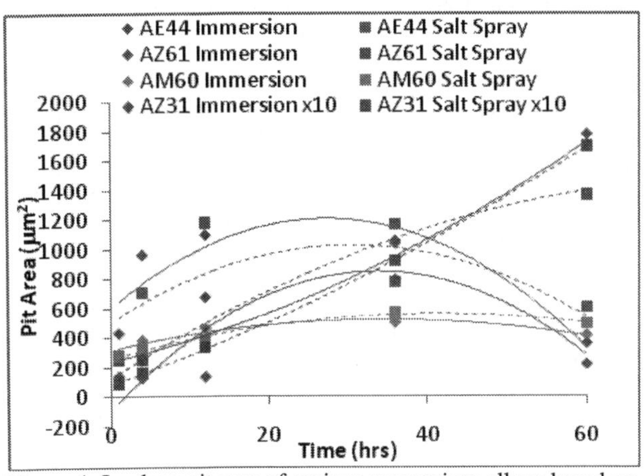

Figure 4: In-plane pit area of various magnesium alloys based on test environment over 60 h. Notice that all surfaces followed second-order polynomial trends. Also notice that the extruded AZ31 surfaces had the largest pit area, which was divided by 10 to ensure all data could be seen. Notice also that the as-cast AM60 surfaces had the smallest pit area. In addition, there was no decrease in pit area on the extruded AZ61 salt spray surface.

Discussion

Total corrosion includes general corrosion, which occurs when water reacts with a magnesium surface to create a $Mg(OH)_2$ film and H_2 gas, and pitting corrosion, which occurs when chloride ions from salt water initiate and maintain pit formation on the magnesium surface [3]. At Mississippi State University, a model is currently being developed that incorporates general corrosion and pitting corrosion, as shown below:

$$\phi = \phi_{gc} + \phi_{IC} \tag{1}$$

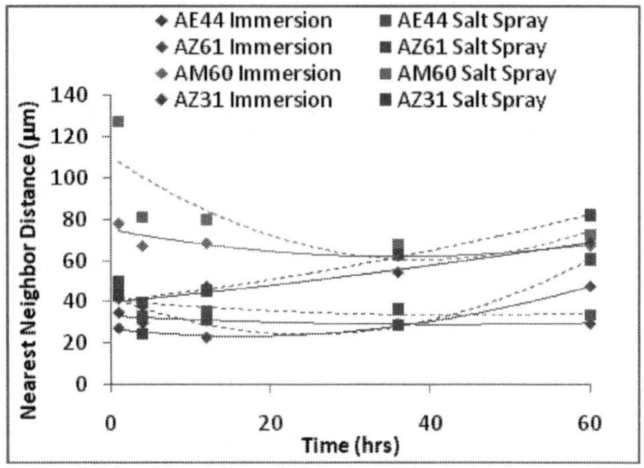

Figure 5: Nearest neighbor distance of various magnesium alloys based on test environment over 60 h. Notice that all surfaces followed second-order polynomial trends. Also notice that the as-cast AM60 surfaces had the largest nearest neighbor distance while the extruded AZ61 surfaces had the smallest nearest neighbor distance.

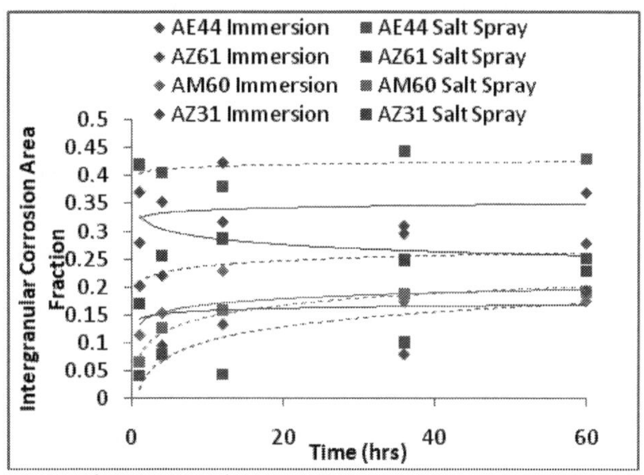

Figure 6: Intergranular corrosion area fraction of various magnesium alloys based on test environment over 60 h. Notice that all surfaces followed logarithmic trends. Also notice that the as-cast AE44 surfaces had the largest intergranular corrosion area fraction.

General corrosion (ϕ_{gc}) is measured by the weight loss and thickness loss of the magnesium coupons, which is only minimally affected by pitting corrosion. Pitting corrosion (ϕ_{IC}), however, is extremely detrimental to materials, as it may not be detected until it penetrates the entire surface since it does not reduce the weight or thickness of the material [3]. Pitting corrosion can be measured by the nucleation, growth, and coalescence of the pits, as shown below:

$$\phi_{IC} = \eta_p v_p c \tag{2}$$

where η is the pit number density, ν is the pit area, and c is a function of the nearest neighbor distance and the intergranular corrosion area fraction (ICAF).

When looking at general corrosion, more weight loss is seen on the immersion surfaces as compared to the salt spray surfaces, except with respect to the as-cast AE44 surfaces (Figure 1). Because the samples in the immersion environment are continuously surrounded by salt containing water, more water can react with the surface, leading to more weight loss as magnesium is removed from the surface. When looking at the salt spray surfaces, which are exposed first to salt containing water, then to 100% humidity, and then to a dry phase, weight loss is not as significant, since there is no continuous exposure to water, meaning there is no continuous general corrosion ongoing. However, more thickness loss is seen on the salt spray surfaces as compared to the immersion surfaces (Figure 2). This is not expected, as one would expect that thickness loss would follow weight loss. The difference in thickness and weight loss, though, can be attributed to the way measurements were taken. Weight loss used a scale, meaning the entire coupon was measured, while thickness loss was taken using calipers, meaning only the edges of the coupons were measured. This difference can account for the difference in thickness loss, as cleaning the samples after each corrosion period removed significant pit debris and salt on the salt spray samples. While both samples hung at 20°, the samples exposed to the cyclical salt spray experienced a collection of pit debris and salt along the edges of the samples, due to the drying phase. This debris led to higher pitting corrosion along the edges, which were measured with the calipers. Because there was no collection of salt or pit debris along the edges of the immersion samples, extra pits could not form meaning the thickness was unaffected by the debris.

General corrosion, however, is only part of the model. The other portion of the model is pitting corrosion, which relates pit number density, pit area, nearest neighbor distance, and intergranular corrosion area fraction (ICAF). The first three values are highly interrelated. Pit number density and pit area are related because at the number of pits increase, the area covered by the pits increase. However, the pits can grow without the pit number increasing, meaning that pit area is not solely related to pit number density. Pit number density and nearest neighbor distance are also related, because as the pit increase in number, the distance between them decreases. Lastly, the pit area and nearest neighbor distance are related because as the pits grow in size, the distance between them decreases.

These relationships are demonstrated in Figures 3-5. As the pit number density increases, so does the pit area, while the nearest neighbor distance decreases. The pit number density begins to decrease prior to pit area decreasing, as the pits can grow in size even when the pit number density decreases. In addition, the nearest neighbor distance decreases even as pit number density decreases, due to the slight growth in pit area. Once pit area begins to decrease, and pit number density continues to decrease, the nearest neighbor distance begins to increase.

The differences seen in pit number density (Figure 3) can be attributed to the environment to which the magnesium is exposed, the form of magnesium, and the type of magnesium alloy. When it comes to the salt spray environment, fewer pits are seen on all surfaces as compared to the immersion surface. This is due to continuous exposure of chloride ions to the immersion surfaces, allowing the chloride ions to continually attack and pit the surfaces. However, chloride ions are only present on the salt spray surfaces for a limited time, meaning that the surfaces can only pit when the chloride ions are present, resulting in lower numbers of pits forming. When comparing the form of magnesium, one can see that the extruded magnesium experienced much higher amounts of pit nucleation than the as-cast magnesium. This is due to the presence of an as-cast skin increasing the corrosion resistance on the as-cast alloys [9, 14]. This skin was removed during the extrusion process, meaning that the extruded magnesium can more easily corrode. Finally, the type of magnesium places a role in the formation of pitting. The higher the amount of aluminum, up to 10%, the higher the corrosion resistance [2]. When comparing the two as-cast materials, the lower pit nucleation corresponded with the higher percentage of aluminum, meaning that aluminum played a higher role in corrosion resistance than either zinc or rare earth elements.

The differences seen in pit area (Figure 4) can also be attributed to the environment to which the magnesium is exposed, the form of magnesium, and the type of magnesium alloy. When looking at the environments, larger pits were seen on the AZ61 and AM60 salt spray surfaces as compared to the immersion surfaces. AE44 and AZ31 showed higher pit areas on the immersion surface as compared to the salt spray surface. The higher pit areas on the salt spray surfaces are due to the presence of pit debris covering the formed pits during the humidity and drying phases. Pit growth is considered autocatalytic, so once it starts, it continues unabated [1]. When pit debris covers the pit, as it does on the salt spray surfaces due to the inability of the humidity and drying phases to remove the pit debris, the pits can continue to grow without general corrosion interfering. When the pit debris is removed, either during the salt spray phase where water is present or during the cleaning process, the larger pits can be seen. While there is a minimal difference in pit area on the AZ31 surfaces, a difference is seen with the AE44 surface, though, because of the shift in corrosion location. When rare earth elements are present, corrosion shifts to the center of the grain and away from the intergranular region [7, 13]. This shift encourages general corrosion and pitting corrosion to "work together", thereby increasing the pit area on the immersion surface. In addition to the environment, though the form of magnesium plays a role in the pit area. The as-cast AM60 material has a smaller pit area on both environments as compared to the extruded AZ61 material. This again can be attributed to the as-cast skin on the AM60 material, which prevents pits from growing due to the small grain size. The extruded AZ61 material does not possess the as-cast skin, meaning that the pits can grow more easily. Again, AE44 does not follow this line, again likely due to the shift in pit formation location. Even though an as-cast skin exists on the AE44 material, the pits form within the magnesium grain. General corrosion works alongside pitting corrosion to degrade the magnesium grains, which grow together, indicating an increase in pit area. The magnesium alloy also appeared to play a role in the pit area, with the smallest pit area occurring on the as-cast AM60 material, the largest pit area occurring on the as-cast AE44 material, and the middle pit area occurring on the extruded AZ61 and AZ31 materials. One could suspect that the presence of manganese affected the growth of the pits differently than the presence of either zinc or rare earth elements, but there is not currently enough alloys to accurately confirm this suspicion.

When it comes to the nearest neighbor distance (Figure 5), as previously mentioned, the pit number density and pit area greatly affect the nearest neighbor distance. The higher the pit number density and the larger the pits, the smaller the nearest neighbor distance. When comparing the environments, since there are more pits formed on the immersion environments, the pits are closer together on the immersion surfaces. When comparing forms of magnesium, the as-cast AE44 surface and the extruded AZ61 surface had similar nearest neighbor distances because of the combination of pit area and pit number density, as did the as-cast AM60 surface and the extruded AZ31 surface. Since the AZ61 surface had a higher pit number density but smaller pit areas as compared to the AE44 surface, the values combine to cause the nearest neighbor distance to appear similar. However, the as-cast AM60 magnesium alloy had the smallest pit number density and the smallest pit areas, the AM60 nearest neighbor distance would be the largest, or furthest apart. In addition, the extruded AZ31 magnesium alloy experienced a decrease in pit number density and an increase in pit area, meaning that the nearest neighbor distance was more influenced by the size of the pits instead of the number of pits. The pit area of AZ31, which was 10x higher than the other pit areas, ensured that the nearest neighbor distance started close and gradually increased, as the large pits incorporated other pits, increasing the distance between the remaining pits.

The intergranular corrosion area fraction (ICAF) is another measure of the coalescence of the pits, because as the pits that form along the intergranular boundary grow together, they eventually grew into each other, forming one long narrow pit. The type of magnesium alloy affected the ICAF much more than either the environment or the form of the magnesium alloy (Figure 6). When comparing the extruded AZ61 alloy with the as-cast AM60 alloy, once can see that the environment affected on the beginning of the ICAF, but by the end of the experiment time, the ICAF had merged between the salt spray and immersion environments. In addition, there was very little difference between the AM60 and AZ61 ICAF. However, there was a significant difference between AM60, AZ61, AZ31, and AE44. The difference is due to the presence of aluminum influencing the corrosion of AZ31 and the rare earth elements and the as-cast skin influencing the corrosion of AE44. For AZ31, there was 3% less aluminum in the magnesium alloy. Since aluminum, increasing to 10%, has been shown to increase the corrosion resistance of magnesium [2], it stands to reason that the lower percentage of aluminum in AZ31 would allow more intergranular corrosion. For AE44, both the alloying elements and the skin contributed to the formation of intergranular corrosion. Since rare earth elements switch corrosion from along the intergranular boundary to the interior of the magnesium grain [7,13] and the as-cast skin results in very small grains, the presence of ICAF means that the grains were degrading and connecting along the intergranular boundary. If there was no as-cast skin, meaning the grains were much larger, there is a chance that the ICAF would have been more in line with the AM60 and AZ61 samples. In addition to the differences caused by the presence of rare earth elements and the as-cast skin, the environment contributed to a difference in ICAF, with the immersion surface experiencing less intergranular corrosion than the salt spray surface. This difference can be contributed to general corrosion, which removed the intergranular boundaries that were left by the pitting and destruction of the magnesium grains. With general corrosion removing, or lowering, the intergranular boundaries, intergranular corrosion may not have been accurately quantified.

Conclusions

Four magnesium alloys in two forms, as-cast AE44, as-cast AM60, extruded AZ61, and extruded AZ31 were examined in two corrosive environments, immersion and salt spray. General corrosion characteristics, weight loss and thickness loss, as well as surface characteristics, pit number density, pit area, nearest neighbor distance, and ICAF, were quantified over 60 hours. The most heavily corroded magnesium alloy, determined by combining general and pitting corrosion, was AZ61, followed by AE44, AZ31, and AM60, respectively. When comparing environments, more pits formed on all surfaces exposed to the immersion environment, while the pits were larger on the salt spray environments.

References

[1] M.G. Fontana, Corrosion Principles, in: M.G. Fontana (Eds.), *Corrosion Engineering*, McGraw-Hill, Boston, 1986, pp. 12-38.

[2] G. Song, A. Atrens, "Understanding Magnesium Corrosion – A Framework for Improved Alloy Performance", *Advanced Engineering Materials* 5 (2003) 837-858.

[3] BA Shaw, Corrosion Resistance of Magnesium Alloys, in: L.J. Korb, ASM (Eds.), ASM Handbook, Vol. 13A: Corrosion, Ninth Ed., ASM International Handbook Committee, Metals Park, 2003, pg. 692.

[4] J.D. Majumdar, R. Galun, B. Mordike, I. Manna, "Effect of laser surface melting on corrosion and wear resistance of a commercial magnesium alloy", *Materials Science and Engineering A*, 361 (2003) 119-129.

[5] C. Blawert, E.D. Morales, W. Dietzel, K.U. Kainer, "Comparison of Corrosion Properties of Squeeze Cast and Thixocast MgZnRE Alloys", *Materials Science Forum*, 488-489 (2005) 697-700.

[6] W. Liu, F. Cao, L. Chang, Z. Zhang, J. Zhang, "Effect of rare earth element Ce and La on corrosion behavior of AM60 magnesium alloy", *Corrosion Science*, 51 (2009) 1334-1343.

[7] W. Liu, F. Cao, L. Zhong, L. Zheng, B. Jia, Z. Zhang, J. Zhang, "Influence of rare earth element Ce and La addition on corrosion behavior of AZ91 magnesium alloy", *Materials and Corrosion*, 60 (2009) 795-803.

[8] Y.L. Song, Y.H. Liu, S.R. Yu, X.Y. Zhu, S.H. Wang, "Effect of neodymium on microstructure and corrosion resistance of AZ91 magnesium alloy", *Journal of Materials Science*, 42 (2007) 4435-4440.

[9] G. Song, "Recent Progress in Corrosion and Protection of Magnesium Alloys", *Advanced Engineering Materials*, 7 (2005) 563-586.

[10] M.C. Zhao, M. Liu, G. Song, A. Atrens, "Influence of pH and chloride ion concentration on the corrosion of Mg alloy ZE41", *Corrosion Science*, 50 (2008) 1939-1953.

[11] G. Song, A. Atrens, X. Wu, B. Zhang, "Corrosion behavior of AZ21, AZ501, and AZ91 in sodium chloride", *Corrosion Science*, 40 (1998) 1769-1791.

[12] Y.L. Song, Y.H. Liu, S.H. Wang, S.R. Yu, X.Y. Zhu, "Effect of cerium addition on microstructure and corrosion resistance of die cast AZ91 magnesium alloy", *Materials and Corrosion*, 58 (2007) 189-192.

[13] N. Birbilis, M.A. Easton, A.D. Sudholz, S.M. Zhu, M.A. Gibson, "On the corrosion of binary magnesium-rare earch alloys", *Corrosion Science*, 51 (2009) 683-689.

[14] G. Song, A. Atrens, M. Dargusch, "Influence of microstructure on the corrosion of diecast AZ91D", *Corrosion Science*, 41 (1998) 249-273.

[15] S.P. White, G.J. Weir, N.J. Laycock, "Calculating chemical concentrations during the initiation of crevice corrosion", *Corrosion Science*, 42 (2000) 605-629.

[16] R.M. Pidaparti, A. Puri, M.J. Palakal, A. Kashyap, "Two-dimensional Corrosion Pit Initiation and Growth Simulation Model," *Computers, Materials, and Continua*, 2 (2005) 65-75.

[17] R.M. Pidaparti, L. Fang, M.J. Palakal, "Computational simulation of multi-pit corrosion process in materials", *Computational Materials Science*, 41 (2008) 255-265.

[18] N.J. Laycock, J.S. Noh, S.P. White, D.P. Krouse, "Computer simulation of pitting potential measurements", *Corrosion Science*, 47 (2005) 3140-3177.

[19] N.J. Laycock, J. Stewart, R.C. Newman, "The initiation of crevice corrosion in stainless steel", *Corrosion Science*, 39 (1997) 1791-1809.

[20] L. Li, X. Li, C. Dong, Y. Huang, "Computational simulation of metastable pitting of stainless steel", *Electrochimica Acta*, 54 (2009) 6389-6395.

[21] T. Johnsen, A. Jossang, T. Jossang, P. Meakin, "An experimental study of the quasi-two-dimensional corrosion of aluminum foils and a comparision with two-dimensional computer simulations", *Physica A*, 242 (1997) 356-276.

[22] B. Malki, B. Baroux, "Computer simulation of the corrosion pit growth", *Corrosion Science*, 47 (2005) 171-182.

[23] J.X. Jia, G. Song, A. Atrens, "Experimental Measurement and Computer Simulation of Galvanic Corrosion of Magnesium Coupled to Steel", *Advanced Engineering Materials*, 9 (2007) 65-74.

[24] ASTM B117 - 07a (2007) Standard Practice for Operating Salt Spray (Fog) Apparatus, Vol. 03.02, 2007.

[25] M.F. Horstemeyer, J. Lathrop, A.M. Gokhale, M. Dighe, "Modeling stress state dependent damage evolution in a cast Al-Si-Mg aluminum alloy", *Theoretical and Applied Fracture Mechanics*, 33 (2000) 31-47.

Magnesium Technology 2011
Edited by: Wim H. Sillekens, Sean R. Agnew, Neale R. Neelameggham, and Suveen N. Mathaudhu
TMS (The Minerals, Metals & Materials Society), 2011

CORROSION PERFORMANCE OF Mg-Ti ALLOYS SYNTHESIZED BY MAGNETRON SPUTTERING

Zhenqing Xu[1], Guang-Ling Song[2*], Daad Haddad[1]
[1]MEDA Engineering and Technical Services, LLC, 17515 W 9 Mile Rd, STE 1075, Southfield, MI 48075, USA
[2]General Motors Research and Development, Mail Code: 480-106-224, 30500 Mound Road, Warren, MI 48090, USA

Keywords: Corrosion, Mg, Ti, Magnetron Sputtering, XPS

Abstract

Mg is difficult to alloy with Ti through a conventional metallurgical approach due to their insolubility in each other and big difference in melting point. However, Mg, if alloyed with Ti, may become corrosion resistant. This hypothesis is verified in this study.

$Mg_{1-x}Ti_x$ alloy thin films (with x=0, 0.2, 0.4, 0.6, 0.8 and 1) were deposited by magnetron sputtering onto a glass substrate. Film compositions were analyzed by electron probe micro-analysis (EPMA). The electrochemical behavior of these alloys was characterized in saturated $Mg(OH)_2$ solutions with and without 0.1M NaCl. The macrostructures of the thin film alloys were compared before and after polarization and immersion measurements. The results showed that the corrosion resistance of the alloy was improved with increasing Ti content. No material loss or corrosion damage was observed for alloys with 80% or more Ti content in both solutions.

Introduction

Mg alloys have found many potential applications recently. Their low density and high strength to weight ratio make Mg alloys attractive to the auto industry in weight reduction and efficiency improvement. However, many of the alloying elements form intermetallic phases with Mg which have a more positive free corrosion potential than the Mg matrix itself [1-3]. Hence, micro-galvanic corrosion attacks usually decrease the overall corrosion resistance of a Mg alloy. It is hypothesized that a precipitation free, single phase, solid solution magnesium alloy will have better corrosion performance.

Mg and Mg alloys present a very negative open circuit potential in a sodium chloride solution and have an active and fast anodic dissolution process [4]. The addition of strong passivation elements like Cr [5] and Ti [6-8] could possibly improve the passivity of a Mg alloy. However, according to the phase diagram, the solubility of Mg in Ti, and vice versa, is very small in crystalline state. Conventional equilibrium metallurgical approach cannot successfully lead to a Mg-Ti solid solution phase due to low solid solubility in each other and very high Ti melting point relative to Mg. Using a non-equilibrium magnetron sputtering deposition technique can substantially extend the solubility of Ti in the Mg-Ti alloy and in this case the large melting point difference between Ti and Mg is no longer an issue.

In this paper, Mg-Ti thin film alloys with different Ti concentrations were deposited and their corrosion behavior was compared, aiming to understand the effect of the Ti solute on the corrosion behavior of Mg.

Experimental

Film Deposition

$Mg_{1-x}Ti_x$ (x=0, 0.2, 0.4, 0.6, 0.8, 1) thin films were deposited onto a round glass disk, 2 cm in diameter, using dc magnetron sputtering in an argon atmosphere at room temperature. The base pressure in the growth chamber was about 5×10^{-8} Torr. The dynamic pressure during film deposition was 2 mTorr. High purity Mg and Ti ingots were used as targets. The thin film was very uniformly formed on the glass substrate. Film composition was controlled by changing the power applied to the Mg and Ti targets. The compositions of the different Mg-Ti thin film samples were confirmed by Electron Probe Microanalysis (EPMA).

Electrochemical Characterization

A glass flat cell and a Solatron 1280 potentiostat system were used for polarization curve measurements. The sample was immersed in the solution at its open circuit potential (OCP) for 5 min before polarization curve measurement. The immersed area was 1 cm^2. A platinum gauze (2.5 cm x 2.5 cm) was used as the counter electrode and a KCl-saturated Ag/AgCl electrode was used as the reference (Ref). Potentiodynamic polarization curve was recorded at a potential scanning rate of 0.1 mV/s from -0.2 V vs. OCP to +1.0 V vs. Ref.

Immersion Test and Surface Film Characterization

Mg-Ti thin film samples were immersed in $Mg(OH)_2$+0.1M NaCl solution for 4 days. Their morphologies after immersion were examined under optical microscope and Scanning Electron Microscope (SEM). A small section in the immersion area was analyzed by X-ray photoelectron spectroscopy (XPS). High resolution XPS has also been performed on Mg 2p, O 1s, and Ti 2p core levels. As is the standard practice in XPS studies, the C1s line corresponding to the C-C bond has been used as the binding energy (BE) reference [9].

Results and Discussion

Deposition and Film Composition

The deposition conditions for growing Mg-Ti thin films and the EPMA composition characterization are listed in Table I. By changing the power applied to the Mg and Ti targets, different growth rates of Mg and Ti can be achieved, forming Mg-Ti thin films with different Ti contents. The EPMA measurements confirm that the deposition parameters used in sputtering quite accurately controlled the thin film compositions. Thickness of these thin films has been measured by cross section SEM. These samples have thickness ranging from 1.2 μm to 1.5 μm.

Table I. EPMA composition characterization of Mg-Ti thin films deposited by magnetron sputtering

Sample	Power (W)		Rate (Å/s)		Calculated Concentration (Using Growth rates)%at		Measured Concentration (EPMA)%at		Measured Thickness (Cross section SEM) μm
	Mg	Ti	Mg	Ti	Mg	Ti	Mg	Ti	
Mg	100	-	4.6	-	100	-	>99	<1	1.2
Mg80Ti20	100	130	4.6	0.9	80	20	79	21	1.3
Mg60Ti40	50	200	2.5	1.3	60	40	59	41	1.3
Mg40Ti60	40	310	2.0	2.1	40	60	42	58	1.5
Mg20Ti80	20	450	1.0	2.9	20	80	19	81	1.4
Ti	-	300	-	2.0	-	100	<1	>99	1.3

Corrosion Behavior

To investigate the effect of Ti on Mg-Ti alloy's electrochemical behavior, potential-dynamic polarization curves were measured in $Mg(OH)_2$ solutions, with and without 0.1M NaCl. Polarization curves of sputter-deposited pure Mg and Ti were also measured for comparison.

Figure 1 shows macroscopic images of the Mg-Ti alloys after polarization in both solutions. The top part of the coupon sample has undergone the polarization measurements in the $Mg(OH)_2$ saturated solution, while the bottom part is in the solution with 0.1M sodium chloride addition. For samples polarized in saturated $Mg(OH)_2$, significant material loss can be observed for pure Mg, Mg80Ti20 and Mg60Ti40. Particularly for pure Mg, no material was preserved after polarization to +1 V vs. Ref. Obviously, a film with increased Ti content was less corroded, indicating that a increase in Ti content can lead to improved corrosion resistance.

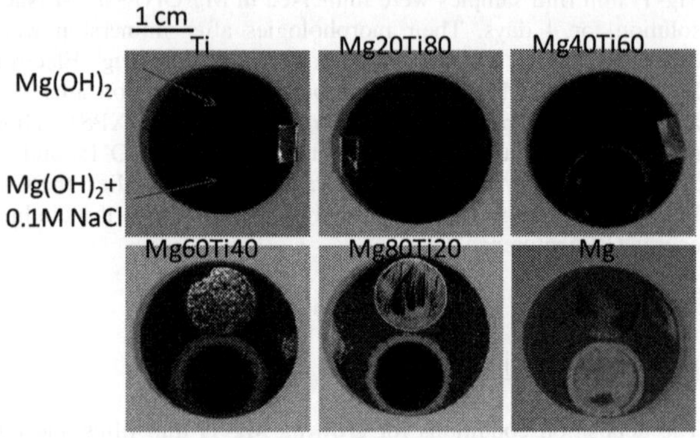

Figure 1. Macroscopic images of Mg-Ti thin films on glass disks after polarization in saturated $Mg(OH)_2$ (top round areas) and $Mg(OH)_2$+0.1 M NaCl (bottom round areas).

With the addition of 0.1M sodium chloride, similar behaviors of these films were observed. Apparent corrosion can be found for all samples other than pure Ti and Mg20Ti80, although much of the materials were preserved for Mg40Ti60, Mg60Ti40 and Mg80Ti20. Crevice corrosion was found at the areas that were in contact with the plastic washer of the electrolyte cell. The severity of the crevice corrosion is in reverse order of the Ti content in the Mg-Ti alloys. The above macroscopic observations show that with increased Ti content, a Mg-Ti alloy thin film becomes more resistant to general corrosion during polarization. Their ability to preserve more material when being anodized with high overvoltage has increased significantly. It is widely known that the corrosion of Ti results in a stable oxide layer (essentially TiO_2) on the surface. The improved passivity of a Mg-Ti alloy with increasing Ti content could result from formation of more TiO_2 on the surface.

Figure 2 shows the polarization curves of Mg-Ti alloys measured in saturated $Mg(OH)_2$ without (a) and with (b) 0.1M NaCl. The influence of the noble Ti on the free corrosion potential can be clearly identified. The higher the Ti content, the more positive is the corrosion potential.

In $Mg(OH)_2$ solution shown in Figure 2a, very little passive tendency is observed for Mg and Mg80Ti20. As the polarization potential is more positive than +0.5 V vs. Ref the current density deceases significantly. This phenomenon cannot be linked to passivation of the material. Rather, considering the previous observation from Figure 1, we can conclude that at a voltage more positive than +0.5 V vs. Ref, significant material was corroded for these two samples and the inert substrate glass was exposed to the solution, resulting in decreased anodic dissolution current densities.

A clear passive regime with constant current can be seen in the anodic region for samples with Ti content larger than 40%. Although Mg60Ti40 has the lowest passivation current density, it loses its passivation when potential is more positive than 0.25 V vs. Ref. Other than Mg60Ti40, the passivation current densities decrease with more Ti addition into Mg-Ti for Mg-Ti alloy when Ti content is higher than 60at%. This higher corrosion resistance can be attributed to structure modification of Mg film by substitution of Mg with Ti and improved stability of the surface film due to Ti incorporation.

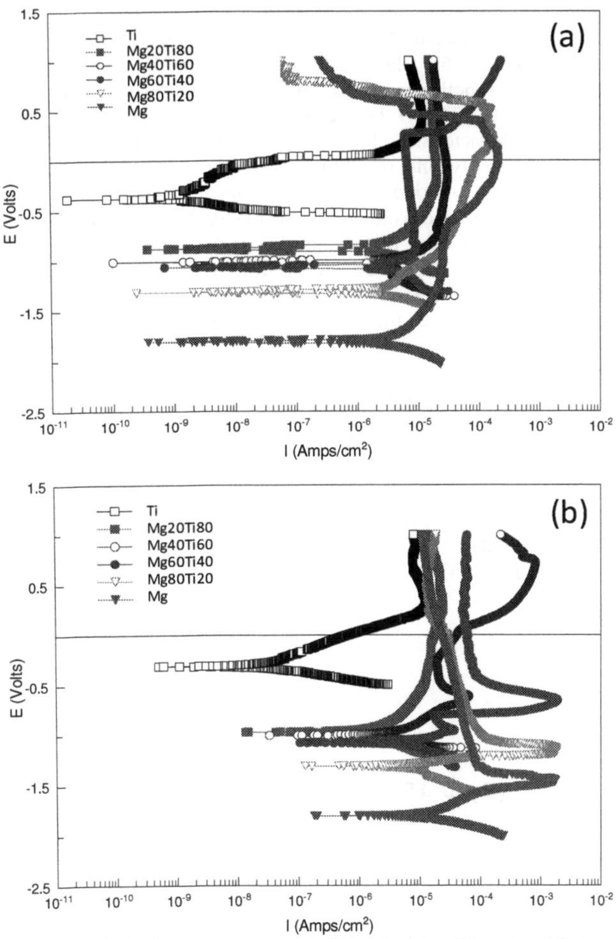

Figure 2. Polarization curves of Mg-Ti thin films in (a) in saturated $Mg(OH)_2$ and (b) saturated $Mg(OH)_2$+0.1 M NaCl;

Similar electrochemical behavior has been observed in solution of $Mg(OH)_2$+0.1M NaCl (Figure 2b). Because of the presence of corrosive Cl^- in the solution, no passivity can be detected for pure Mg and Mg80Ti20. Both Mg60Ti40 and Mg40Ti60 have a small passive region while the latter one has a higher passivation potential and a broader passive potential region from -0.6 to -0.2 V vs. Ref. A dramatic anodic dissolution current density decrease was recorded for Mg-Ti thin films with Ti content less than 40%. A possible scenario is that most of the material was corroded at the crevice corrosion region (shown as a corroded ring in Figure 1) and became electrically insulating, leading to a diminished current density shown in Figure 2b. Mg20Ti80 and pure Ti stay in the passive region even when the polarization potential is at +1 V vs. Ref.

Immersion Test

Mg-Ti thin films morphologies were recorded using an optical microscope for different immersion times in the saturated $Mg(OH)_2$+0.1M NaCl solution (Figure 3). The top part of each image is where the sample was immersed. It was found that after 1 hour of immersion pure Mg was corroded significantly with pitting and cracking evident all over the immersed area. No material loss or cracking can be observed for Mg80Ti20; only a slight color change is found. It was previously shown in Figure 2 that under polarization Mg60Ti40 was more corrosion resistant than Mg80Ti20. However, an interesting finding is that after 1 hour immersion Mg60Ti40 experienced more material loss from the substrate than Mg80Ti20. For Mg-Ti alloys with over 40 at.%Ti, no corrosion occurred on their surface after immersion.

After 4 hour immersion (Figure 3b), all Mg was corroded away for pure Mg thin film. There is no significant material loss for Mg80Ti20. A gap between the top and bottom parts is found, suggesting that some of the film was dissolved along the waterline. Mg60Ti40 has lost considerable amount of the film after 4 hours of immersion. No corrosion can be detected on all other samples.

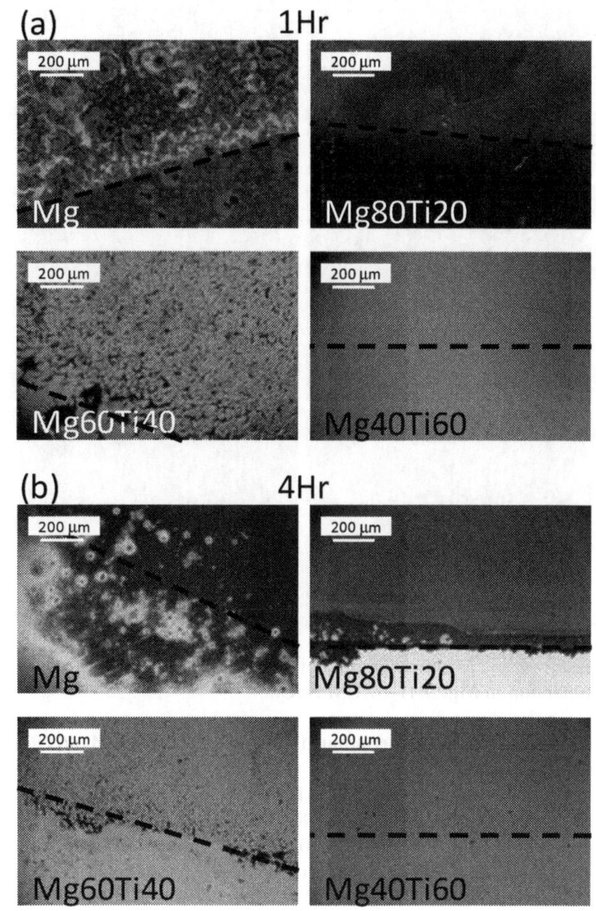

Figure 3. Macroscopic images of Ti-Mg thin film alloys after immersion in $Mg(OH)_2$+0.1M NaCl for (a) 1hr and (b) 4 hr. The top part of the graph is the immersed area.

Further analysis of the structure of these Mg-Ti thin films after immersion test is presented by SEM in Figure 4. Under high magnification cracks can be found in Mg80Ti20, Mg60Ti40 and Mg40Ti60 samples after 4 hour immersion. It seems that the most severe film rupture and cracking is found in Mg60Ti40 (Figure 4b). The density of micro-cracks in Mg80Ti20 is the highest; however, the film seems to adhere very well to the glass substrate. There is no film peeling or delamination. The SEM observation confirms the previous macroscopic finding that Mg60Ti40 has more material loss than Mg80Ti20 after 4 hours of immersion test. Although polarization results suggest that Mg60Ti40 shows more passive nature than Mg80Ti20 (Figure 2), more concentrated stress presented in the film could lead to peeling and delamination after the sample is taken out from the solution and exposed to air. Further analysis such as X-ray diffraction (XRD) and atomic force microscopy (AFM) are being conducted to investigate why stress is more uniformly presented in Mg80Ti20. The grain size, orientation or mechanical properties of the Mg-Ti thin films may all play a role in the stress distribution. Moreover, the difference in strength and brittleness of the compounds could also results in different film cracking patterns.

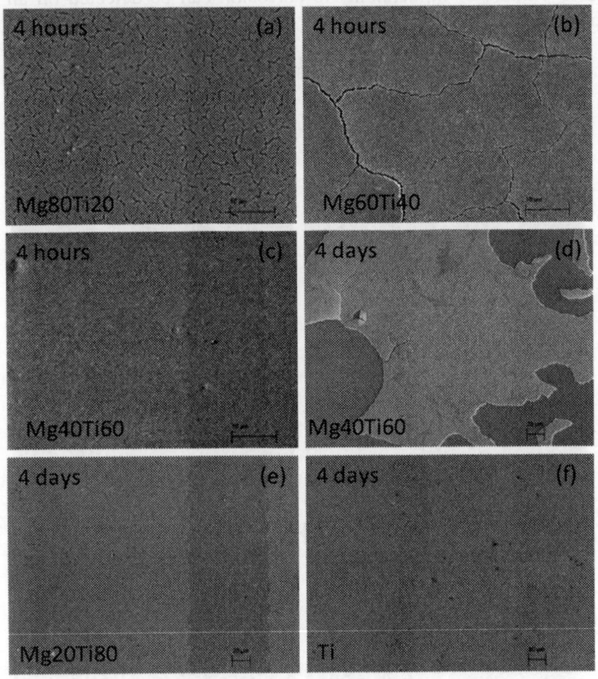

Figure 4. SEM micrographs of Ti-Mg thin film alloys after immersion in $Mg(OH)_2$+0.1 M NaCl.

Immersion test was continued for 4 days for Mg40Ti60, Mg20Ti80 and pure Ti. There is still no cracking or material loss observed for pure Ti and Mg20Ti80 (Figure 4e and f). A small amount of Mg40Ti60 was still left on the substrate (Figure 4d) while Mg80Ti20 and Mg60Ti20 have lost all of the deposited films. Overall, the improvement of corrosion resistance is evident when the Ti content is increased. This beneficial effect observed for Mg-Ti alloys could be partially attributed to the excellent corrosion resistance of Ti. Corrosion of Mg occurs at breaks in a partially protected surface film. The passive film $Mg(OH)_2$ is mechanically weak and unstable and not able to prevent further corrosion of Mg. The formation of continuous TiO_2 passive layer should play an important role in increasing the corrosion resistance of Mg-Ti alloys.

XPS Characterization

XPS measurement was conducted on two selected samples, Mg20Ti80 and Mg80Ti20 after 4 hour immersion test. The survey spectra revealed the presence of Mg 2p, O 1s and Ti 2p. These characteristic peaks correspond to the electron configuration of the electrons within the atoms. High resolution Mg 2p XPS spectra (Figure 5a) show that one broad peak at 50.8 eV is presented for both specimens, suggesting the existence of MgO and $Mg(OH)_2$. The O 1s spectra presented in Figure 5b show different characteristics for these two samples. The Mg20Ti80 sample has only one main O 1s peak presented at 531.0 eV, corresponding to oxide. The O 1s composite peak of Mg80Ti20 sample can be deconvoluted to two big peaks, corresponding to oxide (531.0 eV) and hydroxide (532.8 eV). The measured high resolution peaks for Ti 2p are shown in Figure 5c. The peak positions 459.0 for Ti oxide is in very good agreement with the literature values given for TiO_2 (458.7 eV) [7].

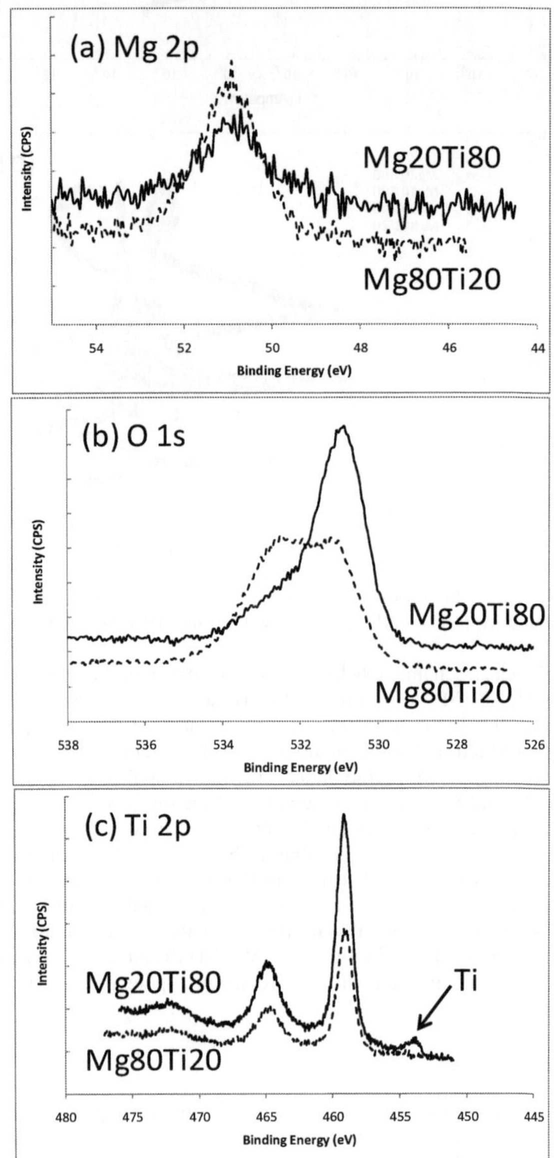

Figure 5. XPS high resolution spectra of Ti-Mg after immersion in $Mg(OH)_2$+0.1M NaCl.

The above findings suggest that the Mg20Ti80 sample surface has a strong oxide layer, possibly TiO_2. It confirms our previous speculation that a stable oxide layer could form on the alloy with higher Ti content and prevent corrosion attack. A small peak is also presented for Mg20Ti80 sample at 453.0, corresponding to Ti in its metallic state. For the Mg80Ti20 sample, Mg is not only presented in its oxide state, but also in the surface film $Mg(OH)_2$. Unfortunately, this film is not very protective [10, 11]. Only a higher concentration of Ti in the film can beneficially enhance the passivity. The mixture of TiO_2 and MgO forms a compact passive film preventing the attack of aggressive Cl^- ions, leading to improved corrosion resistance of a Mg-Ti alloy with a higher Ti content.

Summary

The corrosion properties of Mg_xTi_{1-x} thin films with x ranging from 0 to 1 were investigated. Incorporation Ti in the Mg lattice has been achieved by dc magnetron sputtering. Polarization measurements showed that alloys with high Ti content had better passivity in solutions of $Mg(OH)_2$ with and without 0.1M NaCl. XPS observation confirms the presence of surface oxide film on the Mg-Ti alloy surface. A high ratio of oxide/hydroxide in the Mg20Ti80 sample surface is responsible for improved corrosion resistance.

Production of supersaturated single-phase compounds of Mg-Ti coatings by sputtering provides a promising approach to produce corrosion resistance magnesium based alloys outside the conventional casting technology. Formation of precipitates during sputtering is strongly reduced. As a result, high purity coatings can be obtained and a reduced corrosion rate can be achieved. Compound composition strongly influences the properties and corrosion behaviors of the alloy films. Therefore, other corrosion resistant materials could also be used to form alloys with Mg to prevent corrosion.

Acknowledgements

The authors would like to thank Mr. Richard Waldo for EPMA measurements and Mr. Nicholas Irish for XPS analysis.

References

[1] A. Pardo, M.C. Merino, A.E. Coy, R. Arrabal, F. Viejo, E. Matykina, Corrosion behaviour of magnesium/aluminium alloys in 3.5 wt.% NaCl, Corrosion Science, 50 (2008) 823-834.

[2] G. Song, Investigation on corrosion of magnesium and its alloys, Journal of Corrosion Science and Engineering, 6 (2003).

[3] G. Song, B. Johannesson, S. Hapugoda, D. StJohn, Galvanic corrosion of magnesium alloy AZ91D in contact with an aluminium alloy, steel and zinc, Corrosion Science, 46 (2004) 955-977.

[4] G. Song, Z. Xu, The surface, microstructure and corrosion of magnesium alloy AZ31 sheet, Electrochimica Acta, 55 (2010) 4148-4161.

[5] M. Vilarigues, L.C. Alves, I.D. Nogueira, N. Franco, A.D. Sequeira, R.C. da Silva, Characterisation of corrosion products in Cr implanted Mg surfaces, Surface & Coatings Technology, 158 (2002) 328-333.

[6] K.R. Baldwin, D.J. Bray, G.D. Howard, R.W. Gardiner, Corrosion behaviour of some vapour deposited magnesium alloys, Materials Science and Technology, 12 (1996) 937-943.

[7] T. Mitchell, S. Diplas, P. Tsakiropoulos, Characterisation of corrosion products formed on PVD in situ mechanically worked Mg-Ti alloys, Journal of Alloys and Compounds, 392 (2005) 127-141.

[8] S. Rousselot, M.P. Bichat, D. Guay, L. Roue, Structure and electrochemical behaviour of metastable Mg50Ti50 alloy prepared by ball milling, Journal of Power Sources, 175 (2008) 621-624.

[9] M. Liu, S. Zanna, H. Ardelean, I. Frateur, P. Schmutz, G. Song, A. Atrens, P. Marcus, A first quantitative XPS study of the surface films formed, by exposure to water, on Mg and on the Mg-Al intermetallics: Al3Mg2 and Mg17Al12, Corrosion Science, 51 (2009) 1115-1127.

[10] G. Song, Recent progress in corrosion and protection of magnesium alloys, Advanced Engineering Materials, 7 (2005) 563-586.

[11] A. Seyeux, M. Liu, P. Schmutz, G. Song, A. Atrens, P. Marcus, ToF-SIMS depth profile of the surface film on pure magnesium formed by immersion in pure water and the identification of magnesium hydride, Corrosion Science, 51 (2009) 1883-1886.

Magnesium Technology 2011
Edited by: Wim H. Sillekens, Sean R. Agnew, Neale R. Neelameggham, and Suveen N. Mathaudhu
TMS (The Minerals, Metals & Materials Society), 2011

STRUCTURE AND MECHANICAL PROPERTIES OF MAGNESIUM-TITANIUM SOLID SOLUTION THIN FILM ALLOYS PREPARED BY MAGNETRON-SPUTTER DEPOSITION

Daad Haddad[1], GuangLing Song[2*], Yang Tse Cheng[3]
[1]MEDA Engineering and Technical Services LLC; 17515 W. Nine Mile Road; Southfield, MI 48075, USA
[2]Chemical Sciences and Materials Systems Lab, GM Global Research and Development; 30500 Mound Road; Warren, MI 48090, USA
[3]Department of Chemical and Materials Engineering, University of Kentucky; 177 F. Paul Anderson Tower; Lexington, KY 40506, USA

Keywords: Magnetron Sputtering, Mg-Ti thin films, Nanoindentation, X-ray diffraction, Atomic Force Microscopy

Abstract

Mg alloys are being considered for wider application in automotive industry. Designing new alloys with improved mechanical properties is important to the development of new Mg alloy parts. Mg-Ti is an interesting alloying system that may have good corrosion resistance due to high passivity of Ti. However it is difficult to form through a conventional metallurgical method due to the mutual insolubility of Mg and Ti and the big difference in their melting point. Nevertheless, if the alloy can be formed, it may have other unexpected physical and chemical performance. Therefore, it is of significance to understand the properties of Mg-Ti alloy produced by non-conventional approach.

In this report, $Mg_{(1-x)}Ti_x$ thin film alloys containing 0, 21, 41, 51, 58, 81 and 100 at.% Ti were deposited by dc magnetron sputtering on Si substrates. The mechanical properties of the thin film alloys were obtained using nanoindentation. Electron probe microanalysis (EPMA) was used to determine the film compositions. X-ray diffraction (XRD) measurements showed that single phase magnesium-titanium solid solutions were obtained across the full range of magnesium and titanium mixtures. The topography and the rms roughness of the different alloys were studied using atomic force microscopy (AFM). The mechanical properties of the $Mg_{(1-x)}Ti_x$ thin films were determined by analyzing the nanoindentation load-displacement curves based on the Oliver-Pharr method. The nanoindentation results show that both the elastic modulus and hardness of the $Mg_{(1-x)}Ti_x$ alloy thin films are higher than those of conventional Mg alloys.

Introduction

Magnesium and magnesium alloys have attracted wide attention as lightweight materials that can be used in automotive, aerospace, and hydrogen storage applications. The alloying of magnesium and titanium is of particular interest due to the improved mechanical and corrosion resistant properties of Ti [1-5]. However, magnesium and titanium have very little mutual solubility according to their equilibrium phase diagram, and thus do not form any stable intermetallic compounds under standard alloying conditions [6, 7]. But recently it was reported that by using non-standard alloying techniques, such as electron-beam deposition and magnetron sputtering, the solid solubility of Mg and Ti can be extended significantly and solid solution $Mg_{(1-x)}Ti_x$ thin film alloys can be formed [8-15]. These $Mg_{(1-x)}Ti_x$ films are expected to have the advantages of being lightweight alloys with high specific strength and better corrosion performance.

In this work, the fabrication of $Mg_{(1-x)}Ti_x$ thin film alloys with different Ti content is presented. The topography and the crystal structure of these films are studied using AFM and XRD measurements. Finally, the mechanical properties of these films are measured using nanoindentation experiments and by analyzing the load-displacement curves based on the Oliver-Pharr method [16]. The effect of the composition and morphology on the mechanical properties of these alloys is discussed.

Mg-Ti Thin Films Deposition

The $Mg_{(1-x)}Ti_x$ ($x \approx 0$, 0.21, 0.41, 0.51, 0.58, 0.81, 1) thin films were grown on 1″ Si (100) substrates using dc magnetron sputtering under argon atmosphere. The silicon substrates were ultrasonically cleaned in acetone and methanol successively for 20 minutes. The cleaned substrates were loaded in the load lock chamber where they were heated to 200°C for 30 minutes under vacuum for outgasing. After cooling down, the substrates were transferred to the growth chamber which has a base pressure of about 5.0×10^{-8} Torr.

The deposition plasma for each constituent material was created utilizing a dc power source and a flow of Ar (14 sccm). The dynamic pressure during the growth of the films was 2 mTorr. Deposition of the different alloys is carried out using a recipe which depends on predetermined deposition rates. Growth rates of the magnesium and titanium sources were measured, under the above specified growth conditions, at different dc powers (50 – 600 W) using deposition monitor with the crystal sensor positioned just above the center of the substrate holder. For each composition, the ratio of Mg to Ti growth rate was calculated as a function of the ratio of Mg to Ti atoms. Using the measured growth rates and the calculated growth rate ratio, the film composition was controlled by changing the power to the Mg and Ti targets. The dc powers applied to each target during the deposition of the different alloys together with EPMA determined compositions are summarized in Table I. There is excellent agreement between the measured and calculated compositions. The substrates were rotated at 20 rpm in order to obtain uniform composition across the substrates. The films are typically 1.2-1.5 μm thick as determined from SEM cross section measurements.

Materials Characterization

The composition of the different Mg-Ti thin film samples was determined using Electron probe microanalysis (EPMA). EPMA measurements were made with a Cameca Instruments, Inc. (Nampa, ID) model SX-100 electron probe. Analysis voltage and current were 15 keV and 10 nA. Analyses were done with a focused beam < 1.0 μm in diameter. Each sample was analyzed at 8 random locations with results averaged. Mg and Ti compositions were calculated from x-ray intensities (k-ratios) with the thin film program GMRFILM [17]. Precision and accuracy are both estimated as +/- 2% relative for all samples.

Ex-situ X-ray diffraction was used to study the structures of the Mg-Ti films. Each sample was examined using Cu K_α radiation in a Bruker AXS General Area Detector Diffractometer System (GADDS). The diffraction images are typically collected for a period of 5 minutes using a 0.5 mm collimator and a sample to detector distance of 60 mm. The primary beam incidence angle is 17 degrees. Each diffraction scan was examined for incomplete Debye rings, which indicate preferred crystallite orientation, crystallite size in the film, and single crystal substrate diffraction peaks. Each diffraction image is then integrated to produce a powder diffraction pattern. The diffraction patterns are then compared to reference data for phase identification. Once phase identification is complete lattice parameters can be calculated.

Topography measurements of the Mg-Ti alloy samples were obtained using Dimension V Scanning Probe Microscope (Veeco). The height measurements were obtained using tapping mode and scanning area of 5.0 μm × 5.0 μm. The AFM measurements were performed in air, at ambient temperature and humidity, using a silicon probe with tip ROC (radius of curvature) < 10 nm. Four topography maps at different positions were acquired for each sample. The reported values of the rms roughness are the mean values of the different measurements.

Nanoindentation experiments were carried out with a Hysitron Triboscope (Hysitron Incorporated, Minneapolis, MN). The load controlled indents were made with a Berkovich indenter. The load function consists of three segments, a 5 seconds loading segment, a 2 seconds holding segment at the maximum load, and a 5 seconds unloading segment. Indents were made to a depth of about 10 percent of the film thickness to minimize substrate effects (typically ~140 nm). Data values are averaged from a series of 16 indents per sample. The hardness and modulus are calculated based on the Oliver-Pharr method [16].

Table I. The dc power applied to the Mg and Ti sources during deposition of $Mg_{(1-x)}Ti_x$ thin films at room temperature along with the corresponding Mg and Ti concentration calculated using measured growth rates and the concentration determined from EPMA measurements

Sample	Power (W)		Calculated Concentration (Using Growth rates) (at.%)		Measured Concentration (EPMA) (at.%)	
	Mg	Ti	Mg	Ti	Mg	Ti
Mg	100	-	100	-	>99	<1
$Mg_{0.80}Ti_{0.20}$	100	130	80	20	79	21
$Mg_{0.60}Ti_{0.40}$	50	200	60	40	59	41
$Mg_{0.50}Ti_{0.50}$	50	300	50	50	49	51
$Mg_{0.40}Ti_{0.60}$	40	310	40	60	42	58
$Mg_{0.20}Ti_{0.80}$	20	450	20	80	19	81
Ti	-	300	-	100	<1	>99

Results and Discussion

Crystallographic Structure

The microstructure of the $Mg_{(1-x)}Ti_x$ thin film alloys was determined by XRD measurements. The XRD patterns of the different Mg-Ti films after normalization are shown in Figure 1. The Mg film shows a very strong diffraction peak at 2θ ~ 34.8° which is due to the **(0002)** plane of hexagonal closed packed (HCP) Mg. The Mg film exhibits a strong preferential crystallographic orientation with the **(0001)** plane growing parallel to the substrate surface. The weak diffraction peaks at higher 2θ correspond to the HCP Mg **(10$\bar{1}$2)** and **(10$\bar{1}$3)** peaks. XRD pattern of the $Mg_{0.79}Ti_{0.21}$ thin film shows a strong diffraction peak at 2θ ~ 35.8° that is due to the **(0002)** plane of HCP solid solution unit cell. This peak is shifted to a higher angle (lower d-spacing) with respect to pure Mg due to the partial substitution of Mg atoms by Ti atoms. Because Ti has relatively smaller molar volume than Mg, partial substitution of Mg by Ti causes the Mg lattice to contract and thus d-spacing to decrease. The peak at higher 2θ is due to **(10$\bar{1}$3)** HCP Mg-Ti solid solution. Mg and $Mg_{0.79}Ti_{0.21}$ samples have a strong [001] fiber texture as reflected by the strong **(0002)** diffraction peak of the HCP Mg-Ti unit cell.

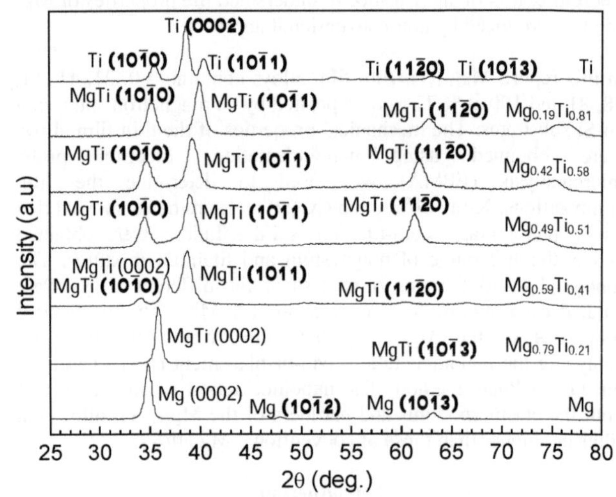

Figure 1. XRD spectra of $Mg_{(1-x)}Ti_x$ thin film alloys.

The film with 41 at.% Ti shows a single phase corresponding to HCP Mg-Ti solid solution. XRD pattern of this film shows three diffraction peaks at lower 2θ (~ 34.0°, 36.6°, and 38.7°) which are due to the **(10$\bar{1}$0)**, **(0002)**, and **(10$\bar{1}$1)** peaks of a contracted HCP Mg lattice with Mg atoms substituted by Ti. The $Mg_{(1-x)}Ti_x$ samples with x ~ 0.51, 0.58 and 0.81 show also a single phase with two diffraction peaks that can be attributed to the **(10$\bar{1}$0)**, and **(10$\bar{1}$1)** diffraction peaks in HCP Mg-Ti lattice. In all these films, the higher 2θ diffraction peaks correspond to the Mg-Ti HCP phase. The pure Ti film, unlike the Mg film, does not show one highly preferred orientation. The XRD spectra of Ti film show the **(10$\bar{1}$0)**,**(0002)**, and **(10$\bar{1}$1)** diffraction peaks of Ti HCP lattice at lower 2θ (35.5°, 38.6°, and 40.4 °).

The above XRD results suggest that all the Mg-Ti thin film alloys are a single hexagonal Mg matrix phase with various contents of Ti in the solid solution. The lattice of the Mg-Ti solid solution contracts with the addition of Ti. In Figure 2, the lattice constants of the $Mg_{(1-x)}Ti_x$ films calculated from XRD spectra are presented as a function of Ti content, x. Also shown in Figure 2 are the lattice constants at different composition calculated using Vegard's Law [18]. Accordingly, the lattice constants of the alloys are linearly interpolated from the values of the pure compounds using the following relation:

$$a_{A_{(1-x)}B_x} = (1 - x)a_A + xa_B \quad (1)$$

The calculated lattice constants from XRD spectra agree with those predicted using Vegard's law confirming that a solid solution of Mg-Ti was obtained.

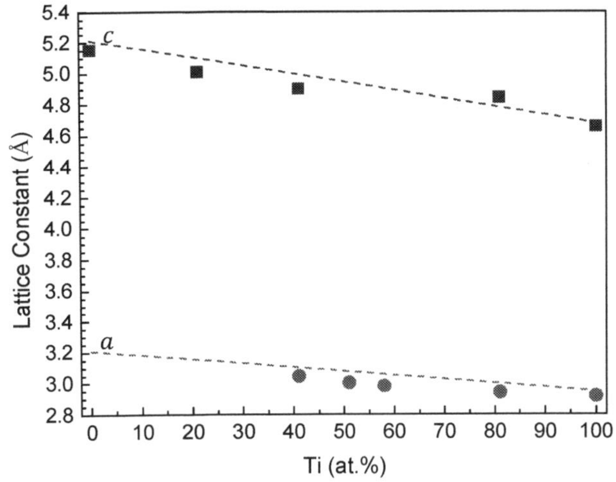

Figure 2. Dependence of lattice constants on the composition of the Mg-Ti samples. Also shown the linear dependence of lattice constants on the alloy composition predicted by Vegard's Law.

Topographic morphology

AFM height images of the Mg-Ti alloys are given in Figure 3. The pure Mg sample shows large hexagonal particles protruding from the film surface at different angles, and making the sample surface very rough. Figure 4 displays the root-mean-square (rms) roughness of the Mg-Ti thin films obtained from AFM height images as a function of Ti concentration. These results show that Mg thin film has a very high rms roughness of ~ 30nm. The AFM image of sample $Mg_{0.79}Ti_{0.21}$ shows that the large hexagonal particles disappeared and are replaced by small irregularly shaped particles that stretch along the sample surface. The refined almost-flat particles in sample $Mg_{0.79}Ti_{0.21}$ leave the surface of this sample significantly smoother with a much smaller rms roughness compared to Mg sample (Figure 4).

As the Ti concentration increases to 41, 51, and 58 at.%, $(10\bar{1}0)$ and $(10\bar{1}1)$ oriented Mg-Ti grains start to emerge in these films. AFM height images of these samples show also angular particles that stick out of the film surface. The appearance of these new $(10\bar{1}0)$ and $(10\bar{1}1)$ oriented Mg-Ti grains in these films affects the surface roughness of these films, which increases to a maximum value for the sample with 58 at.% Ti (Figure 4). The topography of the Ti-rich sample $Mg_{0.19}Ti_{0.81}$ shows long narrow rectangular stick-like particles. These particles grow evenly along the surface of the film and do not protrude out of the surface making the surface of this film smooth with small surface roughness compared to the films with 51 and 58 at.% Ti.

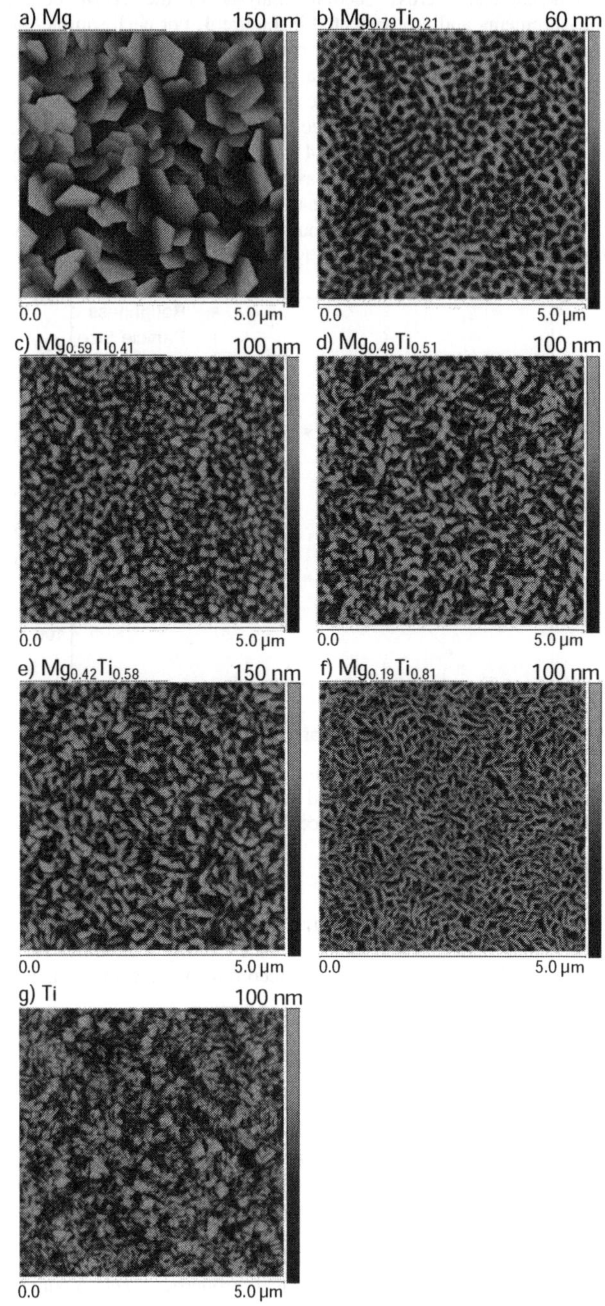

Figure 3. AFM topography images of the different Mg-Ti alloy thin films: a) Mg, b) $Mg_{0.79}Ti_{0.21}$, c) $Mg_{0.59}Ti_{0.41}$, d) $Mg_{0.49}Ti_{0.51}$, e) $Mg_{0.42}Ti_{0.58}$, f) $Mg_{0.19}Ti_{0.81}$, and g) Ti.

Finally, the pure Ti topography shows the fine rectangular stick-like shape as in $Mg_{0.19}Ti_{0.81}$ but in smaller size in addition to some coarse larger particles making the surface roughness of this film slightly higher than that of the $Mg_{0.19}Ti_{0.81}$ sample

The in-plane particle sizes of the different Mg-Ti samples were calculated using cross section analysis of the AFM height measurements and are presented in Figure 4. For each sample the measured lengths of 15 different particles along the short and long axes were averaged to give a mean value of the particle size of that sample. Mg sample shows large particle size of about ~ (750 ± 120) nm. The particle size value decreases to ~ 200 nm for the samples with Ti content of 21, 41 and 51 at.%, and then slightly increases to ~ (250 ± 80) nm for the $Mg_{0.42}Ti_{0.58}$ sample. For the Ti-rich sample (x = 0.81) the particle size decreases again to ~ (175 ± 90) nm reaching a minimum value of ~ (140 ± 40) nm for the pure Ti sample.

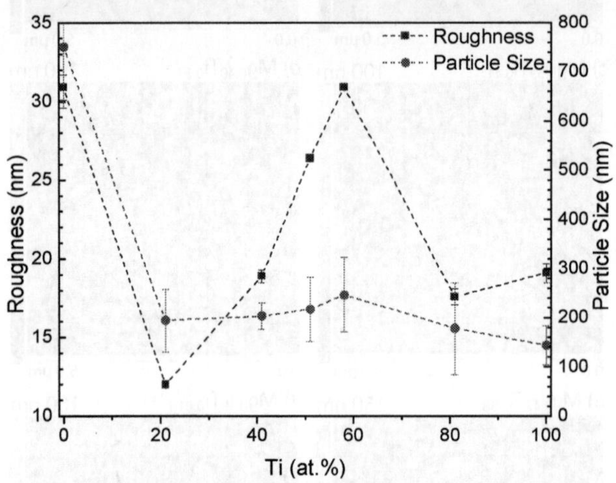

Figure 4. Root-mean-square (rms) roughness and particle size of Mg-Ti alloy thin films, obtained from AFM height images and displayed as a function of Ti concentration.

Micro-Mechanical Properties

The commonly used Oliver-Pharr method was used to analyze the indentation data and to calculate the Young's modulus and hardness of the Mg-Ti samples. According to this method, the hardness is defined as:

$$H = P_{max}/A_c \quad (2)$$

where P_{max} is the maximum indentation load and A_c is the projected contact area of the indenter tip at that load. The contact area of the indenter is calculated as a function of contact depth, h_c, using a series of indents at various loads (various h_c) performed on a sample with a known elastic modulus. Although this method does not account for the resulting indentation shape (e.g., pile-up) or the elastic mismatch between the film and the substrate it serves as a tool for estimating mechanical properties. The Young's modulus is obtained from:

$$E = 1/\gamma\left(\sqrt{\pi}/2\right)\left(S/\sqrt{A_c}\right) \quad (3)$$

where γ is a correction factor that depends on the indenter geometry and S is the stiffness of the unloading curve given by the slope of the initial portion of the unloading curve.

A total of 16 indents are performed on each sample and the load was chosen so that the contact depth was ~ 140 nm to minimize substrate and roughness effects. The calculated mechanical properties of each sample are then averaged and the mean values of modulus and hardness are plotted in Figure 5 along with the standard deviations. The measured Young's modulus and hardness of pure Mg and Ti thin film samples clearly show, as expected, higher Young's modulus and hardness values of Ti sample over Mg sample. It is reasonable to expect the Young's modulus and hardness of the Mg-Ti alloys to increase monotonically with the addition of Ti. However, the results show that $Mg_{(1-x)}Ti_x$ samples with x = 0.21, and 0.81 have both similar Young's modulus and hardness. Actually, sample $Mg_{0.79}Ti_{0.21}$ shows very large improvement in mechanical properties compared to pure Mg sample. But, as Ti content increases to 41, 51, and 58 at.% the measured Young's modulus and hardness of the corresponding samples are lower than those of the Mg-Ti sample with 21 at.% Ti but higher than those of pure Mg sample.

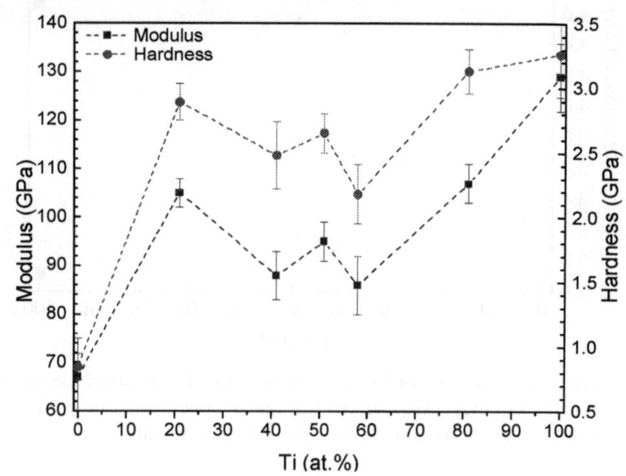

Figure 5. Calculated Young's modulus and hardness of the Mg-Ti alloy samples plotted as a function of Ti content.

These results can be understood once the effect of roughness on the measured modulus and hardness values is taken into account [19, 20]. This effect originates from the initial contact between the indenter tip and the rough surface when some flattening can occur causing the measured h_c and thus A_c to be less accurate. As a result the measured modulus and hardness of the Mg-Ti samples are affected by both the addition of Ti content which should result in improving mechanical properties and the increased surface roughness of the samples which would reduce the values of the modulus and hardness. The $Mg_{0.79}Ti_{0.21}$ shows a very small surface roughness compared to Mg sample, and as a result its mechanical properties are strongly improved due to both Ti addition and lower surface roughness. The surface roughness of Mg-Ti samples with 41, 51, and 58 at.% Ti increases and reaches a maximum value for the sample with 58 at.% Ti. As a result, the modulus and hardness of these samples are lower than the sample with 21 at.% Ti even though they are more Ti-rich. As Ti content increases to 81 at.%, the surface roughness of the $Mg_{0.19}Ti_{0.81}$ decreases. The effect of increased Ti content on the mechanical properties of this film is apparent by the increased modulus and hardness values.

Conclusion

Thin films of $Mg_{(1-x)}Ti_x$ alloys with different Ti content were deposited by magnetron co-sputtering techniques on Si substrates. XRD measurements showed that these alloys have a HCP structure. Moreover, the calculated lattice parameters of the $Mg_{(1-x)}Ti_x$ alloys follow Vegard's rule, indicating that single phase compounds were formed. AFM surface measurements showed that pure Mg thin film has the highest rms roughness while the alloy with ~ 20 at.% Ti has the smoothest surface. The in-plane particle size calculated from AFM height images was the largest (~ 750 nm) for the pure Mg film and fluctuates about a value of ~ 200 nm for the other films. Nanoindentation measurements showed that the alloy thin films have higher Young's modulus and hardness compared to pure Mg. In particular, the relatively smooth $Mg_{(1-x)}Ti_x$ alloy thin film with x ~ 0.20 displays the same improved mechanical properties as the alloy with x~ 0.80 and thus has the potential of being used as a strong and hard protective coating.

Acknowledgements

The authors thank M. Tessema for assistance with XRD measurements and R. Waldo for EPMA measurements.

References

1. K. R. Baldwin, D. J. Bray, G. D. Howard, and R. W. Gardiner, "Corrosion behaviour of some vapour deposited magnesium alloys," *Materials Science and Technology*, 12 (1996), 929-944.

2. Y. Bohne, D. M. Seeger, C. Blawert, W. Dietzel, S. Mändl, and B. Rauschenbach, "Influence of ion energy on properties of Mg alloy thin films formed by ion beam sputter deposition," *Surface & Coatings Technology*, 200 (2006), 6527-6532.

3. T. Mitchell, S. Diplas, and P. Tsakiropoulos, "Characterisation of corrosion products formed on PVD in situ mechanically worked Mg-Ti alloys," *Journal of Alloys and Compounds*, 392 (2005), 127-141.

4. S. Rousselot, M. P. Bichat, D. Guay, and L. Roue, "Structure and electrochemical behaviour of metastable $Mg_{50}Ti_{50}$ alloy prepared by ball milling," *Journal of Power Sources*, 175 (2008), 621-624.

5. Y. T. Cheng, M. W. Verbrugge, M. P. Balogh, D. E. Rodak, and M. Lukitsch, "Magnesium-Titanium solid solution alloys," *United States Patent*, US 7651732B2, Jan. 26, 2010.

6. J. L. Murray, "The Mg-Ti (Magnesium-Titanium) System," *Bulletin of Alloy Phase Diagrams*, 7 (1986), 245-248.

7. W. P. Kalisvaart, H. J. Wondergem, F. Bakker, and P. H. L. Notten, "Mg-Ti based materials for electrochemical hydrogen storage," *Journal of Materials Research*, 22 (2007), 1640-1649.

8. R. A. H. Niessen and P. H. L. Notten, "Electrochemical Hydrogen Storage Characteristics of thin film MgX (X = Sc, Ti, V, Cr) Compounds," *Electrochemical and Solid-State Letters*, 8 (2005), A534-A538.

9. P. Vermeulen, H. J. Wondergem, P. C. J. Graat, D. M. Borsa, H. Schreuders, B. Dam, R. Griessen and P. H. L. Notten, "In situ electrochemical XRD study of (de)hydrogenation of Mg_yTi_{100-y} thin films," *Journal of Materials Chemistry*, 18 (2008), 3680-3687.

10. B. Farangis, P. Nachimuthu, T. J. Richardson, J. L. Slack, B. K. Meyer, R. C. C. Perera, and M. D. Rubin, "Structural and electronic properties of magnesium-3D Transition metal switchable mirrors," *Solid State Ionics*, 165 (2003), 309-314.

11. D. M. Borsa, A. Baldi, M. Pasturel, H. Schreuders, B. Dam, R. Griessen, P. Vermeulen, and P. H. L. Notten, "Mg-Ti-H thin films for smart solar collectors," *Applied Physics Leters*, 88 (2006), 241910-1 – 241910-3.

12. T. J. Richardson, B. Farangis, J. L. Slack, P. Nachimuthu, R. Perera , N. Tamura, and M. Rubin, "X-ray absorption spectroscopy of transition metal magnesium hydride thin films," *Journal of Alloys and Compounds*, 356-357 (2003), 204-207.

13. P. Vermeulen, R. A. H. Niessen, and P. H. L. Notten, "Hydrogen storage in Metastable $Mg_yTi_{(1-y)}$ thin films," *Electrochemistry Communications*, 8 (2006), 27-32.

14. P. Vermeulen, R. A. H. Niessen, D. M. Borsa, B. Dam, R. Griessen, and P. H. L. Notten, "Effect of the deposition technique on the metallurgy and hydrogen storage characteristics of metastable $Mg_yTi_{(1-y)}$ thin films," *Electrochemical and Solid-State Letters*, 9 (2006), A520-A523.

15. G. Song, and D. Haddad, "The topography of magnetron sputter-deposited Mg-Ti alloy thin films," *Materials Chemistry and Physics*, accepted to be published.

16. W. C. Oliver and G. M. Pharr, "An improved technique for determining hardness and elastic modulus using load and displacement sensing indentation experiments," *Journal of Materials Research*, 7 (1992), 1564-1583.

17. R. A. Waldo, M. C. Militello, and S. W. Gaarenstroom, "Quantitative thin-film analysis with an energy-dispersive x-ray detector," *Surface and Interface Analysis*, 20 (1993), 111-114.

18. L. Vegard, "Die Konstitution der Mischkristalle und die raumfüllung der atome," *Zeitschrift für Physik*, 5 (1921), 17-26.

19. M. S. Bobji, S. K. Biswas, and J. B. Pethica, "Effect of roughness on the measurement of nanohardness –A computer simulation study," *Applied Physics Letters*, 71 (1997), 1059-1061.

20. J.Y. Kim, J.J. Lee, Y.H. Lee, J.I. Jang, and D. Kwon, "Surface roughness effect in instrumented indentation: A simple contact depth model and its verification," *Journal of Materials Research*, 21 (2006), 2975-2978.

Magnesium Technology 2011
Edited by: Wim H. Sillekens, Sean R. Agnew, Neale R. Neelameggham, and Suveen N. Mathaudhu
TMS (The Minerals, Metals & Materials Society), 2011

Effect of Adding SiO_2-Al_2O_3 Sol into Anodizing Bath on Corrosion Resistance of Oxidation Film on Magnesium Alloy

Huicong Liu, Liqun Zhu, Weiping Li
Key Laboratory of Aerospace Materials and Performance (Ministry of Education),
School of Materials Science and Technology, Beihang University,
37# Xueyuan Road, Haidian District, Beijing, 100191, China

Keywords: magnesium alloy, modified oxidation film, SiO_2-Al_2O_3 sol, anodic oxidation

Abstract

Due to the widely use in automobile and construction field, AZ91D magnesium alloy need to be protected more effectively for its high chemical activity. In this paper, three kinds of films were formed on magnesium alloy. The first kind of film, named as anodic oxidation film, was prepared by anodic oxidation in the alkaline solution. The processes for preparing the second kind of film, named as multiple film, involved coating sol-gel on the samples and heat-treating before anodic oxidation. The third kind of film was prepared by anodic oxidation in the alkaline oxidation solution containning 5% (vol) SiO_2-Al_2O_3 sol, named as modified oxidation film. The corrosion resistance of the three different films was investigated. The results showed that the modified oxidation film had the highest corrosion resistance due to the largest thickness and most dense surface morphology. Sol was discussed to react during the film forming process, which leaded to the difference between modified oxidation film and anodic oxidation film.

Introduction

Magnesium alloy has been widely used due to its advantages such as low density, high strength-to-density ratio, nice electromagnetic shielding, etc[1-2]. However, magnesium alloy is very active and its oxidation film formed in air is very loose and porous[3], therefore, magnesium alloy is very apt to be corroded in corrosive environment.

In order to protect magnesium alloy, many methods have been used, such as chemical conversion[4-6], anodic oxidation[7-11], micro-arc oxidation[12-13], electroless plating[14], plating[15-16] and so on. Among these methods, anodic oxidation has been studied by many researchers and has made some progress. However, the forming oxidation film on magnesium alloy has many pores[17-18], which will lead to the film easy to be corroded after some certain time.

In recent years, sol-gel dip coating method has been used to increase corrosion resistance of magnesium alloy, aluminum alloy and nickel plating on carbon steel[18-20], and sol plays a role in sealing holes on surface coating.

Generally speaking, sol-gel film is very thin and several dip coating cycles are needed to obain expected thickness. In order to get excellent film through more simple process, sol was added into anodizing bath and was expected to plays a positive role in anodizing process.

In our previous study, electrochemical oxidation characteristic and growth characterization of anodic film on AZ91D magnesium alloy under the effect of silica sol[8,9,17] were investigated, and the effect of sol composition on anodic oxidation film on magnesium alloys[21] was studied. The results showed that the addition of silica sol in Na_2SiO_3 solution could decrease the surface energy and the conductivity of the solution, and could increase the anodic film thickness and improve the uniformity of the anodic film on AZ91D magnesium alloy.

In this paper, modified oxidation film was prepared by adding sol into alkaline oxidation solution, and its corrosion resistance was investigated in comparison with multiple film and anodic oxidation film. The purpose is to form a more protective film on magnesium alloy surface using more simple process.

Experimental Procedure

(1) Materials

The raw material was AZ91D molten magnesium alloy, and samples were cut into the size of 40mm×25mm×2mm. The chemical composition(mass percentage) includes: 8.5 ~ 9.5 % Al, 0.17 ~ 0.4% Mn, 0.45 ~ 1.90% Zn, 0.05 % Si, 0.25% Cu, 0.001 % Ni, 0.004% Fe and Mg is the balance.

(2) Preparation of films on magnesium alloy

Before the films were prepared, the magnesium alloy samples were polished using sand paper and then degreased in acetone in an ultrasonic device.

The prepared sol was SiO_2-Al_2O_3 composite product (volume rate is 70:30). The reactant used to prepare sol was methyltriethoxysilane and aluminum iso-propoxide. The solvent was absolute ethanol and acetone. The reaction time was 6~8 hours, and the temperature was 60~90°C. The power supply for anodic oxidation process was an autotransformer (voltage and current output are 0~250V and 0~4.0A, respectively).

Three kinds of films were made in this paper.

The process for preparing the first kind of film was anodic oxidation in the solution containing 150g/L NaOH, 35 g/L Na_3PO_4 and 35 g/L NaF. The anodic oxidation voltage was 65V and the duration was 30 min. This kind of film is named as "anodic oxidation film".

The second kind of film was made by coating sol on the samples and heat-treating before anodic oxidation. The parameters for anodic oxidation process were the same to that of anodic oxidation film. This kind of film is named as "multiple film".

The third kind of film was prepared by anodic oxidation, the oxidation solution was prepared by adding 5%(vol) SiO_2-Al_2O_3 sol into the anodic oxidation solution. The anodic oxidation voltage and duration were the same to that of anodic oxidation film. And this kind of film is named as "modified oxidation film".

(3) The device and methods for measurement

The thickness of the films was measured by E110B whirlpool testing instrument made by FISCHER company in west Germany. The surface morphology was analyzed by S-530 scanning electron microscope and Hirox KH-3000 digital microscope. The composition of films was examined by Link ISIS instrument made in Japan.

The corrosion resistance was studied in 5 wt% sodium chloride solution, and the techniques used included salt-solution dipping, anodic polarization curve measurement and galvanic corrosion test. The anodic polarization curves were recorded on a CHI604A electrochemical analyzer. In salt-solution dipping test, the anti-corrosion time was defined as the time of the first corrosion spot to be found. The galvanic corrosion test was done on ZRA-1 galvanic corrosion instrument. Galvanic corrosion current was examined relative to LY12CZ aluminum alloy.

Results and discussion

(1) Corrosion resistance of modified oxidation film

As shown in figureⅠ, the modified oxidation film has much higher corrosion resistance than that of anodic oxidation film and multiple film. FigureⅠ shows that the potential of the samples coated with modified oxidation film is more positive than those of anodic oxidation film and multiple film at the same current value, and the largest potential difference for the examined value is about 850mV, which indicates that the samples treated with modified oxidation film have small tendency to be corroded than the other two kinds of films in the same condition.

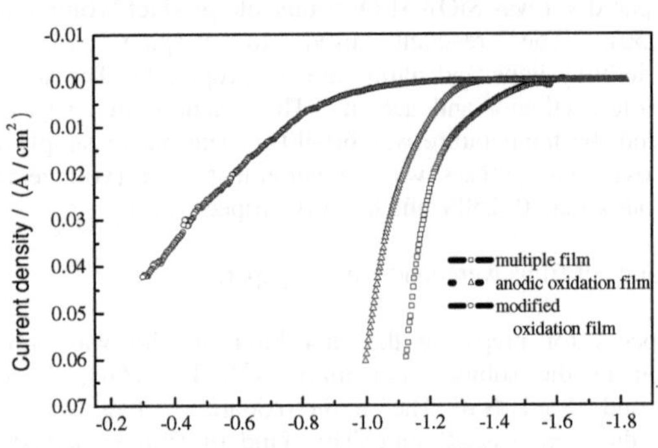

FigⅠ Anodic polarization curves of three kinds of films

The result of salt-solution dipping test is presented in tableⅠ, which shows the corrosion resistance comparison of the three kinds of films. It can be seen that compared with anodic oxidation film and multiple film, modified oxidation film has the lest corrosion pit number after dipped in the 5wt% NaCl solution for 12h in the same area, and can stand the longest time before being corroded. In other word, it owns the highest corrosion resistance.

TableⅠ Corrosion Resistance of the Three Kinds of Films

Films	Anodic oxidation film	Multiple film	Modified oxidation film
Corrosion pit number ($/10cm^2$)	8~10	9~11	1~2
Time for pit emerging (h)	4~6	3~5	12~14

FigureⅡ is the result of galvanic corrosion test for three kinds of films. It can be seen that in the same corrosive condition, modified oxidation film has the lowest current value in comparision with the other two kinds of films. From figureⅢ, there are fewer corrosion pits on the surface of modified oxidation film than that of anodic oxidation film, which further comfirms the results in figureⅠ and Ⅱ.

From the three kinds of corrosion test, the same conclusion can be achieved that the modified oxidation film has the highest corrosion resistance.

(2) Discussion

As shown in tableⅡ, the modified oxidation film has the largest thickness in comparision with anodic oxidation film and multiple film. And the thickness of multiple film is a little smaller than that of anodic oxidation film. FigureⅣ shows the cross section morphology of modified oxidation film, which matches well with the thickness of this film in tableⅡ.

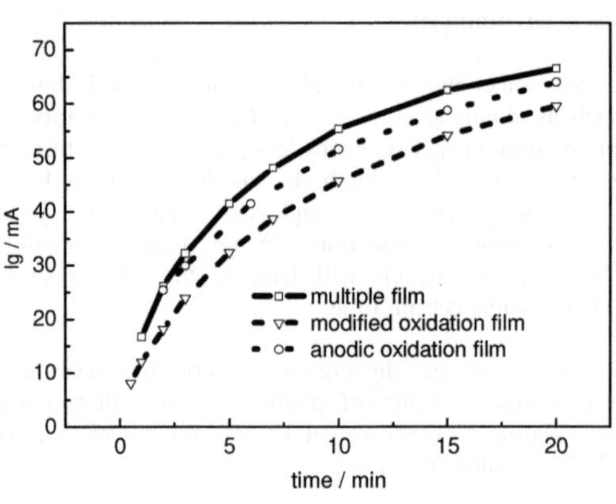

FigⅡ Corrosion current trends for three kinds of films

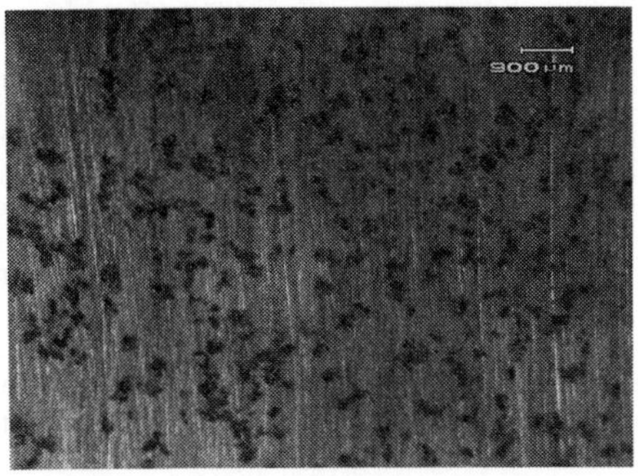

a Anodic oxidation film

b Modified oxidation film

FigIII Micro morphologies of magnesium alloys samples after galvanic corrosion test with aluminum alloy for 15min

TableII Thickness of the Three Kinds of Films

Film	Anodic oxidation film	Multiple film	Modified oxidation film
Thickness(μm)	4.0~4.9	3.4~4.0	29.8~36.2

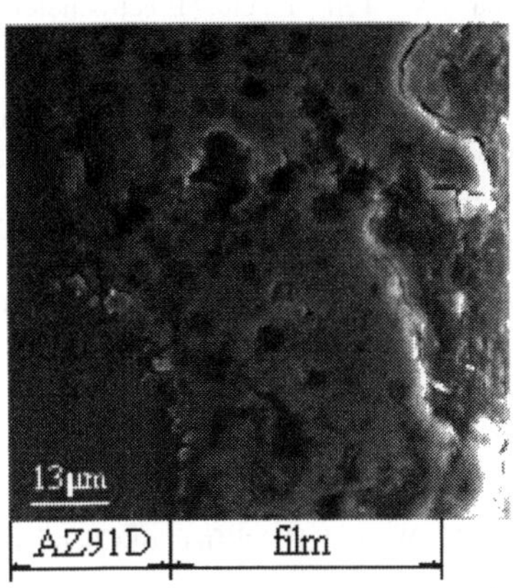

FigIV Cross section morphology of modified oxidation film

The multiple film is thinner, perhaps because there is a breaking and solving effect on the formed sol-gel film on the surface during the anodic oxidation process. Because the sol film is rather thin, it can be equally broken down when the applied voltage is high enough for the following anodic oxidation process, then sol would go into the oxidation solution. As a result, the composition of the multiple film has little difference with that of anodic oxidation film, as shown in tableII . Due to the breaking and solving effect, the anodic oxidation process would be affected, which leads to a slightly thinner film.

After adding 5% (vol) SiO_2-Al_2O_3 sol into anodic oxidation solution, the thickness of oxidation film increased greatly and is much larger than those of anodic oxidation film or multiple film. This may be due to that the sol in the anodic oxidation solution greatly affected the oxidation process, and increased the growing rate of oxidation film. This can be verified in tableIII. It can be seen that Si and Al elements were found in the modified oxidation film, while the Al element content is lower than that of anodic oxidation film or multiple film. The reason is not very clear yet, and will be studied in detail in future work.

TableIII Element Composition Comparison of the Films (Atom%)

Films	O	Mg	Al	Si
anodic oxidation film	27.1~29.2	65.5~68.5	4.1~5.7	——
multiple film	26.8~27.7	65.9~68.3	5.0~6.3	——
modified oxidation film	29.8~33.6	63.5~64.9	2.3~3.2	1.8~2.4

In the process of anodic oxidation in the solution containing sol, sol transfers to magnesium alloy surface. Thanks to the exothermic action of sparks discharging, the sol network is destroyed. The electriferous groups such as :Si: , :Al· , :Si − R (alkyl), :Al − R (alkyl), :Si − O, :Al − O will form and adsorb on the surface, which participate in the oxidation reaction and forming film, and hence SiO_2 and Al_2O_3 formed. These reactions will increase the growth rate of oxidation film on magnesium alloy, so there is an obvious increase in film thickness. So the film can act as barrier effect and increase corrosion resistance of modified oxidation film. Although the comparision experiments have not been done with the addition of inorganic salt, the effect of sol on forming modified oxidation film is obvious.

Comparing the morphology of modified oxidation film(figureV c) with those of anodic oxidation film(figure V a) and multiple film(figureV b), it can be seen that there are some relatively big holes and some holes connect with each other on the anodic oxidation film and multiple film. These holes will obviously reduce the corrosion resistance of magnesium alloy. In comparision with the anodic oxidation film and multiple film, the modified oxidation film is uniform and dense and no big holes on the film are observed, which can effectively increase the corrosion resistance of oxidation film.

With respect to anodic oxidation process, a common viewpoint is that at the initial stage of oxidation, when voltage is higher than the broken voltage of the oxidation film formed in air, sparkle discharging phenomenon would appear. Since the instantaneous temperature induced from sparkle discharging is very high, magnesium and other alloy composition would be partially melted and then oxidation film would be formed under the cooling effect of solution. As oxidation duration prolongs, voltage to break the sample surface film persistently increased, and the repeating break of formed film would lead to the increase in film thickness, and thus oxidation process could go on. A great deal of heat from sparkle discharging was absorbed by solution, and metal oxide formed on sample surface was cooled, which leads to the shrinkage of formed film and hence porous morphology of anodic oxidation film was obtained.

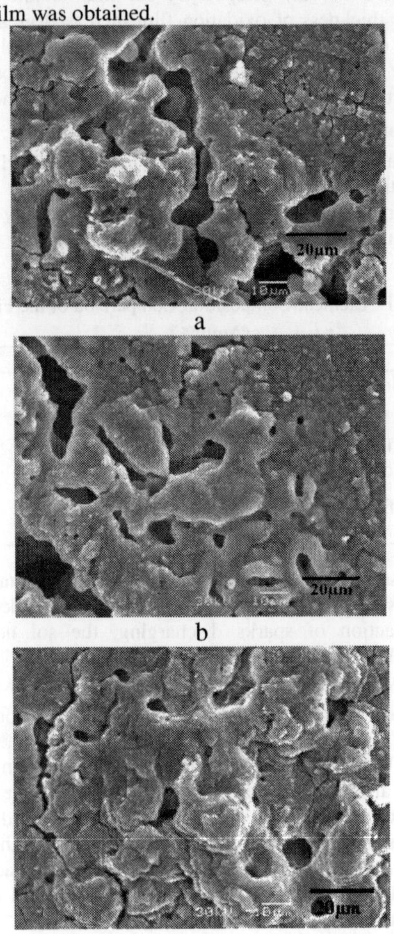

FigV Micro morphologies of the three kinds of films

The effect of sol on the thickness and density of the oxidation film may be explained as follows. The sol in anodic oxidation solution will decrease the conductivity of solution, as reported in reference 17. At the early stage of anodic oxidation, the spark discharging voltage increased, higher than that in anodic oxidation solution, then the growth rate of oxidation film increased and the rate of holes in the film also increased. When anodic oxidation went on, SiO_2-Al_2O_3 sol in the anodizing solution reacted on the interface of anodic oxidation film, and some SiO_2-Al_2O_3 sol composition entered into the film. So the film was more dense and there were fewer big holes. With oxidation duration increasing, the anodic oxidation process continued, and due to sol composition act on the interface of the film or sealed some defaults, the film growed gradually and the thickness of the film increased. The more dense and uniform modified oxidation film will get the better corrosion resistance than anodic oxidation film and multiple film. And from figureIV, there are no penetrable holes in the film, this denser and thicker film would increase the corrosion resistance of magnesium alloy.

Conclusion

(1) After adding SiO_2-Al_2O_3 sol into anodizing bath, modified oxidation film has the largest thickness of about 34μm and the highest corrosion resistance relative to anodic oxidation film and multiple film. In the 5wt% NaCl solution, time for pits emerging of modified oxidation film was about three times longer than that of anodic oxidation film.

(2) The modified oxidation film had the largest thickness, was the most dense and had the least pinholes on the surface among the three kinds of films, and no penetrable holes were observed from the cross section morphology. These were the reasons for the highest corrosion resistance.

(3) Sol played an important role in film forming process, some electriferous groups were formed under the condition of high temperature and electric intensity. These groups adsorbed on the sample surface and reacted to form film. This process leaded to a more thick and dense modified oxidation film in comparision with anodic oxidation film.

Acknowledgement

This work is supported by the Fundamental Research Funds for th e Central Universities(YWF-10-02-036), the Cheung Kong Schola rs and Innovative Research Team Program in University from Mi nistry of Education (Grant No. IRT0805) and the aerial science fund (2008ZE51064).

References

[1] Yan A.J., Zhu X.M., Teng Y., etc. "Electrochemical Corrosion Behavior of MAO Film on AZ31 Magnesium Alloy"[J], *Journal of Dalian Jiaotong University*, 2008, 29(3):45-48.

[2] R. Arrabal, E. Matykina, T. Hashimoto, etc. "Characterization of AC PEO Coatings on Magnesium Alloys"[J], *Surface and Coatings Technology*, 2009, 203(16):2207-2220.

[3] Liu Y.G., Zhang W., Li J.Q.. "Microarc Electrodeposition of Ceramic Coatings on Double Electrodes of AZ91D Magnesium Alloy by AC Pulse Method"[J], *Journal of university of science and technology Beijing*, 2004 , 26(1) : 73-77.

[4] J.K. Lin, J.Y. Uan. "Formation of Mg,Al-hydrotalcite Conversion Coating on Mg Alloy in Aqueous HCO_3^-/CO_3^{2-} and Corresponding Protection Against Corrosion by the Coating"[J], *Corrosion Science*, 2009, 51(5): 1181-1188.

[5] Huo H.W., Li Y., Wang F.H.. "Effect of Chemical Conversion Film Plus Electroless Nickel Plating on Corrosion Resistance of Magnesium Alloys"[J], *The Chinese Journal of Nonferrous Metals*, 2004, 14(2):267-272.

[6] Zhao M., Wu S.S., Luo J.R., etc. "The Present Status and Prospect of Chromium Free Surface Treatment for Magnesium Alloys"[J], *Foundry*, 2003, 52(7):462-465.

[7] A. Yabuki, M. Sakai. "Anodic Films Formed on Magnesium in Organic, Silicate-containing Electrolytes"[J], *Corrosion Science*, 2009, 51(4): 793-798.

[8] Li W.P., Zhu L.Q., Li Y.H.. "Electrochemical Oxidation Characteristic of AZ91D Magnesium Alloy under the Action of Silica Sol"[J], *Surface and Coatings Technology*, 2006, 201(3-4):1085-1092.
[9] W.P. Li, L.Q. Zhu, Y.H. Li, etc. "Growth Characterization of Anodic Film on AZ91D Magnesium Alloy in an Electrolyte of Na_2SiO_3 and KF"[J], *Journal of University of Science and Technology Beijing*, Mineral, Metallurgy, Material, 2006, 13(5): 450-455.
[10] H. Asoh, S. Ono. "Anodizing of Magnesium in Amine-ethylene Glycol Electrolyte"[J], *Materials Science Forum*, 2003(419-422):957-962.
[11] L.Y. Chai, X. Yu, Z.H. Yang, etc. "Anodizing of Magnesium Alloy AZ31 in Alkaline Solutions with Silicate under Continuous Sparking"[J], *Corrosion Science*, 2008, 50(12): 3274-3279.
[12] P. Shi, W.F. Ng, M.H. Wong, etc. "Improvement of Corrosion Resistance of Pure Magnesium in Hanks' Solution by Microarc Oxidation with Sol–Gel TiO_2 Sealing"[J], *Journal of Alloys and Compounds*, 2009, 469(1-2): 286-292.
[13] Liu Y.G., Zhang W., Li J.Q.. "Microarc Electrodeposition of Ceramic Films on Double Electrodes of AZ91D Magnesium Alloy by Symmetrical AC Pulse Method"[J], *Surface Engineering*, 2003, 19(5):345-350.
[14] Huo H.W., Li Y., Wang F.H.. "Electroless Nickel Plating on AZ91D Magnesium Alloys"[J], *Journal of Chinese Society for Corrosion and Protection*, 2002, 22(1):14-17.
[15] Cao W.B., Ren C.X., Guan S.K., etc. "Research Progress in the Electric Ni-P on Magnesium Alloys"[J], Water Conservancy and Electric Power Machinery, 2003, 25(4):28-31.
[16] J.F. Zhang, C.W. Yan, F.H. Wang. "Electrodeposition of Al–Mn Alloy on AZ31B Magnesium Alloy in Molten Salts"[J], *Applied Surface Science*, 2009, 255(9):4926-4932.
[17] W.P. Li, L.Q. Zhu, H.C. Liu. "Effects of Silicate Concentration on Anodic Films Formed on AZ91D Magnesium Alloy in Solution Containing Silica Sol"[J], *Surface and Coatings Technology*, 2006, 201(6): 2505-2511.
[18] Cai Q.Z., Wang D., Luo H.H., etc. Sealing of Micro-arc Oxidation Coating on Magnesium Alloy by SiO_2 Sol Sealing Agent[J], *Special Casting & Nonferrous Alloys*, 2006, 26(10):612-615.
[19] Zhou Q., He C.L., Cai Q.K., etc. Sealing of Anodized Films on Al Alloy with Boehmite Sol[J], *The Chinese Journal of Nonferrous Metals*, 2007, 17(8): 1386-1390.
[20] Liu H.C., Zhu L.q., Du Y.B.. High Temperature Oxidation Resistance of Thin Film Made by Sol-gel Method[J], *Trans Mater Heat Treat*, 2004, 25(4):77~80.
[21] Zhu L.Q., Liu H.C. "The Effect of Sol Ingredient to Anodic Oxidation Film on Magnesium Alloys"[J], *Journal of functional materials*, 2005, 36(6):923-926.

Magnesium Technology 2011
Edited by: Wim H. Sillekens, Sean R. Agnew, Neale R. Neelameggham, and Suveen N. Mathaudhu
TMS (The Minerals, Metals & Materials Society), 2011

Monotonic and Fatigue Behavior of Mg Alloy Friction Stir Spot Welds: An International Benchmark Test in the "Magnesium Front End Research and Development" Project

H. Badarinarayan[1], S.B. Behravesh[2], S.D. Bhole[3], D.L. Chen[3], J. Grantham[4], M.F. Horstemeyer[4], H. Jahed[2], J.B. Jordon[5], S. Lambert[2], H.A. Patel[3], X. Su[6], and Y. Yang[7]

[1] Automotive Products Research Laboratory, Hitachi America Limited, USA
[2] Mechanical and Mechatronics Engineering Department, University of Waterloo, Canada
[3]Department of Mechanical and Industrial Engineering, Ryerson University, Canada
[4]Center for Advanced Vehicular Systems (CAVS), Mississippi State University, USA
[5]Department of Mechanical Engineering, The University of Alabama, USA
[6]Ford Motor Company, USA
[7]Institute for Metals Research, China

Keywords: Fatigue, Friction stir spot welds; Magnesium alloys; Resistance spot weld

Abstract

This paper presents the experimental results of benchmark coupon testing of monotonic and cyclic conditions on friction stir spot welded coupons of Mg AZ31alloy. The results presented here are a product of a collaborative multinational research effort involving research teams from Canada, China, and the United States. Fatigue tests were conducted in load control at R=0.1 at two different maximum loads: 1kN and 3kN. Good agreement was found between the participating labs regarding the number of cycles to failure. Differences in the failure modes were observed for the two different loading conditions tested. At the higher load, fatigue failure was caused by interfacial fracture. However, at the lower load, fatigue cracks formed perpendicular to the loading direction, which led to full width separation. For additional comparison, the monotonic and cyclic results of the friction stir spot welds are compared to resistance spot welded coupons of similar nugget size.

Introduction

The continued push for more fuel efficient automobiles designs is motivation for ongoing research in lightweight metals. Through lightweight designs comes the need to explore alternatives to traditional metals like aluminum and steel currently used in the manufacturing of automobiles. As such, a collaborative multinational research effort involving researchers from Canada, China, and the United States and joined by Chrysler, Ford, and General Motors, is underway with the goal of developing the ability to build a front end of an automobile constructed of magnesium alloys. Magnesium alloys, with a density of 1.74g/cm^3, weigh less than a quarter of steel and two-third of aluminum, and with abundant reserves on the earth, are attractive substitutes for steel and aluminum for vehicle body structures [1–4]. Replacing the largely steel structure of a vehicle's front end with magnesium can also move the vehicle center of gravity away from the front, improving vehicle drivability.

This research project, called the Magnesium Front End Research and Development program, or MFERD, has already yielded several fatigue characterizations of wrought magnesium alloys [cf. 5-7]. In order to fully meet the project's objectives, characterization of mechanical properties of potential joining techniques is also needed. As such, friction stir spot welding is targeted as a possible joining technique in the fabrication of a front end of an automobile using magnesium alloys.

Friction stir welding has steadily been gaining more wide spread use when high integrity and strength are required. Friction stir spot weld is a recent variant of friction stir welding and is an attractive welding technique due to the solid state nature of the process and the lack of stress relieving that is typically needed. A recent literature review of the friction stir spot friction process can be found in [8]. The solid state nature and the ability of joining dissimilar metals have made friction stir spot welding an attractive welding process.

The fatigue behavior of a friction stir spot weld (FSSW) is highly dependent on process parameters employed to create the weld [9,10]. These parameters include speed, depth of plunge, dwell time, and tool configuration. Up to this point, a majority of studies of the nature of fatigue of FSSW's have been almost exclusively focused on aluminum alloys [9-12]. With regards to FSSW's made of aluminum alloys, the failure modes of quasi-static and cyclic loads of FSSW coupons vary based on the load level [9]. Differences have been observed in the fracture path for quasi-static loads compared to fatigue loads for aluminum alloys [9]. In this study[9-10], FSSW coupons failed by interfacial fracture under quasi-static loading, whereas, under cyclic loading, fatigue cracks initiated and grew from several locations including the interfacial tip and outside the weld zone. In addition, fatigue failure modes of aluminum FSSW made using various tooling were also observed to differ based on the shape of the tooling used to make the weld [10]. Also, fatigue of FSSW's of dissimilar aluminum alloys were observed to have different failure modes compared to FSSW's with identical alloys for the top and bottom sheets [11]. The fatigue failure modes of dissimilar metals (aluminum and steel) were also observed to vary compared with joints made of all aluminum alloys [12]. While the FSSW joining technique of magnesium alloys is documented [13-16], limited published literature exists on the fatigue properties of FSSW coupons of magnesium alloys. Mallick and Agarwal [17] were the first to quantify the fatigue behavior of FSSW's made of a magnesium alloy. However, their characterization of the failure mechanisms under cyclic loading was limited in its presentation.

Resistance Spot Weld (RSW), on the other hand, is currently the most common joining process in the automotive industry and hence is the manufacturers' preferred joining technique. Thus, as have been performed for steel [18-22] and aluminum alloys [23-27], characterization of FSSW and RSW techniques and comparing their monotonic and cyclic behavior, provide the motivation for these types of studies. As such, the purpose of this research is the characterization of FSSW'ed sheets of AZ31B magnesium alloy through a round-robin experimental program and also evaluating its mechanical behavior by a comparative analysis to RSW of the same alloy.

Materials

Magnesium AZ31B alloy sheets of 2.0mm thickness are chosen for the present study. Coupons were welded in lap configuration. The individual sheet dimensions were: length 100mm, width 38mm and were welded on an overlap area of 38 × 38 mm. The FSSW tool was made from standard tool steel (H13) material, having a shoulder with diameter 12 mm, pin length of 3.2 mm and left hand threads (M5). The shoulder was a concave profile with the angle of concavity of 10 deg. The welding process parameters were: tool rotation speed 750 RPM, tool plunge speed of 20mm/min, shoulder plunge depth of 0.1mm and a dwell time of 2.5 sec. In regards to the bonded area in FSSW, the welding process produces an annulus shape, and the average inner and outer diameters in this study were found to be 6.5mm and 9.7mm, respectively. During the welding process, the interface between the upper and lower sheet is formed into a hook like shape due to the penetration of the FSSW tool into the bottom sheet.

For comparison purposes, resistant spot weld (RSW) coupons were made of the same alloy as for the FSSW. The thickness of the sheet as well as the specimen configuration and dimensions were similar to the FSSW. Additionally, the weld nugget was approximately the same. The RSW parameters were: welding current of 34 kA, welding time of 8 cycles (8/60 sec), electrode force of 4 kN, and holding time of 30 cycles (0.5 sec). Figure 1 shows general dimensions for FSSW and RSW coupons employed in this study.

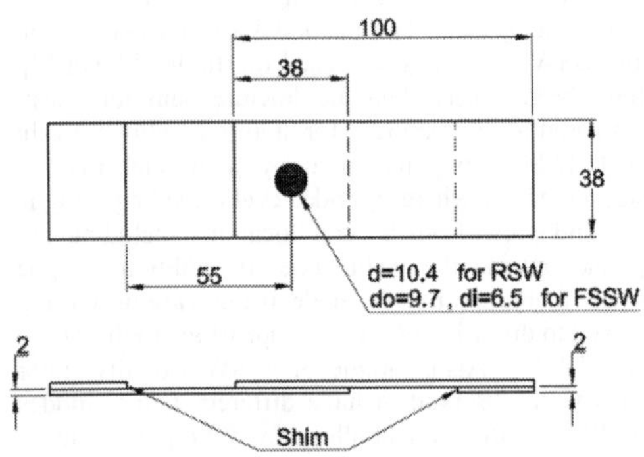

Figure 1. Configuration of friction stir and resistant spot welds single-weld lap-shear coupons. Dimensions are in mm.

Experiments

Mississippi State University (MSU), Ryerson University (RU), University of Waterloo (UW), and the Institute of Metal Research (IMR), all participated in round-robin fatigue testing of lap-shear FSSW coupons. Each institution conducted six (6) load control tests consisting of three (3) specimens conducted at each of the two (2) P_{max} load levels: 3kN and 1kN. The tests were run at $R = \frac{P_{min}}{P_{max}} = 0.1$, and a frequency of 5 Hz. All tests were conducted at ambient temperature and humidity. The test set-up also included certain specifications for mounting including careful attention to grip alignment, grip distance, and use of shims or offsets. The grip-to-grip distance was maintained at 110 mm at each of the labs. In addition, UW also tested RSW coupons under the same conditions as stated for the FSSW coupons.

Results and Discussion

Monotonic Tests

Monotonic tests were performed on a servo-hydraulic load frame under uniaxial displacement rate of 1 mm/min. The test results show that the FSSW coupons exhibited an average ultimate tensile-shear load (UTSL) of 4650±10N. All specimens tested failed in a consistent manner with a partially interfacial failure mode, as shown in Figure 2.

Figure 2. Failure mode in friction stir spot welded coupons under monotonic tensile-shear loading.

Similar to studies on steel FSSW [22] and aluminum FSSW [27], comparing the static and cyclic behavior of emerging FSSW joining technique and commonly used RSW is of interest. Recent research [28] has studied the monotonic and fatigue behavior of RSWs of AZ31B-H24 Mg alloy, and showed that the highest UTSL is achieved using the welding current of 34 kA, and the welding time of 8 cycles. The same welding parameters were utilized in this study, and a solid circular nugget with an average diameter of 10.4 mm was obtained which is close to outer diameter of FSSW (9.7 mm). It should be noted that although FSSW and RSW are of a similar outer diameter, the bonded area are very different (40 mm^2 in FSSW and 85 mm^2 in RSW), due to the different shapes of bonded region. However, comparing the mechanical behavior of these specimens is sound from the application perspective, as the area of coupons contributing in the joints is almost the same in FSSW and RSW specimens. Monotonic testing of tensile-shear RSW specimens yielded an average UTSL of 7620±48N with interfacial failure mode, as shown in Figure 3.

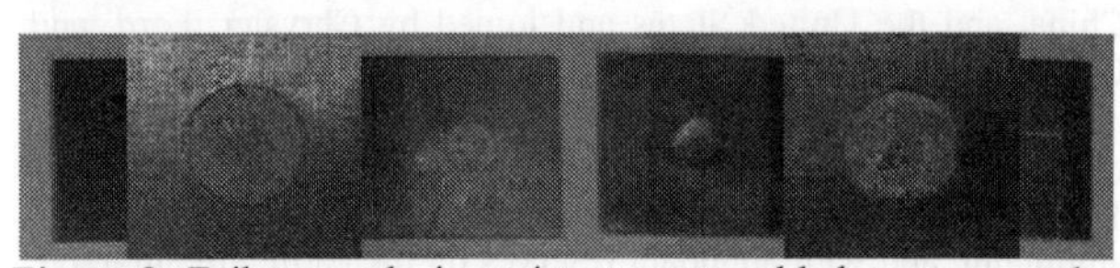

Figure 3. Failure mode in resistant spot welded coupons under monotonic tensile-shear loading.

Load-displacement curves, shown in 4, illustrate that the RSW coupons have higher UTSL compared to FSSW coupons. Higher UTSL of RSW specimens is mainly attributed to the smaller bonded area in FSSW specimens. As mentioned before, RSWs have a solid circular, and FSSWs have an annulus-shaped weld region. Therefore, even for the same outer diameter, a higher UTSL is expected for RSWs.

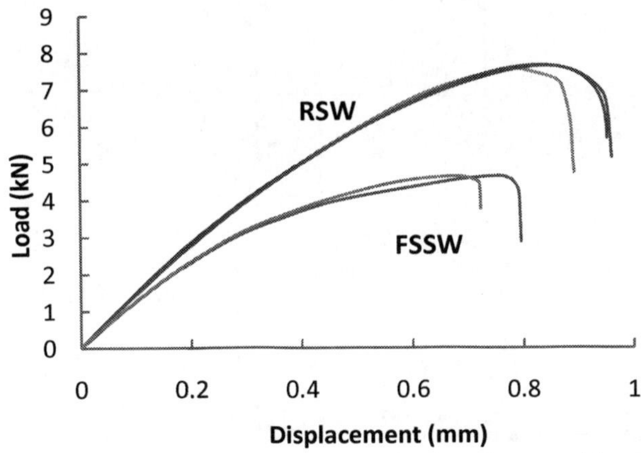

Figure 4. Load-displacement curves for friction stir spot weld and resistance spot weld tensile-shear specimens under quasi-static loading.

<ins>Fatigue Tests</ins>

Figure 5 displays the round-robin fatigue results of the FSSW from the four institutions. The first observation based on the fatigue life results was the overall good correlation between all of the laboratories. Fatigue failure for this particular study was defined as complete separation of the lap-joint. However, there were some outliers in both sets of load levels tested. One of the tests MSU conducted at the 3kN load level failed much earlier than the rest of the tests compared to the other institutions. At the 1kN level, several specimens tested at IMR failed much later compared to the rest of the group of coupons. However, the variation in the fatigue results overall is fairly consistent compared to other published round-robin fatigue testing programs. It is important to point out that at the higher fatigue load level (3kN), the load is above the elastic limit based on the monotonic load-displacement curve shown in Figure 4. At the 1kN load level, the load is within the elastic range. The number of cycles to failure for the FSSW coupons presented here are in the same order of magnitude of similar FSSW [9-12,17]

For further comparisons, the fatigue life of the FSSW coupons are compared to the fatigue life of the RSW. Fatigue results show that the RSW exhibited better fatigue life at the lower load level (1kN) compared to the FSSW coupons. However at the higher load level (3kN), the FSSW coupons exhibited better fatigue resistance compared to the RSW coupons. For the lower load level (1kN), where we have the same failure modes in FSSW and RSW specimens (coupon width separation, Fig. 7.a), higher fatigue life of RSW could be attributed to larger nugget size which causes lower stress concentration and hence smaller hot spot stress and retardation of crack initiation. However, at the higher load level (3kN), the better fatigue performance of FSSW is due to the different failure modes in FSSW and RSW specimens. FSSW specimens failed in interfacial mode (Fig. 7.b), while coupon failure perpendicular to the loading direction (Fig. 7.a) was observed in RSW specimens. The reason why the fatigue strength in interfacial mode is higher than coupon failure is, in both cases, the mode I stress intensity factor (K_I) is the main factor for fatigue crack propagation. The load component contributing to K_I in interfacial failure mode is normal to the coupon surface at the joint, and the load component contributing to K_I in coupon separation is parallel to the coupon surface or normal to the coupon thickness at the joint. And because the load component parallel to the coupon is larger than the one normal to the coupon surface (as bending rotation in coupon at the joint is less than 45 deg), the coupon separation mode is more critical than interfacial failure. However, this still needs more investigations.

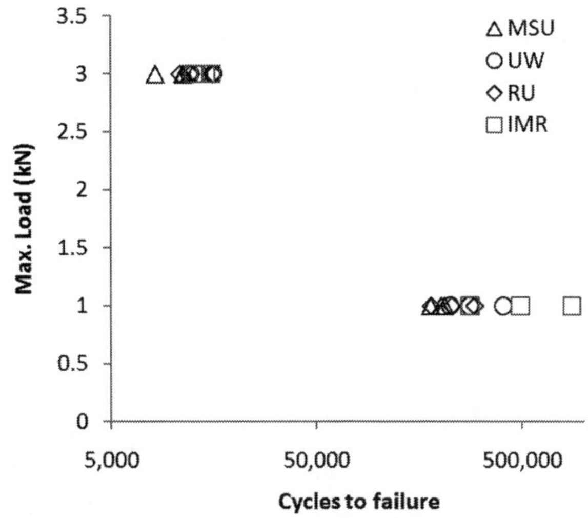

Figure 5. Comparison of Mg AZ31 friction stir spot welds fatigue results from the four universities: Mississippi State University (MSU), Ryerson University (RU), University of Waterloo (UW), and Institute for Metal Research (IMR). Fatigue tests were conducted in load control at R=0.1, at a frequency of 5 Hz and at room temperature.

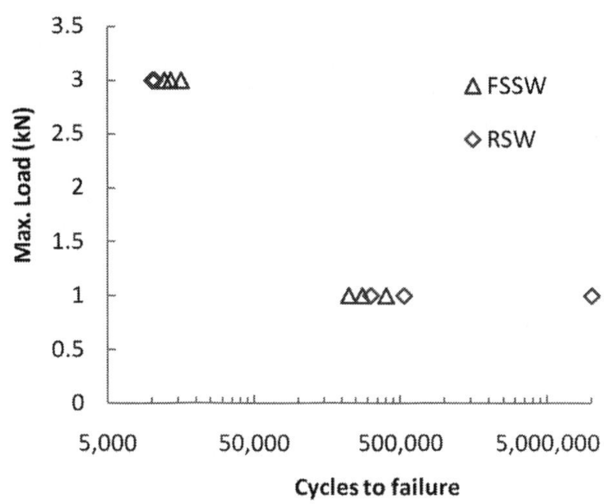

Figure 6. Comparison of fatigue results of the magnesium AZ31 alloy friction stir spot welds to the resistance spot welds tested at University of Waterloo (UW). Fatigue tests were conducted in load control at R=0.1, at a frequency of 5 Hz and at room temperature.

Fractography

The failure modes under cyclic loading were observed to vary for the different load levels tested in the round-robin testing program. All four laboratories reported that the failure mode at the lower cyclic load level of 1 kN was different compared to the higher cyclic load level of 3 kN. That is, the failure at the lower load level occurred perpendicular to the loading direction, while the failure at the higher load level exhibited interfacial fracture, as shown in Figure 7(a) and (b), respectively. The interfacial failure observed for the 3 kN load level likely occurred because as the crack propagated circumferentially around the nugget, the shear/tensile stress in the remaining net area of the nugget increased with each advancement of the crack front. Once the crack had propagated around approximately half of the nugget diameter, the shear/tensile stresses acting on the net area were such that the remaining cross section failed under shear/tensile overload. The other type of fatigue failure occurred at the load level of 1kN. Once the crack had propagated circumferentially around the nugget, the crack then propagated outward through the sheet material.

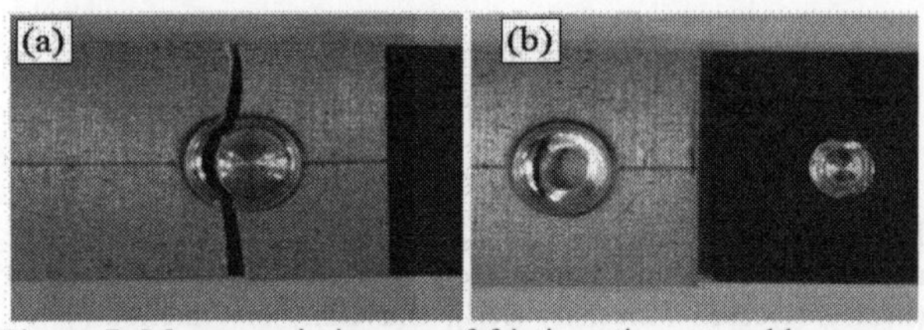

Figure 7. Macroscopic images of friction stir spot weld coupons of magnesium AZ31 alloy samples fatigued at a load level of (a) P_{max}=1 kN, and (b) P_{max}=3 kN (R=0.1, 5 Hz, sine waveform, room temperature).

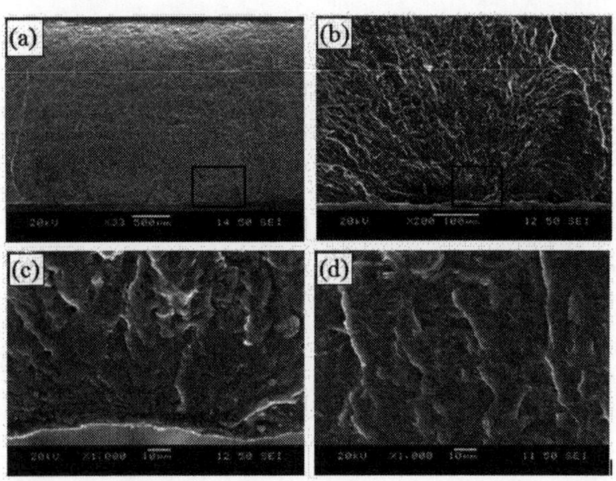

Figure 8. Scanning electron microscope images of fracture surfaces of friction stir spot weld coupons of magnesium AZ31 alloy fatigued at a load level of P_{max}=1 kN, (a) low magnification image near the center of the sample, (b) initiation site in the boxed region in (a), (c) magnified view of the boxed region in (b), and (d) crack propagation area near the center of the sample thickness.

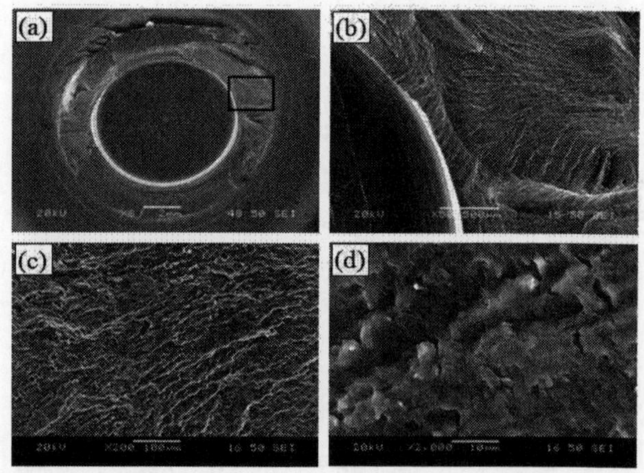

Figure 9. Scanning electron microscope images of fracture surfaces of friction stir spot weld coupons of magnesium AZ31 alloy fatigued at a load level of P_{max}=3 kN, (a) overall view of interfacial fracture of the sample, (b) initiation site in the boxed region in (a), (c) magnified view near the initiation site, and (d) crack propagation area at a higher magnification.

Fracture surfaces of the fatigued specimens were examined under scanning electron microscope (SEM). Figure 8(a-d) and Figure 9(a-d) show the typical SEM images of the coupons tested at P_{max}=1 kN and P_{max}=3 kN, respectively. The low magnification image shown in Figure 8(a) was taken near the center of the sample fractured at P_{max}=1 kN, while 9(a) showed an overall view of interfacial fracture of the sample tested at P_{max}=3 kN. Figure 8(b) and Figure 9(b) showed the boxed region in Figure 8(a) and 9(a), respectively, indicating the crack initiation site. While the failure mode was different (see Figure 7), the fatigue crack initiation at both load levels occurred from the surface. Figure 8(c) and 9(c) showed the higher magnification images near the fatigue crack initiation sites for both load levels. Figure 9(d) shows the fatigue crack propagation area with some striation-like features perpendicular to the crack propagation direction.

Conclusions

A summary of the main conclusions of this work are as follows:

1. The resistant spot weld lap-shear coupons exhibited better monotonic strength compared to the friction stir spot weld coupons with similar specimen dimensions and outer nugget diameter.
2. Friction stir spot weld and resistant spot weld coupons both failed in the interfacial mode under monotonic loading.
3. The fatigue results from the four different testing labs demonstrated consistent fatigue results on the friction stir spot weld lap-shear coupons.
4. Different failure modes were observed for the friction stir spot weld coupons for the two cyclic load levels tested. At the high cyclic load level (3kN), the coupons failed by interfacial fracture. At the lower cyclic load level (1kN), the coupons failed by full width separation.
5. Striations-like features were observed on fracture surfaces of specimens tested at 3kN.

6. The fatigue life the friction stir spot weld coupons were observed to compare closely to the fatigue life of the resistant spot weld coupons. At the higher cyclic load (3kN), the friction stir spot weld coupons exhibited better fatigue life, compared to the resistant spot weld coupons. However, at the lower cyclic load (1kN), the resistant spot weld coupons exhibited better fatigue resistance compared the friction stir spot weld coupons.

Acknowledgments

The authors would like to recognize Richard Osborne, James Quinn, Alan Luo, John Allison, and Robert McCune for their encouragement of this study. This material is based upon work supported by the Department of Energy and the National Energy Technology Laboratory under Award Number No. DE-FC26-02OR22910, AUTO21 Network of Centers of Excellence, Natural Sciences and Engineering Research Council of Canada (NSERC), Premier's Research Excellence Award (PREA). Such support does not constitute an endorsement by the Department of Energy of the views expressed herein. This paper was prepared as an account of work sponsored by an agency of the United States Government. Neither the United States Government nor any agency thereof, nor any of their employees, makes any warranty, express or implied, or assumes any legal liability or responsibility for the accuracy, completeness, or usefulness of any information, apparatus, product, or process disclosed, or represents that its use would not infringe privately owned rights. Reference herein to any specific commercial product, process, or service by trade name, trademark, manufacturer, or otherwise does not necessarily constitute or imply its endorsement, recommendation, or favoring by the United States Government or any agency thereof. The views and opinions of authors expressed herein do not necessarily state or reflect those of the United States Government or any agency thereof.

References

1. K.U. Kainer, Magnesium Alloys and Technology, Wiley-VCH, Cambridge, United Kingdom, 2003.
2. Luo, A.A., "Magnesium: Current and potential automotive applications" JOM. 54 (2): 42–48, 2002.
3. Luo, A.A., Journal of Materials and Manufacturing. SAE Transactions, Warrendale, PA, 411–21, 2005.
4. Friedrich, H.E., and Mordike, B.L., Magnesium Technology—Metallurgy, Design Data, Applications, Springer-Verlag, Berlin, Germany, 2006.
5. S. Begum, D.L. Chen, S. Xu, Alan A. Luo, Met. Mater. Trans. A 39A (2008) 3014
6 C.L. Fan, D.L. Chen, A. A. Luo, Mater Sci. Eng. A 519 (2009) 38
7. J.D. Bernard, J.B. Jordon, M.F. Horstemeyer, H. El Kadiri, J. Baird, David Lamb, Alan A. Luo, "Structure-property relations of cyclic damage in a wrought magnesium alloy," Scripta Materialia, 63 (2010) Viewpoint set no. 47, 751-756.
8. T. Pan, 2007, "Friction stir spot welding (FSSW) – a literature review," Society of Automotive Engineers, Warrendale (PA), SAE Technical Paper No. 2007-01-1702.
9. Lin, P.-C., Pan, J, Pan T., 2008, "Failure Modes and Fatigue Life Estimations of Spot Friction Welds in Lap-Shear Specimens of Aluminum 6111-T4 Sheets, Part 1: Welds Made By a Concave Tool," Int. J. Fatigue, 30(1) pp.74–89.
10. Lin, P.-C., Pan, J., Pan, T., 2008, "Failure Modes and Fatigue Life Estimations of Spot Friction Welds in Lap-Shear Specimens of Aluminum 6111-T4 Sheets. Part 2: Welds Made By a Flat Tool," Int. J. Fatigue, 30(1), pp. 90–105.
11. Tran, V.-X., Pan, J., Pan, T., 2010, "Fatigue behavior of spot friction welds in lap-shear and cross-tension specimens of dissimilar aluminum sheets," Int. J. Fatigue, 32(7), pp. 1022-1041
12. Tran, V.-X., Pan, J., 2010, "Fatigue behavior of dissimilar spot friction welds in lap-shear and cross-tension specimens of aluminum and steel sheets," Int. J. Fatigue, 32(7), pp. 1167-1179
13. Su, P., Gerlich A., and North, T.H., 2005,"Friction stir spot welding of aluminum and magnesium alloy sheets, Society of Automotive Engineers," Warrendale (PA) (2005) [SAE Technical Paper No. 2005-01-1255.
14. Gerlich, A., Su, P., and North, T.H., 2005, "Tool penetration during friction stir spot welding of Al and Mg alloys," J Mater Sci 40 pp. 6473–6481.
15. Pan, T.-Y., Santella, M., Mallick, P.K., Frederick, A., Schwartz, W.J., 2006, "A feasibility study on spot friction welding of magnesium alloy AZ31," In: Proceedings of 63rd annual world magnesium conference, Beijing, China, May 21–24; pp. 179–86.
16. Agarwal, L., Mallick, P.K., and Kang, H.T., 2008, "Spot friction welding of Mg-Mg, Al-Al and Mg-Al alloys", Society of Automotive Engineers, Warrendale (PA), SAE Technical Paper No. 2008-01-0144.
17. Mallick, P.K. and Agarwal, L., 2009, "Fatigue of spot friction welded joints of Mg–Mg, Al–Al and Al–Mg alloys," Society of Automotive Engineers, Warrendale (PA) SAE Technical Paper No. 2009-01-0024.
18. Aota K., Ikeuchi K., "Development of friction stir spot welding using rotating tool without probe and its application to low-carbon steel plates", Welding International, Vol. 23, No. 8, August 2009, pp. 572–580.
19. Ohashi R., Fujimoto M., Mironov S., Sato Y.S., Kokawa H., "Effect of contamination on microstructure in friction stir spot welded DP590 steel", Science and Technology of Welding and Joining, 2009, Vol. 14, No. 3, pp. 221-227.
20. Person N.L., "Tensile-Shear and Fatigue Properties of Resistance and MIG Spot Welds of Some Al Auto Body Sheet Alloys", SAE Paper No. 750463, 1975.
21. Chao Y.J., "Ultimate Strength and Failure Mechanism of Resistance Spot Weld Subjected to Tensile, Shear, or Combined Tensile/Shear Loads", Journal of Engineering Materials and Technology, Vol. 125, APRIL 2003, pp. 125-132.
22. Khan M.I., Kuntz M.L., Su P., Gerlich A., North T. and Zhou Y., "Resistance and friction stir spot welding of DP600: a comparative study", Science and Technology of Welding and Joining, 2007, Vol. 12, No. 2, pp. 175-182.
23. Karthikeyan R., Balasubramanian V., "Predictions of the optimized friction stir spot welding process parameters for joining AA2024 aluminum alloy using RSM", The International Journal of Advanced Manufacturing Technology, April 2010.
24. Thoppul S.D., Gibson R.F., "Mechanical characterization of spot friction stir welded joints in aluminum alloys by combined experimental/numerical approaches", Materials Characterization, November 2009, Vol. 60, No. 11, pp. 1342-1351.
25. Gean A., Westgate S.A., Kucza J.C., Ehrstrom J.C., "Static and Fatigue Behavior of Spot-Welded 5182-0 Aluminum Alloy Sheet", Welding Journal, March 1999, pp. 80s-86s.
26. Hassanifard S., Zehsaz M., Tohgo K., "The Effects of Electrode Force on the Mechanical Behaviour of Resistance Spot-Welded 5083-O Aluminium Alloy Joints", Strain, 2009.
27. Uematsu Y., Tokaji K., "Comparison of fatigue behaviour between resistance spot and friction stir spot welded aluminium

alloy sheets", Science and Technology of Welding and Joining, 2009, Vol. 14, No. 1, pp. 62-71.
28. Behravesh B., Liu L., Jahed H., Lambert S., Glinka G., Zhou Y. , "Effect of Nugget Size on Tensile and Fatigue Strength of Spot Welded AZ31 Magnesium Alloy", SAE Technical Paper, 2010, SAE No. 2010-01-0411.

Magnesium Technology 2011

Addendum

Magnesium Technology 2011
Edited by: Wim H. Sillekens, Sean R. Agnew, Neale R. Neelameggham, and Suveen N. Mathaudhu
TMS (The Minerals, Metals & Materials Society), 2011

CONTROLLING THE BIODEGRADATION RATE OF MAGNESIUM-BASED IMPLANTS THROUGH SURFACE NANOCRYSTALLIZATION INDUCED BY CRYOGENIC MACHINING

Z. Pu[1], D. A. Puleo[2], O.W. Dillon, Jr. [1], I.S. Jawahir[1]
[1]Department of Mechanical Engineering, Center for Manufacturing, University of Kentucky; Lexington, KY 40506, USA
[2]Center for Biomedical Engineering, Wenner-Gren Lab, University of Kentucky; Lexington, KY 40506, USA

Keywords: biodegradable implants, nanocrystallized grain, cryogenic machining, magnesium alloys

Abstract

Magnesium alloys are emerging as a new class of biodegradable implant materials for internal bone fixation. They provide good temporary fixation and do not need to be removed after healing occurs, providing the relief to the patients and reducing the healthcare costs. However, premature failure of these implants often occurs due to the high biodegradation rate caused by low corrosion resistance of magnesium alloys in physiological environments. To control biodegradation/corrosion of magnesium alloys, grain refinement on the surface was achieved through machining-induced severe plastic deformation. Liquid nitrogen was used during machining to suppress grain growth. White layers, which consist of nanocrystallized grain structures, are reported herein for the first time in magnesium alloys. By controlling the machining conditions, white layers with various thicknesses were fabricated. *In vitro* corrosion tests proved that different machining conditions can significantly change the biodegradation rate of magnesium alloys.

Introduction

In the U.S. alone, physician visits for orthopedic surgery reached 48,066,000 in 2006 [1]. Nine out of the twenty five most common orthopedic surgeries involve repair of bone fractures [2]. Internal bone fixation implants, such as bone plates and screws, are widely used to provide temporary fixation for fractured bones. Stainless steels and titanium alloys are two major biomaterials currently used for these implants. However, their excessively stronger mechanical properties compared to bones may lead to stress shielding. The corrosion and fatigue of these materials will inevitably generate metallic ions and particles that may activate adverse tissue reactions. To avoid further reactions after bone healing, these implants need to be removed during a second surgery, which adds additional morbidity (pain, refracture, etc.) to the patients and increases healthcare costs.

While various approaches are being investigated to increase the bio-inertness of traditional implant materials, magnesium alloys are emerging as a novel biodegradable material in which the relatively fast corrosion phenomenon is used as a unique advantage for temporary fixation implants. The potential of magnesium alloys as a biodegradable implant material was explored by several researchers in the first half of the twentieth century. The results of these research investigations were summarized by Staiger et al. [3]. No systematic reaction occurred and little inflammation was observed in these human trials. A marked stimulatory effect for bone healing was also reported. However, the premature failure of magnesium-based implants due to the poor corrosion resistance in physiological environments and gas bubbles generated due to the high corrosion rate impeded further investigation until recently.

In 2005, Witte et al. found results similar to the early researchers through *in vivo* study of magnesium alloys. A stimulatory effect for bone growth was also reported [4]. The formation of a biomimetic layer comprised of magnesium and calcium phosphate at the implant/bone interface was found to be the cause for accelerated bone formation [5, 6].

Despite their attractive features, little progress has been achieved in controlling the biodegradation rate of magnesium alloys. Alloying and coating are two approaches widely studied [7, 8]. However, alloying may introduce elements which may lead to adverse biological reactions. Stability of the coating under cyclic loading in physiological conditions is a great challenge while the complexity of coating techniques may significantly increase the cost of implant.

Mechanical processing of magnesium alloys provides an alternative approach to control the biodegradation rate. Hot rolled magnesium AZ 31 samples were reported to have a marked reduction in biodegradation rate compared with squeeze cast samples [9]. The reduction was attributed to grain refinement from 450 μm to 20 μm. However, further grain refinement by equal channel angular pressing (ECAP) to 2.5 μm did not decrease the biodegradation rate. With the same material, Alvarez-Lopez et al. [10] found that samples with 4.5 μm grain size processed by ECAP and followed by rolling had better corrosion resistance than the initial samples with 25.7 μm grain size. Deep rolled magnesium MgCa3.0 samples were reported to have a pronounced reduction in biodegradation rate [11]. Machining with different cutting speeds also leads to different corrosion rates [11].

Surface and subsurface integrity in machined products is emerging as the new focus in machining research. The performance of the components can be significantly modified by machining through changes in surface integrity factors, such as microstructure, hardness and residual stresses [12]. Significant grain refinement occurs at the machined surface/sub-surface through severe plastic deformation. Nanocrystallized grains of about 5 - 20 nm in size were reported in the white layer of AISI 52100 steel after machining [13]. Ultrafine grains about 175 nm were formed on the machined surface of copper [14]. Due to grain growth caused by the large amount of heat generated during machining, the nanocrystallized grains can be found only at the top surface of the machined component and the cutting speed is limited to very low range. A novel technique based on machining was developed to fabricate thick nanocrystallized layers with the help of the liquid nitrogen cooling [15]. The results proved that severe plastic deformation under cryogenic conditions can successfully introduce nanocrystallized grain structures to the surface and sub-surface layers.

However, only a few studies on controlling the grain refinement through advanced process control have been reported. Also, the relationship between grain refinement, especially in the nanocrystalline range, and biodegradation rates in magnesium alloys is still unknown. Therefore, the aim of the present work was to investigate the microstructural changes under different machining conditions and their influence on biodegradation rate of magnesium alloys incubated in simulated body fluid (SBF).

Experimental Work

Work Material

The work material studied was the commercial AZ31 B-H24 magnesium alloy. *In vivo* tests showed the potential of magnesium AZ31 alloy as a bone implant was significant [16, 17]. The work material was received in the form of 3 mm thick sheet. Disc specimens were made from the sheet and subsequently subjected to orthogonal machining.

Machining Experiments

The machining experiments were conducted on a Mazak Quick Turn-10 Turning Center equipped with an Air Products liquid nitrogen delivery system. The experimental setup is shown in Figure 1.

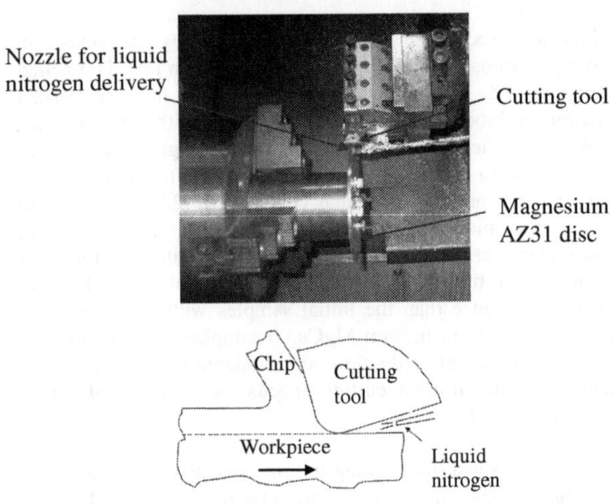

Figure 1. Machining setup with an Air Products liquid nitrogen delivery system

The machining conditions controlled during the experiments were cooling methods and the edge radius of the cutting tool. For dry machining, no coolant was used. For cryogenic machining, liquid nitrogen was applied to the machined surface from the clearance side of the cutting tool. The cutting tools used were uncoated carbide C5/C6 inserts from Kennametal. These cutting tools were ground to three different edge radii. The actual edge radius before machining was measured using a ZYGO New View 5300 measurement system which was based on white light interferometry. A sample measurement is shown in Figure 2.

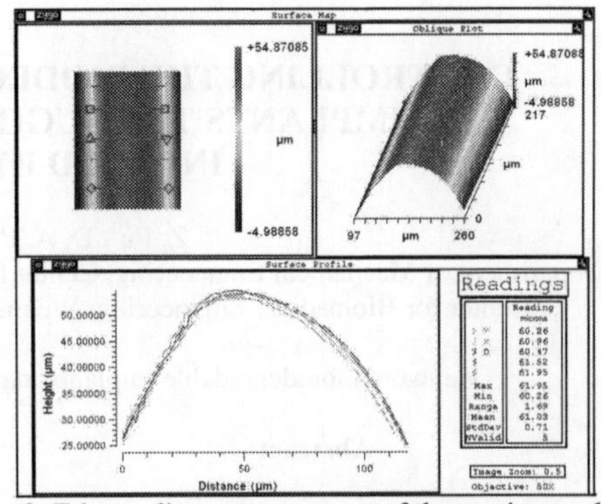

Figure 2. Edge radius measurement of the cutting tool using ZYGO New View 5300

The matrix for the machining experiments is shown in Table I. For each machining experiment, a KISTLER 3-Component Tool Dynamometer was used to measure the cutting forces.

Table I. Matrix for the machining experiments

No.	Cooling Method	Edge Radius (µm)	Cutting Speed (m/min)	Feed Rate (mm/rev)
1	Dry	30	100	0.1
2	Cryogenic	30	100	0.1
3	Cryogenic	68	100	0.1
4	Cryogenic	74	100	0.1

In vitro Corrosion Test

To mimic the human body environment, a simulated body fluid (SBF) was prepared: 8.0 g/l NaCl, 0.4 g/l KCl, 0.14 g/l $CaCl_2$, 0.35 g/l $NaHCO_3$, 1.0 g/l $C_6H_{12}O_6$ (D-glucose), 0.2 g/l $MgSO_4 \cdot 7H_2O$, 0.1 g/l $KH_2PO_4 \cdot H_2O$ and 0.06 g/l $Na_2HPO_4 \cdot 7H_2O$. The pH of the SBF was adjusted to 7.4. The solution was kept in an incubator to maintain the temperature at 37 ± 1 °C. To evaluate the biodegradation rates, hydrogen evolution method [18] was used to continually monitor the corrosion process for 7 days. To reduce the effects of pH increase and accumulation of corrosion products on corrosion rate, large solution volume/surface area (SV/SA) ratio (SV/SA=433) was used [19].

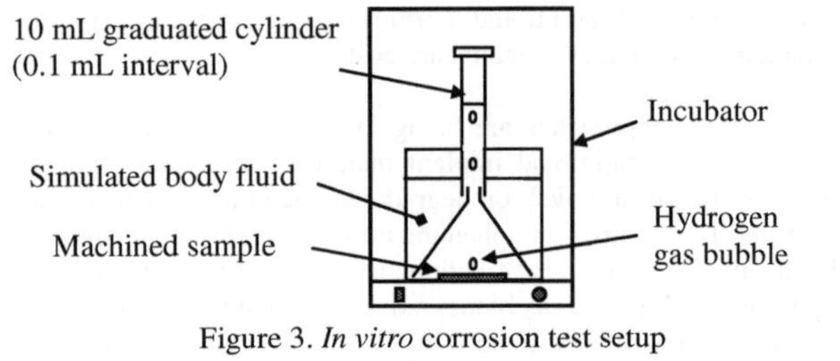

Figure 3. *In vitro* corrosion test setup

Characterization Method

Metallurgical samples were cut from the machined discs. After cold mounting, grinding and polishing, acetic picric solution was used as the etchant to reveal the grain structure. Optical and scanning electron microscopes were used to observe the microstructure of the magnesium alloys. An atomic force microscope (AFM) was used to explore the possible structure of the top layer of the machined samples. For AFM characterization, the samples were observed after grinding and polishing but without etching. The chemical composition of the top layer was determined by energy dispersive spectroscopy (EDS).

Results and Discussion

Microstructure

The initial microstructure before machining experiment is shown in Figure 4. The grain boundaries are clearly visible throughout the sample.

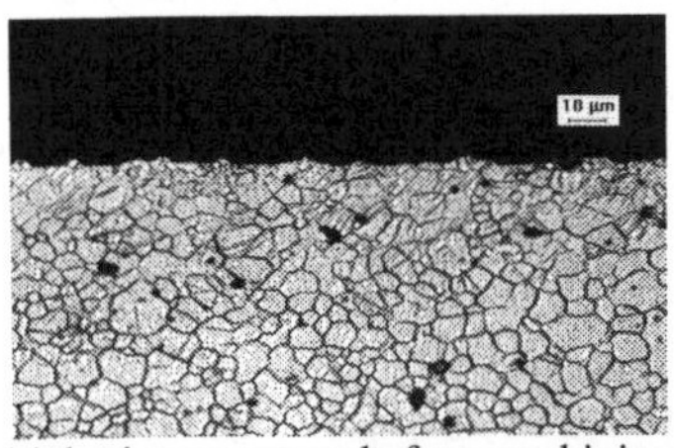

Figure 4. Initial microstructure before machining experiment

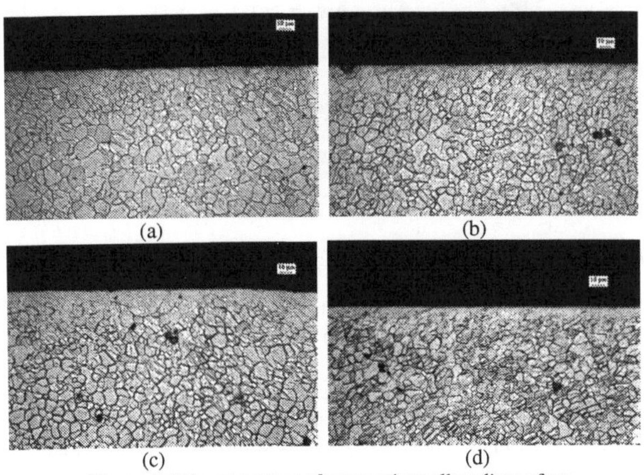

Figure 5. Microstructure of magnesium alloy discs after machining: (a) dry machining, edge radius = 30 μm, (b) cryogenic machining, edge radius = 30 μm, (c) cryogenic machining, edge radius = 68 μm, (d) cryogenic machining, edge radius = 74 μm.

The microstructures of the samples machined under different conditions are presented in Figure 5. Although the grain structures of the bulk material were similar under all machining conditions, significant differences were apparent in the surface layers of different machined samples. For dry machined samples (Figure 5(a)), grain boundaries are clearly visible on the surface. For cryogenic machined samples (Figure 5(b), (c) and (d)), surface layers of different thicknesses, where grain boundaries are not discernable, were formed. This layer of indiscernible grain structure was also reported in other materials after machining, especially in steels, where the term "white layer" was frequently used [13]. While significant research has been done in white layer formation in steels, white layer in magnesium alloys is reported here for the first time.

The influence of machining conditions on white layer formation is clearly seen in Figure 6. Dry machining of magnesium alloys using a tool with 30 μm edge radius did not lead to white layer formation. However, using the same edge radius, a white layer of about 7 μm thickness was formed under cryogenic machining conditions. Under the same cooling condition, the edge radius of the tool played a remarkable role in white layer formation. The thickness of the white layer was increased to about 15 μm when the edge radius was increased to 68 μm. However, further increase in edge radius to 74 μm reduced the white layer thickness.

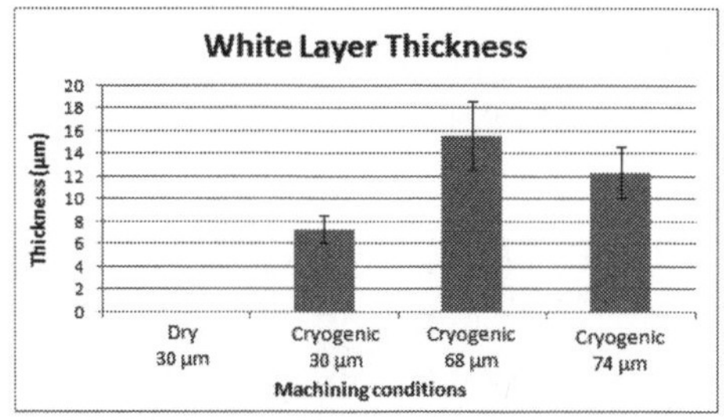

Figure 6. White layer thickness under different machining conditions

Force Analysis

Cutting force and radial force data during machining were analyzed to explore the influencing factors in white layer formation. Both the cutting force and radial force were stable during the 30 second cutting time, indicating little tool-wear and tool/chip adhesion. The cutting forces for all the four experiments remained at about 180 N. However, a significant influence of tool edge radius and cooling method on radial force was present. Figure 7 shows the radial force recorded for the machining experiments. In the cryogenic group, the radial force became larger with increasing edge radius of the tool. A large increase was observed when the edge radius was increased from 30 μm to 68 μm, which corresponds to the large increase in white layer thickness. The radial force was further increased using the tool with 74 μm edge radius. Higher stresses introduced by the large edge radius lead to more deformation twinning within the grains compared with 30 μm and 68 μm. However, the white layer thickness decreased. This may indicate that white layer formation was dependent on both mechanical and thermal effects, which agrees with the research in white layer formation in steels [20].

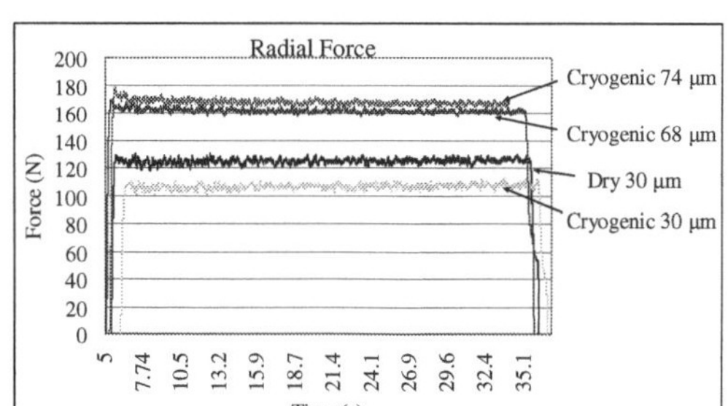

Figure 7. Radial force measurement under different machining conditions

EDS Analysis

To explore the chemical composition of the white layer, energy dispersive spectroscopy (EDS) was used. Figure 8 shows the results of the EDS analysis. Only a little chemical difference between the bulk material and the white layer was found.

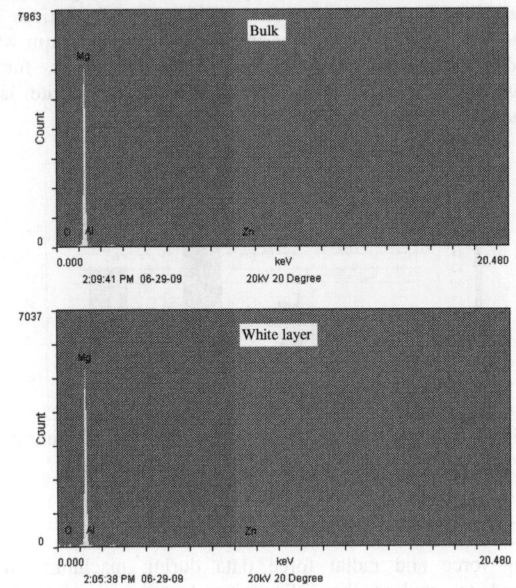

Figure 8. EDS analysis of the bulk material and the white layer

SEM Characterization

Nanocrystallized grain structures were found in white layers in steels [13]. A thick nanocrystallized layer was also found on copper processed by a surface mechanical grinding treatment (SMGT) under liquid nitrogen cooling [15]. Due to the similarity that both cryogenic SMGT and cryogenic machining in producing severe plastic deformation, it is expected that the white layer formed on cryogenically machined magnesium AZ31 samples consisted of nanocrystallized grain structures. To explore this assumption, scanning electron microscope (SEM) and atomic force microscope (AFM) were used to observe the white layer formed on Sample No. 3 (Table I).

Figure 9(a) shows that grain boundaries are clear except the top layer of the sample, and twinning was present more than 100 μm away from the machined surface. Figure 9b shows that grain boundaries in the white layer were not visible even under ×5000 magnification.

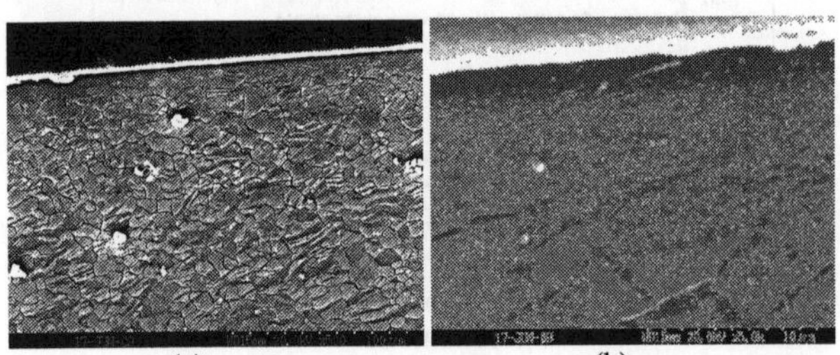

(a) (b)

Figure 9. SEM pictures of Sample No. 3: (a) ×500 (b) ×5000.

AFM Characterization

The ability of atomic force microscope (AFM) to measure grain size was studied by several researchers [21, 22]. Phase imaging tapping-mode AFM was reported to successfully provide grain boundary details and the accuracy of the grain size measurements was comparable to TEM measurement with ±10% depending on AFM calibration accuracy. After grinding and polishing, the top portion of the white layer on Sample No. 3 was observed using tapping-mode AFM. The phase image is shown in Figure 10. While no structures were discernable in optical microscope or SEM pictures, some grain-like features were present in the AFM phase image. The mean size of the features was about 45 nm and all the features were smaller than 100 nm. Based on the AFM picture and the literature review on nanocrystallized grains in white layers of steels and copper, a preliminary conclusion can be made that the white layer on the machined surface of magnesium alloys consisted of nanocrystallized grain structures. Further investigation using transmission electron microscope (TEM) will be conducted to verify this conclusion.

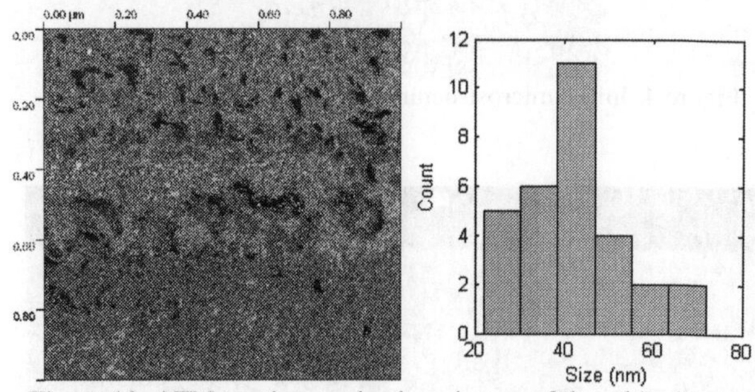

Figure 10. AFM tapping mode phase image of the white layer

In vitro Corrosion Test

The influence of different machining conditions on the biodegradation rate of magnesium alloys was investigated by the *in vitro* corrosion tests. Figure 11 shows accumulative hydrogen evolution per unit area from magnesium AZ31 discs machined under different conditions. The machined samples were left in air for two weeks before the *in vitro* corrosion test, which simulated the passivation stage of implant preparation; a passivated layer was formed on the machined surface due to the oxidation of magnesium in air. This passivated layer was attributed to the different shape of the biodegradation curves compared with other studies, where a very high biodegradation rate occurred at the beginning of the corrosion test.

Preliminary corrosion test results clearly demonstrated the influence of machining conditions on corrosion resistance of magnesium alloys in physiological environment. The slowest corrosion rate occurred on the machined sample with the thickest white layer. However, the sequence of corrosion rates did not agree with the sequence of white layer thickness. Other factors, like deformation twins, may also contribute to the variation in corrosion rate. Other corrosion analyses, such as weight loss and electrochemical methods, will be conducted to verify the results from this preliminary experiment.

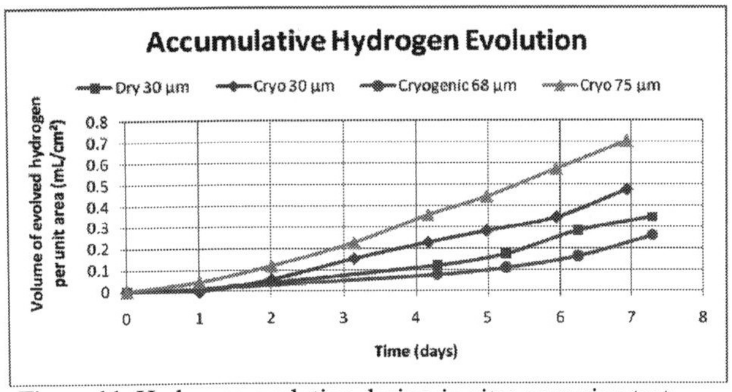

Figure 11. Hydrogen evolution during *in vitro* corrosion test

An interesting phenomenon during the corrosion test was found between the cryogenic machined sample using tool with 68 µm edge radius and the dry machined sample using tool with 30 µm edge radius. The former had the thickest white layer (about 15 µm), while the latter did not have white layer. During the corrosion process, relatively large black spots caused by pitting were formed on the sample without white layer. However, for the sample with the thickest white layer, the whole surface gradually turned dark over time without formation of large black spots. This difference suggested that the sample machined under cryogenic condition underwent more homogeneous corrosion than the one under dry condition. This homogeneity is expected to be beneficial in orthopedic implant applications. The most attractive benefit is the formation of a uniform implant/tissue reaction layer, which may lead to better osseointegration. Also, stress concentration may be avoided due to the homogeneity of the pitting.

Conclusion

The present study shows that the biodegradation rate of the magnesium alloys can be effectively altered by controlling the machining conditions. White layer in magnesium alloys is reported for the first time. The slowest corrosion rate occurred on the cryogenically machined sample with the thickest white layer. Also, corrosion on this sample was found to be more homogeneous compared with the dry machined sample that did not have a white layer.

The AFM phase image suggests that the white layer consisted of nanocrystallized grain structures. The remarkable grain refinement was the combined result of dynamic recrystallization through machining-induced severe plastic deformation and effective suppression of grain growth by liquid nitrogen cooling.

The edge radius of the cutting tool has an important influence on the thickness of the white layer. A combined thermal-mechanical effect for white layer formation was detected.

The preliminary results from the present study reveal a great opportunity to control biodegradation rate of magnesium alloys through advanced process control techniques. Existing knowledge on surface integrity of machined products and various predictive modeling techniques can significantly facilitate the development of a machining-based process to control the biodegradation rate of magnesium alloys. *In vitro* corrosion tests can correlate the surface integrity factors with the corresponding biodegradation rate. In the end, magnesium-based implants with specific biodegradation rate customized to individual medical demands can be manufactured.

References

1. D.K. Cherry, E. Hing, D.A. Woodwell, E.A. Rechtsteiner, "National Ambulatory Medical Care Survey: 2006 Summary" (National Health Statistics Reports, Number 3, August 6, 2008).

2. W.E. Garrett, Jr., M.F. Swiontkowski, J.N. Weinstein, J. Callaghan, R.N. Rosier, D.J. Berry, J. Harrast, G.P. Derosa, "American Board of Orthopaedic Surgery Practice of the Orthopedic Surgeon: Part-II, Certification Examination Case Mix", *The Journal of Bone and Joint Surgery,* 2006, no. 88:660-667.

3. Mark P. Staiger, Alexis M. Pietak, Jerawala Huadma, George Dias, "Magnesium and its alloys as orthopedic biomaterials: A review", *Biomaterials,* 2006, no. 27: 1728–1734.

4. Witte F, Kaese V, Haferkamp H, Switzer E, Meyer-Lindenberg A,Wirth CJ, H. Windhagen, "In vivo corrosion of four magnesium alloys and the associated bone response", *Biomaterials,* 2005, no.26:3557–3563.

5. Liping Xu, Guoning Yu, Erlin Zhang, Feng Pan, Ke Yang, "In vivo corrosion behavior of Mg-Mn-Zn alloy for bone implant application", *Journal of Biomedical Materials Research Part A* , 2007, 83A: 703-711.

6. Zhang E, Xu L, Yu G, Pan F, Yang K, "*In vivo* evaluation of biodegradable magnesium alloy bone implant in the first 6 months implantation", *Journal of Biomedical Materials Research Part A,* 2008,90A:882 – 893.

7. Guangling Song, "Control of biodegradation of biocompatible magnesium alloys", *Corrosion Science,* 2007, no. 49: 1696–1701.

8. Cuilian Wen, Shaokang Guan, Li Peng, Chenxing Ren, Xiang Wang, Zhonghua Hu, "Characterization and degradation behavior of AZ31 alloy surface modified by bone-like hydroxyapatite for implant applications", *Applied Surface Science,* 2009, no.255: 6433–6438.

9. H. Wang, Y. Estrin, Z. Zuberova, "Bio-corrosion of a magnesium alloy with different processing histories", *Materials Letters,* 2008, no.62: 2476-2479.

10. M. Alvarez-Lopez, María Dolores Pereda, J.A. del Valle, M. Fernandez-Lorenzo, M.C. Garcia-Alonso, O.A. Ruano and M.L. Escudero, "Corrosion behaviour of AZ31 magnesium alloy with different grain sizes in simulated biological fluids", 2009, *Acta Biomaterialia.* Article in Press, Corrected Proof.

11. B. Denkena, A. Lucas, "Biocompatible Magnesium Alloys as Absorbable Implant Materials –Adjusted Surface and Subsurface Properties by Machining Processes", 2007, *Annals of the CIRP,* no.56: 113-116.

12. R. M'Saoubi, J.C. Outeiro, H. Chandrasekaran, O.W. Dillon Jr., I.S. Jawahir, "A review of surface integrity in machining and its impact on functional performance and life of machined

products", *International Journal of Sustainable Manufacturing* , 2008, no.1:203~236.

13. A. Ramesh, S.N. Melkote, L.F. Allard, L. Riester, T.R. Watkins, "Analysis of white layers formed in hard turning of AISI 52100 steel", *Materials Science and Engineering A*, 2005, no.390: 88–97.

14. R. Calistes, S. Swaminathan, T.G. Murthy, C. Huang, C. Saldana, M.R. Shankar and S. Chandrasekar, "Controlling gradation of surface strains and nanostructuring by large-strain machining", *Scripta Materialia*, 2009, no.60:17–20.

15. W.L. Li, N.R. Tao, K. Lu, "Fabrication of a gradient nano-micro-structured surface layer on bulk copper by means of a surface mechanical grinding treatment", *Scripta Materialia*, 2008, no.59: 546–549.

16. Yaohua He, Hairong Tao, Yan Zhang, Yao Jiang, Shaoxiang Zhang, Changli Zhao, Jianan Li, Beilei Zhang, Yang Song and Xiaonong Zhang, "Biocompatibility of bio-Mg-Zn alloy within bone with heart, liver, kidney and spleen", *Chinese Science Bulletin*, 2009, no. 54: 484-491.

17. Duygulu O,Kaya RA,Oktay G,Kaya AA, "Investigation on the potential of magnesium alloy AZ31 as a bone implant", *Material Science Forum,* 2007, no. 546-549: 421-424.

18. Song, G., Atrens, A., St John, D. H., "An hydrogen evolution method for the estimation of the corrosion rate of magnesium alloys", *Proceeding of Magnesium Technology 2001,* TMS Annual Meeting. New Orleans, LA. February 11-15, 2001.

19. Lei Yang, Erlin Zhang, "Biocorrosion behavior of magnesium alloy in different simulated fluids for biomedical application", *Materials Science and Engineering C,* 2009, no. 29:1691–1696.

20. Sangil Han, "Mechanisms and Modeling of White Layer Formation in Orthogonal Machining of Hardened and Unhardened Steels" (Ph.D. thesis, Georgia Institute of Technology, 2006)

21. C. H. Pang, P. Hing, A. See, "Application of phase-imaging tapping-mode atomic-force microscopy to investigate the grain growth and surface morphology of TiSi2", *The Journal of Vacuum Science and Technology B*, 2002, no. 20: 1866-1869.

22. Alexander Luce, "Atomic Force Microscopy Grain Structure Characterization of Perpendicular Magnetic Recording Media" (2007 REU Research Accomplishments, National Nanotechnology Infrastructure Network, pp. 134-135).

AUTHOR INDEX
Magnesium Technology 2011

I

J

K

L

M

N

X

Y

Z

SUBJECT INDEX
Magnesium Technology 2011

Q

R

S

T

U

V

W

X

Y

Z

MIX
Papier aus verantwortungsvollen Quellen
Paper from responsible sources
FSC® C105338

If you have any concerns about our products,
you can contact us on
ProductSafety@springernature.com

In case Publisher is established outside the EU,
the EU authorized representative is:
Springer Nature Customer Service Center GmbH
Europaplatz 3, 69115 Heidelberg, Germany

Printed by Libri Plureos GmbH
in Hamburg, Germany